W9-CKI-568

GEOCHIMICA ET COSMOCHIMICA ACTA

SUPPLEMENT 2

PROCEEDINGS
OF THE
SECOND LUNAR SCIENCE CONFERENCE

Houston, Texas, January 11–14, 1971

GEOCHIMICA ET COSMOCHIMICA ACTA

Journal of The Geochemical Society and The Meteoritical Society

SUPPLEMENT 2

PROCEEDINGS OF THE SECOND LUNAR SCIENCE CONFERENCE

Houston, Texas, January 11–14, 1971

Sponsored by
The Lunar Science Institute

Edited by

A. A. LEVINSON

University of Calgary, Calgary, Alberta, Canada

VOLUME 2

CHEMICAL AND ISOTOPE ANALYSES

———

ORGANIC CHEMISTRY

THE MIT PRESS

Cambridge, Massachusetts, and London, England

Set in Monotype Times Roman by
William Clowes and Sons Ltd.
Printed and bound by The Colonial Press Inc.
in the United States of America.

ISBN 0 262 12054 2 (hardcover)

Library of Congress catalog card number: 78–165075

Supplement 2
GEOCHIMICA ET COSMOCHIMICA ACTA
Journal of The Geochemical Society and The Meteoritical Society

Volume 2 — Contents

ORGANIC GEOCHEMISTRY

GEOCHIMICA ET COSMOCHIMICA ACTA

SUPPLEMENT 2

PROCEEDINGS
OF THE
SECOND LUNAR SCIENCE CONFERENCE

Houston, Texas, January 11–14, 1971

Proceedings of the Second Lunar Science Conference, Volume 2, pp. 987–998.
The M.I.T. Press, 1971.

Model history of the lunar surface

H. C. UREY, K. MARTI, J. W. HAWKINS, and M. K. LIU
University of California at San Diego, La Jolla, California 92037

(*Received 22 February 1971; accepted in revised form 30 March 1971*)

Abstract—A model for the early history of the lunar surface is proposed which tries to take into account the relevant experimental and theoretical evidence: mascons, chemical composition, model and isochron ages, seismic and electrical properties. It is suggested that the lead data indicate a two-stage evolution of the lunar surface materials, and that in the remelting process, about half of the radiogenic lead was lost from the crystalline rocks, probably to the soil. Results from heat balance calculations indicate that remelting of basaltic and silicic materials 3.2–4.0 aeons ago is possible if an insulating layer of about 1 km dust, produced mainly by the big collisions that produced the circular mare, is assumed.

INTRODUCTION

THE PRESENCE of substantial mass concentrations (mascons), predominantly in the circular maria and in regions which may be obscured older circular maria, and of negative gravitational anomalies in the Albategnius and Ptolemaeus craters, together with evidence for chemical differentiation and quite nonterrestrial-type seismic phenomena, make it necessary to postulate quite an involved early lunar history. The Rb–Sr model ages and the lead ages of the soil indicate that a differentiation from solar system material occurred about 4.5 aeons ago, and that various remelting periods occurred 3.2 to 4.0 aeons ago. Though the titaniferous basalt does resemble terrestrial basalts to some degree, the differences indicate that the process of production may be quite different in some essential ways. Also, the (anorthositic) highlands, a rock fragment of the coarse soil 10085 and the 12013 rock, indicate very marked differentiations which occurred early in lunar history. Our model is similar to but definitely differs from that of WOOD (1970).

MASCONS

The immense mass concentrations have been supported by the moon from the time of their formation to the present, and this requires a rigid moon during this time. MULLER and SJOGREN (1970) estimate the excess mass of the Imbrium mascon as 1.6×10^{21} grams. To illustrate the magnitude of this mascon, we may assume a 10% difference in mean density between the mascon and its surroundings, and a density of the mascon of about 3.7 g/cm^3. Then the thickness of the mascon, assuming it to be uniform over the inner circle of Mare Imbrium, i.e., 680 km in diameter, would be some 12 km, and the excess pressure below this mass would be 70 bars. If it covers a smaller area, the excess pressure would be greater. If the density is greater, the thickness is less, but excess pressures remain the same. A similar situation exists in Mare Serenitatis and lesser masses and pressures in the case of other mascons. Any lava flow or any other transport of higher density material of this magnitude would be large indeed.

The seismic and electrical properties and the presence of these excess masses taken together indicate that the moon is more rigid than the earth. It has been relatively cool during the time that these excess masses have existed, and is relatively cool as compared with the earth now (SONETT *et al.*, 1971). A slow accumulation of the moon from small solid objects or an accumulation in a lunar gas sphere could supply these conditions. An escape from the earth during the formation of the earth's core or an accumulation from a vapor cloud in orbits near the earth (RINGWOOD and ESSENE, 1970), however, would produce a very hot moon.

If the concentrations of uranium and thorium in the moon, as a whole, are the same as those of the meteorites, and if the potassium concentration is about the same as that estimated for the earth or that of the type III carbonaceous chondrites, namely 250 to 360 ppm, no difficulty in maintaining a rigid moon is encountered, providing it begins its history at low temperatures.

INITIAL MELTING AND DIFFERENTIATION

The surface of the moon is covered with silicate materials which must have secured their chemical composition in a melting and solidification process. The existence of the mascons requires that this melting process was confined to the surface region. Several sources of heat may have been available: (1) the gas sphere, contracting due to radiation of energy to space, would rise to high temperatures near its center; (2) during capture by the earth, the surface regions would be heated by tidal effects; or (3) intense magnetic fields from the sun sweeping over the surface would increase the temperature of an already fairly hot surface. In this melting, some reduction of iron occurred, and this metallic iron carried the siderophile elements down to the bottom of the melted surface layer (UREY and MARTI, 1968). Also, iron sulfide melted and removed the chalcophiles. This material may represent the electrically conducting layer (SONETT *et al.*, 1971). Possibly the more volatile elements were lost also at this time, though these elements may in part be found in materials not yet investigated. As the external heat source declined, temperatures fell slowly, minerals crystallized, and differentiation occurred, due to gravity-controlled mineral separation. The early formed pyroxene and olivine sank, the Ca-plagioclase floated to the surface, and the residual titaniferous basalt liquids solidified last as intermediate layers at this time, since their model ages are near 4.5 aeons. Also, Luny rock 12085 and rock 12013, which we shall refer to as silicic rock, crystallized very close to the basalt at this time, since their model ages are near 4.5×10^9 years.

RINGWOOD and ESSENE (1970) have concluded that the differentiation as outlined above cannot take place at low pressures, and this places another important limitation on the conditions for the early lunar history. However, the origin of the moon in a massive gas sphere (UREY, 1958, 1966) can supply the pressure required (OSTIC, 1965). Such an origin of heating and cooling would supply a uniform source of heat and pressure over the entire surface of the moon, and as the gas sphere dissipated, slow cooling would permit differentiated silicate layers. Tens of millions or more years could be available for this purpose. Though reduction of iron oxides to iron would occur and thus the siderophile elements would be removed to some lower level, it is probable that this would not be complete in a layer some 100 to 200 km thick (UREY, 1966).

COOLING RATES

At this time, the outer parts became sufficiently rigid to support the large craters of the lunar surface, and in particular the negative gravitational anomalies in Albategnius and Ptolemaeus. It should be noted that this cooling occurred when the radioactive nuclides were most abundant.

For the cooling process, we have made a simple model. The temperature of the differentiated layers, T_a, is assumed to be uniformly at about 1300°C initially, and the top surface is cooled to the equilibrium temperature, T_e, between radiation loss and solar heating. Since solid crystalline rocks are assumed to exist, we have used 0.008 cm^2/sec for the thermal diffusivity. Although the radioactive elements were certainly present at this time, we have not included them in the calculations because they do not affect our result significantly. The time dependent temperature T, at a depth x, is obtained (CARSLAW and JAEGER, 1959) from

$$T = T_e + (T_i - T_e)\frac{x}{L} + \frac{2}{\Pi}\sum_{n=1}^{\infty}\frac{T_i\cos n\pi - T_e}{n}\sin\frac{n\pi x}{L}\exp\left(\frac{-\kappa n^2\pi^2 t}{L^2}\right) + \frac{2T_a}{\pi}\sum_{n=1}^{\infty}\frac{1-\cos n\pi}{n}\sin\frac{n\pi x}{L}\exp\left(\frac{-\kappa n^2\pi^2 t}{L^2}\right)$$

where $T_i = 800°C$ is the assumed temperature below the differentiated layer of the moon, L is the thickness of the differentiated layer, and κ is the thermal diffusivity. The temperatures are calculated at two depths of 10 and 20 km in layers of 100 and 200 km thickness, respectively, of solid crystalline rock. The lapse of time for the temperature to fall from 1300°C to 500°C is typically 20 to 50 million years (see Fig. 1A).

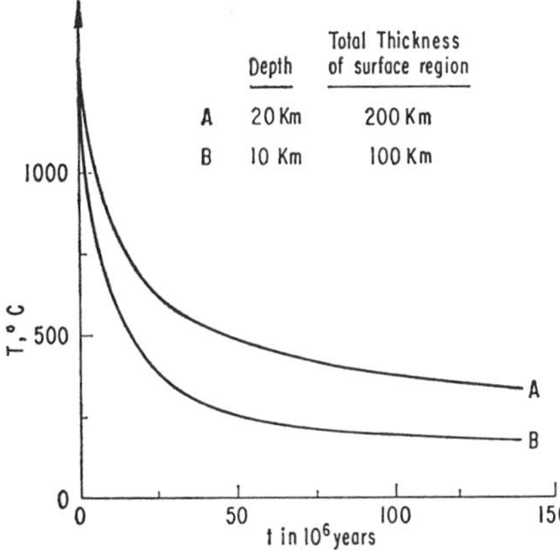

Fig. 1A. Cooling curves for the surface region.

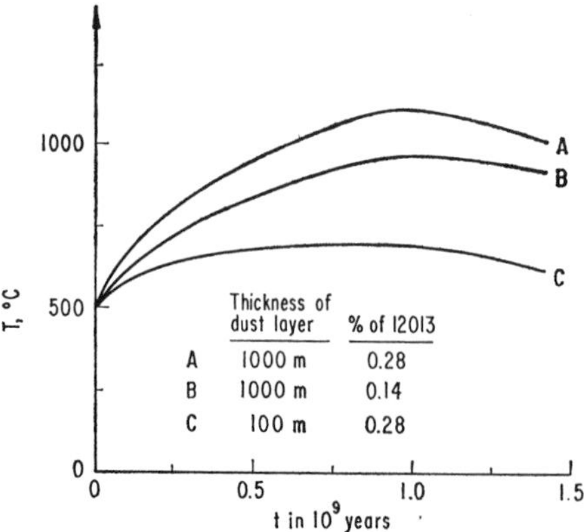

Fig. 1B. Reheating of the region of the basalt layer under three choices of thickness of the dust layer and the fraction of silicic rock.

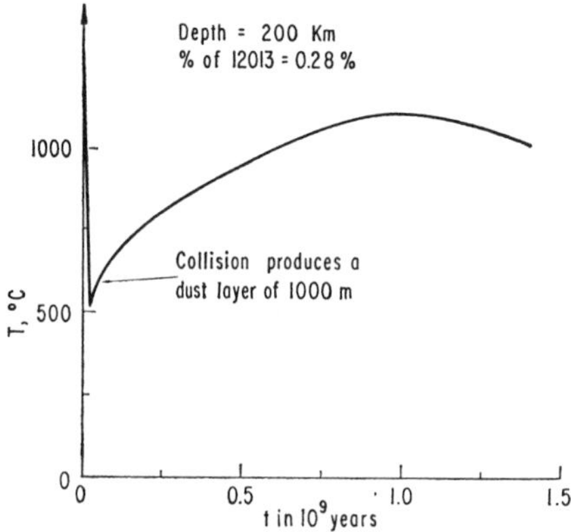

Fig. 1C. Combined cooling and reheating of the basalt layer for one choice of physical parameters.

Collisions and Seismic Results

An early intense bombardment of the moon occurred after this time and left the moon with a cratered surface. The collisions which produced the maria were part of this bombardment, and these collisions may have produced thick layers of dust. The collisions produced in some areas of the moon a very powdered and rubbled surface layer of much lower thermal conductivity.

The greatly broken up and fragmented material of the moon's outer layers probably account for the observed seismic effects. LATHAM *et al.* (1970) and GOLD and SOTER (1970) have discussed these observations, and apparently a thick layer of highly fragmented material is required to account for these observations. It is surprising that this applies to the mare regions where a sheet of solid material might have been expected. Possibly these smooth areas are ash flows with some sizeable rocks included. Also, these fragmented surface layers may account for the wave-like character of the mare rings produced by collisions, that is, the dust and rubble behave like the liquid postulated by VAN DORN (1969) when the energies and mass are very large as in these collisions.

REMELTING

Table 1 summarizes some data bearing on the history of some Apollo 11 and Apollo 12 rocks and soils. The chemical compositions of the lunar surface materials, together with the model ages of some 4.4 to 4.7 aeons, show that an initial melting process must have occurred which produced the titaniferous basalts, the anorthosites, Luny 1 rock 12085 and rock 12013. Model ages show that this occurred early in lunar history by a general surface melting process. It is evident that the rubidium concentrations vary markedly and that the concentrations of some other elements, e.g., potassium, uranium, thorium, and barium, vary in a somewhat similar way indicating that a melting process has produced this variation. On the other hand, the strontium concentrations vary much less and are in fact nearly constant in all Apollo 11 samples. The Rb^{87}–Sr^{87} isochron ages dating the remelting processes vary between 3.2 and 4.0 aeons, and are distinctly younger than the model ages except in the case of the Apollo 11 A rocks. With this exception, it is evidently impossible to produce these rocks by a partial melting process, since the model age requires that the remelting in each case occurred in a closed system (see Table 1; also PAPANASTASSIOU *et al.*, 1970). It is necessary to assume that the individual layers were enclosed in surroundings that did not melt, and that the radioactive element concentrations within each layer supplied the necessary energy to produce melting. This is a very special arrangement which differs markedly from terrestrial situations where deep seated partial melting processes produce the surface igneous rocks.

The Apollo 11 A rocks present a special problem. The isochron and model ages require an increase in the rubidium/strontium ratio such as would result from partial melting processes. Their gross similarity to the Apollo 11 B rocks suggests that they may have started as this B type material, and in the melting process were enriched with trace elements, possibly extracted from the silicic type material, but have left behind about 80% of the radiogenic Sr^{87}. In order to supply the average initial Sr^{87}/Sr^{86} of these rocks, namely, 0.699315 (average of the initial (Sr^{87}/Sr^{86}) for Apollo 11 A rocks) in the time from 4.6 to 3.65 aeons ago, it would be necessary to have a (Rb^{87}/Sr^{86}) ratio of 0.0240, which is about twice that of the B rocks. COMPSTON *et al.* (1971), Murthy *et al.* (1971), and CLIFF *et al.* (1971) have presented similar data on the Apollo 11 and Apollo 12 ages. The COMPSTON *et al.* (1971) data are summarized in the last three columns of Table 1. Though some rocks may not represent exact closed systems, most of the rocks appear to be in this class, and for the most part are recognized as having model ages of approximately 4.5 aeons.

We have assumed a four-layer model to account for the reheating process, namely, a dust layer, an anorthositic layer, a layer consisting of silicic rocks and titaniferous basalt, and a "dunitic" layer. The abundances of K, Th, and U of the Apollo 11 B rocks are used for the titaniferous basalt layer, since we are suggesting that the Apollo 11 A rocks have picked up certain trace elements. We have used the rock 12013 data

Table 1. ^{87}Rb–^{87}Sr ages and average chemical

Element (ppm) or Age (aeons)	Apollo 11					
	Av. A	Av. B	Soil	Breccia	*	†
K	2752[a]	593[a]	1200[a]	1200[a]		
Th	3.404[c]	0.751[c]	2.092[c]	2.574[c]		
U	0.864[c]	0.209[c]	0.544[c]	0.674[c]		
Ba	326[a]	105[a]	175[a]	175[a]		
Rb	5.65[e]	0.70[e]	2.62[e]	3.67[e]	16.7[e]	
Sr	168[e]	166[e]	163[e]	163[e]	205[e]	
Isochron Age	3.65[e]	3.67[e]				4.25[e]
Model Age	3.88[e]	4.7[e]	4.59[e]	4.27[e]	4.45[e]	

* 10085 Luny 1.
† 10085 Leucocratic.
[a] GAST and HUBBARD (1970).
[b] WAKITA and SCHMITT (1970).
[c] TATSUMOTO (1970).
[d] TATSUMOTO *et al.* (1971).
[e] PAPANASTASSIOU *et al.* (1970). In calculating model ages from these data, Babi has been used.

for the silicic layer. We assume that K, U, Th abundances in the anorthositic and dunitic layers are not significant. The proportions of these layers have been chosen in accord with calculations of UREY and MACDONALD (1971), which are not markedly different from those of others, as follows: anorthosite, 9%; basalt, 1.4%; and "dunite," 89%. The silicic layer is taken both as one-tenth and one-fifth of the basaltic layer. The concentrations assumed for silicic and basaltic layers would add up to approximate chondritic concentrations for uranium and thorium. Since our data indicate a lower age (t_1) for the first melting than do the model lead ages of the soil, we have used 4.5 aeons for the time of the first melting. For fragmented rock, a thermal diffusivity of 0.002 cm^2/sec has been used, this being the mean of the data given by HORAI *et al.* (1970) for the Apollo 11 A and C rocks at 1000°K. The thermal conductivity for dust is taken as 10^{-5} cm^2/sec, which is somewhat larger than that reported by BASTIN *et al.* (1970), since some compaction must be present, and, in fact, our value may be too small for thick layers of dust. The temperature T_i for layer i is obtained from

$$\frac{\partial T_i}{\partial t} = \kappa_i \frac{\partial^2 T_i}{\partial x^2} + \sum_{j=1}^{4} q_{ij} \exp(-\lambda_j t), \qquad i = 1, 2, 3, 4$$

where the q_{ij} are the radioactive heating rates and j denotes the four radioactive isotopes K^{40}, U^{235}, U^{238}, and Th^{232}. Continuities of temperatures and heat fluxes are specified at the interfaces of each layer. Laplace transform and numerical inversion techniques used by LIU and WILLIAMS (1971) are employed. The results of calculations for the reheating of the crystalline rocks are shown in Fig. 1B for 100 and 1000 m thick dust layers. It can be seen that the remelting of silicic and basaltic layers is certainly possible if the dust layer is 1000 m thick and 0.28% of rock 12013 type material is used. Figure 1C shows a combination of the cooling and heating phases. Combination of the time required for the cooling and reheating produces just the observed time lag between the model and isochron ages of the crystalline rocks. Various modifications of the conditions will lead to different detailed timing of results.

composition of Apollo 11 and Apollo 12 samples.

Apollo 12					
12013	Av. 12002 12051	12070 Soil	Rocks[j]	Soil[j]	Breccia[j]
14000[b]		1900[i]	530	2600	1030
22 ± 8[b]		5.52[j]			
7.9 ± 2.7[b]		1.69[i]			
2560[f]		390[i]	64	435	125
47.4[f]	0.94[g]	6.34[g]	0.94	7.65	2.15
120[f]	114[g]	126[g]	113	153	116
4.0[h]	3.30[g]		2.8–3.6		
4.52[h]	4.52[g]	4.37[g]	4.4	4.5	4.22

[f] SCHNETZLER *et al.* (1970). Values for the "chip" are used.
[g] PAPANASTASSIOU and WASSERBURG (1970).
[h] ALBEE *et al.* (1970).
[i] WÄNKE *et al.* (1971).
[j] Data in these three columns are calculated or taken from COMPSTON *et al.* (1971). An initial Sr^{87}/Sr^{86} is taken as 0.69920.

It is evident that the combination of low thermal conductivities and higher radioactivities characteristic of the early times of lunar history lead to possible melting processes quite different from those of available earth history. This model for remelting appears to be just marginally able to account for the observations, but since ash flows exist in very limited areas of the moon, this is the required situation.

LEAD AGES

The lead ages have proved to be more difficult to understand (GAST *et al.*, 1970; SILVER, 1970; and TATSUMOTO, 1970) but they may be explained as being only partially closed systems relative to the addition of U and Th or retention of lead in the remelting process. It is assumed here that in the primitive melting process, most of the lead was lost, and that in the remelting process, only a fraction, δ, of the lead produced by the uranium and thorium in the rocks was retained. Then

$$\left(\frac{Pb^{206}}{Pb^{204}}\right) - \left(\frac{Pb^{206}}{Pb^{204}}\right)_I = \left(\frac{U^{238}}{Pb^{204}}\right)(\exp(\lambda_{238}t_1) - \exp(\lambda_{238}t_2))\delta + \frac{U^{238}}{Pb^{204}}(\exp(\lambda_{238}t_2) - 1)$$

gives the relation between the present ratios of (Pb^{206}/Pb^{204}) and (U^{238}/Pb^{204}) and the initial ratio $(Pb^{206}/Pb^{204})_I$ and the times of the first and second meltings, t_1 and t_2. Similar equations can be written for Pb^{207} and Pb^{208}. Assuming that t_2 is given by the Rb^{87}–Sr^{87} isochron age or the K^{40}–Ar^{40} ages as obtained by the Ar^{39}/Ar^{40} techniques, we have two unknowns, and we can solve these equations in pairs for these unknowns. Unfortunately, good rubidium/strontium isochron ages and good lead data appear to be available for only a limited number of rocks. We have used the equations for Pb^{206} and Pb^{207} for these calculations because both have U parents. The data used in securing the results given in Table 2 come from the work of TATSUMOTO (1970) and TATSUMOTO *et al.* (1971) for the uranium–lead data, of TURNER (1970, 1971) for the potassium-argon data, and of PAPANASTASSIOU *et al.* (1970), PAPANASTASSIOU and WASSERBURG (1970), and COMPSTON *et al.* (1971) for the rubidium-strontium data. The data for

Table 2. Solidification and remelting ages of Apollo 11 and Apollo 12 rocks.

Rock Sample	t_2 (aeons)	t_1 (aeons)	δ
10003a*	3.92	6.24	0.022
10003b	3.92	4.59	0.284
10057a	3.63	4.44	0.596
10057b	3.63	4.47	0.554
10071a	3.68	4.44	0.583
10071b	3.68	4.44	0.611
12021	3.28	4.61	0.284
12038	3.21	4.07	0.742
12052	3.24	4.33	0.478
12063 (average a & b)	3.23	4.46	0.427
12000 high average†	3.24	4.47	0.399
12000 low average†	3.24	4.53	0.378
Average‡		4.44	0.485

* Tatsumoto estimates the error in lead and uranium ratios is 6%. This accounts for the high value of t_1 and low δ.
† TATSUMOTO *et al.* data for Apollo 12 rocks 12021, 12052, and 12063 and the 12009, 12022, 12035, 12038, and 12064 were averaged and labeled high and low, respectively.
‡ Average 4.44 omitting 6.24, and 0.485 omitting 0.022.

12021 comes from CLIFF *et al.* (1971). The t_1 ages fall mostly between 4.4 and 4.6 aeons with a very high age for the 10003a sample with reported lead errors of 6%. It appears that the second meltings are not as closed systems for the lead as is the case for the rubidium-strontium in the Apollo 11 B and the Apollo 12 rocks. About half of the radiogenic lead produced in the time between t_1 and t_2 was lost from all these rocks during the remelting process, probably by evaporation to the soil and breccias as has been suggested by SILVER (1970), COMPSTON *et al.* (1971) and as specifically argued by HUEY *et al.* (1971). With a reduction level equivalent to metallic iron, lead oxide will be reduced to metallic lead at 1500°K, and the vapor pressure is about 0.025 atm., and thus, lead would be volatile. The use of the strontium age and the lead analyses gives no consistent initial date for rock 12013, and, hence, no firm statement in regard to closed system melting can be made for this rock from the lead evidence. If averages of all five analyses of TATSUMOTO *et al.* (1971) for this rock are used, an age of 4.6 aeons for t_1 may be secured if an uncertainty in the (Pb^{206}/Pb^{204}) datum of 4% is allowed for, i.e., this ratio should be 1338 instead of 1288. The latter is the average of data running from 415 to 2059. The lead in this rock seems to have moved away from its progenitors to variable degrees. Also, the remaining fractions of the radiogenic leads and of strontium must have been lost to some other material such as possibly the soil, and, in this case, the model ages of the soil may be too large.

CONCLUSIONS: THE MODEL

Figure 2 shows six successive stages which we envison as a model that might supply the conditions as outlined in preceding sections of this paper. *Stage 1.* The moon accumulated in a solid state at a sufficiently low temperature to enable it to remain sufficiently strong to support the mascons from the time they were formed until the present time. *Stage 2.* External sources of heat melted a layer 100–200 km deep, and

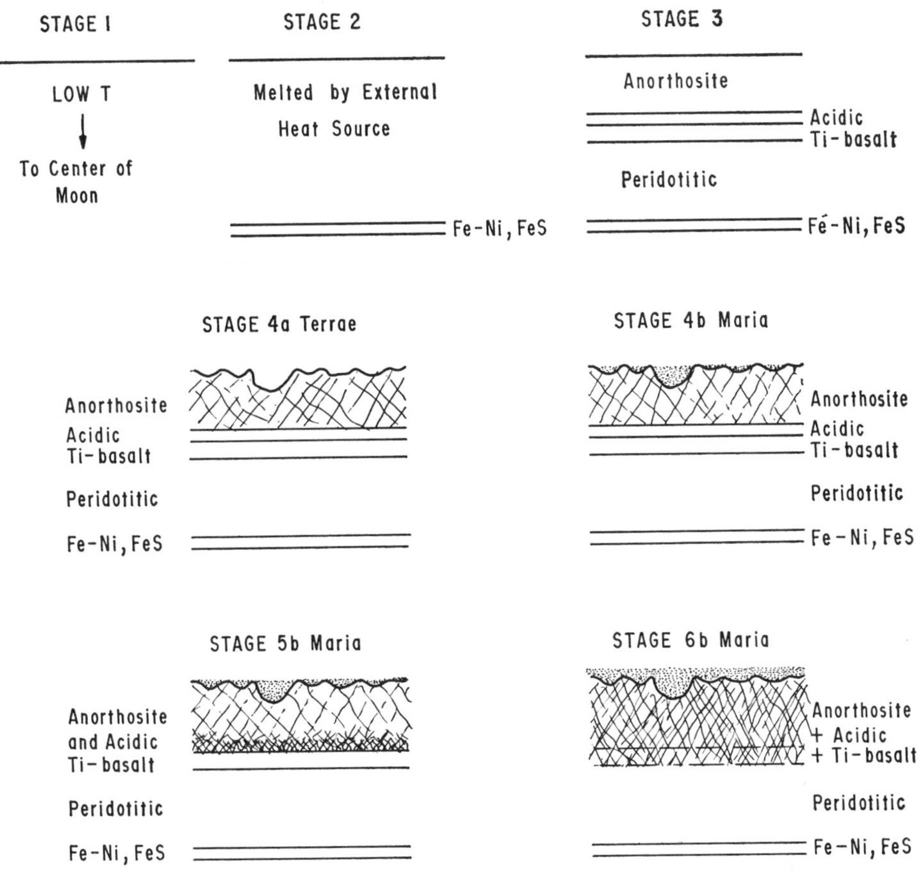

Fig. 2. Six stages in the surface development of the moon. The spaces are not proportioned to scale. The silicic rock is drawn as a definite layer to facilitate numerical calculations. It is probably interspersed in the anorthosite. (Acidic is used for silicic rock.)

some metallic iron-nickel and iron-sulfide sank carrying siderophile and chalcophile elements with them. The electrically higher conducting layer some 200 km below the lunar surface, as observed by Sonett *et al.* (1971), may be this metallic layer. *Stage 3.* The heat source declined and the surface layer differentiated during solidification, producing anorthosite, silicic material, titaniferous basalt, and "peridotitic" layers. *Stage 4a.* Terrae. Intense collisions occurred over the entire moon, but they were less intense in the terrae regions since they were less affected by the great collisions which produced the maria. Little change occurred in these terrae regions due to the intense mare collisions. *Stage 4b.* Maria. The shallow maria regions were bombarded in a similar degree as the terrae, but the great collisions that produced the circular mare broke up the anorthositic layer completely and deposited an insulating dust layer over parts of the surface. *Stage 5b.* Maria. The additional insulation caused the highly radioactive silicic rock to heat up and to melt 4.0 aeons ago, and this material moved upwards into cracks forming silicic dikes in the anorthositic material. The radiogenic lead of the rocks was volatilized and partially lost and separated from the uranium and thorium. *Stage 6b.* Maria. The blanket of dust on the surface and the heat generated partly by the silicic rock remelted the titaniferous basalt 3.3 to 3.65 aeons ago, and probably mixed with some gas, it erupted onto the surface producing the primitive soil and rocks. The lead was partially volatilized and escaped from the rocks probably going into the primitive soil. The surface soil later has acquired additional constituents due to collisions of other particles with highlands and rocks.

Other explanations for the remelting stage are possible. Collisional melting would be an excellent way to account for closed system melting. In this case, the mare collisions must be spread over long periods of time, and, hence, the colliding objects must be stored somewhere in the solar system. Possibly a catastrophic collision in the asteroidal belt could have supplied objects which arrived at the moon and earth during an aeon of time. Or possibly the accumulation of the earth required a long period of time and objects were swept up during some aeon of time. We have not favored these explanations, but they should be developed further.

Subsequent history has consisted of occasional collisions with objects of varying sizes and the development of fairly high temperatures in the deep interior. Some minor and local volcanic activity may exist today on the moon, resulting probably from very deep sources through an outer shell of considerable thickness which is able to support the mascons.

We are not able at this writing to account for the observed magnetic phenomena, which still appear to be difficult to understand.

Acknowledgments—We wish to acknowledge the help we have received from R. N. Clayton, G. W. Latham, S. K. Runcorn, N. Scheinin, C. Sonett, F. A. Williams, and J. A. Wood. We acknowledge support from NASA contract NAS 9–8107.

REFERENCES

ALBEE A. L., BURNETT D. S., CHODOS A. A., HAINES E. L., HUNEKE J. C., PAPANASTASSIOU D. A., PODOSEK F. A., PRICE G. R., and WASSERBURG G. J. (1970) Mineralogic and isotopic investigations on lunar rock 12013. *Earth Planet. Sci. Lett.* **9**, 137–163.

BASTIN J. A., CLEGG P. E., and FIELDER G. (1970) Infrared and thermal properties of lunar rock. *Proc. Apollo 11 Lunar Sci. Conf., Geochim. Cosmochim. Acta* Suppl. 1, Vol. 3, pp. 1987–1991. Pergamon.

CARSLAW H. S., and JAEGER J. C. (1959) *Conduction of heat in solids*, pp. 99–100, Clarendon.

CLIFF R. A., LEE-HU C., and WETHERILL G. W. (1971) Rb–Sr and U, Th–Pb measurements on Apollo 12 Material. Second Lunar Science Conference (unpublished proceedings).

COMPSTON W., BERRY H., VERNON M. J., and CHAPPELL B. W. (1971) Rubidium–strontium chronology and chemistry of lunar material from the Ocean of Storms. Second Lunar Science Conference (unpublished proceedings).

GAST P. W., and HUBBARD N. J. (1970) Abundance of alkali metals, alkaline and rare earths and strontium-87/strontium-86 ratios in lunar samples. *Science* **167**, 485–487.

GAST P. W., HUBBARD N. J., and WIESMANN H. (1970) Chemical composition and petrogenesis of basalts from Tranquillity Base. *Proc. Apollo 11 Lunar Sci. Conf., Geochim. Cosmochim. Acta* Suppl. 1, Vol 2, pp. 1143–1163. Pergamon.

GOLD T., and SOTER S. (1970) Apollo 12 seismic signal: Indication of a deep layer of powder. *Science* **169**, 1071–1075.

HORAI K., SIMMONS G., KANAMORI H., and WONES D. (1970) Thermal diffusivity, conductivity and thermal inertia of Apollo 11 lunar material. *Proc. Apollo 11 Lunar Sci. Conf., Geochim. Cosmochim. Acta* Suppl. 1, Vol. 3, pp. 2243–2249. Pergamon.

HUEY J. M., IHOCHI H., BLACK L. P., OSTIC R G., and KOHMAN T. P. (1971) Lead isotopes and volatile transfer in the lunar soil. Second Lunar Science Conference (unpublished proceedings).

LATHAM G. V., EWING M., PRESS F., SUTTON G., DORMAN J., NAKAMURA Y., TOKSOZ N., WIGGINS, R., DEER J., and DUENNEBIER F. (1970) Passive seismic experiment. *Science* **167**, 455–457.

LATHAM G. V., EWING M., PRESS F., SUTTON G., DORMAN J., NAKAMURA Y., TOKSOZ N., WIGGINS R., DEER J., and DUENNEBIER F. (1970) Apollo 11 passive seismic experiment. *Proc. Apollo 11 Lunar Sci. Conf., Geochim. Cosmochim. Acta* Suppl. 1, Vol. 3, pp. 2309–2320. Pergamon.

LATHAM G. V., EWING M., PRESS F., SUTTON G., DORMAN J., NAKAMURA Y., MEISSNER R., TOKSOZ N., DUENNEBIER F., KOVACH R., and LAMMLEIN D. (1971) Results from the Apollo 12 passive seismic experiment. Second Lunar Science Conference (unpublished proceedings).

LIU M. K., and WILLIAMS F. A. (1971) Heat conduction calculation for a model of the surface of the moon. *Int. J. Heat Mass Transfer* (in press).

MULLER P. M., and SJOGREN W. L. (1968) Mascons-lunar mass concentrations. *Science* **161**, 680–684.

MULLER P. M., and SJOGREN W. L. (1969) Consistency of lunar orbiter residuals with trajectory and local gravity effects. Plenary Meeting COSPAR, Prague, May 12.

MULLER P. M., and SJOGREN W. L. (1970) Private communication.

MURTHY V. R., EVENSEN N. M., JAHN B., and COSCIO M. R., JR., (1971) Rb–Sr isotopic relations and elemental abundances of K, Rb, Sr, and Ba in Apollo 11 and Apollo 12 samples. Second Lunar Science Conference (unpublished proceedings).

OSTIC R. G. (1965) Physical condition in gaseous spheres. *Mon. Not. R. Astro. Soc.*, **131**, 191–197.

PAPANASTASSIOU D. A., and WASSERBURG G. J. (1970) Rb–Sr ages from the Ocean of Storms. *Earth Planet. Sci. Lett.* **8**, 269–278.

PAPANASTASSIOU D. A., WASSERBURG G. J., and BURNETT D. S. (1970) Rb–Sr ages of lunar rocks from the Sea of Tranquillity. *Earth and Planet. Sci. Lett.* **8**, 1–19.

RINGWOOD A. E., and ESSENE E. (1970) Petrogenesis of Apollo 11 basalts, internal constitution and origin of the moon. *Proc. Apollo 11 Lunar Sci. Conf., Geochim. Cosmochim. Acta* Suppl. 1, Vol. 1, pp. 769–799. Pergamon.

SCHNETZLER C. C., PHILPOTTS J. A., and BOTTINO M. L. (1970) Li, K, Rb, Sr, Ba, and rare-earth concentrations, and Rb–Sr age of lunar rock 12013. *Earth Planet. Sci. Lett.* **9**, 185–192.

SILVER L. T. (1970) Uranium–thorium–lead isotopes in some Tranquillity Base samples and their implications for lunar history. *Proc. Apollo 11 Lunar Sci. Conf., Geochim. Cosmochim. Acta* Suppl. 1, Vol. 2, pp. 1533–1574. Pergamon.

SONETT C. P., SMITH B. F., COLBURN D. S., SCHUBERT G., SCHWARTZ K., DYAL P., and PARKIN C. W. (1971) The lunar electrical conductivity profile: Mantle-core stratification, near surface thermal gradient, heat flux and composition. Second Lunar Science Conference (unpublished proceedings).

TATSUMOTO M., and ROSHOLT J. N. (1970) Age of the moon: An isotopic study of uranium–thorium–lead systematics of lunar samples. *Science* **167**, 461–463.

TATSUMOTO M. (1970) Age of the moon: An isotopic study of U–Th–Pb systematics of Apollo 11 lunar samples. *Proc. Apollo 11 Lunar Sci. Conf., Geochim. Cosmochim. Acta* Suppl 1, Vol. 2, pp. 1595–1612. Pergamon.

TATSUMOTO M., KNIGHT R. J., and DOE B. R. (1971) U–Th–Pb systematics of Apollo 12 lunar samples. Second Lunar Science Conference (unpublished proceedings).

TURNER G. (1970) Argon-40/Argon-39 dating of lunar rock samples. *Proc. Apollo 11 Lunar Sci. Conf., Geochim. Acta* Suppl. 1, Vol. 2, pp. 1665–1684. Pergamon.

TURNER G. (1971) ^{40}Ar–^{39}Ar ages from the lunar maria. Second Lunar Science Conference (unpublished proceedings).

UREY H. C. (1958) The early history of the solar system as indicated by the meteorites. Hugo Müller Lecture, March, *Proc. Chem. Soc. of London*, 67–78.

UREY H. C. (1966) Chemical evidence relative to the origin of the solar system. *Mon. Not. R. Astro. Soc.* **131**, 199–223.

UREY H. C., and MARTI K. (1968) Surveyor results and the composition of the moon. *Science* **161**, 1030–1032.

UREY H. C., and MACDONALD G. J. F. (1971) *Physics and Astronomy of the Moon.* (Editor Z. Kopal), p. 263, Academic Press.

VAN DORN W. G. (1969) Lunar maria: Structure and evolution. *Science* **165**, 693–695.

WAKITA H., and SCHMITT R. A. (1970) Elemental abundances in seven fragments from lunar rock 12013. *Earth Planet. Sci. Lett.* **9**, 169–176.

WÄNKE H., WLOTZKA F., TESCHKE F., BADDENHAUSEN H., SPETTEL B., BALACESCU A., QUIJANO-RICO M., JAGOUTZ E., and RIEDER R. (1971) Major and trace elements in Apollo 12 samples and studies on lunar metallic iron particles. Second Lunar Science Conference (unpublished proceedings).

WOOD J. A. (1970) Petrology of the lunar soil and geophysical implications. *J. Geophys. Res.* **75**, 6497–6513.

Proceedings of the Second Lunar Science Conference, Volume 2, pp. 999–1020.
The M.I.T. Press, 1971.

Chemical composition and origin of nonmare lunar basalts

NORMAN J. HUBBARD and PAUL W. GAST
NASA Manned Spacecraft Center
Houston, Texas 77058

(*Received 26 February 1971; accepted in revised form 7 April 1971*)

Abstract—A group of nonmare lunar basalts found at the Apollo 11 and Apollo 12 sites differs from mare lunar basalts by having chemical compositions quite similar to tholeiitic and alkalic earth basalts with high Al_2O_3 concentrations. These nonmare basalts differ from earth basalts by having much lower Na_2O concentrations, high concentrations of Yb and Lu and huge negative Eu anomalies. Such nonmare basalts are abundant (30–70% of the total) in the Apollo 12 regolith and less abundant ($\sim$10% or less) in the Apollo 11 regolith. The Apollo 12 regolith consists almost entirely of mare and nonmare basalts. The Apollo 11 regolith consists of mare and nonmare basalts, plus $\sim$20% of anorthosite. The nonmare basalts have $\sim$4.5 b.y. Rb–Sr model ages. Their old model ages, coupled with their high concentration of Rb, U, and Th, are responsible for the old model ages of the Apollo 11 and Apollo 12 soils. The chemical composition of the nonmare basalts suggests that the moon was extensively differentiated before 4.4 b.y. ago.

INTRODUCTION

GEOLOGIC SAMPLES FROM THE first two manned lunar landings and the first samples returned by an automated station (Luna 16) indicate that the mare regolith surfaces of the moon are probably underlain by ferro-magnesium igneous rocks that have many of the chemical, mineralogical, and textural characteristics of terrestrial basalts. Even though they can be readily distinguished from all terrestrial basalts by their high Fe and low Na content (certain trace element characteristics are even more distinctive), they are quite properly designated by the term, basalt. If, for example, the Apollo 12 boulders and pebbles had been associated with a thick basaltic flow or a dike in a terrestrial volcanic terrain, it is quite unlikely that they would have been designated by any other generic term. The chemical compositions of the mare basalts will not be examined in detail in this paper. However, several salient chemical characteristics of both the mare basalts from the moon and selected terrestrial basalts will be reviewed so that subsequent discussion may be placed in context.

The principal objective of this paper is the further characterization of the soil and individual fragments taken from the soil at the Apollo 12 site. In an earlier paper (HUBBARD *et al.*, 1971) we have proposed that a significant portion of the soil at the Apollo 12 site was derived from beneath the Oceanus Procellarum mare flows or from outside the mare regions. We inferred there that one particular class of fragments, a material which we designated KREEP (from its relatively high K, REE, and P content) was a common component of the lunar highlands. This hypothesis will be further elucidated in this paper.

CHEMICAL CHARACTERISTICS OF MARE BASALTS

The igneous rocks from the Apollo 11, Apollo 12, and Luna 16 sites are readily distinguished from nearly all terrestrial basalts by their chemical composition. Some of the differences are illustrated in Figs. 1–4, where the range of concentrations of CaO, Na_2O, FeO, and TiO_2 in terrestrial and lunar basalts are compared. The compilation of nearly 2,000 analyses of terrestrial basaltic rocks prepared by Manson (1967) is here used as a characterization of terrestrial basalts. It is quite clear that the total iron content and sodium content of lunar mare basalts is unique. Conversely, the calcium content for lunar basalts is essentially identical to that of terrestrial basalts.

The abundance of calcium and aluminum in basaltic rocks is of considerable interest because the distribution of these two elements between co-existing mafic phases is significantly dependent on pressure. At low pressures, most calcium and aluminum in peridotite assemblages occur as plagioclase (see, for example, Green and Ringwood, 1967), i.e., Ca and Al occur in a single phase. Conversely, at intermediate and high pressures, calcium and aluminum occur in different phases, particularly where spinel is stable. In this situation, calcium occurs in diopside quite separated from aluminum. At this point, we should note that Ahrens and Michaelis (1969) have shown that calcium and aluminum are remarkably coherent in almost all types of meteorites. It has been suggested (Gast, 1968a, b; Ahrens and Michaelis, 1969) that calcium and aluminum should be equally fractionated from magnesium, iron, and silicon in chemical processes controlled by volatility of the elements. From this we may infer that it is unlikely that the primitive Ca/Al ratio of a planet should be different from that of chondrites. In particular, we suggest that the primitive mantles of the moon and the earth have or have had Ca/Al ratios similar to those of meteorites. Given this assumption, we can interpret deviations from this ratio in basaltic rocks in terms of magma forming processes. The CaO and Al_2O_3 contents of two very distinct groups of basalts, that is the oceanic ridge basalts and two groups of nepheline normative basalts, are shown in Fig. 5. The Ca and Al contents of these rocks are very clearly separated by the meteoritic Ca/Al ratio.

Data for both the Apollo 11 and Apollo 12 igneous rocks are also plotted in this diagram. The Apollo 11 and Apollo 12 basalts are readily distinguished from each other. Furthermore, both the variation within each group and the fractionation with respect to the chondritic Ca/Al ratio indicate that the lunar igneous liquids are separated from a primitive composition by at least one step of igneous fractionation. Two theoretical fractionation lines are shown in Fig. 5. The first, showing the effect of addition or subtraction of calcium rich plagioclase, falls along the trend observed for oceanic ridge basalts. Kay *et al.* (1970) have made a similar observation and suggest that removal of plagioclase can account for much of the variation in the aluminum content of these basalts. The fractionation line for aluminum rich diopside is almost at right angles to that shown for plagioclase. From the position of this line, one notes that the variation in Ca and Al contents of the undersaturated basalts can be explained by the proportion of calcium rich and aluminum poor minerals that go to form the liquids from which these rocks crystallized. The residue left behind by removing these liquids clearly should be aluminum rich. Lunar basalts from both Apollo 11 and Apollo 12 are only slightly enriched in Ca over Al, except for KREEP, where Al is much greater than Ca. It is, however, quite likely that clinopyroxene was an important phase in the formation of the lunar mare basalt liquids, whereas plagio-

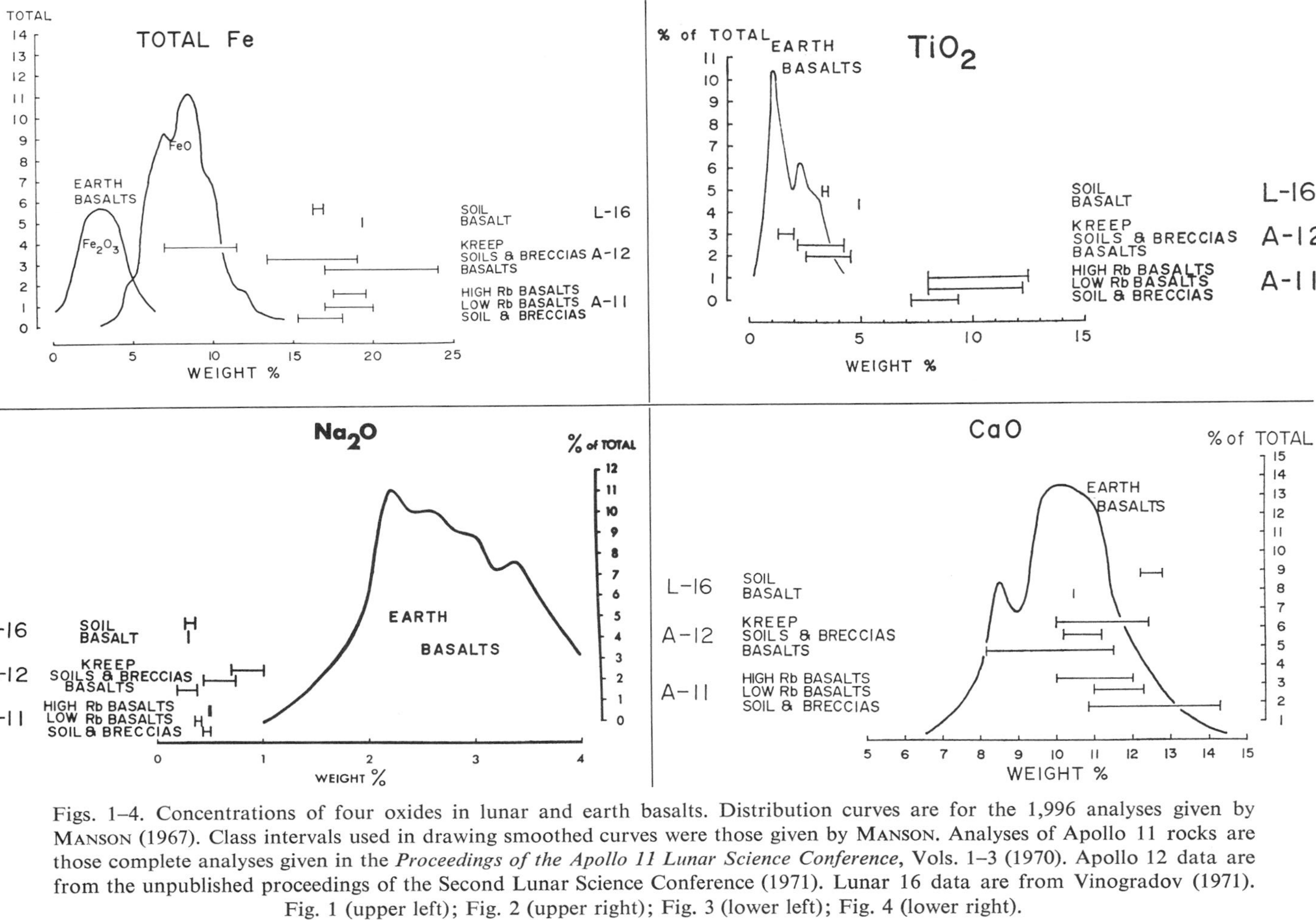

Figs. 1–4. Concentrations of four oxides in lunar and earth basalts. Distribution curves are for the 1,996 analyses given by MANSON (1967). Class intervals used in drawing smoothed curves were those given by MANSON. Analyses of Apollo 11 rocks are those complete analyses given in the *Proceedings of the Apollo 11 Lunar Science Conference*, Vols. 1–3 (1970). Apollo 12 data are from the unpublished proceedings of the Second Lunar Science Conference (1971). Lunar 16 data are from Vinogradov (1971). Fig. 1 (upper left); Fig. 2 (upper right); Fig. 3 (lower left); Fig. 4 (lower right).

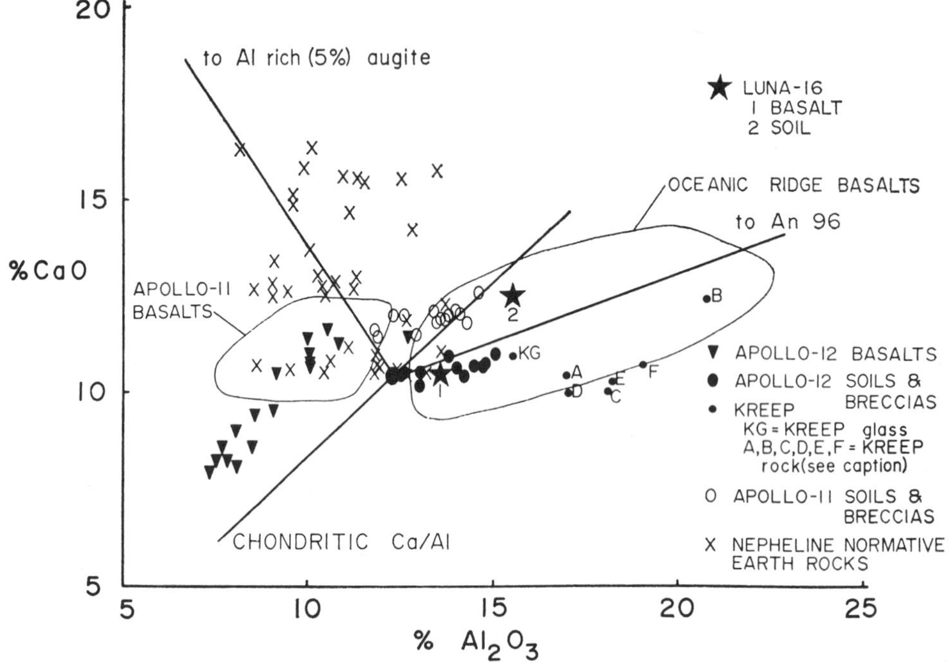

Fig. 5. CaO and Al_2O_3 contents of lunar and terrestrial basalts. Sources for the lunar basalts are those given in Figs. 1–4. Oceanic ridge basalt data is a summary of most published analyses available by September 1970. Undersaturated rocks (nepheline normative) include the analyses given by SPENCER (1969), a series of unpublished analyses from the post-erosional basalts of Oahu and Kauai (unpublished data), and averages for several rock types given by NOCKOLDS (1954).

clase is also a likely important phase in the nonmare (KREEP) basalts. In any of the above cases, more than one Ca or Al phase may be involved in the formation of a basaltic liquid. The differences cited here indicate the degree to which one phase dominated over other phases during melting or during fractional crystallization. Thus, the data cited here cannot be used to exclude the presence of specific Ca and Al bearing phases—for example, plagioclase—that may be in contact with the lunar liquids.

Abundances of REE and other lithophile trace elements further distinguish terrestrial and lunar basalts. In fact, these same elements also distinguish different groups of terrestrial and lunar basalts from each other. Chondrite normalized abundance patterns for U, Ba, and REE in lunar basaltic rocks are shown in Fig. 6. We have used only stable isotope dilution data for REE and Ba reported by PHILPOTTS and SCHNETZLER (1970), GAST and HUBBARD (1970), and in this paper in constructing this figure so that intralaboratory systematic errors are not attributed to the lunar samples. Uranium data are also from isotope dilution analyses. The chondrite concentrations used for normalization of the REE were determined by identical procedures.

Several representative abundance patterns for a series of terrestrial basalts are

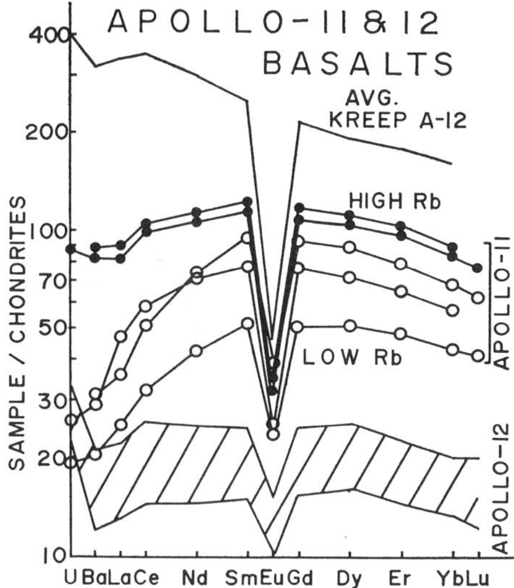

Fig. 6. Rare earth, Ba, and U data for lunar basalts. Rare earth and Ba data are from this report and PHILPOTTS *et al.* (1971). Uranium data are from TATSUMOTO and KNIGHT; SILVER; and CLIFF *et al.* (all 1971). Chondrite values used for normalization have been newly determined by us using the same spike solutions and methods and for the lunar data, except for U, where we have used a semiarbitrary value of 0.01 ppm.

shown in Fig. 7. The data shown in these illustrations illustrate a dramatic difference between terrestrial and lunar basalts; that is, the fractionation of light vs. heavy rare earths is much greater in many terrestrial basalts. The abundance of the heavy REE is relatively constant in terrestrial samples. In contrast, the concentration of heavy REE varies markedly in lunar basalts (Fig. 6). Secondly, we note that the fractionation of Eu from neighboring REE is small or unobserved in terrestrial basalts. In contrast, lunar basaltic samples show consistent fractionations of both Sr and Eu from the trivalent REE. The magnitude of the Eu and Sr anomalies in the lunar basalts increases regularly with increasing total REE content (Fig. 6).

The major element compositions, combined with lithophile element abundance patterns, distinguish three chemical groups among the mare basalts. They are: (1) the high Fe, intermediate Ti, low K, Ba, La, and U basalts found at the Apollo 12 and Lunar 16 sites; (2) the high Fe, high Ti, low K, Ba, La, and U basalts from the Apollo 11 site; and (3) the high Fe, high Ti, high K, Ba, La, and U basalts from the Apollo 11 site. The distinctly different abundance patterns of these three types are clearly evidence in Fig. 6.

POSSIBLE NONMARE ROCKS FOUND IN THE LUNAR REGOLITH

We have assumed that the hand specimen-size samples of basaltic rocks returned from the Apollo 11 and Apollo 12 sites and the pebble that apparently terminated the drilling at the Lunar 16 site (Vinogradov, 1971) were derived from an underlying

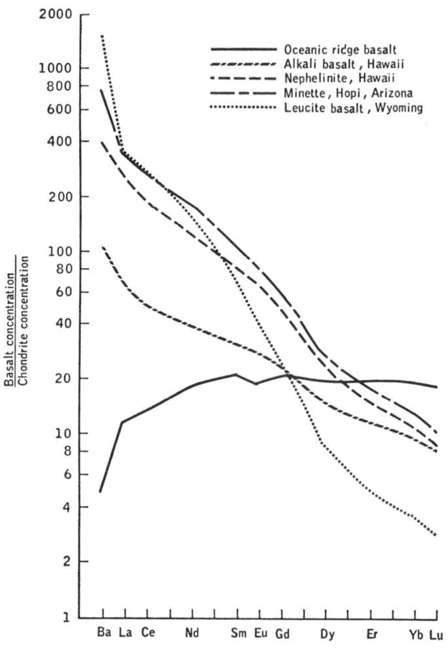

Fig. 7. Chondrite normalized abundances of REE and Ba. Data are taken from KAY (written communication) and KAY *et al.* (1970). There is a regular progression from a horizontal (La and Ba depleted) pattern for the oceanic ridge basalts to highly fractionated, Ba-La enriched, and heavy REE depleted patterns for undersaturated rocks. The latter rocks are generally rich in trace elements, e.g., U, Pb, Nb, and P_2O_5, when compared with the oceanic ridge basalts.

basaltic bedrock. Morphological and statistical studies of many small and medium sized lunar craters (OBERBECK and QUAIDE, 1968) strongly support this assumption. The cratering process that produced the lunar regolith has provided a mechanism for introducing material from outside the local environment into the regolith. It is clear, for example, from most whole-moon photographs that some material associated with ray craters must have been transported for distances up to several hundred kilometers. It is furthermore quite possible that much of the ray material from craters like Kepler and Copernicus could have been derived from a premare lunar crust.

The exact thickness of mare basalts at the Apollo 12 site is not well known. However, the incomplete submergence of moderate size craters suggests that some mare regions have rather thin covers of basalt flows—perhaps less than several hundred meters thick. In these regions, postmare craters that are only a few kilometers in diameter may penetrate the mare flows and eject submare material onto the mare surface. The distribution of premare material in the postmare regolith is thus dependent on the distribution of large craters and on the variation in thickness of the mare flows. In this case, it seems quite likely that the proportion of premare regolith fluctuates greatly from site to site. Nevertheless, the probability of incurring premare materials in any given postmare regolith sample appears to be significant.

Comparison of the soil or regolith chemical composition with that of the igneous rocks at the three sites where mare regolith samples have been investigated provides strong evidence that the soil contains material that has not been derived from the underlying basalt. The characteristics of this foreign component in the Apollo 11 soil have been investigated by mass balance calculations (see, for example, GOLES *et al.*, 1970, and COMPSTON *et al.*, 1970) and by direct examination of individual particles in the regolith (see, for example, WOOD *et al.*, 1970, and ALBEE and CHODOS, 1970).

Preliminary chemical information for the rocks and soil samples at the Apollo 12 site (LSPET, 1970) clearly demonstrated that the Apollo 12 soil and breccia must contain significant proportions of materials quite different from the presumed bedrock. This observation was amplified in the detailed studies of rock 12013 (All authors, *Earth Planet. Sci. Lett.* **9**, no. 2, 1970). It was suggested that the characteristics of the dark portion of this rock were those of the material that might account for the differences between the soil and the igneous rocks (HUBBARD *et al.*, 1970, and SCHNETZLER *et al.*, 1970). Isotope data on soil and igneous rocks (PAPANASTASSIOU and WASSERBURG, 1970) further demonstrated the existence of a foreign component in the Apollo 12 soil. FUCHS (1970) and MEYER and HUBBARD (1970) noted that the Apollo 12 soil contained material in many ways similar to the dark portion of rock 12013. Subsequently, it was suggested that the chemical composition of many rather common fragments in the Apollo 12 soil were, indeed, similar to those of the dark portion of rock 12013 (HUBBARD *et al.*, 1971).

The morphology and mineralogy of these fragments have been described in some detail by a large number of investigators (KING *et al.*; MCKAY *et al.*; MEYER *et al.*; QUAIDE *et al.*; WOOD *et al.*; and ANDERSON *et al.*; all 1971). These fragments are distinguished by the occurrence of a calcium-poor pyroxene which has, in several cases, been identified as an orthopyroxene. The Fe content ranges from that of hypersthene to ferro-hypersthene. The hypersthene bearing fragments range from brecciated holocrystalline to fragments that are almost entirely glass. The totally glassy fragments can be distinguished by their yellow-brown color and ropy convoluted morphology. Most of the soil at the Apollo 12 site contains from 20 to 30% of the brecciated crystalline fragments. The abundance of the glassy fragments is rather variable. In one sample, 12033 coarse fines, ropy yellow-brown glass makes up about 50% of the sample. The abundance of the hypersthene-bearing fragments also varies with depth in the double core tube from this site (MCKAY *et al.*, 1971). The chemical composition of both individual glass and crystalline fragments have been made by most of the investigators listed above. Averages reported by individual investigators are summarized in Table 1. The range of compositions observed for the yellow-brown glasses and hypersthene-bearing crystalline fragments are quite similar to those of some rather common terrestrial basalts, as shown in Table 1.

Among the major elements, the low Na content of the fragments is the only feature that distinguishes them chemically from these, and all other, terrestrial basalts. The composition of the fragments summarized in Table 1 is clearly distinguished from that of the mare basalts by their high Al/Ca ratio. This is illustrated in Fig. 5. In addition, both the yellow-brown glassy fragments and the hypersthene bearing lithic fragments have consistently lower iron contents than the mare igneous rocks. The iron contents (cf. Fig. 4) of these fragments are, in fact, quite comparable to those of terrestrial basalts. The textures of the hypersthene-bearing materials clearly indicate that virtually all of the crystalline fragments have undergone brecciation and welding. Quantitative mineralogical compositions are difficult to obtain from many of the fragments that have been observed. WOOD *et al.* (1971) give an average composition—57% plagioclase, 36% pyroxene plus olivine, and 6% glass.

MCKAY *et al.* (1971) have analyzed 250 individual particles ranging from 1 to 75 μ in diameter adhering to larger hypersthene bearing fragments and derive an average composition consisting of 42% plagioclase, 17% clinopyroxene, 13% orthopyroxene, and 17% glass. The mineralogical composition of the hypersthene-bearing fragments and associated glass fragments is thus quite different from that of the mare basalts both in the abundance of orthopyroxene and in the predominance of plagioclase over mafic minerals. In the absence of general terrestrial, mineralogical, or genetic criteria that can describe the distinctive features of such a diverse set of particles, we have adopted the acronym, KREEP, to designate the plagioclase hypersthene basalts and chemically similar glass fragments. It is clear from the abundant descriptions and major element analyses now available that the crystalline fragments designated norite by WOOD *et al.* (1971) and the ropy yellow-brown high alkali glasses share the distinctive chemical characteristics that were used to define KREEP (HUBBARD *et al.*, 1971). We suggest that the dark portions of rock 12013 and Luny Rock 1 also fall within this classification.

We recognize the possibility that there may be recognizable chemical differences between the various materials that are here grouped together. For example, the major element analyses for individual brecciated fragments shown in Table 1 con-

Table 1. Chemical comparison of lunar and earth basalts.

	Average Lunar Basalts (Wet Chemical and XRF Data)			Assorted KREEP Materials (Microprobe Data)										
	Apollo 11* Low Rb	Apollo 11* High Rb	Apollo 12	Average Brown Glass	Hypersthene: Bearing Brecciated Crystalline						Oceanic† Alkali Basalt	Hawaiian‡ Tholeiite	Hawaiian‡ Alkalic Basalt	Myvatn§ Iceland
					A	B	C	D	E	F				
SiO_2	40.58	40.30	45.03	49.0	48.4	46.6	49.3	45	48.7	48	48.2	49.4	46.5	48.8
Al_2O_3	10.19	8.72	9.17	15.0	16.9	20.7	18.1	17	19.7	19	16.5	14.0	14.6	16.0
TiO_2	10.23	11.49	3.36	2.0	1.6	1.7	1.7	2	1.3	1.6	2.9	2.5	3.0	1.2
FeO	19.03	19.55	21.60	11.5	11.6	6.9	9.3	9	9.3	9	11.3	11.2	12.0	9.5
MgO	6.92	7.64	10.89	8.0	9.8	9.5	7.6	9.5	7.4	8	5.3	8.4	8.2	7.0
CaO	11.55	10.45	9.81	11.0	0.4	12.4	10.1	10	10.5	10.5	9.1	10.3	10.3	12.5
Na_2O	0.41	0.52	0.29	0.8	0.9	0.7	1.0	1.0	1.0	1.0	3.7	2.1	2.9	2.1
P_2O_5	0.09	0.18	0.10	0.6	0.5	0.2	1.5	1.0	0.9	0.8	0.48	0.4	0.8	0.8
K_2O	0.08	0.30	0.058	0.8	0.5	0.3	1.1	0.8	0.9	0.9	1.9	0.3	0.4	0.3

* Analysts are Chappell, Peck, Maxwell, Engel, Scoon, and Wiik. Data are from the Moon Issue, Vol. 167, No. 3918 (1970).

(A) Average of 7 analyses of "Norite," Keil *et al.* (1971a);‖
(B) Average of 8 analyses of anorthositic basalt, Keil *et al.* (1971a);‖
(C) Average of 6 analyses of "Norite," Wood *et al.* (1971);
(D) Average of 7 analyses of gray mottled fragments, Anderson *et al.* (1971);
(E) Average of 11 analyses of brecciated crystalline KREEP, Meyer *et al.* (1971);
(F) Overall average of brecciated crystalline KREEP.

† ENGEL and ENGEL (1970);
‡ MACDONALD and KATSURA (1964);
§ SIGVALDASON (1969).
‖ These analysis are grouped differently in Keil et al. (1971b).

sistently have higher Al_2O_3 contents than the average for all glass fragments. The dispersion in Al_2O_3 contents in the holocrystalline fragments also appears to exceed that of the glass fragments. It may be argued that this indicates that the glass may be contaminated with some mare basalt. Alternatively, the glasses may be derived from a more restricted range of nonmare basalts. We suggest that such differences are best described by considering distinguishable groups of fragments as subsets of a larger set of objects identified by the KREEP chemistry. The mixing models discussed below suggest that the subset consisting of glass fragments is a better approximation of the end member chemical composition than the Al rich holocrystalline fragments.

If further studies support the proposition that the KREEP chemical characteristics are those of characteristic highland intrusive or extrusive rock, we would suggest the term highland basalt in place of KREEP to designate this material.

Lithophile Trace Elements in Soil Samples and Kreep Fragments

The concentrations of Ba, Sr, K, Rb, Na, and the REE in Apollo 12 soil samples 12070, 12040, and 12033 (coarser than 1 mm) and breccia sample 12073 have been determined by the stable isotope dilution method. These elements have also been determined in a group of millimeter-sized fragments selected from soil samples 12003 and 12033. Results of these analyses are given in Table 2 along with previously published results for dark material from rock 12013. Chondrite normalized abundance patterns for the soil and fragment samples listed in Table 2, along with data for related samples analyzed by other investigators, are shown in Fig. 9. The individual fragments analyzed in this study were selected from a larger collection which had previously been analyzed for their major element composition and characterized in terms of texture and mineral composition. They are typical of the ropy glass and hypersthene bearing fragments in the Apollo 12 soil. We conclude from the data shown in Fig. 9 and Table 2 that these fragments consistently have unusually high abundances of Rb, Ba, and REE along with the high abundance of K, P, and Na that was noted previously and that the abundance of these lithophile elements in the KREEP fragments and in the dark portion of rock 12013 are essentially identical. The similarity of the abundance patterns in individual fragments and in soil samples from the Apollo 12 site indicates that a significant portion of this soil could have been derived from material similar in composition to the KREEP fragments. In addition, if these fragments were also derived from a preexisting nonmare basalt, as we have proposed, they indicate that highly differentiated igneous liquids could have been quite common on the pre-mare lunar surface.

Many of the trace element characteristics of this liquid are similar to those of many terrestrial basalts. For example, the REE are fractionated from each other so that the light REE (Ce through Sm) are much enriched relative to the heavy REE as is the case for alkalic, Hawaiian tholeiitic, and many Icelandic basalts. (See Table 1 for major element compositions of representative terrestrial basalts.) However, when viewed in the context of all lunar materials, the REE abundances in KREEP show that the heavy REE vary by more than a factor of 10 in lunar basalts. In contrast, the concentration of these elements in terrestrial basalts varies by less than a factor

Table 2.

Element	12002,124 Basalt 251 mg	12020,32 Basalt 259 mg	12022,38 Basalt 188 mg	12035,16 Basalt 141 mg	12051,40 Basalt 203 mg	12053,32 Basalt 202 mg	12073,28 Fragment Basalt 8.0 mg	12075,11 Basalt 208 mg	12044,10 Fines Soil 141 mg	12070,04 Fines Soil 200 mg
La ppm	6.02			3.87	6.53	7.32	—	6.34	—	30.3
Ce ppm	17.0	16.1	17.4	11.5	19.2	20.9	22.4	16.1	91.6	90.0
Nd ppm	12.3	12.0	14.4	8.91	15.4	15.3	17.2	11.6	54.8	55.1
Sm ppm	4.24	4.50	5.38	3.22	5.68	5.25	5.85	3.94	16.7	16.4
Eu ppm	0.853	0.839	1.26	0.751	1.23	1.10	1.10	0.828	1.72	1.74
Gd ppm	5.65	5.43	7.71	4.30	7.89	—	7.8	5.30	19.2	—
Dy ppm	6.34	6.13	9.37	5.07	9.05	8.01	8.70	6.22	21.2	21.7
Er ppm	3.89	3.75	5.42	3.09	5.57	4.88	5.29	3.73	13.3	13.5
Yb ppm	3.78	3.69	5.69	3.04	5.46	4.71	5.18	3.71	12.9	13.0
Lu ppm	—			0.423	—	0.667	0.724	0.508	1.67	1.78
Ba ppm	67.2	64.4		47.2	73.6	84.4	83.1	63.9	380	404
Sr ppm	101	93.6	143	84.3	148	130	105	94.3	157	150
$Sr^{87/86}$*	0.7010		0.7002	0.7001				0.7011	0.7073	0.7078
	±0.0003		±0.0003	±0.0003				±0.0003	±0.0003	±0.0003
Rb ppm	1.04	0.997	0.738	0.689	0.909	1.24	1.89	0.993	6.08	6.46
K ppm	476	468	536	363	530	583	753	457	2,110	2,150
Na %	0.16	0.14	0.18	—	0.23	0.20	—	—	0.36	0.34
K/Rb	458	469	728	527	583	470	398	460	347	333

* Errors listed are approximately 2σ limits.

of 3. In fact, the high abundance of Yb and Lu in KREEP clearly distinguishes this basalt from any terrestrial basalt. The depletion of Eu and Sr relative to Sm and Gd in the lunar basalts is also in strong contrast with the abundance of these elements in terrestrial basalts. In fact, the Eu anomalies in these particular basalts are greater than those observed for any other lunar sample. It is also of interest that the ratio of Sr/Eu varies regularly with the degree of Sr and Eu separation from the trivalent ions (Fig. 8). This ratio varies from about 115 to 60 in going from the Apollo 12 mare basalts to KREEP. We infer from this that the distribution coefficients between the main Sr and Eu containing mineral in the moon and the basaltic liquids in contact with this mineral favor Sr^{++} over Eu^{++} in the mineral; that is, if this mineral is assumed to be plagioclase, $K^{plag/liq}$ is greater for Sr^{++} than Eu^{++}.

We also note that La and Ba were consistently depleted relative to a smooth extrapolation from the lighter REE (Figs. 6 and 9). The Ba and La depletions range from 20 to 40 percent. This anomaly could result from several factors. It may, in part, be due to the semi-arbitrary selection of a chondritic Ba concentration. It may also be further evidence that plagioclase was at one time in equilibrium with these liquids. Previous calculations (GAST and HUBBARD, 1970) suggest that plausible distribution coefficients for plagioclase and pyroxene predict such a fractionation for Ba. Finally, we also note that Ba is strongly enriched in the silica rich immiscible liquid reported in lunar basalts by ROEDDER and WEIBLEN (1970). Separation of such an immiscible phase from a parent liquid could also account for the depressed Ba abundance. At the present time, we have no explanation of the depressed La abundance shown in Figs. 6 and 9.

Analytical data.

12033,70 Coarse‡ Fines Soil 132 mg	12073,28 Breccia 126 mg	12033,73MH Fragment Breccia 3.6 mg	12003,B1 KREEP glass 4.3 mg	12003,S.A. KREEP crystalline 3.8 mg	12003,A. 8.3 mg	12033,97,1A KREEP glass 6.5 mg	12033,97,2A KREEP agglutinate 11.8 mg	12013,10-05 Dark Material 41.5 mg	CCS-3 Chondrite composite 500 mg
69.7	48.4	55.0	—	134	108	105	93.4	98.7	0.325
180	127	151	265	360	275	265	243	267	0.798
108	76.7	89.2	155	230	155	149	138	156	0.567
31.0	22.4	26.0	44.5	61.7	42.4	40.5	38.2	44.7	0.186
2.79	2.17	3.46	3.67	3.40	2.59	3.11	3.11	3.38	0.0692
35.7	26.9	30.0	50.8	73.2	49.0	50.0	45.1	—	0.255
39.2	28.6	31.5	57.0	74.8	55.3	—	—	55.8	0.305
25.1	18.9	19.7	36.0	43.9	—	—	—	35.1	0.209
24.0	17.4	19.5	35.9	39.1	—	—	—	34.4	0.231
—	—	2.76	—	—	—	—	—	4.66	0.0349
858	571	723	1,365	851	602	1,207	1,110	1,162	3.8
185	163	233	216	192	168	195	198	199	11.0
0.7126 ±0.0003	0.7094 ±0.0003	0.707 ±0.0005 −0.0015	0.716 ±0.001 −0.003	0.714 ±0.0008	0.7106 ±0.0017	0.7166 ±0.0007	0.7139 ±0.0004	0.710 ±0.0005	
14.4	9.26	8.79	20.0	14.6	11.4	18.3	17.1	9.80	2.30
4,697	3,180	2,910	7.030	4,740	3,600	6,490	5,490	3,830	805
0.61	—	—	—	0.73	0.52	0.66	0.64	0.96	—
330	347	335	356	325	316	355	321	391	350

‡ 132-mg split of 0.98-g homogenized sample of ground-up coarse fines (1–2 mm)

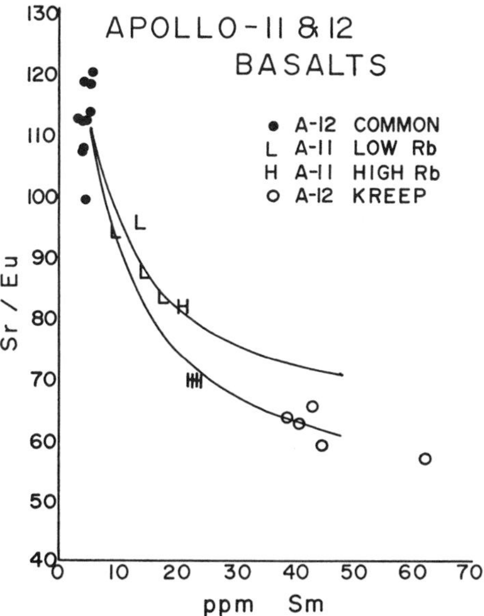

Fig. 8. Variation of Sr/Eu ratio vs. Sm concentration. Data are from Philpotts *et al.* (1971), Gast *et al.* (1970), and this report. The line drawn through the data is from one of several single stage partial melting model calculations, which all give similar lines. Discussion of models is later in the paper.

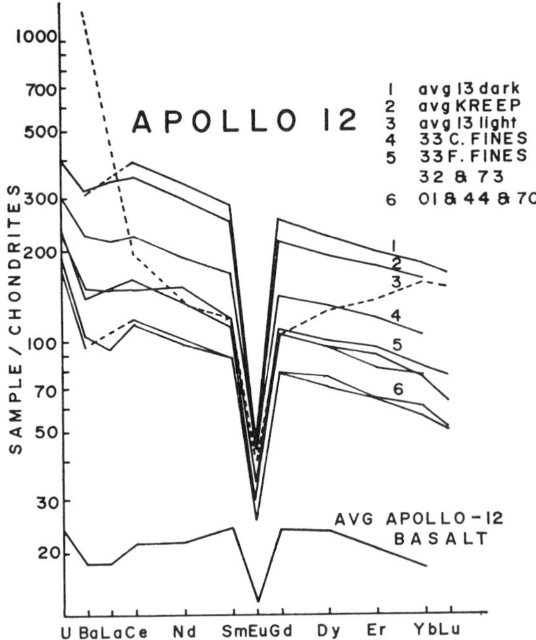

Fig. 9. Summary of rare earth, Ba, and U data for Apollo 12 samples. Data are from this report, PHILPOTTS *et al.*; HASKIN *et al.*; TATSUMOTO and KNIGHT; SILVER; and CLIFF *et al.* (all 1971).

We have already noted that KREEP fragments consistently have high abundances of K. We note here, in addition, that the K/Rb ratio is consistently lower than that of the mare basalts. It ranges from 315 to 391 with an average value of 343. This is essentially identical with that of chondrites. Finally, we note that the Nb concentration of soil sample 12033, which contains about 70 wt.% of KREEP, according to a mixing model presented below, is much higher than that of the mare basalts. Nb is also strongly enriched in alkali-olivine basalts (HUCKENHOLZ, 1965) relative to oceanic ridge basalts (CANN, 1971). The Cr content of soil sample 12033 (0.32%) and breccia 12034 (0.25%) is significantly lower than that of other lunar samples. We infer from this that the KREEP basalts have distinctly lower Cr contents than the mare basalts. Nevertheless, the Cr content is still 3 to 4 times that of most terrestrial basalts.

ISOTOPIC CHARACTERISTICS OF THE REGOLITH

The isotopic composition of Pb (TATSUMOTO and ROSHOLT, 1970; SILVER, 1970) and Sr (PAPANASTASSIOU *et al.*, 1970; COMPSTON *et al.*, 1970) and the "ages" derived from these data for the Apollo 11 soil were the first compelling evidence that a detectable portion of the post-mare regolith consisted of materials much older than those presumed to underlie the regolith. Similar data for the Apollo 12 soils (PAPANASTASSIOU and WASSERBURG, 1970) confirmed the earlier conclusion. The unknown portion of the soil presumably responsible for the apparent old age of the soil was designated by the term "magic component" by PAPANASTASSIOU *et al.* (1970).

The question now arises whether or not a component similar to KREEP can explain the isotopic characteristics of either or both of the Apollo 11 or Apollo 12 soil samples. The occurrence of Luny Rock 1 and other such fragments (WOOD *et al.*, 1971) in the Apollo 11 soil and the KREEP-like characteristics of this rock support this hypothesis. We have determined $Sr^{87/86}$ and Rb^{87}/Sr^{86} ratios for each of the fragments described above as well as several soil samples. These data along with measurements made by other investigators are plotted on an isochron diagram in Fig. 10. The isotopic composition of these fragments is similar to that of Luny Rock 1. They yield an apparent age of ~4.5 b.y. At this point, we should also note that the isotopic results reported for rock 12013 (LUNATIC ASYLUM, 1970) show quite clearly that the dark portions of this rock must have existed before 4.0 b.y. ago; that is, they predate the crystallization of the Apollo 11 and Apollo 12 mare basalts. The apparent or model ages of the KREEP basalt fragments cannot be unambiguously interpreted as the time of formation of a simple igneous rock which subsequently became the source of the brecciated fragments and ropy yellow-brown glass. A more complex 2- or 3-stage evolution can explain the observed isotope characteristics equally well. However, any alternative evolution which implies a younger age for the source of the KREEP fragments also requires that the source of the KREEP basalt must have had a higher Rb/Sr ratio than the present basalt. If we assume that the most primitive lunar materials had no Eu and Sr anomalies, we see immediately that such a more complex evolution requires: first, the generation with an extremely high Rb/Sr ratio, perhaps associated with material like the light portion of rock 12013, and then the

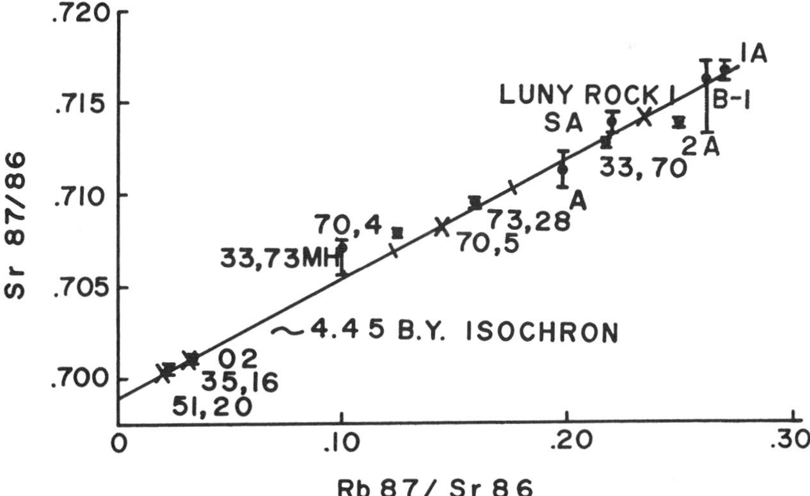

Fig. 10. A partial summary of Rb–Sr data for Apollo 12 basalts, soils, and breccias, as well as six members of the KREEP class of basaltic lunar rocks (Luny Rock 1, 1A, 2A, B1, S.A., and A.). Data in X are from PAPANASTASSIOU and WASSERBURG (1970) and PAPANASTASSIOU *et al.* (1970).

derivation of KREEP-like basalts from this material. Thus, the postulate of a more complex history simply strengthens the conclusion that highly differentiated materials existed early in the lunar history.

Isotopic data on rock 12013 leads to an identical conclusion (LUNATIC ASYLUM, 1971). We suggest that the simplest explanation of the isotopic data from both the KREEP fragments and rock 12013 is that the silicious and basaltic material that gives apparent ages approaching 4.5 b.y. were derived independently from each other or that the silicious material was produced by segregating an immiscible liquid from a 4.5 b.y. old basalt which may have had the characteristics of KREEP. We suggest that it is very unlikely that basaltic materials were derived from materials with highly silicious (granitic) compositions.

In summary, we suggest that the materials from both the Apollo 11 and Apollo 12 sites that are described by the chemical characteristics of KREEP are all derived from 4.3 to 4.5 b.y. old igneous rocks. The occurrence of this material at two widely separated sites on the moon suggests that these igneous rocks are not associated with a single isolated body. We suggest that such material will be found in abundance at non-mare sites on the moon.

MIXING MODELS

Mixing models to explain lunar regolith can be made on the basis of chemical and isotopic data or on the basis of morphological and mineralogical features of constituent fragments. They can also be constructed from hypothetical or derived end members. Ultimately, any mixing model not constructed entirely from observational data must be tested by such data.

Soil and breccia samples from Apollo 11 and Apollo 12 have been extensively studied so that their various components have been identified both as to kind and abundance. Possible components are (1) the anorthosites from Apollo 11; (2) the non-mare basaltic fragments (KREEP) in Apollo 12 and their Apollo 11 counterparts (Luny Rock 1 type fragments); (3) the high Si, K, Rb, and Ba white material from 12013 and a few small fragments of similar material from Apollo 12 soils; (4) the mare basalts from Apollo 11 and Apollo 12; and (5) meteoritic materials (the discussion that follows considers only those elements explained by lunar-lunar mixing models, not those explained by lunar-meteorite mixing models).

The most abundant components of the Apollo 12 regolith are mare basalt fragments and KREEP fragments (ANDERSON *et al.*, KING *et al.*, MCKAY *et al.*, MEYER *et al.*, QUAIDE *et al.*, WOOD *et al.*, all 1971). Anorthosite fragments are relatively rare in Apollo 12, and silicious materials are almost entirely restricted to 12013. Apollo 11 contains moderately abundant anorthosite fragments and similar, although smaller, amounts of Luny Rock 1 type fragments. No fragments of silicious material have been reported in Apollo 11 soils and breccia.

Apollo 12

The chemical and isotopic mixing model (HUBBARD *et al.*, 1971) gives relative abundances of mare basalts and KREEP fragments ranging from ~70% mare basalts and ~70% KREEP to ~30% mare basalts and ~70% KREEP. (See Fig. 11 for a summary of chemical differences in Apollo 12 materials.) The relative abundances of mare and nonmare (KREEP) materials in various Apollo 12 soil and breccia samples are reported in detail by MEYER *et al.* and MCKAY *et al.* (both 1971) and show that the relative abundances derived from the chemical model are quite similar to those observed. A detailed consideration of the internal consistency of the chemical mixing model indicates that a second non-mare material, the silicious component of 12013, is present at about the 1% level. The amount of the silicious component is severely limited in the Apollo 11 and Apollo 12 regoliths by its very high Ba, Rb, and K concentrations and very unusual REE abundance pattern (Fig. 9). In contrast, the REE, Ba, and U abundance patterns of KREEP are identical to those of the Apollo 12 soil and breccia samples (Fig. 9).

Apollo 11

The Apollo 11 soil is not so easily matched by mixing models. The Apollo 12 regolith is rather undisturbed by reworking processes whereas the Apollo 11 regolith is extensively reworked such that large amounts of hybrid material have been formed from initially distinct components.

The chemical composition of the mare basalt component in Apollo 11 is difficult to estimate because there are two quite distinct chemical groups. Realistic mixing models are difficult to make if large amounts cf Apollo 11 high Rb basalts are included in the average mare basalt component. Recent exposure age data (EBERHARDT *et al.*,

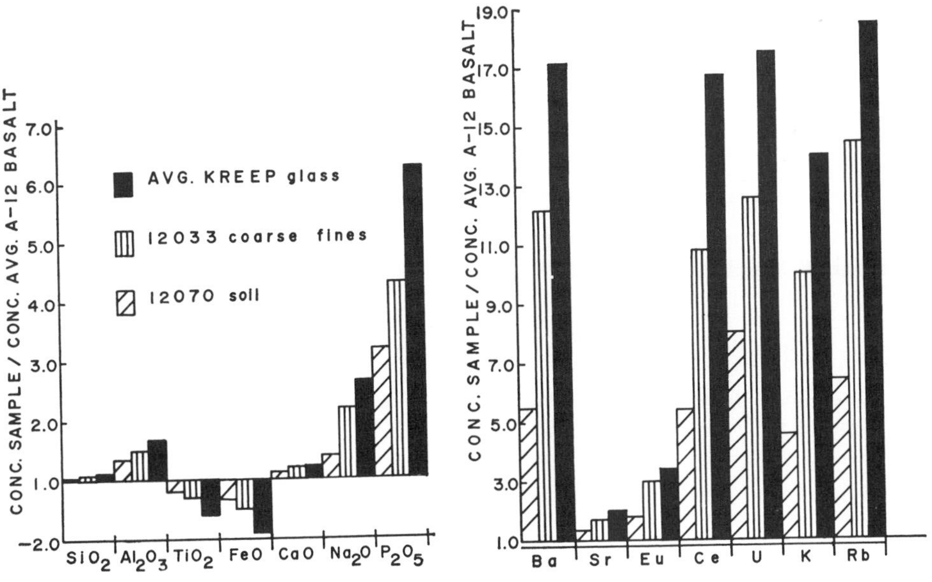

Fig. 11. Some general chemical features of Apollo 12 samples 12070, 12033 coarse fines, and KREEP glass relative to average Apollo 12 basalt. Data are for the same sources as Apollo 12 data in Figs. 1–4.

1970) point to a different history of the high-Rb basalts and suggest a nonlocal source. A satisfactory mixing model for Apollo 11 soil consists of ~70% average low Rb basalt, ~20% anorthosite and ~10% KREEP. This model gives essentially the analyzed concentrations of REE, Sr, Rb, U, Al_2O_3, and TiO_2 and has the advantage over previous models in that it uses only analyzed components that are known to be present.

CONCLUSIONS AND EARLY LUNAR HISTORY

The soil from two mare sites on the moon contain strong indirect evidence that highly differentiated (with respect to the probable average composition of the planet) silica-rich and basaltic rocks occur in the highlands or in the submare parts of the moon.

The detailed petrogenesis of these materials cannot be inferred without a more specific knowledge of their occurrence on the lunar surface. It is possible, for example, that nearly all materials described here had their ultimate origin in the Imbrium basin; that is, they are part of the Fra Mauro ejecta blanket. However, even without the knowledge of the structural situation of the source material for the regolith fragments, some general conclusions regarding the highlands and early history of the moon can be inferred from the presumed chemical composition of this source. The high CaO and Al_2O_3 contents (see Fig. 5) clearly suggest that a liquid with very high normative plagioclase contents occurred in or on the lunar crust. The first, or most abundant, liquidus mineral in this liquid probably was plagioclase. These liquids thus provide a source for rocks that are enriched in plagioclase; that is, gabbroic anorthosites and anorthosites.

The unusual abundance of large ion lithophile elements (K, Rb, Ba, REE, U, Th, etc.) suggests that KREEP liquids were produced by small degrees of partial melting followed by subsequent segregation and ascent of the liquid. Extensive fractional crystallization involving pyroxene or olivine is an unlikely explanation for the strong enrichment in large ion lithophile elements in that it would not produce liquids with the observed Mg/Fe + Mg ratios or Cr contents. The lithophile element abundance patterns; that is, both the enrichment in light REE over heavy REE and the extreme depletion in Eu and Sr, are predicted characteristics of such a liquid. Previous calculations (GAST and HUBBARD, 1970) show that moderately fractionated REE patterns may result from very small degrees of partial melting of a plagioclase-clinopyroxene-orthopyroxene assemblage when the distribution coefficients for these elements between liquid and clinopyroxene are assumed to be considerably greater (i.e., favor the liquid) than those inferred from phenocryst data. In the case previously reported, the slope of the rare earth pattern tends to approach that of the equilibrium liquid clinopyroxene pattern when the extent of partial melting is less than 2%. The large Eu anomalies observed for the KREEP basalt are even more difficult to produce by fractional crystallization than those observed in the Apollo 11 high Rb basalts. We have shown earlier that partial melting of a plagioclase bearing source can explain these anomalies in a plausible way.

When taken all together, the increased range of Sm/Eu ratios and the regular variation in Sr/Eu ratios observed here, the range in total REE content and the

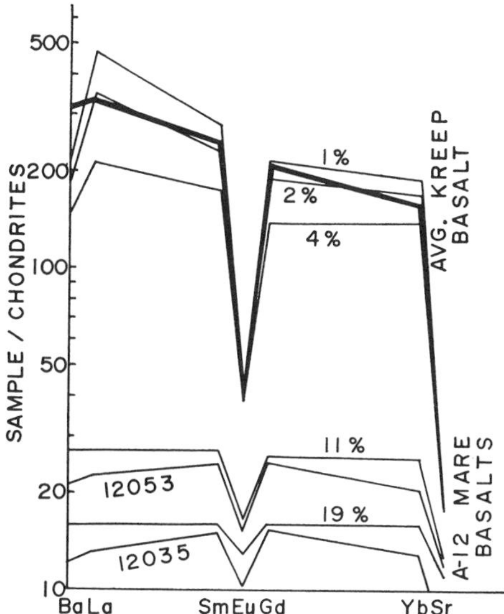

Fig. 12. Comparison of chondrite normalized REE, Ba, and Sr data for KREEP and Apollo 12 mare basalts with the results of partial melting calculations. The KREEP and Apollo 12 mare basalts have been matched by using different starting materials for each basalt type. Note that Apollo 12 mare basalts like 12053 are matched by the model while the model Eu and Sr anomalies are too small to match olivine rich mare basalts like 12035. See Table 3 for parameters and boundary conditions used.

variation in relative abundance of the trivalent REE greatly restrain the range of parameters in allowed partial melting models. We have investigated these restraints by extending the calculations that we have given previously (GAST and HUBBARD, 1970). The distribution coefficients used in these calculations are somewhat modified from those chosen in our previous model. They are summarized in Table 3 along with other parameters, etc., of the present model. The values shown there for clinopyroxene and liquid for the trivalent rare earths assume a somewhat less extreme fractionation than the values which we used previously. The values used here for the distribution of Eu between plagioclase and basaltic liquid are similar to those experimentally determined (RINGWOOD, personal communication, 1970). The distribution of Sr between coexisting calcic plagioclase and basalt is estimated from unpublished studies of plagioclase phenocrysts in oceanic basaltic rocks (SHIH, personal communication, 1971).

We have investigated the effect of changing both the phase composition and the degree of partial melting in order to determine if a given starting material can produce liquids with the observed range of REE, Ba, and Sr concentrations and relative abundances. These calculations indicate that (1) phase compositions with less than 5 wt. % of plagioclase failed to produce liquids with the observed characteristics; in particular, such liquids will not simultaneously have significant Eu anomalies and unfractionated trivalent REE abundances; (2) even with higher plagioclase contents,

Table 3. Parameters and boundary conditions for partial melting model.

	Distribution Coefficients $K^{liq/min}$			
	cpx	plag	opx	oliv
Element				
Ba	120	8	70	500
La	36	30	600	500
Sm	9	30	200	500
Eu	80	1.2 (KREEP) 0.8 (mare)	160	500
Gd	6	30	105	500
Yb	6	30	36	500
Sr	100	0.55	80	500
	Composition of Sources			
	cpx	plag	opx	oliv
KREEP source (~9X chondrites for REE, Sr, etc.)	15%	25%	30%	30%
Apollo 12 mare basalt source (~3X chondrites for REE, Sr, etc.)	17.5%	10%	32.5%	40%
	Composition of Liquids			
	cpx	plag	opx	oliv
KREEP liquid (1–2% liquid produced)	15%	50%	20%	15%
Apollo 12 mare basalt liquid (~10% liquid produced)	30%	20%	35%	15%

Boundary Conditions

A. Chondritic relative abundances for REE and Sr
B. Reasonable $K^{liq/plag}$ values for Sr and Eu
C. Must match observed range of REE, Sr, and Ba concentrations
D. Moderately realistic source and liquid bulk compositions

a given phase composition and bulk composition can produce the observed range of concentrations and relative abundances only if we assume that the initial bulk materials that underwent melting had a significant negative Eu and Sr anomaly; and (3) none of the liquids observed can be produced from starting materials with chondritic concentrations of the REE, Ba, and Sr. With the assumed distribution coefficients, these elements must be enriched by at least a factor of 3, relative to chondrites. If the relative abundances of the REE, Ba, and Sr in the primitive moon were those of chondrites, i.e., had no primitive Eu anomaly, we must furthermore conclude that the observed range of concentrations and relative abundances in the various lunar basalts can be produced only if the degree of enrichment of these elements relative to chondrites in the source of these basaltic liquids varies by at least a factor of 2 or 3; specifically, we find that the Apollo 12 mare basalts may be produced from a source with about three times the chondritic concentration of these elements, while the KREEP basalts require a source with six to nine times the chondritic concentration of these elements.

The large variation in trace element concentrations thus leads to the conclusion that even the upper part of the moon is chemically heterogeneous. It seems quite likely that these heterogeneities are not limited to large ion refractory trace elements. Other refractory elements such as Ca and Al should vary in concentration along with the trace elements discussed here. WETHERILL (1968) and RINGWOOD and ESSENE (1970a) have shown that the Ca and Al content of the deeper lunar interior must be limited by the density of eclogites stabilized by high pressure. This constraint, combined with the evidence cited here, suggests that the Ca and Al content of the lunar interior must decrease markedly as a function of depth. The enrichment in large ion refractory elements can also be inferred from a simple mass balance calculation. If we assume, for example, that the crust of the moon contains the equivalent of a 1 kilometer shell of KREEP basalt, we require a shell at least 80–100 kilometers thick enriched by a factor of 5 (relative to chondrites) in the elements noted above in order to derive such a differentiated crust.

In conclusion, we suggest that the KREEP basalts were derived from the upper 100 kilometers of the moon shortly after it accreted. We suggest that this shell was heated to about the melting temperature of basalt by the release of gravitational energy during the accretion of the last 10–20% of the mass of the moon. Alternatively, this portion could have accreted at temperatures just below the melting point (900–1100°C.) and melted by internal heat due to radioactive decay during the first 100 or 200 million years of lunar history. Such early melting would be limited to the refractory element-enriched outer layer of the moon inferred here.

Mare basalts were probably produced by more extensive melting at depths in excess of 100 kilometers (see, for example, RINGWOOD and ESSENE, 1970b). Perhaps the different depth of origin may explain the differences in Fe content and Ti content observed for the two basalt types. The heat source for the mare basalts must have been, in large part, the decay of long-lived radionuclides. The production of the basaltic liquids inferred here must have resulted in a substantial upward transfer of radioactive nuclides from the upper $\frac{1}{4}$–$\frac{1}{2}$ of the mass of the moon very early in lunar history. This early differentiation may preclude later heating of this part of the lunar interior to the point where basaltic liquids were formed.

Acknowledgments—We particularly thank Charles Meyer for stimulating a cooperative study of the KREEP fragments; Ernest Schonfeld for his assistance in computer calculations of partial melting models; Larry Nyquist for his aid in high precision Sr isotope analyses, and Henry Wiesmann for maintaining a smoothly functioning clean chemical laboratory. We greatly appreciate the assistance of Donna Sanders in typing and assembling this manuscript.

REFERENCES

AGRELL S. O., LONG J. V. P., and REED S. J. B. (1971) Glasses from Apollo 11 and Apollo 12 soils and microbreccias. Second Lunar Science Conference (unpublished proceedings).

AHRENS L. H. and VON MICHAELIS H. (1969) The composition of stony meteorites III. Some inter-element relationships. *Earth Planet. Sci. Lett.* **5**, 395–400.

ALBEE A. L. and CHODOS A. A. (1970) Microprobe investigations on Apollo 11 samples. *Proc. Apollo 11 Lunar Sci. Conf., Geochim. Cosmochim. Acta* Suppl. 1, Vol. 1, pp. 135–157. Pergamon.

ANDERSON A. T., Jr., NEWTON R. C., and SMITH J. V. (1971) Apollo 12 mineralogy and petrology: Light-colored fragments; minor-element concentrations; petrologic development of the Moon. Second Lunar Science Conference (unpublished proceedings).

CANN J. R. (1971) Rb, Sr, Y, Zr, and Nb in some ocean floor basaltic rocks. *Earth Planet. Sci. Lett.* **10**, 7–11.

CLIFF R. A., LEE-HU C., and WETHERILL G. W. (1971) Rb–Sr and U, Th–Pb measurements on Apollo 12 material. Second Lunar Science Conference (unpublished proceedings).

COMPSTON W., BERRY J., VERNON M. J., CHAPPELL B. W., and KAYE M. J. (1971) Rubidium-strontium chronology and chemistry of lunar material from the Ocean of Storms. Second Lunar Science Conference (unpublished proceedings).

COMPSTON W., CHAPPELL B. W., ARRIENS P. A., and VERNON M. J. (1970) The chemistry and age of Apollo 11 lunar material. *Proc. Apollo 11 Lunar Sci. Conf., Geochim. Cosmochim. Acta* Suppl. 1, Vol. 2, pp. 1007–1027. Pergamon.

EBERHARDT P., GEISS J., GRAF H., GRÖGLER H., KRÄHENBÜHL U., SCHWALLER H., SCHWARZMÜLLER J., and STETTLER A. (1970) Correlation between rock type and irradiation history of Apollo 11 igneous rocks. *Earth Planet. Sci. Lett.* **10**, 67–72.

ENGEL A. E. J. and ENGEL C. G. (1970) Mafic and ultramafic rocks. In *The Sea: Ideas and Observations on Progress in the Study of the Seas* (editor A. E. Maxwell), Vol. 4, Wiley-Interscience.

ENGEL C. G. and ENGEL A. E. J. (1971) Major element composition of three Apollo 12 rocks and some petrogenic considerations. Second lunar Science Conference (unpublished proceedings).

FREDRIKSSON K., NELEN J., ANDERSEN C. A., and HINTHORNE J. R. (1971) A summary of phase analysis on Apollo 12 samples. Second Lunar Science Conference (unpublished proceedings).

FUCHS L. H. (1970) Orthopyroxene-plagioclase fragments in the lunar soil from Apollo 12. *Science* **169**, 866–868.

GAST P. W. (1968a) Implications of the Surveyor V chemical analysis. *Science* **159**, 897–899.

GAST P. W. (1968b) Upper mantle chemistry and evolution of the earth's crust. In *History of the Earth's Crust* (editor R. Phinney), pp. 15–27, Princeton University Press.

GAST P. W. and HUBBARD N. J. (1970) Rare earth abundances in soil and rocks from the Ocean of Storms. *Earth Planet. Sci. Lett.* **10**, 94–100.

GAST P. W., HUBBARD N. J., and WIESMANN H. (1970) Chemical composition and petrogenesis of basalts from Tranquillity Base. *Proc. Apollo 11 Lunar Sci. Conf., Geochim. Acta* Suppl. 1, Vol. 2, pp. 1143–1163. Pergamon.

GOLES G. G., DUNCAN A. R., OSAWA M., MARTIN M. R., BEYER R. L., LINDSTROM D. J., and RANDLE K. (1971) Analyses of Apollo 12 specimens and a mixing model for Apollo XII "soils." Second Lunar Science Conference (unpublished proceedings).

GOLES G. G., RANDLE K., OSAWA M., LINDSTROM D. J., JÉROME D. Y., STEINBORN T. L., BEYER R. L., MARTIN M. R., and MCKAY S. M. (1970) Interpretations and speculations on elemental abundances in lunar samples. *Proc. Apollo 11 Lunar Sci. Conf., Geochim. Cosmochim. Acta* Suppl. 1, Vol. 2, pp. 1177–1194. Pergamon.

GREEN D. H. and RINGWOOD A. E. (1967) The genesis of basalt magmas. *Contrib. Mineral. Petrol.* **15**, 103–190.

HASKIN L. A., ALLEN R. O., HELMKE P. A., ANDERSON M. R., KOROTEV R. L., and ZWEIFEL K. A. (1971) Rare earths and other trace elements in Apollo 12 lunar materials. Second Lunar Science Conference (unpublished proceedings).

HUBBARD N. J., GAST P. W., and WIESMANN H. (1970) Rare earth, alkaline and alkali metal and $^{87/86}$Sr data for subsamples of lunar sample 12013. *Earth Planet. Sci. Lett.* **9**, 181–184.

HUBBARD N. J., MEYER C., Jr., GAST P. W., and WIESMANN H. (1971) The composition and derivation of Apollo 12 soils. *Earth Planet. Sci. Lett.* 10, 341–350.

HUCKENHOLZ H. G. (1965) Die Verteilung des Niobs in den Gesteinen und Mineralen der Alkalibasalt-Assoziation der Hocheifel. *Geochim. Cosmochim. Acta* **29**, 807–820.

KAY R., HUBBARD N. J., and GAST P. W. (1970) Chemical characteristics and origin of oceanic ridge volcanic rocks. *J. Geophys. Res.* **75**, 1585–1613.

KEIL K., PRINZ M., and BUNCH T. E. (1971a) Mineralogical aspects of Apollo 12 rocks. Second Lunar Science Conference (unpublished proceedings).

KEIL K., PRINZ M., and BUNCH T. E. (1971b) Mineralogy, petrology, and chemistry of some Apollo 12 samples. Second Lunar Science Conference (unpublished proceedings).

KING E. A., BUTLER J. C., and CARMAN M. F. (1971) The lunar regolith as sampled by Apollo 11 and Apollo 12: Grain size analyses, modal analyses, origins of particles. Second Lunar Science Conference (unpublished proceedings).

LSPET (Lunar Sample Preliminary Examination Team (1970) Preliminary examination of lunar samples from Apollo 12. *Science* **167**, 1325–1339.

LUNATIC ASYLUM (1970a) Ages, irradiation and chemical composition of lunar rocks from the Sea of Tranquillity. *Science* **167**, 463–466.

LUNATIC ASYLUM (1970b) Mineralogic and isotopic investigations on lunar rock 12013. *Earth Planet. Sci. Lett.* **9**, 137–163.

MACDONALD G. A. and KATSURA T. (1964) Chemical composition of Hawaiian lavas. *J. Petrol.* **5**, 82–133.

MANSON V. (1967) Geochemistry of basaltic rocks: major elements. In *Basalts: The Poldervaart Treatise on Rocks of Basaltic Composition* (editors H. H. Hess and A. Poldervaart), Vol. 1, pp. 215–271, Interscience.

MAXWELL J. A. and WIIK H. B. (1971) Chemical composition of Apollo 12 lunar samples 12004, 12033, 12051, 12052, and 12065. Second Lunar Science Conference (unpublished proceedings).

MCKAY D., MORRISON D., LINDSAY J., and LADLE G. (1971) Apollo 12 soil and breccia. Second Lunar Science Conference (unpublished proceedings).

MEYER C., Jr., AITKEN F. K., BRETT P. R., MCKAY D. S., and MORRISON D. A. (1971) Rock fragments and glasses rich in K, REE, P in Apollo 12 soils: Their mineralogy and origin. Second Lunar Science Conference (unpublished proceedings).

MEYER C., Jr., and HUBBARD N. J. (1970) High potassium, high phosphorous glass an an important rock type in the Apollo 12 soil samples. *Meteoritics* **5**, 210–211.

NOCKOLDS S. R. (1954) Average chemical compositions of some igneous rocks. *Bull. Geol. Soc. Amer.* **65**, 1007–1032.

OBERBECK V. R. (1971) A mechanism for the production of lunar crater rays. *The Moon* (in press).

OBERBECK V. R. and QUAIDE W. L. (1968) Genetic implications of lunar regolith thickness variations. *Icarus* **9**, 446–465.

PAPANASTASSIOU D. A. and WASSERBURG G. J. (1970) Rb–Sr ages from the Ocean of Storms. *Eath Planet. Sci. Lett.* **8**, 269–278.

PAPANASTASSIOU D. A., WASSERBURG G. J., and BURNETT D. S. (1970) Rb–Sr ages of lunar rocks from the Sea of Tranquillity. *Earth Planet. Sci. Lett.* **8**, 1–19.

PHILPOTTS J. A. and SCHNETZLER C. C. (1970) Apollo 11 lunar samples: K, Rb, Sr, Ba, and rare-earth concentrations in some rocks and separated phases. *Proc. Apollo 11 Lunar Sci. Conf., Geochim. Cosmochim. Acta* Suppl. 1, Vol. 2, pp. 1471–1486. Pergamon.

PHILPOTTS J. A., SCHNETZLER C. C., BOTTINO M. L., and FULLAGAR P. D. (1971) Li, K, Rb, Sr, Ba and rare-earth concentrations and $^{87}Sr/^{86}Sr$ in some Apollo 12 soils, rocks and separated phases. Second Lunar Science Conference (unpublished proceedings).

QUAIDE W., OBERBECK V., BUNCH T. and POLKOWSKI G. (1971) Investigations of the natural history of the regolith at the Apollo 12 site. Second Lunar Science Conference (unpublished proceedings).

RINGWOOD A. E. and ESSENE E. (1970a) Petrogenesis of lunar basalts and the internal constitution and origin of the moon. *Science* **167**, 607–610.

RINGWOOD A. E. and ESSENE E. (1970b) Petrogenesis of Apollo 11 basalts, internal constitution and origin of the moon. *Proc. Apollo 11 Lunar Sci. Conf., Geochim. Cosmochim. Acta* Suppl. 1, Vol. 1, pp. 769–799. Pergamon.

ROEDDER E. and WEIBLEN P. W. (1971) Lunar petrology of silicate melt inclusions, Apollo 11 and Apollo 12, and terrestrial equivalents. Second Lunar Science Conference (unpublished proceedings).

ROEDDER E. and WEIBLEN P. W. (1970) Lunar petrology of silicate melt inclusions, Apollo 11 rocks. *Proc. Apollo 11 Lunar Sci. Conf., Geochim. Cosmochim. Acta* Suppl. 1, Vol. 1, pp. 801–837. Pergamon.

ROSE H. J., Jr., CUTTITTA F., ANNELL C. S., CARRON M. K., CHRISTIAN R. P., DWORNIK E. J., HELZ A. W., and LIGON D. T., Jr. (1971) Semimicroanalysis of Apollo 12 samples. Second Lunar Science Conference (unpublished proceedings).

SCHNETZLER C. C., PHILPOTTS J. A., and BOTTINO M. L. (1970) Li, K, Rb, Sr, Ba, and rare-earth concentrations, and Rb–Sr age of lunar rock 12013. *Earth Planet. Sci. Lett.* **9**, 185–192.

SCOON J. J. (1971) Quantitative chemical analyses of lunar samples 12040,36 and 12064,38. Second Lunar Science Conference (unpublished proceedings).

SIGVALDASON G. E. (1969) Chemistry of basalts from the Icelandic rift zone. *Contrib. Mineral. Petrol.* **20**, 357–370.

SILVER L. T. (1970) Uranium–thorium–lead isotope relations in lunar materials. *Science* **167,** 468–471.

SILVER L. T. (1971) U–Th–Pb isotope relations in Apollo 11 and Apollo 12 lunar samples. Second Lunar Science Conference (unpublished proceedings).

SPENCER A. B. (1969) *Alkalic Igneous Rocks of Balcones Province, Texas*, Vol. 10, pp. 272–306.

TATSUMOTO M. and KNIGHT R. J. (1971) U–Th–Pb systematics of Apollo 12 lunar samples. Second Lunar Science Conference (unpublished proceedings).

TATSUMOTO M. and ROSHOLT J. N. (1970) Age of the moon: An isotopic study of uranium–thorium–lead systematics of lunar samples. *Science* **167**, 461–463.

VINOGRADOV A. P. (1971) Preliminary data on lunar ground brought to earth by automatic probe Luna-16. Second Lunar Science Conference (unpublished proceedings).

WAKITA H. and SCHMITT R. A. (1970) Lunar anorthosites: Rare-earth and other elemental abundances. *Science* **170**, 969–974.

WETHERILL G. W. (1968) Lunar interior: Constraint on basaltic composition. *Science* **160**, 1256.

WILLIS J. P., AHRENS L. H., DANCHIN R. V., ERLANK A. J., GURNEY J. J., HOFMEYR P. K., MCCARTHY T. S., and ORREN M. J. (1971) Some interelement relationships between lunar rocks and fines, and stony meteorites. Second Lunar Science Conference (unpublished proceedings).

WOOD J. A., DICKEY J. S., Jr., MARVIN U. B., and POWELL B. N. (1970) Lunar anorthosites and a geophysical model of the moon. *Proc. Apollo 11 Lunar Sci. Conf., Geochim. Cosmochim. Acta* Suppl. 1, Vol. 1, pp. 965–988. Pergamon.

WOOD J. A., MARVIN U., REID J. B., TAYLOR G. J., BOWER J. F., POWELL B. N., and DICKEY J. S., Jr. (1971) Relative proportions of rock types, and nature of the light-colored lithic fragments in Apollo 12 soil samples. Second Lunar Science Conference (unpublished proceedings).

Proceedings of the Second Lunar Science Conference, Volume 2, pp. 1021–1036.
The M.I.T. Press, 1971.

Volatile and siderophile elements in lunar rocks: Comparison with terrestrial and meteoritic basalts

EDWARD ANDERS, R. GANAPATHY, REID R. KEAYS,*
J. C. LAUL, and JOHN W. MORGAN
Enrico Fermi Institute and Department of Chemistry,
University of Chicago, Chicago, Illinois 60637

(*Received 23 February 1971; accepted in revised form 15 April 1971*)

Abstract—Fifteen trace elements were measured by neutron activation analysis in four Apollo 11 and nine Apollo 12 rocks. Among siderophile elements, Ag and Au are strongly depleted in lunar rocks compared to eucrites and terrestrial basalts, while Ir is not. This difference contradicts the fission hypothesis, and may reflect more strongly reducing conditions during metal-silicate segregation on the Moon. The volatile elements Cs, Rb, In, Tl, Cd, Zn, Br, and Bi are depleted 10- to 100-fold in both lunar basalts and eucrites, relative to terrestrial basalts. These fractionations apparently happened in the solar nebula. The Cs/U ratios suggest that the Earth, Moon, and eucrites incorporated 11%, 1.5%, and 0.5% low-temperature material during their formation. From the abundances of Tl, Bi, In, and Pb^{204}, nominal mean accretion temperatures are estimated: Earth 560°K, Moon 620°K, and eucrites 470°K. The Earth–Moon difference probably reflects a 100-fold higher accretion rate of volatile-rich material in the final stages of accretion, rather than a true temperature difference.

INTRODUCTION

WE HAVE MEASURED 15 trace elements in Apollo 12 lunar samples by radiochemical neutron activation analysis. The present paper reports all our data on crystalline rocks, which provide information on large-scale chemical fractionation processes during and after the Moon's formation. The companion paper by LAUL *et al.* (1971) gives results on soils and breccias, which provide information on meteorite influx and the evolution of the regolith. Both papers also contain unpublished results on Apollo 11 samples that were not included in our Apollo 11 paper (GANAPATHY *et al.*, 1970a). Together, these three papers give a complete account of all our Apollo 11 and Apollo 12 data, including those already published in brief Reports or Letters (GANAPATHY *et al.*, 1970b; LAUL *et al.*, 1970; MORGAN *et al.*, 1971).

The roster of elements has changed somewhat since our Apollo 11 work: Cu and Pd were dropped while Se and Te were added.

EXPERIMENTAL TECHNIQUES

Samples

Twenty-one ~0.1-gram samples of four Apollo 11 and nine Apollo 12 rocks were analyzed. Some of the replicates were ordinary repeats for the purpose of checking doubtful numbers. Others were separate analyses of morphologically distinct portions of heterogeneous rocks, e.g., the glassy coating and crystalline interior of 12017, or the bewildering chiaroscuro of 12013. A sample of Columbia River basalt BCR–1 was included with every irradiation.

* Present address: Department of Geology, University of Melbourne, Parkville, Victoria, 3052, Australia.

Table 1. Abundances in Apollo 11

Sample	Type	Ir (ppb)	Au (ppb)	Zn (ppm)	Cd (ppb)	Ag (ppb)	Bi (ppb)
10017,87	AB	0.020	0.72	18	68	16	1.15
10047,65	B Fragm.	0.005	0.029	1.8	3.4	1.89	0.16
10047,32	B Powder	*0.24*	*0.33*	*5.76*	*255*	*24.7*	*2.15*
10057,41	A	0.009	0.013	1.75	3.5	—	—
10057,41	A	0.023	0.017	1.71	3.15	0.69	0.27
10072,23	A	0.022	0.10	1.72	14	—	—
10072,23	A	*4.02*	0.14	1.81	6.5	*17.3*	0.73
12002,126	B	*0.62*	0.024	0.70	1.4	0.81	*1.4*
12008,11	AB Ilm.	0.06	0.074	1.04	1.3	1.17	0.26
12017,17	AB Int.	0.20	0.072	1.02	1.1	1.45	0.25
12017,20	AB Glass	2.64	1.23	3.30	5.4	2.81	0.59
12020,39	A	0.03	0.036	0.74	1.1	0.98	0.15
12038,57	A Powder	0.04	*0.36*	*2.09*	*5.2*	*80*	0.76
12040,28	B	0.17	0.012	0.78	3.3	0.41	0.45[b]
12051,45	AB	0.09	0.008	0.52	1.2	0.82	0.53
12051,45	AB	*0.54*	0.007	0.54	1.1	0.80	—
12065,35	AB	0.078	0.013	0.93	—	1.37	0.30
12065,35	AB	≤0.05	0.011	0.67	1.2	—	—
Ave. Apollo. 12[c]	AB	0.095	0.034	0.80	1.51	1.00	0.33
12013,10,06	Ign. Brec.	4.6	3.10	4.10	34	*31.4*	0.63
12013,10,15	Ign. Brec.	0.047	0.21	2.06	91	0.88	0.46
12013,10,18	Ign. Brec.	0.72	0.46	1.94	44	0.68	0.55
12013,10,35	Ign. Brec.	0.19	*13.2*	3.04	40	3.80	0.52
12013,10,37 + 24	Ign. Brec.	0.50	*2.37*	2.85	64	3.49	1.04
12013,10,41	Ign. Brec.	*10.0*	0.45	2.21	300	0.44	≤0.9
12013,10,44	Ign. Brec.	0.84	0.34	2.44	24	1.17	0.55
Ave. 12013[d]		0.434	0.353	2.77	71	1.65	0.60
BCR-1	Terr. Bas	0.012	0.84	118	124	31.5	46
BCR-1	Terr. Bas.	*4.0*	0.41	116	127	26.3	49
BCR-1	Terr. Bas.	≤0.06	—	130	120	25.5	46
BCR-1	Terr. Bas.	≤0.05	0.35	108	140	—	—
BCR-1	Terr. Bas.	0.007	0.44	108	134	26.3	48
Ave. BCR-1		≤0.03	0.51	116	129	28	47

[a] Doubtful values are shown in italics. They include all of the higher In and Ag values (contamination from vacuum gaskets in sample containers and LRL); most results on 10047,32 and 12038,57 (powders prepared at LRL), and a few unusually high Ir and Au values, where contamination during cutting or handling was suspected.

[b] Error from counting statistics was ±0.027 for this sample.

[c] Excluding Ir, Au, Zn, Cd, Ag, Bi, Tl, and Br in 12017,20, which appear to be largely of meteoritic origin.

[d] Excluding Ir and Au in 12013,10,06, which appear to be largely of meteoritic origin.

Procedure

Our procedure, judged unpublishable by editorial fiat, is available as NAPS Document 01194 from ASIS National Auxiliary Publication Service, c/o CCM Information, 909 Third Avenue, New York, New York 10022; at a charge of $2 for microfiche and $5 for photocopy. It is also available free of charge from the authors as preprint EFI-70-30, as long as the supply lasts.

RESULTS

The data are shown in Table 1. On the whole, very little compositional variation is evident. Sample 12017,20, representing the glassy coating of this rock, shows unusually high abundances of most elements. This apparently implies addition of meteoritic material (MORGAN *et al.*, 1971), and derivation from a more alkali-rich rock than 12017 (GOLES *et al.*, 1971). Sample 12038,57, a powder prepared at LRL, is

and Apollo 12 Lunar Rocks[a]

Tl (ppb)	Br (ppb)	Te[e] (ppb)	Se (ppb)	Ga (ppm)	Rb (ppm)	Cs (ppb)	In (ppb)	Co (ppm)	Ref.
6.16	190	—	215	5.1	6.6	186	*70*	31	
0.28	29	—	250	—	1.54	60	2.8	12	
0.57	*102*	13	—	5.4	1.25	45	*109*	14	1
—	—	—	181	—	—	—	—	30	
1.09	25	8	—	3.5	3.68	159	3.2	27	1
—	—	—	188	—	—	—	—	50	
0.92	36	—	—	4.7	5.98	159	*179*	27	1
0.25	10	10	141	2.4	0.97	39	1.9	—	2
0.34	11	50	139	3.2	0.62	28	*28*	60	
0.34	16	20	156	2.9	1.06	45	2.0	32	3
0.88	49	70	182	3.3	2.30	87	5.1	44	3
0.61	16	—	114	—	0.85	39	2.7	68	
0.76	10	—	120	4.3	0.41	15	*77*	25	
0.06	4	—	58	1.9	0.29	13	0.4	54	
0.36	16	10	201	4.6	1.03	40	2.0	—	2
0.37	16	16	204	4.3	1.06	42	1.2	—	2
0.53	22	—	200	3.8	1.15	67	2.2	43	
—	—	—	180	—	1.05	49	—	42	
0.36	13	18	144	3.3	0.96	41	1.8	47	
1.9	114	30	99	6.4	13.5	820	*380*	31	4
18	26	30	54	6.3	66.5	2670	*32*	16	4
18	48	10	30	5.9	49.4	2280	2.9	21	4
7.2	54	30	75	6.0	26.7	1260	3.9	27	4
18	47	10	68	6.0	50.0	2370	2.8	34	4
34	40	30	420	5.9	98.7	4030	6.6	34	4
4.8	36	30	40	6.0	20.7	1010	3.1	21	4
12.8	58	25	94	6.1	42	1900	3.7	26	
278	—		95	22	45	890	91	36	
290	31		94	21	46	925	96	35	4
282	60		85	22	46	900	88	37	
—	—		95	—	47	910	—	—	
270	—		90	22	47	930	91	38	
280	46		92	22	46	910	92	37	

[e] Tellurium was determined via 117d Te^{123m}. Some of the samples were recycled when Se^{75} contamination (up to 25% the total counting rate) was observed. Other samples were checked for selenium contamination by γ–γ coincidence counting, and corrected when necessary. However, owing to the very low activity of the samples, small amounts of Se^{75} may have escaped detection; thus some of the lower Te values may be systematically too high by as much as 20 to 40%.

References: (1) GANAPATHY *et al.* (1970a); (2) GANAPATHY *et al.* (1970b); (3) MORGAN *et al.* (1971); (4) LAUL *et al.* (1970).

highest in Au, Zn, Cd, Ag, Tl, and In. We suspect that these high values are due to contamination during grinding, as already noted for 10047 (GANAPATHY *et al.*, 1970a).

Among the remaining samples, 12040 tends to be lowest. Otherwise there are no clearcut groupings, or correlations with proposed classification schemes (WARNER, 1970; GOLES *et al.*, 1971; BIGGAR *et al.*, 1971; WÄNKE *et al.*, 1971). This does not in any way cast doubts on the validity of these schemes; it merely reflects the narrow range of variation of our data.

Comparison with data by other authors

Iridium, Cadmium, Selenium. Our results agree with those of BAEDECKER *et al.* (1971) and BRUNFELT *et al.* (1971).

Gold. Our results for crystalline rocks are 1–3 orders of magnitude lower than those of BRUNFELT *et al.* (1971) and SMALES *et al.* (1971). Our confidence in our own results is strengthened by the internal agreement of four pairs of replicates, the agreement of our BCR–1 results with literature values, and the consistent difference between rocks on the one hand and soils and breccias on the other. This difference leads to a meteoritic component of reasonable composition and magnitude (LAUL *et al.*, 1971).

Zinc. Our results for rocks 12040, 12051, and 12065 are systematically lower by factors of 2–10 than those of BOUCHET *et al.* (1971), COMPSTON *et al.* (1971), MORRISON *et al.* (1971), and SMALES *et al.* (1971). However, our results for soils and breccias and for Apollo 11 rock 10017 agree with literature values.

Silver and Indium. Contamination from ALSRC and LRL vacuum gaskets seems to be less endemic than in Apollo 11 samples. The few available Ag values by other authors are higher than ours by about a factor of 10. Indium values by BAEDECKER *et al.* (1971) and WAKITA *et al.* (1971) generally agree with ours; those by BRUNFELT *et al.* (1971) are mostly higher.

Bismuth and Thallium. No analyses by other authors are available for comparison.

Bromine. Only one direct cross-check is available, for rock 12065. The BOUCHET *et al.* (1971) value is fivefold higher than ours. Their remaining analyses range from 80 to 120 ppb, higher than the values of Reed and Jovanovic (1971), 22 to 53 ppb, which in turn are higher than ours, 4 to 22 ppb.

Tellurium. Our values agree fairly well with those of Reed and Jovanovic (1971), though some of their upper limits are an order of magnitude lower than our values.

Gallium, Rubidium, Cesium, Cobalt. These elements were measured by many authors. Their results agree well with ours.

DISCUSSION

Comparison with achondrites and terrestrial basalts

The geochemical behavior of most of our elements is not well enough understood for a quantitative petrologic interpretation. We shall therefore limit ourselves to an empirical comparison of lunar basalts with their terrestrial and meteoritic counterparts (Figs. 1 and 2). Elements are arranged roughly in the order of decreasing abundance relative to Cl chondrites. Lunar data are represented by filled symbols, meteoritic data by open symbols, and average terrestrial basalt by crosses. The meteoritic data are taken from LAUL *et al.* (unpublished work), except Rb (PAPANASTASSIOU and WASSERBURG, 1969). Among the 5 classes of Ca-rich achondrites plotted here, eucrites and shergottites are most appropriate for such a comparison because they closely resemble lunar basalts, as noted by many authors. Howardites, though still similar in bulk chemistry, often contain chondritic and other extraneous material, and thus are less suitable for comparison (MAZOR and ANDERS, 1967; JÉROME, 1970). Angrites and nakhlites are represented by only one meteorite each and resemble lunar basalts less closely than do the other classes. But since they are usually classified as "basaltic" achondrites, we have included them in this figure.

All of these meteorites probably come from the asteroid belt, and thus reflect basalt formation processes in smaller bodies (<400 km radius) and at greater distances from the Sun (2–3 a.u.). To be sure, an asteroidal origin of Ca-rich achondrites is inferred largely by default: O^{18}/O^{16} ratios and other arguments imply that at least the eucrites are not lunar (TAYLOR and EPSTEIN, 1970; HEYMANN *et al.*, 1968); differentiated composition suggests that they are not cometary; and the now-established absence of lunar rocks in the world's meteorite collections apparently rules out an origin from larger bodies (Mars, Mercury, etc.) because ejection of rocks by impact is apparently very improbable even for the Moon whose escape velocity is as low as

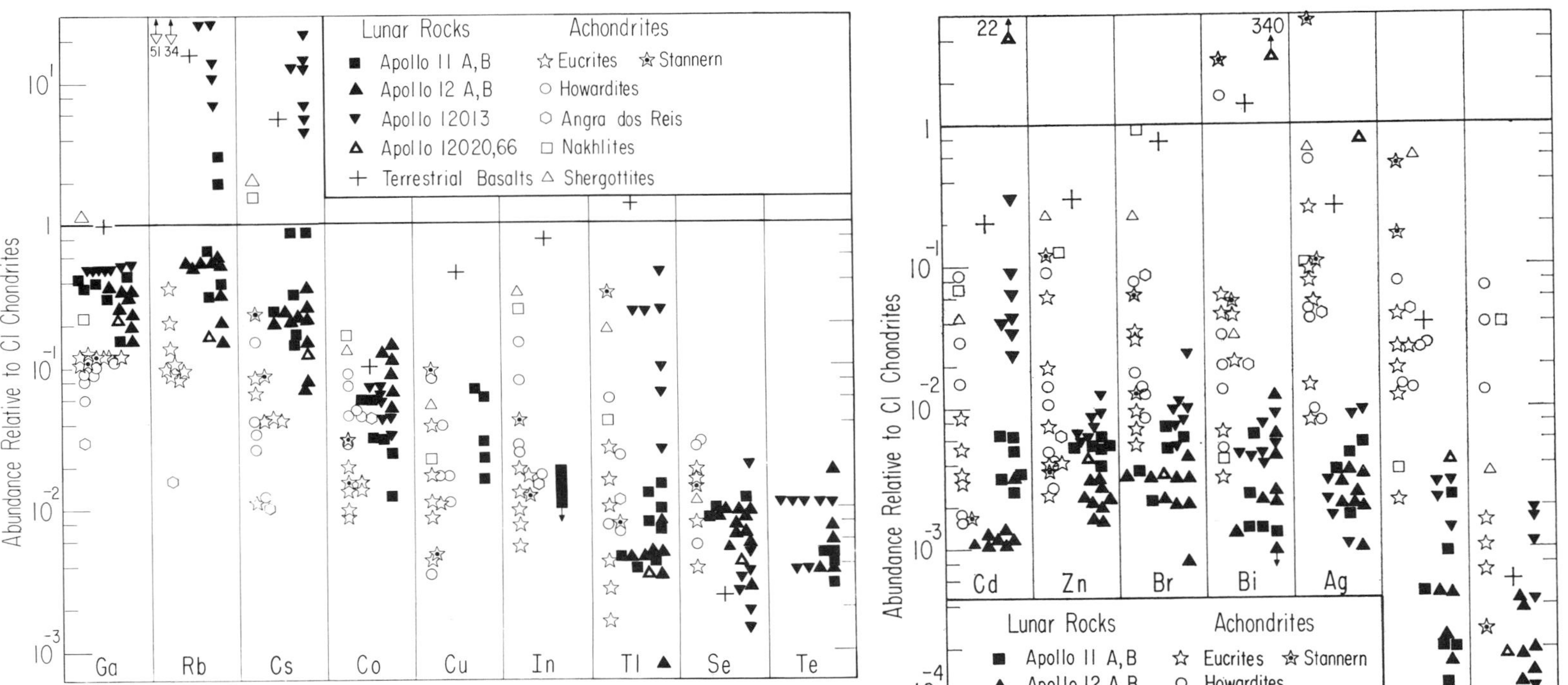

Figs. 1 and 2. Comparison of lunar, terrestrial, and meteoritic basalts. Volatile elements (Cs, Rb, In, Tl, Cd, Zn, Br, Bi) are depleted in lunar basalts by 1–2 orders of magnitude, relative to terrestrial continental basalts (TAYLOR, 1964). Only a few rare and atypical samples (rock 12013; core 12028,66) occasionally deviate from this trend. Eucrites often show comparable depletions. Among siderophile elements, Ag and Au are strongly depleted in lunar rocks relative to both eucrites and terrestrial basalts, while Ir (and presumably Se, Te) are not.

2.4 km/sec. The discovery that the asteroid Vesta has a basaltic surface (MCCORD *et al.*, 1970) further strengthens the case for an asteroidal origin of Ca-rich achondrites.

Breccia 12013, with 7 analyses (inverted triangles), is clearly overrepresented in Figs. 1 and 2. Its "dark" component, typified most nearly by our sample 06 which is lowest in Cs, Rb, Tl, and Cd, seems to comprise 10–60% of Apollo 12 soils and thus is an important rock type at least in that area of the Moon (HUBBARD *et al.*, 1971; LAUL *et al.*, 1971). But the "light," granitic component prominent in the other 6 samples seems to represent an extreme differentiate of lower abundance. We shall therefore use only the 12013,10,06 point in those cases (Cs, Rb, Tl, Cd) where the remaining 12013 samples lie well above the commonplace lunar basalts, and count their average as a single analysis in all other cases. We shall also ignore the spectacularly high Cd, Bi, Ag values of core sample 12028,66, as such material apparently comprises less than 10^{-4} of the Apollo 11 and Apollo 12 regolith (GANAPATHY *et al.*, 1970b). Au and Ir in sample 12013,10,06 probably are largely meteoritic and will also be disregarded.

Siderophile elements

Let us compare five siderophile elements: Ag, Au, Ir, Re, and Ni (Table 2). The

Table 2. Mean Abundances of Siderophile Elements Relative to Type I Carbonaceous Chondrites

	Ag	Au	Ir	Re	Ni	Ref.
Apollo 11, 12 basalts	3×10^{-3}	6×10^{-4}	2×10^{-4}	2×10^{-3}	3×10^{-3}	a, b, c, d
Eucrites	8×10^{-2}	4×10^{-2}	9×10^{-4}	2×10^{-3}	1.3×10^{-3}	e, f, g
Terrestrial basalts (Continental)	3×10^{-1}	1×10^{-2}	6×10^{-4}	1×10^{-2}	1×10^{-2}	h, i, j, k

(a) GANAPATHY *et al.* (1970a): Ag, Au, Ir; (b) This work: Ag, Au, Ir; (c) HERR *et al.* (1971); Re; (d) Misc. authors, Proceedings Apollo 11 and Apollo 12 Conferences: Ni; (e) LAUL *et al.* (unpublished work): Ag, Au, Ir; (f) LOVERING and BUTTERFIELD (1970): Re; (g) NICHIPORUK *et al.* (1967): Ni; (h) TAYLOR (1964): Ag, Ni; (i) FLANAGAN (1969); FLEISCHER (1969); and this work: Au; (j) CROCKET *et al.* (1968): Ir; (k) MORGAN and LOVERING (1967): Re.

last three show similar depletions in all three types of basalt, while the first two show much greater depletion in the Moon, by nearly a factor of 100. The difference for Au which is crucial to our arguments seems well substantiated (Fig. 3). On a purely empirical basis, one can argue that the contrast between the lunar and terrestrial pattern speaks against the fission hypothesis, as already noted in our earlier paper (GANAPATHY *et al.*, 1970a). If the Moon had lost its siderophiles while still part of the Earth, it should show a terrestrial siderophile pattern.

The reason for the differences in pattern is not easy to determine, however. One seemingly obvious explanation is volatility. The elements that show disproportionate depletion on the Moon, Ag and Au, happen to be the most volatile ones of the five. If the metal particles that accreted into the Moon had ceased to equilibrate with the nebula at high temperatures (>1300°K), Au and Ag would be left behind in the gas (LARIMER, 1967). However, it does not seem possible to account for both the siderophile and volatile element depletion by volatility alone, without invoking special, rather contrived, assumptions.

Another possibility is a second stage of metal extraction on the Moon, i.e., reduction of some Fe^{2+} to Fe, followed by gravitational segregation of this secondary

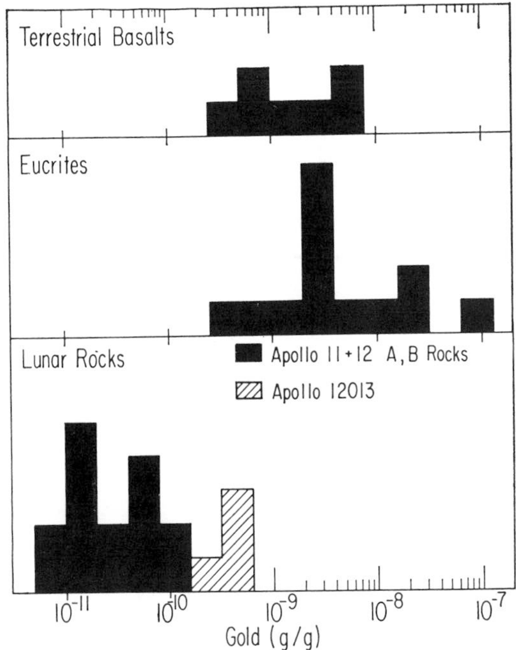

Fig. 3. Gold content of lunar basalts lies below that of all terrestrial and meteoritic basalts investigated to date.

metal. If this second extraction happened under the same redox conditions as the first, it should have depleted each element by a factor proportional to its metal/silicate distribution coefficient (which in turn is proportional to the depletion factors in Table 2). Iridium, with the greatest depletion factor, should then have been depleted the most, while Au, Re, and Ni, with essentially identical depletion factors in terrestrial basalts, should have been depleted in lunar basalts to a uniform degree, contrary to observation.

A third possibility is chemical complexing of Ag and Au in the Earth, which would cause them to favor the silicate phase. GANAPATHY *et al.* (1970a) had suggested complexing by water, but this does not explain why the bone-dry eucrites, angrites, and other achondrites were equally successful in retaining Ag and Au. A shift of the partition equilibria by pressure must be rejected for the same reason, as pressures in the achondrite parent bodies were surely smaller than those in the Earth and Moon.

The most likely explanation of the Ag and Au depletion is a difference in redox conditions during metal-silicate segregation. Under more strongly reducing conditions, these elements may be expected to concentrate in the metal to a greater degree. The lower Ag and Au abundance in lunar basalts compared to terrestrial basalts may thus be attributed to more complete reduction to the metal. This is consistent with the lower oxygen fugacity estimated for lunar basalts ($f_{O_2} = 10^{-14}$ to 10^{-16} vs. 10^{-10} bars; SMITH *et al.*, 1970).

If this explanation is correct, eucrites (and probably all Ca-rich achondrites; see Fig. 2) apparently formed at an oxygen fugacity closer to the terrestrial than to the lunar value. Regrettably, no direct estimates of f_{O_2} for achondrites are available that could be used to check this prediction.

Volatile elements

All of the remaining elements in Figs. 1 and 2 are volatile, having condensation temperatures of $\sim$1100°K to $\sim$400°K from a solar gas at 10^{-2} to 10^{-4} atm (LARIMER, 1967; ANDERS, 1971). Except for Se (and probably Te, for which no comparison data exist) they are consistently less abundant in the Moon and eucrites than in the Earth. The most volatile of these (In, Tl, and Bi) have been used as "cosmothermometers" to estimate the accretion temperature of the Earth and chondrite parent bodies (UREY, 1954; LARIMER and ANDERS, 1967; KEAYS *et al.*, 1971; ANDERS, 1971). It would obviously be of interest to make a similar estimate for the Moon and eucrites. (We have stated in our Apollo 11 paper why we think that the depletion in volatiles happened in the solar nebula rather than on the lunar surface, and shall not repeat the arguments here.)

In terms of the two-component model of WOOD (1962a) and LARIMER and ANDERS (1967, 1970), each planet consists of a low-temperature fraction that condensed most of its volatiles and a remelted, high-temperature fraction that lost them. (We have no direct evidence that this model is applicable to the Moon. However, as it accounts fairly well for the fractionation of 55 depleted elements in meteorites, its extension to the Moon would seem justified by Occam's principle.) Let us first estimate the amount of the low-temperature fraction in the Moon and then its accretion temperature. In order to obtain figures representative of the whole planet, we shall use elements that are largely concentrated in the crust.

Amount of Low-Temperature Material. We can estimate this from the Cs/U ratio, because Cs, being rather volatile, should occur only in the low-temperature fraction, while U, being refractory, should occur in both the low- and high-temperature fractions. Both elements concentrate almost quantitatively in the crust during planetary differentiation. They are not readily separated from one another by igneous processes, judging from the near-constancy of Cs/U, K/U, and K/Cs ratios in a wide range of rocks. Thus the Cs/U ratio in lunar basalts should be a good approximation of the whole-moon ratio.

Table 3 gives Cs/U ratios for the Earth, Moon, and eucrites, all normalized to Cl chondrites. If these ratios are taken at face value, they suggest that the amount of low-temperature fraction decreases from 11% in the Earth to 0.5% in the eucrite parent bodies. Of course, this is based on the tacit assumption that the high-temperature fraction had cosmic composition. Were it enriched in U and other refractories by some factor (e.g., 5, GAST, 1971), then the percentages of low-temperature fraction in Table 3 would have to be raised by the same factor.

Accretion Temperatures. A nominal accretion (or condensation) temperature of the low-temperature fraction can be estimated from the abundance of a suitably volatile element. In the range of partial condensation, its abundance is a sensitive function of temperature. We shall use Tl for this purpose because it correlates with Cs over nearly 3 orders of magnitude (Fig. 4), and hence appears to be similarly concentrated in the crust (GANAPATHY *et al.*, 1970a). Since Cs indicates the amount of low-temperature

Table 3. Abundance of volatile elements.

	$\frac{Cs^a}{U}$	$\frac{Tl^a}{Cs}$	Accretion Temperatures °K[b] Tl	Bi	Pb^{204}	In	Mean	Ref.[e]
Moon (Apollo 11, Apollo 12)	0.015	0.018	601	658	613	(582)	624	1
Earth (Cont. basalts)	0.11	0.25	530	567	572	561	558	1
Eucrites[c]	0.005	0.18	454	470		482	469	2
L-Chondrites (Abund. Ave.)[d]			513	521		514	516	3
L-Chondrites (Temp. Ave.)[d]			539	542		533	538	3

[a] Relative to Cl chondrites, on the basis of the following mean abundances: Cs = 186 ppb, U = 12 ppb, Tl = 74 ppb.

[b] At a total nebular pressure of 10^{-2} atm for Earth and Moon, and 10^{-4} atm for meteorites.

[c] Excluding an atypical sample of Stannern that is strongly enriched in a variety of trace elements (Figs. 1, 2).

[d] *Abundance average:* average composition of L-chondrite parent body was estimated by calculating mean abundance in each petrologic type and combining these into a grand average, weighted by relative frequency of each petrologic type (Van Schmus, 1969). A nominal accretion temperature was then calculated from this "grand average." *Temperature average:* individual accretion temperatures were calculated for each meteorite for which data were available. These temperatures were first averaged for each petrologic type, and then combined into a grand average after weighting by relative frequency of each petrologic type.

[e] (1) This work; (2) LAUL *et al.* (unpublished work); (3) KEAYS *et al.* (1971).

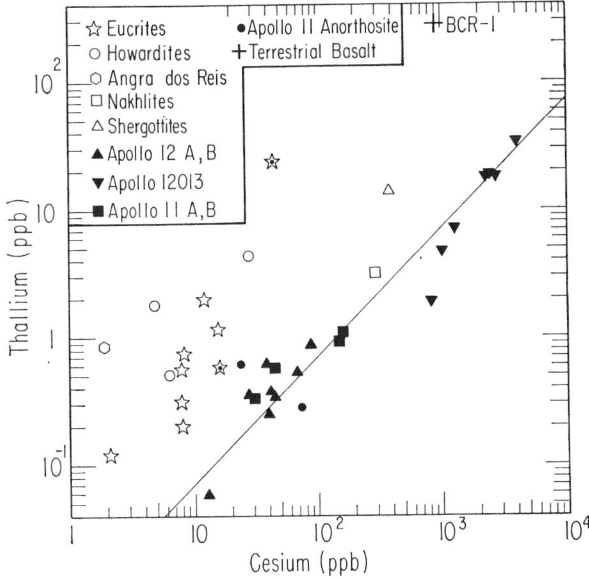

Fig. 4. Tl and Cs correlate in lunar rocks over 3 orders of magnitude. The Tl/Cs ratio is lower than that of terrestrial and most meteoritic basalts.

fraction, the Tl/Cs ratios (Table 3, column 3) give the abundance of Tl in the low-temperature fraction.

Evidently lunar basalts have a much lower Tl/Cs ratio than do eucrites and terrestrial basalts. The constancy of the ratio from the most alkali-poor lunar basalts and anorthosites to the alkali-rich rock 12013 (Fig. 4) suggests that the Tl/Cs ratio is a meaningful average for the whole Moon. We can use this ratio to estimate a mean accretion temperature. Similar estimates can be obtained from Bi, In, and Pb^{204}

abundances, but these will be less reliable since it is less certain that these elements were quantitatively concentrated in the crust during planetary differentiation.

In order to relate these abundances to temperature, we need to know the condensation curves of these elements. The best available estimates (Fig. 5) come from the work of Larimer (unpublished data, 1970, see KEAYS *et al.*, 1971). They involve guesses of certain thermodynamic quantities (solubilities of trace metals in nickel-iron) and may hence be in error by as much as 50°K.

Condensation temperatures also depend on the total pressure in the nebula. For the meteorites, representing a heliocentric distance of 2–3 a.u., a value of $10^{-4\pm2}$ atm has been estimated (LARIMER and ANDERS, 1970). For the Earth and Moon, no independent estimates are available, but most published models of the solar nebula predict pressures some 1–2 orders of magnitude higher at 1 a.u. than at 2–3 a.u. Accordingly, we shall assume a nebular pressure of 10^{-4} atm for the meteorites and 10^{-2} atm for the Earth and Moon. (It seems fair to use the same value for the Earth and Moon, because even the capture theory in its latest version postulates formation of the Moon at very nearly 1.0 a.u.; SINGER, 1969.)

Temperatures calculated in this manner are given in columns 4 to 8 of Table 3. We see that the four cosmothermometers agree rather well for each body. This suggests that the condensation curves are not grossly in error, and that the assumption

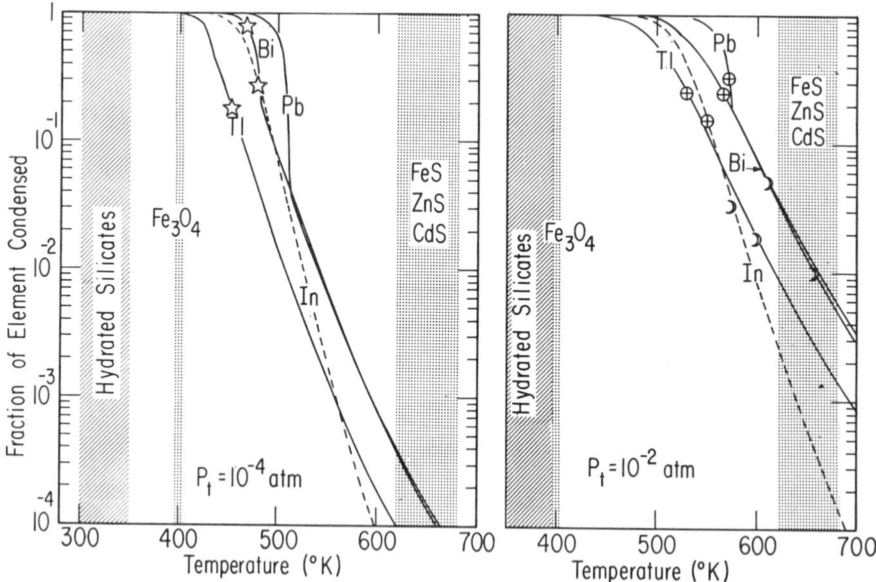

Fig. 5. Nominal accretion temperatures of the Earth, Moon, and eucrites can be estimated by plotting abundances of volatile elements on the condensation curves of these elements (Larimer, 1967, and unpublished work cited by KEAYS *et al.*, 1971). The two pressures used are probably representative for the solar nebula at 2–3 a.u. and 1 a.u., respectively. Shaded areas indicate temperature ranges for the reactions Fe → FeS, Fe → Fe_3O_4, and anhydrous silicates → hydrated silicates. (The latter is a rough estimate.)

of quantitative concentration in the crust is nearly as valid for Bi, In, and Pb as for Tl and Cs. Abundances would need to be changed by no more than a factor of 5 to establish complete concordancy.

Interpretation. Some caution is needed in interpreting these temperatures. Values for chondrites are probably most meaningful, being based on decimeter-sized samples of relatively undifferentiated matter that evidently accreted over a short span of time. But the remaining values are based on differentiated material, which at best represents planet-wide average abundances of the thermometric elements. Temperatures probably fell during accretion of the planets, and accretion rates also varied. Since the abundance of volatile elements in the accreting dust was not a linear function of temperature, but more nearly a logarithmic function, the mean abundance of a volatile element does not give the true mean temperature, but a value unduly weighted toward lower temperatures, when the bulk of the volatile elements accreted (TUREKIAN and CLARK, 1969). To facilitate comparisons, two values are therefore given for the L-chondrite parent body (Table 3): one representing a *true* mean temperature based on a weighted average of individual chondrite temperatures, and the other, a *nominal* average temperature based on the average abundance of thermometric elements in the body. The latter is appropriate for direct comparisons with the Earth-Moon-eucrite values.

Temperatures for the Moon are consistently higher than those for the Earth. A possible explanation is that the Moon originated closer to the Sun, but this can at most account for a small part of the difference. The temperature distribution in the inner solar nebula must have been quite flat, no steeper than $T \propto r^{-1/2}$, judging from the small difference in the accretion temperatures of the Earth (1.00 a.u.) and meteorites (2–3 a.u.). The gradient between 1 and 2 a.u. could not have exceeded 150°, even if all parameters are stretched to the limit. Since the capture theory (SINGER, 1969) constrains the Moon to an initial orbit at about 1.00 ± 0.05 a.u., a temperature difference of less than 15° would be expected between Earth and Moon. Apparently the major part of the difference is due to some other cause.

It may be more realistic to ascribe the Moon's volatile depletion to a lower growth rate in the terminal stages of accretion, when temperatures were low enough for volatiles to condense extensively on residual dust. GANAPATHY *et al.* (1970a) had suggested that such a low growth rate might reflect accretion in a satellite orbit, the Moon's motion about the Earth leading to high encounter velocities and hence low capture efficiencies for dust in circular heliocentric orbits. SINGER and BANDERMANN (1970) have reexamined this problem and found that the effect would work in the opposite direction, as long as gas pressures were too low to damp the motion of accreting particles. Under these conditions, dust would concentrate in the Earth's neighborhood, raising rather than lowering the Moon's accretion rate.

SINGER and BANDERMANN note that their treatment is valid as long as n, the gas density in molecules/cm^3, is less than $10^{14}\, a$, where a is the particle radius in cm. Let us see whether this condition was satisfied under the expected accretion conditions in the inner solar system. It seems likely on experimental and theoretical grounds that the radius of the volatile-bearing grains was on the order of 10^{-4} to 10^{-5} cm. This is the observed particle size in carbonaceous chondrites and the matrix of unequilibrated chondrites (WOOD, 1962b; KERRIDGE, 1964, 1970), and also the maximum grain radius for diffusional equilibration of dust grains in a cosmic gas at 500° to 600°K (Fig. 5 of LARIMER, 1967). At this grain radius, gas effects can be neglected only at $P < 10^{-9}$ to 10^{-10} atm; this is several orders of magnitude less than the estimated pressure in the inner solar nebula, 10^{-2} to 10^{-4} atm, and well below the minimum values consistent with the meteoritic data (10^{-6} atm) or with the requirement that sufficient mass remain in the outer solar system to build Jupiter and Saturn (10^{-5} atm). Apparently a realistic treatment of the problem must take the effects of the gas into account. This has not yet been attempted. Meanwhile, the GANAPATHY *et al.* and SINGER–BANDERMANN treatments may be regarded as limiting cases.

We can also try to approach the problem empirically, now that data for eucrites have become available. Do the abundance data form a simple monotonic sequence, or do they show an anomaly for the Moon that might be attributed to formation as a satellite rather than an independent body? The results are somewhat contradictory. The Cs/U ratio, indicating the amount of low-temperature material, falls monotonically with radius, from 0.11 for the Earth to 5×10^{-3} for the eucrites (Table 3). As yet there is no theory that predicts the ratio of remelted to unremelted material in planets as a function of size and heliocentric distance, and hence we have no way of telling whether the Moon's position in this sequence is anomalous. The Tl/Cs ratio, on the other hand, shows no correlation with radius, the Moon's being the lowest.

The trend persists when the comparison is extended to other classes of achondrites (angrites, nakhlites, shergottites; Fig. 4). Only one measurement is available for each class, but except for Nakhla, they again fall distinctly above the lunar ratio. A trivial reason for this difference may be the more distant origin of the achondrites. The high Tl/Cs ratio may merely reflect lower temperatures in the asteroid belt.

Thus we have no reliable frame of reference, empirical or theoretical, against which to judge the Earth–Moon difference in volatile content. All we can do is estimate the magnitude of this difference from our latest data, for comparison with future models.

As noted above, the bulk of the volatile elements was probably brought in toward the end of accretion, when temperatures in the nebula had fallen low enough for Bi, Tl, etc. to condense. The volatile contents of Earth and Moon thus reflect relative accretion rates in the final stages of accretion.

The overall content of volatile elements is probably best expressed by the Tl/U ratio (normalized to Cl chondrites). It is 2.7×10^{-4} for the Moon and 2.7×10^{-2} for the Earth. If both planets have equal U abundances (per unit mass) and are wholly differentiated, then these ratios imply $Tl_{Earth}/Tl_{Moon.} = 100$. If only the outer parts of the Moon melted (UREY *et al.*, 1971), corresponding to a mass fraction α, then $Tl_E/Tl_M = 100/\alpha$, because the melted layer contains all the Moon's Tl but only α its U. Most authors nowadays agree that $\alpha \geq 0.3$, and thus the abundance ratio Tl_E/Tl_M must lie between 100 and 300. (The abundances were calculated per unit mass of the differentiated portion of the planet, not unit area as done by SINGER and BANDERMANN, 1970.) The Moon's depletion in water must be at least as great as that in Tl. Its initial water content apparently was no greater than 370 g/cm² (ANDERS, 1970).

The Moon's low volatile content may actually be the cause of its singular depletion in some siderophile elements, such as Ag and Au. We concluded earlier that the depletion of Ag and Au was due to more strongly reducing conditions in the Moon. In terms of the two-component model (LARIMER and ANDERS, 1967, 1970), the redox state of a planet depends mainly on the *amount* of low-temperature material, and on the equilibration *temperature* of this material with the nebular gas. As temperature decreases, the material first becomes richer in Fe^{2+}, then Fe^{3+} and H_2O (LARIMER and ANDERS, 1967; LARIMER, 1968; ANDERS, 1971). Thus the redox state will be roughly correlated with Tl content, which in turn reflects both the amount and equilibration temperature of low-temperature material. The Moon, with the lowest Tl/U ratio, should be the most highly reduced body in Table 3, and apparently is.

Rocks 12013 and 12008

We have already discussed our results on rock 12013 in a separate publication (LAUL *et al.*, 1970). As far as our 15 elements are concerned, this rock fits the trends of the ordinary Apollo 11 and Apollo 12 basalts. It shows a typical lunar abundance pattern, quite unlike the terrestrial or tektite pattern. Its description as "tektite glass" by O'KEEFE (1970) therefore seems inappropriate, as already noted by KING *et al.*

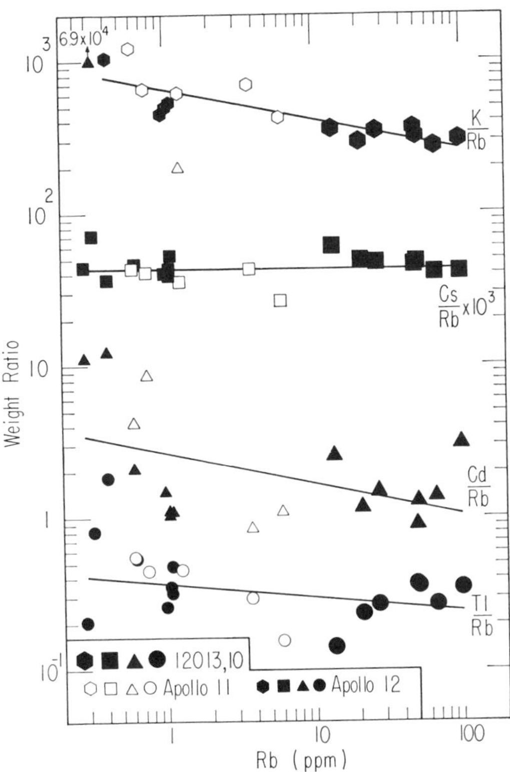

Fig. 6. Abundance ratios for alkalis, Tl, and Cd remain nearly constant, as the concentration varies over a range of 300. Straight lines are least-squares fits.

(1970). Interelement correlation plots for 12013 have been published in our earlier paper. A novel and interesting trend, first brought to light by 12013, is the correlation between Cd and alkalis in lunar rocks, which seems to have no parallel on Earth (Fig. 6). From the leaching experiment on Apollo 11 soil (LAUL *et al.*, 1971), we suspect that Cd, U, Th, and Pb, though correlated with Ca and alkalis in bulk analyses, are located in an acid-soluble minor phase associated with feldspar, not in feldspar itself. Alkalis and Ca, which largely reside in feldspars or glasses, are much less soluble in HNO_3 than are the above elements.

The titanium-rich, glassy rock 12008 shows no noteworthy compositional trends, at least insofar as the elements measured by us are concerned. A possible exception is its Co content, which is somewhat higher than that of the other rocks in Table 1.

Acknowledgments—We are indebted to Rudy Banovich, Mary Joan Hasche, Annie Pierce, and Frank Quinn for assistance in various phases of this work. Financial support came mainly from NASA Contract NAS 9–7887, supplemented by AEC Contract AT(11–1)–382. Page charges, not an allowable expense under NAS 9–7887, were provided by NASA Grant NGL 14–001–167.

REFERENCES

ANDERS E. (1970) Water on the moon? *Science* **169**, 1309–1310.

ANDERS E. (1971) Meteorites and the early solar system. *Ann. Rev. Astron. Astrophys.* **9** (in press).

BAEDECKER P. A., SCHAUDY R., ELZIE J. L., KIMBERLIN J., and WASSON J. T. (1971) Trace element studies of rocks and soils from Oceanus Procellarum and Mare Tranquillitatis. Second Lunar Science Conference (unpublished proceedings).

BIGGAR G. M., O'HARA M. J., and PICKETT A. (1971) Crystallization history of protohypersthene basalts from the Ocean of Storms. Second Lunar Science Conference (unpublished proceedings).

BOUCHET M., KAPLAN G., VOUDON A., and BERTOLETTI M. J. (1971) Spark mass spectrometric analysis of major and minor elements in six lunar samples—Age determinations of four rocks. Second Lunar Science Conference (unpublished proceedings).

BRUNFELT A. O., HEIER K. S., and STEINNES E. (1971) Determination of 40 elements in Apollo 12 materials by neutron activation analysis. Second Lunar Science Conference (unpublished proceedings).

COMPSTON W., BERRY H., VERNON M. J., CHAPPELL B. W., and KAYE M. (1971) Rubidium-strontium chronology and chemistry of lunar material from the Ocean of Storms. Second Lunar Science Conference (unpublished proceedings).

CROCKET J. H., KEAYS R. R., and HSIEH S. (1968) Determination of some precious metals by neutron activation analysis. *J. Radioanal. Chem.* **1**, 487–507.

FLANAGAN F. J. (1969) U.S. Geological Survey standards, II. First compilation of data for the new USGS rocks. *Geochim. Cosmochim. Acta* **33**, 80–120.

FLEISCHER M. (1969) U.S. Geological Survey standards, I. Additional data on rocks G–1 and W–1, 1965–1967. *Geochim. Cosmochim. Acta* **33**, 65–78.

GANAPATHY R., KEAYS R. R., LAUL J. C., and ANDERS E. (1970a) Trace elements in Apollo 11 lunar rocks: Implications for meteorite influx and origin of moon. *Proc. Apollo 11 Lunar Sci. Conf., Geochim. Cosmochim. Acta* Suppl. 1, Vol. 2, pp. 1117–1142. Pergamon.

GANAPATHY R., KEAYS R. R., and ANDERS E. (1970b) Apollo 12 lunar samples: Trace element analysis of a core and uniformity of the regolith. *Science* **170**, 533–535.

GAST P. W. (1971) The chemical composition of the earth, the moon, and chondritic meteorites. To be published in Proceedings of a Symposium in Honor of Francis Birch. McGraw-Hill.

GOLES G. G., DUNCAN A. R., OSAWA M., MARTIN M. R., BEYER R. L., LINDSTROM D. J., and RANDLE K. (1971) Analyses of Apollo 12 specimens and a mixing model for Apollo 12 "soils." Second Lunar Science Conference (unpublished proceedings).

HERR W., HERPERS U., MICHEL R., ABDEL RASSOUL A., and WOELFLE R. (1971) Spallogenic

^{53}Mn(T = 2 × 10^6 a) and search for Re isotopic anomalies in lunar surface material by means of neutron bombardment. Second Lunar Science Conference (unpublished proceedings).

HEYMANN D., MAZOR E., and ANDERS E. (1968) Ages of calcium-rich achondrites, I. Eucrites. *Geochim. Cosmochim. Acta* **32**, 1241–1268.

HUBBARD N. J., GAST P. W., and MEYER C. (1971) The origin of the lunar soil based on REE, K, Rb, Ba, Sr, P and Sr $^{87/86}$ data. Second Lunar Science Conference (unpublished proceedings).

JÉROME D. Y. (1970) Composition and origin of some achondritic meteorites. Ph.D. Dissertation, University of Oregon.

KEAYS R. R., GANAPATHY R., and ANDERS E. (1971) Chemical fractionations in meteorites, IV. Abundances of fourteen trace elements in L-chondrites: Implications for cosmothermometry. *Geochim. Cosmochim. Acta* **35**, 337–363.

KERRIDGE J. F. (1964) Low-temperature minerals from the fine-grained matrix of some carbonaceous meteorites. *Ann. N.Y. Acad. Sci.* **119**, 41–53.

KERRIDGE J. F. (1970) Some observations on the nature of magnetite in the Orgueil meteorite. *Earth Planet. Sci. Lett.* **9**, 299–306.

KING E. A., Jr., MARTIN R., and NANCE W. B. (1970) Tektite glass *not* in Apollo 12 sample. *Science* **170**, 199–200.

LARIMER J. W. (1967) Chemical fractionations in meteorites, I. Condensation of the elements. *Geochim. Cosmochim. Acta* **31**, 1215–1238.

LARIMER J. W. (1968) Experimental studies on the system Fe–MgO–SiO_2–O_2 and their bearing on the petrology of chondritic meteorites. *Geochim. Cosmochim. Acta* **32**, 1187–1207.

LARIMER J. W. and ANDERS E. (1967) Chemical fractionations in meteorites, II. Abundance patterns and their interpretation. *Geochim. Cosmochim. Acta* **31**, 1239–1270.

LARIMER J. W. and ANDERS E. (1970) Chemical fractionations in meteorites, III. Major element fractionations in chondrites. *Geochim. Cosmochim. Acta* **34**, 367–388.

LAUL J. C., KEAYS R. R., GANAPATHY R., and ANDERS E. (1970) Abundance of 14 trace elements in lunar rock 12013,10. *Earth Planet. Sci. Lett.* **9**, 211–215.

LAUL J. C., MORGAN J. W., GANAPATHY R., and ANDERS E. (1971) Meteoritic material in lunar samples: Characterization from trace elements. *Proc. Second Lunar Sci. Conf.*, Vol. 2, pp. 1139–1158. M.I.T. Press.

LOVERING J. F. and BUTTERFIELD D. (1970) Neutron activation analysis of rhenium and osmium in Apollo 11 lunar material. *Proc. Apollo 11 Lunar Sci. Conf., Geochim. Cosmochim. Acta* Suppl. 1, Vol. 2, pp. 1351–1355. Pergamon.

MAZOR E. and ANDERS E. (1967) Primordial gases in the Jodzie howardite and the origin of gas-rich meteorites. *Geochim. Cosmochim. Acta* **31**, 1441–1456.

MCCORD T. B., ADAMS J. B., and JOHNSON T. V. (1970) Asteroid Vesta: Spectral reflectivity and compositional implications. *Science* **168**, 1445–1447.

MORGAN J. W., LAUL J. C., GANAPATHY R., and ANDERS E. (1971) Glazed lunar rocks: Origin by impact. *Science* **172**, 556–557.

MORGAN J. W. and LOVERING J. F. (1967) Rhenium and osmium abundances in some igneous and metamorphic rocks. *Earth Planet. Sci. Lett.* **3**, 219–224.

MORRISON G. H., GERARD J. T., POTTER N. M., GANGADHARAM E. V., ROTHENBERG A. M., and BURDO R. A. (1971) Elemental abundances of lunar soil and rocks from Apollo 12. Second Lunar Science Conference (unpublished proceedings).

NICHIPORUK W., CHODOS A., HELIN E., and BROWN H. (1967) Determination of iron, nickel, cobalt, calcium, chromium and manganese in stony meteorites by X-ray fluorescence. *Geochim. Cosmochim. Acta* **31**, 1911–1930.

O'KEEFE J. A. (1970) Tektite glass in Apollo 12 sample. *Science* **168**, 1209–1210.

PAPANASTASSIOU D. A. and WASSERBURG G. J. (1969) Initial strontium isotopic abundances and the resolution of small time differences in the formation of planetary objects. *Earth Planet. Sci. Lett.* **5**, 361–376.

REED G. W. and JOVANOVIC S. (1971) The halogens and other trace elements in Apollo 12 soil and rocks; Halides, platinum metals and mercury on surfaces. Second Lunar Science Conference (unpublished proceedings).

Singer S. F. (1969) Was the moon captured and what were the consequences? Paper presented at the Natl. Fall Meeting of the Amer. Geophys. Union, San Francisco, December 17.

Singer S. F. and Bandermann L. W. (1970) Where was the moon formed? *Science* **170**, 438–439.

Smales A. A., Mapper D., Webb M. S. W., Webster R. K., Wilson J. D., and Hislop J. S. (1971) Elemental composition of lunar surface material. (Part 2) Second Lunar Science Conference (unpublished proceedings).

Smith J. V., Anderson A. T., Newton R. C., Olsen E. J., Wylie P. J., Crewe A. V., Isaacson M. S., and Johnson D. (1970) Petrologic history of the moon inferred from petrography, mineralogy, and petrogenesis of Apollo 11 rocks. *Proc. Apollo 11 Lunar Sci. Conf., Geochim. Cosmochim. Acta* Suppl. 1, Vol. 1, pp. 897–925. Pergamon.

Taylor H. P., Jr. and Epstein S. (1970) O^{18}/O^{16} ratios of Apollo 11 lunar rocks and minerals. *Proc. Apollo 11 Lunar Sci. Conf., Geochim. Cosmochim. Acta* Suppl. 1, Vol. 2, pp. 1613–1626. Pergamon.

Taylor S. R. (1964) Abundance of chemical elements in the continental crust: A new table. *Geochim. Cosmochim. Acta* **28**, 1273–1285.

Turekian K. K. and Clark S. P., Jr. (1969) Inhomogeneous accumulation of the earth from the primitive solar nebula. *Earth Planet. Sci. Lett.* **6**, 346–348.

Urey H. C. (1954) On the dissipation of gas and volatilized elements from protoplanets. *Astrophys. J.* Suppl. 1, 147–173.

Urey H. C., Marti K., Hawkins J. W., and Liu M. K. (1971) Model history of the lunar surface. Second Lunar Science Conference (unpublished proceedings).

Van Schmus W. R. (1969) The mineralogy and petrology of chondritic meteorites. *Earth-Sci. Rev.* **5**, 145–184.

Wakita H., Rey P., and Schmitt R. A. (1971) Abundances of the 14 rare earth elements plus 22 major, minor, and trace elements in ten Apollo 12 rocks and soil samples. Second Lunar Science Conference (unpublished proceedings).

Wänke H., Wlotzka F., Teschke R., Baddenhausen H., Spettel B., Balacescu A., Quijano-Rico M., Jagoutz E., and Rieder R. (1971) Major and trace elements in Apollo 12 samples and studies on lunar metallic iron particles. Second Lunar Science Conference (unpublished proceedings).

Warner J. L. (1970) Apollo 12 crystalline rocks—A preliminary classification. Submitted to *Earth Planet. Sci. Lett.*

Wood J. A. (1962a) Chondrules and the origin of the terrestrial planets. *Nature* **194**, 127–130.

Wood J. A. (1962b) Metamorphism in chondrites. *Geochim. Cosmochim. Acta* **26**, 739–749.

Proceedings of the Second Lunar Science Conference, Volume 2, pp. 1037–1061.
The M.I.T. Press, 1971.

Trace element studies of rocks and soils from Oceanus Procellarum and Mare Tranquillitatis

PHILIP A. BAEDECKER, RUDOLF SCHAUDY, JOHN L. ELZIE,
JEROME KIMBERLIN, and JOHN T. WASSON
Department of Chemistry and Institute of Geophysics and Planetary Physics,
University of California, Los Angeles, California 90024

(Received 22 February 1971; accepted in revised form 5 April 1971)

Abstract—Neutron-activation data on Zn, Ga, Ge, Cd, In, and Ir are reported for six rocks and two soils from the Apollo 12 mission. A comparison of these data and similar data for Apollo 11 samples indicates extralunar components in the 12070 and 10084 soils of about 1.0 and 1.1% expressed in terms of an assumed composition which is the same as the water-free portion of C1 chondrites. A relationship between the integrated flux of extralunar material and the increase in concentration of such material in the fines portion of the lunar regolith is derived. According to this relationship, the integrated flux at the Apollo 11 site is 25% higher than that at the Apollo 12 site, which indicates that the influx of extralunar material has been decreasing with time. A re-analysis of the data on the Apollo 11 fines shows them to be consistent with a minor depletion in ilmenite and inclusion of a 20% component resembling Wood's anorthositic gabbro. Terrestrial basalts show relatively small variations in their contents of Zn, Ga, Ge, In, and possibly Cd, independent of their geographical origin. Apollo 12 rocks have concentrations of Zn, Ge, Cd, In, and possibly Ir which are lower by factors of 60 or more relative to terrestrial basalts. A mechanism is proposed for the late accretion of volatile-rich materials, including comets, in which a primitive terrestrial atmosphere is invoked to explain the significantly higher concentrations of such substances on the earth.

INTRODUCTION

OUR PREVIOUS WORK on Apollo 11 samples involved the determination of Ga, Ge, In, Ir, and Au (WASSON and BAEDECKER, 1970). In the present study we have added Zn and Cd, which proved to have unexpectedly high concentrations in Apollo 11 fines and breccias, and we have dropped Au, which provided little information in addition to that from our Ge and Ir determinations, and which was also measured by several other investigative teams. Our six elements were chosen in order to provide a maximum of information about two different questions: (1) the nature of the extralunar component, and its usefulness in evaluating the formation of the lunar regolith; and (2) the distribution of elements of similar ionic properties in lunar crystalline rocks, and the interpretation of this data in terms of plausible geochemical and cosmochemical processes. Although our elements are not ideally suited for such questions, we have also attempted to assess (3) the nature of the extraneous lunar component in the soil. Through the aid of the Manned Spacecraft Center curator and the Lunar Sample Analysis Planning Team, we have obtained a set of six rocks from the Apollo 12 mission which range from the Mg-rich, Al-poor to the Mg-poor, Al-rich extremes described in the LSPET (1970) report. In addition, we have investigated samples of two soils, the typical local (12070) and unusual light gray (12033) fines.

Since a number of our geochemical interpretations depend on a comparison of data on lunar, terrestrial, and meteoritic samples, we have followed the practice of WASSON and BAEDECKER (1970) and included the latter two types of samples in the same neutron activation runs as our lunar samples. In such a fashion we greatly reduce the magnitude of errors introduced in such comparisons. In this paper we have added 14 terrestrial samples to the nine studied by WASSON and BAEDECKER (1970). We have analyzed all samples in duplicate, in the belief that the resulting greater precision would reveal effects which would otherwise have been missed.

EXPERIMENTAL TECHNIQUES

Samples and sample preparation

Approximately one gram each of rock samples 12009, 12014, and 12038, and soil samples 12033 and 12070 were provided for analysis. Two 500-mg aliquots of a powder prepared by grinding a 10 g piece of rock 12063 were analyzed in this work, the remainder of the rock being used for Rb–Sr dating and lead isotope studies by CLIFF *et al.* (1971). The procedures followed in preparing these samples for analysis were identical to those described in WASSON and BAEDECKER (1970).

In our studies on the Apollo 11 series of rock samples, large variations in the observed In content of the material which we analyzed indicated that the samples had been contaminated prior to their receipt in our laboratory, most probably by In–Ag seals on the rock boxes. Larger samples of rocks 12002 and 12022 weighing 3.5 and 3.2 g, respectively, were provided for our studies, in order that the surfaces of the rock chips could be removed prior to analysis, and thus minimize the possibility of surface contamination. An S. S. White airbrasive apparatus using 50 μ SiC as the abrasive powder was used for removing surface material. An analysis of the abrasive powder for the trace elements of interest showed concentrations of 0.48, 0.14, 0.00016, 0.00033, 1.89, and 0.027 ppm for Ga, Ge, In, Ir, Zn, and Cd respectively. Ge and Cd were present in higher concentrations in the SiC powder than was expected in the lunar material, Zn and Ir at approximately the same level of concentration, and Ga and In were significantly lower. If the lunar samples were contaminated by an 0.1% addition of SiC, this could cause as much as a 1% error in the results for Ge and Cd. We believe that the actual contamination is smaller. Approximately 280 mg of rock 12022 was abraded from the 3.2 g chip, and the sample washed in reagent-grade acetone using a sonic cleaner prior to crushing and splitting. Rock sample 12002 was split into one large fragment weighing 1.83 g and several smaller fragments. Two powdered samples of 500 mg each were prepared by crushing some of the uncleaned smaller fragments. The 1.83 g chip was treated with the same sandblasting procedure as rock 12022, 480 mg being removed in the process. The remainder of the sample was crushed, and chips totaling 550 mg in weight were powdered and analyzed. As will be discussed in more detail later, our attempts to remove In contamination were moderately successful.

Our sample of the light gray fines, 12033, was packaged at the Lunar Receiving Laboratory in a stainless steel bolt-top container. Upon opening the container, the sample was found to be contaminated with metal turnings from the threads of the bolt-top. The fines were sieved through a 176 μ screen to remove the turnings, and only the -176 μ fraction was analyzed. The sieved fines were examined under a microscope; no metal turnings could be detected.

Sample packaging, flux monitor preparation and irradiation procedures were identical to those described in WASSON and BAEDECKER (1970). During one of our irradiations an experiment was performed to show whether systematic errors might arise from possible volatilization of the deposited salts from the flux-monitor foils as a result of heating during the irradiation. Each flux-monitor foil was wrapped with a second piece of high-purity aluminum foil; following irradiation the outer foils were processed in the same manner as the flux monitors. An irradiated blank sample of high-purity Al was also processed. The same level of activity for all six radionuclides was observed in the cover foils as was found in the blank. This indicates that any loss of activity from the flux monitor foils due to volatilization is negligible. Although one might argue that volatilized monitor compounds would not necessarily recondense on the second foil during the irradiation, we believe that this is where the largest fraction should be found when any vapor condenses upon cooling at the end of the irradiation.

Analytical and radiometric procedures

The six elements Zn, Ga, Ge, Cd, In, and Ir were determined simultaneously in each rock sample. Since description of the procedures used in the determination of Ga, Ge, In, and Ir was provided in WASSON and BAEDECKER (1970), only our procedures for Zn and Cd will be discussed here. Following sample dissolution, Ga, Ge, In, and Ir were precipitated as hydroxides. Zn and Cd are only partially carried down in the precipitation step, and the supernatant was retained for subsequent chemical processing. The hydroxide precipitate was dissolved in HBr; and Ga, Ge, and In were recovered by solvent extraction. The Ir-containing solution is passed through an anion exchange column, which completely retained the Ir. The column was washed with 6N HNO_3 to elute any Zn and Cd which may have been retained by the column, all column effluents combined with the supernatant from the initial hydroxide precipitation, and the solution evaporated to dryness. The resulting salt cake was dissolved in H_2O, and the solution scavenged of Cr and Sc activities by a hydroxide precipitation step. Zn and Cd were then precipitated from weakly basic solution as the sulfides. The sulfide precipitate was dissolved in concentrated HCl, sulfide was removed by evaporation, and Zn and Cd were separated following an anion exchange procedure outlined by KALLMAN *et al.* (1956). Zn was removed from the column by eluting first with a solution 2M in NaOH and 0.34M in NaCl, and secondly with H_2O. The Cd was eluted with 1M HNO_3, the eluate evaporated to dryness and the residue dissolved in 3N HCl. Cu and Cr carriers were added to the Cd-containing eluate and Cu precipitated as the sulfide with H_2S. Cd was then precipitated as the sulfide by lowering the pH of the solution, and the precipitation cycle was repeated. Both Zn and Cd were recovered from their respective solutions by precipitation as the sulfide. The precipitates were dissolved in HCl, the resulting solutions counted, and the chemical yields determined by atomic absorption.

Gamma counting, using a 3 × 3 in. NaI well detector, was used to measure the Zn and Cd activities (the 0.335 MeV photopeak of the ^{115}In daughter of ^{115}Cd and the 1.114 MeV photopeak of ^{65}Zn).

Precision and accuracy

In order to test the precision of our procedures we have analyzed several replicates of standard rocks BR and W–1. Based on the replicate analyses reported for W–1 (see Table 2) and by WASSON and BAEDECKER (1970) we calculated relative standard deviations of $\pm$3.5, 3.4, 20, 8, 7, and 25, respectively for the elements Zn, Ga, Ge, Cd, In, and Ir. The precision for Zn and In was poorer by a factor of 2 in our analyses of BR, based on fewer determinations. The precision of our results for Ge appear to be poorer for these standard rocks than in most other terrestrial and lunar basalts we have investigated. We have analyzed 21 replicates of the Waianae, Hawaii basalt. If we assume that these all have the same Ge content, we can calculate a relative standard deviation of 7%, which should be an upper limit on the true value for our procedure.

The accuracy of our results can best be assessed by comparing them with the data obtained by other workers on the same rocks. A comparison of our data on the USGS series of standard rocks has indicated no systematic differences for Ga, Ge, In, and Ir which would point to systematic errors in our results. Our mean Zn value is about 10% higher than the average of some 20 analyses of W–1, which may indicate a systematic error in our Zn results. A check of our Zn flux monitor solution failed to explain this error. To date no reliable mean Cd value for W–1 is available for comparison.

RESULTS

The results of our analyses on the Apollo 12 returned lunar samples are presented in Table 1. (In addition we have previously unpublished Ir results on two Apollo 11

Table 1. Replicate and mean concentrations of six trace

Sample	Type	Zn (ppm) replicates	Zn (ppm) mean	Ga (ppm) replicates	Ga (ppm) mean	Ge (ppb) replicates	Ge (ppb) mean
12009,24	A	2,0, 1.3†	1.8	3.2, 3.2	3.2	≤42, ≤40	≤41
12038,24	A	1.3, 34‡	1.3	5.2, 5.2	5.2	≤6, ≤30	≤18
12063,51	A	1.9, 2.7	2.3	4.1, 4.6	4.3	≤30, ≤10	≤20
12002,121	B	1.1, 1.2 1.3	1.2	2.9, 3.0, 2.8	2.9	86, 10, ≤18	≤57
12014,12	B	1.0, 1.0	1.0	2.9, 2.9	2.9	18, ≤7	≤13
12022,51	B	1.2, 3.3	2.0§	3.8, 3.9	3.9	≤10, 50	≤30
12033,94	D	7.2, 7.0	7.1	5.4, 5.4	5.4	221, 225	223
12070,84	D	9.0, 8.8	8.9	4.4, 4.6	4.5	349, 361	355

* Not corrected for uranium fission contribution.
† Value of less than usual accuracy. Given one-half weight in the determination of the mean.
‡ Value shows evidence of contamination. Not included in the determination of the mean.

samples; the 10084,26 fines for which we obtained additional results of 9.7 and 8.2 ppb to those previously reported, and the type B crystalline rock 10058,30 on which we obtained a single result of 0.09 ppb).

In general the problem with In contamination encountered in the Apollo 11 samples appears to be less pervasive on the Apollo 12 samples. The In concentrations observed on the samples of 12022 and 12002 which were treated by the sandblasting procedure described above, and the untreated specimens of 12009, 12038, and 12063 appear to indicate an In concentration of 0.5–2 ppb in lunar crystalline rocks, slightly lower than the lower limit of 3 ppb obtained from the Apollo 11 data. However, rock 12014, the untreated specimen of 12002, and the soil samples 12033 and 12070 have very high In contents, which we believe indicates In contamination.

The amounts of Ge, Cd, and In detected in lunar crystalline rocks are near the limits of detection for our method. Our Ge results for the lunar crystalline rocks are reported as upper limits. We are at present unable to account for the poor reproducibility of our Cd results. The high results reported in samples 12009, 12063, 12002, 12022, and 12070 are believed to be erroneous. The precision of our results for terrestrial rocks is considerably better. Because of the high U and low Cd abundances in lunar rocks, corrections for fission-produced ^{115}Cd have been applied to our data using the U values of BRUNFELT *et al.* (1971), CLIFF *et al.* (1971), MORRISON *et al.* (1971), O'KELLEY *et al.* (1971), RANCITELLI *et al.* (1971), ROSHOLT and TATSUMOTO (1971), SILVER (1971), and TAYLOR *et al.* (1971). For rock 12014 no U data are as yet available, therefore no corrections have been carried out.

Our previous Ga results reported in WASSON and BAEDECKER (1970) should be increased by a factor of 1.05 to provide agreement with our latest Ga results, due to a recalibration of our monitor solution. Our most recent Ga results appear to be too high by 5–10% when compared with these corrected results. We have applied a uniform correction factor of 0.95 to these values, and are searching for a discrepancy. There appears to be some systematic difference between the Apollo 12 crystalline rocks and those recovered from Mare Tranquillitatis. A comparison of our data and that of GANAPATHY *et al.* (1970b) with that of GANAPATHY *et al.* (1970a) for the Apollo 11 rocks indicates that both Zn and Cd appear to be less abundant in the Apollo 12

elements in six rocks and two soils from Oceanus Procellarum

Cd (ppb) replicates	mean	In (ppb) replicates	mean	Ir (ppb) replicates	mean
2.2, ≤38‡	2.2	4.2‡, 1.6	1.6	0.12, 0.05	0.08
4.1, 8.1	6.1	1.0, 1.3	1.1	0.57‡, 0.09	0.09
1.1, ≤21‡	1.1	0.9, 1.4	1.1	≤0.03, 0.06	≤0.04
50‡, 2.0	2.0	5.0‡, 14.2‡, 0.62‖	0.6	0.04, 0.06	0.05
≤2.0					
4.4, 4.0	≤4.2*	133, 201‡	≤133¶	0.09, 0.07	0.08
≤50‡, 6.4	6.4	1.1‖, 2.4‖	1.6§	0.09, 0.49‡	0.09
56, 72	64	56, 38	≤47¶	3.9, 3.9	3.9
456‡, 195	195	131, 248‡	≤131¶	8.1, 7.4	7.7

§ Geometric mean.
‖ Surface of rock sample removed with SiC abrasive powder prior to analysis.
¶ All analyses show evidence of contamination. Reported mean is probably too high.

samples. Whereas Ga was observed to be nearly constant in abundance in the Apollo 11 basalts (Wasson and Baedecker, 1970), a much wider range of Ga concentrations was found in the Apollo 12 rocks. This is consistent with the wider range in major element composition observed by LSPET (1970).

The results on a number of terrestrial rock samples which were analyzed simultaneously with the lunar rocks are summarized in Table 2. For most rocks we report means of analyses of two or more aliquots of the same powder. Results of single analyses are italicized. In general the individual results of replicate analyses fall within the limits of precision previously estimated based on our data for BR and W–1. We have extended the work reported by Wasson and Baedecker (1970) on the Waianae volcano to include three additional specimens from the lower member of this volcano, and have reanalyzed the three specimens from the lower member which were included in our previous study. In addition to our work on the Waianae volcano, we have also analyzed 12 rock samples from the Skaergaard, Greenland intrusion. The results of our analyses on the chilled margin gabbro (which is believed to represent the parent liquid) are presented in Table 2. A detailed discussion and presentation of our data on the Skaergaard will be published elsewhere. Two samples of peridotite were analyzed, PCC–1 and a rock from Salt Lake, Hawaii. Both rock samples were poorer in Zn, Ga, Ge, Cd, and In than all basalt samples analyzed, and Ir was found to be markedly enriched in both rocks. Ga, Cd, and In were found to be higher by a factor of 2 in the Salt Lake sample as compared to PCC–1. The Brown Point pyroxene gabbro is a Ti-rich facies belonging to a mass of coarse gabbroic anorthosite from the Adirondack Mountains. The abundances of all elements are similar to those in other terrestrial igneous rocks, with the exception of Ga, which is lower.

The Extralunar Component

The moon's cratered surface shows ample evidence of accretion of extralunar materials, and it is to be expected that evidence of this extralunar component should be found in the finely comminuted portion of the lunar regolith. Evidence for such a

Table 2. Zn, Ga, Ge, Cd, In and Ir concentrations in terrestrial basalts and other terrestrial rocks.*

Sample Location	Type	No.	Source‡	Zn (ppm)	Ga (ppm)	Ge (ppm)	Cd (ppb)	In (ppb)	Ir (ppb)
Waianae† Hawaii	Tholeiitic basalt	C–7	GAM	115	19.5	1.62	250	78	0.48
Waianae Hawaii	Tholeiitic basalt	C–6	GAM	117	18.2	1.56	114	72	0.57
Waianae Hawaii	Tholeiitic basalt	C–11	GAM	108	21.0	1.65	100	80	0.50
Waianae Hawaii	Tholeiitic basalt	C–5	GAM	123	22.5	1.68	172	88	0.38
Waianae Hawaii	Tholeiitic basalt	C–40	GAM	109	19.9	1.63	173	84	0.58
Waianae Hawaii	Tholeiitic basalt	C–20	GAM	111	21.8	1.75	105	89	0.34
Mid-Atl. Ridge	Tholeiitic basalt	1–3	FA	88	17.0	1.58	149	67	0.18
Mid-Atl. Ridge	Tholeiitic basalt	56–2/3	FA	82	16.0	*1.54*	125	59	0.08
Skaergaard Greenland	Marginal gabbro	EG4507	EAV	82	22.2	1.26	42	41	0.26
Bridal Veil Oregon	Basalt	BCR–1	USGS	133	21.7	1.55	*82*	94	≤0.12
Centerville Virginia	Diabase	W–1	USGS	92	17.9	1.74	155	64	0.32
Essey-la-Côte France	Basalt	BR	CRPG	167	19.0	1.67	124	65	0.04
Mellenbach Germany	Basalt	BM	ZGI	132	18.7	1.89	76	61	0.38
Westerly R.I.	Granite	G–2	USGS	93	23.3	1.36	27	30	0.07
Silver Plume Colorado	Grano-diorite	GSP–1	USGS	107	24.3	1.74	*64*	55	≤0.16
Guano Valley Oregon	Andesite	AGV–1	USGS	96	21.7	1.36	*75*	45	≤0.20
Brown Point New York	Pyroxene gabbro	114	GG	149	*13.7*	1.98	181	*87*	*0.07*
Twin Sisters Washington	Dunite	DTS–1	USGS	53	0.45	0.97	*10*	2.5	0.56
Cozadero California	Peridotite	PCC–1	USGS	45	*0.66*	1.07	*17*	3.4	5.7
Salt Lake Honolulu	Peridotite	HK571–01204d	IK	52	2.27	1.14	44	8.2	3.3

* Italicized values represent the result of a single determination.

† Waianae basalts listed in order of decreasing crystallization index.

‡ The sources are abbreviated as follows: FA: F. Aumento, Dalhousie University; GG: G. Goles, University of Oregon (described by A. F. Buddington, 1939); IK: I. Kushiro, University of Tokyo; GAM: G. A. Macdonald, University of Hawaii; EAV: E. A. Vincent, Oxford University; CRPG: Centre des Recherches Petrographiques et Geochimiques, Nancy, France; USGS: United States Geological Survey; ZGI: Zentrales Geologisches Institut, Berlin, East Germany.

component in Apollo 11 soils and breccias was reported by the LSPET (1969) team, and later by several other investigators. The most comprehensive and conclusive study of Apollo 11 samples is that of GANAPATHY *et al.* (1970a), who report that 12 elements are clearly enriched in the fines and breccias relative to the local rocks. They argued that the extra amounts of these siderophile and volatile elements were consistent with a 1.9% addition of material resembling C1 chondrites to a matrix composed of a roughly equal mixture of the two local rock types at the landing site.

We have studied two soils (but no breccias) from the Apollo 12 site. Sample 12033 is light gray soil, and contains substantially smaller amounts of siderophilic and volatile elements than does 12070. The 12070 soil is similar in composition (GANAPATHY *et al.*, 1970b) and color (LSPET, 1970) to most of the fine material in the 12028 core. We have attempted to define the magnitude of the extralunar component in 12070 and in the more typical portions of 12028 by a method similar to that used by GANAPATHY *et al.* (1970a) but differing in important details. Like these authors, we have estimated the concentrations of the elements of interest in the "lunar" matrix of the fines by taking an average of the available data on local rocks, and have subtracted these values from the average concentrations found in the 12070 and 12028 fines. Such data for 11 elements are summarized in Table 3. Following WASSON and BAEDECKER (1970) we have normalized the results by dividing by C1 chondrite concentrations on a water-free basis. This has the important dividend that the resulting values are better estimates of the fraction of C1-like material actually present in the fines. We list estimated errors in the resulting values (which mainly depend on the scatter in the data on the fines, since in most cases the additional contribution from the matrix is negligible), and have shown these errors as bars in Fig. 1a. This facilitates the assessment of the correctness of the assumption that the extralunar material closely resembles the composition of C1 chondrites. We have also reevaluated the Apollo 11 data on fines and breccias in the manner described above. These are summarized in Table 3 and Figs. 1b and 1c.

As can be seen from the last three lines of Table 3, Ge and Ir are the first and second most precisely determined of the 11 elements listed. Interestingly enough, the magnitudes of the extralunar component estimated on the basis of Ir are about 1.6 times greater than those estimated on the basis of the Ge data. We have drawn bands across Figs. 1a, 1b, and 1c which are just wide enough to include the Ge and Ir points. If we take the midpoints of these bands to be the best estimates of the magnitudes of the extralunar component, and the widths of the bands as estimates of the errors, we arrive at values of 1.04 ± 0.24, 1.14 ± 0.25 and $1.07 \pm 0.25\%$ for the Apollo 12 and Apollo 11 fines and the Apollo 11 breccias, respectively. The latter must be taken *cum grano salis*, since the breccias have not all originated in the same localized area of time and space.

The error bars of the other nine elements touch the defined bands, with the following exceptions: Cd in 12D; Zn and Cd in 11D and 11C. The significance of these exceptions will be discussed later. The remainder of the data support the conclusion of GANAPATHY *et al.* (1970a) that the extralunar component has a mean composition similar to that of C1, C2, or E3–4 chondrites, and distinctly different from that expected from any other known class of meteorites. The agreement is not surprising, since C1 chondrites are believed to most nearly resemble the nonvolatile portion of mean solar-system material (see ANDERS, 1971), and may be the most abundant form of meteoritic material falling on the earth (SHOEMAKER and LOWERY, 1967) or the moon. The composition of interplanetary dust is not known, but it is reasonable to assume that that portion which is produced by cometary attrition is similar in composition to the C1 chondrites.

That the observed Ir/Ge ratio in the extralunar component is 1.6 times greater

Table 3. Summary of data needed to estimate the magnitude of

	Ni (ppm)	Ir (ppb)	Pd (ppb)	Au (ppb)	Ge (ppb)
11C	204 ± 20	8.5 ± 1.0	10 ± 3	2.4 ± 0.5	360 ± 40
11D	190 ± 20	8.9 ± 1.0	9.8 ± 3.0	2.0 ± 0.5	390 ± 40
11AB	5.8	0.1†	3.3	0.071	20†
12D	200 ± 20	8.3 ± 1.0		2.2 ± 0.5	355 ± 40
12AB	60 ± 40	0.1		0.016	≤20
C1‡	13300	634	625	208	42000
$\left(\frac{\text{11D–AB}}{\text{C1}}\right)100$	1.38 ± 0.15	1.39 ± 0.16	1.04 ± 0.51	0.91 ± 0.24	0.89 ± 0.09
$\left(\frac{\text{11C–AB}}{\text{C1}}\right)100$	1.49 ± 0.15	1.32 ± 0.16	1.07 ± 0.53	1.11 ± 0.24	0.82 ± 0.09
$\left(\frac{\text{12D–AB}}{\text{C1}}\right)100$	1.05 ± 0.34	1.29 ± 0.16		1.06 ± 0.24	0.80 ± 0.09

* Data taken from the following papers:

Apollo 11—ANNELL and HELZ (1970) [Ni]; COMPSTON *et al.* (1970) [Ni]; GANAPATHY *et al.* (1970a) [Ir, Pd, Au, Ag, Zn, Cd, Br, Bi, Tl]; HASKIN *et al.* (1970) [Au, Ag, Br]; MORRISON *et al.* (1970) [Zn]; REED and JOVANOVIC (1970) [Br]; and WASSON and BAEDECKER (1970) [Ir, Au, Ge].

Apollo 12—GANAPATHY *et al.* (1970b) [Ir, Au, Ag, Zn, Cd, Br, Bi, Tl]; LAUL *et al.* (1970a) [Ir, Au, Ag, Zn, Cd, Br, Bi, Tl]; this work [Ir, Ge, Zn, Cd].

Carbonaceous chondrites—CROCKET *et al.* (1967) [Pd, Au]; EHMANN *et al.* (1970) [Ir, Au]; FOUCHÉ and SMALES (1967) [Ge]; GOLES *et al.* (1967) [Br]; GREENLAND (1967) [Pd, Ag, Zn, Cd]; GREENLAND and

than that found in C1 chondrites is an interesting and possibly important point. The errors we give in the last three lines of Table 3 are meant to be one standard deviation, and do *not* include an estimate of the error in the C1 chondrite concentrations. Using the errors as given, and assuming errors of about 10% and 5% in the C1 chondrite concentrations of Ir and Ge, respectively, we estimate a probability of less than 0.1 that the apparent difference in the Ir/Ge ratios between the extralunar component of the fines and the C1 chondrites reflects experimental error.* Thus it is fairly likely that the Ir/Ge ratio of the extralunar material is really higher than that of C1 chondrites. Germanium is one of the most volatile of the siderophile elements, and the Ir/Ge ratio in the ordinary chondrites is about 4 times higher than the same ratio for C1 chondrites. Also, this ratio varies by over 5 orders of magnitude in iron meteorites, and small additions of certain types of iron meteorite material to the soil could have a profound effect on the Ir/Ge ratio.

There is also an apparent tendency for volatile elements (other than Bi, to which we tend to assign less weight—see the following paragraph) to be slightly higher in Figs. 1a, b, c than would be predicted on the basis of the siderophilic-element contents of the soils and breccias. Concentrations of these elements are lower in "recrystallized" chondrites, and apparently in the bulk earth and bulk moon. If these "lost"

* The concentration of Ir in Cl chondrites is an arithmetic average of one analysis of Ivuna and two analyses of Orgeuil reported by EHMANN *et al.* (1970). Atomic abundances (relative to Si) based on these data are very similar to those found by the same group for C2 and H chondrites. The data are based on nondestructive neutron activation analysis. The data of CROCKET *et al.* (1967), which are based on radiochemical neutron activation, show Ir abundances in C1 chondrites which are only about 60% as large as those in C2 chondrites. It appears that the latter workers may have achieved incomplete exchange between carrier and the activated radionuclides, perhaps as a result of the highly reducing character of the C1 chondrites. Their data on the enstatite chondrites also appear too low, perhaps for similar reasons.

the meteoritic component at the Apollo 11 and Apollo 12 sites.*

Ag (ppb)	Zn (ppm)	Cd (ppb)	Br (ppb)	Bi (ppb)	Tl (ppb)
18 ± 10	29 ± 5	92 ± 14	160 ± 50	2.2 ± 0.7	2.8 ± 0.7
10 ± 5	21 ± 5	44 ± 10	107 ± 30	1.6 ± 0.7	2.2 ± 0.7
1.5	1.64	4.6	22	0.33	0.67
5.4 ± 2.5	7.0 ± 1.1	49 ± 8	125 ± 30	2.0 ± 0.6	1.9 ± 0.6
0.80	1.4 ± 0.7	1.6	13.2	0.8 ± 0.4	0.31
487	421	1260	6870	149	90
1.7 ± 1.0	4.5 ± 1.2	3.1 ± 0.8	1.24 ± 0.44	0.87 ± 0.47	1.7 ± 0.8
3.5 ± 2.1	6.4 ± 1.2	6.9 ± 1.1	2.01 ± 0.73	1.28 ± 0.47	2.3 ± 0.9
0.94 ± 0.51	1.3 ± 0.3	3.7 ± 0.6	1.63 ± 0.44	0.81 ± 0.54	1.8 ± 0.7

LOVERING (1965) [Ni, Ge, Zn]; LAUL *et al.* (1970b) [Tl]; LAUL *et al.* (1970c) [Bi]; LIEBERMAN and EHMANN (1967) [Br]; REED and ALLEN (1966) [Br]; REED *et al.* (1960) [Bi, Tl]; SCHMITT *et al.* (1963) [Cd]; and WIIK (1969) [Ni, Zn].

† The 11AB Ge and Ir values of WASSON and BAEDECKER (1970) and GANAPATHY *et al.* (1970a) showed some scatter. Since their higher values may include some contamination, we have used values for 11AB which are typical of the 12AB rocks.

‡ Cl values raised by 1.25 to place them on water-free basis.

volatiles condensed out in some colder portion of the solar system, it is possible that such a material enriched in these elements accounts for a measurable fraction of the extralunar component.

Although it may be ultimately possible to define a mean composition for the extralunar component which differs in some details from that of C1 chondrites, a number of other elements must be determined with precision comparable to those available for Ge and Ir before detailed testing of hypothetical mixtures of different types of meteorites and/or other cosmic matter can be undertaken.

ANDERS *et al.* (1971) have made the novel suggestion that local variations (on a scale of tens or hundreds of meters) in the extralunar component can be found at the Apollo 12 site. Such evidence, if correct, would be extremely important for evaluating the source of the extralunar component, and would help to assess the relative contributions from interplanetary dust and fractionated meteorites to the extralunar component. Interplanetary dust, if present, should fall at comparable rates and have comparable compositions at all geographical locations on the moon. However, we can find no evidence in the tabulated data of ANDERS *et al.* (1971) to support their proposal, with the exception of their Bi concentrations. Their Bi values show large fluctuations both in soils and crystalline rocks, however, and we question whether an appreciable fraction of these variations may not result from experimental error.

As noted above, Cd is enriched in all soils and breccias by a factor of 3–5 relative to that expected on the basis of the estimated amounts of C1-like material; Zn is enriched by comparable factors in the Apollo 11 samples, but is not present in amounts in excess of that predicted in Apollo 12 samples. GANAPATHY *et al.* (1970a) attribute the excess amounts of these elements in Apollo 11 samples to a hypothetical Cd- and Zn-rich lunar rock which was not included in the limited suite of rocks returned by the astronauts. Alternatively, one could attribute at least a portion of

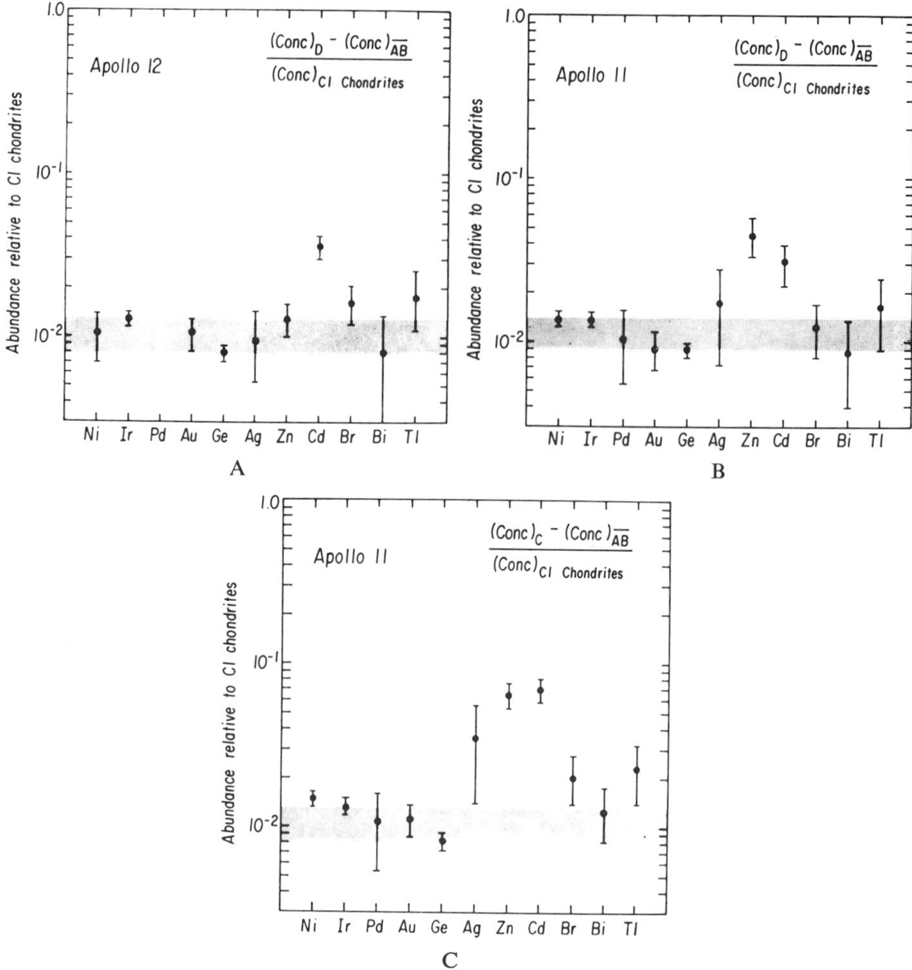

Fig. 1. Abundance pattern of 11 trace elements in the extralunar component obtained by subtracting average abundances for crystalline rocks from those of the soils and breccias and normalizing to C1 chondrites. The error bars represent estimates of one standard deviation based on the replicate analyses of the soil and breccia samples. (A) Apollo 12 soils. The band represents an extralunar component amounting to a 1.04 ± 0.24% addition of a C1-chondrite-like material to soil derived from local crystalline rocks. (B) Apollo 11 soils. The band represents an addition of 1.14 ± 0.25% of a C1-chondrite-like material to soil derived from local crystalline rocks. (C) Apollo 11 breccias. The band represents an addition of 1.07 ± 0.25% of a C1-chondrite-like material to soil derived from local crystalline rocks.

this material to the hypothetical volatile-rich extralunar material mentioned above. The great difference in the Cd/Zn ratios between the Tranquillitatis and Procellarum sites appears to lend support to the "unusual lunar rock" explanation.

The concentration of the extralunar component in the fines portion of the lunar regolith can be considered to be a function of the time elapsed since the regolith started to form. In the discussion to follow, we will assume that the particle-size distribution of the extralunar material has remained constant during this period. This assumption is supported by the fact that the same cometary source is probably responsible for (1) the fine interplanetary dust, which ÖPIK (1969) believes provides most of the mass influx to the moon; and (2) the Apollo asteroid and live comet influx, which Öpik estimates causes most of the primary cratering. We shall also assume that the regolith turns over as a result of a bombardment which is a small fraction of that suffered by the surface since its formation. The lowest curve in Fig. 2 of SHOEMAKER *et al.* (1970) indicates that the bombardment of the surface at the Mare Tranquillitatis site has been sufficient to turn the surface over six times. The regolith thickness increases with increasing integrated flux of extralunar material. However, as the thickness increases, the lower limit of the projectile mass which can penetrate the regolith also increases, and the rate of regolith growth decreases. The relationship between the regolith thickness and integrated flux can be evaluated on the basis of data given by OBERBECK and QUAIDE (1968). They have estimated thicknesses of the lunar regolith at four different locations on the basis of Lunar Orbiter photographs. We have evaluated the extralunar flux from data given in their Fig. 7a (note that the points labeled 50 and 500 m on the abscissa of that figure should be 31.6 and 316 m, respectively). SHOEMAKER *et al.* (1970) give a power law of the form $F = \chi c^{\lambda}$, where F is the cumulative number of craters/km^2 with diameters greater than c (in meters); λ is -2.93, and χ is a measure of the cumulative flux of extralunar material. In order to avoid the effects of crater obliteration by repetitive bombardment, we have fitted this equation to the large-crater-diameter portion of the Oberbeck and Quaide curves, where crater erosion processes have been relatively unimportant. Doing so, we have estimated values of relative cumulative fluxes for the four locations. The use of a value of -2.93 for λ in the power function is rather arbitrary in our treatment of the data of Oberbeck and Quaide. The actual slope of the lower portion of their curves appears to be somewhat higher. However, the actual value of λ is not of great importance providing that the actual projectile (and crater) size distributions at the four different locations were the same.

The relationship between relative cumulative flux and regolith depth is demonstrated by Fig. 2, where the logarithms of the two variables are plotted. They appear to be related in exponential fashion, and a line is drawn in Fig. 2 which corresponds to the relationship

$$\log D_R = 0.585 \log \chi + 0.466, \quad (1)$$

where D_R is the regolith depth. According to this relationship, a factor of 10 increase in the integrated meteorite flux brings about an increase in the regolith depth by a factor of about 3.8.

Of more direct interest for us is the relationship between the infall of extralunar material and the concentration of such material in the fines portion of the regolith. The concentration in the fines of a particular element resulting from the extralunar component will be proportional to the relative influx, and inversely proportional to the regolith depth: i.e., $[X] \propto \chi/D_R$, where $[X]$ is the net concentration of a "extralunar tracer" element such as Ir. Here we assume that the trace element is

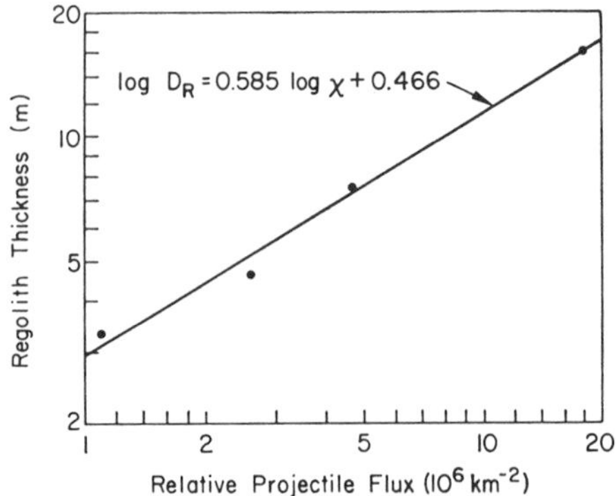

Fig. 2. A plot of the logarithm of the regolith depth (D_R) vs. the logarithm of the cumulative flux constant (χ) for the four regolith types studies by OBERBECK and QUAIDE (1968) using the data presented in Table 4.

homogeneously distributed within the regolith. It then follows that

$$\log [X] = \log \chi - \log D_R + k', \tag{2}$$

where k' is a unitless constant. Combining eqs. (1) and (2) we obtain

$$\log [X] = 0.415 \log \chi + k, \tag{3}$$

where k is a constant which can be calculated for each element of interest at each lunar landing site. For purposes of comparing the concentrations of a given element at two different landing sites (designated by subscripts 1 and 2) we can eliminate this constant and write the equation

$$2.41 \log \frac{[X]_2}{[X]_1} = \log \frac{\chi_2}{\chi_1}. \tag{4}$$

We have calculated that the extralunar component at the Apollo 11 site corresponds to about an addition of 1.14% C1 chondrite-like material, and that that at the Apollo 12 site is about 1.04%. The ratio of these two values is about 1.10. The error in this ratio is relatively small (1 σ of about 4%) since it involves only the experimental precision of our combined Ir and Ge measurements, and does not have to allow for systematic errors or the correctness of the assumption that the extralunar component has a composition resembling C1 chondrites. Using eq. (4), we calculate that the extralunar influx at the Mare Tranquillitatis site was about 1.25 times greater than that at the Oceanus Procellarum site. (This factor would decrease to 1.17 if the value for λ used in the power function to fit the Oberbeck and Quaide curves is set equal to -4.3.)

Rubidium-strontium ages at these sites are 3.65 Gyr (PAPANASTASSIOU *et al.*, 1970) and 3.31 Gyr (PAPANASTASSIOU and WASSERBURG, 1970), respectively. That the ratio of these two ages (1.10) is substantially smaller than the ratio of integrated extralunar fluxes at the two sites indicates that the flux of extralunar objects has been decreasing with the passage of time. A similar conclusion has recently been reached by SHOEMAKER and SODERBLOM (private communication) on the basis of crater counts and crater morphology. GANAPATHY *et al.* (1970a) have estimated an infall rate of 3.8×10^{-9} g cm^{-2} yr^{-1} of C1 chondrite-like material (including H_2O) at the Apollo 11 site. Assuming the same regolith thickness at that site (4.5 m) and a thickness 1.14 times less at the Apollo 12 site (an estimate from eq. (1)), the same packed density (1.8 g/cm^3), fraction of large rocks (10%) and solar wind fraction (2%), we estimate influxes of 2.8×10^{-9} g cm^{-2} yr^{-1} and 2.5×10^{-9} g cm^{-2} yr^{-1} of material with C1-like composition (including H_2O). Because of the evidence for a flux decreasing with time, 2.5×10^{-9} g cm^{-2} yr^{-1} must be regarded as an upper limit on the recent flux of such material on to the lunar surface.

Although we have estimated an average extralunar component of the Apollo 11 breccias, they should each be considered individually, since neither their time nor locus of formation is known. Our earlier data (WASSON and BAEDECKER, 1970) indicate that the extralunar components of breccia samples 10021, 10046, and 10048 are very similar to those of the 10084 fines, and are consistent with formation at a local site relatively late in the history of the regolith. Breccia 10060 has an extra-lunar component about 0.65 that of the other samples, and breccia 10056 contains less than 0.05 times the C1-like component of the more typical samples. In fact, WASSON and BAEDECKER (1970) pointed out that composition-wise, 10056 was a type B crystalline rock.

MCKAY *et al.* (1970) have proposed that the lunar breccias are formed by sintering of material heated and transported during the base-surge phase of crater formation, and not by shock-lithification. They state that compositional evidence provides support for such an origin. We do not agree with the latter conclusion; rather, the compositional evidence defines conditions that must be met by any mechanism for breccia formation. The amount of extralunar component in 10021, 10046, and 10048 is too close to that of the local soil for this to be fortuitous; it is very likely that these consist of lithified (or sintered) *preexisting* regolith. Breccia 10060 is most likely a mixture of 65% regolith and 35% comminuted, deeper-lying bedrock. Breccia 10056 was investigated by MCKAY *et al.* (1970), who confirmed that it was petrologically a breccia. Assuming that the 10056 sample we investigated was not mislabelled, it should consist entirely of comminuted bedrock, and is probably a monomict breccia. Breccias 10056 and 10060 deserve further study, since they are likely to contain features contrasting with those of the more typical breccias.

The unusual light gray fines 12033 has a meteoritic component which is about 0.59 times that in the 12070 and 12028 soils. BAEDECKER *et al.* (1971) proposed that this meteoritic component has resulted from either (1) a 59:41 mixture of local soil with an Al-, P-, and K-rich, Fe- and Ti-poor exotic component from elsewhere on the moon; or (2) a fortuitous contribution of 0.59% C1 chondrite-like material to an Al-rich, Fe-poor rock type during the cratering event. They feel that the former possibility is more persuasive, and assuming such a picture they calculated the composition of an "exotic" component from major element data on both 12033 and the more typical soil 12070. Their calculated composition is similar in K and P content to the KREEP (high K, rare earth element, and P) component of HUBBARD *et al.* (1971). These workers report that the 12033 soil contains 65% KREEP, and that

12070 contains 25% KREEP. If it is assumed that the exotic component defined by BAEDECKER *et al.* (1971) is KREEP, and amounts to 25% of the 12070 fines, the 59:41 12070/exotic ratio given by these authors indicates a 56% KREEP component in 12033, in relatively good agreement with the estimate of HUBBARD *et al.* (1971). As indicated by BAEDECKER *et al.* (1971), the KREEP component is probably of more recent origin than the Copernican event, and we favor the Fra Mauro formation as the source of this material.

ANORTHOSITIC AND OTHER LUNAR COMPONENTS IN THE SOIL

WOOD *et al.* (1970a) found that about 5% of the coarse fines from the Apollo 12 site were of anorthositic composition. Independently WOOD *et al.* (1970b) and WASSON and BAEDECKER (1970) estimated from the composition of the "fine" fines and breccias that these materials contain 20% of material of anorthositic composition. That this fine material has more of this component than the coarser material is not surprising, since the finer material has higher ejection velocities in cratering events. GOLES (1970) and GOLES *et al.* (1970) have argued for a "cryptic" component distinct from the anorthosite, chiefly on the basis of variation diagrams where Al, Si, Ca, and Mg contents of crystalline rocks, soils and breccias are plotted versus their Ti contents. Although it seems quite certain that lunar materials other than anorthositic fragments and local rocks have contributed to the Apollo 11 soil, we fail to see the evidence for appreciable amounts of these materials in these authors' variation diagrams. Their choice of TiO_2 as the independent variable was unfortunate, since the relatively dense ilmenite phase may be depleted in the lunar surface materials by settling (COMPSTON *et al.*, 1970). Thus, the Ti values of the soil and, to a lesser degree, the breccias, should be regarded as lower limits. When one allows for this by an error bar (+1% Ti in length) and adds appropriate error bars to the other plotted points, there is no difficulty in finding straight lines which intercept the crystalline rocks, breccias, soils, and anorthositic gabbro on the variation diagrams of GOLES (1970). We find no reason to believe that other compositionally distinct lunar materials account for as much as 5% of the mass of the Apollo 11 fines and breccias.

Like the Apollo 11 soils, the soil samples recovered from Oceanus Procellarum are chemically distinct from the crystalline rocks, being enriched in Al and depleted in Ti, and the existence of an exotic component in the lunar soil was indicated by the original LSPET (1970) data. As mentioned in the previous section, HUBBARD *et al.* (1971) have characterized this component as being a high-K, REE, P rock which they call KREEP, and independently WOOD *et al.* (1971) have characterized the same component, which they term *norite*. It is difficult to estimate the abundance of most of our elements in the KREEP component due to their enrichment in the soil samples from the extralunar component. However, we can estimate the Ga concentration in the KREEP component, since the presence of the meteoritic component has only a minor effect on the observed Ga concentration of the soil. In Fig. 3 we have plotted Ga versus the Al, Ca, and Mg contents of the rocks and soils (the data on the latter elements are from ANNELL *et al.*, 1971; COMPSTON *et al.*, 1971; ENGEL and ENGEL, 1971; LSPET, 1970; ROSE *et al.*, 1971; and WILLIS *et al.*, 1971). The regression

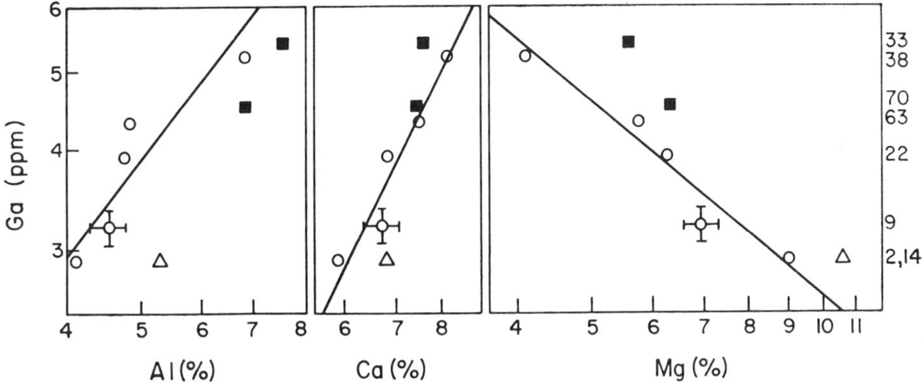

Fig. 3. Gallium data for six crystalline rocks and two soil samples plotted vs. their Al, Ca, and Mg contents. The regression lines were based on the data points for the crystalline rocks represented by open circles (the open triangle is rock 12014, for which the major element data is of low quality). Correlations between each pair of variables are significant at confidence levels of >95%.

lines are calculated using York's least squares fitting procedure, with a ±5% error assigned to all variables. The data point for 12014 was excluded from the fitting procedure, since the only major element data available on this sample was that of LSPET (1970), which we considered to be unreliable. The LSPET analyses generally yielded Al, Ca, and Mg results which were systematically higher than the values obtained for the same rock samples by wet chemical or X-ray fluorescence techniques. The Ga contents of the crystalline rocks are positively correlated with Al and Ca and negatively correlated with Mg at confidence levels greater than 95%.

Hubbard *et al.* (1971) estimate that soils 12070 and 12033 contain 25 and 65% KREEP, respectively, based on their Rb, Sr, and rare earth element data. The microprobe data of Wood *et al.* (1971) shows the KREEP component to be enriched in Al and Ca with respect to the local crystalline rocks, with a wide range of Mg concentrations (2.5–6.3% Mg). Figure 3 points to an apparent enrichment of Ga in the KREEP component, and suggests a greater enrichment of Ga than Ca. From the available data on soils 12070 and 12033 and an assumed composition of 25% and 65% KREEP in 12033 and 12070, respectively, one can calculate average compositions for the crystalline rocks and KREEP components. The average crystalline rock Ga content calculated is 3.9 ppm, 0.1 ppm higher than the mean Ga concentration of our suite of rocks, and the Ga content estimated for the KREEP component is 6.2 ppm.

The presence of the KREEP component in the fines and breccias is probably of minor import for the calculations used to evaluate the extralunar component in the previous section. In the following section we show that the concentrations of the elements we have studied show little variation between different types of mafic and acidic igneous rocks. Thus, a negligible error should be introduced in subtracting out a lunar contribution which is based only on the observed concentrations in the local rocks. A second potential problem has to do with the fact that the extralunar component is probably considerably greater in the highlands regolith than in the maria basins. Although the anorthositic and KREEP materials are probably of highlands origin, the material presently in the maria regolith has been ejected at high velocities from relatively large cratering explosions. The highest velocity ejecta is associated with the central "plume" during crater formation, and probably consists of material from well below the regolith. It should therefore contain nearly no extralunar material.

COMPOSITIONAL COMPARISON OF LUNAR AND TERRESTRIAL IGNEOUS ROCKS

We have attacked the problem of understanding the geochemical behavior of our selection of trace elements from two different standpoints. First, we have initiated studies of suites of cogenetic terrestrial samples, in order to observe whether fractionation trends are present, and hopefully, to gain some insights into the processes which have produced these fractionations. Second, it was thought that a comparison of bulk concentrations in similar materials from the earth and moon might reveal some trends associated with the separate origins of these two planets and assist in formulating and testing hypotheses regarding formational and planet-wide geochemical processes.

One of the first type of studies, an investigation of a rock suite from the Skaergaard Intrusion, will be published elsewhere. Data on another suite of related rocks from the Waianae, Hawaii, volcano, are given in Table 2. WASSON and BAEDECKER (1970) measured five trace elements in three rocks from the lower member and two rocks from the upper member of the Waianae Volcano, as these members are defined by MACDONALD and KATSURA (1964). Since rocks from the different members are probably not cogenetic, we have obtained an additional three specimens of the lower-member rocks originally collected by Macdonald. We have selected our samples to show as much chemical variation as possible while still plotting in the main trend on alkali-iron-magnesium triangular plots and alkali-titania two-parameter plots. Figure 4 shows Zn, Ga, Ge, and In data plotted for these rocks versus the crystallization index (CI) of POLDERVAART and PARKER (1964). The CI is roughly proportional to the normative contents of Mg and Ca silicate minerals. Presumably, a lower CI indicates a later-forming differentiate. Also plotted are the In, Ge, and corrected Ga values reported by WASSON and BAEDECKER (1970) for the two rocks from the Waianae upper member.

The data on the six rocks are listed in order of increasing crystallization index in Table 2. As can be seen there, the Cd and Ir data tend to scatter, with no trends evident. For this reason we have not plotted them in Fig. 4. The scatter in Ir reflects relatively large experimental errors in our data, which are near the detection limit for this element. The Cd data are quite consistent between splits of the same powder, but vary widely from specimen to specimen. We attribute this variation to sampling problems. Apparently the Cd is concentrated in a minor phase which is very inhomogeneously distributed, and which contains only minor amounts of Zn, Ga, Ge, and In.

Figure 4 shows that Zn and Ge remain remarkably constant throughout the Waianae suite, although no Zn data are available on rocks from the upper member. The concentration of Ge also remains constant through all of the Skaergaard differen-

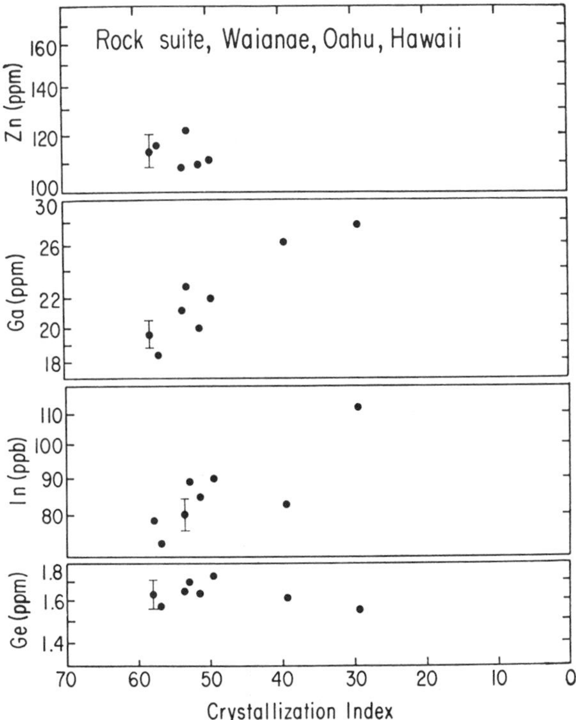

Fig. 4. Plots of our Zn, Ga, In, and Ge data vs. crystallization index for a suite of Waianae, Hawaii, rocks. The error bars correspond to 95% confidence limits on the means of duplicate determinations.

tiates before increasing in the last few percent to solidify. As will be discussed in more detail below, Ge also remains remarkably constant throughout all igneous samples which we have studied. The trends of increasing Ga and In with decreasing CI which were observed in the mixed Waianae suite of WASSON and BAEDECKER (1970) are also observable in the suite from the lower member only. The increase in concentration of each element is small (10–15%) within this relatively small range in CI. Gallium is thought to substitute for Al (III) in igneous rocks (GOLDSCHMIDT, 1954), and In for Fe (II) (GOLDSCHMIDT, 1954; SHAW, 1952). Our data show a positive correlation between Ga and Al (the correlation coefficient is 0.70) but no correlation between In and FeO. Although no trend in the Ir data is observed for the samples from the lower member of the Waianae suite, the average Ir content appears to be higher than that in the rocks from the upper member (WASSON and BAEDECKER, 1970), perhaps reflecting a tendency for Ir to concentrate in early differentiates. The high Ir concentrations observed for the peridotite samples analyzed may reflect a still higher level of abundance of Ir in the mantle materials from which the basalts were derived.

In addition to the above samples we have investigated seven other rocks of basaltic composition. These include two basalts dredged from the Mid-Atlantic Ridge, the chilled marginal gabbro of the Skaergaard Intrusion, and four standard rocks:

BCR-1 and W-1 of the USGS; BR of the CRPG, Nancy; and BM of the ZGI Berlin. All elements (except Ir, for which the data are of lower quality) show remarkably small variations despite the fact that these basaltic rocks are from widely varying locations. Indium varies by a factor of 2.3, Zn by a factor of 2, and Ga and Ge by factors of 1.4. Cd varies by about a factor of 6; however, if we discard the highest and lowest values as reflecting sampling inhomogeneities, the spread is reduced to about a factor of 2.

During the past few years major advances have been made in our understanding of the origin of basalts. Both high-pressure thermodynamic studies (e.g., ITO and KENNEDY, 1968; O'HARA, 1965) and trace-element distribution studies (SCHILLING and WINCHESTER, 1967; GAST, 1968; GRIFFIN and MURTHY, 1969; SCHNETZLER and PHILPOTTS, 1970) have shown that basalts have probably formed by partial melting of upper mantle materials at depths of 50–100 km, with relatively little differentiation of the magma during transport to the surface. The upper mantle source material probably has had the highest trace-element content near the rising portions of convective cells under midocean ridges, with depletion of certain elements under the continents. The relatively low trace-element contents of oceanic abyssal basalts are thought to reflect extensive (20–30%) partial melting of the source materials, whereas the alkalic basalts were formed by smaller amounts of partial melting.

Our data are entirely consistent with such a picture. The midocean basalts have the lowest concentrations of all elements except Cd, where our data scatter. We find it remarkable, however, that the concentrations of most elements in continental basalts are so similar to the oceanic ones. Either the depletion in these elements in continental upper-mantle materials is nicely compensated for by lesser degrees of partial melting, or the compositions of the source materials and the degrees of partial melting are comparable at the two locations. Fractionation during transport and final emplacement would appear to be minor. We plan to determine our suite of elements in alkalic basalts and in ultramafics in addition to those discussed below in an attempt to draw more detailed conclusions about basalt formation. Also listed in Table 2 are data on four other igneous rocks, the USGS standard granite, granodiorite, and andesite, and Buddington's pyroxene gabbro, which has been studied by other lunar investigators. Inclusion of the first three together with the basalts causes the ranges in concentration to increase only in the cases of In (to 3.1) and Cd (to about 9) in excess over the elemental concentration ranges listed in the previous paragraph. Inclusion of the pyroxene gabbro increases the Ga and Ge ranges to 1.7 and 1.5, respectively. That basic and acidic igneous rocks are so similar clearly shows that they have originated in relatively simple processes (as opposed to multiple-plate fractionation processes, which should cause much larger concentration ranges to be observed).

The last three rocks listed in Table 2 are ultramafics—the USGS standard dunite and peridotite, and a Hawaiian peridotite described by KUSHIRO and KUNO (1963). These yield concentrations of Zn and Cd which are lower by factors of about 2 than the lowest values in the mafic and acidic rocks. Gallium and In are lower in the ultramafics by factors of 5 to 30. Germanium is only slightly lower in the ultramafics. Iridium in the dunite is similar to the mafic values, but is about a factor of 10 higher in the two peridotite samples. The differences between the peridotites and the mafic

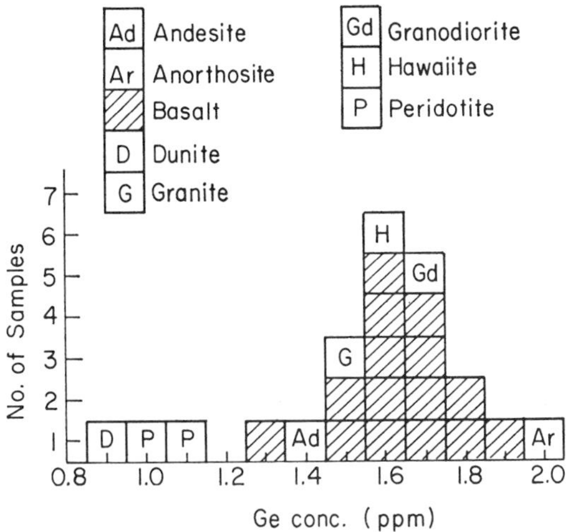

Fig. 5. A histogram of Ge data on terrestrial rocks (as presented in Table 2).

rocks probably reflect the direction and general magnitude of the distribution of these elements between the basaltic magmas and the parent material (i.e., during partial melting of upper mantle materials, Ir tends to remain in the unmelted solid). The remarkably small range of Ge in terrestrial igneous rocks is illustrated in Fig. 5, a histogram showing all of our data on terrestrial rocks. Apparently Ge is distributed about equally between all major phases in these rocks, and for those rock types which have formed by partial melting about equally between the molten and solid phases. Germanium is clearly not a good "tracer" element for studying igneous fractionation processes.

As noted above, our Ir data show quite a bit of scatter on the terrestrial samples studied. In some cases, however, the data on one type of rock from a single locality are relatively consistent. For example, the Ir content of the Waianae basalts derived from the lower member is quite well defined at about 0.47 ± 0.10 ppb and those of the W–1 diabase at 0.32 ± 0.05 ppb. The BCR–1 and BR basalts gave values which are about an order of magnitude lower than the above values. Because of the large spread, it is not possible to speak of an average Ir content of terrestrial basaltic rocks. Other igneous rocks show Ir concentration ranges similar to those encountered in the basalts. Only the peridotites show relatively consistent and high values, as mentioned above, but our data are too few to establish a world-wide average for these rocks.

In Table 4 are listed average concentrations in terrestrial and lunar basalts for the five elements other than Ir. Gallium is 5.2 times lower in the Apollo 12 rocks than in average terrestrial basalts; the factor for Apollo 11 basalts is about 4.5, after correcting the WASSON and BAEDECKER (1970) data for the 5% systematic error noted in the experimental section. The other elements are depleted by factors of 60 or more, and could, within the scatter of the data, all be depleted by the same factor (of about 70–80). The difference between Ga and the other elements is striking. The lunar Ga abundances (relative to Si) are slightly lower than those in the ordinary chondrites (TANDON and WASSON, 1968; FOUCHÉ and SMALES, 1967), and slightly higher than

Table 4. A comparison of the trace-element composition of terrestrial and lunar (Oceanus Procellarum) rocks.

	Zn (ppm)	Ga (ppm)	Ge (ppb)	Cd (ppb)	In (ppb)
Terrestrial basaltic rocks	114	19.5	1620	118	68
Lunar basaltic rocks	1.6	3.7	$\leq$20	$\leq$2	1.2
Ratio terr/lun	71	5.2	$\gtrsim$80	$\geq$60	57

the values for terrestrial peridotites (Table 2). In the LL-chondrites and mesosiderites, which have oxidation states comparable to lunar samples (as estimated by the Fe/Mg ratios of ferromagnesian silicates coexisting with metallic iron), Ga is about equally distributed between the magnetic (metallic) and nonmagnetic (silicate) separates (FOUCHÉ and SMALES, 1967; VAN ALSTINE *et al.*, 1970). Thus, the Ga content of lunar surface materials should not be depleted by more than a factor of about 2 by the separation of a metallic phase.

The depletion of Ge may be ascribed to its extraction by a metal phase during the formation of the lunar basalts, since Ge is concentrated in the metal of LL-chondrites and mesosiderites (FOUCHÉ and SMALES, 1967; VAN ALSTINE *et al.*, 1970). The other siderophile elements such as Ir would also have been lost through such a process. Thus, the fact that the Ge/Ga ratio is about 20 times higher in terrestrial rocks indicates that the extraction of siderophiles into a metal phase was not so efficient on the earth as on the moon. A higher terrestrial oxygen fugacity may also have resulted in a higher silicate/metal distribution ratio for Ge.

Two different geochemical affinities can be invoked to explain the depletion of the other three elements in lunar rocks. RINGWOOD and ESSENE (1970) and GANAPATHY *et al.* (1970a) classify them as volatile elements, and believe that their low abundances (together with those of Rb, Pb, etc.) may be characteristic of the (basaltic) source regions, and already established during the accretion of the moon. They and others (e.g., TERA *et al.*, 1970) quite reasonably reject the possibility that these elements were boiled off during extrusion, since the Sr^{87}/Sr^{86} intercept indicates that the Rb–Sr fractionation took place about 4.6 Gyr ago, 1.3 Gyr before the extrusion of the Oceanus Procellarum basalts. Alternatively, Zn, Cd, and In may be chalcophile, and may have been extracted with a settling FeS phase (or, more likely, an FeS–Fe eutectic) during a molten stage in the moon's history. Unfortunately, we do not know either (1) how strong their chalcophile properties are, nor (2) how effectively they would be scavenged by an iron-troilite eutectic. It appears to be simplest to attribute the low abundance of the three elements to volatility, since WASSON (1971) has shown that this is the property which best explains the similar lunar/terrestrial and ordinary chondrite/C1 chondrite abundance patterns.

Two models have been proposed to explain the low volatile-element concentrations of lunar rocks relative to terrestrial basalts. RINGWOOD and ESSENE (1970) attribute the low abundance of voltaile elements on the moon to an incomplete recondensation of materials evaporated from a late, high-temperature stage during the formation of the earth. GANAPATHY *et al.* (1970a) invoke a model developed by ANDERS (1968) and TUREKIAN and CLARK (1969) in which the accretion of the moon and earth occurred during a period of falling temperatures, with condensation and accretion of volatiles as a very late event during the formation of these planets. If, when this occurred, (1) the earth and moon had nearly their present masses, and (2) the relative capture velocities were low, it is reasonable that the earth should have swept up much more of the material than the moon. GANAPATHY *et al.* (1970a) argue for an enhancement of this effect resulting from the geocentric orbital velocity of the moon.

Although WASSON (1971) has argued that the lunar/terrestrial concentration ratios are more in keeping with a "selective accretion" model of the latter type than with a "partial condensation" model as proposed by RINGWOOD and ESSENE (1970), SINGER and BANDERMANN (1970) have shown that the maximum enrichment attainable by this mechanism is a factor of 22 more mass per unit area, or a factor of only 3.6 more mass per unit volume, if the volatile elements are distributed through masses on the earth and moon which are proportional to the total masses of these planets. Since four of the terrestrial lunar concentration ratios listed in Table 4 are ≥ 60, it appears that the GANAPATHY *et al.* (1970a) model must be rejected.

WASSON (1971) has proposed an alternative manner in which a late, volatile-rich material could be selectively accreted by the earth relative to the moon, independent of the sign of the temperature-time differential. (Because of the inherently low probability for the capture of the moon had it formed elsewhere in the solar system (see, e.g., WISE, 1969), it is assumed that it and the earth formed at about the same distance from the sun.) It seems reasonable that the initial accretion of material to form the earth and the moon chiefly involved volatile-poor material in roughly circular heliocentric orbits, and that this process, once started, proceeded quite rapidly. At the end of this stage, the earth and the moon had achieved masses very similar to their present ones. It is further assumed that the earth had formed and retained an atmosphere of mass comparable to or greater than its present one, whereas the moon had no appreciable atmosphere. Accretion continued, but at a much slower rate. If now, the major source of accreted material were comets which were occasionally perturbed by Jupiter into orbits which intersected those of the earth and moon, the selective accretion of volatile-rich material to the earth can be understood. According to SINGER and BANDERMANN (1970), material with cometary geocentric velocities is captured at a higher rate (per unit mass) on the moon than on the earth (or at a rate slightly less than a factor of 2 in favor of the earth if expressed in units of mass per unit area). However, the retention efficiency for such material would have been much greater on the earth than on the moon. The earth's atmosphere would have given the earth a retentivity of nearly 1.0 for such material, whereas collisions of high-velocity, volatile-rich comets with the moon may very well have resulted in the retention of only a very small fraction of the original mass. Loss would have occurred both because of the vaporization of the lost material to atomic velocities greater than the escape velocity, as well as the entrainment of unvaporized particles by the relatively dense cloud of vapor which would result from an explosively heated dirty-snowball-type comet (WHIPPLE, 1950). It is not critical whether or not the moon was gravitationally attached to the earth when this process occurred. If the mixing volumes involved masses of the earth and moon which were proportional to their total masses, the retention efficiency would have to be 300 times lower on the moon than on the earth in order to explain a terrestrial/lunar concentration of 100 for volatile elements. More details are given by WASSON (1971).

Acknowledgments—We are greatly indebted to J. Kaufman, R. Glimp, H. Mahoney, and W. Simpson for assistance. Terrestrial samples were kindly provided by F. Aumento, H. de la Roche, F. Flanagan, G. Goles, I. Kushiro, G. A. Macdonald, K. Schmidt, and E. A. Vincent. We also wish to thank L. H. Ahrens, B. W. Chappell, W. Compston, F. Cuttitta, E. Engel, G. Goles, J. A. Maxwell, L. C. Peck, H. J. Rose, Jr., R. A. Schmitt, and J. H. Scoon for kindly providing us with their data prior to the Apollo 12 Lunar Science Conference. Neutron irradiations at the UCLA and Ames Laboratory reactors were capably handled by J. Brower, A. F. Voight, and their associates. This research has been supported by NASA contract NAS 9–8096 and NSF grant GA 15731.

REFERENCES

ANDERS E. (1968) Chemical processes in the early solar system, as inferred from meteorites. *Acc. Chem. Res.* **1**, 289–298.

ANDERS E. (1971) How well do we know cosmic abundances? *Geochim. Cosmochim. Acta* **35** (in press).

ANDERS E., LAUL J. C., KEAYS R. R., GANAPATHY R., and MORGAN J. W. (1971) Elements depleted on lunar surface: Implications for origin of moon and meteorite influx. Second Lunar Science Conference (unpublished proceedings).

ANNELL C. S. and HELZ A. W. (1970) Emission spectrographic determination of trace elements in lunar samples from Apollo 11. *Proc. Apollo 11 Lunar Sci. Conf., Geochim. Cosmochim. Acta* Suppl. 1, Vol. 2, pp. 991–994. Pergamon.

ANNELL C. S., CARRON M. K., CHRISTIAN R. P., CUTTITTA F., DWORNIK E. J., HELZ A. W., LIGON D. T., Jr., and ROSE H. J., Jr. (1971) Chemical and spectrographic analyses of lunar samples from Apollo 12 mission. Second Lunar Science Conference (unpublished proceedings).

BAEDECKER P. A. and WASSON J. T. (1970) Gallium, germanium, indium and iridium in lunar samples. *Science* **167**, 503–505.

BAEDECKER P. A., CUTTITTA F., ROSE H. J., Jr., SCHAUDY R., and WASSON J. T. (1971) On the origin of lunar soil 12033. *Earth Planet. Sci. Lett.* (in press).

BRUNFELT A. O., HEIER K. S., and STEINNES E. (1971) Determination of 40 elements in Apollo 12 materials by neutron activation analysis. Second Lunar Science Conference (unpublished proceedings).

BUDDINGTON A. F. (1939) Adirondack igneous rocks and their metamorphism. *Geol. Soc. Amer. Mem.* **7**, 36.

CLIFF R. A., LEE-HU C., and WETHERILL G. W. (1971) Rb–Sr and U–Th–Pb measurements on Apollo 12 materials. Second Lunar Science Conference (unpublished proceedings).

COMPSTON W., CHAPPELL B. W., ARRIENS P. A., and VERNON M. J. (1970) The chemistry and age of Apollo 11 lunar material. *Proc. Apollo 11 Lunar Sci. Conf., Geochim. Cosmochim. Acta* Suppl. 1, Vol. 2, pp. 1007–1027. Pergamon.

COMPSTON W., BERRY H., VERNON M. J., CHAPPELL, B. W., and KAYE M. J. (1971) Rubidium-strontium chronology and chemistry of lunar material from the Ocean of Storms. Second Lunar Science Conference (unpublished proceedings).

CROCKET J. H., KEAYS R. R., and HSIEH S. (1967) Precious metal abundances in some carbonaceous and enstatite chondrites. *Geochim. Cosmochim. Acta* **31**, 1615–1623.

EHMANN W. D., BAEDECKER P. A., and MCKOWN D. M. (1970) Gold and iridium in meteorites and some selected rocks. *Geochim. Cosmochim. Acta* **34**, 493–507.

ENGEL C. G. and ENGEL A. E. J. (1971) Major element composition of three Apollo 12 rocks and some petrogenic considerations. Second Lunar Science Conference (unpublished proceedings).

FOUCHÉ K. F. and SMALES A. A. (1967) The distribution of trace elements in chondritic meteorites. I. Gallium, germanium, and indium. *Chem. Geol.* **2**, 5–33.

GANAPATHY R., KEAYS R. R., LAUL J. C., and ANDERS E. (1970a) Trace elements in Apollo 11 lunar rocks: Implications of meteorite influx and origin of moon. *Proc. Apollo 11 Lunar Sci. Conf., Geochim. Cosmochim. Acta* Suppl. 1, Vol. 2, pp. 1117–1142. Pergamon.

GANAPATHY R., KEAYS R. R., and ANDERS E. (1970b) Apollo 12 lunar samples: Trace element analysis of a core and the uniformity of the regolith. *Science* **170**, 533–535.

GAST P. W. (1968) Trace element fractionation and the origin of tholeiitic and alkaline magma types. *Geochim. Cosmochim. Acta* **32**, 1057–1086.

GOLDSCHMIDT V. M. (1954) In *Geochemistry* (editor A. Muir) Oxford University Press. 730 pp.

GOLES G. G. (1970) Comments on the genesis and evolution of Apollo XI "soil," preprint.

GOLES G. G., GREENLAND L. P., and JEROME D. Y. (1967) Abundances of chlorine, bromine and iodine in meteorites. *Geochim. Cosmochim. Acta* **31**, 1771–1787.

GOLES G. G., RANDLE K., OSAWA M., LINDSTROM D. J., JEROME D. Y., STEINBORN T. L., BEYER R. L., MARTIN M. R., and MCKAY S. M. (1970) Interpretations and speculations on elemental abundances in lunar samples. *Proc. Apollo 11 Lunar Sci. Conf., Geochim. Cosmochim. Acta* Suppl. 1, Vol. 2, pp. 1177–1194. Pergamon.

GREENLAND L. (1967) The abundances of selenium, tellurium, silver, palladium, cadmium, and zinc in chondritic meteorites. *Geochim. Cosmochim. Acta* **31**, 849–860.

GREENLAND L. and LOVERING J. F. (1965) Minor and trace element abundances in chondritic meteorites. *Geochim. Cosmochim. Acta* **29**, 821–858.

GRIFFIN W. L. and MURTHY V. R. (1969) Distribution of K, Rb, Sr and Ba in some minerals relevant to basalt genesis. *Geochim. Cosmochim. Acta* **33**, 1389–1414.

HASKIN L. A., ALLEN R. O., HELMKE P. O., PASTER T. P., ANDERSON M. R., KOROTEV R. L., and ZWEIFEL K. A. (1970) Rare earths and other trace elements in Apollo 11 lunar samples. *Proc. Apollo 11 Lunar Sci. Conf., Geochim. Cosmochim. Acta* Suppl. 1, Vol. 2, pp. 1213–1231. Pergamon.

HUBBARD N. J., GAST P. W., and MEYER C. (1971) The origin of the lunar soil based on REE, K, Rb, Ba, Sr, P, and $Sr^{87/86}$ data. Second Lunar Science Conference (unpublished proceedings).

ITO K. and KENNEDY G. C. (1968) Melting and phase relations in the plane tholeiite-lherzolite-nepheline basenite to 40 kilobars with geological implications. *Contrib. Mineral. Petrol.* **19**, 177–211.

KALLMANN S., STEELE C. G., and CHU N. Y. (1956) Determination of cadmium and zinc. Separation from other elements and each other by anion exchange. *Anal. Chem.* **28**, 230–233.

KUSHIRO I. and KUNO H. (1963) Origin of primary basalt magmas and classification of basaltic rocks. *J. Petrol.* **4**, 75–89.

LAUL J. C., KEAYS R. R., GANAPATHY R., and ANDERS E. (1970a) Abundance of 14 trace elements in lunar rock 12013,10. *Earth Planet. Sci. Lett.* **9**, 211–215.

LAUL J. C., PELLY I., and LIPSCHUTZ M. E. (1970b) Thallium contents of chondrites. *Geochim. Cosmochim. Acta* **34**, 909–920.

LAUL J. C., CASE D. R., SCHMIDT-BLEEK F., and LIPSCHUTZ M. E. (1970c) Bismuth contents of chondrites. *Geochim. Cosmochim. Acta* **34**, 89–103.

LIEBERMAN K. W. and EHMANN W. D. (1967) Determination of bromine in stony meteorites by neutron activation. *J. Geophys. Res.* **72**, 6279–6287.

LSPET (Lunar Sample Preliminary Examination Team) (1969) Preliminary examination of lunar samples from Apollo 11. *Science* **165**, 1211–1227.

LSPET (Lunar Sample Preliminary Examination Team) (1970) Preliminary examination of lunar samples from Apollo 12. *Science* **167**, 1325–1339.

MACDONALD G. A. and KATSURA T. (1964) Chemical composition of Hawaiian lavas. *J. Petrol.* **5**, 82–133.

MCKAY D. S., GREENWOOD W. R., and MORRISON D. A. (1970) Origin of small lunar particles and breccia from the Apollo 11 site. *Proc. Apollo 11 Lunar Sci. Conf., Geochim. Cosmochim. Acta* Suppl. 1, Vol. 1, pp. 673–694. Pergamon.

MORRISON G. H., GERARD L. T., KASHUBA A. T., GANGADHARAM E. V., ROTHENBERG A. M., POTTER N. M., and MILLER G. B. (1970) Elemental abundances of lunar soil and rocks. *Proc. Apollo 11 Lunar Sci. Conf., Geochim. Cosmochim. Acta* Suppl. 1, Vol. 2, pp. 1383–1392. Pergamon.

MORRISON G. H., GERARD J. T., POTTER N. M., GANGADHARAM E. V., ROTHENBERG A. M., and BURDO R. A. (1971) Elemental abundances of lunar soil and rocks from Apollo 12. Second Lunar Science Conference (unpublished proceedings).

OBERBECK V. R. and QUAIDE W. L. (1968) Genetic implications of lunar regolith thickness variations. *Icarus* **9**, 446–465.

O'HARA M. J. (1965) Primary magmas and the origin of basalts. *Scot. J. Geol.* **1**, 1–40.

O'KELLEY G. D., ELDRIDGE J. S., SCHONFELD E., and BELL P. R. (1971) Comparative radionuclide concentrations and ages of Apollo 11 and Apollo 12 samples from nondestructive gamma-ray spectrometry. Second Lunar Science Conference (unpublished proceedings).

ÖPIK E. J. (1969) The moon's surface. *Ann. Rev. Astron. Astrophys.* **7**, 473–526.

PAPANASTASSIOU D. A., WASSERBURG G. J., and BURNETT D. S. (1970) Rb–Sr ages of lunar rocks from the Sea of Tranquillity. *Earth Planet. Sci. Lett.* **8**, 1–19.

PAPANASTASSIOU D. A. and WASSERBURG G. J. (1970) Rb–Sr ages from the Ocean of Storms. *Earth Planet. Sci. Lett.* **8**, 269–278.

POLDERVAART A. and PARKER A. B. (1964) The crystallization index as a parameter of igneous differentiation in binary variation diagrams. *Amer. J. Sci.* **262**, 281–289.

RANCITELLI L. A., PERKINS R. W., FELIX W. D., and WOGMAN N. A. (1971) Cosmogenic and primordial radionuclide measurements in Apollo 12 lunar samples by nondestructive analysis. Second Lunar Science Conference (unpublished proceedings).

REED G. W., Jr. and ALLEN R. O., Jr. (1966) Halogens in chondrites. *Geochim. Cosmochim. Acta* **30**, 779–800.

REED G. W. and JOVANOVIC S. (1970) Halogens, mercury, lithium and osmium in Apollo 11 samples. *Proc. Apollo 11 Lunar Sci. Conf., Geochim. Cosmochim. Acta* Suppl. 1, Vol. 2, pp. 1487–1492. Pergamon.

REED G. W., KIGOSHI K., and TURKEVICH A. (1960) Determinations of concentrations of heavy elements in meteorites by activation analysis. *Geochim. Cosmochim. Acta* **20**, 122–140.

RINGWOOD A. E. and ESSENE E. (1970) Petrogenesis of Apollo 11 basalts, internal constitution and origin of the moon. *Proc. Apollo 11 Lunar Sci. Conf., Geochim. Cosmochim. Acta* Suppl. 1, Vol. 2, pp. 769–799. Pergamon.

ROSE H. R., Jr., CUTTITTA F., ANNELL C. S., CARRON M. K., CHRISTIAN R. P., DWORNIK E. J., HELZ A. W., and LIGON D. T., Jr. (1971) Semimicroanalysis of Apollo 12 samples. Second Lunar Science Conference (unpublished proceedings).

ROSHOLT J. N. and TATSUMOTO M. (1971) Isotopic composition of thorium and uranium in Apollo 12 samples. Second Lunar Science Conference (unpublished proceedings).

SCHILLING J. G. and WINCHESTER J. W. (1967) Rare earth fractionation and magmatic processes. In *Mantles of Earth and Terrestrial Planets* (editor S. K. Runcorn) Interscience, 267–283.

SCHMITT R. A., SMITH R. H., and OLEHY D. A. (1963) Cadmium abundances in meteoritic and terrestrial matter. *Geochim. Cosmochim. Acta* **27**, 1077–1088.

SCHNETZLER C. C. and PHILPOTTS J. A. (1970) Partition coefficients of rare earth elements between igneous matrix material and rock forming mineral phenocrysts, II. *Geochim. Cosmochim. Acta* **34**, 331–340.

SHAW D. M. (1952) The geochemistry of indium. *Geochem. Cosmochim. Acta* **2**, 185–206.

SHOEMAKER E. M. and LOWERY C. J. (1967) Airwaves associated with large fireballs, and the frequency distribution of energy of large meteoroids. (Abstract) *Meteoritics* **3**, 123–124.

SHOEMAKER E. M., HAIT M. H., SWANN G. A., SCHLEICHER D. L., SCHABER G. G., SUTTON R. L., DAHLEM D. H., GODDARD E. N., and WATERS A. C. (1970) Origin of the lunar regolith at Tranquillity Base. *Proc. Apollo 11 Lunar Sci. Conf., Geochim. Cosmochim. Acta* Suppl. 1, Vol. 3, pp. 2399–2412. Pergamon.

SILVER L. T. (1971) U–Th–Pb isotope relations in Apollo 11 and Apollo 12 lunar samples. Second Lunar Science Conference (unpublished proceedings).

SINGER S. F. and BANDERMANN L. W. (1970) Where was the moon formed? *Science* **170**, 438–439.

TANDON S. N. and WASSON J. T. (1968) Gallium, germanium, indium and iridium variations in a suite of L-group chondrites. *Geochim. Cosmochim. Acta* **32**, 1087–1109.

TAYLOR S. R., KAYE M., GRAHAM A., RUDOWSKI R., and MUIR P. (1971) Trace element chemistry of lunar samples from the Ocean of Storms. Second Lunar Science Conference (unpublished proceedings).

TERA F., EUGSTER O., BURNETT D. S., and WASSERBURG G. J. (1970) Comparative study of Li, Na, K, Rb, Cs, Ca, Sr, and Ba abundances in achondrites and in Apollo 11 lunar samples. *Proc. Apollo 11 Lunar Sci. Conf., Geochim. Cosmochim. Acta* Suppl. 1, Vol. 2, pp. 1637–1657. Pergamon.

TUREKIAN K. K. and CLARK S. P. (1969) Inhomogeneous accumulation of the earth from the primitive solar nebula. *Earth Planet. Sci. Lett.* **6**, 346–348.

VAN ALSTINE D. R., SCHAUDY R., and WASSON J. T. (1970) Everything you always wanted to know about mesosiderites but were afraid to ask. (Abstract) *Meteoritics* **5**, 226.

WASSON J. T. (1971) Volatile elements on the earth and moon. *Earth Planet Sci. Lett.*, submitted.

WASSON J. T. and BAEDECKER P. A. (1970) Ga, Ge, In, Ir, and Au in lunar, terrestrial and meteoritic basalts. *Proc. Apollo 12 Lunar Sci. Conf., Geochim. Cosmochim. Acta* Suppl. 1, Vol. 2, pp. 1741–1750. Pergamon.

WHIPPLE F. L. (1950) A comet model. I. The acceleration of Comet Encke. *Astrophys. J.* **111**, 375–394.

WIIK H. P. (1969) On regular discontinuities in the composition of meteorites. *Comen. Phys. Mathemat.* **34**, 135–145.

WILLIS J. P., AHRENS L. H., DANCHIN R. V., ERLAND A. J., GURNEY J. J., HOFMEYR P. K., MCCARTHY T. S., and ORREN M. J. (1971) Some interelement relationships between lunar rocks and fines, and stony meteorites. Second Lunar Science Conference (unpublished proceedings).

WISE D. U. (1969) Origin of the moon from the earth: Some new mechanisms and comparisons. *J. Geophys. Res.* **74**, 6034–6045.

WOOD J. A., DICKEY J. S., Jr., MARVIN U. B., and POWELL B. N. (1970a) Lunar anorthosites. *Science* **167**, 602–604.

WOOD J. A., DICKEY J. S., Jr., MARVIN U. B., and POWELL B. N. (1970b) Lunar anorthosites and a geophysical model of the moon. *Proc. Apollo 11 Lunar Sci. Conf., Geochim. Cosmochim. Acta* Suppl. 1, Vol. 1, pp. 965–998. Pergamon.

WOOD J. A., MARVIN U. B., REID J. B., TAYLOR G. J., BOWER J. F., POWELL B. N., and DICKEY J. S., Jr. (1971) Relative proportions of rock types, and nature of the light-colored lithic fragments in Apollo 12 soil samples. Second Lunar Science Conference (unpublished proceedings).

Proceedings of the Second Lunar Science Conference, Volume 2, pp. 1063–1081.
The M.I.T. Press, 1971.

Analyses of Apollo 12 specimens: Compositional variations, differentiation processes, and lunar soil mixing models

GORDON G. GOLES, ANDREW R. DUNCAN, DAVID J. LINDSTROM, MARILYN R. MARTIN, ROBERT L. BEYER, MASUMI OSAWA,* KEITH RANDLE,† LINDA T. MEEK, TERRY L. STEINBORN, and SHEILA M. MCKAY
Center for Volcanology, University of Oregon, Eugene, Oregon 97403

(*Received 22 February 1971; accepted in revised form 29 March 1971*)

Abstract—Instrumental neutron activation analyses were used to determine abundances of 25 elements in 13 samples of 10 Apollo 12 specimens. Apollo 12 crystalline rocks may be divided into two compositional groups which are most clearly defined by Ca, Mg, and Co abundances. Simple differentiation models, such as fractional crystallization, cannot explain the compositional distinctions observed. Rare-earth abundances may be explained by a two-stage process involving both partitioning of these elements into phosphates during crystallization of a primitive magma and subsequent partial melting. The slight relative enrichment in light rare-earth elements observed in some Apollo 12 specimens may be related to slight fractionation of zircon, baddeleyite, or Zr–Ti silicates from magmas ancestral to components of those specimens. Mixing models for Apollo 11 and Apollo 12 soils closely match observed compositions, which suggests that we have identified the nature and relative proportions of the clastic components which have been the principal contributors to Apollo 11 and Apollo 12 mare soils.

INTRODUCTION

STUDIES OF Apollo 11 samples by instrumental activation analysis techniques demonstrated the value of these techniques for obtaining geochemical data which support useful interpretations, while conserving lunar materials (e.g., GOLES *et al.*, 1970). Here we report results of similar studies of Apollo 12 samples, interpret compositional groups in Apollo 12 crystalline rocks, interpret rare-earth abundances, and discuss mixing models for Apollo 11 and Apollo 12 soils.

ANALYTICAL DATA

Samples of Apollo 12 crystalline rocks 12002, 12018, 12021, 12040, 12065, a mixture of crystalline rock and glass from 12017, breccia 12034 with glass coating, breccia 12010, and soils 12044 and 12070 were analyzed for abundances of 25 elements. The sample of 12017 sent to us has unknown proportions of glass and rock and is compositionally unusual. Until the two materials are separated, no attempt will be made to interpret the results on that sample. In contrast, the mixture of breccia 12034 and its glass coating is very similar in composition to breccia 12010. Consequently, we assume that the glass coating on breccia 12034 is not very different in composition from the breccia itself, and that taken together they may be averaged with breccia 12010 in making the interpretations discussed below.

Generally our procedures for instrumental neutron activation analyses of samples (usually about 1 gm in mass) closely follow those of GORDON *et al.* (1968). Ti, Al, V, Ca, Mn, and Na were

* Present address: Department of Chemistry, Kanazawa University, Kanazawa, Japan.
† Present address: Scottish Research Reactor Centre, East Kilbride, Glasgow, Scotland.

determined from short-lived isotopes in a separate irradiation (SCHMITT et al., 1970). Si analyses were done by T. W. Osborn, Oregon State University, Corvallis, by 14 MeV neutron activation (VOGT and EHMANN, 1965).

Abundances found in Apollo 12 crystalline rocks are given in Table 1A, those for soils and breccias appear in Table 1B. Monitor solutions were used both as absolute comparison standards and as an aid in subtracting interfering gamma-ray photopeaks. Extensive use was made of USGS standard rocks W-1 and BCR-1, both for checking the accuracy of our determinations and, in some cases, as absolute standards. Elemental abundances for these standards rocks, as determined in the present work, are given in Table 1C with those values assumed as standards given in italics.

Data reduction was done by computerized peak searching, followed by calculation of photopeak areas and elemental abundances. Plotted spectra were used to correct backgrounds and identify interferences in regions of the spectra too complex for the computer program to interpret properly. Also, hand calculation of photopeak areas was necessary for photopeaks too small to be found by our computer procedure.

Estimates of analytical precisions, expressed as relative deviations, were made primarily from the counting statistics and range from 1 to 50%. Fe, Co, Sc, Cr, and Na determinations have estimated precisions of 1–2%. La, Sm, Eu, Yb, Lu, Si, Al, Mn, and Hf determinations have estimated precisions of 2–5%, and Ca, Ti, Tb, Ce, and Ho determinations have estimated precisions of 5–10%. Ba, Nd, Zr, Ta, and V abundances approached the limits of sensitivity of the procedure, and precision in this group is strongly dependent on the absolute elemental abundances in the various samples. Thus Ba precisions range from 15% in the breccia 12010 to 50% in sample 12040A; Nd precisions range from 3 to 45% in the same rocks; Zr precisions range from 10% in breccias to 50% in 12021B; Ta precisions range from 2% in breccia 12010 to 25% in rock 12002; and V precisions range from 10% in sample 12040B to 20% in breccia 12010. Contents of K, Rb, and Cs were below detection limits for our techniques.

Extremely high Cr contents of the crystalline rocks posed several new problems. The ^{233}Pa

Table 1A. Abundances in Apollo 12 crystalline rocks.

Element (ppm except where indicated)	12002,129	12018,40	12021,56 A	12021,56 B	12040,17 A	12040,17 B	12040,17 C	12065,38
Si (%)	21.0	20.4	21.5	22.0	21.0	20.3	20.7	21.0
Ti (%)	1.35	1.45	2.10	2.12	1.38	1.36	1.37	1.86
Al (%)	4.28	3.66	5.29	5.57	3.69	3.53	3.82	4.87
Fe (%)	16.4	16.02	14.9	14.9	17.88	15.28	16.02	15.3
Ca (%)	4.8	5.2	7.1	7.5	5.6	5.9	5.3	7.9
Na	1510	1460	1940	2030	1540	1260	1420	1840
Ba	80	70	120	130	40	120	50	90
La	5.34	5.30	7.29	7.46	5.43	3.49	4.21	6.9
Ce	12	14	19	20	17	13	14	17
Nd	13	14	13	16	9	12	12	16
Sm	3.94	3.89	5.43	5.68	4.30	2.87	3.46	5.02
Eu	0.80	0.797	1.12	1.24	0.93	0.715	0.813	1.01
Tb	1.00	1.13	1.41	1.38	1.26	0.81	0.92	1.30
Ho	1.3	—	1.7	1.8	—	—	—	1.8
Yb	3.45	3.50	4.81	4.77	3.66	2.69	2.94	4.15
Lu	0.459	0.602	0.675	0.69	0.629	0.525	0.540	0.64
Th	0.91	1.12	1.7	1.5	1.2	0.9	1.1	1.3
Hf	2.49	2.48	4.03	4.09	2.76	2.06	2.34	3.58
Zr	< 100	60	—	100	150	80	70	—
Ta	0.28	0.48	0.40	0.41	0.40	0.40	0.33	0.39
Mn	1880	1900	1890	1950	1890	1890	1880	2000
Co	65.8	58.0	29.6	27.7	65.3	61.1	61.7	38.8
Sc	38.3	38.9	51.0	48.3	36.7	38.4	37.5	50.0
V	—	190	—	130	—	230	150	180
Cr	5620	3690	2060	1870	3560	4270	3650	3090

Table 1B. Abundances in Apollo 12 soils, breccias, and rocks with glass coatings.

Element (ppm except where indicated)	Soils 12044,19	Soils 12070,71	Breccia 12010,23	Breccia chip with glass coating 12034,17	Crystalline rock chip with glass coating 12017,21
Si (%)	—	—	20.8	21.4	—
Ti (%)	1.6	1.64	1.4	1.37	—
Al (%)	6.4	6.50	6.32	7.73	—
Fe (%)	13.02	12.6	10.55	10.33	15.16
Ca (%)	7.0	6.5	6.6	8.1	—
Na	3220	3240	6200	5800	2540
Ba	380	340	640	720	150
La	33.1	32.1	75.6	60.4	17.1
Ce	94	87	226	176.7	49
Nd	63	40	112	92	40
Sm	16.2	15.8	34.9	28.3	9.65
Eu	1.69	1.71	2.58	2.69	1.36
Tb	3.79	3.69	9.60	8.12	—
Ho	4.2	4.3	—	—	—
Yb	12.9	11.7	27.8	21.7	8.08
Lu	1.81	1.68	4.05	3.14	1.26
Th	7.4	6.7	15.65	11.53	2.19
Hf	12.9	11.3	25.2	20.4	6.70
Zr	350	370	550	630	180
Ta	1.90	2.05	3.72	3.09	0.86
Mn	1570	1550	1450	1290	—
Co	45.5	44.7	33.4	30.4	45.0
Sc	37.9	36.8	32.8	30.0	47.4
V	—	—	90	—	—
Cr	2590	2480	1870	1840	3460

312 KeV peak, used to determine Th, was a small shoulder on the huge ^{51}Cr 320 KeV peak; nevertheless, we used plotted spectra to aid in estimating Th abundances with precisions ranging from 1% in breccias up to 15% for samples with especially high Cr abundances, such as rock 12002. The high Cr activity introduces a large Compton edge and backscatter peak which form much of the background in the 140–180 KeV region of the spectra. The backscatter peak is centered at 143 KeV, under the ^{141}Ce–^{59}Fe (142–145 KeV) multiple photopeak. The shape of the backscatter peak was estimated from a plot of a pure Cr monitor spectrum and was graphically subtracted from the composite peak, yielding Ce values with precision estimates depending on Cr content, generally 5–10% but as high as 25% in the extreme case of rock 12002. Only the ^{175}Yb 283 KeV peak was used in Yb determinations since the ^{169}Yb 177 KeV peak appeared on the steep Compton slope from the Cr photopeak.

Compositional Groups in Apollo 12 Crystalline Rocks

By considering analytical data, Goles *et al.* (1971) and Cuttitta (in oral presentation of papers by Annell *et al.*, 1971, and Rose *et al.*, 1971) suggested that the Apollo 12 crystalline rocks may be divided into at least two distinctive compositional groups. By considering hypothetical differentiation paths involving olivine and spinel fractionation, Compston *et al.* (1971) inferred the existence of at least six compositional groups among the Apollo 12 crystalline rocks. Dr. J. L. Warner has kindly provided us with a magnetic tape file (hereafter referred to as "the Apollo 12 data-base") containing most of the analytical data presented at the Second Lunar Science Conference. Thus we can now verify with more comprehensive data the existence of the two groups we had previously proposed (Goles *et al.*, 1971), and define the compositional criteria characteristic of each group.

Table 1C. Abundances in USGS standard rocks (ppm except where otherwise noted).

Element	W-1	BCR-1
Ti (%)	0.65	1.33
Al (%)	7.58	6.96
Fe (%)	*7.80*	9.36
Ca (%)	4.7	7.7
Na (%)	1.58	2.36
Ba	161	655
La	10.8	*24.5*
Ce	23	52
Nd	18	30.1
Sm	3.5	*7.0*
Eu	1.07	1.95
Tb	0.61	*1.15*
Ho	—	1.15
Yb	2.1	*3.5*
Lu	0.360	0.535
Th	2.35	6.02
Hf	2.3	4.57
Ta	*0.67*	0.99
Mn	1300	1380
Co	*47.0*	37.9
Sc	*34.0*	30.8
V	458	285
Cr	*110*	10.7

The values given above were largely determined during the analyses of Apollo 12 specimens whose results are reported in Tables 1A and 1B. Those values given in italics above were used for calibration purposes and were *not* determined during this work.

Selected analytical data for abundances of 17 elements in 18 Apollo 12 crystalline rocks were taken from the Apollo 12 data-base. We have not attempted to classify at this time the remaining crystalline rocks in the Apollo 12 collection, either because there are insufficient analytical data for them, or because the available data are of unknown reliability.

Distribution patterns for abundances of the 17 elements in selected data for the crystalline rocks are of two general types. In the first type, elemental abundances are distributed either as two distinct groups with no intermediate values (Ca), or exhibit strongly bimodal distributions (Mg, Al, Co, Na, and Sc) (Fig. 1). The two groups of Apollo 12 crystalline rocks defined by the distinct groups of Ca abundances also form the modal groups for the other elements. The second type of distribution is either weakly bimodal or unimodal and is typified by rare-earth elements (REE), Fe, and Ti. Most elements whose abundances have the second type of distribution show little or no overlap of the data (even though the *distribution* is continuous) for the two groups of crystalline rocks as defined above. However a few elements (e.g., Rb, Cs) show an almost complete overlap of data from the two groups, suggesting that those elements have been differentiated by processes unlike the ones responsible for most of the compositional variations we observe.

By our classification, group 1 rocks have higher Ca, Al, Na, Sc, and REE abundances, and lower Mg, Co, and Fe abundances, than have group 2 rocks. The elements

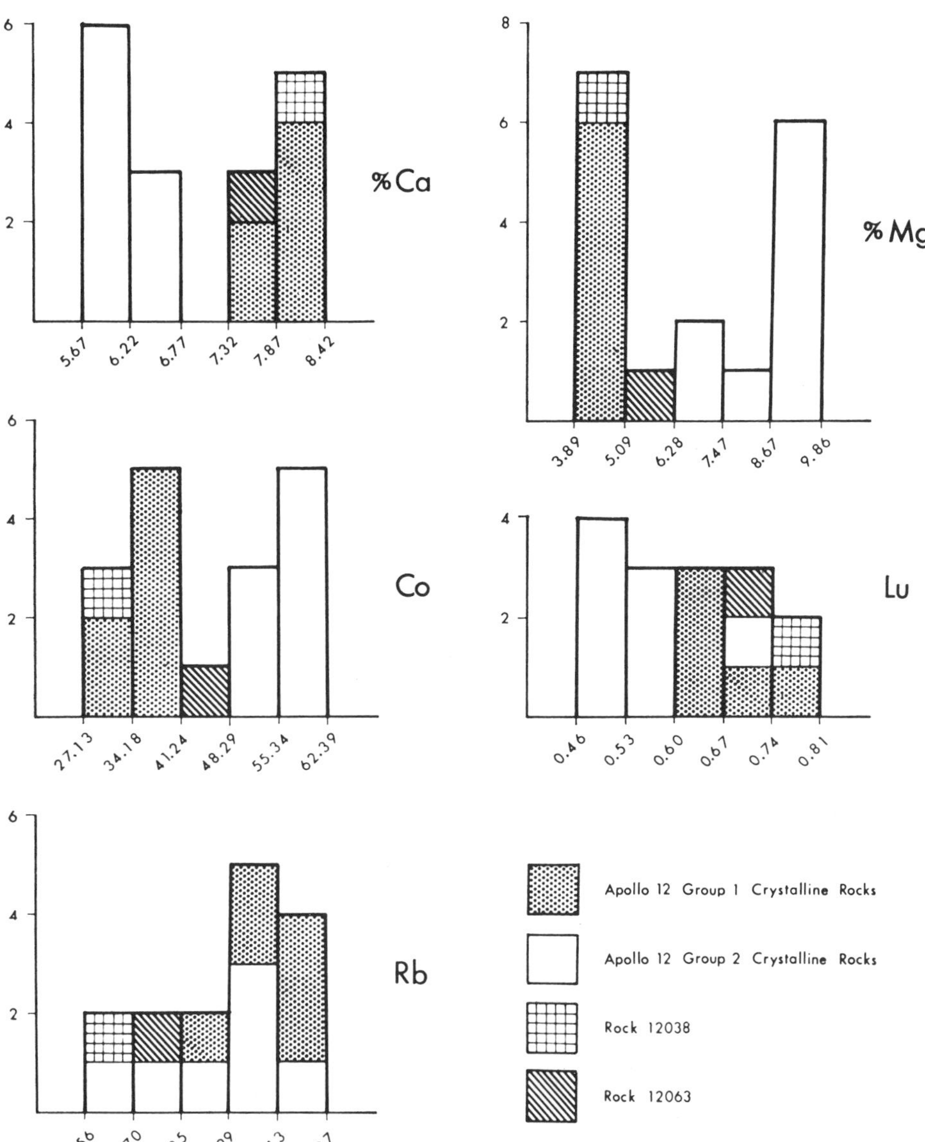

Fig. 1. Histograms of abundances of Ca, Mg, Co, Lu, and Rb in Apollo 12 crystalline rocks, using selected data from the Apollo 12 data-base. A computer program designed to choose the most meaningful (constant) compositional intervals was used; these intervals are given only to two decimal places in this figure. See text for discussion of the different types of distributions exhibited here.

of most use for classifying rocks into one of these two groups are Ca, Mg, and Co.

On the basis of this classification scheme we have placed 16 of the 18 selected Apollo 12 crystalline rocks into one or other of these two groups as follows: Group 1—12017 (tentative classification), 12021, 12051, 12053, 12064, and 12065; Group 2—12002, 12004, 12008, 12009, 12018, 12020, 12022, 12035, 12040, and 12075. Rock 12063 appears to have elemental abundances which are intermediate between those of groups 1 and 2, and rock 12038 has distinctly higher Al, Na, and REE abundances than have any of the other 18 selected crystalline rocks. It is certainly possible that the elemental abundance distribution patterns observed may be caused by strongly biased sampling of the crystalline rock types present at the Apollo 12 landing site, but in spite of that possibility (which we consider to be unlikely), allocation of many of the Apollo 12 crystalline rocks to one or the other of these two groups is valuable for many purposes (such as the computation of soil mixing models, as discussed below). It should be noted that divisions of the Apollo 12 crystalline rocks according to the compositional criteria proposed above do not correspond with divisions made according to the mainly textural criteria proposed by WARNER and ANDERSON (1971).

FRACTIONAL CRYSTALLIZATION AND PARTIAL MELTING

As discussed above, two distinct compositional groups are present in the Apollo 12 crystalline rocks. Previously (GOLES *et al.*, 1971), we suggested a crude fractional crystallization model for relating these groups to one another, and COMPSTON *et al.* (1971) presented a much more elaborate discussion of a similar fractionation model. Fractional crystallization models of this kind may be tested by using partition coefficients and some appropriate theory such as that which yields the Rayleigh fractionation law, as discussed by GAST (1968) and by others. We have assumed a partition coefficient fractionation model based on surficial equilibrium in order to test whether crystallization of the kinds and amounts of minerals inferred from comparisons of bulk compositions gives results which are self-consistent. In principle, this approach provides a powerful test of a fractionation model since it requires not only that the end results of the putative fractionation be in agreement with observation but also that the path taken by the model be in agreement with well-established physical-chemical laws.

In particular, COMPSTON *et al.* (1971) presented a model relating rocks 12009, 12018, 12020, and 12057 by addition or removal of olivine (and, in one case, spinel). We have taken the amounts of olivine postulated by those authors and have used, for consistency, only their own analytical data to construct a partition coefficient fractionation model, in which we allow the partition coefficients themselves to vary as may be required to fit the observed compositions and the proposed fractionation paths. Rock 12018 is proposed by COMPSTON *et al.* to be related to rock 12009 by the addition of 17.1% olivine ($Fo_{70.2}$) (which we shall term process A) and rock 12020 is proposed to be related to rock 12009 by the addition of 11.2% olivine ($Fo_{69.4}$) (which we shall term process B). Processes A and B give olivine-liquid partition coefficients of 0.45 and 1.0 respectively for Ba, and 0.42 and 1.0 respectively for Rb, all of which are much higher than expected. Furthermore, the estimated olivine-liquid partition coefficients differ between the two processes by about a factor of two, which is very difficult to justify because the olivines in the two processes are postulated to have almost identical major element compositions and to have grown under very similar conditions. Some of the data of COMPSTON *et al.* give physically-unrealistic partition coefficients (e.g., −0.8 for La fractionation and −2.2 for Ce fractionation in process A; negative partition coefficients have no physical significance), but these values probably reflect analytical errors. It seems most unlikely that the same cause accounts for the inconsistencies with Ba and especially Rb partition coefficients noted above.

We have also tried to fit a simple partition coefficient model to the two compositional groups observed in Apollo 12 crystalline rocks. No path has been found that involves fractionation of any combination of olivine, spinel, clinopyroxene, orthopyroxene, and plagioclase, and that is adequate to explain the compositional differences between the two groups. Consequently, it seems likely that both fractional crystallization and partial melting were effective in modifying the composition of Apollo 12 crystalline rocks. Perhaps some degree of contamination by easily mobilized components of regolith or wall-rock material also contributed to the compositional variations observed. In any case, it is clear that simple models in which only one process of differentiation is considered are unlikely to yield both the observed compositions and chemically reasonable paths by which those compositions may be attained. We are continuing to explore differentiation models, in hopes of setting plausible limits on the extents to which the processes mentioned above might have contributed to the observed compositional variations.

RARE-EARTH ABUNDANCES AND THEIR INTERPRETATION

Abundances of REE are more variable in Apollo 12 specimens taken all together than in Apollo 11 specimens. Compared to the REE abundances in Apollo 11 rocks, soils, and breccias, REE abundances in Apollo 12 crystalline rocks are noticeably lower, and those in the Apollo 12 soils and breccias are markedly higher. The range of variability in REE abundances within the Apollo 12 crystalline rocks is comparable to that within one flow unit of terrestrial basalts (HASKIN *et al.*, 1971; OSAWA and GOLES, 1970).

Representative chondrite-normalized REE distribution patterns for the two Apollo 12 crystalline rock groups and for Apollo 12 soils and breccias are plotted in Fig. 2. The crystalline rocks, represented by 12002 and 12065, have relatively flat distribution patterns, with sizeable negative Eu anomalies, much like the Apollo 11 samples. The Apollo 12 soils and breccias (12070 and 12010) are slightly enriched in light REE relative to heavy REE, and show larger negative Eu anomalies than do the crystalline rocks.

We have chosen to indicate the relative Eu depletion by the ratio of Eu to Sm, which is essentially constant for all Apollo 12 crystalline rocks ($\sim$0.20) and which in these rocks is distinctly higher than the Eu/Sm ratio in Apollo 12 soils and breccias ($\sim$0.10) as determined from our data. Data given in the Apollo 12 data-base confirm this observation. Apollo 11 samples have Eu to Sm ratios of about 0.15 and the ratios form a continuous distribution, in contrast to the bimodal distribution found for Apollo 12 crystalline rocks, breccias, and soils. No obvious distinction can be made

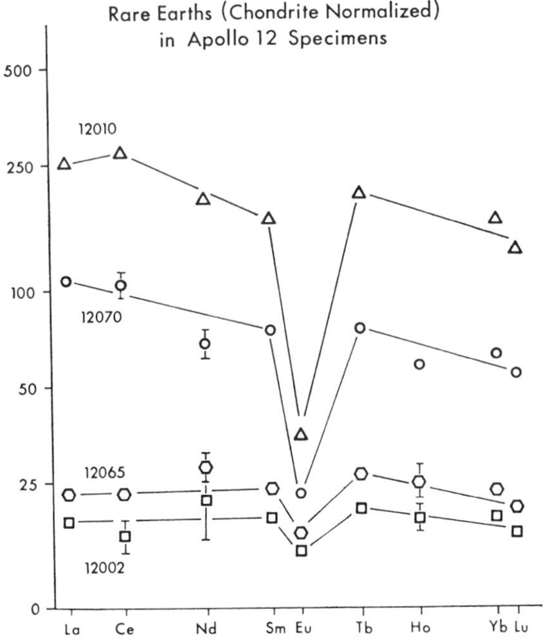

Fig. 2. Chondrite-normalized rare-earth abundances of typical Apollo 12 specimens. Sample 12002 represents group 2 crystalline rocks, 12065 represents group 1 crystalline rocks, 12070 represents well-mixed soil, and 12010 represents breccias.

between Apollo 11 crystalline rocks, soils, and breccias using Eu/Sm data from the Apollo 11 data-base (a magnetic tape file of Apollo 11 data taken from *Proc. Apollo 11 Lunar Sci. Conf., Geochim. Cosmochim. Acta* Suppl. 1, 1970, and provided by courtesy of Dr. J. L. Warner).

In general, lunar samples with a low Eu/Sm ratio (a marked Eu depletion) have higher REE abundances than do samples with a higher Eu/Sm ratio (a smaller Eu depletion). A correlation of this kind is not observed within the groups of Apollo 12 crystalline rocks, but is seen for all Apollo 12 samples taken together. Similar correlations are apparent for Apollo 11 crystalline rocks, soils, and breccias, considered either as separate groups or all together. If plagioclase fractionation has played a role in generating these Eu anomalies, it appears to have affected Apollo 11 crystalline rocks more than Apollo 12 crystalline rocks. The suggestion of TAYLOR *et al.* (1971) that increasing Eu depletion in Apollo 12 *crystalline* rocks is correlated with increasing total REE content is not supported either by our data, or by those for all samples of crystalline rocks in selected data from the Apollo 12 data-base.

Consider the mechanisms which may generate the high abundances of REE in lunar specimens, and the slight, but definite, relative enrichment of light REE in some of the Apollo 12 specimens. Materials representing residual siliceous liquids present at a late stage in the crystallization of lunar magmas have been observed to be rich in REE (as indeed would be expected both from data on minerals separated from lunar crystalline rocks and from partition coefficients for REE into the early-forming minerals of terrestrial igneous rocks), and perhaps fractionation of such residual liquids would generate REE-enriched specimens. This mechanism may have been important in the genesis of unusual rocks such as 12013, but in general it does not seem plausible that small amounts of residual liquids have moved from their point(s) of origin and have been subsequently incorporated in quite similar amounts into all the lunar basalts sampled as large fragments at the Apollo 11 and Apollo 12 landing sites.

Rare-earth-rich phosphate phases have been found in all types of Apollo 11 and Apollo 12 specimens (ALBEE and CHODOS, 1970; FUCHS, 1970 a,b; DRAKE *et al.*, 1970; KEIL *et al.*, 1971). Previously (GOLES *et al.*, 1971) we suggested that most of the REE in lunar rocks would be located in such phosphate phases. Crystal-liquid partition coefficients determined by NAGASAWA (1970) for REE in terrestrial apatites are very large, lending support to this suggestion. Differentiation processes involving such phosphates are thus likely to be important in explaining REE distribution patterns of lunar materials.

Rare-earth-rich phosphates seem to be late-crystallizing minerals in lunar magmas, and so a one-stage fractionation model for lunar crystalline rocks relying on these phosphates would be subject to the same difficulties as is the fractionation of residual liquids discussed above. However, we may construct a two-stage model. Using NAGASAWA's (1970) partition coefficients, it is easy to show that phosphates crystallizing from a liquid with approximately chondritic REE relative abundances (but not necessarily chondritic absolute abundances) would incorporate large amounts of REE whose relative abundance pattern would be nearly flat, but with a bow upward in the middle of the REE series and a large Eu depletion. If at some later time, rocks containing phosphates with that REE pattern were partially melted, REE abundances of the resulting magma would be dominated by the phosphate pattern, and hence the magma would have a REE pattern very similar to those observed in almost all lunar specimens. Minor plagioclase fractionation, presumably at very shallow levels, might then be required to generate the variety of Eu/Sm ratios observed in lunar specimens.

Relative enrichment in light REE is characteristic of KREEP glass (HUBBARD *et al.*, 1971 a,b) and of materials rich in a KREEP-like component (see our discussion of mixing models below). During partial melting which generated ancestral magmas of the KREEP components, clinopyroxene may have been left behind in such a way as to produce a slight enrichment in light REE in the melt. However, there is another, very efficient, mechanism for generating relative enrichments in light REE. Zircon, baddeleyite, and Zr–Ti silicates (mineral A, etc.; CAMERON, 1970; RAMDOHR and EL GORESEY, 1970; BROWN *et al.*, 1971; FRICK *et al.*, 1971) also incorporate large amounts of REE. If the partition coefficients for REE in lunar zircons and Zr–Ti silicates are similar to those of terrestrial zircons (NAGASAWA, 1970), and if these phases are, in at least some cases, early-crystallizing minerals which could be fractionated from their parental magmas, it would be easy to produce relative enrichments in the light REE.

Mechanisms of REE fractionation such as those suggested above should now be tested by construction of explicit models of fractionation paths. Nevertheless, they seem to be plausible in the light of available data.

MIXING MODELS FOR APOLLO 12 SOILS

Possible origins of the lunar regolith, and in particular the lunar soil, have been discussed by many authors (e.g., DUKE *et al.*, 1970; FREDRIKSSON *et al.*, 1970; MCKAY *et al.*, 1970; QUAIDE and BUNCH, 1970; SHORT, 1970; WOOD *et al.*, 1970a; SHOEMAKER *et al.*, 1970; GOLES, 1971b), and it is apparent that comminution of various rock types and addition or generation of a glassy component have both played a major role in the history of the soil. When different rock types which have essentially similar mineralogy undergo comminution simultaneously, it is frequently very difficult to estimate the proportions of the parents from a study of their ground-up remains. There are also statistical and sampling problems involved in making estimates of the proportions of different rock types in lunar soils from counts of identifiable particles within a restricted size range. The compositions of some glass types present in lunar soils have been extensively studied (e.g., the KREEP glasses), but compositional data for many apparently distinct glass types are either meagre or lacking altogether. Owing to these limitations on interpretation of data derived from studies of individual particles in the lunar soil, examination of compositional data for bulk samples of lunar soils and interpretation of these data by means of compositional mixing models may usefully complement such studies.

In order to evaluate the mixing models discussed below we made use of a computer program modified after one originally written by W. Bryan, L. Finger, and F. Chayes. The technique used ensures that the solutions attained are best fits according to a least-squares criterion, and is essentially the same as that described by BRYAN *et al.* (1969). However, since the compositional data we used had a range of 6 orders of magnitude, and since we wished to give trace and major elements equal weight in determining the solution, all data were normalized by dividing each mixing component, element by element, by elemental abundances in the constituent whose least-squares approximation was sought (e.g., the soil).

Four mixing models for Apollo 12 soils will be discussed. The compositions of all components used are given in Table 2, and the proportions of the components in each mix, together with percentage residuals for each element, are given in Table 3.

Model 1 is essentially a recalculation of the model presented by GOLES *et al.* (1971), except that slight alterations have been made in our own data, and a typical KREEP composition (HUBBARD *et al.*, 1971a) and an average composition for type-one carbonaceous chondrites were added as components. Model 1 differs from the other models presented in that only our own data for elemental abundances in Apollo 12 samples have been used. This particular model thus has the advantage of avoiding purely analytical inconsistencies by using data determined entirely in one laboratory, but suffers the disadvantage of including only a small number of samples of each Apollo 12 compositional type as its components. The fit for Model 1 is extremely good, with each residual being less than the estimated precision of our analyses for the element concerned. The largest residual is −7.9%, for Ce. Anorthosite and KREEP were included as components since both of these types of materials have been reported in all Apollo 12 soils, though in varying proportions (e.g., WOOD *et al.*, 1971; QUAIDE *et al.*, 1971; MCKAY *et al.*, 1971). The proportion of anorthosite (4.0%) in Model 1 is reasonable when compared to data from WOOD *et al.* (1971, Fig. 1).

From the data of WOOD *et al.* (1971) our estimate of 7.1% KREEP for Model 1 appears reasonable but is substantially lower than that inferred from a mixing model discussed by HUBBARD *et al.* (1971b). However, the splits of Apollo 12 breccias analyzed by us are clearly richer in KREEP than is the "average" Apollo 12 breccia (an average of so few samples of such highly variable rocks is obviously not very reliable), and we attribute the apparently low KREEP proportion in Model 1 to sampling errors. Note that we include breccias as an independent component of our mixing models, even though it is evident that these clastic rocks are not primary in the sense that our other components are. We have attempted to construct mixing models for breccias themselves in terms of primary components, with indifferent success, and feel that at present it is preferable to use observed breccia compositions to approximate the composition of an important set of components of the soils. Since the breccias are not simple mixtures of Apollo 12 crystalline rocks, KREEP, and anorthosite, they apparently supply components of which we have now only limited or no explicit knowledge. ANDERS *et al.* (1971) and GANAPATHY *et al.* (1970) consider that a chondritic component, with a composition similar to that of carbonaceous chondrites, is present in Apollo 12 soils. The Model 1 estimate of the chondritic proportion of the soil as 1.5% is almost entirely dependent on relative Co abundances in the different components, and is consequently in remarkably good agreement with GANAPATHY *et al.*'s (1970) estimate of 1.8% of a chondritic component. (The abundances of the elements they analyzed appear to be much better suited to this calculation than are Co abundances.)

Type-two carbonaceous chondrites give essentially the same results as do type-one carbonaceous chondrites when used as a component in our mixing models, and are probably more consistent with the mineralogical evidence discussed by JEDWAB and HERBOSCH (1970).

The compositions of Apollo 12 components in Models 2, 3, and 4 are derived from the Apollo 12 data-base as discussed earlier. The "average" soil used in Model 2 includes data for samples 12001, 12025 (surface sample only), 12044, and 12070. These four samples are very similar, both with regard to their chemical composition, and from studies of physical properties of their constituent particles (MCKAY *et al.*, 1971; QUAIDE *et al.*, 1971), and we consider them to be representative of comparatively well-mixed samples of the surface soil at the Apollo 12 landing site (see also QUAIDE *et al.*, 1971). Many other Apollo 12 soils (including core-tube strata) have markedly different compositions, both from each other, and from the soils discussed above. Mixing models for two of these, 12032 and 12033, are given as Models 3 and 4.

Model 2 shows appreciably larger residuals for many elements than does Model 1. While many of these larger residuals may be due to variations related to systematic analytical differences between laboratories, we feel that the large residuals for Na (−20.1%), Rb (−21.7%), and Cs (−31.4%) are significant. The proportions of anorthosite in this model appear to be reasonable, and our estimate of the KREEP component (25.3%) agrees almost exactly with the estimate (25%) of HUBBARD *et al.* (1971b) for soil 12070. The estimated chondritic component is markedly low (0.3%) compared to the estimate (1.8%) by GANAPATHY *et al.* (1970).

Model 3, for soil 12032, generates residuals like those of Model 2, and we consider

Table 2. Compositions of components used in mixing models of

	11–AV–A	11–AV–B	11–AV–D	12–AV–GP1	12–AV–GP2	12–AV–C	12–AV–D	12–G–GP1
%Si	18.9	18.8	19.7	21.6	20.6	21.6	21.5	—
%Ti	7.13	6.40	4.43	2.18	1.61	1.97	1.68	1.99
%Al	4.12	5.28	7.25	5.42	4.26	6.67	6.75	5.15
%Ca	7.53	8.26	8.67	8.07	6.23	7.73	7.54	7.60
%Fe	14.8	14.4	12.4	15.4	16.4	12.2	12.7	15.1
%Mg	4.58	4.23	4.75	4.47	8.37	5.02	6.20	—
Na	3630	3010	3180	2120	1550	4990	3310	1920
Rb	5.83	0.81	2.76	1.11	0.962	5.94	6.47	—
Cs	0.169	0.032	0.103	0.06	0.061	0.49	0.42	—
Sr	171	178	164	130	98.3	140	145	—
La	25.5	10.5	15.0	6.63	5.60	62.9	33.0	6.48
Ce	75.7	37.0	50.0	19.4	16.50	174	89.6	28.3
Nd	64.1	35.5	37.9	15.4	12.0	76.8	55.6	15.3
Sm	20.4	13.8	13.2	5.40	4.29	28.8	16.3	5.29
Eu	2.18	2.11	1.78	1.15	0.91	2.53	1.74	1.10
Gd	27.4	17.9	15.7	7.35	5.93	30.1	19.6	—
Dy	32.4	22.9	19.8	8.44	7.05	31.8	23.1	—
Er	18.9	12.1	11.9	6.19	3.86	19.9	13.4	—
Yb	18.9	13.1	11.2	4.71	3.64	22.4	12.1	4.47
Lu	2.69	1.87	1.56	0.66	0.55	3.19	1.80	0.66
Mn	1770	2090	1570	2090	2120	2090	1740	1960
Co	27.1	14.0	29.2	34.0	57.9	35.3	41.9	33.8
Sc	78.0	85.0	60.2	56.9	43.5	33.2	37.9	49.9
Cr	2240	1680	1860	2730	3800	2120	2650	2530

Column Headings: 11–AV–A—Average of Apollo 11 A-type rocks, selected data from Apollo 11 data-base. Samples 10017, 10022, 10024, 10049, 10057, 10069, 10071, 10072. 11–AV–B—Average of Apollo 11 B-type rocks, selected data from Apollo 11 data-base. Samples 10003, 10020, 10044, 10045, 10047, 10050, 10058, 10062. 11–AV–D—Average of Apollo 11 soil sample 10084, selected data from Apollo 11 data-base. 12–AV–GP1—Average of Apollo 12 group 1 crystalline rocks, selected data from Apollo 12 data-base. Samples 12017, 12021, 12051, 12053, 12064, 12065. 12–AV–GP2—Average of Apollo 12 group 2 crystalline rocks, selected data from Apollo 12 data-base. Samples 12002, 12004, 12008, 12009, 12018, 12020, 12022, 12035, 12040, 12075. 12–AV–C—Average of Apollo 12 breccias, selected data from Apollo 12 data-base. Samples 12010, 12034, 12073. 12–AV–D—Average of four well-mixed Apollo 12 soils, selected data from Apollo 12 data-base. Samples 12001, 12025 (surface sample only), 12044, 12070. 12–G–GP1—Average of Apollo 12 group 1 crystalline rocks, data from this paper only. Samples 12021, 12065. 12–G–GP2—Average

that most of the higher residuals could be attributed to combination of data from a number of laboratories. However again the residuals for Na (−40.3%) and Rb (−25.6%) seem to be too large for explanation in this fashion. Model 4, for soil 12033, not only has high residuals for Na (−38.9%) and Cs (+25.5%), but also has higher residuals than have appeared in previous models for a number of other elements (e.g., Eu, −18.6%; Gd, +10.8%; Sc, +15.7%). Either additional components are present in 12033 soil compared with those present in soil 12032 and in 12070-type soil (12-AV-D), or one or more of the components present in 12033 soil is of distinctly different composition from the corresponding component(s) in the other two soils.

The model estimates of anorthosite proportions in 12032 and 12033 appear quite reasonable, and the proportion of KREEP in 12033 (58.5%) is close to the estimate (65%) by HUBBARD et al. (1971b). Owing to the comparatively large residuals for some elements in Models 2, 3, and 4, apparent variations in the proportions of Apollo 12 group 1 and group 2 crystalline rocks which are present in these soils should be interpreted with caution, and we do not believe that they merit specific discussion at this stage.

Table 3 (ppm except where otherwise noted).

12–G–GP2	12–G–C	12–G–D	12032	12033	Anorthosite	KREEP	CC–1
—	—	—	21.7	22.1	21.5	22.5	10.8
1.39	1.39	1.60	1.53	1.47	0.18	1.10	0.05
3.87	7.03	6.35	7.19	7.60	17.9	8.25	0.95
5.20	7.35	6.75	7.57	7.64	12.5	7.8	1.08
16.3	10.4	12.8	11.8	11.0	2.2	8.30	17.1
—	—	—	5.97	5.45	1.03	4.70	9.39
1460	6000	3230	4460	4900	2820	3700	5110
—	—	—	9.21	10.7	0.64	15.3	2.38
—	—	—	0.37	0.40	0.05	0.82	0.18
—	—	—	161	176	180	208	8.2
4.84	68.0	29.7	47.0	62.2	1.58	101	0.19
25.5	201	99	122	162	3.54	263	0.60
12.7	102	52	73.0	108	2.55	153	0.42
3.79	31.6	16.0	21.2	33.7	0.69	43.8	0.13
0.81	2.64	1.70	2.20	2.82	0.89	3.23	0.53
—	—	—	24.9	35.7	0.84	50.6	0.24
—	—	—	27.3	34.3	0.80	56.0	0.22
—	—	—	16.6	22.6	0.51	35.6	0.15
3.35	24.8	12.3	16.1	21.7	0.47	35.2	0.13
0.54	3.60	1.75	2.28	2.67	0.074	4.66	0.02
1890	1370	1560	1580	1470	290	1210	1880
62.2	31.9	45.1	38.3	32.3	6.4	35	480
38.2	31.4	37.4	36.0	27.0	5.5	25	5.1
4380	1860	2540	2530	2190	400	1150	2430

of Apollo 12 group 2 crystalline rocks, data from this paper only. Samples 12002, 12018, 12040. 12–G–C—Average of Apollo 12 breccias, data from this paper only. Samples 12010, 12034. 12–G–D—Average of two well-mixed Apollo 12 soils, data from this paper only. Samples 12044, 12070. 12032—Average composition of soil 12032, selected data from Apollo 12 data-base. 12033—Average composition of soil 12033, selected data from Apollo 12 data-base. Anorthosite—Gabbroic anorthosite from Apollo 11 regolith, data from Wood *et al.* (1970a) and Wakita and Schmitt (1970a). KREEP—Average KREEP glass, data from Meyer *et al.* (1971) and Hubbard *et al.* (1971a,b) supplemented by data on the dark portion of rock 12013 from Wakita and Schmitt (1970b) and Laul *et al.* (1970). CC–1—Average type-one carbonaceous chondrite, data from Schmitt *et al.* (1964), Schmitt *et al.* (1971), Keil (1969), Kaushal *et al.* (1970), and Goles (1971a).

The generally larger residuals for Na, Rb, and Cs in Models 2, 3, and 4 suggest that a high-alkali component, distinct from the KREEP group of materials, may be a component of the Apollo 12 soils. If this component had alkali abundances similar to those of the light-colored portion of 12013, less than 1% of such a component would be needed in the soil to correct the observed alkali residuals. The "silicic component" reported by McKay *et al.* (1971) in Apollo 12 soils may well be this minor, alkali-rich component needed in the mixing models. Alternatively, the Na, Rb, and Cs abundances in KREEP materials themselves, or in high-KREEP breccias, may be markedly variable.

Hubbard *et al.* (1971b, Fig. 4) showed that mixtures of Apollo 12 crystalline rocks and KREEP fall very close to a 4.4 billion year isochron (the model age for soil) and are thus consistent with the available Sr isotopic data. At this time, since we cannot identify how much, if any, of a putative high-alkali component (other than KREEP) is present in the Apollo 12 soils, our models necessarily contain the same major components as those of Hubbard *et al.* (1971b) and should therefore be consistent with the Sr isotopic data.

Table 3. Mixing models for Apollo 11 and Apollo 12 soils. Proportions of components (%).

	1	2	3	4	5
11–AV–A					25.78
11–AV–B					35.22
11–AV–D					⋆
12–AV–GP1		21.20	27.58	13.51	
12–AV–GP2		36.44	25.55	19.45	15.15
12–AV–C		9.30			
12–AV–D		⋆			
12–G–GP1	34.85				
12–G–GP2	21.12				
12–G–C	21.42				
12–G–D	⋆				
12032			⋆		
12033				⋆	
Anorthosite	4.02	7.43	5.94	8.51	18.85
KREEP	7.09	25.31	40.92	58.53	3.81
CC–1	1.51	0.31			1.19

Differences between observed and calculated abundances (%)

	1	2	3	4	5
Si	—	−0.4	(0.1)*	(0.6)*	(−0.1)*
Ti	−5.6	−9.3	−3.7	−13.8	−0.4
Al	−3.3	0.1	−2.3	4.1	0.2
Ca	5.6	0.9	2.4	3.8	−2.0
Fe	0.8	−0.1	(1.8)	(5.8)*	(−0.6)*
Mg	—	−7.1	−10.3	−7.0	−6.9
Na	2.6	−20.2	−40.3	−38.9	−6.8
Rb	—	−21.7	−25.6	−12.7	−3.4
Cs	—	−31.4	0.2	25.5	2.2
Sr	—	−1.8	(2.8)*	(1.0)*	(0.0)*
La	7.3	6.0	−4.9	−1.6	1.8
Ce	−7.9	4.0	−3.8	−1.2	−8.5
Nd	−1.9	(−3.4)*	(4.0)*	(12.8)*	(−1.8)*
Sm	−1.8	1.4	−3.1	−19.1	−4.8
Eu	−2.6	−2.7	−12.6	−18.6	−2.3
Gd	—	(−1.0)*	(2.2)*	(10.8)*	(4.2)*
Dy	—	(−6.5)*	(0.6)*	(−3.1)*	(−0.2)*
Er	—	(1.5)*	(−4.2)*	(0.6)*	(−6.1)*
Yb	2.2	10.4	3.5	1.3	2.4
Lu	3.4	1.2	(2.2)*	(−9.9)*	(4.3)*
Mn	5.0	2.0	(3.1)*	(3.1)*	(4.4)*
Co	−0.4	1.2	1.5	14.2	−1.0
Sc	0.0	−0.4	3.8	15.7	2.7
Cr	−0.6	−6.0	−24.5	−17.1	1.8

⋆ Designates the component whose approximation is sought by the mixing model.

* Differences for these elements were calculated using the observed mixing proportions, but observed abundances were not used to generate the model.

N.B. Composition and explanation of components is given in Table 2.

MIXING MODEL FOR APOLLO 11 SOIL

GOLES et al. (1970) presented a rudimentary mixing model to show that a large fraction of the Apollo 11 soil consisted of a "cryptic component" which was not represented in the collection of crystalline rocks sampled at the Apollo 11 landing site. WOOD et al. (1970a,b) observed and analyzed anorthositic clasts in a collection of lithic fragments (ranging from 1 to 5 mm in the least dimension) from Apollo 11 soil and suggested that the composition of the Apollo 11 breccias could be approximated by a mixture of 20% gabbroic anorthosite and 80% basalt. However, GOLES (1971b) showed that Apollo 11 soil was not a simple mixture of anorthositic clasts and fragments of basaltic rocks like those collected at the Apollo 11 site, and thus that at least one additional component was necessary. Now that more lunar samples have been collected and analyzed, we have constructed a mixing model which describes the composition of the Apollo 11 soil in more detail. Our results are presented as Model 5 in Table 3.

The Apollo 11 A-type rocks comprise only about 42% of the A + B contribution, in contrast to the minimum ratio of 4 to 1, A to B rocks, as proposed by TERA *et al.* (1970) to account for the composition of Apollo 11 breccias, and as used by GOLES (1971b) for the Apollo 11 soil. The discrepancy originates in the high (and previously unsuspected) alkali contents of the KREEP-like component. In Model 5 the sum of the contributions from A and B type rocks (61%) to the soil is close to the maximum value allowed by the Ti content of the soil (GOLES *et al.*, 1970). The 19% contribution of gabbroic anorthosite is a factor of four greater than the proportion of anorthositic particles found by WOOD *et al.* (1970a) in one size fraction of Apollo 11 soil. Either the anorthositic material is largely concentrated in the fine fractions of the soil as is suggested by their relatively higher Al contents (FRONDEL *et al.*, 1970; GOLES *et al.*, 1970), or many fragments of anorthositic composition in the coarser fraction studied by WOOD *et al.* were not identified as such, perhaps because they were glassy.

The choice of group 2 Apollo 12 crystalline rocks as a component in Model 5 is not completely arbitrary, but depends on three observations: (1) As discussed above, at least one component with a relatively mafic composition which is distinct from that of Apollo 11 crystalline rocks is required by the composition of Apollo 11 soil. This component can of course have a wide range of compositions. The only compositional data available to us for such an additional component were those for Apollo 12 crystalline rocks. (2) When group 1 Apollo 12 rocks are used instead of group 2 rocks, the residuals became relatively large. (3) When mixing models are computed with both rock types included, the proportion of group 2 rocks is very much greater than that of group 1 rocks. We do not mean to imply that fragments of these rocks were actually transported from the Apollo 12 site to the Apollo 11 site, but merely suggest that about 15% of the Apollo 11 soil is apparently composed of fragments of rocks compositionally similar to these group 2 rocks. The fact that nearly all of the differences in calculated vs. observed elemental abundances are within analytical error (Table 3) suggests that the composition of the Apollo 12 group 2 rocks is a good approximation to the composition of a principal component of Apollo 11 soil.

The carbonaceous chondritic contribution (discussed above) amounts to 1.2% for Model 5, in fair agreement with the estimate of GANAPATHY *et al.* (1970) of 1.7% for the chondritic contribution to Apollo 11 soil. A component of chondritic composition is included primarily for completeness and, with the exception of Co, makes only small contributions to the fit obtained. A representative KREEP component was also included in model 5, since KREEP glasses have been found in Apollo 11 soils (CHAO *et al.*, 1970; REID *et al.*, 1970) and since Luny Rock 1 of ALBEE and CHODOS (1970) has been identified as a KREEP sample by MEYER *et al.* (1971). The amount of KREEP obtained in Model 5 (3.8%) is small, but critical to the overall fit, for without it the REE and alkali elements tend to force so much A and B type rocks into the mixture that the residuals for major elements become very large.

CONCLUSIONS

Although we have analyzed only a limited number of samples from two lunar sites, it is already possible to state some useful conclusions. Differentiation processes which generated lunar crystalline rocks cannot be described by a single mechanism such as fractional crystallization or partial melting. Rather, a combination of such mechanisms is required by the observed compositional differences between crystalline rocks. Rare earths are especially informative in this context and suggest that a two-stage (or more complex) process is required to explain the observed range of REE abundances. Soil mixing models can be devised which are in gratifying agreement with observed compositions and with the limitations placed on soil components by other workers. These mixing models suggest that we have already sampled and analyzed the major kinds of materials which contributed to mare regoliths at the Apollo 11 and Apollo 12 landing sites.

Acknowledgments—We thank T. W. Osborn, III for determining Si in many of our samples. Dr. W. B. Bryan of Woods Hole Oceanographic Institution provided the original version of the computer program used in constructing our mixing models. G. J. Allison aided greatly in drafting figures, as did J. E. Duncan in typing drafts of the text. This work was supported in large part by NASA Contract NAS 9–7961.

REFERENCES

ALBEE A. L., and CHODOS A. A. (1970) Microprobe investigations on Apollo 11 samples. *Proc. Apollo 11 Lunar Sci. Conf., Geochim. Cosmochim. Acta* Suppl. 1, Vol. 1, pp. 135–158. Pergamon.

ANDERS E., LAUL J. C., KEAYS R. R., GANAPATHY R., and MORGAN J. W. (1971) Elements depleted on lunar surface: Implications for origin of moon and meteorite influx rate. Second Lunar Science Conference (unpublished proceedings).

ANNELL C. S., CARRON M. K., CHRISTIAN R. P., CUTTITTA F., DWORNIK E. J., HELZ A. W., LIGON D. T., and ROSE H. J. (1971) Chemical and spectrographic analyses of lunar samples from Apollo 12 mission. Second Lunar Science Conference (unpublished proceedings).

BROWN G. M., EMELEUS C. H., HOLLAND J. G., PECKETT A., and PHILLIPS R. (1971) Mineral chemistry of contrasted Apollo 12 basalt-types and comparisons with Apollo 11. Second Lunar Science Conference (unpublished proceedings).

BRYAN W. B., FINGER L. W., and CHAYES F. (1969) Estimating proportions in petrographic mixing equations by least-squares approximation. *Science* **163**, 926–927.

CAMERON E. N. (1970) Opaque minerals in certain lunar rocks from Apollo 11. *Proc. Apollo 11 Lunar Sci. Conf., Geochim. Cosmochim. Acta* Suppl. 1, Vol. 1, 221–246. Pergamon.

CHAO E. C. T., BOREMANN J. A., MINKIN J. A., JAMES O. B., and DESBOROUGH G. A. (1970) Lunar glasses of impact origin: Physical and chemical characteristics and geologic implications. *J. Geophys. Res.* **75**, 7445–7479.

COMPSTON W., BERRY H., VERNON M. J., CHAPPELL B. W., and KAYE M. J. (1971) Rb–Sr chronology and chemistry of lunar material from the Ocean of Storms. Second Lunar Science Conference (unpublished proceedings).

DRAKE M. J., MCCALLUM I. S., MCKAY G. A., and WEILL D. F. (1970) Mineralogy and petrology of Apollo 12 sample no. 12013: A progress report. *Earth Planet. Sci. Lett.* **9**, 103–123.

DUKE M. B., WOO C. C., SELLERS G. A., BIRD M. L., and FINKELMAN R. B. (1970) Genesis of lunar soil at Tranquillity base. *Proc. Apollo 11 Lunar Sci. Conf., Geochim. Cosmochim. Acta* Suppl. 1, Vol. 1, pp. 347–361. Pergamon.

FREDRIKSSON K., NELEN J., and MELSON W. G. (1970) Petrography and origin of lunar breccias and glasses. *Proc. Apollo 11 Lunar Sci. Conf., Geochim. Cosmochim. Acta* Suppl. 1, Vol. 1, pp. 419–432. Pergamon.

FRICK C., HUGHES T. C., LOVERING J. F., REID A. F., WARE N. G., and WACK D. A. (1971) Electron probe, fission track and activation analysis of lunar samples. Second Lunar Science Conference (unpublished proceedings).

FRONDEL C., KLEIN C., JR., ITO J., and DRAKE J. C. (1970) Mineralogical and chemical studies of Apollo 11 lunar fines and selected rocks. *Proc. Apollo 11 Lunar Sci. Conf., Geochim. Cosmochim. Acta* Suppl. 1, Vol. 1, pp. 445–474. Pergamon.

FUCHS L. H. (1970a) Fluorapatite and other accessory minerals in Apollo 11 rocks. *Proc. Apollo 11 Lunar Sci. Conf., Geochim. Cosmochim. Acta* Suppl. 1, Vol. 1, pp. 475–479. Pergamon.

FUCHS L. H. (1970b) Orthopyroxene–plagioclase fragments in the lunar soil from Apollo 12. *Science* **169**, 866–868.

GANAPATHY R., KEAYS R. R., and ANDERS E. (1970) Apollo 12 lunar samples: Trace element analysis of a core and the uniformity of regolith. *Science* **170**, 533–535.

GAST P. W. (1968) Trace element fractionation and the origin of tholeiitic and alkaline magma types. *Geochim. Cosmochim. Acta* **32**, 1057–1086.

GOLES G. G. (1971a) Caesium. In *Elemental Abundances in Meteorites* (editor B. Mason) (New York: Gordon and Breach, in press).

GOLES G. G. (1971b) Comments on the genesis and evolution of Apollo XI "soil". *Lithos* **4**, 71–81.

GOLES G. G., RANDLE K., OSAWA M., LINDSTROM D. J., JEROME D. Y., STEINBORN T. L., BEYER R. L., MARTIN M. R., and MCKAY S. M. (1970) Interpretations and speculations on elemental abundances in lunar samples. *Proc. Apollo 11 Lunar Sci. Conf., Geochim. Cosmochim. Acta* Suppl. 1, Vol. 2, pp. 1177–1194. Pergamon.

GOLES G. G., DUNCAN A. R., OSAWA M., MARTIN M. R., BEYER R. L., LINDSTROM D. J., and RANDLE K. (1971) Analyses of Apollo 12 specimens and a mixing model for Apollo 12 "soils." Second Lunar Science Conference (unpublished proceedings).

GORDON G. E., RANDLE K., GOLES G. G., CORLISS J. B., BEESON M. H., and OXLEY S. S. (1968) Instrumental activation analysis of standard rocks with high-resolution γ-ray detectors. *Geochim. Cosmochim. Acta* **32**, 369–396.

HASKIN L. A., ALLEN R. O., HELMKE P. A., ANDERSON M. R., KOROTEV R. L., and ZWEIFEL K. A. K. A. (1971) Rare earths and other trace elements in Apollo 12 lunar materials. Second Science Conference (unpublished proceedings).

HUBBARD N. J., MEYER C., GAST P. W., and WIESMANN H. (1971a) The composition and derivation of Apollo 12 soils. *Earth Planet. Sci. Lett.* **10**, 341–350.

HUBBARD N. J., GAST P. W., and MEYER C. (1971b) The origin of the lunar soil based on REE, K, Rb, Ba, Sr, P, and $Sr^{87/86}$ data. Second Lunar Science Conference (unpublished proceedings).

JEDWAB J., and HERBOSCH A. (1970) Tentative estimation of the contribution of Type 1 carbonaceous meteorites to the lunar soil. *Proc. Apollo 11 Lunar Sci. Conf., Geochim. Cosmochim. Acta* Suppl. 1, Vol. 1, pp. 551–559. Pergamon.

KAUSHAL S. K., and WETHERILL G. W. (1970) Rubidium-87–strontium-87 age of carbonaceous chondrites. *J. Geophys. Res.* **75**, 463–468.

KEIL K. (1969) Meteorite composition. In *Handbook of Geochemistry* (V. 1) (editor K. H. Wedepohl), 78–115. Springer-Verlag.

KEIL K., PRINZ M., and BUNCH T. E. (1971) Mineralogical and petrological aspects of Apollo 12 rocks. Second Lunar Science Conference (unpublished proceedings).

LAUL J. C., KEAYS R. R., GANAPATHY R., and ANDERS E. (1970) Abundance of 14 trace elements in lunar rock 12013,10. *Earth Planet. Sci. Lett.* **9**, 211–215.

MCKAY D. S., GREENWOOD W. R., and MORRISON D. A. (1970) Origin of small lunar particles and breccia from the Apollo 11 site. *Proc. Apollo 11 Lunar Sci. Conf., Geochim. Cosmochim. Acta* Suppl. 1, Vol. 1, pp. 673–694. Pergamon.

McKAY D. S., MORRISON D., LINDSAY J., and LADLE G. (1971) Apollo 12 soil and breccia. Second Lunar Science Conference (unpublished proceedings).

MEYER C., AITKEN F. K., BRETT R., McKAY D. S., and MORRISON D. A. (1971) Rock fragments and glasses rich in K, REE, P in Apollo 12 soils: Their mineralogy and origin. Second Lunar Science Conference (unpublished proceedings).

NAGASAWA H. (1970) Rare-earth concentrations in zircons and apatites and their host dacites and granites. *Earth Planet. Sci. Lett.* **9**, 359–364.

OSAWA M., and GOLES G. G. (1970) Trace element abundances in Columbia River Basalts. *Proceedings of the Second Columbia River Basalt Symposium* (ed. Gilmore E. H. and Stradling D.) pp. 55–71. Eastern Washington State College Press.

QUAIDE W., and BUNCH T. (1970) Impact metamorphism of lunar surface materials. *Proc. Apollo 11 Lunar Sci. Conf., Geochim. Cosmochim. Acta* Suppl. 1, Vol. 1, pp. 711–729. Pergamon.

QUAIDE W., OBERBECK V., BUNCH T., and POLKOWSKI G. (1971) Investigations of the natural history of the regolith at the Apollo 12 site. Second Lunar Science Conference (unpublished proceedings).

RAMDOHR P., and EL GORESEY A. (1970) Opaque minerals of lunar rocks and dust from Mare Tranquillitatis. *Science* **167**, 615–618.

REID A. M., FRAZIER J. Z., FUJITA H., and EVERSON J. E. (1970) Chemical composition of the major phases in Apollo 11 lunar samples. Scripps Institution of Oceanography Report 70–4.

ROSE H. J., CUTTITTA F., ANNELL C. S., CARRON M. K., CHRISTIAN R. P., DWORNIK E. J., HELZ A. W., and LIGON D. T. (1971) Semi-micro analysis of Apollo 12 samples. Second Lunar Science Conference (unpublished proceedings).

SCHMITT R. A., SMITH R. H., and OLEHY D. A. (1964) Rare-earth, yttrium and scandium abundances in meteoritic and terrestrial matter, II. *Geochim. Cosmochim. Acta* **28**, 67–86.

SCHMITT R. A., LINN T. A., and WAKITA H. (1970) The determination of fourteen common elements in rocks via sequential instrumental activation analysis. *Radiochim. Acta* **13**, 200–212.

SCHMITT R. A., GOLES G. G., and SMITH R. H. (1971) *Elemental Abundances in Stone Meteorites.* (To be published).

SHOEMAKER E. M., HAIT M. H., SWANN G. A., SCHLEICHER D. L., SCHABER G. G., SUTTON R. L., DAHLEM D. H., GODDARD E. N., and WATERS A. C. (1970) Origin of the lunar regolith at Tranquillity Base. *Proc. Apollo 11 Lunar Sci. Conf., Geochim. Cosmochim. Acta* Suppl. 1, Vol. 3, pp. 2399–2412. Pergamon.

SHORT N. M. (1970) Evidence and implications of shock metamorphism in lunar samples. *Proc. Apollo 11 Lunar Sci. Conf., Geochim. Cosmochim. Acta* Suppl. 1, Vol. 1, pp. 865–871. Pergamon.

TAYLOR S. R., KAYE M., GRAHAM A., RUDOWSKI R., and MUIR P. (1971) Trace element chemistry of lunar samples from the Ocean of Storms. Second Lunar Science Conference (unpublished proceedings).

TERA F., EUGSTER O., BURNETT D. S., and WASSERBURG G. J. (1970) Comparative study of Li, Na, K, Rb, Cs, Ca, Sr, and Ba abundances in achondrites and in Apollo 11 lunar samples, *Proc. Apollo 11 Lunar Sci. Conf., Geochim. Cosmochim. Acta* Suppl. 1, Vol. 2, pp. 1637–1657. Pergamon.

VOGT, J. R., and EHMANN W. D. (1965) The nondestructive determination of silicon and oxygen in meteorites by fast neutron activation analysis. *Proceedings of the 1965 International Conference on Modern Trends in Activation Analysis*, p. 82.

WAKITA H., and SCHMITT R. A. (1970a) Lunar anorthosites: Rare-earth and other elemental abundances. *Science* **170**, 969–974.

WAKITA H., and SCHMITT R. A. (1970b) Elemental abundances in seven fragments from lunar rock 12013. *Earth Planet. Sci. Lett.* **9**, 169–176.

WARNER J. L., and ANDERSON D. H. (1971) Lunar crystalline rocks: Petrology, geology, and origin. Second Lunar Science Conference (unpublished proceedings).

WOOD J. A., DICKEY J. S., JR., MARVIN U. B., and POWELL B. N. (1970a) Lunar anorthosites and a geophysical model of the moon. *Proc. Apollo 11 Lunar Sci. Conf., Geochim. Cosmochim. Acta* Suppl. 1, Vol. 1, pp. 965–988. Pergamon.

Wood J. A., Marvin U. B., Powell B. N., and Dickey J. S., Jr. (1970b) Mineralogy and petrology of the Apollo 11 lunar sample. Smithsonian Astrophysical Observatory, Spec. Rep. No. 307.

Wood J. A., Marvin U., Reid J. B., Taylor G. J., Bower J. F., Powell B. N., and Dickey J. S., Jr. (1971) Relative proportions of rock types and nature of the light-colored lithic fragments in Apollo 12 soil samples. Second Lunar Science Conference (unpublished proceedings).

Proceedings of the Second Lunar Science Conference, Volume 2, pp. 1083–1099.
The M.I.T. Press, 1971.

Trace element chemistry of lunar samples from the Ocean of Storms

S. R. TAYLOR, R. RUDOWSKI, PATRICIA MUIR, and A. GRAHAM
Department of Geophysics and Geochemistry, Australian National University, Canberra, Australia
and
MAUREEN KAYE
Department of Geology, Australian National University, Canberra, Australia

(*Received 22 February 1971; accepted in revised form 23 March 1971*)

Abstract—Variations in element abundances in Apollo 12 rocks are due both to varying degrees of partial melting and to limited fractional crystallisation. Trends in element concentrations occur for many samples in a NW–SE direction across the Apollo 12 traverse area. Elements concentrated in early crystal phases (Ni, Co, Cr, V) decrease, and those concentrated in residual phases (Th, U, REE, Y, Nb, Ba, Zr, Hf) increase southeastwards. These variations are consistent with derivation of the rocks from a shallow fractionated sequence, with the lower basic portions exposed in the NW quadrant as throwout from Middle Crescent Crater (WARNER and ANDERSON, 1971). Absolute variations in REE abundances without change of relative patterns are interpreted as due mainly to varying degrees of partial melting. The decrease in Eu depletion with decrease in REE total abundances, observed in Apollo 11 and 12, is interpreted as due to increase in the amount of partial melting, Eu (and Sr) being present in major phases in the lunar interior, while the trivalent REE occur in minor accessory phases. The minor Eu depletion in 12038 is probably due to a small degree of partial melting followed by plagioclase accumulation. The parallel geochemical behaviour of Eu and Sr indicates that Eu is not depleted in the whole moon.

It is necessary to add to the Apollo 12 rock compositions, a component similar to the dark portion of 12013, to produce the Apollo 12 soil compositions. The soil is not well mixed at the Apollo 12 site. Differences between Apollo 11 and 12 rocks indicate that the lunar interior is heterogeneous on a small scale. The lunar lavas are enriched in refractory elements and depleted in volatile and siderophile elements by factors of 100 or more in comparison with primitive solar system nebula abundances. These differences are common to both Apollo 11 and Apollo 12 sites and are sufficiently large to indicate that the entire moon shows similar compositional characteristics, indicating that the lunar material was heated to high temperatures at or prior to accretion. A lunar origin for tektites is rejected.

INTRODUCTION

THIS PAPER discusses the abundances of 35 elements in rock samples 12002, 12022, 12052, 12063, and 12038, in fines sample 12070, and in 10 mineral separates.

ANALYTICAL METHODS AND DATA

Thirty-five elements were determined by spark-source mass spectrometry and emission spectrography. The methods used and calibration standards used will be reported in detail elsewhere. The mass spectrometric method was similar to that described by TAYLOR (1965) and the emission spectrographic method, to that given by TAYLOR *et al.* (1969). A new procedure for processing spark source mass spectrographic photoplate data has been developed. The Ilford Q-2 photoplate response to incident mass is limited to an intensity range of about 50 over the linear portion of the emulsion calibration curve. Intensity ranges of greater than 1000 have to be recorded to cover the range in isotopic concentrations. Precise measurement of relative exposures is difficult because of variations in source geometry and it is necessary to provide a measurable internal standard for each exposure over a wide range.

To overcome the inherent limited photoplate response which makes the use of a single isotope line impossible, lutetium which has two isotopes 175 and 176 with a mass ratio of 37.61 has been selected. At a concentration of 50 ppm Lu_2O_3, Lu^{175} is readable on the short exposures and Lu^{176} is measurable on the larger exposures. Lutetium meets nearly all the conditions for an ideal internal standard for spark source mass spectrographic analysis of silicates (TAYLOR, 1965). Lu^{176} is corrected for interference by Hf^{176}, Yb^{176}, and natural Lu^{176} as follows: I Lu^{176} = I Lu^{176} meas. − (0.89 I Yb^{171} + 0.19 I Hf^{178} + 0.026 I Tm^{169}). The Tm correction, which is very small, is used as a measure of natural Lu^{176}. The intensity values are derived by a modification of the self-calibration method of AHRENS and TAYLOR (1961). Gamma for the photoplate is derived by using isotope pairs (eg., Yb^{173}, Yb^{174}; Gd^{155}, Gd^{158}) which provide a fixed-intensity scale, enabling the photographic density versus intensity relationship to be established. The value is commonly 50°. A Seidel function (log do/d − 1) which extends the linear portion of the photoplate calibration curve, is used to obtain relative intensities, by extrapolation of each density reading, using the derived gamma value, to a standard density of 25. Intensities for the individual nuclide lines are ratioed to the Lu intensity value for the same exposure. Calibration of intensity versus element concentration was carried out using USGS international rock standards and data from the Apollo 11 soil sample 10084, using data quoted by LEVINSON and TAYLOR (1971).

Mineral separations were carried out in the following manner. Samples were ground in new agate mortars. Following ultrasonic treatment to disperse aggregates, density fractions were separated by centrifuging with bromoform. Metal particles were removed from alcohol suspensions using plastic wrapped magnets. The density fractions were divided into size fractions by sieving through silk cloth to achieve maximum efficiency in handpicking. Mineral fractions were >98% pure. From the one gram samples supplied, it was possible to obtain clean 30 mg amounts of the major phases.

Analytical data are given for the rocks in Tables 1A and 1B and for the minerals in Tables 2A and 2B. Precision is dependent on measurement of the isotope/Lu intensity ratios and is ±10%. Accuracy is dependent on the values used for the calibration standards, and may be checked against values obtained by other analysts on the same samples. Space does not permit a detailed assessment, which will be published elsewhere, but agreement is generally satisfactory. Minor corrections to the data issued in preprint form at the Second Lunar Science Conference include deletion of the lead data in which laboratory contamination of about one ppm occurred during processing. The U and Th data for all rocks was a factor of two high due to an arithmetical error in data processing. Data for Ba, Sr, Cr, V, Sc, Ni, Co, Cu, and Li were obtained by emission spectrography. All other elements were determined by spark source mass spectrography.

SAMPLE DISTRIBUTION

Four of the rock samples (12002, 12022, 12052, and 12038) lie along a roughly N–S line across the Apollo 12 traverse area, from north of Head Crater to Bench Crater. Sample No. 12022 may be from the mound south of ALSEP. The fifth sample, 12063, is less precisely located near Halo Crater in the southernmost part of the traverse area (WARNER and ANDERSON, 1971).

TRENDS IN ELEMENT CONCENTRATIONS

The rocks collected by the astronauts during the Apollo 12 traverse are random samples to a very high degree. The regolith is complex and not well mixed (see later). The rocks lying at present on the surface are derived from all craters which penetrate

Table 1A. Trace element abundances in Apollo 12 lunar samples. All data in parts per million (ppm) except Cs (ppb).

	12002,139	12022,46	12052,47	12063,66	12038,27	12070,93
Cs ppb	60	30	30	25	20	300
Ba ppm	57	60	65	60	125	340
Eu ppm	0.84	1.4	0.90	1.6	2.6	1.9
Sr ppm	95	140	110	140	180	135
Ba/Sr	0.60	0.43	0.59	0.43	0.69	2.5
Th ppm	0.65	0.75	1.15	0.70	1.0	5.6
U ppm	0.15	0.17	0.30	0.15	0.25	1.6
Zr ppm	110	135	130	140	200	500
Hf ppm	3.5	5.2	4.6	5.4	6.5	14
Sn ppm	0.23	0.39	0.16	0.12	0.10	0.30
Nb ppm	4.5	6.0	6.0	4.0	7.8	25
Mo ppm	0.05	—	0.03	0.04	0.05	0.03
W ppm	—	—	—	—	—	0.5
Th/U	4.3	4.4	3.9	4.5	4.0	3.5
Zr/Hf	31	26	28	26	30	35
Cr ppm	5000	3300	3300	2500	2100	2800
V ppm	190	150	160	130	120	110
Sc ppm	41	55	54	60	50	40
Ni ppm	100	42	21	23	1.0	180
Co ppm	67	52	42	40	29	45
Cu ppm	5	5	5	5	5	5
Ni/Co	1.49	0.80	0.50	0.58	0.02	4.0
Cr/V	29	22	21	19	17	25
V/Ni	1.90	3.6	7.6	5.7	50	0.60
Cr/Ni	50	79	157	109	2100	15.6

Table 1B. Rare earth elements (and yttrium) in Apollo 12 samples. All data in parts per million (ppm).

	12002,139	12022,46	12052,47	12063,66	12038,27	12070,69
La	5.9	6.3	6.3	9.8	14	32
Ce	16	19	21	30	36	76
Pr	2.5	3.0	2.6	4.2	5.3	12
Nd	14	16	17	20	26	54
Sm	5.4	6.4	6.3	7.6	9.4	19
Eu	0.84	1.4	0.90	1.6	2.6	1.9
Gd	7.5	9.7	9.0	12	13	25
Tb	1.5	1.8	2.0	2.2	2.9	3.8
Dy	6.5	12	9.6	12	15	23
Ho	1.9	3.0	2.7	2.9	3.9	6
Er	4.6	8.2	7.2	9.0	9.9	15
Tm	0.8	1.3	1.1	1.5	1.6	2.4
Yb	4.2	7.2	5.3	6.8	9.0	14
Lu	—	—	—	—	—	—
$\sum$REE	72	95	91	120	149	283
Y	39	64	48	65	68	120
$\sum$REE + Y	111	159	139	185	217	403
La/Yb	1.40	0.88	1.19	1.44	1.55	2.29
Sm/Eu	6.4	4.6	7.0	4.8	3.6	10

Table 2A. Trace element data in pyroxenes from Apollo 12 samples.

	Clino-pyroxene 12002,139	Clino-pyroxene 12052,47	Clino-pyroxene 12063,66	Pigeonite 12038,27	Augite 12038,27	Clino-pyroxene 12070,69
Ba ppm	30	40	40	12	40	70
Eu ppm	0.60	1.4	0.77	0.35	0.80	0.72
Sr ppm	15	17	7	8	13	10
La ppm	3.0	3.6	4.7	1.3	5.0	7.8
Ce ppm	7.4	11.5	12	4.7	16	17
Pr ppm	1.0	1.9	2.1	0.72	2.5	3.4
Nd ppm	5.2	9.5	12	3.7	12	16
Sm ppm	2.4	5.8	5.1	1.50	4.5	6.2
Eu ppm	0.60	1.4	0.77	0.35	0.80	0.72
Gd ppm	4.1	7.7	9.4	2.4	7.1	6.6
Tb ppm	0.87	1.8	2.0	0.53	1.5	2.1
Dy ppm	4.5	7.3	12	2.6	8.9	10.5
Ho ppm	1.3	2.1	3.0	0.68	2.2	3.0
Er ppm	4.0	7.4	8.5	2.3	6.1	8.4
Tm ppm	—	—	—	—	0.92	—
Yb ppm	4.3	8.6	8.4	2.3	5.4	8.1
Lu ppm	—	—	—	—	—	—
$\sum$REE ppm	39	69	85	23	74	85
Y ppm	30	30	55	—	40	70
$\sum$REE + Y ppm	69	99	140	—	114	155
Th ppm	0.46	0.62	0.51	—	0.16	1.34
U ppm	0.14	—	0.15	—	—	0.35
Zr ppm	70	70	95	—	100	130
Hf ppm	2.4	2	3.2	—	3.0	3.6
Nb ppm	1.1	1.6	0.70	0.30	0.70	3.3
Th/U	3.3	—	3.4	—	—	3.8
Zr/Hf	29	35	30	—	33	36
Cr ppm	4000	4300	2700	3900	2800	3300
V ppm	250	270	150	250	140	180
Sc ppm	80	85	90	70	85	90
Ni ppm	20	—	—	—	—	25
Co ppm	37	30	34	34	35	34
Ni/Co	0.54	—	—	—	—	0.74
Cr/V	16	16	18	16	20	18
V/Ni	12.5	—	—	—	—	7.2
Cr/Ni	200	—	—	—	—	130

the regolith. These include small local craters (e.g., Sharp), large local craters (Surveyor, Middle Crescent) which excavate more deeply, and contributions from distant unidentifiable sources. Smaller craters (e.g., Block) reexcavate ejecta from larger craters. In view of the exceedingly complex origins possible for individual rocks, it is accordingly of interest if any regularity (however faint) appears in element trends across the traverse area.

The LSPET (1970) examination showed that the rock samples displayed systematic variations in element abundances when arranged in order of Mg content. This variation corresponds in general to N–S or NW–SE directions across the traverse area. Figure 1 shows the variation in the large cations (Ba), the large high-valency cations (Th, U, Zr, Sr, Nb), and the rare earth elements (REE + Y). Note that they generally

Table 2B. Trace element data in plagioclases and olivines from lunar samples. All data in ppm except Cs (ppb).

	Plagioclase 12063,66	Plagioclase 12038,27	Olivine 12002,139	Olivine 12070,69
Cs ppb	20	40	—	—
Ba	56	80	16	40
Pb	.34	.38	—	—
Eu	8.3	7.0	.098	.39
Sr	350	350	—	—
Ba/Sr	.16	.23	—	—
La	2.5	2.8	1.6	2.5
Ce	6.1	6.0	4.2	5.4
Pr	.82	.62	.36	1.1
Nd	4.0	2.5	1.3	5.0
Sm	1.6	.82	.39	2.5
Eu	8.3	7.0	.098	.39
Gd	2.1	1.2	.65	3.5
Tb	.49	.17	.11	.80
Dy	2.1	.99	.55	4.9
Ho	.56	.23	.16	1.5
Er	1.4	.51	.33	3.8
Tm	—	—	—	—
Yb	1.2	.40	.38	4.5
Lu	—	—	—	—
$\sum$ REE	31	23	10	36
Y	—	—	—	—
Zr	—	—	—	65
Hf	—	—	—	1.7
Zr/Hf	—	—	—	38
Cr	90	85	3300	3400
V	—	—	120	150
Sc	—	—	14	40
Ni	—	—	150	70
Co	—	—	95	55
Cu	—	—	—	—
Ni/Co	—	—	1.6	1.3
Cr/V	—	—	28	23
V/Ni	—	—	.80	2.1
Cr/Ni	—	—	22	49

increase in concentration from left to right, which corresponds to a southerly or SE direction across the traverse area. Note the parallel behaviour of Eu and Sr. Nickel, Co, Cr, and V, which enter sixfold sites in the main rockforming minerals, show contrasted decreases in the same direction (Fig. 2). When these trends are plotted against MgO content, they exhibit regular trends (Fig. 3). There are exceptions to the trends discussed here. For example, samples 12021, 12035, and 12040 do not fit. This is to be expected because of the complex origins and heterogeneity of the regolith and the random nature of the sampling.

Possible explanations for the variations observed are partial melting, fractional crystallisation, or a combination of both causes. The derivation of the lavas by partial melting from the lunar interior is considered established, both from the petrological and experimental petrology studies and, decisively, by the large spread in ages of maria filling shown by Apollo 11 and Apollo 12 rocks (LEVINSON and TAYLOR, 1971).

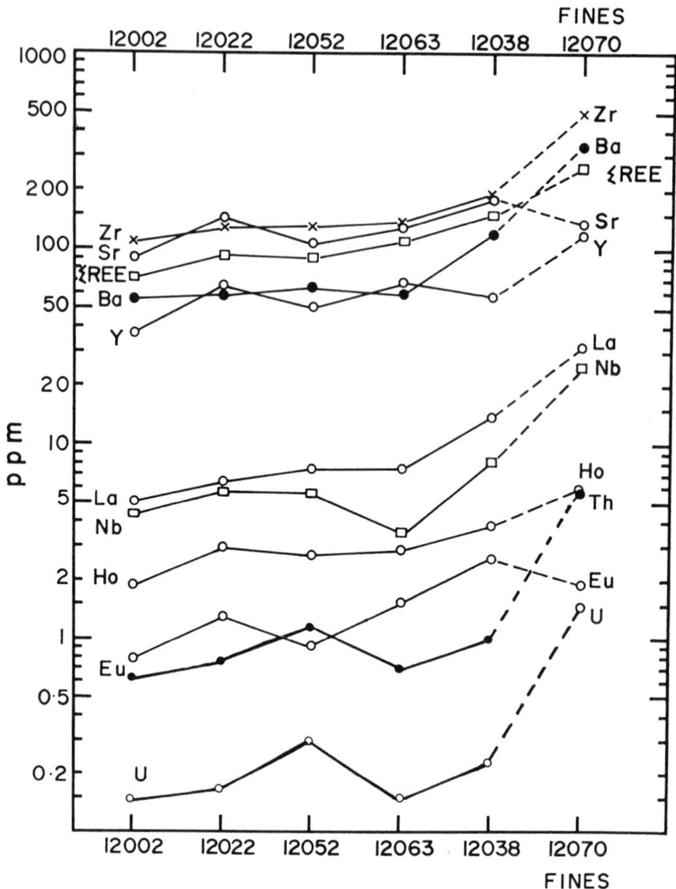

Fig. 1. Variation in abundance of large cations in N–S direction across Apollo 12 sample area. The most northerly sample (12002) is at left.

However, limited fractional crystallisation also appears to have operated, and it is difficult to disentangle these two processes. The changes in composition and the wide variety of rocks from a small area are consistent with processes of near surface fractionation acting on a limited number of primary melts. Removal of olivine (Kushiro *et al.*, 1971) can account for most of the variation among the ferromagnesian trace elements. Addition or subtractions of the interstitial material carrying the rare earth elements Th, U, Zr, Hf, Nb, etc. can account for much of the variation in these elements.

In this connection it is important to note the scale of the variations. A majority of the rock samples appear to be derived from the smaller craters and hence are of local origin. The collection area is about 500 × 500 meters. Within this small area, most of the rocks have differing compositions. If these differences in the local rocks are primary ones, due to partial melting, then they have been preserved without mixing from the zone of melting (>200 km depth) and over unknown but probably large

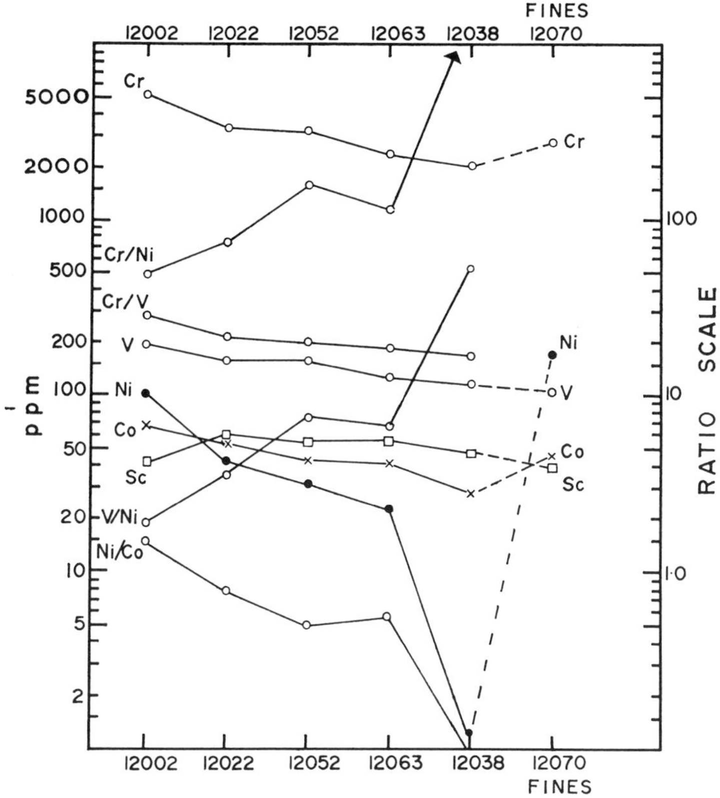

Fig. 2. Variation in abundance of Cr, V, Sc, Ni, and Co in a N–S direction across the Apollo 12 sample area.

lateral distances (10 < 1000 km). Rocks thrown out from distant (> 10 km) locations on to the Apollo 12 site do not suffer from this restriction, but the rocks discussed here appear to be local on geological grounds (SCHMITT and SUTTON, 1971). However, it should be recalled that the very low lunar lava viscosity may result in very thin flows or intrusive units. Hence large primary differences in composition may occur over smaller distances than expected on the basis of terrestrial experience.

The data presented here can be fitted into the stratigraphic scheme proposed by WARNER and ANDERSON (1971). They suggested a two-layer model for the upper portion of the mare fill underlying the regolith. An upper differentiated unit, represented by samples 12063 and possibly 12038, overlies a lower more basic unit (samples 12002, 12022, 12052). The lower unit, enriched in Ni, Co, Cr, and V, was excavated by Middle Crescent Crater, northwest of the traverse area, and forms an outthrown apron across the northwest portion of the traverse area. Further southeast, the original sequence is preserved, and rocks from the upper unit, enriched in Ba, REE, Th, U, Zr, Nb, etc. and depleted in Ni, Co, Cr, and V were thrown out, for example from Bench Crater. Although this is a much simplified picture (the rocks represent a

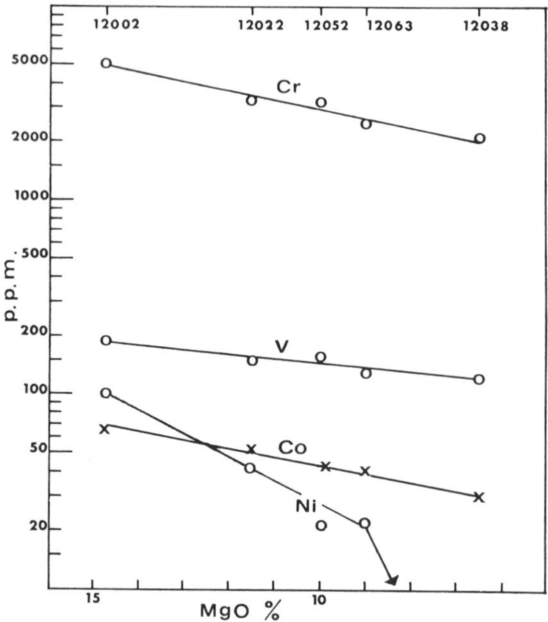

Fig. 3. Relation of Cr, V, Co and Ni concentrations to MgO content.

continuum in composition) it is generally consistent with the geology of the site (SCHMITT and SUTTON, 1971; WARNER and ANDERSON, 1971).

RARE-EARTH ELEMENTS

The rare earth abundances are plotted in Fig. 4, ratioed, element by element, to the chondritic abundances. In the lower diagram, the *rock* data are shown, with Apollo 11 *rock* data (LEVINSON and TAYLOR, 1971) for comparison. The absolute abundances vary, but the relative patterns do not alter, except that in the Apollo 12 rocks the relative Eu depletion, expressed as Sm/Eu ratios, is less. The rock data thus show a general decrease in Eu depletion with decreasing abundance of total REE.

The center diagram in Fig. 4 shows that the Apollo 12 fines contain higher absolute abundances of the rare earths, a greater Eu depletion, and are relatively enriched in the larger REE (La–Sm), compared to the Apollo 11 fines. Such differences are consistent with higher amounts of "cryptic" component similar to the dark portion of rock 12013. The top diagram in Fig. 4 shows the similar relative patterns in Nuevo Laredo achondrite and the most fractionated rock of the series (12038).

Two significant trends appear from a study of the rare earth patterns in Apollo 11 and Apollo 12 rocks.

(1) The total abundances vary by large factors, but the relative patterns show only slight changes. These shifts in the rare earth abundances, without alteration of the relative patterns, argue against unaccompanied mineral separation of phases such as plagioclase. Olivine removal will not effect the overall abundance patterns notably.

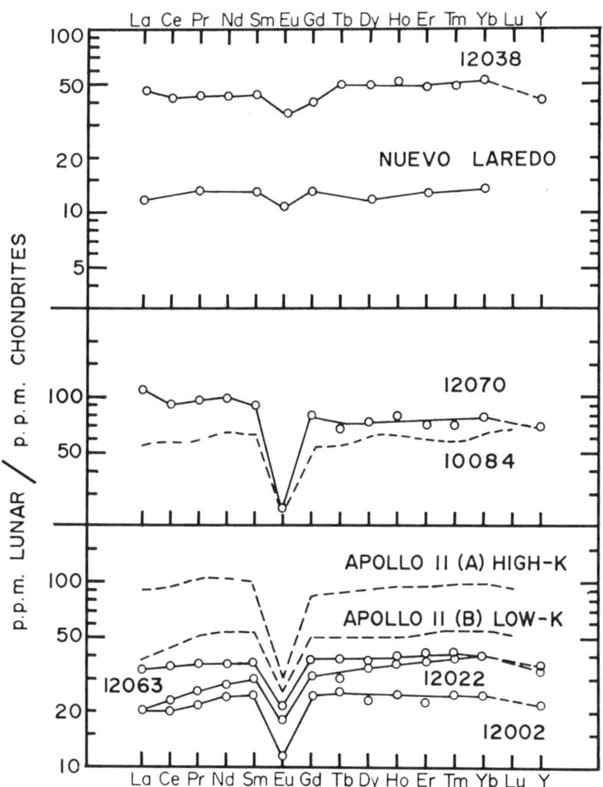

Fig. 4. Variation in rare earth elements (+Y) relative to abundances in chondritic meteorites. *Top*: Sample 12038 and Nuevo Laredo achondrite (GAST and HUBBARD, 1970). *Center*: Apollo 12 fines (12070) compared to Apollo 11 fines (10084). Note absolute enrichment in Apollo 12 fines and relative enrichment in large REE (La–Sm). *Bottom*: Relative REE abundance patterns of Apollo 12 rocks compared with Apollo 11-high-K and low-K rocks. Note the lower abundances in Apollo 12 and the lesser depletion of Eu. There is little change in relative patterns (except Eu).

Removal of some clinopyroxene could be tolerated without changing the relative total rock patterns (Fig. 5) but the amounts which need to be removed cannot account for the variations observed between the Apollo 11 high-K and low-K and Apollo 12 rocks. These variations are more simply explained by varying degrees of partial melting.

(2) Except for Nos. 12038 and 12002 the general case is that Eu is more depleted in those rocks with the highest total abundance of REE. Five causes have been invoked to explain the Eu depletion in Apollo 11 and Apollo 12 rocks. (a) Plagioclase removal by near surface fractionation (GAST and HUBBARD, 1970). Excessive amounts (50–90%) need to be removed, and plagioclase crystallises late in the cooling history as shown by many petrological studies (see summary by LEVINSON and TAYLOR, 1971). (b) Partial melting in the interior, with plagioclase as a liquidus phase (HASKIN *et al.*, 1970; PHILPOTTS and SCHNETZLER, 1970). The experimental evidence (RINGWOOD and

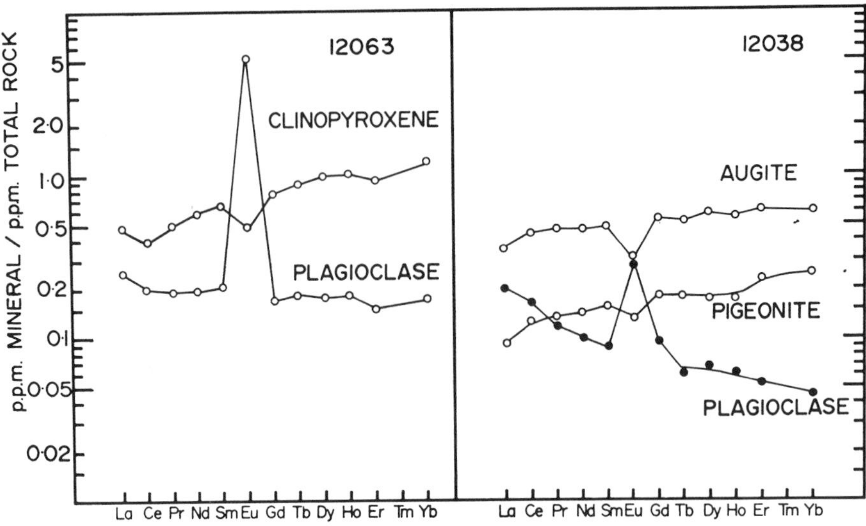

Fig. 5. Relative REE abundances in mineral separates relative to total rock abundances.

Essene, 1970) shows that plagioclase is not stable at the depth of formation of the lavas. (c) In a variation of the above two hypotheses, a primitive fractionation concentrated Eu in the uplands and depleted the source regions of the lavas. However, fractionation to produce upland material probably occurred due to melting of surficial layers and did not extend to depths of 200–500 km from which the maria lavas come. (d) The moon as a whole is depleted in Eu. This would imply that Eu was lost along with volatile and siderophile elements during accretion. However, Eu and Sr show parallel behaviour (Fig. 6). If the moon is depleted in Eu, then it is also depleted in Sr. But Sr is notably involatile, and the moon is enriched in such refractory elements. Accordingly Eu, along with Sr and the other REE is probably not depleted in the moon. Philpotts *et al.*, (1971) reached similar conclusions from a study of Eu and Ba abundances. (e) Eu (and Sr) are selectively retained in the interior during partial melting. This supposes that the trivalent REE are contained in minor phases (e.g., Whitlockite) and Eu (and Sr) in a major phase. Support for this view is obtained from the close geochemical association of Eu and Sr, which has been noted, and which is due to the similarity in ionic radius between divalent Eu and Sr. Small degrees of partial melting involving the minor and interstitial phases will then produce a high concentration of REE with a large Eu depletion (Green *et al.*, 1971). Increasing amount of partial melting will involve the major phases and dilute the total REE content and the relative Eu depletion will decrease, as observed (Fig. 4). We prefer this explanation. This implies that partial melting takes place on a scale involving individual phases, and that equilibrium melting of all phases is not achieved. Partial melting in the lunar interior can be predicted to be very localised, at least in the initial stages, on account of the lack of water and other volatiles.

Sample 12038 represents a special case. It has a larger total REE concentration

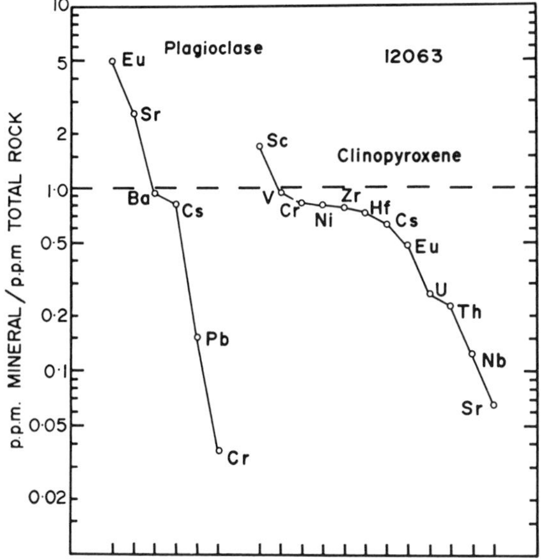

Fig. 6. Relative element abundances in minerals from rock 12063.

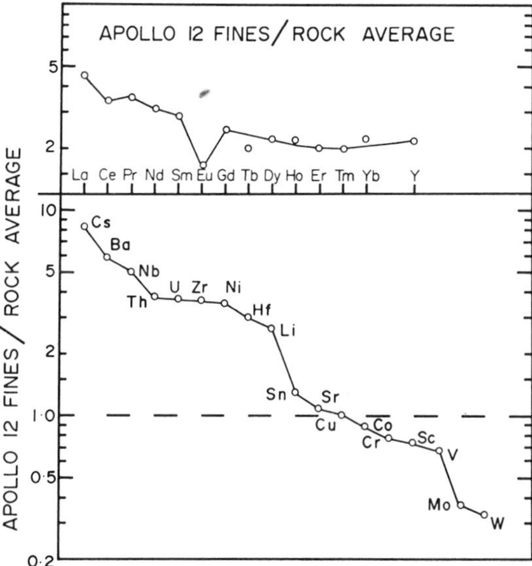

Fig. 7. Comparison of element abundances in Apollo 12 fine material (12070) compared to average of Apollo 12 rocks (excluding 12038). Note absolute enrichment of REE and relative enrichment of large REE (La–Sm). Note enrichment of large cations (Cs, Ba) and Th, U, Zr, Hf, Nb.

than the other Apollo 12 rocks (but is lower than Apollo 11 low-K, Fig. 11). The REE pattern (top, Fig. 4) may be due to (i) A small degree of partial melting, followed by near surface accumulation of plagioclase (Haskin *et al.*, 1971) to reduce the Eu depletion. (ii) Fractional crystallisation, involving olivine removal and accumulation of mesostasis to enrich total REE, with plagioclase accumulation to reduce the Eu depletion. (iii) Partial melting in a differing source region, with differing ratios of phases containing trivalent REE and Eu.

Abundance Comparisons

Comparison of Apollo 12 fine material and rocks

As discussed by LSPET (1970) it is clear that the fine material at the Apollo 12 site (12070) cannot be made up from the typical rocks collected. Figure 7 compares the abundances in the fines with those in the average rock (excluding No. 12038). The REE abundances (top section, Fig. 7) are enriched absolutely (by a factor of 2 for the smaller REE Gd–Lu + Y) and relatively for the larger elements (La–Sm) by factors of up to 4 times. Similar patterns are shown by other trace elements (lower section, Fig. 7). The fines are enriched in the large cations (Cs, Ba), and the large high valency cations (Th, U, Zr, Hf, Nb) compared with the rocks. These are the elements concentrated in the residual material or in minor and accessory phases. The elements entering the main rock-forming minerals are not strongly enriched or depleted. Thus the "extra" component in the fines has to be strongly enriched in elements concentrated in residual phases, or in the dark portion of rock no. 12013 (Wakita and Schmitt, 1970). Material balance calculations at the Apollo 12 site must be made with caution, because of the stratified nature of the regolith and the presence of the light grey fine layer (12033). The overall mixing appears less thorough than at the Apollo 11 site. Contributory factors include the presence of large craters (e.g., Middle Crescent, Surveyor) near or within the traverse area, so that the regolith is heterogeneous and much less well mixed than at Tranquillity Base.

Comparison of Apollo 12 and Apollo 11 fines

Similar features to those noted in the previous section are observed. The 12 fines (12070) are enriched absolutely in the small REE and relatively in the large REE (Fig. 8). They are also enriched in the large cations, the large high-valency cations, and in V and Co reflecting the higher abundance of these elements in the Apollo 12 rocks. A small component of material similar in composition to the dark portion of 12013, would account for many of these differences, but the comparisons should not be pushed too far because of the less well mixed nature of the 12 fines.

Comparison of Apollo 12 with Apollo 11 rocks

Figures 9 and 10 show the relative abundances in Apollo 12 rock average (excluding 12038), Apollo 11–type A–high-K and Apollo 11–type B–low-K rocks. In comparison

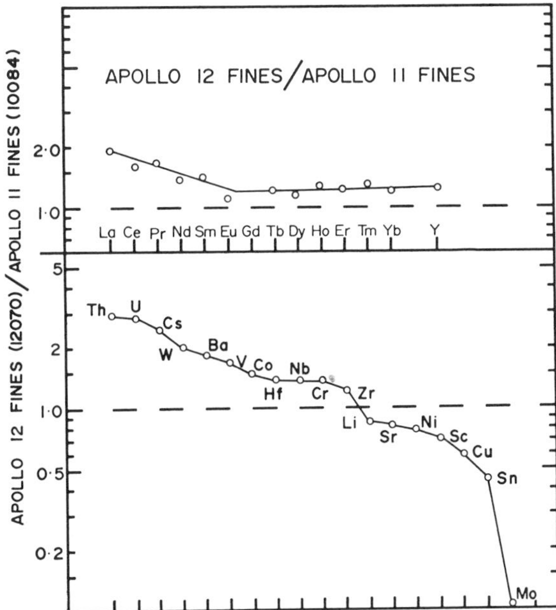

Fig. 8. Comparison of element abundances in Apollo 12 fines (12070) compared to Apollo 11 fines (10084). Note the absolute enrichment of REE and relative enrichment of large REE, and the enrichment of the large cations (Ba, Cs) large high-valency cations (Zr, Hf, Th, U, Nb) and Cr, V, and Co in Apollo 12 fine material.

with the high-K rocks, the Apollo 12 rocks are enriched in those elements entering ferromagnesian minerals (Ni, V, Co, Cr) and depleted in large cations (Cs, Ba), large. high valency cations (Th, U, Zr, Hf, Nb) Li and REE. The same relative enrichments and depletions are observed, but to a lesser degree in the comparison with the Apollo 11–low-K rocks (Fig. 10). Figure 11 compares sample 12038 with the Apollo 11–high-K rocks. Although this rock is the most highly fractionated at the 12 site (excluding 12013) it is clearly depleted in REE, large cations and large high-valency cations relative to Apollo 11–high-K rocks. Thus although volcanic activity is producing rocks of rather similar major element chemistry at both Apollo 11 and Apollo 12 sites, the many differences in detail in the trace element abundances most probably reflect real inhomogeneities on a small scale in the lunar interior.

Comparison with cosmic abundances

A plot of the Apollo 12 rocks and fine material relative to abundances in type 1 carbonaceous chondrites, here equated with primitive solar nebula abundances, is given in Fig. 12. A striking dependence on volatility and siderophile character is apparent. The involatile elements (REE, Th, U, Zr, Hf, Nb, Ba) are rather uniformly concentrated in the fines by a factor of about 100 over the chondritic abundances. The vertical bars give the spread in the abundances in the rock samples. These show similar but less extreme enrichments over the chondritic abundances for the involatile

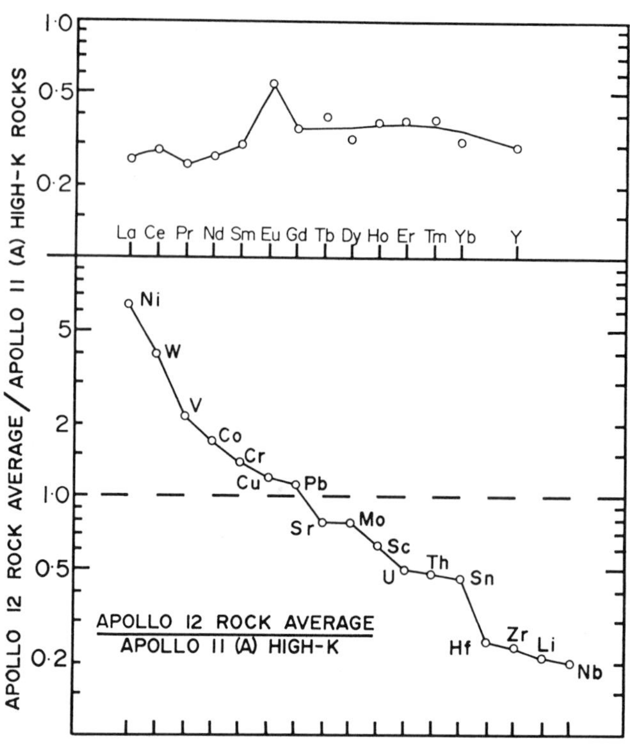

Fig. 9. Comparison of the compositions of average Apollo 12 rock (excluding 12038) with Apollo 11–(A)–high-K rocks. Note depletion of REE, large cations, large high-valency cations, and enrichment of Ni, Co, Cr + V in Apollo 12 rocks.

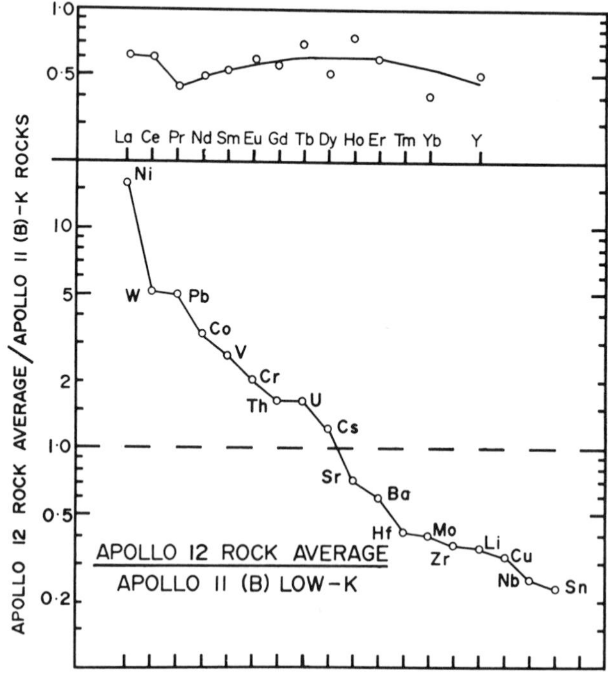

Fig. 10. Comparison of the composition of the average Apollo 12 rocks (excluding 12038) with Apollo 11–(B)–low-K rocks. Note depletion of REE and enrichment of Ni, Co, V, and Cr in Apollo 12 rocks.

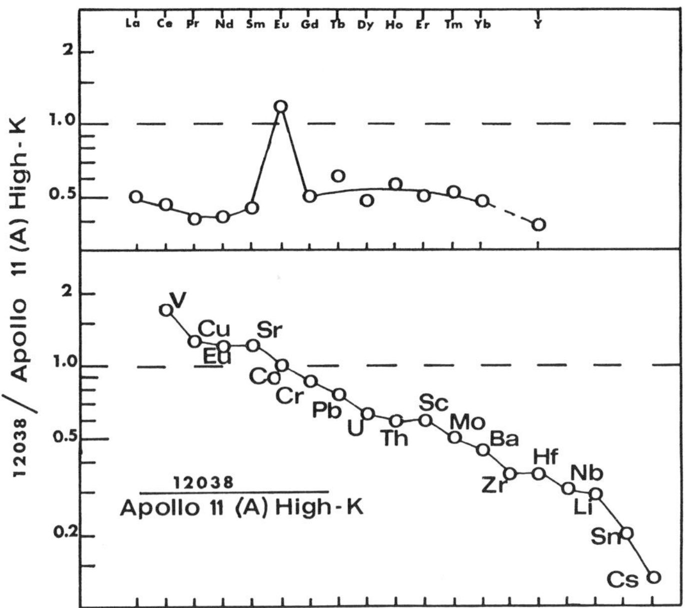

Fig. 11. Comparison of the composition of sample 12038 with Apollo 11–(A)–high-K rocks.

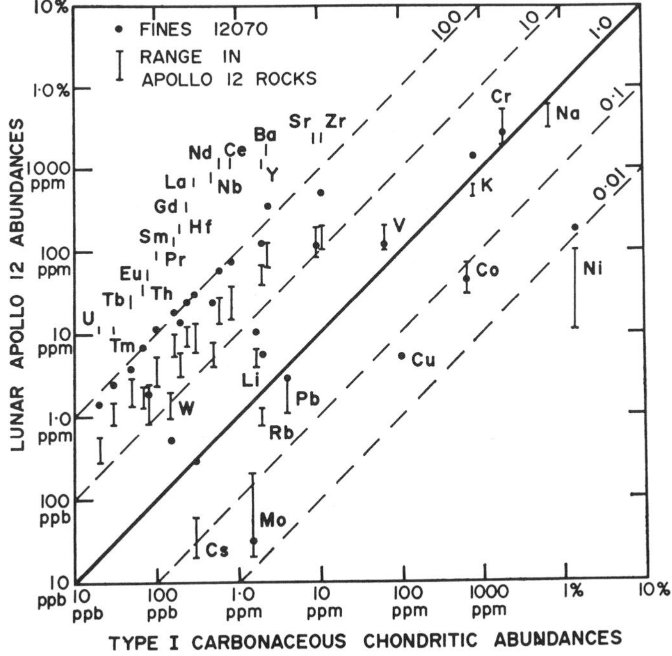

Fig. 12. Comparison of element abundances in Apollo 12 lunar samples with abundances in type 1 carbonaceous chondrites. Vertical bars give range in composition in rocks. Dots are concentrations in fines. Points lying along the 45° diagonal line indicate equality in composition. Points lying to the left of the diagonal line are enriched and those lying to the right are depleted in Apollo 12 samples relative to chondritic abundances. Dashed lines indicate relative enrichment factors of 100 and 10 and depletion factors of 0.1 and 0.01. Note the uniform enrichment of many involatile elements.

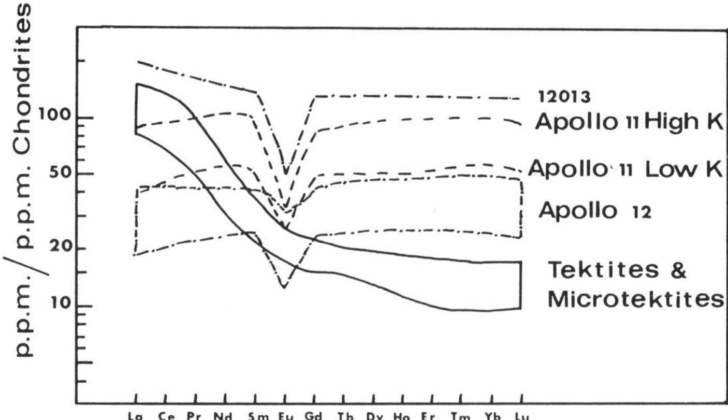

Fig. 13. Rare earth abundances in tektites and microtektites (FREY *et al.*, 1970) compared with those in Apollo 11 and Apollo 12 rocks, including sample 12013 (WAKITA and SCHMITT, 1970). Note the relative enrichment of large REE (La–Sm) in tektites and the absence of Eu anomalies, compared to the lunar rocks.

elements. The alkalies show progressive depletions in the order Li > (Na, K) > Rb > Cs in order of increasing volatility. The siderophile elements show depletions by factors of up to 100. The overall pattern is similar for both rocks and fine material, at Apollo 11 and Apollo 12 sites. The isotopic evidence indicates that loss of lead and Rb occurred well before the eruption of the lavas (LEVINSON and TAYLOR, 1971). This element abundance pattern, with its uniform enrichments and depletions is probably a primary feature, reflected to varying degrees in all lunar rocks, and consistent with the accretion of the moon from material which was already strongly fractionated relative to the composition of the primitive solar nebula.

Tektites

The relative rare earth patterns in tektites and micro-tektites (FREY *et al.*, 1970) are compared in Fig. 13 with those for the Apollo 11 and Apollo 12 rocks, including rock 12013. The differences are clear. Most notable are the relatively strong enrichment in the large REE in tektites and the absence of an Eu anomaly. Taken in conjunction with the other differences in chemistry (e.g. Cr, Ba, Y, Zr) the higher $\delta\, O^{18}$ values (9–11.5 for tektites compared with 4–7 for lunar rocks), the order of magnitude age differences, the Pb isotopic differences and much higher lunar $\mu(U^{238}/Pb^{204})$ values, the evidence against a lunar origin for tektites is overwhelming (TAYLOR, 1971).

REFERENCES

AHRENS L. H. and TAYLOR S. R. (1961) *Spectrochemical Analysis* (Addison-Wesley).

FREY F. A., SPOONER C. M., and BAEDECKER P. A. (1970) Microtektites and tektites: A chemical comparison. *Science*, **170**, 845–847.

GAST P. and HUBBARD N. W. (1970) Rare earth abundances in soil and rocks from the Ocean of Storms. *Earth Planet. Sci. Lett.* **10**, 94.

GAST P., HUBBARD N. J., and WIESMAN H. (1970) Chemical composition and petrogenesis of basalts from Tranquillity Base. *Proc. Apollo 11 Lunar Sci. Conf., Geochim. Cosmochim. Acta* Suppl. 1, Vol. 2, pp. 1143–1163. Pergamon.

GREEN D. H., RINGWOOD A. E., WARE N. G., HIBBERSON W. O., MAJOR A., and KISS E. (1971) Experimental petrology and petrogenesis of Apollo 12 basalts. Second Lunar Science Conference (unpublished proceedings).

HASKIN L. A., ALLEN R. O., HELMKE P. A., PASTER T. P., ANDERSON M. R., KOROTEV R. L., and ZWEIFEL K. A. (1970). Rare earths and other trace elements in Apollo 11 lunar samples. *Proc. Apollo 11 Lunar Sci. Conf., Geochim. Cosmochim. Acta* Suppl. 1, Vol. 2, pp. 1213–1231. Pergamon.

HASKIN L. A., ALLEN R. O., HELMKE P. A., ANDERSON M. R., and KOROTEV R. L. (1971) Rare earths and other trace elements in Apollo 12 lunar materials. Second Lunar Science Conference (unpublished proceedings).

KUSHIRO I., NAKAMURA Y., HAMAMURA H., and AKIMOTO S. (1971) Origin of chemical and mineralogical variations of the Apollo 12 crystalline rocks. Second Lunar Science Conference (unpublished proceedings).

LEVINSON A. A. and TAYLOR S. R. (1971) *Moon Rocks and Minerals.* Pergamon.

LSPET (Lunar Sample Preliminary Examination Team) (1970) Preliminary examination of lunar samples from Apollo 12. *Science* **167**, 1325–1339.

PHILPOTTS J. A. and SCHNETZLER C. C. (1970) Apollo 11 lunar samples: K, Rb, Sr, Ba and rare earth concentrations in some rocks and separated phases. *Proc. Apollo 11 Lunar Sci. Conf., Geochim. Cosmochim. Acta* Suppl. 1, Vol. 2, pp. 1471–1486. Pergamon.

PHILPOTTS J. A., SCHNETZLER C. C., BOTTINO M. L., and FULLAGAR P. D. (1971) Li, K, Rb, Sr, Ba and rare earth concentrations and $^{87}Sr/^{86}Sr$ in some Apollo 12 soils, rocks, and separated phases. Second Lunar Science Conference (unpublished proceedings).

RINGWOOD A. E. and ESSENE E. (1970) Petrogensis of Apollo 11 basalts, internal constitution and origin of the moon. *Proc. Apollo 11 Lunar Sci. Conf., Geochim. Cosmochin. Acta* Suppl. 1, Vol. 1, pp. 769–799. Pergamon.

SCHMITT H. H. and SUTTON R. L. (1971) Stratigraphic sequence for samples returned by Apollo missions 11 and 12. Second Lunar Science Conference (unpublished proceedings).

TAYLOR S. R. (1965) Geochemical analysis by spark source mass spectrography. *Geochim. Cosmochim. Acta* **29**, 1243–1261.

TAYLOR S. R. (1971) Tektites and the moon. *Comments on Earth Sciences: Geophysics* (In press).

TAYLOR S. R., KAYE M., WHITE A. J. R., DUNCAN A. R., and EWART A. (1969) Genetic significance of Co, Cr, Ni, Sc, and V content of andesites. *Geochim. Cosmochim. Acta* **33**, 275–286.

WAKITA H. and SCHMITT R. A. (1970) Elemental abundances in seven fragments from lunar rock 12013. *Earth Planet. Sci. Lett.* **9**, 169–176.

WARNER J. L. and ANDERSON D. H. (1971) Lunar crystalline rocks: petrology, geology, and origin. Second Lunar Science Conference (unpublished proceedings).

Proceedings of the Second Lunar Science Conference, Volume 2, pp. 1101–1122.
The M.I.T. Press, 1971.

Alkali, alkaline earth, and rare-earth element concentrations in some Apollo 12 soils, rocks, and separated phases

C. C. SCHNETZLER and JOHN A. PHILPOTTS
Planetology Branch, Goddard Space Flight Center, Greenbelt, Maryland 20771

(*Received 22 February 1971; accepted in revised form 31 March 1971*)

Abstract—Concentrations of Li, K, Rb, Sr, Ba, and rare-earth elements have been determined by isotope dilution in a number of soils, core samples, breccia, igneous rocks, and separated mineral fractions. Soil and core samples have similar relative concentrations of these trace elements, but absolute concentrations vary by up to a factor of 2. These compositions can be matched by various mixtures (approximately 2:1 to 4:1) of normal Apollo 12 igneous rocks and the dark-colored portion of 12013 (KREEP). The majority of Apollo 12 igneous rocks appear to be related by olivine crystallization-cumulation control; other samples (e.g., 12038, 12022, 12063) might be related to these "normal" rocks by plagioclase and/or pyroxene cumulation. Mineral data suggest that 12035 and 12040 augite, pigeonite, and plagioclase approached quasi-equilibration; in contrast, 12021 pigeonite is phenocrystic. Eu anomalies continue to be best explained as due to a plagioclase effect, rather than volatilization or phosphate fusion control. Model ages and initial Sr ratios, together with data indicating mineral cumulation in some rocks, suggest differentiation occurred at 4.5 b.y. for the Apollo 12 and low-alkali Apollo 11 rocks; their mineral isochron ages represent subsequent local, closed-system melting.

INTRODUCTION

THIS PAPER PRESENTS data on the concentration of alkali elements Li, K, and Rb, alkaline earths Sr and Ba, and rare earth elements (REE) Ce, Nd, Sm, Eu, Gd, Dy, Er, Yb, and Lu in a number of Apollo 12 igneous rocks and separated minerals, soils, breccia, and core samples. It is essentially a continuation of our study of the moon, initiated by the Apollo 11 measurements (PHILPOTTS and SCHNETZLER, 1970a). Our study of Apollo 12 rock 12013 has been reported recently (SCHNETZLER *et al.*, 1970), and will not be discussed here except in as far as it relates to the petrogenesis of the other rocks and soils.

All our analyses reported in this paper were by mass spectrometric isotope dilution (PHILPOTTS and SCHNETZLER, 1970a; SCHNETZLER *et al.*, 1967). A few changes have been made in our procedure since the Apollo 11 analyses; Li has been added to the list of elements routinely analyzed, Sr is now determined using a ^{84}Sr spike so that the Sr isotopic composition can also be calculated, and Gd is usually determined from single rhenium filament GdO^+ spectra.

From a comparison of Apollo 11 results, GAST *et al.* (1970) suggested that our Nd and Sm results were systematically high. Extensive recalibrations of our spikes have verified their suspicion, and both Nd and Sm concentrations reported by our laboratory before the summer of 1970 (including our Apollo 11 results) should be reduced by 11%. As suggested by GAST *et al.*, variable amounts of CO_2 and H_2O absorption by the REE oxides used as standards caused erroneously high values for these two elements. A comparison of our results with the isotope dilution results of the MSC group (Gast *et al.*, 1970; Hubbard *et al.*, 1970) for 10084, BCR–1 and 12070 now show agreement for all the REE analyzed in common within ±3%, with the exception of Yb. The MSC Yb values are about 9% higher than ours. Also, although the average Lu difference is within ±3%, the individual differences of the three samples are too scattered to reveal any systematic bias, in contrast to the situation for the other REE.

A detailed interlaboratory comparison of our results with those of the 14 other laboratories who analyzed REE in Apollo 12 materials is beyond the scope of this paper. Systematic bias is difficult to discern for many of the neutron activation and spark-source data (in contrast with the isotope dilution results) due primarily to the less precise nature of these methods. Utilizing those elements analyzed in common in powdered samples (to eliminate sample inhomogeneity seen in "whole-rock" analyses, particularily when small chips of coarse-grained rocks were used), the results of Haskin *et al.* (1971), Wakita *et al.* (1971), and Goles *et al.* (1971) agree with the isotope dilution results (this work; Hubbard *et al.*, 1971; Smales, 1971) to within ±10%, and to within ±5% for Ce, Sm, and Eu.

Soils, Breccia, Core

We have analyzed 3 samples of fine soil material (12001, 12032, and 12070), a breccia (12073) and five samples from the double core tube (12025/12028). The results are given in Table 1. All of these samples have generally similar compositions, with the exception of core sample 12028,65, which is from a depth (uncorrected) of 13.2 to 14.4 cm. It is from the coarse layer, noted visually in the core as "comprised mostly of olivine and olivine-rich gabbro" (LSPET, 1970). Our trace element concentrations for this sample are a factor of about two to ten times lower than other regolith samples. The relative concentrations are quite similar to those observed in the Apollo 12 igneous rocks and approximate those in 12035 (the rock with the lowest abundances in these trace elements; see Fig. 4). Data on other trace elements (Ganapathy *et al.*, 1970) in this level of the core show this same relationship, in general—i.e., this level resembles the Apollo 12 igneous rocks and not the other soil or core samples. However, Ganapathy *et al.* (1970) found the concentrations of Cd and Bi to be about three orders of magnitude higher in this layer than any other Apollo 12 rock or soil; we find no strikingly anomalous concentrations. From our data this layer appears to be a crumbled rock, similar to the most olivine rich variety found at the site, and which has not been mixed with any "cryptic" component (see below).

The other fine soil and core samples we have analyzed are quite similar to each other, in a relative sense, but show some differences in absolute concentrations. This can be seen in Fig. 1, where we have normalized the concentrations to those in soil 12070. Thus, the regolith is neither vertically nor laterally chemically homogeneous. This observation can be extended to other soils from data reported at the Second Lunar Science Conference. Sample 12037 appears to be depleted in the same elements we analyzed, compared to 12070, by about a factor of 0.8 (Wänke *et al.*, 1971); 12044 is indistinguishable from 12070 in the abundances of these elements (Hubbard *et al.*, 1971); 12042 is about 20% higher in REE than 12070 (Haskin *et al.*, 1971), similar to core samples 12028,89 and 12028,120 (Fig. 1); and REE in 12033 are about

Table 1. Abundances, in parts per million by weight, in Apollo 12 soils, breccia, and core samples.

	Soils			Breccia
	12001	12032	12070*	12073
Li	17.8	23.4	18.0 ± 0.08%	25.8
K	1980	3030	1960 ± 0.1%	3162
Rb	6.48	9.24	6.47 ± 0.5%	10.2
Sr	145.5	161.4	144.2 ± 0.1%	164.8
Ba	370	529	373 ± 0.5%	565
Ce	87.2	117	89.4 ± 1.5%	120
Nd	55.1	73.0	55.4 ± 0.2%	76.8
Sm	16.1	20.7	16.0 ± 0.3%	21.6
Eu	1.78	2.12	1.77 ± 1.0%	2.19
Gd	19.4	24.9	20.1 ± 0.7%	26.8
Dy	22.1	28.0	22.3 ± 0.09%	28.9
Er	12.6	15.9	13.0 ± 0.5%	17.4
Yb	12.0	15.2	11.9 ± 0.4%	16.0
Lu	1.81	2.24	1.84 ± 0.05%	2.43
Wt (mg)	241	195	197 198	131

* Avg. of two independent analyses. ± are % deviation from average.

	Core					12032
	12025,71	12028,65	12028,89	12028,120	12028,144	Cos. Frag.
Depth, cm†	1.7–2.5	13.2–14.4	18.9–19.7	31.2–32.2	37.2–38.2	
Li	18.1	7.69	21.3	21.2	22.4	18.56
K	2120	280	2430	2495	2750	593
Rb	6.84	0.613	7.84	7.93	8.96	0.498
Sr	144.4	80.9	155.2	152.7	154.9	124.7
Ba	389	44.9	442	463	518	202
Ce	90.2	10.2	112	109	121	20.6
Nd	57.2	7.82	70.1	68.4	78.2	11.3
Sm	16.5	2.70	19.6	19.4	22.2	3.00
Eu	1.74	0.729	2.03	1.97	2.025	1.53
Gd	20.8	3.68	23.2	23.4	27.6	3.96
Dy	21.2	4.68	25.5	26.1	30.0	4.72
Er	13.1	2.83	14.9	15.5	17.1	3.65
Yb	12.0	2.77	14.0	14.2	16.1	4.16
Lu	1.86	0.421	2.08	2.19	2.42	0.722
Wt. (mgs)	110	96	92	103	113	23

† Depths of core samples are distances from top of core tube (LSPET, 1970). True depths, due to compression in the core tube, are estimated by CARRIER et al. (1971) and are given in parentheses: 1.7–2.5 (2.0–2.8), 13.2–14.4 (17.8–19.4), 18.9–19.7 (25.8–26.9), 31.2–32.2 (44.5–46.1), 37.2–38.2 (55.6–57.2)

30 to 50% higher than in 12070 (HASKIN *et al.*, 1971), slightly more enriched than core sample 12028,144 (Fig. 1). REE have not been reported in soils 12041 and 12030; however, on the basis of major element analyses (FRONDEL *et al.*, 1971) 12041 appears to be quite similar to 12070, while 12030 appears to be intermediate in composition between 12042 and 12032. Breccia 12034 is almost a factor of two higher in K, Rb, Ba, and REE than 12070 (WAKITA *et al.*, 1971).

No unambiguous picture emerges from the geographic locations (LSPET, 1970) and compositions of these soils. Samples 12001 and 12070, probably collected within a few meters of each other on the north rim of Surveyor Crater, are essentially identical

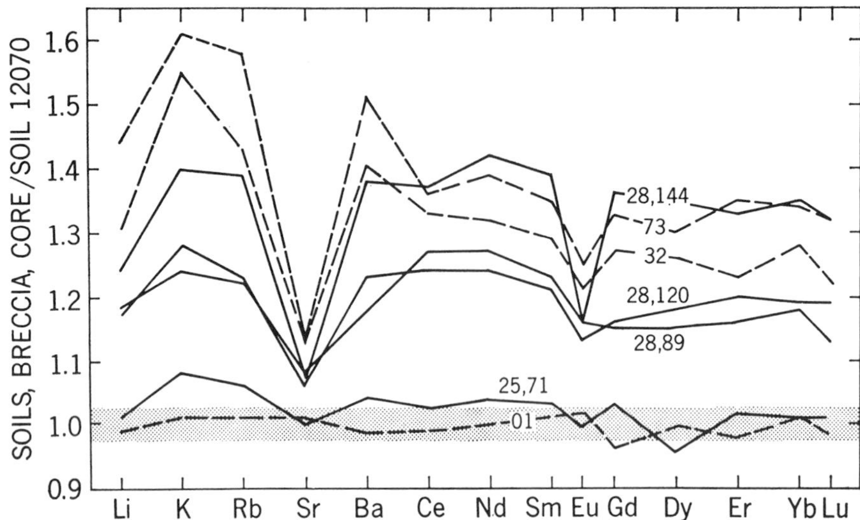

Fig. 1. Trace element concentrations in Apollo 12 soils, breccia, and core, normalized to those in soil 12070. Soils (12001 and 12032) and breccia (12073) shown by dashed lines, core samples by solid lines. Shaded area is ±2.5%. Data from Table 1.

to 12044, collected about 200 meters away on the south rim of the crater, and to the top 1 to 2 cm of the double core (12025,71), taken about 70 meters to the SSW of 12044. Additionally, sample 12041, collected between Bench and Halo craters, about 150 meters to the west of 12044, appears to be similar in composition to samples 12044, 12070, 12001, and the top of the core, although the agreement is much less well documented. The enriched soils seem to lie on the west side of the collecting area, around Head and Bench craters. However, samples 12037 and 12032 were collected within 40 meters of each other on the northwest and north rim of Bench crater, and the former is depleted in these trace elements by about 20%, compared to the soils to the east, and the latter is enriched by about 30%. Also the type and degree of enrichment that is found in breccia 12073 and soils 12033, 12032, 12042, and perhaps 12030 can be found in depth within the core from the southeastern part of the collecting area.

Cryptic Component of the Soil

Although there are considerable absolute variations in the compositions of the soils and rocks, the trace element compositions of these two types of materials are quite distinct (LSPET, 1970). This can be seen in Fig. 2, which shows the level of normalized Ba and REE concentrations of all 7 soils and 4 core samples for which data have been reported, and the level for the twelve "normal sequence" igneous rocks discussed below (see Fig. 4). It is obvious the soils are not simply pulverized samples of the rocks, nor, conversely, are the rocks merely remelted regolith. The soils at the Apollo 12 site, and to a lesser extent the Apollo 11 site, must contain a component (or components) which is not observed in the igneous rocks. This material has been called, by various authors, the magic, cryptic, or transferred component.

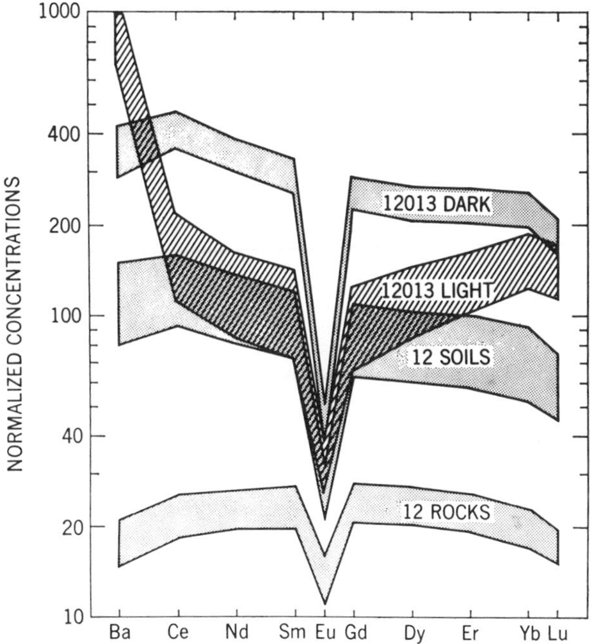

Fig. 2. Chondrite normalized Ba and REE concentration ranges observed for 12013 light and dark-colored portions, Apollo 12 soils and the Apollo 12 "normal sequence" igneous rocks. 12013 data from SCHNETZLER *et al.* (1970), HUBBARD *et al.* (1970), and WAKITA and SCHMITT (1970a). Soils data from Table 1, WÄNKE *et al.* (1971), HUBBARD *et al.* (1971), and HASKIN *et al.* (1971). See caption of Fig. 4 for the origin of data on igneous rocks.

In our study of the unique breccia 12013, we noted that (1) the light and dark colored portions of the rock were characterized by two distinct trace element patterns (Fig. 2); (2) the dark-colored portion appeared to be a primary late-stage differentiate liquid; (3) the dark-colored portion satisfied the requirements, at least for the elements we determined, of the cryptic component that must be added to the rocks to produce the soil; and (4) the required mixture to produce the alkali, alkaline earth, and RE elements in soil 12070 was about 20% dark portion and 80% normal igneous rock (SCHNETZLER *et al.*, 1970). The subsequent publication of other chemical data on rock 12013 and the release of large amounts of data on igneous rocks and soils at the Second Lunar Science Conference allows us to further test our simple two-component mixing model. Figure 3 shows the concentrations, normalized to soil 12070, of a number of elements in a typical dark portion of 12013, a normal Apollo 12 igneous rock, and a hypothetical 20%–80% mixture of these two components. Li, K, Rb, Cs, Ba, REE (except Eu), Th, U, and Zr concentrations differ by more than an order of magnitude between 12013 dark portion and normal igneous rocks. The 1:4 mixture, however, has concentrations within a factor of two of those observed in soil 12070. Considering the variations observed in 12013 dark portion, igneous rocks and soils, and possible analytical errors, we feel this "across the board" (i.e., including some less sensitive

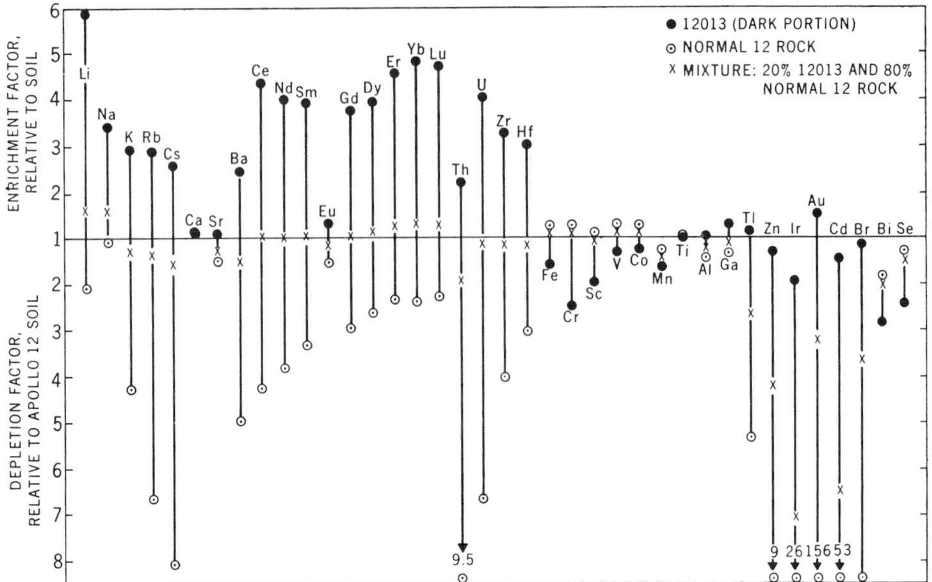

Fig. 3. Concentrations in dark-colored portion of 12013, normal Apollo 12 igneous rock, and 1:4 mixture, all relative to those in soil 12070. Data for 12013 from SCHNETZLER *et al.* (1970), WAKITA and SCHMITT (1970a), and LAUL *et al.* (1970). Data for soil 12070 and normal igneous rocks from LSPET (1970), GANAPATHY *et al.* (1970), WAKATA *et al.* (1971), and Tables 1 and 2.

elements) agreement is excellent, and strongly supports the original model we proposed on much more limited data. Tl, Zn, Ir, Au, Cd, and Br also have large differences in concentrations in 12013 dark portion and igneous rocks, but the 1:4 mixture concentrations do not match soil 12070. These are the elements which GANAPATHY *et al.* (1970) propose have been added to the soil by a small (~1%) meteoritic component. The mixing model we propose, therefore, still requires that these elements be primarily from a third, meteoritic component, but the addition of 20% 12013 dark portion does not materially change the estimated amount of meteoritic component.

The variations in soil and core compositions, discussed above, can be readily explained by the mixing of different proportions of 12013 dark portion and igneous rocks. Thus, while 12070, 12001, and 12044 require a 20%–80% mixture, 12032 requires a 30%–70% mixture; 12033 (the soil with the greatest enrichment of many lithophilic elements) requires a 35%–65% mixture, and 12037 (the soil which seems to have the lowest concentrations of these elements) requires a 15%–85% mixture. Breccia 12034 requires about a 50%–50% mixture, whereas breccia 12010 (GOLES *et al.*, 1971) requires approximately a 65%–35% mixture.

Several other models for the cryptic component of the soil have been proposed. On the basis of Rb, Cs, and U abundances, Laul and co-workers (1970) proposed whole-rock 12013, whereas, primarily on the basis of radiogenic components, Lunatic Asylum (1970) proposed the "white end member" of 12013. Mass-weighted average values of a number of REE analyses of light, dark, and mixed portions of 12013

(WAKITA and SCHMITT, 1970a; HUBBARD *et al.*, 1970; SCHNETZLER *et al.*, 1970), the analysis of the saw dust from the cutting of 12013 (HUBBARD *et al.*, 1970), and gamma ray whole-rock analyses for K, U, and Th (O'KELLEY *et al.*, 1971) show that the whole-rock concentrations are dominated by the light-colored end member. Thus, these two proposals (LAUL *et al.*, 1970; Lunatic Asylum, 1970) are essentially the same. Ba and REE concentrations in the rocks, soils, and light and dark portions of 12013 (Fig. 2) show clearly that the light-colored portion (and therefore also the whole rock) cannot represent the cryptic component of the soil. Whereas Nd, Sm, and Gd in the light-colored portion have essentially the same abundances as the soils, the heavy REE Er, Yb, and Lu have concentrations about a factor of two higher than the soil and Ba is about an order of magnitude higher than the soil. Thus, it is not possible to balance simultaneously Ba, light REE, and heavy REE by adding some percentage of light portion of 12013 to the rocks.

GOLES *et al.* (1971) have proposed a complicated mixing model to produce soil 12070, utilizing about 40% of breccia 12010; 28% and 27% of two groups (?) of Apollo 12 igneous rocks; and 5% anorthosite. This model is probably not fundamentally different from our model in terms of the component used to supply the high concentrations of large lithophilic cations. In their model, this is the breccia 12010, which they point out is not a "primary" lunar material, but is itself a mixture. In fact, they further point out that the breccia appears to be composed of about 60% of a material similar to the dark-colored portion of 12013. Because their model requires 40% breccia, it essentially agrees with our calculation of about 20% 12013 dark portion in soil 12070.

Several papers were presented at the Second Lunar Science Conference by investigators from the Manned Spacecraft Center (MSC) which proposed as the cryptic component a material they called KREEP (HUBBARD *et al.*, 1971; MEYER *et al.*, 1971; MCKAY *et al.*, 1971). This type of material has been noted by a number of other investigators as being a rather common component of the soils (e.g., ENGELHARDT *et al.*, 1971; FREDRIKSSON *et al.*, 1971; GLASS, 1971). The MSC groups have shown KREEP to be essentially the same as the dark portion of 12013 (see especially HUBBARD *et al.*, 1971, Fig. 1). Indeed, MEYER *et al.* (1971) state that KREEP and the dark portion of 12013 "represent a discrete magma type." The estimates of HUBBARD *et al.* (1971) on the proportion of KREEP in the soils differ from our estimates based on the dark portion of 12013 (for example, 25% of KREEP in 12070 instead of our estimate of 20% of 12013 dark). The concentrations of REE used to arrive at these estimates are quite variable in absolute, but not relative, abundance in this KREEP–12013 dark material, and they picked a lower REE abundance material for their end member than we did. The important point, however, is that this unusual material is a significant portion of all the soils, and in some soils reaches dominant proportions. Thus, our model based on the dark portions of 12013, the model of GOLES *et al.* (1971) based on breccia 12010 and that of the MSC groups based on KREEP are not significantly different—we are all proposing the same type of material which is abundant in all the soils.

MEYER *et al.* (1971) state that KREEP is chemically unrelated to the Apollo 11 and Apollo 12 mare basalts, and is foreign to the Apollo 12 site. Fragments of intermediate composition seem to be lacking or sparse and the texture appears quite distinctive. Also the age determined from 12013 is 4.0 b.y. or older. However, the chemical composition appears to be that of a late stage differentiate of the normal Apollo 11 and Apollo 12 igneous rocks. Also, the large abundance of this material in all of the soils suggests to us a local origin. Thus, we feel the suggestion that this material is Copernican ray material (e.g., MEYER *et al.*, 1970) should be viewed with some caution.

IGNEOUS ROCKS

Our K, Rb, Sr, Ba, and REE concentration data for the igneous rocks (Table 2) and the data of other investigators indicate some grouping. The majority of the rocks have very similar relative concentrations of the trace elements and only a small range in absolute concentrations (factor of about 1.3). Normalized trace element concentrations for these rocks fall within the shaded area in Fig. 4; in this paper, we term such rocks "normal." Our samples 12018 and 12040 fall in the lower part of this area

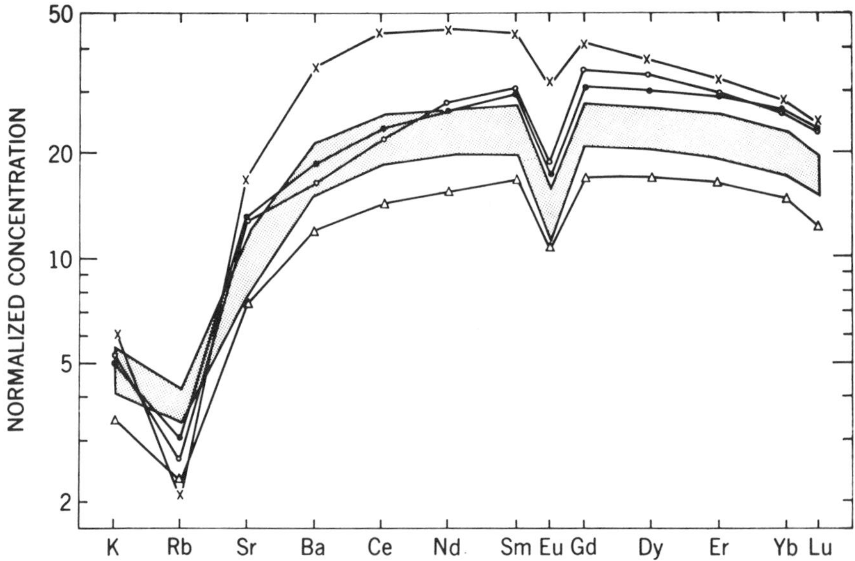

Fig. 4. Chondrite normalized concentrations (K and Rb times 10) in Apollo 12 igneous rocks. Twelve rocks have similar relative trace element patterns which fall in the shaded area. They are, from generally higher to lower abundances: 12052, 53, 64, 21, 65, 04, 18, 02, 09, 40, 75, and 20. Individual patterns of rocks which show small differences from the shaded area pattern are for 12038 (×), 12063 and 12022 (○), 12051 (●), and 12035 (△). Data for 12038, 52, 21, 18, and 40 from Table 2; for 12051, 53, 02, 75, and 35 from HUBBARD *et al.* (1971); for 12063, 22, 64, 09 and 20 REE from HASKIN *et al.* (1971), and other elements from CLIFF *et al.* (1971), MURTHY *et al.* (1971), HUBBARD *et al.* (1971), O'KELLEY *et al.* (1971), COMPSTON *et al.* (1971), and SMALES (1971); 12065 from SMALES (1971) and GOLES *et al.* (1971); 12004 from WAKITA *et al.* (1971), MURTHY *et al.* (1971) and O'KELLEY *et al.* (1971). Where interlaboratory bias was obvious from analyses of 12070 and other powdered samples, the data were normalized to a common 12070 value.

Table 2. Abundances, in parts per million by weight, in Apollo 12 igneous rocks and separated minerals

	12018	12021				12035		
	W. Rock	Matrix	Plagioclase	Low-Ca Augite	Pigeonite	Plagioclase	Low-Ca Augite	Pigeonite
Li	7.51	8.37	14.4	6.79	5.51	8.95	5.50	7.65
K	429	529	890	288	20.6	418	94.5	70.4
Rb	1.04	1.14	1.16	0.938	0.104	0.210	0.282	0.174
Sr	89.3	128.5	323	38.0	3.51	367	13.2	9.11
Ba	60.3	71.1	97.0	45.0	4.99	43.7	13.1	13.1
Ce	15.9	19.8	12.7	16.4	1.50	1.81	4.67	3.63
Nd	11.8	14.4	8.77	13.9	1.40	1.19	4.14	2.79
Sm	3.91	4.84	2.57	4.82	0.570	0.298	1.68	0.904
Eu	0.834	1.116	2.31	0.517	0.042	2.24	0.171	0.109
Gd	5.55	6.59	3.12	7.09	1.20	0.392	2.78	1.25
Dy	6.54	7.86	2.96	8.44	1.77	0.245	3.60	1.52
Er	3.80	4.53	1.50	4.87	1.33	0.122	2.28	1.02
Yb	3.42	4.12	1.13	4.56	1.42	0.120	2.12	1.09
Lu	0.52	0.64	0.162	0.70	0.259	0.0193	0.336	0.191
Wt. (mgs)	175	178	122	145	101	99	175	120
Eu^{2+}/Eu^{3+}		11.2	18	0.98	2.1	155	0.88	1.02

	12038	12040				
	W. Rock	W. Rock	Plagioclase	Low-Ca Augite	Pigeonite	Opaque
Li	10.8	6.67	9.07	5.60	7.71	2.49
K	634	411	526	74.3	74.1	384
Rb	0.604	1.000	0.306	0.234	0.191	1.11
Sr	190	85.5	454	11.1	8.37	24.0
Ba	130	57.2	55.3	10.8	11.6	66.1
Ce	35.0	15.3	2.95	4.65	3.67	15.3
Nd	26.3	12.0	2.02	4.46	2.99	11.5
Sm	8.02	4.03	0.540	1.97	1.04	3.76
Eu	2.19	0.796	2.60	0.177	0.109	0.370
Gd	10.6	5.60	0.755	3.22	1.52	5.29
Dy	11.1	6.36	0.602	4.37	1.89	5.99
Er	6.12	3.71	0.288	2.76	1.27	4.17
Yb	5.26	3.38	0.248	2.75	1.40	5.21
Lu	0.814	0.521	0.038	0.424	0.234	1.02
Wt. (mgs)	124	148	165	363	350	73
Eu^{2+}/Eu^{3+}			86	0.54	0.77	0.59

	12052	12013 (Density Fractions)	
	Whole Rock	12013,15C S.G. = 2.93–3.21	12013,15E S.G. > 3.50
Li	8.04	105	101
K	534	14,500	768
Rb	1.26	48.8	4.69
Sr	116	96.8	8.3
Ba	75.0	2110	285
Ce	18.8	80.3	76.7
Nd	14.7	34.2	(43)
Sm	4.91	10.2	11.52
Eu	1.04	1.54	< 0.3
Gd	6.87	—	17.4
Dy	7.74	25.2	34.0
Er	4.55	21.2	30.9
Yb	4.32	30.4	46.5
Lu	0.651	4.94	8.15
Wt. (mg)	204	8.3	1.8

along with 12020, 12002, 12075, 12004, and 12009; our samples 12052 and 12021 fall in the upper part, along with 12065 and 12053. There might be some subgrouping of samples within this normal group, but if present it is obscured by analytical errors, interlaboratory bias, sampling, or other factors. It may be of interest to note that all of this normal group of rocks with the exception of 12040 (granular) are of WARNER and ANDERSON's (1971) textural type 1—that is, they are porphyritic. Conversely, only one porphyritic type, 12022, falls outside the normal zone.

Major element data has proved informative in deciphering these normal rocks. As demonstrated particularly well by KUSHIRO *et al.* (1971) but also by other investigators (e.g., GREEN *et al.*, 1971; BIGGAR *et al.*, 1971), chemical differences between many of the Apollo 12 rocks can be accounted for largely in terms of the addition (accumulation into) or subtraction (crystallization out of) of olivine. Additional data reinforce the case for olivine. This is illustrated in Fig. 5 for Mg and Al; other elements also appear to be consistent. The data points are averages (a few anomalous values were discarded) of all available analyses. The porphyritic rocks, which include all the "normal" rocks except 12040 (but also including abnormal 12022), are represented by the filled circles in Fig. 5. The solid line drawn through these data points extrapolates to an abscissa intercept corresponding to the composition of olivine $\sim Fo_{74}$, which,

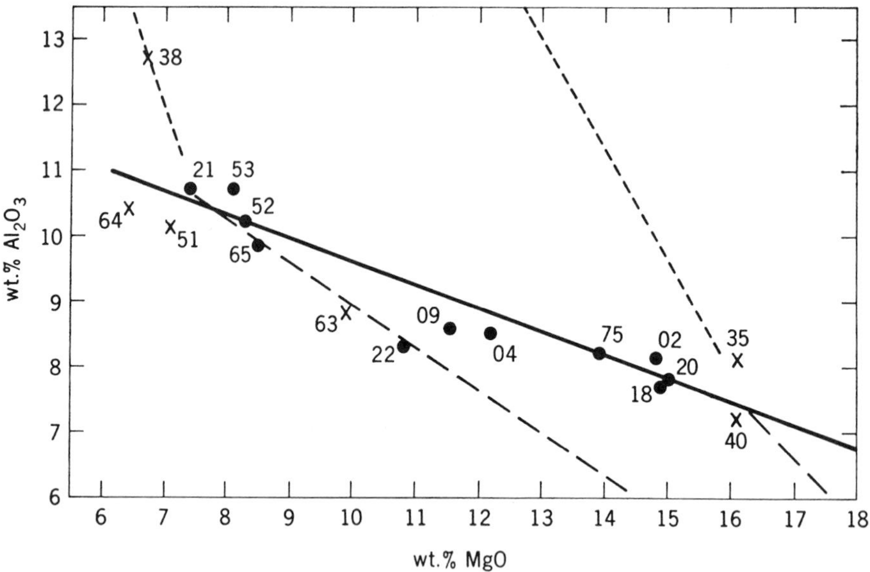

Fig. 5. Correlation of Al_2O_3 with MgO in Apollo 12 igneous rocks. WARNER and ANDERSON's (1971) Type 1 (●) and Type 2 (×) classification is used. Solid line matches olivine control, heavy dashed lines are pigeonite control, and light dashed lines are plagioclase control. Data are from COMPSTON *et al.* (1971), EHMANN and MORGAN (1971), WAKITA *et al.* (1971), ENGEL and ENGEL (1971), ROSE *et al.* (1971), ANNELL *et al.* (1971), WÄNKE *et al.* (1971), BIGGAR *et al.* (1971), GOLES *et al.* (1971), BOUCHET *et al.* (1971), BRUNFELT *et al.* (1971), MORRISON *et al.* (1971), MAXWELL and WIIK (1971), and SMALES (1971).

interestingly enough, is fairly typical of the most magnesian olivine found in many of these rocks. Scatter of the normal rocks off the olivine line might be due to the addition or subtraction of other phases. BIGGAR *et al.* (1971) have suggested that 12021, 12052, and 12065 of the normal rocks considered here, are related by pigeonite control; this seems possible as indicated by the heavy dash lines which extrapolate from the high and low MgO rocks to the composition of the magnesian pigeonite in 12021. Similarly a supposed effect of plagioclase may be gauged from the light dash lines which extropolate to a composition typical of feldspar in the Apollo 12 igneous rocks.

It is apparent, however, that scatter of the normal rock values off the olivine control line is small compared to their spread along it. Indeed the ranges in the given compositions indicate that much if not all of this scatter off the olivine line could be an artifact. The spread along the olivine line can be accounted for by the addition to and subtraction from an intermediate rock of about 15% olivine; this would also nicely explain the range in absolute trace elements concentrations (Fig. 4) shown by the normal igneous rocks. KUSHIRO *et al.* (1971) have argued that olivine control favors a crystallization-cumulation relationship, as opposed to a partial melting relationship, among these samples, because a residue composed only of olivine is not palatable. The disappearance of olivine from the liquidus of some of the less magnesian samples at only moderate pressures (GREEN *et al.*, 1971; BIGGAR *et al.*, 1971; KUSHIRO *et al.*, 1971) also argues against a difference in degree of partial fusion relationship. The trace-element data, the *straight-line* olivine control, and the propinquity when collected, would seem to indicate that the Apollo 12 normal igneous rocks are probably consanguineous, closely related samples of an olivine crystallization-cumulation sequence. Apparently at odds with this view, however, is the less than desirable agreement of initial $^{87}Sr/^{86}Sr$ and ages amongst these samples (see below).

We consider next the rocks that do not fall into the normal trace element group. Perhaps the most abnormal in both major and trace element chemistry of the igneous rocks considered here is 12038. It is an ophitic rock (WARNER and ANDERSON, 1971) rich in feldspar and correspondingly has a distinctly higher aluminum content (Fig. 5). BIGGAR *et al.* (1971) have postulated that 12038 represents a cotectic liquid related to some of the mare magnesian rocks (12018, 12020, 12040) by olivine control. This cannot be so; the Rb content, for example, of 12038 is lower than that of these other rocks (Fig. 4). A more likely relationship is that 12038 formed from a later liquid and contains an excess of feldspar (cf. HASKIN *et al.*, 1971). If there is a relationship, a later liquid is indicated by the higher trace element concentrations (Fig. 4) except for Rb; these large cations, of course, generally show enrichment in the liquid with differentiation. Feldspar addition is indicated by the higher aluminum content and by trace element tendencies characteristic of feldspar, such as lower Rb, higher K/Rb, lower Rb/Sr, and higher concentrations of Eu relative to the other REE (e.g., PHILPOTTS and SCHNETZLER, 1970b; SCHNETZLER and PHILPOTTS, 1970). The aluminum increase can be accounted for by the addition of about 10% plagioclase; considerably more feldspar than this is required to modify the trace element characteristics however. Also, the major element data relate 12038 by simple feldspar addition along the tie line to the trace element enriched, low magnesium type of normal rock, whereas the higher trace element content of 12038 seems to require a later liquid and this would be expected to fall to the left of the normal rocks in Fig. 5 judging from the mineral control-lines. These problems may be resolved if it is assumed that considerably more than 10% plagioclase was added but also that significant amounts of a magnesian, low aluminum phase (clinopyroxene is preferred) were also added. 12038 in this model could be either a later liquid containing considerable excess phenocrysts of plagioclase and clinopyroxene, or simply a cumulate from such a liquid. Fig. 5 of PHILPOTTS and SCHNETZLER (1970a) illustrates how K, Rb, Sr, and Eu would tend to be dominated by feldspar in such a mixture and the other REE by clinopyroxene. The present ophitic texture of 12038 is not in accord with the model; perhaps remelting is required. Other models relating 12038 to the normal igneous rocks are possible but considerably less likely.

Sample 12022 is anomalous in both major and trace element chemistry. It has been postulated as a discrete magma type (GREEN *et al.*, 1971; KUSHIRO *et al.*, 1971; BIGGAR *et al.*, 1971). Sample 12063 is quite similar in chemical composition to 12022 although analyses of both samples display wide, possibly significant, scatter. The trace element concentrations in these two samples are, in general, indistinguishable. Therefore treating them as a group appears to be justified in spite of their differing textures (12022 porphyritic, 12063 ophitic—WARNER and ANDERSON, 1971). They are characterized by higher heavier REE concentrations than the normal type and steeper slope to the pattern (Fig. 4). These characteristics could be explained in terms of adding excess clinopyroxene to a later liquid following on from the normal sequence. The somewhat higher K/Rb and lower Rb/Sr are not adequately explained in this model judging from data obtained to date on lunar pyroxenes (Fig. 6), although partition coefficients measured for terrestrial clinopyroxenes (PHILPOTTS and SCHNETZLER,

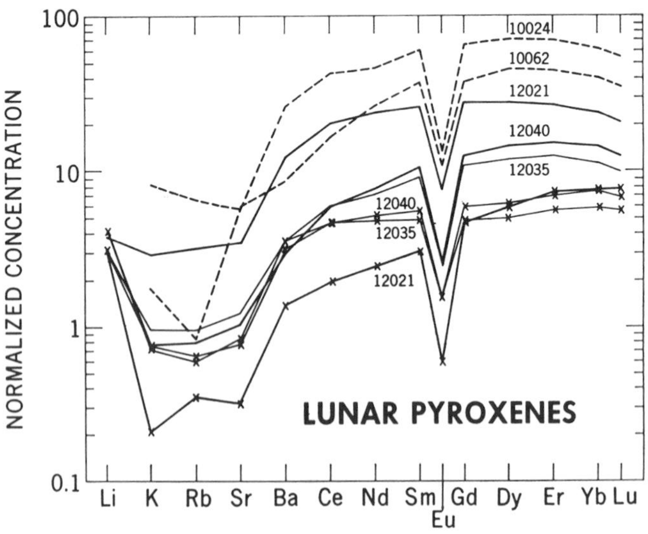

Fig. 6. Chondrite normalized concentrations (K and Rb times 10) in lunar pyroxenes. Apollo 11 samples from PHILPOTTS and SCHNETZLER (1970a). Apollo 12 samples, Table 2, include low-Ca augites (solid line) and pigeonites (×).

1970b) would fit; perhaps some excess feldspar is also required in these samples. The major element chemistry of 12022 and 12063 is not inconsistent with the idea of adding clinopyroxene and some olivine to a later liquid, but the higher titanium content and perhaps the lower silicon content are not accounted for. A separate magma may be the best explanation. The REE pattern of 12022–12063, although lower, is similar to that of the low alkali type of Apollo 11 basalt, for which we favor a cumulative origin (PHILPOTTS and SCHNETZLER, 1970a). Sample 12051 and possibly 12064 (lacking Rb, Sr, and Ba data at time of writing), both of which are ophitic (WARNER and ANDERSON, 1971), appear to be approximately intermediate both in trace elements (Fig. 4) and major elements (e.g., Fig. 5) between 12022–12063 and a later liquid of the normal type.

The remaining sample, 12035, granular in texture, is not especially anomalous. The trace element concentrations are similar in a relative sense to those of the normal rocks but somewhat lower in absolute amount (Fig. 4). This can be explained in terms of excess olivine. The somewhat higher K/Rb, lower Rb/Sr, and smaller Eu anomaly might indicate a small excess of plagioclase, in accord with the higher aluminum content (Fig. 5). Trace element concentrations, although somewhat lower, are very similar in 12035 and 12040 plagioclase (Fig. 7), augite, and pigeonite (Fig. 6). A close genetic link between 12035 and 12040, and the other normal rocks, is indicated.

MINERALS

Trace element data for minerals separated out of 12021, 12035, and 12040 are given in Table 2. Analyses of Apollo 12 minerals by TAYLOR *et al.* (1971) appear to be in general agreement. Our separates have been identified by electron microprobe

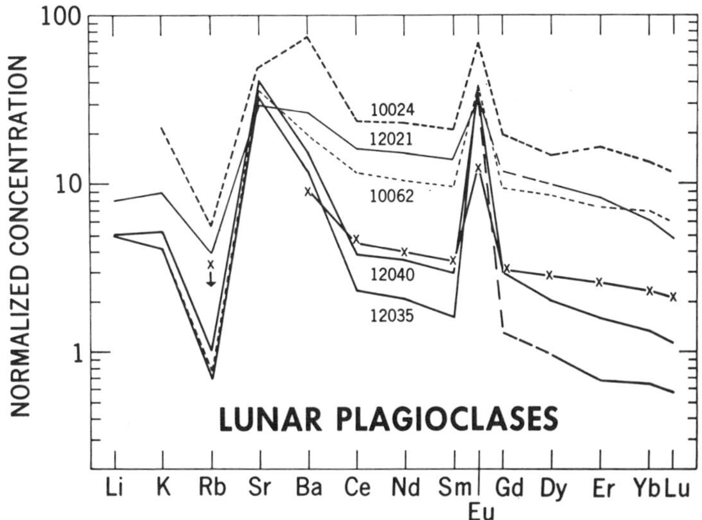

Fig. 7. Chondrite normalized concentrations (K and Rb times 10) in lunar plagioclases. Apollo 11 plagioclases from PHILPOTTS and SCHNETZLER (1970a), Apollo 12 plagioclase from Table 2, and anorthosite fragment (×) from WAKITA and SCHMITT (1970b).

analysis. The trace element data indicate that essentially all of the Sr and Eu in the rocks can be accounted for in terms of modally adjusted major mineral concentrations. In fact, these two elements are contained (along with aluminum) largely in the plagioclase. The close coherence of Sr and Eu in the (reduced) lunar samples is expected (e.g., PHILPOTTS, 1970). Concentrations of the other trace elements in the separates are not sufficient to build up the measured whole-rock values. It appears that roughly half the total content of these elements are in accessory phases and/or the mesostasis. This is in accord with the Apollo 11 mineral results (PHILPOTTS and SCHNETZLER, 1970a; GOLES *et al.*, 1970).

The opaque separate from 12040 (Table 2) has trace element concentrations similar to those in our two Apollo 11 opaque separates (PHILPOTTS and SCHNETZLER, 1970a), particularly when the data are whole-rock normalized. The opaque patterns do not resemble those of other major phases or of the whole rocks. This fact, plus the consistent pattern, suggest that we are not analyzing random contamination but rather a particular phase with a characteristic trace element pattern.

The Apollo 12 clinopyroxene separates (Table 2, Fig. 6) have whole-rock normalized trace element concentrations similar to those of Apollo 11 augites. Each of our three Apollo 12 samples yielded both augite and pigeonite. Samples 12035 and 12040 show similar absolute concentrations and partitioning for augite and pigeonite; plagioclase separates from these two rocks also have similar trace element patterns (Table 2, Fig. 7), differing somewhat from each other in absolute values in the same direction as to their whole-rock patterns (Fig. 4). In contrast, 12021 pigeonite has lower trace element concentrations than do the 12035 and 12040 pigeonites, whereas 12021 augite and plagioclase have higher concentrations than do the respective minerals in the other two rocks, far higher than could be accounted for solely in terms of whole-rock trace element differences. The rock textures illuminate this matter. Rocks 12035 and 12040 are both relatively coarse-grained and granular. This and the trace element data suggest not only that these samples are closely related to each other but also that internal equilibration has been approached. Rock 12021 has quite a different aspect, i.e., large phenocrysts of pigeonite in a relatively fine-grained matrix. The sample is so obviously porphyritic that we did not perform a whole-rock analysis but instead separated the coarse pigeonite and its groundmass, the latter presumably being representative of a liquid. The lower trace element concentrations of 12021 pigeonite are therefore understood in terms of this phase being phenocryst in quasi-equilibrium with 12021 matrix which in a relative sense at least has trace element contents similar to those in 12035 and 12040 whole rocks. The trace element distribution coefficients between 12021 pigeonite and groundmass are consistent with those measured between "orthopyroxene" phenocrysts and matrix materials (SCHNETZLER and PHILPOTTS, 1970). The augite and plagioclase from 12021 groundmass represent late crystallizing phases, hence their higher concentrations. The fact that this matrix was "quenched" is probably largely responsible for the lack of full separation (partitioning), at least for our large cations (e.g., Eu), of elements between the augite and plagioclase. Other effects such as temperature and phase composition are probably not as important.

The Apollo 12 plagioclases show similar patterns to Apollo 11 plagioclases (Fig. 7). Some of the differences in absolute concentrations are due to whole-rock differences, and some, along with much of the variously developed Eu anomalies, are due to the effect described above. It is of interest that the Apollo 11 feldspathic fragment analyzed by WAKITA and SCHMITT (1970b) bears a marked similarity in trace element signature to the basaltic feldspars (Fig. 7). The most obvious interpretation of these trace element data is that this fragment does indeed represent such feldspar. If this fragment represents highland anorthosite as purported, then cumulation of the anorthosite from a liquid with a REE pattern similar to those in the basalts is indicated. We have analyzed a light colored fragment from the 12032 coarse fines (Table 1). It consists of plagioclase (An_{96}) and low-Ca pyroxene, probably orthopyroxene (CaO $\sim$ 1.0%, MgO $\sim$ 28%). The K, Rb, Sr, Ba, and light rare-earth concentrations in this fragment resemble the feldspar values and are probably dominated by the plagioclase. The small positive Eu anomaly is also due to the feldspar. Concentrations of the heavier REE increase with atomic number; this is probably due to the pyroxene. The possible relationships of this fragment to other lunar rocks must remain obscure until further samples and minerals are analyzed.

The mineral rare-earth data permit the calculation of Eu^{2+}/Eu^{3+} ratios of the phases during crystallization (PHILPOTTS, 1970). Inasmuch as the calculation is based upon the assumption of equivalence of Sr and Eu^{2+} partitioning, errors are likely to be accentuated for phases of similar Sr content crystallizing under reducing conditions. To obviate such errors the Eu^{2+}/Eu^{3+} values given in the table are based upon plagioclase to other-phase data only. This yields a number of values for the feldspar ratio and these agree with each other to within a few percent. The 12021 pigeonite value was not obtained in this manner but by the more appropriate comparison with bulk matrix. There is good agreement within the same type of phase in Eu^{2+}/Eu^{3+} in the various lunar rocks including the two Apollo 11 samples (PHILPOTTS, 1970). Eu^{2+}/Eu^{3+} in feldspars range from 20 to 200, in pyroxenes from 0.5 to 2, and in opaques from 0.6 to 1. In accord with the discussion above, the spread in Eu^{2+}/Eu^{3+} between feldspar and pyroxene is smaller for 12021 matrix than for 12035 or 12040. The mineral and estimated whole-rock Eu^{2+}/Eu^{3+} values put the lunar basalts in a position intermediate to terrestrial basalts and achondritic meteorites; this is presumably a sequence of increasingly reduced conditions of crystallization.

BASALT Eu ANOMALIES

A characteristic feature of lunar basalts is a distinct negative Eu anomaly. Several explanations of this feature have been advanced. It has been suggested that Eu was lost through volatilization either during the formation of the moon (NGUYEN-LONG-DEN and YOKOYAMA, 1970) or from the surface of lava lakes (O'HARA and BIGGAR, 1970). A specific problem with the former is that a wide variety of solar system materials such as the bulk earth and many meteorites, including examples of (reduced) achondrites, show no anomalous Eu concentration; it therefore seems unlikely that initial lunar material would have an anomaly. A specific objection to the lava lake hypothesis is that the model ages do not seem to permit loss of volatile Rb relative to Sr during extrusion of the basalts; this would appear to argue against any great loss of other volatile species. Objections common to both hypotheses are that there is no evidence that conditions were reducing enough to produce volatile metallic Eu, and no evidence of anomalous concentrations of other rare-earth elements that could be reduced and volatilized, such as Yb and Sm. Finally it would be noted that the size of the Eu anomaly displays a considerable range in bulk lunar materials (Eu/Sm from 0.7 to 0.1 of the chondritic value) whereas some ratios of volatile to involatile elements show less variation in spite of equal or greater ranges in individual element concentration; thus K/Ba in lunar samples is 7 ± 3 (about $\frac{1}{50}$ of its chondritic value) over two orders of magnitude in element concentrations. This is not readily explainable in terms of volatilization, but it fits in well with what is known about mineral-melt differentiation (e.g., PHILPOTTS and SCHNETZLER, 1970b; SCHNETZLER and PHILPOTTS, 1970).

An alternative explanation of the basalt anomalies involves leaving Eu behind in the residue, relative to the other rare earths (e.g., GOLES *et al.*, 1970; RINGWOOD and ESSENE, 1970; TAYLOR *et al.*, 1971; GREEN *et al.*, 1971). An immediate problem is that no phase in terrestrial rocks (e.g., SCHNETZLER and PHILPOTTS, 1970) or in lunar rocks (e.g., Figs. 6 and 7) has been demonstrated unequivocally to take excess Eu relative to a melt with the exception of feldspar, but it is just to avoid feldspar involvement that the present hypothesis was advanced (see below). Of course, there may be such phases at depth in the moon but this is considered to be unlikely; the most likely, although quite improbable, would seem to be orthopyroxene. At any rate, to avoid these problems GREEN *et al.* (1971) have turned to surface or local equilibrium melting of low Eu phosphates in the basalt source.

This model is rejected for the following reasons. (1) Because nonequilibrium fusion is invoked, the appropriateness of the experimental (supposed equilibrium) studies, which dictated the search for a new model in the first place, is suspect. There is no doubt that major and trace element variations are largely decoupled in igneous processes, but this decoupling cannot be of the arbitrary kind used for a number of years by the Australian school (e.g., wall rock reaction, etc.). Incoherence, like beauty, is in the eyes of the beholder. (2) Recent work questions the premise that phosphates would have large negative Eu anomalies relative to associated mafic phases, even under reducing conditions (NAGASAWA, 1970; unpublished data). (3) The Apollo 11 high-alkali type, which would represent the smallest degree of partial melting in simple fusion models (e.g., 3%, RINGWOOD and ESSENE, 1970), contain about 0.1% phosphate; hence, disregarding complications, the source rock would contain roughly 30 ppm phosphate. One may wonder whether this amount of phosphate would occur in the source material as a discrete phase or would be dissolved in major minerals. And further if a discrete phosphate phase did exist, how it could be brought together with other components of the disequilibrium melt and brought to the surface, all without equilibrating with the major phases. (4) If Apollo 11 high alkali rocks and the Apollo 12 rocks are construed as the result of different degrees of partial fusion of similar source materials, then the large size of the Eu anomaly in the Apollo 12 rocks, which would represent considerably greater degree of fusion involving major minerals, does not fit this model for reasonable partition coefficients. (5) If the lunar rocks are formed by nonequilibrium fusion at their Rb–Sr mineral isochron ages, then model ages (see below) for Apollo 11 low-alkali type and Apollo 12 rocks indicate no change in Rb/Sr in spite of the fact that relative to the rare earths, much of the Sr (geochemically similar to Eu^{2+}) was left in the residue. Further, different Rb/Sr ratios are required in the source materials. (6) Finally, but perhaps most important, is the close coherence of Ba and rare-earths in the lunar rocks (e.g., Fig. 4) in marked contrast to their abundance ratios in phosphates. Thus, analyses of Apollo 12 apatites and whitlockites (BANCROFT *et al.*, 1971; BROWN *et al.*, 1971; KEIL *et al.*, 1971) do not report any K, let alone Ba. This is accord with terrestrial apatite data. For example, an apatite with a matrix partition coefficient for Ce of 52 (NAGASAWA, 1970) yields a Ba partition coefficient that is a factor of at least 2500 times lower (our unpublished data). Seemingly, the only way around this problem would be to postulate an additional trace phase that contains the Ba, but this cannot be K-feldspar (positive Eu anomaly) or zircon (high heavy rare earths—NAGASAWA, 1970); further, when more than one phase is involved the chances of coherence decrease. For these reasons the phosphate fusion model is rejected. A seemingly better disequilibrium fusion model involves eclogite but this too has its drawbacks.

The best explanation of the lunar basalt Eu anomalies still seems to be plagioclase, either in the fusion residue, or crystallizing out of the magma. The major objection to this model is the assertion of some experimental petrologists that plagioclase is not a liquidus phase in these samples (e.g., GREEN *et al.*, 1971). Others, however, report liquidus plagioclase for some of the basalts. The contention of BIGGAR *et al.* (1971), for example, that the Apollo 12 basalts represent cotectic liquids after extreme fractional crystallization of pyroxene and plagioclase, fits the simple explanation of the Eu anomaly very well. RINGWOOD and ESSENE (1970) have also argued against feldspar crystallization on the basis of the constancy of Ba/Yb in lunar basalts. Apollo 11 high-alkali and the Apollo 12 samples do exhibit fair constancy. The higher Ba/Yb ratio of feldspars is apparent in Fig. 7. However, measured pyroxene partition coefficients (PHILPOTTS and SCHNETZLER, 1970b; SCHNETZLER and PHILPOTTS, 1970) indicate considerably lower Ba/Yb in pyroxenes relative to their liquids (see also 12021 pigeonite, Fig. 6). Thus, constancy of Ba/Yb could be used as an argument for feldspar involvement in order to balance the pyroxene effect.

Perhaps some of the difficulties apparently involved in reconciling some of the experimental data with a feldspar-produced Eu anomaly arise from a possibly erroneous interpretation of the rocks as liquids in a simple, single stage process. Thus, if the low-alkali Apollo 11 rocks represent cumulates as we believe, then it is immaterial whether feldspar appears on their liquidi or does not, inasmuch as these rocks would not have been liquids. Our own genetic prejudices are set out below.

TIME OF DIFFERENTIATION

An interesting feature of the Apollo 12 rocks and the low-alkali type of Apollo 11 rocks is that they have 4.5 b.y. model Rb–Sr ages (e.g., PAPANASTASSIOU *et al.*, 1970; PAPANASTASSIOU and WASSERBURG, 1970; COMPSTON *et al.*, 1971; MURTHY *et al.*, 1971). This means that starting with reasonable "primordial" initial strontium isotopic composition, the Rb/Sr ratios of these samples need not (and probably did not) have changed over the past 4.5 b.y., except for changes due to radioactive decay. However, some event did homogenize the Sr in these samples at their mineral isochron ages of typically 3.3 and 3.7 b.y., respectively. This event might have been partial fusion in which a large fraction of the Rb and Sr went into the melt. A problem arises, however, if it is accepted that 12022, 12038, and 12063 (see above), and the Apollo 11 low-alkali type rock (Philpotts and Schnetzler, 1970a) are (partial) cumulates of pyroxene and plagioclase as indicated by the trace element data. In general, equilibrated pyroxene and plagioclase have quite different Rb/Sr ratios which, in turn, differ from the liquid Rb/Sr; fractional crystallization or, particularly, cumulation of these minerals in various proportions, therefore, yield rocks with wide ranges in Rb/Sr. It would seem to be too much of a coincidence to expect all such lunar cumulates to have the same Rb/Sr ratios as their parental liquids and source materials. It is therefore suggested that these cumulates formed about 4.5 b.y. ago.

Perhaps the best case that can be made for this argument is that of sample 12038. Major and trace element data indicate that this sample contains considerable excess (cumulative) feldspar (see above); the low Rb/Sr ratio is then readily explained. The 3.3 b.y. initial $^{87}Sr/^{86}Sr$ in this rock, which is lower than that of any other Apollo 12 sample (Compston *et al.*, 1971), is also explained if cumulation occurred at 4.5 b.y.; if differentiation occurred at 3.3 b.y. then the correlation of feldspar content, low Rb/Sr, and low initial $^{87}Sr/^{86}Sr$ is fortuitous. The suggestion is therefore made, in agreement with the suggestion of Urey *et al.* (1971), that lunar rocks with 4.5 b.y. model Rb–Sr ages differentiated approximately 4.5 b.y. ago. Supporting the case for early differentiation is sample 12013 with a component age of 4 b.y. (Lunatic Asylum, 1970; Schnetzler *et al.*, 1970) but also with initial $^{87}Sr/^{86}Sr$ high enough to require the prior existence of high Rb/Sr material for several b.y.

This proposal requires no change in Rb/Sr during the events dated by the mineral isochrons. Metamorphism appears to be ruled out by the rock textures (e.g., mineral zoning, quench features, etc.) and the high temperatures of mineral equilibration indicated by the oxygen isotope data (e.g., Onuma *et al.*, 1971; Epstein and Taylor, 1971). Closed-system whole-rock melting is therefore suggested. This might be a local phenomenon, perhaps involved in, or triggered by, meteorite impact. Remelting, of course, would obliterate pre-existing (e.g., cumulate) textures. Finally, this 4.5 b.y. differentiation model makes it possible to relate rocks of different mineral isochron initial Sr or even different mineral ages, where relationships are indicated by major or trace element data.

Lunar Evolution Model

An attempt at a serious synthesis at the present time might be premature, particularly in view of interesting preliminary reports concerning the Apollo 14 material. The following should be viewed as an eclectic framework on which to hang future observations.

The moon accreted $\sim$4.6 b.y. ago, largely in a cool state. The accreting material was characterized by such features as lower relative volatile element contents, compared to the earth, and terrestrial rather than achondritic oxygen isotopic compositions (e.g., ONUMA *et al.*, 1971; EPSTEIN and TAYLOR, 1971). Melting and differentiation started during or soon after accretion. This was probably restricted to the outer couple of hundred kilometers as indicated by low lunar interior temperatures (SONETT *et al.*, 1971) and the existence of mascons (UREY *et al.*, 1971). The differentiation involved basaltic rocks from which considerable feldspar was lost (Eu anomalies); a (highland) crust formed of separated feldspar and late liquids as indicated in part by the Surveyor 7 Tycho analysis (PATTERSON *et al.*, 1969) and by the "anorthositic" fragments in Apollo soils (e.g., WOOD 1970, 1971). Rocks with high Rb/Sr ratios as in 12013 dark portion (alias KREEP) were formed and emplaced at the surface. The mare basins and mare basalts formed at this time ($\sim$4.5 b.y.) including the Apollo 12 basalts and Apollo 11 low-alkali cumulates.

Such a model obviates RINGWOOD and ESSENE's (1970) arguments for a deep seated origin, based upon a thick cooled crust at 3.7 to 3.3 b.y. Cooling of the outer portions of the moon progressed rapidly. At 4 b.y. 12013 underwent some process, such as metamorphism which partially redistributed Sr. At 3.7 b.y. the low-alkali Apollo 11 (ilmenite) cumulates underwent closed-system melting and emplacement at or near the surface. At the same time the Apollo 11 high-alkali basalts were formed by partial fusion of rocks similar to the low-alkali type but having higher Rb/Sr; the Rb/Sr was further increased during the partial fusion event. At 3.3 b.y. the bulk of the Apollo 12 rocks were formed by local closed-system whole-rock fusion. Subsequent other small scale whole-rock or partial fusion events have taken place. The lunar soil has formed by impact of surface rocks.

Acknowledgments—S. Schuhmann and P. Shadid were of great assistance in the analytical portion of this work. We are also indebted to W. Kouns for aid in the mineral separations, and F. Wood for aid in the microprobe analyses.

REFERENCES

ANNELL C. S., CARRON M. K., CHRISTIAN R. P., CUTTITTA F., DWORNIK E. J., HELZ A. W., LIGON D. T., and ROSE H. J., JR. (1971) Chemical and spectrographic analyses of lunar samples from Apollo 12 mission. Second Lunar Science Conference (unpublished proceedings).

BANCROFT G. M., BROWN M. G., GAY P., MUIR I. D., and WILLIAMS P. G. L. (1971) Mineralogical and petrographic investigation of some Apollo 12 samples. Second Lunar Science Conference (unpublished proceedings).

BIGGAR G. M., O'HARA M. J., and PECKETT A. (1971) Origin, eruption, and crystallization of protohypersthene basalts from the Ocean of Storms. Second Lunar Science Conference (unpublished proceedings).

BOUCHET M., KAPLAN G., VOUDON A., and BERTOLETTI M.-J. (1971) Spark mass spectrometric analysis of major and minor elements in six lunar samples. Second Lunar Science Conference (unpublished proceedings).

BROUN G. M., EMELENS C. H., HOLLAND J. G., PECKETT A., and PHILLIPS R. (1971) Mineral chemistry of contrasted Apollo 12 basalt-types and comparisons with Apollo 11. Second Lunar Science Conference (unpublished proceedings).

BRUNFELT A. O., HEIER K. S., and STEINNES E. (1971) Determination of 40 elements in Apollo 12 materials by neturon activation analysis. Second Lunar Science Conference (unpublished proceedings).

CARRIER W. D. III, JOHNSON S. W., WERNER R. A., and SCHMIDT R. (1971) Disturbance in samples recovered with the Apollo core tubes. Second Lunar Science Conference (unpublished proceedings).

CLIFF R. A., LEE-HU C., and WETHERILL G. W. (1971) Rb–Sr and U, Th–Pb Measurements on Apollo 12 material. Second Lunar Science Conference (unpublished proceedings).

COMPSTON W., BERRY H., VERNON M. J., CHAPPELL B. W., and KAYE M. J. (1971) Rb–Sr chronology and chemistry of lunar material from the Ocean of Storms. Second Lunar Science Conference (unpublished proceedings).

EHMANN W. D. and MORGAN J. W. (1971) Major element abundances in Apollo 12 rocks and fines by 14 MeV neutron activation. Second Lunar Science Conference (unpublished proceedings).

ENGEL C. G. and ENGEL A. E. J. (1971) Major element composition of three Apollo 12 rocks and some petrogenic considerations. Second Lunar Science Conference (unpublished proceedings).

ENGELHARDT W. V., ARNDT J., MULLER W. F., and STOFFLER D. (1971) Shock metamorphism and origin of the regolith at the Apollo 11 and Apollo 12 landing sites. Second Lunar Science Conference (unpublished proceedings).

EPSTEIN S. and TAYLOR H. P. JR (1971) O^{18}/O^{16}, Si^{30}/Si^{28}, D/H, and C^{13}/C^{12} ratios in lunar samples. Second Lunar Science Conference (unpublished proceedings).

FREDRIKSSON K., NELEN J., ANDERSEN C. A., and HINTHORNE J. R. (1971) A summary of phase analysis on Apollo 12 samples; ion microprobe mass analysis. Second Lunar Science Conference (unpublished proceedings).

FRONDEL C., KLEIN C., and ITO J. (1971) Mineralogical and chemical data on Apollo 12 lunar fines. Second Lunar Science Conference (unpublished proceedings).

GANAPATHY R., KEAYS R. R., and ANDERS E. (1970) Apollo 12 lunar samples: Trace element analysis of a core and the uniformity of the regolith. *Science* **170**, 533–535.

GAST P. W., HUBBARD N. J., and WIESMANN H. (1970) Chemical composition and petrogenesis of basalts from Tranquillity Base. *Proc. Apollo 11 Lunar Sci. Conf., Geochim. Cosmochim. Acta* Suppl. 1, Vol. 2, pp. 1143–1163. Pergamon.

GLASS B. P. (1971) Investigation of glass recovered from Apollo 12 sample 12057. Second Lunar Science Conference (unpublished proceedings).

GOLES G. G., RANDLE K., OSAWA M., LINDSTROM D. J., JEROME D. Y., STEINBORN T. L., BEYER R. L., MARTIN M. R., and MCKAY S. M. (1970) Interpretations and speculations on elemental abundances in lunar samples. *Apollo 11 Lunar Sci. Conf., Geochim. Cosmochim. Acta* Suppl. 1, Vol. 2, pp. 1177–1194. Pergamon.

GOLES G. G., DUNCAN A. R., OSAWA M., MARTIN M. R., BEYER R. L., LINDSTROM D. J., and RANDLE K. (1971) Analyses of Apollo 12 specimens and a mixing model for Apollo 12 "soils." Second Lunar Science Conference (unpublished proceedings).

GREEN D. H., RINGWOOD A. E., WARE N. G., HIBBERSON W. O., MAJOR A., and KISS E. (1971) Experimental petrology and petrogenesis of Apollo 12 basalts. Second Lunar Science Conference (unpublished proceedings).

HASKIN L. A., ALLEN R. O., HELMKE P. A., ANDERSON M. R., KOROTEV R. L., and ZWEIFEL (1971) Rare earths and other trace elements in Apollo 12 lunar materials. Second Lunar Science Conference (unpublished proceedings).

HUBBARD N. J., GAST P. W., and WIESMANN H. (1970) Rare earth, alkaline and alkali metal, and $^{87/86}Sr$ data for subsamples of lunar sample 12013. *Earth Planet. Sci. Lett.* **9**, 181–184.

HUBBARD N. J., MEYER C., GAST P. W., and WIESMANN H. (1971) The composition and derivation of Apollo 12 soils. *Earth Planet. Sci. Lett.* **10**, 341–350.

KEIL K. and PRINZ M. (1971) Mineralogical and petrological aspects of Apollo 12 rocks. Second Lunar Science Conference (unpublished proceedings).

KUSHIRO I., NAKAMURA Y., HARAMURA H., and AKIMOTO S. (1971) Origin of chemical and mineralogical variations of the Apollo 12 crystalline rocks. Second Lunar Science Conference (unpublished proceedings).

LAUL J. C., KEAYS R. R., GANAPATHY R., and ANDERS E. (1970) Abundance of 14 trace elements in lunar rock 12013. *Earth Planet. Sci. Lett.* **9**, 211–215.

LSPET (Lunar Sample Preliminary Examination Team) (1971) Preliminary examination of the lunar samples from Apollo 12. *Science* **167**, 1325–1339.

LUNATIC ASYLUM (1970) Mineralogic and isotopic investigations on lunar rock 12013. *Earth Planet. Sci. Lett.* **9**, 137–163.

MAXWELL J. A. and WIIK H. B. (1971) Chemical composition of Apollo 12 lunar samples 12004, 12033, 12051, 12052, and 12065. Second Lunar Science Conference (unpublished proceedings).

MCKAY D., MORRISON D., LINDSAY J., and LADLE G. (1971) Apollo 12 soil and breccia. Second Lunar Science Conference (unpublished proceedings).

MEYER C. JR., AITKEN F. K., BRETT R., MCKAY D. S., and MORRISON D. A. (1971) Rock fragments and glasses rich in K, REE, P in Apollo 12 soils: Their mineralogy and origin. Second Lunar Science Conference (unpublished proceedings).

MORRISON G. H., GERARD J. T., POTTER N. M., GANGADHARAM E. V., ROTHENBERG A. M., and BURDO R. A. (1971) Elemental abundances of lunar soil and rocks from Apollo 12. Second Lunar Science Conference (unpublished proceedings).

MURTHY V. R., EVENSEN N. M., JAHN B. M., and COSCIO M. R. (1971) Rb–Sr isotopic relations and elemental abundances of K, Rb, Sr, and Ba in Apollo 11 and Apollo 12 samples. Second Lunar Science Conference (unpublished proceedings).

NAGASAWA H. (1970) Rare earth concentrations in zircons and apatites and their host dacites and granites. *Earth Planet. Sci. Lett.* **9**, 359–364.

NGUYEN-LONG-DEN and YOKOYAMA, Y. (1970) Depletion of Eu in lunar rocks. *Meteoritics* **5**, 214–215 (abstract).

O'KELLEY G. D., ELDRIDGE J. S., SCHONFELD E., and BELL P. R. (1971) Comparative radionuclide concentrations and ages of Apollo 11 and Apollo 12 samples from nondestructive gamma-ray spectrometry. Second Lunar Science Conference (unpublished proceedings).

O'HARA M. J. and BIGGAR G. M. (1970) Volatilization-fractionation in lunar lava lakes. *Geol. Soc. Am.* abstracts, Vol. 2, No. 7, 639–640 (abstract).

ONUMA N., CLAYTON R. N., and MAYEDA T. K. (1971) Oxygen isotope fractionation in Apollo 12 rocks and soils. Second Lunar Science Conference (unpublished proceedings).

PAPANASTASSIOU D. A. and WASSERBURG G. J. (1970) Rb–Sr ages from the Ocean of Storms. *Earth Planet. Sci. Lett.* **8**, 269–278.

PAPANASTASSIOU D. A., WASSERBURG G. J., and BURNETT D. S. (1970) Rb–Sr ages of lunar rocks from the Sea of Tranquillity. *Earth Planet. Sci. Lett.* **8**, 1–19.

PATTERSON J. H., FRANZGROTE E. J., TURKEVICH A. L., ANDERSON W. A., ECONOMOU T. E., GRIFFIN H. E., GROTCH S. L., and SOWINSKI K. P. (1969) Alpha-scattering experiment on Surveyor 7: Comparison with Surveyors 5 and 6. *J. Geophys. Res.* **74**, 6120–6148.

PHILPOTTS J. A. (1970) Redox estimation from a calculation of Eu^{2+} and Eu^{3+} concentrations in natural phases. *Earth Planet. Sci. Lett.* **9**, 257–268.

PHILPOTTS J. A. and SCHNETZLER C. C. (1970a) Apollo 11 lunar samples: K, Rb, Sr, Ba, and rare-earth concentrations in some rocks and separated phases. *Proc. Apollo 11 Lunar Sci. Conf., Geochim. Cosmochim. Acta* Suppl. 1, Vol. 2, pp. 1471–1486. Pergamon.

PHILPOTTS J. A. and SCHNETZLER C. C. (1970b) Phenocryst-matrix partition coefficients for K, Rb, Sr, and Ba, with applications to anorthosite and basalt genesis. *Geochim. Cosmochim. Acta* **34**, 307–322.

RINGWOOD A. E. and ESSENE E. (1970) Petrogenesis of Apollo 11 basalts, internal constitution and origin of the moon. *Proc. Apollo 11 Lunar Sci. Conf. Geochim. Cosmochim. Acta* Suppl. 1, Vol. 1, pp. 769–799. Pergamon.

ROSE H. J. JR., CUTTITTA F., ANNELL C. S., CARRON M. K., CHRISTIAN R. P., DWORNIK E. J., HELZ A. W., and LIGON D. T. JR. (1971) Semimicroanalysis of Apollo 12 samples. Second Lunar Science Conference (unpublished proceedings).

SCHNETZLER C. C., THOMAS H. H., and PHILPOTTS J. A. (1967) The determination of rare-earth elements in rocks and minerals by mass spectrometric, stable isotope dilution technique. *Anal. Chem.* **39**, 1888–1890.

SCHNETZLER C. C. and PHILPOTTS J. A. (1970) Partition coefficients of rare-earth elements between igneous matrix material and rock-forming mineral phenocrysts, II. *Geochim. et Cosmochim. Acta* **34**, 331–340.

SCHNETZLER C. C., PHILPOTTS J. A., and BOTTINO M. L. (1970) Li, K, Rb, Sr, Ba, and rare-earth concentrations, and Rb–Sr age of lunar rock 12013. *Earth Planet. Sci. Lett.* **9**, 185–192.

SMALES A. A. (1971) Elemental composition of lunar surface material (part 2). Second Lunar Science Conference (unpublished proceedings).

SONETT C. P., SMITH B. F., COLBURN D. S., SCHUBERT G., SCHWARTZ K., DYAL P., and PARKIN C. W. (1971) The lunar electrical conductivity profile: Mantle-core stratification, near surface thermal gradient, heat flux, and composition. Second Lunar Science Conference (unpublished proceedings).

TAYLOR S. R., KAYE M., GRAHAM A., RUDOWSKI R., and MUIR P. (1971) Trace element chemistry of lunar samples from the Ocean of Storms. Second Lunar Science Conference (unpublished proceedings).

UREY H. C., MARTI K., HAWKINS J. W., and LIN M. L. (1971) Model history of the lunar surface. Second Lunar Science Conference (unpublished proceedings).

WAKITA H. and SCHMITT R. A. (1970a) Elemental abundances in seven fragments from lunar rock 12013. *Earth Planet. Sci. Lett.* **9**, 169–176.

WAKITA H. and SCHMITT R. A. (1970b) Lunar anorthosites: Rare-earth and other elemental abundances. *Science* **170**, 969–974.

WAKITA H., REY P., and SCHMITT R. A. (1971) Abundances of the 14 rare earth elements plus 22 major, minor, and trace elements in 10 Apollo 12 rocks and soil samples. Second Lunar Science Conference (unpublished proceedings).

WÄNKE H., WLOTZKA F., TESCHKE F., BADDENHAUSEN H., SPETTEL B., BALACESEU A., QUIJANO-RICO M., JAGOUTZ E., and RIEDER R. (1971) Major and trace elements in Apollo 12 samples and studies on lunar metallic iron particles. Second Lunar Science Conference (unpublished proceedings).

WARNER J. L. and ANDERSON D. H. (1971) Lunar crystalline rocks: Petrology, geology, and origin. Second Lunar Science Conference (unpublished proceedings).

WOOD J. A., DICKEY J. S. JR., MARVIN U. B., and POWELL B. N. (1970) Lunar anorthosites and a geophysical model of the moon. *Proc. Apollo 11 Lunar Sci. Conf., Geochim. Cosmochim.* Acta Suppl. 1, Vol. 1, pp. 965–988. Pergamon.

WOOD J. A., MARVIN U., REID J. B., TAYLOR G. J., BOWER J. F., POWELL B. N., and DICKEY J. S. JR. (1971) Relative proportions of rock types, and nature of the light-colored lithic fragments in Apollo 12 soil samples. Second Lunar Science Conference (unpublished proceedings).

Proceedings of the Second Lunar Science Conference, Volume 2, pp. 1123–1138.
The M.I.T. Press, 1971.

Some interelement relationships between lunar rocks and fines, and stony meteorites

J. P. Willis, L. H. Ahrens, R. V. Danchin, A. J. Erlank,
J. J. Gurney, P. K. Hofmeyr, T. S. McCarthy, and M. J. Orren
Department of Geochemistry, University of Cape Town. Rondebosch, C.P., South Africa

(*Received 24 February 1971; accepted in revised form 2 April 1971*)

Abstract—The concentrations of 27 elements have been determined in three Apollo 12 rocks and two fines. Together with published analyses of Apollo 11 materials, these data are used to discuss elemental variations within and between Apollo 11 and Apollo 12 rocks and fines. The variations within the Apollo 12 rocks, and between Apollo 11 and Apollo 12 rocks are considered to be due mainly to differing degrees of partial melting. The Apollo 12 fines are enriched, relative to both Apollo 12 rocks and Apollo 11 fines, particularly in K, Rb, Ba, P, Zr, Nb, Y, La, Yb, and Be. These enrichments are considered to be due to admixture of foreign material analogous to KREEP and rock 12013, and estimates are made of the Zr, Nb, and Be contents of such materials. These calculations support previous conclusions that the lunar interior is enriched in refractory elements.

Emphasis has been placed on chemical relationships between stony meteorites and the lunar surface, and in these discussions we have paid particular attention to the relationships between Ca and Al, Sr and Eu, and Mg and Cr. The Ca/Al ratio in lunar fines is closely similar to that found in stony meteorites. The Sr–Eu relationship between stony meteorites and lunar rocks and fines is remarkably well-developed, whereas the Mg–Cr relationship in these materials further supports the similarity previously postulated for the lunar surface and basaltic achondrites.

Introduction

The purposes of this paper are five-fold: (1) presentation of our analytical techniques and data as determined either by X-ray fluorescence or optical emission spectroscopy in three Apollo 12 rocks (12002,113; 12053,24; 12063,52) and two Apollo 12 fines (12032,38 and 12070,88); (2) a general discussion of the compositional variations observed within and between lunar rocks and fines, with special reference to certain interelement relationships; (3) a critical evaluation of the Ca-Al relationship in stony meteorites and all types of lunar materials, particularly lunar fines; (4) a study of the relationship between geochemically closely coherent pairs of refractory elements, with particular reference to Sr and Eu, in stony meteorites and lunar materials; and (5) a comparison of Cr abundances between terrestrial and extra-terrestrial matter (stony meteorites and lunar rocks and fines); and a brief examination of Mg–Cr relationship between basaltic achondrites (howardites and eucrites) and lunar rocks and fines.

Analytical Techniques and Data

As this is the first time we are reporting data for lunar materials, brief details of our analytical techniques are set out below.

Sample preparation

The amounts of lunar material received varied between 2 g and 6 g. The samples were first crushed by hand in a carbon steel pestle and mortar to pass a 60 mesh nylon screen, and then ground in an automatic agate mortar to approximately 300 mesh size.

X-ray fluorescence

A Philips PW 1220 2 kW semiautomatic X-ray spectrometer was used for all determinations, except Ba, which was determined on a PW 1540 manual spectrometer (WILLIS *et al.*, 1969). The total sample, in powder form, was used for the determination of Rb, Sr, Y, Zr, and Nb, using techniques similar to those discussed by CHERRY *et al.* (1970). The mass absorption coefficient was also determined on the powder by the method of REYNOLDS (1963). 1 g of powder was then briquetted and used for the determination of Zn, Cu, Ni, Ba, La, S, and Na. Zn and S were determined first to minimize any possible contamination from the vacuum oil.

Another 1 g of powder was dried at 120°C overnight in a preheated Vitreosil crucible, and then heated at 1000°C to constant weight (~12 hours). Three 0.28 g portions of the heated material were fused with Johnson Matthey Spectroflux no. 105 using the method of NORRISH and HUTTON (1969) for the analysis of Fe, Mn, Cr, Ti, Ca, K, P, Si, Al, and Mg.

Optical emission spectroscopy

Of the remaining heated material, 0.1 g was mixed with specpure graphite powder, containing Pd as internal standard, in the ratio 1 part sample to 2 parts graphite, for the determination of Be, Sc, V, Co, and Yb. A Jarrell-Ash 3.4 m plane grating spectrograph (5 Å/mm dispersion in the first order) was used for the analyses, with a conventional DC arc and a modified Stallwood jet with an argon–oxygen mixture. Spectral lines used were Be 3131, Sc 3613, V 3276, Co 3453, Yb 3289, Pd 3258, and Pd 3065 Å. Background corrections were made on all lines.

Standards

For XRF determinations either G-1 or W-1 was used as reference standard for the trace elements. PCC–1 and DTS–1 were used for Cr_2O_3 and a departmental standard for S. The rock standards BR, T-1, S-1, GSP-1, AGV-1, PCC-1, and DTS-1 were used for major element analysis, together with artificial standards for Fe and Ti. For optical emission analysis a number of international rock standards were used to establish working curves.

Table 1. Major element data (%) for Apollo 12 rocks and fines by X-ray fluorescence spectrometry

	s.d.*	Rocks 12002,113	12053,24	12063,52	Average	Fines 12032,38	12070,88	Average
SiO_2	0.30	43.56	46.21	43.48	44.42	46.58	45.74	46.16
TiO_2	0.02	2.60	3.32	5.00	3.64	2.56	2.79	2.68
Al_2O_3	0.12	7.87	10.14	9.27	9.09	13.59	12.67	13.13
FeO†	0.10	21.66	19.77	21.26	20.90	15.11	16.52	15.82
MnO	0.002	0.283	0.280	0.280	0.281	0.207	0.222	0.215
MgO	0.10	14.88	8.17	9.56	10.87	9.89	10.42	10.16
CaO	0.05	8.26	11.01	10.49	9.92	10.53	10.45	10.49
Na_2O	0.01	0.23	0.26	0.31	0.27	0.56	0.39	0.48
K_2O	0.003	0.051	0.065	0.061	0.059	0.363	0.241	0.302
P_2O_5	0.01	0.11	0.14	0.14	0.13	0.38	0.30	0.34
Cr_2O_3	0.01	0.96	0.49	0.44	0.63	0.37	0.42	0.40
S	0.001	0.062	0.078	0.090	0.077	0.069	0.075	0.072
Subtotal		100.526	99.933	100.381		100.209	100.238	
O = S		0.031	0.039	0.045		0.035	0.038	
Total		100.49	99.89	100.33		100.17	100.20	

* Average experimental standard deviation (1 σ)
† Total Fe as FeO

Table 2. Trace element data (ppm) for Apollo 12 rocks and fines

	s.d.*	Rocks 12002,113	12053,24	12063,52	Average	s.d.*	Fines 12032,38	12070,88	Average
				X-ray fluorescence spectrometry					
S	8	622	785	905	771	8	688	752	720
K	30	420	540	510	490	30	3010	2000	2510
Cr	100	6570	3350	3010	4310	100	2530	2870	2700
Ni	1.2	64	10	20	31	1.4	117	202	˙160
Cu	0.9	4.6	26†	8.0	13	0.9	29†	8.0	19
Zn	0.6	1.5	1.2	4.5	2.4	0.5	22†	9.7	16
Rb	0.2	1.25	1.1	0.8	1.05	0.2	8.9	6.4	7.7
Sr	1.4	86	111	145	114	1.3	154	136	145
Y	1.6	39	52	65	52	1.8	197	149	173
Zr	1.5	106	138	133	126	2.2	741	549	645
Nb	1.6	8.5	10.0	7.9	8.8	1.5	58	45	52
Ba	2	66	86	67	73	4	531	373	452
La		<6	<6	<6	<6	2.6	51	33	42
				Emission spectrography					
Be	0.2	1.0	1.7	1.5	1.4	0.7	4.3	4.0	4.2
Sc	9	45	50	55	50	8	39	39	39.0
V	7	175	148	135	153	8	96	112	104
Co	5	62	30	36	43	5	31	40	36
Yb	0.6	4.3	3.8	5.7	4.6	1.5	16	12	14

* s.d. Average experimental standard deviation.
† Contamination?

A study of the data from the Second Lunar Science Conference (unpublished proceedings) indicates a probable systematic error for P_2O_5 in our lunar rock analyses. This error may also be present in our soil analyses. The samples had already been returned to NASA and could not be reanalyzed. However, a check on the Apollo 11 rock 10017,70 indicates that the rock analyses reported here may be high by $\sim$0.09% P_2O_5, due to blank problems. A study of the Cr data in the unpublished proceedings, together with that for PCC-1 and DTS-1, (Flanagan, 1969), indicates an urgent need for the accurate standardization of Cr below the 1% level. Cl data presented at the Second Lunar Science Conference have been omitted from this paper, as the samples are thought to have been contaminated by the high local sea salt content of the air. All analytical data are presented in Tables 1 and 2, together with average experimental standard deviations which are derived from replicate determinations and take into account all sources of imprecision. Some interelement ratios are presented in Table 3.

General Discussion on Lunar Rocks and Fines

Lunar rocks

As indicated by the preliminary data (LSPET 1970) the Apollo 12 rocks show a larger variation in major element composition than the Apollo 11 rocks and this is evident from the data given in Table 1, even though we have only analyzed three rocks. From the petrographic data available to us, and as shown by the variations in Mg and Ni, it might be inferred that these three rocks are members of a differentiated suite which has undergone olivine fractionation. However, detailed inspection of the data does not support this contention. For example, a substantial amount of olivine would have to be removed from rock 12002 (14.9% MgO) to produce rock 12063 (9.6% MgO), yet Ba, which is not readily accommodated in the olivine structure, is present in virtually identical amounts in the two samples. This would imply removal also of

Table 3. Some interelement ratios in lunar rocks and fines.

	Rocks				Fines		
	12002,113	12053,24	12063,52	Average	12032,38	12070,88	Average
K/Rb	336	491	638	467	338	313	327
K/Ba	6.36	6.28	7.61	6.71	5.67	5.36	5.54
Ba/Rb	53	78	84	72	60	58	59
Zr/Nb	12.5	13.8	16.8	14.4	12.7	12.3	12.5
Ba/Zr	0.62	0.62	0.50	0.58	0.72	0.68	0.70
K/Sr	4.9	4.9	3.5	4.3	19.5	14.7	17.3
K/Zr	3.96	3.91	3.83	3.88	4.06	3.64	3.88
Rb/Sr	0.014	0.010	0.006	0.010	0.058	0.047	0.052
Ba/Sr	0.77	0.77	0.46	0.66	3.45	2.74	3.09
$Rb/Zr \times 10^2$	1.18	0.80	0.60	0.86	1.20	1.17	1.19

interstitial material (mesostasis) and the existence of open system conditions. The variation of other elements (e.g., Sr, Y, and Nb) is also not in accord with that expected from olivine fractionation alone (or spinel or pyroxene fractionation) and hence we infer that varying degrees of partial melting must also be invoked to explain the observed compositional variation in the Apollo 12 rocks. Critical interelement ratios (Table 3) do not suggest that different source areas are involved in the production of the three rocks analyzed here. We have insufficient data and samples to further evaluate these possibilities.

Compared to the Apollo 11 low K (Rb) rock suite (COMPSTON *et al.*, 1970; GAST *et al.*, 1970), and allowing for compositional variations and analytical error, the Apollo 12 rocks are enriched in Si, Mg, Ni, V, Co, and Cr, and depleted in Ti, Zr, S, Sc, Y, Nb, La, Yb, and possibly Be, indicating the more mafic character of the Apollo 12 rocks. Considered collectively, the data suggest a greater degree of partial melting during the production of the liquids giving rise to the Apollo 12 rocks, in view of the relative similarity in major element composition when compared to the larger differences present for dispersed elements such as Nb and Zr in the Apollo 11 and Apollo 12 rocks. The case for varying degrees of partial melting giving rise to the observed compositions, as opposed to advanced near-surface crystal fractionation (O'HARA, *et al.*, 1970), has been cogently argued by RINGWOOD (1970) and GAST AND HUBBARD (1970). Different source areas may also be partly responsible as K/Rb ratios for example are lower in the Apollo 12 basalts than in the low K Apollo 11 basalts (Fig. 1). However, K/Ba ratios are indistinguishable in the two suites (Fig. 2).

Lunar fines

The two samples of fines analyzed (Tables 1 and 2) have fairly similar major element compositions, but differ appreciably for several trace elements, for example K, Ba, Rb, Zr, and Nb. This demonstrates that the regolith at Apollo 12 is not well-mixed. It may be noted however that several interelement ratios (e.g., Ba/Rb, Ba/Zr, Zr/Nb) are remarkably similar (Table 3) and this merits consideration in any mixing process which is proposed to account for the lunar fines.

Compared with data for the Apollo 11 fines 10084 (COMPSTON *et al.*, 1970; ANNELL and HELZ 1970; MORRISON *et al.*, 1970; SMALES *et al.*, 1970) the Apollo 12 fines are depleted in Ti and Sc (ilmenite?) and enriched in Mg and V (olivine from the more

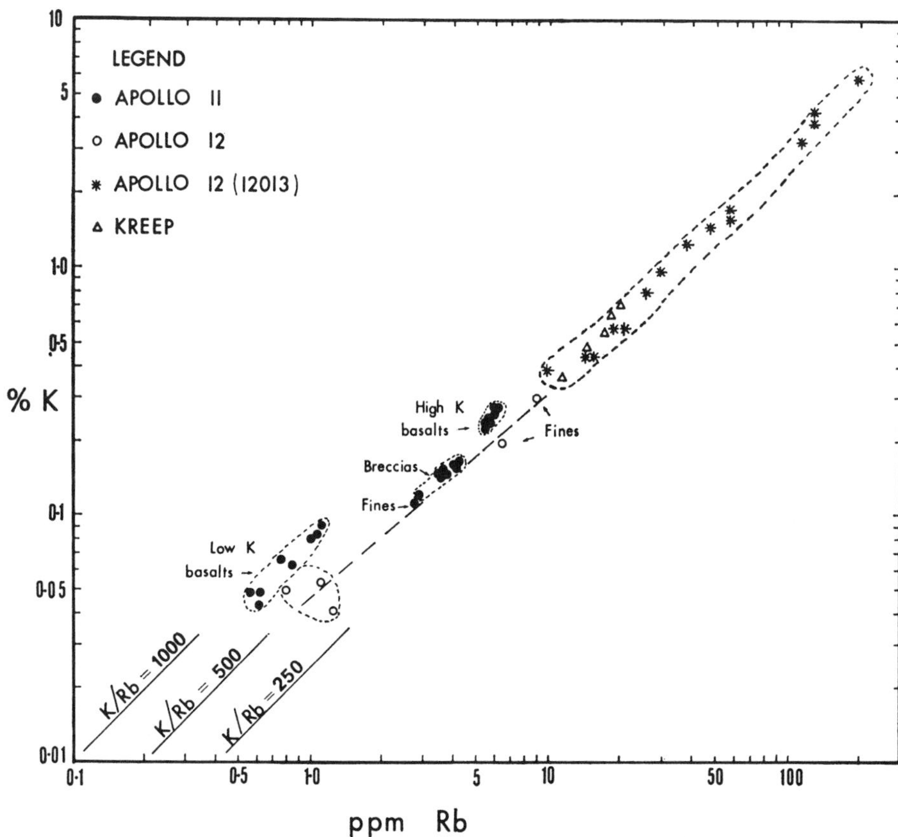

Fig. 1. The K–Rb relationship in lunar materials. Apollo 12 data for rocks and fines are from this paper. Apollo 11 rocks, breccia, and fines data are from Compston *et al.* (1970), Gopalan *et al.* (1970), Murthy *et al.* (1970), Gast *et al.* (1970), Philpotts and Schnetzler (1970), Tera *et al.* (1970), Smales *et al.* (1970), and Hubbard *et al.* (1970). Apollo 12 rock 12013 data from Gast *et al.* (1970) and Schnetzler *et al.* (1970). Apollo 12 KREEP data from Hubbard *et al.* (1971b).

Mg rich Apollo 12 rocks?) and K, Rb, Ba, P, Zr, Nb, Y, La, Yb, (La > Yb), and possibly Be. Where reliable data are available it may be shown that some interelement ratios involving the last named elements also vary. For example it can be seen from Fig. 1 that the Apollo 12 fines have lower K/Rb ratios than Apollo 11 10084. These features reflect varying admixture of foreign material and this is discussed below.

Lunar rocks versus fines

Several authors have concluded that the Apollo 11 fines cannot be solely derived from the range of Apollo 11 rocks which have been analyzed. In particular, Tera *et al.* (1970) showed that an extra component, with lower K/Rb and K/Ba ratios than those observed in the rocks, was necessary to explain the composition of the fines. LSPET (1970) also demonstrated that the Apollo 12 fines cannot be directly derived from the

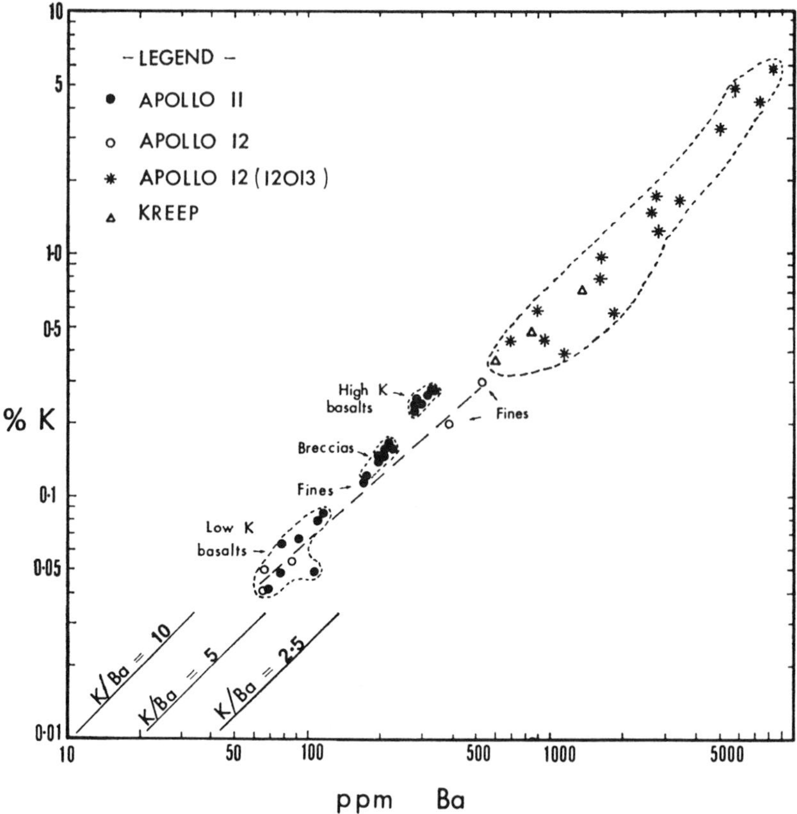

Fig. 2. The K–Ba relationship in lunar materials. Sources of data as for Fig. 1.

Apollo 12 rocks and this is confirmed by the data presented here. It is instructive to consider the ratio of fines/rocks for the averages given in Tables 1 and 2 for those elements where distinctions are reasonably clear-cut.

Ratio Fines/Rocks	*Elements*
0.6–0.9	Fe, Mn, Cr, V, Sc
1.0–2.0	Al, Na, Sr
>2.0	K, Rb, Ba, P, Y, La, Yb, Zr, Nb, Be, Ni, Zn

The first group of elements (ratios < 1) shows that the fines have less ferromagnesian and opaque minerals than the rocks. The second group indicates the presence of more plagioclase or anorthosite in the fines. Of those elements greatly enriched in the fines, the increased abundance of Ni and Zn is considered to be due to small amounts of meteoritic infall; the others are all characteristically enriched in residual liquids (dispersed or incompatible elements) and hence their increased abundance must be due to admixture of more differentiated rock types, or rocks derived by smaller degrees of partial melting.

SCHNETZLER *et al.* (1970) and HUBBARD *et al.* (1970) have suggested that the dark (mafic) portions of the unique rock 12013 represent the "cryptic" or "magic" component of Apollo 12 fines. More recently HUBBARD *et al.* (1971a, b) have drawn attention to the presence of high K, REE, P glass (KREEP) in the Apollo 12 fines and demonstrated that these fines are quantitatively matched by mixtures of average Apollo 12 basalt and KREEP. K/Rb and K/Ba plots shown in Figs. 1 and 2 confirm that either the mafic portions of rock 12013 or KREEP may represent the missing cryptic component in the Apollo 12 soils, if they are derived from the Apollo 12 basalts. The mixing lines in these diagrams are drawn through the average of our three rock analyses and the average of our two analyses of fines. The K/Ba plot also shows that it is possible to derive the Apollo 11 fines from a mixture of either the Apollo 11 low K rocks on the one hand and the mafic portions of rock 12013 or KREEP on the other. However, the K/Rb plot demonstrates that rock 12013 or KREEP cannot be the precise missing component added to either of the Apollo 11 basalt types to produce the Apollo 11 fines; these can be derived from the Apollo 12 rocks. If the mafic portions of rock 12013 or KREEP are wide-spread components of the lunar surface (fines), the implication is that neither of the Apollo 11 rock types are major components (or are present) in the Apollo 11 soil. It is pertinent to recall that the Apollo 11 rocks are younger than the model age given for Apollo 11 fines (COMPSTON *et al.*, 1970) and it has yet to be established that the basal fragments in the Apollo 11 fines (WOOD *et al.*, 1970) are derived from either of the two Apollo 11 basalt types which have been analyzed.

The abundances of Zr and Nb are substantially higher (by a factor of about five) in the Apollo 12 fines when compared to the Apollo 12 rocks (Table 2). As no data are available for individual fragments of rock 12013 or KREEP glass it is of interest to calculate estimated abundances in these materials. This could be done singly for either element using the same approach and mixing proportions given for various Apollo 12 fines by HUBBARD *et al.*, (1971a, b), but this requires the mixing proportions to be correct and necessitates the use of an average composition for the rocks (Zr varies by about 30% in the three samples analyzed). A more sophisticated approach is possible because the K/Zr and Zr/Nb ratios given in Table 3 are relatively constant for both rocks and fines and it follows that the ratios in the missing components must be similar also, hence problems due to assumptions of average rock compositions and mixing proportions are avoided. The KREEP 1 glass used by HUBBARD *et al.* (1971a, b) in their mixing calculations contains 7030 ppm K. Using the ratios given in Table 3 for Apollo 12 fines (K/Zr = 3.9, Zr/Nb = 12.5), KREEP 1 glass should contain 1800 ppm Zr and 144 ppm Nb. A similar approach may be used for other KREEP glasses or rock 12013 fractions. It may be noted that a single sample of rock 12013 analyzed by LSPET (1970) yielded concentrations of 2200 ppm Zr and 170 ppm Nb (Zr/Nb = 12.9).

We emphasize that Zr and Nb concentrations of this order are, together with the rare-earth contents (HUBBARD *et al.*, 1971b), substantially higher than those observed in terrestrial basaltic rocks. A wide variety of oceanic island basalts analyzed in this laboratory (KABLE *et al.*, 1971) generally contain less than 400 ppm Zr and 80 ppm Nb (Zr/Nb 4–10). It is difficult to envisage the derivation of a basaltic liquid (HUBBARD *et al.*, 1971a give 48% SiO_2 for average KREEP glass) with the high contents of Zr and Nb estimated above for KREEP from a chondritic type parent ($\sim$10 ppm Zr, $\approx$1 ppm Nb, ERLANK and WILLIS, 1964). Even if all Zr and Nb partition into the melt during melting, it is necessary to assume about 0.5% partial melting which we consider to be unrealistic. The alternative is that these two refractory elements are enriched in lunar basalt source regions and this is in agreement with the conclusion reached by GAST and HUBBARD (1970) for other refractory elements.

The enrichment of Be in the Apollo 12 lunar fines (Table 2) is interesting as this has not been previously reported. No suggestion for this may be found in the Apollo 11 data presented by ANNELL and HELZ (1970) and MORRISON *et al.* (1970); however, their Be results show a fairly large scatter. We assume that the excess Be in the Apollo 12 fines is contributed by the KREEP component and using a K–Be plot we estimate that KREEP 1 glass will contain of the order of 10 ppm Be. This again attests to the differentiated nature of this component. It is also noteworthy that the Be concentration in eucrites Pasamonte and Sioux County is of the order of 0.3 ppm (SILL and WILLIS, 1962).

The features discussed above indicate that a small amount of meteoritic component and a "differentiated" component (related to rock 12013 and KREEP) are required to partly account for the composition of lunar fines. Mention has also been made of an anorthitic component as the Apollo 11 and Apollo 12 fines contain more Al than any rocks from these missions (Fig. 3) and as the Apollo 11 and Apollo 12 fines have also been shown to contain anorthositic fragments. The special significance of this observation is discussed under Ca–Al below.

The Ca–Al relationship in stony meteorites and lunar materials

The close relationship between Ca and Al in chondrites and basaltic achondrites is now well established (AHRENS and VON MICHAELIS, 1969; AHRENS, 1970a; AHRENS, 1971). The average Ca/Al ratio is 1.08 (mesosiderites have a characteristically lower ratio, average = 0.86). It should be noted (1) that most minerals which have Ca and Al as major constituents have ratios far removed from these values (see Fig. 3 for anorthite and clinopyroxene ratios); (2) that despite a large variation in the overall concentration of the principal host minerals for Ca and Al, clinopyroxene and plagioclase (anorthitic in basaltic achondrites and sodic maskelynite glass in chondrites) the ratio remains virtually the same; and (3) even if the mineralogy is unusual, e.g., the presence of many Ca and Al phases in Allende (C III chondrite) including gehlenite and Ca–Al rich glass (see Fig. 3), the ratio for the meteorite as a whole remains unchanged.

We have experienced some difficulty in compiling available Ca and Al data on Apollo 11 and Apollo 12 lunar rocks and fines, as the data are often not in satisfactory agreement (data for sample 12070 presented at the Second Annual Lunar Science Conference show a spread in the Ca/Al ratio of 0.89–1.13). For this reason we have used only the data of AGRELL *et al.* (1970), MAXWELL *et al.* (1970), MAXWELL and WIIK (1971), COMPSTON *et al.* (1970, 1971), MASON *et al.* (1971), SCOON (1971), and our own data. This choice in no way implies that all other data are in error (inaccurate or imprecise), but we consider from our experience that the data from these

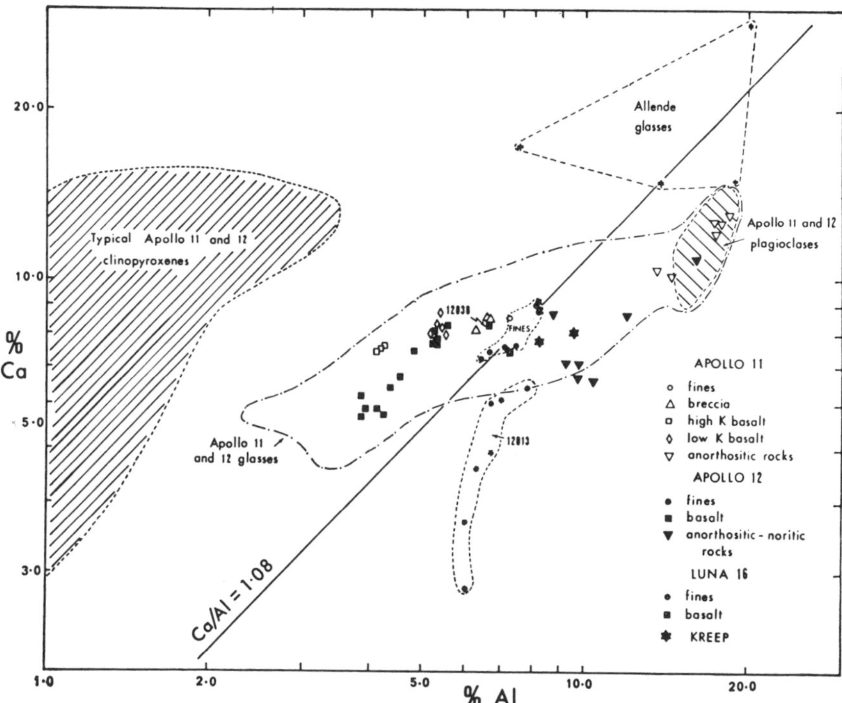

Fig. 3. The Ca–Al relationship in lunar materials. Note the clustering of the lunar fines about the stony meteorite trend line. Sources of data: Dence *et al.* (1971), Keil *et al.* (1971), Greene *et al.* (1971), Agrell *et al.* (1970), Frondel *et al.* (1970), Ware and Lovering (1970), Douglas *et al.* (1970), Keil *et al.* (1970), Duke *et al.* (1970), Chao *et al.* (1970), Adler *et al.* (1970), Brown *et al.* (1970), Hargraves *et al.* (1970), Drake *et al.* (1970), Carter *et al.* (1971), Compston *et al.* (1970), Maxwell *et al.* (1970), Compston *et al.* (1971), Maxwell and Wiik (1971), Scoon (1971), Mason *et al.* (1971), Wood *et al.* (1970), Bunch *et al.* (1970), Wakita and Schmitt (1970), Hubbard *et al.* (1971a), Wood *et al.* (1971), Vinogradov (1971), Marvin *et al.* (1970), and this paper.

laboratories is internally consistent and reliable. The only available Luna 16 data are from Vinogradov (1971). Rock 12013 data are from Wakita and Schmitt (1970), KREEP glass data from Hubbard *et al.* (1971a), and the anorthosite-norite data from Wood *et al.* (1971). Our requirements for mineral and impact glass data are not so stringent and are derived from several authors. Allende glass data may be found in Marvin *et al.* (1970). The assembled Ca–Al plot is shown in Fig. 3.

It is immediately apparent that, apart from a single sample of Luna 16 basalt, the only samples which cluster around the stony meteorite trend line are lunar fines. Ca/Al values are as follows:

Locality	*Ca/Al*
Apollo 11 (10084)	1.19
Apollo 12 (5 samples)	1.02–1.13
Luna 16 (4 samples)	1.08–1.12

Allowing for analytical error there is probably a real spread of at least 10% in the Ca/Al ratio showing that the lunar fines are not perfectly mixed (note also variation of Ca and Al absolute abundances in Fig. 3). Nevertheless, considering that these samples are from three separate localities and bearing in mind the overall consistency of the Ca/Al ratio, the similarity of the Ca/Al ratio in lunar fines and stony meteorites can hardly be fortuitous. In considering the significance of this feature, the following comments may be briefly considered.

(1) The lunar fines are mixtures of Apollo 11 and Apollo 12 basalts, anorthositic rocks (WOOD *et al.*, 1970, 1971; and KEIL *et al.*, 1970), KREEP (HUBBARD *et al.*, 1971a, b) and possibly material similar to the mafic portion of rock 12013 (note that KREEP has higher Ca and Al abundances than any of the fractions in rock 12013). The exact proportions of each are difficult to evaluate, but any reasonable combination implies a rather uniform mixing process.

(2) The lunar fines are derived from rocks with lower Ca/Al ratios, similar to the Luna 16 basalt, with smaller amounts of anorthosite or KREEP admixture. Such rocks have not been reported from the Apollo 11 or Apollo 12 sites (rock 12038 has the lowest Ca/Al observed), but many glasses have ratios close to the meteoritic value. Glasses comprise 50% of the <1 mm fraction of the Apollo 11 fines (WOOD *et al.*, 1970), and it appears that the most abundant glass has a Ca/Al ratio of 1.16 (FRONDEL *et al.*, 1970). Further discussion on the mixing history of the lunar fines will not be attempted here, but we interpret the above observations to indicate that the lunar fines may, at least for the major elements, represent fairly well mixed averages of the lunar surface, and, by implication, may reveal the Ca/Al *ratio* of the lunar interior. If so, this places restrictions on the composition of the source regions of lunar basalts. RINGWOOD (1970) considers that these regions are comprised essentially of clinopyroxene and orthopyroxene, and gives analyses for synthetic clinopyroxene and orthopyroxene liquidus phases which have ratios of 2.5 and 1.3, respectively. Any mixture of these two minerals therefore cannot generate the required ratio. Hence the lunar interior must either be composed of pyroxenes which are more aluminous than inferred by Ringwood, or else another aluminous phase, such as plagioclase or spinel, must be present (the presence of plagioclase in the lunar interior is not supported by experimental evidence, see RINGWOOD, 1970).

The Ca/Al ratio in the earth's mantle, as estimated from kimberlitic garnet peridotites (CARSWELL and DAWSON, 1970), and the sun (ALLER, 1968) appear to be similar (Ca/Al ~ 1.0) and thus close to the stony meteorite average. One form of Allende glass has a ratio of 1.08 (Fig. 3) and we suggest that this ratio is characteristic of the solar system.

In order to account for the uniformity of the Ca/Al ratio and the trends of other refractory elements, it may be instructive to bear in mind that a variety of species exist, AlO, CaO, TiO, ZrO, and others, all of them spectrum emitters, at the high temperatures prevailing in hot nebulae (the vaporization experiments on lunar samples by DE MARIA *et al.* (1971) appear to be pertinent in this respect). If a given nebular mass is quenched or isolated in some way, the relative proportions of these elements (which may or may not be similar to so-called cosmic proportions) will presumably be independent of the mineralogy of the final condensed object (moon, meteorite etc.), provided that all processes acting on this object subsequent to condensation are of an isochemical nature. Rather, the mineralogy in the first place is dependent on the relative proportions of elements in the condensate and subsequently on the post-condensation history. The pair Ca–Al provide a classic example in this respect. Thus the Ca/Al ratio of Allende is in fact determined by the relative proportions of Ca and Al in the condensate rather than by its present complex mineralogy. We consider that greater attention should be given to preaccretion chemistry in order to understand more fully the problems of lunar and solar system petrology and geochemistry.

The comments above on the Ca/Al ratio in condensing nebulae apply also to Sr and Eu, a pair of elements which we will now consider.

The Relationship between Closely Coherent Pairs of Refractory Elements

Although the chemistry of the refractory group of elements (Al, Ca, Sc, Ti, Sr, Y, Zr, Nb, Ba, rare earths, Hf, Ta, Th, and U) varies considerably when taken as a whole, some pairs or larger groups are chemically and geochemically very similar, e.g., Zr and Hf, trivalent rare earths, Nb and Ta, Th and U, and Sr and Eu^{II}. In our opinion, a critical and highly quantitative comparison of the relationships between such pairs or larger groups of these geochemically coherent refractory elements in stony meteorites and lunar materials would be highly significant for cosmochemistry. Much has been written on the rare earths, but data on the other geochemically coherent groups is scant or unsatisfactory, except for Sr and Eu^{II}, and our discussion here will be restricted to these two elements.

In order to reduce analytical uncertainty to a minimum, and as it is desirable to have both elements determined in the same meteorite or lunar specimen, the data of one laboratory only will be considered (Philpotts and Schnetzler, 1970; Schnetzler and Philpotts, 1968; and Philpotts and Schnetzler, 1971). The Sr–Eu relationship (Fig. 4) is remarkably well developed, particularly with respect to the stony meteorites. (We have not plotted data for Serra de Mage and Moore County, as these eucrites are generally considered unusual). The slope of the fitted line is less than unity (rate of increase of Eu is slightly greater than that of Sr) and if substantiated by further data this relationship may be of particular significance with respect to small differences in bond type (Sr $\rightarrow$ O and Eu $\rightarrow$ O) in various compounds, whether we are emphasizing processes in cooling nebulae or in melts (partial melting or differentiation). The importance of *small* differences in bond type in geochemical processes has been stressed by Ahrens (1964).

The significance of the presence of high rare earth content glass (KREEP) in Apollo 12 soil has been discussed by Hubbard *et al.* (1971b). Highly differentiated or unusual material such as KREEP and rock 12013 will, however, not be considered here.

Cr Abundances and the Mg–Cr Relationship

The characteristically high Cr concentration in *all* stony meteorites and lunar rocks and fines relative to terrestrial rocks of approximately the same bulk chemistry has been stressed before (Ahrens and Danchin, 1970), and Ringwood (1970) has

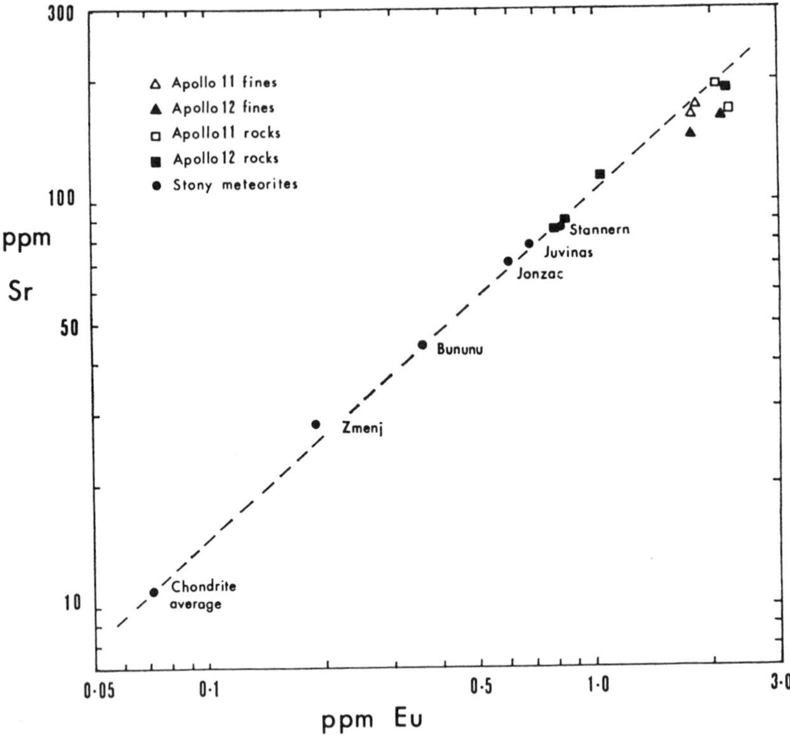

Fig. 4. The relationship between Sr and Eu in some stony meteorites and Apollo 11 and Apollo 12 rocks and fines. The regularity, particularly with respect to the meteorites, is well developed.

emphasized the importance of this element in lunar petrology. Here we wish, in the first place, to examine the Cr abundance relationship more fully and more quantitatively than before. On the basis of several hundred Cr determinations in terrestrial basalts (MANSON, 1967) a histogram (not shown) was constructed using a linear scale. This histogram displayed extreme positive skewness and is similar in appearance to that for Cr in Ontario diabases (Fig. 20 of AHRENS, 1954). Although the spread of the Cr concentration is very large, a striking characteristic of this element in most types of terrestrial rock, it is quite clear that the terrestrial levels are much lower than those typical of chondrites, basaltic achondrites, and lunar materials. It is noteworthy that the Cr level in rock 12013 is remarkably high ($\sim$1000 ppm or more; WAKITA and SCHMITT, 1970) for this type of material. In fact, Cr may be one of the most effective index elements for distinguishing terrestrial from much of extraterrestrial matter in the solar system.

A sympathetic relationship between Mg and Cr in howardites, eucrites, and lunar rock 12038 has been noted (Fig. 1 of AHRENS 1970). The problem of a thorough quantitative study of such a relationship is complicated, mainly, it seems, because of considerable uncertainty in the analytical data. For example, Cr levels in fines 12070 range from $\sim$2000 ppm to $\sim$3000 ppm. Mg is far more uniform. As a result of this

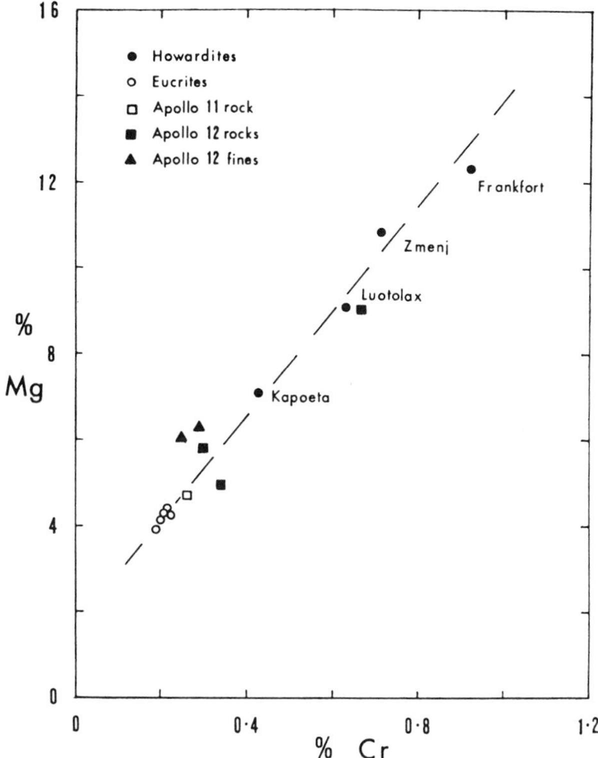

Fig. 5. The Mg–Cr relationship in basaltic achondrites and some lunar materials.

problem we have adopted a rather arbitrary procedure and for Mg and Cr in basaltic achondrites have used the data of AHRENS (1970b) and for lunar materials used only our own data. A graphical representation of the Mg–Cr relationship is shown in Fig. 5. In a general way, the lunar data fit the basaltic achondrite trend line. Lunar sample 12002 (high olivine) is of considerable interest as both Mg and Cr are very high. High Mg and Cr concentrations have also been observed in other samples.

Acknowledgments—We wish to acknowledge the patience and cooperation of Mrs. E. H. Tydeman and Miss D. C. Zoutendyk in the preparation of this manuscript. We are grateful to both the C.S.I.R., Pretoria and the U.C.T. Staff Research Fund for their encouragement and generous financial support.

REFERENCES

ADLER I., WALTER L. S., LOWMAN P. D., GLASS B. P., FRENCH B. M., PHILPOTTS J. A., HEINRICH K. J. F., and GOLDSTEIN J. I. (1970) Electron microprobe analysis of lunar samples. *Science* **167**, 590–592.

AGRELL S. O., SCOON J. H., MUIR I. D., LONG J. V. P., MCCONNELL J. D. C., and PECKETT A. (1970) Mineralogy and petrology of some lunar samples. *Science* **167**, 583–586.

AHRENS L. H. (1954) The lognormal distribution of the elements. *Geochim. Cosmochim. Acta* **5**, 49–73.

AHRENS L. H. (1964) The significance of the chemical bond for controlling the geochemical distribution of the elements. Part 1. In *Physics and Chemistry of the Earth* (editors L. H. Ahrens, F. Press, and S. K. Runcorn) Vol. 5, Chap. 1, pp. 1–54, Pergamon.

AHRENS L. H. (1970a) The composition of stony meteorites (IX) Abundance trends of the refractory elements in chondrites, basaltic achondrites and Apollo 11 fines. *Earth Planet. Sci. Lett.* **10**, 1–6.

AHRENS L. H. (1970b) The composition of stony meteorites (VI): (a) The Mg–Cr relationship in some basaltic achondrites, Apollo 11 soil and Apollo 12 rock 12038. (b) The Ba-rare earth relationship in stony meteorites and Apollo 11 rocks and soil. *Earth Planet. Sci. Lett.* **9**, 336–340.

AHRENS L. H. (1971) The Ca–Al relationship in stony meteorites and some lunar materials. In *Vinogradov 75th Birthday Anniversary Volume* (in press).

AHRENS L. H. and DANCHIN R. V. (1970) Chemical composition of the lunar surface. *Science* **167**, 87.

AHRENS L. H. and VON MICHAELIS H. (1969) The composition of stony meteorites (III). Some interelement relationships. *Earth Planet. Sci. Lett.* **5**, 395–400.

ALLER L. H. (1968) The chemical composition of the sun and the solar system. *Proc. ASA* **1**, 133–135.

ANNELL C. S. and HELZ A. W. (1970) Emission spectrographic determination of trace elements in lunar samples from Apollo 11. *Proc. Apollo 11 Lunar Sci. Conf., Geochim. Cosmochim. Acta* Suppl. 1, Vol. 2, pp. 991–994. Pergamon.

BROWN G. M., EMELEUS C. H., HOLLAND J. G., and PHILLIPS R. (1970) Petrographic, mineralogic and X-ray fluorescence analysis of lunar igneous-type rocks and spherules. *Science* **167**, 599–601.

BUNCH T. E., KEIL K., and PRINZ M. (1970) Electron microprobe analyses of pyroxenes, plagioclases and ilmenites from Apollo 11 lunar samples. UNM Institute of Meteoritics, Special Publication No. 1.

CARSWELL D. A. and DAWSON J. B. (1970) Garnet periodititе xenoliths in South African Kimberlite pipes and their petrogenesis. *Contrib. Mineral. Petrol.* **25**, 163–184.

CARTER N. L., FERNANDEZ L. A., AVE'LALLEMANT H. G., and LEUNG I. S. (1971) Pyroxenes and olivines in crystalline rocks from Ocean of Storms. Second Lunar Science Conference (unpublished proceedings).

CHAO E. C. T., JAMES O. B., MINKIN J. A., BOREMAN J. A., JACKSON E. D., and RALEIGH C. B. (1970) Petrology of unshocked crystalline rocks and shock effects in lunar rocks and minerals. *Science* **167**, 644–647.

CHERRY R. D., HOBBS J. B. M., ERLANK A. J., and WILLIS J. P. (1970) Thorium, uranium, potassium, lead, strontium and rubidium in silicate rocks by gamma-spectroscopy and/or X-ray fluorescence. *Canadian Spectr.* **15**, 1–8.

COMPSTON W., BERRY H., VERNON M. J., CHAPPELL B. W., and KAYE M. J. (1971) Rubidium-strontium chronology and chemistry of lunar material from the Ocean of Storms. Second Lunar Science Conference (unpublished proceedings).

COMPSTON W., CHAPPELL B. W., ARRIENS P. A., and VERNON M. J. (1970) The chemistry and age of Apollo 11 lunar material. *Proc. Apollo 11 Lunar Sci. Conf., Geochim. Cosmochim. Acta* Suppl. 1, Vol. 2, pp. 1007–1027. Pergamon.

DE MARIA G., BALDUCCI G., GUIDO M., and PIACENTE V. (1971) Mass spectrometric investigation of the vaporization process of lunar samples. Second Lunar Science Conference (unpublished proceedings).

DENCE M. R., DOUGLAS J. A. V., PLANT A. G., and TRAILL R. J. (1971) Mineralogy and petrology of some Apollo 12 samples. Second Lunar Science Conference (unpublished proceedings).

DOUGLAS J. A. V., DENCE M. R., PLANT A. G., and TRAILL R. J. (1970) Mineralogy and deformation in some lunar samples. *Science* **167**, 594–597.

DRAKE M. J., MCCALLUM I. S., MCKAY G. A., and WEILL D. F. (1970) Mineralogy and petrology of Apollo 12 sample No. 12013: A progress report. *Earth. Planet. Sci. Lett.* **9**, 103–123.

DUKE M. B., WOO C. C., BIRD M. L., SELLERS G. A., and FINKELMAN R. B. (1970) Lunar Soil: Size distribution and mineralogical constituents. *Science* **167**, 648–650.

ERLANK A. J. and WILLIS J. P. (1964) The zirconium content of chondrites and the zirconium-hafnium dilemma. *Geochim. Cosmochim. Acta* **28**, 1715–1728.

FLANAGAN F. J. (1969) U.S. Geological Survey Standards, II. First compilation of data for the new U.S.G.S. rocks. *Geochim. Cosmochim. Acta* **33**, 81–120.

FRONDEL C., KLEIN C., ITO J., and DRAKE J. C. (1970) Mineralogy and composition of lunar fines and selected rocks. *Science* **167**, 681–683.

FRONDEL C., KLEIN C., ITO J., and DRAKE J. C. (1970) Mineralogical and chemical studies of Apollo 11 lunar fines and selected rocks. *Proc. Apollo 11 Lunar Sci. Conf., Geochim. Cosmochim. Acta* Suppl. 1, Vol. 1, pp. 445–475. Pergamon.

GAST P. W. and HUBBARD N. J. (1970) Rare-earth abundances in soil and rocks from the Ocean of Storms. *Earth Planet. Sci. Lett.* **10**, 94–101.

GAST P. W., HUBBARD N. J., and WIESMANN E. H. (1970) Chemical composition and petrogenesis of basalts from Tranquillity Base. *Proc. Apollo 11 Lunar Sci. Conf., Geochim. Cosmochim. Acta* Suppl. 1, Vol. 2, pp. 1143–1163. Pergamon.

GOPALAN K., KAUSHALL S., LEE-HU C., and WETHERILL G. W. (1970) Rb–Sr and U, Th–Pb ages of lunar materials. *Proc. Apollo 11 Lunar Sci. Conf., Geochim. Cosmochim. Acta* Suppl. 1, Vol. 2, pp.1195–1205. Pergamon.

GREENE C. H., PYE D. L., STEVENS H. J., RASE D. E., and KAY H. F. (1971) Compositions, homogeneity, densities and thermal history of lunar glass particles. Second Lunar Science Conference (unpublished proceedings).

HARGRAVES R. B., HOLLISTER L. S., and OTALORA G. (1970) Compositional zoning and its significance in pyroxenes from three coarse-grained lunar samples. *Science* **167**, 631–633.

HUBBARD N. J., GAST P. W., and MEYER C. (1971b). The origin of the lunar soil based on REE, K, Rb, Ba, Sr, P, and $Sr^{87/86}$ data. Second Lunar Science Conference (unpublished proceedings).

HUBBARD N. J., GAST P. W., and WIESMANN H. (1970). Rare-earth, alkaline and alkali metal, and $Sr^{87/86}$ data for subsamples of lunar sample 12013. *Earth Planet. Sci. Lett.* **9**, 181–184.

HUBBARD N. J., GAST P. W., and WIESMANN H. (1971a) The composition and derivation of Apollo 12 soils. *Earth Planet. Sci. Lett.* (in press).

KABLE E. J. D., ERLANK A. J., and CHERRY R. D. (1971) Geochemical features of lavas from Marion and Prince Edward Islands. In *Scientific Results of the South African Marion and Prince Edward Islands Scientific Expedition 1965/66* (editor E. M. van Zinderen Bakker), Balkema.

KEIL K., BUNCH T. E., and PRINZ M. (1970) Mineralogy and composition of Apollo 11 lunar samples. *Proc. Apollo 11 Lunar Sci. Conf., Geochim. Cosmochim. Acta* Suppl. 1, Vol. 1, pp. 561–598. Pergamon.

KEIL K., PRINZ M., and BUNCH T. E. (1970) Mineral chemistry of lunar samples. *Science* **167**, 597–599.

KEIL K., PRINZ M., and BUNCH T. E. (1971) Mineralogical and petrological aspects of Apollo 12 rocks. Second Lunar Science Conference (unpublished proceedings).

LSPET (Lunar Sample Preliminary Examination Team) Preliminary examination of the lunar samples from Apollo 12. *Science* **167**, 1325–1339.

MANSON V. (1967) Geochemistry of basaltic rocks: Major elements. In *Basalts. The Poldevaart treatise on rocks of basaltic composition* (editors H. H. Hess and A. Poldevaart) Vol. 1, pp. 215–270, Interscience.

MARVIN U. B., WOOD J. A., and DICKEY J. S. (1970) Ca–Al rich phases in the Allende meteorite. *Earth Planet. Sci. Lett.* **7**, 346–350.

MASON B., MELSON W. G., HENDERSON E. P., JAROSEWICH E., and NELEN J. (1971) Mineralogy and petrography of some Apollo 12 samples. Second Lunar Science Conference (unpublished proceedings).

MAXWELL J. A., PECK L. C., and WIIK H. B. (1970) Chemical composition of Apollo 11 lunar samples 10017, 10020, 10072, and 10084. *Proc. Apollo Lunar Sci. Conf., Geochim. Cosmochim. Acta* Suppl. 1, Vol. 2, pp. 1369–1374. Pergamon.

MAXWELL J. A. and WIIK H. B. (1971) Chemical composition of Apollo 12 lunar samples 12004, 12033, 12051, 12052, and 12065. Second Lunar Science Conference (unpublished proceedings).

MORRISON G. H., GERARD J. T., KASHUBA T., GANGADHARAM E. V., ROTHENBERG A. M., POTTER N. M., and MILLER G. B. (1970) Elemental abundances of lunar soil and rocks. *Proc. Apollo 11 Lunar Sci. Conf., Geochim. Cosmochim. Acta* Suppl. 1, Vol. 2, pp. 1383–1392. Pergamon.

Murthy V. R., Evensen N. M., and Coscio M. R. (1970) Distribution of K, Rb, Sr and Ba, and Rb–Sr isotopic relations in Apollo 11 lunar samples. *Proc. Apollo 11 Lunar Sci. Conf., Geochim. Cosmochim. Acta* Suppl. 1, Vol. 2, pp. 1393–1406. Pergamon.

Norrish K. and Hutton J. T. (1969) An accurate X-ray spectrographic method for the analysis of a wide range of geological samples. *Geochim. Cosmochim. Acta* **33**, 431–453.

O'Hara M. J., Biggar G. M., Richardson S. W., Ford C. E., and Jamieson B. G. (1970) The nature of seas, mascons, and the lunar interior in the light of experimental studies. *Proc. Apollo 11 Lunar Sci. Conf., Geochim. Cosmochim. Acta* Suppl. 1, Vol. 1, pp. 695–710. Pergamon.

Philpotts J. A. and Schnetzler C. C. (1970) Apollo 11 lunar samples: K, Rb, Sr, Ba, and rare-earth concentrations in some rocks and separated phases. *Proc. Apollo 11 Lunar Sci. Conf., Geochim. Cosmochim. Acta* Suppl. 1, Vol. 2, pp. 1471–1486. Pergamon.

Philpotts J. A., Schnetzler C. C., Bottino M. L., and Fullager P. D. (1971) Li, K, Rb, Ba, and rare-earth concentrations and $^{87}Sr/^{86}Sr$ in some Apollo 12 soils, rocks, and separated phases. Second Lunar Science Conference (unpublished proceedings).

Reynolds R. C. (1963) Matrix corrections in trace element analysis by X-ray fluorescence. Estimation of the mass absorption coefficient by Compton scattering. *Amer. Mineral.* **48**, 1133–1143.

Ringwood A. E. (1970) Petrogenesis of Apollo 11 basalts and implications for lunar origin. *J. Geophys. Res.* **75**, 6453–6479.

Schnetzler C. C. and Philpotts J. A. (1968) Genesis of the calcium-rich achondrites in light of rare-earth and barium concentrations. *Proc. Vienna Symposium on Meteorite Research* (editor P. W. Millman), D. Reidel.

Schnetzler C. C., Philpotts J. A., and Bottino M. L. (1970) Li, K, Rb, Sr, Ba, and rare-earth concentrations, and Rb–Sr age of lunar rock 12013. *Earth Planet. Sci. Lett.* **9**, 185–192.

Scoon J. H. (1971) Quantitative chemical analyses of lunar samples 12040,36 and 12064,33. Second Lunar Science Conference (unpublished proceedings).

Sill C. W. and Willis C. P. (1962) The beryllium content of some meteorites. *Geochim. Cosmochim. Acta* **26**, 1209.

Smales A. A., Mapper D., Webb M. S. W., Webster R. K., and Wilson J. D. (1970) Elemental composition of lunar surface material. *Proc. Apollo 11 Lunar Sci. Conf., Geochim. Cosmochim. Acta* Suppl. 1, Vol. 2, pp. 1575–1581. Pergamon.

Tera F., Eugster O., Burnett D. S., and Wasserburg G. J. (1970) Comparative study of Li, Na, K, Rb, Cs, Ca, Sr, and Ba abundances in achondrites and in Apollo 11 lunar samples. *Proc. Apollo 11 Lunar Sci. Conf., Geochim. Cosmochim. Acta* Suppl. 1, Vol. 2, pp. 1637–1657. Pergamon.

Vinogradov A. P. (1971) Preliminary data on lunar ground brought to earth by automatic probe "Luna 16". Second Lunar Science Conference (unpublished proceedings).

Wakita H. and Schmitt R. A. (1970) Elemental abundances in seven fragments from lunar rock 12013. *Earth Planet. Sci. Lett.* **9**, 169–176.

Ware N. G. and Lovering J. F. (1970) Electron-microprobe analyses of phases in lunar samples. *Science* **167**, 517–520.

Wiik H. B. (1969) On regular discontinuities in the composition of meteorites. *Comm. Phys.-Math.* **34**, 135–145.

Willis J. P., Fesq H. W., Kable E. J. D., and Berg G. W. (1969) The determination of barium in rocks by X-ray fluorescence spectrometry. *Canadian Spectr.* **14**, 3–11.

Wood J. A., Marvin U. B., Powell B. N., and Dickey J. S. (1970) Mineralogy and petrology of the Apollo 11 lunar sample. Smithsonian Astrophys. Observatory Special Report 307.

Wood J. A., Marvin U. B., Reid J. B., Taylor G. J., Bower J. F., Powell B. N., and Dickey J. S. (1971) Relative proportion of rock types and nature of the light-coloured lithic fragments in Apollo 12 soil samples. Second Lunar Science Conference (unpublished proceedings).

Proceedings of the Second Lunar Science Conference, Volume 2, pp. 1139–1158.
The M.I.T. Press, 1971.

Meteoritic material in lunar samples: Characterization from trace elements

J. C. LAUL, JOHN W. MORGAN, R. GANAPATHY, and EDWARD ANDERS
Enrico Fermi Institute and Department of Chemistry,
University of Chicago, Chicago, Illinois 60637

(*Received 26 February 1971; accepted in revised form 31 March 1971*)

Abstract—Fifteen trace elements were measured by neutron activation analysis in 5 soils, 5 core samples, and 2 breccias from the Apollo 12 site and in 2 samples of anorthositic fragments from Apollo 11 soil. Contingency soil 12070, collected some 50 meters from Surveyor Crater, showed a primitive meteoritic component of about 1.9% C 1 equivalent, very similar to that of Apollo 11 soil for which a meteoritic influx rate of 3.8×10^{-9} g cm^{-2} yr^{-1} was estimated. Soils collected from crater rims and the 2 breccias contain less than 1% meteoritic material of decidedly fractionated type (ordinary chondrite or iron). Many of the trace elements in crater rim soils, including some of apparently meteoritic origin, are correlated with the KREEP content, and extrapolation to 100% KREEP indicates that the impact at the source of this material (Copernicus?) may have been due to a fractionated meteorite. Anorthosites also appear to contain a meteoritic component of fractionated composition. The glassy coating of rock 12017 is of impact origin, and has been contaminated by about 0.5% C 1 type material, or alternatively, by a mixture of 10 to 15% local soil, similar to 12070, and 0.06 to 0.3% fractionated meteoritic material. Four of the five core samples contain amounts of meteoritic material similar to 12070 soil, with a pattern transitional between the primitive and fractionated type. The coarse-grained 13-cm core layer contains no apparent meteoritic component, but shows extreme enrichment in Bi and Cd (38 and 22 ppm). Neighboring layers, 11 cm above and 6 cm below, have only 10^{-4} to 10^{-5} the Bi and Cd content of the 13-cm horizon, which suggests negligible mixing since the deposition of this horizon, 10^7 to 10^8 years ago. The predicted "mean turnover time" for this depth is only $\sim 10^6$ years, indicating that turnover times may not be meaningful on a small scale. Apparently the dominant process in the regolith is blanketing by successive impacts, not "gardening." A sequential leaching of Apollo 11 soil with HNO_3 dissolved meteoritic elements preferentially. Apparently they mainly reside in small grains or on surfaces. There may be an additional labile component of indigenous Zn and Cd.

INTRODUCTION

COSMOCHEMICAL PROCESSES strongly fractionate certain siderophile and volatile trace elements (LARIMER and ANDERS, 1967), giving them a disproportionate importance in meteoritic and planetary studies. In lunar investigations the abundances of these elements can: (1) place restrictions upon the conditions under which the Moon was formed; (2) indicate some of the chemical processes which have taken place during and after its accretion; and (3) give insight into the way in which the lunar surface has been modified by meteoritic impact.

The first two topics are discussed in a companion paper, covering our studies on crystalline rocks (ANDERS *et al.*, 1971). In the present communication we are concerned exclusively with breccias and soils. Whereas our previous study (GANAPATHY *et al.*, 1970a) dealt with only one Apollo 11 soil and two breccias from a single location, the present paper contains results for five soils, two breccias and five core tube samples, collected over a range of a few hundred meters. We also carried out an acid leaching experiment on Apollo 11 soil (SILVER, 1970) in order to learn more about the location of the meteoritic component.

EXPERIMENTAL

Samples

Breccias were coarsely crushed in an agate mortar before analysis; fines were analyzed as received except for two handpicked anorthosite fractions. The U.S. Geological Survey standard basalt BCR-1 was included in every irradiation as a control. Sample weights were typically ~0.1 g, except for scarce material where as little as 0.02 g was used.

Procedure

Details of the radiochemical neutron activation procedure are unpublished, but may be obtained from the sources given in our companion paper (ANDERS *et al.*, 1971).

Leaching experiment

After irradiation, 0.48 g of fines 10084,49 were sequentially leached in concentrated HNO_3 by the procedure of Silver (1970, Experiment 6). Each of the 3 leach fractions was digested with carriers for several hours in $HCl–HNO_3$ to ensure equilibration of active species, particularly Ir (KIMBERLIN *et al.*, 1968). The residue was fused with $NaOH–Na_2O_2$in the presence of carriers as in our normal procedure. No further residue was observed after this treatment. Crude estimates for Fe (via Fe^{59}), Ni (via Co^{58}) and Ca (via Ca^{47}) were made for all four fractions, and also for two anorthosite samples. These values are probably good to about 10%, except for Ca in the leach residue where ~30% would be more realistic.

RESULTS

Results for Apollo 12 and Apollo 11 material are given in Tables 1 and 2 respectively.

Analytical accuracy and precision

Comparison of our soil and breccia results with those of other authors leads generally to the same conclusions reached in our companion paper (ANDERS *et al.*, 1971) for crystalline rocks. However, results for Au and Zn in soils and breccias agree well, the poorer agreement in crystalline rocks probably being due to the lower abundances. The agreement of Cd values in soils is not satisfactory, on the other hand. Our results scatter less than others and may be more reliable.

To reduce contamination we have avoided conventional sampling techniques, and the precision of our measurements is of some concern. We have estimated our analytical precision from replicate determinations of elements in 5 aliquots of BCR-1 and this may be compared with the actual precision of replicate analyses of lunar samples. Nineteen replicates are available for such comparison: duplicates of two Apollo 11 and two Apollo 12 crystalline rocks (we exclude the contaminated 10047 powder), and 11 aliquots of one Apollo 11 and 3 Apollo 12 fines samples. For each sample, replicate analyses were normalized to the mean value, and pooled to derive separate standard deviations for rocks and soils.

The results of this (statistically dubious) procedure are shown graphically in Fig. 1, where the percent standard deviation (for a single determination), for each element in lunar rocks or fines, is plotted against the same parameter for BCR-1. Points lying above the diagonal indicate poorer precision in the lunar materials and the distance from the diagonal may be regarded as rule-of-thumb measure of the relative sampling error. The noble metals Ir and Au lie below the diagonal,

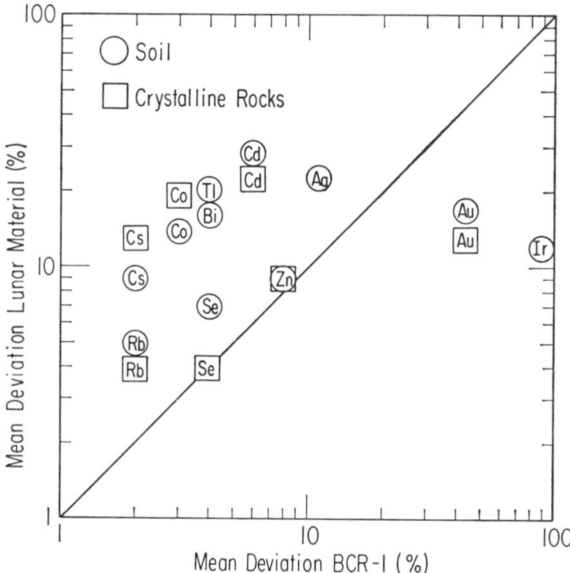

Fig. 1. Comparison of relative analytical precision for replicate analyses of U.S.G.S. standard basalt BCR-1 with that for lunar rocks and fines. The standard deviations for a single determination, calculated from pooled replicate analyses of 4 lunar crystalline rocks and 4 samples of lunar fines, are plotted against the same parameter for 5 replicate analyses of BCR-1.

Points lying above the diagonal indicate an additional sampling error for lunar material. Apparently Ir and Au are more uniformly distributed in lunar material than in terrestrial basalts.

indicating that for these elements the sampling problem is worse for BCR-1 than for the lunar rocks.

The sampling problem in lunar material appears to be particularly serious for Cd, and may indicate the presence of a very minor phase, rich in this element, which is sporadically distributed. Bismuth and Tl, too, have poorer precision in lunar material, partly a result of analytical difficulties in measuring the rather low levels present. For these two elements, counting statistics alone may account for errors of up to 10%, particularly in crystalline rocks. On the other hand, Co and Cs are rather easily determined at the levels found in lunar rocks and fines. Their scatter, therefore, indicates a real sampling problem, leading to additional errors of 10 to 20%.

Discussion

Calculation of the meteoritic component

In our previous work (Keays *et al.*, 1970; Ganapathy *et al.*, 1970a) we assumed that the soil consisted largely of material similar to average crystalline rocks, plus a small meteoritic component. It was clear, however, that certain elements, Rb, Cs, Zn, and particularly Cd were too abundant to be accounted for by this simple model, and it was necessary to postulate the presence of a third component rich in these elements.

The abundance of an element i in a soil sample is then

$$[i]_S = (1 - \alpha - \beta)[i]_{AB} + \alpha[i]_K + \beta[i]_M \qquad (1)$$

Table 1. Abundances in Apollo

Sample	Ir (ppb)	Au (ppb)	Zn (ppm)	Cd (ppb)	Ag (ppb)	Bi (ppb)
Fines						
12025,72 Core 1.7 cm[b]	5.9	2.5	6.1	70	28	3.5
12028,66 Core 13.2 cm[b,c]	0.08	0.63	1.5	22000	301	38500
12028,90 Core 18.9 cm[b,c]	8.1	1.7	5.1	53	*140*	2.1
12028,121 Core 21.2 cm[b,c]	8.7	2.1	5.4	48	3.6	1.2
12028,145 Core 37.2 cm[b,c]	9.2	2.0	4.3	49	7.2	1.4
12032,33	3.9	1.5	6.5	26	3.4	0.63
12032,33	4.6	1.4	5.0	16		
12032,33	3.5	1.3	4.7	10	3.0	0.41
Mean	4.0	1.4	5.4	17	3.2	0.52
12033,20	3.9	1.6	5.1	27	7.5	0.30
12033,20	4.4	1.4	4.9	22		
12033,20	3.4	1.4	5.1	29	17	0.40
Mean	3.9	1.5	5.0	26	12	0.35
12037,25	3.5	1.6	6.3		5.1	0.72
12037,25	5.0	1.5	5.9	35		
Mean	4.3	1.5	6.1	35	5.1	0.72
12057,78	5.0	5.7	6.9	34	*43*	1.6
12070,69[c]	8.5	2.4	6.9	45	*46*	2.4
Breccias						
12010,22	2.7	0.82	8.0	34	3.7	0.32
12073,37	4.5	2.0	6.5	19	2.7	0.46
Ave. 12013,10[e]	0.43	0.35	2.8	71	1.7	0.60
Ave. Crystalline Rocks[e]	0.10	0.03	0.8	2	1.0	0.33
Non-Meteoritic Component[f]	0.15	0.08	1.1	12	1.1	0.37

[a] Doubtful values are shown in italics. They include all of the higher In and Ag values where contamination from vacuum gaskets in sample containers and LRL was suspected.

[b] Depths given for core samples represent nominal distances from top of core tube (LSPET, 1970). These are smaller than the true depths, owing to compression during sampling (CARRIER *et al.*, 1970). For comparison, the estimated true depths (cm) are given in parentheses: 1.7–2.5 (2.0–2.8); 13.2–14.4 (17.8–19.4); 18.9–19.7 (25.8–26.9); 31.2–32.2 (44.5–46.2); 37.2–38.2 (55.6–57.2).

[c] GANAPATHY *et al.* (1970b).

[d] Tellurium was determined via 117d Te^{123m}. Some of the samples were recycled when Se^{75} contamination

Table 2. Abundances

Sample	Ir (ppb)	Au (ppb)	Zn (ppm)	Cd (ppb)	Ag (ppb)	Bi (ppb)	Tl (ppb)	Br (ppb)
Fines (<1 mm)								
Acid Leach[a]								
10084,49-4	1.7	0.76	6.8	31	2.4	0.90	0.29	
10084,49-5	1.7	0.77	3.1	6.6	1.8	0.22	0.27	
10084,49-6	0.8	0.52	3.1	4.9	1.8	0.20	0.30	
10084,49-7	3.9	0.72	6.2	6.8	3.0	0.24	0.74	
10084,49-S	8.0	2.8	19	49	9.0	1.6	1.6	
Bulk								
10084,49	7.0	2.7	18	51	8.6	1.7	1.7	90
10084,49[c]	7.6	4.2	21	53	8.9	1.6	1.7	87
10084,49[c]	6.9	2.0	21	35	8.6	1.4	1.5	77
Mean	7.2	3.0	20	46	8.7	1.6	1.6	84
Anorthosite								
10084,49,1-1[d]	3.4	1.1	2.0	17	1.7	0.59	0.28	47
10085,107[e]	3.6	1.5	1.4	9	0.49	0.42	0.61	42

[a] Fractions from conc. HNO_3 leach, after SILVER (1970) 4 = 15 min. at 25°C; 5 = 30 min at 70°C; 6 = 7 hr at 70°C; 7 = residue; S = sum of fractions 4–7.

[b] Large error in this value (see text).

[c] Reference: GANAPATHY *et al.* (1970a).

12 fines and breccias[a]

Tl (ppb)	Br (ppb)	Te[d] (ppb)	Se (ppb)	Ga (ppm)	Rb (ppm)	Cs (ppb)	In (ppb)	Co (ppm)
2.2	140	130	215	3.9	6.0	250	*77*	39
0.26	16	10	86	2.7	0.32	23	*42*	
0.51	141	80	230	5.2	8.6	350	*290*	
2.2	124	30	247	5.0	9.0	360	9	
2.3	116	90	237	5.2	10.8	340	*26*	
2.5	180	20	200	5.1	7.6	400	9	34
			190		7.5	370		43
1.3							11	49
1.9	180	20	195	5.1	7.6	390	10	42
2.1	130	40	160	4.7	8.0	380	*24*	26
			190		8.6	420		41
1.7							*58*	35
1.9	130	40	180	4.7	8.3	400		34
5.2	140	40	190	3.4	5.0	200	*27*	49
			220		5.1	280		60
5.2	140	40	210	3.4	5.1	240		55
1.7	100	60	190	3.5	5.1	220	*1280*	125
1.6	120	100	260	4.3	6.3	250	*220*	
4.3	130		190	4.1	8.8	260	7.0	38
2.8	130	100	190	4.5	9.2	390	6.5	35
12.8	58	25	94	6.1	42	1900	3.7	26
0.36	13	18	144	3.3	0.96	41	1.8	47
2.2	20	19	137	3.7	7.1	320	2.1	44

(up to 25% the total counting rate) was observed. Other samples were checked for selenium contamination by γ–γ coincidence counting, and corrected when necessary. However, owing to the very low activity of the samples, small amounts of Se^{75} may have escaped detection; thus some of the lower Te values may be systematically too high by as much as 20 to 40%.

[e] ANDERS *et al.* (1971).

[f] Typical value (85% crystalline rock, 15% 12013,10). Actual values calculated for individual samples vary slightly.

in Apollo 11 fines

Te (ppb)	Se (ppb)	Ga (ppm)	Rb (ppm)	Cs (ppb)	In (ppb)	Co (ppm)	Ni (ppm)	Ca (%)	Fe (%)
		0.35	0.08	3.4	*590*	5.9	45	0.3	0.5
		0.74	0.15	5.9	8.8	7.7	59	1.9	1.3
		1.3	0.25	8.2	0.64	4.5	44	2.4	1.3
	140	2.5	2.5	86	1.3	13.4	43	6.0[b]	9.4
		4.9	3.0	103	*600*	31	191	10.7[b]	12.5
	330	4.8	2.8	97	*3500*	29	150	8.6	12.3
		5.4	3.3	98	*520*	27			
		5.2	3.1	94	*770*	28			
	330	5.1	3.1	96		28	150	8.6	12.3
	90	4.6	1.8	74	2.7	14	90	11.8	6.5
	29	3.2	0.70	24	6.6	12	75	12.7	3.1

[d] 21.4 mg of light lithic fragments (~40), hand-picked by Dr. Julie Morgan from a >100 mesh fraction of 10084,49 soil.

[e] 5 anorthosite fragments totalling 41.1 mg from 1–10 mm fraction of fines; hand-picked at LRL.

where the subscripts S, AB, K, and M refer to the sample, crystalline rock, alkali-rich and meteoritic components and α and β are the fractional abundance of the alkali-rich and the meteoritic components. In practice, β is at most ~ 0.02 and is $\ll (1 - \alpha)$. For calculation of the meteoritic component we use only those elements which are enriched at least twofold over their indigenous abundances, i.e.,

$$[i]_S \geq 2\{(1 - \alpha - \beta)[i]_{AB} + \alpha[i]_K\} \tag{2}$$

If β is neglected in the term $(1 - \alpha - \beta)$, at worst an error of $\leq 1\%$ will be introduced into the estimate of β.

The meteoritic component is then

$$\beta \approx \frac{[i]_S - (1 - \alpha)[i]_{AB} - \alpha[i]_K}{[i]_M} \tag{3}$$

We evaluate α from the Rb and Cs abundances. The chondritic values for these elements are sufficiently close to those for crystalline rocks, and the meteoritic fraction is so small that we can put

$$(1 - \alpha - \beta)[i]_{AB} + \beta[i]_M \approx (1 - \alpha)[i]_{AB} \tag{4}$$

and

$$\alpha \approx \frac{[i]_S - [i]_{AB}}{[i]_K - [i]_{AB}} \tag{5}$$

The identity of the alkali-rich component presents a problem. Rare earth element (REE) studies of Apollo 12 soil (HUBBARD *et al.*, 1971) have shown the presence of a component high in K, REE, and P, acronymically terms KREEP. This material is similar in composition to the dark component in breccia 12013 (for example, 12013,10, 06, LAUL *et al.*, 1970). However, it appears to be a specific rock type and probably a widespread one. KREEP cannot be the only alkali-rich component in lunar soil. A treatment of trace element trends in soils containing known amounts of KREEP (MORGAN *et al.*, 1971b) indicates that even a KREEP-free soil may contain an excess of alkali metals and Cd over known crystalline rocks. Breccia 12013,10 the most silica-rich, granitic sample #41 is high in Rb and Cs, as well as Cd and Tl. A small admixture of such granitic material in the soil could account for much of the excess Cd.

Until more is known of the siderophile and volatile trace element distribution in KREEP, and until we have incorporated into our analysis an unequivocal indicator for the presence of this material, we have chosen to use a mass-weighted average of 12013,10 (LAUL *et al.*, 1970) as the alkali-rich component. The average Rb and Cs values of 12013,10 correspond to 67% KREEP (12013,10,06) and 33% silica-rich 12013,10,41. (We recognize that other elements, e.g., REE, may give significantly different proportions. The biggest correction, however, is to Cd; as this element correlates with the alkali metals, the calculation is comparatively insensitive to the actual KREEP fraction.) We have listed in Table 1 the mean abundances for Apollo 12 crystalline rocks (ANDERS *et al.*, 1971) and breccia 12013,10 (LAUL *et al.*, 1970) used in the calculation of the nonmeteoritic component. A typical composition for the nonmeteoritic component is also shown (85% Apollo 12 crystalline rocks and 15% 12013,10).

The calculated meteoritic component is normalized to type 1 carbonaceous chondrites (eq. 3), to provide a convenient frame of reference for chemical characterization of the meteoritic component. Fractionated meteorite types (iron meteorites, ordinary chondrites) deviate systematically from the uniform C 1 pattern. The comparative sensitivities of elements for the detection and identification of meteoritic types may be illustrated by plotting the increase in abundance on addition of 1 weight percent of meteoritic material to average Apollo 12 crystalline rock (plus 15% 12013,10; Fig. 2).

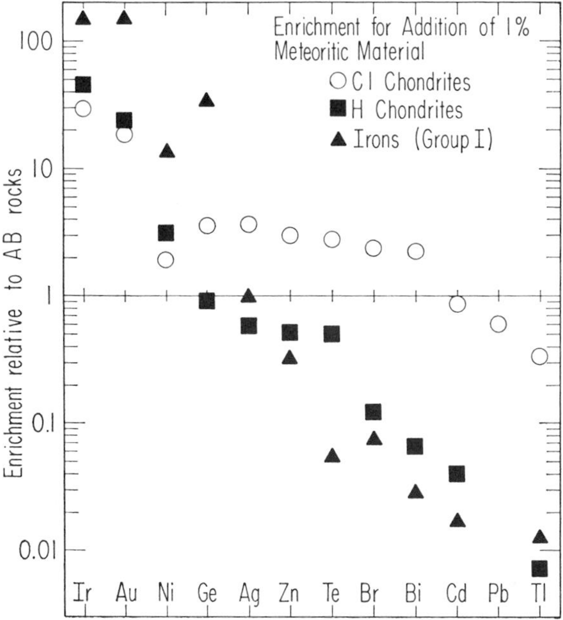

Fig. 2. Enrichment which would result from the addition of 1% meteoritic material to a typical non-meteoritic component (Table 1) calculated by

$$\frac{\beta[i]_M}{(1 - \alpha - \beta)[i]_{AB} + \alpha[i]_K}$$

where $\alpha = 0.15$ and $\beta = 0.01$. If a twofold enrichment is regarded as significant, then the number of diagnostic elements is 9 for C 1 chondrites, 3 for H-chondrites, and 4 for group I iron meteorites.

Assuming as in our earlier work that enrichments by a factor of 2 or more are significant, we see that 9 of these elements are diagnostic for C 1 chondrites, whereas only 3 or 4 (Au, Ir, Ni, and possibly Ge) are suitable indicators for fractionated meteorites. Not all of these indicators are equally reliable, however. Ag, Zn, Br, Cd, Pb, and Tl show rather large variations from rock to rock, and are sometimes overabundant in soils, presumably due to the presence of rock types rich in these elements. Tl and Zn, in particular, are not useful indicators of meteoritic material at the Apollo 12 site. The presence of alkali- and Tl-rich rock types makes the correction for indigenous Tl larger and less well determined than for Apollo 11 soil. Our most trustworthy indicators of meteoritic material therefore are Ir and Au whose abundances in lunar rocks are very low, and Bi whose abundance, though higher, is very uniform. Ni, Ge, and Te are also suitable, but we do not measure the first two and have only rather approximate data for the third.

Apollo 12 meteoritic component

Lunar soils and breccias are generally enriched in elements thought to be largely of meteoritic origin. We have estimated the meteoritic component from the data given in Table 1 and using eqs. 3 and 5.

The results for the contingency soil sample 12070 closely resemble those of Apollo 11 soil 10084 (Fig. 3). Both show a meteoritic component of ~1.9% with a decidedly primitive meteoritic pattern. Among the "less reliable" indicator elements, Zn, Cd, and Br are overabundant, while Tl is underabundant, presumably due to an overcorrection for indigenous Tl. Other Apollo 12 soils reveal somewhat different patterns, depending upon location and depth.

An apparently primitive meteoritic component, equivalent to ~0.5% C 1 material, has been found in the glassy coating of rock 12017 (MORGAN *et al.*, 1971a; see this reference or ANDERS *et al.*, 1971 for analytical results for 12017). However the trace

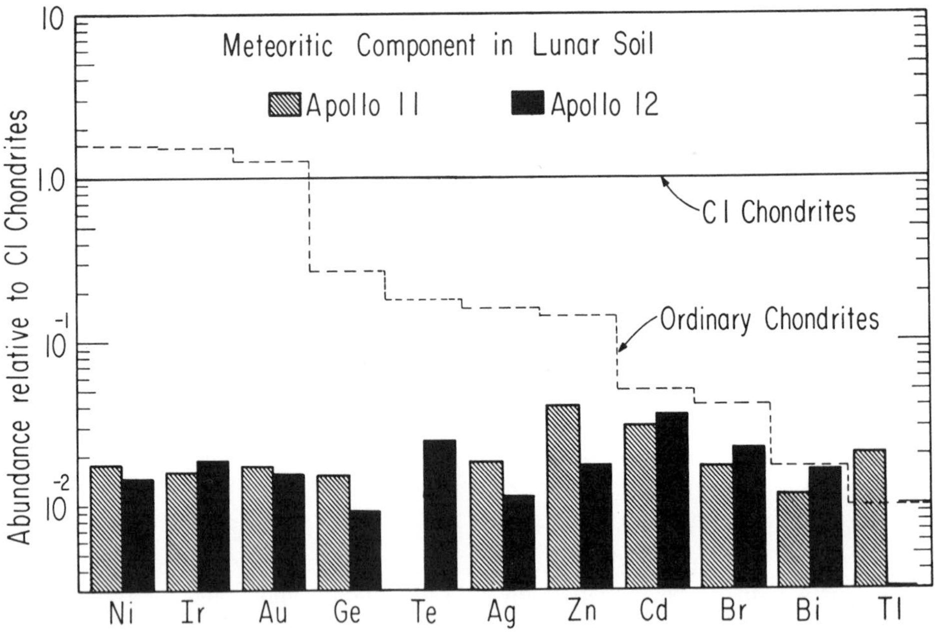

Fig. 3. Comparison of the meteoritic component pattern of Apollo 12 soil, 12070, collected from a relatively undisturbed area, with that of Apollo 11 fines. The relatively flat distribution in both cases indicates a meteoritic component consisting largely of carbonaceous-chondrite-like material, present to about 1.9%. Ordinary chondrite material would have given a staircase pattern parallel to the dashed curve. Apollo 11 values from GANAPATHY *et al.* (1970a); Apollo 12 values from Table 1, except for Ni (LSPET, 1970) and Ge (BAEDECKER *et al.*, 1971).

element composition also could be matched by a mixture of 10 to 15% of local soil, similar to 12070, and 12017 crystalline rock plus a small fractionated meteoritic component (0.3% ordinary chondrite or 0.06% group I iron).

Breccia 12010 possesses a fractionated meteoritic pattern, defined mainly by the Bi abundance which does not exceed that of average crystalline rock. The Ir and Au abundances are compatible with a meteoritic component of 0.5% ordinary chondrite or 0.1% group I iron meteorite. The preposterously high excesses of Zn, Cd, Tl, and Br illustrate the unreliability of these elements as indicators of meteoritic material. Apparently breccia 12010 contains a rock type rich in these elements.

The meteoritic components of the two anorthosites cannot be determined precisely, as we do not know the indigenous abundance corrections. These may not be large for many elements if the anorthosites are early-formed crystal cumulates. In particular, they are probably negligible for Ir and Au which occur in meteoritic ratio and are grossly overabundant compared to other lunar rocks. For other elements, we can obtain upper limits for the meteoritic component by making no correction for the indigenous content. The results are given in Table 3, together with those for breccia 12010 and 12017 glass.

Table 3. Apparent meteoritic components in four special samples.

Sample	Type	Percent Equivalent Cl Material[a]							
		Ir	Au	Zn	Cd	Ag	Bi	Tl	Br
12010,22[c]	C	0.59	0.49	(2.1)	(3.2)	0.69	~0	(5.3)	(2.4)
12017,20[b]	Glass	0.55	0.77	0.69	0.43	0.35	0.30	0.73	0.69
10084,49,1-1[c]	Anorth.	~0.77	~0.71	<0.61	<1.68	<0.42	<0.52	<0.38	<0.98
10085,107-1[c]	Anorth.	~0.79	~0.98	<0.42	<0.89	<0.13	<0.37	<0.82	<0.87

[a] Values in parentheses are too high to be meteoritic, and probably imply presence of a volatile-rich rock type in the breccia.
[b] MORGAN *et al.* (1971a).
[c] Not corrected for indigenous element.

Particularly interesting is the very low Ag value in 10085,107-1, which establishes firmly that these fragments exhibit a fractionated meteorite pattern. If we compare Ag/Ir ratios, 10085,107-1 yields a value of <0.16, compared to 0.16 for ordinary chondrites and 0.052 for group I iron meteorites. Since we do not know the indigenous Ag contribution, it is not possible to distinguish between these fractionated meteorite types. Abundances for the remaining indicator elements cluster between 0.3 and 0.9% C 1 equivalent, which at first glance suggests a primitive pattern. However, these are upper limits, uncorrected for indigenous abundances. If these abundances were similar to those in mare basalts, most of the apparent excesses would vanish.

The two anorthosites differ somewhat in their nonmeteoritic composition. The coarser sample 10085,107-1 is similar in Fe, Ca, and Rb to material of the same type analyzed by WAKITA and SCHMITT (1970). The finer 10084,49,1-1 sample is lower in Ca and higher in Fe and alkalis. Some of the grains of 10084,49,1-1 had brown glass fused to them, which may be the source of excess Fe. Differences in Cd, Ag, and Tl content between the two anorthosites may be a reflection of the major element chemistry.

Where precise corrections for indigenous abundances can be made, most informative of our suite of elements for the investigation of the meteoritic component are Ir,

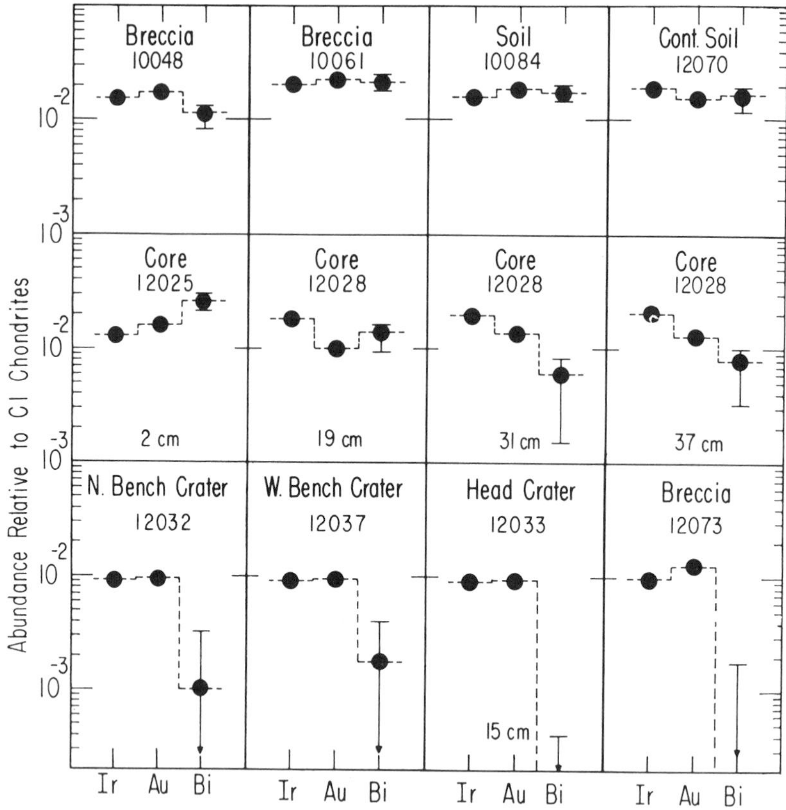

Fig. 4. Abundances of three meteoritic elements in lunar soils and breccias, corrected for indigenous Ir, Au, and Bi. The first four samples show a flat pattern, characteristic of primitive meteorites. Apparently micrometeorites of C 1-like composition dominate the meteoritic component in these samples. The last 4 samples show a stepwise pattern characteristic of fractionated meteorites. Three of these are soil samples collected on or near crater rims; presumably these craters were produced by ordinary chondrites or iron meteorites. The core samples show patterns which are transitional between primitive and fractionated types, and probably represent mixtures rather than specific meteorite types. The relatively high Bi in 12025,72 may be a slight contaminant from 12028,66, which is greatly enriched in this element. The core depths for each sample are shown; the true depths in the lunar surface (CARRIER *et al.*, 1971) are shown in footnote b of Table 1.

Au, and Bi. The calculated meteoritic components for Apollo 12 soils and breccia 12073 for these three elements are shown in Fig. 4. Apollo 11 breccias and soil are shown for comparison. All four samples in the top row of Fig. 4 show a primitive meteoritic pattern, at a level corresponding to $\sim 2\%$ C 1 material or equivalent. The four samples in the bottom row (soils 12032, 12037, 12033, and breccia 12073) show a step pattern characteristic of fractionated meteorites (irons or ordinary chondrites). The enrichment of Bi is much less than that of Ir and Au, and may in fact be zero within the uncertainty of the correction for indigenous lunar Bi. The amount of meteoritic material is only about one-half that in the other samples: 0.9% to 1% C 1 equivalent (corresponding to 0.86% L-chondrites, 0.60% H-chondrites, or 0.17% group I irons).

All three soil samples that showed a fractionated meteorite pattern were collected on crater rims, whereas those showing a primitive pattern were taken at greater distances. Apparently the soil far away from craters is heavily contaminated by the steady influx of micrometeorite material of C 1 composition. Material on crater rims is less contaminated, and still shows the abundance pattern of the projectile. The effect disappears at distances of only a few tenths of a crater diameter from the rim.

The abundances in Fig. 4 are not sufficient to establish the exact nature of the meteoritic material in the crater rim soils and breccia 12073. One approach which may yield a further clue is to assume that the pattern for each soil is the result of a single large impact, and to apply cratering theory (Öpik, 1961). The ratio of eroded mass to projectile mass, M/μ, depends on impact velocity w, the crushing strength s, and density ρ of the target rock, and a dimensionless factor k that allows for the extra momentum of backfiring:

$$M/\mu = kw(\rho/s)^{1/2} \tag{6}$$

The parameter k is a slowly varying function of impact velocity, projectile density, etc.; it generally falls between 2 and 5 and can be computed exactly from relations given by Öpik. For average lunar rocks, $\rho = 3.3$ g cm^{-3}. The crushing strength is estimated as 9×10^8 dynes cm^{-2} (Öpik, 1969).

From eq. (6) we find that the percentages of meteoritic material in the crater rim soils and breccia 12073 correspond to impact velocities of 8, 11, and 24 km/sec for the three types of projectiles. A velocity as high as 24 km/sec is fairly improbable for an iron meteorite (Millman, 1969; Öpik, 1961), although a single iron meteor with $V_G = 30$ km/sec has been observed (Ceplecha, 1966). On the other hand, velocities of 8 and 11 km/sec for chondrites are close to the mean geocentric velocity for stony meteorites, 14 km/sec (Millman, 1969, and earlier references cited therein).

The middle row of Fig. 4 shows samples of a transitional type. The two upper core samples 12025,72 and 12028,90 also approximate a C 1 type distribution (although there are reservations about these samples which will be discussed in a later section). The lower two core samples show a meteoritic Bi level only about one-half that for C 1, and may be regarded as transitional to a fractionated meteoritic pattern. However, on the basis of the relatively unfractionated noble metals, all the core samples have about 2% C 1 equivalent meteoritic material.

KREEP-related meteoritic component

The light gray crater-rim soils 12032, 12033, and 12037 contain substantial amounts of KREEP (Hubbard *et al.*, 1971). Some authors have proposed that this material represents ejecta from the crater Copernicus (LSPET, 1970; Meyer and Hubbard, 1970; Dollfus *et al.*, 1971) while others have favored the Fra Mauro formation (Quaide *et al.*, 1971). Whatever its source, if KREEP is of impact origin, it must contain the remains of the projectile responsible for the impact. We therefore attempted to identify the meteoritic component associated with KREEP.

The KREEP content of 12033 is 65% (HUBBARD *et al.*, 1971) and we have estimated that of 12032 and 12037 from published REE values (HUBBARD *et al.*, 1971; WAKITA *et al.*, 1971; HASKIN *et al.*, 1971; PHILPOTTS *et al.*, 1971). With KREEP content as the independent variable we have calculated regression lines for the elemental abundances in these three soils, and have estimated the abundances in pure KREEP and in KREEP-free soil from the regression coefficients. From the extrapolated elemental abundances for KREEP we have determined upper limits for the associated meteoritic component. The details of our procedure and a full discussion of results will be given elsewhere (MORGAN *et al.*, 1971b). However, the conclusions emerge that: (1) KREEP carries records of an impact at its source (Copernicus?) and a local impact; and (2) both were by *fractionated* meteorites.

Abundance of primitive and fractionated meteorites in space

In our Apollo 11 paper (GANAPATHY *et al.*, 1970a), it appeared that the meteoritic influx to the Moon was largely of carbonaceous chondrite material. This supported a widely held view that carbonaceous chondrites comprise a major part of the terrestrial influx, but are underrepresented in collections owing to destruction in the atmosphere (MAZOR and ANDERS, 1967; SHOEMAKER and LOWERY, 1967; MCCROSKY, 1968; WHIPPLE, 1966). Of the 6 discrete impacts on the Moon which have now been chemically characterized, only one (the glass coating of 12017) seems likely to be of primitive chondritic material. Therefore, on admittedly poor statistics, it appears that the carbonaceous chondrite contribution is largely as micrometeorites, presumably debris from spontaneously disintegrating comets. Among larger objects, fractionated meteorites (ordinary chondrites, irons) predominate.

If crater-forming events are largely due to fractionated meteorite types, we may roughly estimate the cumulative mass influx rate of fractionated material from Fig. 5 of GANAPATHY *et al.* (1970a). The curve for asteroidal objects (impact velocity = 15 km/sec) is appropriate for fractionated meteorites. Even if we assume that *all* of the meteoritic component in soil is due to micrometeorites of C 1-like composition, the cumulative mass influx rate of fractionated projectiles up to 10 km diameter (forming craters up to 100 km diameter) is 50 to 100 times higher. The effective throwout range is $15 + 0.3B_0$, where B_0 is the crater diameter in km (ÖPIK, 1969) so that large craters add relatively little meteoritic material to the regolith outside their immediate vicinity.

Double core 12025 and 12028

Five samples from the core were analyzed. Results of four of these, all from 12028, have previously been published (GANAPATHY *et al.*, 1970b). The variation of trace element abundance with depth is shown in Fig. 5.

The nominal 13-cm layer (Unit VI; LSPET, 1970) is spectacularly enriched in Bi and Cd, by factors of 10^4 to 10^5. The implications of this observation have already been discussed (GANAPATHY *et al.*, 1970b). One significant conclusion is that the turnover rate of the regolith at the 12028 site must be orders of magnitude slower than previously estimated. From the data presented graphically by GAULT (1970), the turnover time at a given depth may be estimated. Using the corrected depth of 17.8–19.2 cm

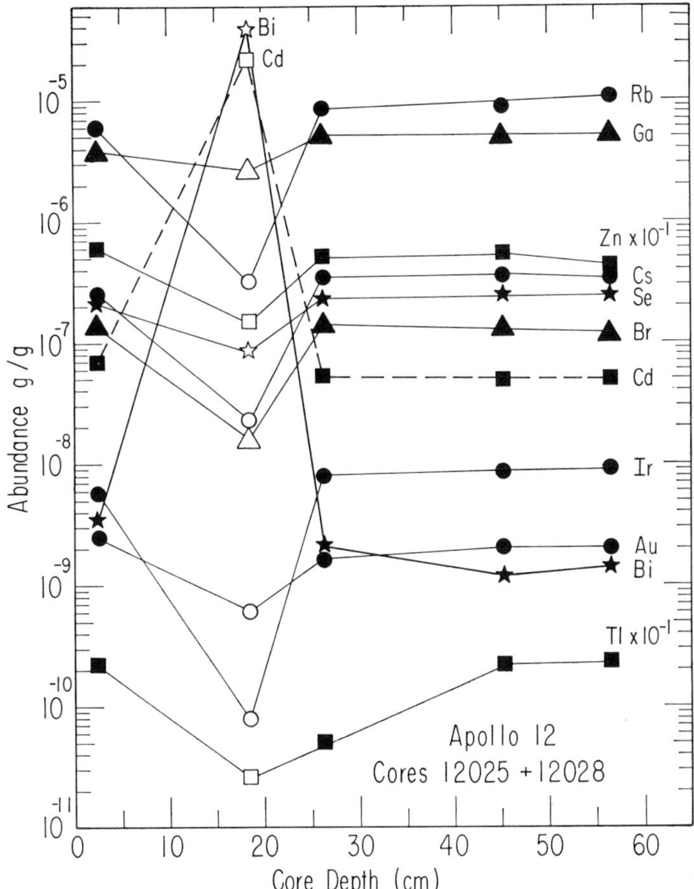

Fig. 5. The distribution of trace elements in the double core 12025/12028 versus true depth (CARRIER *et al.*, 1971). The abundances vary rather uniformly except for the coarse olivine-rich layer 12028,66 which shows great enrichment of Bi and Cd and smaller, but significant variations in other elements. The lack of mixing is clearly demonstrated.

(CARRIER *et al.*, 1971) for our sample of Unit VI, we estimate (for a Naumann-Hawkins flux) a turnover time of about 8×10^5 yr.

There is however only very slight evidence of any vertical spreading of Bi or Cd from the marker horizon Unit VI. For example, if all the Bi in sample 12025,72 (Unit IX) were derived from Unit VI, this would represent only 0.01% mixing at a distance of some 16 cm. We cannot rule out the possibility that the high Bi and Cd contents are not typical of Unit VI as a whole, but that the enrichment observed in our sample represents the chance inclusion of one grain of very Bi- and Cd-rich material. However, the presence of at least 10 distinct layers in the core 12025/28 (four of these in the top 18 cm alone!) independently rules out any substantial mixing since deposition of these layers. The actual deposition time is not known, but is probably on the order of several times 10^7 yr or greater. Unit VI has a surface exposure age of 1.5×10^7 yr (CROZAZ *et al.*, 1971), and if this represents a typical interval between deposition of successive layers, the four overlying layers probably were deposited over a period on the order of 5×10^7 yr. An even higher age is obtained if one assumes that the regolith grows linearly with time, at a rate of $\sim 10^{-9}$ m yr^{-1}. In any case, it appears that the turnover time estimates of GAULT (1970) are in error, or are not applicable to our case.

Two reasons why these turnover estimates do not coincide with our observations occur to us. The first possibility is that there is a fundamental flaw in the cratering law used. It has been stressed (ÖPIK, 1961) that in meteorite impact, the momentum of the projectile, rather than the kinetic energy as assumed by GAULT (1970), determines crater volume. This alone can apparently lead to errors of 1 to 2 orders of magnitude in estimates of crater volume. However, even discrepancies of this order are not large enough to account for the negligible mixing observed and the short turnover times estimated. The second possibility is that the concept of "mean turnover time" is not very meaningful on the scale at which the double core was sampled.

It is axiomatic that the lunar regolith increases in mean thickness with time. However, the presence of craters clearly indicates that the increase cannot be uniform on a small scale. We suggest therefore that the growth of the regolith in small areas (of the order of meters^2, perhaps) may be regarded as a "biased random walk." The regolith in a particular small area is built up from overlapping ejecta blankets, probably in increments of the order of cm generally. Rarely there is a direct hit by a large object which effectively removes the accumulated regolith, and the process begins again. Since the area over which a perceptible layer of ejecta is deposited (say to a mean thickness of $\geq \frac{1}{10}$ or $\frac{1}{20}$ of the mean crater depth) is roughly 10 to 20 times greater than the area of the crater, addition of material is clearly more probable than removal, by a similar factor. Addition therefore proceeds as many small increments; but removal by a few large decrements. In this way many successive layers can be built up, before the whole stratigraphy is blasted away. It appears that the turnover rates estimated by GAULT (1970) may represent an average for km^2 areas, but on a small scale there is considerable "graininess" and strata may remain undisturbed for considerable periods. In any case, it seems that the term "gardening" is a misnomer, as applied to the lunar regolith. It implies a mixing process in which vertical and lateral displacements are of equal magnitude, with excavation and blanketing proceeding concurrently. In the lunar regolith, on the other hand, mixing proceeds mainly by large lateral displacements. Excavation and blanketing proceed separately, never concurrently in the same area. A similar concept was suggested independently by McKay *et al.* (1971).

The source of the enrichment in Unit VI and the mineral host of Bi and Cd is as yet unknown. It may be significant that the Bi-to-Cd atomic ratio is close to unity. In a fine-grained residual phase of a fragment from this unit, SELLERS *et al.* (1971) detected an unidentified phosphate mineral, of which they were able to account for ~90% of the constituents. Could Bi and Cd be among the missing components?

We find no evidence of a significant meteoritic component in Unit VI. The very low Ir abundance immediately excludes all chondrites and all but a very few iron meteorites. The low abundance of Au would exclude remaining irons at levels of ~0.05% or less. Because all the elements enriched in achondrites are also enriched in lunar surface rocks, we would be unable to detect an achondritic impacting body. If Unit VI is the result of an impact in which significant mixing of projectile and target occurred, then it appears that a fractionated meteorite was responsible.

Other double core samples resemble typical soils in showing distinct enrichments of several elements of probable meteoritic origin. The abundances of Ir and Au in the core samples are quite uniform with depth, and agree well with those in contingency sample 12070. On the other hand there is apparently a decrease in the abundances of Bi, Cd, and Zn with depth. This may be exaggerated by slight contamination of 12025,72, and perhaps 12028,90, by Unit VI (during the coring operation?). However, if we substitute sample 12070 as a representative sample of the top few cm of soil the trend is still observed. Superficially this could be interpreted as a change with time in the ratio of primitive to fractionated meteorites. This may, however, be only on a local scale, and may reflect a disproportionate amount of highly fractionated meteoritic material introduced by one event early in the core history. For example, there are significant amounts of KREEP distributed in the lower core (McKay *et al.*, 1971). An alternative is that the volatile trace element variation reflects shorter surface residence times for the lower part of the core. However, noble gas measurements (Funkhouser *et al.*, 1971) and particle track studies (Comstock *et al.*, 1971) reveal no systematic differences with depth.

10084 leaching experiment

Silver (1970) describes an experiment in which a sample of 10084 fines was leached by concentrated HNO_3 under increasingly rigorous conditions. Lead isotopic measurements and U and Th analyses were made on three leach fractions and the residue. From this and other experiments Silver inferred the presence of a labile Pb fraction in the soil, and suggested that the enrichment of volatile elements in lunar fines is due to a volatile transfer process. Albedo measurements by Hapke *et al.* (1971) also indicate that vapor phase deposition may be an important process on the lunar surface. We repeated Silver's leaching experiment for our suite of elements (Table 2, Fig. 6). An uncertainty in interpretation is introduced by recoil effects following neutron capture. These effects are probably more serious for elements with several oxidation states, such as Ir, than for most of our other elements. The low solubility of the major phases is indicated by the small amounts of alkali metals leached. Surprisingly, Fe and particularly Ca appear to be more soluble. Probably most of the soluble Fe and Ca reside in glasses which contain 10–20% CaO but are generally poor in alkalis (Fredriksson *et al.*, 1970).

The meteoritic elements are found in all leach fractions as shown in Table 4, where the abundances from Table 2 are normalized to percentages of C 1 abundances. No correction is made for any indigenous component.

An interesting pattern emerges, if the results are compared with Ni, which is almost entirely meteoritic. About $\frac{1}{3}$ of the Ni in the residue would account for the Ni contribution from crystalline rocks. Thallium, which in lunar rocks is strongly correlated with K, Rb, and Cs (Ganapathy *et al.*, 1970a; Anders *et al.*, 1971), appears in the leach fractions to approximately the same extent as Ni, indicating the presence of meteoritic Tl. That remaining in the residue is comparable to the abundance in crystalline rocks plus a meteoritic component of the size indicated by Ni. The same pattern is apparent for Ag and Au.

Rubidium and Cs in the leach fractions must be regarded as being largely indigenous. The soluble Rb and Cs may be present in small amounts in glasses, or as a vapor-deposited surface coating.

Iridium is leached to roughly the same extent as Ni in the first two fractions, but is underabundant in the third. The abundance in the residue is very much higher than that found in crystal-

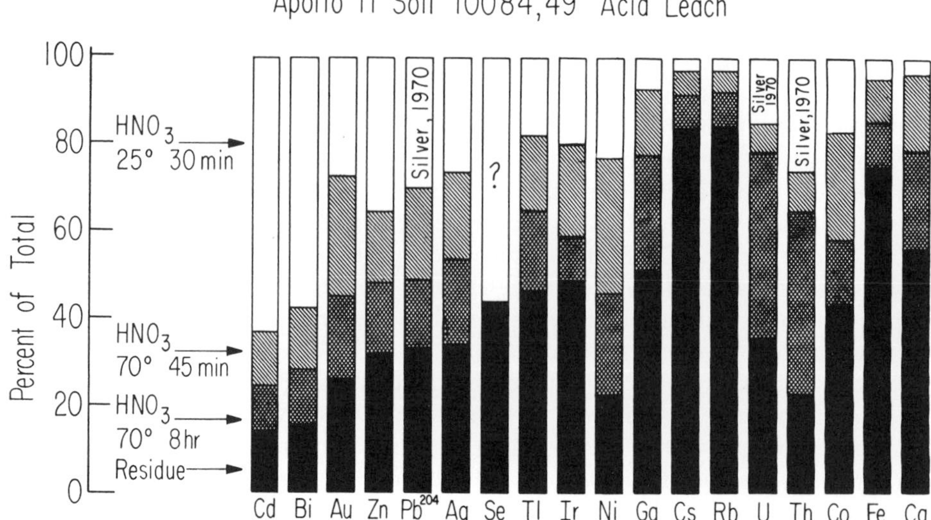

Fig. 6. Apollo 11 soil 10084,49 acid leaching experiment. The procedure follows that of SILVER (1970) and is described in detail in the text. Elements of largely meteoritic origin are leached significantly whereas those which are largely indigenous are not. Uranium and Th are exceptions, probably because they reside in an acid soluble indigenous component. The large amounts of Cd and Zn in the first leach solution may also indicate the presence of an acid soluble indigenous phase.

Table 4. Nitric acid leaching of Apollo 11 soil 10084,49.

Leach Fraction[a]	Abundance (Percent Equivalent C 1 Material)								
	Ir	Au	Zn	Cd	Ag	Bi	Tl	Co	Ni
4	0.37	0.51	2.1	3.1	0.61	0.79	0.39	1.23	0.44
5	0.39	0.51	0.94	0.65	0.45	0.19	0.37	1.60	0.57
6	0.18	0.35	0.94	0.49	0.45	0.18	0.41	0.98	0.43
7	0.88	0.48	1.88	0.67	0.78	0.21	1.00	2.78	0.42
Apollo 11 Basalts[b]	0.004	0.09	1.13	1.34	1.14	0.39	2.05	5.0	0.15
Apollo 11 Basalts[c]	0.004	0.04	0.50	0.56	0.40	0.25	0.80	4.7	0.15

[a] See footnote to Table 2 for description of leach fractions. Abundances represent 100 (ppm relative to original soil sample)/(ppm in C 1 chondrites). No correction has been applied for indigenous trace element content. The approximate magnitude of such a correction is indicated by the "Apollo 11 Basalts" (Means) in the last two lines of the table.

[b] Including trace-element-rich crystalline rock 10017.

[c] Mean, excluding 10017.

line rocks. Some of the Ir may reside in an acid resistant phase or in metallic particles protected by silicate glass coatings, as observed by AGRELL *et al.* (1970) and others.

Metal grains from 10084 have been analyzed by WÄNKE *et al.* (1971), and all of the Ni, Au, and Ir appear to be in the metal phase. Unfortunately a satisfactory mass balance is lacking and the contribution of the metal fraction to the Ir and Au abundance of the bulk soil exceeds that found in the direct bulk soil analysis by a factor of between 2 and 3. This is said to be a problem of sampling; the grains used were atypically large, and apparently richer than normal in Au and Ir (WÄNKE *et al.*, 1971). On this basis it seems that WÄNKE *et al.* are not justified in asserting "100% conversion of the siderophilic elements of C 1 material into metal seems not very plausible." Clearly the uncertainties of the data would not distinguish between 100% and, say, 70% conversion.

Nevertheless, much of the Au and Ir does appear to reside in metallic particles as indicated by the results of our leaching experiment.

The autoreduction of FeO by C in carbonaceous chondrites is a well-known process (e.g., RINGWOOD, 1966) and appears to be a feasible mechanism for production of metal during hypervelocity impact of micrometeorites. Each micrometeorite particle will be vaporized (or at least melted) on impact. On cooling part or all of the Fe should condense as metal, and with it virtually all of the siderophiles. As these processes typically involve only 10^{-6} to 10^{-12} g, one would expect the metal produced in the first instance to be very small grains or even surface coatings. Subsequent impact acting upon these could produce many features observed in lunar soil. Metal drops intermingled with troilite and silicates are common in Apollo soils, but recognizable meteoritic metal is observed rarely (AGRELL *et al.*, 1970). Fractionated meteorites undoubtedly make an important contribution to lunar soil, as we have shown in this paper. However, the concentration of siderophilic elements in metallic grains does not, of itself, rule out a C 1-like component.

Perhaps we should restate our position on the observed primitive type component in lunar soil. Although we normalize to C 1 abundances, we do not necessarily interpret a primitive or C 1-like pattern to imply the presence of C 1 chondrites exactly as we know them on Earth. These have been exposed to liquid H_2O in the meteorite parent body, which has apparently destroyed much of their FeS. If metal had been present it would have been destroyed too, under conditions which were sufficiently oxidizing to convert S^{2-} to SO_4^{2-}. It is quite conceivable that "dry" uncorroded C 1-type material may exist in, for example, comets. Primitive solar system material was apparently very close to C 1 composition (ANDERS, 1971), and hence this composition may be expected to turn up in all those corners of the solar system that were not affected by major chemical fractionations. In addition, it would be very difficult to distinguish an E 4 enstatite chondrite, say, from a C 1 pattern.

In the first leach fraction Zn and Cd are highly enriched relative to Ni. Cadmium is present in the other two acid fractions to about the same extent as other meteoritic elements and in the residue at a level comparable with crystalline rocks. On the other hand, Zn is high in the other leach fractions and in the residue. It appears that the rocks from which the regolith formed were higher in Zn than the mean of our crystalline rock measurements might indicate. The pattern for Bi is *qualitatively* similar to Cd. However, the *total* content in the three leach fractions is similar to that of Ni, and the abundance in the residue agrees very well with that of crystalline rocks.

The higher acid-solubility of the meteoritic elements is not surprising, because much of the meteoritic material appears to be present as small particles or surface coatings. We have previously shown that these elements were enriched in the finest (−325 mesh) size fraction of the soil (GANAPATHY *et al.*, 1970a). Moreover, some part of the meteoritic material must be present as vapor-deposited surface coatings on larger grains. In Silver's (1970) experiments it was found that Th and U were leached rather easily. Neither of these elements is enriched in meteorites. However they mainly reside in minor acid-soluble phases. The lability of Zn and Cd might be explained by a small amount of rather easily soluble Zn–Cd mineral (sulfide?) present in the lunar soil. Terrestrially Zn and Cd are generally associated in sulfide ores.

CONCLUSION

The agreement between Apollo 11 soil and the Apollo 12 contingency sample indicates that the meteoritic influx and the average regolith mixing depths have been similar in both regions. The results for the core indicate that local mixing is orders of magnitude slower than implied by mean turnover rates. We propose a "biased random walk" model which may be more applicable to small areas of the regolith. The meteoritic component in most soils probably contains a contribution from micrometeorites of primitive composition, which may be largely cometary debris. Not knowing the thickness of the regolith, we cannot calculate an absolute infall rate at the Apollo 12 site. However, the similarity to Apollo 11 soil, for which a value has been calculated

(GANAPATHY *et al.*, 1970a), suggests that it too will be close to 3.8×10^{-9} g cm^{-2} yr^{-1}. This in turn agrees with estimates for the terrestrial influx, $(1.2 \pm 0.6) \times 10^{-8}$ g cm^{-2} yr^{-1} (BARKER and ANDERS, 1968).

Although the micrometeorite influx appears to be largely primitive chondritic material, it now seems the larger impacting bodies are fractionated meteorites. At present we cannot distinguish precisely between iron meteorites and fractionated chondritic types. However, there are elements which could be used for this purpose. With improved resolution of meteoritic types we may be able to identify the debris of planetesimals which impacted upon the Moon's surface during its early history.

We can now be certain that anorthositic material and KREEP are not a major source of Bi and Tl. Therefore the analysis in the companion paper (ANDERS *et al.*, 1971) based on Bi/Cs and Tl/Cs ratios probably gives a fairly reliable estimate of the abundances of these elements in the whole Moon.

Acknowledgments—We are indebted to Dr. Reid R. Keays for the first of the replicate analyses for Ir, Zn, Cd, Ag, and Bi in 12032, 12033, and 12037 and to Dr. Julie Morgan for handpicking anorthositic fragments 10084,49,1-1. Valuable assistance in various phases of this work was given by Rudy Banovich, Mary Joan Hasche, Annie Pierce, and Frank Quinn. Financial support came mainly from NASA Contract NAS 9–7887, supplemented by AEC Contract AT(11–1)–382, and page charges were provided by NASA Grant NGL 14–001–167.

REFERENCES

AGRELL S. O., SCOON J. H., MUIR I. D., LONG J. V. P., MCCONNELL J. D., and PECKETT A. (1970) Observations on the chemistry, mineralogy, and petrology of some Apollo 11 lunar samples. *Proc. Apollo 11 Lunar Sci. Conf., Geochim. Cosmochim. Acta* Suppl. 1, Vol. 1, pp. 93–128. Pergamon.

ANDERS E. (1971) How well do we know cosmic abundances? *Geochim. Cosmochim. Acta*, in press.

ANDERS E., GANAPATHY R., KEAYS R. R., LAUL J. C., and MORGAN J. W. (1971) Volatile and siderophile elements in lunar rocks: Comparison with terrestrial and meteoritic basalts. Second Lunar Science Conference (unpublished proceedings).

BAEDECKER P. A., SCHAUDY R., ELZIE J. L., KIMBERLIN J., and WASSON J. T. (1971) Trace element studies of rocks and soils from Oceanus Procellarum and Mare Tranquillitatis. Second Lunar Science Conference (unpublished proceedings).

BARKER J. L., JR., and ANDERS E. (1968) Accretion rate of cosmic matter from iridium and osmium contents of deep-sea sediments. *Geochim. Cosmochim. Acta* **32**, 627–645.

CARRIER III W. D., JOHNSON S. W., WERNER R. A., and SCHMIDT R. (1971) Disturbance in samples recovered with the Apollo core tubes. Second Lunar Science Conference (unpublished proceedings).

CEPLECHA Z. (1966) Complete data on iron meteoroid (Meteor 36221). *Bull. Astron. Inst. Czech.* **17**, 195–206.

COMSTOCK G. M., EVWARAYE A. O., FLEISCHER R. L., and HART H. R., JR. (1971) The particle track record of the Ocean of Storms. Second Lunar Science Conference (unpublished proceedings).

CROZAZ G., WALKER R., and WOOLUM D. (1971) Cosmic ray studies of "recent" dynamic processes on the surface of the moon. Second Lunar Science Conference (unpublished proceedings).

DOLLFUS A., GEAKE J. E., and TITULAER C. (1971) Polarimetric and photometric properties of Apollo lunar samples. Second Lunar Science Conference (unpublished proceedings).

FREDRIKSSON K., NELEN J., and MELSON W. G. (1970) Petrography and origin of lunar breccias and glasses. *Proc. Apollo 11 Lunar Sci. Conf., Geochim. Cosmochim. Acta* Suppl. 1, Vol. 1, pp. 419–432. Pergamon.

FUNKHOUSER J., BOGARD D., and SCHAEFFER O. A. (1971) Noble gas analyses of core tube samples from Mare Tranquillitatis and Oceanus Procellarum. Second Lunar Science Conference (unpublished proceedings).

GANAPATHY R., KEAYS R. R., LAUL J. C., and ANDERS E. (1970a) Trace elements in Apollo 11 lunar rocks: Implications for meteorite influx and origin of moon. *Proc. Apollo 11 Lunar Sci. Conf., Geochim. Cosmochim. Acta* Suppl. 1, Vol. 2, pp. 1117–1142. Pergamon.

GANAPATHY R., KEAYS R. R., and ANDERS E. (1970b) Apollo 12 lunar samples: Trace element analysis of a core and the uniformity of the regolith. *Science* **170**, 533–535.

GAULT D. E. (1970) Saturation and equilibrium conditions for impact cratering on the lunar surface: Criteria and implications. *Radio Sci.* **5**, 273–291.

HAPKE B. W., CASSIDY W. A., and WELLS E. N. (1971) The albedo of the moon: Evidence for vapor-phase deposition processes on the lunar surface. Second Lunar Science Conference (unpublished proceedings).

HASKIN L. A., ALLEN R. O., HELMKE P. A., ANDERSON M. R., KOROTEV R. L., and ZWEIFEL K. A. (1971) Rare earths and other trace elements in Apollo 12 lunar materials. Second Lunar Science Conference (unpublished proceedings).

HUBBARD N. J., GAST P. W., and MEYER C. (1971) The origin of the lunar soil based on REE, K, Rb, Ba, Sr, P, and $Sr^{87/86}$ data. Second Lunar Science Conference (unpublished proceedings).

KEAYS R. R., GANAPATHY R., LAUL J. C., ANDERS E., HERZOG G. F., and JEFFERY P. M. (1970) Trace elements and radioactivity in lunar rocks: Implications for meteorite infall, solar-wind flux, and formation conditions of moon. *Science* **167**, 490–493.

KIMBERLIN J., CHAROONRATANA C., and WASSON J. T. (1968) Neutron activation determination of iridium in meteorites. *Radiochim. Acta* **10**, 69–76.

LARIMER J. W., and ANDERS E. (1967) Chemical fractionations in meteorites—II. Abundance patterns and their interpretation. *Geochim. Cosmochim. Acta* **31**, 1239–1270.

LAUL J. C., KEAYS R. R., GANAPATHY R., and ANDERS E. (1970) Abundance of 14 trace elements in lunar rock 12013,10. *Earth Planet. Sci. Lett.* **9**, 211–215.

LSPET (Lunar Sample Preliminary Examination Team) (1970) Preliminary examination of lunar samples from Apollo 12. *Science* **167**, 1325–1339.

MAZOR E., and ANDERS E. (1967) Primordial gases in the Jodzie howardite and the origin of gas-rich meteorites. *Geochim. Cosmochim. Acta* **31**, 1441–1456.

MCCROSKY R. E. (1968) Distributions of large meteoric bodies. *Smithson. Astrophys. Obs. Spec. Rep.* No. 280.

MCKAY D., MORRISON D., LINDSAY J., and LADLE G. (1971) Apollo 12 soil and breccia. Second Lunar Science Conference (unpublished proceedings).

MEYER C., JR., and HUBBARD N. J. (1970) High potassium, high phosphorus glass as an important rock type in the Apollo 12 soil samples. *Meteoritics* **5**, 210–211.

MILLMAN P. M. (1969) Astronomical information on meteorite orbits. In *Meteorite Research* (editor P. M. Millman), Chap. 45, pp. 541–551. Reidel.

MORGAN J. W., LAUL J. C., GANAPATHY R., and ANDERS E. (1971a) Glazed lunar rocks: Origin by impact. *Science*, in press.

MORGAN J. W., GANAPATHY R., LAUL J. C., and ANDERS E. (1971b) Lunar crater Copernicus: Possible nature of impact. Manuscript in preparation.

ÖPIK E. J. (1961) Notes on the theory of impact craters. *Proc. Geophys. Lab.-Lawrence Radiation Lab. Cratering Symp.*, Washington, D.C., March 28–29, Vol. 2, Paper S, 1–28. UCRL Report 6438.

ÖPIK E. J. (1969) The moon's surface. *Ann. Rev. Astron. Astrophys.* **7**, 473–526.

PHILPOTTS J. A., SCHNETZLER C. C., BOTTINO M. L., and FULLAGER P. D. (1971) Li, K, Rb, Sr, Ba, and rare earth concentrations and $^{87}Sr/^{86}Sr$ in some Apollo 12 soils, rocks, and separated phases. Second Lunar Science Conference (unpublished proceedings).

QUAIDE W., OBERBECK V., BUNCH T., and POLKOWSKI G. (1971) Investigations of the natural history of the regolith at the Apollo 12 site. Second Lunar Science Conference (unpublished proceedings).

RINGWOOD A. E. (1966) Genesis of chondritic meteorites. *Rev. Geophys.* **4**, 113–175.

SELLERS G. A., WOO C. C., BIRD M. L., and DUKE M. B. (1971) Descriptions of the composition and grain-size characteristics of fines from the Apollo XII double-core tube. Second Lunar Science Conference (unpublished proceedings).

SHOEMAKER E. M., and LOWERY C. J. (1967) Airwaves associated with large fireballs and the frequency distribution of energy of large meteoroids. *Meteoritics* **3**, 123–124.

SILVER L. T. (1970) Uranium-thorium-lead isotopes in some Tranquillity Base samples and their implications for lunar history. *Proc. Apollo 11 Lunar Science Conf., Geochim. Cosmochim. Acta* Suppl. 1, Vol. 2, pp. 1533–1574. Pergamon.

WAKITA H., and SCHMITT R. A. (1970) Lunar anorthosites: Rare-earth and other elemental abundances. *Science* **170**, 969–974.

WAKITA H., REY P., and SCHMITT R. A. (1971) Abundances of the 14 rare-earth elements plus 22 major, minor, and trace elements in ten Apollo 12 rocks and soil samples. Second Lunar Science Conference (unpublished proceedings).

WÄNKE H., WLOTZKA F., TESCHKE F., BADDENHAUSEN H., SPETTEL B., BALACESCU A., QUIJANO-RICO M., JAGOUTZ E., and RIEDER R. (1971) Major and trace elements in Apollo XII samples and studies on lunar metallic iron particles. Second Lunar Science Conference (unpublished proceedings).

WHIPPLE F. L. (1966) Before Type I carbonaceous chondrites. Paper presented at 29th Meeting of the Meteoritical Society, Washington, D.C., Nov. 3–5.

Proceedings of the Second Lunar Science Conference, Volume 2, pp. 1159–1168.
The M.I.T. Press, 1971.

Abundances of the primordial radionuclides K, Th, and U in Apollo 12 lunar samples by nondestructive gamma-ray spectrometry: Implications for origin of lunar soils

G. DAVIS O'KELLEY and JAMES S. ELDRIDGE
Oak Ridge National Laboratory, Oak Ridge, Tennessee 37830

and

ERNEST SCHONFELD and P. R. BELL*
Manned Spacecraft Center, Houston, Texas 77058

(Received 22 February 1971; accepted in revised form 29 March 1971)

Abstract—Gamma-ray spectrometers with low background were used to determine concentrations of K, Th, and U in crystalline rocks 12002, 12004, 12021, 12039, 12051, 12052, 12053, 12054, 12062, 12064, and 12065; in breccias 12013, 12034, and 12073; and in fines samples 12032 and 12070. The crystalline rocks studied resemble the low-potassium rocks of Apollo 11 and have average concentrations of K, Th, and U of 520, 0.97, and 0.25 ppm, respectively, with remarkably small deviations from the average values. Concentration ratios K/U for all Apollo 11 and Apollo 12 materials vary only from 1200 to 3200 for a range of 50 in K or U concentration. With the body of data now available on samples from Apollo 11 and Apollo 12, it appears that there are five groups of materials with characteristic K/U ratios.

The Apollo 12 fines and breccia together are relatively uniform, but, because of their much higher concentrations of radioactive elements, they could not have been formed directly from the crystalline rocks examined. Rock 12013 has the highest concentrations of K, Th, and U, but the lowest ratio Th/U of any lunar sample we have studied. Our systematics of primordial radioelement concentrations predict some of the properties of the foreign component of the lunar soil and breccia. It is suggested that the dark portion of 12013 contains a high concentration of this foreign material.

Concentrations of K, Th, and U obtained in this study were combined with gas analysis data from other workers to obtain K–^{40}Ar ages from 2.34 to 3.37 aeons and U, Th–^{4}He ages from 1.22 to 2.69 aeons.

INTRODUCTION

DATA ON THE abundances of the primordial radioelements K, Th, and U are essential to many studies of geological processes. Such information is of great importance in formulating theories of planetary evolution where, because of their radioactivity, the contribution of primordial radionuclides to the thermal balance of the planet must be considered. The concentration of K can serve as a useful indicator for alkali-metal chemistry. Further, since K is a relatively volatile element and Th and U are refractory, concentrations of these elements may be invoked to establish limits on a temperature profile for a planetary body or other geological system under study.

Gamma-ray spectrometry has proved to be a very useful method for determination of K via its radioactive isotope ^{40}K, and U and Th via their respective decay products. Studies on lunar samples from Apollo 11 demonstrated that nondestructive, gamma-ray spectrometry measurements could be made with good accuracy on samples which

* Present address: Oak Ridge National Laboratory, Oak Ridge, Tennessee 37830.

varied in size from a few grams to kilograms (O'KELLEY *et al.*, 1970a, 1970b; PERKINS *et al.*, 1970; HERZOG and HERMAN, 1970; WRIGLEY and QUAIDE, 1970).

Because of the widespread geochemical interest in concentrations of K, Th, and U, and because the experimental method is rapid and nondestructive, it was deemed useful to scan a large number of lunar samples by gamma-ray spectrometry. The measurements reported here show interesting geochemical trends for the Apollo 12 surface materials which are compared to a less extensive survey of Apollo 11 materials. In addition, this survey yielded the first evidence that rock 12013 was dramatically different in its chemistry from all lunar samples previously examined.

Preliminary data on some of the samples examined in this study during the Apollo 12 mission were reported in LSPET (1970). Since the publication of the preliminary study, the data analyses have been refined and additional samples have been analyzed.

EXPERIMENTAL METHODS

The gamma scintillation coincidence spectrometer which was used at the Lunar Receiving Laboratory (LRL) for analysis of most of the samples reported in this study contained two NaI(Tl) detectors, each 23 cm in diameter and 13 cm long, with 10 cm pure NaI light guides. This spectrometer system and associated data reduction techniques were described previously by O'KELLEY *et al.* (1970b). Stainless steel cans of 0.8 mm wall thickness, 16 cm in diameter and either 5.6 or 7.6 cm in height were used as sample containers during the quarantine period; later, thin teflon bags were used to protect the samples during measurements after quarantine restrictions were removed.

A second detector system, identical to that at the LRL and located at Oak Ridge National Laboratory (ORNL), was used to analyze two of the samples (12002,20 and 12013,11) reported in this study. The ORNL data acquisition system used a 4096-channel coincidence analyzer and was operated in a 64 × 64 channel matrix configuration with both energy axes adjusted to 60 keV/channel to cover an energy range of 0–3.8 MeV in each detector. An IBM 360/91 program was used to pre-process the matrix data and separate the spectra into singles spectra from the two detectors, a summed singles spectrum from the two detectors, and gamma-gamma coincidence spectra from selected regions of the matrix folded in a manner similar to that which we described for the LRL system (O'KELLEY *et al.*, 1970b). The series of single-parameter spectra obtained in this way were plotted and then were analyzed by use of ALPHA-M, a computer program for quantitative radionuclide determination by the method of least squares (SCHONFELD, 1967). Analyses carried out with the ORNL system were in excellent agreement with those made with the LRL system.

For analyses of data recorded during the preliminary study and reported in LSPET (1970), calibration of the LRL coincidence spectrometer was established by recording a library of spectra from cylindrical radioactive standards prepared by dispersing known amounts of radioactivity in quantities of iron powder. Absorption factors to correct the data for the presence of the stainless steel containers were computed. These corrections were especially important for the cosmogenic radionuclide data, relatively small for the U and Th determinations, and least significant for the ^{40}K data. The data reported here were obtained by using, when required, a new library of spectra recorded with the standard sources placed inside the stainless steel containers actually used.

Spectrum libraries employed in the analysis of samples 12002,0; 12002,20; 12013,11; 12032,16; 12034,0; 12070,0; and 12073,0 were obtained from replicas which accurately reproduced the electronic and bulk densities of the lunar samples. Procedures for preparation of the cylindrical standards and the replicas were described earlier by O'KELLEY *et al.* (1970a, 1970b).

Error statements given in Table 1 are quite conservative estimates of the overall uncertainties, including counting statistics, geometrical factors, and calibration errors. A detailed analysis of a low-level synthetic lunar sample containing eight components showed that our methods were capable of absolute uncertainties $\leq 2.5\%$ for samples of the same degree of complexity as the samples studied here (O'KELLEY *et al.*, 1970b).

Table. 1. Concentrations of primordial radionuclides in lunar samples*

Sample	Weight (g)	K (ppm)	Th (ppm)	U (ppm)	K/U Mass ratio	Th/U Mass ratio
Crystalline rocks						
12002,0	1529	450 ± 20	0.89 ± 0.06	0.23 ± 0.02	1960 ± 192	3.87 ± 0.42
12002,20	260	425 ± 21	0.73 ± 0.04	0.21 ± 0.02	2020 ± 216	3.50 ± 0.39
12002,30	46	440 ± 20	0.86 ± 0.06	0.22 ± 0.02	2000 ± 212	3.91 ± 0.45
12004,1	502	469 ± 33	0.92 ± 0.09	0.24 ± 0.03	1960 ± 282	3.83 ± 0.61
12021,0	1877	500 ± 50	0.98 ± 0.10	0.26 ± 0.03	1920 ± 292	3.77 ± 0.53
12039,0	255	673 ± 40	1.20 ± 0.06	0.31 ± 0.03	2170 ± 246	3.87 ± 0.42
12051,0	1660	530 ± 50	1.00 ± 0.10	0.26 ± 0.03	2040 ± 304	3.85 ± 0.59
12052,1	201	540 ± 20	1.03 ± 0.06	0.27 ± 0.02	2000 ± 166	3.82 ± 0.36
12053,0	879	535 ± 40	1.06 ± 0.11	0.28 ± 0.03	1910 ± 244	3.79 ± 0.57
12054,0	687	530 ± 35	0.79 ± 0.08	0.22 ± 0.03	2410 ± 365	3.59 ± 0.61
12062,0	739	510 ± 35	0.83 ± 0.09	0.22 ± 0.03	2320 ± 354	3.78 ± 0.66
12064,0	1214	520 ± 35	0.87 ± 0.09	0.23 ± 0.02	2260 ± 248	3.78 ± 0.51
12065,0	2109	510 ± 50	1.06 ± 0.11	0.27 ± 0.03	1890 ± 280	3.92 ± 0.54
Average, crystalline rocks					2066	3.79 ± 0.14
Breccia						
12034,0	155	4560 ± 130	13.1 ± 0.3	3.4 ± 0.2	1341 ± 88	3.85 ± 0.24
12073,0	405	2960 ± 90	8.45 ± 0.10	2.19 ± 0.08	1352 ± 68	3.86 ± 0.15
Fines						
12032,16	89.6	3100 ± 100	8.8 ± 0.2	2.35 ± 0.07	1320 ± 58	3.75 ± 0.14
12070,0	354	2030 ± 120	6.25 ± 0.50	1.65 ± 0.16	1230 ± 140	3.79 ± 0.48
Average, fines and breccia					1311	3.81 ± 0.14

* Standardization for assay of K, Th, and U assumed terrestrial isotopic abundances and equilibrium of Th and U decay series.

Results and Discussion

Analysis of rocks and fines

Concentrations of K, Th, and U in 11 crystalline rocks, two breccias, and two samples of fines are listed in Table 1. The large number of samples studied permits some detailed comparisons to be made between the concentrations of Table 1 and comparable measurements carried out on Apollo 11 samples by several gamma-ray spectrometry teams (O'Kelley *et al.* 1970a, 1970b; Perkins *et al.* 1970; Herzog and Herman, 1970; Wrigley and Quaide, 1970).

The crystalline rocks of Apollo 11 appeared to fall into two chemical groups with different alkali metal concentrations. Our Apollo 11 work showed that one chemical group had average K, Th, and U concentrations of about 500, 1.0, and 0.25 ppm, respectively; the other group had respective average K, Th, and U concentrations of about 2400, 3.3, and 0.8 ppm. Similar chemical differences were noted by Gast *et al.* (1970) and by Compston *et al.* (1970).

All of the crystalline rocks of Apollo 12 which we have examined resemble the low-potassium group of Apollo 11 and have average concentrations of K, Th, and U of about 520, 0.97, and 0.25 ppm, respectively. These Apollo 12 samples show remarkably small deviations from the average values. Warner and Anderson (1971) have suggested that the crystalline rocks of Oceanus Procellarum can be classified into petrographic groups and some attempts have been made to classify these Apollo 12 basalts into various chemical groups. However, the chemical differences are quite subtle (e.g., GOLES *et al.*, 1971) and more data are needed to compare the petrographic and chemical classifications.

The fines and breccia from the Ocean of Storms are quite different from the crystalline rocks in several respects. The concentrations of the radioactive elements K, Th, and U are much higher and more variable in the fines and breccia than in the crystalline rocks. The average ratio K/U for fines and breccia is only about 1300 (see Table 1), compared with an average value of about 2100 for the crystalline rocks. It is important to note that the mass ratio Th/U is approximately 3.8 for all materials listed in Table 1, in good agreement with average Th/U ratios for Tranquillity Base materials and terrestrial materials. The concentrations of K, U, and Th in the fines and breccia from the Apollo 11 site were intermediate between the respective concentrations for the two chemical classes of crystalline rocks. It appeared on this basis that the fines and breccia might have been derived from the crystalline rocks at the Sea of Tranquillity; however, more detailed chemical studies later showed that Apollo 11 soil contained other components differing slightly from the crystalline rocks in composition (WOOD *et al.*, 1970; WASSON *et al.*, 1970). Clearly, the crystalline rocks whose concentrations are given in Table 1 could not alone have formed the Apollo 12 fines and breccia.

Preliminary observations that the crystalline lunar rocks are strongly depleted in volatile elements (LSPET 1969, 1970) have been substantiated by detailed studies (e.g., GANAPATHY *et al.*, 1970a, 1970b). This depletion of volatile elements is thought to be characteristic of the whole moon. Studies of the primordial radioactive elements are particularly useful in this connection, because the ratios of K (a volatile element) to U (a refractory element) for common terrestrial rock types are remarkably constant over a wide range of K concentration. Thus, the K/U ratio appears to be a result of early chemical fractionation and is not affected by later igneous processes. It was noted by WASSERBURG *et al.*, (1964) that the terrestrial average K/U ratio of 10^4 was distinct from K/U ratios for meteorites.

In Fig. 1 we show a plot of K/U mass ratios as a function of potassium concentration, normalized to 18 wt. % silicon. Data on lunar samples are from our gamma spectrometry studies and from other chemical analyses reported in the literature. It is clear that eucrites, carbonaceous chondrites, and ordinary chondrites fall into groups which are distinct from one another and from either lunar or terrestrial materials. It may be noted that on these systematics the tektites fall within the terrestrial group. The meteorites fall on a common "trend line" for which the product of the K and U concentrations is constant. In contrast, both the earth and the moon possess constant K/U ratios characteristic of each planet.

The values of K/U for lunar material appear to be separable into five distinct groups, as shown in Fig. 1: (a) Apollo 12 rocks + low-K Apollo 11 rocks; (b) Apollo 11 fines and breccia; (c) high-K Apollo 11 rocks; (d) Apollo 12 fines and breccia; and (e) rock 12013. The relationships between these "islands" in the K/U systematics suggest some interesting possibilities. If trend lines were drawn in Fig. 1 through the Apollo 11 points and through the Apollo 12 data (excluding rock 12013), the intersection falls in the group of points for the crystalline rocks of Apollo 12 and the low-K rocks of Apollo 11. It may be significant that this intersection falls near the extra-

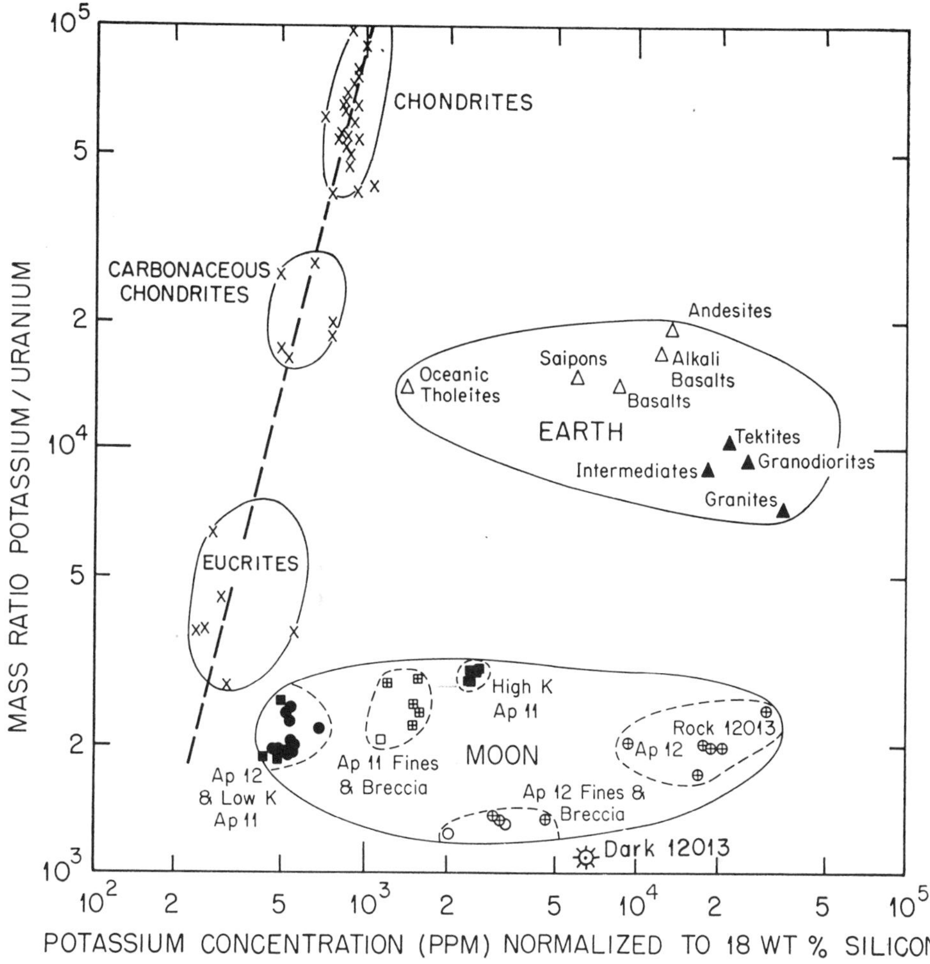

Fig. 1. Plot of mass ratios K/U as a function of K concentration normalized to 18 wt. % Si. The distinct grouping of meteorites and terrestrial and lunar materials is apparent.

polated trend line for meteorites. The divergence of the trend lines for Apollo 11 and Apollo 12 materials from each other suggests that the samples at the two landing sites are the products of two separate magmatic processes which originated from material with similar K and U concentrations. This conclusion was reached by Fanale and Nash (1971) who analyzed preliminary K and U data by a different graphical method. This analysis also implies that the white portion of rock 12013 was involved in a third magmatic event. The starting material for all three events is depleted in K but enriched in U with respect to chondrites.

Rock 12013 and the origin of Apollo 12 fines and breccia

Sample 12013 proved to be one of the most interesting samples studied. Our gamma-ray spectrometry measurements during the preliminary examination showed clearly that the specimen was dramatically different from all lunar samples previously examined (LSPET, 1970). The concentrations of K and U were about 40 times the average concentrations of the crystalline rocks of Apollo 12 and about 10 times the concentrations of the high-K rocks of Apollo 11. In Table 2 we show a refined analysis of our data on 12013,0 and an additional measurement on the 66-g piece (12013,11) which remained after samples were cut for destructive analysis. Despite the inhomogeneity of this breccia on a microscopic scale, the primordial radionuclide concentrations from our two measurements on large rock fragments agree very well with each other. The Th/U ratio is distinctly different, namely, 3.3 compared with 3.8 for other Apollo 12 material.

It is interesting to speculate on the possible role of rock 12013 or one of its components in forming the Apollo 12 fines and breccia. Our K/U systematics in Fig. 1 suggest that the required material to mix with the crystalline rocks must have a low ratio K/U of about 1000 and a K concentration of about 5000–7000 ppm. Further, since our Th/U ratios of Table 1 show that the fines, breccia, and crystalline rocks all have a common Th/U ratio of 3.8, not all of the constituents of 12013 can be common to the lunar soil.

Studies on 12013 have disclosed a general chemical similarity between the light and gray material, which is distinct from the dark material. Chemical analyses on these two classes of components were averaged with the results shown in Table 2. The average K concentration of 6500 ppm for the dark material and 25,500 ppm for the light and gray is in fair agreement with our measured concentration of about 20,000 ppm for the whole rock. The K concentration of the light material is too high to be the missing component. Further, WAKITA and SCHMITT (1970) report a range of 2.3–3.2 for the Th/U ratio, which also militates against the light material. However, TATSUMOTO (1970) found dark material from 12013,10 which had a U concentration of 5.7 ppm and a "normal" Th/U mass ratio of 3.7. Thus, it appears that a component of

Table 2. Concentrations of K, Th, and U in lunar rock 12013.

Sample	Weight (g)	K (ppm)	Th (ppm)	U (ppm)	K/U Mass ratio	Th/U Mass ratio
12013,0*	80.0	20,400 ± 600	34.2 ± 0.8	10.3 ± 0.5	1980 ± 112	3.32 ± 0.18
12013,11*	66.2	21,100 ± 600	32.2 ± 1.4	9.8 ± 1.0	2160 ± 228	3.29 ± 0.36
12013,10 (dark)	—	~6500†		~5.7‡	~1100	3.7‡
12013,10 (light)	—	25,500§		~13.0‖	~2000	(2.3–3.2)‖

* Standardization for assay of K, Th, and U assumed terrestrial isotopic abundances and equilibrium of Th and U decay series.

† Average of analyses by SCHNETZLER *et al.* (1970), HUBBARD *et al.* (1970) and TURNER (1970).

‡ Estimated from TATSUMOTO (1970).

§ Average of analyses by ALEXANDER (1970), SCHNETZLER *et al.* (1970), HUBBARD *et al.* (1970), TURNER (1970), and WAKITA and SCHMITT (1970).

‖ WAKITA and SCHMITT (1970).

12013 with a K concentration of 6500 ppm, a K/U ratio of ~1100, and a Th/U ratio of 3.8 does exist and may be the component required to form the lunar fines and breccia from the crystalline rocks. This point has been added to the K/U systematics of Fig. 1.

These results strongly support the suggestions of HUBBARD *et al.* (1970) and SCHNETZLER *et al.* (1970) that the dark component of 12013 may be a major component of the lunar soil. However, as recently proposed by HUBBARD *et al.* (1971), the dark material probably contains in turn the exotic basaltic material high in K, rare earth elements and P (KREEP), which appears to comprise a large percentage of the lunar soil. If the KREEP material is found to be high in uranium as discussed above, it would serve the purpose of our K/U systematics equally well.

In Fig. 2 we show two-component mixing diagrams which are in accord with the requirements of our K/U systematics and with the properties of the dark component of 12013. The K mixing line is fitted to our average K concentration of 520 ppm for lunar basalts and to 6500 ppm for the foreign component of the soil. The U mixing line is required to intercept the left ordinate at our average concentration for lunar basalts of 0.25 ppm; however, the intercept on the right ordinate was determined by the best fit of a straight line to all data points. On this basis our best estimate of the U concentration in the foreign component is 5.2 ppm. The agreement obtained for five sets of data is very good for such a simple model. Although concentrations of Al, Mg, and Fe in the crystalline rocks vary over a rather wide range, HUBBARD *et al.* (1971) found that the two-component mixing model for Apollo 12 soils also could explain adequately the variations in the concentrations of the major elements Si, Al, Mg, Ti, and Fe.

Gas retention ages

The elemental analyses for K, Th, and U reported in Table 1 may be combined with rare gas concentrations to estimate crystallization ages of rocks. In Table 3 we show

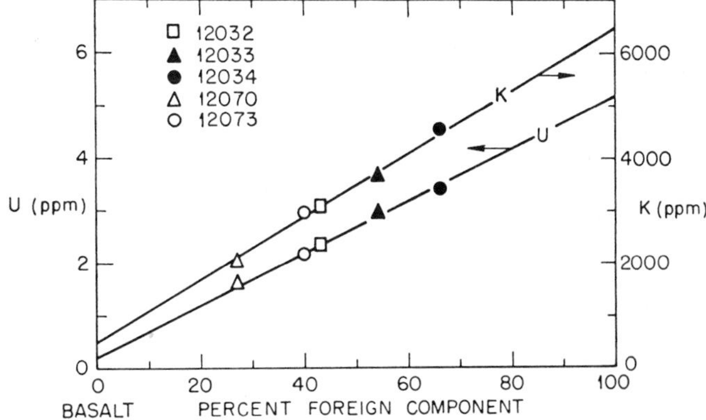

Fig. 2. Two-component mixing lines for K and U concentrations in lunar soils and breccia. Samples (and per cent foreign component) are: soils 12032 (43%), 12033 (54%), 12070 (27%); and breccia 12034 (66%), 12073 (40%).

Table 3. Estimation of gas retention ages of lunar samples.

Sample	K–^{40}Ar (10^9y)	U, Th–^{4}He (10^9y)
12002	2.55*	2.32*
12004	2.92*	2.69*
12021	3.22†	
12052	2.34†	
12062	2.61‡	1.22§
12064	3.10*	2.69*
12065	2.79*	1.95*
12013	3.37‡	1.47‡

* Rare gas concentration from Hintenberger *et al.* (1971); K, Th, U concentrations from Table 1.

† Rare gas concentrations from Marti and Lugmair (1971); K, Th, U concentrations from Table 1.

‡ Rare gas concentrations from LSPET (1970) and from Schaeffer *et al.* (1970); K, Th, U concentrations from Table 1.

§ Rare gas concentrations from LSPET (1970); Th and U concentrations from Table 1.

gas retention ages for 8 samples. Although the K–^{40}Ar and U, Th–^{4}He ages are approximately concordant for 12002, 12004, 12021, and 12064, the U, Th–^{4}He ages are consistently younger than the K–^{40}Ar ages. The low U, Th–^{4}He ages for 12062, 12065, and 12013 suggest that these rocks have experienced significant heating accompanied by preferential loss of He.

Gas retention ages of large samples are generally found to give somewhat shorter ages than those obtained by other methods. However, the K–^{40}Ar ages of Table 3 for 12004, 12021, and 12064 show evidence of little or no loss of radiogenic argon. These ages compare well with the range for all Apollo 12 rocks reported by Albee *et al.* (1971) to lie between 3.15 and 3.35 b.y. Our value for 12064 of 3.10 b.y. lies within experimental error of this range. Our value of 3.22 b.y. for 12021 is in good agreement with 3.28 b.y. obtained by Cliff *et al.* (1971). Similarly, our value of 2.92 b.y. for 12004 agrees with 3.00 b.y. by Murthy *et al.* (1971). None of the Apollo 12 rocks we have studied have exhibited an age as high as the 3.97 b.y. we determined for Apollo 11 rock 10003 (O'Kelley *et al.*, 1970b).

Acknowledgments—The authors gratefully acknowledge contributions to the work reported here by R. S. Clark, J. E. Keith, V. A. McKay, K. J. Northcutt, W. R. Portenier, M. K. Robbins, R. T. Roseberry, and R. E. Wintenberg. We thank the management and staff of the Lunar Receiving Laboratory for their hospitality. Discussions with P. W. Gast helped formulate our ideas. This research was carried out under Union Carbide's contract with the U.S. Atomic Energy Commission through interagency agreements with the National Aeronautics and Space Administration.

References

Albee A. L., Burnett D. S., Chodos A. A., Haines E. L., Huneke J. C., Podosek F. A., Papanastassiou D. A., Price G., Tera F., and Wasserburg G. J. (1971). Rb–Sr ages, chemical abundance patterns and history of lunar rocks. Second Lunar Science Conference (unpublished proceedings).

Alexander E. C., Jr. (1970). Rare gases from stepwise heating of lunar rock 12013. *Earth Planet. Sci. Lett.* **9**, 201–208.

Compston W., Chappell B. W., Arriens P. A., and Vernon M. J. (1970). The chemistry and age of Apollo 11 lunar material. *Proc. Apollo 11 Lunar Sci. Conf., Geochim. Cosmochim. Acta* Suppl. 1, Vol. 2, pp. 1007–1029. Pergamon.

CLIFF R. A., LEE-HU C., and WETHERILL G. W. (1971). Rb–Sr and U, Th–Pb measurements on Apollo 12 material. Second Lunar Science Conference (unpublished proceedings).

FANALE F. P. and NASH D. B. (1971). Potassium–uranium systematics of Apollo 11 and Apollo 12 samples: Implications for lunar material history. *Science* **171**, 282–284.

GANAPATHY R., KEAYS R. R., LAUL J. C., and ANDERS E. (1970a). Trace elements in Apollo 11 lunar rocks: Implications for meteorite influx and origin of moon. *Proc. Apollo 11 Lunar Sci. Conf., Geochim. Cosmochim. Acta* Suppl. 1, Vol. 2, pp. 117–1143. Pergamon.

GANAPATHY R., KEAYS R. R., and ANDERS E. (1970b). Apollo 12 lunar samples: Trace element analysis of a core and the uniformity of the regolith. *Science* **170**, 533–535.

GAST P. W., HUBBARD N. J., and WIESMANN H. (1970). Chemical composition and petrogenesis of basalts from Tranquillity Base. *Proc. Apollo 11 Lunar Sci. Conf., Geochim. Cosmochim. Acta* Suppl. 1, Vol. 2, pp. 1143–1165. Pergamon.

GOLES G. G., DUNCAN A. R., OSAWA M., MARTIN M. R., BEYER R. L., LINDSTROM D. J., and RANDLE K. (1971). Analyses of Apollo 12 specimens and a mixing model for Apollo 12 soils. Second Lunar Science Conference (unpublished proceedings).

HERZOG G. F. and HERMAN G. F. (1970). Na^{22}, Al^{26}, Th, and U in Apollo 11 lunar samples. *Proc. Apollo 11 Lunar Sci. Conf., Geochim. Cosmochim. Acta* Suppl. 1, Vol. 2, pp. 1239–1247. Pergamon.

HINTENBERGER H., WEBER H., and TAKAOKA N. (1971). Concentrations and isotopic abundances of the rare gases in lunar matter. Second Lunar Science Conference (unpublished proceedings).

HUBBARD N. J., GAST P. W., and WIESMANN H. (1970). Rare earth, alkaline and alkali metal and $^{87/86}Sr$ data for subsamples of lunar sample 12013. *Earth Planet. Sci. Lett.* **9**, 181–184.

HUBBARD N. J., MEYER C., GAST P. W., and WIESMANN H. (1971). The composition and derivation of Apollo 12 soils. *Earth Planet. Sci. Lett.* **10**, 341–350.

LSPET (Lunar Sample Preliminary Examination Team) (1970). Preliminary examination of lunar samples from Apollo 11. *Science* **165**, 1211–1227.

LSPET (Lunar Science Preliminary Examination Team) (1970). Preliminary examination of lunar samples from Apollo 12. *Science* **167**, 1325–1339.

MARTI K. and LUGMAIR G. W. (1971). Kr^{81}–Kr and K–Ar^{40} ages, cosmic-ray spallation products and neutron effects in Apollo 11 and Apollo 12 lunar samples. Second Lunar Science Conference (unpublished proceedings).

MURTHY V. R., EVENSEN N. M., JAHN B., and COSCIO M. R. (1971). Rb–Sr isotopic relations and elemental abundances of K, Rb, Sr and Ba in Apollo 11 and Apollo 12 samples. Second Lunar Science Conference (unpublished proceedings).

O'KELLEY G. D., ELDRIDGE J. S., SCHONFELD E., and BELL P. R. (1970a). Elemental compositions and ages of lunar samples by non-destructive gamma-ray spectrometry. *Science* **167**, 580–582.

O'KELLEY G. D., ELDRIDGE J. S., SCHONFELD E., and BELL P. R. (1970b). Primordial radionuclide abundances, solar-proton and cosmic-ray effects and ages of Apollo 11 lunar samples by non-destructive gamma-ray spectrometry. *Proc. Apollo 11 Lunar Sci. Conf., Geochim. Cosmochim. Acta* Suppl. 1, Vol. 2, pp. 1407–1423. Pergamon.

PERKINS R. W., RANCITELLI L. A., COOPER J. A., KAYE J. H., and WOGMAN N. A. (1970). Cosmogenic and primordial radionuclide measurements in Apollo 11 lunar samples by nondestructive analysis. *Proc. Apollo 11 Lunar Sci. Conf., Geochim. Cosmochim. Acta* Suppl. 1, Vol. 2, pp. 1455–1471. Pergamon.

SCHONFELD E. (1967). ALPHA M— an improved computer program for determining radioisotopes by least-squares resolution of the gamma-ray spectra. *Nucl. Instrum. Methods* **52**, 177–178.

SCHAEFFER O. A., FUNKHOUSER J. G., BOGARD D. D., and ZÄRINGER J. (1970). Potassium–argon ages of lunar rocks from Mare Tranquillitatis and Oceanus Procellarum. *Science* **170**, 161–162.

SCHNETZLER C. C., PHILPOTTS J. A., and BOTTINO M. L. (1970). Li, K, Rb, Sr, Ba and rare-earth concentrations, and Rb–Sr age of lunar rock 12013. *Earth Planet. Sci. Lett.* **9**, 185–192.

TATSUMOTO M. (1970). U–Th–Pb age of Apollo 12 rock 12013. *Earth Planet. Sci. Lett.* **9**, 193–200.

TURNER G. (1970). ^{40}Ar–^{39}Ar age determination of lunar rock 12013. *Earth Planet. Sci. Letters* **9**, 177–181.

WAKITA H. and SCHMITT R. A. (1970). Elemental abundances in seven fragments from lunar rock 12013. *Earth Planet. Sci. Lett.* **9**, 169–176.

WARNER J. L. and ANDERSON D. L. (1971). Lunar crystalline rocks: petrology, geology and origin. Second Lunar Science Conference (unpublished proceedings).

WASSERBURG G. J., MACDONALD G. J. F., HOYLE F., and FOWLER W. A. (1964). Relative contributions of uranium, thorium, and potassium to heat production in the earth. *Science* **143**, 465–467.

WASSON J. T. and BAEDECKER P. A. (1970). Ga, Ge, In, Ir, and Au in lunar, terrestrial, and meteoritic basalts. *Proc. Apollo 11 Lunar Sci. Conf., Geochim. Cosmochim. Acta* Suppl. 1, Vol. 2, pp. 1741–1750. Pergamon.

WOOD J. A., DICKEY J. S., Jr., MARVIN U. B., and POWELL B. N. (1970). Lunar anorthosites and a geophysical model of the moon. *Proc. Apollo 11 Lunar Sci. Conf., Geochim. Cosmochim. Acta* Suppl. 1, Vol. 1, pp. 965–988. Pergamon.

WRIGLEY R. C. and QUAIDE W. L. (1970). Al^{26} and Na^{22} in lunar surface materials: Implications for depth distribution studies. *Proc. Apollo 11 Lunar Sci. Conf., Geochim. Cosmochim. Acta* Suppl. 1, Vol. 2, pp. 1751–1757. Pergamon.

Proceedings of the Second Lunar Science Conference, Volume 2, pp. 1169–1185.
The M.I.T. Press, 1971.

Elemental abundances of lunar soil and rocks from Apollo 12

G. H. MORRISON, J. T. GERARD, N. M. POTTER,
E. V. GANGADHARAM, A. M. ROTHENBERG, and R. A. BURDO
Department of Chemistry,
Cornell University, Ithaca, N.Y. 14850

(*Received 12 February 1971; accepted 5 March 1971*)

Abstract—Results are presented for the multielement analysis of samples of lunar soil (12032, 12042, and 12070), breccia (12073) and igneous rocks (12004, 12021, 12051, and 12053). U.S.G.S. standard W-1 was used as a comparative standard and 56 elements were determined so far using spark source mass spectrometry and neutron activation analysis. The chemistry of these samples show that the igneous rocks may fall into two groups. The soil samples from Apollo 12 have higher trace elemental abundances than those from Apollo 11, while the igneous rocks have lower trace elemental abundances than those from Apollo 11. The Apollo 12 breccia sample very closely resembles the soil with respect to elemental abundances. However, the soil and breccia trace elemental abundances are considerably higher than in the igneous rocks. Again, comparison with meteoritic and terrestrial abundances revealed depletion of volatile elements and enrichment of the refractory elements. Titanium is enriched, but not as much as in Apollo 11. The negative Eu anomaly is more pronounced in the soil and breccia but less so in the igneous rocks this time. The lunar material was compared to basaltic achondrites and terrestrial basalts with the closest comparison being to basaltic achondrites. There is an overall similarity of the material returned from the two Apollo sites but yet sufficient differences to suggest detailed geochemical processes special to each site.

INTRODUCTION

A DETAILED elemental abundance study of the Apollo 11 material by this laboratory has been published by MORRISON *et al.* (1970a, 1970b). The present study reports multielement abundance data for three lunar soils, one breccia, and four igneous rocks from Apollo 12 as presented by MORRISON *et al.* (1971), with full consideration given to the experience gained in analyzing the Apollo 11 samples. Information has been obtained so far for 56 elements using spark source mass spectrography, instrumental neutron activation analysis, and irradiations followed by radiochemical separations and high resolution gamma spectrometry. The details of the procedure have been previously described by MORRISON *et al.* (1969a, 1969b, 1970b). Based on an evaluation of the Apollo 11 and 12 soil samples 10084 and 12070, for which abundant data of other investigators are available for comparison, the overall accuracy of the mass spectrographic and neutron activation methods is estimated at 10%. Eleven elements (in Table 1) are reported separately as approximate values because of analytical problems associated with their determination. These include the volatile nature of most of these elements or their occurrence at the detection limit. Despite these limitations we believe that some of these elements are of comparable accuracy as that reported above.

Results and Discussion

The results of the analyses of the eight Apollo 12 samples are given in Table 1. Included are our mass spectrometric (MS) and neutron activation (NAA) results, the average of other investigators reported at the Second Lunar Science Conference, and the range of these values. The mean values should be viewed critically, particularly where the range is large or where the number of values reported is limited. The wide range of values for some elements in the rock samples may reflect sampling differences. In other cases systematic errors on the part of a few investigators tended to bias the average. Where valid comparisons are possible, as for example soil sample 12070 with a large number of values reported, our results are in excellent agreement with the mean values.

Internal comparison of our data reveals the following major observations. The variation in elemental composition of the four igneous rocks analyzed is relatively small, suggesting that they are genetically related; rock 12004, however, exhibits significantly higher Mg, Cr, and Co concentrations. The breccia is similar in composition to the soils analyzed, indicating its closer relationship with the latter than with the igneous rocks. The three soil samples show minor variations in chemistry.

Trace elements are enriched approximately fivefold in the soils and breccia over those in the igneous rocks except for V and Sc which are depleted (Fig. 1). A similar comparison for the major elements in Fig. 2 shows that Ti, Fe, and Mn are higher in

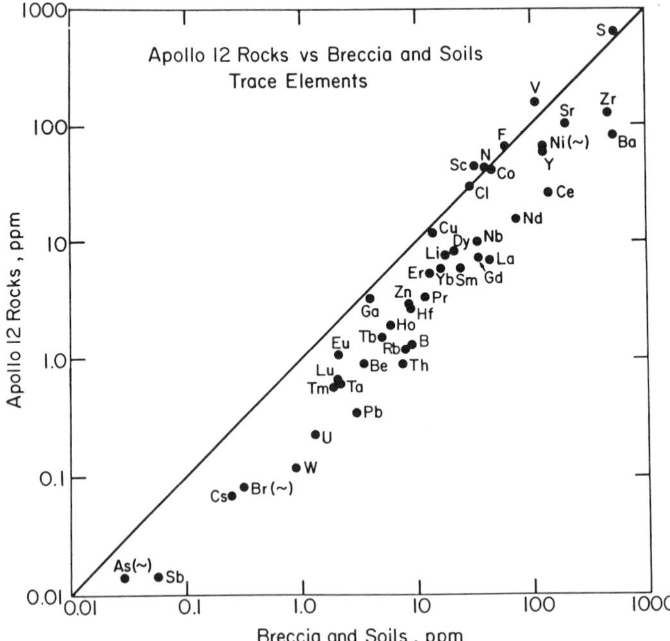

Fig. 1. Comparison of average trace element abundances of igneous rocks vs. soils and breccia in Apollo 12.

Table 1. Elemental abundances in the Apollo 12 lunar samples studied using spark source mass spectroscopy (MS) and neutron activation analysis (NAA). The results are grouped according to NASA sample type: A, fine-grained crystalline; B, medium-grained crystalline; C, breccia; and D, soil

Element	Method	Type A Rocks		Type AB Rock	Type B Rock	Type C Rock	Type D Soils		
		12004,40	12053,25	12051,42	12021,83	12073,35	12042,46	12070,07	12032,35
		Major elements, weight percent							
Al	NAA	4.7	6.0	5.9	5.9	7.0	7.0	6.9	8.3
	Aver.*	4.7 (6)	5.36 (2)	5.3 (4)	5.7 (6)	7.37 (1)	6.9 (1)	6.7 (10)	7.3 (3)
	Range*	4.24–5.8	5.33, 5.39	5.11–5.5	5.43–5.9	7.37	6.9	6.50–6.82	7.10–7.5
Ti	NAA	1.6	1.6	2.2	2.3	1.3	1.4	1.5	1.1
	Aver.	1.7 (4)	2.1 (3)	2.8 (4)	2.1 (5)	1.3 (1)	1.6 (1)	1.7 (10)	1.6 (2)
	Range	1.53–1.9	2.0–2.2	2.6–3.2	1.93–2.24	1.3	1.6	1.6–2.1	1.54, 1.7
Fe	NAA	16.0	14.3	15.6	15.6	11.2	13.7	12.2	11.7
	Aver.	16.2 (2)	15.5 (2)	15.6 (4)	14.9 (6)	11.3 (1)	12.6 (1)	12.5 (11)	11.9 (3)
	Range	15.1–17.11	15.37, 15.6	14.7–16.3	14.2–15.1	11.3	12.6	11.5–13.2	11.76–12.4
Mg	NAA	8.4	4.9	4.6	4.4	4.6	4.9	6.8	6.6
	Aver.	7.2 (6)	4.95 (2)	4.2 (4)	4.46 (4)	5.58 (1)	6.3 (1)	6.4 (8)	6.85 (3)
	Range	6.5–7.63	4.86, 4.92	3.7–4.5	4.27–4.6	5.58	6.3	5.7–7.9	5.96–8.1
Ca	NAA	8.0	6.9	7.3	9.1	7.6	8.7	6.8	7.4
	Aver.	6.7 (5)	7.4 (3)	8.17 (4)	7.9 (4)	8.0 (1)	7.6 (1)	7.3 (9)	7.8 (2)
	Range	5.9–7.7	6.7–7.87	7.8–8.6	7.3–8.2	8.0	7.6	6.5–7.6	7.52, 8.1
Na	NAA	0.16	0.19	0.24	0.21	0.48	0.26	0.32	0.44
	Aver.	0.17 (5)	0.21 (3)	0.23 (4)	0.21 (5)	0.446 (1)	0.40 (1)	0.33 (11)	0.43 (2)
	Range	0.147–0.22	0.1929–0.23	0.208–0.236	0.185–0.223	0.446	0.40	0.289–0.371	0.4155, 0.4465
K	NAA	0.047	0.050	0.051	0.059	0.32	0.14	0.19	0.31
	Aver.	0.0476 (6)	0.053 (4)	0.052 (6)	0.051 (6)	0.32 (3)	0.21 (1)	0.189 (14)	0.31 (4)
	Range	0.037–0.06	0.050–0.0582	0.03–0.0664	0.043–0.058	0.3162–0.33	0.21	0.12–0.217	0.298–0.35
Mn	NAA	0.22	0.22	0.21	0.22	0.14	0.17	0.19	0.17
	Aver.	0.21 (5)	0.219 (2)	0.21 (3)	0.21 (6)	0.159 (1)	0.176 (2)	0.17 (11)	0.157
	Range	0.19–0.225	0.216, 0.222	0.206–0.217	0.192–0.217	0.159	0.162, 0.190	0.155–0.18	0.1525, 0. 1604
Cr	NAA	0.48	0.22	0.16	0.25	0.22	0.27	0.28	0.24
	Aver.	0.45 (5)	0.34 (2)	0.21 (3)	0.26 (5)	0.226 (1)	0.27 (2)	0.27 (13)	0.245 (2)
	Range	0.375–0.490	0.335, 0.348	0.178–0.246	0.1965–0.318	0.226	0.250, 0.293	0.208–0.700	0.237, 0.25
		Trace elements, parts per million							
Li	MS	—	7.6	7.4	—	—	—	18	—
	Aver.	11 (2)	6.1 (2)	—	7.9 (4)	—	—	19 (6)	—
	Range	11, 11	4.8, 7.3	—	6.6–8.71	—	—	12–28	—
Be	MS	—	0.9	0.9	—	—	—	3.5	—
	Aver.	—	1.7 (1)	—	1.7 (2)	—	6.8 (1)	3.5 (4)	4.3 (1)

Table 1 (continued)

Element	Method	Type A Rocks		Type AB Rock	Type B Rock	Type C Rock	Type D Soils		
		12004,40	12053,25	12051,42	12021,83	12073,35	12042,46	12070,07	12032,35
	Range	—	1.7	—	1.5–1.9	—	6.8	3.21–4.0	4.3
B	MS	—	1.4	1.2	—	—	—	9	—
	Aver.	1.9 (2)	1.4 (1)	—	—	—	—	7 (2)	—
	Range	0.8, 3	1.4	—	—	—	—	2.3, 11	—
Sc	NAA	44	46	48	45	28	34	35	32
	MS	—	42	41	—	—	—	38	—
	Aver.	43 (2)	53 (2)	60 (1)	51 (4)	36.2 (1)	38 (1)	39 (10)	37 (2)
	Range	43, 43.8	50, 56.4	60	48–53.5	36.2	38	36–43	36, 39
V	NAA	180	170	100	190	83	150	72	130
	MS	—	140	120	—	—	—	100	—
	Aver.	172 (4)	148 (1)	116 (2)	160 (6)	—	108 (1)	110 (10)	103 (2)
	Range	100–230	148	102–130	130–205	—	108	49–143	96, 110
Co	NAA	60	36	35	37	39	52	44	39
	MS	—	36	35	—	—	—	43	—
	Aver.	50 (5)	35 (2)	37 (3)	34 (4)	37 (2)	52 (1)	43 (12)	37 (3)
	Range	31–63	30, 39.1	33–42	28.6–38	36, 38.2	52	17–52	31–42
Cu	NAA	6.2	24	9.1	9.7	6.8	28	7.7	6.5
	Aver.	11.5 (4)	17 (2)	11 (2)	11.7 (3)	5.7 (1)	10 (1)	8.1 (9)	29 (1)
	Range	6.9–22	7.1, 26	6, 16	8.2–14	5.7	10	5–12.5	29
Zn	NAA	4.2	2.1	1.6	4.2	6.8	8.5	8.7	8.4
	Aver.	5.5 (2)	—	0.53 (1)	3.2 (3)	6.49 (1)	7.2 (1)	8.2 (9)	13.9 (2)
	Range	3, 8	—	0.53	1.2–4.3	6.49	7.2	6–15.5	5.71, 22
Ga	NAA	2.5	2.9	3.9	3.4	4.4	4.0	3.8	4.3
	Aver.	3.2 (3)	4.1 (1)	3.7 (2)	4.6 (3)	4.7 (2)	4.4 (1)	3.9 (8)	5.1 (1)
	Range	1.2–4.5	4.1	2.9, 4.5	3.6–5.4	4.3, 5.1	4.4	2.5–4.9	5.1
Rb	NAA	≤1	≤3	≤3	≤1	9.2	9.9	5.8	7.5
	MS	—	1.2	1.2	—	—	—	5.9	—
	Aver.	1.20 (5)	1.3 (3)	0.987 (4)	1.19 (4)	10.0 (4)	5.5 (1)	6.8 (13)	8.5 (4)
	Range	0.7–2	1.1–1.7	0.909–1.05	1.14–1.25	9.24–11.3	5.5	5.6–9.5	7.5–9.24
Sr	MS	—	120	150	—	—	—	130	—
	Aver.	93.4 (5)	127 (3)	145 (3)	102 (4)	174 (3)	110 (1)	144 (11)	158 (2)
	Range	72–115	111–140	140–148	73–128.5	163–190	110	120–190	154, 161.4
Y	MS	—	54	65	—	—	—	130	—
	Aver.	52 (4)	52 (1)	51 (3)	50 (2)	—	128 (1)	135 (8)	180 (2)
	Range	36–87	52	42–64	48, 53	—	128	110–180	164, 197
Zr	NAA	120	120	130	140	390	390	460	590
	MS	—	110	130	—	—	—	420	—
	Aver.	130 (2)	138 (1)	149 (2)	115 (3)	—	482 (1)	480 (8)	631 (3)

	Range	110, 150	138	128, 170	100–133	—	482	370–600	512–741
Nb	MS	—	9.4	11	—	—	—	35	—
	Aver.	9.5 (2)	10.0 (1)	7 (1)	13 (2)	—	26 (1)	33.5 (6)	58 (1)
	Range	7, 12	10.0	7	11, 16	—	26	25–45	58
Sb	NAA	0.010	0.008	0.004	0.032	0.003	0.035	0.009	0.130
	Aver.	—	—	—	0.04 (1)	—	—	—	—
	Range	< 0.7	—	—	0.04	—	—	0.04, 1.27	—
Cs	NAA	≤0.2	≤0.1	≤0.1	≤0.2	0.3	0.2	0.2	0.3
	MS	—	0.07	—	—	—	—	0.23	—
	Aver.	0.07 (2)	0.10 (1)	0.045 (2)	0.068 (1)	0.44 (2)	—	0.30 (5)	0.37 (2)
	Range	0.05, 0.09	0.10	0.041, 0.05	0.068	0.385, 0.5	—	0.24–0.39	0.36, 0. 38
Ba	NAA	69	86	96	72	650	510	330	690
	MS	—	89	95	—	—	—	330	—
	Aver.	64 (5)	97 (3)	83 (3)	80 (5)	509 (3)	445 (1)	365 (13)	487 (3)
	Range	51.8–79	84.4–120	73.6–100	44–125	390–571	445	270–423	400–531
La	NAA	6.4	6.4	7.4	7.2	50	42	36	52
	MS	—	6.9	7.6	—	—	—	36	—
	Aver.	5.5 (3)	7.1 (3)	5.7 (4)	6.6 (3)	48.9 (2)	43 (2)	36 (14)	49 (2)
	Range	5–6.0	7.07–7.26	5–6.53	6.18–7.38	48.4, 49.8	38.3, 48	22–82	49, 51
Ce	NAA	21	27	28	27	120	110	97	130
	MS	—	19	24	—	—	—	99	—
	Aver.	19 (3)	23 (3)	17.3 (3)	19 (4)	126 (3)	116 (1)	83 (13)	116 (2)
	Range	15–26	19.0–30	15.1–19.2	17.1–20	120–131	116	50.7–115	114, 117
Pr	MS	—	3.5	3.1	—	—	—	12	—
	Aver.	3.3 (3)	2.8 (1)	2.0 (2)	—	—	—	14 (5)	15 (1)
	Range	1.9–5.5	2.8	1.87, 3.2	—	—	—	6.26–26	15
Nd	NAA	13	17	19	16	86	67	59	97
	MS	—	19	19	—	—	—	68	—
	Aver.	14.0 (3)	15 (2)	14 (3)	15 (3)	77 (2)	82 (1)	53 (9)	74 (2)
	Range	12.9–16	15, 15.3	10.7–15.8	14.4–16	76.7, 77	82	32.3–74	73, 75.7
Sm	NAA	5.4	5.4	6.9	6.0	25	24	19	28
	MS	—	5.9	6.7	—	—	—	19	—
	Aver.	4.2 (3)	5.2 (3)	5.59 (2)	5.2 (4)	22 (3)	20.5 (1)	17.1 (11)	21.0 (2)
	Range	3.2–5.0	4.7–5.5	5.50, 5.68	4.8–5.55	21.4–22.4	20.5	14.7–19	20.7, 21.7
Eu	NAA	0.89	0.96	1.3	1.1	2.4	2.3	1.7	2.3
	MS	—	0.93	1.2	—	—	—	1.8	—
	Aver.	1.0 (3)	0.94 (3)	1.25 (3)	1.1 (4)	2.3 (3)	2.11 (1)	1.6 (11)	2.15 (2)
	Range	0.82–1.4	0.67–1.10	0.76–1.28	1.045–1.18	2.17–2.44	2.11	0.88–1.8	2.12, 2.18
Gd	NAA	7.4	6.7	6.8	8.0	31	30	23	49
	MS	—	6.0	6.7	—	—	—	17	—
	Aver.	5.0 (3)	7.6 (2)	6.6 (3)	6.7 (2)	26.9 (2)	23.4 (1)	17 (7)	24.8 (2)
	Range	4.7–5.2	7.53, 7.6	5.21–7.89	6.59, 6.9	26.8, 26.9	23.4	12.8–25	24.8, 24.9
Tb	NAA	1.4	1.3	1.8	1.4	5.3	4.7	4.1	6.1

Table 1 (continued)

Element	Method	Type A Rocks		Type AB Rock	Type B Rock	Type C Rock	Type D Soils		
		12004,40	12053,25	12051,42	12021,83	12073,35	12042,46	12070,07	12032,35
	MS	—	1.3	1.5	—	—	—	3.8	—
	Aver.	1.3 (3)	1.5 (2)	1.35 (1)	1.2 (2)	6.2 (1)	4.03 (1)	4.0 (9)	4.3 (1)
	Range	0.97–2	1.33, 1.73	1.35	1.14, 1.25	6.2	4.03	1.99–6	4.3
Dy	MS	—	7.5	8.9	—	—	—	21	—
	Aver.	6.7 (3)	7.7 (3)	7.3 (3)	8.5 (3)	31 (1)	26.9 (1)	22 (10)	27 (2)
	Range	5.5–9	6.53–8.44	5.58–9.05	7.86–9.3	31	26.9	13.7–29	26.5, 28
Ho	NAA	1.3	1.8	2.4	2.1	6.3	6.1	4.8	6.7
	MS	—	2.0	2.3	—	—	—	4.9	—
	Aver.	1.5 (2)	1.8 (2)	1.92 (3)	1.6 (2)	7.58 (1)	4.65 (1)	4.9 (8)	6.5 (1)
	Range	1.4, 1.5	1.49, 2.1	1.38–2.27	1.4, 1.75	7.58	4.65	2.65–6.7	6.5
Er	MS	—	4.6	5.8	—	—	—	13	—
	Aver.	3.9 (2)	5.0 (3)	5.5 (2)	5.5 (3)	—	15 (1)	13 (8)	16.6 (2)
	Range	3.84, 3.9	4.8–5.4	5.4, 5.57	4–7.9	—	15	8.4–16	15.9, 17.3
Tm	NAA	0.51	0.53	0.69	0.53	2.0	1.8	1.6	2.2
	MS	—	0.66	0.68	—	—	—	1.5	—
	Aver.	0.66 (1)	—	0.64 (2)	—	—	—	1.8 (3)	2.8 (1)
	Range	0.66	—	0.49, 0.78	—	—	—	1.07–2.4	2.8
Yb	NAA	5.7	5.9	6.1	5.3	15	18	13	18
	MS	—	4.1	5.6	—	—	—	11	—
	Aver.	5.0 (4)	4.5 (4)	5.3 (4)	5.5 (5)	16.5 (3)	14.4 (1)	12.3 (13)	16.1 (2)
	Range	3.17–7	3.92–5.2	3.38–7.5	4.12–9.5	16.0–17.4	14.4	6.7–20	15.2, 17.1
Lu	NAA	0.61	0.61	0.73	0.72	1.9	2.1	1.8	2.5
	MS	—	0.67	0.72	—	—	—	1.4	—
	Aver.	0.75 (3)	0.63 (3)	0.73 (1)	0.82 (5)	2.30 (2)	—	1.7 (9)	2.28 (2)
	Range	0.44–1.3	0.614–0.662	0.73	0.523–1.56	2.17, 2.43	—	1.30–1.9	2.24, 2.32
Hf†	NAA	3.9	3.4	4.4	4.6	12	11	13	15
	MS	—	3.3	3.9	—	—	—	12	—
	Aver.	4.3 (3)	4.0 (1)	3.3 (1)	3.7 (2)	21.7 (1)	—	13 (7)	18 (1)
	Range	2.7–5.1	4.0	3.3	3.3, 4.06	21.7	—	11.2–17.2	18
Ta	NAA	0.5	0.6	0.5	0.7	2.7	1.4	3.3	1.8
	Aver.	0.33 (1)	0.45 (1)	—	0.41 (2)	2.1 (1)	—	1.6 (3)	—
	Range	0.33	0.45	—	0.41, 0.42	2.1	—	1.41–2.05	—
W	NAA	0.07	0.16	0.09	0.15	0.86	0.97	0.73	1.0
	Aver.	0.14 (1)	0.12 (1)	—	0.23 (1)	1.21 (1)	—	0.63 (3)	—
	Range	0.14	0.12	—	0.23	1.21	—	0.5–0.74	—
Th	NAA	0.79	0.82	0.95	1.1	10	6.1	6.9	9.6
	Aver.	0.77 (4)	1.1 (3)	1.00 (3)	1.17 (4)	8.3 (2)	—	6.1 (11)	8.5 (2)
	Range	0.4–0.92	0.85–1.34	0.901–1.1	0.932–1.6	8.17, 8.45	—	4.4–6.86	8.1, 8.8

U	NAA	0.19	0.24	0.20	0.27	2.2	1.0	1.4	1.5
	Aver.	0.27 (4)	0.26 (4)	0.25 (2)	0.24 (4)	2.25 (2)	—	1.6 (10)	2.35 (1)
	Range	0.238–0.303	0.20–0.322	0.234, 0.260	0.19–0.261	2.19, 2.32	—	1–1.95	2.35
					Approximate Values, ppm‡				
N	MS	—	36	54	—	—	—	40	—
	Aver.	—	—	—	—	130 (1)	—	—	48 (1)
	Range	—	—	—	—	130	—	—	48
F	MS	—	51	85	—	—	—	60	—
	Aver.	113 (1)	—	—	—	—	—	152 (2)	—
	Range	113	—	—	—	—	—	63, 241	—
P	MS	—	320	420	—	—	—	1200	—
	Aver.	166 (2)	599 (1)	345 (2)	390 (2)	—	1440 (1)	1200 (4)	1691 (1)
	Range	112, 220	599	310, 380	390, 390	—	1440	644–1440	1691
S	MS	—	540	780	—	—	—	530	—
	Aver.	896 (2)	785 (1)	950 (2)	—	—	—	1020 (3)	688 (1)
	Range	700, 1092	785	900, 1000	—	—	—	752–1200	688
Cl	MS	—	30	32	—	—	—	30	—
	Aver.	30 (2)	61 (3)	—	8.1 (2)	—	—	32 (4)	77 (1)
	Range	29, 31.5	9.4–146	—	4.7, 11.5	—	—	24–106	77
Ni	NAA	≤20	39	66	87	120	≤50	210	100
	MS	—	42	43	—	—	—	200	—
	Aver.	67 (4)	19 (2)	6 (1)	15 (2)	230 (1)	—	190 (8)	117 (1)
	Range	52–80	10, 28	6	13, 16	230	—	164–222	117
As	NAA	0.007	0.02	0.02	0.01	0.03	0.03	0.1	0.002
	Aver.	— (2)	0.01 (1)	—	0.14 (2)	0.026 (1)	—	0.24 (3)	—
	Range	0.004, 0.1	0.01	—	0.09, 0.18	0.026	—	0.022–0.57	—
Br	NAA	0.04	0.1	0.1	0.09	0.13	0.2	0.3	0.5
	Aver.	0.08 (1)	0.019 (1)	0.016 (1)	—	0.13 (1)	—	0.20 (3)	0.18 (1)
	Range	0.08	0.019	0.016	—	0.13	—	0.13–0.33	0.18
Mo	NAA	≤0.3	≤0.8	≤0.5	≤0.5	≤0.6	≤0.2	≤0.5	≤0.7
	MS	—	0.3	0.2	—	—	—	0.3	—
	Aver.	0.6 (1)	—	—	—	—	—	0.45 (2)	—
	Range	0.6	—	—	—	—	—	0.3, 0.6	—
Hg	NAA	< 0.03	< 0.06	< 0.06	< 0.05	< 0.06	< 0.03	< 0.06	< 0.06
	Aver.	—	0.0058 (1)	—	0.002 (1)	—	—	0.0021 (1)	—
	Range	—	0.0058	—	0.002	—	—	0.0021	—
Pb	MS	—	0.3	0.4	—	—	—	3	—
	Aver.	1.7 (2)	0.672 (1)	—	0.419 (1)	—	—	2.6 (3)	—
	Range	0.848, 2.5	0.672	—	0.419	—	—	1.1–3.51	—

* The averages and ranges were derived from the data reported by the following authors in their unpublished proceedings at the Second Lunar Science Conference. The numbers in the parentheses indicate the number of laboratories included in the average. ANDERS *et al.* (1971), ANNELL *et al.* (1971), BAEDECKER *et al.* (1971), BOUCHET *et al.* (1971), BRUNFELT *et al.* (1971), COMPSTON *et al.* (1971), EHMANN and MORGAN (1971), ENGEL and ENGEL (1971), FRICK *et al.* (1971),

Gast and Hubbard (1971), Goles *et al.* (1971), Haskin *et al.* (1971), Hess *et al.* (1971), Hubbard *et al.* (1971), Lipshutz *et al.* (1971), Maxwell and Wiik (1971), Moore *et al.* (1971), Murthy *et al.* (1971), O'Kelley *et al.* (1971), Philpotts *et al.* (1971), Rancitelli *et al.* (1971), Reed and Javanoic (1971), Rose *et al.* (1971), Sievers *et al.* (1971), Silver (1971), Smales *et al.* (1971), Tatsumoto (1971), Taylor (1971), Travesi and Palomares (1971), Wikita *et al.* (1971), Wänke *et al.* (1971), Willis *et al.* (1971).

† Hf is the only element whose values have been changed from those presented at the Apollo 12 Lunar Science Conference. The values were calculated using a corrected value for the Hf standard employed.

‡ These elements have analytical problems associated with their determination and therefore are reported only as approximate values.

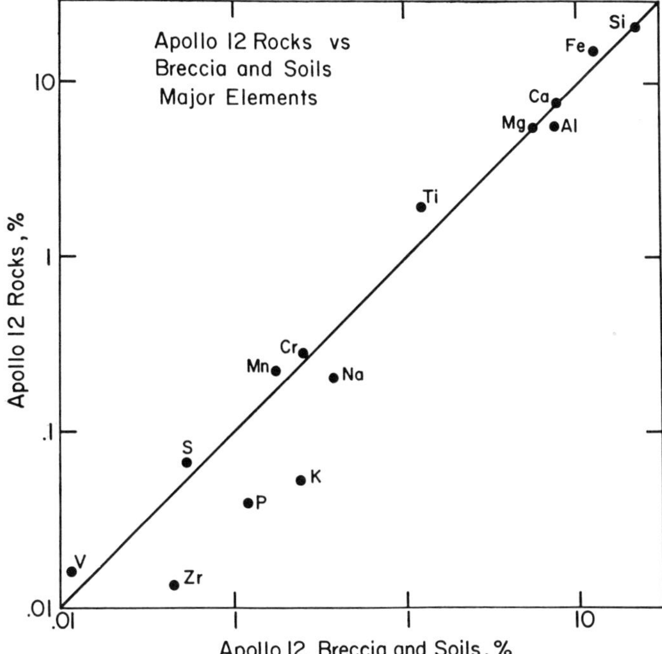

Fig. 2. Comparison of average major element abundances of igneous rocks vs. soils and breccia in Apollo 12.

the igneous rocks while Al, Na, K, and Zr are higher in the breccia and soils. Silicon, Mg, Ca, and Cr show little variation. The observations made on the eight samples reported here are in general agreement with the findings of LSPET (1970) and the Second Lunar Science Conference.

Several investigators have reported the possible existence of two chemical groups of igneous rocks based on the limited number of samples they received. Goles *et al.* (1971) classified their rocks into group 1, containing lower Ti, Al, Ca, Na, rare earths, Hf, and Sc and higher Co than their group 2 rocks. Wakita *et al.* (1971) also suggested two groups based on lower Ti, Ca, Na, and rare earths and higher V in one group relative to the other. Ehmann and Morgan (1971) suggested the possibility of two groups based on the behavior of Al. Based on the behavior of Mg, Cr, Co, and Al, our analyses of the four igneous rocks received shows rock 12004 to be chemically distinguishable from the other three.

When the total data on all igneous rocks are examined, however, a continuous range of concentrations are observed for all of the above elements. Chemical grouping of Apollo 12 rocks, therefore, is not as obvious as that in Apollo 11 igneous rocks, which fell into two chemically distinct groups based on the behavior of many elements. The work of Compston *et al.* (1971) is certainly of importance at this point, since chemical grouping of Apollo 12 rocks is somewhat indecisive. Based on strontium isotopic composition, they found at least five groups of igneous rocks which corresponded with six chemical groups based on the behavior of Rb, Sr, Y, and Zr.

Comparison of the elemental abundance patterns of the materials from the two mare sites sampled by Apollo 12 and Apollo 11 reveals significant differences. Fig. 3 compares average trace element abundances of the igneous rocks in Apollo 12 and Apollo 11, where the Apollo 11 data represent the average of all igneous rocks reported on in the *Moon Issue of Science* (1970). All trace elements except Co and V are lower in the Apollo 12 rocks. A similar comparison for the major elements in Fig. 4 shows considerable depletion of Ti, Zr, K, Na, P, and S in the Apollo 12 rocks.

Trace element abundances of Apollo 12 and Apollo 11 soils are compared in Fig. 5. All trace elements except Ga, Zn, Sc, Ni, and N are enriched in the Apollo 12 soils. In general, this is the opposite of the relationship observed in the igneous rocks. Amongst the major and minor elements Ti and S are lower in the Apollo 12 soils while most of the others are slightly enriched as shown in Fig. 6.

These features reflect to a large extent the behavior of the incompatible elements, i.e., U, Th, Zr, Ba, rare earths, etc. According to RINGWOOD and ESSENE (1970) the behavior of these incompatible elements can be explained by partial melting from parental material within the lunar interior. This has been used by COMPSTON *et al.* (1970) to explain the two groups of igneous rocks in Apollo 11 where differences in their trace-element and radiogenic isotope chemistry suggest two magmas formed by different degrees of partial melting. The Apollo 12 rocks appear to be derived from partial melting from another source region or regions with at least five episodes of

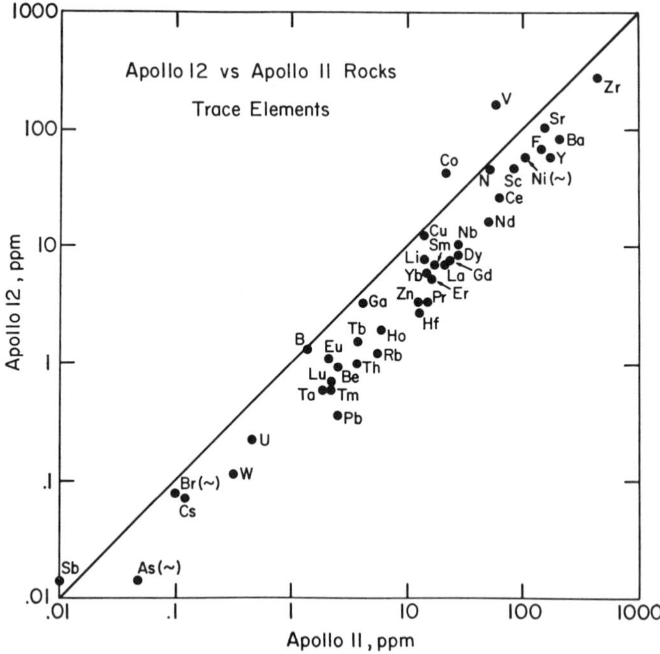

Fig. 3. Comparison of average trace element abundances of igneous rocks in Apollo 12 and Apollo 11.

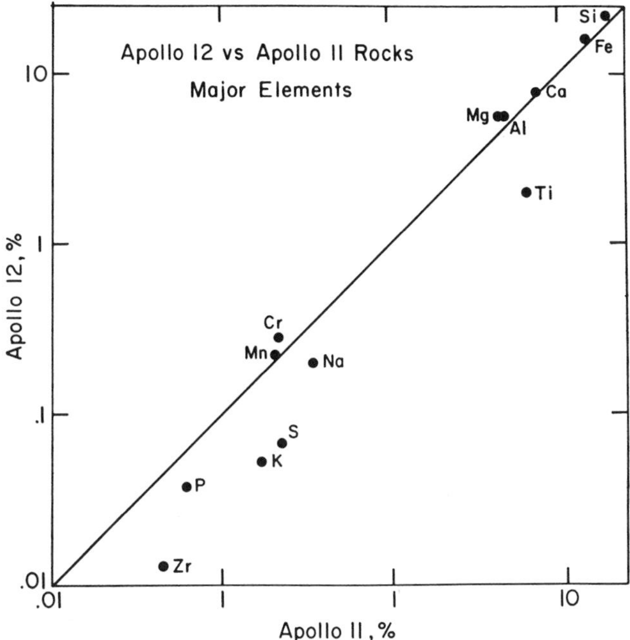

Fig. 4. Comparison of average major element abundances of igneous rocks in Apollo 12 and Apollo 11.

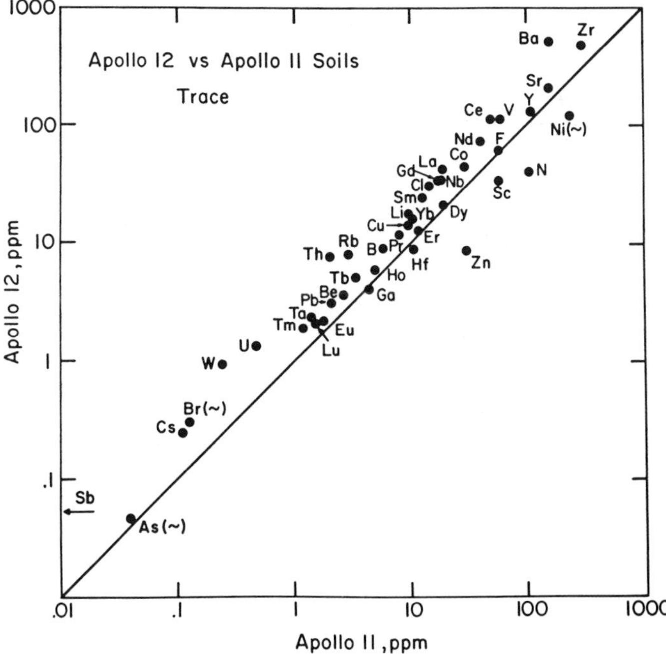

Fig. 5. Comparison of average trace element abundances of soils in Apollo 12 and Apollo 11.

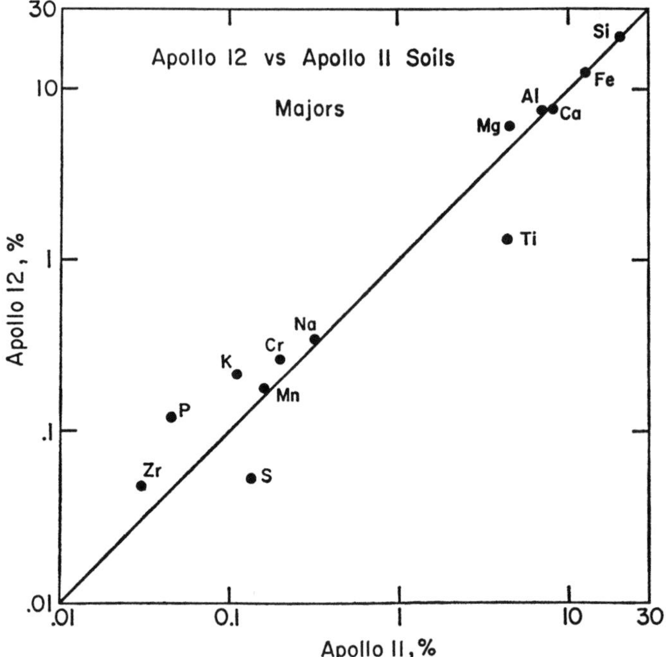

Fig. 6. Comparison of average major element abundances of soils in Apollo 12 and Apollo 11.

partial melting according to COMPSTON *et al.* (1971). The degree of partial melting is greater in the Apollo 12 episodes than in the Apollo 11.

It is observed from the Apollo 12 analyses that these same elements are concentrated approximately fivefold in the soils and breccia over the igneous rocks. The soils and breccia could not have been formed directly from the crystalline rocks unless there is yet another suite of rocks which have not been sampled. One rock which suggests such a possibility is the composite and complex rock 12013, which has lithic components that are late stage magmatic differentiates (*Earth Planet. Sci. Lett.* special issue on lunar sample 12013 (1970)). Alternatively, partial melting followed by crystal differentiation could have further enriched the incompatible elements in the residual liquid which subsequently solidified and comminuted to yield the soil material.

The rare-earth elements in the Apollo 12 igneous rocks exhibit a lower absolute abundance than in the Apollo 11 rocks. The degree of this depletion is of the order of 2.5 to 3. In contrast, the breccia and soils of Apollo 12 contain a higher absolute abundance than the corresponding Apollo 11 material. Fig. 7 shows the rare-earth abundances of Apollo 12 materials normalized to chondrites. A negative Eu anomaly is observed similar to that found in Apollo 11 material. The degree of Eu depletion in the soils and breccia is considerably more than in the igneous rocks. In addition, the breccia and soils show a greater tendency to concentrate the lighter rare earths relative

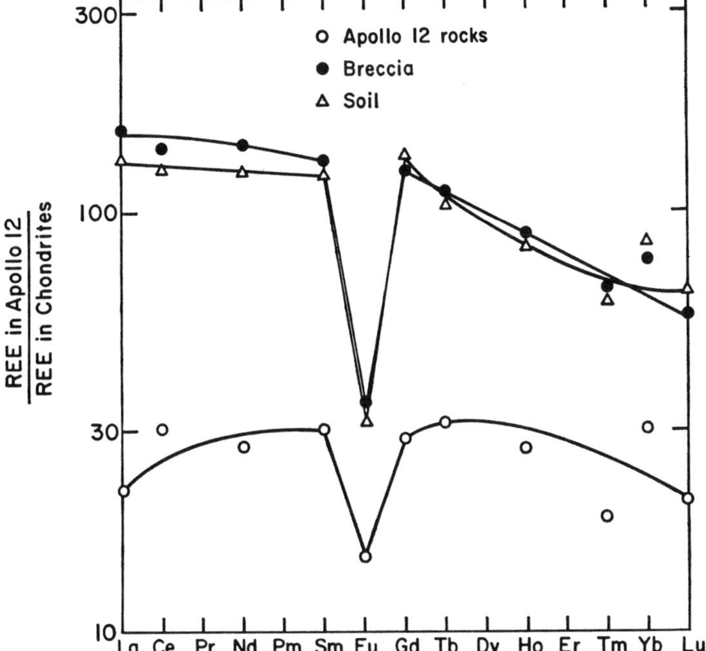

Fig. 7. Relative rare-earth element abundances of Apollo 12 materials.

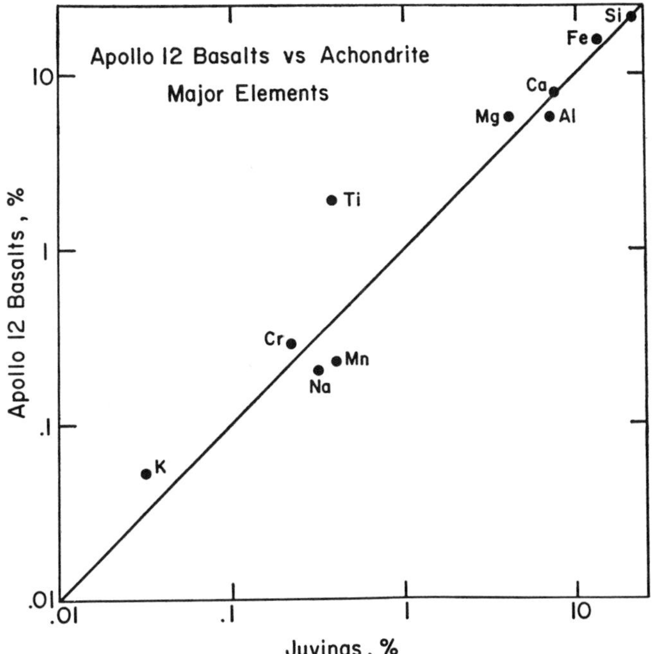

Fig. 8. Comparison of average major element abundances of igneous rocks in Apollo 12 vs. basaltic achondrite.

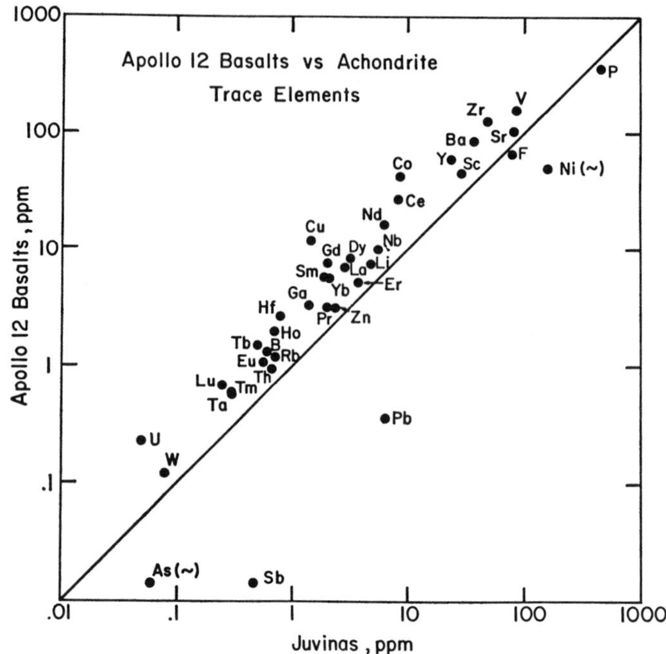

Fig. 9. Comparison of average trace element abundances of igneous rocks in Apollo 12 vs. basaltic achondrite.

to the heavier ones than in the case of the igneous rocks of Apollo 12 or indeed of Apollo 11 material in general. (Figure 2, MORRISON *et al.* (1970b)).

When the average Apollo 12 lunar abundance data is compared with data from appropriate meteoritic and terrestrial materials, the Apollo 12 basalts like those of Apollo 11 are seen to closely resemble basaltic achondrites, for example Juvinas, in most major elements with the notable exception of Ti, which is highly enriched in the lunar rocks (Fig. 8). The trace elements in the lunar basalts are uniformly enriched relative to Juvinas, as shown in Fig. 9, but the degree of enrichment is less than in Apollo 11 basalts.

Comparison of the trace element abundances in Apollo 12 basalts with those in oceanic tholeiites shows that though there is considerable scatter about a 1:1 trend line, a larger number of elements appear enriched in the tholeiites (Fig. 10). This is in contrast to the situation with Apollo 11 basalts where, although there is a more even distribution, the lunar basalts are more enriched.

In summary, there is an overall similarity of the material returned from the two Apollo mare sites; however, there are sufficient differences to suggest detailed geochemical processes special to each site.

Acknowledgments—The authors gratefully acknowledge the support of this work by the National Aeronautics and Space Administration under grant NAS 9–9986, the National Science Foundation under grant GP–6471X, and the Advanced Research Projects Agency through the Cornell Materials Science Center.

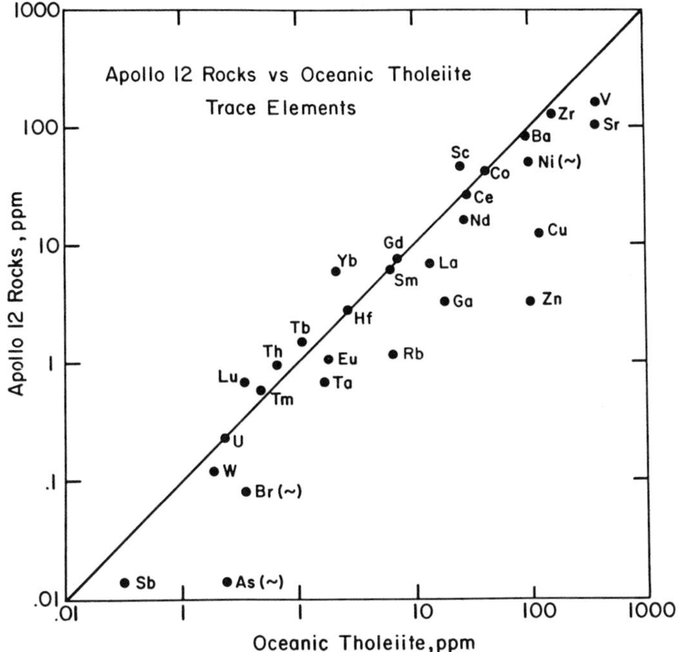

Fig. 10. Comparison of average trace element abundances of igneous rocks in Apollo 12 vs. oceanic tholeiite.

References

ANDERS E., LAUL J. C., KEAYS R. R., GANAPATHY R., and MORGAN J. W. (1971) Elements depleted on lunar surface: Implications for origin of moon and meteorites influx rate. Second Lunar Science Conference (unpublished proceedings).

ANNELL C. S., CARRON M. K., CHRISTIAN R. P., CUTTITTA F., DWORNIK E. J., HELZ A. W., LIGON D. T., JR., and ROSE H. J., JR. (1971) Chemical and spectrographic analysis of lunar samples from Apollo 12 mission. Second Lunar Science Conference (unpublished proceedings).

BAEDECKER P. A., SCHANDY R., ELZIE J. L., KIMBERLIN J., and WASSON J. T. (1971) Trace element studies of rocks and soils from Oceanus Procellarum and Mare Tranquillitatis. Second Lunar Science Conference (unpublished proceeding).

BOUCHET M., KAPLAN G., VOUDON A., and BERTOLETTI M. (1971) Spark Mass spectrometric analysis of major and minor elements in six lunar samples; age determinations of four rocks. Second Lunar Science Conference (unpublished proceedings).

BRUNFELT A. O., HEIER K. S., and STEINNES E. (1971) Determination of 40 elements in Apollo 12 materials by neutron activation analysis. Second Lunar Science Conference (unpublished proceedings).

COMPSTON W., BERRY H., VERNON M. J., CHAPPELL B. W., and KAYE M. J. (1971) Rubidium–Strontium chronology and chemistry of lunar material from the Ocean of Storms. Second Lunar Science Conference (unpublished proceedings).

COMPSTON W., CHAPPELL B. W., ARRIENS P. A., and VERNON M. J. (1970) The chemistry and age of Apollo 11 lunar material. *Proc. Apollo 11 Lunar Sci. Conf., Geochim. Cosmochim. Acta* Suppl. 1, Vol. 2, pp. 1007–1027. Pergamon.

EHMANN W. D., and MORGAN W. (1971) Major element abundances in Apollo 12 rocks and fines by 14 MeV neutron activation. Second Lunar Science Conference (unpublished proceedings).

ENGEL C. G., and ENGEL A. E. J. (1971) Major element composition of three Apollo 12 rocks and some petrogenic considerations. Second Lunar Science Conference (unpublished proceedings).

FRICK C., HUGHES T. C., LOVERING J. F., REID A. F., WARE N. G., and WARK D. A. (1971) Electron probe, fission track, and activation analysis of lunar samples. Second Lunar Science Conference (unpublished proceedings).

GAST P. W., and HUBBARD N. J. (1971) Rare-earth abundances in soil and rocks from the Ocean of Storms. Second Lunar Science Conference (unpublished proceedings).

GOLES G. G., DUNCAN A. R., OSAWA M., MARTIN M. R., BEYER R. L., LINDSTROM D. L., and RANDLE K. (1971) Analyses of Apollo 12 specimens and a mixing model for Apollo 12 soils. Second Lunar Science Conference (unpublished proceedings).

HASKIN L. A., ALLEN R. O., HELMKE P. A., ANDERSON M. R., KOROTEV R. L., and ZWEIFEL K. A. (1971) Rare earths and other trace elements in Apollo 12 lunar materials. Second Lunar Science Conference (unpublished proceedings).

HESS F. D., PALMER D. F., and BISCHOFF J. L. (1971) Relations of some lunar rocks and fines: Evidence by radiochemical analysis of rare-earth elements. Second Lunar Science Conference (unpublished proceedings).

HUBBARD N. J., GAST P. W., and MEYER C. (1971) The origin of the lunar soil based on REE, K, Rb, Ba, Sr, P, and $^{87}Sr/^{86}Sr$ data. Second Lunar Science Conference (unpublished proceedings).

LIPSCHUTZ M. E., BALSIGER H., and PELLY I. Z. (1971) Vanadium isotopic composition and contents in lunar rocks and dust from the Ocean of Storms. Second Lunar Science Conference (unpublished proceedings).

LSPET (Lunar Sample Preliminary Examination Team) (1970) Preliminary examination of the lunar samples from Apollo 12. *Science* **167**, 1325–1339.

MAXWELL J. A., and WIIK H. B. (1971) Chemical composition of Apollo 12 lunar samples 12004, 12033, 12051, 12052, and 12065. Second Lunar Science Conference (unpublished proceedings).

MOORE C. B., LEWIS C. F., DELLES F. M., GOOLEY R. C., and GIBSON E. K., JR. (1971) Total carbon and nitrogen abundances in Apollo 12 lunar samples. Second Lunar Science Conference (unpublished proceedings).

MORRISON G. H., GERARD J. T., KASHUBA A. T., GANGADHARAM E. V., ROTHENBERG A. M., POTTER N. M., and MILLER G. B. (1970a) Multielement analysis of lunar soil and rocks. *Science* **167**, 505–507.

MORRISON G. H., GERARD J. T., KASHUBA A. T., GANGADHARAM E. V., ROTHENBERG A. M., POTTER N. M., and MILLER G. B. (1970b) Elemental abundances of lunar soil and rocks. *Proc. Apollo 11 Lunar Sci. Conf., Geochim. Cosmochim. Acta* Suppl. 1, Vol. 2, pp. 1383–1392. Pergamon.

MORRISON G. H., GERARD J. T., POTTER N. M., GANGADHARAM E. V., ROTHENBERG A. M., and BURDO R. A. (1971) Elemental abundances of lunar soil and rocks from Apollo 12. Second Lunar Science Conference (unpublished proceedings).

MORRISON G. H., GERARD J. T., TRAVESI A., CURRIE R. L., PETERSON S. F., and POTTER N. M. (1969a) Multielement neutron activation analyses of rock using chemical group separations and high resolution gamma spectrometry. *Anal. Chem.* **41**, 1633–1637.

MORRISON G. H., and KASHUBA A. T. (1969b) Multielement analysis of basaltic rock using spark source mass spectrometry. *Anal. Chem.* **41**, 1842–1846.

MURTHY V. R., EVENSEN N. M., JAHN B., and COSCIO M. R., JR. (1971) Rb–Sr isotopic relations and elemental abundances of K, Rb, Sr, and Ba in Apollo 11 and Apollo 12 samples. Second Lunar Science Conference (unpublished proceedings).

O'KELLEY G. D., ELDRIDGE J. S., SCHONFELD E., and BELL P. R. (1971) Comparative radionuclide concentrations and ages of Apollo 11 and Apollo 12 samples from nondestructive gamma-ray spectrometry. Second Lunar Science Conference (unpublished proceedings).

PHILPOTTS J. A., SCHNETZLER C. C., BOTTINO M. L., and FULLAGAR P. D. (1971) Li, K, Rb, Sr, Ba, and rare earth concentrations and $^{87}Sr/^{86}Sr$ in some Apollo 12 soils, rocks, and separated phases. Second Lunar Science Conference (unpublished proceedings).

RANCITELLI L. A., PERKINS R. W., FELIX W. D., and WAGMAN N. A. (1971) Cosmogenic and primordial radionuclide measurements in Apollo 12 lunar samples by nondestructive analysis. Second Lunar Science Conference (unpublished proceedings).

REED G. W., and JOVANOVIC S. (1971) The halogens and other trace elements in Apollo 12 soils and rocks; halides, platinum metals and mercury on surfaces. Second Lunar Science Conference (unpublished proceedings).

RINGWOOD A. E., and ESSENE E. (1970) Petrogenesis of Apollo 11 basalts, internal constitution and origin of the moon. *Proc. Apollo 11 Lunar Sci. Conf., Geochim. Cosmochim. Acta* Suppl. 1, Vol. 1, pp. 769–799. Pergamon.

ROSE H. J., JR., CUTTITTA F., AMNELL C. S., CARRON M. K., CHRISTIAN R. P., DWORNIK E. J., HELZ A. W., and LIGON D. J., JR. (1971) Semi-micro analysis of Apollo 12 samples. Second Lunar Science Conference (unpublished proceedings).

SIEVERS R. E., EISENTRAUT K. J., JOHNSON D. G., RICHARDSON M. F., and WOLF W. R. (1971) Variations in beryllium and chromium concentrations in lunar fines compared with crystalline rocks. Second Lunar Science Conference (unpublished proceedings).

SILVER L. T. (1971) U–Th–Pb isotope relations in Apollo 11 and 12 lunar samples. Second Lunar Science Conference (unpublished proceedings).

SMALES A. A., MAPPER D., WEBB M. S. W., WEBSTER R. K., WILSON J. D., and HISLOP J. S. (1971) Elemental composition of lunar surface material. Second Lunar Science Conference (unpublished proceedings).

Special Issue on Lunar Sample 12013 (1970) *Earth Planet. Sci. Lett.* **9**.

TATSUMOTO M., KNIGHT R. J., and DOE B. R. (1971) U–Th–Pb systematics of Apollo 12 lunar samples. Second Lunar Science Conference (unpublished proceedings).

TAYLOR S. R., KAYE M., GRAHAM A., RUDOWSKI R., and MUIR P. (1971) Trace element chemistry of lunar samples from the Ocean of Storms. Second Lunar Science Conference (unpublished proceedings)

The Moon Issue (1970), *Science* **167**.

TRAVESI A., and PALOMARES J. (1971) Multielement neutron activation analysis of trace elements. Second Lunar Science Conference (unpublished proceedings).

WAKITA H., REY P., and SCHMITT R. A. (1971) Abundances of the 14 rare-earth elements plus 22 major, minor, and trace elements in ten Apollo 12 rocks and soil samples. Second Lunar Science Conference (unpublished proceedings).

WÄNKE H., WLOTZKA F., TESCHKE F., BADDENHAUSEN H., SPETTEL B., BALACESCU A., QUIJANO-RICO M., JAGOUTZ E., and RIEDER R. (1971) Major and trace elements in Apollo 12 samples and studies on lunar metallic iron particles. Second Lunar Science Conference (unpublished proceedings).

WILLIS J. P., AHRENS L. H., DANCHIN R. V., ERLANK A. J., GURNEY J. J., HOFMEYR P. K., MCCARTHY T. S., and ORREN M. J. (1971) Some interelement relationships between lunar rocks and fines, and stony meteorites. Second Lunar Science Conference (unpublished proceedings).

Proceedings of the Second Lunar Science Conference, Volume 2, pp. 1187–1208.
The M.I.T. Press, 1971.

Apollo 12 samples: Chemical composition and its relation to sample locations and exposure ages, the two component origin of the various soil samples and studies on lunar metallic particles

H. WÄNKE, F. WLOTZKA, H. BADDENHAUSEN, A. BALACESCU, B. SPETTEL, F. TESCHKE, E. JAGOUTZ, H. KRUSE, M. QUIJANO-RICO, and R. RIEDER
Max-Planck-Institut für Chemie (Otto-Hahn-Institut), Mainz, Germany

(*Received 24 February 1971; accepted in revised form 25 March 1971*)

Abstract—Concentrations of 47 major and trace elements were determined in 14 samples of lunar igneous rocks, soils, and breccias. It was observed that the igneous rocks belong to two chemically well-defined groups. We call them high- and low-Mg rocks. By splitting the igneous rocks into these two groups, clusters show up in the distribution of the cosmic ray exposure ages. Nearly all high-Mg rocks were collected in the plain area close to the LM during the first EVA while low-Mg rocks were all collected inside or close to the rims of craters. Five of the rocks from the rim of Surveyor Crater are chemically very similar and have, within limits of error, the same cosmic ray exposure ages of about 180 million years. (Age data is from BOGARD *et al.* (1971) and HINTENBERGER *et al.* (1971)). This might indicate that Surveyor Crater is only 180 million years old.

As in the case of the Apollo 11 samples, the samples from Apollo 12 also show the negative Eu-anomaly. Furthermore, the REE abundance pattern is different for soil samples and breccias, high-Mg rocks and low-Mg rocks.

The various soil samples from Apollo 12 have different chemical compositions. Computer calculations showed that the various soils and breccias are a mixture of only two components but with varying mixing ratios. One of the components is represented by the local basaltic rocks while the other component must be "granitic." The admixture of the "granitic" component must have taken place within the last 100 million years.

Metal particles were separated from the fines 12001. Individual grains were analyzed for Ni and Co by the microprobe and for Fe, Ni, Co, Cu, Ga, Au, Ir, W, and As by neutron activation techniques. The results allow us to distinguish a group of primary particles with meteoritic bulk and trace element composition from a group of secondary particles with low abundances of siderophile trace elements. The W content was found to be very high (up to 500 ppm) in some particles and highly variable. Some of the "meteoritic" metal grains contain high amounts of W which they may have acquired on the lunar surface. About 50% of the total amount of Au and Ir in the lunar fines is located in the metal grains. Metal from meteorites, from rock 12053, and from breccia 12073 were studied for comparison.

INTRODUCTION AND PROCEDURE

THE CONCENTRATIONS of 47 major and trace elements have been determined in 13 different samples from Oceanus Procellarum (9 rocks, 3 fines, and one breccia) and one additional rock sample from Mare Tranquillitatis. The data for almost all the elements were obtained by combined and successive instrumental activation analysis with fast and thermal neutrons and radiochemical neutron activation analysis of one and the same sample of about 0.5 grams. Only boron was determined on aliquots of about 20 mg by a fluorimetric method. In some cases Li and F were also determined on separate aliquots of about 100 mg via radiochemical neutron activation. For the determination of 44 elements on one and the same sample, the samples were processed as follows (for details see WÄNKE *et al.* (1970a) and RIEDER and WÄNKE (1969)).

Irradiation I: Fast neutrons (14 MeV, ^3H (d, n)^4He) 5 × 10 sec each. ^{16}N and ^{28}Al were counted for the determination of O and Si.

Irradiation II: Fast neutrons (14 MeV, ^3H (d, n)^4He) 3 × 5 min each. ^{56}Mn, ^{28}Al, ^{27}Mg, and ^{24}Na were counted with a large volume Ge (Li) detector for the determination of Fe, Mg, and Al. Si was used as an internal standard.

Irradiation III: Thermal neutrons—TRIGA research reactor. 6 hrs with 7 × 10^{11} n/cm^2 sec. During a cooling period of 3 weeks activities of ^{24}Na, ^{42}K, ^{46}Sc, ^{47}Sc (from Ti and Ca), ^{48}Sc (from Ti), ^{51}Cr, ^{56}Mn, ^{59}Fe, ^{60}Co, ^{140}La, ^{141}Ce, ^{147}Nd, ^{153}Sm, ^{160}Tb, ^{175}Yb, ^{177}Lu, ^{181}Hf, and ^{182}Ta were counted several times with a large volume Ge (Li) detector for the determination of these elements.

Irradiation IV: Thermal neutrons—TRIGA research reactor, 6 hrs with 7 × 10^{11} n/cm^2 sec. The samples were decomposed under vacuum with Na_2O_2 and radiochemically processed to separate and determine the following elements: Ni (via ^{58}Co), Cu, Ga, Ge, Rb, Sr, Pd, In, Cs, Ba, Pr, Gd, Dy, Ho, Er, W, Ir, Au, Th (via ^{233}Pa), and U (via ^{135}Xe).

RESULTS

The results are shown in Table 1; values for the accuracy of individual elements are also given. The quality of the determination can be estimated in two ways. First, as O is determined directly and all major, semimajor, and most of the minor elements were also determined, one can add up to 100% with confidence; the sum is indeed in all cases very close to 100%; the two samples 12002 and 12063 where we get 98.31% and 102.05% are exceptions due to less accurate Ca-measurements caused by a failure of the counting equipment. Second, in Table 2 we have compared our values for the fines 12070 with those of all other authors. The deviation of our values from the mean values of all investigators is generally lower than the standard deviation of the mean values.

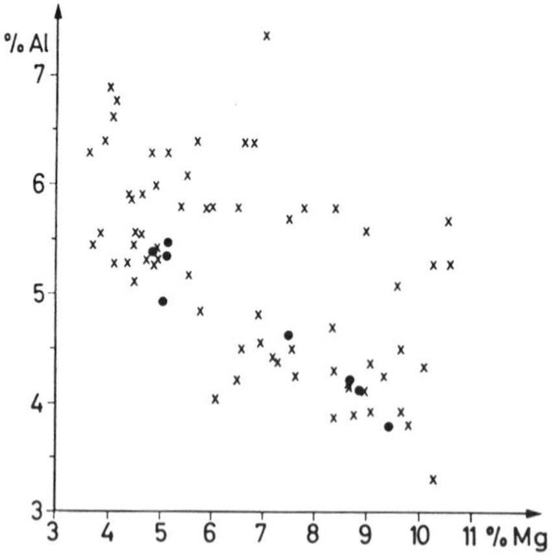

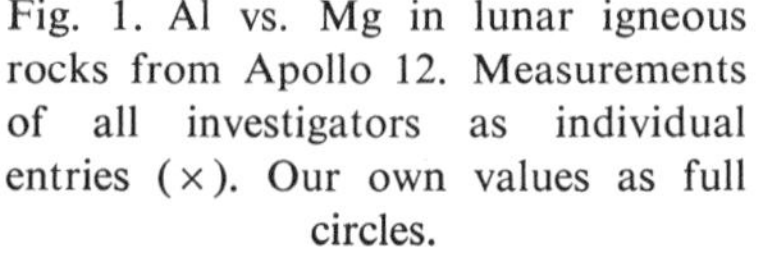
Fig. 1. Al vs. Mg in lunar igneous rocks from Apollo 12. Measurements of all investigators as individual entries (×). Our own values as full circles.

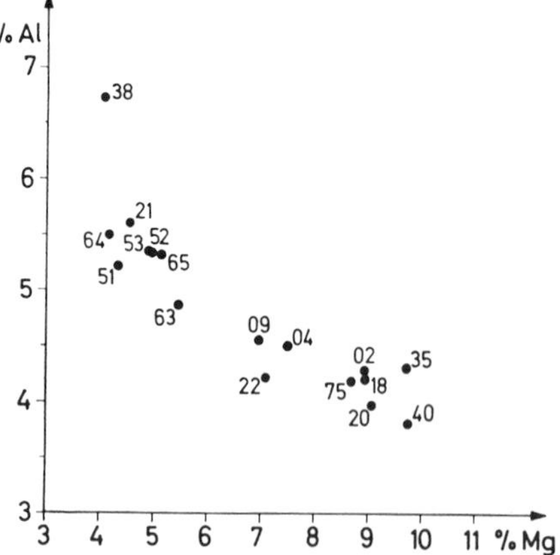

Fig. 2. Al vs. Mg in lunar igneous rocks from Apollo 12. Average values of all investigators (see Table 3).

The rock samples form two rather discrete groups. A low-Mg group (~5% Mg) and a high-Mg group (~8–9% Mg). The increase in Mg is correlated with the decrease in Al. The correlation as well as the two groups are difficult to see in a diagram in which all determinations reported by the various investigators are included (Fig. 1) (for references see Table 2). For the calculation of average values for the various rock samples we have excluded all individual determinations which are far off. The average values obtained in this way for Al and Mg are plotted in Fig. 2. In this case the Al–Mg correlation is comparable to that found by using only our own data (full circles in Fig. 1).

Our values for O are generally higher than that reported by EHMANN and MORGAN (1971). The difference is small but consistent as was the case for Apollo 11 samples. EHMANN and MORGAN (1970) have suggested that the difference might be due to absorbed water, as our samples are handled in normal laboratory air. We believe that absorbed water can only explain part of the difference observed. We have also calculated O values from data of the elements as oxides, and the values come very close to our O concentrations.

The two groups of rocks can be distinguished not only from the concentrations of Mg and Al, but also from the concentration of nearly every element. The differences are especially pronounced for Ni, Co, Cr, Na, K, Ca, Sr, Ba, and REE. The ratio of the average concentrations in high-Mg rocks to that in low-Mg rocks for a number of elements calculated from our data is listed in Table 3.

The rare earth elements of all samples investigated show the negative Eu-anomaly (Fig. 3). This anomaly is somewhat smaller, as it was for the samples from Mare Tranquillitatis. The REE abundance curve looks somewhat different for the fines and the rock samples, indicating a different REE abundance pattern for other parts of the Moon. A considerable difference in the REE abundance pattern for high-Mg rocks (for example 12004) and low-Mg rocks is also noticeable.

Local distribution of high- and low-Mg rocks at the Apollo 12 landing site and a discussion of their cosmic-ray exposure ages

If we take a Mg concentration of 6% as the dividing line between high- and low-Mg rocks we get the following distribution: high-Mg rocks—12002, 12004, 12009, 12012, 12014, 12015, 12018, 12020, 12022, 12035, 12040, and 12075; low-Mg rocks—12021, 12038, 12051, 12052, 12053, 12062, 12063, 12064, and 12065.

According to the map (Fig. 4) prepared by SUTTON and SCHABER (1971), we see that rock 12021 (a low-Mg rock) was picked up much closer to Head Crater than most of the other rocks (all high Mg-rocks) collected during the first EVA.

Among the rocks from the second EVA, eight low-Mg rocks and only two high-Mg rocks have been found so far. Nearly all rock samples from the second EVA were collected on crater rims. In particular, four of the five rocks from Surveyor Crater, for which data on chemical composition exist, were collected inside the rim of Surveyor Crater; and the last one just outside the rim. All five of these rocks belong to the low-Mg group. According to rare gas measurements by BOGARD *et al.* (1971) and HINTENBERGER *et al.* (1971) five of the rocks from the Surveyor Crater have, within the limits of error, a cosmic ray exposure age of 180 m.y. (the individual values

Table 1. Major and trace elements in lunar

	Apollo 12			Apollo 11			
		Fines		Breccia			
	12001,114	12037,26	12070,74	12073,31	10049	12002,120	12004,33
O %	42.6	42.4	42.6	43.3	41.0	41.7	41.9
Mg %	5.92	6.99	5.84	5.58	4.4	8.90	7.49
Al %	6.65	5.90	6.72	7.37	4.5	4.13	4.62
Si %	21.6	21.6	21.5	22.1	20.0	20.8	21.5
Ca %	6.1	7.4	7.6	8.0	6.3	—	5.9
Ti %	1.6	1.5	1.6	1.3	4.8	—	1.53
Fe %	13.1	14.3	12.7	11.4	14.1	17.2	16.4
B ppm			2.3		—	—	0.8
Li ppm			12.0		—	—	11.0
Na ppm	3200	2760	3090	4460	3600	1560	1470
Cl ppm			24.4		—	—	31.5
K ppm	2100	1700	1900	3300	2280	450	410
Sc ppm	38.1	39.0	37.3	36.2	80.9	42.6	43.8
Cr ppm	2430	3230	2270	2260	1960	6570	4100
Mn ppm	1710	1910	1590	1590	1600	2230	1990
Co ppm	38.3	47.1	41.5	38.2	24.0	68.7	47.9
Ni ppm	310	180	200	230	—	150	80
Cu ppm	7.2	4.5	7.2	5.7	—	5.5	6.9
Ga ppm	4.2	4.4	3.3	5.1	4.3		3.8
Ge ppb	200	—	210	110	≤1		100
As ppb	27	—	22	26	—		4
Rb ppm	23	6.1	8.7	11.3	—		0.9
Sr ppm	130	90	140	190	180		72
Pd ppb	9.0	13.5	6.5	10.2	—		≤1
In ppb	92	109	486	12.5	16		10.4
Cs ppb	530	310	390	500	—		90
Ba ppm	460	190	390	390	202		79
La ppm	32.4	24.5	33.0	49.8	27.4	5.65	5.43
Ce ppm	87	—	86	131	118	—	15
Pr ppm	10.8	9.1	10.6	—	—	—	1.9
Nd ppm	72	—	—	—	59.1	—	12.9
Sm ppm	15.0		14.7	21.4	12.8	3.3	3.2
Eu ppm	1.80	1.57	1.80	2.44	2.11	0.96	0.82
Gd ppm	19.4	16.0	15.7	—	—	—	4.7
Tb ppm	3.78	3.22	4.0	6.2	5.46	1.10	0.97
Dy ppm	22.6	17.8	20.2	31.0	—	5.8	5.5
Ho ppm	5.0	4.13	5.2	7.58	—	—	1.40
Er ppm	14.5	12.7	15.8	—	—	—	3.84
Yb ppm	11.0	8.9	10.6	16.0	16.4	3.33	3.17
Lu ppm	1.56	1.35	1.52	2.17	2.45	0.55	0.44
Hf ppm	13.3	10.9	15.6	21.7	—	3.6	5.1
Ta ppm	1.4	1.04	1.46	2.1	2.0	0.54	0.33
W ppm	0.63	0.45	0.74	1.21	—	—	0.14
Ir ppb	11	5	7.5	8.8	—		(33)
Au ppb	2.6	1.5	1.8	2.7	(1.1)		(4.0)
Th ppm	5.50	3.52	5.52	8.17	4.03		0.82
U ppm	1.67	0.72	1.69	2.32	0.814		0.238
%	98.31	100.95	99.35	100.01	—	—	100.14

Samples 12002, 12020, 12064, and 12065 are exchange samples from H. Hintenberger. These samples were analyzed in order to obtain data on the K content; the data for the other elements listed were obtained as byproducts. For the Ir and Au values in brackets we suspect

are 170, 160, 205, 182, and 177 m.y.). For one of these rocks (12054) there are no data on its chemical composition, but it would be highly desirable to have those data. The sixth rock from Surveyor Crater (12063) has an exposure age of only 65 m.y., but a close look on the map shows that this rock was collected on the rim of Block Crater, a younger crater within Surveyor Crater.

fines, breccias, and igneous rocks.

			Apollo 12 Rocks				
12018,41	12020,37	12052,94	12053,20	12063,69	12064,15	12065,29	accuracy %
42.0	42.0	42.2	42.1	41.4	—	41.4	1
8.69	9.44	5.10	4.86	5.05	—	5.12	4
4.22	3.81	5.35	5.39	4.94	—	5.47	2
21.1	20.5	22.1	22.1	20.9	—	21.9	1
4.4	5.8	8.5	6.7	9.3	10.2	7.7	10
1.59	1.58	1.5	2.2	2.8	2.6	2.1	10
16.5	16.4	15.2	15.6	16.7	15.1	15.4	3
0.9		1.1	1.4	5.2	—		20
4.5		4.7	4.8	5.9	—		20
1430	1290	1760	1960	2150	1980	1800	5
17.8		28.0	27.0	17.2	—		20
410	380	560	500	630	670	660	5
41.7	45.4	50.6	56.4	62.9	63.1	56.5	5
3730	4560	3490	3480	2580	2160	3560	5
1980	2170	2180	2220	2200	2280	2290	5
51.6	61.0	38.4	39.1	39.1	27.2	38.8	5
100		39	28	49	—		20
5.5	6.9	39	7.1	12.9	6.6	7.8	10
3.2		3.9	4.1	5.3	—		10
100		60	100	100	—		30
10		6	10	53	—		15
1.5		—	1.7	1.1	—		20
96		110	140	130	—		10
3		≤1	≤1	≤1	—		30
1.9		7.8	0.86	2.3	—		10
90			100	80	—		20
84		70	120	140—			15
5.68	4.82	6.52	7.07	6.88	6.33	6.68	5
21	—	21	30	19	20	24	15
1.75	—	—	2.8	3.1	—	—	10
14.7	—	—	—	—	—	24	30
3.8	3.4	4.5	4.7	6.6	5.5	4.5	10
0.84	0.82	1.08	1.34	1.62	1.30	1.06	5
4.7	—	—	7.6	11.0	—	—	10
1.0	0.91	1.35	1.73	2.65	1.75	1.58	10
5.8	5.68	7.44	6.53	10.3	9.48	7.64	5
1.11	1.07	1.34	1.49	2.46	1.87	1.11	10
4.2	—	—	5.4	7.35	—	—	10
3.45	2.91	3.70	3.92	5.71	5.25	3.78	5
0.40	0.42	0.58	0.62	0.84	0.67	0.59	10
3.4	3.8	3.8	4.0	6.3	3.9	3.9	10
0.36	0.45	0.44	0.45	0.47	0.33	0.51	15
0.15		0.15	0.12	0.14	—		10
(0.5)		(3.6)	(0.24)	(1.3)	—		20
(3.6)		(0.9)	(2.2)	(2.9)	—		15
0.85	0.71	1.28	0.87	0.82	—		10
0.252		0.356	0.242	0.236	—		10
98.76	100.37	100.75	99.77	102.05	—	99.82	

contamination before we received the samples. Accuracy of the Ca values for samples 12001 and 12063 is 20%.

We suggest the following model: The impact which made Surveyor Crater took place 180 m.y. ago, and it excavated rocks of rather uniform chemical composition (12051, 12054, 12062, 12064, 12065), while 115 m.y. later, i.e., 65 m.y. ago, Block Crater was created inside Surveyor Crater and more rocks were brought to the surface having, of course, the same chemical composition as all the rocks in this crater; one

Table 2. Major and trace elements in sample 12070, comparison of date from all investigators.

Element	This work	Mean	Deviation of this work to mean	Standard deviation all authors	Other authors
	%		%	%	
O	42.6	42.0	1.4		41.4[4].
Mg	5.84	6.15	5.0	5.6	6.2[11], 6.3[10], 5.8[10], 5.7[9], 6.8[6], 6.14[3], 6.3[4], 6.29[1].
Al	6.72	6.71	0.2	2.6	6.8[8], 6.8[11], 6.9[10], 6.7[10], 6.7[9], 6.9[6], 6.64[5], 6.50[2], 6.74[5], 6.65[1], 6.59[4], 6.61[3].
Si	21.5	21.45	0.2	2.5	20.3[8], 21.4[11], 21.4[10], 21.5[10], 22.2[9], 21.29[1], 21.43[3], 22.0[4].
Ca	7.6	7.32	3.8	5.0	7.5[8], 7.5[11], 7.4[10], 7.6[10], 7.4[9], 6.8[6], 6.5[2], 7.45[1], 7.47[3].
Ti	1.6	1.66	3.6	3.9	1.7[8], 1.70[11], 1.67[10], 1.69[10], 1.5[6], 1.67[5], 1.75[5], 1.64[2], 1.67[1], 1.68[3].
Fe	12.7	12.45	2.0	4.3	12.4[8], 12.7[11], 11.7[15], 11.6[15], 12.8[10], 12.8[10], 13.2[9], 12.2[6], 12.5[5], 12.1[5], 12.6[2], 12.83[1], 13.06[3], 11.5[4].
	ppm		%	%	
B	2.3	7.4		—	11[9], 9[6].
Li	12.0	15.5	22.6	15.8	17[11], 15[10], 15[10], 18[6], 13[16], 18.4[16].
Na	3090	3326	7.1	7.2	3430[8], 3710[11], 3561[10], 3487[10], 3300[9], 3200[6], 3190[5], 3260[5], 3240[2], 2893[1], 3190[3], 3690[12].
Cl	24.4	27.6	11.6	36.6	30[6], 21[5], 27[5], 17[16], 46[16].
K	1900	1961	3.1	9.1	2075[11], 1776[23], 1500[15], 1909[10], 1990[21], 1900[6], 1970[5], 1890[5], 2056[24], 2150[20], 1980[1], 2242[3], 2170[12], 1909[10].
Sc	37.3	39.5	5.6	7.2	36[8], 42[11], 41.9[15], 43.8[15], 38[10], 44[10], 35[6], 38[6], 39.3[5], 38.3[5], 40[18], 36.8[2], 39[1], 43[12].
Cr	2270	2676	15.2	11.5	2430[8], 3060[11], 3000[23], 3224[10], 2430[10], 2800[6], 2570[5], 2620[5], 2800[18], 2480[2], 2870[1], 2080[3], 2800[12], 2700[22].
Mn	1590	1669	4.7	5.2	1610[8], 1650[11], 1670[10], 1540[10], 1730[5], 1750[5], 1550[2], 1735[1], 1782[3], 1750[12].
Co	41.5	43.2	3.9	7.1	44[8], 39.1[15], 39.6[15], 49[10], 49[10], 44[6], 43[6], 41.3[5], 40.2[5], 45[18], 44.7[2], 40[1], 45[3], 43[12].
Ni	200	202	1.0	6.5	222[11], 215[10], 210[6], 200[6], 180[18], 202[1], 186[3], 200[12].
Cu	7.2	7.1	1.4	16.7	8.8[15], 7[9], 7.7[6], 6.7[5], 5[18], 8[1], 6[3].
Ga	3.3	4.0	17.5	22.1	4.9[11], 2.5[15], 4.8[10], 5.0[10], 3.8[6], 4.1[5], 4.0[5], 4.5[28], 2.5[3], 4.3[29].
Ge	0.210	0.283		—	0.355[28].
As	0.022	—		—	0.15[9], 0.1[6], 0.57[5], 0.58[5].
Rb	8.7	6.71	29.7	17.2	5.8[8], 6.2[11], 6.25[23], 5.2[10], 5.9[10], 9.5[9], 5.8[6], 5.9[6], 8.3[5], 8.1[5], 6.59[24], 6.46[20], 6.4[1], 6.33[3], 6.4[12], 6.6[12], 6.3[29].
Sr	140	142	1.4	13.7	123[11], 140[23], 115[10], 125[10], 145[9], 160[6], 135[18], 150[20], 136[1], 143.3[3], 190[12].
Pd	0.0065	—		—	
In	0.486	0.415	17.1	74.5	0.360[5], 0.880[5], 0.131[28], 0.218[29].
Cs	0.390	0.286	36.8	20.7	0.24[8], 0.2[6], 0.23[6], 0.29[5], 0.31[5], 0.3[18], 0.290[12], 0.360[12], 0.248[29].
Ba	390	364	7.1	12.4	270[8], 423[11], 383[23], 420[10], 430[10], 330[6], 330[6], 304[5], 340[18], 404[20], 340[2], 373[1], 350[3], 390[12], 350[12].
La	33.0	33.4	1.2	16.9	37.7[8], 36.0[8], 40[11], 21.5[15], 22.6[15], 46[10], 39[10], 34.9[17], 31.6[17], 36[6], 36[6], 33.2[5], 33.7[5], 26.6[19], 32[18], 30.3[20] 32.1[2], 33[1], 29[3], 38[12].
Ce	86	82.9	3.7	14.7	87.2[8], 83[15], 78[15], 86.3[17], 80.2[17], 97[6], 99[6], 84[5], 90[5], 50.7[19], 76[18], 90[20], 87.7[2], 62[3], 90[12].
Pr	10.6	10.7	0.9	24.2	12.6[8], 12[6], 6.26[19], 12[18].
Sm	14.7	16.7	12.0	8.8	16.6[8], 16.0[8], 15[9], 15[15], 18.1[17], 16.3[17], 19[6], 19[6], 16.1[5], 15.8[5], 19[18], 16.4[20], 15.8[2], 18[12], 16.7[12].
Eu	1.80	1.75	2.9	3.9	1.73[8], 1.67[8], 1.79[17], 1.7[6], 1.8[6], 1.77[5], 1.84[5], 1.9[18], 1.73[20], 1.71[2], 1.7[12], 1.7[12], 1.67[17].
Gd	15.7	18.9	16.9	25.9	19.0[8], 18[9], 17.6[17], 22[17], 23[6], 17[6], 12.8[19], 25[18].
Tb	4.0	3.2	25.0	22.5	3.3[8], 2.3[15], 2.2[15], 3.5[15], 3.7[17], 3.6[17], 4.1[6], 3.8[6], 2.69[5], 2.50[5], 1.99[19], 3.8[18].
Dy	20.2	22.5	10.2	16.4	20.5[8], 26[9], 24.3[17], 23.3[17], 21[6], 22.7[5], 24.1[5], 13.7[19], 23[18], 21.7[20], 29[12].

Table 2 (continued)

Element	This work	Mean	Deviation of this work to mean	Standard deviation all authors	Other authors
			%	%	
	ppm				
Ho	5.2	4.9	6.1	9.9	4.9[8], 4.9[17], 4. 9[17], 4.8[6], 4.9[6], 4.3[5], 5.2[5], 6[18], 4.3[2].
Er	15.8	14.4	9.7	11.6	12.9[8], 18[17], 14[17], 13[6], 13.1[5], 14.2[5], 15[18], 13.5[20].
Yb	10.6	11.7	9.4	18.9	11.4[8], 12.0[8], 14[11], 7.5[15], 6.0[15], 14[10], 12[10], 12.9[17], 12.2[17], 13[6], 11[6], 12.9[5], 12.6[5], 8.3[19], 14[18], 13.0[20], 10.7[2], 13[1], 13.6[12].
Lu	1.52	1.7	10.6	11.5	1.67[8], 1.66[8], 1.90[17], 1.75[17], 1.8[6], 1.4[6], 1.90[5], 1.90[5], 1.30[19], 1.78[20], 1.68[2], 1.9[12].
Hf	15.6	12.2	27.9	22.2	12[8], 9.8[15], 12.6[15], 17.2[9], 8.5[6], 7.9[6], 13.4[5], 11.6[5], 14[18], 11.3[2], 12[12].
Ta	1.46	1.93	24.4	42.3	3.3[6], 1.53[5], 1.30[3], 2.05[2].
W	0.74	0.65	13.9	29.7	0.73[6], 0.64[5], 0.64[5], 0.5[18].
Ir	0.0075	0.0065	15.4	30.6	0.00436[5], 0.00433[5], 0.0077[28], 0.0085[29].
Au	0.0018	0.0035	48.6	48.5	0.0045[5], 0.0054[5], 0.00239[29].
Th	5.52	6.00	8.0	13.6	6.6[3], 6.52[26], 6.9[6], 6.358[25], 6.02[27], 6.6[9], 5.6[18], 4.6[5], 4.2[5], 6.73[21], 5.99[24], 6.7[2], 5.6[8].
U	1.69	1.62	4.3	7.0	1.6[3], 1.70[26], 1.4[6], 1.651[25], 1.64[27], 1.6[18], 1.5[5], 1.6[5], 1.7[21], 1.85[24], 1.63[24], 1.5[16].

1 WILLIS *et al.* (1971).
2 GOLES *et al.* (1971).
3 COMPSTON *et al.* (1971).
4 EHMANN and MORGAN (1971).
5 BRUNFELT *et al.* (1971).
6 MORRISON *et al.* (1971).
7 MAXWELL and WIIK (1971).
8 WAKITA *et al.* (1971).
9 BOUCHET and KAPLAN (1971).
10 ROSE *et al.* (1971).
11 ANNELL *et al.* (1971).
12 SMALES *et al.* (1971).
13 SCOON (1971).
14 PECK see BIGGAR *et al.* (1971).
15 TRAVESI and PALOMARES (1971).
16 REED and JOVANOVIC (1971).
17 HASKIN *et al.* (1971).
18 TAYLOR *et al.* (1971).
19 HESS *et al.* (1971).
20 HUBBARD *et al.* (1971).
21 RANCITELLI *et al.* (1971).
22 SIEVERS *et al.* (1971).
23 MURTHY *et al.* (1971).
24 CLIFF *et al.* (1971).
25 SILVER (1971).
26 FIELDS *et al.* (1971).
27 ROSHOLT and TATSUMOTO (1971).
28 BAEDECKER *et al.* (1971a).
29 ANDERS *et al.* (1971).

Table 3. Average element ratios between high- and low-Mg rocks. Data from Table 1.

Elements	High-Mg group / Low-Mg group	Elements	High-Mg group / Low-Mg group
Ni	2.85	Sc	0.75
Mg	1.72	Ba	0.74
Cr	1.62	Cs	0.74
Co	1.59	K	0.70
Fe	1.06	Ti	0.70
Ta	1.00	Yb	0.69
O	1.00	Dy	0.68
Si	0.97	Ho	0.67
Mn	0.94	Lu	0.66
Hf	0.88	Sr	0.66
U	0.87	Eu	0.64
Th	0.84	Sm	0.64
Na	0.84	Ca	0.60
La	0.81	Tb	0.54
Al	0.80	Gd	0.51
Ga	0.79		

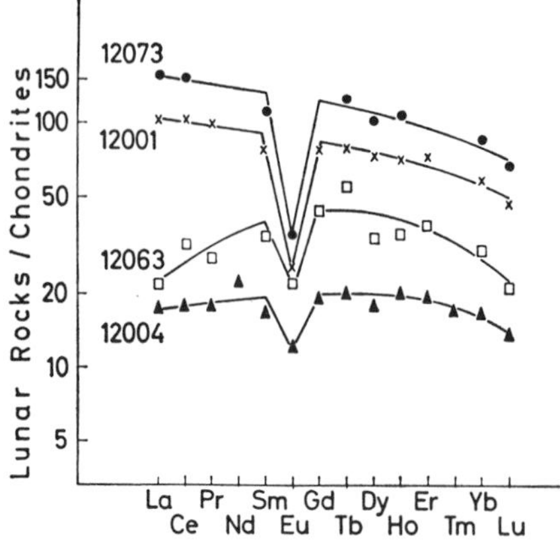

Fig. 3. REE abundance normalized to chondritic values. Data for chondrites are taken from WAKITA *et al.* (1970). Note the difference in the abundance pattern for soil 12001 and breccia 12073 to the rock samples 12063 and 12004. Low-Mg rocks like 12063 and high-Mg rocks like 12004 also differ somewhat in their REE pattern.

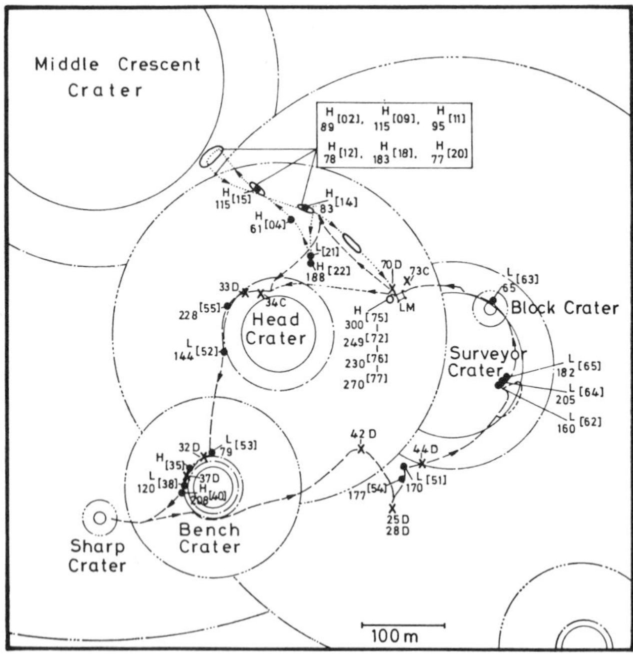

Fig. 4. Map showing sample location according to SUTTON and SCHABER (1971). H and L refer to high-resp. low-Mg rocks. The figures left of the brackets refer to the cosmic ray exposure ages in million years calculated from concentrations of ^{3}He reported by BOGARD *et al.* (1971) and HINTENBERGER *et al.* (1971).

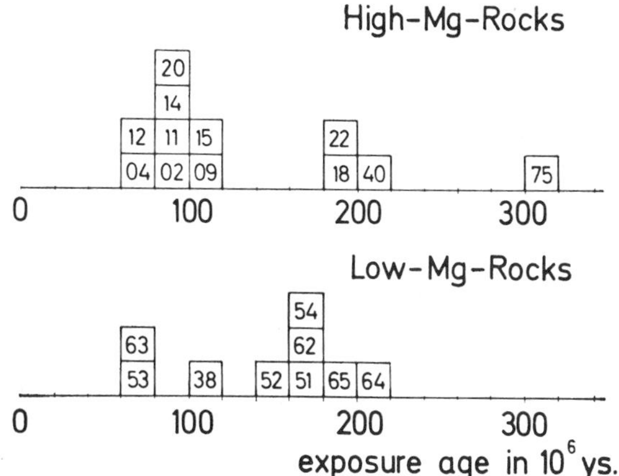

Fig. 5. Cosmic ray exposure ages of igneous rocks from Apollo 12. Clusters show up rather distinctly after splitting into high- and low-Mg rocks. A similar observation was made by EBERHARDT *et al.* (1970) for Apollo 11 rocks. This grouping of exposure ages indicates an ejection from a few single craters which must have corresponding ages. Values for the exposure ages from ^{3}He data reported by BOGARD *et al.* (1971) and HINTENBERGER *et al.* (1971).

of which is rock 12063. The model depends strongly on the validity of the sample locations given by SUTTON and SCHABER (1971).

The origin of the high-Mg group rocks from the first EVA cannot yet be unambiguously explained. According to their exposure ages these rocks were brought to the surface in at least three, perhaps four, different events.

The cosmic ray exposure ages (Fig. 5) of the high-Mg rocks cluster around 90 m.y. (8 rocks), 190 m.y. (3 rocks) and 260 m.y. (1 rock, 12075). The rocks 12072, 12076, and 12077, all collected within a few meters, have also an exposure age of 260 m.y. but no chemical analyses have been made. Only the group with 190 m.y. could come from the event which might have created the Surveyor Crater. If this should be correct, high- and low-Mg rocks must then originate from different depths. The stratigraphy and mechanism of ejection must then be responsible for the fact that the low-Mg rocks are to be found inside or on the rim of the craters and the high-Mg rocks outside the crater rims.

The classification of the rocks into two groups according to their chemical composition may not be correct; it could well be that the two groups are due to the sampling, as it just happened that we have collected and analyzed rocks from two sections with a continuously varying Al/Mg ratio (i.e., rocks with a varying proportion of feldspar). In this respect we would like to point out again the fact that the majority of the low-Mg group rocks was found within or on the rim of Surveyor Crater. According to the map (Fig. 4) prepared by SUTTON and SCHABER (1971), Head Crater should be younger than Surveyor Crater and Bench Crater even younger. If then the 90 m.y. cluster of high-Mg group rocks was thrown out from a nearby crater, one might speculate the age of either Bench Crater or Head Crater being 90 m.y. In the first case, Head Crater must then have an age of 90 to 180 m.y. Rock 12052 found just on

Table 4. Comparison of the measured values with the computed

		1	2	3	4	5	6
						Samples used for the calcu-	
		high-Mg rocks		Soil	Soil	Soil	Soil
		12018	12004	12037	12044	12070	12042
Si%	M	21.10	21.50	21.60	23.20	21.44	21.40
	C	21.50	21.51	21.64	21.71	21.71	21.72
	D	−2	0	0	6	−1	−1
Fe%	M	16.00	16.50	14.20	13.60	12.46	13.20
	C	16.17	16.14	14.31	13.33	13.32	13.17
	D	−1	2	−1	2	−7	0
Mg%	M	8.69	7.49	6.99	7.30	6.15	5.60
	C	7.89	7.88	7.10	6.69	6.69	6.62
	D	9	−5	−2	8	−9	−18
Al%	M	4.22	4.62	5.90	6.50	6.71	7.00
	C	4.62	4.63	5.76	6.37	6.37	6.46
	D	−9	0	2	2	5	8
Ca%	M	4.40	5.90	7.40	7.00	7.32	8.20
	C	5.69	5.70	6.55	7.00	7.00	7.07
	D	−29	3	12	0	4	14
Na ppm	M	1430	1470	2760	3220	3330	3010
	C	1530	1540	2580	3140	3140	3220
	D	−7	−5	7	3	6	−7
K ppm	M	410	410	1700	2110	1960	1740
	C	400	410	1460	2030	2030	2110
	D	2	−1	14	4	−3	−22
Cs ppm	M	0.057	0.070	0.253	—	0.285	0.200
	C	0.068	0.069	0.193	0.259	0.260	0.270
	D	−19	1	24	—	9	−35
Sr ppm	M	79.0	89.0	90.0	—	142.0	141.0
	C	79.2	79.7	117.0	137.0	137.1	140.1
	D	0	10	−30	—	3	1
Ba ppm	M	84	79	190	380	364	(478)
	C	64	66	252	353	353	368
	D	24	16	−33	7	3	—
Sc ppm	M	41.7	43.8	39.0	38.0	39.5	36.0
	C	43.2	43.1	39.4	37.5	37.5	37.2
	D	−4	2	−1	1	5	−3
Cr ppm	M	3730	4100	3230	2590	2680	2700
	C	3760	3750	3210	2910	2910	2870
	D	−1	8	1	−13	−9	−6
Mn ppm	M	1980	1990	1910	1570	1669	1660
	C	1978	1975	1802	1710	1709	1695
	D	0	1	6	−9	−2	−2
Zr ppm	M	83	127	310	350	459	436
	C	119	121	293	385	385	400
	D	−43	5	5	−10	16	8
La ppm	M	5.7	5.4	24.5	33.1	33.4	45.0
	C	5.4	5.7	24.5	34.6	34.6	36.2
	D	4	−5	0	−5	−4	20
Eu ppm	M	0.84	0.82	1.54	1.71	1.75	2.30
	C	0.83	0.84	1.49	1.84	1.84	1.90
	D	1	−3	3	−8	−5	18
Yb ppm	M	3.4	3.2	8.9	12.3	11.7	16.0
	C	3.0	3.1	9.2	12.5	12.5	13.0
	D	12	2	−4	−1	−7	19
Hf ppm	M	3.4	5.1	10.9	12.9	12.2	(7.6)
	C	3.7	3.7	10.1	13.6	13.6	14.1
	D	−8	27	7	−5	−11	—
Ta ppm	M	0.39	0.42	1.04	1.90	1.93	1.40
	C	0.44	0.45	1.16	1.54	1.54	1.60
	D	−13	−7	−12	19	20	−14
W ppm	M	0.150	0.140	0.450	—	0.65	0.97
	C	0.125	0.130	0.526	0.74	0.73	0.77
	D	17	7	−17	—	−14	21
U ppm	M	0.25	0.24	(0.72)	—	1.64	(1)
	C	0.25	0.26	1.09	1.53	1.54	1.60
	D	1	−9	—	—	6	—

Average element concentrations using data of all investigators for the Apollo 12 soils and breccias (M) compared with the computed values (C). D is the difference in % between measured and calculated values. Note that data for KREEP and rock 12013 have not been used for the calculated element concentrations. As a measure for the accuracy (M) we have listed the standard deviation from the mean values of all authors

ones according to a two-component model from least-squares fits.

7	8	9	10	11	12	13	14
lation of element-lines				Standard deviation for 12070 %	KREEP	Breccia	
Soil 12032	Breccia 12073	Soil 12033	Breccia 12034			12013 Total	12013 Dark
21.50	22.10	22.03	21.40	2.5	22.50	27.90	25.20
21.82	21.83	21.88	21.91		22.00	22.01	22.02
-1	1	1	-2		2	21	13
11.80	11.20	11.20	10.30	4.3	8.30	8.30	8.50
11.77	11.58	10.89	10.52		9.20	9.05	8.91
0	-3	3	-2		-11	-9	-5
6.40	5.58	5.69	5.80	5.6	4.70	3.60	5.40
6.03	5.95	5.66	5.50		4.95	4.88	4.82
6	-7	1	5		-5	-36	11
7.30	7.37	7.47	8.30	2.6	8.30	6.80	8.00
7.33	7.44	7.87	8.09		8.90	9.00	9.09
0	-1	-5	2		-7	-32	-14
7.50	7.80	7.60	8.15	5.0	7.80	5.10	6.40
7.72	7.80	8.12	8.29		8.90	8.97	9.03
-3	0	-7	-2		-14	-76	-41
4340	4460	4490	4460	7.2	3700	9300	10000
4020	4130	4520	4730		5470	5570	5640
7	7	-1	-6		-48	40	44
3190	3300	3530	3500	9.1	5460	13700	3800
2910	3030	3420	3630		4380	4470	4550
9	8	3	-4		20	67	-20
0.359	0.395	0.420	0.480	20.7	—	—	—
0.364	0.377	0.424	0.449		0.537	0.548	0.557
-2	4	-1	7		—	—	—
217.0	178.0	156.0	—	13.7	216.0	—	199.0
168.7	172.6	186.6	194.1		220.8	224.1	226.9
22	3	-20	—		-2	—	-14
540	537	613	(470)	12.4	906	2300	1162
511	531	601	638		772	788	802
5	1	2	—		15	66	31
36.0	32.0	32.0	31.0	7.2	—	27.0	—
34.3	33.9	32.6	31.8		29.2	28.8	28.6
5	-6	-2	-3		—	-7	—
2430	2230	2260	2370	11.5	—	1410	—
2450	2390	2190	2080		1690	1640	1600
-1	-7	3	12		—	-16	—
1609	1495	1424	1525	5.2	—	1200	—
1563	1545	1480	1445		1321	1306	1293
3	-3	-4	5		—	-9	—
657	390	669	635	14.5	—	—	—
531	550	614	649		772	787	800
19	-41	8	-2		—	—	—
50.0	49.0	61.0	67.5	16.9	110.0	66.0	98.7
50.6	52.5	59.6	63.4		76.8	78.5	79.9
-1	-7	2	6		30	-19	19
2.25	2.40	3.00	2.80	3.9	3.25	2.00	3.38
2.39	2.46	2.71	2.84		3.30	3.36	3.41
-6	-3	10	-1		-2	-68	-1
17.0	16.0	22.2	23.7	18.9	37.5	30.0	34.4
17.7	18.4	20.6	21.9		26.3	26.8	27.3
-4	-15	7	8		30	11	21
18.0	21.7	20.0	27.0	22.2	—	—	—
19.0	19.7	22.1	23.4		28.0	28.5	29.0
-6	9	-10	13		—	—	—
1.80	2.40	—	—	42.3	—	—	—
2.14	2.22	2.48	2.63		3.13	3.20	3.25
-19	8	—	—		—	—	—
1.00	1.21	—	—	29.7	—	—	—
1.07	1.12	1.26	1.34		1.63	1.66	1.69
-7	8	—	—		—	—	—
(1.5)	2.32	2.70	(5.84)	7.0	—	7.90	—
2.24	2.33	2.64	2.80		3.40	3.47	3.53
—	0	2	—		—	56	—

for sample 12070 (see Table 2). Values in brackets were omitted for the computation. In the case of rock samples, soils, and breccias 12034 and 12073 see Table 2 for references of the data used. Data for KREEP and sample 12013 from HUBBARD *et al.* (1971) and MEYER *et al.* (1971).

the rim of Head Crater shows an exposure age of 144 m.y. (BOGARD *et al.*, 1971). The age values for the craters estimated above are surprisingly low. We can of course, not prove with certainty that the exposure age clusters of the igneous rocks correspond to the ages of the craters in which the rocks were found.

The chemistry and the origin of the various soil samples and breccias from the Apollo 12 landing site

Samples 12032, 12033, 12037, 12042, 12044, and 12070 represent six different fines from separated parts of the lunar surface at the Apollo 12 landing site. One can suspect that the two breccias are rather closely related to the soil samples. We have then altogether eight samples which by definition are well mixed and are, in some respects, superior to all rock samples which are small in number and statistically less significant than the billions of grains in each soil sample. In Table 4 we have put together what we think to be the best averages for the concentration of the major elements and some characteristic trace elements of all soil samples and the two breccias. With the assumption that all these samples represent a mixture of only two components but with different and unknown mixing ratios, we computed the individual mixing lines by the least-squares method, finding the best fit to the 21 element lines for each of the eight samples using a CDC 3300. Our first attempt proved unambiguously that the assumption of a two-component mixture was correct. We then dropped about seven individual values out of the 168 values (i.e., 8 × 21) which showed the highest deviations; all of them were single determinations of elements which are not easy to determine (see Table 4). The improvement was small but was of some importance for our next step, in which we tried to fit rock samples into the diagram calculated from the data of soils and breccias only. The two high-Mg rocks 12004 and 12018, using mainly our own data for these two rocks, were found to be especially suitable. The fit to the element lines calculated from soils and breccias was very satisfactory. We then added the data of these two rocks to the data set of the soils and breccias and made another computation. The element lines calculated in this way differed only slightly from the previous ones. The calculated concentrations and the deviations from the measured figures are listed in Table 4. The agreement is in most cases better than the uncertainty of the determinations, in spite of the fact that the variation of some elements exceeds a factor of ten.

We then took the composition of the samples KREEP, 12013 total, and 12013 dark given by HUBBARD *et al.* (1971) and computed their position in our mixing diagram (Fig. 6), but without correcting the element lines in respect to the data of these additional samples. The agreement with the expected concentration and the actual ones for KREEP and 12013 (total and dark) is considerable worse as for the other samples as it can be seen from Table 4 and Fig. 6.

While there cannot be much doubt that the high-Mg rocks come very close to the first end member of our mixing model, KREEP or 12013 (in part or total) is probably only a first approximation to the second end member. Let us assume for the moment that this approximation is correct; we then derive for the concentration of this end member material in the various samples the following values: 12037 (25%), 12044 and 12070 (40%), 12042 (42%), 12032 (62%), 12073 (65%), 12033 (74%), and 12034 (80%).

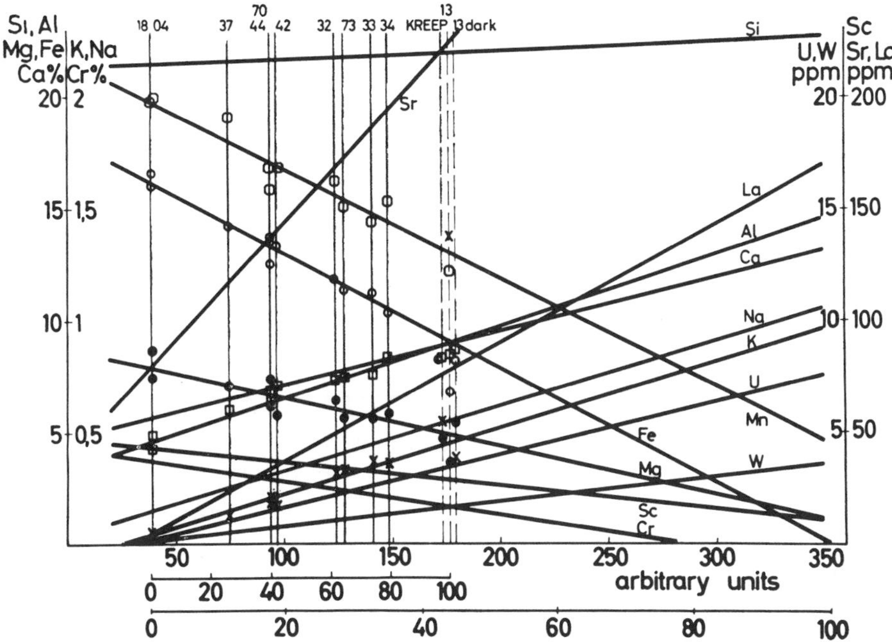

Fig. 6. Mixing diagram. Element lines were plotted for only 13 elements out of 21 and data points for only 5 elements for the sake of clarity (for details see Table 5). Note that the element lines and the sample positions along the x-axis were calculated via a least-squares fit from the data for basaltic rocks, soils, and the breccias 12034 and 12073. Data for KREEP (Hubbard *et al.*, 1971) and samples of 12013 were not used for the computation of element lines but the position of these samples along the x-axis was calculated via a least-squares fit.

We shall see later on that the admixture of "granitic" type material probably does not exceed a value of 50% for soil 12033; we believe, therefore, that the end member lies even further to the right than for rock 12013, which is a breccia itself. In any case, the end member must lie somewhere between rock 12013 and the point on the x-axis of Fig. 6 where Fe becomes zero.

One of the six soil samples, namely 12033, has attracted considerable interest. This sample is much lighter in color than the other soil samples. It was dug out from a depth of about 15 cm just inside the northwest rim of Head Crater. The light grey soil was overlain by a darker soil. A layer similar in appearance was found at the bottom of the double core tube south of Surveyor Crater (LSPET, 1970).

As can be seen from Fig. 6, the two breccias, 12034 and 12073, are compositionally very similar to soil 12033. Breccia 12034 was also found inside Head Crater at the same place where soil 12033 was taken. Breccia 12073 was taken during the first EVA beside the LM. Both samples 12033 and 12034 are distinctly different from all the other soil samples and breccias in that the light solar wind rare gases are practically absent. According to the data given by Funkhouser *et al.* (1971) the depletion reaches more than a factor of ten for helium when compared with average fines (for example, 12070), but decreases for the heavier rare gases, being a factor of two for xenon. This distribution suggests a partial loss of rare gases. From the concentrations of strongly siderophilic and volatile elements, BAEDECKER *et al.* (1971b) have shown that soil 12033 contains only 60% of the meteoritic component present in soil 12070, the total amount being only about 1% in the case of 12070. BAEDECKER *et al.* (1971b) have suggested that 12033 consists of a mixture of 60% of local soil and 40% of an exotic component. Our computation shows that this exotic component exists in all soil samples and breccias, and only the mixing ratio differs. This exotic component is identical with what we have called above the "granitic" component. As nearly all U, Th, and K found in 12033 as well as in 12034 and 12073 comes from the "granitic" component, this component must have suffered a nearly complete loss of radiogenic He and Ar at the time, or just before, the mixing took place. This would explain the low concentration of ^{4}He and ^{40}Ar in sample 12033 as well as in 12034 and 12073 (FUNKHOUSER *et al.*, 1971; BLOCH *et al.*, 1971). At the time of the mixing, the local component with a normal solar-wind rare gas concentration lost considerable amounts of the light solar-wind rare gases. The concentrations of cosmic ray-produced rare gases in the three samples (perhaps with the exception of breccia 12073) amount only to about 50% of that in 12070. Taking the evidence from solar wind, radiogenic, and cosmic ray-produced rare gases into account, one reaches the conclusion that the "granitic" component which amounts to about 50%, was completely outgassed prior to, or at, the time of mixing and that the local component which represents the other 50% of these three samples lost part of its solar wind gas but nearly none of the spallogenic and radiogenic rare gases during the mixing process. The mixing must have occurred rather recently—i.e., within the last 100 m.y.—as no appreciable amounts of either spallogenic or radiogenic rare gases seem to have been produced after the mixing. In any case, it seems obvious that soil 12033 must be rather young; otherwise, the difference observed in chemical composition as well as in the rare gas content is unexplained.

Our computation shows that 12070 itself is a mixture of the soil derived from the local basaltic rocks and a "granitic" component; only the mixing ratio differs. This in principle is not in conflict with the model of BAEDECKER *et al.* (1971b).

The "granitic" component must have been thrown to the Apollo 12 landing site rather recently and in amounts large enough to cover the area with tens of cm of it. (In any case, the event must have occurred after Head Crater was made.) A connection with the ejecta of Copernicus has already been disproved by BAEDECKER *et al.* (1971b) since Copernicus must be much older.

It is interesting that the soil from Apollo 11 (i.e., sample 10084) does not fit in the mixing diagram for Apollo 12 samples. A similar computation for Apollo 11 soil points toward an addition of an anorthositic component, as already proposed by WOOD *et al.* (1970), which is chemically different from the "granitic" component required for the Apollo 12 soils. Anorthosite can only supply an iron- and Mg-poor and Al- and Ca-rich admixture but not the alkaline, rare-earth, and other elements in the required amounts.

Metal particles

During our investigation of the Apollo 11 lunar samples, we separated some trace amounts of metallic iron from fines 10084. With instrumental neutron activation we studied their Ni and Co content and some trace elements. The most puzzling result was its high W content (WÄNKE *et al.* (1970b). We extended this study to the Apollo 12 samples. The total amount of metallic iron in sample 12001 was measured at 0.13%, considerably less than the 0.6% found in sample 10084. Most of the metal from 12001 was found in the $>150\ \mu$ (65% by weight) sieve fraction (Fig. 7). We also separated 4 mg metallic iron from a 40 g sample of rock 12053, which we pulverized for our measurements of cosmic ray induced radioisotopes. This metal was much finer (mostly $<35\ \mu$) than that from the fines.

Table 5 shows the results for Ni and Co and for the trace elements Cu, Ga, Au, Ir, and W. As we expected an inhomogenous distribution of W, we divided our metal sample out of 10084 into 5 parts, fractions f1 through f5. Indeed, W varies in these five fractions between 3 and 43 ppm while the other elements vary not more than by a factor of two.

The metal from 12001 shows an even higher W content: 74 to 123 ppm, with an average of 110 ppm. The metal from breccia 12073 shows similar high contents. The metal from the rock 12053 has an average of 33 ppm W. For comparison we also measured metal from various types of meteorites and found similar high W contents only in the metal from the eucrite Juvinas (Table 5).

Table 6 shows the calculation of the metal contribution to the Au, Ir, and W content of the bulk fines. It is evident that about half of the Au and Ir in the fines 12001 is located in the metal grains. It is interesting to note that in the case of W, the same amount (about 150 ppb) is contributed from the metal to both fines 12001 and 10084, despite the five-fold higher W content of the 12001 metal compared to that from 10084.

The analyses of individual metal grains (Table 7) show that the higher W content of the bulk metal from 12001 is not due to a higher W content of all the grains, but to a higher proportion of W-rich grains. In the metal fraction of 12001, nearly all the W is found in about one-fourth of the grains with ~ 300 ppm W, while in 10084, metal grains with this W content are about a factor of five rarer.

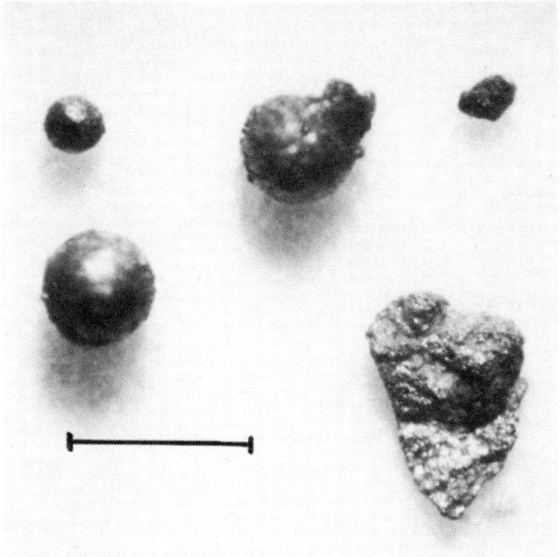

Fig. 7. Metal particles from fines 12001. Upper row: No. 8n, 8o, 8q. Lower row: No. 7 and 6 (numbers refer to Table 8). Scale bar: 0.5 mm.

Table 5. Composition of metal from different meteorites and lunar samples divided into several grain size fractions. From 70 g of the fines 12001 we obtained 40 mg of metal > 150 μ, 15 mg 35–150 μ and 3 mg < 35 μ.

fraction	Metal of	(sample weight)	Ni %	Co %	Cu ppm	Ga ppm	Au ppm	Ir ppm	W ppm
	Pultusk	(8.41 mg)	9.1	0.46	433	18	1.18	4.25	1.1
	Norton County	(14.1 mg)	9.1	0.29	170	42	1.38	0.55	0.3
	Juvinas 1	(1.91 mg)	2.9	0.62	150	9	0.76	0.9	21
	Juvinas 2	(0.022 mg)	3.5	1.0	1840	8.6	0.25	≤2.5	23.6
	Juvinas 3	(0.020 mg)	15.3	1.1	6700	36	1.16	≤6	31.5
	Juvinas 4	(0.041 mg)	7.2	0.8	1960	7.4	0.67	≤8	19.0
	Juvinas 5	(0.034 mg)	1.12	1.24	15.3	≤4	≤0.2	1.8	17.5
f	10084-18 (fines)		4.7	0.52	340	11	0.83	2.6	24
	↓ subdivided:								
f1	20 rounded grains	(0.77 mg)	6.20	0.45	—	—	1.68	2.3	19
f2	20 irregul. shaped gr.	(1.68 mg)	5.22	0.63	—	—	1.01	2.0	43
f3	20 irregul. shaped gr.	(1.17 mg)	4.08	0.34	—	—	0.83	2.4	16
f4	"fine" grains	(2.75 mg)	3.69	0.48	—	—	0.74	1.6	15
f5	3 "large" grains	(1.20 mg)	3.63	0.71	—	—	0.75	1.4	3.2
	12001-114 (fines)								
c	< 35 μ	(1.22 mg)	4.10	0.90	751	8.7	0.45	2.76	74
b	35–150 μ	(1.72 mg)	4.29	0.77	363	4.47	0.55	1.86	102
a	> 150 μ (16 grains)	(0.86 mg)	7.62	0.59	225	18	1.27	7.3	123
	12053-2 (rock)								
d	M 15	(4.32 mg)	1.89	1.36	675	9.0	0.1	0.5	33
	M 25 a	(0.163 mg)	2.2	1.8	—	3.7	0.26	≤2.7	26.1
	M 25 b	(0.177 mg)	—	1.5	605	10.4	≤0.1	≤0.8	16
	M 25 c	(0.146 mg)	—	1.3	610	≤3.1	≤0.1	≤1.1	243
	12073-31 (breccia)								
e	5 fine grains	(0.033 mg)	4.5	0.45	201	(60)	6.4	45	134
	↓								
	1 grain	(0.004 mg)	2.6	0.34	300	8	19	389	50

Table 6. Concentration of metal in different lunar samples and their Ni, Au, Ir, and W content.

	Metal content %	Ni %	Au ppm	Ir ppm	W ppm
Metal from rock 12053,2	0.036	1.89	0.1	0.5	33
Metal from fines 12001,114	0.124	6.4	1.0	5.5	110
Metal from fines 10084,18	0.60	4.7	0.83	2.6	24
12001,114		ppm	ppb	ppb	ppb
Bulk sample		314	2.6	11	630
Contribution of metal fraction to bulk composition		80	1.2	6.8	136
10084,18					
Bulk sample		280	2.1	6.9	220
Contribution of metal fraction to bulk composition		280	5.0	15.6	144

Calculated is the amount of each element contributed by the metal fraction to the bulk composition. The high results obtained for Au and Ir for sample 10084 may be due to the fact that we measured a coarse grained sample, which may contain higher concentrations of the elements than the average metal.

Table 7. Composition of individual metal grains from lunar fines.

fraction	Sample number	Sample weight in 10^{-6} g	Ni %	Co %	W ppm	Ir ppm	Au ppm	Cu ppm	Ga ppm	As ppm
				Metal grains from 12001						
	2	260	5.3	0.41	15.6	7.0	1.65	654	19	
	3	360	5.2	0.25	1.8	4.3	0.89	463	4	
	6	66.6	0.1	0.16	560	0.9	0.1	124	3	
	7	95.9	8.5	0.46	6	5.5	1.3	43	13	
a	8a	23.9	5.8	0.78	560	18	2.3	185	58	
	8b	82.8	4.7	0.36	25	5	0.3	20	8.5	
	8c	135	5.1	0.63	360	10	2.0	270	7	
	8d	55.3	5.0	0.55	<20	<10	1.9	100	86	
	8e	43.5	5.0	1.07	320	17	1.9	364	16	
	8f	37.9	n.d.	1.16	78	7	2.4	270	46	
	8g	52.5	4.9	0.79	365	<10	2.1	300	26	
	8h	114	5.0	0.36	<10	5.6	0.63	345	28	
	8l	29.8	4.7	0.26	<20	~5	~0.5	100	68	
	8m	38.6	7.7	0.88	90	18	5	393	<20	
	8n	14.7	5.5	0.44	<20	<10	<2	258	<20	
	8o	71.7	5.1	0.68	<20	34	0.5	86	45	
				Metal grains from 10084						
f2	10a	110	3.2	0.46	36	6	1.0	207	25	44
	10b	54.5	0.4	0.59	<6	<8	<0.3	140	20	<8
	10c	137	5.2	0.51	<20	4.5	0.8	218	9.6	13
	10d	138	4.9	0.53	36	8	0.7	299	12.7	<3
	10e	247	12.6	1.16	279	11.5	2.9	396	5.0	34
	10f	120	4.5	0.51	<15	6	1.3	94	10	17
	10g	76	5.0	0.40	<20	8.2	0.9	206	7.6	25
	10h	55	4.6	0.57	<6	5	1.6	14	32	20
	10i	177	3.6	1.22	<5	<2	<0.1	248	28	<4
	10k	51.5	5.1	0.20	<5	<5	~0.3	68	<7	<7
	10m	54.5	4.9	0.90	<20	14	2.3	891	35	60
	10n	24	5.0	0.45	29	<7	2.2	369	<16	53
	10o	51	n.d.	0.96	<25	<10	<0.3	370	<17	<10
	10p	59	4.8	0.58	<10	<5	1.6	617	<16	<10
	10q	27	8.3	0.62	170	7	2.1	1600	<25	<20
	10r	46	n.d.	0.45	<20	<4	1.0	41	35	<30
	10s	29	0.0	0.34	<20	<5	<0.3	195	<20	<20
f3	121	76	n.d.	0.43	<10	<5	1.3	330	<20	24
	122	83	n.d.	0.43	45	12	1.6	229	<30	42
	123	62	n.d.	0.13	110	6	0.4	140	<20	<10
	125	22	n.d.	0.53	<20	<5	1.1	226		<20
	127	40	n.d.	0.35	<20	5	1.7	261		<10
	128	29	n.d.	0.27	<20	<3	<0.3	150		<30
	129	171	n.d.	0.19	28	0.6	0.07	100	13.2	<2
f5	5a	500	7.2	0.7	3	0.8	1.6	130	14	
	5b	300	0.9	0.9	6	<0.3	<0.1	250	32	
	5c	130	8.0	0.5	2	2.5	1.4	180	12	

In Fig. 8 the Au vs. Co content of individual metal grains is plotted. They show a good correlation. A similar correlation is found for Ir and Co. Cu and Co are only slightly correlated, while Ga and Co show no correlation. The average values for the atomic ratios Co/Au = 12500 and Co/Ir = 2050 derived from the correlation diagram are very close to the cosmic ratios, whereas the Co/W ratios in the 10084 metal (680) and in the 12001 metal (190) are much lower than the cosmic ratios given by CAMERON, 1967 (14400) or SUESS and UREY, 1956 (3700). All those grains which do not fit into the Co/Au or Co/Ir correlation, i.e., those high in Co and Ni but low in Au and Ir

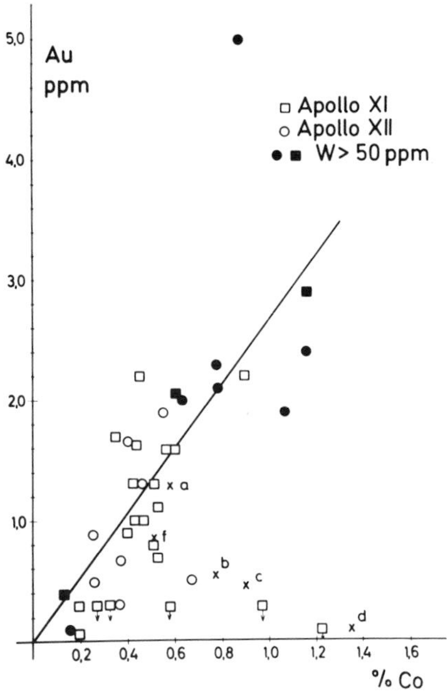

Fig. 8. Au vs. Co, measured in individual metal grains via neutron activation. Points a, b, and c give measurements obtained on bulk samples of grain size $> 150\mu$, 35–150μ, and $< 35\mu$ respectively; d is metal from rock 12053; f metal from fines 10084. Points with a downward arrow give upper limits for Au only.

represent probably particles from the local basaltic rocks. Following REID *et al.* (1970) we call them secondary metal grains. On the other hand, an extralunar origin seems evident for all grains with the cosmic Co/Au or Co/Ir ratios; some of them are highly enriched in W (primary metal grains).

This metal may have taken up W from the lunar silicates. The degassing of the "granitic" component of soil 12033 indicates an example for a reheating event in which such an equilibration may occur. Also GOLDSTEIN and YAKOWITZ (1971) inferred a prolonged high temperature stage on the Moon at about 500°C of the order of years from the composition and structure of two of their lunar "meteoritic" metal particles. As another source, metal from achondrites may be considered which would fit better in respect to W, but their metal content is too low to account for an appreciable metal component in the lunar fines. There may be other extralunar sources of W rich metal, but mostly likely is an origin in lunar igneous rocks high in W. Metal from one of the local rocks (12053) contains only one-third of the W (33 ppm) compared to the metal from the fines. Under the reducing conditions of the lunar surface, W will become siderophile and enter metal grains in variable amounts, depending on the source rock and the degree of reduction, in a manner similar to that of Ni and Co (REID *et al.*, 1970).

On Earth granitic rocks are known to contain more W than basalts (KRAUSKOPF, 1970), thus rocks like sample 12013 may be good candidates for the production of W

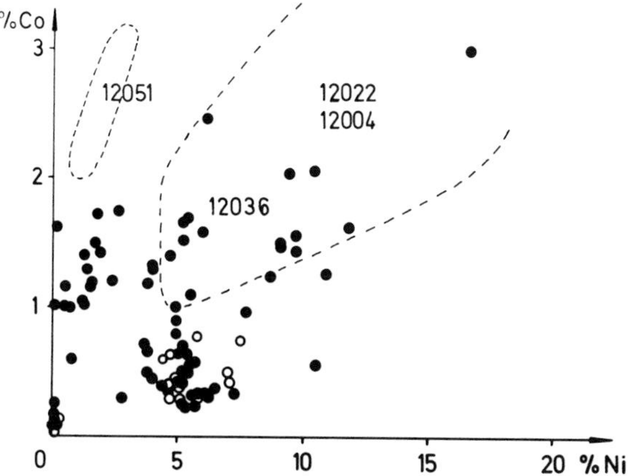

Fig. 9. Ni vs. Co from microprobe measurements in individual metal grains of size fraction 35–150μ from 12001 fines. Open circles represent grains containing troilite inclusions. The dashed lines enclose the measurements made on metal grains from the Apollo 12 rocks 12051 and 12036 (KEIL *et al.*, 1971) and 12022 and 12004 (REID *et al.*, 1970).

rich metal. The existence of a "granitic" component rich in W on the Moon has been demonstrated above in our mixing model (see Fig. 6). In this connection it is interesting that VINOGRADOV (1971) reports W contents up to 30 ppm in the lunar soil brought back by Luna-16.

About 100 grains of the 35–150 μ metal fraction from 12001 were mounted and polished for metallographic and microprobe study. Most of them are clean metal, some have drop-like troilite inclusions, and a few display the dendritic intergrowth of metal and troilite known from the Ramsdorf chondrite (BEGEMANN and WLOTZKA, 1969) and other lunar metal particles (GOLDSTEIN *et al.*, 1970; WÄNKE *et al.*, 1970b; MASON *et al.*, 1970; and GOLDSTEIN and YAKOWITZ, 1971).

Their Ni and Co contents were measured with a microprobe and are plotted in Fig. 9. A cluster of points forms at about 6% Ni and 0.5% Co, which is typical for meteoritic metal and probably represents the meteoritic component of the metal. Most of the troilite containing grains are found at this cluster. Another group lies at 0.1% Ni, 0.1–0.3% Co, values typical of the Apollo 11 rocks. Some of these also contain troilite. The other points scatter widely, but it is interesting that the areas occupied by the metal grains of some of the Apollo 12 rocks (REID *et al.*, 1970; KEIL *et al.*, 1971) are either not represented in the fines at all (rock 12051) or are only partly so (rock 12022).

Acknowledgments—We are grateful to NASA for making available the lunar material for this investigation. The samples were irradiated in TRIGA-research reactor of the Institut für Anorganische Chemie und Kernchemie der Universität Mainz. We thank Prof. F. Begemann and Prof. J. T. Wasson for many valuable discussions. We wish to thank the staff of the TRIGA-research reactor and the staff of our Institute, in particular, Mr. P. Deibele, Mr. H. Döpke, Mr. H. Engler, and Miss H. Prager. The financial support by the Bundesministerium für Bildung und Wissenschaft is gratefully acknowledged.

References

Anders E., Laul J. C., Keays R. R., Ganapathy R., and Morgan J. W. (1971) Elements depleted on lunar surface: Implications for origin of moon and meteorite influx rate. Second Lunar Science Conference (unpublished proceedings).

Annell C. S., Carron M. K., Christian R. P., Cuttitta F., Dwornik E. J., Helz A. W., Ligon D. T., Jr., and Rose H. J., Jr. (1971) Chemical and spectrographic analyses of lunar samples from Apollo 12 mission. Second Lunar Science Conference (unpublished proceedings).

Baedecker P. A., Schaudy R., Elzie J. L., Kimberlin J., and Wasson J. T. (1971a) Trace element studies of rocks and soils from Oceanus Procellarum and Mare Tranquillitatis. Second Lunar Science Conference (unpublished proceedings).

Baedecker P. A., Cuttitta F., Rose H. J., Jr., Schaudy R., and Wasson J. T. (1971b) On the origin of lunar soil 12033. *Earth Planet. Sci. Lett.* **10**, 361–364.

Begemann F. and Wlotzka F. (1969) Shock induced thermal metamorphism and mechanical deformation in the Ramsdorf chondrite. *Geochim. Cosmochim. Acta* **33**, 1351–1370.

Biggar G. M., O'Hara M. J., and Peckett A. (1971) Origin, eruption and crystallization of protohypersthene basalts from the Ocean of Storms. Second Lunar Science Conference (unpublished proceedings).

Bloch M., Fechtig H., Funkhouser J., Gentner W., Jessberger E., Kirsten T., Müller O., Neukum G., Schneider E., Steinbrunn F., and Zähringer J. (1971) Location and variation of rare gases in Apollo 12 lunar samples. Second Lunar Science Conference (unpublished proceedings).

Bogard D. D., Funkhouser J. G., Schaeffer O. A., and Zähringer J. (1971) Noble gas abundances in lunar material—II. Cosmic ray spallation products and radiation ages from Mare Tranquillitatis and Oceanus Procellarum. Preprint.

Bouchet M., and Kaplan G. (1971) Spark mass spectrometric analysis of major and minor elements in six lunar samples. Age determinations of four rocks. Second Lunar Science Conference (unpublished proceedings).

Brunfelt A. O., Heier K. S., and Steinnes E. (1971) Determination of 40 elements in Apollo 12 materials by neutron activation analysis. Second Lunar Science Conference (unpublished proceedings).

Cameron A. G. W. (1968) A new table of abundances of the elements in the solar system. In *Origin and Distribution of the Elements* (editor L. H. Ahrens), pp. 125–143. Pergamon.

Cliff R. A., Lee-Hu C., and Wetherill G. W. (1971) Rb–Sr and U, Th–Pb measurements on Apollo 12 material. Second Lunar Science Conference (unpublished proceedings).

Compston W., Berry H., Vernon M. J., Chappell B. W., and Kaye M. J. (1971) Rubidium–strontium chronology and chemistry of lunar material from the ocean of storms. Second Lunar Science Conference (unpublished proceedings).

Eberhardt P., Geiss J., Graf H., Grögler N., Krähenbühl U., Schwaller H., Schwarzmüller J., and Stettler A. (1970) Correlation between rock type and irradiation history of Apollo 11 igneous rocks. *Earth Planet. Sci. Lett.* **10**, 67–72.

Ehmann W. D., and Morgan J. W. (1970) Oxygen, silicon, and aluminum in Apollo 11 rocks and fines by 14 MeV neutron activation. *Proc. Apollo 11 Lunar Sci. Conf., Geochim. Cosmochim. Acta* Suppl. 1, Vol. 2, pp. 1071–1079. Pergamon.

Ehmann W. D., and Morgan J. W. (1971) Major element abundance in Apollo 12 rocks and fines by 14 MeV neutron activation. Second Lunar Science Conference (unpublished proceedings).

Fields P. R., Diamond H., Metta D. N., Stevens C. M., and Rokop D. J. (1971) Isotopic abundances of actinide elements in Apollo 12 samples. Second Lunar Science Conference (unpublished proceedings).

Funkhouser J. G., Schaeffer O. A., Bogard D. D., and Zähringer J. (1971) Noble gas abundances in lunar material—I. Solar wind implanted gases in Mare Tranquillitatis and Oceanus Procellarum. Preprint.

Goldstein J. I., Henderson E. P., and Yakowitz H. (1970) Investigation of lunar metal particles. *Proc. Apollo 11 Lunar Sci. Conf., Geochim. Cosmochim. Acta* Suppl. 1, Vol. 1, pp. 499–512. Pergamon.

GOLDSTEIN J. I. and YAKOWITZ H. (1971) Metallic inclusions and metal particles in the Apollo 12 lunar soil. Second Lunar Science Conference (unpublished proceedings).

GOLES G. G., DUNCAN A. R., OSAWA M., MARTIN M. R., BEYER R. L., LINDSTROM D. J., and RANDLE K. (1971) Analyses of Apollo 12 specimens and a mixing model for Apollo 12 soils. Second Lunar Science Conference (unpublished proceedings).

HASKIN L. A., ALLEN R. O., HELMKE P. A., ANDERSON M. R., KOROTEV R. L., and ZWEIFEL K. A. (1971) Rare earths and other trace elements in Apollo 12 lunar materials. Second Lunar Science Conference (unpublished proceedings).

HESS F. D., PALMER D. F., and BISCHOFF J. L. (1971) Relations of some lunar rocks and fines: Evidence by radiochemical analysis of rare earth elements. Second Lunar Science Conference (unpublished proceedings).

HINTENBERGER H., WEBER H., and TAKAOKA N. (1971) Concentrations and isotopic abundances of the rare gases in lunar matter. Second Lunar Science Conference (unpublished proceedings).

HUBBARD N. J., GAST P. W., and MEYER C. (1971) The origin of the lunar soil based on REE, K, Rb, Ba, Sr, P, and $Sr^{87/86}$ data. Second Lunar Science Conference (unpublished proceedings).

KEIL K., PRINZ M., and BUNCH T. E. (1971) Mineralogical and petrological aspects of Apollo 12 rocks. Second Lunar Science Conference (unpublished proceedings).

KRAUSKOPF K. B. (1970) Tungsten. In *Handbook of Geochimistry* (editor K. H. Wedepohl), Vol. II–2, Chap. 74. Springer.

LSPET (Lunar Sample Preliminary Examination Team) Preliminary examination of the lunar samples from Apollo 12. *Science* **167**, 1325–1339.

MASON B., FREDRIKSSON K., HENDERSON E. P., JAROSEWICH E., MELSON W. G., TOWE K. M., and WHITE J. S., JR. (1970) Mineralogy and petrography of lunar samples. *Proc. Apollo 11 Lunar Sci. Conf., Geochim. Cosmochim. Acta* Suppl. 1, Vol. 1. pp. 655–660. Pergamon.

MAXWELL J. A., and WIIK H. B. (1971) Chemical composition of Apollo 12 lunar samples 12004, 12033, 12051, 12052, and 12065. Second Lunar Science Conference (unpublished proceedings).

MEYER C., JR., AITKEN F. K., BRETT R., MCKAY D. S., and MORRISON D. A. (1971) Rock fragments and glasses rich in K, REE, P in Apollo 12 soils: Their mineralogy and origin. Second Lunar Science Conference (unpublished proceedings).

MORRISON G. H., GERARD J. T., POTTER N. M., GANGADHARAM E. V., ROTHENBERG A. M., and BURDO R. A. (1971) Elemental abundances of lunar soil and rocks from Apollo 12. Second Lunar Science Conference (unpublished proceedings).

MURTHY V., EVENSEN N. M., JAHN B., and COSCIO M. R., JR. (1971) Rb–Sr isotopic relations and elemental abundances of K, Rb, Sr, and Ba in Apollo 11 and Apollo 12 samples. Second Lunar Science Conference (unpublished proceedings).

PECK L. C. values given in paper by BIGGAR *et al.* (1971).

RANCITELLI L. A., PERKINS R. W., FELIX W. D., and WOGMAN N. A. (1971) Cosmogenic and primordial radionuclide measurements in Apollo 12 lunar samples by non-destructive analysis. Second Lunar Science Conference (unpublished proceedings).

REED G. W. and JOVANOVIC S. (1971) The halogenes and other trace elements in Apollo 12 soil and rocks. Second Lunar Science Conference (unpublished proceedings).

REID A. M., MEYER C., JR., HARMON R. S., and BRETT R. (1970) Metal grains in Apollo 12 igneous rocks. *Earth Planet. Sci. Lett.* **9**, 1–5.

RIEDER R., and WÄNKE H. (1969) Study of trace element abundance in meteorites by neutron activation. In *Meteorite Research* (editor P. M. Millman), pp. 75–86. Reidel.

ROSHOLT J. N. and TATSUMOTO M. (1971) Isotopic composition of thorium and uranium in Apollo 12 samples. Second Lunar Science Conference (unpublished proceedings).

ROSE H. J., JR., CUTTITTA F., ANNELL C. S., CARRON M. K., CHRISTIAN R. P., DWORNIK E. J., HELZ A. W., and LIGON D. T., JR. (1971) Semimicroanalysis of Apollo 12 samples. Second Lunar Science Conference (unpublished proceedings).

SCOON J. H. (1971) Quantitative chemical analyses of lunar samples 12040,36 and 12064,33. Second Lunar Science Conference (unpublished proceedings).

SIEVERS R. E., EISENTRAUT K. J., JOHNSON D. G., RICHARDSON M. F., WOLF W. R., ROSS W. D., FREW N. J., and ISENHOUR T. L. (1971) Variations in beryllium and chromium concentrations in lunar fines compared with crystalline rocks. Second Lunar Science Conference (unpublished proceedings).

SILVER L. T. (1971) U–Th–Pb isotope relations in Apollo 11 and Apollo 12 lunar samples. Second Lunar Science Conference (unpublished proceedings).

SMALES A. A., HISLOP J. S., MAPPER D., WEBB M. S. W., WEBSTER R. K., and WILSON J. D. (1971) Further studies on the elemental composition of lunar surface materials (part 2). Second Lunar Science Conference (unpublished proceedings).

SUESS H. E., and UREY H. C. (1956) Abundances of the elements. *Rev. Mod. Phys.* **28**, 53–74.

SUTTON R. L., and SCHABER G. G. (1971) Lunar locations and orientations of rock samples from Apollo missions 11 and 12. Second Lunar Science Conference (unpublished proceedings).

TAYLOR S. R., KAYE M., GRAHAM A., RUDOWSKI R., and MUIR P. (1971) Trace element chemistry of lunar samples from the Ocean of Storms. Second Lunar Science Conference (unpublished proceedings).

TRAVESI A., and PALOMARES J. (1971) Multielement neutron activation analysis of trace elements in lunar fines. Second Lunar Science Conference (unpublished proceedings).

VINOGRADOV A. P. (1971) Preliminary data on lunar ground brought to Earth by automatic probe Luna-16. Second Lunar Science Conference (unpublished proceedings).

WÄNKE H., WLOTZKA F., JAGOUTZ E., and BEGEMANN F. (1970b) Composition and structure of metallic iron particles in lunar "fines." *Proc. Apollo 11 Lunar Sci. Conf., Geochim. Cosmochim. Acta* Suppl. 1, Vol. 1, pp. 931–935. Pergamon.

WÄNKE H., RIEDER R., BADDENHAUSEN H., SPETTEL B., TESCHKE F., QUIJANO-RICO M., and BALACESCU A. (1970a) Major and trace elements in lunar material. *Proc. Apollo 11 Lunar Sci. Conf., Geochim. Cosmochim. Acta* Suppl. 1, Vol. 2, pp. 1719–1727. Pergamon.

WAKITA H., SCHMITT R. A., and REY P. (1970) Elemental abundances of major, minor, and trace elements in Apollo 11 lunar rocks, soil and core samples. *Proc. Apollo 11 Lunar Sci. Conf., Geochim. Cosmochim. Acta* Suppl. 1, Vol. 2, pp. 1685–1717. Pergamon.

WAKITA H., REY P., and SCHMITT R. A. (1971) Abundances of the 14 rare earth elements plus 22 major, minor, and trace elements in ten Apollo 12 rock and soil samples. Second Lunar Science Conference (unpublished proceedings).

WILLIS J. P., AHRENS L. H., DANCHIN R. V., ERLANK A. J., GURNEY J. J., HOFMEYR P. K., MC-CARTHY T. S., and ORREN M. J. (1971) Some inter-element relationships between lunar rocks and fines, and stony meteorites. Second Lunar Science Conference (unpublished proceedings).

WOOD J. A., DICKEY J. S., JR., MARVIN U. B., and POWELL B. N. (1970) Lunar anorthosites and a geophysical model of the moon. *Proc. Apollo 11 Lunar Sci. Conf., Geochim. Cosmochim. Acta* Suppl. 1, Vol. 1, pp. 965–988. Pergamon.

Proceedings of the Second Lunar Science Conference, Volume 2, pp. 1209–1215.
The M.I.T. Press, 1971.

Comparison of the analytical results from the Surveyor, Apollo, and Luna missions

ANTHONY L. TURKEVICH
Enrico Fermi Institute and Chemistry Department,
University of Chicago, Chicago, Illinois 60637

(*Received 15 February 1971; accepted 5 March 1971*)

Abstract—The principal chemical element composition and inferred mineralogy of the powdered lunar surface material at five mare and one terra sites on the moon are compared. The mare compositions are all similar to one another and comparable to that of terrestrial ocean ridge basalts except in having higher titanium and much lower sodium contents than the latter. These analyses suggest that most, if not all, lunar maria have this chemical composition and are derived from rocks with an average density of 3.19 g cm^{-3}. Mare Tranquillitatis differs from the other three maria in having twice the titanium content of the others.

The chemical composition of the single highland site studied (Surveyor 7) is distinctly different from that of any of the maria in having much lower amounts of titanium and iron and larger amounts of aluminum and calcium. The inferred mineralogy is 45 mole percent high anorthite plagioclase and the parent rocks have an estimated density of 2.94 g cm^{-3}. The Surveyor 7 chemical composition is the principal contributor to present estimates of the overall chemical composition of the lunar surface.

ANALYTICAL INFORMATION for the principal chemical elements is available at present from six sites on the moon. The location of these sites, their characteristics, and the nature of the analytical technique used, are summarized in Table 1. The phrase "principal chemical elements" is used here to denote the elements usually present in rocks in amounts greater than about 0.3 atom percent. They comprise the elements O, Na, Mg, Al, Si, Ca, Ti, and Fe. These elements constitute about 99% of the atoms and therefore determine the gross chemical nature of the lunar surface. The analyses compared in this report are those of the powdered material at a given site. It is felt

Table 1. Locations on the Moon from which chemical composition data are available

Mission	Date	Selenographic Coordinates Long.	Lat.	Ref.	Geographical Area	Type of Area	Type of Analysis
Surveyor 5	1967, Sept. 11	23.20°E	1.42°N	(a)	Mare Tranquillitatis	Mare	in situ α scattering
Surveyor 6	1967, Nov. 10	1.37°W	0.46°N	(a)	Sinus Medii	Mare	in situ α scattering
Surveyor 7	1968, Jan. 10	11.44°W	40.97°S	(a)	outside Crater Tycho	Terra	in situ α scattering
Apollo 11	1969, July 20	23.43°E	0.69°N	(b)	Mare Tranquillitatis	Mare	returned samples
Apollo 12	1969, Nov. 19	23.34°W	2.45°S	(c)	Oceanus Procellarum	Mare	returned samples
Luna 16	1970, Sept. 19	56.18°E	0.41°S	(d)	Mare Foecunditatis	Mare	returned samples

(a) JAFFE (1969); (b) LSPET (1969); (c) LSPET (1970); (d) VINOGRADOV (1971).

that, in the case of the principal chemical elements, the chemical composition of this material represents the average of a given area of the moon better than the composition of individual rocks.

Although the Surveyor analyses, obtained by the remote-control alpha-scattering technique, are usually appreciably less precise than those on returned lunar samples, they represent half of the sites on the moon from which information is available. Their accuracy is adequate to establish the gross rock type and even some of its special characteristics. In addition, the Surveyor 7 chemical analysis is at present the only one from a highland site on the moon.

Finally, at the present stage of lunar and planetary investigations, a comparison between remote control analytical results and those obtained on returned samples can help make proper decisions about future exploration of the moon, asteroids, and planets. In this connection it should be borne in mind that the capabilities of remote control analyses by various techniques are continually improving. The Surveyor missions represent the state of technology of one particular technique almost ten years ago. Not only could the accuracy for the principal chemical elements be improved today, but there are possibilities also of measuring some of the minor constituents.

Table 2 presents the results of chemical analyses of surface lunar fines for the principal chemical elements at the five lunar *mare* sites that have been studied so far. Estimates of the accuracies of the Surveyor analyses are indicated; the accuracies of the analyses of the returned samples are probably all less than 3% at the same confidence level. Although the Surveyor analyses for Ca, Ti, and Fe are, strictly speaking, for groups of elements with about the same atomic weight, assignment of the values primarily to the elements indicated appears justified on geochemical grounds.

The comparison between the Surveyor 5 and Apollo 11 analyses, columns 2 and 3 of Table 2, is particularly to be noted since both apply to Mare Tranquillitatis, at sites

Table 2. Concentrations of principal chemical elements in lunar mare material (fines) (Percent by Atom)

	Surveyor* 5	Apollo 11	Surveyor* 6	Apollo 12	Luna 16
Element	(a)	(b)	(a)	(c)	(d)
O	61.1 ± 1.0	59.87	59.3 ± 1.6	59.9	(60.15)
Na	0.47 ± 0.15	0.33	0.6 ± 0.24	0.30	0.37
Mg	2.8 ± 1.5	4.57	3.7 ± 1.6	6.8	4.99
Al	6.4 ± 0.4	6.30	6.5 ± 0.4	6.3	6.95
Si	17.1 ± 1.2	16.31	18.5 ± 1.4	16.0	15.97
Ca*	5.5 ± 0.7	4.92	5.2 ± 0.9	4.1	4.99
Ti*	2.0 ± 0.5	2.19	1.0 ± 0.8	0.9	0.98
Fe*	3.8 ± 0.4	5.12	3.9 ± 0.6	5.4	5.39

* For the Surveyor analyses, the elemental symbols Ca, Ti, and Fe represent a range of elements (see text and ECONOMOU, 1970). The principal contributors are expected to be the elements listed. The quoted errors of the Surveyor analyses are estimates at the 90% confidence level. In the case of Surveyor 7, an amount of fluorine equal to 0.29 ± 0.12 atom percent was also found.

(a) FRANZGROTE (1970); (b) Averages of the more accurate of many lunar soil analyses reported in LEVINSON (1970); (c) LSPET (1970); (d) VINOGRADOV (1971).

less than 30 km apart. It is seen that the agreement is just about within the Surveyor analytical errors except in the case of Fe, where the Surveyor answer is some 25% lower than the Apollo result. The analytical results for all the maria lead to a similar geochemical picture—that of a silicate rock with Al and Ca each comparable to or greater than Mg, low Na (0.3 to 0.6 atom percent) and relatively high Ti (0.9 to 2.2 atom percent). The Fe content is also relatively high at 3.8 to 5.4 atom percent. Although there may be some indication of systematic biases in the alpha scattering method leading to high values of Si and low values of Mg and Fe, the variations in mare compositions illustrated in Table 2 are relatively small.

Since the first chemical analysis of the lunar surface on the Surveyor 5 mission (TURKEVICH *et al.*, 1967), the compositions of the lunar maria have been compared with those of terrestrial basaltic rocks. The more extensive data available now make possible a more detailed comparison. This is done in Fig. 1, where the amounts of Mg, Al, Ca, Fe, Ti, and Na in the lunar maria are compared with the amounts of these elements in terrestrial ocean ridge basaltic rocks (KAY *et al.*, 1970). The comparison is not made for O or Si since these are not too sensitive indicators of rock type. There is, however, an indication that the amount of Si in the lunar maria may be one to two atom percent lower than in terrestrial basaltic rocks.

Fig. 1 shows that the Al, Mg, and Ca contents of the lunar maria are well within the ranges observed by KAY *et al.* In the case of Fe, also, there is overlap, although the overlap is provided primarily by the alpha-scattering results, both of which are lower than those on returned lunar samples. Thus, for the six chemical elements present in amounts greater than 3 atom percent, the chemical composition of the lunar maria is quite well represented by that of terrestrial ocean ridge basaltic rocks.

The striking differences between the maria compositions and those of terrestrial basaltic rocks are in the high Ti and low Na contents of the former. Both these aspects are well illustrated in Fig. 1. Even larger differences in the amounts of minor constituents have, of course, been noted (see e.g., LSPET, 1969, 1970).

The principal variation in composition among the different maria is in the Ti content, which is twice as high in Mare Tranquillitatis than in the other three lunar maria studied. The apparently even larger variation in Mg content appears to be at least partially due to the larger errors in the Surveyor analyses for this element. On the other hand, the Al content at all five sites is remarkably constant at 6.6 ± 0.3 atom percent.

Considering the fact that the four maria sampled are widely separated on the moon—although all are close to the equator—these analyses suggest that most, if not all lunar maria have this gross chemical composition. An average lunar mare chemical composition calculated on the assumption that each analysis is representative of its mare (the Surveyor 5 and Apollo 11 data for Mare Tranquillitatis were averaged) is given in column 2 of Table 3.

Column 3 of Table 3 gives the average lunar soil composition determined by the alpha-scattering technique on the Surveyor 7 mission to a highland site outside the crater Tycho (PATTERSON *et al.*, 1970). It is seen that the composition here is distinctly different from that of the maria. There is about a factor of three fewer atoms of elements heavier than Ca (i.e., much less Fe and very little, if any, Ti), and about a 40%

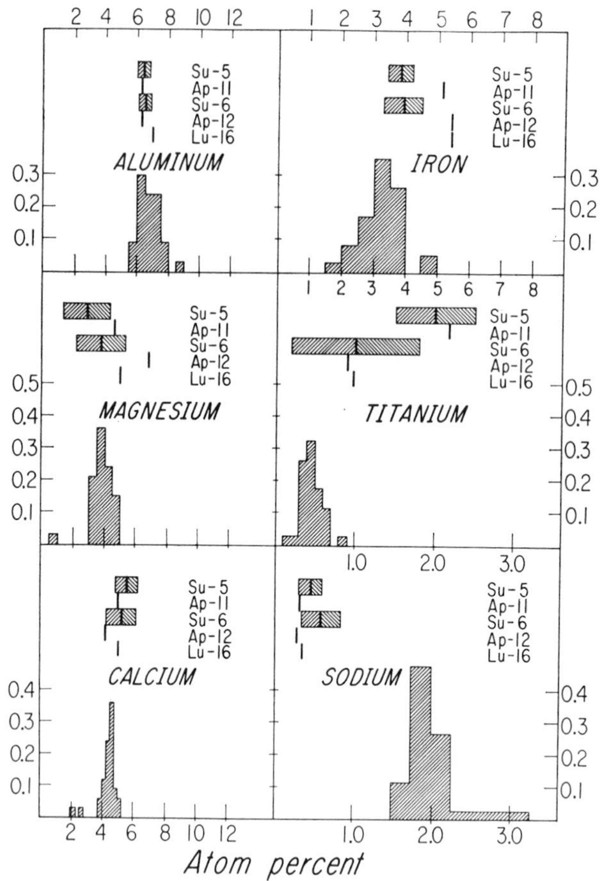

Fig. 1. Comparison of the amounts of some of the principal chemical elements in lunar mare material with the amounts in terrestrial ocean ridge basaltic rocks. The abscissae are in atom percent. The histograms are the distributions in the amounts of the element as determined by KAY, HUBBARD, and GAST (1970). The shaded areas about the Surveyor analyses represent the 90% confidence range of these analyses.

increase in Al and Ca contents. The amount of Si is, within the Surveyor errors, the same as in the maria.

The applicability of this analysis at one highland site on the moon to the other highland areas is at present only speculative, even though fragments of similar composition have been found in the Apollo 11 soil (e.g., SMITH *et al.*, 1970; WOOD, 1970) and in some of the Apollo 12 breccia rocks (e.g., ANDERSON *et al.*, 1971) and these have been attributed to highland material. If it is assumed that the Surveyor 7 analyses are representative of all the lunar highlands, it has been pointed out (TURKEVICH *et al.*, 1968b; PHINNEY *et al.*, 1969; PATTERSON *et al.*, 1969) that the differences in chemical composition between the lunar terrae and maria may be the explanation for the albedo differences of the major topographical features of the moon. They also

Table 3. Comparison of the chemical composition of lunar maria and terrae

	(Percent by atom)		
Element	Average Mare	Terra (Surveyor 7)*	Average Lunar Surface†
O	60.0	61.8 ± 1.0	61.4
Na	0.4	0.5 ± 0.2	0.5
Mg	4.8	3.6 ± 1.6	3.8
Al	6.5	9.2 ± 0.4	8.7
Si	16.8	16.3 ± 1.2	16.4
Ca	4.9	6.9 ± 0.6	6.5
Ti	1.3	0 ± 0.4	0.3
Fe	4.8	1.6 ± 0.4	2.2

* The errors are estimates at the 90% confidence level (see PATTERSON *et al.*, 1970).

† See text.

imply lower density for the highlands as compared to the maria, with a suggestion that isostatic adjustment has occurred in many regions of the moon. The chemical composition of the highlands as represented by the Surveyor 7 analyses, also suggests the existence of a crust on the moon, chemically as well as physically different from the main body of the moon (PATTERSON *et al.*, 1970).

The average chemical composition of the lunar surface, on the assumption that the mare composition of column 2 of Table 3 is applicable to 20% of the surface and the Surveyor 7 composition to the rest, is given in column 4 of Table 3.

Although an elemental chemical composition cannot provide the detailed information about rock type and mineralogy that can be obtained from examination of returned samples, some data of this type can be reasonably inferred from chemical analyses. This was first done using the early Surveyor analyses by TURKEVICH *et al.* (1968a) and PHINNEY *et al.* (1969). A normative mineral composition is derived following an order that minimizes the uncertainties due to the errors of the Surveyor analyses. The amount of ilmenite is determined by the Ti content. The amount of Al, with its relatively small error, determines the amount of plagioclase feldspar ($CaAl_2Si_2O_8$ and $NaAlSi_3O_8$ with the Na content determining the relative amounts of the two). The remaining atoms of Ca, Mg, Fe, and Si are assigned to pyroxenes and olivine or quartz. It is in the detailed assignment of the different types of pyroxenes and in the amounts of olivine or of silica minerals that the relatively large errors of the alpha-scattering method for Mg and Si play the largest role. The results of the above procedure are shown in Table 4 for the six lunar sites for which analyses are available.

The table illustrates, from a mineralogical standpoint, the similarities and differences among the different maria. The higher ilmenite content at Mare Tranquillitatis than at the other maria follows directly from the higher Ti content there. The constancy of the plagioclase contribution at 24.5 ± 2.4 mole percent is a consequence of the constant Al contents of the maria. The anorthite fraction of the plagioclase is constant at 88%, although here the inability of the alpha scattering method to distinguish between Ca and K must be remembered. Likewise, the detailed composition of the pyroxenes and the apparent considerable amount of silica minerals in the results from

Table 4. Normative mineral composition inferred from chemical analyses (mole percent)

		Surveyor 5		Apollo 11		Surveyor 6		Apollo 12		Luna 16		Surveyor 7	
Ilmenite $FeTiO_3$	il	13.1		14.7		6.8		6.8		7.6		—	
Plagioclase Feldspars		22.8		22.2		24.5		24.9		28.4		45.0	
$NaAlSi_3O_8$	ab		3.1		2.2		4.1		2.3		2.9		4.6
$CaAl_2Si_2O_8$	an		19.7		20.0		20.4		22.6		25.5		40.4
Pyroxenes		46.5		63.1		59.9		43.6		42.3		39.4	
$MgSiO_3$	en		16.4		12.9		15.0		8.3		13.2		22.9
$CaSiO_3$	wo		18.3		30.6		25.2		20.3		15.5		11.0
$FeSiO_3$	fs		11.8		19.6		19.7		15.0		13.6		5.5
Olivine		—		—		—		24.8		21.8		15.6	
Mg_2SiO_4	fo		—		—		—		15.4		11.6		11.0
Fe_2SiO_4	fa		—		—		—		9.4		10.2		4.6
Quartz SiO_2	qtz	17.7		—		8.8		—		—		—	
Estimated density ($gm\ cm^{-3}$)		3.17		3.27		3.13		3.17		3.22		2.94	

Average mare rock density: $\bar{\rho} = 3.19$ gm cm^{-3}.

Surveyors 5 and 6 are less certain because of the rather large errors (and anticorrelation) attached to the Si and Mg results. The gross mineralogical characteristics inferred from the treatment of Table 4 have actually been observed in the case of the powdered lunar material returned to earth by the Apollo 11 and Apollo 12 and Luna 16 missions (LSPET, 1969; LSPET, 1970; VINOGRADOV, 1971).

The last column of Table 4 indicates that the material outside the crater Tycho is representable by negligible ilmenite, and a 50% increase in plagioclase content (again highly anorthitic) over lunar mare material. There appears to be no silica excess; in fact, some olivine content is suggested.

The bottom row of Table 4 gives the particle densities for the material making up the lunar fines at the different sites. They are calculated according to the procedure of PHINNEY *et al.* (1969), and represent estimates of the densities of the *average* rock from which these lunar fines were derived. These average densities at all five mare sites are relatively constant at 3.19 ± 0.06 gm cm^{-3}. The density of the corresponding terra material, as represented by the Surveyor 7 analyses, is significantly lower, 2.94 gm cm^{-3}. These numbers are in good agreement with the estimates made by PHINNEY *et al.* (1969) from the preliminary results of the Surveyor missions, namely 3.20 ± 0.05 and 3.00 ± 0.05 gm cm^{-3}.

A result of special interest from the Surveyor 7 chemical analysis is the apparent presence of 0.3 atom percent of F. Since the amount of P that would be required to have this F in the form of apatite was definitely excluded by the analyses, this F may be in the form of CaF_2 or as a partial replacement for O in the silicates. It will be of interest to see if any F is detected in those particles of mare material that are considered as coming from highlands.

Acknowledgment—This work was supported by NASA Contract NAS-9-7883. The author also acknowledges the support of a U.S. National Science Foundation Fellowship and the hospitality of the Institut de Physique Nucléaire, Facultés des Sciences, Orsay, France.

References

Anderson A. T., Jr., Newton R. C., and Smith J. V. (1971) Apollo 12 mineralogy and petrology: Light-colored fragments; minor element concentrations: Suggested detailed petrological development of Moon. Second Lunar Science Conference (unpublished proceedings).

Economou T. E., Turkevich A. L., Sowinski K. P., Patterson, J. H., and Franzgrote E. J. (1970) The alpha-scattering technique of chemical analysis. *J. Geophys. Res.* **75**, 6514–6523.

Franzgrote E. J., Patterson J. H., Turkevich A. L., Economou T. E., and Sowinski K. P. (1970) Chemical composition of the lunar surface in Sinus Medii. *Science* **167**, 376–379.

Jaffe L. D. (1969) The Surveyor lunar landings. *Science* **164**, 774–788.

Kay R., Hubbard N. J., and Gast P. W. (1970) Chemical characteristics and origin of oceanic ridge volcanic rocks. *J. Geophys. Res.* **75**, 1585–1613.

Levinson A. A. (1970) editor, *Proc. Apollo 11 Lunar Sci. Conf., Geochim. Cosmochim. Acta* Suppl. I. Pergamon.

LSPET (Lunar Sample Preliminary Examination Team) (1969) Preliminary examination of lunar samples from Apollo 11. *Science* **165**, 1211–1227.

LSPET (Lunar Sample Preliminary Examination Team) (1970) Preliminary examination of lunar samples from Apollo 12. *Science* **167**, 1325–1339.

Patterson J. H., Franzgrote E. J., Turkevich A. L., Anderson W. A., Economou T. E., Griffin H. E., Grotch S. L., and Sowinski K. P. (1969) Alpha scattering experiment on Surveyor 7: Comparison with Surveyors 5 and 6. *J. Geophys. Res.* **74**, 6120–6148.

Patterson J. H., Turkevich A. L., Franzgrote E. J., Economou T. E., and Sowinski K. P. (1970) Chemical composition of the lunar surface in a terra region near the crater Tycho. *Science* **163**, 825–828.

Phinney R. A., O'Keefe J. A., Adams J. B., Gault D. E., Kuiper G. P., Masursky H., Collins R. J., and Shoemaker E. M. (1969) Implications of the Surveyor 7 results. *J. Geophys. Res.* **74**, 6053–6080.

Smith J. V., Anderson A. T., Newton R. C., Olsen E. J., Wyllie P. J., Crewe A. V., Isaacson M. S., and Johnson D. (1970) Petrological history of the moon inferred from petrography, mineralogy, and petrogenesis of Apollo 11 rocks. *Proc. Apollo 11 Lunar Sci. Conf., Geochim. Cosmochim. Acta* Suppl. I, Vol. 1, pp. 897–925. Pergamon.

Turkevich A. L., Franzgrote E. J., and Patterson J. H. (1967) Chemical analysis of the moon at the Surveyor 5 landing site: Preliminary results. *Science* **158**, 635–637.

Turkevich A. L., Patterson J. H., and Franzgrote E. J. (1968a) Chemical analysis of the moon at the Surveyor 6 landing site. *Science* **160**, 1108–1110.

Turkevich A. L., Franzgrote E. J., and Patterson J. H. (1968b) Chemical analysis of the moon at the Surveyor 7 landing site: Preliminary results. *Science* **162**, 117–118.

Vinogradov A. P. (1971) Preliminary data on lunar ground brought to earth by automatic probe "Luna-16." Second Lunar Science Conference (unpublished proceedings) January 14. See also (1970) *Nature* **228**, 492.

Wood J. A. (1970) Petrology of the lunar soil and geophysical implications. *J. Geophys. Res.* **75**, 6497–6513.

Proceedings of the Second Lunar Science Conference, Volume 2, pp. 1217–1229.
The M.I.T. Press, 1971.

Elemental composition of some Apollo 12 lunar rocks and soils*

F. CUTTITTA, H. J. ROSE, JR., C. S. ANNELL, M. K. CARRON, R. P. CHRISTIAN, E. J. DWORNIK, L. P. GREENLAND, A. W. HELZ, and D. T. LIGON, JR.
U.S. Geological Survey, Washington, D.C. 20242

(*Received 23 February 1971; accepted 16 March 1971*)

Abstract—The major, minor, and trace element composition of fourteen lunar samples (6 soils and 8 igneous rocks) collected at the Ocean of Storms site by the Apollo 12 Mission has been determined by combined semimicro chemical, X-ray fluorescence, and spectrographic techniques. As with the Apollo 11 samples, these lunar materials have higher concentrations of refractory elements, lower contents of volative elements, and unusually high reducing capacities compared to their terrestrial analogs. The Apollo 12 rocks are more mafic in character than the Apollo 11 materials; variations in trace elements noted among samples from the Ocean of Storms reflect their greater diversity in texture and modal mineralogy. There is evidence for a bimodal distribution in Ni, Co, and Cr; this would reflect fractional crystallization sequences of mafic magmas. Systematic relationships noted for Mg with Si, Ni, Cr, Ba, as well as the Fe/Ni, Ni/Co, and Ba/V ratios, suggest that Mg may be the most significant parameter in processes associated with the formation of the Apollo 12 crystalline rocks.

INTRODUCTION

FOURTEEN SAMPLES (6 soils and 8 igneous rocks) collected by the Apollo 12 Mission at the Oceanus Procellarum site (LSPET 1970 and Apollo 12: Preliminary Science Report, 1970) were analyzed. The very small quantities allocated for two samples (12009,42 and 12035,15) were sufficient only for emission spectrographic analysis of the trace elements, and no data for major elements could be generated for these two samples. Twelve of the lunar samples were analyzed for major, minor, and trace elements by a combination of semimicro chemical, X-ray fluorescence, and dc arc emission spectrographic techniques. Analytical data for major and minor elements are given in Tables 1, 2, and 3, and include both chemical and X-ray determinations. Trace element determinations were made for 44 elements, of which only 19 were detected by the spectrographic dc arc methods used (Tables 4, 5, and 6). If present, the remaining 25 elements are below the concentrations indicated in Table 4.

The igneous rocks all have similar mineral assemblages, varying only in proportions of pyroxene, olivine, plagioclase, and ilmenite. Compared to Apollo 11 soils (Apollo 11: Preliminary Science Report, 1969) the Apollo 12 fines (and in particular, soil 12033) are lighter, reflecting different proportions of pyroxene ($\approx$40%), plagioclase ($\approx$30%), glass ($\approx$20%), and olivine (5–10%). Minor quantities (a few percent each) of ilmenite, tridymite, cristobalite, nickel-iron phase, and several unidentified phases have been reported (Apollo 12: Preliminary Science Report, 1970).

* Publication authorized by the Director, U.S. Geological Survey.

Table 1. Major and minor element composition of some Apollo 12 lunar igneous rocks (in wgt. %).

Element Oxide	12018,34	12020,42	12075,18	12021,97	12021,78	12038,28
SiO_2	43.6	44.6	44.8	46.5	46.7	47.1
Al_2O_3	8.28	8.00	7.87	10.5	11.1	12.8
Fe_2O_3	0.00	0.00	0.00	0.00	0.00	0.00
FeO	20.6	20.7	20.7	19.4	19.1	17.4
MgO	15.1	14.4	14.4	7.48	7.38	6.80
CaO	8.54	8.53	8.53	11.3	11.5	11.4
Na_2O	0.26	0.23	0.23	0.30	0.30	0.64
K_2O	0.04	0.06	0.07	0.06	0.05	0.07
H_2O^-	0.10	0.00	0.00	0.00	0.05	0.00
Total Ti as TiO_2	2.60	2.56	2.55	3.51	3.45	3.17
P_2O_5	0.07	0.08	0.08	0.09	0.09	0.17
MnO	0.28	0.27	0.27	0.27	0.27	0.24
Cr_2O_3	0.69	0.60	0.60	0.41	0.38	0.31
TOTAL	100.16	100.03	100.1	99.82	100.37	100.10
Total Fe as Fe_2O_3	22.9	23.0	23.0	21.5	21.2	19.4
Total reducing capacity as FeO	20.7	20.7	20.7	20.0	19.2	17.7
ΔRC[1]	0.1	0.0	0.0	0.6	0.1	0.3

[1] ΔRC = Total reducing capacity measured for the lunar samples less the reducing capacity attributable to the FeO content of the lunar samples.

Table 2. Major and minor element composition of Apollo 12 lunar soils (in wgt. %).

Element Oxide	12033,59	12033,70	12042,40	12070,89	12070,90	12070,143b
SiO_2	47.2	48.2	45.7	45.8	45.7	46.0
Al_2O_3	14.3	15.1	13.0	12.9	13.0	12.7
Fe_2O_3	0.00	0.00	0.00	0.00	0.00	0.00
FeO	14.2	12.9	16.2	16.3	16.4	16.7
MgO	9.28	8.45	10.4	10.2	10.5	9.56
CaO	10.6	10.6	10.6	10.5	10.4	10.5
Na_2O	0.66	0.87	0.54	0.50	0.48	0.47
K_2O	0.41	0.54	0.25	0.25	0.23	0.23
H_2O^-	0.00	0.00	0.00	0.00	0.00	0.00
Total Ti as TiO_2	2.48	2.33	2.71	2.83	2.78	2.82
P_2O_5	0.52	0.55	0.33	0.33	0.32	0.32
MnO	0.19	0.18	0.24	0.22	0.23	0.22
Cr_2O_3	0.32	0.37	0.39	0.44	0.42	0.43
TOTAL	100.16	100.09	100.36	100.27	100.46	100.03
Total Fe as Fe_2O_3	15.8	14.3	18.0	18.1	18.2	18.5
Total reducing capacity as FeO	14.5	13.1	17.8	17.9	17.9	17.9
ΔRC[1]	0.3	0.2	1.6	1.6	1.5	1.2

[1] ΔRC = Total reducing capacity measured for the lunar samples less the reducing capacity attributable to the FeO content of the lunar soils.

DESCRIPTION OF THE SAMPLES

Brief descriptions of the Apollo 12 lunar samples, as received in this laboratory, follow: Most samples generally weighed less than 1 gram each. In general, we are in agreement with the classifications in the Apollo 12: Preliminary Science Report (1970).

Selected sample 12009 (split 42)

This rock contains euhedral olivine crystals up to 1 mm in size in a fine-grained matrix. Rock

Table 3. Comparison of range and average values (wgt. %) of major and minor elements of igneous type tocks with lunar soils.

	Igneous Rocks		Soils	
Element Oxide	Range	Average	Range	Average
SiO_2	43.6 –47.1	45.6	45.7 –48.2	46.4
Al_2O_3	7.87–12.8	9.76	12.7 –15.1	13.5
Fe_2O_3	0.00– 0.00	0.00	0.00– 0.00	0.00
FeO	17.4 –20.7	19.7	12.9 –16.7	15.5
MgO	6.80–15.1	10.9	8.45–10.5	9.73
CaO	8.53–11.5	9.97	10.4 –10.6	10.5
Na_2O	0.23– 0.64	0.33	0.47– 0.87	0.59
K_2O	0.04– 0.07	0.05_8	0.23– 0.54	0.32
H_2O^-	0.00– 0.10	0.02_5	0.00– 0.00	0.00
Total Ti as TiO_2	2.55– 3.51	2.97	2.33– 2.83	2.66
P_2O_5	0.07– 0.17	0.09_7	0.32– 0.55	0.40
MnO	0.24– 0.28	0.27	0.18– 0.24	0.21
Cr_2O_3	0.31– 0.69	0.50	0.32– 0.44	0.40
TOTAL		100.15		100.21
Total Fe as Fe_2O_3	19.4 –23.0	21.8	14.3 –18.5	17.2
Total reducing capacity as FeO	17.7 –20.7	19.8	13.1 –17.9	16.5
ΔRC	0.0 – 0.6	0.37	0.2 – 1.6	1.1

Table 4. Trace element abundances in some Apollo 12 lunar igneous rocks (in ppm).

Constituent	12035,15	12018,34	12020,42	12075,18	12009,42	12021,97	12021,78	12038,28
Zn	<4	ND	<4	<4	4	4.3	4.0	<4
Cu	6.5	8.1	9.0	8.5	14	13	14	10
Ga	4.2	5.0	4.8	4.8	5.0	4.7	5.4	5.9
Rb	1.2	1.2	1.4	1.4	1.4	1.2	1.4	<1
Li	5.8	6.0	5.7	5.7	6.4	8.1	6.6	8.5
Mn	2140	2000	2150	2020	2120	2150	2160	1850
Cr	4540	4600	4330	4000	4390	2850	3180	2210
Co	71	70	64	69	59	38	36	34
Ni	101	84	77	72	62	16	13	6.7
Ba	51	67	61	58	71	71	88	142
Sr	65	58	65	59	75	84	73	158
V	204	158	155	158	166	130	147	126
Be	1.6	1.6	1.4	ND	1.4	1.5	1.9	<1
Nb	<10	<10	13	16	<10	11	16	12
Sc	42	42	39	37	44	51	48	50
La	ND	ND	<20	<20	<20	<20	ND	<20
Y	36	37	37	38	42	53	48	71
Yb	4.7	5.4	5.1	5.1	5.3	5.4	5.7	6.3
Zr	88	99	119	95	114	133	112	186

Samples tabulated in the order of decreasing MgO and Ni contents.

The following elements were looked for but not detected (ND) in the analyzed samples. If present, they would be in concentrations below those (in ppm) indicated in parenthesis: Ag (0.2), As (4), Au (0.2), B (10), Bi (1), Cd (8), Ce (100), Cs (1), Ge (1), Hf (20), Hg (8), In (1), Mo (2), Nd (100), P (2000), Pb (1), Pt (3), Re (30), Sb (100), Sn (10), Ta (100), Te (300), Th (100), Tl (1), U (500), and W (200).

color is a greenish-gray, the olivine phenocrysts are clear yellowish-green. No vugs or vesicles are evident in this small sample. This sample is classified as an olivine basalt.

Selected sample 12018 (split 34)

The sample consists of two fragments weighing approximately 1.5 and 0.5 grams. The rock is medium-grained and vuggy, and is characterized by abundant olivine in equant grains up to 0.6 mm in size. The vugs contain well-developed crystals of clear to cloudy plagioclase and reddish-brown

Table 5. Trace element abundances in Apollo 12 lunar soils (in ppm).

Constituent	12033,59	12033,70	12042,40	12070,89	12070,90	12070,143b
Zn	<4	4.6	7.5	8	7.6	8.2
Cu	8.7	8.0	10	12	14	11
Ga	5.2	4.8	4.4	4.9	4.8	5.0
Rb	10.3	16	5.5	6.2	5.2	5.9
Li	23	24	16	17	15	15
Mn	1430	1150	1620	1650	1670	1540
Cr	2960	2680	2930	3060	3220	2430
Co	36	106	52	52	49	49
Ni	137	210	235	222	215	150
Ba	667	990	445	423	420	430
Sr	137	178	110	123	115	125
V	106	80	108	121	110	114
Be	8.0	5.3	6.8	3.3	4.3	3.2
Nb	38	74	26	29	29	30
Sc	38	38	38	42	38	44
La	61	88	48	40	46	39
Y	190	245	128	145	133	142
Yb	20	23	14	14	14	12
Zr	645	790	482	498	462	410

Table 6. Comparison of range and average values (ppm) of trace element content of igneous type rocks with lunar soils.

Element	Basalt and Gabbro		Lunar Soils	
	Range	Average	Range	Average
Zn	ND–4.3	<4	<4–8.2	6.7
Cu	6.5–14	10.4	8.0–14	10.6
Ga	4.2–5.9	5.0	4.4–5.2	4.9
Rb	<1–1.4	1.2	5.2–16	8.2
Li	5.7–8.5	6.6	15–24	18
Mn	1850–2160	2074	1150–1670	1510
Cr	2210–4600	3763	2430–3220	2880
Co	34–71	55	36–106	58
Ni	6.7–101	54	137–235	195
Ba	51–142	76	420–990	563
Sr	58–158	80	110–178	131
V	126–204	156	80–121	107
Be	ND–1.9	1.4	3.2–8.0	5.2
Nb	<10–16	12	26–74	38
Sc	37–51	44	38–44	40
La	ND–<20	<20	39–88	54
Y	36–71	45	128–245	164
Yb	4.7–6.3	5.4	12–23	16
Zr	88–186	118	410–790	548

pyroxene with some intergrowths attaining a length greater than 1 mm. Ilmenite is present in the vugs in tabular crystals about 0.1 mm. long. The rock is classified as an olivine dolerite.

Selected sample 12020 (split 42)

This specimen is fine-grained, grayish-brown in color, and has many vugs. The vugs contain well-developed and large (up to 2 mm) crystals of clear to cloudy plagioclase, honey- to deep-brown clinopyroxene, yellowish-green olivine, and to a lesser extent, ilmenite. The rock is classified as an olivine basalt.

Selected sample 12021 (splits 78 and 97)

The two sample splits were pulverized at the LRL with ultimate grain size ranging from powdery material less than 0.01 μM to 0.5 mm fragments of the predominant minerals; pigeonite, augite, plagioclase, and ilmenite. Olivine is present in trace amounts. The pigeonite is clear yellow, the augite, reddish-brown to deep red-brown with many inclusions that appear to be ilmenite. The rock is classified as a pigeonitic dolerite.

Documented sample 12033 (splits 59 and 70)

The two sample splits of lunar soil are light gray, chiefly fine-grained and powdery, and with basaltic fragments. The samples contained crystals of plagioclase, pigeonite, augite, olivine, and ilmenite, aggregates of cristobalite, glass spherules, and partially fused aggregates up to 2.0 mm in size. The coarser glass fragments are cinder-like and range in color from near white through shades of brown to black.

Glass spherules up to 300 μM in diameter occur sparingly. The fine portion of the sample contains relatively few < 100 μM spherules. No metallic particles were found.

Documented sample 12035 (split 15)

This specimen is coarse-grained and vuggy, greenish gray, and contains abundant olivine, pyroxene, and plagioclase. The two latter minerals are commonly intergrown in vugs and attain lengths up to 3 mm. Also evident in the vugs are octahedra of a spinel-type black opaque mineral. The rock has been classified by the LSPET (1970) team as a troctolite.

Documented sample 12038 (split 28)

This rock is fine-grained and dark gray, and has vugs containing acicular plagioclase intergrown with pyroxene crystals. An opaque mineral, probably ilmenite, is also present in tabular form in vugs. This rock is a basalt.

Documented sample 12042 (split 40)

This soil sample is a dark gray cohesive powder with fragments approximately 2 mm long present in the coarser portion. These include rock fragments (breccia-like and medium-grained basaltic), augite and olivine crystals, and aggregates showing all degrees of fusion from slight glazing to frothy glasses. Color ranges from light gray to black. Glass spherules are rare in both the coarse and fine portions of this sample. No metallic particles were observed.

Contingency sample 12070 (splits 89, 90, and 143b)

Three splits of this lunar soil were received. These samples are almost black and contain nearly the same assortment of characteristic particles in the coarser fraction as indicated in the description of soils 12033 and 12042. Glassy spherules, however, are more abundant and range in size up to 600 μM. They show all degrees of sphericity and vary in color from light bottle green to black. Some are dense, others vesicular. Spherules < 100 μM in diameter are much more abundant in the finer portion of this sample than in the two lunar soils previously described. The coarser aggregates are partially fused and glazed, some being as large as 3 mm on edge. The black color of the sample probably results from the presence of black glassy particles. No metallic fragments were found in this sample.

Contingency sample 12075 (split 18)

This specimen of olivine basalt was received as a small chip weighing less than 0.6 g. The specimen was mineralogically quite similar to selected sample 12020 (42). Subsequent elemental analysis showed an almost identical composition in terms of major, minor, and trace elements.

Analytical Methods

The lunar samples were weighed, photographed, and thoroughly examined mineralogically under clean-room conditions prior to undertaking the elemental analyses. Each whole rock sample was carefully disaggregated and pulverized in a boron carbide mortar in preparation for chemical, optical-emission spectrographic, and X-ray fluorescence analysis using semimicro techniques. All analytical work was monitored with analyzed geological materials such as U.S. Geological Survey standard rocks, andesite AGV-1; basalt, BCR-1; diabase, W-1; dunite, DTS-1; granites G-1 and G-2; granodiorite, GSP-1; and peridotite, PCC-1; as well as with synthetically prepared standards.

Chemical analyses

Using the methods shown in parentheses, the following constituents were determined spectrophotometrically: SiO_2 (molybdenum blue), Al_2O_3 (Ca alizarin red S), total iron (*o*-phenanthroline), total titanium (tiron), P_2O_5 (molybdenum blue), and MnO (catalyzed persulfate oxidation). CaO and MgO were determined complexometrically using versene as the titrant. Na_2O, K_2O, and Cr_2O_3 determinations as well as redeterminations of CaO, MgO, and MnO were made by atomic absorption. Determinations of H_2O^- were made by dehydration of the samples at 110°C for 1 hour. The total reducing capacity of the crystalline rocks and soil samples was determined by dissolution of the samples in the presence of excess NH_4VO_3 and subsequent back titration of excess vanadium (V). FeO was determined volumetrically after catalytic oxidation of any other elements that might be present in a reduced state (GRIMALDI *et al.*, 1943).

X-ray fluorescence

The samples and standards were prepared using the fusion-heavy absorber procedure of ROSE *et al.* (1963). Eighty mg of sample, 80 mg of La_2O_3, and 500 mg of $Li_2B_4O_7$ were mixed in a high-purity graphite crucible with a Teflon stirring rod. The mixture was fused at 1100°C for 20 min. The resultant fused bead was cooled and then brought to a weight of 800 mg with cellulose powder. The bead was cracked in a mixer-grinder capsule and the fragments and cellulose powder ground together for 15 min. in a mixer-grinder to less than 350 mesh. The ground powder was then pressed at 50,000 psi into a one-inch disc supported by 2 g of cellulose powder as a backing. Si, Al, Fe, Ca, K, Ti, P, Mn, and Cr were determined with conventional air-path and vacuum spectrometers.

Optical-emission spectrographic analysis

For the spectrographic analysis of the lunar soils, the sample was quartered twice, and one quarter of the original sample was ground in an agate mortar. A 200 mg portion of each finely ground sample was mixed with 50 mg of graphite powder by grinding together in an agate mortar, and the mixtures were stored in polyethylene capsules.

Three methods of dc arc emission spectroscopy were used. *Method 1.* A 15-A arc in air vaporized a 25 mg sample-graphite mixture to completion. First order spectra from 2300 to 4800 Å were photographed using a 3.4-m Ebert spectrograph. The spectra were examined for 38 elements. *Method 2.* A 25-A arc in an atmosphere of argon was used selectively to volatilize and determine nine elements: Ag, Au, Bi, Cd, Ge, In, Pb, Tl, and Zn. To do this, a 25-mg sample-graphite mixture was spectrochemically buffered with 30 mg of Na_2CO_3. Second order spectra in the 2400–3650 Å region were recorded with a 3-m Eagle spectrograph (ANNELL, 1967). *Method 3.* A 12.5-mg portion of the sample-graphite mixture, buffered with 20-mg of K_2CO_3, was vaporized in a 15-A arc in air for the determination of Cs, Rb, and Li. The 3-m Eagle spectrograph was used to record first order spectra in the 6500–9000 Å region (ANNELL, 1964).

Based on the report by the Preliminary Examination Team for the general composition of the Apollo 12 samples (LSPET, 1970), a matrix of high Fe, Ti, and Si containing proportionate amounts of Al, Mg, Ca, Na, Mn, and Cr as oxides or carbonates was prepared and sintered. This matrix was used for dilution of other standards and mixtures to provide spectra and interelement reactions comparable to those for the lunar samples.

Analytical accuracy

The results reported for major, minor, and trace elements are averages of replicate analyses, the major and minor constituents being determined by both chemical and X-ray techniques.

The percent standard deviation for all determinations (replicates) reported in Table 1 are as follows: SiO_2, 0.22; Al_2O_3, 0.11; FeO, 0.10; MgO, 0.10; CaO, 0.05; Na_2O, 0.01; K_2O, 0.005; TiO_2, 0.015; P_2O_5, 0.01; Mn, 0.002; and Cr, 0.01.

For those elements determined by emission spectroscopy (Method 1), a coefficient of variation of $\pm 15\%$ of the amount present is assigned (BASTRON *et al.*, 1960). Methods 2 and 3, specifically designed for a selective group of elements, both have a coefficient of variation of $\pm 10\%$ of the amount present.

COMPOSITIONAL CHARACTER OF APOLLO 12 LUNAR SAMPLES

Elemental variations

Data for the major, minor, and trace elements for the igneous rocks and soils are shown in Tables 1–6. The soils are tabulated in numerical sequence while the igneous rocks are in the order of decreasing Mg and Ni contents. As in the case of the Apollo 11 lunar samples, examination of the analytical data for the major and minor elements shows a number of similarities and differences when comparing the Apollo 12 crystalline rocks with the lunar soils. Al_2O_3, Na_2O, K_2O, P_2O_5, excess reducing capacity values, and the trace elements Zn, Rb, Li, Ni, Ba, Sr, Be, Nb, La, Y, Yb, and Zr are higher in the lunar fines. The FeO, MnO, and V concentrations are discernibly higher in the Apollo 12 igneous rocks. The ranges and average concentrations of Na_2O do not help to differentiate igneous rocks from lunar soils in either the Apollo 11 or Apollo 12 samples. However, the Apollo 12 samples are characterized by a much higher K_2O concentration in the soil fines (0.23–0.41%) as compared to the igneous rocks ($\approx 0.05\%$). The analyzed crystalline rocks have minor but significant internal variations in composition, particularly in MgO, which varies by almost twofold. There is also evidence (Table 1) of an inverse correlation between MgO and SiO_2 and a direct relationship of MgO with Cr_2O_3. This constituent (MgO) appears to be a significant formative parameter among the major components of the Apollo 12 crystalline rocks. Compared to the Apollo 11 igneous rocks (ROSE *et al.*, 1970), the Apollo 12 samples have lower K, Rb, and TiO_2 concentrations, and higher Fe, Mg, and Ni contents. These are in accord with the more mafic character of the Apollo 12 rocks compared to the Apollo 11 rocks.

Crystalline rocks

The results in Table 1 show two noteworthy compositional features, i.e., the similarity between samples 12020 and 12075 (and to a lesser degree of rock 12018) and the uniqueness of basalt 12038 compared to the other Apollo 12 crystalline rocks. Contingency sample 12075 was collected from a group of small rocks aligned roughly NE between the two small Head and Surveyor craters some 15 m NW of the LM. Selected sample 12020 was collected late in the first EVA period near the two mounds located approximately 120 m and 160 m NW of the LM and in the vicinity of the ALSEP. The remarkable similarity in major, minor, and trace element composition for these two rocks indicates that they are ejecta fragments from a common source material.

Compared to the other Apollo 12 crystalline rocks, basalt 12038 has higher SiO_2, Al_2O_3, Na_2O, and P_2O_5 contents, and lower concentrations of FeO, MgO, MnO, and Cr_2O_3. The trace elements Co, Ni, and V (Table 4) are markedly depleted in sample 12038, whereas Ba, Sr, Y, Yb, and Zr are enriched. Such elemental distributions imply two possibilities: that rock 12038 and the other Apollo 12 rocks are part of the same fractional crystallization sequence, or that the Apollo 12 crystalline rocks represent members of two separate but similar magmatic events.

Lunar soil 12033 (splits 59 and 70)

The two splits of soil 12033 are notably different in composition from both the local rocks at the Apollo 12 site and all other lunar materials returned to date. It consists of light-to-medium gray fines collected from a trench dug on the northwest rim of the Head Crater (LSPET, 1970 and Apollo 12: Preliminary Science Report, 1970) approximately 15 cm below the surface. It was overlain by darker soil and was also found near the bottom of the core tube (LRL Catalog, 1970). The trace element content of soil 12033 is considerably different from that of the other four analyzed lunar soils (Tables 2 and 5), having higher concentrations of Rb, Li, Ba, Sr, Nb, La, Y, Yb, and Zr; a higher ratio of Cr/Ni; and lower Cu and Zn contents. Compared to the other lunar soils, sample 12033 is also characterized by its major element content, having higher concentrations of SiO_2, Al_2O_3, Na_2O, K_2O, and P_2O_5 and lower FeO, TiO_2, MnO, and Cr_2O_3 contents. Despite these generalities, there are some internal inconsistencies among the two analyzed splits of soil 12033, i.e., compared to split 70, split 59 has higher Mn, Cr, V, and Be concentrations and lower Rb, Co, Ni, Ba, Sr, Nb, La, Y, and Zr contents. These differences may reflect variations in the mineral composition of the two different particle-size fractions (split 59 $<$ 1 mm and split 70 between 1 and 2 mm). The excess reducing capacity (ΔRC) for sample 12033 is far less ($+0.2$ to $+0.3\%$) than that of other lunar soils; the values were as high as $+4\%$ for the Apollo 11 soils (ROSE, *et al.*, 1970), and about $+1.5\%$ for the Apollo 12 samples (Table 2). The excess reducing capacity for soil 12033 is, in fact, quite similar to that found for the lunar igneous rocks (Table 1).

This soil sample has been tentatively considered by LSPET (1970) to be a crystal-vitric ash with similarities to terrestrial volcanic ash. They suggest that it may be part of the ray emanating from Copernicus. However, BAEDECKER *et al.* (1971) found that sample 12033 contains 0.6 as much meteoritic component as the more typical Apollo 12 lunar soils. They believe that it must be quite young compared to both the 12070 soil and the Apollo 11 soil 10084 because of the lesser amount of meteoritic component and the retention of its identity.

KEAYS, *et al.* (1970) pointed out that the enrichment of a number of siderophilic (Ni and Ir) and volatile trace (Ge and Zn) elements in the lunar soils and breccias is attributable to an influx of extralunar material having a composition similar to that of the C-1 chondrites. BAEDECKER *et al.* (1971) reported that in examining their data for sample 12033 for evidence of the meteoritic component, they noted these elements to be consistently lower in concentration. They conclude from this that the amount of meteoritic component is either the residuum from the extralunar object which

Table 7. Some elemental ratios of the Apollo 12 returned lunar samples.

Sample No.		Fe/Ni	Rb/Sr	Ni/Co	Cr/V	Cr/Ni	Ba/V	Ba/Sr
					Igneous rocks			
12035,15	Troctolite	—	0.018	1.42	22.3	45	0.25	0.78
12018,34	Olivine dolerite	1908	0.021	1.20	29.1	55	0.42	1.16
12020,42	Olivine basalt	2091	0.021	1.20	27.9	56	0.40	0.94
12075,18	Olivine basalt	2236	0.023	1.04	25.3	56	0.37	0.98
12009,42	Olivine basalt (feldspar)	—	0.018	1.05	26.5	71	0.43	0.95
12021,97	Pigeonite dolerite	9429	0.014	0.42	21.9	178	0.55	0.85
12021,78	Pigeonite dolerite	11,415	0.019	0.36	21.6	245	0.60	1.21
12038,28	Basalt	20,196	<0.006	0.20	17.5	330	1.13	0.90
					Soils			
12033,59		807	0.075	3.81	27.9	22	6.3	4.9
12033,70		477	0.089	1.98	33.5	13	12.4	5.6
12042,40		536	0.050	4.52	27.2	13	4.1	4.0
12070,89		571	0.050	4.26	25.3	14	3.5	3.4
12070,90		596	0.045	4.39	29.3	15	3.8	3.7
12070,143b		794	0.047	3.98	21.8	14	4.0	3.4

Igneous rocks listed in the order of decreasing MgO and Ni contents.
Soil samples listed in numerical sequence.

produced the 12033 material as crater ejecta, or that the 12033 soil is a mixture of an unknown component with the local soil in approximately 4:6 proportions.

Significant element ratios

A comparison of some geochemically significant element ratios for the Apollo 12 rocks and soils is presented in Table 7. Several of these element ratios also distinguish the crystalline rocks from the lunar soils. The Fe/Ni ratios for the crystalline rocks are much higher (roughly 1,900 to 20,000) than the ratios of 500–800 in the soils. Similarly, the Cr/Ni ratios for the Apollo 12 samples are much higher in the rocks. Both the Fe/Ni and Cr/Ni ratios show an inverse relationship to the MgO content. The data in Table 7 also show that the Rb/Sr, Ba/V, Ni/Co, and Ba/Sr ratios are lower in the rocks and higher in the fines. The uniqueness of soil sample 12033 and basalt rock 12038 compared to the other Apollo 12 samples is once again indicated in their element ratios. The ratios (Table 7) show evidence of a bimodal distribution that is also discernible in both the major and trace element data for the crystalline rocks, i.e., the Fe/Ni, Ni/Co, Cr/Ni, and Ba/V ratios.

Bimodal distribution

One of the unusual features of the trace element distribution in the Apollo 12 crystalline rocks is the bimodal distribution of the elements readily discernible in Table 4 for Ni (<20 and >60 ppm), Co (<40 and >60 ppm), and Cr (<3200 and >4000 ppm), and the Fe/Ni, Ni/Co, Cr/Ni, and Ba/V ratios (Table 7). The distribution of most of the major, minor, and trace elements within the two modes is generally similar to that of known differentiated sequences of terrestrial mafic magmas. This bimodal distribution suggests that these rocks could have resulted from either a number of similar magmatic sequences or that it reflects the lack of a sufficient number of samples to better establish their relationship in a fractional crystallization sequence.

Excess reducing capacity

We have reported (ROSE *et al.*, 1970) that the total reducing capacity (TRC) for the samples from the Apollo 11 Tranquillity Base was considerably higher than the values determined for total iron as FeO. The ΔRC values (TRC minus FeO) expressed as FeO for the Apollo 11 samples ranged from +0.8 to +4.1%. The ΔRC values for the Apollo 12 samples are lower, ranging from 0.0 to +1.6%, with soil samples 12042 and 12070 having characteristically higher ΔRC values averaging +1.5%. In contrast to the Apollo 11 and Apollo 12 lunar soils, soil sample 12033 has a ΔRC (+0.2 to +0.3%) comparable to those of the Apollo 12 rocks, which range from 0.0 to +0.6 and average +0.37% ΔRC.

The high vacuum and absence of water on the lunar surface are ideal conditions for a reducing environment. In addition, because neither the metallic Fe nor the S content can account for the excessive reducing capacity, the ΔRC strongly suggests the presence of another reducing constituent such as Ti(III) and/or the invalidity of assuming a terrestrial element-oxygen stoichiometry as generally reported in terrestrial rock analyses. The analytical data (Tables 1 and 2) show that the ΔRC and TiO_2 values are much lower in the Apollo 12 samples as compared to the values reported for Apollo 11 (ROSE *et al.*, 1970). The exact source for the excessive reducing capacity cannot be stated unequivocably and requires further study.

Oxygen determinations by instrumental 14 MeV neutron activation (EHMANN and MORGAN, 1971) have been made on 7 Apollo 12 rocks and 4 soils. Their results show that the lunar materials have an apparent oxygen deficiency of approximately 1.6% as compared to the simple element-oxygen stoichiometry assumed in terrestrial rock summations. Their finding of an oxygen depletion supports our observations on the excess reducing capacity and high summations reported here and for the Apollo 11 breccias and soils (EHMANN and MORGAN, 1970; ROSE *et al.*, 1970).

Some geochemical implications

A notable feature of the chemistry of the crystalline rocks is the great variation of Ni. This was observed previously by LSPET (1970), where it was interpreted as a fractional crystallization sequence. Three of the samples are very low in Ni (<20 ppm) and the other five contain more than 60 ppm Ni. LSPET reported a continuous series of Ni contents almost spanning the range of our values. Nonetheless, the appearance of two groups in our data is probably not fortuitous. Cr, Co, and Mg are depleted whereas Ba, Sr, Y, and Zr are enriched in the low-Ni rocks; V and Fe show some depletion and Sc, Yb, Si, Al, Ca, Na, Ti, and P are enriched somewhat in the low-Ni rocks. Zn, Cu, Ga, Mn, Be, Nb, and, most significantly, K and Rb are little different in the two Ni groups. The distribution of most of the major and trace elements between the two Ni groups is generally similar to that of a fractional crystallization sequence. The mineralogic variation reported by LSPET (1970) implies extensive crystal/liquid fractionation processes. The chemical data, however, imply only a relatively limited differentiation. Ni, Co, and Cr are usually rapidly depleted in the early stages of crystallization but, in these samples, Co and Cr vary only by a factor of two, and only Ni is greatly fractionated. Elements that are concentrated in residual liquids, such as K, Rb, P, and Ba, should increase in concentration by a factor of two after about 80% crystallization (GREENLAND, 1970); of these elements, K, Rb, and P show little variation and only Ba increases by a factor of 2. The preferential incorporation of Sr relative to Ba in plagioclase leads to a marked increase in the Ba/Sr ratio of residual liquids; in these rocks, however, the Ba/Sr ratio remains constant. In fact, most (not all) of the observed chemical variations can be attributed to a varying (plagioclase + pyroxene)/olivine ratio with crystallization from a magma of nearly constant composition.

The mineralogic and chemical data can be reconciled by assuming that our rocks represent perhaps as much as 80% crystallization of the original magma. Further differentiation of the remaining liquid could yield a composition similar to that of the alkali-rich rock 12013 described by LSPET (1970). However, the high Ni and Cr content of this rock precludes its being a direct product of this residual liquid. Other evidence suggestive of the existence of a residual liquid not represented in our rocks can be found in the chemistry of the Apollo 12 lunar soils.

The lunar soils are markedly enriched in K, Rb, P, Ba, and to a slightly lesser extent, in rare earths, Y, and Zr, relative to the crystalline rocks. Notable enrichments of Nb, Be, and Li are also present. All these elements would be expected to be concentrated to a greater or lesser degree in the residual magma whose existence was inferred above. Of the elements strongly depleted in the residual magma, Cr is about the same, Co is slightly higher, and Ni is considerably higher in the soils than in the low-Ni rocks. This distribution pattern suggests that the soils are derived primarily from mafic rocks such as those analyzed here, a subordinant component of fractionated rocks from the inferred residual liquid, and a very small meteoritic component to provide the excess Ni and Co.

If all the Ni in our most Ni-rich soil (12042) is derived from a meteoritic component containing 1.5% Ni (typical of chondrites), this component represents $<1.6\%$ of the soil. Estimating the fractionated component is more difficult: it could not represent more than 50% without depleting Cr and Co (the meteoritic contribution of Co must be <15 ppm) to a much greater extent than is observed; on the other hand, a 10% addition implies $\approx 5\%$ K in the fractionated component which appears implausibly high. A 30% contribution would imply $\approx 1.5\%$ K (about the same as 12013) and still be compatible with the observed Cr and Co content.

SUMMARY AND CONCLUSIONS

The major conclusions derived from examination and evaluation of the analytical data are as follows:

(1) Lunar materials from both the Apollo 11 and Apollo 12 sites have higher refractory and lower volatile element concentrations than comparable terrestrial rocks.

(2) The Apollo 12 rocks have higher concentrations of FeO, MgO, and Ni, and

lower concentrations of K_2O, TiO_2, Rb, and Zr than the Apollo 11 lunar materials—the lower TiO_2 content reflecting their lower ilmenite content.

(3) The Apollo 12 rocks are more mafic than the Apollo 11 materials, with systematic variations occurring among their SiO_2, MgO, Ni, Co, and Ba contents, and their Ni/Co, Cr/Ni, and Ba/V ratios. Such data are quite similar to those found in the enrichment or depletion of elements in terrestrial rocks as a result of differentiation processes. The Apollo 12 rocks are either segments of a single fractionation sequence or members of separate but similar magmatic events.

(4) Magnesium appears to be a significant parameter among the major components of the Apollo 12 crystalline rocks.

(5) Compared to the Apollo 11 samples, trace elements variations in Apollo 12 igneous rocks reflect their greater diversity in texture and modal mineralogy.

(6) The excess reducing capacity (ΔRC) of the Ocean of Storms samples is less than that of the Tranquillity Base lunar materials. The high ΔRC of the lunar materials from both sites strongly suggests that the element-oxygen stoichiometry generally accepted for terrestrial rocks is invalid for the lunar materials and/or the presence, in addition to iron, of a major element capable of existing in more than one valence state, i.e., Ti(III) vs. Ti(IV).

(7) Documented soil sample 12033 is compositionally different from the other lunar fines and local rocks at the Apollo 12 site. The analytical evidence suggests that this soil contains either a meteoritic component from an extralunar object which produced the 12033 material as crater ejecta, or that the 12033 soil is a mixture of a cryptic component with the local soil.

(8) The remarkable compositional similarity between crystalline rocks 12020 and 12075 indicates that they are ejecta fragments from a common, basaltic source material.

(9) The trace element content of the crystalline rocks has a bimodal distribution for the elements Ni, Co, and Cr and the Fe/Ni, Ni/Co, and Ba/V ratios. These distribution modes may reflect either a lack of sufficient samples identifiable with a single fractional crystallization sequence, or that these crystalline rocks are parts of a number of similar magmatic sequences.

Acknowledgments—This research was undertaken in behalf of NASA under Contract No. 8–078049, Order No. T–75447 and Contract No. 9–170021, Order No. T–74398.

REFERENCES

ANNELL C. S. (1964) A spectrographic method for the determination of cesium, rubidium, and lithium in tektites. U.S. Geol. Survey Prof. Paper 501–B, B148–B151.

ANNELL C. S. (1967) Spectrographic determination of volatile elements in silicates and carbonates of geologic interest, using an argon d-c arc. U.S. Geol. Survey Prof. Paper 575–C, C132–C136.

APOLLO 11 Preliminary Science Report (1969) NASA SP–214, 41–84, 123–142.

APOLLO 12 Preliminary Science Report (1970) NASA SP–235, 113–182, 189–216.

BAEDECKER P. A., CUTTITTA F., ROSE H. J., JR., SCHAUDY R., and WASSON J. T. (1971) On the origin of lunar soil 12033. *Earth Planet. Sci. Lett.* **10** (3), 361–364.

BASTRON H., BARNETT P. R., and MURATA K. J. (1960) Method for the quantitative analysis of rocks, minerals, ores, and other materials by a powder d-c arc technique. *U.S. Geol. Survey Bull.* **1084–G**, 165–182.

EHMANN W. D. and MORGAN J. W. (1970) Oxygen, silicon, and aluminum in Apollo 11 rocks and fines by 14 MeV neutron activation. *Proc. Apollo 11 Lunar Sci. Conf., Geochim. Cosmochim. Acta* Suppl. 1, Vol. 2, pp. 1071–1079. Pergamon.

EHMANN W. D. and MORGAN J. W. (1971) Major element abundances in Apollo 12 rocks and fines by 14 MeV neutron activation. Proc. Second Lunar Science Conference (unpublished proceedings).

GREENLAND L. P. (1970) An equation for trace element distribution during magmatic crystallization. *Amer. Mineral.* **55**, 455–465.

GRIMALDI F. S., STEVENS R. E., and CARRON M. K. (1943) Determination of iron. *Anal. Ed., Indust. Engrg. Chem.* **15**, 387–388.

KEYS R. R., GANAPATHY R., LAUL J. C., ANDERS E., HERZOG G. F., and JEFFREY P. M. (1970) Trace elements and radioactivity in lunar rocks: Implications for meteorite infall, solar-wind flux and formation conditions of moon. *Science*, **167**, 490–493.

LRL (Lunar Sample Information Catalog Apollo 12), MSC–01512, January 12, 1970.

LSPET (Lunar Sample Preliminary Examination Team) (1970) Preliminary examination of lunar samples from Apollo 12. *Science* **167**, 1325–1339.

ROSE H. J., JR., ADLER I., and FLANAGAN F. J. (1963) X-ray fluorescence analysis of the light elements in rocks and minerals. *Applied Spec.* **17**, (4), 81–85.

ROSE H. J., JR., CUTTITTA F., DWORNIK E. J., CARRON M. K., CHRISTIAN R. P., LINDSAY J. R., LIGON D. T., JR., and LARSON R. R. (1970) Semimicro X-ray fluorescence analysis of lunar samples. *Proc. Apollo 11 Lunar Sci. Conf., Geochim. Cosmochim. Acta* Suppl. 1, Vol. 2, pp. 1493–1497. Pergamon.

Proceedings of the Second Lunar Science Conference, Volume 2, pp. 1231–1236.
The M.I.T. Press, 1971.

Bulk elemental composition of Apollo 12 samples: Five igneous and one breccia rocks and four soils

H. WAKITA and R. A. SCHMITT
Department of Chemistry and the Radiation Center,
Oregon State University, Corvallis, Oregon 97331

(*Received 12 February 1971; accepted 4 March 1971*)

Abstract—Bulk compositions have been determined by instrumental and radiochemical neutron activation analysis (INAA and RNAA). In general, the chemical compositions of Apollo 12 porphyritic and granular-ophitic basalts may be distinguished by their TiO_2, Al_2O_3, MgO, CaO, Na_2O, and Cr_2O_3 contents. Igneous rocks have significantly lower Al_2O_3, Na_2O, and K_2O and higher FeO contents than either soils or breccia rocks. The striking similarity in major, minor, and trace elemental abundances for breccia rock 12034 and soil 12033, both obtained from a 15-cm deep trench, suggest a direct relationship between these two lunar samples. The chemical compositions of igneous and breccia rocks and soils are different at both the Sea of Tranquillity and Ocean of Storms sampling sites.

INTRODUCTION

ABUNDANCES OF ten major and minor elements have been determined by INAA and RNAA and bremsstrahlung activation analysis, prior to the determination of 14 REE (rare-earth elements) by RNAA (SCHMITT *et al.*, 1970; WAKITA *et al.*, 1970). In addition, elemental abundances of twenty-six minor and trace elements (i.e., 14 REE, Y, Cd, In, Rb, Cs, V, Sc, Co, Zr, Hf, Th, and Ba) in the same samples have also been determined by us and are reported in a separate paper of this volume.

RESULTS AND DISCUSSION

Results are shown in Table 1 and Fig. 1. In general, the elemental abundance data of this work are in good agreement with the data reported by LSPET (1970) and recent bulk data presented at the Apollo 12 Lunar Science Conference (i.e., FRONDEL *et al.*, 1971; ROSE *et al.*, 1971; ANNELL *et al.*, 1971; COMPSTON *et al.*, 1971; ENGEL and ENGEL, 1971; MAXWELL and WIIK, 1971; and others). On the average, our SiO_2 and MgO values are $\sim 4\%$ low and $\sim 6\%$ high, respectively, compared to those obtained by chemical techniques.

Lunar soils

Soil samples were collected at the Ocean of Storms from different sites, separated by only a few meters to a few hundred meters (LSPET, 1970). Analytical data show significant differences between these soils. The observed differences in elemental abundances for different soils may be accounted for by mixing and pulverizing varying amounts of either breccia, anorthositic, and igneous rocks (GOLES *et al.*, 1971), or mixing "KREEP," noritic, or 12013 dark-phase matter with igenous rocks (HUBBARD

Table 1. Abundances of major and minor elements in Apollo 12

Element (%)	Crystalline Rocks									
	Porphyritic Basalts (Type 1)[b]						Granular and Ophitic Basalts (Type 2)[b]			
	12004-45		12020-24		12075-17		12051-44		12063-70	
	0.439 g	0.582 g	0.444 g	0.408 g	0.496 g	0.523 g	0.449 g	0.535 g	.0481 g	0.479 g
SiO_2[d]	44.3	—	42.2	—	41.1	—	47.1	—	—	42.8
TiO_2	3.3	3.2	2.7	3.0	3.0	2.6	4.8	5.0	5.0	5.0
Al_2O_3	8.5	8.5	8.5	8.3	8.1	7.9	10.6	10.0	9.1	9.1
FeO[f]	21.5	—	21.8	—	21.6	—	21.0	—	—	21.2
MgO[g]	10.9	—	16.1	—	13.9	—	6.1	—	—	11.4
CaO	9.9	9.8	8.7	8.8	8.4	8.5	12.3	11.7	10.2	10.8
Na_2O	0.226	0.230	0.220	0.213	0.220	0.208	0.318	0.317	0.275	0.279
K_2O[h]	—	0.055	—	0.069	—	0.100	—	0.065	0.064	—
MnO	0.274	0.249	0.253	0.260	0.257	0.257	0.261	0.273	0.284	0.265
Cr_2O_3	0.708	—	0.612	—	0.622	—	0.307	—	—	0.410
Total (%)[e]	100 ± 3		101 ± 3		97 ± 3		103 ± 3			101 ± 3

[a] One standard deviation due to counting statistics and other errors for single determinations are approximately ±5% for Si, Ti, Fe, Mg, and Ca; ±2 − 3% for Al, Na, Mn, and Cr; and ±20% for K.
[b] Classification is by WARNER and ANDERSON (1971).
[c] Sample taken from 15-cm deep trench.
[d] Si abundances are determined by T. W. OSBORN III.
[e] The average total summation $\pm 1\sigma$ is 100 ± 2% for ten specimens.

et al., 1971; WOOD *et al.*, 1971; PHILPOTTS *et al.*, 1971). Some fine soils such as 12070 are quite homogeneous in chemical composition, as evidenced by the excellent agreement of the data obtained by many investigators.

On the other hand, duplicate analyses of the light-colored soil 12033, taken from a 15-cm deep trench, show wide variations in trace elemental abundances that are not immediately apparent in bulk elemental composition (WAKITA *et al.*, 1971). When comparing the reported data for 12033 of this study, LSPET (1970), MAXWELL and WIIK (1971), FRONDEL *et al.* (1971), ROSE *et al.* (1971), and COMPSTON *et al.* (1971), small bulk compositional differences appear to be real for ~0.5–1 gm samples.

The average Fe, Ca, Al, Mn, and Cr abundances in the four Apollo 12 soils and in the Apollo 11 soil are almost identical (Fig. 1). Relative to Apollo 11 10084 soil, Ti is reduced by a factor of ≈3 in Apollo 12 soils. Mg is enriched by a factor ≈1.4 in Apollo 12 soils relative to Apollo 11 soil, indicating a more mafic character for Apollo 12 soils. Na and K concentrations vary by factors of 2 and 4, respectively, in the four Apollo 12 soil samples. Na is higher by a factor of 1–2 and K is also higher by a factor of 1–4 in the Apollo 12 soils relative to Apollo 11 soil.

A generally linear correlation exists between K and Rb and between Rb and Cs in Apollo 11 and 12 igneous, breccia, and soil samples with K abundances ranging from 0.03 to 1.66% (e.g., ALBEE *et al.*, 1971 and WAKITA and SCHMITT, 1970). However, the lack of Na and K correlation may be attributable to the residence of Na in plagioclase. K, Rb, and Cs probably are found in interstitial phases of lunar rocks and soil fragments (TERA *et al.*, 1970).

five crystalline rocks, one breccia rock, and four soil samples[a]

Breccia Rock		Soils						
12034-18[c]		12032-36		12033-71[c] (1–2 mm)		12037-40 (2–4 mm)		12070-77
0.199 g	0.267 g	0.292 g	0.265 g	0.246 g	0.202 g	0.221 g	0.239 g	0.573 g
45.8	—	—	40.3	46.9	—	41.8	—	43.5
2.7	2.8	2.8	3.0	2.3	2.8	3.2	3.3	2.8
15.7	15.7	14.1	14.0	17.3	16.8	14.1	12.1	12.8
13.3	—	—	15.3	11.7	—	16.3	—	16.0
9.6	—	—	13.4	10.6	—	10.8	—	13.1
11.2	11.6	12.2	10.6	12.3	11.9	11.9	10.2	10.5
0.788	0.755	0.601	0.602	0.834	0.805	0.455	0.461	0.462
—	0.529	0.421	—	—	0.469	—	0.102	—
0.167	0.173	0.195	0.199	0.154	0.168	0.207	0.227	0.208
0.271	—	—	0.346	0.235	—	0.352	—	0.355
100 ± 3			98 ± 3	102 ± 3		99 ± 3		100 ± 3

[f] Fe abundances are relative to the average Fe abundances in the U.S.G.S. standard rocks W-1, GSP-1, BCR-1, and DTS-1.

[g] Mg abundances determined from only a single instrumental photonuclear activation. Mg values of this work for 12070 and 12033 soils agree within error with corresponding Mg abundances of LSPET (1970).

[h] For purpose of chemical yield determinations in the RNAA procedure, K abundances have been normalized to K = 4400 ppm for 12034, as measured by Gast and Hubbard (1970).

Breccia rock

Breccia rock 12034 and soil 12033 were taken from a 15-cm deep trench near the northwest rim of Head Crater (LSPET, 1970). Abundance ratios (12034/12033) of the average major and minor elements are as follows: Si (0.98), Ti (1.08), Al (0.92), Ca (0.94), Na (0.94), and Mn (1.06). Furthermore, supported by similar abundance ratios of nearly unity for other lithophilic trace elements (Wakita *et al.*, 1971), it appears that either this 12034 breccia rock was compacted from the soil in the regolith that was identical to 12033 soil, and then both rock 12034 and 12033 soil were ejected onto the regolith surface (now 15 cm below the present surface) by some cratering event; or that these 12033 soil particles are the disintegration fragments from a precursor rock to the 12034 rock. Since abundances of trace elements in Apollo 11 breccia rocks and soil differed by $>10\%$, many investigators have concluded that the Apollo 11 breccia rocks were not compacted solely from the returned Apollo 11 soil and very likely also had a cryptic component (e.g., see Goles *et al.*, 1970).

Crystalline rocks

Apollo 12 crystalline rocks have been petrographically classified into two magma types; i.e., porphyritic basalts (type 1) and the granular and ophitic basalts (type 2) (Warner and Anderson, 1971). Our data for five Apollo 12 crystalline rocks support this classification. As seen in Table 1 and Fig. 1, the three porphyritic basalts are depleted in Ti, Al, Ca, and Na, and enriched in Mg and Cr, compared to two granular and ophitic basalts. For Na, accurately determined by INAA, the average value of

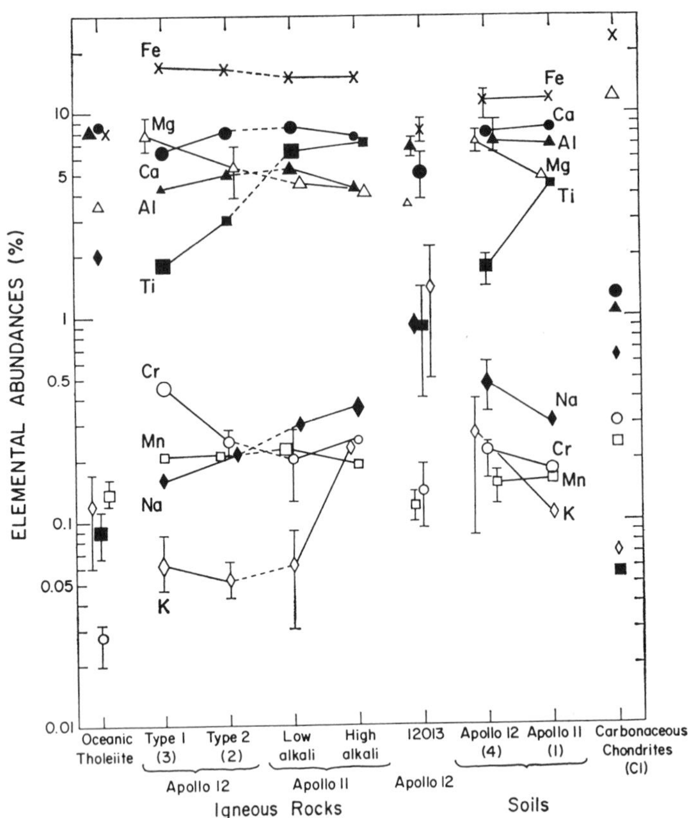

Fig. 1. Abundances with ranges for five Apollo 12 rocks and four soil samples are from this work. Values with 1σ deviation for Apollo 11 rocks and soil are from eight different groups, whose values are summarized in Table 4 by WAKITA *et al.* (1970). The K value and its range were taken from TERA *et al.* (1970). Abundances and 1σ deviation for 12013 rock are from WAKITA and SCHMITT (1970), except for Mg, which is from LSPET (1970). Carbonaceous chondritic values are from MASON (1963), SCHMITT *et al.* (1971), and LOVELAND *et al.* (1969). Most ocean tholeiitic values and mean deviations are taken from ENGEL and ENGEL (1965); Cr, Mn, and K and their ranges are from HART (1969) and CORLISS (1970).

0.22% Na_2O (1630 ppm Na) in type 1 rocks is significantly lower than 0.30% Na_2O (2200 ppm Na) obtained in type 2 rocks. For K, however, there is no clear difference between these two types of rock. Rb and Cs are also essentially identical in these two igneous rock types (WAKITA *et al.*, 1971). Furthermore, some trace elements, viz. V and Co, are enriched by a factor of ~1.5 in type 1 rocks (WAKITA *et al.*, 1971). These trends may be accounted for by either addition or removal of olivine in a magma lake during magmatic differentiation (COMPSON *et al.*, 1971). Different amounts of interstitial phases may change the abundances of the coherently incompatible elements; e.g., K, Rb, Cs, REE, etc.

Comparing Apollo 12 igneous rocks with Apollo 12 soils (Table 1 and Fig. 1), we observe significant differences in the bulk as well as in the minor and trace elemental compositions. From this we can conclude that Apollo 12 soils were not derived solely from pulverization of Apollo 12 crystalline rocks but also from fragmentation of differing amounts of extraneous materials. Comparing bulk compositions of Apollo 12 type 2 igneous rocks with Apollo 11 low alkali rocks (Fig. 1), we observe that Ti is depleted by a factor of ~2 in Apollo 12 type 2 rocks; Na is also depleted in type 2 rocks, but K (also Rb and Cs) are about the same. WARNER and ANDERSON (1971) have also classified Apollo 11 low alkali rocks as granular and ophitic basalts (type 2 rocks). However, since bulk chemical data of this study and also minor and trace elemental studies (e.g., see WAKITA *et al.*, 1970, 1971) do not support this classification strictly, we suggest that slightly different degrees of partial melting on a common parent rock may have been responsible for the observed differences between Apollo 11 low alkali and Apollo 12 type 2 rocks. Finally, no simple abundance correlations exist between any lunar samples and either common terrestrial basalts or primordial chondrites.

In summary, an evaluation of elemental abundances of the bulk major and minor elements of this study corroborates the observations by the LSPET (1970) and many other investigators, that (1) significantly different degrees of partial melting or fractional crystallization must be invoked for the generation of the basaltic rocks from which the soils were at least partially derived at both the Sea of Tranquillity and Ocean of Storms landing sites; and (2) major, minor, and trace elemental abundance data of this work divide Apollo 12 igneous rocks into at least two distinct magma types (WARNER and ANDERSON, 1971). However, some exceptions to this simple classification are observed (GOLES *et al.*, 1971). Moreover, Rb–Sr investigations indicate that more than two primary basaltic magmas existed at the Ocean of Storms site (COMPSTON *et al.*, 1971).

Acknowledgments—This study was supported by NASA contract 9–8097 and NASA grant NGL 38–002–020. We are grateful to the Oregon State University TRIGA reactor group for neutron activations and to T. W. OSBORN III for Si analyses by 14-MeV neutron activation. J. MACKENZIE of Gulf General Atomic provided assistance in the bremsstrahlung activations for Mg.

REFERENCES

ALBEE, A. L., BURNETT D. S., CHODOS A. A., HAINES E. L., HUNEKE J. C., PODOSEK F. A., PAPANASTASSIOU D. A., PRICE G., TERA F., and WASSERBURG G. J. (1971) Rb–Sr ages, chemical abundance patterns and history of lunar rocks, Second Lunar Science Conference (unpublished procedings).

ANNELL C. S., CARRON M. K., CHRISTIAN R. P., CUTTITTA F., DWORNIK E. J., HELZ A. W., LIGON D. T., JR., and ROSE H. J., JR. (1970) Chemical and spectrographic analyses of lunar samples from Apollo 12 mission. Second Lunar Science Conference (unpublished proceedings).

COMPSTON W., BERRY H., VERNON M. J., CHAPPELL B. W., and KAYE M. J. (1971) Rubidium–strontium chronology and chemistry of lunar material from the Ocean of Storms. Second Lunar Science Conference (unpublished proceedings).

CORLISS J. (1970) Mid-ocean ridge basalts. Ph.D. thesis, University of California, San Diego.

ENGEL A. E. J., ENGEL C. G., and HAVENS R. G. (1956) Chemical characteristics of oceanic basalts and the upper mantle. *Geol. Soc. Amer. Bull.* **76**, 719–734.

ENGEL C. G., and ENGEL A. E. J. (1971) Major elements composition of three Apollo 12 rocks and some petrogenic considerations. Second Lunar Science Conference (unpublished proceedings).

FRONDEL C., DRAKE J. C., ITO J., and KLEIN C. (1971) Studies of Apollo 12 lunar rocks and fines. Part II. Mineralogical and chemical data on Apollo lunar fines. Second Lunar Science Conference (unpublished proceedings).

GAST P. W., and HUBBARD N. J. (1970) Private communication.

GOLES G. G., DUNCAN A. R., OSAWA M., MARTIN M. R., BEYER R. L., LINSTROM D. J., and RANDEL K. (1971) Analyses of Apollo XII specimens and a mixing model for Apollo XII "soils." Second Lunar Science Conference (unpublished proceedings).

GOLES G. G., RANDLE K., OSAWA M., LINDSTROM D. J., JÉROME D. Y., STEINBORN T. L., BEYER R. L., MARTIN M. R., and MCKAY S. M. (1970) Interpretation and speculation on elemental abundances in lunar samples, *Proc. Apollo 11 Lunar Sci. Conf. Geochim. Cosmochim. Acta* Suppl. 1, Vol. 2, pp. 1177–1194. Pergamon.

HART S. R. (1969) K, Rb, and Cs contents with K/Rb, K/Cs ratios of fresh and altered submarine basalts. *Earth Planet. Sci. Lett.* **6**, 295–303.

HUBBARD N. J., MEYER C., GAST P. W., and WIESMANN H. (1971) The Composition and derivation of Apollo 12 soils. *Earth Planet. Sci. Lett.* **10**, 341–350.

LOVELAND W., SCHMITT R. A., and FISHER D. E. (1969) Aluminum abundances in stony meteorites. *Geochim. Coshochim. Acta* **33**, 375–385.

LSPET (Lunar Sample Preliminary Examination Team) (1970) Preliminary examination of the lunar samples from Apollo 12. *Science* **167**, 1325–1339.

MASON B. (1963) Carbonaceous chondrites. *Space Sci. Rev.* **1**, 621–646.

MAXWELL J. A., and WIIK H. B. (1971) Chemical composition of Apollo 12 lunar samples 12004, 12033, 12051, 12052, and 12065. Second Lunar Science Conference (unpublished proceedings),

PHILPOTTS J. A., SCHNETZLER C. C., BOTTINO M. L., and FULLAGAR P. D. (1971) Li, K, Rb, Sr, Ba, and Rare Earth concentrations and $^{87}Sr/^{86}Sr$ in some Apollo 12 soils, rocks and separated phases. Second Lunar Science Conference (unpublished proceedings).

ROSE H. J., JR., CUTTITTA F., ANNELL C. S., CARRON M. K., CHRISTIAN R. P., DWORNIK E. J., HELZ A. W., and LIGON D. T., JR. (1971) Semi-micro analysis of Apollo 12 samples. Second Lunar Science Conference (unpublished proceedings).

SCHMITT R. A., LINN T. A., JR., and WAKITA H. (1970) The determination of fourteen common elements in rocks via sequential instrumental activation analysis. *Radiochim. Acta* **13**, 200–212.

SCHMITT R. A., GOLES G. G., and SMITH R. H. (1971) Elemental abundances in stone meteorites. In preparation.

TERA F., EUGSTER O., BURNETT D. S., and WASSERBURG G. J. (1970) Comparative study of Li, Na, K, Rb, Cs, Sr, and Ba abundances in achondrites and in Apollo 11 lunar samples. *Proc. Apollo 11 Lunar Sci. Conf., Geochim. Cosmochim. Acta* Suppl. 1, Vol. 2, pp. 1637–1657. Pergamon.

WAKITA H., SCHMITT R. A., and REY P. (1970) Elemental abundances of major, minor, and trace elements in Apollo 11 lunar rocks, soil, and core samples. *Proc. Apollo 11 Lunar Sci. Conf., Geochim. Cosmochim. Acta* Suppl. 1, Vol. 2, pp. 1685–1717. Pergamon.

WAKITA H., and SCHMITT R. A. (1970) Elemental abundances in seven fragments from lunar rock 12013. *Earth Planet. Sci. Lett.* **9**, 169–176.

WAKITA H., REY P., and SCHMITT R. A. (1971) Abundances of the 14 rare earth elements plus 22 major, minor, and trace elements in ten Apollo 12 rock and soil samples. Second Lunar Science Conference (unpublished proceedings).

WARNER J. L., and ANDERSON D. H. (1971) Lunar crystalline rocks: Petrology, geology, and origin. Second Lunar Science Conference (unpublished proceedings).

WOOD J. A., MARVIN U., REID J. B., TAYLOR G. J., BOWER J. F., POWELL B. N., and DICKEY J. S., JR. (1971) Relative proportions of rock types, and nature of the light-colored lithic fragments in Apollo 12 Soil Samples. Second Lunar Science Conference (unpublished proceedings).

Proceedings of the Second Lunar Science Conference, Volume 2, pp. 1237–1245.
The M.I.T. Press, 1971.

Major element abundances in Apollo 12 rocks and fines by 14 MeV neutron activation

WILLIAM D. EHMANN and JOHN W. MORGAN*
Department of Chemistry, University of Kentucky, Lexington, Kentucky 40506

(*Received 18 February 1971; accepted 5 March 1971*)

Abstract—Abundances of O, Si, Al, Mg, and Fe in seven rocks and four samples of fines collected by the Apollo 12 mission have been determined by instrumental 14 MeV neutron activation. These five elements constitute approximately 90% of the mass of these materials. The abundances for type A, type AB, type B, and type D, respectively, are O—39.3, 39.4, 40.4, 42.0%; Si—20.7, 21.2, 21.4, 23.0%; Al—4.3, 5.1, 4.8, 6.7%; Mg—6.9, 4.5, 4.5 to 10.1, 6.6%; Fe—15.5, 14.7, 15.1, 12.6%. Evidence is presented for the existence of several chemical groups among the crystalline rocks. The O–Si correlation found for the Apollo 12 crystalline rocks is similar to that observed for the Apollo 11 crystalline rocks, and both differ significantly from that observed for terrestrial igneous rocks. An apparent O depletion in the Apollo 12 materials relative to that calculated on the basis of simple stoichiometry has been observed.

INTRODUCTION

ACTIVATION ANALYSIS employing 14 MeV neutrons has been used to determine the abundances of O, Si, Al, Mg, and Fe in rocks and fines collected during the Apollo 12 mission. This method of analysis has been shown to yield accurate and precise abundance data while retaining the advantage of speed. The method is essentially nondestructive, except for the very minor levels of radioactivity induced in the samples and radiation damage. We regard this as the most reliable method available for the direct determination of O in bulk samples.

Analyses of Apollo 11 lunar rocks and fines using these techniques have been reported by EHMANN and MORGAN (1970). The more important observations based on our analyses of Apollo 11 materials may be summarized as follows:

(1) The crystalline rocks fall into two distinct chemical groups based on their Al abundances. These chemical groups are not identical to the groups defined by crystal texture.

(2) The breccias and fines are enriched in O, Si, and Al, as compared to the crystalline rocks and must contain components distinct in their composition from the crystalline rocks.

(3) The breccias and fines exhibit an apparent O deficiency of approximately 1.6% O, as compared to the calculated O abundance required for simple stoichiometry based on total silicate analyses. This may be due to reduction and subsequent O depletion of the fines and breccias due to the action of solar wind hydrogen. The crystalline rocks exhibit little or no O depletion calculated on the same basis.

---

* Present address: Enrico Fermi Institute for Nuclear Studies, University of Chicago, Chicago, Illinois 60637.

(4) Si and O are strongly correlated in the Apollo 11 rocks. The regression line for the crystalline rocks is represented by the equation: O% = 0.98 Si% + 20.8. This regression line has a slope significantly different than that reported by EUGSTER (1969) for terrestrial igneous rocks. The low O/Si ratio in the lunar rocks is in part due to the presence of ilmenite (31.6% O) and reduced species such as Ti(III), but the difference in the slopes of the terrestrial and lunar regression lines is not simply explained.

In this paper the results of the analyses on Apollo 12 rocks and fines are examined in light of these observations on Apollo 11 materials.

ANALYTICAL METHOD

Apparatus

14 MeV neutrons were produced by a Kaman Nuclear model A–1250 Cockcroft–Walton generator, and the neutron yield monitored by a low geometry enriched BF_3 detector. The pneumatic single-sample transfer system, using dry N_2 as propellant gas, and the sequential programming circuit are essentially those described by VOGT *et al.* (1965). Minor modifications were made to interface with a data acquisition system based on a Nuclear Data ND 2201 4096 channel analyzer. Gamma-ray activity measurements were made using a 10 cm × 10 cm well-type NaI (Tl) detector.

Preparation of samples

Samples of lunar rocks were received under a double-dry nitrogen seal as irregular-shaped chips, ranging in weight between 0.48 and 2.1 g. The design of our usual polyethylene rabbits (EHMANN and MCKOWN, 1968) was modified to take $\frac{3}{8}$ in.i.d. capsules. Before use, all polyethylene parts for the rabbit assembly were immersed in absolute ethanol and agitated ultrasonically for 20 minutes. They were dried in a jet of high purity dry N_2, transferred to a vacuum desiccator, and dried under vacuum. Samples were transferred to inner capsules in a glove box under dry N_2. After sealing, the capsule was positioned in the outer container with polyethylene spacers, so that the apparent center of mass was centered in the neutron beam position. The outer container was flushed with dry N_2 before it too was sealed.

Preparation of standards

The encapsulation of the standards was identical to that described for the samples. Fused optical quartz L-1 (pieces of a broken quartz lens donated by Dr. W. Blackburn of the University of Kentucky) was used as a standard for Si and O. Potassium dichromate NBS #136b was used, as received, for an additional standard for the O determinations. In addition to powdered standards, chunks of L-1 quartz, similar in shape to the rock chips, were prepared. Initially, these were used together with powdered standards for Si and O analyses, until it was established that geometry effects introduced no detectable systematic bias. For O, a two-way comparison was possible, with powdered quartz and NBS #136b $K_2Cr_2O_7$. The quartz chunks agreed with both of these to an accuracy of about 0.4 relative per cent.

Opal glass (NBS #91) and potassium feldspar (NBS #70a) were dried at 110°C and used for Al standards. The abundances of this element in these materials bracket those reported in the lunar samples. In addition, these standard materials contain Si as a major constituent and this element produces a primary interference in the Al determination. A correction for this interference was made specifically for each analysis using the measured Si content of the appropriate samples. The apparent Al abundance of the standards was also adjusted accordingly. Any residual error in the estimation of this correction will be largely self-canceling for samples and standards possessing similar Al/Si ratios.

Standards for Mg and Fe were NBS #88a dolomitic limestone (dried at 110°C for two hours) and reagent grade ferrous ammonium sulfate, respectively.

Procedures

Detailed procedures for the determination of Si and O have previously been given by MORGAN and EHMANN (1970a). Procedures for Al, Mg, and Fe have been outlined in MORGAN and EHMANN (1971), along with a detailed discussion of the analytical principles and potential interferences. These procedures have been shown to yield precise and accurate data for these elements, based on analyses of selected U.S.G.S. standard rocks.

Special care was taken to handle all lunar rocks and fines under a high-purity dry N_2 atmosphere. Therefore, it is felt the O data presented here closely reflect the true abundance of O under lunar conditions.

Results

The results of the analyses of seven Apollo 12 crystalline rocks and four samples of fines are presented in Table 1. The error limits for the individual rocks are standard deviations of the means based typically on eight replicate analyses for O, seven replicate analyses for Si, and three replicate analyses for Al, Mg, and Fe. The error limits for the group means are standard deviations of the

Table 1. Major elements in Apollo 12 rocks and fines*

Rock	Type	%O	%Si	%Al	%Mg	%Fe†	Total
12004,32	A	39.9	21.2	4.38	7.3	15.1	87.9
12063,60	A	38.7	20.2	4.23	6.5	15.9	85.5
MEAN	A	39.3	20.7	4.31	6.9	15.5	86.7
12051,46	AB	39.4	21.2	5.11	4.5	14.7	84.9
12021,81	B	40.0	21.7	5.54	4.6	14.2	86.0
12022,53	B	38.9	19.6	4.04	6.1	15.1	83.7
12035,04	B	41.1	21.7	4.35	10.1	16.0	93.3
12064,34	B	41.4	22.4	5.46	4.5	15.1	88.9
MEAN	AB & B	40.2	21.3	4.90	6.0	15.0	87.4
12032,34	D	41.9	23.9	7.29	6.5	12.4	92.0
12044,18	D	42.0	23.2	6.51	7.3	14.2	93.0
12057,79	D	42.8	23.0	6.58	6.4	12.3	91.1
12070,70	D	41.4	22.0	6.59	6.3	11.5	87.8
MEAN	D	42.0	23.0	6.74	6.6	12.6	91.0

* Standard deviations of the means for the replicate analyses of the individual samples were ±0.2% O, 0.2% Si, 0.06% Al, 0.3% Mg, and 0.5% Fe.

† The accuracy of these Fe data is probably no better than 5 relative percent due to the several correction factors involved in the calculations and the low activity levels obtained. Irradiation conditions were optimized for the O, Si, and Al determinations and the sensitivity for Mg and Fe could be considerably improved.

Table 2. O, Si, and Al in Apollo 12 rock 12013*
(MORGAN and EHMANN, 1970b)

Chip Number 12013	Sample Weight (mg)	O%	%Si	%Al
10,06	88	44.2 ± 0.3	25.1 ± 0.1	7.1 ± 0.3
10,15	99	44.9 ± 0.5	29.6 ± 0.2	6.6 ± 0.1
10,18	66	46.9 ± 0.4	29.9 ± 0.1	6.3 ± 0.2
10,41	32	47.3 ± 0.4	33.0 ± 0.4	5.1 ± 0.3
10,44	42	44.1 ± 0.3	26.5 ± 0.1	6.3 ± 0.6
MEAN (MASS WEIGHTED)		45.2 ± 0.6	28.4 ± 1.3	6.5 ± 0.3

* Error limits are standard deviations of the means of 6–8 replicate analyses for O and Si, and 3 replicate analyses for Al. The error limits for the mass weighted means are standard deviations of the means based on the dispersion of the individual chip abundances.

mean, based on the distribution of the individual values within each group. Previously published results (MORGAN and EHMANN, 1970b) on five chips of Apollo 12 rock 12013 are summarized in Table 2. All abundances are given in units of weight per cent.

DISCUSSION

Chemical groupings in Apollo 12 rocks

The Apollo 11 crystalline rocks were found to fall into two distinct groups based on their Al abundances (EHMANN and MORGAN, 1970). As can be seen from the data in Table 1, at least two major chemical groups can also be delineated for the Apollo 12 crystalline rocks. Rocks 12004, 12022, and 12063 exhibit lower Al abundances and higher Mg and Fe abundances than do rocks 12021, 12051, and 12064. The unusual mineralogy of rock 12035, a troctolite, is reflected in the high abundances of Mg and Fe observed. LSPET (1970) reports a composition of 15% pyroxene, 40% olivine, and 45% plagioclase for this rock and it clearly does not fall into either of the two major chemical groups defined above. The chemistry of rock 12013, as given in Table 2, also places it in a unique position among the materials collected by the Apollo 12 mission (MORGAN and EHMANN, 1970b). WÄNKE *et al.* (1971), COMPSTON *et al.* (1971), and others have also reported the existence of distinct groups among the Apollo 12 crystalline rocks based on chemical or isotopic data.

As was the case for the Apollo 11 crystalline rocks, the two major chemical groups for the Apollo 12 crystalline rocks do not correspond to groups defined by crystal texture alone. The abundances within chemical groups are summarized in Table 3. The unusual rocks 12035 and 12013, as well as the breccias and fines, are listed separately.

SUTTON and SCHABER (1971) have attempted to determine the lunar locations of rock samples from the Apollo 12 mission. Based on their descriptions two out of three of our Apollo 12 group 1 samples (low Al, high Mg, and Fe) lie in the region defined by a triangle drawn through Middle Crescent, Bench, and Head craters. This region is

Table 3. Comparison of Apollo 11 and Apollo 12 abundances within the apparent chemical groups*

	%O	%Si	%Al	%Mg	%Fe
Apollo 11					
Group 1 Crystalline Rocks (10022, 10024, 10069, 10071)	39.0 ± 0.6	18.8 ± 0.3	4.1 ± 0.1	—	—
Group 2 Crystalline Rocks (10063, 10047, 10050, 10058, 10062)	39.3 ± 0.5	18.7 ± 0.4	5.3 ± 0.1	—	—
Breccias (18 type C rocks)	41.1 ± 0.2	19.7 ± 0.2	6.6 ± 0.1	—	—
Fines (3 aliquants of 10084)	40.8 ± 0.7	20.2 ± 0.1	7.2 ± 0.1	—	12.4 ± 0.1
Apollo 12					
Group 1 Crystalline Rocks (12004, 12022, 12063)	39.2 ± 0.2	20.3 ± 0.4	4.2 ± 0.1	6.6 ± 0.3	15.4 ± 0.3
Group 2 Crystalline Rocks (12021, 12051, 12064)	40.3 ± 0.6	21.8 ± 0.3	5.0 ± 0.3	4.5 ± 0.1	14.7 ± 0.3
Rock 12035	41.1 ± 0.2	21.7 ± 0.1	4.4 ± 0.1	10.1 ± 0.3	16.0 ± 0.6
Rock 12013 (5 chips)	45.2 ± 0.3	28.4 ± 0.6	6.5 ± 0.1	—	—
Fines (12032, 12044, 12057, 12070)	42.0 ± 0.3	23.0 ± 0.4	6.7 ± 0.2	6.6 ± 0.3	12.6 ± 0.6

* Error limits are standard deviations of the means for the individual abundances within each group.

to the northwest of the LM landing site. All three of the group 2 samples (high Al, low Mg, and Fe) lie close to Surveyor Crater. However, since some of these locations are uncertain, we feel it is not possible to assign the chemical groups to the ejecta of any specific crater, based on the information currently available.

Element correlations in Apollo 12 rocks

The Apollo 12 crystalline rocks and fines exhibit a significant (99% confidence level) positive correlation for Al and Si: Al% = 0.79 Si% − 11.8. This may be interpreted as an increase in feldspar with differentiation and a further enrichment in the fines due to the addition of anorthosite. The Al–Si regression line (Fig. 1) for the Apollo 12 rocks and fines connects directly with the Al–Si regression line for rock 12013, but Al becomes negatively correlated with Si in the 12013 chips (MORGAN and EHMANN, 1970b). This negative correlation could be explained by intrusion of a compacted soil by a liquid rich in Si (plus some potassium feldspar). Figure 1 also illustrates clearly the two distinct major chemical groups in the Apollo 11 and Apollo 12 crystalline rocks, based on Al abundances.

O and Si are found to exhibit a significant (99% confidence level) positive correlation (Fig. 2) in the Apollo 12 crystalline rocks according to the equation:

$$O\% = 0.96\ Si\% + 19.6.$$

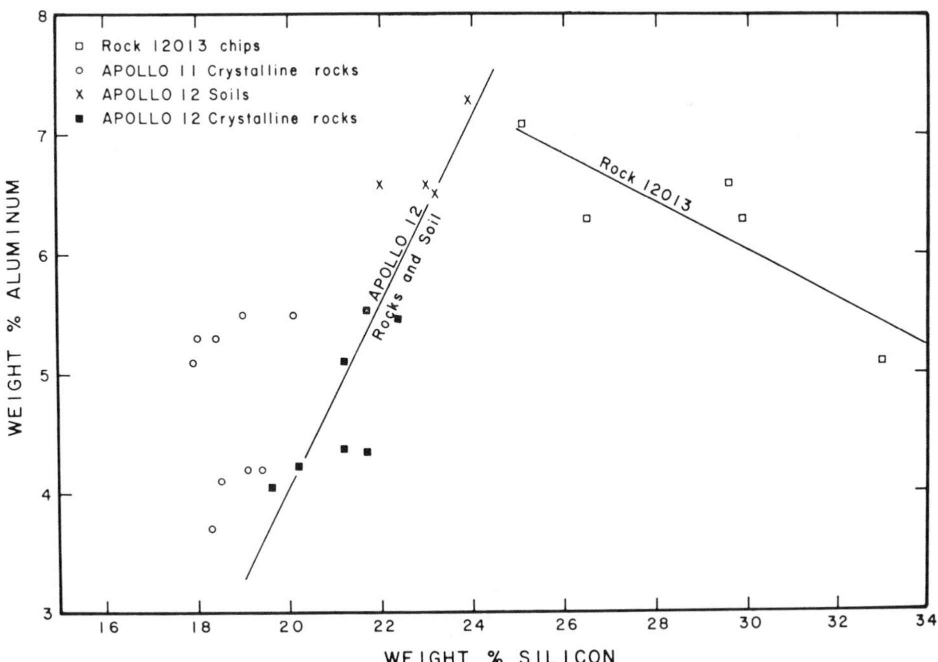

Fig. 1. Relation of Si and Al in lunar materials.

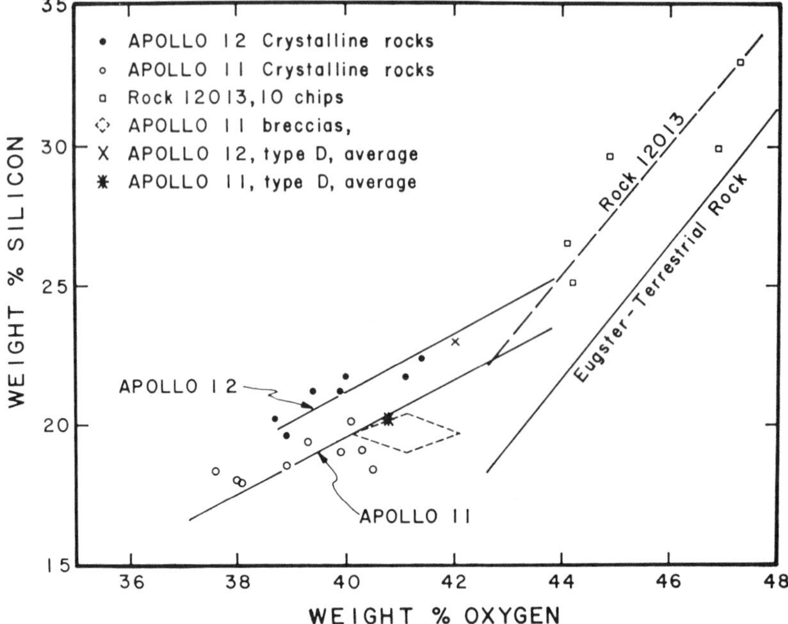

Fig. 2. Relation of Si and O in lunar materials. The line for terrestrial igneous rocks is based on Eugster (1969).

This regression line is nearly identical to that found by Ehmann and Morgan (1970) for the Apollo 11 crystalline rocks:

$$O\% = 0.98\ Si\% + 20.8.$$

In both cases the slope of the regression line is significantly different than that given by Eugster (1969) for terrestrial igneous rocks:

$$O\% = 0.415\ Si\% + 35.0.$$

The Apollo 12 rocks are generally higher in both O and Si than are the corresponding Apollo 11 rocks. This is probably a reflection of the generally lower ilmenite content of the Apollo 12 samples. The presence of ilmenite in the lunar samples results in a displacement of the O–Si regression line to the O deficient side of the line for the terrestrial rocks (ilmenite contains only 31.6% O), but does not in itself explain the significant difference in the slopes of the lunar and terrestrial regression lines.

In contrast, Morgan and Ehmann (1970b) have shown that analyses of five chips of rock 12013 result in a O–Si regression line that closely parallels the terrestrial line:

$$O\% = 0.43\ Si\% + 33.0,$$

but represents rocks poorer by some 2% O. The differences exhibited among these several regressions must in part be related to the amount of O available during the course of the crystallization of the rocks. Sato and Helz (1971) have reported minimum bulk rock oxygen fugacity values of the order of 10^{-12} to 10^{-14} for the Apollo

12 rocks. These values are several orders of magnitude smaller than found for terrestrial basalts. The data suggest the existence of a highly reducing system at the time of crystallization of the lunar basalts.

Fines

The Apollo 12 data presented here suggest that the fines are not derived exclusively from the crystalline rocks sampled, but represent rather complex mixtures. Material of anorthositic composition would be required in the fines to account for the higher Al abundance in the fines (6.7% Al), as compared to a simple 50–50% mixture of Type A and B crystalline rocks (4.6% Al). Using the data presented here, an approximately 17% admixture of average anorthositic material (16.2% Al) similar to that reported by WOOD *et al.* (1970) would be required to account for the Al abundance in the fines. GANAPATHY *et al.* (1970) have estimated a 1.7% admixture of meteoritic matter (Type I carbonaceous chondrite) in the Apollo 12 fines, and they have noted that an additional admixture of 10% of material similar to rock 12013 would account for the high Rb, Cs, and U abundances in the fines. The calculation of the anorthositic contribution as made above based on Al is rather insensitive to the amount of 12013 material present (6.5% Al) and the small amount of meteoritic matter. However, a mixture of 2% meteoritic matter, 71% average A and B rock, 17% anorthositic material, and 10% rock 12013 material does yield a composition for the fines very similar to that found experimentally for the major elements we determined. Additional speculations concerning the composition of the fines must be based on detailed mineralogical studies and their trace element contents.

An apparent O deficiency of approximately 1.6 $\pm$ 0.7% O was noted in the Apollo 11 fines and breccias, as compared to the amount of O required for simple stoichiometry based on total silicate analyses (EHMANN and MORGAN, 1970). Several groups reported high summations in their total silicate analyses, especially in the fines and breccias. The work of MAXWELL *et al.* (1970) indicates that the O depletion based on stoichiometry is minimal or possibly nonexistent in the Apollo 11 crystalline rocks. EHMANN and MORGAN (1970) suggested that the O depletion in the fines and breccias might be due to reduction by solar wind H. This suggestion is consistent with the observed high abundance of solar-wind rare gases in the fines and breccias, as compared to the crystalline rocks.

In our paper (EHMANN and MORGAN, 1971) we calculated an O depletion in the Apollo 12 fines of some 2.6% O, compared to a small depletion of only 0.7% O in the Apollo 12 crystalline rocks. This calculation was based in part on preliminary analyses of LSPET (1970). We have now recalculated the apparent O depletion based on our O data and the new Apollo 12 data presented at the Second Lunar Science Conference (MAXWELL and WIIK, 1971; SCOON, 1971; KLEIN *et al.*, 1971; ANNELL *et al.*, 1971; ENGEL and ENGEL, 1971). These calculations yielded an apparent O depletion with respect to stoichiometry of 1.6 $\pm$ 0.9% O with no significant difference noted between the Apollo 12 fines and crystalline rocks. ANNELL *et al.* (1971) report an "excess reducing capacity" equivalent to 1.6% FeO in the 12070 fines and little or no "excess reducing capacity" in several crystalline rocks.

HAPKE *et al.* (1971) have carried out thermogravimetric analyses in an oxygen atmosphere on Apollo 11 fines and crystalline rocks. They found that the fines increased their weight by approximately 2%, whereas a sample of pulverized crystalline rock gained only 1%. If all the Fe in the samples were oxidized to Fe(III), the weight gain due to the uptake of oxygen would be only 1%. They suggest the excess weight gain by the fines could be due to the uptake of oxygen by a nonstoichiometric film on the particles of the fines which may have been vapor-deposited. The source of this vapor is suggested as sputtering by the solar wind and/or vaporization by meteorite impact. HAPKE *et al.* also suggest the opacity of these nonstoichiometric films contributes to the low albedo of the lunar fines.

As stated previously, our Apollo 12 O data does suggest an O depletion in the fines, but in contrast to the Apollo 11 data this depletion is no greater in the fines than in the Apollo 12 crystalline rocks. Since these calculated depletions involve subtraction of large numbers to obtain small differences, the apparent O depletion should be reevaluated when additional total silicate analyses become available.

Acknowledgments—Financial support for this work was provided by the University of Kentucky Research Foundation and NASA Contract NAS 9–8017. The assistance of Mr. David E. Gillum and Mrs. Susan Smith in processing the data and in manuscript preparation is gratefully acknowledged.

REFERENCES

ANNELL C. S., CARRON M. K., CHRISTIAN R. P., CUTTITTA F., DWORNIK E. J., HELZ A. W., LIGON D. T., JR., and ROSE H. J., JR. (1971) Chemical and spectrographic analyses of lunar samples from Apollo 12 mission. Second Lunar Science Conference (unpublished proceedings).

COMPSTON W., BERRY H., VERNON M. J., CHAPPELL B. W., and KAYE M. J. (1971) Rubidium–strontium chronology and chemistry of lunar material from the Ocean of Storms. Second Lunar Science Conference (unpublished proceedings).

EHMANN W. D., and MCKOWN D. M. (1968) Heat-sealed polyethylene sample containers for neutron activation analysis. *Anal. Chem.* **40**, 1758.

EHMANN W. D., and MORGAN J. W. (1970) Oxygen, silicon, and aluminum in Apollo 11 rocks and fines by 14 MeV neutron activation. *Proc. Apollo 11 Lunar Sci. Conf. Geochim. Cosmochim. Acta* Suppl. I, Vol. 2, pp. 1071–1079. Pergamon.

EHMANN W. D., and MORGAN J. W. (1971) Major element abundances in Apollo 12 rocks and fines by 14 MeV neutron activation. Second Lunar Science Conference (unpublished proceedings).

ENGEL C. G., and ENGEL A. E. J. (1971) Major element composition of three Apollo 12 rocks and some petrogenic considerations. Second Lunar Science Conference (unpublished proceedings).

EUGSTER H. P. (1969) Oxygen abundance in common igneous rocks. In *Handbook of Geochemistry* (editor K. H. Wedepohl), Vol. II, part I, Chap. 8, p. E-1.

GANAPATHY R., KEAYS R. R., and ANDERS E. (1970) Apollo 12 lunar samples: Trace element analysis of a core and the uniformity of the regolith. *Science* **170**, 533–535.

HAPKE B. W., CASSIDY W. A., and WELLS E. N. (1971) The albedo of the Moon: Evidence for vapor-phase deposition processes on the lunar surface. Second Lunar Science Conference (unpublished proceedings).

KLEIN C., DRAKE J. C., ITO J., and FRONDEL C. (1971) Studies of Apollo 12 lunar rocks and fines. Second Lunar Science Conference (unpublished proceedings).

LSPET (Lunar Sample Preliminary Examination Team) (1970) Preliminary examination of the lunar samples from Apollo 12. *Science* **167**, 1325–1339.

MAXWELL J. A., ABBEY S., and CHAMP W. H. (1970) Chemical composition of lunar material. *Science* **167**, 530–531.

MAXWELL J. A., and WIIK H. B. (1971) Chemical composition of Apollo 12 lunar samples 12004, 12033, 12051, 12052, and 12065. Second Lunar Science Conference (unpublished proceedings).

MORGAN J. W., and EHMANN W. D. (1970a) Precise determination of oxygen and silicon in chondritic meteorites by 14 MeV neutron activation using a single transfer system. *Anal. Chim. Acta* **49**, 287–299.

MORGAN J. W., and EHMANN W. D. (1970b) Lunar rock 12013; O, Si, Al, and Fe abundances. *Earth Planet. Sci. Lett.* **9**, 164–168.

MORGAN J. W., and EHMANN W. D. (1971) 14 MeV neutron activation analysis of rocks and meteorites. *Proceedings of the NATO Advanced Study Institute-Activation Analysis in Geochemistry and Cosmochemistry*, Kjeller, Norway, September 7–12, 1970, Universitetsforlaget, Oslo. In press.

SATO M., and HELZ R. T. (1971) Oxygen fugacity studies of Apollo 12 basalts by the solid-electrolyte method. Second Lunar Science Conference (unpublished proceedings).

SCOON J. H. (1971) Quantitative chemical analyses of lunar samples 12040,36 and 12064,33. Second Lunar Science Conference (unpublished proceedings).

SUTTON R. L., and SCHABER G. G. (1971) Lunar locations and orientations of rock samples from Apollo missions 11 and 12. Second Lunar Science Conference (unpublished proceedings).

VOGT J. R., EHMANN W. D., and MCELLISTREM M. T. (1965) An automated system for rapid and precise fast neutron activation analysis. *J. Appl. Radiat. Isotop.* **16**, 573–580.

WÄNKE H., WLOTZKA F., TESCHKE F., BADDENHAUSEN H., SPETTEL B., BALACESCU A., QUIJANO-RICO M., JAGOUTZ E., and RIEDER R. (1971) Major and trace elements in Apollo 12 samples and studies on lunar metallic iron particles. Second Lunar Science Conference (unpublished proceedings).

WOOD J. A., DICKEY J. S., JR., MARVIN U. B., and POWELL B. M. (1970) Lunar anorthosites. *Science* **167**, 602–604.

Proceedings of the Second Lunar Science Conference, Volume 2, pp. 1247–1252.
The M.I.T. Press, 1971.

Spark mass spectrometric analysis of major and minor elements in six lunar samples

MICHEL BOUCHET, GRÉGOIRE KAPLAN, ANTOINETTE VOUDON, and MARIE-JEANNE BERTOLETTI
Laboratoire d'Analyses Physiques, 64 Serres-Castet, France

(*Received 19 February 1971; accepted in revised form 31 March 1971*)

Abstract—Investigations for 72 major and trace elements were carried out by spark mass spectrometry on four rocks and one lunar soil sample from the Apollo 12, and one lunar soil sample from the Apollo 11 mission. Of these 72 elements, 59 were successfully analyzed. All these samples were compared with a standard basalt, B.R. Trace elements are more abundant in the Apollo 12 soil sample than in the other samples. It was also possible to determine $^{207}Pb/^{206}Pb$ ratios by spark mass spectrometry and to establish maximum model ages of $4.25 \pm 0.15 \times 10^9$ years for four rocks.

INTRODUCTION

ANALYSIS OF TWO lunar soils, Apollo 11 10084,142 and Apollo 12 12070,92, and four Apollo 12 rocks 12018,43; 12004,28; 12063,53; and 12065,41 were performed with spark source mass spectrography. For the major elements, emission spectroscopy was used only for analysis of the two lunar soil samples in order to use these two soils as standards for rock analysis by mass spectrography. The French standard basalt B.R., from the University of Nancy, France (ROUBAULT *et al.*, 1966) was again used as a multielement standard for all major and minor elements in all lunar samples, soils included. Recommended values of B.R. basalt are shown in Table 1.

ANALYTICAL METHOD

About 250 mg of each sample was crushed and half of the powder was mixed and briquetted with an equal amount of graphite (Spektralkohle pulver Ringsdorffwerke GMBH Badgodsberg Mehlem Qualität RWA Serien Nr XII/125/910); the other half was mixed and briquetted with aluminum powder (8201 w—Aluminum powder 99.999% 90–150 μ—Batch n° 40,512 KOCH-LIGHT Lab. England). In this way the readings were not affected by carbon and aluminum interferences. Iron in the sample was used as an internal standard. To compare lunar samples with the basalt B.R. a special process was employed with the spark mass spectrograph. Each photoplate can receive 32 exposures (16 for the sample, 16 for the standard) and each sample, once analyzed, was used as a standard for the following one as shown in Table 2. This complete operation was performed once with each graphite and aluminum medium. By this method, it was possible to avoid the eventual defects of the photoplates.

Each sample was analyzed 12 times (6 times with carbon and 6 times with aluminum). The values were calculated by arithmetical average for each medium; the sensitivity coefficients were different for graphite and for aluminum. After comparison with the standard analyses under the same conditions, the arithmetical average of the two methods was calculated.

To establish standards for the major elements, it was found more advisable to use another method. Readings were obtained with as little as 100 mg of the two soil samples. In that way, two

Table 1. Compared values for the different elements in the 6 lunar samples and in Basalt B.R.

Element	Rocks				Soils		
	12018,43	12004,28	12063,53	12065,41	10084,142 Apollo 11	12070,92 Apollo 12	B.R.
	Major elements percentage by weight						
	Spark Mass Spectrometry				Emission Spectroscopy‡		
Si	18.7	18.9	19.6	18.1	19.7	22.2	18.00
Al	6.4	5.8	6.1	7.4	7.3	6.7	5.47
Fe	16.6	19.0	16.9	17.1	12.3	13.2	9.01
Mn	0.47	0.19	0.20	0.31	0.16	0.18	0.16
Mg	6.8	6.5	5.5	7.0	4.8	5.7	7.97
Ca	7.4	7.7	8.6	6.9	8.5	7.4	9.94
Na	0.37	0.22	0.44	0.30	0.33	0.33	2.28
K	0.058	0.037	0.066	0.041	0.12	0.12	1.14
Ti	1.7	2.7	2.5	2.5	5.4	2.1	1.96
Cr	0.48	0.49	0.31	0.40	0.50	0.70	0.042
	Trace Elements (parts per million)						
Li	9	11	15	7	12	28	12
B	2	3	2	3	8	11	8
F	150	113	84	78	96	63	1050
P	206	112	203	155	213	644	4460
S	1520	1090	1510	1660	1750	1100	462
Cl	35	29	48	29	84	48	343
Sc	70	73	123	78	62	63	28
V	96	100	54	51	22	49	240
Co	42	31	23	20	14	17	50
Ni	90	62	46	32	163	164	270
Cu	7	9	12	8	9	7	72
Zn	6	8	16	8	22	14	160
Ga	9	4.5	7	5.5	8.5	15	26
Ge	0.5	0.5	0.8	0.6	1	1.1	1
As†	0.05	0.1	0.09	0.04	0.1	0.15	
Se†	0.2	0.3	0.4	0.5	0.5	1.1	
Br	0.08	0.08	0.12	0.11	0.14	0.15	0.8
Rb	3	2	3	3	8	9.5	45
Sr	188	115	137	95	186	145	1300
Y	106	87	128	74	173	180	34
Zr	290	204	174	190	301	872	240
Nb	15	12	12	16	23	42	80
Mo	0.7	0.6	0.9	0.6	1	0.6	3
Ru†	0.9	0.7	0.5	0.7	0.6	1.1	
Rh†	0.15	0.3	0.1	0.3	0.1	0.4	
Cd†	0.8	1	0.9	1	1.6	1.9	
In†	0.6	0.7	0.9	0.8	1.7	2	
Cs	0.3	0.3	0.3	0.4	0.8	1.7	2
Ba	200	165	146	153	410	672	1050
La	29	18	25	22	37	82	90
Ce	26	26	28	31	60	115	130
Pr†	4	5.5	8	5.8	17	26	
Nd	12	16	54	22	44	74	17
Sm	3	5	9	5	12	15	7.4
Eu†	1.2	1.4	3	1.3	3	3.6	
Gd	4	5	9	6	15	18	7
Tb†	1.4	2	5	2.5	9.0	7.3	
Dy	11	9	16	12	20	26	9
Ho†	0.3	0.2	0.7	3	11	15.7	
Er†	4.6	8.6	19.7	7.8	33	29	
Yb	8	7	16	13	21	20	4
Lu†	0.6	1.3	2.7	1.1	3.6	4.2	
Hf†	3	5	10	6.2	12.9	17.2	
Pb	3	2.5	2	5	4	8	16
Th	1.2	1.7	2.9	2.4	3.9	6.6	12.6
U	0.25	0.29	0.47	0.62	0.67	1	1.8
U*	0.66	0.33	0.39	0.26	0.63	1.12	
Th*	3.29	1.64	1.94	1.30	3.15	5.60	

* These values were obtained via α autoradiography by Professor R. Coppens, University of Nancy (CRR).

† Certain elements were not visible at the level of detection used for the basalt B.R. On the other hand, they were noticed in the lunar samples and, in their case, the coefficient of correction was taken as equal to unity.

‡ These analyses were performed by K. Govindaraju (CRPG), Nancy, France.

Table 2. Disposition of pairs of samples on each photoplate. This matrix was made once with a graphite medium and once with an aluminum medium.

	R1	R2	R3	R4	P12	P11	B
R1		(R1 R2)	(R1 R3)	(R1 R4)	(R1 P12)	(R1 P11)	(R1 B)
R2			(R2 R3)	(R2 R4)	(R2 P12)	(R2 P11)	(R2 B)
R3				(R3 R4)	(R3 P12)	(R3 P11)	(R3 B)
R4					(R4 P12)	(R4 P11)	(R4 B)
P12						(P12 P11)	(P12 B)
P11							(P11 B)
B							

R1 = sample 12018,43; R2 = sample 12004,28; R3 = sample 12063,53; R4 = sample 12065,41; P12 = sample 12070,92; P11 = sample 10084,142; B = sample basalt B.R. (Nancy).

Table 3. $^{207}Pb/^{206}Pb$ ratios in four Apollo 12 lunar rock samples.

Sample	$^{207}Pb/^{206}Pb$	Maximum Model Ages, 10^9 years
12004,28	0.451	4.15 ± 0.20
12018,43	0.504	4.32 ± 0.08
12063,53	0.458	4.17 ± 0.17
12065,41	0.517	4.35 ± 0.26

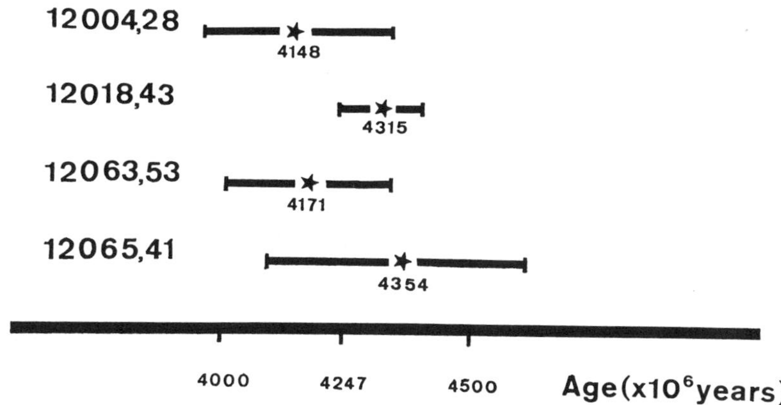

Fig. 1. Reproductability of age measurements with lead isotopes ^{206}Pb and ^{207}Pb by spark mass spectrometry.

more standards for the major elements were obtained. This technique permitted good analytical results to be obtained from only 40 mg samples per rock. The accuracy and precision of this method depends on the accuracy of recommended values for basalt B.R. and the reproductibility of the spark mass spectrometric analyses.

Despite the fact that exposures of 300 nC and, in certain cases, of 1000 nC were used, some elements were beyond the limits of sensitivity: Ag, Sn, Sb, I, W, Re, Os, Ir, Pt, Au, Hg, Tl, and Bi. Other elements, such as Pd, Te, and Tm, may be affected by interferences. Because Ta was the principal element for the mass spectrometer source as well as for the instruments used in the preparation of the samples, this metal was not subjected to analysis.

The mass spectrograph has an electrical detection system with automatic peak switching, thus making it possible to compute isotopic ratios for model age determinations. Only lead isotopes could be measured by this technique. Interferences were seen for ^{204}Pb, but coherent readings were obtainable for ^{207}Pb/^{206}Pb ratios (Table 3). Because graphite was used as an electrode, a mass-204 interference may possibly be attributable to seventeen ^{12}C masses. Similar effects were not detected for the 206, 207, and 208 masses. The lead blank was beyond the limit of detection (10 ppb). The reproductability of this method is shown in Fig. 1.

RESULTS AND DISCUSSIONS

Figure 2 gives a comparison between the chemical composition of Apollo 12 rocks and soils. Abundances in Apollo 12 soil 12070 are compared to the average values in four Apollo 12 rocks. It is seen that certain elements, especially the REE (rare-earth elements), are systematically more abundant in the soil.

A comparison of the major and minor elements in lunar rocks and soils, in the basalt B.R. and in the meteorite Juvinas (a monomict pigeonite-plagioclase achondrite) is given in Fig. 3. Great similarity between the four rock samples is observed. Likewise it is evident that the composition of Apollo 11 lunar soil 10084 is closer to these four rocks than to the composition of the Apollo 12 soil 12070. On the other hand, the latter is closer to the meteorite Juvinas. Finally, the basalt B.R. is sufficiently

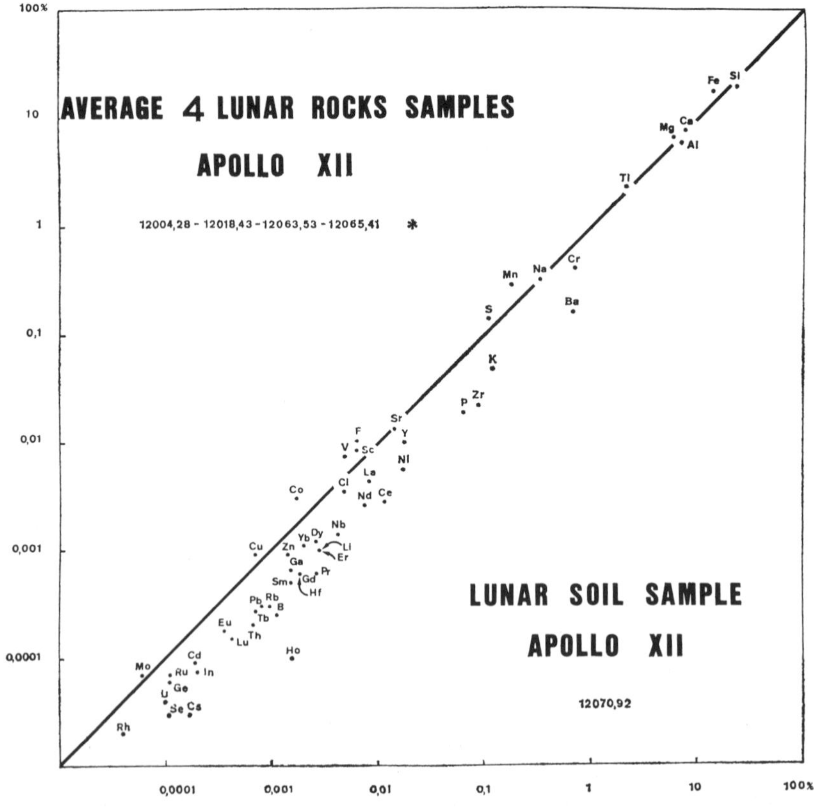

Fig. 2. Comparison of abundances in Apollo 12 soil and four rocks.

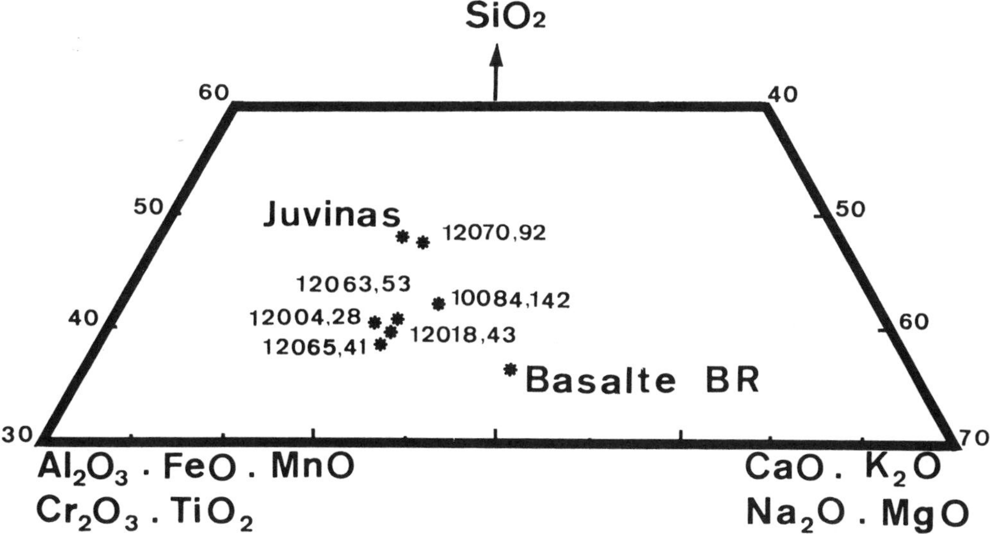

Fig. 3. A comparison of the major elemental composition in Apollo 12 four rocks and soil, Apollo 11 soil, Basalt B.R., and the achondritic meteorite Juvinas.

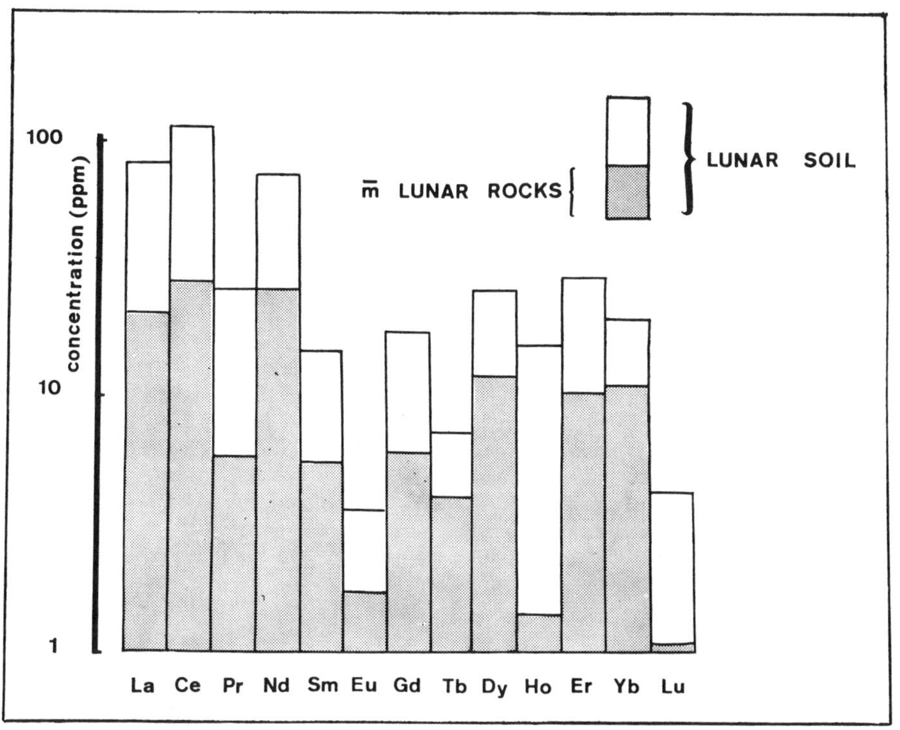

Fig. 4. Comparison between abundances of the rare earths in Apollo 12 soil and the average of four Apollo 12 rocks.

similar to all the samples, thereby justifying its use as a comparison standard. The graph of Fig. 4, a comparison of REE abundances in Apollo 12 soil and four rocks, reveals a systematic REE abundance increase in the lunar soil relative to the average abundances in the four rocks.

From Table 1 we note that the average Fe value of the four rocks is much greater than the Fe content of the Apollo 11 and Apollo 12 soils and of the basalt B.R. On the contrary, K and Na are much less abundant in the six lunar samples than in the basalt B.R. The content of other elements such as Si, Al, Mg, and Ti are rather similar in the four rocks and the basalt B.R. However the distribution of the major elements in the Apollo 12 soil is more similar to eucrite meteorites, such as Juvinas. This suggests that this lunar soil can contain an important component of this particular meteoritic class.

With respect to the trace elements, it is noticeable that there exist certain elements which distinguish the soil samples from the rock samples, such as Ni, Rb, Ba, Nb, and the REE. Probably these systematic differences may be attributed to different origins for some of the constituents of the soil and rocks.

On account of the very low ^{204}Pb content found in lunar material, the $^{207}Pb/^{206}Pb$ ratios are given without correction for common lead (TATSUMOTO and ROSHOLT, 1970; SILVER, 1970). Therefore, the model ages of this work represent only maximum ages. The corrections for coefficients of sensitivity were not sufficiently reliable for Pb, Th, and U. For example, Pb abundances of this work are generally 3 to 6 times higher than Pb values obtained by SILVER (1971) and TATSUMOTO *et al.* (1971); U and Th values of this work are also 1 to 4 times higher than those obtained by the above two groups. Therefore, accurate Pb–U and Pb–Th ages cannot be determined with confidence by using our data.

Acknowledgments—This study was supported by funds of the Laboratoire d'Analyses Physiques. The authors are grateful to NASA for supplying the lunar sample, to M. Legrand, Director of Société Générale, Pau, for safeguarding the samples, and to A. Bélisson, R. Dumez, A. Méon, and H. Montoechio for technical and scientific assistance.

REFERENCES

ROUBAULT M., DE LA ROCHE H., and GOVINDARAJU K. (1966) *Sci. de la Terre* **11**, 105–121.

SILVER L. T. (1970) Uranium–thorium–lead isotope relations in lunar material. *Science* **167**, 468–471.

SILVER L. T. (1971) U–Th–Pb Isotope Relations in Apollo 11 and Apollo 12 lunar samples. Second Lunar Science Conference (unpublished proceedings).

TATSUMOTO M., KNIGHT R. J., and DOE B. R. (1971) U–Th–Pb systematics of Apollo 12 lunar sample. Second Lunar Science Conference (unpublished proceedings).

TATSUMOTO M., and ROSHOLT J. N. (1970) Age of the Moon, an isotopic study of Uranium–Thorium–lead systematics of lunar samples. *Science* **167**, 461–463.

Proceedings of the Second Lunar Science Conference, Volume 2, pp. 1253–1258.
The M.I.T. Press, 1971.

Elemental composition of lunar surface material (part 2)

A. A. SMALES, D. MAPPER, M. S. W. WEBB, R. K. WEBSTER, J. D. WILSON, and J. S. HISLOP

Analytical Sciences Division, Atomic Energy Research Establishment, Harwell, England

(*Received 9 February 1971; accepted in revised form 29 March 1971*)

Abstract—Results are given for the elemental abundances in further samples of Apollo 11 materials and in three samples from the Apollo 12 mission. Comments are made on similarities and differences between Apollo 11 and Apollo 12 materials and between the latter and the meteorite Angra dos Reis.

INTRODUCTION

WE PRESENT HERE results so far obtained on samples 10057,52, 10059,34, 12022,49, 12065,51, and 12070,78, together with results for some further elements on 10060,26 and 10084,9,10. The methods used were activation analysis (AA); instrumental neutron activation analysis (INAA); radiochemical neutron activation analysis (RNAA); high-energy gamma activation analysis (GAA), and mass spectrometric isotope dilution analysis (MSID); supplemented by spark source mass spectrography (SSM) emission spectrography (ES), and X-ray fluorescence spectrometry (XRF). Their estimated accuracies are given in our previous paper (SMALES *et al.* 1970).

Results for minor and trace elements using AA and MSID are given in Table 1, those for major elements in Table 2, and some supplementary results by ES in Table 3. We have examined all our samples by SSM and have established upper limits for many more elements similar to those in Table 4 in our previous paper (SMALES *et al.* 1970).

DISCUSSION OF RESULTS

(1) The similarity of the two rocks 12022 and 12065, and the difference between them and the fines sample 12070 is readily apparent from Table 1. Apart from Sc, Cr, and Mn the fines sample is much higher in minor and trace element contents.

(2) The 12070 fines sample is significantly higher in K, Rb, Cs, Ba, REE, and Zr, but significantly lower in Ti and Zn, than the 10084 fines.

(3) Some of the values reported for the "volatile or depleted" elements of GANAPATHY *et al.* (1970) for types A and B rocks of Apollo 11 are given in Table 4 together with our results on the Apollo 12 rocks 12022 and 12065, listed by NASA (1970) as types B and AB, respectively. We have included As and Sb results by HASKIN *et al.* (1970) and a Ge result by WASSON and BAEDECKER (1970) on Apollo 11 rocks to compare with ours on Apollo 12, these elements being amongst the "normally depleted" elements of LARIMER (1967). It is clear that the pattern of behavior of all the elements listed is similar in the Apollo 11 and Apollo 12 rocks of the A and B types, though As seems significantly lower in the Apollo 12 rocks. It is perhaps worth recording here that our Ge, Ga, and In determinations on rock 12065 were by

Table 1. Minor and trace elements determined by activation analysis and/or mass spectrometric isotope dilution analysis.

		Method*	Apollo 11 10057	Apollo 11 10059	Apollo 12 12022	Apollo 12 12065	Apollo 12 12070
			Parts per million by weight				
9	F	4	90	90	40	40	
11	Na	1	4040	3740	1680	1790	3690
19	K	5	2740	1600	510	520	2170
21	Sc	1	86	68	59	60	43
24	Cr	1	2250	2070	—	3400	2800
25	Mn	1	1810	1720	2410	2050	1750
27	Co	3	(~30[2])		48.5	40	43
29	Cu	3				8	
30	Zn	3			2	2	11
37	Rb	3, 5	5.1, 6.3	3.2, 4.2	0.7, 0.8	1.0, 1.2	6.4, 6.6
38	Sr	4, 5	—, 190	—, 160	120, 170	110, 115	—, 190
47	Ag	3	0.006	0.009	0.011		0.026
55	Cs	3, 5	0.20, 0.21	0.12, —	0.04, 0.044	0.054, 0.06	0.29, 0.36
56	Ba	3, 5	260, 310	206, 210	54, 55	68, 67	390, 350
57	La	2	~28	~19	~5.5	~8	~38
58	Ce	5	(77[2])	66	17	16	90
60	Nd	5	(63[2])	51	15	14	57
62	Sm	1, 5	22, 21	15, 14	6, 5.4	5, 4.7	18, 16.7
63	Eu	1, 5	2.1, 1.9	2.1, 2.1	1.0, 1.3	0.9, 0.9	1.7, 1.7
65	Tb	2	~7	~4.5	~2.4	~2.3	~6
66	Dy	1	36	25	13	12	29
67	Ho	2	—	—	~3.1	~2.3	~6.7
70	Yb	1	19	13	4.9	4.7	13.6
71	Lu	2	~2.7	~1.9	~0.7	~0.6	~1.9
72	Hf	2	~17		~4	~4	~12

* Methods: (1) INAA ± 5%, (2) INAA ± 20%, (3) RNAA ± 5%, (4) GAA ± 10%, (5) MSID ± 5% where two methods are quoted results are given, respectively.

Additional information by method 3 on sample 12065: Ga 2.8, Ge 0.02, As 0.003, Pd <0.001, In 0.0024 Sb 0.009, Re <0.001, Au 0.0003 ppm. On sample 10084: Co 28.6, Rb 3.0, Cs 0.11, La 16, Zn 23. On Sample 10060: Co 27.4, Rb 4.2, La 21, Zn 27.

Table 2. Major elements in lunar rocks expressed as % oxide

	Apollo 11 10057 XRF	10057 INAA	10057 MSID	10059 XRF	10059 INAA	10059 MSID	10060 XRF	Apollo 12 12022 XRF	12022 INAA	12022 MSID	12022 GAA	12065 XRF	12065 INAA	12065 MSID	12065 GAA
SiO_2	41.4			42.4			41.8	42.3				47.1			
Al_2O_3	8.1	7.6		12.6	12.5		11.5	8.2	8.5			9.8			
FeO	19.1			16.4			16.7	21.6			21.7	20.0			20.3
MgO	7.4			8.1			7.6	11.9				9.2			
CaO	10.4			11.5			11.2	9.0			9.2	10.1			10.9
Na_2O		0.54			0.51				0.23				0.24		
K_2O			0.33			0.19				0.061				0.063	
TiO_2	10.9			8.3			8.7	4.9			4.8	3.1			3.4
MnO		0.23			0.22				0.31				0.26		
Cr_2O_3		0.35			0.32								0.50		

Insufficient sample was available for XRF method on fines 12070. Results for Na, K, Mn, Cr, and trace elements on that sample are available in Table 1.

radio-chemical neutron activation on a single chip which was surface etched with acid after the irradiation, the etch liquor being discarded.

(4) In the fines, not only are the levels of alkali metals higher in 12070 than in 10084, but there has clearly been less differentiation of the heavier alkali metals in 12070, as can be seen from Table 5, where ratios are shown. Table 6 shows that the Apollo 12 rocks fit into the "low-Rb" classification of GAST *et al.* (1970) for Apollo

Table 3. Some supplementary results by emission spectrography*

		Apollo 11 10057	Apollo 11 10059	Apollo 12 12022	Apollo 12 12065	Apollo 12 12070
		Parts per million by weight				
28	Ni	< 20	300	50	20	200
23	V	50	50	120	150	110
29	Cu			5		
39	Y	250	190	60	45	130
40	Zr	770	630	200	180	600

* Estimated accuracy for ES. ± 20%.

Table 4. "Depleted" elements in Apollo 11 and Apollo 12 rocks.

	Apollo 11 Samples "Average" values for some type A and type B rocks*	Apollo 12 Samples Results for Rock 22 and/or Rock 65
Cu ppm	9.33	8.2
Zn ppm	7.41	2
Co ppm	20.9	48.5, 40
Pd ppb	3.40	< 1
As ppb	68†	3
Sb ppb	6.4†	9
Ge ppm	≤0.08‡	0.02
Ga ppm	4.48	2.8
In ppb	2.9	2.4
Cs ppb	121	40–60
Rb ppm	3.58	0.8, 1.1

* From Ganapathy *et al.* (1970).
† From Haskin *et al.* (1970).
‡ From Wasson and Baedecker (1970).

Table 5. Alkali metal ratios in Apollo 11 and Apollo 12 fines.

	Apollo 11 Fines, 10084	Apollo 12 Fines, 12070
Na/K	2.67	1.70
Na/Rb	1067	568
Na/Cs	29,091	11,182
K/Rb	400	334
K/Cs	10,909	6,576
Rb/Cs	27.3	19.7

Table 6. Alkali metal contents of Apollo 11 and Apollo 12 rocks.

	Apollo 11 samples 10057, type A	Apollo 11 samples 10058*, type B	Apollo 11 samples 10060, type C	Apollo 12 samples 12022	Apollo 12 samples 12065
Na, ppm	4040	2900	3770	1680	1790
K	2740	877	1700	506	520
Rb	5.7	0.98	4.2	0.8	1.1
Cs	0.20	0.027	0.17	0.04	0.06

* Gast *et al.* (1970).

Table 7. Alkali metal ratios in Apollo 11 and Apollo 12 rocks.

	Apollo 11 samples			Apollo 12 samples	
	10057, type A	10058*, type B	10060, type C	12022	12065
Na/K	1.47	3.31	2.22	3.32	3.44
Na/Rb	709	2959	898	2100	1630
Na/Cs	20,200	107,400	22,180	42,000	29,830
K/Rb	481	895	405	633	473
K/Cs	13,700	32,480	10,000	12,650	8670
Rb/Cs	28.5	36.3	24.7	20.0	18.3

* Using results from GAST *et al.* (1970).

Table 8. Some REE ratios in Apollo 11 and Apollo 12 materials

	Apollo 11*				Apollo 12		
	A	B	C	D	12022	12065	12070
La/Sm	1.0	0.57	1.1	1.1	0.95	1.6	2.2
Sm/Eu	9.5	6.5	7.4	7.1	4.5	5.2	10.2
Lu/Sm	0.11	0.13	0.12	0.12	0.12	0.12	0.11
Yb/Lu	7.4	7.1	7.2	7.1	7.0	7.8	7.2

* From GAST *et al.* (1970), WAKITA *et al.* (1970).

Table 9. Elemental similarities between Angra dos Reis and Apollo 12 rocks.

	Angra dos Reis	12022	12065
Al%	5.7	4.4	5.2
Ti%	1.7	2.9	1.9
Sc ppm	57	59	60
Sr	133*	145	115
La	10	6	8
Ce	21†	17	16
Nd	20†	15	14
Sm	8	6	5
Eu	2.0	1.0	0.9
Tb	2	2	2
Dy	12	13	12
Ho	3.5	3	2
Yb	6	5	5
Lu	0.7	0.7	0.6

* TERA *et al.* (1970).
† SCHNETZLER and PHILPOTTS (1969).

11 rocks, generally of type B. However, the alkali metal ratios in Table 7 show that less differentiation of the heavier alkali metals has occurred in the Apollo 12 rocks, as compared with the Apollo 11 type B rocks. Indeed, in the Apollo 12 rocks, ratios between K, Rb, and Cs are nearer those for types A, C, and D of Apollo 11.

(5) As with Apollo 11 materials, there is a negative Eu anomaly in the Apollo 12 samples though not so pronounced in the rocks as in the fines. In Apollo 11 the REE levels were generally higher in type A than in type C, which in turn was higher than type B, with type C levels being similar to or higher than type D; we have only three samples of Apollo 12 material, but, as can be seen from Table 1, while the levels in 12070 (type D) are much higher than those for 12022 (type B), in line with

the Apollo 11 materials, 12070 is also much higher than 12065 (type AB). Table 8 lists some of the elemental abundance ratios used for example by GAST *et al.* (1970) and WAKITA *et al.* (1970) to discuss REE behavior in Apollo 11 materials, together with corresponding ratios from our Apollo 12 samples. The following points emerge:

(a) La/Sm: The value for 12070 (type D) is much higher than that for Apollo 11 type D; that for the 12022 (type B) rock is similar to those of Apollo 11 types A, C, D but much higher than that of Apollo 11 type B; while that for the 12065 (type AB) is higher than for any Apollo 11 type.

(b) Sm/Eu: The value for 12070 (type D) is much higher than for Apollo 11 types B, C, or D and even higher than for Apollo 11 type A, whereas the values for both Apollo 12 rocks are lower than any of the Apollo 11 types, even type B.

(c) Lu/Sm and Yb/Lu: The values in all three Apollo 12 samples are much the same as for all types of Apollo 11.

From these considerations it seems that there is some difference of behavior of the light REE in the Apollo 12 materials from that in Apollo 11.

(6) As was the case with Apollo 11 materials, we have compared the elemental abundances of the Apollo 12 samples with those available for the unique achondrite Angra dos Reis. It has not been our intention to suggest that Angra dos Reis might have derived from the moon, but rather that any similarity in patterns of elemental abundance might be used to assist in the thinking about the crystallization or melting processes involved in the lunar surface rock formation. Table 9 lists values from our previous paper (SMALES *et al.* 1970), for 14 elements in the meteorite and in the two Apollo 12 rocks available to us. The similarity is quite remarkable. However, the differences in levels of calcium and the alkali metals, noted in our previous paper, still persist for the Apollo 12 materials.

Acknowledgments—We thank M. W. Cooper, R. Hallett, T. C. Hughes, R. P. Kay, M. Perkins, A. G. Pratchett, C. A. J. McInnes, P. F. Ralph, and D. R. Williams for experimental assistance.

REFERENCES

GANAPATHY R., KEAYS R. R., LAUL J. C., and ANDERS E. (1970) Trace elements in Apollo 11 lunar rocks: Implications for meteorite influx and origin of moon. *Proc. Apollo 11 Lunar Sci. Conf., Geochim. Cosmochim. Acta* Suppl. 1, Vol. 2, pp. 1117–1142. Pergamon.

GAST P. W., HUBBARD N. J., and WIESMANN H. (1970). Chemical composition and petrogenesis of basalts from Tranquillity Base. *Proc. Apollo 11 Lunar Sci. Conf., Geochim. Cosmochim. Acta* Suppl. 1, Vol. 2, pp. 1143–1163. Pergamon.

HASKIN L. A., ALLEN R. O., HELMKE P. A., PASTER T. P., ANDERSON M. R., KOROTEV R. L., and ZWEIFEL K. A. (1970) Rare earths and other trace elements in Apollo 11 lunar samples. *Proc. Apollo 11 Lunar Sci. Conf., Geochim. Cosmochim. Acta* Suppl. 1, Vol. 2, pp. 1213–1231. Pergamon.

LARIMER J. W. (1967) Chemical fractionations in meteorites—I. Condensation of the elements. *Geochim. Cosmochim. Acta* **31**, 1215–1238.

NASA (1970) Apollo 12 preliminary science report. NASA SP–235. Houston.

SCHNETZLER C. C., and PHILPOTTS J. A. (1969) Genesis of the calcium-rich achondrites in light of rare earth and barium concentrations. In *Meteorite Research* (editor P. M. Millman), pp. 206–216. Reidel.

SMALES A. A., MAPPER D., WEBB M. S. W., WEBSTER R. K., and WILSON J. D. (1970) Elemental composition of lunar surface material. *Science* **167**, 509–512.

TERA F., EUGSTER O., BURNETT D. S., and WASSERBURG G. J. (1970) Comparative study of Li, Na, K, Rb, Cs, Ca, Sr, and Ba abundances in achondrites and in Apollo 11 lunar samples.

Proc. Apollo 11 Lunar Sci. Conf., Geochim. Cosmochim. Acta Suppl. 1, Vol. 2, pp. 1637–1657. Pergamon.

WAKITA H., SCHMITT R. A., and REY P. (1970) Elemental abundances of major, minor, and trace elements in Apollo 11 lunar rocks, soil, and core samples. *Proc. Apollo 11 Lunar Sci. Conf., Geochim. Cosmochim. Acta* Suppl. 1, Vol. 2, pp. 1685–1717. Pergamon.

WASSON J. T., and BAEDECKER P. A. (1970) Ga, Ge, In, Ir, and Au in lunar, terrestrial, and meteoritic basalts. *Proc. Apollo 11 Lunar Sci. Conf., Geochim. Cosmochim. Acta* Suppl. 1, Vol, 2, pp. 1741–1750. Pergamon.

Proceedings of the Second Lunar Science Conference, Volume 2, pp. 1259–1260.
The M.I.T. Press, 1971.

Chemical analyses of lunar samples 12040 and 12064

J. H. Scoon
Department of Mineralogy and Petrology, University of Cambridge, Cambridge, England

(*Received 22 February 1971; accepted 29 March 1971*)

Abstract—Two samples, 12040,36 and 12064,38 of the Apollo 12 collection, were analyzed by classical chemical methods for major elements. The results of these analyses and C.I.P.W. norms are given.

Titanium is much less abundant than in the Apollo 11 samples. Water is again absent, and iron is present almost entirely in the ferrous state.

The compositions of lunar samples 12040,36 and 12064,38 together with their C.I.P.W. norms are given in Table 1. Classical methods of chemical analysis were used, as described by Scoon (Agrell *et al.*, 1970), except that 1 gm of PbO replaced the $PbCrO_4$ previously used as a flux in the determination of total water by the Penfield method. No attempt was made to estimate metallic iron, as the determination is complicated by the presence of sulphur, and possibly of some titanium in the titanous state. In addition there is the possibility that some of the chromium may be present in the chromous state.

In sample 12040,36, the presence of 0.01% Fe_2O_3 is reported, whereas in 12064,38 the ferrous iron determination is in slight excess of that of the total iron; this discrepancy being due to the presence of reducing substances other than FeO. The TiO_2 values reported are considerably lower than those of the rocks of the Apollo 11

Table 1. Chemical analyses of lunar rocks (wt. %) and C.I.P.W. norms.

	Analyses 12040,36	12064,38		C.I.P.W. Norms 12040,36	12064,38
SiO_2	43.89	46.41	Q	—	2.99
Al_2O_3	7.41	10.50	Or	0.24	0.24
Fe_2O_3	0.01	nil	Ab	1.69	2.54
FeO	20.83	19.95	An	19.18	27.15
MnO	0.26	0.27	Di	15.92	25.87
MgO	16.10	6.38	CEn	4.27	4.54
CaO	7.87	11.71	Fs	3.57	8.56
Na_2O	0.20	0.30	CWO	8.08	12.78
K_2O	0.04	0.07	Hy	31.28	32.71
H_2O^+	nil	nil	En	17.02	11.33
H_2O^-	nil	nil	Fs	14.26	21.38
TiO_2	2.74	4.14	Ol	25.28	
P_2O_5	0.07	0.04	Fo	13.14	
Cr_2O_3	0.70	0.38	Fa	12.14	
S	0.06	0.07	Cm	1.03	0.56
F	nil	nil	M^+	0.01	—
	100.18	100.22	Il	5.20	7.85
Less S = O	0.02	0.02	Ap	0.17	0.09
	100.16	100.20			
Total iron as Fe_2O_3	23.16	21.89			

collection, as are the values for the alkali metals. No fluorine could be detected using the Sen Gupta modification of the colorimetric method of Huang and Johns (SEN GUPTA, 1968).

REFERENCES

AGRELL S. O., SCOON J. H., MUIR I. D., LONG J. V. P. L., MCCONNELL J. D. C., and PECKETT A. (1970) Observations on the chemistry, mineralogy, and petrology of some Apollo 11 lunar samples. *Proc. Apollo 11 Lunar Sci. Conf., Geochim. Cosmochim. Acta* Suppl. 1, Vol. 1, pp. 93–128. Pergamon.

SEN GUPTA, J. G. (1968) Determination of fluorine in silicate and phosphate rocks, micas and stony meteorites. *Anal. Chim. Acta* **42**, 119–125.

Proceedings of the Second Lunar Science Conference, Volume 2, pp. 1261–1276.
The M.I.T. Press, 1971.

The halogens and other trace elements in Apollo 12 samples and the implications of halides, platinum metals, and mercury on surfaces

G. W. Reed and S. Jovanovic
Chemistry Division, Argonne National Laboratory, Argonne, Illinois 60439

(*Received 18 February 1971; accepted in revised form 25 March 1971*)

Abstract—Evidence is presented for chemical and physical processes which affected trace elements and which may give clues to phenomena that occurred and in some cases are still occurring on and in the Moon. Water-soluble halides are found in rock and soil samples. The distribution of Cl and Br between soluble and insoluble phases in lunar material appears to parallel more closely that found for terrestrial basalts than that in meteorites. Os is depleted in exterior relative to interior samples of rocks, but Ru abundances do not vary. There is an increase in the concentration of surface-related Hg with depth in cores and this increase is relevant in determining the orientation of rocks on the lunar surface. Elemental Hg and volatile compounds of Hg and Os, if present, may be lost from the Moon by a nonthermal mechanism.

Introduction

Fluorine, Cl, Br, and I; Te; the platinum metals, Ru and Os; Li; U; and Hg have been measured in soils and a breccia; in interior and exterior samples of rocks; at various depths in cores; and in some grain fractions from soils. Most of these measurements have been on Apollo 12 samples but some were made on soil samples and on samples from a core from Apollo 11. Cl, Br, and I were determined on the same sample aliquant (50–100 mg) along with Te, U, and Li. F and Br were occasionally measured in samples from which Hg had been removed by volatilization and in which Ru, Os, and U were measured. In our experiments surface-related Hg is volatilized and water-soluble phases are extracted. A regular increase in the amount of surface-related Hg with depth in cores has been observed. This observation makes the behavior of the halogens of interest since they and many of their compounds are volatile. Chemical identification of the water-soluble halogen compounds we detect may prove to be difficult because of the solubility of micrograins of abundant phases. However, comparison of these water-leachable components with those measured in meteorites and terrestrial basalts, in particular, permits speculation as to their origin.

Experimental Data

The details of our neutron- and photon-activation analysis procedures have been reported (Reed and Jovanovic, 1969). Some modifications in timing and in evaluating data permitted better resolution of the problems arising from fission-produced I and Ru isotopes. Ru, previously not reported, was determined directly by measuring 2.9-day ^{97}Ru or indirectly from 40-day ^{103}Ru corrected for the contribution from U fission. The treatment of the I measurements is discussed later. Our standard procedure is to perform all sample handling in a N_2 dry-box.

Table 1. Halogens, tellurium, mercury, uranium, lithium, ruthenium, and osmium contents in lunar samples. portions,

Sample	F (ppm)	Cl (ppm)	Br (ppm)	I (ppb)
Soil				
10084,1		$24_r + 21_l$	$0.035_r + 0.003_l$	$nd_r + 74_l$
10084,1	—			
10084,2, < 44 μm	$254_r + 72_l$*	$21_r + 18_l$	$0.31_r + 0.26_l$	$nd_r + 754_l$
10084,2, 74–150 μm	900*	19	0.55	142
10084,2, 44–74 μm	—			
10084,2, > 150 μm	—			
12070,61	241*	17	0.49	nd
12070,61		$26_r + 20_l$	$0.072_r + 0.019_l$	$nd_r + 35_l$
12070,61, 74–150 μm		25	0.49	nd
12070,61, 1–2 mm		$28_r + 6.8_l$	$0.14_r + 0.1_l$	$nd_r + 58_l$
12070,61, < 74 μm	—			
12070,61, > 150 μm	—			
12033,18	300*	41	0.45	nd
12033,18		$17_r + 2.6_l$	$0.47_r + 0.022_l$	nd
12033,18	—			
12034,23, breccia		$46_r + 4.6_l$	$0.13_r + 0.03_l$	$nd_r + 10_l$
Core				
10004,27, 0.0 cm	225 ± 82			
10004,27, 3.3 cm		38	0.048	nd
10004,27, 9.9 cm		17	0.54	nd
10004,27, 13.2cm	520 ± 255			
12025,74, 1.5–2.5 cm	45		0.034	
12028,215, 20–21 cm	101		0.38	
12028,216, 38–39 cm	—		0.089	
Rock				
12052,18, i‡		$1.9_r + 1.7_l$	$0.019_r + 0.02_l$	$nd_r + 8.7_l$
12052,18, i	18		0.053	
12052,49, e†		$1.4_r + \leq 0.4_l$	$0.016_r + 0.008_l$	$nd_r + 17_l$
12052,49, e	31		0.022	
12053,31, i		6.8	0.017	nd
12053,31, i		$4.4_r + 7.6_l$	$0.019_r + 0.006_l$	$nd_r + 56_l$
12053,31, i	—			
12053,31, i	—			
12053,53, e		9.4	0.015	nd
12053,53, e	—			
12022,109, i		8.7	≤0.039	508
12022,109, i	—			
12022,69, e		$2.6_r + 1.5_l$	$0.022_r + 0.019_l$	$nd_r + 14_l$
12022,69, e	13		0.043	
12021,84, i		$3.3_r + 1.4_l$	nd	$nd_r + 19_l$
12021,84, i	—			
10017,22		$11_r + 2.1_l$	$0.17_r + 0.027_l$	$nd_r + 4.7_l$

Results

Fluorine

Apollo 12 soils 12070 and 12033 have fluorine contents in the 150–300 ppm range found in Apollo 11 soil and rock. Within the rather large errors of the measurements, results on Apollo 11 core samples from depths of 0 cm and 13.3 cm agree with our other Apollo 11 results. This general agreement is not observed for the Apollo 12 core; samples from the 1–2 cm and 20–21 cm depths have only 45 and 101 ppm F,

All data on a line are for a single aliquant; subscripts l and r refer to water extractable and nonextractable respectively.

Te (ppb)	Hg (ppb)	U (ppm)	Li (ppm)	Ru (ppb)	Os (ppb)
—		1.1_r + nd_l	13		
	2.8	1.2		—	13
—		—	$14_r + 0.2_l$		
1410 ± 200		0.65	6.6		
	2.5	—		—	—
	0.016	—		—	—
≤423		1.5	13		
nd		$2.5_r + 0.012_l$	$16_r + 2.4_l$		
≤28		2.6	24		
≤9.8		$1.2_r + 0.013_l$	$34_r + \leq 0.4_l$		
	3.3	—		—	—
	0.9	—		—	—
—		1.6	31		
—		1.5_r + 0.028	$42_r + 0.77_l$		
	2.6	1.5		—	3.6
—		$5.8_r + 0.035_l$	$17_r + \leq 1.6_l$		
	3.0	—		—	8.0
nd		3.2	20		
(<100)		12	18		
	39	1.2		—	24
	10	1.1		71	9.0
	17	2.9		65	3.2
	20	1.6		5.5	11
≤1.8		$0.31_r + 0.0077_l$			
	3.0	—		9.1	0.26
11 ± 5		$0.23_r + 0.01_l$			
	9.0	—		6.6	≤0.10
≤10		0.35	8.7		
≤2.6		$0.21_r + 0.0007_l$	4.6		
	1.6	0.12		5.1	—
	3.8	0.13		2130	3.0
16 ± 8		0.29	8.6		
	12	—		12	0.38
≤1.2		$0.16_r + 0.001_l$	0.92		
	8.0	—		17	1.2
—		$0.22_r + 0.0014_l$	$15_r + \leq 0.33_l$		
	7.6	—		18	0.13
—		$0.19_r + 0.006_l$	$7.5_r + 0.48_l$		
	2.0	(0.2)		(25)	0.95
117 ± 60		$0.23_r + 0.2_l$	$16_r + \leq 0.3_l$		

nd = not detected; * separate aliquant; ( ) uncertain results; † exterior; ‡ interior; — not determined; § Errors in the Te contents range as high as 50%.

respectively. This departure is further exaggerated in rocks 12052 and 12022; they have F contents of 18, 31, and 13 ppm, respectively. The new results are listed in Table 1.

Smales *et al.* (1970) reported a lower F content in Apollo 11 soil than we observed. This discrepancy could be and probably is a sampling variation. In order to check our own results, we measured the F content of W–1. We found 187 ppm, which is lower, but not unreasonably so, than the 200–300 ppm expected (REED and JOVANOVIC, 1969).

Discrepancies also appear in the results for F in Apollo 12. We, along with SMALES *et al.* (1971), find low F contents in rocks but in this case the SMALES *et al.* (1971) value of 40 ppm for a sample of rock 12022 is higher than our 13 ppm. BOUCHET *et al.* (1971) measured different rocks as well as soil 12070 and found 78–150 ppm F in the rocks compared with 63 in the soil. It does not seem likely that the discrepancies between our results and those of BOUCHET *et al.* (1971) can be due entirely to sampling. These investigators' results on Cl and Br in rocks, discussed later, also are higher than ours; the differences appear to be systematic.

Most reported P_2O_5 contents in Apollo 12 soils are about 0.3–0.4%, but those in rocks are near 0.08% (ROSE *et al.*, 1971; ENGEL and ENGEL, 1971; MORRISON *et al.*, 1971; MAXWELL and WIIK, 1971; and WILLIS *et al.*, 1971). Our F data for soils 12070 and 12033 give F/P_2O_5 ratios consistent with those for fluorapatite. The rocks, on the other hand, have lower F/P_2O_5 ratios. The KREEP component which MEYER *et al.* (1971) think constitute as much as 30% of 12070 and 65% of 12033 soils contains a REE-rich phosphate which is not fluorapatite. Thus accounting for most of the phosphorous as fluorapatite does not appear to be justified. Instead it may be necessary to attribute some of the F to another mineral.

Chlorine

Chlorine contents average 39 ppm in Apollo 12 soils and breccia and ~6 ppm in rocks; there is no overlap of rock and soil results (Table 1). With a single exception, our reported Cl concentrations in Apollo 11 samples fall between 8 and 16 ppm. Two new Apollo 11 soil sample measurements fall outside this range, those on 10084,1 and on the 3.3-cm deep sample from core 10004 (Table 1).

BRUNFELT *et al.* (1971) also observed a higher Cl concentration in soil than in rocks. Two other groups, WÄNKE *et al.* (1971) and BOUCHET *et al.* (1971) found Cl contents in rocks as high as those in the soil. There is reasonable agreement between all the groups on the Cl content of the 12070 soil.

Bromine

Bromine contents in Apollo 12 rocks fall in a concentration range of 0.017–0.042 ppm. The soil sample Br contents spread between 0.09–0.5 ppm. The sampled rocks certainly cannot supply the higher Br contents observed in soils. The soil contents are either derived from other rocks or from an outside source such as meteorites. Meteoritic Cl contents appear to be as variable as Br contents (REED, 1971) so one might expect that if Cl and Br are derived from meteorites, the Cl in soils will show as great a spread as the Br; but the spread in the Cl results is smaller (17–50 ppm) than that of Br. Some data on water soluble Br discussed below are also pertinent to these considerations.

The value of 0.09 ppm Br for a 12070 aliquant agrees with results of ANDERS *et al.* (1971) and BOUCHET *et al.* (1971). However they do not report values as high (~0.49 ppm) as we found in another aliquant of 12070 and in 12033. The former group also found less Br in rocks than in soil whereas the latter did not; however, different rocks were measured.

Iodine

Neutron activation of lunar soil and rocks leads to the production of copious amounts of fission I radioactivities with half-lives of 52 minutes to 8 days. The 25 min ^{128}I radioactivity produced from natural iodine must be resolved from these radioactivities. Fortunately, when the samples were leached with hot water, a separation of I radioactivities occurred.

Sample and hot H_2O —
- Leach: 25 min—^{128}I
  - 2.3 hr—^{132}I
  - Small amt. of longer-lived (~15 hr) decay
- Residue dissolved by peroxide fusion:
  - 52 min—^{134}I
  - 15 hr—^{133}I and ^{135}I
  - 8 day—^{131}I

This separation is probably due to the difference in oxidation states of ^{128}I and ^{132}I and the higher-mass fission-produced ^{133}I and ^{135}I (Hall and Walton, 1961).

We examined a few cases in which the amount of ^{128}I activity found in the leach should have been resolvable, if present, from a 52-min ^{134}I component in the residue. None was detected; thus, less than this amount of ^{128}I activity could have been present in the residue.

The most reliable data on natural iodine are those determined in leach solutions and these results, listed in Table 1, should be considered lower limits. In a few instances the iodine in the total sample was observed (12022,109 and the 74–150 μm fraction from 10084,2 as well as 10017,22 and 10072,24 reported previously by Reed and Jovanovic, 1970). No case for enrichment of iodine in soil and breccia over that in rocks can be made on the basis of the data; both have I contents ranging from about 10 to 60 ppb. These values are within the concentration range (9–68 ppb) reported in ordinary chondrites (Reed, 1971) and somewhat less than that reported for eucrites (Clark *et al.*, 1967).

Alexander *et al.* (1971) have measured I and Br in Apollo 11 rocks 10044 and 10057 by irradiating with neutrons for long periods and then measuring the halogens via rare-gas daughters. This method measures only the gases produced in retentive sites (Hohenberg and Reynolds, 1969 and Alexander *et al.*, 1971). Iodine contents in meteorites were found to be lower by an order of magnitude or more when measured via neutron-capture derived ^{128}Xe than when measured by techniques similar to ours.

Accepting the leachable I as an estimate of the I contents in lunar samples, we note that these and the contents of < 1 ppb reported by Alexander *et al.* (1971) parallel the pattern in meteorites. A similar trend is noted in the results for Br and for Kr derived from Br. In this case however Br is not only in water-soluble sites but also at sites near surfaces from which (n, γ)-produced Kr may escape.

Tellurium

Because of the high U content of lunar samples, the amount of ^{131}I contributed by ^{130}Te is largely masked by fission produced ^{131}I. No ^{131}I radioactivity was detected in the leach solutions. Our counting technique (β-γ coincidence) could detect less than one ppb Te as ^{130}Te. As can be seen in Table 1, most Te contents are expressed as limits and in the few cases where numbers can be reported, errors of 50% apply. Apollo 11 and 12 soils and rocks generally contain between 1 and 30 ppb Te with a couple of exceptions (10017,22 and the 74–150 μm fraction of 10084,2).

Te contents of $\sim$10 ppb and 100 ppb are reported by ANDERS *et al.* (1971) in three rocks and in 12070 soil, respectively.

Te contents in ordinary chondrites are several hundred ppb and in carbonaceous chondrites, several ppm (GOLES and ANDERS, 1962; REED and ALLEN, 1966). Ca-rich achondrites contain 100–300 ppb Te (CLARK *et al.*, 1967). Te appears to be depleted in all the lunar samples relative to meteorites with the possible exception of 10084,2 soil.

Interhalogen comparisons

Both F and Cl are lower in Apollo 12 rocks than in the soil but the data are too sparse to permit more than the most tentative speculation. The F/Cl ratio is 11 (spread of 5–17) for the two soil samples and rock 12052; rock 12022 has a lower ratio of $\sim$3.2. BOUCHET *et al.* (1971) also get F/Cl ratios near 3 in rocks but the F and Cl contents they report are factors of 2–6 higher than we find. Most Apollo 11 rocks and soil have constant Cl and F contents and a F/Cl ratio of 22. Basaltic terrestrial rocks have lower ratios (1–8) (HUANG and JOHNS, 1967; GREENLAND and LOVERING, 1966) as do basaltic achondrites (2–4) and ordinary and carbonaceous chondrites (0.7–2) (REED, 1971). The authors above noted an increase in the F/Cl ratio in rocks which were late differentiates; on this basis Apollo 12 rock and soil represent earlier differentiates than does Apollo 11 material. This designation is consistent with the LSPET (1970) characterization of Apollo 12 rocks as being more mafic than Apollo 11. In this regard, total halogen contents as well as those of other strongly lithophilic elements such as U and Li (Table 1; see also LSPET (1970) for other examples) are lower in Apollo 12 rocks than in the soil or in Apollo 11 rock and soil. Thus, Apollo 12 rocks are more basic than the soil which must then have been in part derived from another source; a component such as KREEP, enriched in strongly lithophilic elements must have contributed to the soil at Procellarum.

The Cl/Br ratios vary. It can be seen in Table 1 that although Apollo 12 samples tend to have higher Cl/Br ratios (100–500) than do Apollo 11 samples ($\sim$50), sample 10084,1 has an even higher ratio (1200) and an aliquant of 12070,61 has a low ratio (34). Even though BOUCHET *et al.* (1971) report larger Cl and Br contents than we find their Cl/Br ratios in Apollo 12 samples are 250–430. So far there is little justification for inferring a Cl–Br correlation in lunar samples on the basis of the total elemental contents.

A significant fraction of the halogens in lunar samples is present as water-soluble compounds. These compounds are found in the rocks and soil and are not necessarily surface correlated as might be expected if they were derived from a volatile phase such as volcanic emanations or impact-produced vapors. A cursory examination of the data for Cl and Br in Table 1 reveals that the <44 μm—10084,2 and 1–2 mm—12070,61 samples have soluble Cl fractions consistent with deposition from a vapor but the Br fractions do not follow this trend. On the other hand, samples 10084,1 and 12070,61 have about the same contents and fractions of soluble Cl as does the fine grained 44 μm—10084,2 sample. The same applies to the coarse grained rock 12021 as compared with some of the finer grained ones. Furthermore, the rocks contain as large a fraction of soluble Cl and Br as do the soils.

The cations with which the soluble halogens are associated cannot be identified. The anticipated masking effect is caused by the solubility of cations, especially Ca but also Mg and Al, from major phases as reported by MASON *et al.* (1970, 1971) and KELLER and HUANG (1971). In a series of heating experiments on rock 12022, DERBY *et al.* (1971) found no measurable amounts of halogens accompanying the alkali metals, Na and K. They report that the K is surficial and present on basaltic fines and on sample surfaces. They later report that Cl is present in material evaporated at temperatures above 850°C but a number of cations were also evaporated.

We shall examine the processes that could have occurred to see if the source(s) of the soluble component can be inferred. Magmatic emanations and impact-produced volatiles have been mentioned. Volatile transfer has been suggested by SILVER (1970, 1971) and by HUEY *et al.* (1971) as a mechanism for introducing an isotopically extraneous Pb and possibly other volatiles into the regolith. Another source might be associated with the magmatic fluids that caused the extensive vesiculation observed in many lunar igneous rocks; still another is a refining process, caused by slow cooling, which could leave grains coated with exsolved salts.

Besides the grain size test applied above for the presence of condensible vapors, the interior and exterior surfaces of rocks should contain different amounts of condensed salts, unless the process occurred at depths where higher temperatures cause increased mobility. Rock 12052 appears to have a relatively higher content of leachable Cl and Br in the interior sample. The evidence is fragmentary but seems to suggest no special surface correlated concentration effect.

When the soluble Cl/soluble Br ratio is examined, it is found that most samples, soil and rock, cluster in a group with a mean ratio of about 68. Three samples give ratios which are significantly higher (1000–7000); they are 10084,1; 12070,61; and 12053,31.

An entirely different distribution is found in the insoluble Cl and Br. All Apollo 11 samples have a ratio, insoluble Cl/insoluble Br, between 25 and 100 whereas this ratio in Apollo 12 samples ranges from 100 to 350. These observations indicate that the conditions of formation of the insoluble Br and Cl containing phases were different at the two landing sites, but that the water soluble Br and Cl may be related and may have a common origin.

An approach to understanding what this origin may have been is to compare the relative amounts, [insoluble]/[soluble], of Cl and Br in lunar materials with those in meteorites, chondrites, and achondrites, and terrestrial basalts. If magmatic processes caused the distribution between water soluble and insoluble phases, then similar patterns may indicate similar conditions. The histograms in Fig. 1 summarize the data we have available. The meteorite data are restricted to ordinary and types I and II carbonaceous chondrites and to the Ca-rich achondrites and are from REED (1971);

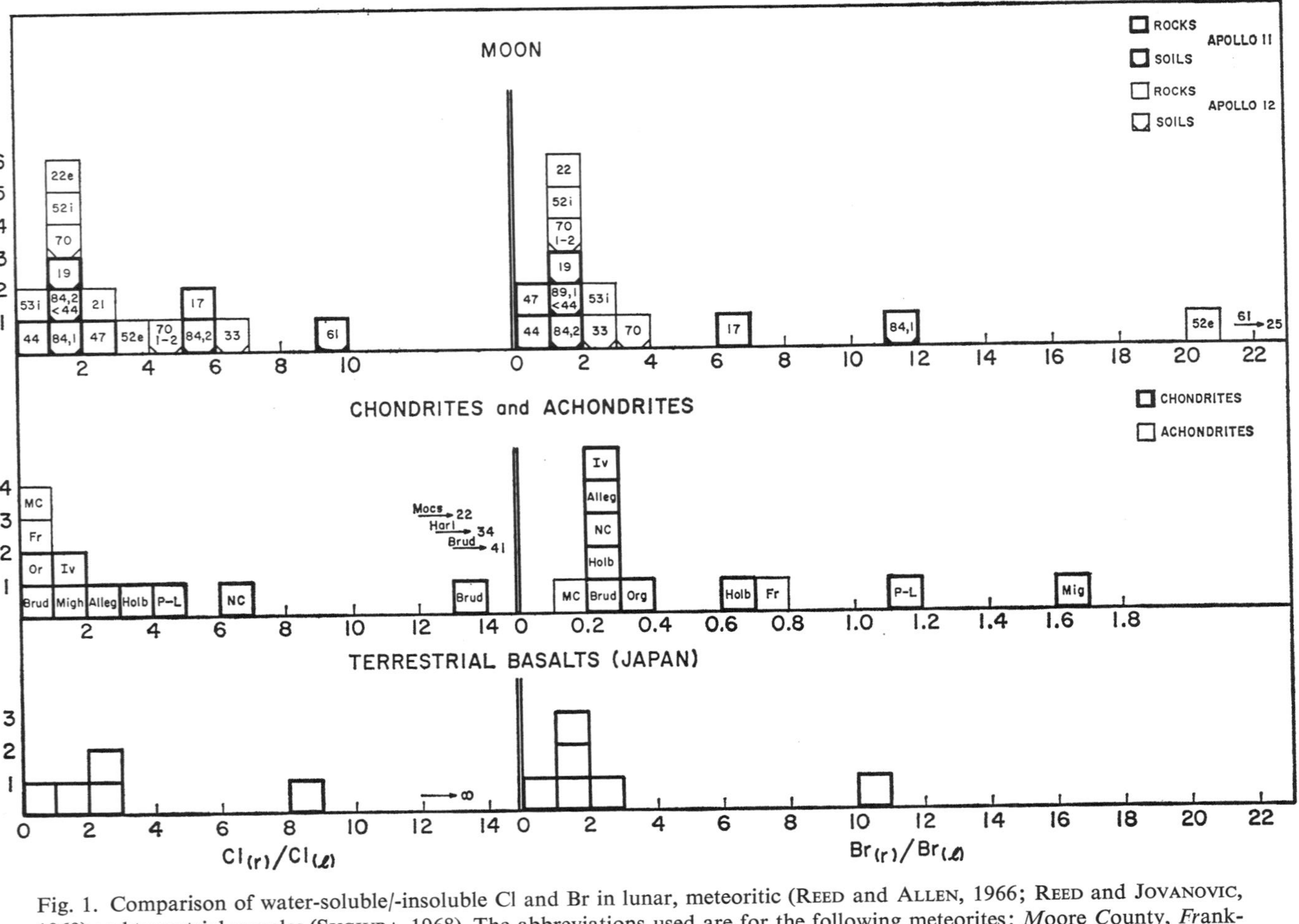

Fig. 1. Comparison of water-soluble/-insoluble Cl and Br in lunar, meteoritic (Reed and Allen, 1966; Reed and Jovanovic, 1969) and terrestrial samples (Sugiura, 1968). The abbreviations used are for the following meteorites: *M*oore *C*ounty, *Fr*ankfurt, *Or*gueil, *Brud*erheim, *Iv*una, *Migh*ei, *Alleg*an, *Holb*rook, *P*antar-*L*, *N*ew *C*oncord, *Harl*eton.

the terrestrial basalt data are from SUGIURA (1968). More data on achondrites and terrestrial basalts are clearly needed but even with the available data, a close parallel appears between the lunar and terrestrial samples. The meteoritic ratios for Br are an order of magnitude lower than those found for the lunar and terrestrial samples.

Terrestrial magmas are wet and the presence of water soluble compounds is not surprising. These compounds may also be accounted for by subsequent contamination of the rocks by ground water. A dry moon will require different explanations.

Uranium

The uranium contents of Apollo 11 rocks and soil are remarkably uniform with most values falling between 0.4 and 0.8 ppm. A factor-of-2 uncertainty due to sampling can account for any differences reported. In contrast the Apollo 12 rocks and soil have very different U contents. Our results on soil and breccia give U contents of 1.5 to 6 ppm and on rocks, 0.16 to 0.35 ppm. These values are consistent with those reported by LSPET (1970). Our most interesting U data, however, are those for *Apollo 11* core, 10004,27 and a separate soil sample, 10084,1. Samples from 3.3, 9.9, and 13.2 cm depths have U contents of 3.2, 12, and 1.2 ppm, respectively. Soil sample 10084,1 contains 1.2 ppm U. These are all higher than any other reported Tranquillity Base U results.

Because of these apparently disparate results we examine our data in the light of results reported by others. We have determined U on 50–100 mg samples by neutron activation, measuring in some experiments fission I and in others fission Ba. Replicate determinations were made on most samples and generally a spread in contents was found. The best value is probably the weighted average. Listed below are averages for the U results listed in Table 1 and the ranges of values reported by other investigators (BOUCHET *et al.*, 1971; BRUNFELT *et al.*, 1971; CLIFF *et al.*, 1971; FIELDS *et al.*, 1971; MORRISON *et al.*, 1971; O'KELLEY *et al.*, 1971; RANCITELLI *et al.*, 1971; ROSHOLT and TATSUMOTO, 1971; SILVER, 1971; TAYLOR *et al.*, 1971; and WÄNKE *et al.*, 1971).

Sample	12033	12034	12070	12052	12053	12021	12022
This work	1.5	5.8	2.0	0.28	0.22	0.20	0.19
Other work	2.67, 3.27	3.4–3.6	1.0–2.1	0.27–0.59	0.24–0.32	0.26, 0.26	0.34, 0.198

The greatest discrepancies are between the results on 12033 and 12034. Our values for 12033, based on both fission I and on fission Ba, are particularly puzzling since this soil contains a large fraction of KREEP component and would be expected to have a higher U content.

Our results on Apollo 11 core and soil 10084,1 are also outside the range of typical results reported by ourselves and others. Results on one of the 10084,1 samples and the 13.2 cm core sample are based on fission Ba. WAKITA *et al.* (1970) have reported constant Th contents (0.8 $\pm$ 0.1 ppm) in samples from core 10005 which was taken about 3 meters from core 10004. Because these Th contents appear to be too low (factor of 3), WAKITA *et al.* (1970) consider their core-sample Th data reliable only on a relative basis. Using a Th/U ratio of $\sim$4 one might expect $\sim$0.2 ppm U if the Th contents are reliable. Even if the U content is 0.55 ppm as reported for soil 10084, values as high as 3.2 and 12 ppm in two of the 10004 core samples appear to be outside of any expected sampling uncertainty. However, extreme sampling errors may arise from the presence of minor phases such as phosphate and zirconium minerals that are high in U (BURNETT *et al.*, 1971; FRICK *et al.*, 1971). As noted, the WAKITA *et al.* (1971) samples are from core 10005 and ours from 10004. The latter, identified as core 2, is faintly stratified according to FRYXELL *et al.* (1970). This core was taken nearer the rim of a small crater (Apollo 11 Preliminary Scientific Report, 1969) and therefore could be sampling an ejecta blanket which may contain a KREEP-type component. Core 10004 then may be similar to such samples as 12033 and 12034 which are from a 15 cm deep trench on the rim of Head Crater.

MEYER *et al.* (1971) have measured the distribution of KREEP glasses in the 12025,28 double core. It is of interest to compare the abundances of the elements we have measured with the KREEP content. The three strongly lithophilic elements F, Br, and U exhibit a regular pattern with the 20–21 cm deep fraction containing the largest contents of these elements. This section contains a larger relative abundance of KREEP glass than does the 38–39 cm deep fraction. Two sections adjacent to our 1.5–2.5 cm deep fraction were low in this component. This correlation does not seem to hold for Cl, Br, and U in soils 12070 and 12033 (Table 1). The former contains only ~25% KREEP component whereas the latter contains ~65% (MEYER *et al.*, 1971). Comparisons with trends or lack thereof found by the other investigators (ALBEE *et al.*, 1971; PHILPOTTS *et al.*, 1971; ANDERS *et al.*, 1971) are not necessarily valid since in general the same core sections were not sampled.

Lithium

LSPET (1970) reports Li contents averaging 5.5 ppm for rocks and 11–25 ppm for soil and breccia. Li contents in our rocks are about 8 ppm. Soil 12070 and breccia 12034 average about 16 ppm Li; soil 12033 has more than twice as much.

Ruthenium-osmium

The only Apollo 12 soil data are those from cores 12025, 12028. There is a large variation, with an average concentration of about 68 ppb Ru for two samples from the upper half of the core and 5.5 ppb Ru for a sample near the bottom. The Ru contents in rocks range from 5 to 18 ppb with one sample having 2.1 *ppm* Ru. This same sample had the highest Os of any of the Apollo 12 rocks (Table 1).

Osmium contents in the soil average ~7 ppb, those in the interior of the rocks, ~1 ppb. However, the Os contents in the exterior rock samples are lower than those in the interior by factors of 2.5 to 10. This loss of Os from surfaces of rocks is surprising. Significantly, Ru, which is chemically similar to Os, was not lost. Both elements form volatile oxides and halides, however those of Os are more stable (BREWER, 1953; HEPWORTH and ROBINSON, 1957; CLAASSEN *et al.*, 1961).

Ru (NICHIPORUK, 1971) and Os (MORGAN, 1971) have about the same cosmic abundance (1.0 and 0.6 ppm, respectively). The Ru/Os ratio in the soil is about 10/1. If the Moon accreted with a Ru/Os ratio of about unity, then the Os in the soil has been fractionated from the Ru.

The Ru/Os ratio in the rocks (for interior samples in the case of Os) is about 10/1. That this is the same ratio found in the soil may be a coincidence.

On closer examination it appears that any assumption of a fractionation between Ru and Os may be fallacious. Comparison of the abundance ratios of Ir, Re, and Os in lunar samples and chondritic meteorites reveals that fractionation between these elements apparently did not occur (Table 2). If meteorite abundance ratios of Os/Re and Ir/Os hold, instead of depletion of Os from the exterior of rocks it appears that Os has become enriched in the interior of rocks (note the data for rocks 12053 and 12021 in columns 3 and 4 of Table 2). Similarly soil, 12070, and all the rocks appear to have been enriched in Ru, since the Os contents of the soil and the exteriors of rocks give the same ratios to Re and Ir as are found in chondrites.

The relative enrichment of both Os and Ru in the soil vs. the rocks of $\sim 5/1$ is consistent with larger concentrations of Re (HERR *et al.*, 1971), Ir (ANDERS *et al.*, 1971) and Os and Re (Apollo 11, LOVERING and BUTTERFIELD, 1970) in soil relative to rocks.

The possible loss of Os from the exterior of the rocks may be explained by the formation of volatile compounds. We have noted (REED *et al.*, 1971) that some of the Hg on the Moon is labile and could be lost by the Manka-Michel mechanism (see below). This argument may be applied to the loss of Os from the Moon. However, this mechanism does not explain the Ru/Os ratio inside the rocks nor the depletion by a factor of 5 of both elements in rocks relative to soil.

Mercury

The contents of Hg in Apollo 12 soils range from 2 to 20 ppb and in rocks from 1.6 to 12 ppb. A gradient with depth in surface-related Hg concentrations (i.e., Hg that is volatilized from samples at a temperature of 130°C or less) was observed in cores (REED *et al.*, 1971). It was concluded that some Hg is labile and can be volatilized from the surface during lunar daytime. If this occurs, it may escape from the Moon by the MANKA-MICHEL (1970) mechanism. Should such escape occur, this Hg must be replenished from an interior reservoir or from outside by the solar wind and/or meteorites. The increasing amount of surface-related Hg with depth suggests that a cold trap exists below the surface. The orientation of rocks on the lunar surface may

Table 2. Ir, Os, Ru, and Re in lunar samples.*

Sample	Ir/Re	Os/Re	Ir/Os	Ru/Os
12033	10	10	1.0	
12070	11	12	0.9	[9(12025/28)]
12053	7	91 (int. Os)†	0.08 (int. Os)	3 (int. Os)
		9 (ext. Os)	0.63 (ext. Os)	32 (ext. Os)
12022		37 (int. Os)		14 (int. Os)
		3.4 (ext. Os)		138 (ext. Os)
12021		11 (int.)		26 (int.)
Chondrites*	7	12	0.6	1.7

* Ir: ANDERS *et al.* (1971), BAEDECKER and EHMANN (1965); Re: HERR *et al.* (1971), MORGAN and LOVERING (1964); Os: LOVERING and BUTTERFIELD (1970).

† int. = interior sample, ext. = exterior sample.

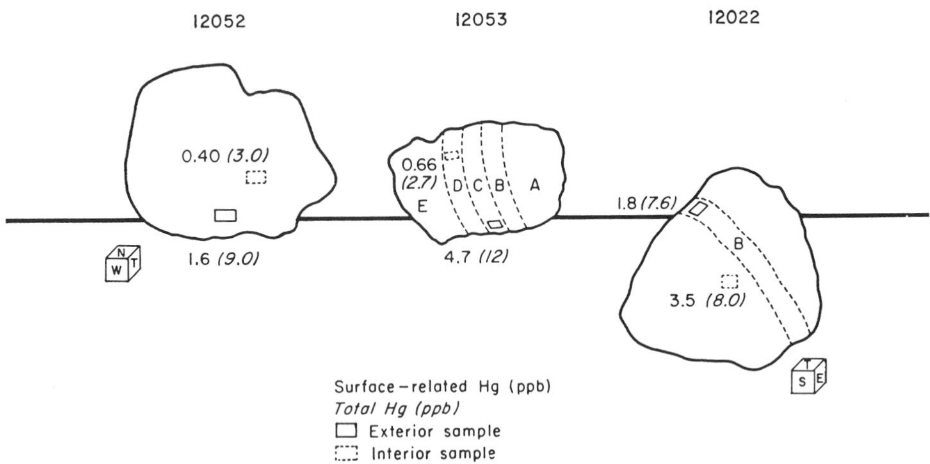

Fig. 2. Orientation of rocks 12052, 12053, and 12022 on the lunar surface based on surface-related Hg concentrations. Total Hg concentrations are given in parentheses. Orientations are as given in the Lunar Receiving Laboratory's cutting drawings of the samples.

be deduced from measurements of surface-related Hg and is illustrated for rocks 12022, 12052, and 12053 in Fig. 2.

New information reported at the Second Lunar Science Conference necessitates a reexamination of the sources and manner of trapping of volatiles in the Moon. MARTI and LUGMAIR (1971) report an increase in ^{40}Ar with depth in 12025,28 core samples. A few other observations that may bear on these problems are: (1) BLOCK *et al.* (1971) report that solar wind gases have diffused to greater than penetration depth. (2) This diffusion could be facilitated by the abundance of microcracks deduced from thermal expansion experiments by BALDRIDGE and SIMMONS (1971). (3) The presence of trapped inert gases in such cracks and other voids is established by the crushing experiments of HEYMANN and YANIV (1971), who observed copious amounts (50%) of trapped heavy rare gases released.

We have reported that surface-related Hg concentrations increase with depth in cores. The *total* Hg extracted from the samples follows this same trend. In fact the total Hg found in rocks and soil seems to be correlated with the Hg on surfaces. The intermediate temperature fractions, 130°–325°C and 130°–475°C, also show a correlation with the <130°C fraction. [Hg is extracted from samples by stepwise heating (REED and JOVANOVIC, 1967) usually at temperatures of 75°, 130°, 175°, 225°, 275°, 325°, 475°, 625°, and 1200°C.] These fractions constitute most of the total Hg. The 325°–475°C and the 475°–1200°C fractions do not show this correlation. The correlation is only approximate as might be expected for material of varying composition, grain size, and porosity. The type and abundance of sites emptied at each temperature will vary from sample to sample. However, an equilibrium in the Hg distribution between the sites seems to be a reasonable assumption. Hg released at the highest temperatures, >450°C, may not have equilibrated since some of this may be originally trapped Hg. In a strict sense all the Hg released below 450°C may be surface related, but only that released below 130°C is in readily emptied sites.

Whereas implantation of volatiles in surface material may be the mechanism for trapping atmospheric and solar wind gases this cannot explain the increase in ^{40}Ar or Hg with depth. These volatiles must diffuse into or out of the subsurface regions where they are concentrated. The presence of Kr and Xe with apparently primordial isotopic compositions in lunar anorthosite and not in lunar basalt led HEYMANN and YANIV (1971) to conclude that these noble gas samples were indeed primordial. Hence an origin of other volatiles within the Moon is probable.

Acknowledgments—We thank the Chemistry Department and the operating staff of the High Flux Reactor, Brookhaven National Laboratory, for making their facilities available. We are especially grateful to J. Hudis and Mrs. E. Rowland for their assistance. The cooperation of the Argonne National Laboratory Linac staff is greatly appreciated. Fruitful discussions with L. Glendenin and K. Flynn are gratefully acknowledged. The assistance of L. Fuchs and K. Jensen is much appreciated. Based on work supported by the U.S. Atomic Energy Commission and by NASA Contract T-76356.

REFERENCES

ALBEE A. L., BURNETT D. S., CHODOS A. A., HAINES E. L., HUNEKE J. C., PODOSEK F. A., PAPANASTASSIOU D. A., PRICE RUSS G., TERA F., and WASSERBURG G. L. (1971) The irradiation history of lunar samples. Second Lunar Science Conference (unpublished proceedings).

ALEXANDER E. C., DAVIS P. K., KAISER W. A., LEWIS R. S., and REYNOLDS J. H. (1971) Some specific rare gas studies in Apollo samples. Second Lunar Science Conference (unpublished proceedings).

ANDERS E., LAUL J. C., KEAYS R. R., GANAPATHY R., and MORGAN J. W. (1971) Elements depleted on lunar samples: Implications for origin of moon and meteorite influx rate. Second Lunar Science Conference (unpublished proceedings).

BALDRIDGE S. and SIMMONS G. (1971) Thermal expansion of lunar rocks. Second Lunar Science Conference (unpublished proceedings).

BAEDECKER P. A. and EHMANN W. D. (1965) The distribution of some noble metals in meteorites and natural materials, *Geochim. Cosmochim. Acta* **29**, 329–342.

BLOCK B., FECHTIG H., FUNKHOUSER J., GENTNER W., JESSBERGER E., KIRSTEN T., MÜLLER O., NEUBUM G., SCHNEIDER E., STEINBRUNN F., and ZÄHRINGER J. (1971) Active and inert gases in lunar material released by crushing at room temperature and by heating at low temperatures. Second Lunar Science Conference (unpublished proceedings).

BOUCHET M., KAPLAN G., VOUDON A., and BERTOLETTE M. (1971) Spark mass spectrometric analysis of major and minor elements. Second Lunar Science Conference (unpublished proceedings).

BREWER L. (1953) Thermodynamic properties of oxides. *Chem. Rev.* **52**, 1–25.

BRUNFELT A. O., HEIER K. S., and STEINNES E. (1971) Determination of 40 elements in Apollo 12 materials by neutron activation analysis. Second Lunar Science Conference (unpublished proceedings).

BURNETT D., MORRISON M., SEITZ M., WALKER R., and YUHAS D. (1971) U–Th distributions and fission track dating of lunar samples. Second Lunar Science Conference (unpublished proceedings).

CLAASSEN H. H., SELIG H., MALM J. G., CHERNICK C. L., and WEINSTOCK B. (1961) Ruthenium hexafluoride. *J. Amer. Chem. Soc.* **83**, 2390–2391.

CLARK R. S., ROWE M. W., GANAPATHY R., and KURODA P. K. (1967) Iodine, uranium, and tellurium contents in meteorites. *Geochim. Cosmochim. Acta* **31**, 1605–1613.

CLIFF R. A., LEE-HU C., and WETHERILL G. W. (1971) Rb–Sr and U, Th–Pb measurements on Apollo 12 material. Second Lunar Science Conference (unpublished proceedings).

Derby J. V., Lewis V. A., Hale D., Legrone H., and Naughton, S. S. (1971) Investigation of lunar erosion by volatilized alkalies. Second Lunar Science Conference (unpublished proceedings).

Engel C. G. and Engel A. E. J. (1971) Major element composition of three Apollo 12 rocks and some petrogenic considerations. Second Lunar Science Conference (unpublished proceedings).

Fields P. R., Diamond H., Metta D. N., Stevens C. M., and Rokop, D. J. (1971) Isotopic abundances of actinide elements in Apollo 12 samples. Second Lunar Science Conference (unpublished proceedings).

Frick C., Hughes T. C., Lovering J. F., Reid A. F., Ware N. G., and Wark, D. A. (1971) Fission track and activation analysis of lunar samples. Second Lunar Science Conference (unpublished proceedings).

Fryxell R., Anderson D., Carrier D., Greenwood W., and Heiken G. (1970) Apollo 11 drive-tube core samples: An initial physical analysis of lunar surface sediment. *Science* **167**, 734–736.

Goles G. G. and Anders E. (1962) Abundances of iodine, tellurium, and uranium in meteorites, I. *Geochim. Cosmochim. Acta* **26**, 723–737.

Greenland L. and Lovering J. F. (1966) Fractionation of fluorine, chlorine, and other trace elements during differentiation of a tholeiitic magma. *Geochim. Cosmochim. Acta* **30**, 963–982.

Hall D. and Walton G. N. (1961) Chemical effects in fission product recoil, recoil-V. *J. Inorg. Nucl. Chem.* **19**, 16–26.

Hepworth M. A. and Robinson P. L. (1957) Simple and complex oxyhalides of Ru and Os. *J. Inorg. Nucl. Chem.* **4**, 24–29.

Herr W., Herpers J., Michel R., Rassoul A. A., and Woelple R. (1971) Spallogenic ^{53}Mn ($T = 2 \times 10^6$ yr) and search for the isotopic anomalies in lunar surface material by means of neutron bombardment. Second Lunar Science Conference (unpublished proceedings).

Heymann D. and Yaniv A. (1971) Inert gases from Apollo 11 and Apollo 12 fines. Second Lunar Science Conference (unpublished proceedings).

Hohenberg C. M., and Reynolds J. H. (1969) Preservation of the iodine-xenon record in meteorites. *J. Geophys. Res.* **74**, 6679–6683.

Huang W. H., and Johns W. D. (1967) The chlorine and fluorine contents of geochemical standards. *Geochim. Cosmochim. Acta* **31**, 597–602.

Huey J. M., Ihochi H., Black L. P., Ostic R. G., and Kohman T. P. (1971) Lead isotopes and volatile transfer in the lunar soil. Second Lunar Science Conference (unpublished proceedings).

Keller W. D. and Huang W. H. (1971) Response of Apollo 12 lunar dust to reagents simulative of those in the weathering environment of the earth. Second Lunar Science Conference (unpublished proceedings).

Lovering J. F. and Butterfield D. (1970) Neutron activation analysis of Re and Os in Apollo 11 lunar material. *Proc. Apollo 11 Lunar Sci. Conf., Geochim. Cosmochim. Acta* Suppl. 1, Vol. 2, pp. 1351–1355. Pergamon.

LSPET (Lunar Sample Preliminary Examination Team) (1970) Preliminary examination of lunar samples from Apollo 12. *Science* **167**, 1325–1339.

Manka R. H. and Michel F. C. (1970) Lunar atmosphere as a source of ^{40}Ar and other lunar surface elements. Second Lunar Science Conference (unpublished proceedings).

Marti K. and Lugmair G. W. (1971) ^{81}Kr–Kr and K–^{40}Ar ages, cosmic-ray spallation products and neutron effects in Apollo 11 and Apollo 12 lunar samples. Second Lunar Science Conference (unpublished proceedings).

Mason B., Fredriksson K., Henderson E. P., Jarosewich E., Melson W. G., Towe K. M., and White J. S., Jr. (1970) Mineralogy and petrography of lunar samples. *Proc. Apollo 11 Lunar Sci. Conf., Geochim. Cosmochim. Acta* Suppl. 1, Vol 2, pp. 1685–1717. Pergamon.

Mason B., Melson W. G., Henderson E. P., Jarosewich E., and Nelen, J. (1971) Mineralogy and petrography of some Apollo 12 samples. Second Lunar Science Conference (unpublished proceedings).

Maxwell J. A. and Wiik H. B. (1971) Chemical composition of Apollo 12 lunar samples 12004, 12033, 12051, 12052, and 12065. Second Lunar Science Conference (unpublished proceedings).

MEYER C., JR., AIKEN F. K., BRETT R., MCKAY D. S., and MORRISON D. A. (1971) Rock fragments and glasses rich in K. REE, P in Apollo 12 soils: Their mineralogy and origin. Second Lunar Science Conference (unpublished proceedings).

MORGAN J. W. (1971) "Osmium," Elemental Abundances in Meteorites (B. Mason, ed.) To be published, Gordon and Breach.

MORGAN J. W. and LOVERING J. F. (1964) Rhenium and osmium abundances in stony meteorites. *Science* **144**, 835–836.

MORRISON G. H., GEROD J. T., PORTER N. M., GANGADHARAM E. V., ROTHENBERG A. M., and BURDO, R. A. (1971) Elemental abundances of lunar soil and rocks from Apollo 12. Second Lunar Science Conference (unpublished proceedings).

NICHIPORUK W. (1971) "Ruthenium," Elemental Abundances in Meteoritic Matter (B. Mason, ed.) To be published, Gordon and Breach.

O'KELLEY G. D., ELDRIDGE J. S., SCHONFELD E., and BELL P. R. (1971) Comparative radionuclide concentrations and ages of Apollo 11 and Apollo 12 samples from nondestructive gamma-ray spectrometry. Second Lunar Science Conference (unpublished proceedings).

PHILPOTTS J. A., SCHNETZLER C. C., BOTTINO M. L., and FULLAGAR P. D. (1971) Li, K, Rb, Sr, Ba, and rare earth concentrations and $^{87}Sr/^{86}Sr$ in some Apollo 12 soils, rock, and separated phases. Second Lunar Science Conference (unpublished proceedings).

Preliminary Science Report, Apollo 11 (1969) NASA, SP–214, p. 52.

RANCITELLI L. A., PERKINS R. W., FELIX W. D., and WOGMAN N. A. (1971) Cosmogenic and primordial radionuclide measurements in Apollo 12 lunar samples by nondestructive analysis. Second Lunar Science Conference (unpublished proceedings).

REED G. W., and ALLEN R. O. (1966) Halogens in chondrites. *Geochim. Cosmochim. Acta* **30**, 779–800.

REED G. W., and JOVANOVIC S. (1967) Mercury in chondrites. *J. Geophys. Res.* **72**, 2219–2228.

REED G. W., and JOVANOVIC S. (1969) Some halogen measurements on achondrites. *Earth Planet. Sci. Lett.* **6**, 316–320.

REED G. W., and JOVANOVIC S. (1970) Halogens, mercury, lithium, and osmium in Apollo 11 samples. *Proc. Apollo 11 Lunar Sci. Conf., Geochim. Cosmochim. Acta* Suppl. 1, Vol. 2, pp. 1487–1492. Pergamon.

REED G. W., GOLEB J. A., and JOVANOVIC S. (1971) Surface-related Hg in lunar samples. *Science* **172**, 258–261.

REED G. W. (1971) "Fluorine," "Chlorine," "Bromine," "Iodine," Elemental Abundances in Meteoritic Matter (B. Mason, ed.) To be published, Gordon and Breach.

ROSE H. J., JR., CUTTITTA F., ANNELL C. S., CARRON M. K., CHRISTIAN R. P., DWORNIK E. J., HELZ A. W., and LIGNON D. T., JR. (1971) Semimicroanalysis of Apollo 12 samples. Second Lunar Science Conference (unpublished proceedings).

ROSHOLT J. N., and TATSUMOTO M. (1971) Isotopic composition of thorium and uranium in Apollo 12 samples. Second Lunar Science Conference (unpublished proceedings).

SILVER L. T. (1970) U–Th–Pb isotopes in some Tranquillity Base samples and their implications for lunar history. *Proc. Apollo 11 Lunar Sci. Conf., Geochim. Cosmochim. Acta* Suppl. 1, Vol. 2, pp. 1533–1574. Pergamon.

SILVER L. T. (1971) U–Th–Pb isotope relations in Apollo 11 and Apollo 12 lunar samples. Second Lunar Science Conference (unpublished proceedings).

SMALES A. A., MAPPER D., WEBB M. S. W., WEBSTER R. K., and WILSON J. D. (1970) Elemental composition of lunar surface material. *Science* **167**, 509–512.

SMALES A. A., MAPPER D., WEBB M. S. W., WEBSTER R. K., WILSON J. D., and HISLOP, J. S. (1971) Elemental composition of lunar surface material (Part 2). Second Lunar Science Conference (unpublished proceedings).

SUGIURA T. (1968) Bromine and chlorine ratios in igneous rocks. *Bull. Chem. Soc. Japan* **41**, 1133–1139.

TATSUMOTO M., KNIGHT R. J., and DOE B. R. (1971) U–Th–Pb systematics of Apollo 12 lunar samples. Second Lunar Science Conference (unpublished proceedings).

TAYLOR S. R., KAYE M., GRAHAM A., RUDOWSKI R., and MUIR P. (1971) Trace element chemistry of lunar samples from the Ocean of Storms. Second Lunar Science Conference (unpublished proceedings).

WAKITA H., SCHMITT R. A., and REY P. (1970) Elemental abundances of major, minor, and trace elements in Apollo 11 lunar rocks. *Proc. Apollo 11 Lunar Sci. Conf., Geochim. Cosmochim. Acta* Suppl. 1, Vol. 2, pp. 1685–1717. Pergamon.

WÄNKE H., TESCHKE F., BADDENHAUSEN H., SPETELL B., BALACESU A., QUIJANO-RICO M., JAGOUTZ E., and RIEDER R. (1971) Major and trace elements in Apollo 12 samples and studies on lunar metallic iron particles. Second Lunar Science Conference (unpublished proceedings).

WILLIS J. P., AHRENS L. H., DANCHIN R. V., ERLANK A. J., GURNEY J. J., HOFMEYER P. K., MCCARTHY T. S., and ORREN M. J. (1971) Some interelement relationships between lunar rocks and soils, and stony meteorites. Second Lunar Science Conference (unpublished proceedings).

Proceedings of the Second Lunar Science Conference, Volume 2, pp. 1277–1280.
The M.I.T. Press, 1971.

Multielement neutron activation analysis of trace elements in lunar fines

A. Travesí, J. Palomares, and J. Adrada
Junta de Energía Nuclear, Ciudad Universitaria, Madrid-3 Spain

(*Received 22 February 1971; accepted in revised form 31 March 1971*)

Abstract—Preliminary results of the trace element contents in Apollo 11 and Apollo 12 lunar fines 10084 and 12070 are presented. A multielement neutron activation analysis scheme using radiochemical group separations and high-resolution gamma-ray spectrometry with Ge(Li) detectors has been used. Quantitative data for more than 25 trace elements are in good agreement with previously reported results for the same materials. Upper detection limits for Sb, Au, and Ho are presented. Apollo 12 fines show an enrichment relative to the Apollo 11 fines in K, Th, La, and the light REE (rare earth elements), but Eu and the heavy REE, Yb and Lu, have similar values in both materials. Cr and Co show only slight enrichment. Sc, Zn, and Ga are depleted in Apollo 12 fines relative to the Apollo 11 fines, and Cu and Hf have similar abundance values in both materials.

Introduction

Abundances and distributions of elements in lunar samples are essential to our knowledge of the outer space chemical composition. Comparison of lunar abundances with average abundances in the earth's crust and meteorites are useful for testing lunar models.

Since abundances of many elements are in the range of parts per million (ppm) or less, the analytical methods for cosmic abundance studies are those with high sensitivity and specificity such as NAA (neutron activation analysis) and mass spectrometry. These methods have been used extensively to determine trace element abundances in gelogical materials and meteorites over the past twenty years. Also both methods have been applied by many investigators for analyses of lunar samples obtained from the Apollo 11 and Apollo 12 missions.

The advent of high resolution Ge(Li) detectors for gamma ray spectrometry has provided NAA with the nondestructive capability of multielement analysis. This technique has been applied by Gordon *et al.* (1968) to the analysis of standard rocks and by Goles *et al.* (1970) to the analysis of Apollo 12 lunar samples for the simultaneous determination of up to 26 elements. However, the use of simple chemical group separations permits full utilization of the desirable features of Ge(Li) detectors without the extra difficulties of separating individual elements. This results in considerable improvement in the NAA technique, as demonstrated by Morrison *et al.* (1969) in the analysis of standard rocks and by Morrison *et al.* (1970) in the analysis of lunar samples.

Experimental Techniques

Ground samples of about 0.2 gm were sealed in polyethylene vials and activated for ten hours at a flux of 5×10^{11} neutrons cm^{-2} sec^{-1} in the vertical activation facilities of the JEN-1 reactor. Standards were activated together with the lunar samples under the same activation conditions.

To minimize standard manipulation, counting time and data handling, some standards were combined and activated as a unit. These generally included the elements that were present in each group of the radiochemical separations. A flux monitor of a special Al–Co alloy was attached to sample and standard vials in order to correct for any flux differences during the activation.

After waiting fifteen hours for decay, the samples were dissolved and radiochemical separations into five groups were carried out, according to the procedure described by Morrison *et al.* (1969). Some additional modifications were introduced for the REE groups in order to eliminate the high ^{51}Cr activity which overlapped some activities under the 0.32 MeV Photopeak. The REE were precipitated as oxalates and dissolved again for counting in the normal way as described below.

All the group fractions and standard mixtures containing 40 ml of solution were counted in plastic vials. Counting started twenty hours after the activations, and each group was counted at several different times in order to optimize determinations for isotopes of different half-lives. Counting was continued for one month after activation. Gamma ray spectra were plotted with a Hewlett-Packard X–Y plotter. Punch paper tape output from the analyzer was fed directly into a Hewlett-Packard digital computer HP–2114 B, with an 8K memory. The "AREAS" program written in Fortran provides the exact location of the peaks, their areas corrected for background, and the detection limits according to the criteria of Currie (1968). Computer programs are also used to obtain experimental half-lives from the data, to obtain individual activities in the case of composite peaks, and to obtain the elemental concentration from comparison of the selected photopeak areas with the activity of a standard of known concentration activated and counted under identical conditions.

To ensure complete recovery of each element, the gamma ray spectrum of each chemical group was examined for every element, and in the few cases where an element was found to be distributed into two groups, the combined activities were used to obtain its abundance.

Results

Two samples of Apollo 11 and Apollo 12 lunar fines, 10084 and 12070, have been analyzed by the procedure described above. At present three determinations of the 10084 sample and only two of the 12070 sample have been completed. More determinations are presently under way. Also determinations of the U.S. Geological Survey Rock andesite AGV–1 has been carried out to assess the methods and procedures.

Table 1 presents the abundances for the 25 trace elements. Abundances given are the average values of all the values obtained in the calculations, including different measurements taken at appropriate decay times of every photopeak from the different radionuclides. Whenever possible, several photopeaks and several radionuclides have been used to determine abundances, as indicated in the second column of Table I. Standard deviations (S.D.) of the average abundances indicated in the table have been obtained from either the statistical error of counting or from averaging all the experimental data; the larger error has been listed.

Table 1 also presents for comparative purposes the average abundances of other published results for the same samples. Apollo 11 data are obtained from the compilation of Warner (1971). Apollo 11 fines data are obtained from Baedecker *et al.* (1971), Brunfelt *et al.* (1971), Gast and Hubbard (1971), Goles *et al.* (1971), Haskin *et al.* (1971), Morrison *et al.* (1971), Rose *et al.* (1971), Smales *et al.* (1971), Taylor *et al.* (1971), Wakita *et al.* (1971), and Willis *et al.* (1971). Standard deviations are obtained by averaging all the available experimental data. When only the average elemental abundance is available, as in the case of some data for Apollo 11 fines, the range of the data is included. The last column of Table 1 gives the abundances

Table 1. Abundances of elements in lunar fines

Element determined	Radionuclide measured	Apollo 11 Fines 10084,140 This Work average S.D.	Apollo 11 Fines 10084,140 Other Authors average S.D.	Apollo 12 Fines 12070,81 This Work average S.D.	Apollo 12 Fines 12070,81 Other Authors average S.D.	Relative Apollo 12–Apollo 11
K	K–42	1150 ± 100	1160 (830–1330)	2000 ± 100	2000 ± 120	1.74
Sc	Sc–46	56.8 ± 4.8	59.7 ± 3.5	41.9 ± 2	38.8 ± 2.6	0.68
Cr	Cr–51	2460 ± 170	1985 (1700–2430)	2870 ± 100	2670 ± 320	1.17
Fe%	Fe–59	12.1 ± 0.6	12.25 (11.3–14.0)	12.1 ± 0.5	12.5 ± 0.4	1.0
Co	Co–60	31 ± 2.7	30.2 ± 5.2	41.3 ± 2.7	44.6 ± 3.5	1.33
Cu	Cu–64	9.1 ± 1.5	9.2 ± 1.0	8.9 ± 0.8	8.4 ± 2.9	0.98
Zn	Zn–69^{m}, Zn–65	21.7 ± 2.7	21.2 ± 1.6	~7 ± 3.0	8.1 ± 1.1	0.33
Ga	Ga–72	4.7 ± 0.2	4.8 ± 0.5	2.8 ± 0.4	4.5 ± 0.4	0.60
Sb	Sb–122	<0.06	0.005	<0.05	0.04	—
Hf	Hf–181	10.9 ± 1.2	9.5 ± 1.3	9.8 ± 2.4	11.4 ± 1	0.90
Th	Pa–233	2.3 ± 0.4	2.3 ± 0.2	5.5 ± 0.3	6.0 ± 0.9	2.4
Ta	Ta–182	<2	1.4 ± 0.2	1.1 ± 0.3	2.0 ± 0.9	—
Au	Au–198	<0.04	0.0025 ± 0.001	<0.01	0.005 ± 0.0006	—
La	La–140	15.6 ± 1.2	17.2 ± 2.7	31.2 ± 2.0	34.5 ± 3.2	2.0
Ce	Ce–141	46.8 ± 2.5	48.9 ± 3.9	80.5 ± 3.5	87.9 ± 6.6	1.73
Pr	Pr–142	6.2 ± 2	7.7 ± 2.2	15 ± 3	12.2 ± 0.4	2.42
Sm	Sm–153	12.2 ± 1.4	13.1 ± 0.7	19 ± 1	17 ± 1.4	1.55
Eu	Eu–152^{m}, Eu–152	1.72 ± 0.15	1.84 ± 0.13	1.78 ± 0.11	1.75 ± 0.07	1.03
Gd	Gd–153	17 ± 3	17 ± 2.5	<30	20.2 ± 3.1	—
Tb	Tb–160	2.7 ± 0.7	3.2 ± 0.5	3.3 ± 0.4	3.4 ± 0.6	1.22
Ho	Ho–166	<8	5.4 ± 0.8	<10	4.9 ± 0.5	—
Tm	Tm–170	1.2 ± 0.3	1.5 ± 0.4	<3	1.8 ± 0.4	—
Yb	Yb–175	10.9 ± 1.1	10.8 ± 1.1	11.2 ± 0.6	12.7 ± 1.0	1.03
Lu	Lu–177	1.45 ± 0.1	1.47 ± 0.13	1.5 ± 0.2	1.76 ± 0.15	1.03

of the Apollo 12 fines relative to the Apollo 11 10084 fines. Some interesting compositional differences between the Apollo 11 and Apollo 12 soils may be summarized. Apollo 12 fines are enriched relative to the Apollo 11 fines in many trace elements by factors of 1.5 to 2.5. For example, K, Th, La, and the light REE are enriched, but some REE such as Eu, Yb, and Lu have similar abundances in both materials. The Cu and Hf values are rather similar in both fines.

The relative enrichment factors of 1.17 for Cr and 1.33 for Co (Table 1, last column) show only a slight enrichment for these elements in the Apollo 12 fines. The elements Sc, Zn, and Ga are depleted by factors from 0.7 to 0.3 in the Apollo 12 fines relative to the Apollo 11 fines.

Acknowledgments—We thank the Spanish Comisión Nacional de Investigación del Espacio for their continuous support of this research project, in the forms of both financial aid and collaboration in all aspects of this investigation. All lunar materials were obtained from the National Aeronautics and Space Administration. U.S.G.S. rocks were kindly supplied by the U.S. Geological Survey through the courtesy of F. J. Flanagan and M. Fleisher.

References

Baedecker P. A., Schaudy R., Elzie J., Kimberlin J., and Wasson J. T. (1971) Trace element studies of rocks and soils from Oceanus Procellarum and Mare Tranquillitatis. Second Lunar Science Conference (unpublished proceedings).

Brunfelt A. O., Heier K. S., and Steinnes E. (1971) Determination of 40 elements in Apollo 12 materials by neutron activation analysis. Second Lunar Science Conference (unpublished proceedings).

Currie L. A. (1968) Limits for qualitative detection and quantitative determination, application to radiochemistry. *Anal. Chem.* **40**, 586–593.

Gast P. W. and Hubbard N. J. (1971) Rare earth abundances in soil and rocks from the Ocean of Storms. Second Lunar Science Conference (unpublished proceedings).

Goles G. G., Osawa M., Randle K., Beyer R. L., Jérome D. Y., Lindstrom D. J., Martin M. R., McKay S. M., and Steinborn T. L. (1970) Instrumental neutron activation analysis of lunar specimens. *Science* **167**, 497–499.

Goles G. G., Duncan A. R., Osawa M., Martin M. R., Beyer R. L., Lindstrom D. J., and Randle K. (1971) Analysis of Apollo 12 specimens and a mixing model for Apollo 12 "soils." Second Lunar Science Conference (unpublished proceedings).

Gordon G. E., Randle K., Goles G. G., Corliss J. B., Beeson M. M., and Oxley S. S. (1963) Instrumental activation analysis of standard rocks with high-resolution γ-ray "detectors." *Geochim. Cosmochim. Acta.* **32**, 369–396.

Haskin L. A., Allen R. O., Helmke P. A., Anderson M. R., Korotev R. C., and Sweifel K. A. (1971) Rare earths and other trace elements in Apollo 12 Lunar materials. Second Lunar Science Conference (unpublished proceedings).

Morrison G. H., Gerard J. T., Travesí A., Currie R. L., Petterson S. F., and Potter N. M. (1969) Multielement neutron activation analysis of rock using chemical group separations and high resolution gamma spectroscopy. *Anal. Chem.* **41**, 1633–1637.

Morrison G. H., Gerard J. T., Kashuba A. T., Gangadharan E. V., Rothemberg A. M., Potter N. M., and Miller G. B. (1970) Multielement analysis of lunar soil and rocks. *Science* **167**, 505–507.

Morrison G. H., Gerard J. T., Potter N. M., Gangadharan E. V., Rothemberg A. M., and Bordo R. A. (1971) Elemental abundances of lunar soil and rocks from Apollo 12. Second Lunar Science Conference (unpublished proceedings).

Rose H. J., Jr., Cuttitta F., Annell C. S., Carron M. K., Christian R. P., Dwornik, E. J., Helz A. W., and Ligon D. T., Jr. (1971) Semimicroanalysis of Apollo 12 samples. Second Lunar Science Conference (unpublished proceedings).

Smales A. A., Mapper D., Webb M. S. W., Webster L. K., Wilson J. D., and Hislop J. S. (1970) Elemental composition of lunar surface material, Part 2. Second Lunar Science Conference (unpublished proceedings).

Taylor S. R., Kaye M., Graham A., Rudowski R., and Muir P. (1971) Trace element chemistry of lunar samples from the Ocean of Storms. Second Lunar Science Conference (unpublished proceedings).

Wakita H., Rey P., and Schmitt R. A. (1971) Abundances of the 14 rare earth elements plus 22 major, minor, and trace elements in ten Apollo 12 rock and soil samples. Second Lunar Science Conference (unpublished proceedings).

Warner J. (1971) A summary of Apollo 11 chemical, age and model data. NASA Publication, Curator's Office Manned Spacecraft Center, Houston.

Willis J. P., Ahrens L. H., Danchin R. V., Erlank A. J., Gurney J. J., Hofmeyr P. K., McCarthy T. S., and Orren M. J. (1971) Some interelement relationships between lunar rocks and fines and stony meteorites. Second Lunar Science Conference (unpublished proceedings).

Proceedings of the Second Lunar Science Conference, Volume 2, pp. 1281–1290.
The M.I.T. Press, 1971.

Determination of 40 elements in Apollo 12 materials by neutron activation analysis

A. O. BRUNFELT and K. S. HEIER
Mineralogisk-Geologisk Museum, Sars gt. 1, Oslo 5, Norway
and
E. STEINNES
Institutt for Atomenergi, 2007 Kjeller, Norway

(*Received 22 February 1971; accepted in revised form 30 March 1971*)

Abstract—The abundances of forty elements in four Apollo 12 lunar rocks (12002, 12018, 12021, and 12038) and one soil sample (12070) have been determined by neutron activation analysis. The rock 12038 is compared to eucrites with respect to a possible common origin. Though there are several chemical similarities, the differences are sufficient to demonstrate that the sample is not necessarily of achondritic origin. Elemental abundances as well as alkali element ratios in the lunar samples are discussed and compared with Apollo 11 material, meteorites, and terrestrial rocks. These comparisons show that the lunar rocks analyzed by us have no identical terrestrial or meteoritic counterpart. In many respects the Apollo 12 rocks are intermediate between Apollo 11 material and eucrites.

INTRODUCTION

NEUTRON ACTIVATION analysis has been applied to the determination of 40 elements in five samples of lunar material from the Apollo 12 mission. Four rocks and one soil were analyzed. The following samples were received:

12070,79, 1.03 g; contingency sample, lunar fines.
12002,132, 0.45 g; 12002,133, 0.23 g; 12002,134, 0.19 g;
12002,135, 0.15 g; rock chips.
12018,42, 1.02 g; rock chip.
12021,77, 2.00 g; crushed rock.
12038,59, 1.00 g; rock fines.

EXPERIMENTAL TECHNIQUES

All samples were ground by hand in agate mortars before analysis.

Chips 12002,133, 134, and 135 were ground and analyzed separately. Insufficient material was left of each chip for the determination of As, Se, Rb, Ag, Sb, Cs, Ir, and Au. These elements were therefore determined in a composite of the three chips. Chip 12002,132 was retained and prepared for analysis afterwards in order to check some of the results obtained on the other chips.

The irradiations were performed in the reactor JEEP–II at a thermal neutron flux of about 1.5×10^{13} n cm^{-2} sec^{-1}. Seven different series of analyses have been run, as indicated in the following: (A) 10 mg sample, 30 sec irradiation, thermal neutrons, counting immediately. (B) Samples from (A), 5 min irradiation, thermal neutrons, counting after 1 hr. (C) 100 mg sample, 2 d irradiation, epithermal neutrons, counting after 3–20 d. (D) Samples from (C), 1 d irradiation, thermal neutrons, counting after 6–30 d. (E) 50 mg samples, 15 min irradiation, thermal neutrons, 20 min delay, radiochemical separation of short-lived activities. (F) 100 mg samples, 20 h irradiation, thermal neutrons, 24 h delay, radiochemical separation of medium half-life activities. (G)

Table 1. Elemental abundances determined by

Element	BCR–1	12070 79		12002 132	12002 133	12002 134	12002 135
Na %	2.42	0.319,	0.326		0.153	0.160	0.158
Al %	7.40	6.64,	6.74		4.72	4.28	4.49
Cl ppm	54	21,	27		—	—	17
K ppm	13300	1970,	1890		438	398	420
Sc ppm	30.7	39.3,	38.3	41.7	45.0	42.1	43.9
Ti %	1.29	1.67,	1.75		1.57	1.50	1.65
V ppm	476	143,	142,		223	212	227
Cr ppm	—	2570,	2620		6780	5650	5450
Mn ppm	1419	1730,	1750		2090	2150	2080
Fe %	9.37	12.5,	12.1	16.5	17.1	16.5	16.0
Co ppm	36.2	41.3,	40.2	58.7	59.8	58.3	58.1
Cu ppm	15.7	6.7,	6.7		5.9	6.5	5.3
Zn ppm	127.4	7.1,	7.6		3.4	2.9	7.2
Ga ppm	22.2	4.1,	4.0		3.1	2.9	2.6
As ppm	0.60	0.57,	0.58		←	< 0.05	→
Se ppb	116	239,	253		←	172	→
Rb ppm	50	8.3,	8.1		←	1.14	→
Sr ppm	312	< 100		< 100	←	< 100	→
Ag ppm	0.031	0.14,	0.29		←	0.03	→
In ppb	103	360,	880		—	—	24
Sb ppm	0.60	0.05,	0.03		←	0.07	→
Cs ppm	0.97	0.29,	0.31		←	0.075	→
Ba ppm	580	321,	304	66	58	65	48
La ppm	23.7	33.2,	33.7		5.8	5.6	5.1
Ce ppm	53	84,	90		22	21	19
Sm ppm	6.52	16.1,	15.8		4.38	5.50	4.32
Eu ppm	1.94	1.77,	1.84		0.91	0.82	0.86
Tb ppm	0.96	3.22		0.84	0.93	1.22	0.79
Dy ppm	5.65	22.7,	24.1		7.5	10.6	6.0
Ho ppm	1.20	4.3,	5.2		—	—	—
Er ppm	4.5	13.1,	14.2		—	—	—
Yb ppm	3.21	12.9,	12.6		7.0*	12.8*	5.9*
Lu ppm	0.535	1.90,	1.90		1.11*	2.07*	0.86*
Hf ppm	4.72	13.4,	11.6	2.9	3.3	3.1	3.1
Ta ppm	0.74	1.53,	1.30		0.35	0.37	0.37
W ppm	0.38	0.64,	0.64		0.14	0.29	0.20
Ir ppb	< 0.1	4.36,	4.33		←	< 0.1	→
Au ppb	0.75	4.5,	5.4		←	11	→
Th ppm	4.98	4.6,	4.2	0.60	2.5*	9.3*	2.3*
U ppm	1.68	1.5,	1.6	0.19	2.1*	8.5*	1.7*

* These samples were apparently contaminated with U, Th, and heavy REE during grinding in agate mortars in our laboratory.

200 mg samples, 7 d irradiation, thermal neutrons, 7 d delay, radiochemical separation of long-lived activities.

The samples were always run in duplicate. The right-hand column of Table 1 shows which elements were determined by the various schemes indicated above. A more detailed description of the analytical schemes will be published elsewhere (BRUNFELT AND STEINNES, 1971). For most measurements 20 cc Ge(Li) detectors were used, except for a few cases where the sensitivity was improved by the use of a 3″ × 3″ NaI(Tl) crystal.

The scheme was tested by replicate analysis of BCR–1 before application on lunar rocks, and the mean values are listed in Table 1. During the lunar rock analyses BCR–1 was used as standard except for chromium and some elements with low abundance, where separate standards were used for each element.

neutron activation in Apollo 12 lunar samples

12018 42		12021 77		12038 59		Analytical method
0.151,	0.152	0.188,	0.182	0.466,	0.471	B
4.59,	4.37	5.97,	5.53	6.67	6.65	A
5.1,	5.2	10.5,	12.5	4.7,	8.5	E†
404,	395	486,	490	531,	500	A
39.4,	40.2	54.6,	52.5	44.6,	42.7	D
1.58,	1.64	1.93,	1.93	1.91,	2.04	A
191,	200	192,	217	154,	144	A
3720,	3940	2570,	2650	2040,	2010	D†
2070,	2050	2090,	2090	1870,	1900	B
16.1,	16.2	14.9,	15.3	14.1,	13.8	C
54.2,	55.0	31.7,	31.2	28.4,	28.0	F
5.0,	4.4	8.1,	8.3	16.6,	10.3	F
2.6,	1.9	1.2,	1.2	6.1,	3.2	F
2.6,	2.8	3.7,	3.5	5.1,	5.5	F
<0.05		0.09,	0.18	0.15,	0.10	F
181,	177	226,	221	182,	174	G†
1.42		1.19,	1.30,	0.43	0.44	G†
79,	77	137		137,	152	C
0.01,	0.01	0.10,	0.13	0.08,	0.06	G†
2.6,	<3.3	640,	470	171,	149	E†
0.21,	0.05	0.04,	0.04	0.01,	0.02	G†
0.047,	0.035	0.062,	0.073	0.014,	0.021	G†
65,	61	46,	42	91,	89	C
5.8		5.9,	6.7	11.6,	11.5	D
12,	18	24,	17	24,	30	D
3.90,	3.90	5.32,	5.55	6.93,	6.87	C
0.76,	0.76	1.05,	1.04	2.04,	2.05	D
0.76,	0.74	1.28,	1.21	1.19,	1.19	C
5.6,	5.3	9.1,	9.5	8.8,	8.0	E
1.33		2.32,	2.22	1.48,	1.48	E†
3.2		7.9*,	8.0*	5.0,	4.9	E
3.5,	3.7	10.2*,	9.4*	4.9,	4.5	D
0.61,	0.56	1.65*,	1.48*	0.75,	0.76	D
3.0,	2.8	3.8,	3.3	5.0,	4.8	C
0.36,	0.37	0.43,	0.41	0.59,	0.63	C
0.115,	0.092	0.26,	0.21	0.108,	0.106	F
<0.1		<0.1		<0.1		G†
1.9,	1.7	2.1,	2.2	42,	3.5	G†
0.48,	0.44	5.2*,	3.8*,	0.29,	0.18	C
0.22,	0.17	3.3*,	3.4*	0.14,	0.10	F

† For these elements separate standards were used; for all the other elements the results are relative to BCR–1, using the values listed in the table.

Results and Discussion

The concentrations of 40 elements were determined; the analytical data are given in Table 1. Duplicate values are presented, in order to demonstrate the reproducibility for the determination of each element. In some cases the difference between duplicates is unexpectedly high, indicating inhomogeneities in the samples (Cu, Zn,and Au in 12038, Sb in 12018, Ag and In in 12070 and 12021). For Ag and In this might be associated with contamination, as mentioned below.

In Table 2, a comparison between the data published for Na, K, Rb, Sr, Ba, and eight REE by Gast and Hubbard (1970), Zn, Ga, Se, Rb, Ag, In, Cs, Ir, and Au by Ganapathy *et al.* (1970a), and those reported in this work for the same elements in

Table 2. Comparison of data reported in this work with those published by Ganapathy *et al.* (1970) and Gast and Hubbard (1970).

	12070 soil			12002 basalt		
Element	Ganapathy *et al.* (1970)	Gast and Hubbard (1970)	This work	Ganapathy *et al.* (1970)	Gast and Hubbard (1970)	This work*
Na %		0.34	0.323		0.16	0.153–0.160
K ppm		2150	1930		476	389–438
Ti %		1.7	1.71		1.7	1.50–1.65
Zn ppm	6.9		7.3	0.70		2.9–7.2
Ga ppm	4.26		4.1	2.40		2.6–3.1
Se ppb	259		246	141		172
Rb ppm	6.3	6.46	8.2	0.97	1.04	1.14
Sr ppm		159	< 100		101	< 100
Ag ppm	0.046		0.14–0.29	0.00081		0.03
In ppb	218		360–880	1.9		24
Cs ppm	0.248		0.31	0.039		0.075
Ba ppm		404	313		67.2	48–65
La ppm		30.1	33.5		6.02	5.1–5.8
Ce ppm		90	87		17.0	19–22
Sm ppm		16.4	16.0		4.24	4.32–5.50
Eu ppm		1.73	1.81		0.853	0.82–0.91
Dy ppm		21.7	23.4		6.34	6.0–10.6
Er ppm		13.5	13.7		3.89	–
Yb ppm		13.0	12.8		3.78	(5.9–12.8)†
Lu ppm		1.78	1.90		—	(0.86–2.07)†
Ir ppb	8.5		4.34	0.62		< 0.1
Au ppb	2.39		4.9	0.024		11

* Range of the values of analyzed chips 133, 134, and 135.
† Probably contaminated.

the lunar soil 12070 and rock 12002, is presented. Except for Rb, Sr, and Ba in 12070, the agreement with the work of Gast and Hubbard is good. The stable isotope dilution data by these workers are probably more accurate for these elements. The high values found for Yb and Lu in 12002,133, 134, and 135 in this work are probably due to contamination during the grinding of the samples in agate mortars in our laboratory. Likewise the high values found for U and Th in 12002,133, 134, and 135 must be due to contamination, as evident from the data obtained independently on chip 12002,132. Sample 12021,77 was apparently contaminated by the same elements.

We shall not enter into a discussion of all our data, but concentrate on a few important aspects.

Comparison of sample 12038 with eucrites

Eucrites and sample 12038 have similar contents of a number of elements, including most of the major elements. In Table 3, our data for 12038,59 are compared with the average eucrite composition (Ahrens and Danchin, 1970). To our knowledge no data on eucrites exist for As, Ag, and W.

Although the concentrations of 16 elements are in agreement within a factor of 2, there are a number of elements that exhibit considerable differences, including the major elements Ti and Mn. Of the elements that differ in concentration by a factor between 2 and 10, all except Cl and Mn have the higher concentration in sample 12038. This group includes such geochemically different elements as Ti, Ba, REE,

Table 3. Comparison between present data for rock 12038,59 and average eucrite composition (AHRENS and DANCHIN, 1971).

A. Elements similar within a factor of 2			*B*. Elements similar to a factor between 2 and 10			*C*. Elements differing by an order of magnitude or more		
Element	12038,59	Eucrites	Element	12038,59	Eucrites	Element	12038,59	Eucrites
Na %	0.469	0.30	Cl ppm	7	22	Se ppb	178	1.6
Al %	6.66	6.7	Ti %	1.98	0.42	In ppb	160†	0.24–1.6
K %	0.052	0.05	Mn %	0.189	0.40			
Sc ppm	44	25	Cu ppm	13	5			
V ppm	149	100	Zn ppm	5	2			
Cr ppm	2030	2000	Ga ppm	5.3	~1			
Fe %	14.0	13.6	Ba ppm	90	35			
Co ppm	28	15	REE	—	—			
Rb ppm	0.44	0.30	Hf	4.9	0.97‡			
Sr ppm	144	80	Ta ppm	0.61	0.12			
Sb ppm	0.02	0.024–0.1						
Cs ppm	0.017	0.012						
Ir ppb	<0.1	0.29*						
Au ppb	3.5	2.5						
Th ppm	0.23	0.42						
U ppm	0.12	0.11						

* Average of three analyses from WASSON and BAEDECKER (1970).
† Probably contaminated.
‡ Recently reported by EHMANN and REBAGAY (1970).

Hf, and Ta on the one hand, and Cu, Zn, and Ga on the other. Only Se and In differ in concentration by an order of magnitude or more. The high concentration of In in 12038 is probably related to contamination as indicated by WASSON and BAEDECKER (1970). The only determination of Se in a eucrite (Nuevo Laredo) is by SCHINDEWOLF (1960) and should be verified.

Though the similarities between sample 12038 and eucrites are striking, the difference noted in the group (*B*) elements of Table 3 demonstrate that this sample is not necessarily of achondritic origin.

The alkali elements

The following statements can be made about the observed alkali element contents: (1) The overall low contents of these elements as demonstrated for Apollo 11 is verified. (2) All the alkali elements except Na are found in higher concentrations in the lunar fines (12070) than in any of the rocks. (3) Rock 12038 has the maximum Na content. (4) The Rb content of 12070,79 (8.2 ppm) is, to our knowledge, the highest published value for lunar fines. Values of 6.3 ppm and 6.46 ppm have, however, been obtained for 12070, respectively by GANAPATHY *et al.* (1970a) using neutron activation analysis and GAST AND HUBBARD (1970) using mass spectrometric isotope dilution. (5) Based on their Rb contents, all rock samples would be placed in Group 2 (low Rb-group) of the COMPSTON *et al.* (1970) classifications for Apollo 11 rocks. This would also agree with their K and Ba contents. (6) The Na/K ratios of all the rocks except 12038 are remarkably similar (Table 4). Na/K is significantly higher for rock 12038 and significantly lower for the fines 12070 than for the other rock samples. The average Na/K ratio of eucrites is 6, but varies between 0.95 (Bereba) and 14.8

Table 4. Alkali element ratios in lunar samples.

	12070 79	12002 133	12002 134	12002 135	12018 42	12021 77	12038 59
Na/K	1.7	3.5	4.0	3.8	3.8	3.8	9.1
K/Rb	235		360		282	390	1,170
K/Cs	6,400		5,500		10,000	7,300	30,300

(Moore County) (data from HEIER, 1970). The value of this ratio in terrestrial basalts is about 2. (7) The K/Rb ratios of the rocks, except 12038, and the fines are rather typical of terrestrial upper crustal ratios and lower than reported for Apollo 11 material by GAST *et al.* (1970) (about 400 to 500 for fines, breccia, and high Rb-rocks; 700 to 900 for low Rb-rocks). (8) The high K/Rb ratio of 12038 is similar to that of eucrites and primitive terrestrial basalts. It is higher than reported for Apollo 11 rocks. (9) The K/C's ratios are considerably lower (except for 12038) than those determined by GAST *et al.* (1970) for Apollo 11 material (16,000 to 32,000 for the rocks and 12,000 for the fines) and those in terrestrial oceanic basalts. The K/Cs ratios of the Apollo 12 samples, as well as the K/Rb ratios, are in the range of upper crustal continental rocks.

Tungsten

As far as we have been able to ascertain, tungsten was determined on the Apollo 11 rocks only by MORRISON *et al.* (1970) (0.13–0.42, av. 0.31 ppm) and WÄNKE *et al.* (1970) (0.24–0.43, av. 0.36 ppm). The present data are within the same concentration range (0.10–0.29 ppm), but the average value 0.17 ppm is significantly lower than for the Apollo 11 rocks.

A comparatively high W concentration (0.64 ppm) was found for the lunar fines 12070. For the analyzed Apollo 11 fines values of 0.22 ppm (WÄNKE *et al.*, 1970) and 0.25 ppm (MORRISON *et al.*, 1970) were reported.

Selenium

Of the three elements S, Se, and Te only Se was determined in the present work. Se determinations on Apollo 11 material were by HASKIN *et al.* (1970), MORRISON *et al.* (1970), and TUREKIAN and KHARKAR (1970). The data of TUREKIAN and KHARKAR (1970) (0.12–0.39 ppm) are similar to those obtained by GANAPATHY *et al.* (1970b) (0.086–0.259 ppm) and by us (0.172–0.246 ppm) on Apollo 12 material.

As already mentioned, these Se concentrations are one order of magnitude higher than the one existing Se determination on a eucrite. Sulphur determinations on Apollo 11 materials by COMPSTON *et al.* (1970), KAPLAN *et al.* (1970), and MAXWELL *et al.* (1970) indicate S concentrations between 0.1 and 0.2 ppm vs. 0.25 to 0.36 in howardites (no determinations of S in eucrites have been reported). Te was found by GANAPATHY *et al.* (1970a, 1970b) to be in the range of 8 to 73 ppb in Apollo 11 and 10 to 100 ppb in Apollo 12 material; this is one to two orders of magnitude above the Te content in eucrites. If we accept the existing data on S, Se, and Te in lunar and achondritic materials, we have to conclude that their S contents are similar, while their Se and Te contents are different by one to two orders of magnitude. It is not possible to state whether similar differences exist within the group As, Sb, and Bi as only Sb data have been reported for achondrites. Our values for Sb in lunar rocks are similar to those for achondrites (HAMAGUCHI *et al.*, 1961; TANNER and EHMANN, 1967).

Chromium

Our results for Cr range between 0.20% in 12038 (equal to the average eucrite value) and 0.67% in one of the chips from sample 12002. Cr was determined by a number of workers to be in the range 0.12 to 0.30% in Apollo 11 material (COMPSTON *et al.*, 1970; GOLES *et al.*, 1970; HASKIN *et al.*, 1970; MAXWELL *et al.*, 1970; SMALES *et al.*, 1970). Rock 12002 is unique in having approximately twice the Cr content of any of the Apollo 11 material. Nevertheless, the most striking thing about Cr in the lunar rocks is its uniformly high concentration compared with terrestrial rocks. This is even more noteworthy as Cr contents in terrestrial rocks are very sensitive to the stage of crystal–liquid differentiation.

Chlorine

Cl was determined in Apollo 11 rocks by HASKIN *et al.* (1970), MORRISON *et al.* (1970), REED and JOVANOVIC (1970), SMALES *et al.* (1970). Excluding the data of MORRISON *et al.* (1970) because of reported possible contamination and of SMALES *et al.* (1970) who only give an upper limit of 70 ppm, the Cl concentration in Apollo 11 material ranges between 3 and 30 ppm in the rocks and fines. HASKIN *et al.* (1970) report 65.4 ppm Cl in the breccias. The range of our data is from 5.2 ppm (12018) to 24 ppm (12070). It should not noted that our established value for Cl in BCR–1 is 54 ppm while HASKIN *et al.* (1970) have reported 30.5 ppm.

Silver, iridium, and gold

Silver was determined in Apollo 11 rocks by GANAPATHY *et al.* (1970a), HASKIN *et al.* (1970), MORRISON *et al.* (1970), and TUREKIAN and KHARKAR (1970). High concentrations of Ag in the Apollo 11 rocks were regarded as questionable because of possible contamination by In–Ag vacuum gaskets in Apollo sample return containers. The same comment may therefore be made about Ag in Apollo 12 samples; discussion of the significance of this element is unwarranted. Our data for Ag are within the range of the Apollo 11 data.

Iridium in Apollo 11 rocks was determined by GANAPATHY *et al.* (1970a) and WASSON and BAEDECKER (1970). GANAPATHY *et al.* (1970b) have also determined Ir in Apollo 12 samples. We find less than 0.1 ppb Ir in the crystalline rocks and higher contents in the fines. This trend agrees with the data of the above authors.

There is some uncertainty about the determinations of gold in the Apollo 11 crystalline rocks. GANAPATHY *et al.* (1970a), HASKIN *et al.* (1970), and WASSON and BAEDECKER (1970) generally agreed with Au contents less than 0.5 ppb. TUREKIAN and KHARKAR (1970) and WÄNKE *et al.* (1970) found Au concentrations between

0.9 and 8.1 ppb. GANAPATHY *et al.* (1970b) have confirmed the low gold level in the lunar rocks by analysis of Apollo 12 samples. Our data on Apollo 12 material are all above 1 ppb with a maximum of 11 ppb in sample 12002. This sample has also a higher Cr content than the other rocks analyzed by us. All investigators found above 1 ppb Au in the Apollo 11 fines, and the concentration of 5 ppb as reported in this work on Apollo 12 fines is less than the 8.5 ppb determined by TUREKIAN and KHARKAR (1970) for Apollo 11.

Rare-earth elements

We have determined ten rare-earth elements (REE). In Fig. 1 the data obtained for the soil 12070 and the rocks 12020 and 12038 normalized to those for a composite sample of chondritic meteorites (HASKIN *et al.*, 1968) are shown. For dysprosium a chondritic abundance value of 0.29 ppm as established by H. WAKITA and R. A. SCHMITT (private communication, 1971) was used.

The lunar fines 12070 shows the same pattern for the elements Sm–Lu as that established for Apollo 11 D-type fines with almost no difference in absolute concentration. The lighter REE appear to be relatively enriched in 12070.

For comparison, the REE patterns for the average abundances of group 1 and group 2 Apollo 11 rocks, according to the chemical classification by COMPSTON *et al.* (1970), are shown in Fig. 1. This classification is supported by textural criteria as

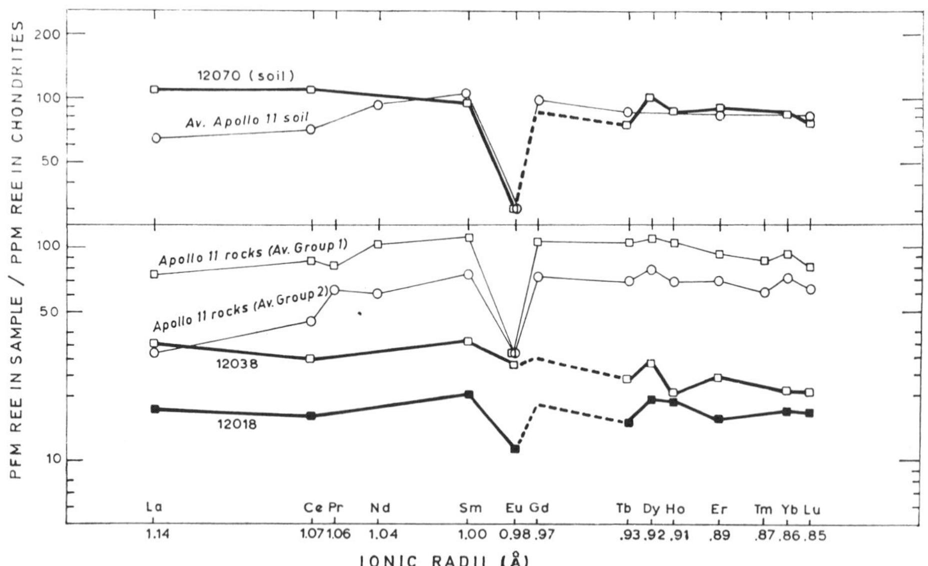

Fig. 1. Chondritic normalized REE abundances in Apollo 12 lunar samples plotted as a function of ionic radii. Gd is interpolated from the Sm and Tb values. The soil sample 12070 is compared with the average Apollo 11 soil. The lunar rocks 12018 and 12038 are compared to the two rock groupings as proposed by COMPSTON *et al.* (1970). The average abundances of the Apollo 11 samples are evaluated from the compilation by WARNER (1971).

discussed by JAMES and JACKSON (1970). The lunar rocks studied in this work have lower REE contents than the Apollo 11 rocks, although they are significantly higher in REE than the chondritic meteorites. The lunar rock 12038, which has the highest plagioclase content, shows the least Eu deficiency (Sm/Eu ratio 3.4).

Acknowledgment—Financial support by the Royal Norwegian Council for Scientific and Industrial Research (Research contract B 1206.3070) is gratefully acknowledged.

REFERENCES

AHRENS L. H. and DANCHIN R. V. (1971) The chemical composition of the basaltic achondrites. In *Physics and Chemistry of the Earth*. Pergamon. (in press)

BRUNFELT A. O. and STEINNES E. (1971) A neutron activation scheme for the determination of 42 elements in lunar material. To be published.

COMPSTON W., CHAPPEL B. W., ARRIENS P. A., and VERNON M. J. (1970) The chemistry and age of Apollo 11 lunar material. *Proc. Apollo 11 Lunar Sci. Conf., Geochim. Cosmochim. Acta* Suppl. 1, Vol. 2, pp. 1007–1027. Pergamon.

EHMANN W. D. and REBAGAY T. V. (1970) Zirconium and hafnium in meteorites by activation analysis. *Geochim. Cosmochim Acta* **34**, 649–658.

GANAPATHY R., KEAYS R. R., and ANDERS E. (1970a) Apollo 12 lunar samples: Trace element analysis of a core and the uniformity of the regolith. *Science* **170**, 533–535.

GANAPATHY R., KEAYS R. R., LAUL J. C., and ANDERS E. (1970b) Trace elements in Apollo 11 lunar rocks: Implications for meteorite influx and origin of moon. *Proc. Apollo 11 Lunar Sci. Conf., Geochim. Cosmochim. Acta* Suppl. 1, Vol. 2, pp. 1117–1142. Pergamon.

GAST P. W. and HUBBARD N. J. (1970) Rare earth abundances in soil and rocks from the Ocean of Storms. *Earth Planet. Sci. Lett.* **9**, 94–101.

GAST P. W., HUBBARD N. J., and WIESMANN H. (1970) Chemical composition and petrogenesis of basalts from Tranquillity Base. *Proc. Apollo 11 Lunar Sci. Conf., Geochim. Cosmochim. Acta* Suppl. 1, Vol. 2, pp. 1143–1163. Pergamon.

GOLES G. G., RANDLE K., OSAWI M., SCHMITT R. A., WAKITA H., EHMANN W. D., and MORGAN J. W. (1970) Elemental abundances by instrumental activation analyses in chips from 27 lunar rocks. *Proc. Apollo 11 Lunar Sci. Conf., Geochim. Cosmochim. Acta* Suppl. 1, Vol. 2, pp. 1165–1176. Pergamon.

HAMAGUCHI H., NAKAI T., and ENDO T. (1961) Determination of arsenic and antimony in stone meteorites by neutron activation. *Nippon Kagaku Zasshi*, **82**, 1485–1489.

HASKIN L. A., ALLEN R. O., HELMKE P. A., PASTER T. P., ANDERSON M. R., KOROTOV R. L., and ZWEIFEL K. A. (1970) Rare-earths and other trace elements in Apollo 11 lunar samples. *Proc. Apollo 11 Lunar Sci. Conf., Geochim. Cosmochim. Acta* Suppl. 1, Vol. 2, pp. 1213–1231. Pergamon.

HASKIN L. A., HASKIN M. A., FREY F. A., and WILDEMAN T. R. (1968) Relative and absolute terrestrial abundances of the rare earth. In *Origin and Distribution of the Elements* (editor L. H. Ahrens), Int. Ser. Mon. Earth Sci., Vol. 30, pp. 889–912. Pergamon.

HEIER K. S. and BILLINGS G. K. (1970) Potassium. In *Handbook of Geochemistry* (editor K. H. Wedepohl), Vol. II/2, Chap. 19 B-E, Springer Verlag, Berlin.

JAMES O. D. and JACKSON E. D. (1970) Petrology of the Apollo 11 ilmenite basalts. *J. Geophys. Res.*, **75**, 5793–5824.

KAPLAN I. R., SMITH J. W., and RUTH E. (1970) Carbon and sulfur concentration and isotopic composition in Apollo 11 lunar samples. *Proc. Apollo 11 Lunar Sci. Conf., Geochim. Cosmochim. Acta* Suppl. 1, Vol. 2, pp. 1317–1329. Pergamon.

MAXWELL J. A., PECK L. C., and WIIK H. B. (1970) Chemical composition of Apollo 11 lunar samples 10017, 10020, 10072 and 10084. *Proc. Apollo 11 Lunar Sci. Conf., Geochim. Cosmochim. Acta* Suppl. 1, Vol. 2, pp. 1369–1374. Pergamon.

MORRISON G. H., GERARD J. T., KASHUBA A. T., GANGADHARAM E. V., ROTHENBERG A. M.,

Potter N. M., and Miller G. B. (1970) Elemental abundances of lunar soil and rocks. *Proc. Apollo 11 Lunar Sci. Conf., Geochim. Cosmochim. Acta* Suppl. 1, Vol. 2, pp. 1383–1392. Pergamon.

Reed G. W., Jr. and Jovanovic S. (1970) Halogens, mercury, lithium and osmium in Apollo 11 samples. *Proc. Apollo 11 Lunar Sci. Conf., Geochim. Cosmochim. Acta* Suppl. 1, Vol. 2, pp. 1487–1492. Pergamon.

Schindewolf U. (1960) Selenium and tellurium content of stony meteorites by neutron activation. *Geochim. Cosmochim. Acta* **19**, 134–138.

Smales A. A., Mapper D., Webb M. S. W., Webster R. K., and Wilson J. D. (1970) Elemental composition of lunar surface material. *Proc. Apollo 11 Lunar Sci. Conf., Geochim. Cosmochim. Acta* Suppl. 1, Vol. 2, pp. 1575–1581. Pergamon.

Tanner J. T. and Ehmann W. D. (1967) The abundance of antimony in meteorites, tektites and rocks by neutron activation analysis, *Geochim. Cosmochim. Acta* **31**, 2007–2026

Turekian K. K. and Kharkar D. P. (1970) Neutron activation analysis of milligram quantities of Apollo 11 lunar rocks and soil. *Proc. Apollo 11 Lunar Sci. Conf., Geochim. Cosmochim. Acta* Suppl. 1, Vol. 2, pp. 1659–1664. Pergamon.

Wänke H., Rieder R., Baddenhausen H., Spettel B., Teschke F., Quijano-Rico M., and Balacescu A. (1970) Major and trace elements in lunar material. *Proc. Apollo 11 Lunar Sci. Conf., Geochim. Cosmochim. Acta* Suppl. 1, Vol. 2, pp. 1719–1727. Pergamon.

Warner J. (1971) A summary of Apollo 11 chemical, age and model data. Extracted from the *Proc. Apollo 11 Lunar Sci. Conf.*, Suppl. *Geochim. Cosmochim. Acta* Curator's Office Manned Spacecraft Center Houston, Texas January 1971.

Wasson J. T. and Baedecker P. A. (1970) Ga, Ge, In, Ir and Au in lunar, terrestrial and meteoritic basalts. *Proc. Apollo 11 Lunar Sci. Conf., Geochim. Cosmochim. Acta* Suppl. 1, Vol. 2, pp. 1741–1750. Pergamon.

Proceedings of the Second Lunar Science Conference, Volume 2, pp. 1291–1300.
The M.I.T. Press, 1971.

Radioanalytical determination of elemental compositions of lunar samples

MILOSLAV VOBECKÝ and JAROSLAV FRÁNA
Nuclear Research Institute of Czechoslovak Academy of Sciences,
Řež near Prague

JAROSLAV BAUER
Institute of Chemical Technology, Prague

and

ZDENĚK ŘANDA, JAROSLAV BENADA, and JAROSLAV KUNCÍŘ
Institute of Mineral Raw Materials, Kutná Hora

(*Received 3 March 1971; accepted in revised form 31 March 1971*)

Abstract—A multielement analysis was made on Apollo 11 and Apollo 12 lunar soil samples 10084,141 and 12070,83 and Apollo 12 rock fragment 12063,73 and their 26 separated mineral fractions and glasses (olivine, pyroxene, ilmenite, feldspar, glass, slag fractions etc.). Instrumental neutron activation analysis (INAA) was used to determine Na, Mg, Al, (Si), K, Ca, Sc, Ti, V, Cr, Mn, Fe, Co, Ni, Sr, Zr, Cs, Ba, La, Ce, Nd, Sm, Eu, Gd, Tb, Dy, Ho, Tm, Yb, Lu, Hf, Ta, W, Ir, Au, Th, and U. Nondispersive X-ray fluorescence analysis was applied to the determination of Ti, Fe, Sr, Y, and Zr in original samples. Separated plagioclase was identified as bytownite with about An_{80} and clinopyroxene as an augite-pigeonite type. Olivine and especially ilmenite are to a great extent inhomogeneous with contents of the main components differing from the predicted values. Lunar fines 12070,83 is enriched in V, Cr, Co, Zr, Hf, Th, Au, and REE in comparison with the fines 10084,141 from Tranquillity Base; on the other hand the Ti content is 2.5 times lower.

INTRODUCTION

LUNAR SOIL SAMPLES 10084,141 (Apollo 11) and 12070,83 (Apollo 12), and a fragment of Apollo 12 olivine basalt 12063,73 were analyzed by INAA and high resolution Ge(Li) gamma-ray spectrometry. Separated fractions of feldspar, ilmenite, olivine, clinopyroxene, and glasses from these samples were analyzed as well.

EXPERIMENTAL TECHNIQUES

About 100 milligram portions of the original samples were used for each analysis. The samples of soil were sifted through a 160 mesh polyamide sieve. The portions having a particle diameter over 0.116 mm were visually separated under a microscope. The residue after the visual separation was separated magnetically. The rock sample, wrapped in a polyethylene foil, was crushed in Plattner's mortar and the same separation procedure as used for soils was applied. Minerals were identified from morphological and optical properties and by X-ray diffraction. Monomineral fractions obtained from the separation were not quite homogeneous because of the presence of numerous small inclusions. In this manner the separated fractions of 1–3 milligrams were taken for analyses; the weight of the glassy fractions was of the order of tens of micrograms.

The samples, sealed in polyethylene foil, were then activated. Besides single-element standards, multielement mixtures were successfully used in which the elements were combined with regard to sample composition, activation and measurement conditions according to KUNCÍŘ *et al.* (1970). Both short activations (2 or 3 min. in the core at 10^{13} n cm^{-2} sec^{-1}) and long activations (20 and 70 hours in a thermal neutron column at 2×10^{12} n cm^{-2} sec^{-1}) were used. By means of short activations the elements Na, Mg, Al, (Si), K, Ca, Ti, V, Mn, Co, Ba, Sm, Eu, Dy, and U were determined. By using longer activation times, determinations of Na, K, Sc, Cr, Fe, Co, Ni, Zr, Cs, Ba, La, Ce, Nd, Sm, Eu, Gd, Tb, Ho, Tm, Yb, Lu, Hf, Ta, W, Ir, Au, and Th were obtained.

The gamma-ray spectra of the activated samples were measured by a 30 cc Ge(Li) detector coupled to a 4096-channel analyzer. The detector resolution (FWHM) for 1332.5 keV photons of ^{60}Co was 2.6 keV. The intensities of analytically usable photopeaks in the sample and standard spectra, together with other calculations, were processed by a computer. Also nondispersive X-ray fluorescence analysis with Si(Li) detector (excitation source 500 mCi^{147} Pm/Al, 415 eV FWHM for K_α of Fe) was useu to determine the contents of Ti, Fe, Sr, Y, and Zr in these basic samples.

Even though the method of short-term activation did not permit the analysis for some macrocomponent elements such as Si, Ca, Mg, and Al with optimum accuracy, especially when small sample portions were weighed, we proceeded to estimate abundances of these elements. The accuracy of the determination of Si by means of the ^{29}Al activity cannot be better than $\pm 30\%$; however, even knowledge of a less accurate value permits approximate corrections to be made for the interfering contributions which are met with when determining Mg and Al. For the final interpretation of analytical results the best SiO_2 values were obtained by subtracting the macroelemental oxides from 100%. For milligram sample weights, $\pm 30\%$ errors were obtained for Mg, Ni, Zr, Cs, Ba, Nd, Gd, and Tm; $\pm 20\%$ errors were obtained for Na, Al, K, Ca, Co, Ce, Dy, Ho, Ta, W, Ir, Au, and U; $\pm 10\%$, for Ti, V, Mn, Fe, Sm, Eu, Tb, Lu, Hf, and Th; and $\pm 5\%$, for Sc, Cr, La, and Yb. The minimum detectable limit was estimated as a 2σ of the Compton continuum in the region of the analyzed photopeak. Ti, Fe, and Zr were determined by means of X-ray fluorescence with about 10% accuracy, whereas the accuracy for Sr and Y is worse.

RESULTS AND DISCUSSION

The results of analyses of lunar fines 10084,141 and separated mineral fractions of feldspar, ilmenite, clinopyroxene, red-brown and green glasses together with three fractions marked 11–7, 11–8 and 11–9 are summarized in Table 1. For the comparison of results the published values of lunar fines 10084 (WÄNKE *et al.*, 1970; WAKITA *et al.*, 1970; TUREKIAN and KHARKAR, 1970; and KEAYS *et al.*, 1970) are introduced in the last column of the table.

Feldspar was separated as colorless clear or white translucent grains of size from 0.1 mm to 1.0 mm with vitreous gloss and significant cleavage. From the values shown in Table 1, it follows that the average composition of feldspars is in good agreement with a ratio of components for bytownite $Ab_{20}An_{80}$. Relatively high content of Fe and Mg in these plagioclase samples could be caused by inclusions of ilmenite and olivine. In the comparison with the original sample of fines the contents of the trace elements (REE, Sc, V, Co, Hf) in the feldspar samples are much lower. Eu is slightly higher comparing with the other REE. The corrected content of Si = 20.1%.

Ilmenite (?), *titanium-rich mineral fraction*. Black opaque, metallic glossy fragments of tabular crystals and irregular grains of the size from 0.1 mm to 1.5 mm were separated. The analysis of this fraction showed great discrepancies from the theoretical composition of ilmenite. This phenomenon might be caused by the great inhomogenity of this mineral (e.g., see ANDERSEN *et al.*, 1970 and HAGGERTY *et al.*, 1970). The increase of Sc and V content in the ilmenite samples is significant compared with the original sample of fines. A similar increase is observed in this fraction for the content of Si at 16.3%.

Clinopyroxene. Light or dark brown fragments of transparent to translucent crystals from 0.1 mm to 1.5 mm were separated for an analysis. The compositions

Table 1. Abundances of major and trace elements in lunar fines 10084,141 and mineral fractions. Comparison with the reported values (in ppm except for those noted by %).

Element	10084,141	feldspar	ilmenite (?)	clinopy-roxene	glass red-brown	glass green	fraction 11–7	11–8	11–9	Reported values (10084)
Na (%)	0.28	0.55	0.31	0.11	0.38	0.23	0.30	0.37	0.33	0.33[a]
Mg (%)	5.4	2.5	4.3	6.7	5.6	3.9	4.2	5.7	5.5	4.73[a]
Al (%)	7.5	17.2	4.6	2.4	11.2	9.3	7.5	8.3	6.6	7.13[a]
Si (%)	(23)	(24)	(28)	(37)		(19)	(20)	(16)	(20)	19.53[a]
K (%)	0.11	<0.06		0.07			0.14	0.13	0.14	0.1107[a]
Ca (%)	7.8	11.7	7.3	8.3		10.5	8.0	8.3	8.0	8.40[a]
Sc	60	10.1	123	133	90	67	63	63	74	61.4[a]
Ti (%)	4.6 4.2*	0.5	9.6	2.6	5.8	3.8	4.4	4.4	5.4	4.36[a]
V	77	18	107	97	154	86	74	78	86	63[b]
Cr (%)	0.18	0.044	0.27	0.27	0.33	0.18	0.22	0.21	0.27	0.2005[a]
Mn (%)	0.16	0.046	0.21	0.23	0.18	0.15	0.16	0.16	0.19	0.1603[a]
Fe (%)	11.2 10.9*	2.8	16.8	14.5	15.0	12.7	11.7	12.1	12.7	12.34[a]
Co	28	8.5	24	22	115	32	35	32	23.4	29.2[a]
Ni	200							<250	250	229[a]
Sr	230*						<380	<300	<250	162[a]
Y	85*									96[b]
Zr	340 340*			400			190	380	410	380[c]
Cs	<0.2			<1			<0.3	<0.3	<0.3	0.122[a]
Ba	160	210		230		180	350	220	180	167[a]
La	20.4	16.4	15.7	12.2	78	113	21	21	20	17.2[a]
Co	53	47		30		47	51	52	53	49.4[a]
Nd	48	45								40.5[a]
Sm	12.1	7.1	19.5	11.2	19.2	11.1	11.9	13.4	13.9	13.5[a]
Eu	1.8	1.8	2.3	1.2	2.2	2.7	1.7	1.9	1.7	1.84[a]
Gd	18						17	17	17	17.6[a]
Tb	2.9	1.4	11.8	3.3		2.4	3.1	3.5	3.4	3.14[a]
Dy	22	9	28	20	27	18	19	18.5	21.3	20.1[a]
Ho	4.3	2.1	5.0	3.3		4.6	3.5	5.5	3.8	5.03[a]
Tm	1.9	1.3	9.3	4.6			2.9	2.2	2.3	1.8[b]
Yb	10.1	5.0	18.3	11.2		8.2	12.9	12.0	12.8	10.9[a]
Lu	1.44	0.56	2.2	1.65		1.05	1.4	1.6	1.5	1.49[a]
Hf	9.2	5.0	11.0	5.5		6.5	7.9	9.3	10.2	10.0[a]
Ta	1.16			1.25			1.3	1.3	1.2	1.40[a]
W	<2	<6	<10	27.5		24	<4	<1.6	<1.4	0.235[a]
Ir	0.0045							<0.03	<0.02	0.00688[d]
Au	<0.02	<0.04	<0.08	0.25		0.65	0.3	<0.02	<0.02	0.00275[a]
Th	1.9	1.7		<1.5			1.2	2.0	1.7	2.35[a]
U							1.4	1.5	2.1	0.37[a]
Weight of sample milligram	1.490 98.762†	1.216	0.108	0.279	0.015	0.153	3.326	7.797	15.072	

The values in parentheses are semiquantitative only.
* Values obtained by the nondispersive X-ray fluorescence analysis.
† For 20 hours irradiation only.
[a] Wänke *et al.* (1970).
[b] Wakita *et al.* (1970).
[c] Turekian and Kharkar (1970).
[d] Keays *et al.* (1970).

correspond to pyroxenes of augite-pigeonite type. The high contents of Au and W in pyroxene samples are meaningful. The content of Sc compared with the original sample is more than doubled. The corrected content of Si 21.5%.

Red-brown glass is represented by a 15 μg fragment. A higher concentration of V, Co, and La is notable. *Green glass* was isolated from the sample in a form of smooth

Table 2. Abundances of major and trace elements in lunar fines 12070,83

				glass particles				
Element	12070,83	olivine (?)	pyroxene	dark violet	black opaque	green	dark brown	black opaque (slag)
Na (%)	0.29	0.41	0.10	0.09	0.30		0.19	0.25
Mg (%)	6.6	7.4	6.4	7.0	5.6	6.7	5.1	6.2
Al (%)	7.1	6.1	2.1	6.6	6.6	13.8	8.3	5.9
Si (%)	(19)	(28)	(21)		(38)		(27)	(37)
K (%)	0.20	1.65	0.06	0.05	0.23		0.42	0.28
Ca (%)	6.8	4.5	6.2	7.3	4.2		8.5	
Sc	43	24	70	47	45	25	48	34
Ti (%)	{1.74 1.7*	0.5	1.5	2.5	1.8		1.0	1.8
V	129	72	220	140	120	100	115	130
Cr (%)	0.27	0.16	0.32	0.24	0.31	0.20	0.35	0.24
Mn (%)	0.18	0.13	0.22	0.18	0.14	0.06	0.08	0.15
Fe (%)	{13.4 13.6*	9.5	14.6	13.3	12.9		12.8	11.5
Co	46	39	35	62	53	< 100	80	39
Ni	200	< 400						
Sr	170*	< 500						
Y	97*							
Zr	{680 605*	< 300						
Cs		2.2						
Ba	390	1000		1000	6200		1000	700
La	45	50	28	32	36		70	32
Ce	103	112		93	94		172	67
Nd	60	< 40						
Sm	13.9	11.0	9.2	16.7	18.4		30	14.3
Eu	1.9	1.2	0.8	2.1	2.2		2.8	1.7
Gd	23	17						
Tb	3.6	3.9		5.5			12.1	4.6
Dy	22.5	21.4	13.4	27	22.0	8.0	35	27
Ho	4.4	5.2	11.9	4.9	5.5		4.3	3.8
Tm	2.7	5.2						9.0
Yb	12.9	25	8.8	13.1	13.2	6	23	16
Lu	1.7	3.4	1.1	2.1	1.9	0.7	2.5	1.1
Hf	12.7	11.7	6.8	11.6	13.3		13.8	10.8
Ta	1.4	5.0						
W		< 2	< 2	< 5	< 5		< 6	< 5
Ir								
Au	0.016	< 0.01	< 0.03	< 0.04	< 0.07		< 0.07	< 0.07
Th	5.9	33		5.3	4.6		10.6	9.3
U	2.3	9.0	0.7		3.2		4.6	
Weight of sample milligram	7.614 100.315†	2.431	1.296	0.050	0.090	0.008	0.041	0.074

The values in parentheses are semiquantitative only.
* Values obtained by the nondispersive X-ray fluorescence analysis.

bright spheres less than 0.2 mm in diameter. Compared with the original sample, higher contents of La and exceptionally high contents of W and Au are observed. A relative decrease of Eu content is smaller comparing with the original sample. The corrected content of Si = 17.5%.

Fraction 11–7 consists of magnetically separated particles and appears as dark irregular grains (scoriaceous glass) and slags, ranging in length from 0.2 mm to 1 mm.

mineral and glass fractions (in ppm except for those noted by %).

fractions						
12–8	12–5	12–6	12–7	12–13	12–14	Element
	0.34	0.40	0.29	0.35	0.32	Na (%)
<3	5.2	5.6	6.2	7.0	7.0	Mg (%)
9.2	7.2	7.6	7.2	6.9	6.9	Al (%)
	(21)	(29)	(25)	(23)	(22)	Si (%)
<2	0.22	0.27	0.53	0.24	0.21	K (%)
	6.9	6.9	3.0	6.8	7.1	Ca (%)
<40	38	38	41	44	48	Sc
1.5	1.8	1.8	1.6	1.6	1.6	Ti (%)
74		112	150	130	135	V
0.16	0.32	0.24	0.27	0.26	0.29	Cr (%)
0.12	0.17	0.16	0.18	0.19	0.18	Mn (%)
<10	12.1	11.4	12.6	12.2	13.5	Fe (%)
	57	41	134	35	52	Co
	<300	340		<200	<200	Ni
	<430	<200	<470	<200	<200	Sr
	460	740	530	740	590	Zr
	<0.4	<0.6	<0.6	<0.2	(0.25)	Cs
5000	410	650	490	400	370	Ba
210	45	45	21	49	43	La
	109	114	85	111	93	Ce
	<60	80	40			Nd
42	19.4	19.7	14.8	18.1	15.8	Sm
14	1.8	2.2	1.6	1.5	1.7	Eu
	22	19		24	26	Gd
	4.6	4.4	3.4	4.4	3.7	Tb
96	28	25	23.7	26	21.3	Dy
23	5.3	5.9	4.7		3.8	Ho
	3.1	3.2	2.6	2.9	2.6	Tm
140	14.3	14.0	11.4	15.8	13.0	Yb
7.3	2.0	1.8	1.6	1.8	1.8	Lu
	13.1	13.6	12.1	14.6	14.3	Hf
	1.3	1.9	1.1	1.6	1.8	Ta
	3.9	<6	<3	<4	4.2	W
	<0.04		<0.04	<0.01	<0.01	Ir
	<0.04	<0.07	<0.04	<0.02	0.06	Au
	5.6	6.2	5.5	6.3	5.4	Th
	2.4	2.6	2.5		2.8	U
0.014	1.793	1.526	1.089	52.968	18.251	Weight of sample milligram

† For 20 hours irradiation only.

The elemental composition is close to the original sample; higher contents are observed for Co, Ba, U and especially Au. *Fraction* 11–8 is a magnetically separated portion with the biggest particles (> 1.0 mm). *Fraction* 11–9 is an oversize residue after the separation of larger grains from analyzed minerals. This fraction and the previous one have similar compositions compared to the original sample.

Table 2 contains the results of analyses of the lunar fines sample 12070,83 and particular mineral fractions separated from this sample. The original sample contains more V, Cr, Co, Zr, Hf, Th, REE, and Au in comparison with the fines from Tranquillity Base; on the contrary, the Ti content is substantially lower. The results are in a good agreement with the data published by LSPET (1970).

Olivine (?) was separated as light green or yellow-green round translucent grains containing numerous microscopic dark inclusions. These inclusions can cause considerable disagreement with the theoretical composition of olivines, especially high content of Al, K, and Ca. For the trace elements in this sample higher concentrations of Ba, Th, U, and some of heavier REE are meaningful. The corrected content of Si = 25.0%.

Pyroxene. Brown fragments of 0.1–1.5 mm crystals were separated. In appearance and chemical composition these pyroxenes are similar to pyroxenes from the sample 10084. Only a higher content of V is remarkable. A higher value for Ho was obtained from only one measurement.

Glasses. Partly dark purple, green, and dark brown glasses in the form of small transparent spheres and ovaloids (dumbbells and partly black opaque spheres) of a slag appearance were separated. Concerning trace elements only a higher concentration of Ba is remarkable, especially in the case of black opaque glass.

Unidentified mineral was marked 12–8. This black, brightly opaque, and flat crystalline fragment of about 14 μg weight resembles ilmenite. The Ti content is only 1.5%, and concentrations of siderophilic elements are also low. On the other hand the contents of Ba and REE are much higher than in the original sample and the other fractions. *Fractions* 12–5 and 12–7 are magnetic portions separated in the stronger magnetic field. In the case of sample 12–7 (1100 Oe) the analysis shows about three times higher content of Co. *Fraction* 12–6 contains black particles of slag. *Fraction* 12–13 is an oversize portion containing particles greater than 0.5 mm. *Fraction* 12–14 are particles with the smallest size (less than 0.1 mm). The major, minor, and trace elemental compositions of fraction 12–14 are rather similar to the abundances found in the other fractions of 12–5, 6, 7, and 13.

The results of analyses of rock fragments 12063,73 and four basic mineral components are shown in Table 3.

The gray-brown, fine-grained rock 12063,73 has been characterized as ilmenite microgabbro with olivine, traces of cristobalite, and a subophitic texture. Compared with a sample of fines the rock contains substantially less Zr, Hf, Th, REE, and Ta. The results obtained from nondispersive X-ray fluorescence analysis are not considered to be quite certain because of poorly defined geometry of measurements; a sample was exposed only from one side. The determination of the main elements by means of a short activation has not been carried out.

Ilmenite (?). Opaque fragments of glossy crystals were separated. It follows from Table 3 that the contents of Fe and Ti are not in accordance with a theoretical composition of ilmenite and, the sample seems to contain about 50% of strange admixtures. They might be both larger grains of minerals intergrown by ilmenite and inner fine inhomogeneities (armalcolite, rutile etc.). It is obvious that higher contents of REE, Zr, Hf, Th, and Ta were found in this fraction.

Pyroxene. Crystals are brown, glossy, and translucent with sizes from 0.2 to

Table 3. Abundances of major and trace elements in lunar rock fragment 12063,73 and mineral fractions separated from it (in ppm except for those noted by %).

Element	12063,73	ilmenite(?)	pyroxene	feldspar	olivine
Na (%)	0.19	0.13	0.08	0.64	0.056
Mg (%)		2.4	7.5	1.8	15
Al (%)		3.5	2.3	15.3	1.0
Si (%)		(17)	(26)	(27)	(23)
K (%)	0.056	0.053	0.052	<0.1	<0.1
Ca (%)		4.6	6.9	10.8	1.5
Sc	61	38	98	97	17.2
Ti (%)		14.0	1.9	0.56	0.77
V		133	277	26	109
Cr (%)	0.29	0.19	0.43	0.38	0.25
Mn (%)		0.24	0.27	0.05	0.23
Fe (%)	{15.1 17.1*	18.2	17.9	2.8	16.2
Co	44	29	45	6.4	70
Ni	<100	<160	<500	<90	<250
Sr		<200	<300	<200	
Y	400*				
Zr	250*	400			
Cs	<0.2	<0.2	<1	<0.1	<0.3
La	7.1	11.2	6.5	1.5	0.9
Ce	19.2	32	21	5.6	<4
Nd	16.5	36	<90	<30	<60
Sm	5.1	9.1	5.9	1.3	0.9
Eu	1.17	1.3	0.85	1.9	0.2
Gd	<6	16	7.0		<4
Tb	1.5	2.2	1.6	0.42	<0.1
Dy		19.6	10.0	9.0	2.2
Ho	1.3	2.9	2.4	0.5	<0.4
Tm	1.1	2.0	1.7	0.54	<0.3
Yb	5.5	8.4	6.6	1.05	1.0
Lu	0.7	1.3	0.86	0.1	0.15
Hf	3.9	8.2	3.9	0.9	0.7
Ta	0.5	1.3	<0.8	<0.4	<0.1
W		<3	<2	<2	<0.7
Ir	<0.004	<0.01	<0.01		
Au	<0.008	<0.02	<0.03	<0.01	<0.01
Th	0.66	1.5	1.0	0.3	0.4
U		1.1		<0.85	3.3
Weight of sample milligram	101.007†	1.559	1.921	2.856	2.066

The values in parentheses are semiquantitative only.
* Values obtained by the nondispersive X-ray fluorescence analysis.
† For 20 hours irradiation only.

0.6 mm, and are similar to pyroxenes from the fines samples. A chemical composition is very similar as well. A crystal isolated from the rock permitted geometrical measurements. This prismatic crystal of a cinnamon brown color has dimensions of 0.27 mm × 0.16 mm. The faces are irregular and the crystal has many inner accessory minerals. On a terminal the twinned obtuse angle is perceivable. It is possible to observe within the crystal numerous cracks and light inclusions that cause irregular extinction effects between crossed nicols. The signals of the planes are very poor. The forms found correspond to terrestrial augites: (010), (100), (110) and (111) (Fig. 1). The crystal is twinned along (100). The values of positional angles are summarized in Table 4. As

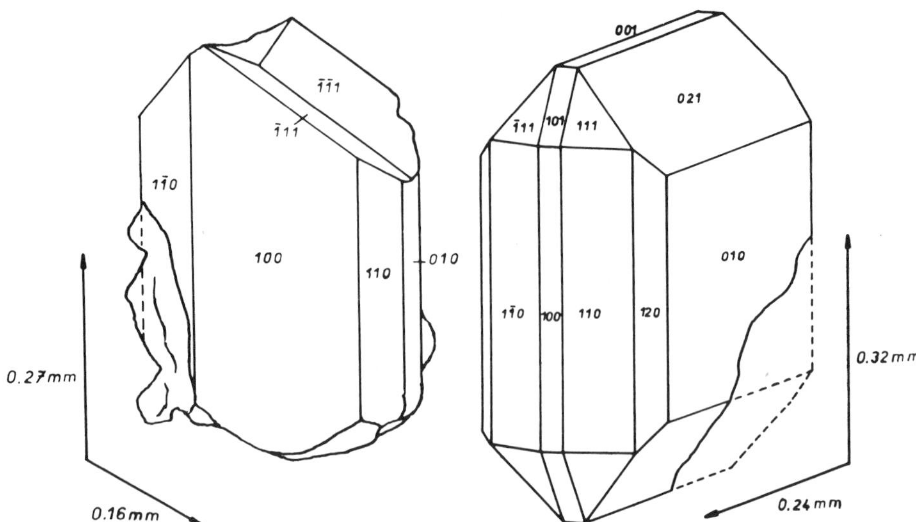

Fig. 1. Augite crystal twinned on {100}, with short prismatic habit, 0.27 mm in length, 0.16 mm in width.

Fig. 2. Idealized olivine crystal with tabular habit isolated from rock sample 12063,73, length 0.32 mm, width 0.24 mm.

the quality of faces was rather poor the ratio of parameters obtained from the experimental data was only approximately determined; $a:b:c = 1.09:1:0.60$, $\beta = 106°$. These constants are in agreement with the values given for terrestrial augites ($a:b:c = 1.092:1:0.589$, $\beta = 105°50'$).

Feldspar. Colorless and whitish fragments of monoclinic feldspar were isolated. These feldspars can be classified as bytownite plagioclase which are similar to feldspar isolated from lunar fines 10084. The trace element contents are lower compared with other minerals isolated from this rock. Relative to Sm, the Eu content is enriched ten times more in these feldspars than is observed in other minerals. It is obviously an isomorphic substitution of alkaline earth elements in feldspars by divalent Eu. The corrected value for Si = 21.7%. A crystal with original faces was found in isolated feldspar material. The crystal is colorless and clear with dimensions 0.48 mm × 0.19 mm of column habit. Light brown pyroxene and black opaque ilmenite intergrow into the crystal of feldspar. It is possible to measure goniometrically only one zone ((001), (010)). On the terminal, there are just refractive irregular faces. Many irregular cracks, which obviate more accurate determinations of optical constants are found within the crystal. According to measurements on the universal stage we can classify the crystal of feldspar as a high temperature plagioclase An_{55} in Baveno twinn.

Olivine. Separated greenish or green grains containing numerous microscopic opaque inclusions are clear and transparent. Due to inclusions, the chemical composition differs from olivine. Contrary to the overall composition of the rock the isolated olivines contain less Sc, REE, and Hf. On the other hand, the content of U is higher than is observed in other minerals.

Table 4. Two-circle goniometer data for olivine and augite isolated from rock sample 12063,73

hkl	Olivine		hkl	Augite	
001	—	0°00′	010	0°00′	90°00′
010	0°00′	90°00′	100	90°00′	90°00′
100	90°00′	90°00′	110	44°30′	90°00′
110	65°36′	90°00′	$\bar{1}$11	26°10′	33°10′
120	47°48′	90°00′			
021	0°00′	48°31′			
101	90°00′	51°29′			
111	65°36′	54°02′			

A light green crystal of olivine from the rock fragment 12063,73 was separated for the goniometric measurement. The crystal has a tubular habit with dimensions 0.32 mm × 0.24 mm. Numerous dark opaque inclusions are apparent in transmitted light in immersion liquid. Enclosed particles are mostly deflated by irregular octahedra; some of them are spherical. An irregularly limited black opaque grain (probably ilmenite) is growing from the broken part of olivine. Olivine gives an irregular undulatory extinction between crossed nicols probably because of the numerous inclusions.

Only some forms of crystals are smooth and straight, and are mostly wrinkled with various accessories which cause relatively poor signals for goniometric measurements. Forms (010), (021), (120) and (110) (Fig. 2) were found to be dominant in this lunar olivine. Only the basal pyramid (111) was found; the faces (001), (101) and (100) are developed only as narrow facets. From the measured positional angles included in Table 4 the ratio of parameters a:b:c = 0.449:1.000:0.564. The experimentally obtained value of the index of refraction $\beta = 1.687$ corresponds roughly to the theoretical composition of olivine $Fo_{80}Fa_{20}$. The lower value of the parameter ratios obtained for this lunar olivine crystal could be caused by the presence of monticellite component ($CaMgSiO_4$, a:b:c = 0.434:1:0.576) according to WINCHELL and WINCHELL (1951).

REFERENCES

ANDERSEN C. A. and HINTHORNE J. R. (1970) Ion microprobe analysis of lunar material from Apollo 11. *Proc. Apollo 11 Lunar Sci. Conf., Geochim. Cosmochim. Acta* Suppl. 1, Vol. 1, pp. 159–167. Pergamon.

HAGGERTY S. E., BOYD F. R., BELL P. M., FINGER L. W., and BRYAN W. B. (1970) Opaque minerals and olivine in lavas and breccias from Mare Tranquillitatis. *Proc. Apollo 11 Lunar Sci. Conf., Geochim. Cosmochim. Acta* Suppl. 1, Vol. 1, pp. 513–538. Pergamon.

KEAYS R. R., GANAPATHY R., LAUL J. C., ANDERS E., HERZOG G. F., and JEFFERY P. M. (1970) Trace elements and radioactivity in lunar rocks: Implications for meteorite infall, solar wind flux, and formation conditions of moon. *Science* **167**, 490–493.

KUNCÍŘ J., BENADA J., ŘANDA Z., and VOBECKÝ M. (1970) Multielement standard for routine instrumental activation analysis of trace elements in rocks and tektites. *J. Radioanal. Chem.* **5**, 369–378.

LSPET (Lunar Sample Preliminary Examination Team) (1970) Preliminary examination of the lunar samples from Apollo 12. *Science* **167**, 1325–1339.

TUREKIAN K. K. and KHARKAR K. P. (1970) Neutron activation analysis of milligram quantities of Apollo 11 rocks and soil. *Proc. Apollo 11 Lunar Sci. Conf., Geochim. Cosmochim. Acta* Suppl. 1, Vol. 2, pp. 1659–1664. Pergamon.

WAKITA H., SCHMITT R. A., and REY P. (1970) Elemental abundances of major, minor and trace elements in Apollo 11 lunar rocks, soil and core samples. *Proc. Apollo 11 Lunar Sci. Conf., Geochim. Cosmochim. Acta* Suppl. 1, Vol. 2, pp. 1685–1717. Pergamon.

WÄNKE H., RIEDER R., BADDENHAUSEN H., SPETTEL B., TESCHKE F., QUIJANO-RICO M., and BALACESCU A. (1970) Major and trace elements in lunar material. *Proc. Apollo 11 Lunar Sci. Conf., Geochim. Cosmochim. Acta* Suppl. 1, Vol. 2, pp. 1719–1727. Pergamon.

WINCHELL A. N. and WINCHELL H. (1951) In *Elements of Optical Mineralogy*, Part II, p. 498, J. Wiley.

Proceedings of the Second Lunar Science Conference, Volume 2, pp. 1301–1305.
The M.I.T. Press, 1971.

Analyses of Apollo 11 and Apollo 12 rocks and soils by neutron activation

D. P. KHARKAR and K. K. TUREKIAN
Department of Geology and Geophysics, Yale University,
New Haven, Connecticut 06520

(*Received 19 March 1971; accepted in revised form 31 March 1971*)

Abstract—New analyses of Apollo 11 and Apollo 12 rocks made by INAA for Na, Mn, Dy, Sm, Lu, La, Yb, Ce, Eu, Tb, Sc, Hf, Cr, Co, Fe, Ti, and Ta, and by X-ray fluorescence for Ca, together with radiochemical results for Au, Ag, and Mo and fission track U determinations for Apollo 12 rocks are reported. The excess of Au and Ag in soil relative to rocks reported by GANAPATHY, KEAYS, and ANDERS is confirmed, and contamination of a LRL processed rock for Au, Ag and Mo is clearly indicated.

INTRODUCTION

THE APOLLO 12 ROCKS and samples of regolith sent to us were analyzed using the methods described before for the Apollo 11 materials (TUREKIAN and KHARKAR, 1970). Several improvements were made, however, to correct some of the difficulties encountered in our earlier work. The problems of resolution and drift in the analyzer were corrected so that proper background and interference corrections could be made. All samples were processed in a laminar flow hood in a limited-access clean room far from the ambient contamination of our general laboratories.

The lunar samples were irradiated with three silicate base standards (W–1 diabase, "standard pot," and BCR–1) along with gold, silver, and molybdenum standards in solution pipetted in microgram amounts into quartz vials. All samples were run both as cellulose bound pellets as described earlier and in quartz vials, each in duplicate. The standards as pellets were also run in duplicate. Samples to be analyzed for Ag, Au, and Mo were those irradiated in quartz vials without binder added in order to eliminate a source of contamination found in the binder material. Values in parts per million for Na (2610), Co (14.06), Cr (115.10), Sc (20.55), Fe (10,170), Ti (7820) are those compared with the "standard pot" (PERLMAN and ASARO, 1969); Mn (1239), Ca (78,473), and U (0.45) are those compared with W–1 diabase (FLEISCHER, 1969; AHRENS *et al.*, 1967); and La (25.20), Ce (54.20), Sm (7.23), Eu (1.97), Dy (6.55), Tb (1.15), Yb (3.48), Lu (0.53), Hf (5.23), and Ta (0.90) are those compared with BCR–1 (NICHOLLS *et al.*, 1967).

X-ray fluorescence assays for Fe, Ti, and Ca were made. The Fe and Ti concentrations so determined were in excellent agreement with the INAA (Instrumental Neutron Activation Analysis) results. The calcium determinations are new in this program.

Because of these improved techniques we reran all the Apollo 11 samples previously reported as well as an additional sample of Apollo 11 "soil" received after

Table 1. Composition of Apollo 11 and Apollo 12 rocks and soils.[a,b] (ppm except as noted)

Sample	Type	Na	Na	%Ca	%Ca	Mn	Mn	%Fe	%Fe	%Ti	%Ti	La	La
10084,57	Soil D	3116		9.44		1587	1590	11.74		4.44		17.61	
		3118	3120	8.49	9.0	1589		12.44	12.1	4.00	4.2	17.33	17.5
10084,139	Soil D	3100		9.26		1588	1600	10.97		4.34		16.13	
		3108	3100	8.62	8.9	1612		11.41	11.2	4.31	4.3	15.88	16.0
10044,27	Vesicular	3630	3630	8.19	9.1	1941	1940	13.35	13.6	4.76	5.1	10.43	10.5
	basalt A			9.98		1935		13.75		5.48		10.52	
10049,33	Vesicular	3800	3780	8.33	7.7	1738	1740	13.64	13.7	6.98	6.7	23.89	24.6
	basalt A	3750		7.07		1732		13.74		6.36		25.28	
10057,79	Vesicular	4050	4050	7.40	7.8	1781	1780	13.94	14.5	7.80	7.3	22.63	23.1
	basalt A			8.28		1775		14.97		6.79		23.56	
10020,27	Gabbro B	2700	2690	9.55	8.7	1958	1960	13.09	13.7	6.35	5.8	6.67	6.6
		2680		7.91		1952		14.27		5.22		6.56	
10056,21	Gabbro B	2800	2810	9.53	8.9	2070	2070	13.06	13.1	4.98	5.1	11.95	11.3
		2827		8.24		2063		13.19		5.16		10.57	
10062,25	Gabbro B	2995	3000	9.09	8.3	2023	2020	14.25	14.0	6.91	6.9	11.21	11.5
		3000		7.53		2016		13.67		6.81		11.73	
10021,33	Breccia C	3470	3470	9.77	9.6	1582	1580	12.14	12.2	5.54	5.3	17.82	17.8
		3472		9.37		1577		12.33		5.04		17.73	
10046,24	Breccia C	3720	3700	8.73	8.8	1636	1630	12.13	12.1	5.29	5.4	18.36	18.6
		3680		8.95		1631		12.13		5.54		18.80	

Sample	Type	Ce	Ce	Sm	Sm	Eu	Eu	Dy	Dy	Tb	Tb	Yb	Yb
10084,57	Soil D	39.41	39.7	9.01	9.0	1.90	2.1	23.22	21.4	2.50	2.5	8.37	8.4
		40.00				2.20		19.54				8.44	
10084,139	Soil D	38.52	38.5	9.91	9.9	1.81	2.0	20.13	20.8	2.47	2.5	8.72	8.5
						2.18		21.45				8.36	
10044,27	Vesicular	37.21	37.6	11.43	11.4	2.83	3.0	27.45	27.6	3.90	3.9	10.87	10.4
	basalt A	38.00				3.18		27.78				9.96	
10049,33	Vesicular	70,67	71.1	15.65	15.7	2.00	2.1	31.61	30.6			14.08	14.2
	basalt A	71.60				2.22		29.49				14.36	
10057,79	Vesicular	68.39	69.6	14.27	14.3	2.27	2.4	33.30	29.9			13.16	13.2
	basalt A	70.70				2.44		26.52					
10020,27	Gabbro B	25.00	24.9	9.50	9.5	1.17	1.4	18.22	17.3	2.60	2.6	6.66	6.5
		24.80				1.67		16.31				6.39	
10056,21	Gabbro B	33.18	32.7	11.06	11.1	3.08	3.1	30.75	31.5			10.67	10.5
		32.24				3.06		32.27				10.23	
10062,25	Gabbro B	37.00	37.6	8.73	8.7	2.34	2.2	24.34	24.6			7.84	7.8
		38.10				1.97		24.87				7.81	
10021,33	Breccia C	48.07	48.3	11.13	11.2	1.90	1.9	22.36	20.9	3.10	3.1	10.08	9.9
		48.50		11.27		1.85		19.35				9.75	
10046,24	Breccia C	52.31	52.3	9.91	9.7	1.95	2.0	22.14	19.9			9.74	10.2
				9.48		2.11		17.61				10.74	

Sample	Type	Lu	Lu	Sc	Sc	Hf	Hf	Ta	Ta	Cr	Cr	Co	Co
10084,57	Soil D	1.64	1.64	68.05	68	10.84	10.8	1.11	1.3	1988	2070	33.88	33
		1.63		68.79				1.49		2152		33.90	
10084,139	Soil D	1.62	1.67	72.30	69	10.24	10.2	1.28	1.4	2423	2130	34.00	34
		1.71		66.03				1.54		1841		34.20	
10044,27	Vesicular	2.06	2.11	100.00	100	13.68	14.0	1.50	1.5	1416	1420	12.84	13
	basalt A	2.15		100.74		14.33		1.55		1419		13.00	
10049,33	Vesicular	2.60	2.58	86.77	86	17.34	17.3	1.97	1.9	2107	2100	22.25	23
	basalt A	2.56		86.15				1.91		2082		22.88	
10057,79	Vesicular	2.45	2.48	87.46	90	17.25	18.1	1.35	1.7	2257	2290	25.90	26
	basalt A	2.50		91.83		18.86		1.94		2328			
10020,27	Gabbro B	1.48	1.45	88.52	91	7.85	8.2	1.08	1.1	2258	2350	17.56	18
		1.42		94.03		8.62		1.14		2449		18.00	
10056,21	Gabbro B	2.17	2.0	99.05	100	14.94	16.3	1.65	1.8	1393	1380	11.54	12
		1.91		100.18		17.62		1.94		1369		11.58	
10062,25	Gabbro B	1.83	1.7	87.32	86	12.68	11.9	1.99	1.7	1691	1660	13.61	13
		1.62		85.28		11.16		1.41		1629		12.03	

Table 1 (continued)

Sample	Type	Lu	Lu	Sc	Sc	Hf	Hf	Ta	Ta	Cr	Cr	Co	Co
10021,33	Breccia C	2.94	2.2	71.82	72	14.00	13.4	1.77	1.8	2068	2100	26.32	27
		1.91		72.86		12.81				2138		28.26	
10046,24	Breccia C	1.90	1.93	72.31	72	12.71	13.4	1.41	1.4	2137	2150	27.60	27
		1.95		72.31		14.09				2165		26.93	

Sample	Type	Na	Na	%Ca	%Ca	Mn	Mn	%Fe	%Fe	%Ti	%Ti	La	La
12070,80	Fines D	3372	3410	8.56	8.4	1654	1640	12.21	12.1	2.27	2.2	33.42	33.4
		3450		8.20		1634		11.96		2.03		33.29	
12044,20	Fines D	3379	3360	8.63	9.1	1718	1710	12.15	12.2	2.58	2.4	32.98	32.7
		3342		9.49		1705		12.30		2.30		32.34	
12001,46	Fines D	3347	3390	8.25	8.4	1619	1620	12.36	12.3	2.18	2.3	31.85	32.0
		3425		8.54		1617		12.31		2.35		32.13	
12021,98	Crushed & homo-genized B	2025	2030	7.94	8.0	1984	2030	14.78	14.7	2.66	2.5	6.36	6.3
		1999		8.09		2078		14.59		2.37		6.16	
12052,54	Rock A	1844	1870	7.99	8.0	1980	1990	14.74	14.7	2.40	2.3	6.64	6.7
		1887				1990		14.68		2.14		6.82	
12009,41	Rock A	1827	1850	7.18	7.0	2040	2090	15.40	15.5	2.17	2.3	5.25	5.5
		1869		6.74		2144		15.49		2.40		5.81	
12020,41	Rock A	1460	1480	5.58	5.6	2061	2070	15.76	15.9	2.08	2.2	5.01	5.0
		1495		5.62		2119		15.94		2.25		5.06	
120063,116	Rock A	2077	2100	7.03	7.1	2040	2100	15.66	15.6	3.02	2.9	5.39	5.5
		2125		7.09		2166		15.56		2.69		5.63	

Sample	Type	Ce	Ce	Sm	Sm	Eu	Eu	Dy	Dy	Tb	Tb	Yb	Yb
12070,80	Fines D	74.48	74	15.00	15.1	1.80	1.80	27.44	26.7	3.26	3.3	12.77	12.8
		74.00		15.10		1.79		26.04		3.25		12.86	
12044,20	Fines D	70.98	71	15.23	15.1	1.80	1.82	27.45	26.8	3.25	3.2	12.46	12.3
		71.87		14.96		1.84		26.04		3.23		12.19	
12001,46	Fines D	71.00	71	15.31	15.2	1.82	1.82	25.22	24.6	3.00	3.1	12.49	12.6
		70.00		15.12		1.81		23.93		3.13		12.77	
12021,98	Crushed & homo-genized B	50.00	49	3.24	3.7	1.39	1.2	14.68	14.3	1.79	1.79	4.34	4.4
		48.00		4.13		0.99		13.93		1.79		4.45	
12052,54	Rock A	45.00	45	3.99	3.5	1.16	1.16	14.61	14.2	1.78	1.78	3.18	3.7
		44.50		3.09				13.86				4.14	
12009,41	Rock A	36.00	38	3.32	3.2	0.86	0.83	15.18	14.8	1.75	1.77	3.28	3.4
		40.00		2.98		0.79		14.41		1.78		3.54	
12020,41	Rock A	40.00	42	2.97	2.8	0.71	0.80	14.58	14.2	1.73	1.74	3.30	3.3
		43.00		2.69		0.89		13.83		1.75		3.25	
120063,116	Rock A	44.50	45	3.99	3.9	1.09	1.2	14.89	14.5	1.53	1.64	4.94	5.3
				3.82		1.22		14.13		1.75		5.61	

Sample	Type	Lu	Lu	Sc	Sc	Hf	Hf	Ta	Ta	Cr	Cr	Co	Co
12707,80	Fines D	1.75	1.75	39.30	40	14.02	14.2	1.83	1.8	2446	2400	43.49	43
				40.08		14.40				2355		41.56	
12044,20	Fines D	1.72	1.76	40.43	41	14.56	14.3	1.97	2.0	2410	2410	41.36	42
		1.79		40.84		14.00				2409		42.03	
12001,46	Fines D	1.66	1.69	38.49	39	13.99	13.7	2.00	2.0	2418	2430	40.64	40
		1.71		38.87		13.44				2443		40.17	
12021,98	Crushed & homogen-ized B	0.82	0.83	55.46	55	4.25	4.2	0.83	0.79	2413	2390	30.57	32
		0.84		55.30		4.23		0.75		2367		33.96	
12052,54	Rock A	0.84	0.89	51.90	53	4.05	4.2	1.01	0.88	3124	3140	38.49	40
		0.93		53.36		4.42		0.75		3161		40.58	
12009,41	Rock A	0.85	0.86	45.76	46	3.65	4.0	0.56	0.55	4336	4330	51.89	52
		0.86		45.78		4.25				4326		52.84	

Table 1 (continued)

Sample	Type	Lu	Lu	Sc	Sc	Hf	Hf	Ta	Ta	Cr	Cr	Co	Co
12020,41	Rock A	0.75 0.86	0.81	44.04 43.30	44	3.60 3.50	3.6	0.44 0.67	0.56	4157 4177	4170	56.84 58.58	58
120063,116	Rock A	0.92 1.09	1.00	57.75 58.57	58	4.01 4.94	4.5	0.63	0.62	2700 2702	2700	40.49 42.30	41

Sample	Type	U	U	Mo + U		(Mo)	Au	(ppb)	Ag	(ppb)
12070,80	Fines D	1.15 1.08	1.12	1.21 1.29	1.25	0.13	3.66 2.80	3.2	78 79	79
12044,20	Fines D	1.36 1.36	1.36				2.50 2.57	2.5	79 —	79
12001,46	Fines D	1.25 1.20	1.23	1.34 1.42	1.38	0.15	3.03 2.67	2.9	42 71	57
12021,98	Crushed & homogenized B	0.31 0.29	0.30	0.81 0.76	0.79	0.49	1.00 0.78	0.89	256 292	274
12052,54	Rock A	0.31 0.27	0.29	0.35 0.33	0.34	0.05	0.07 0.08	0.08	<1 <1	<1
12009,41	Rock A	0.25 0.23	0.24	0.38 0.35	0.37	0.13	0.05 0.06	0.06	<1 <1	<1
12020,41	Rock A	0.26	0.26	0.39 0.36	0.38	0.13	0.09 0.08	0.09	<1 <1	<1
12063,116	Rock A	0.20	0.20	0.28 0.24	0.26	0.08	0.05 0.05	0.05	<1 <1	<1

[a] The coefficients of variation based on the pooled estimates of duplicate analyses are as follows: Na 2.2%; Mn 2.3%; Dy 4.7%; Sm 8.7%; Lu 5.7%; Sc 1.1%; Hf 7.7%; Cr 1.0%; Co 3.5%; Fe 0.9%; La 2.7%; Yb 6.1%; Ce 2.9%; Eu 8.8%; Tb 2.4%; Ti 9.0%; Ca 3.0%; Au 12%; Ag 19%; Ta 11%; U 5.0%. Errors on duplications do not include uncertainties in the standards. Generally the total counting errors from all sources (samples, standards, blanks) are about the same as the errors based on duplicates.

[b] The uranium concentrations for the Apollo 11 materials, as reported in our previous study, are valid here. They are (in ppm): 10084,57 (0.29); 10044,27 (0.41); 10049,33 (0.74); 10057,79 (0.47); 10020,27 (0.22); 10056,21 (0.35); 10062,25 (0.28); 10021,33 (0.39); 10046,24 (0.69).

the initial work. These results are also presented, and they supercede our earlier work.

The results of our analyses are presented in Table 1.

The uranium results for the Apollo 11 samples are the same as those in TUREKIAN and KHARKAR (1970). Because there was an indication that our Apollo 11 samples, opened and processed in our earliest work, may have become contaminated, we did not think it advisable to determine Au, Ag, and Mo in these samples again. Since the Apollo 12 samples were processed in much more stringently clean areas, we believe results on Au, Ag, and Mo are not subject to major contamination effects.

Under any conditions it is clear that rock 12021,98 crushed and homogenized at the Lunar Receiving Laboratory is considerably higher in Au, Ag, and Mo relative to the other rocks, indicating the probability of contamination there when extensive processing has occurred.

Silver and gold have also been reported for Apollo 12 rocks and soils by GANAPATHY *et al* (1970). For the one common sample analyzed, soil 12070, they report 2.4 ppb Au to our 3.2 ppb, and 46 ppb Ag to our 78 ppb. We consider this to be reasonable agreement considering that there is some doubt of homogeneity in soil samples for these trace elements. Our averages for A and B rocks (ignoring the LRL processed sample) are Au: 0.067 ppb compared to 0.013 for GANAPATHY *et al.*; and

< 1 ppb Ag compared to their 0.81 ppb average. Our average soil values are 2.9 ppb Au compared to GANAPATHY *et al.*; 2.0 ppb and 71 ppb Ag compared to their 5.4 ppb (ignoring their high values, of which 12070 is one). Although differences exist between our two sets of analyses the data clearly indicate an excess of Ag and Au in lunar soil relative to lunar rock. The comparisons for Mo are less sure, but the data show no recognizable differences among the rocks and soils analyzed except for the highly contaminated 12021 processed at the LRL.

The differences and similarities between the Apollo 11 and Apollo 12 rocks and soils have been well stated by LSPET (1970). Our results are totally compatible with these results. We defer a detailed discussion of lunar surface rock and soil composition in terms of models of origin to a later time.

Acknowledgments—We wish to thank Dr. E. Perry for the X-ray fluorescence work reported here, Dr. L. Chan for the assistance in the uranium determinations, and Mr. L. Grossman and Mr. Joel Hasbrouck (an undergraduate working with Mr. Grossman) for their assistance in the computerized analysis of the analyzer data. The research was supported by the National Aeronautical and Space Administration under contract NAS-9-8032.

REFERENCES

AHRENS L. H., CHERRY R. D., and ERLANK A. J. (1967) Observations on the Th-U relationship in zircons from granitic rocks and from kimberlites. *Geochim. Cosmochim. Acta* **31**, 2379–2387.

FLEISCHER M. (1969) U.S.G.S. Standards—I. Additional data on rocks G-1 and W-1, 1965–1967. *Geochim. Cosmochim. Acta* **33**, 65–79.

GANAPATHY R., KEAYS R. R., and ANDERS E. (1970) Apollo 12 lunar samples: Trace element analysis of a core and the uniformity of the regolith. *Science* **170**, 533–535.

LSPET (Lunar Sample Preliminary Investigation Team) (1970) Preliminary examination of the lunar samples from Apollo 12. *Science* **167**, 1325–1339.

NICHOLLS G. D., GRAHAM A. L., WILLIAMS E., and WOOD M. (1967) Precision and accuracy in trace element analysis of geological materials, using solid source mass spectrography. *Anal. Chem.* **39**, 584–590.

PERLMAN I. and ASARO F. (1969) Pottery analysis by neutron activation. *Archaeometry*, **11**, 21–52.

TUREKIAN K. K. and KHARKAR D. P. (1970) Neutron activation analysis of milligram quantities of Apollo 11 lunar rocks and soil. *Proc. Apollo 11 Lunar Sci. Conf., Geochim. Cosmochim. Acta* Suppl. 1, Vol. 2, pp. 1659–1664. Pergamon.

Proceedings of the Second Lunar Science Conference, Volume 2, pp. 1307–1317.
The M.I.T. Press, 1971.

Rare-earth elements in Apollo 12 lunar materials

L. A. HASKIN, P. A. HELMKE, R. O. ALLEN, M. R. ANDERSON,
R. L. KOROTEV, and K. A. ZWEIFEL
Department of Chemistry, University of Wisconsin, Madison, Wisconsin 53706

(*Received 22 February 1971; accepted in revised form 9 April 1971*)

Abstract—With respect to their relative rare-earth element (REE) abundances, all except 3 of the Apollo 12 igneous rocks studied so far belong to a single group. Variations in REE concentrations among the members of that group are comparable to those found in small samples from a single, 10 meter thick flow of Icelandic basalt. Differences in REE concentrations and other characteristics among the members of this group appear to be related to olivine crystallization from a common initial liquid. Rocks 12022 and 12063 have identical relative REE abundances that can be distinguished from those found for rocks of the larger group by their relatively lower concentrations for the light REE. Rock 12038 has relatively higher light REE abundances than do the other rocks. All Apollo 12 samples have large negative Eu anomalies like those found in Apollo 11 materials.

NINE APOLLO 12 IGNEOUS ROCKS and 4 samples of "fines" were analyzed for the rare-earth elements (REE). These data were obtained by the same procedures of neutron activation that were used for our analyses of Apollo 11 samples (HASKIN *et al.*, 1970). Details of the procedures are found in the paper by DENECHAUD *et al.* (1970).

The results of our analyses are given in Tables 1 and 2. The uncertainties in the values for 12009 (column 1, Table 1), if converted to percentages, would be typical of the percent uncertainties for the data from the other samples. These uncertainties are standard deviations based on counting statistics, which we have found to be the controlling factor in our precision. For each sample, there is an additional, systematic uncertainty of approximately one percent for the entire group of REE.

To a first approximation, the abundances of the REE in the Apollo 12 igneous rocks and soils resemble those found in the Apollo 11 materials (e.g., HASKIN *et al.*, 1970). The relative abundances have a fairly flat distribution in comparison with the

Table 1. Rare-earth concentrations in Apollo 12 igneous rocks (ppm).

	12009,40	12020,35	12021,82	12022,54	12038,58	12053,28	12063,111	12064,16	12075,14
La	6.1 ± 1.2	5.19	6.16	5.81	11.8	7.13	6.24	6.76	6.33
Ce	16.8 ± .5	14.5	17.1	16.7	29.1	19.0	17.8	17.5	17.0
Nd	16 ± 5	13	16	19	22	15	16	16	13
Sm	4.53 ± .03	3.92	4.75	6.31	7.57	5.51	6.48	5.51	4.41
Eu	0.94 ± .01	0.76	1.055	1.32	1.969	1.05	1.36	1.161	0.91
Gd	5.2 ± .8	5.3	6.9	9.2	10.1	7.53	9.4	7.2	6.6
Tb	1.11 ± .07	1.02	1.14	1.56	1.61	1.33	1.66	1.27	1.06
Dy	7.13 ± .06	7.1	8.30	10.8	9.73	8.44	11.3	9.03	7.26
Ho	1.4 ± .1	1.34	1.36	1.87	1.99	2.1	2.0	1.72	1.37
Er	3.6 ± 1.2	3.8	4	5.8	5	4.8	5.3	6	5
Yb	3.74 ± .09	3.43	4.20	5.34	4.80	4.36	5.4	4.59	3.76
Lu	0.551 ± .007	0.510	0.523	0.767	0.689	0.614	0.79	0.67	0.53

Table 2. Rare-earth concentrations in Apollo 12 fines (ppm).

	12033,44	12042,45	12070,72	12070,145*
La	48.5 ± .6	36.8	34.9 ± .5	31.6 ± .6
Ce	127 ± 2	111	86.3 ± 1.3	80.2 ± 1.3
Nd	90 ± 20	79	57 ± 9	60 ± 15
Sm	21.5 ± .2	19.7	18.1 ± .3	16.3 ± .1
Eu	2.31 ± .02	2.03	1.79 ± .04	1.67 ± .01
Gd	28 ± 2	22.5	17.6 ± 1.2	22 ± 2
Tb	4.6 ± .2	3.87	3.7 ± .1	3.6 ± .1
Dy	31.8 ± .2	25.8	24.3 ± .3	23.3 ± .6
Ho	6.1 ± .3	4.46	4.90 ± .05	4.9 ± .2
Er	20 ± 6	14	18 ± 3	14 ± 2
Yb	17.4 ± .5	13.8	12.9 ± .4	12.2 ± .3
Lu	2.43 ± .03	2.09	1.90 ± .03	1.75 ± .02

* Coarse fines

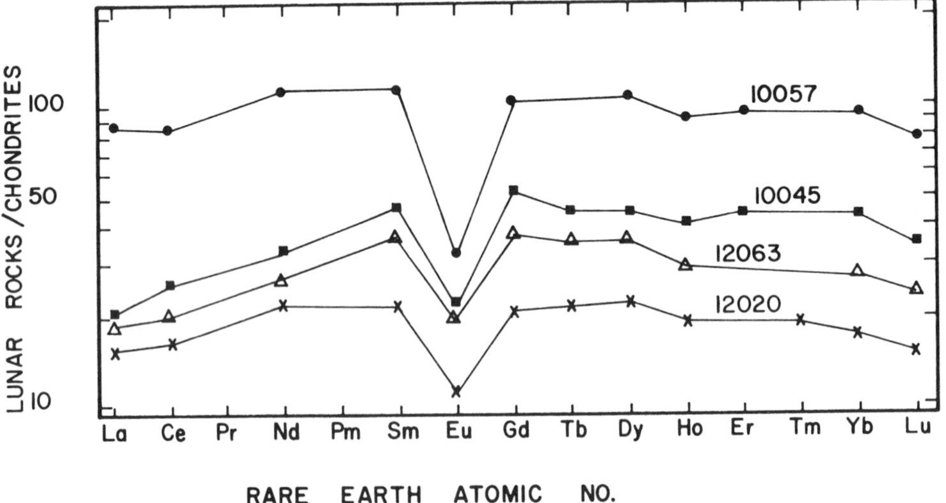

Fig. 1. Comparison diagram for Apollo 12 rocks 12020 and 12063 which represent the extremes of REE concentrations among the igneous rocks we analyzed (except for 12038). The Apollo 11 type B rock with the lowest REE concentrations (10045) and the type A rock with the highest concentrations (10057) are shown for comparison.

REE distribution in chondritic meteorites. There are large, negative Eu anomalies that increase in size with increasing total REE concentration. These features can be seen in the comparison diagrams in Figs. 1, 2, and 6.

Of the nine igneous rocks we have analyzed, all except one (12038) are closely similar to each other in REE concentrations. Of the 8 similar rocks, 12020 has the lowest and 12063 the highest REE concentrations. The REE abundances for these two rocks and, hence, the range in abundances for the 8 rocks, are shown in Fig. 1. This range is expanded slightly when data from other investigators for other igneous rocks are considered. The range of REE concentrations in all Apollo 12 igneous rocks studied so far is narrower than that found for the Apollo 11 rocks or for the type B

Apollo 11 rocks alone. The REE concentrations in the Apollo 12 igneous rocks (12038 excepted) are lower than the lowest of the type B Apollo 11 rocks. (Data from the papers by GAST and HUBBARD, 1971; PHILPOTTS *et al.*, 1971; WAKITA *et al.*, 1971; BRUNFELT *et al.*, 1971; and GOLES *et al.* 1971 are used in all these comparisons.) Data for Apollo 11 rocks 10057 (type A) and 10045 (type B), which represent the extremes for Apollo 11 igneous rocks, are also shown in Fig. 1 for comparison.

The range in concentration for La in the Apollo 12 igneous rocks (12038, 12022, and 12063 excepted, for reasons discussed below) is +18% and −35% from the mean value of 6.0 ppm. The ranges (same rocks excepted) are for Sm +24, −31%, for Eu +26, −21%, for Yb +41, −23%, and for the ratio of Sm to Eu ±10% from the mean values. The range of variation for the ratio of Sm to Eu is the same as that found among samples cored from a single flow of homogeneous Icelandic Basalt (HASKIN *et al.*, 1971). The ranges in concentrations for La, Sm, Eu, and Yb exceed the ranges in the Icelandic basalt only by factors of about 2 to 3. They are considerably less than the ranges of concentrations found for basalts cored from successive lava flows from a single volcano (Steens Mountain, P. HELMKE and L. HASKIN, to be published).

The ranges of variation in REE concentrations for those rocks that contain olivine phenocrysts (12022, 12004, 12009, 12018, 12020, 12035, 12040, 12075) are about the same as found for the cores of the Icelandic basalt. The ranges of variation for those rocks with no olivine phenocrysts (12021, 12051, 12052, 12053, 12064, 12065) are considerably smaller than those found for the Icelandic basalt. On the basis of the REE data, the group of rocks containing olivine phenocrysts could be from one single flow, those containing no olivine phenocrysts from another. In fact, the members of both groups could be from the same flow. Their relative REE abundance distributions are all nearly indistinguishable.

The relative REE abundances for the rocks excluded from the above groups can be readily distinguished from those for the rest of the rocks. The REE distributions for 12022 and 12063 are identical and differ from those for the rest of the rocks by being depleted in La (35%), Ce (30%), and Nd (15%). Rock 12038 has REE concentrations similar to those for the rest of the Apollo 12 igneous rocks for the elements Gd through Lu, but is relatively enriched in the lighter REE. Thus, 3 distinguishable relative REE abundance distributions are present, and are shown in Figs. 1 and 2.

The narrow range of variation among these rocks is further emphasized in Fig. 3. A graph of the ratio of Sm concentrations to Eu concentrations versus ppm Sm was found to give a straight line for Apollo 11 igneous rocks (HASKIN *et al.*, 1970). The theoretical justification for this kind of graph was discussed briefly in the same paper. We are aware that there is danger in interpreting graphs of x/y versus x inasmuch as such graphs always show at least some correlation, and a very good one whenever y is nearly constant. The variation in Eu concentrations among the Apollo 11 rocks is not great, but was large enough to establish the relationship shown in Fig. 3. The line is a least squares fit to the Apollo 11 data. Such a straight-line relationship could be exhibited by data for rocks that were chilled liquids from a single sequence of partial melting or fractional crystallization. (The absence of any overlap in the ranges for several elements, e.g., K, Rb, Ba, U, Th, between the A and B types of Apollo 11

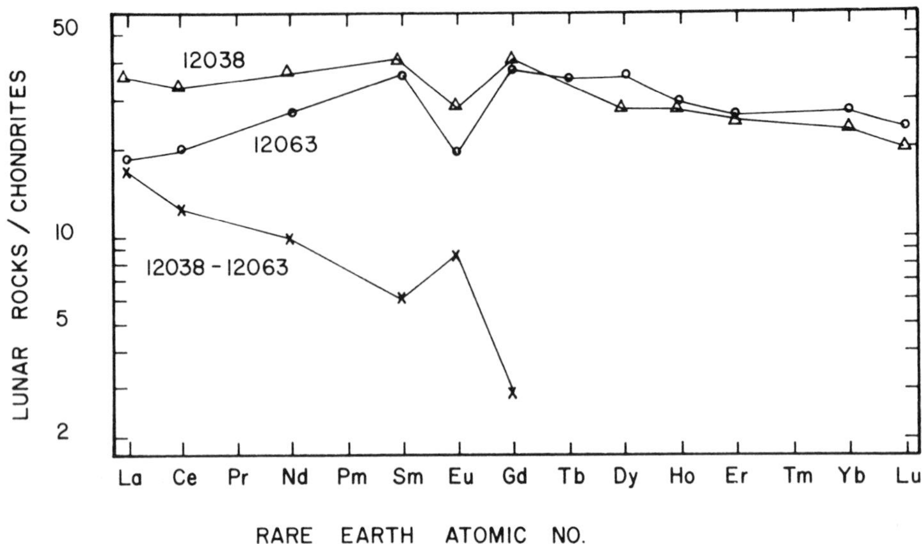

Fig. 2. Comparison diagram for rocks 12038 and 12063 and the difference in REE concentrations between them. The difference has a REE distribution similar to that characteristic of plagioclase.

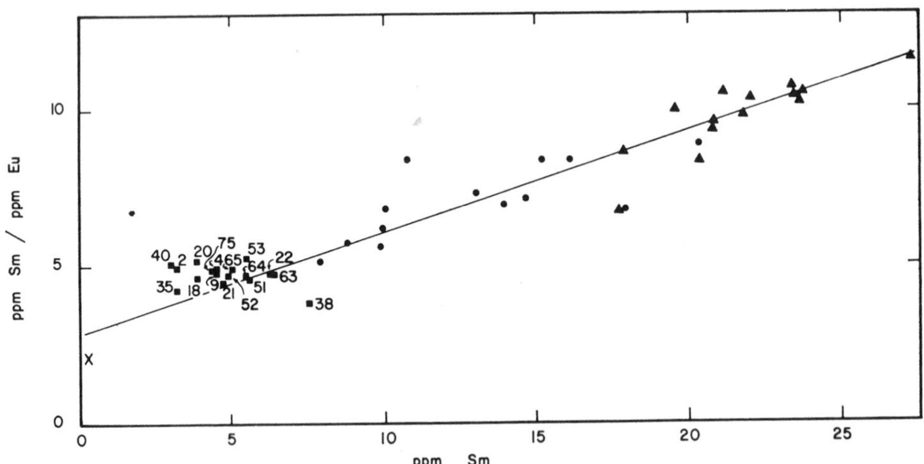

Fig. 3. Graph of ratios of concentration of Sm to Eu versus concentrations of Sm. Triangle symbols are Apollo 11 type A rocks, circle symbols are Apollo 11 type B rocks. The line corresponds to a least squares fit to the Apollo 11 data. The X represents chondritic meteorites. Square symbols denote Apollo 12 igneous rocks, which are numbered.

basalts argues against such a simple relationship for the members of those two rock types.)

There is no hint in Fig. 3 of a similar trend of increase in the ratio of Sm to Eu with increasing concentration of Sm for the Apollo 12 igneous rocks. Nevertheless, most of the rocks yield values that lie close to the line drawn for the Apollo 11 samples. Except for rock 12038, the values for the Apollo 12 rocks lie on or to the left of the line. All of the Apollo 12 rocks except 12038 have essentially the same value for the ratio of Sm to Eu. Such variation among the Apollo 12 rocks as appears in Fig. 3 arises mainly from variation in Sm concentrations. There is no a priori reason to presume that the Apollo 12 rocks were derived from a liquid whose REE values would fall on the line for Apollo 11 rocks but, if they did, why might the values for the rocks themselves not lie on the line? Dilution of such a liquid with a mineral of low REE concentrations would appropriately lower the Sm concentrations for the rocks from that of the initial liquid without changing appreciably the value of the ratio of Sm to Eu. Any mineral with solid-liquid distribution coefficients of the order of 0.1 or lower would produce such an effect. Olivine has suitably low distribution coefficients (e.g., PHILPOTTS and SCHNETZLER, 1970; HASKIN *et al.*, 1970) and occurs as phenocrysts in many of the Apollo 12 basalts (KLEIN *et al.*, 1971; WALTER *et al.*, 1971; BRETT *et al.*, 1971; ANDERSON *et al.*, 1971; DENCE *et al.*, 1971; HOLLISTER *et al.*, 1971; BROWN *et al.*, 1971).

Modal mineral data for the rocks are required to test the above hypothesis, namely, do the Apollo 12 rocks correspond to different extents of dilution of a single liquid with crystals of olivine settled from that liquid as it cooled? Modal data are hard to find and, in any case, might not be appropriate for the chips of rock analyzed for the REE because the distributions of olivine phenocrysts on a 1 g scale in the rocks are not uniform. Qualitatively, data from some of the rocks described as containing significant quantities of olivine phenocrysts do lie farthest to the left of the line in Fig. 3, with data from some rocks that contain few or no phenocrysts somewhat farther to the right. However, rocks 12022 and 12063 clearly contain olivine (BRETT *et al.*, 1971; HOLLISTER *et al.*, 1971) and their REE values lie very near the line.

Some petrologists have also suggested that the rocks might come from a single initial liquid that underwent crystal settling of olivine on a small scale. The bulk chemical compositions of the rocks would then depend on the extent of olivine crystallization (e.g., COMPSTON *et al.*, 1971; KUSHIRO *et al.*, 1971; ANDERSON *et al.*, 1971; GREEN *et al.*, 1971; BIGGAR *et al.*, 1971; WALTER *et al.*, 1971). KUSHIRO *et al.* (1971) have used a variation diagram of SiO_2 versus Al_2O_3 to show the effect of olivine crystallization on the final compositions of the rocks. Figure 4 is such a diagram, based on the analytical data of COMPSTON *et al.* (1971); ENGEL and ENGEL (1971); ANNELL *et al.* (1971); WILLIS *et al.* (1971); MAXWELL and WIIK (1971); SMALES (1971); SCOON (1971); WÄNKE *et al.* (1971); and EHMANN and MORGAN (1971). The data, except those for rocks 12022, 12063, 12038, and 12035, define a line whose intercept on the SiO_2 axis is at 39.1% SiO_2. In effect, this intercept gives the value for the SiO_2 content of the hypothetical rock in the series that would contain no Al. In fact, for the Apollo 12 rock series, only the mineral olivine contains silica but no Al. The composition of the olivine corresponding to 39% SiO_2 is Fo_{79}, just the value found

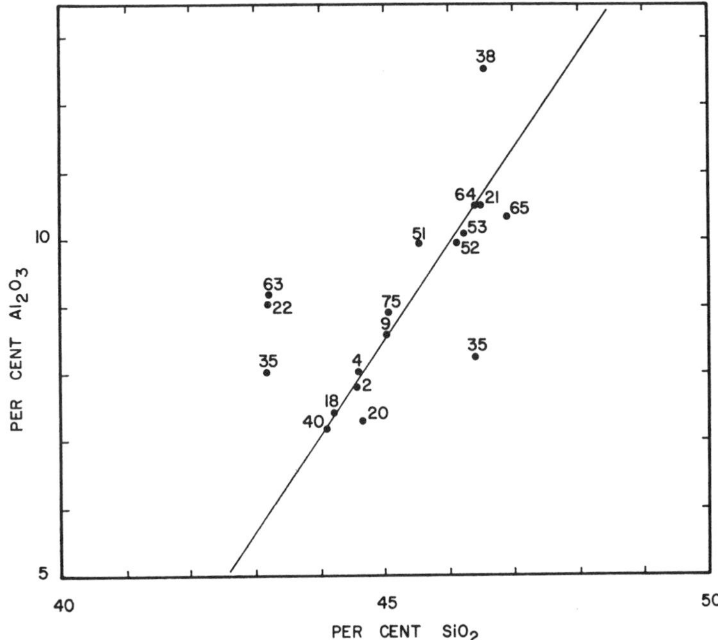

Fig. 4. Variation diagram for SiO_2 and Al_2O_3 which illustrates the olivine control line for the fractional crystallization of a parent liquid. For discussion see KUSHIRO *et al.* (1971).

experimentally to make up the cores of the olivine phenocrysts in the rocks. (The value of SiO_2 used for rock 12035, COMPSTON *et al.*, 1971, puts the point for that rock too far to the left of the line in Fig. 4. The value given by EHMANN and MORGAN, 1971, puts the point an equal distance to the right of the line. Possibly a value for the SiO_2 content that would make the point fall nearer to the line would be more representative of the rock than either of those available.)

Further justification for considering rocks 12022 and 12063 as part of some different series, and 12038 as having still another source is found in Fig. 5. In this figure, the concentrations of a typical REE (Sm) are plotted against the concentrations of a trace element (Co) that may have been enriched in crystallizing olivine compared with its concentration in the liquid. Abundances for Co were taken from the papers of ANNELL *et al.* (1971), BRUNFELT *et al.* (1971), COMPSTON *et al.* (1971), GOLES *et al.* (1971), SMALES (1971), WAKITA *et al.* (1971), and WÄNKE *et al.* (1971). Values for all rocks except 12022, 12063, and 12038 are considered, their relative olivine contents (as inferred from Figs. 4 and 5) correlate qualitatively quite well with the distance to the left of the line in Fig. 3 at which the corresponding points for the REE data are found.

The REE data for rocks 12022 and 12063 give points very near the line in Fig. 3. Possibly, they should not, if abundant olivine phenocrysts are present and if the line represents the composition of their parent liquid. They may, however, correspond to

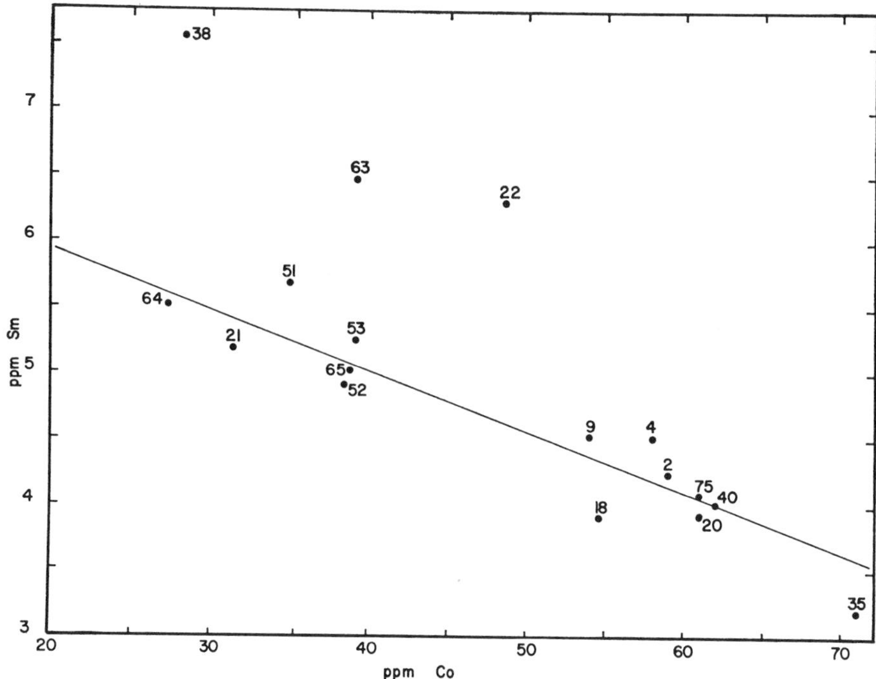

Fig. 5. Variation diagram for Sm and Co.

olivine phenocrysts plus all of the residual liquid from which the phenocrysts crystallized.

The REE data for rock 12038 plot to the right of the line in Fig. 3. For that rock to have derived from a liquid whose REE concentrations would give a point on the line, the ratio of Sm to Eu would have to be decreased more rapidly than the concentration of Sm was changed. Qualitatively, this might occur if some crystals of plagioclase feldspar were added to the liquid or became dissolved in it. Feldspars are selectively enriched in Eu relative to the other REE during crystal fractionation (e.g., SCHNETZLER and PHILPOTTS, 1970; HASKIN *et al.*, 1970). Figure 2 shows the REE distribution for 12038, and the difference between the REE distributions for 12038 and for 12063, both of which have similar concentrations of the heavy REE. The differences give points on the comparison diagram that are similar to those found for feldspars (SCHNETZLER and PHILPOTTS, 1970). There is no independent evidence for modification of an initial liquid by feldspar to produce rock 12038.

The significance of the relationship shown in Fig. 3 and discussed in detail in an earlier paper (HASKIN *et al.*, 1970) remains uncertain. There is no compelling reason to believe that all liquids that produce rocks in the lunar maria should yield points on the line in Fig. 3, but the data so far obtained apparently lie on or near it. Clearly, it is not reasonable to suggest that the Apollo 12 igneous rocks are derived from chilled liquids from the same sequence of partial melting or fractional crystallization that might have produced the distant Apollo 11 basalts. It is plausible, however, to suggest that the Apollo 12 igneous rocks and the type A and type B Apollo 11 rocks all derived by similar processes (e.g., partial melting) from starting materials that were identical or very similar in gross composition and character and in REE concentrations. In the cases of the Apollo 11 rocks, the liquids that were derived from the starting materials would have chilled with little further modification. The liquid that produced most of the Apollo 12 rocks would correspond to a greater extent of partial melting of the starting material and would have been modified by fractional crystallization before the rocks solidified.

In general, the data for REE in Apollo 12 igneous rocks support the conclusions we made from our studies of Apollo 11 igneous rocks. Values for liquid-solid distribution coefficients for the REE are not known well enough to allow us to distinguish between partial melting and fractional crystallization for the production of the liquids. The need for large amounts of feldspar remains as does the difficulty that feldspar apparently does not lie on the liquidus under the conditions of temperature and pressure at which the liquids could have formed (Ringwood and Essene, 1970). Nevertheless, the possibility remains that the highlands contain the large amounts of feldspar needed to account for the Eu deficiencies in the rocks of the lunar seas (Wood *et al.*, 1970; Wakita and Schmitt, 1970).

The Apollo 12 "fines" (Table 2, Fig. 6) are not as close in REE concentrations to the Apollo 12 igneous rocks as the Apollo 11 "fines" were to the Apollo 11 igneous rocks. The necessity for an "exotic" component in the soil with high REE concentrations and the possible identity of that component has been discussed by other investigators (e.g., Hubbard *et al.*, 1971; Philpotts *et al.*, 1971; Goles *et al.*, 1971; Wakita *et al.*, 1971; Meyer *et al.*, 1971). The highest REE concentrations found in our samples of Apollo 12 "fines" were in sample 12033, the light-colored soil taken from the trench on the rim of head crater. The lowest concentrations were found in sample 12070 from the contingency sample. The REE concentrations for the 12070

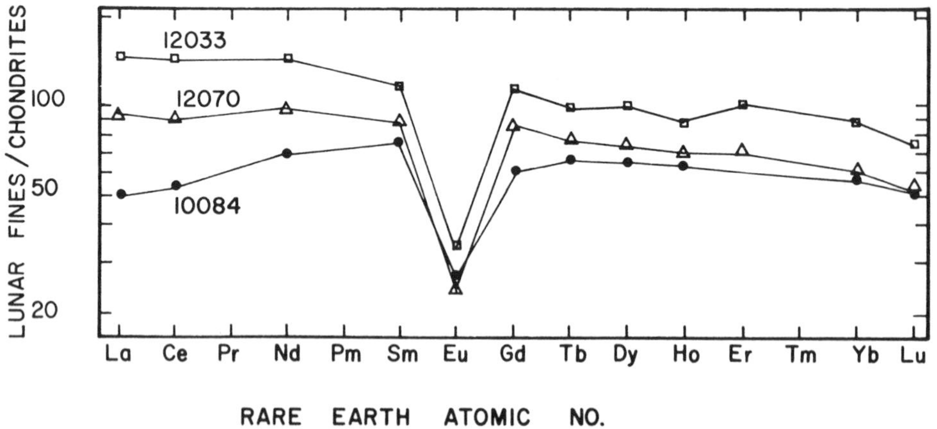

Fig. 6. Comparison diagram for 12070 and 12033 which represent the extremes of REE concentrations found in the Apollo 12 "fines". Data from the Apollo 11 soil (10084) are included for comparison.

coarse "fines" are essentially the same as those for the 12070 fine "fines." Compared with the Apollo 11 fines (10084) the Apollo 12 fines have higher REE concentrations, are relatively enriched in the lightest REE (e.g., La, 80%) and are slightly depleted in Eu (by about 10%, based on the interpolated value). These are, qualitatively, the same features that distinguish the average REE abundances for the Earth's crust from that found in the chondritic meteorites (e.g., HASKIN *et al.*, 1968).

The very high REE concentrations for the soils in both lunar maria (the Sea of Tranquillity and the Ocean of Storms) and the comparatively lower REE concentrations for many of the igneous rocks require that some abundant source of material of high REE content exist. Should materials with these required high REE contents be plentiful in the lunar highlands, which are a large and obvious potential source for material for the lunar fines, then the difficulties of having average absolute REE concentrations for the moon as a whole that are the same as those found in the chondritic meteorites (HASKIN *et al.*, 1970) are compounded far beyond redemption.

Acknowledgments—We are grateful to the Apollo 12 astronauts and supporting personnel whose efforts made these lunar samples available, and to the crew of the University of Wisconsin nuclear reactor for the neutron irradiations. This work was supported in part by the National Aeronautics and Space Administration under contract NAS–9–7975.

REFERENCES

ANDERSON A. T., JR., NEWTON R. C., and SMITH J. V. (1971) Apollo 12 mineralogy and petrology: Light-colored fragments, trace-element concentrations; petrologic development of the Moon. Second Lunar Science Conference (unpublished proceedings).

ANNELL C. S., CARRON M. K., CHRISTIAN R. P., CUTTITTA F., DWORNIK E. J., HELZ A. W., LIGON D. T., JR., and ROSE H. J., JR. (1971) Chemical and spectrographic analysis of lunar samples from the Apollo 12 mission. Second Lunar Science Conference (unpublished proceedings).

BIGGAR G. M., O'HARA M. J., and PECKETT A. (1971) Origin, eruption, and crystallization of protohypersthene basalts from the Ocean of Storms. Second Lunar Science Conference (unpublished proceedings).

BRETT R., BUTLER P., JR., MEYER C., JR., REID A. M., TAKEDA H., qnd WILLIAMS R. J. (1971) Apollo 12 igneous rocks 12004, 12008, 12009, and 12022: A mineralogical and petrological study. Second Lunar Science Conference (unpublished proceedings).

BROWN G. M., EMELEUS C. H., HOLLAND J. G., PECKETT A., and PHILLIPS R. (1971) Mineral chemistry of contrasted Apollo 12 basalt-types and comparisons with Apollo 11. Second Lunar Science Conference (unpublished proceedings).

BRUNFELT A. O., HEIER K. S., and STEINNES E. (1971) Determination of 40 elements in Apollo 12 materials by neutron activation analysis. Second Lunar Science Conference (unpublished proceedings).

COMPSTON W., BERRY H., VERNON M. J., CHAPPELL B. W., and KAYE M. J. (1971) Rubidium-strontium chronology and chemistry of lunar material from the Ocean of Storms. Second Lunar Science Conference (unpublished proceedings).

DENCE I. R., DOUGLAS J. A. V., PLANT A. G., and TRAILL R. J. (1971) Mineralogy and petrology of some Apollo 12 samples. Second Lunar Science Conference (unpublished proceedings).

DENECHAUD E. B., HELMKE P. A., and HASKIN L. A. (1970) Analysis for the rare earth elements by neutron activation and Ge (Li) spectrometry. *J. Radioanal. Chem.* **6**, 97–113.

EHMANN W. D. and MORGAN J. W. (1971) Major element abundances in Apollo 12 rocks and fines by 14 MeV neutron activation. Second Lunar Science Conference (unpublished procedings).

ENGEL C. G. and ENGEL A. E. J. (1971) Major element composition of three Apollo 12 rocks and some petrogenic considerations. Second Lunar Science Conference (unpublished proceedings).

Gast P. W. and Hubbard N. J. (1971) Rare earth abundances in soil and rocks from the Ocean of Storms. Second Lunar Science Conference (unpublished proceedings).

Goles G. G., Duncan A. R., Osawa M., Martin M. R., Beyer R. L., Lindstrom D. J., and Randle K. (1971) Analyses of Apollo 12 specimens and a mixing model for Apollo XII "soils." Second Lunar Science Conference (unpublished proceedings).

Green D. H., Ringwood A. E., Ware N. G., Hibberson W. O., Major A., and Kiss E. (1971) Experimental petrology and petrogenesis of Apollo 12 basalts. Second Lunar Science Conference (unpublished proceedings).

Haskin L. A., Haskin M. A., Frey F. A., and Wildeman T. R. (1968) Relative and absolute terrestrial abundances of the rare earths. In *Origin and Distribution of the Elements* (editor L. H. Ahrens), pp. 889–912, Pergamon.

Haskin L. A., Allen R. O., Helmke P. A., Paster T. P., Anderson M. R., Korotev R. L., and Zweifel K. A. (1970) Rare earths and other trace elements in Apollo 11 lunar samples. *Proc. Apollo 11 Lunar Sci. Conf., Geochim. Cosmochim. Acta* Suppl. 1, Vol. 2, pp. 1213–1231. Pergamon.

Haskin L. A., Helmke P. A., Paster T. P., and Allen R. O. (1971) Rare earths in meteoritic, terrestrial, and lunar matter. *Proc. NATO Conf. Activation Analysis in Geochemistry, Oslo* (in press).

Hollister L., Trzcienski W., Hargraves R., and Kulick C. (1971) Crystallization histories of samples 12063 and 12065. Second Lunar Science Conference (unpublished proceedings).

Hubbard N. J., Gast P. W., and Meyer C. (1971) The origin of the lunar soil based on REE, K, Rb, Ba, Sr, P, and $^{87}Sr/^{86}Sr$ data. Second Lunar Science Conference (unpublished proceedings).

Klein C., Drake J. C., Ito J., and Frondel C. (1971) Mineralogical, petrological, and chemical features of four lunar gabbros. Second Lunar Science Conference (unpublished proceedings).

Kushiro I., Nakamura Y., Haramura H., and Akimoto S. (1971) Origin of chemical and mineralogical variations of the Apollo 12 crystalline rocks. Second Lunar Science Conference (unpublished proceedings).

Maxwell J. A. and Wiik H. B. (1971) Chemical compositions of Apollo 12 lunar samples. Second Lunar Science Conference (unpublished proceedings).

Meyer C. Jr., Aitken F. K., Brett R., McKay D. S., and Morrison P. A. (1971) Rock fragments and glasses rich in K, REE, P in Apollo 12 soils: Their mineralogy and origin. Second Lunar Science Conference (unpublished proceedings).

Philpotts J. A., Schnetzler C. C., Bottino M. S., and Fullagar P. D. (1971) Li, K, Rb, Sr, Ba and rare earth concentrations and $^{87}Sr/^{86}Sr$ in some Apollo 12 soils, rocks and separated phases. Second Lunar Science Conference (unpublished proceedings).

Ringwood A. E. and Essene E. J. (1970) Petrogenesis of Apollo 11 basalts, internal constitution and origin of the moon. *Proc. Apollo 11 Lunar Sci. Conf., Geochim. Cosmochim. Acta* Suppl. 1, Vol. 1, pp. 769–799. Pergamon.

Schnetzler C. C. and Philpotts J. A. (1970) Partition coefficients of rare-earth elements between igneous matrix material and rock-forming mineral phenocrysts—II. *Geochim. Cosmochim. Acta* **34**, 331–340.

Scoon J. H. (1971) Quantitative chemical analysis of lunar samples 12040 and 12064. Second Lunar Science Conference (unpublished proceedings).

Smales A. A. (1971) Elemental composition of lunar surface material. Second Lunar Science Conference (unpublished proceedings).

Wakita H. and Schmitt R. A. (1970) Rare-earth and other elemental abundances in lunar anorthosites. *Science* **170**, 969–974.

Wakita H., Rey P., and Schmitt R. A. (1971) Abundances of the 14 rare-earth elements plus 22 major, minor, and trace elements in ten Apollo 12 rock and soil samples. Second Lunar Science Conference (unpublished proceedings).

Walter L. S., French B. M., Ghose S., Heinrich K. F. J., Spijkerman J. J., Lowman P. D., Jr., Doan A. S., and Adler I. (1971) Mineralogical studies of Apollo 12 samples. Second Lunar Science Conference (unpublished proceedings).

Wänke H., Wlotzka F., Teschke F., Baddenhausen H., Spettel B., Balacescu A., Quijano-Rico M., Jagoutz E., and Rieder R. (1971) Major and trace elements in Apollo 12 samples and studies on lunar metallic iron particles. Second Lunar Science Conference (unpublished proceedings).

Willis J. P., Ahrens L. H., Danchin R. V., Erlank A. J., Gurney J. J., Hofmeyr P. K., McCarthy T. S., and Orren M. J. (1971) Some interelement relationships between lunar rocks and fines, and stony meteorites. Second Lunar Science Conference (unpublished proceedings).

Wood J. A., Dickey J. S., Jr., Marvin U. B., and Powell B. N. (1970) Lunar anorthosites and a geophysical model of the moon. *Proc. Apollo 11 Lunar Sci. Conf., Geochim. Cosmochim. Acta* Suppl. 1, Vol. 1, pp. 965–988. Pergamon.

Proceedings of the Second Lunar Science Conference, Volume 2, pp. 1319–1329.
The M.I.T. Press, 1971.

Abundances of the 14 rare-earth elements and 12 other trace elements in Apollo 12 samples: Five igneous and one breccia rocks and four soils

H. WAKITA, P. REY, and R. A. SCHMITT
Department of Chemistry and the Radiation Center,
Oregon State University, Corvallis, Oregon 97331

(*Received 12 February 1971; accepted 5 March 1971*)

Abstract—Abundances of fourteen REE (rare-earth elements), Y, Cd, In, Rb, Cs, V, Sc, Co, Zr, Hf, Th, and Ba have been determined by radiochemical and instrumental neutron activation analysis (RNAA and INAA). The chondritic normalized REE distribution patterns, REE abundance ranges, and the degrees of Eu depletion observed in Apollo 12 rock and soil samples are significantly different from each other and also from Apollo 11 rock and soil samples. Total REE + Y abundances in four Apollo 12 soils vary from 390 to 720 ppm and are higher by a factor of 4–7 compared to Apollo 12 crystalline rocks. The REE in a 15-cm deep soil 12033 (this work) are significantly more abundant and display about the same chondritic normalized distribution patterns compared to REE abundances and patterns measured in five core samples (1–37 cm) by PHILPOTTS *et al.* (1971). These soil analyses stress the striking vertical and lateral inhomogeneity of the regolith at the Apollo 12 site. Sm/Eu ratios in Apollo 12 soils and basalts are 9.7 $\pm$ 0.3 and 4.8 $\pm$ 0.2, respectively, which may be compared to average ratios of 7.4 $\pm$ 0.4 in Apollo 11 soil and 9.4 $\pm$ 0.6 and 6.5 $\pm$ 0.7 in Apollo 11 high and low alkali rocks, respectively (Chondritic Sm/Eu = 2.67). Indium abundances are 2 $\pm$ 1 ppb in two Apollo 12 soils and four basalts, and are similar to the lowest in abundances found in Apollo 11 rocks. Rubidium and Cs are enriched by factors of 1–4 and 2–4, respectively, in Apollo 12 soils relative to Apollo 11 soil. In general, trace elemental abundances of Apollo 12 porphyritic and granular-ophitic basalts may be distinguished by their REE, Y, Co, V, Hf, and Ba abundances. The striking similarity in major, minor, and trace elemental abundances for breccia rock 12034 and soil 12033, both obtained from a 15-cm deep trench, suggest a direct relationship between these two samples.

INTRODUCTION

BULK COMPOSITIONS for the same specimens have also been determined by WAKITA *et al.* (1971) and are discussed in a separate paper of this volume. Each specimen was split into two roughly equal portions. After determinations of V (Al, Ti, Ca, Na, and Mn) by INAA in each split, half of the sample splits were subjected to RNAA for the fourteen REE, Y, Cd, In, (K), Rb, and Cs. For the other half of sample splits, Mg was determined via 23-MeV bremsstrahlung activation analysis, and finally Sc, Co, Zr, Hf, Th, Ba, (Fe), (Cr), La, Sm, Eu, Yb, and Lu were analyzed via INAA. For the soil sample 12070, assumed to be homogeneous at the 0.5 g size, INAA were done on half of the sample; the other split was directly subjected to RNAA. Analytical methods have been previously described by REY *et al.* (1970), SCHMITT *et al.* (1970), and WAKITA *et al.* (1970).

Results and Discussion

Rare earths and yttrium

In detail, the REE distribution patterns, their abundance ranges, and the degrees of Eu depletion observed in Apollo 12 samples are significantly different from each other. The Apollo 12 rock and soil samples have highly variable REE distribution patterns and total REE + Y abundance ranges, compared to Apollo 11 samples, which suggest more complex histories for the Ocean of Storms sites. Relative to REE abundance in Apollo 11 soil, the light REE in Apollo 12 soils are progressively riched from Sm to La. Similar heavy REE patterns are observed in both Apollo 11 and Apollo 12 soils. The abundances of all REE + Y in four Apollo 12 soils vary from 390 to 720 ppm, compared to 300 ppm in Apollo 11 soil. These total values are about 80–140 and 55 times more abundant, respectively, than REE + Y in chondrites.

Comparing the REE abundance data of seven different Apollo 12 soils, four samples from this work and three from other investigators (Gast and Hubbard, 1971; Goles *et al.*, 1971; Haskin *et al.*, 1971; Philpotts *et al.*, 1971; and others), we note that Apollo 12 soils are chemically inhomogeneous. The soils 12001, 12044, and 12070 *and* the soils 12032 and 12042 have relatively tight abundance ranges of 16.5 ± 0.7 and 21.2 ± 0.6 Sm ppm, respectively. On the other hand, two determinations of the Sm abundance in the soil 12033 vary from 21.5 to 31 ppm (Haskin *et al.*, 1971 and Wakita *et al.*, 1971). Furthermore, significant differences in homogeneity up to a factor of two were observed in the soil 12037 (2–4 mm fragments); for example, La varies from 26.2 to 51.3 ppm and Sm, from 13.0 to 23.5 ppm. The abundances of other trace elements and common bulk elements are also significantly different for duplicate ~0.2 g sample analyses. Soil samples 12033 and 12037, both coarse fines, were distinctly inhomogeneous under visual inspection.

The chondritic normalized REE distribution patterns are quite similar for all four Apollo 12 soils and one breccia rock. Above all, the REE distribution patterns and the REE and other trace elemental abundances in breccia 12034 and soil 12033, were found to be almost identical (Table 1 and Fig. 1). Both specimens were retrieved from a 15-cm deep trench at the northwest rim of Head Crater (LSPET, 1970). Consistent with similar major and minor elemental abundances in both samples (Wakita *et al.*, 1971), we conclude that either (a), this 12034 breccia rock was compacted from soil in the regolith that was identical to the 12033 soil, and then both the 12034 breccia and the 12033 soil were ejected onto the surface in a cratering event, or (b) it is also possible that the 12033 soil fragments are the disintegration fragments of a precursor rock to 12034.

For Apollo 12 crystalline rocks, the REE distribution patterns for the heavier REE (Gd–Lu) are quite similar to those of Apollo 11 rocks and both Apollo 11 and 12 soils. Excluding the degree of Eu depletion, overall similarity is observed between the Apollo 11 and Apollo 12 igneous rocks (Fig. 1). The total REE + Y abundances (90–140 ppm) in Apollo 12 crystalline rocks (Table 1) are approximately 17–27 times higher than chondritic abundances. Relative to Apollo 12 soils, the igneous rocks are depleted in REE by a factor of 4–7.

Warner and Anderson (1971) have petrologically classified the Apollo 12 crystalline rocks into two magma types, porphyritic basalts (type 1) and granular and

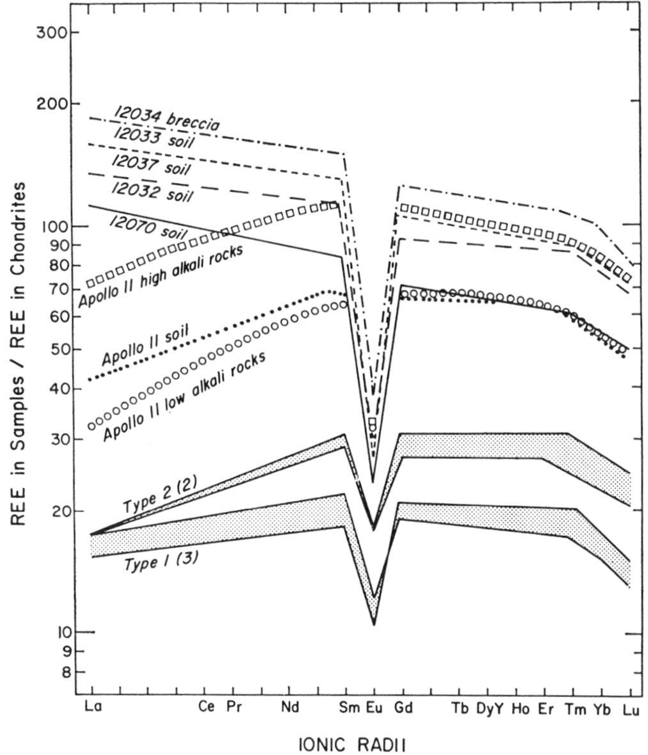

Fig. 1. REE abundances in Apollo 12 samples are taken from this work; values of Apollo 11 samples are taken from WAKITA *et al.* (1970). For normalization, the chondritic abundances were determined by WAKITA and ZELLMER (1970) in composite sample of 12 chondrites (La 0.34, Ce 0.91, Pr 0.121, Nd 0.64, Sm 0.195, Eu 0.073, Gd 0.26, Tb 0.047, Dy 0.30, Ho 0.078, Er 0.20, Tm 0.032, Yb 0.22, Lu 0.034, and Y 1.7; $\sum$ REE + Y = 5.2 ppm); REE radii by WHITTAKER and MUNTUS (1970).

ophitic basalts (type 2). Major and minor elemental abundance data (WAKITA *et al.*, 1971) and trace element data of this work strongly support their classification.

The Eu anomalies in Apollo 12 samples have a broader range than those observed in Apollo 11 rocks and soil. We have compiled all the Apollo 11 and 12 data on Sm/Eu ratios versus Sm abundances, which are approximately proportional to the total REE abundances (Fig. 2). All data were taken from this study and other works (see footnote to Fig. 2). A higher Sm/Eu ratio indicates greater Eu depletion. In general, this diagram shows that higher total REE abundances are associated with a greater Eu depletion (HASKIN *et al.*, 1970). The seven Apollo 12 soils have rather constant Sm/Eu ratios, even though their Sm abundances range within a factor of two, i.e., from 16 to 31 ppm. The average Sm/Eu ratios of 9.7 ± 0.3 (72 ± 2% Eu depletion relative to chondritic meteorites) for Apollo 12 soils is strikingly similar to that of the Apollo 11 high alkali rocks. The average Sm/Eu ratio of 6.5 ± 0.7 (or 59 ± 6% relative Eu depletion) found in Apollo 11 low alkali rocks is rather similar to

Table 1. Abundances (ppm) of the fourteen REE and Y, and Rb, Cs, Sc, V, Co, Zr, Hf,

Element (ppm)	Crystalline Rocks									
	Porphyritic basalts (Type 1)[b]						Granular and ophitic basalts (Type 2)[b]			
	12004,45		12020,24		12075,17		12051,44		12063,70	
	0.439 g	0.582 g[d]	0.444 g	0.408 g[d]	0.496 g	0.523 g[d]	0.449 g	0.535 g[d]	0.418 g	0.479 g[d]
Rb	—	0.7 ± .3	—	1.0 ± .4	—	1.4 ± .7	—	1.0 ± .4	0.4 ± .2	—
Cs	—	0.05	—	0.05	—	0.08	—	0.05	0.05	—
Sc	43	—	42	—	43	—	60	—	—	55
V	210	220	180	200	210	190	130	—	130	140
Co	58	—	61	—	61	—	35	—	—	46
Cd (ppb)	—	<7	—	<5	—	<9	—	<30	18	—
In (ppb)	—	2.0	—	12	—	1.0	—	2.3	3.4	—
Zr	—	—	—	—	—	—	—	—	—	—
Hf	2.7	—	2.4	—	2.7	—	3.3	—	—	3.6
Th[e]	0.4 ±0.3	—	0.7 ±0.2	—	0.7 ±0.2	—	1.1 ±0.3	—	—	0.7 ±0.2
Ba	60 ±40	—	25 ±20	—	30 ±20	—	100 ±40	—	—	60 ±30
La	6.0	5.9	5.9	5.6	5.8	5.3	5.9	5.9	6.0	5.9
Ce	—	16.3	—	16.1	—	14.9	—	17.8	17.5	—
Pr	—	2.6	—	2.2	—	2.2	—	3.2	2.9	—
Nd	—	13.2	—	12.0	—	10.9	—	15.8	17.1	—
Sm	4.70	4.32	4.08	4.10	4.10	3.68	5.50	5.49	6.05	5.60
Eu	0.96	0.87	0.79	0.76	0.86	0.84	1.27	1.30	1.30	1.43
Gd	—	5.2	—	5.4	—	5.1	—	6.7	8.2	—
Tb	—	1.04	—	0.96	—	0.85	—	1.35	1.50	—
Dy	—	5.7	—	5.2	—	5.7	—	7.2	8.9	—
Ho	—	1.50	—	1.55	—	1.46	—	2.10	2.24	—
Er	—	3.9	—	3.5	—	3.4	—	5.4	6.0	—
Tm	—	0.66	—	0.59	—	0.54	—	0.78	0.96	—
Yb	3.7	3.7	3.8	3.7	3.6	3.4	5.0	4.9	5.6	5.4
Lu	0.52	0.51	0.54	0.52	—	0.44	0.75	0.71	0.79	0.80
Y	—	36	—	34	—	30	—	42	55	—
$\sum$ REE + Y	—	102	—	96	—	89	—	121	140	—

[a] One standard deviation due to counting statistics and other errors for single determinations are approximately ±5% for Co, Sc, the fourteen REE and Y; ±10% for Cs, V, Hf, Th, Cd, and In; and ±30 — 50% for Z.
[b] Classification is by WARNER and ANDERSON (1971).
[c] Rock and soil were taken from a 15-cm deep trench.

those observed in Apollo 11 breccia rocks or soil, and very different from those of the Apollo 11 high alkali rocks or Apollo 12 rocks and soils.

For Apollo 12 igneous rocks, the average Sm/Eu ratio 4.8 ± 0.2 (45 ± 2% depletion) is approximately half of the ratio found in the Apollo 12 soils and Apollo 11 high alkali rocks. Because rock 12038 has a distinctly lower Sm/Eu ratio and a higher relative light REE pattern (e.g., see HASKIN *et al.*, 1971) than Apollo 12 igneous rocks and also shows anomalous chalcophilic elemental abundance (ANDERS *et al.*, 1971), we have excluded this rock from group averages. In more detail, the Sm/Eu ratios in the Apollo 12 crystalline rocks may be grouped in two small islands with average Sm/Eu ratios of the 4.9 ± 0.2 and 4.5 ± 0.2 for porphyritic basalts (type 1) and granular-ophitic basalts (type 2), respectively. These two population means are barely statistically different. If these two Sm/Eu groupings are real, we suggest that at least two individual flows were responsible for the Apollo 12 crystalline rocks. OSAWA and

Th, and Ba in Apollo 12 five crystalline rocks, one breccia rock, and four soil samples[a]

Breccia Rock		Soils							
12034,18[c]		12032,36		12033,71[c] (1–2 mm)		12037,40 (2–4 mm)		12070,77	
0.199 g	0.267 g[d]	0.292 g	0.265 g[d]	0.246 g	0.202 g[d]	0.211 g	0.239 g[d]	0.451 g	0.573 g[d]
—	11 ± 2	8.5 ± .4	—	—	12 ± 1	—	3.3 ± 1.0	5.8 ± .6	—
—	0.48	0.36	—	—	0.42	—	0.22	0.24	—
31	—	—	36	27	—	40	—	—	36
90	110	110	110	80	90	100	160	—	130
35	—	—	42	30	—	40	—	—	44
—	< 15	53	—	—	8	—	< 10	(460)	—
—	35	(130)	—	—	2.0	—	3.0	(470)	—
640 ± 130	—	—	640 ± 130	510 ± 100	—	310 ± 160	—	—	370 ± 110
27	—	—	18	20	—	9	—	—	12
13.2 ± 0.7	—	—	8.1 ± 0.4	13.5 ± 0.5	—	5.1 ± 0.3	—	—	5.6 ± 0.3
470 ± 50	—	—	400 ± 50	600 ± 60	—	200 ± 50	—	—	270 ± 50
75	60	46	48	75	62	26.2	51.3	37.7	36
—	172	114	—	—	180	—	136	87.2	—
—	20.4	15.0	—	—	21.0	—	19.0	12.6	—
—	100	75.7	—	—	91	—	80	51.0	—
32.0	29.4	21.6	21.8	32	29	13.0	23.5	16.6	16.0
2.95	2.65	2.19	2.18	3.08	2.89	1.52	1.96	1.73	1.67
—	33.3	24.8	—	—	33.0	—	30.0	19.0	—
—	5.51	4.3	—	—	5.40	—	4.6	3.3	—
—	34.0	26.5	—	—	32.0	—	28.0	20.5	—
—	8.8	6.5	—	—	8.8	—	7.9	4.9	—
—	21.7	17.3	—	—	20.1	—	17.4	12.9	—
—	3.6	2.8	—	—	2.9	—	2.6	2.0	—
25.0	22.4	16.5	17.6	26.3	21.0	11.3	17.8	11.4	12.0
3.30	2.74	2.34	2.30	3.20	2.61	1.49	2.43	1.67	1.66
—	186	164	—	—	211	—	156	110	—
—	703	540	—	—	723	—	579	393	—

[d] Abundances were obtained via radiochemical neutron activation analysis except for V, via instrumental neutron activation analysis.

[e] Th abundances are relative to the average Th abundances in the U.S.G.S. standard rocks W-1, GSP-1, and BCR-1.

GOLES (1970), in a study of Columbia River Plateau basalt, have found that La, Sm, and Yb abundances of nine distinct and layered flows of Picture Gorge basalt are significantly uniform, with population standard deviations of 8 to 12%, i.e., the deviation of a REE abundance in one flow from the average in the nine flows. If we assume a maximum REE abundance dispersion of 10% in a given flow and a dispersion of 5% in the Sm/Eu ratios within a given flow, the porphyritic basalts could result from one flow and the granular and ophitic basalts, from two flows. However, HASKIN *et al.* (1971) suggest that all the Sm/Eu ratios and total Sm spread in Apollo 12 basalts may have been indigenous with one flow, which is consistent with the relatively wide REE contents observed in a single Icelandic basaltic flow.

The similarity observed in the REE distribution patterns in both Apollo 12 rock types suggests different amounts of olivine addition or removal during magmatic differentiation of a common magma (e.g., COMPSTON *et al.*, 1971 and HASKIN *et al.*,

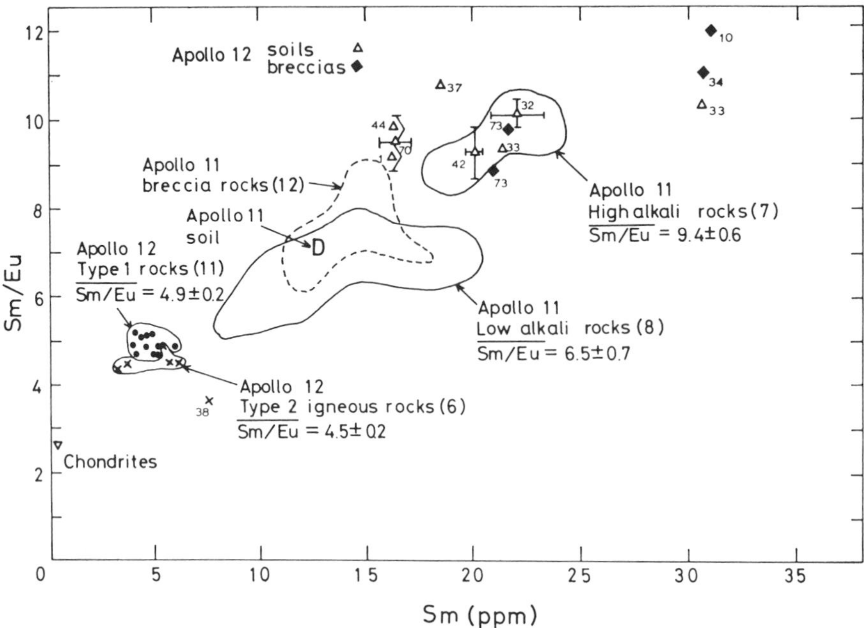

Fig. 2. Sm/Eu ratios versus Sm (ppm) are plotted for Apollo 11 and Apollo 12 rock and soil samples. Values of Apollo 11 samples are taken from GAST *et al.* (1970), GOLES *et al.* (1970), HASKIN *et al.* (1970), PHILPOTTS and SCHNETZLER (1970), and WAKITA *et al.* (1970). Values of Apollo 12 samples are taken from BRUNFELT *et al.* (1971), GAST and HUBBARD (1971), GOLES *et al.* (1971), HASKIN *et al.* (1971), MORRISON *et al.* (1971), PHILPOTTS *et al.* (1971), SMALES *et al.* (1971), and WAKITA *et al.* (1971). For Apollo 12 igneous rock, type 1 rocks are porphyritic basalts and type 2 rocks are granular and ophitic basalts by WARNER and ANDERSON (1971). For the Apollo 12 soils, two analyses were done for 12044 and 12042, three for 12032, and seven for 12070. Also each datum (● or ×) for Apollo 12 igneous rocks generally represents the average of two or more analyses done by two or more groups.

1971). The average of 100 ppm total REE + Y abundance found in type 1 rocks may be obtained by diluting type 2 rock magma with approximately 20% of olivine.

The observed Eu depletion mechanism could be caused by plagioclase precipitation and separation or other processes such as partial melting.

An intriguing mechanism has been explored by GREEN *et al.* (1971), who experimentally showed that calcium phosphate in equilibrium with diopside is enriched in Gd^{3+} compared to Sr^{2+} and presumably Eu^{2+}. They concluded that a small degree of partial melting would melt calcium phosphate mineral first and thereby yield magmas with negative Eu anomalies.

If indeed the Rb–Sr age of rock 12004 (COMPSTON *et al.*, 1971) is 0.4–0.5 × 10^9 yr. younger than the average Apollo 12 igneous rocks, then the remarkable similarity in REE distribution patterns and Eu depletion (similar Sm/Eu ratios) and very similar chemical compositions in 12004 and other Apollo 12 porphyritic basalts (e.g., see Table 1) suggests that nearly the same percentages of partial melting must have occurred for generation of two magmas separated in time by 0.4–0.5 $\times$ 10^9 yr. This seems very improbable to us.

Other trace elements

Indium abundances of 2 and 3 ppb were found in two Apollo 12 soils 12033 and 12037, which are nearly equal to the "uncontaminated" level of 1 to 2 ppb observed in a few Apollo 11 basalts and in many Apollo 12 basalts. GANAPATHY *et al.* (1970) and BAEDECKER *et al.* (1971) have reported much higher values for 12033 and 12037 soils. The 2–3 ppb In in two Apollo 12 soils of this work are not inconsistent with the KEAYS *et al.* (1970) hypothesis that 1–2% of type 1 carbonaceous chondritic like matter was added to the lunar soil.

The alkali elemental abundances vary within a factor range of 2–4 in four Apollo 12 soils. Rb and Cs are enriched by factors of 1–4 and 2–4, respectively, in Apollo 12 soils relative to Apollo 11 soils. Abundances of the alkali elements in Apollo 12 rocks are more like those in Apollo 11 low alkali rocks. Although some groups of rocks may deviate slightly from a direct correlation, it seems quite striking that K and Rb or K and Cs are coherently related in all Apollo 11 and Apollo 12 samples over a factor of $\sim$40 in abundance. The average K/Rb ratio of $\approx$400 in Apollo 11 and Apollo 12 samples is compared with the 430 and 350 found in terrestrial basic rocks and chondrites (ERLANK, 1968).

Abundance ratios of Zr and Hf in Apollo 11 and Apollo 12 samples of this work and others are summarized in Table 2. The observed 20 to 30% difference in ratios between Apollo 11 and 12 samples seems to be statistically real. These differences in the Zr/Hf ratio, probably the most geochemically coherent pair of chemical elements, may be attributable to very different fractionation histories at the Sea of Tranquillity and Ocean of Storms.

Table 2. Summary of Zr/Hf ratios in Apollo 11 and Apollo 12 lunar soils and rocks

	Apollo 11[a]		Apollo 12[b]	
	Number of soils or rocks	*Avg. Zr/Hf*	*Number of soils or rocks*	*Avg. Zr/Hf*
Soils	3 Zr and 5 Hf analyses on 10084	31 ± 4	7 Zr and 4 Hf analyses on 12070	42 ± 5
			4 difference soils, 12032, 12033, 12042, and 12070	40 ± 4
Breccia	8	32 ± 6	3	28 ± 6
Igneous	6-High alkali	30 ± 8	10 Porphyritic	35 ± 6
Rocks	4-Low alkali	26 ± 3	3 Granular and ophitic	37 ± 5

[a] Only neutron activation results for Hf and X-ray fluorescence and emission spec. results for Zr were used. Data was taken from ANNELL and HELZ (1970), COMPSTON *et al.* (1970), GOLES *et al.* (1970), MORRISON *et al.* (1970), SMALES *et al.* (1970), WAKITA *et al.* (1970), and WÄNKE *et al.* (1970). One standard deviation from the mean are given as ± values.

[b] Again neutron activation results for Hf were largely used and X-ray fluor. and emission spec. results for Zr were used. Hf values by TAYLOR *et al.* (1971) via emission spec. and spark source mass spec. and Hf value by MORRISON *et al.* (1971) via neutron activation were multiplied by factors of 0.86 and 1.4, respectively, in order to normalize their values to other neutron activation results for 12070 soil. Data was taken from ANNELL *et al.* (1971), BRUNFELT *et al.* (1971), COMPSTON *et al.* (1971), GOLES *et al.* (1971), MAXWELL and WIIK (1971), MORRISON *et al.* (1971), ROSE *et al.* (1971), SMALES *et al.* (1971), TAYLOR *et al.* (1971), WAKITA *et al.* (1971), and WILLIS *et al.* (1971).

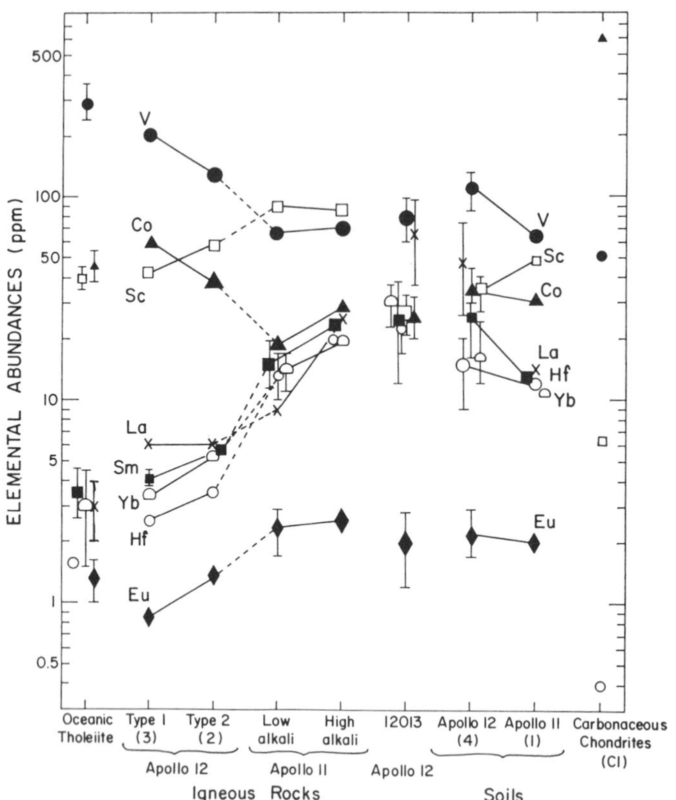

Fig. 3. Abundances with ranges for five Apollo 12 igneous rocks and four soil samples are from this work. Values and 1σ deviations for Apollo 11 rocks and soil are from many groups whose work is summarized in Tables 5 and 7 of WAKITA *et al.* (1970). Abundances with 1σ deviations in 12013 rock are from WAKITA and SCHMITT (1970), carbonaceous Cl chondritic abundances (water free basis) are taken from NICHPORUK and BINGHAM (1970) for V; Co and Sc from SCHMITT *et al.* (1971); and Hf, from EHMANN and REBAGAY (1970). Carbonaceous Cl chondritic abundances that are not shown in the figure are below 0.3 ppm. V abundances with mean deviations in oceanic basalts are from ENGEL *et al.* (1965); Co and Sc with ranges, from CORLISS (1970); REE with 1σ deviations, by KAY *et al.* (1970); and Hf in a single basalt by REBAGAY (1969). "Type 1" are porphyritic basalts and "Type 2" are granular and ophitic basalts by WARNER and ANDERSON (1971).

Briefly, abundances of other trace elements, i.e., Co, V, Sc, etc. in Apollo 11 and Apollo 12 samples, oceanic tholeiites and carbonaceous Cl chondrites are shown in Fig. 3. Abundances of V, Co, Sc, Hf, and REE distinguish Apollo 12 igneous rocks into two types. Space limitation does not permit further discussion.

Acknowledgments—This study was supported by NASA Contract NAS 9–8097 and NASA Grant 38–002–020. We acknowledge the assistance of the Oregon State University TRIGA reactor group for neutron activations and D. G. COLES, T. COOPER, D. HILL, and T. W. OSBORN III for assistance in some phases of RNAA.

References

Anders E., Laul J. C., Keays R. R., Ganapathy R., and Morgan J. W. (1971) Elements depleted on lunar surface: Implications for origin of moon and meteorite influx rate. Second Lunar Science Conference (unpublished proceedings).

Annell C. S., and Helz A. W. (1970) Emission spectrographic determination of trace elements in lunar samples from Apollo 11. *Proc. Apollo 11 Lunar Sci. Conf., Geochim. Cosmochim. Acta* Suppl. 1, Vol. 2, pp. 991–994. Pergamon.

Annell C. S., Carron M. K., Christian R. P., Cuttitta F., Dwornik E. J., Helz A. W., Ligon D. T., Jr., and Rose H. J., Jr. (1971) Chemical and spectrographic analyses of lunar samples from Apollo 12 mission. Second Lunar Science Conference (unpublished proceedings).

Baedecker P. A., Schaudy R., Elzie J. L., Kimberlin J., and Wasson J. T. (1971) Trace element studies of rocks and soils from Oceanus Procellarum and Mare Tranquillitatis. Second Lunar Science Conference (unpublished proceedings).

Brunfelt A. D., Heier K. S., and Steinnes E. (1971) Determination of 40 elements in Apollo 12 materials by neutron activation analysis. Second Lunar Science Conference (unpublished proceedings).

Compston W., Chappell B. W., Arriens P. A., and Vernon M. J. (1970) The chemistry and age of Apollo 11 lunar material. *Proc. Apollo 11 Lunar Sci. Conf., Geochim. Cosmochim. Acta* Suppl. 1, Vol. 2, pp. 1007–1027, Pergamon.

Compston W., Berry H., Vernon M. J., Chappell B. W., and Kaye M. J. (1971) Rubidium–strontium chronology and chemistry of lunar material from the Ocean of Storms. Second Lunar Science Conference (unpublished proceedings).

Corliss J. B. (1970) Mid-ocean ridge basalts. Ph.D. thesis, University of California, San Diego.

Ehmann W. D., and Rebagay T. V. (1970) Zirconium and hafnium in meteorites by activation analysis. *Geochim. Cosmochim. Acta* **34**, 649–658.

Engel A. E. J., Engel C. G., and Havens R. G. (1965) Chemical characteristics of oceanic basalts and the upper mantle. *Geol. Soc. Amer. Bull.* **76**, 719–734.

Erlank A. J. (1968) The terrestrial abundance relationship between potassium and rubidium. In *Origin and Distribution of the Elements* (editor L. H. Ahrens), pp. 871–888. Pergamon Press.

Ganapathy R., Keays R. R., Laul J. C., and Anders E. (1970) Trace elements in Apollo 11 lunar rocks: Implications for meteorite influx and origin of moon. *Proc. Apollo 11 Lunar Sci. Conf., Geochim. Cosmochim. Acta* Suppl. 1, Vol. 2, pp. 1117–1142, Pergamon.

Gast P. W., Hubbard N. J., and Wiesmann H. (1970) Chemical composition and petrogenesis of basalts from Tranquillity Base. *Proc. Apollo 11 Lunar Sci. Conf., Geochim. Cosmochim. Acta* Suppl. 1, Vol. 2, pp. 1143–1163. Pergamon.

Gast P. W., and Hubbard N. J. (1971) Rare earth abundances in soil and rocks from the Ocean of Storms. *Earth Planet. Sci. Lett.* **10**, 341–350.

Goles G. G., Randle K., Osawa M., Schmitt R. A., Wakita H., Ehmann W. D., and Morgan J. W. (1970) Elemental abundances by instrumental activation analyses in chips from 27 lunar rocks. *Proc. Apollo 11 Lunar Sci. Conf., Geochim. Cosmochim. Acta* Suppl. 1, Vol. 2, pp. 1165–1176. Pergamon.

Goles G. G., Duncan A. R., Osawa M., Martin M. R., Beyer R. L., Lindstrom D. J., and Randle K. (1971) Analyses of Apollo XII specimens and a mixing model for Apollo XII "soils." Second Lunar Science Conference (unpublished proceedings).

Green D. H., Ringwood A. E., Ware N. G., Hibberson W. O., Major A., and Kiss E. (1971) Experimental petrology and petrogenesis of Apollo 12 basalts. Second Lunar Science Conference (unpublished proceedings).

Haskin L. A., Allen R. O., Helmke P. A., Paster T. P., Anderson M. R., Korotev R. L., and Zweifel K. A. (1970) Rare earth and other trace elements in Apollo 11 lunar samples. *Proc. Apollo 11 Lunar Sci. Conf., Geochim. Cosmochim. Acta* Suppl. 1, Vol. 2, pp. 1213–1231. Pergamon.

Haskin L. A., Allen R. O., Helmke P. A., Anderson M. R., Korotev R. L., and Zweifel K. A. (1971) Rare earth and other trace elements in Apollo 12 lunar materials. Second Lunar Science Conference (unpublished proceedings).

KAY R., HUBBARD N. J., and GAST P. W. (1970) Chemical characteristics and origin of oceanic ridge volcanic rocks. *J. Geophy. Res.* **75**, 1585–1613.

KEAYS R. R., GANAPATHY R., LAUL J. C., ANDERS E., HERZOG G. F., and JEFFREY P. M. (1970) Trace elements and radioactivity in lunar rocks: Implications for meteorite infall, solar-wind flux, and formation conditions of moon. *Science* **30**, 490–493.

LSPET (Lunar Sample Preliminary Examination Team) (1970) Preliminary examination of the lunar samples from Apollo 12. *Science* **167**, 1325–1339.

MAXWELL J. A., and WIIK H. B. (1971) Chemical composition of Apollo 12 lunar samples 12004, 12033, 12051, 12052, and 12065. Second Lunar Science Conference (unpublished proceedings).

MORRISON G. H., GERARD J. T., KASHUBA A. T., GANGADHARAM E. V., ROTHENBERG A. M., POTTER N. V., and MILLER G. B. (1970) Elemental abundances of lunar soil and rocks. *Proc. Apollo 11 Lunar Sci. Conf., Geochim. Cosmochim. Acta* Suppl. 1, Vol. 2, pp. 1383–1392. Pergamon.

MORRISON G. H., GERARD J. T., POTTER N. M., GANGADHARAM E. V., ROTHENBERG A. M., and BURDO R. A. (1971) Elemental abundances of lunar soil from Apollo 12. Second Lunar Science Conference (unpublished proceedings).

NICHPORUK W., and BINGHAM E. (1970) Vanadium and copper in chondrites. *Meteoritics* **5**, 115–130.

OSAWA M., and GOLES G. G. (1970) Trace element abundances in Columbia River basalts. *Proceedings of the Second Columbia River Basalt Symposium* (editors E. H. Gilmour and E. Stradling) Eastern Washington State Press (Cheney, Wash.) pp. 55–71.

PHILPOTTS J. A., and SCHNETZLER C. C. (1970) Apollo 11 lunar samples: K, Rb, Sr, Ba, and rare-earth concentrations in some rocks and separated phases. *Proc. Apollo 11 Lunar Sci. Conf., Geochim. Cosmochim. Acta* Suppl. 1, Vol. 2, pp. 1471–1486, Pergamon.

PHILPOTTS J. A., SCHNETZLER C. C., BOTTIND M. L., and FULLAGAR P. D. (1971) Li, K, Rb, Sr, Ba, and rare earth concentrations and $^{87}Sr/^{86}Sr$ in some Apollo 12 soils, rocks, and separated phases. Second Lunar Science Conference (unpublished proceedings).

REBAGAY T. V. (1969) The determination of zirconium and Hafnium in meteorites and terrestrial materials by activation analysis. Ph.D. thesis, University of Kentucky.

REY P., WAKITA H., and SCHMITT R. A. (1970) Radiochemical activation analysis of In, Cd, and the 14 rare earth elements and Y in rocks. *Anal. Chim. Acta* **51**, 163–178.

ROSE H. J., JR., CUTTITTA F., ANNELL C. S., CARRON M. K., CHRISTIAN R. P., DWORNIK E. J., HELZ A. W., and LIGON D. T., JR. (1971) Semi-micro analysis of Apollo 12 samples. Second Lunar Science Conference (unpublished proceedings).

SCHMITT R. A., LINN T. A., JR., and WAKITA H. (1970) The determination of fourteen common elements in rocks via sequential instrumental activation analysis. *Radiochim. Acta* **13**, 200–212.

SCHMITT R. A., GOLES G. G., and SMITH R. H. (1971) Elemental abundances in stone meteorites. (in preparation)

SMALES A. A., MAPPER D., WEBB M. S. W., WEBSTER R. K., and WILSON J. D. (1970) Elemental composition of lunar surface material. *Proc. Apollo 11 Lunar Sci. Conf., Geochim. Cosmochim. Acta* Suppl. 1, Vol. 2, pp. 1575–1581. Pergamon.

SMALES A. A., MAPPER D., WEBB M. S. W., WEBSTER R. K., WILSON J. D., and HISLOP J. S. (1971) Elemental composition of lunar surface material (Part 2). Second Lunar Science Conference (unpublished proceedings).

TAYLOR S. R., KAYE M., GRAHAM A., RUDOWSKI R., and MUIR P. (1971) Trace elements chemistry of lunar samples from the Ocean of Storms. Second Lunar Science Conference (unpublished proceedings).

WAKITA H., and SCHMITT R. A. (1970) Elemental abundances in seven fragments from lunar rock 12013. *Earth Planet. Sci. Lett.* **9**, 169–176.

WAKITA H., SCHMITT R. A., and REY P. (1970) Elemental abundances of major, minor, and trace elements in Apollo 11 lunar rocks, soil, and core samples. *Proc. Apollo 11 Lunar Sci. Conf., Geochim. Cosmochim. Acta* Suppl. 1, Vol. 2, pp. 1685–1717, Pergamon.

WAKITA H., and ZELLMER D. (1970) unpublished work.

WAKITA H., REY P., and SCHMITT R. A. (1971) Abundances of the 14 rare earth elements plus 22 major, minor, and trace elements in ten Apollo 12 rock and soil samples. Second Lunar Science Conference (unpublished proceedings).

WÄNKE H., RIEDER R., BADDENHAUSEN H., SPETTEL B., TESCHKE F., QUIJANO RICO M., and BALACESCU A. (1970) Major and trace elements in lunar material. *Proc. Apollo 11 Lunar Sci. Conf., Geochim. Cosmochim. Acta* Suppl. 1, Vol. 2, pp. 1719–1727. Pergamon.

WARNER J. L., and ANDERSON D. H. (1971) Lunar crystalline rocks: Petrology, geology, and origin. Second Lunar Science Conference (unpublished proceedings).

WHITTAKER E. J. W., and MUNTUS R. (1970) Ionic radii for use in geochemistry. *Geochim. Cosmochim. Acta* **34**, 945–956.

WILLIS J. P., AHRENS L. H., DANCHIN R. V., ERLANK A. J., GURNEY J. J., HOFMEYER P. K., McCARTHY T. S., and ORREN M. J. (1971) Some interelement relationships between lunar rocks and fines, and stony meteorites. Second Lunar Science Conference (unpublished proceedings).

Proceedings of the Second Lunar Science Conference, Volume 2, pp. 1331–1335.
The M.I.T. Press, 1971.

Rhenium and osmium abundance determinations and meteoritic contamination levels in Apollo 11 and Apollo 12 lunar samples

J. F. LOVERING and T. C. HUGHES
School of Geology, University of Melbourne, Parkville, Victoria 3052, Australia

(*Received 24 February 1971; accepted in revised form 2 April 1971*)

Abstract—Re and Os abundance levels in Apollo 12 basaltic rocks (0.014–0.026 ppb Re; 0.17–1.1 ppb Os) are strongly depleted relative to chondritic meteorites. These data would imply some early stage metal phase separation and segregation if the parent rock in the moon's interior has ever been chemically related to primitive chondritic material.

The fines and breccias studied from the Apollo 11 site show relative Re and Os enrichments (and Os/Re ratios) corresponding to 0.12% iron meteorite or 1.3% chondritic meteorite contamination levels.The Apollo 12 fines and breccia studied show evidence of contamination levels of 0.07% for iron meteorites or 0.7% for chondritic meteorites. These levels are about one-half those observed at the Apollo 11 site. Our data do not enable us to differentiate between an iron meteorite *versus* a chondritic meteorite origin for the contaminating component. However the low iron meteorite levels calculated are well within estimates of iron meteoritic fragment abundances generally observed in fines and breccias from both Apollo 11 and Apollo 12 sites.

INTRODUCTION

LOVERING and BUTTERFIELD (1970) have already reported some preliminary data on Re and Os abundances in Apollo 11 volcanic rocks, breccias, and fines. These data suggested that if the Apollo 11 breccias and fines had been formed from the Apollo 11 volcanic rocks then these impact generated materials contain possible chondritic contamination levels up to 2%. However their data indicated extremely variable Os/Re ratios in the breccias and fines samples analyzed which were not consistent with the relatively constant Os/Re ratio of 11.5 found in chondrites. The present study repeats the Os and Re abundance measurements previously made on some Apollo 11 materials and provides new data on the abundances in volcanic rocks, fines and a breccia from the Apollo 12 site.

EXPERIMENTAL PROCEDURES

The sampling and neutron activation methods developed previously for the analysis of Re and Os in lunar materials by LOVERING and BUTTERFIELD (1970) were used in the present study except that the silica vial containing the irradiated samples was no longer dissolved along with the sample. This step had previously added particularly high Os blanks to the very low-level Os abundances measured in some of the lunar volcanic rocks. New work has shown that neither Re nor Os are lost during transfer of irradiated sample to fusion crucible. The analytical methods were checked by re-analysing three terrestrial rocks previously analyzed by MORGAN and LOVERING (1967b). The results (Table 1) are in good agreement with their data.

Table 1. Rhenium and osmium abundances in three terrestrial rocks and in Apollo 11 breccias and fines.

Rock Type	Rock No.	Re* (ppb)	Os* (ppb)	Os/Re
Terrestrial Rocks				
		0.086 ± 0.009	1.5 ± 0.1	17
Kimberlite (Kimberley Mine)	R275	0.110 ± 0.010	2.3 ± 0.1	21
Basalt	U.S.G.S. BCR-1	0.80 ± 0.03	≤0.11	≤0.14
		0.86 ± 0.03	≤0.21	≤0.24
Peridotite	U.S.G.S. PCC-1	0.102 ± 0.007	3.1 ± 0.2	31
		0.037 ± 0.005	6.1 ± 0.3	164
Apollo 11 materials				
Breccia (C)	10019,12	0.73 ± 0.03	7.8 ± 0.35	10.7
	10046,20	0.67 ± 0.03	7.6 ± 0.33	11.3
Fines (D)	10084,53	0.85 ± 0.03	8.1 ± 0.35	9.5
Average Apollo 11 breccias and fines	—	0.75	7.8	10.4
Average 12 chondrites (MORGAN and LOVERING (1967a)	—	57	660	11.5
Average 28 iron meteorites HIRT *et al.* (1963)	—	590	7060	11.9

* Errors quoted refer to sample statistical counting errors only.

DATA AND DISCUSSION

Apollo 11 material

New analyses are reported in Table 1 on single aliquots of samples for which preliminary analyses were reported previously by LOVERING and BUTTERFIELD (1970). The materials re-analyzed were two breccias (10019,12, 10046,20) and one sample of fines (10084,53). The new data show a much greater consistency for both Re and Os (and Os/Re ratio) between all three samples than the previous results and largely on these grounds are to be preferred.

The Apollo 11 secondary materials (i.e., breccias and fines) have Re and Os abundances considerably enriched relative to lunar basalt levels as typified by the Apollo 12 basalts (Table 2). Since the breccias and fines are composed largely (but not exclusively) of lunar basalt material then they also contain significant amounts of some other material which is relatively enriched in both Re and Os. The most likely contaminating material with enriched Re and Os would be of meteoritic origin (Table 1). It is significant in this regard that the Os/Re ratios measured in the breccias and fines are very different from lunar basalt ratios and very close to meteoritic ratios. If the average Re and Os levels in the Apollo 12 basaltic rocks reported below are assumed to be typical also of the Apollo 11 basalts, then the breccias and soils have Re and Os abundances which are consistent with either chondritic contamination levels of about 1.3 weight per cent or iron meteorite contamination levels of about 0.12%.

Our data do not enable us to decide between the alternative chondrite versus iron

Table 2. Rhenium and osmium abundances in Apollo 12 materials.

Rock Type	Rock	Re (ppb)	Os (ppb)	Os/Re
Volcanic Rocks A				
Olivine basalt	12020,28	0.019 ± 0.003	0.53 ± 0.05	28
		0.022 ± 0.003	0.54 ± 0.05	25
	12020,40	0.014 ± 0.003	0.19 ± 0.03	14
		0.019 ± 0.003	0.22 ± 0.03	12
Volcanic Rocks AB				
Olivine basalt	12051,48	0.022 ± 0.005	0.59 ± 0.09	27
		0.026 ± 0.005	0.84 ± 0.09	32
Pigeonite porphyry	12065,40	0.025 ± 0.004	0.74 ± 0.08	30
		0.021 ± 0.004	0.64 ± 0.08	31
Volcanic Rocks B				
Olivine dolerite	12002,130	0.015 ± 0.004	1.1 ± 0.1*	—
		0.016 ± 0.004	0.5 ± 0.08	31
Olivine dolerite	12022,55	0.021 ± 0.003	0.17 ± 0.03	8
		0.018 ± 0.003	0.24 ± 0.03	13
Average Volcanic Rock		0.020	0.47	24
Fines (D)	12032,23	0.35 ± 0.03	3.9 ± 0.3	11
		0.46 ± 0.03	4.2 ± 0.3	9
	12033,38	0.33 ± 0.03	3.7 ± 0.3	11
		0.46 ± 0.03	4.2 ± 0.3	9
	12037,27	0.34 ± 0.03	4.2 ± 0.3	12
		0.34 ± 0.03	3.4 ± 0.3	10
	12070,26	0.61 ± 0.02	7.2 ± 0.2	12
		0.57 ± 0.02	6.8 ± 0.2	12
	Average Fines	0.43	4.7	11
Breccia (C)	12073,33	0.41 ± 0.02	7.8 ± 0.3	19
		0.52 ± 0.02	7.2 ± 0.2	14

* Value not included in Os average for volcanic rocks.

meteorite hypotheses for the source of meteoritic contamination in the Apollo 11 breccias and fines. However probable iron meteorite fragments have been reported by many workers (e.g., LOVERING and WARE, 1970) in Apollo 11 breccias and fines. Precise data are not available on the observed level of iron meteorite contamination but preliminary estimates suggest that levels of about 0.1% are quite reasonable. Under these circumstances the Re and Os data reported here for Apollo 11 breccias and fines can be adequately, and most simply, explained on the basis of about 0.1% iron meteorite infall. It is possible that an additional chondritic infall contribution may also be present but it would need to occur at a level very much less than 1% to explain the Re and Os data obtained in this study.

Apollo 12 material

Volcanic rocks. Re and Os analyses on samples of Type A, Type AB, and Type B volcanic rocks from the Apollo 12 site are reported in Table 2. Re abundances measured on aliquots of each rock show good reproducibility and between samples the observed abundance range is relatively small (0.014–0.026 ppb Re). Os abundance ranges between samples is somewhat larger (0.17–1.1 ppb Os) and analyses on aliquots of single samples show differences of up to a factor of 2. These data indicate that Re is more homogeneously distributed through the rocks than is Os and this would imply that Re and Os occur in separate phases in the rocks.

Two samples of the same rock 12020 also show significant differences in Os content with the sample taken from the exterior of the rock (12020,28) containing more than twice as much Os than the sample 12020,40 taken from the interior of the rock. No other samples were available to test whether there is a real concentration of osmium towards the surfaces of individual rock fragments and the observed difference may simply be explained on the basis of a fortuitous inhomogeneity in Os distribution in the rock. The matter is complicated by some evidence reported by REED and JOVANOVIC (1971) for apparent Os *loss* from Apollo 12 rock exteriors but their evidence must be investigated further.

The Re and Os abundances measured in the Apollo 12 basaltic rocks are severely depleted with respect to chondritic abundances (Table 1), as are terrestrial basalts and "basaltic" achondrites (see LOVERING and BUTTERFIELD, 1970). As both Re and Os are strongly siderophilic elements then evolution of the lower basalts from chondritic material must have involved an efficient separation and segregation of metal phase at some stage in such a model for the chemical evolution of the moon.

Fines and breccias. Re and Os abundances in Apollo 12 fines (Table 2) show a high degree of consistency both between different samples and aliquots of the same sample indicating unusual homogeneity with regard to Re and Os distribution. The high coherence between both elements is consistent with the hypothesis that they are both present in a single constituent in the fines. By analogy with the Apollo 11 fines and breccias it seems that this constituent is most likely meteoritic material. The observation that the observed Os/Re ratios are very close to meteorite ratios (see Table 1) is consistent with this hypothesis.

The level of meteoritic contamination can be calculated if it is assumed that silicate material of lunar origin in the fines has Re and Os abundance levels close to the average reported in Table 2 for Apollo 12 volcanic rocks. On this basis the average meteoritic contamination levels in the fines studied would be around 0.07 weight per cent if iron meteorite material is the contaminant or 0.7% if chondrites are considered. These levels are about one-half the meteorite contamination levels calculated above for the Apollo 11 fines and breccias.

Our data do not allow us to choose between the alternatives of a chondrite *versus* an iron meteorite contaminating source for the Apollo 12 fines but, in view of the observed occurrence of rare iron meteorite fragments in these fines, it seems very reasonable to propose that in our samples an iron meteorite source of about 0.07% is the main contaminant with some additional chondritic contaminant possible at a level much less than 0.7%.

The one Apollo 12 breccia sample (12073,33) analyzed in this study (Table 2) showed Re and Os levels virtually identical with the fines sample 12070,26 so that similar levels of meteoritic contamination are observed in this breccia. Our data were also consistent with the hypothesis that the breccias are simply compact soils from the same locality.

The higher Re content reported previously by LOVERING and BUTTERFIELD (1970) for another aliquot of 10084,53 must either be the result of contamination or, more likely, the result of sample inhomogeneities. The lower value reported here is consistent with a "normal" Os/Re ratio of 9.5 and is preferred to the higher value previously reported. There is also an apparent inconsistency between the Re value of 0.85 ppb reported by us for 10084,53 and the values 7.85 and 7.25 ppb reported by HERR *et al.* (1971) for 10084,21. Our analysis was carried out on an aliquot of the total sample supplied which contained *all* size ranges <1 mm. HERR *et al.* (1971) reported that their analysis was carried out on a separated fine fraction (≤ 100 microns) only. Since the siderophilic elements are enriched in the finest fractions of the lunar fines (R. KEAYS, personal communication) then it is only to be expected that the analyses by HERR *et al.* (1971) would be anomalously high.

ANDERS *et al.* (1971) have also estimated the meteoritic contamination levels in the same suite of fines and breccia analyzed in this study. From their data they concluded that the fines 12032, 12037 and breccia 12073 show a contamination pattern characteristic of "differentiated" meteorites (i.e., irons or ordinary chondrites). These levels of contamination corresponded to 0.86% L chondrite, 0.60% H chondrite or 0.17% iron meteorite component. These estimates are within a factor of 3 of our estimates. However they found the fines 12070 from the contingency sample to have about 1.8% of a "primitive" (i.e., C 1 carbonaceous chondrite) component. Our data on 12070 were indistinguishable from the other samples of fines studied but since C 1 carbonaceous chondrites are not enriched in Re and Os it is not possible for us to verify their conclusion of variable meteoritic contamination levels and variable types of contaminating meteoritic materials in the soils of the Apollo 12 site.

Acknowledgments—We wish to thank H. Berry and M. Cowan (Australian National University) and A. Robertson (University of Melbourne) for assistance with this project. Officers of the Australian Atomic Energy Commission also provided special assistance with thermal neutron irradiations. The work was supported in part by grants from the Australian Research Grants Committee, the Australian Institute of Nuclear Science and Engineering and the Department of Education and Science.

REFERENCES

ANDERS E., LAUL J. C., KEAYS R. R., GANAPATHY R., and MORGAN J. W. (1971) Elements depleted on lunar surface. Implications for origin of moon and meteorite influx rate. Second Lunar Science Conference (unpublished proceedings).

HERR W., HERPERS U., MICHEL R., RASSOUL A. A., and WOELFLE R. (1971) Spallogenic ^{53}Mn ($T = 2 \times 10^6$a) and search for Re isotopic anomalies in lunar surface material by means of neutron bombardment. Second Lunar Science Conference (unpublished proceedings).

HIRT G., HERR W., and HOFFMEISTER W. (1963) Age determination by the rhenium-osmium method. In *Radioactive Dating*, pp. 35–43. IAEA.

LOVERING J. F. and BUTTERFIELD D. (1970) Neutron activation analysis of rhenium and osmium in Apollo 11 lunar material. *Proc. Apollo 11 Lunar Science Conf., Geochim. Cosmochim. Acta*, Suppl. 1, Vol. 2, pp. 1351–1353. Pergamon.

LOVERING J. F. and WARE N. G. (1971) Electron probe microanalyses of minerals and glasses in Apollo 11 lunar samples. *Proc. Apollo 11 Lunar Science Conf., Geochim. Cosmochim. Acta*, Suppl. 1, Vol. 1, pp. 633–654. Pergamon.

MORGAN J. W. and LOVERING J. F. (1967a) Rhenium and osmium abundances in chondritic meteorites. *Geochim. Cosmochim Acta*, **31**, 1893–1909.

MORGAN J. W. and LOVERING J. F. (1967b) Rhenium and osmium abundances in some igneous and metamorphic rocks. *Earth Planet. Sci. Lett.* **3**, 219–224.

REID G. W. and JOVANOVIC S. (1971) The halogens and other trace elements in Apollo 12 soil and rocks. Second Lunar Science Conference (unpublished proceedings).

Proceedings of the Second Lunar Science Conference, Volume 2, pp. 1337–1341.
The M.I.T. Press, 1971.

Search for rhenium isotopic anomalies in lunar surface material by neutron bombardment

W. HERR, U. HERPERS, R. MICHEL, A. A. ABDEL RASSOUL, and R. WOELFLE*
Institut für Kernchemie der Universitaet Koeln, Germany.

(*Received 22 February 1971; accepted in revised form 31 March 1971*)

Abstract—Three Apollo 12 rocks, 12021, 12022, and 12053, and two lunar regolith samples, 10084 and 12070, were analyzed for their Re and its isotopic composition by means of neutron activation, considering the decay of ^{186}Re and of ^{188}Re. An extensive chemical purification was necessary to measure the very low concentration of the natural β-decaying radioelement rhenium, ranging in the rocks from 0.03 to 0.09 ppb. Relative to the Re content in Apollo 12 rocks, the Apollo 12 soil samples contained more Re by a factor of ~ 10, and relative to that of Apollo 11 soil, by a factor of ~ 100. To the first approximation the isotopic abundance was found to be "normal." However, because of the rather poor statistics due to the short halflife of ^{188}Re, the measurements do not provide conclusive evidence of an isotopic effect. There may be a slight depletion of ^{185}Re in basaltic rocks. A depletion of ^{187}Re in soil is not observed by the contribution of Re in solar wind, in which ^{187}Re might have decayed by "bound state" β-decay.

INTRODUCTION

IN SPITE OF THE fact that the cosmochemistry of Re is not very well known, the natural β^- radioactivity of ^{187}Re has grown in interest, since we were able to show its usefulness in dating special ore deposits, iron meteorites, etc. (HERR *et al.*, 1961). A large number of Re-containing minerals have been dated (HERR *et al.*, 1967). The difficulties and discrepancies in the determination of the decay constant and the extremely low β-energy ($E_\beta = 2.6$ keV, HUSTER and VERBEEK, 1967) led to a discussion about the existence of a "bound state" β-decay. GILBERT (1958) calculated the probability of this mode of decay, and experiments made by BRODZINSKI and CONWAY (1965) gave a ratio of "bound" to "continuum state" decay $= 0.5 \pm 0.3$. In these considerations the important question concerning the temperature dependence of the ^{187}Re decay constant therefore arose, because the degree of ionization would be a determining factor. The possibility of dating nucleosynthesis events was also pointed out by CLAYTON (1964). This author suggests that the ^{187}Re halflife can be shortened by a factor of 10 at temperatures of 10^8 to 10^9 °K (CLAYTON, 1969). Such temperatures exist in the interior of the sun. Therefore, measurement of Re and its isotopic composition in the solar wind might provide valuable information about its decay properties.

The presence of Re in solar wind is expected to be about 10^{-13} atoms cm$^{-2}\cdot$sec^{-1}, and this Re flux gives about 10^{-8} g cm^{-2} in 10^9 yr. The quantitative determination of Re and its isotopic abundance turned out to be an intricate analytical task. In lunar fines 10084 from the Apollo 11 mission, we obtained a figure of 11 ppb Re for the regolith (HERR *et al.*, 1970). We extended this work, now measuring both of the

* Institut für Radiochemie der Kernforschungsanlage Juelich.

induced radioisotopes ^{186}Re (T = 3.75 d) and ^{188}Re (T = 16.7 h). Re was analyzed in three rock samples (12021,57; 12022,35, and 12053,39) and two soil samples (10084,21 and 12070,87). LOVERING and BUTTERFIELD (1970) found the same Re content in Apollo 11 regolith 10084, and a somewhat lower content in rocks.

EXPERIMENTAL TECHNIQUES

Samples of 50 to 100 mg were sealed in quartz tubes and activated for 2 days ($\Phi = 6 \times 10^{13}$ cm^{-2} sec^{-1}) in the FRJ–2 reactor (DIDO–type) at Juelich together with some Re standards ($KReO_4$ on Al foil). The measurements were done by γ spectrometry with a 20 cc Ge(Li) detector using the most intense peaks of 137 keV for ^{186}Re and 155 keV for ^{188}Re. Surface contamination was serious because the small Re quantities were of the order of 10^{-13} g. There are many interfering radionuclides in the 130 to 160 keV energy region. Chemical purifications of high quality are necessary. After transport and cooling times of 6 to 8 hours the samples were fused in a Ni crucible with NaOH + Na_2O_2 and suitable $KReO_4$ and other noble element carriers. After solution of the melt, a $Fe(OH)_3$ scavenging preceded the Re precipitation by tetraphenylarsoniumchloride under addition of EDTA. The precipitate was reduced by H_2 in a quartz system (600°C), then oxidized by O_2 to Re_2O_7 (400 to 500°C) and sublimated. The NaOH solution was adjusted to 0.2 N HCl and passed through a Dowex–1 ion-exchanger column. Re was eluated with 0.2 N HCl/5% NH_4SCN mixture. ^{99}Tc, ^{99}Mo, and other possible interferences were thereby eliminated. After destroying SCN^- ions, a second iron precipitation was done followed by a second Re tetraphenylarsonium-chloride precipitation in NH_4OH solution. The Re standards were treated in the same way. The chemical yield was about 70%. The combination of dry distillation and ion exchange procedures resulted in a sample of high purity. The radiochemical purity was also checked by a precise energy determination of the counted peaks (Fig. 1). The resolution of the detector was 1.65% at 137 keV. The induced halflifes were measured for each peak separately. The literature values of the ^{188}Re halflife vary between 16.7 h and 20.0 h. The value of T = 16.7 h (Flammersfeld, 1953) was used for this investigation.

RESULTS AND DISCUSSION

In the first row of Table 1 the elemental abundances of Re in the lunar material are presented. The abundances are obtained by calculating each of the two Re isotopes by way of its radioisotope and then taking the sum. Exceptions are marked by asterisks.

Table 1. Rhenium content and isotopic composition of lunar material.

Type	lunar rock			lunar soil	
Sample No	12021,57	12022,35	12053,39	12070,87	10084,21 ($\leq$100 μm)
Re (ppb)	0.094 ± 0.005	0.044 ± 0.002	0.033 ± 0.002	0.67 ± 0.01	7.85 ± 0.02
	0.077 ± 0.007*	0.041 ± 0.002	0.030 ± 0.003	0.66 ± 0.01	7.25 ± 0.06
	0.078 ± 0.007*	—	0.026 ± 0.002	—	11.20 ± 0.40**
^{185}Re Isotope	33.4 ± 5.5	33.9 ± 1.9	31.0 ± 3.6	35.6 ± 0.97	37.0 ± 0.12
(%)	—	35.2 ± 1.9	35.3 ± 2.7	34.7 ± 0.97	38.4 ± 1.5
^{187}Re Isotope	66.6 ± 3.8	66.1 ± 3.1	69.0 ± 12.3	64.4 ± 1.4	63.0 ± 0.21
(%)	—	64.8 ± 3.1	64.7 ± 8.3	65.3 ± 1.4	61.6 ± 1.9

* These Re values are determined only by ^{186}Re (T = 3.75 d). For the isotopic abundance of terrestrial Re, data from WHITE and CAMERON (1948) are taken; ^{185}Re = 37.07 (±0.17%) ^{187}Re = 62.93 (±0.11%); however the errors given here represent the deviations in our own experiments!

** See earlier work, HERR *et al.* (1970).

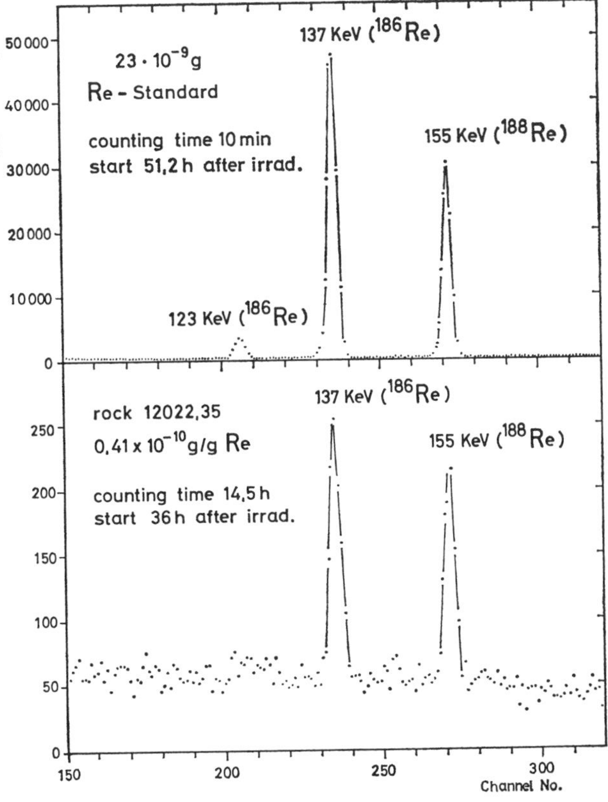

Fig. 1. Typical Re γ-ray spectra.

The errors in the Re elemental abundances are calculated from the uncertainty in each separate isotope measurement.

It can be seen that the Mare Tranquillitatis regolith is about one hundred times enriched in Re ($\sim$7 ppb) compared with basaltic rocks, which contain only 0.03 to 0.08 ppb. The Apollo 12 soil has only $\sim$0.6 ppb Re. These differences may result from small grains of meteoritic origin. HERR *et al.* (1961) reported Re values for irons in the 0.2 ppm region, and MORGAN and LOVERING (1967) found an average of 0.05 ppm in chondrites. As already mentioned, the Re content for soil 10084 (LOVERING and BUTTERFIELD, 1970) are concordant with our Apollo 11 results. However, in a more recent publication (FRICK *et al.*, 1971) these authors have revised their earlier Re values for Apollo 11 fines (10084) and now report a rhenium content of 0.85 ppb. They state that their new data show a much greater consistency for both the elements Re and Os, and for the Re/Os ratio, respectively.

With regard to our Apollo 11 soil results we noticed that this material was heterogeneous. We made 3 independent Re determinations that ranged from 7 to 11 ppb. A prior contamination could not be totally excluded. The isotopic analysis of this specific soil sample seems to come nearest to "normal" terrestrial rhenium. We do not think that the discrepancy is due to the analytical technique, since FRICK *et al.*

(1971) and we have both found similar Re contents for 12070 "soil." Further examinations and a possible exchange of the respective lunar soil sample 10084 should help to clarify the situation.

In rows 2 and 3 of Table 1 the Re isotope abundances are given in percent of Re element. The errors are one standard deviation due to counting statistics. We plotted the sample to standard activity ratios of ^{186}Re against the activity ratios of ^{188}Re (Fig. 2). The 45° line represents the isotopic composition of terrestrial Re. Evidently

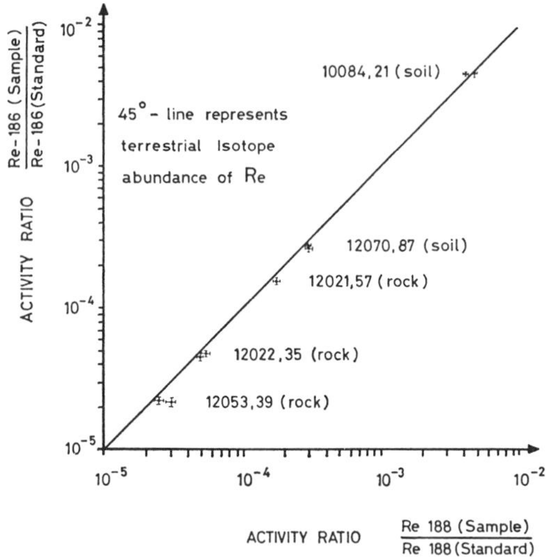

Fig. 2. Determination of the Re isotopic composition in lunar samples.

Re isotope effects, if they are real, are rather small. The accuracy of the ratio measurements decreases with decreasing Re content. Because Re analysis in the 10^{-12} to 10^{-13} g range is difficult, statistical errors are too large to prove the existence of anomalies. However, the rocks seem to be slightly depleted in ^{185}Re, whereas the regolith samples seem more "normal." There is a necessity for further improved techniques and, still more important, for analysis of surface samples of lunar rocks.

The ideas of Clayton were discussed by MORGAN (1970), who suggested β-counting for the measurements of isotopic effects. It is our opinion that γ spectrometry should be superior to β counting. KEAYS *et al.* (1970) have attributed the enrichment of trace elements such as Ir, Au, etc., in the soil to meteoritic origin (even cometary origin has to be considered), and possibly to "solar wind" influx. The rather large differences in the Re content of the regolith and rocks led us to similar considerations.

For further Re work, we are interested in following up the report by REED and JOVANOVIC (1971), who observed an Os depletion in rock surfaces compared to rock interiors and soils. The Re/Os ratio is of general importance (HERR *et al.*, 1961; MORGAN and LOVERING, 1967).

REFERENCES

BRODZINSKI R. L. and CONWAY D. C. (1965) Decay of ^{187}rhenium. *Phys. Rev.* **138B**, 1368–1371.

CLAYTON D. D. (1964) Cosmoradiogenic chronologies of nucleosynthesis. *Astrophys. J.* **139**, 637–663.

CLAYTON D. D. (1969) Isotopic composition of cosmic importance. *Nature* **224**, 56–57.

FLAMMERSFELD A. (1953) ^{188m}Re, ein neues Kernisomer von $T = 18.7$ m Halbwertszeit. *Z. Naturforsch.* **8a**, 217–218.

FRICK C., HUGHES T. C., LOVERING J. F., REID A. F., WARE N. G., and WARK D. A. (1971) Electron probe, fission track, and activation analysis of lunar samples. Second Lunar Science Conference (unpublished proceedings).

GILBERT N. (1958) Étude théorique de la désintégration β^- du rhenium 187. *Comp. Rend.* **247**, 868–871.

HERR W., HOFFMEISTER W., HIRT B., GEISS J., and HOUTERMANS F. G. (1961) Versuch zur Datierung von Eisenmeteoriten nach der Re/Os-Methode. *Z. Naturforsch.* **16a**, 1053–1058.

HERR W., WOELFLE R., EBERHARDT P., and KOPP E. (1967) Development and recent application of the Re/Os dating method. In *Radioactive Dating and Methods of Low-Level Counting*, pp. 499–508, International Atomic Energy Agency, Vienna.

HERR W., HERPERS U., HESS B., SKERRA B., and WOELFLE R. (1970) Determination of manganese-53 by neutron activation and other miscellaneous studies on lunar dust. *Science* **167**, 747–749.

HUSTER E. and VERBEEK H. (1967) Das β-Spektrum des natürlichen Rhenium-187, *Z. Phys.* **203**, 435–442.

KEAYS R. R., GANAPATHY R., LAUL J. C., ANDERS E., HERZOG G. F., and JEFFERY P. M. (1970) Trace elements and radioactivity in lunar rocks: Implications for meteorite infall, solar wind flux and formation conditions of moon. *Science* **167**, 490–493.

LOVERING J. F. and BUTTERFIELD D. (1970) Neutron activation analysis of rhenium and osmium in Apollo 11 lunar material. *Proc. Apollo 11 Lunar Sci. Conf., Geochim. Cosmochim. Acta* Suppl. 1, Vol 2, pp. 1351–1355. Pergamon.

MORGAN J. W. and LOVERING J. F. (1967) Rhenium and osmium abundances in chondritic meteorites. *Geochim. Cosmochim. Acta* **31**, 1893–1909.

MORGAN J. W. (1970) Anomalous rhenium isotopic ratio in the solar wind: Detection at the nanogram level. *Nature* **225**, 1037–1038.

REED G. W. and JOVANOVIC S. (1971) The halogens and other trace elements in Apollo 12 soil and rocks. Halides, platinum metals and mercury on surfaces. Second Lunar Science Conference (unpublished proceedings).

WHITE J. R. and CAMERON A. E. (1948) The natural abundance of isotopes of the stable elements. *Phys. Rev.* **74**, 991–1000.

Proceedings of the Second Lunar Science Conference, Volume 2, pp. 1343–1350.
The M.I.T. Press, 1971.

Total carbon and nitrogen abundances in Apollo 12 lunar samples

C. B. Moore, C. F. Lewis, J. W. Larimer
F. M. Delles, R. C. Gooley, and W. Nichiporuk
Arizona State University, Tempe, Arizona 85281

and

E. K. Gibson, Jr.
NASA Manned Spacecraft Center, Houston, Texas 77058

(*Received 22 February 1971; accepted in revised form 25 March 1971*)

Abstract—Total carbon and nitrogen abundances in Apollo 12 samples are generally similar to those from Apollo 11. A major difference occurs in several samples of lunar fines. These fines from crater ejecta are relatively light in color and have total carbon in the range of 23–65 μg/g and nitrogen contents of 46–96 μg/g. Regular Apollo 12 lunar fines range from 115 to 180 μg/g carbon and from 85 to 140 μg/g nitrogen. Eight samples from a core tube sample had carbon and nitrogen contents within the above limits and showed a random distribution with depth. A dark-colored breccia had a carbon content of 120 μg/g and a light-colored breccia 65 μg/g. Seven Apollo 12 basalts ranged from 16 to 45 μg/g in total carbon while two basalts had 43 and 44 μg/ of nitrogen.

The Apollo 12 results support the model that the carbon and nitrogen in fines are composed of indigenous lunar material together with a major solar wind component.

Introduction

Total carbon and nitrogen contents of Apollo 12 lunar samples were determined utilizing analytical techniques developed for these elements in meteorites (Moore *et al.*, 1965, 1966, 1967, 1969a, b, 1970a, b; Gibson and Moore, 1970). For the determination of total carbon, samples were burned in a flowing oxygen atmosphere at over 1600°C to form CO_2. After necessary purification and trapping of the effluent gases, the CO_2 was detected utilizing a LECO 589–400 gas chromatographic low-carbon analyzer. For the nitrogen determination samples were heated to about 2400°C in a graphite crucible in a helium atmosphere to reduce all nitrogen compounds to N_2 which was detected in a LECO Nitrox-6 gas chromatographic analyzer. Differential thermal conductivity is utilized as the detection method in both systems. National Bureau of Standards low carbon (101e) and nitrogen (33d) steel standards were used to construct standard analytical curves for both determinations. The combustion-gas chromatographic detection technique determines total carbon and nitrogen but does not discriminate their chemical state in the analyzed samples.

Investigations of carbon in Apollo 11 samples indicated that it is concentrated in lunar fines and breccias with respect to the lunar basalts (Moore *et al.*, 1970a, b). Studies of the chemical species of carbon released from the lunar materials by pyrolysis, leaching, and extraction techniques indicated that it was mainly present as inorganic carbon. Carbon dioxide was reported to be the most abundant carbon containing phase released by vacuum pyrolysis below 500 C. (Friedman *et al.*, 1970; Chang *et al.*, 1970; Abell *et al.*, 1970a; Oro *et al.*, 1970; Murphy *et al.*, 1970). Other gaseous carbon phases released include CO and CH_4 (Chang *et al.*, 1970). Hydrolysis studies using HC1 and DC1 released indigenous CH_4 and deuterated hydrocarbons probably produced from carbides (CHANG *et al.*, 1970; ABELL *et al.*, 1970). Residual carbon in the extracted fines was attributed to elemental carbon (KAPLAN and SMITH, 1970; FRIEDMAN *et al.*, 1970; EPSTEIN and TAYLOR, 1970). There is little conclusive evidence for the presence of significant concentrations of hydrocarbon compounds more complex than methane. The total carbon abundances reported by all of the above authors were very similar; ranging from 100 to 350 μg/g in the lunar fines and breccias to 15 to 150 μg/g in the lunar basalts. Studies by MOORE *et al.*, (1970) and KAPLAN and SMITH (1970) indicated that the finest lunar regolith material had significantly higher carbon contents than the coarser grained fines. Detailed studies of the total nitrogen contents in Apollo 11 samples were fewer in number. MOORE *et al.* (1970) and MORRISON *et al.* (1970) reported similar results for Apollo 11 basalts, fines, and breccias. Generally the relative distributions of carbon and nitrogen were similar, but it is interesting to note that both groups found basalt samples with nitrogen contents as high or higher than those in the lunar fines and breccia (100–150 μg/g).

The carbon and nitrogen distributions in the fines and breccias are a mixture of indigenous lunar carbon and nitrogen with major additions of solar wind material.

RESULTS

The results of the Apollo 12 analyses are given in Table 1 for total carbon and in Table 2 for total nitrogen. Total carbon analyses indicated with an asterisk (*) were run at the Lunar Receiving Laboratory as a part of the preliminary examination (LSPET, 1970). In Table 3 the total carbon and nitrogen contents of selected samples from the double core tube are listed. For each lunar sample the weights of material used for individual analyses are listed together with the concentration of carbon or nitrogen detected. No attempt was made to homogenize individual lunar samples. This minimized the possibility of contamination and also provided a test for sample inhomogeneity. Rock samples were crushed with a single stroke in a clean diamond mortar. To guard against contamination they were not sieved or run through a mechanical splitter. The precision indicated as $\pm$ is taken from the 90% confidence levels on the standard analytical curves.

The carbon and nitrogen abundances in the Apollo 12 samples are generally similar to those found by the authors in the Apollo 11 samples (MOORE *et al.*, 1970a, b). The lunar fines have higher abundances of both elements than do the rock samples. Figure 1 in which carbon is plotted against nitrogen for the Apollo 11 and Apollo 12 samples indicates that there is a direct correlation between them within individual samples.

The fines of "D" samples analyzed appear to be divided into two groups. Those that are dark black in color and similar in appearance to the Apollo 11 fines have carbon contents that range from 115 to 180 μg/g and nitrogen contents of 85 to 140 μg/g. The ranges for the Apollo 11 samples were 142 to 226 μg/g for carbon and 102 to 153 μg/g for nitrogen. In addition to these samples there were three lighter colored fines samples that had significantly lower carbon contents. In these samples the

Table 1. Total carbon in Apollo 12 lunar samples

Sample number	Sample wt. (g)	Total carbon (μg/g)	Weighted mean (μg/g)
D Samples—Fines			
12001,33	0.1214	130 ± 15	
	0.1718	100 ± 10	
	(0.2932)	—	120 ± 15
12003,1*	0.1571	185 ± 20	
	0.1128	195 ± 25	
	0.1480	120 ± 10	
	0.1605	200 ± 10	
	(0.5784)	—	180 ± 15
12003,2	0.1474	130 ± 15	
	0.0564	150 ± 20	
	(0.2038)	—	135 ± 15
12023,11	0.1521	140 ± 12	
	0.1254	155 ± 16	
	(0.2773)	—	150 ± 15
12024,7*	0.1395	115 ± 20	115 ± 20
12032.1*	0.3813	25 ± 4	25 ± 4
12032,13	0.2225	60 ± 5	
	0.3090	60 ± 5	
	(0.5315)	—	60 ± 5
12033,1*	0.6014	23 ± 2	23 ± 2
12033,7	0.1994	60 ± 5	
	0.1121	36 ± 20	
	0.3265	50 ± 4	
	(0.6380)	—	50 ± 5
12037,10	0.1198	80 ± 10	
	0.2077	30 ± 10	
	0.2869	80 ± 8	
	(0.6144)	—	65 ± 10
12042,1*	0.2452	125 ± 10	
	0.3409	130 ± 10	
	(0.5861)	—	130 ± 10
12042,12	0.0886	165 ± 20	
	0.1297	100 ± 10	
	(0.2183)	—	130 ± 15
C Samples—Breccia			
12034,2*	0.2446	65 ± 12	65 ± 12
12057,1* (dark colored breccia chip from D sample)	0.2194	120 ± 10	120 ± 10
B Samples—Coarse-grained rocks			
12002,176	0.1430	14 ± 7	
	0.1695	18 ± 6	
	(0.3125)	—	16 ± 6
12022,80	0.1275	24 ± 8	
	0.1895	70 ± 9	
	0.2209	30 ± 8	
	(0.5379)	—	40 ± 8
12040,2* (rock chip from D sample)	0.2245	45 ± 15	45 ± 15
12044,2,1* (rock chip from D sample)	0.3071	44 ± 5	44 ± 5
A Samples—Fine-grained rocks			
12052,4*	0.4487	25 ± 4	
	0.3326	34 ± 10	
	0.1261	65 ± 15	
	(0.9074)	—	34 ± 10
12063,1*	0.2755	35 ± 10	35 ± 10
12065,1*	0.4920	29 ± 4	
	0.1438	38 ± 10	
	(0.6358)	—	31 ± 6

* Samples run at LRL during preliminary examination.

Table 2. Total nitrogen in Apollo 12 lunar samples

Sample number	Sample wt. (g)	Total nitrogen (μg/g)	Weighted mean (μg/g)
D Samples—Fines			
12001,33	0.1124	110 ± 5	110 ± 5
12003,2	0.0923	82 ± 5	
	0.1379	87 ± 5	
			85 ± 5
12023,11	0.0961	120 ± 5	120 ± 5
12032,13	0.1639	48 ± 5	48 ± 5
12033,7	0.1581	46 ± 5	46 ± 5
12037,10	0.1073	96 ± 5	96 ± 5
12042,12	0.1432	99 ± 5	
		170 ± 5	
			130 ± 5
B Samples—Coarse-grained rocks			
12002,176	0.1482	43 ± 5	43 ± 5
12022,80	0.1434	44 ± 5	44 ± 5

Table 3. Total carbon and nitrogen in splits from double core tube sample

Sample number	Depth (cm)	Sample wt. (g)	Total carbon (μg/g)	Sample wt. (g)	Total nitrogen (μg/g)
12025,58	8–9	0.0648	150 ± 30	0.0714	120 ± 7
12028,165	11–12	0.0890	140 ± 30	0.0500	140 ± 7
12028,181	16.4–17.4	(0.0640)	(250 ± 30)		
		(0.0575)	(70 ± 25)		
		0.1215	160 ± 30	0.0766	100 ± 7
12028,170	20.8–21.8	0.0681	150 ± 20	0.0669	76 ± 7
12028,174	25.4–26.1	0.0661	160 ± 30	0.0634	90 ± 7
12028,177	28.8–30.0	0.0890	100 ± 30	0.0607	69 ± 7
12028,114	30.0–30.6	0.0722	140 ± 20	0.0618	86 ± 7
12028,186	36.7–37.2	0.0662	170 ± 30	0.0553	96 ± 7

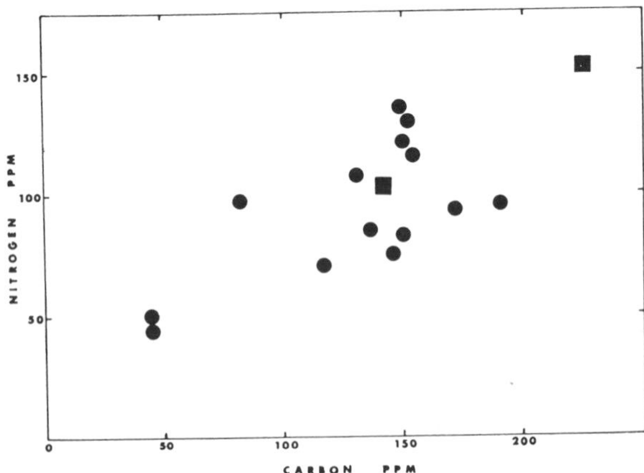

Fig. 1. Correlation of total carbon and nitrogen in lunar fines ■ Apollo 11; ● Apollo 12

carbon contents ranged from 23 to 65 μg/g. The nitrogen ranged from 46 to 96 μg/g.

The small samples taken from selected depths in double core tube (12025, 12028) are surprisingly constant in their carbon and nitrogen contents. Small fluctuations between samples are most likely the result of variations in the size distribution of the fines. It has been shown by MOORE *et al.* (1970a, b) and KAPLAN and SMITH (1970a, b) that the finest regolith material has a higher carbon content than the coarser fraction so that slight variations in the size distribution in the random samples could produce a noticeable difference in their carbon contents. This type of variation is evident in the sample 12028,181 in which two individual analyses show a larger variation than that between individual separate double core tube samples. All of the analyzed fines samples from the double core tube appear to fall in the dark or normal fines group.

The dark colored breccia 12057 is similar to the dark fines in both carbon and nitrogen. The light colored breccia 12034 is lower in both of these elements and hence similar to the lighter colored fines.

The fine-grained and coarse-grained basalts studied ranged from 16 to 45 μg/g in total carbon and from 43 to 44 μg/g in nitrogen. These rocks are similar to the related Apollo 11 basalts. The relatively high nitrogen content detected in an A type fine-grained Apollo 11 rock was not confirmed in this study. The two rocks made available for nitrogen analysis were both of the B type which in Apollo 11 also showed a relatively low nitrogen content.

DISCUSSION

The total carbon values reported in this paper are similar to those reported by other investigators (KAPLAN and PETROWSKI, 1971; EPSTEIN and TAYLOR, 1971; CHANG *et al.*, 1971). HINTENBERGER *et al.* (1971) have reported a major difference with respect to total nitrogen. Their investigations indicated values of $<1 \times 10^{-3}$ to $<3 \times 10^{-3}$ cm^3 N_2 STP/g or about 1 to 4 μg/g in the lunar basalts 10003, 10049, and 10072 and a range of 0.061 to 0.072 cm^3 N_2 STP/g or 76 to 90 μg/g in the bulk lunar fines 10084, 10087, and 12070. They suggest a possible contamination of lunar samples by atmospheric nitrogen. Other determinations of nitrogen include values of 36 and 54 μg/g in two lunar basalts (12053 and 12051) and 40 μg/g in lunar fines (12070) found by MORRISON *et al.* (1971). CHANG *et al.* (1971) found an indication of 40 μg/g in 12023 fines and 63 μg/g in 10009 breccia of chemically bound nitrogen released during an hydrolysis reaction. On the basis of the available data, the reason for the reported differences is difficult to evaluate.

The analytical results for the Apollo 12 samples reported here tend to support the model for the chemical evolution of the lunar regolith developed on the basis of our Apollo 11 data (MOORE *et al.*, 1970b). This model is developed in an attempt to explain the apparent paradox in the abundance and distribution of the volatiles C, N and S in the lunar regolith. For illustrative purposes, let us assume that the regolith is comprised of about equal parts of Type A and B rocks, 20% anorthosite to preserve a balance in Al content, and 2% carbonaceous chondrite material (KEAYS *et al.*, 1970).

The expected amounts of C, N, and S in such a model of the regolith are: 760 ppm C, 110 ppm N, and 2600 ppm S as compared to the observed amounts of about 150 ppm C, 130 ppm N, and 1000 ppm S. In fact, unlike carbon and nitrogen, sulfur is less abundant in the lunar fines and breccias than in the lunar basalts (KAPLAN and SMITH, 1970; PECK and SMITH, 1970; AGRELL *et al.*, 1970).

In order to explain these observations, it is hypothesized that during the conversion of the rocks to fines, presumably by high-energy meteorite impacts, there has been a net loss of these volatiles from the lunar surface. Thus, all volatiles in the meteoritic material, including not only C, N, and S but H_2O and the noble gases as well, along with a significant fraction of the indigenous volatiles, are lost during the impact event. We estimate the fractional loss of indigenous material on the basis of the glass content of the regolith to be about 35% in Apollo 11 fines and 25% in Apollo 12 fines. Solar wind bombardment adds carbon, nitrogen, and sulfur to the lunar fines in solar proportions (H/C/N/S = 50,000/16/6/1, MÜLLER, 1968), but since the solar abundance of sulfur is less than that of carbon or nitrogen and the indigenous sulfur content of the rocks high, the relative addition of sulfur is not significant.

Table 4 illustrates an elemental balance sheet for carbon, nitrogen, and sulfur. The calculations are based on the assumption that (1) the average proton flux on the lunar surface has been 2×10^8 cm^{-2} sec^{-1} (TILLES, 1965), (2) for Apollo 11 35% of the indigenous volatiles from the lunar rock and accumulated solar wind component in the lunar regolith are lost by impact and for Apollo 12 the loss factor is 25%, (3) the depths of the regolith are 4.6 and 2.5 m respectively, and (4) both have an age of 3.6×10^9 years.

The agreement of the calculated and measured values may be considered satisfactory in view of the problems related to sampling and, in the case of sulfur, the small amount of data. A similar but simpler calculation in which the calculated solar wind contributions for nitrogen and sulfur are normalized to the measured total carbon abundances gave similarly satisfactory results. If, in fact, future studies indicate that a major revision is needed in the total nitrogen contents of lunar rocks, this model will require modification.

Since the publication of the Apollo 11 results, several investigators have performed

Table 4. Calculated vs. observed concentrations of C, N, S in lunar fines (μg/g).

	Calculated Contributions			Measured	
	Lunar basalt	Solar Wind	Theoretical fines	Fines	References
Apollo 11					
C	45	110	155	140–225	a
N	65	35	100	100–150	a
S	950	15	960	650–1400	b, c, d
Apollo 12					
C	25	125	150	115–180	e
N	35	50	85	85–130	e
S	600	20	600	600–1000	f, g

References:
(a) MOORE *et al.* (1970); (b) KAPLAN and SMITH (1970); (c) PECK and SMITH (1970); (d) AGRELL *et al.* (1970); (e) this paper; (f) KAPLAN and PETROWSKI (1971); (g) MAXWELL and WIIK (9171).

additional experiments to support the evidence for a solar wind contribution to the elemental makeup of the lunar regolith. We concur with Epstein and Taylor (1971) who point out that some of the isotopic variations in carbon, oxygen, and silicon might be due to solar wind contributions or fractional evaporation. Several investigators notably Chang *et al.* (1971) and Eglinton *et al.* (1971) have performed interesting experiments with the hydrolysis of lunar fines with HCl, DCl, NaOH, and NaOD. Their studies indicate a release of both CH_4 and deuterated hydrocarbons. The CH_4 is attributed to either indigenous methane or solar wind material. The deuterated hydrocarbons are attributed to carbides from a meteoritic source although they may also be formed from individual labile carbon atoms introduced by solar wind. As noted above, Chang *et al.* (1971) also have detected ammonia released by this type of hydrolysis attack. These data, together with rare gas studies, show that at least some of the volatiles in the lunar regolith are of solar wind origin.

If the above model is valid, the light colored fines samples (12032, 12033, and 12037) have not accumulated as large a fraction of solar wind carbon and nitrogen as the dark fines. This suggests that they have not been exposed at the lunar surface for a long enough period of time to mature in terms of solar wind component.

References

Abell P. I., Draffan G. H., Eglinton G., Hayes J. M., Maxwell J. R., and Pillinger C. T. (1970) Organic analysis of the returned lunar sample. *Science* **167**, 757–759.

Agrell S. O., Scoon J. H., Muir I. D., Jong J. V. P., McConnell J. D. C., and Peckett A. (1970) Mineralogy and petrology of some lunar samples. *Science* **167**, 583–586.

Burlingame A. L., Calvin M., Han J., Henderson W., Reen W., and Simoneit B. R. (1970) Lunar organic compounds: Search and characterization. *Science* **167**, 751–752.

Chang S., Smith J. W., Kaplan I., Lawless J., Kvenvolden K. A., and Ponnamperuma C. (1970) Carbon compounds in lunar fines from Mare Tranquillitatis. IV. Evidence for oxides and carbides. *Proc. Apollo 11 Lunar Sci. Conf., Geochim. Cosmochim. Acta* Suppl. 1, Vol. 2, pp. 1857–1870. Pergamon.

Chang S., Kvenvolden K., Lawless J., Ponnamperuma C., and Kaplan I. R. (1971) Carbon in an Apollo 12 sample: Concentration, isotopic composition, pyrolysis products, and evidence for indigenous carbides and methane. Second Lunar Science Conference (unpublished proceedings).

Eglinton G., Abell P. I., Cadogan P. H., Maxwell J. R., and Pillinger C. T. (1971) Survey of lunar carbon compounds. Second Lunar Science Conference (unpublished proceedings).

Epstein S. and Taylor H. P., Jr. (1970) The concentration and isotopic composition of hydrogen, carbon and silicon in Apollo 11 lunar rocks and minerals. *Proc. Apollo 11 Lunar Sci. Conf., Geochim. Cosmochim. Acta* Suppl. 1, Vol. 2, pp. 1085–1096. Pergamon.

Epstein S. and Taylor H. P. (1971) O^{18}/O^{16}, Si^{30}/Si^{28}, D/H, and C^{13}/C^{12} ratios in lunar samples. Second Lunar Science Conference (unpublished proceedings).

Friedman I., Gleason J. D., and Hardcastle K. (1970) Water, hydrogen, deuterium, carbon and ^{13}C content of selected lunar material. *Proc. Apollo 11 Lunar Sci. Conf., Geochim. Cosmochim. Acta* Suppl. 1, Vol. 2, pp. 1103–1110. Pergamon.

Gibson E. K. and Moore C. B. (1970) Inert carrier-gas fusion determination of total nitrogen in rocks and meteorites. *Anal. Chem.* **42**, 461–464.

Hintenberger H., Weber H. W., Voshage H., Wänke H., Begemann F., Vilscek E., and Wlotzka F. (1970) Rare gases, hydrogen, and nitrogen: Concentrations and isotopic composition in lunar material. *Science* **167**, 543–545.

Hintenberger H., Voshage H., and Specht S. (1971) Heat extraction and mass spectrometric analysis of gases from lunar samples. Second Lunar Science Conference (unpublished proceedings).

JEDWAB J., HERBOSCH A., WALLAST R., NAESSENS G., and VAN GEEN-PEERS N. (1970) Search for magentite in lunar rocks and fines. *Science* **167**, 618–619.

KAPLAN I. R. and SMITH J. W. (1970) Concentration and isotopic composition of carbon and sulfur in Apollo 11 lunar samples. *Science* **167**, 541–543.

KAPLAN I. R. and PETROWSKI C. (1971) Carbon and sulfur isotopic studies on Apollo 12 lunar samples. Second Lunar Conference (unpublished proceedings).

KEAYS R. R., GANAPATHY R., LAUL J. C., ANDERS E., HERZOG G. F., and JEFFERY, P. M. (1970) Trace elements and radioactivity in lunar rocks: Implications for meteorite infall, solar wind flux and formation conditions of moon. *Science* **167**, 490–493.

LSPET (Lunar Sample Preliminary Examination Team) (1970) Preliminary examination of lunar samples from Apollo 12. *Science* **167**, 1325–1339.

MASON B. (1963) The carbonaceous chondrites. *Space Sci. Rev.* **1**, 621–646.

MAXWELL J. A. and WIIK H. B. (1971) Chemical composition of Apollo 12 lunar samples 12004, 12033, 12051, 12052, and 12065. Second Lunar Science Conference (unpublished proceedings).

MOORE C. B. and LEWIS C. F. (1965) Carbon abundances in chondritic meteorites. *Science* **149**, 317–318.

MOORE C. B. and LEWIS C. F. (1966) The distribution of total carbon content in enstatite chondrites. *Earth Planet. Sci. Lett.* **1**, 376–378.

MOORE C. B. and LEWIS C. F. (1967) Total carbon content of ordinary chondrites. *J. Geophys. Res.* **72**, 6289–6292.

MOORE C. B. and GIBSON E. K. (1969a) Nitrogen in chondritic meteorites. *Science* **163**, 174–176.

MOORE C. B., GIBSON E. K., and KEIL K. (1969b) Nitrogen abundances in enstatite chondrites. *Earth Planet. Sci. Lett.* **6**, 457–460.

MOORE C. B., LEWIS C. F., GIBSON E. K., and NICHIPORUK W. (1970b) Total carbon and nitrogen abundances in lunar samples. *Science* **167**, 496–497.

MOORE C. B., GIBSON E. K., LARIMER J. W., LEWIS C. F., and NICHIPORUK W. (1970b) Total carbon and nitrogen abundances in Apollo 11 lunar samples and selected achondrites and basalts. *Proc. Apollo 11 lunar Sci. Conf., Geochim. Cosmochim. Acta* Suppl. 1, Vol. 2, pp. 1375–1382. Pergamon.

MORRISON G. H., GERARD J. T., KASHUBA A. T., GANGADHARAM E. V., ROTHENBERG A. M., POTTER N. M., and MILLER G. B. (1970) Elemental abundances of lunar soil and rocks. *Proc. Apollo 11 Lunar Sci. Conf., Geochim. Cosmochim. Acta* Suppl. 1, Vol. 2, pp. 1383–1392.

MORRISON G. H., GERARD J. T., POTTER N. M., GANGADHARAM E. V., ROTHENBERG A. M., and BURDO R. A. (1971) Elemental abundances of lunar soil and rocks from Apollo 12. Second Lunar Science Conference (unpublished proceedings).

MÜLLER E. (1968) The solar abundances. In *Origin and Distribution of the Elements*, (editor L. H. Ahrens), Vol. 30, Sec. 2, pp. 155–176. Pergamon.

MURPHY M. E., MODZELESKI V. E., NAGY B., SCOTT W. M., YOUNG M., DREW C. M., HAMILTON P. B., and UREY H. C. (1970) Analysis of Apollo 11 lunar samples by chromatography and mass spectrometry: Pyrolysis products, hydrocarbons, sulfur amino acids. *Proc. Apollo 11 Lunar Sci. Conf., Geochim. Cosmochim. Acta* Suppl. 1, Vol. 2, pp. 1879–1890. Pergamon.

ORÓ J., UPDEGROVE W. S., GIBERT J., MCREYNOLDS J., GIL-AV E., IBANEZ J., ZLATKIS A., FLORY D. A., LEVY R. L., and WOLF C. J. (1970) Organogenic elements and compounds in Type C and D lunar samples from Apollo 11. *Proc. Apollo 11 Lunar Sci. Conf., Geochim. Cosmochim. Acta* Suppl. 1, Vol. 2, pp. 1901–1920. Pergamon.

PECK L. C. and SMITH V. C. (1970) Analysis of moon samples Apollo 11 mission. *Science* **167**, 532.

TILLES D. (1965) Atmospheric noble gases: Solar wind bombardment of extraterrestrial dust as a possible source mechanism. *Science* **148**, 1085–1088.

WOOD J. A., DICKEY J. S., MARVIN U. B., and POWELL B. N. (1970) Lunar anorthosites. *Science* **167**, 602–604.

Proceedings of the Second Lunar Science Conference, Volume 2, pp. 1351–1366.
The M.I.T. Press, 1971.

Thermal analysis-inorganic gas release studies of lunar samples

EVERETT K. GIBSON, JR.
National Aeronautics and Space Administration, Manned Spacecraft Center,
Houston, Texas 77058

and

SUZANNE M. JOHNSON
Lockheed Electronics Corporation, Houston, Texas 77058

(*Received 23 February 1971; accepted in revised form 24 March 1971*)

Abstract—Thermal analysis–quadrupole mass spectrometric analyses have been made on Apollo 11 and Apollo 12 lunar samples and synthetic lunar analogs. The samples were heated to 1400°C under vacuum, and the released gaseous species were determined along with their abundances, temperature ranges, and sequence of release. The volatiles measured included H_2, He, H_2O, N_2, CO, O_2, H_2S, CO_2, and SO_2. The concentrations of the alkali elements Li, K, Na, and Rb along with selected rare-earth elements were measured via isotopic dilution and atomic absorption analysis for selected sample residues. The K, Na, and Rb concentrations were found to be depleted by three to five fold during the vacuum pyrolysis while the lithium and rare-earth elements concentrations were not changed. Weight loss measurements and initial melting point determinations were obtained for the samples investigated. The evolved gaseous constituents appear to be derived from: (1) chemical reactions between solid phases, (2) solar wind, and (3) gaseous components from gas-rich inclusions and vesicles. The gas evolution patterns obtained from synthetic lunar analogs composed of troilite, cohenite, and a silicate matrix duplicated the gas release patterns for the lunar samples at temperatures above 500°C. Nitrogen, carbon monoxide, carbon dioxide, and possibly water vapor are the major components found within the gas-rich inclusions present in the Apollo 11 and Apollo 12 lunar samples.

INTRODUCTION

A LARGE NUMBER of the Apollo 11 and Apollo 12 holocrystalline rocks have vesicles, indicating that a gas phase was present during their formation (LSPET, 1969, 1970). The gas release most likely occurred because of the decrease of the overburden pressure on the molten material during the eruption and subsequent crystallization of the lunar basalts. When this occurred, the gaseous species dissolved within the melt began to exsolve from the magma. The vesicles were then formed. The actual composition of the gas that originally exsolved from the lunar magma is unknown but can be tentatively identified by using phase equilibria considerations. WELLMAN (1970) has shown that the assemblage iron–troilite–ilmenite found in the lunar basalts places restrictions upon both the p_{O_2} and p_{S_2} of the lunar basalts and the composition of the gas which formed the vesicles, provided equilibrium conditions prevailed. He also showed that only a small number of gaseous species could have pressures exceeding 1% of the vapor pressure of the constituent elements in the melting range of the lunar basalts and 0.01 to 100 atmospheres total pressure. Among the gaseous species possible are H_2, H_2O, N_2, CH_4, CO, CO_2, COS, and the noble gases. All of these gaseous species have been found in the lunar samples (ABELL *et al.*, 1970a, 1970b; CHANG *et al.*, 1970, 1971; BURLINGAME *et al.*, 1970; EGLINTON *et al.*, 1971; ORÓ *et al.*, 1970; MURPHY *et al.*, 1970). FRIEDMAN *et al.* (1970), CHANG *et al.* (1971), ABELL *et al.* (1970a), ORÓ *et al.* (1970, 1971), and MURPHY *et al.* (1970) reported that carbon dioxide was the most abundant carbon containing phase released during vacuum pyrolysis of the lunar soil below 500°C. BURLINGAME *et al.* (1970) reported that carbon monoxide was the major phase released during heating up to 500°C. No other research group confirmed their results regarding the abundances of CO and CO_2 during pyrolysis of soil samples to 500°C.

Gas inclusions in the lunar rocks have been identified by ROEDDER and WEIBLEN (1970). Small amounts of volatile constituents, possibly in part H_2O or CO_2, were present in the lunar magmas. These gases are still trapped in the silicate melt inclusions and as occasional gas inclusions. ROEDDER and WEIBLEN (1970) reported finding small amounts of noncondensable gases in some of the bubbles in glass samples from the lunar soil though many bubbles were empty.

The study of the volatiles in the returned lunar samples is important for the understanding of past thermal conditions on the lunar surface. The redox state of the planetary body can also be inferred from the volatile gaseous species released by heating the lunar samples under vacuum. The released volatiles may be from several sources: (1) indigenous to the moon, (2) meteoritic or cometary contributions, (3) solar wind, (4) remnants from a primordial lunar atmosphere, (5) outgassing products from the interior of the moon, (6) pyrolysis reaction products from mineral phases in the samples, and (7) contamination from the LM, astronaut activity, or sample processing in the LRL.

The soils returned by Apollo 11 and Apollo 12 contained crystalline rock and mineral fragments, glass, and a meteoritic component. The glasses found in the soil, which are depleted in volatile content, are the melting products resulting from meteoritic or cometary impact. The indigenous volatiles from the lunar rocks and solar wind components in the lunar regolith are lost during impact melting processes. The meteoritic volatiles appear to be lost during the impacting event.

The purpose of our investigation was to determine the gaseous species released from Apollo 11 and Apollo 12 soils, as well as from crystalline rock fragments and breccias during vacuum pyrolysis of the samples at programmed heating rates. This type of experiment is important in order to determine which volatiles are present and their relative abundance, as well as the temperature release range and sequence of release for these volatiles.

EXPERIMENT PROCEDURES

The lunar samples were analyzed utilizing a Mettler recording vacuum thermoanalyzer interfaced with a Finnigan 1015S/L quadrupole mass spectrometer (GIBSON and JOHNSON, 1971). The mass spectrometer source was placed directly in the reaction chamber. With this arrangement the evolved gaseous species could be analyzed without requiring gas transfer procedures.

Samples used in this study were placed in a previously outgassed 16 mm diameter platinum crucible and evacuated to 2×10^{-6} torr. The sample weight, temperature, and pressure were continuously recorded. The sensitivity of the analytical balance used for weight loss studies was

0.05 mg. The samples were heated from room temperature to 1400°C at a heating rate of 4°C/min. Sample temperatures were measured with calibrated platinum/platinum-10% rhodium thermocouples located at the base of the sample crucibles. Spectra were obtained every 12.5°C during the heating cycle. Reproducible background spectra were obtained during the bakeout procedures before sample analysis began and were later subtracted from the obtained spectra.

The lunar soil samples were neither crushed nor homogenized before analyses, under the assumption that previous sample splitting during sample description and processing had homogenized the soil samples for the size fractions allocated. The sample sizes utilized for the soil analyses were 242 mg for the Apollo 11 sample and 208 mg for the Apollo 12 soil sample. The samples of the breccia and crystalline rock used in this work were from 2.5 gm samples which had previously been homogenized (< 60 mesh) for analyses. The sample size used for the analysis of the Apollo 11 breccia was 202 mg. Because of the expected lower volatile contents of the crystalline rocks, larger sample sizes (400 and 500 mg) were used.

After the thermal analysis-mass spectrometric study was completed, the sample residues from the breccia and crystalline rock analysis were analyzed for their potassium, rubidium, sodium, lithium, and selected rare-earth elements concentrations by isotopic dilution and atomic absorption techniques. The analytical procedures employed have been described previously by GAST *et al.* (1970).

In order to determine the possible origins and sources of the released gaseous species, synthetic lunar analogs with small amounts of troilite, graphite, and cohenite were prepared. The matrix of this lunar analog was composed of olivine (40 wt. %), pyroxene (40 wt. %), and anorthite (20 wt. %). The mixture was crushed to less than 150 mesh. Various components (i.e., troilite, cohenite, graphite) were added and their gas release patters determined in an attempt to duplicate the release patterns of the lunar samples.

EXPERIMENTAL RESULTS

The samples analyzed included: (1) Apollo 11 soil 10086,16; (2) Apollo 11 breccia 10073,23; (3) Apollo 11 crystalline rock 10017,57 (surface chip); (4) Apollo 12 soil 12023,9; and (5) Apollo 12 crystalline rock 12022,38 (interior chip). The gas release patterns for the soils and crystalline rocks are shown in Figs. 1–4. The gas release patterns obtained for the Apollo 11 soil and breccia and the Apollo 12 soil samples show only minor differences. The gas release curves give evidence for solar wind components, chemical reaction products, and gaseous components from vesicles and/or gas-rich inclusions. The gas release patterns for the synthetic lunar analogs, with troilite, cohenite, and graphite added, are given in Fig. 5.

LUNAR SOIL

Carbon dioxide is the major carbon containing gaseous phase released below 500°C in both Apollo 11 and Apollo 12 soils. Low temperature carbon dioxide has been previously identified by ABELL *et al.* (1970a), ORÓ *et al.* (1970), and CHANG *et al.* (1970, 1971). The source of the low temperature carbon dioxide is not fully understood. ORÓ *et al.* (1970) attribute the CO_2 to the thermal decomposition of carbonates. Since the only carbonate reportedly found in the lunar fines is aragonite (GAY *et al.*, 1970), and its occurrence is still questionable, another source must be found for the low temperature carbon dioxide. The exact nature of the carbon found in the lunar fines has not been elucidated. MOORE *et al.* (1970) have proposed that the majority of the carbon found in the lunar fines can be accounted for from the solar wind. The carbon

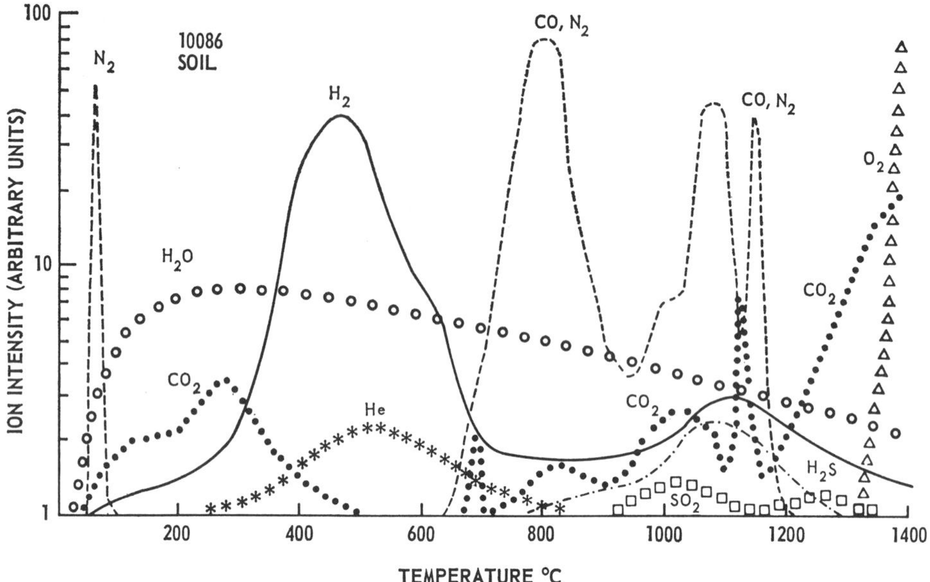

Fig. 1. Gas release pattern for Apollo 11 soil 10086,16. Sample weight 242.94 mg. Heating rate 4°C/minute.

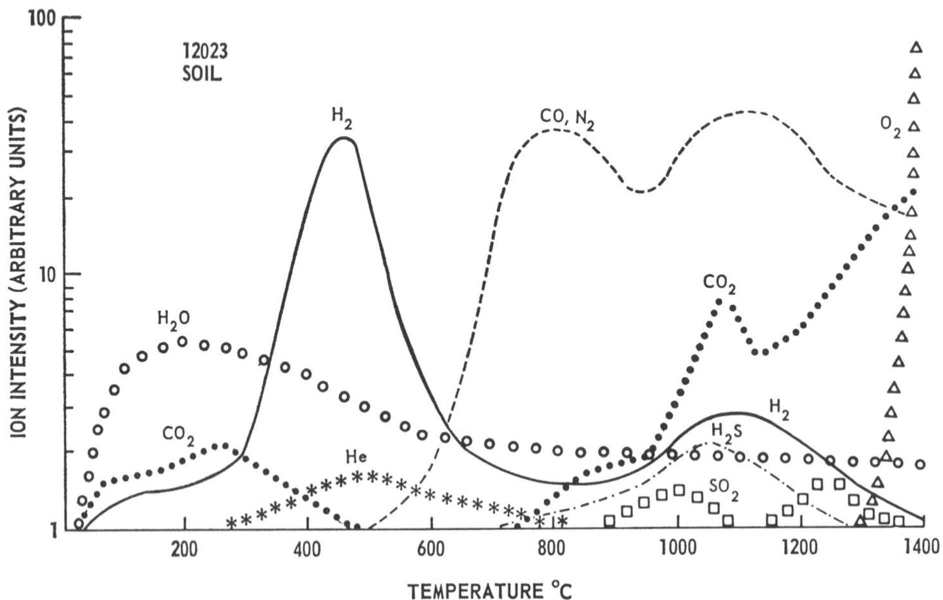

Fig. 2. Gas release pattern for Apollo 12 soil 12023,9. Sample weight 208.50 mg. Heating rate 4°C/minute.

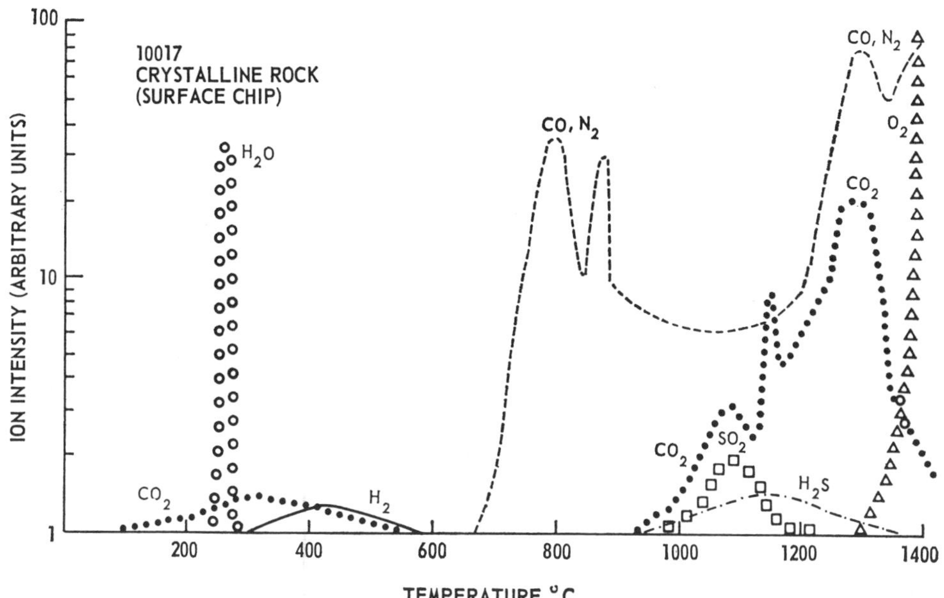

Fig. 3. Gas release pattern for a surface chip of Apollo 11 crystalline rock 10017,57. Sample weight 400.45 mg. Heating rate 4°C/minute.

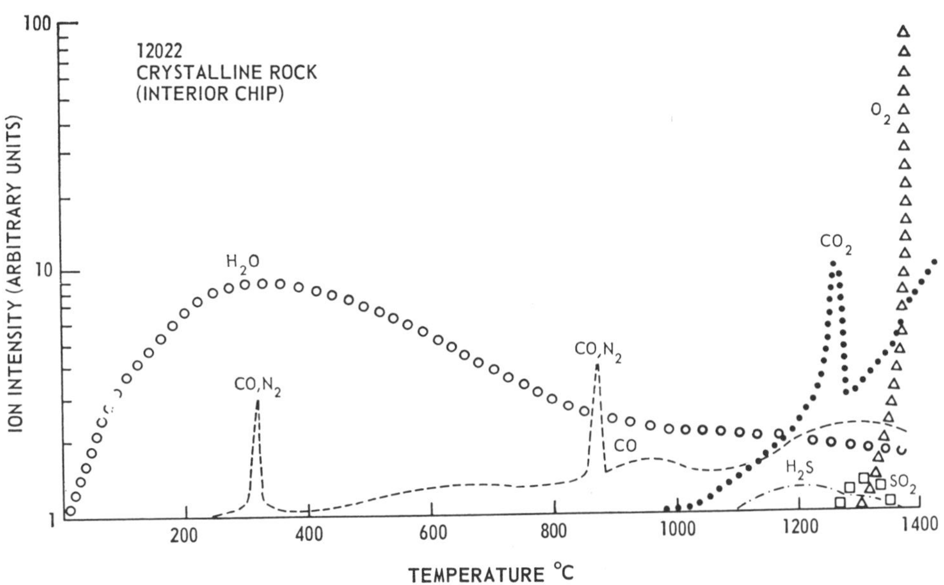

Fig. 4. Gas release pattern for an interior piece of Apollo 12 crystalline rock 12022,38. Sample weight 500.65 mg. Heating rate 4°C/minute.

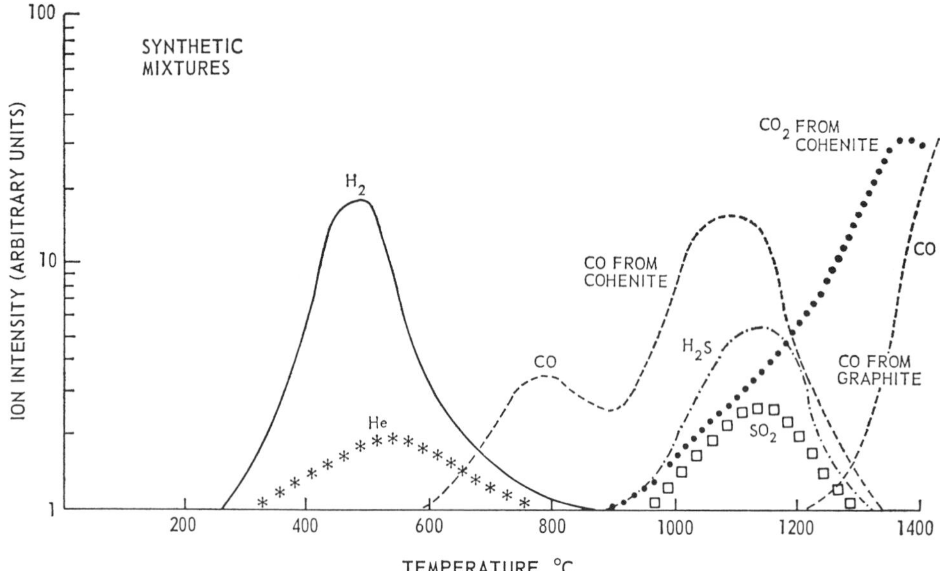

Fig. 5. Gas release patterns for synthetic lunar analogs. Matrix composed of olivine, pyroxene, and anorthite, with $<5\%$ troilite, cohenite, and graphite added. The H_2 and He release patterns are taken from LORD, 1968. The CO released below 1200°C is from the reaction of cohenite with the silicate matrix. All of the CO_2 observed is from cohenite. The CO released above 1300°C is from the reaction products of of graphite and the silicate matrix. The H_2S and SO_2 are reaction products of troilite and the silicate matrix.

from the solar wind might remain as elemental carbon or combine with the oxygen atoms of the silicates and produce carbon dioxide. The CO_2 could then be easily released by heating below 500°C if it were only weakly bonded within the silicates.

The solar wind hydrogen and helium are released between 300°–700°C. The release temperatures for both the hydrogen and helium are identical with those found by PEPIN *et al.* (1970) and LORD (1968). Additional hydrogen is evolved also between 1000°–1300°C. This high temperature released hydrogen is believed to be a chemical reaction product. The H/He ratio found in this investigation for the Apollo 11 fines is 7.3 and for the Apollo 12 fines is 9.5. STOENNER *et al.* (1970) reported a H/He atom ratio of 7.8 measured from the Apollo 11 lunar soils. HANNEMAN (1970) reported the H/He ratio for Apollo 11 soil to be 8.5. The presently accepted solar H/He atom ratio is 17 (LAMBERT, 1967). Diffusion of the hydrogen from the samples may account for the lower experimentally obtained H/He ratio.

During the gas release study of the lunar fines, water was released over a wide temperature range. The manner in which it was released indicates that it was bonded to the lunar sample in some manner. Our previous experience with the manner in which the water is released during gas evolution analysis indicates that the water was possibly chemisorbed on the samples. If the water had only been physically absorbed on the surface of the lunar particles, it would have been released very rapidly at temperatures below 150°–170°C. Possibly the water was produced by the irradiation of silicates by protons (ZELLER *et al.*, 1970). Experimental investigations are presently underway in our laboratory to test this mechanism of water formation on lunar samples.

At temperatures above 700°C the most abundant gaseous species released is mass 28, carbon monoxide and/or nitrogen. The carbon monoxide and/or nitrogen is evolved in two distinct regions, (1) between 700°–900°C and (2) between 1000°–1200°C. The source of the released carbon monoxide is believed to be a reaction product of carbon containing phases such as cohenite with the silicate minerals. Cohenite [$(Fe, Ni)_3C$] and other carbon containing phases have been identified in the lunar fines (ANDERSON *et al.*, 1970; FRONDEL *et al.*, 1970; CARTER and MACGREGOR, 1970; CHANG *et al.*, 1970, 1971). When the synthetic lunar analogs, with small amounts of cohenite (<5 wt. %), are heated, carbon monoxide is released in two distinct regions between 700°–1200°C. The release of carbon monoxide between 700°–900°C is characteristic for carbonaceous chondrite components (GIBSON and JOHNSON, 1971). The carbon monoxide released *does not* appear to be from the reaction of graphite and silicates. Synthetic lunar analogs, with graphite (≤ 200 mesh) added, failed to evolve carbon monoxide below 1350°C (Fig. 5). Therefore, the suggestion of BURLINGAME *et al.* (1970) that the reaction $FeO + C \rightarrow Fe + CO$ contributed to the observed carbon monoxide does not seem feasible. ARRHENIUS *et al.* (1970) reported a single graphite grain, but it is still questionable if its origin is lunar. None of the other detailed mineralogical and petrologic studies have found additional graphite.

The sudden release of carbon dioxide at 675°C and 1125°C from the Apollo 11 fines possibly confirms ROEDDER and WEIBLEN'S (1970) observation of noncondensable carbon dioxide and other gas phases within inclusions. The release of the carbon dioxide is therefore believed to arise from the rupturing of gas filled vesicles. The carbon dioxide release is also correlated with the observation of ORÓ *et al.* (1970) that CO_2 is released when the lunar fines are treated with acid. However, the CO_2 produced with the acid treatment could possibly be only the release of the solar wind produced CO_2. The CO_2 evolved above 1200°C is from reaction of cohenite with the silicate matrix. Carbon dioxide is similarly produced from the reaction of the synthetic lunar analogs and meteoritic cohenite mixtures (Fig. 5). We find it quite surprising that the inclusions which apparently are rupturing contain only a single gas phase, CO_2. Because of the low oxygen fugacity in the moon, the CO_2 should be unstable relative to H_2 and CO at magmatic temperatures. Solar wind derived gaseous components may account for the observed CO_2.

Previous investigators (ABELL *et al.*, 1970b; CHANG *et al.*, 1971; HENDERSON *et al.*, 1971; ORÓ *et al.*, 1971; NAGY *et al.*, 1970) have reported the presence of methane in both the Apollo 11 and Apollo 12 lunar soils. During our investigation we were not able to see the methane because of the type of analytical system used.

The hydrogen sulfide and sulfur dioxide released between 900°–1300°C are chemical reaction species produced between troilite and the silicates in the soil. However, the possibility that the SO_2 and H_2S have a solar wind origin cannot be completely ruled out. We believe that the quantities of H_2S and SO_2 which could be produced from the solar wind would be very small in comparison to the amounts released from the chemical reactions of troilite and the silicates. Troilite was identified during the initial examination of the lunar samples (LSPET, 1969, 1970). Identical gas release patterns can be generated with synthetic lunar analogs and meteoritic troilite mixtures (Fig. 5). The source of the H_2S is the reaction $FeS + 2H \rightarrow Fe + H_2S$. The source of the SO_2 may be from reactions of the type: $FeS + 2FeSiO_3 \rightarrow SO_2 + 3Fe + 2SiO_2$.

The oxygen released at temperatures above 1300°C is a reaction product of iron from the melt with the platinum crucibles. As the iron dissolves in the platinum crucible, oxygen is liberated by the reaction: $FeO + Pt = [Fe, Pt] + \frac{1}{2}O_2$.

LUNAR BRECCIAS

The results of the gas release study of the Apollo 11 breccia 10073 are identical to those found from investigation of the lunar soils. However, the amounts of gases released are slightly decreased. This can be accounted for by the amount of lithic fragments in the breccias and fines. The low temperature carbon dioxide is present in the breccia along with the solar wind hydrogen and helium components. The H/He ratio found for the breccia was 4.8, lower than the values of 7.3 and 9.5 found for the Apollo 11 and Apollo 12 soil samples. Carbon monoxide and/or N_2 is the major gaseous component released at temperatures above 700°C. The sudden release of carbon dioxide during the melting of the lunar breccia and soils, as noted by the carbon dioxide "spikes" in Figs. 1 and 2 in the temperature region of 1100°–1125°C, is of considerable importance. The carbon dioxide released is believed to result from rupturing of gas-rich inclusions found in the soils and breccia components. These inclusions have been previously observed by ROEDDER and WEIBLEN (1970). The amount of carbon dioxide released was very small, in the ppb range, but nevertheless it is similar to the quantities released from gas inclusions found in terrestrial olivine grains or glass inclusions (ROEDDER, 1965).

CRYSTALLINE ROCKS

The gas evolution analysis of the crystalline rocks 10017 and 12022 are shown in Fig. 3 and 4. Notable differences occur in the gas release patterns for the crystalline rocks as compared to the soils and breccia. The low temperature carbon dioxide, hydrogen, and helium component are present in sample 10017 but are absent from 12022. The concentrations of the CO_2, H_2, and He in the crystalline rocks are lower than in the soils and breccia by a factor of 8 to 10. The low abundances of solar wind hydrogen in the surface chip 10017 and the absence (below the detection limit) of solar wind hydrogen found in the interior chip 12022 allows one to speculate on the origin of other possible solar wind derived products. Carbon dioxide was released from the surface chip 10017 but was not detected in the interior chip 12022 at temperatures below 500°C. Therefore, we conclude that the carbon dioxide released below 500°C from the surface rock chip, breccia, and soils is derived from the solar wind. The isotopic ratios for the low temperatures carbon dioxide should be indicative of the possible solar wind ratio. CHANG *et al.* (1971) reported a $\delta^{13}C$ (per mil) value between 10.0 and 17.7 for the carbon released below 500°C whereas the carbon released at high temperatures had $\delta^{13}C$ isotopic composition between +7.4 and +22. They concluded that the presence of isotopically light, indigenous CH_4 probably caused the consistently low $\delta^{13}C$ values below 500°C. We interpret the low values to be from solar wind derived carbon.

Carbon monoxide and/or nitrogen is the major volatile component(s) released at temperatures above 700°C, again similar to the soils and breccia. The nature of the occurrence of carbon in the lunar rocks is difficult to determine. CHANG *et al.* (1970) presented evidence for the presence of carbides in the Apollo 11 rocks. WELLMAN (1970) has shown that carbon monoxide should be the major carbon containing phase present during the crystallization of the holocrystalline lunar rocks. The carbon monoxide should be approximately four times more abundant than carbon dioxide. During the gas release study of basaltic rock 12022 (Fig. 4) a gas phase of mass 28 was sporadically released, at 310°C and 850°C. The gas phase could be carbon monoxide and/or nitrogen but it is difficult to determine which gas because of the low resolution mass spectrometer utilized in this investigation. MOORE *et al.* (1970) presented evidence that nitrogen was unevenly distributed in the Apollo 11 fine grained crystalline rocks. The CO and/or N_2 could perhaps result from rupturing of gas-rich inclusions found in the fine grained rocks. BLOCH *et al.* (1971) noted that nitrogen was released from lunar crystalline rocks by crushing.

Sulfur dioxide and hydrogen sulfide are evolved between 1000°–1300°C from the crystalline rocks during vacuum pyrolysis. These two volatile species are reaction products between sulfide containing phases (e.g., troilite) and the silicate phases of the crystalline rocks (Fig. 5). Troilite has been observed in sample 10017 by ADLER *et al.* (1970), while LSPET (1970) and BRETT *et al.* (1971) identified troilite in 12022.

The amount of volatiles released from the crystalline rock is less than that which is released in the analysis of soils. Similar trends have been previously observed for the analysis of the volatile elements carbon and nitrogen (MOORE *et al.*, 1970). The decreased volatile content of the crystalline rock results from the absence of solar wind and meteoritic components which are found in the lunar soils and breccias. (GANAPATHY *et al.*, 1970).

The analysis of the holocrystalline rock 10017 indicates the presence of water vapor released from the lunar sample (Fig. 3). The water vapor is believed to have been released from the rupturing of a gas-rich vesicles. The quantity of water vapor released from the sample is in the order of 1 to 10 ppm. FRIEDMAN *et al.* (1970) found 25 ppm H_2O in an analysis of sample 10017 while MAXWELL *et al.* (1970) reported 100 ppm H_2O in a separate analysis of the same sample. BROWN *et al.* (1970) reported that a thin section of 10017,50 contained 4.8% vesicles which were more abundant than in the coarser-grained rocks. The vesicles reported were usually circular in section but did not contain any hydrous mineral phases. ROEDDER and WEIBLEN (1970) have previously described the gas-rich inclusions found in the Apollo 11 holocrystalline rocks.

The release of the water vapor, as shown in Fig. 3, at 262°–275°C would be very sudden when the gas or vapor filled vesicles and inclusions, ruptured, and could occur at any elevated temperature depending upon the pressure inside the vesicle and the tensile strength of the walls of the silicate or glass matrix in which the inclusions are found. The gaseous components found within the inclusions would be identical to those originally present during the crystallization of the minerals or glass phases. Consequently, the analysis of these primary inclusions will yield an insight into the chemical composition of the gas phase which was present early in the moon's history.

DREVER *et al.* (1970) reported the presence of a phyllosilicate on the surface of rock 10017 while GAY *et al.* (1970) and TRAILL *et al.* (1970) reported the presence of amphiboles in other lunar samples. The possibility that the water vapor which we observed during the pyrolysis of sample 10017 resulted from the decomposition of a phyllosilicate or amphibole is unlikely in light of the recent work of CHARLES *et al.* (1971). CHARLES *et al.* have examined the stability of the amphiboles previously found in lunar samples and found that the hydroxyl ions normally associated with terrestrial amphiboles have been replaced by the halogens, fluorine and chlorine. Their experimental results indicated that at 600°C only about 0.05 bars H_2O were required to produce the amphibole found in the vugs of selected lunar samples. Therefore, CHARLES *et al.* (1971) concluded that the water content of lunar magmas was very small.

The maximum lunar surface temperature of 140°C and the high vacuum of the moon would rule out the possiblity of finding volatiles being released below this temperature. Any volatiles released from lunar samples below ~140°–150°C would be contaminants or strongly adsorbed gaseous species introduced during the terrestrial handling and processing of the lunar samples. Therefore, it appears that the water vapor found in sample 10017 is indigenous.

WEIGHT LOSS STUDIES

During the gas release studies the weight of each sample was simultaneously recorded with a sensitivity of 0.05 mg. The results of the weight loss measurements are given in Table 1. Initial examination of the lunar samples (LSPET, 1969, 1970) noted the complete absence of low temperature mineral phases such as hydrates, sulfates, or carbonates commonly found in terrestrial rock and soils. The results of our investigation confirm the "dryness" of the lunar samples. The Apollo 11 soil sample lost 1.41 wt. % while the Apollo 12 soil lost 2.13 wt. %. The Apollo 11 and 12 crystalline rocks along with the breccia lost 1.02, 0.51, and 0.95 wt. %, respectively when heated to 1400°C under vacuum. The weight loss results reported in this work are

TABLE 1. Summary of weight loss for lunar samples

Sample	Weight mg	Percent weight-loss to 1000°C	Percent weight-loss to 1400°C	Initial melting °C
Apollo 11 Soil 10086,16	242.96	0.13	1.41	1120 ± 10
Apollo 12 Soil 12023,9	208.50	0.20	2.13	1125 ± 10
Apollo 11 Breccia 10073,23	201.87	0.15	0.95	1125, 1230 ± 10
Apollo 11 Rock A 10017,57	400.45	0.073	1.02	1100 ± 10
Apollo 12 Rock 12022,38	500.65	0.038	0.51	1140 ± 10

similar to those reported by HANNEMAN (1970). However, a portion of Hanneman's investigation was carried out under oxidizing conditions while this investigation was carried out under vacuum. HANNEMAN (1970) reported that the specimens exhibited a net weight gain due to oxidation; while the gases were being evolved upon heating through the range 600°–1200°C. We interpret these results to mean that the lunar soil is only partially oxidized and that its oxidation rate is fairly rapid.

Heating of the lunar soils in a platinum crucible under vacuum produced an initial meling of the Apollo 11 soil at 1120°C ± 10°C while initial melting of the Apollo 12 soil occurred at 1125°C. The breccia examined showed two distinct regions of melting. Initial melting occurred at 1125°C ± 10°C and a second region of melting beginning at 1230°C ± 10°C. The lower melting point is identical with the soil component while the higher melting point possibly results from the lithic fragments found in the breccia (e.g., plagioclase feldspar or ilmenite). The initial melting or solidus of the Apollo 11 crystalline rock 10017 was found to be around 1100°C ± 10°C, slightly lower than the soils and breccia samples. The initial melting of the 12022 crystalline rock occurred at 1140°C ± 10°C.

THERMAL VOLATILIZATION OF SELECTED ELEMENTS

O'HARA *et al.* (1970) noted that the potassium content of a synthetic lunar analog was decreased when placed in a vacuum-furnace at 1200°C for extended periods of time. Therefore, selected sample residues obtained after the gas release studies of the lunar basalts and breccias were analyzed for their potassium, rubidium, sodium, lithium, and selected rare-earth element contents. The experimental results, obtained via isotopic dilution and atomic absorption analysis, are given in Table 2. The analytical results indicate that the potassium, sodium, and rubidium contents of the samples were reduced by factors of 3 to 5 while the concentration of refractory element lithium and the rare-earth elements remained essentially constant. The samples had been subjected to a 4°C/min heating rate to 1400°C under a vacuum of 2×10^{-6} torr. This experiment shows that the volatile elements potassium, sodium, and rubidium can be depleted by partial volatilization of the lunar samples. However, the europium anomaly, observed by GAST *et al.* (1970) and other investigators, does not appear to result from volatilization, as demonstrated by these experimental results. The vacuum on the lunar surface is greater than that employed during our experimental study;

Table 2. Concentrations of selected elements from volatilization study

Element	Breccia 10073		Basalt 10017		Basalt 12022	
	initial	after 1400°C	initial	after 1400°C	initial	after 1400°C
ppm						
K	1200	385	2610	641	536	179.3
Rb	2.84	0.69	5.63	1.05	0.738	0.173
Na	3500	—	3800	500	1824	673
Li	—	—	18.7	19.3	9.51	11.8
Ce	46.5	—	77.3	—	17.4	17.7
Nd	35.4	—	59.5	—	14.8	15.0
Sm	12.4	—	20.9	—	5.58	5.59
Eu	1.70	—	2.14	—	1.28	1.28
Gd	15.9	—	27.4	—	7.90	7.85

therefore, we can only speculate about the exact nature of the alterations of the chemical composition of the lunar basalts.

Analyses of the Apollo 11 rocks by GANAPATHY *et al.* (1970) have shown that the lunar rocks are depleted by factors of 3 to 100 for certain relative volatile elements as compared to terrestrial basalts. RINGWOOD and ESSENE (1970) have explained this depletion on the basis of assuming that the depletion is a primary feature of the lunar basalt-source region. O'HARA *et al.* (1970) have suggested that the volatile element depletion resulted from "boiling-off" during extrusion of the basalts. RINGWOOD and ESSENE (1970) have pointed out the problems associated with the thermal volatilization of selected elements (e.g., sodium) and the observed trends for the Apollo 11 basalts. If thermal volatilization had been active, the Apollo 11 plagioclases should have been zoned with K and Na contents decreasing outward from the core. However, the Apollo 11 basalts are remarkably constant in their sodium contents despite the wide range of cooling histories (ESSENE *et al.*, 1970). Conceivably, the low-potassium lunar basalts possibly have been through a period of greater selective thermal volatilization during their crystallization history than for the high-potassium basalts; but more likely, the differences occur because of different source regions for the lunar basalts.

CONCLUSIONS

ANDERS (1970) recently reviewed the possibilities for finding water on the moon in the light of a model used for the accretion of meteorites and terrestrial planets. He suggested that the depletion of volatile elements would follow a priori if the moon had originally accreted as a satellite of the earth, and that this would imply a very low initial water content. The volatile elements and compounds, such as carbon, nitrogen, sulfur, and water, which are associated with the carbonaceous chondrites are all either depleted in the lunar rocks and soil or can be accounted for as having a solar wind origin (MOORE *et al.*, 1970). The trace element studies of the Apollo 11 and Apollo 12 soils and breccias contain around 1.9% meteoritic material, believed to be of carbonaceous chondrite composition (GANAPATHY *et al.*, 1970; ANDERS *et al.*, 1971). However, the volatiles sulfur and water, normally associated with carbonaceous chondrites, are either absent or strongly depleted in the lunar soils and breccias. MOORE *et al.* (1970) pointed out that the impacting velocities of carbonaceous chondrites on the surface of the moon are probably great enough to completely volatilize the sulfur and other easily volatilized materials such as water.

EPSTEIN and TAYLOR (1970, 1971) and FRIEDMAN *et al.* (1970) suggested that most of the water in the lunar samples is of terrestrial origin and only a small fraction is indigenous lunar water. They think that the lunar water was either primary or from oxidized solar wind. Our experimental results indicate that there is evidence of water from both sources in selected lunar (basaltic) crystalline samples. The discovery of possible gas-rich vesicles and inclusions containing water during this investigation, along with the observations of ROEDDER and WEIBLEN (1970), possibly confirm the presence of primary water in the lunar samples.

ZELLER *et al.* (1970) have reported that chemical alterations of the lunar surface materials might result from proton irradiation of the lunar surface. Water and hydroxyl ions can be produced on the surface of silicates which have been irradiated with protons. The solar wind protons can remove the reaction products at the surface by sputtering processes, while those products formed deeper by the higher energy solar flare and cosmic protons can be retained. Hydrocarbons are also a possible product from the proton irradiation of the surfaces if carbon-rich materials are present (e.g., carbonaceous chondrite material). The methane and ethane observed by ABELL *et al.* (1970b), NAGY *et al.* (1970, 1971), EGLINTON *et al.* (1971) and ORÓ *et al.* (1971) might have originated by (1) the proton irradiation of carbonaceous chondrite matter or indigenous organic matter found within the lunar fines or (2) formation from the solar wind by hydrogenation of the solar wind carbon with the hydrogen bombardment from the solar wind.

A portion of the water and other volatiles (e.g., CO, CO_2, N_2, etc.) originally present on the moon in the form of water vapor and other gases trapped within gas-rich vesicles and inclusions may still be present today, provided that the numerous cratering events have not caused rupturing of the vesicles. The water and other volatiles such as carbon dioxide, carbon monoxide, hydrogen sulfide, sulfur dioxide, methane, nitrogen, etc. which were released into the lunar exosphere during past lunar volcanism have been removed from the moon by two mechanisms. The one-sixth lunar gravity, as compared to the Earth's gravitational attraction, would give gaseous constituents a definite lunar half-life before they would escape from the moon. The gas pressure on the moon is so small that its exosphere begins essentially at the lunar surface; therefore, the escape velocity becomes one of the criteria used to determine whether a gaseous molecule will remain in the lunar atmosphere or escape into space. The lunar escape velocity is 2.38 km/sec. At the lunar noon temperature of about 390°K any gaseous phase whose molecular weight is below 43 would escape from the moon while those gases with higher molecular weight would be retained.

However, a second and more important phenomenon occurs on the moon which removes all of the gaseous species present in the lunar atmosphere. Solar radiation in the form of a solar wind from the sun will ionize any gaseous species in the lunar exosphere. As soon as the ionization occurs, the electrostatic mechanism proposed by ÖPIK and SINGER (1960) and SINGER (1961) will remove the ionized gases. The speed of removal is identical with the speed of ionization. With the extremely thin atmosphere which could have originally been present on the moon after the various periods of lunar volcanism, a virtually total dispersal of the "lunar atmosphere" by electrostatic action (solar wind) would be accomplished in a time-span of 10^6–10^7 seconds (i.e., 10–100 days) (KOPAL, 1969). Therefore, no gaseous species should be found, except in possible trace amounts, in the lunar atmosphere today. However, small amounts have been detected by the lunar atmosphere measuring experiments left on the surface of the moon during the Apollo 12 mission (FREEMAN *et al.*, 1971). The measurements recorded to date indicate that gaseous species are released during seismic activity, both man-made and lunar in origin. The components present, as inferred from these measurements, are principally the rare gases associated with solar wind products. Solar wind derived gaseous species such as helium, neon, argon, and krypton are found to be enriched in the outer surfaces of lunar materials. Thus, the inorganic gaseous species commonly associated with terrestrial volcanism will be found only in the gas-rich inclusions and vesicles of the lunar rocks and soil samples.

The gaseous species found in this work indicate that nitrogen, carbon monoxide, carbon dioxide, and possibly water are the major components found within the gas-rich inclusions present in lunar samples. The remaining gaseous species reported are derived from the pyrolysis of the lunar samples and are the chemical reaction products from reactions of the various components such as troilite, cohenite, and the silicates found in the lunar samples.

Acknowledgments—The authors gratefully acknowledge the assistance of Drs. Paul W. Gast and J. Oró for allowing joint analyses of the samples used in this study. We thank Henry Wiesmann for carrying out the chemical separations and atomic absorption analyses. Dr. Norman J. Hubbard performed the isotopic dilution analyses and his assistance is acknowledged. The meteoritic samples of cohenite and troilite were obtained through the courtesy of Dr. C. B. Moore, Arizona State University.

REFERENCES

ABELL P. I., DRAFFAN C. H., EGLINTON G., HAYES J. M., MAXWELL J. R., and PILLINGER C. T. (1970a) Organic analysis of the returned Apollo 11 lunar samples. *Proc. Apollo 11 Lunar Sci. Conf., Geochim. Cosmochim. Acta* Suppl. 1, Vol. 2, pp. 1757–1773. Pergamon.

ABELL P. I., EGLINTON G., MAXWELL J. R., and PILLINGER C. T. (1970b) Indigenous lunar methane and ethane. *Nature* **226**, 251–252.

ADLER I., WALTER L. S., LOWMAN P. D., GLASS B. P., FRENCH B. M., PHILPOTTS J. A., HEINRICH K. J. F., and GOLDSTEIN J. I. (1970) Electron microprobe analysis of lunar samples. *Science* **167**, 590–592.

ANDERS E. (1970) Water on the moon. *Science* **169**, 1309–1310.

ANDERS E., LAUL J. C., KEAYS R. R., GANAPATHY R., and MORGAN J. W. (1971) Elements depleted on lunar surface: Implications for origin of moon and meteorite influx rate. Second Lunar Science Conference (unpublished proceedings).

ANDERSON A. T., JR., CREWE A. V., GOLDSMITH J. R., MOORE P. B., NEWTON J. C., OLSEN E. J., SMITH J. V., and WYLLIE P. J. (1970) Petrologic history of the moon suggested by petrography, mineralogy, and crystallography. *Science* **167**, 587–590.

ARRHENIUS G., ASURMAA S., DREVER J. I., EVERSON J. E., FITZGERALD R. W., FRAZER J. Z., FUJITA H., HANOR J. S., LAL D., LIANG S. S., MACDOUGALL D., REID A. M., SINKAS J., and WILKENING L. (1970) Phase chemistry, structure, and radiation effects in lunar samples. *Science* **167**, 659–661.

BLOCH M., FECHTIG H., FUNKHOUSER J., GENTNER W., JESSBERGER E., KIRSTEN I., MÜLLER O., NEUKUM G., SCHNEIDER E., STEINBRUNN F., and ZÄHRINGER J. (1971) Active and inert gases in lunar material released by crushing at room temperature and by heating at low temperatures. Second Lunar Science Conference (unpublished proceedings).

BRETT R., BUTLER JR. P., MEYER JR. C., REID A. M., TAKEDA H., and WILLIAMS R. J. (1971) Apollo 12 igneous rocks 12004, 12008, 12009, and 12022: Metal grains and their relation to the crystallization history. Second Lunar Science Conference (unpublished proceedings).

BROWN G. M., EMELEUS C. H., HOLLAND J. G., and PHILLIPS R. (1970) Mineralogical, chemical and petrological features of Apollo 11 rocks and their relationship to igneous processes. *Proc. Apollo 11 Lunar Sci. Conf., Geochim. Cosmochim. Acta* Suppl. 1, Vol. 1, pp. 195–219. Pergamon.

BURLINGAME A. L., CALVIN M., HAN J., HENDERSON W., REED W., and SIMONEIT B. R. (1970) Study of carbon compounds in Apollo 11 lunar samples. *Geochim. Cosmochim. Acta* Suppl. 1, Vol. 2, pp. 1779–1791. Pergamon.

CARTER J. L. and MACGREGOR I. D. (1970) Mineralogy, petrology and surface features of some Apollo 11 samples. *Proc. Apollo 11 Lunar Sci. Conf., Geochim. Cosmochim. Acta* Suppl. 1, Vol. 1, pp. 247–265. Pergamon.

Chang S., Smith J. W., Kaplan I., Lawless J., Kvenvolden K. A., and Ponnamperuma C. (1970) Carbon compounds in lunar fines from Mare Tranquillitatis-IV. Evidence for oxides and carbides. *Proc. Apollo 11 Lunar Sci. Conf., Geochim. Cosmochim. Acta* Suppl. 1, Vol. 2, pp. 1857–1869. Pergamon.

Chang S., Kvenvolden K., Lawless J., Ponnamperuma C., and Kaplan I. R. (1971) Carbon in an Apollo 12 sample: Concentration, isotopic composition, pyrolysis products, and evidence for indigenous carbides and methane. *Science* **171**, 474–477.

Charles R. W., Hewitt D. A., and Wones D. R. (1971) H_2O in lunar processes: Stability of amphibole in lunar samples 10058 and 12013. Second Lunar Science Conference (unpublished proceedings).

Drever J. I., Fitzgerald R. W., Liang S. S., and Arrhenius G. (1970) Phyllosilicates in Apollo 11 samples. *Proc. Apollo 11 Lunar Sci. Conf., Geochim. Cosmochim. Acta* Suppl. 1, Vol. 1, pp. 341–345. Pergamon.

Eglinton G., Abell P. I., Cadogan P. H., Maxwell J. R., and Pillinger C. T. (1971) Survey of lunar carbon compounds. Second Lunar Science Conference (unpublished proceedings).

Epstein S. and Taylor H. P., Jr. (1970) The concentration and isotopic composition of hydrogen, carbon, and silicon in Apollo 11 lunar rocks and minerals. *Proc. Apollo 11 Lunar Sci. Conf., Geochim. Cosmochim. Acta* Suppl. 1, Vol. 2, pp. 1085–1096. Pergamon.

Epstein S. and Taylor H. P., Jr. (1971) C^{18}/O^{16}, Si^{30}/Si^{28}, D/H, and C^{13}/C^{12} ratios in lunar samples. Second Lunar Science Conference (unpublished proceedings).

Essene E., Ringwood A. E., and Ware N. (1970) Petrology of lunar rocks from the Apollo 11 landing site. *Proc. Apollo 11 Lunar Sci. Conf., Geochim. Cosmochim. Acta* Suppl. 1, Vol. 1 pp. 385–397. Pergamon.

Freeman J. W., Jr. Hills H. K., and Fenner M. A. (1971) Some results from the Apollo XI superthermal ion detector. Second Lunar Science Conference (unpublished proceedings)

Friedman I., Gleason J. D., and Hardcastle K. G. (170) Water, hydrogen, deuterium, carbon and C^{13} content of selected lunar material. *Proc. Apollo 11 Lunar Sci. Conf., Geochim. Cosmochim. Acta* Suppl. 1, Vol. 2, pp. 1103–1109. Pergamon.

Frondel C., Klein Jr. C., Ito J., and Drake J. C. (1970) Mineralogical and chemical studies of Apollo 11 lunar fines and selected rocks. *Proc. Apollo 11 Lunar Sci. Conf., Geochim. Cosmochim. Acta* Suppl. 1, Vol. 1, pp. 445–474. Pergamon.

Ganapathy R., Keays R. R., Laul J. C., and Anders E. (1970) Trace elements in Apollo 11 lunar rocks: Implications for meteorite influx and origin of the moon. *Proc. Apollo 11 Lunar Sci. Conf., Geochim. Cosmochim. Acta* Suppl. 1, Vol. 2, pp. 1117–1142. Pergamon.

Gast P. W., Hubbard N. J., and Wiesmann H. (1970) Chemical composition and petrogenesis of basalts from Tranquillity Base. *Proc. Apollo 11 Lunar Sci. Conf., Geochim. Cosmochim. Acta* Suppl. 1, Vol. 2, pp. 1143–1163. Pergamon.

Gay P., Bancroft G. M., and Bown M. G. (1970) Diffraction and Mössbauer studies of minerals from lunar soils and rocks. *Proc. Apollo 11 Lunar Sci. Conf., Geochim. Cosmochim. Acta* Suppl. 1, Vol. 1, pp. 481–497. Pergamon.

Gibson E. K., Jr. and Johnson S. M. (1971) Thermogravimetric-quadrupole mass spectrometric analysis of geochemical samples. *Analytical Chemistry* (submitted).

Hanneman R. E. (1970) Thermal and gas evolution behavior of Apollo 11 samples. *Proc. Apollo 11 Lunar Sci. Conf., Geochim. Cosmochim. Acta* Suppl. 1, Vol 2, pp. 1207–1211. Pergamon.

Henderson W., Kray W. C., Newman W. A., Reed W. E., Burlingame A. L., Simoneit B. R., and Calvin M. (1971) Study of carbon compounds in Apollo 11 and Apollo 12 returned lunar samples. Second Lunar Science Conference (unpublished proceedings).

Kopal Z. (1969) *The Moon*, D. Reidel, Dordrecht-Holland, 525 pp.

Lambert D. L. (1967) Abundance of helium in the sun. *Nature* **215**, 43–44.

Lord H. C. (1968) Hydrogen and helium ion implantation into olivine and enstatite: Retention coefficients, saturation concentrations, and temperature release profiles. *J. Geophys. Res.* **73**, 5271–5280.

LSPET (Lunar Sample Preliminary Examination Team) (1969) Preliminary examination of lunar samples from Apollo 11. *Science* **165**, 1211–1227.

LSPET (Lunar Scicnce Preliminary Examination Team) (1970) Preliminary examination of lunar samples from Apollo 12. *Science* **167**, 1325–1339.

MAXWELL J. A., PECK L. C., and WIIK H. B. (1970) Chemical composition of Apollo 11 lunar samples 10017, 10020, 10072, and 10084. *Proc. Apollo 11 Lunar Sci. Conf., Geochim. Cosmochim. Acta* Suppl. 1, Vol. 2, pp. 1369–1374. Pergamon.

MOORE C. B., GIBSON E. K., LARIMER J. W., LEWIS C. F., and NICHIPORUK W. (1970) Total carbon and nitrogen abundances in Apollo 11 lunar samples and selected achondrites and basalts. *Proc. Apollo 11 Lunar Sci. Conf., Geochim. Cosmochim. Acta* Suppl. 1, Vol. 2, pp. 1375–1382. Pergamon.

MURPHY R. C., PRETI G., NAFISSI-V M. M., and BIEMANN K. (1970) Search for organic material in lunar fines by mass spectrometry. *Proc. Apollo 11 Lunar Sci. Conf., Geochim. Cosmochim. Acta* Suppl. 1, Vol. 2, pp. 1891–1900. Pergamon.

NAGY B., SCOTT W. M., MODZELESKI V., NAGY L. A., DREW C. M., MCEWAN W. S., THOMAS J. E., HAMILTON P. B., and UREY H. C. (1970) Carbon compounds in Apollo 11 Lunar Samples. *Nature* **225**, 1028–1032.

NAGY B., DREW C. M., HAMILTON P. B., MODZELESKI V. E., NAGY L. A., UREY H. C. (1971) The organic compounds in the Apollo 12 lunar samples. Second Lunar Science Conference (unpublished proceedings).

O'HARA M. J., BIGGAR G. M., RICHARDSON S. W., FORD C. E., and JAMIESON B. G. (1970) The nature of seas, mascons, and the lunar interior in the light of experimental studies. *Proc. Apollo 11 Lunar Sci. Conf., Geochim. Cosmochim. Acta* Suppl. 1, Vol. 1, pp. 695–710. Pergamon.

ÖPIK E. J. and SINGER S. F. (1960) Escape of gases from the moon. *J. Geophys. Res.* **65**, 3065–3070.

ORÓ J., UPDEGROVE W. S., GIBERT J., MCREYNOLDS J., GIL-AV E., IBANEX J., ZLATKIS A., FLORY D. A., LEVY R. L., and WOLF C. J. (1970) Organogenic elements and compounds in type C and D lunar samples from Apollo 11. *Proc. Apollo 11 Lunar Sci. Conf., Geochim. Cosmochim. Acta* Suppl. 1, Vol. 2, pp. 1901–1920. Pergamon.

ORÓ J., FLORY D. A., GIBERT J. M., MCREYNOLDS J., LICHTENSTEIN H. A., WIKSTROM S., and GIBSON E. K., JR. (1971) Abundances and distribution of organogenic elements and compounds in Apollo 12 lunar samples. Second Lunar Science Conference (unpublished proceedings).

PEPIN R. O., NYQUIST L. E., PHINNEY D., and BLACK D. C. (1970) Rare gases in Apollo 11 lunar materials. *Proc. Apollo 11 Lunar Sci. Conf., Geochim. Cosmochim. Acta* Suppl. 1, Vol. 2, pp. 1435–1454. Pergamon.

RINGWOOD A. E. and ESSENE E. (1970) Petrogenesis of Apollo 11 basalts, internal constitution and origin of the moon. *Proc. Apollo 11 Lunar Sci. Conf., Geochim. Cosmochim. Acta* Suppl. 1, Vol. 1, pp. 769–799. Pergamon.

ROEDDER E. (1965) Liquid CO_2 inclusions in olivine-bearing nodules and phenocrysts from basalts. *Amer. Mineral.* **50**, 1746–1782.

ROEDDER E. and WEIBLEN P. W. (1970) Lunar petrology of silicate melt inclusions, Apollo 11 rocks. *Proc. Apollo 11 Lunar Sci. Conf., Geochim. Cosmochim. Acta* Suppl. 1, Vol. 1, pp. 801–837. Pergamon.

SINGER S. F. (1961) Atmosphere near the moon. *Astronaut. Acta* **7**, 135–140.

STOENNER R. W., LYMAN W. J., and DAVIS R., JR. (1970) Cosmic-ray production or rare-gas radioactivities and tritium in lunar material. *Proc. Apollo 11 Lunar Sci. Conf., Geochim. Cosmochim. Acta* Suppl. 1, Vol. 2, pp. 1583–1594. Pergamon.

TRAILL R. J., PLANT A. G., and DOUGLAS J. A. V. (1970) Garnet: First occurrence in the lunar rocks. *Science* **169**, 981–982.

WELLMAN T. R. (1970) Gaseous species in equilibrium with the Apollo 11 holocrystalline rocks during their crystallization. *Nature* **225**, 716–717.

ZELLER E. J., DRESCHHOFF G., and KEVAN L. (1970) Chemical alterations resulting from proton irradiation of the lunar surface. *Modern Geology* **1**, 141–148

Proceedings of the Second Lunar Science Conference, Volume 2, pp. 1367–1380.
The M.I.T. Press, 1971.

Mass spectrometric investigation of the vaporization process of Apollo 12 lunar samples

G. De Maria, G. Balducci, M. Guido, and V. Piacente
Istituto di Chimica Fisica ed Elettrochimica, Università di Roma, Rome, Italy

(*Received 22 February 1971; accepted in revised form 30 March 1971*)

Abstract—The vaporization processes of lunar samples 12022 and 12065 were investigated by means of the Knudsen cell-mass spectrometric technique, using two different instruments.

In addition to monatomic species Na, K, Fe, Mg, Ca, Al, Cr, Mn, and O, the molecules O_2, SiO, SiO_2, FeO, AlO, Al_2O, TiO, and TiO_2 were detected in a temperature range up to 2500°K. Partial pressures were measured at different temperatures in the vapor phase over the condensed system, according to the vaporization procedure adopted ("progressive depletion" or "fixed temperature procedure.")

Activity measurements for iron were made utilizing a multiple-rotating Knudsen cell technique. The thermodynamic treatment of the equilibrium reactions involving the observed species was based mainly on the third-law procedure. The general internal consistency of the data and the agreement with literature values, shows that the thermodynamic equilibrium conditions were attained in the vapor phase. A discussion concerning the characterization of the mode of vaporization of both samples is also presented.

Introduction

Characterization of the vaporization of lunar samples is of particular interest not only in treating a variety of astrochemical and cosmochemical problems, but also in connection with problems related to the exploitation of resources of lunar soil (De Maria and Piacente, 1969).

It has long been recognized that condensation from a primeval nebula led to processes resulting in the formation of the solid bodies we now observe in the solar system (Schatzmann, 1966; Lord III, 1965). Therefore knowledge of molecular dissociative equilibria in the gas-phase over condensed materials which could be considered to be somewhat representative of the primary crust of the planets, should prove to be useful in treating the physical-chemical aspects of condensation processes of the primeval nebula (Tsuji, 1964; Morris, 1967).

Experimental Techniques

The vaporization of selected lunar specimens was carried out by Knudsen effusion-mass spectrometric technique. Two mass spectrometers were utilized for these experiments. The first one was an Inghram-type Nuclide Analysis HT model (NAA), 60° magnetic-sector, 12″ radius of curvature, vertically mounted. The second one was a high-temperature mass spectrometer, Bendix Co. time-of-flight, model 3015 (BC).

General characteristics of the method and the experimental procedure utilized have been described previously (Inghram and Drowart, 1960; De Maria, 1966; Bowles, 1969).

The Knudsen cell region, the ionization chamber and the analyzer tube or flight-tube are kept evacuated by means of a differential pumping system. A movable slit placed between the Knudsen cell region and the ionization chamber makes it possible to distinguish species effusing from the Knudsen cell from those in the background.

Both radiation heating and electron bombardment were used to vaporize the sample from the Knudsen cell, reaching temperatures as high as 2500°K. Temperatures were measured using calibrated Leeds and Northrup optical pyrometers, sighting a bottom black body hole in the NAA instrument and the effusion hole in the BC instrument, through bottom and top viewing window, respectively. A tungsten, tungsten–rhenium thermocouple enclosed in the central crucible support leg, was used together with the optical pyrometer in the BC instrument. The agreement between the two readings was generally satisfactory.

Procedure and Experimental Results

Vaporization of Apollo 12 lunar specimens 12022,57 and 12065,36 was carried out using rhenium containers having a knife-edge effusion hole 2.3 mm in diameter. The choice of the container material was dictated by previous vaporization studies of olivine (De Maria *et al.*) and terrestrial basalts (De Maria and Piacente, 1969). On the whole 680.2 mg of sample 12022,57 and 928.6 mg of 12065,36 were utilized in 6 quantitative vaporization experiments carried out with both instruments. The experiments were performed utilizing alternate procedures which consisted either of studying the vaporization of the sample progressively, namely by changing the temperature step by step ("progressive depletion" procedure) or of heating the sample to a predetermined temperature ("fixed temperature" procedure). The latter mode was controlled by the limits imposed by the validity of the Knudsen effusion conditions. This procedure was mainly used in the initial part of the BC experiments.

In addition to the monatomic species Na, K, Fe, Mg, Ca, Al, Cr, Mn, and O, the molecular species O_2, SiO, SiO_2, FeO, AlO, Al_2O, TiO, and TiO_2 were detected during the vaporization of the samples. Ions were identified by ionization potentials, intensity profiles in the molecular beam and isotopic distribution. Absolute partial pressures were determined utilizing the relation

$$P = \frac{I_i^+ T}{S\sigma\gamma a_i} \tag{1}$$

where P is the partial pressure, I_i^+ the ion intensity of the isotopic peak i, a_i its abundance, T the temperature of the effusion cell, S the sensitivity factor as deduced by an instrument calibration, σ the relative ionization cross-section at the operating electron energy, and γ the electron multiplier efficiency. The σ values at the maximum of the ionization efficiency curve were taken from Mann (Mann, 1967). These values were reduced to the operating electron energies utilizing experimental ionization efficiency curves. Ion intensities were measured at each temperature at 30 or 70 eV.

To convert ion intensities into pressures, the silver calibration (Inghram and Drowart, 1960), and, whenever possible, the O/O_2 calibration, were utilized. The latter procedure is based on direct measurements of the ratio between the O^+ and O_2^+ ion intensities and the well known (Brix and Herzberg, 1954; JANAF, 1965) equilibrium constant for the dissociation of O_2. In making this calibration, corrections for ionization cross sections, fragmentation patterns and secondary electron yelds were taken into account. The agreement between the silver calibration and the O/O_2 calibration was satisfactory at high temperatures, while at lower temperatures, owing to the contribution of the O^+ fragment to the O^+ parent ion, the O/O_2 calibration proved to be less reliable.

As the sample vaporized a depletion in the volatile components occurred. This process caused a change in the intensities of species at constant temperature to values which, in some cases, dropped below the sensitivity-detection limits of the instruments. We therefore considered only the "initial points," namely the values measured in a temperature range corresponding to the conditions under which presumably no appreciable variation of the concentration of the relative components in the condensed phase occurred.

The treatment of the experimental results given in this paper, is mainly restricted to such data.

Sample 12022,57

The chemical and mineralogical composition of the 12022 specimen is reported elsewhere (Smales, 1971); 167.6 mg and 162.6 mg were utilized in two series of experiments carried out with BC instrument, while 350 mg were vaporized quantitatively in one series performed with the NAA instrument. The behavior of the system is described below.

Alkaline components

Na^+ and K^+ were the first ion species observed at temperatures about 1250°K. Owing to the low content in the sample (0.36 and 0.068 wt. % of Na_2O and K_2O, respectively) depletion of these components occurred very soon, so that only a few initial points could be measured. Molecular oxygen was also detected during the evaporation of these components, showing a partial pressure far below the stoichiometric amount corresponding to the process: $Me_2O(s) \rightarrow 2Me(g) + \frac{1}{2}O_2(g)$. This behavior could be partially attributed to a certain degree of interaction of oxygen with the rhenium container. This interpretation is supported by the results obtained in the vaporization of sample 12065, where passivated rhenium cells were used and a higher ratio O_2^+/Me^+ was measured. Moreover, ReO_2 and ReO_3 gaseous molecules were observed at higher temperatures. Partial pressures of the alkali metal species and their apparent temperature dependences are reported in Figs. 1 and 2. Partial pressure values for oxygen in a temperature range corresponding approximately to the vaporization of the alkali metals are reported in Table 1.

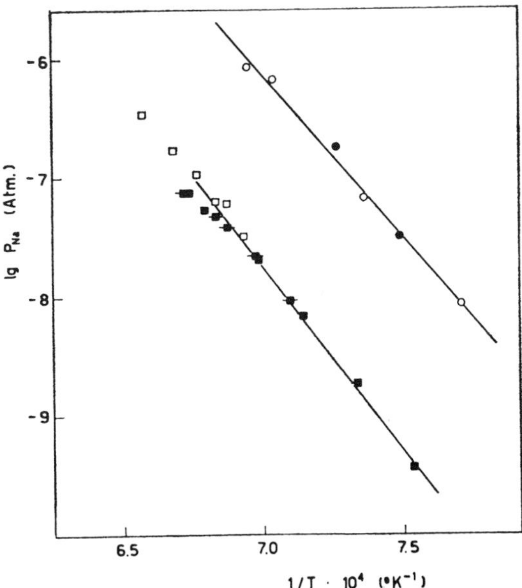

Fig. 1. Temperature dependence of Na partial pressures over lunar samples 12022 and 12065. The circles are referred to the sample 12022 while the squares are relative to the sample 12065. Open points are referred to experiments carried out with BC instrument; full points are relative to NAA experiments. Barred points are relative to different series of data. This notation is followed through the text.

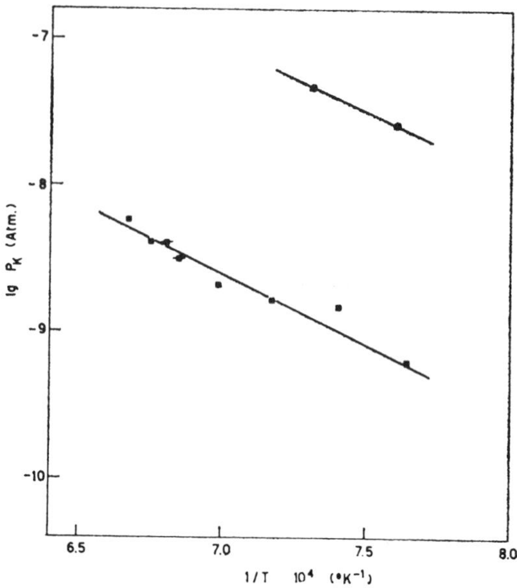

Fig. 2. Temperature dependence of K partial pressures over samples 12022 and 12065. For the symbolism used see Fig. 1.

Table 1. Partial pressures of O_2 over sample 12022 and 12065.[a]

Sample	T (°K)	*P* (atm)
12022	1396	5.54×10^{-9}
	1475	4.96×10^{-8}
12065	1433	3.78×10^{-9}
	1482	1.22×10^{-8}
	1499	2.11×10^{-8}
	1471	1.27×10^{-8}
	1459	8.78×10^{-9}

[a] Measurements carried out with NAA instrument.

Iron

Fe^+ ion peaks were first observed around 1300°K. This element vaporizes at nearly unit activity and the temperature dependence, as shown in Fig. 3, is very close to that for pure iron reported in the literature (Hultgren *et al.*, 1967).

This could be explained in part as due to the presence of free iron in the sample. However nearly unit activity of iron was observed in the vaporization study of olivine (De Maria *et al.*) and it was explained as a result of the process (Mueller, 1965):

$$Fe_2SiO_4(c) \longrightarrow Fe(c) + O_2(g) + SiO_2(c) \qquad (2)$$

More complex reactions might occur in the lunar sample, but the net process has to be regarded as being very similar to reaction (2).

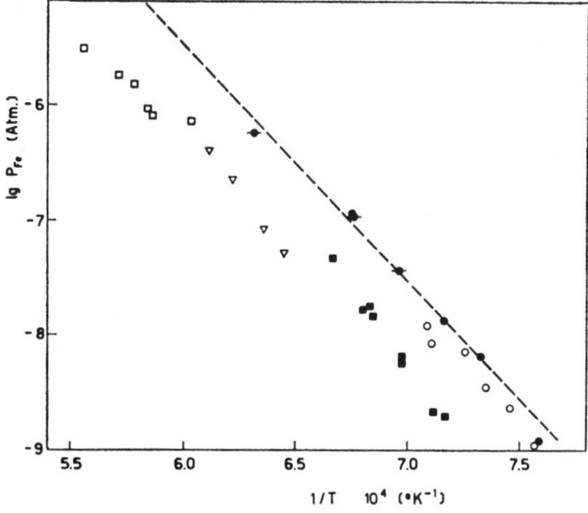

Fig. 3. Temperature dependence of Fe partial pressures over samples 12022 and 12065. The dashed line represents the vapor pressure of pure iron (HULTGREN, 1967). The Δ points are relative to sample 12065 data, obtained using the method of multiple-rotating cell. For the symbolism used see Fig. 1.

Table 2. Partial pressures of Mn and Cr over sample 12022.[a]

T (°K)	P_{Mn} (atm)	P_{Cr} (atm)
1583	7.66×10^{-9}	—
1609	1.32×10^{-8}	—
1725	8.47×10^{-8}	1.03×10^{-8}
1794	—	6.2×10^{-8}

[a] Measurements carried out using NAA instrument.

Chromium and Manganese

The "initial points" vapor pressure data are reported in Table 2. These elements are to be considered as minor constituents, and display a very low activity (10^{-5} to 10^{-6}).

Magnesium and Calcium

These components were observed as atomic species. The partial pressure values corresponding to a few indicative "initial points" are reported in Table 3 and Fig. 4 for Mg and Ca, respectively. They exhibit low values of activity. The vapor pressure data here reported are of the same order of magnitude as the decomposition pressures over the pure oxides (ALTMAN, 1963; ACKERMANN and THORN, 1961).

The measurement of calcium ion intensities was made difficult by a strong re-evaporation effect. For magnesium, a decrease of activity was noticed as the initial observation of the species was made. This behavior renders questionable the attribution of the reported data to "initial points."

Table 3. Partial pressures of Mg over samples 12022 and 12065.

Sample	T (°K)	P (atm)	
		NAA	BC
12022	1650	—	1.6×10^{-8}
	1694	5.7×10^{-8}	—
	1815	3.96×10^{-6}	—
12065[a]	1747	1.65×10^{-9}	—
	1854	9.4×10^{-9}	—
	1776	2.40×10^{-9}	—
	1857	8.85×10^{-8}	—
	1927	3.20×10^{-8}	—

[a] These points probably are not to be considered "initial points."

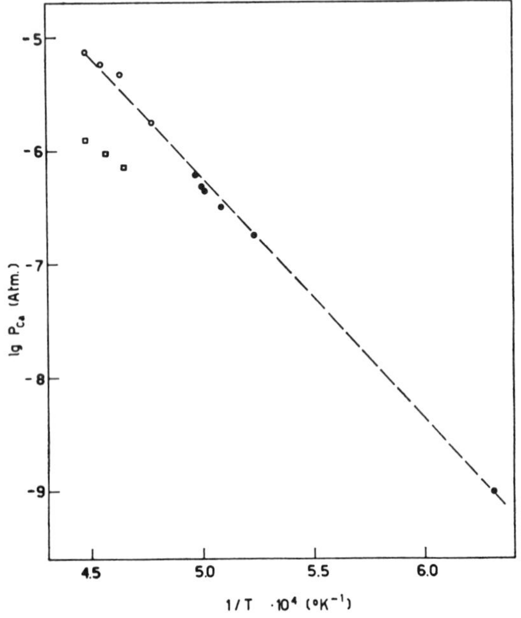

Fig. 4. Pressure values of Ca over 12022 and 12065 lunar samples.

Silicon

SiO and SiO_2 were the silicon-containing molecular species observed in the vapor. SiO was first observed at temperatures around 1650°K.

The relative proportion of these species was affected by the vaporization procedure carried out. In particular, the fixed temperature procedure, mainly utilized in the BC experiments, tends to create less reducing conditions, which determine as a consequence the shift of the equilibrium towards the preferential formation of SiO_2 species. The thermodynamic treatment of the data is reported in the following paragraph "Equilibrium reactions." Partial pressure values of SiO relative to the initial points

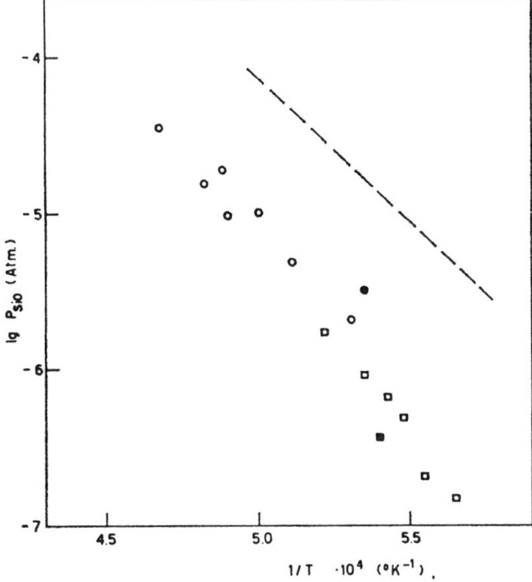

Fig. 5. Partial pressures of SiO over 12022 and 12065 lunar samples. The dotted line represents the SiO partial pressure over cristobalite (PORTER *et al.*, 1955).

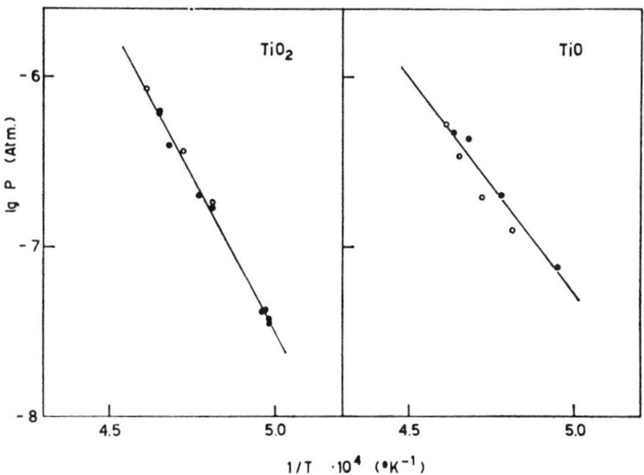

Fig. 6. Partial pressure of TiO and TiO_2 over 12022 and 12065 lunar samples.

are reported in Fig. 5. They are about a factor of 10 less than the partial pressure of the same species over SiO_2 in form of cristobalite (PORTER *et al.*, 1955).

Titanium

TiO and TiO_2 were the only titanium-containing molecular species observed in the vaporization of the sample. Their partial pressures in the temperature range 2000°–2170°K were comparable. The relative data are reported in Fig. 6. This behavior could be explained on the ground of the particular thermodynamic conditions created by the simultaneous vaporization of residual silicon-oxygen condensed component. This constituent creates oxidizing conditions in the system, determining therefore the shift of the equilibrium: $TiO(g) + O(g) \leftrightharpoons TiO_2(g)$ towards the formation of the higher oxidation state species.

Aluminum

Al, Al_2O, and AlO were the species observed in the vapor for this constituent. Partial pressure values are reported in Fig. 7 for Al and Table 4 for AlO and Al_2O.

It should be emphasized that the partial pressures of these species increased as the vaporization of the other less refractory constituents occurred. Therefore in this initial part of the experiments the reproducibility of the ion intensity with temperature was poor. An analogous situation occurred in the final part of the vaporization process, where a marked decrease of the ion intensity was observed, likely due to the sample exhaustion. For this reason, only the data taken in the intermediate range were deemed reliable and therefore included in the tables. AlO points measured with NAA instrument were taken mostly in the final range, so that they could not be included in the tables. Atomic oxygen was detected in this temperature range and the relative pressure data are reported in Table 5.

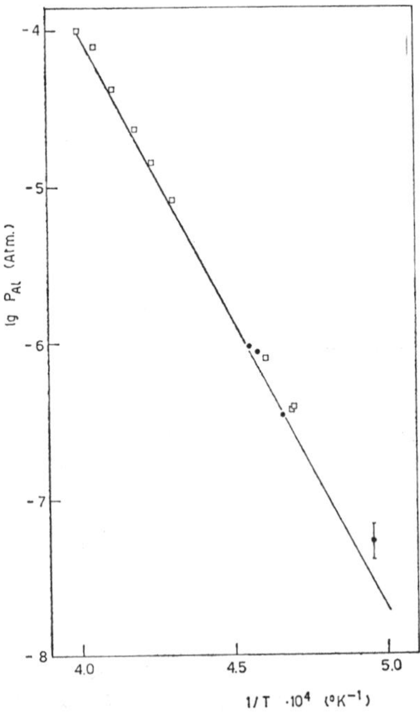

Fig. 7. Partial pressure of A1 over 12022 and 12065 lunar samples.

Table 4. Partial pressures of AlO and Al_2O over samples 12022 and 12065.

Sample	T (°K)	P_{Al_2O} (atm) NAA	P_{Al_2O} (atm) BC	P_{AlO} (atm) BC
12022	2060	5.07×10^{-9}	—	—
	2104	1.13×10^{-8}	—	—
	2125	1.57×10^{-8}	—	—
	2146	2.24×10^{-8}	—	—
	2180	3.65×10^{-8}	—	—
	2183	3.82×10^{-8}	—	—
	2194	4.01×10^{-8}	—	—
	2290	1.55×10^{-7}	—	—
	2270	1.04×10^{-7}	—	—
12065	2320	—	—	1.05×10^{-6}
	2360	—	6.90×10^{-7}	1.30×10^{-6}
	2395	—	9.88×10^{-7}	2.22×10^{-6}
	2425	—	1.90×10^{-6}	3.79×10^{-6}
	2465	—	3.39×10^{-6}	6.84×10^{-6}
	2500	—	4.73×10^{-6}	8.96×10^{-6}

Table 5. Partial pressures of O over samples 12022 and 12065 in the range of vaporization of aluminum containing species.

Sample	T(°K)	P (atm) NAA	P (atm) BC
12022	2015	9.2×10^{-8}	—
	2060	2.2×10^{-7}	—
	2104	3.2×10^{-7}	—
	2146	6.5×10^{-7}	—
	2162	9.0×10^{-7}	—
	2194	8.54×10^{-7}	—
	2270	2.0×10^{-6}	—
12065	2320	—	1.63×10^{-6}
	2360	—	2.91×10^{-6}
	2395	—	3.80×10^{-6}
	2425	—	7.40×10^{-6}
	2465	—	1.13×10^{-5}
	2500	—	1.42×10^{-5}

Sample 12065

305 and 150 mg of this sample were used in two experiments carried out with the BC instrument, while in the NAA experiments 304 mg were employed. The data relative to this sample are reported in Figs. 1–7 and Tables 1–5, together with the data relative to the sample 12022. The investigation of sample 12065 was carried out following a procedure similar to that described for sample 12022. The similarity of the general characteristics of the mode of vaporization of the two samples emerges from an inspection of the reported diagrams. Some differences in the partial pressures of a few species (Na, K, Fe, Ca, and Mg) can be pointed out. The differences might be attributed to differences in the chemical composition of the samples. Nevertheless, apart from the case of Na and K, the differences could appear less marked if one takes into account an uncertainty of at least a factor 2 in the pressure values. This uncertainty is mainly due to the errors connected to the calibration procedure.

In discussing the results one should also bear in mind the fact that the use of passivated rhenium crucibles may give conditions which could favor the formation of species in the higher oxidation state (TiO_2, SiO_2).

Equilibrium Reactions

Equilibrium reactions involving atomic and molecular gaseous species examined are reported in Table 6 together with the relative ΔH_0° values. For each equilibrium reaction, the basic data utilized in the calculations were the partial pressures of the gaseous species measured simultaneously at various temperatures in the entire course of the vaporization experiments. The data were obtained from both samples in experiments carried out with both instruments.

The thermodynamic treatment of the pressure data was performed using third-law and where possible, second-law procedures (Inghram and Drowart, 1960). The necessary free-energy and heat-content functions of molecular and atomic species were taken from JANAF tables (JANAF 1965/67 and Hultgren (Hultgren *et al.*, 1967), respectively. The errors quoted for second-law ΔH_0°'s are the standard deviations while in the third-law ΔH_0° they are inclusive of the standard deviations and estimated errors connected to the previously discussed calibration procedure.

As an example of equilibrium calculation, third-law data for Reactions (1) and (2) are reported in Tables 7 and 8.

An inspection of data reported in Table 6 shows that ΔH_0° values obtained in different experiments with two instruments are in agreement.

This proves that thermodynamic equilibrium conditions were attained in the vapor over the condensed material.

The most significative test for the validity of this assumption is the comparison of ΔH_0° values for the reaction: $AlO \rightleftharpoons Al + O$ for which a direct measurement of the equilibrium constant is available from literature (Drowart *et al.*, 1960). Our value of 112 ± 3 kcal/mole compares fairly well with the reported value of 115 ± 5 kcal/mole.

In the light of these results, data concerning the remaining equilibria could be utilized to derive not well established thermodynamic quantities for molecular

Table 6. Third-law heats of equilibrium reactions in the vapor phase over samples 12022 and 12065.

Reaction	ΔH_0° (kcal/mole) BC	NAA
(1) $SiO_{2(g)} \rightleftharpoons SiO_{(g)} + \frac{1}{2}O_{2(g)}$	47 ± 2	46 ± 2
(2) $AlO_{(g)} \rightleftharpoons Al_{(g)} + O_{(g)}$	112 ± 3	—
(3) $Al_2O_{(g)} + O_{(g)} \rightleftharpoons 2AlO_{(g)}$	15.6 ± 2	—
(4) $FeO_{(g)} \rightleftharpoons Fe_{(g)} + \frac{1}{2}O_{2(g)}$	39 ± 2	38 ± 2
	38 ± 3[a]	—
(5) $TiO_{2(g)} \rightleftharpoons TiO_{(g)} + \frac{1}{2}O_{2(g)}$	83 ± 2	—
(6) $TiO_{2(g)} \rightleftharpoons TiO_{(g)} + O_{(g)}$	—	142 ± 3
		143 ± 8[a]

[a] Second-law value.

Table 7. Third-law heat of reaction: $SiO_{2(g)} = SiO_{(g)} + \frac{1}{2}O_{2(g)}$ in kilocalories/mole.

Sample	T (°K)	$-R \ln K_p$	$-\Delta(\text{fef})$	ΔH_0°
12022	2060	4.09	19.025	47.6
	2080	3.80	19.02	47.5
	2120	3.07	19.00	46.8
	2150	3.07	18.99	47.4
	2170	2.70	18.98	47.0
12065	1990	4.62	19.05	47.1
	2035	4.30	19.03	47.5
	2060	3.98	19.02	47.4
				$\Delta\bar{H}_0^\circ = 47.3 \pm 0.3$[a]
	1850	5.99_5	19.105	46.4[b]

[a] The error quoted is the standard deviation.
[b] Experiment carried out with NAA instrument.

Table 8. Third-law heat of reaction: $AlO_{(g)} = Al_{(g)} + O_{(g)}$ in kcal/mole.[a]

T (°K)	K_p	$-R \ln K_p$	$-\Delta(\text{fef})$	ΔH_0°
2360	3.19×10^{-5}	20.56	27.26	112.8
2395	4.02×10^{-5}	20.01	27.30	113.4
2425	8.03×10^{-5}	18.73	27.32	111.6
2465	1.33×10^{-4}	17.74	27.35	111.1
2500	1.58×10^{-4}	17.39	27.38	111.9
				$\Delta\bar{H}_0^\circ = 112.2 \pm 0.9$[b]

[a] Experiments carried out on sample 12065 with BC instrument.
[b] The errors quoted is the standard deviation.

species involved such as D_0° (FeO). The discussion of the values obtained in relation to those available in literature is beyond the spirit of this work and are reported elsewhere (BALDUCCI *et al.*).

ACTIVITY DETERMINATION BY MULTIPLE-ROTATING KNUDSEN CELL

Accurate measurements of iron activity were made using a multiple rotating Knudsen cell assembly. The description of the method has been reported elsewhere DE MARIA and PIACENTE). For this experiment a system of three Knudsen cells inserted in a graphite block was employed.

The crucibles were made out of three different materials: rhenium, iron, and magnesium oxide.

The rhenium crucible was loaded with 106 mgrs. of 12065 lunar sample, the MgO crucible with Na_2O and the iron crucible with iron chips. The lid of this latter crucible was protected with a tantalum external cap. The effusion holes of all the crucibles had a knife-edge orifice, 1 mm in diameter. The calibration of the system was carried out following the procedure described previously.

Activity measurements of Na could not be performed, owing to the high vapor pressure of the standard reference system, while positive results could be obtained for iron. The measured values are plotted in Fig. 3.

Previous experiments carried out using platinum crucibles were unsuccessful, due to the formation of a foam which overflowed the container, running down the external walls.

Conclusions

A general picture of the vaporization behavior of sample 12022 is represented in Fig. 8.

The complexity of the system investigated and some arbitrary peculiarity of the vaporization procedure seem to play a less severe role than one, at first sight, could be brought to think. Even though the significance of the so called "initial points" appears somewhat questionable, especially for the components Mg and SiO, nevertheless the reproducibility of the data obtained by two independent groups of researchers, using two different instruments and pressure calibration procedures, demonstrates a substantial validity of the results.

On this ground one can adequately characterize the mode of vaporization of Apollo 12 lunar samples as following: (a) In a first stage the preferential vaporization of alkali components and an almost contemporary formation of pure iron occurs. This stage is also characterized by release of molecular oxygen. (b) At temperatures around 1600°K the vaporization of alkaline-earth components start to be observed. Silicon oxide species are observed at a higher temperature range around 1800°K. (c) The residue contains the more refractory components, namely titanium and

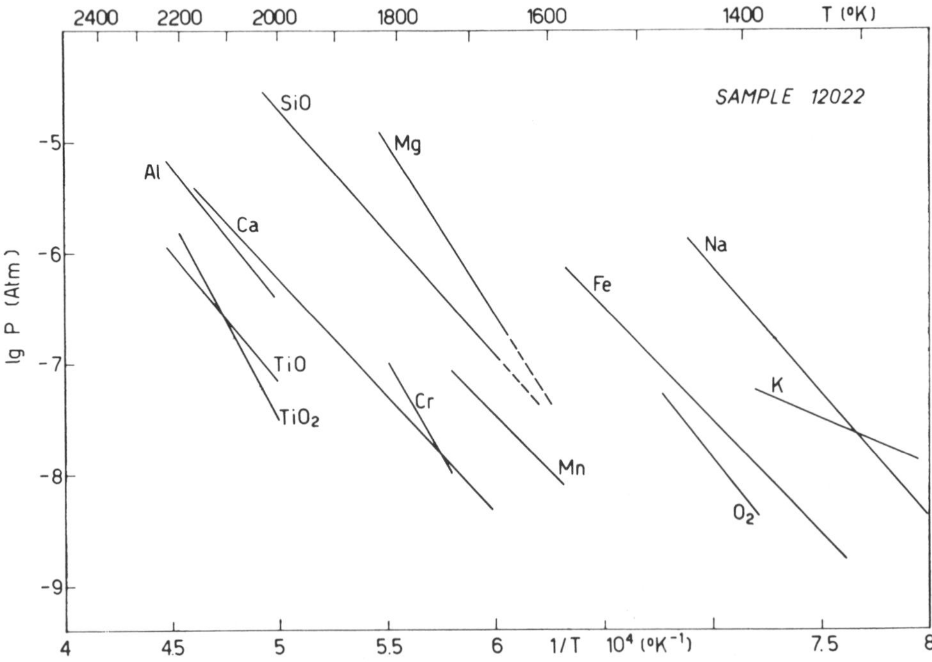

Fig. 8. General figure of the vaporization behavior of 12022 lunar sample.

aluminum, whose gaseous oxide species start to be observed in an appreciable amount around 1900°–2000°K.

The decrease of activity observed after a more or less extensive period of recording the "initial points" could be attributed either to depletion of the residue or to possible reactions in the condensed phase.

To the extent to which the Apollo 12 lunar rocks can be considered representative of the primordial lunar surface (or the surfaces of its predecessor planetesimals), the species SiO, SiO_2, Al_2O, AlO, TiO, TiO_2, and FeO are to be considered significant gaseous molecular constituents of the primeval nebula involved in the moon formation.

More extensive experiments, especially of the type "fixed temperature procedure" should be carried out with different compositions of lunar samples. Investigation of the nature of the residual condensed-phase will usefully complement the vaporization studies.

Acknowledgments—We wish to acknowledge the valuable help of Mr. Di Egidio in the course of the experiments. This work has been supported by the Consiglio Nazionale delle Ricerche through the Centro di Studio per la Termodinamica Chimica alle Alte Temperature and the Commissione Spaziale Italiana.

We are grateful to NASA for making available a generous supply of lunar material for our investigation.

References

Ackermann R. J. and Thorn R. J. (1961) *Progress in Ceramic Sciences*, Vol. 1, p. 49, Pergamon.

Altmann R. L. (1963), Vaporization of magnesium oxide and its reaction with alumina. *J. Phys. Chem.* **67**, 366–369.

Balducci G., De Maria G., Guido M., and Piacente V., *J. Chem. Phys.* (in press).

Bowles R. (1969), High temperature studies of inorganic solids. In *Time of Flight Mass Spectrometry* (editors D. Price and J. E. Williams), pp. 211–226, Pergamon.

Brix P. and Herzberg G. (1954) Fine structure of the Schumann–Runge bands near the convergence limit and the dissociation energy of the oxygen molecule. *Can. J. Phys.* **32**, 110–135.

De Maria G. (1966) Condensed-phase and gas-phase equilibria at high temperature. In *Simposio di Dinamica delle Reazioni Chimiche*, Vol. 5, CNR, Rome.

De Maria G. and Piacente V. (1969) Mass spectrometric study of rock like lunar surface material. *Atti Accad. Naz. Lincei* **47**, 525–536.

De Maria G., Piacente V., and Carapezza M. Thermodynamic study of vaporization process of olivine (in preparation).

De Maria G. and Piacente V. Activity measurements by multiple rotating Knudsen cell-mass spectrometric method. *Bull. Soc. Chim. Belg.* (in publication).

Drowart J., De Maria G., Burns R. P., and Inghram M. G. (1960). Thermodynamic study of Al_2O using a mass spectrometer. *J. Chem. Phys.* **32**, 1366–1372.

Hultgren R., Orr R. L., and Kelley K. K. (1967) *Supplement to selected values of thermodynamic properties of metals and alloys.* Univ. of California.

Inghram M. G. and Drowart J. (1960). *Mass Spectrometry Applied to High Temperature Chemistry, in High Temperature Technology.* McGraw-Hill.

JANAF (1965) Thermochemical Tables, Publ. 168–370, U.S. Department of Commerce.

JANAF (1965–67) Thermochemical Tables, Pb 168–370–371–372.

Lord III H. C. (1965) Molecular Equilibria and Condensation in a Solar Nebula and Cool Stellar Atmospheres. *Icarus* **4**, 279–288.

Mann J. B. (1967) Ionization cross sections of the elements calculated from mean-square radii of atomic orbitals. *J. Chem. Phys.*, **46**, 1646–1651.

MORRIS S. and WYLLER A. A. (1967) Molecular dissociative equilibria in carbon stars. *Astrophys. J.* **150**, 877–907.

MUELLER R. F. (1965). System Fe–MgO–SiO_2–O_2 with applications to terrestrial rocks and meteorites. *Geochim. Cosmochim. Acta* **29**, 967–976.

PORTER R. F., CHUPKA W. A., and INGHRAM M. G. (1955). Mass spectrometric study of gaseous species in the Si–SiO_2 system. *J. Chem. Phys.* **23**, 216–217.

SCHATZMAN E. (1966) *The Origin and Evolution of the Universe.* Hutchinson and Co.

SMALES A. A., MAPPER D., WEBB M. S. W., WEBSTER R. K., WILSON J. D., and HISLOP J. S. (1971). Elemental composition of lunar surface material (Part 2). Second Lunar Science Conference (unpublished proceedings).

TSUJI T. (1964). Abundance of molecules in stellar atmospheres. *Proc. Japan Acad.* **40**, 99–104.

Proceedings of the Second Lunar Science Conference, Volume 2, pp. 1381–1396.
The M.I.T. Press, 1971.

Active and inert gases in Apollo 12 and Apollo 11 samples released by crushing at room temperature and by heating at low temperatures

J. FUNKHOUSER,* E. JESSBERGER, O. MÜLLER, and J. ZÄHRINGER
Max-Planck-Institut für Kernphysik,
69 Heidelberg, Germany

(*Received 23 February 1971; accepted in revised form 31 March 1971*)

Abstract—Active and inert gases were released from lunar material, several meteorites, and a terrestrial basalt by crushing at room temperature and by heating at 225° and 300°C. Independent analyses were performed by mass spectrometry and gas chromatography. Density measurements determined before and after crushing allowed the internal gas pressure in vesicles and vugs to be calculated.

The composition of gases released from lunar crystalline rocks by crushing is variable and consists, in most cases, of H_2, N_2, CH_4 and the noble gases, all in ppb quantities. Gas pressures were below 1 mm Hg indicating incomplete retention of the vesiculating volatiles. The gases released by crushing lunar breccias consist of ppm quantities of solar wind helium and ppb amounts of H_2, N_2, H_2S, CH_4, C_2H_6, C_2H_4, C_3H_8. The noble gas elemental and isotopic ratios generally conform to those in bulk samples of the rocks and breccia with the exception of enhanced $^{20}Ne/^{36}Ar$ ratios in the breccia. The efficiency of noble gas release by crushing is approximately 10^{-3}.

The chemical and physical state of carbon monoxide in lunar rocks and breccia is enigmatic as it is not detected by crushing at room temperature, yet it is common to all samples heated as low as 225°C. A search for volatile carbonyls in lunar fines yielded negative results. Water, CO_2 and NH_3 were also not detected during crushing lunar samples.

INTRODUCTION

THE OCCURRENCE of vesicles in some of the crystalline rocks returned from Mare Tranquillitatis and Oceanus Procellarum indicates that magmatic gases were once present. The chemical and mineralogical evidence of severe fractionation of volatile elements and crystallisation under very low oxygen fugacity and anhydrous conditions places certain constraints on the composition of the magmatic gas phase. In lunar fines and breccia, on the other hand, the solar wind has contributed both directly and indirectly relatively large amounts of volatiles. Many investigators have released gases such as CO, CO_2, H_2, N_2, noble gases, H_2O, and low-molecular weight hydrocarbons by both low-temperature heating and acid dissolution. However, many of the observed gases may be produced during the extraction from reactions with contaminants and reagents, e.g., $C + H_2O \rightarrow CO + H_2$ or $Fe_3C + 4H^+ \rightarrow CH_4 + 3Fe^{++}$. In fact, the amount of carbon occurring as carbides has been ingeniously estimated by measuring hydrocarbons produced during deuterium exchange (EGLINTON *et al.*, 1971; CHANG *et al.*, 1971).

We have released trapped gases from selected lunar samples by gentle crushing

*Permanent address: State University of New York, Stony Brook, N.Y. 11790

at room temperature and have analyzed the constituents both by gas chromatography and mass spectrometry. Comminution by grinding has been used recently by ABELL *et al.* (1970a, b), CHANG *et al.* (1971), and EGLINTON *et al.* (1971) to study trapped gases in lunar material, and by BELSKY and KAPLAN (1970) to determine trapped hydrocarbons in meteorites. We have also carried out some low-temperature heating experiments in order to learn more about the source of the gases released by crushing and their chemistry.

EXPERIMENTAL PROCEDURE

Gas release by crushing was performed on aliquot chips of samples using different techniques. Two separate crushers (Fig. 1) and analytical systems were employed. In one, a stainless steel sample crusher was connected directly to a double column, double detector VARIAN MAT gas chromatograph, Model 1532, 2B, (MÜLLER and GENTNER, 1968). Samples of 100–500 mg were crushed under highly pure helium which served as a carrier gas. The gases released were evenly split into a 6-foot molecular sieve column and a 6-foot Porapak Q column, each equipped with a very sensitive He-detector. The ion current from both detectors was registered with a two-pen recorder (see Fig. 2).

In the mass spectrometer system, a similar crusher was connected directly to a rapid scanning mass filter (VARIAN MAT AMP 3) and indirectly to a 60°, 6-inch magnetic deflection mass spectrometer (JESSBERGER and GENTNER, 1971). The gases released by crushing were immediately measured with the mass filter. The pressure was not measured during analysis as an ion gauge would perturb the gas composition. The noble gases were subsequently separated over cold

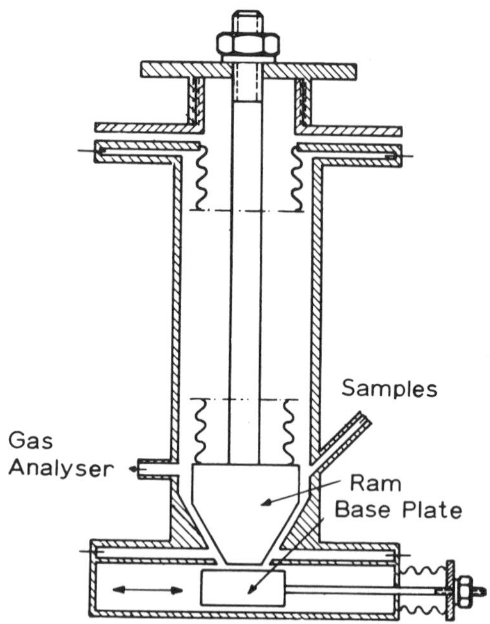

Fig. 1. Sample crusher. Constructed of stainless steel with Viton gaskets (for gas chromatograph) or gold gaskets (for mass spectrometer). An extensible bellows actuated by a hand screw depresses the hardened steel ram against a base plate, also hardened steel. Crushed samples can be removed from base plate without breaking vacuum.

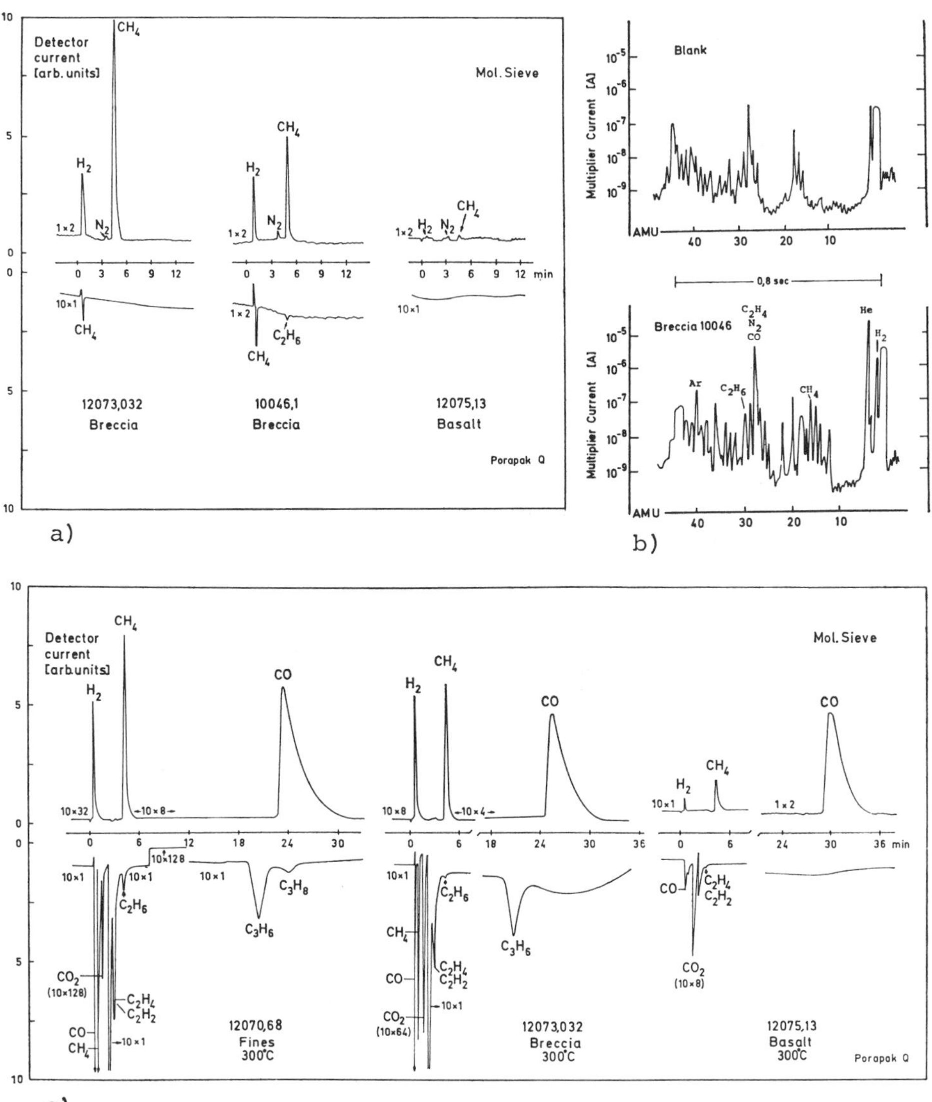

Fig. 2. (a) Gas chromatograms of gases released from lunar breccias 12073,032, 10046,1 and basalt 12075,13 by crushing at room temperature. Detector voltage 400 V (molecular sieve), 280 and 320 V (Porapak Q). Column temperature 35°C. Helium flow 30 cm^3/min. Length of both columns 6 feet. 1×2, 10×1 = range × attenuator. (b) Mass filter spectra of gases released in a blank run and from lunar breccia 10046,1 by crushing at room temperature. The resolution of the mass filter is 40. Note multiplier current is logscale. (c) Gas chromatograms of gases released from lunar fines 12070,68, breccia 12073,032 and basalt 12075,13 by heating in vacuum sealed glass vials at 300°C for 15 hours. Conditions same as Figure 2a.

Table 1. Summary of gases released from lunar, meteoritic, and terrestrial material by crushing at room temperature.

Sample (no. of analyses)	Gas Content in Units of 10^{-8} cc STP/g											
	H_2	H_2O	H_2S	CO	CO_2	N_2	NH_3	CH_4	C_2H_6	He	Ar	Total measured
10057 (8) vesicular basalt	≤3–90	—	—	—	—	21–220	—	≤2–200	—	30–48	12–24	70–300
12021 (3) gabbro	≤4–60	—	—	—	—	≤4–32	—	≤1–14	—	29–64	2–6	70–140
12063 (4) microgabbro	≤2–180	—	—	—	—	12–110	—	10–27	—	10–87	2–8	70–320
12075 (3) basalt	≤2–970	—	—	—	—	≤2–45	—	≤5–32	—	43–45	2–6	420–1100
10046 (5) breccia	1600–2900	—	4–18	—	—	1100–1800	—	320–2200	41–99[1]	77000–250000	140–390	82000–260000
10046 (1) light fragment	≤17	n.d.	n.d.	—	—	180	n.d.	34	—	n.d.	n.d.	230
10061 (1) breccia	2200	—	8	—	—	2100	—	2000	110[2]	76000	200	85000
12073 (1) breccia	2700	n.d.	n.d.	—	—	87	n.d.	2500	—	n.d.	n.d.	5300
Bruderheim (2)	150–260	—	—	—	—	22–40	—	32–36	—	44	13	210–390
Allende (2)	240–560	—	—	—	12–88	240–1400	—	150–200	8	27	12	2100–2400
Abee (2)	3300–36000	6500	2800	—	25–210	7200–40000	—	1090–12000	215–1000[3]	2200	340	12000–100000
Kapoeta (2)	≤2–250	—	—	—	—	≤4–130	—	≤2–31	—	74	—	480
Olivine basalt (3)	1100–13000	8000–31000	330–450	200–800	25–1800	690–2000	—	40–570	—	21–41	33–42	19000–65000
Duran (Vycor) glass (3)	10–950	—	—	—	—	10–180	—	10–50	—	35–43	≤0.2	30–1200
	Detection Limits in Units of 10^{-8} cc STP											
Mass spectrometry	0.5	1	1	5	0.5	1.5	5	0.5	1	0.5	0.1	
Gas chromatography	1	n.d.	n.d.	5	1	1	n.d.	1	1	n.d.	n.d.	

n.d. = not determined; — = below detection limits; also in the above units: (1) 10 C_2H_4, 7–16 C_3H_8; (2) 51 C_3H_8; (3) 120 C_2H_4, 87 C_3H_8, 46 C_3H_6.

charcoal and their abundance and isotopic ratios determined statically with the magnetic deflection spectrometer. The sensitivity of each method of analysis for different gases can be seen from Table 1. Typical gas chromatograms and mass spectra are illustrated in Fig. 2.

Crushing experiments were performed on lunar crystalline rocks and breccia, and on meteorites and a terrestrial basalt to provide a basis for comparison. These results are listed in Table 1 as a range of measured values. It should be emphasized that the manner in which the gases were released from our samples involved more a breaking open of each sample, rather than a continuous grinding and pulverizing comminution process. Samples were loaded as individual pieces. Typically, a lunar basalt was crushed such that ~40% of the original sample was greater than 0.5 mm grain size and only ~10% less than 0.05 mm. The softer breccias were more efficiently broken.

Due to the extremely small quantities of gases evolved from our samples, contamination from adsorbed terrestrial components and products synthesized in the apparatus during crushing and analysis is especially serious. Methane is commonly produced in steel crushers merely by grinding. We employed gentle crushing in order to minimize such reactions, while laboratory glass was crushed to determine blank values caused both by adsorbed air and products desorbed. The results illustrate, however, the difficulty in obtaining blank values from supposedly inert substances. The gases evolved from the glass are probably trapped (MULFINGER, 1966). Blank values were best assessed from the lower limits of the gas content in some of the rock samples (Table 1), and are considered negligible with respect to the conclusions drawn from our results. The isotopic composition of argon also indicated that little atmospheric contamination had occurred (Table 2).

Table 2. Isotopic composition of noble gases released by crushing.

Sample (No. of analyses)	^{4}He 10^{-8} cc STP/g	$^{4}He/^{3}He$	$^{4}He/^{20}Ne$	$^{20}Ne/^{22}Ne$	$^{22}Ne/^{21}Ne$	$^{20}Ne/^{36}Ar$	$^{36}Ar/^{38}Ar$	$^{40}Ar/^{36}Ar$	$^{4}He/^{40}Ar$
10057 (6)	30–48	—	—	—	—	—	—	280–2000	1.5–4
total [1]*	91000	1570	368	9.9	4.3	6.0	2.9	97	22.8
10022 (1)	92	1450	840	—	—	25	—	510	42
total [1]	63000	180	350	3	1.3	3.4	1.16	100	10.5
12021 (2)	29–64	1100–1500	37–85	—	—	17–38	—	100–140	10–13.5
12063 (3)	10–87	900–1800	100–180	—	—	11.4	—	114–150	4–19
total [1]	11300	154	170	4.21	1.50	4.1	1.44	65	11
12075 (2)	43–45	800	105–135	—	—	2–21	—	40–108	7–20
total [1]	76000	250	95	7.40	2.48	5.5	2.92	7.5	70
Bruderheim (1)	44	1000	205	—	—	12	4.4	700	3.5
total [2]	500	10.5	55	0.8	1.1	6.1	1.0	740	0.45
Allende (1)	27	—	—	—	—	—	5.7	230	2.2
total [3]	2600	350	480	3.3	1.35	0.32	5.1	115	1.3
Kapoeta (1)	74	—	98	—	—	—	—	—	—
total [4] avg. light	10200	1660	196	10.4	5.7	13.3	4.1	514	5.6
10046	105000	2800	35.8	13.0	31.6	32.8	5.40	2.41	480
10046 a†	118000	3750	69.0	13.3	31.2	24.2	5.40	2.29	710
b	87500	2800	62.5	13.4	32.3	24.2	5.34	2.48	540
c	48800	2750	61.5	13.7	29.5	26.8	4.85	2.70	600
10046 a	66500	2700	50.0	13.5	30.8	24.6	5.35	2.33	530
b	11000	nd	nd	nd	nd	nd	nd	2.40	565
total [5]	$2.6 \cdot 10^{7}$	2940	59.3	12.8	30.6	6.7	5.61	2.50	162
10061 a	14900	2900	41.2	13.4	32.3	21.2	5.45	2.2	395
b	61000	nd	34	nd	nd	23	nd	2.1	370
total [5]	$2.55 \cdot 10^{7}$	2980	50.8	12.5	30.1	5.85	5.13	1.98	150

(n.d.) = not determined; — = below detection limits.

* references for total gas amounts [1] BOGARD *et al.* (1970); [2] KIRSTEN *et al.* (1963); [3] FIREMAN and GOEBEL (1970); [4] MÜLLER and ZÄHRINGER (1966); [5] HINTENBERGER *et al.* (1971b).

† letters refer to incremental crushing steps.

In addition, the gases released from several crystalline rocks that were predegassed at 120°C for 8 hours under vacuum showed no differences within the range of results obtained from non-heated specimens.

The use of a multiple sample loading system for the crusher of the mass spectrometry analytical system allowed us to crush a number of different samples before breaking vacuum and cleaning out the crusher; however, it also created occasional problems of cross-contamination. For example, the sample of Abee (Table 1) was crushed after a lunar breccia and was hopelessly contaminated with noble gases released from a few residual breccia grains on the base plate.

Replicate density measurements of lunar material, using a 1 cm^3 capillary pycnometer and tetrahydro-naphthalene as liquid, were performed on several crystalline rock samples and a breccia before and after crushing. From the volume difference, void space of the specimens was determined. The gas pressure in internal cavities was then calculated from total gas amount and void space. The results are presented in Table 3.

Before each density measurement was initiated, air adsorbed on surfaces and in cracks of the sample was removed by subjecting the pycnometer to ultrasonic vibration while vacuum pumping.

Also X-ray photography was used to detect the presence of bubbles and voids before sample splitting. By this method voids as small as 0.2 mm in diameter can be detected as test photographs showed. Figure 3 demonstrates X-ray photographs of Apollo 12 rock samples 12021, 12063, and 12075. These rocks appear to be very dense as no voids are visible. For comparison, an X-ray picture of a vesicular nuclear explosion glass sample from the test site near Trinity, New Mexico, is included showing the location of internal bubbles clearly. Fig. 3 also shows a normal photograph of vesicular rock 10057. Visual estimates of vesicle volume on the surface of this rock proved to be larger by over a factor of ten compared to void space determined by density (see Table 3). This implies that most bubbles of this rock are not closed, but open to the atmosphere.

Low temperature heating experiments of lunar crystalline rocks, breccia, fines, stone meteorites, and a terrestrial basalt were performed. Samples of about 5 mg weight were preheated at 150°C in glass vials under vacuum, sealed, and heated at 225° and also 300°C for 15 hours, respectively. The gases released from the samples were measured by breaking the vials in the crushing apparatus connected to the gas chromatograph. The results of the heating experiments are compiled in Table 4. Blanks were determined from processing in the same manner empty vials and vials containing quartz grains. Only negligible amounts of atmospheric gases were found.

RESULTS AND DISCUSSION FROM CRUSHING

Lunar rocks

The gases released by crushing lunar crystalline rocks are hydrogen, helium, nitrogen, methane, and lesser amounts of argon (Table 1). The quantity, as well as

Table 3. Specific gravity measurements of lunar rocks before and after crushing. Accuracy: ±1‰. Total gas pressure in voids was calculated from void space and total gas amount including rare gases.

Sample	Weight (mg) before crushing	Specific gravity (g/cm^3) before crushing	Specific gravity (g/cm^3) after crushing	Void space (mm^3)	Total gas pressure (mm Hg)
10057,22 vesicular basalt	308.2	3.404	3.415	0.30	0.4
10057,80,1	373.9	3.392	3.415	0.7_2	1.2
10057,80,2	377.8	3.391	n.d.		
10057,80,3	154.5	3.390	3.457	0.9_0	0.4
12063,112 microgabbro	494.2	3.401	3.419	0.7_3	0.4
12075,13 basalt	446.6	3.375	3.394	0.7_3	0.04
12021,16 gabbro	249.6	3.376	3.410	0.7_4	0.03
10046,1 breccia	84.2	3.205	3.256	0.4_1	300

n.d. = not determined.

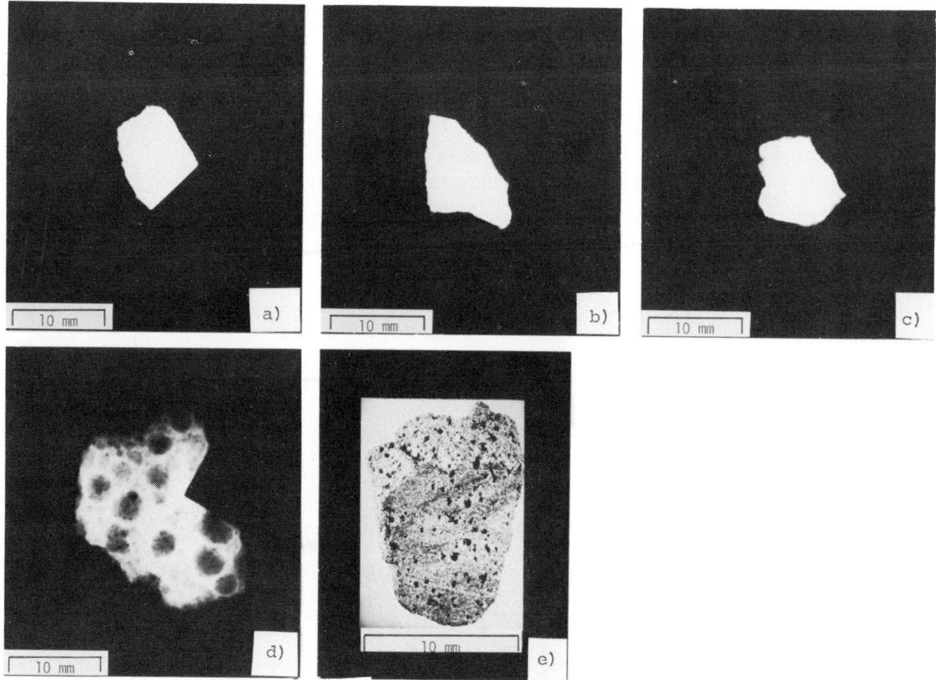

Fig. 3 (a–c) X-ray photographs of three crystalline lunar rock samples 12021, 12063, and 12075 before crushing. No voids are visible. (d) An X-ray picture of a vesicular nuclear explosion glass sample from the test site near Trinity, New Mexico, is included for comparison. Vesicles are clearly visible (dark areas). (e) Normal photograph of vesicular basalt 10057.

Table 4. Results of low-temperature heating experiments. Gases released at 225° and 300°C from lunar material, meteorites, and a terrestrial basalt sealed in vacuum vials. Sample weights about 5 mg. Gas amounts are given in 10^{-8} cc STP/5 mg sample.

Sample	°C	H_2	CO	CO_2	N_2	CH_4	C_2H_6	$C_2H_4(C_2H_2)$	C_3H_6	C_3H_8	H_2O
12063,112	225	10	110	440	—	3	—	—	—	—	—
microgabbro	300	22	170	600	—	8	—	3	—	—	—
12075,13	225	5	120	380	—	3	—	—	—	—	—
basalt	300	9	140	640	—	10	—	—	—	—	—
10046,1	225	340	410	1400	—	8	—	—	—	—	—
breccia	300	910	720	2400	—	52	3	30	43	—	—
12073,032	225	87	550	1000	—	17	—	—	—	—	—
breccia	300	920	2700	4600	—	130	15	80	150	—	—
10084,31	225	100	140	430	—	11	—	—	—	—	—
fines	300	1200	860	2300	—	50	5	11	15	—	—
12070,68	225	320	440	1600	—	9	—	—	—	—	—
fines	300	3500	5800	6600	—	350	50	160	140	16	—
Bruderheim	225	170	620	2600	—	11	1	3	—	—	—
L6	300	490	830	5400	—	38	3	19	22	—	—
Allende	225	370	1700	3700	—	50	—	—	—	—	—
C3	300	410	210	8600	—	250	11	—	120	—	—
Abee	225	1100	47	2400	—	36	—	—	—	—	—
E4	300	15200	3300	8800	23	140	10	3	47	—	detected
Olivine-basalt	225	230	44	—	—	2	—	—	—	—	—
terrestrial	300	210	240	630	—	14	1	11	—	—	—

— = below detection limits.

the composition of the gases, varies widely from rock to rock and even within a single specimen as illustrated for a typical sample analysis in Table 5. This probably reflects the opening of different volumes of internal surfaces and cavities where these gases are trapped as well as the degree of mobility of a gas in moving to and from such sites.

The internal pressure of the gas in vesicles and vugs of the crystalline rocks is less than 1 mm Hg (Table 3), which corresponds to mere centimeters overburden. It seems unlikely that these rocks were extruded that close to the lunar surface, so we must conclude that either the vesicles were formed primarily by condensable and/or reactive gases or, more likely, that the gaseous phase has leaked out along grain boundaries and microfractures. We have noted that terrestrial basalts also have lost most of their vesiculating gases, yet vesicles or inclusions within single mineral grains or glasses do retain their constituents (ROEDDER, 1965; MÜLLER and GENTNER, 1968; JESSBERGER and GENTNER, 1971).

By comparing the amount of noble gases released by crushing to the total released by melting (Table 2), we find an extraction efficiency of about 10^{-3}. Proportionately more radiogenic Ar than radiogenic He is released upon crushing, but this is not surprising considering that K and U/Th are located in different sites in the rocks and also that the more mobile He atoms would be preferentially lost to the lunar atmosphere from cavities and grain boundaries. Spallogenic gases liberated upon crushing were in most cases too close to the detection limits of the mass spectrometer to be measured accurately. The higher $^{40}Ar/^{36}Ar$ ratios compared to those from the total gas analysis reflects the greater mobility of the radiogenic gases vs spallogenic gases and are consistent with results from step-wise temperature extractions of the crystalline rocks (BOGARD *et al.*, 1970). No evidence for a trapped primordial noble gas component is apparent.

If we assume the same extraction efficiency for nitrogen as is found for the radiogenic gases, then the crystalline rocks must contain on the order of 10^{-3} to 10^{-4} cc STP/g molecular nitrogen (ca. 0.1–1 ppm N), which is within the limits estimated by HINTENBERGER *et al.* (1970, 1971a). MOORE *et al.* (1970, 1971), on the other hand, report total N values of 40–110 ppm in Apollo 11 and Apollo 12 rocks. If indeed this amount of nitrogen is indigenous to lunar crystalline rocks, then the moot question of its mode of occurrence is raised.

The extraction efficiency for hydrogen appears to be about 10–100 times lower than that of Ar or He assuming a total bulk rock hydrogen content of 0.03–0.8 cc

Table 5. Replicate analyses of gas released from lunar basalt 12063,112 by crushing at room temperature.

Sample Wt. mg	Method	Gas Contents in Units of 10^{-8} cc STP/g					
		H_2	N_2	CH_4	He	Ar	$^{40}Ar/^{36}Ar$
51	MS	180	98	27	10	2.9	—
138	MS	180	12	10	30	1.6	110
65	MS	≤ 2	110	14	87	7.9	150
494	GC	≤ 4	55	11	n.d.	n.d.	n.d.

n.d. = not determined.

STP/g (D'AMICO *et al.*, 1970; STOENNER *et al.*, 1970; FRIEDMAN *et al.*, 1971; HINTENBERGER *et al.*, 1971a). Possibly hydrogen is chemisorbed or tied up in hydrides.

Methane is released in quantities ranging from 5 to 200 $\times$ 10^{-8} cc STP/g. With an extraction efficiency of 10^{-3}, this corresponds to a maximum of $\sim$1 ppm C, which is that amount found by EGLINTON *et al.* (1971) by acid etching of an Apollo 12 rock. No other hydrocarbons have been detected in the crystalline rocks in our experiments, even during low temperature heating (Table 4). The total carbon content of a number of Apollo 12 rocks ranges from 8 to 45 ppm (FRIEDMAN *et al.*, 1971; KAPLAN and PETROWSKI, 1971; MOORE *et al.*, 1971). Very little is in the form of hydrolyzable carbides (EGLINTON *et al.*, 1971; KAPLAN and PETROWSKI, 1971). Thus, much of the carbon in the lunar rocks is unaccounted for. We have found no trace of CO (or CO_2) in our crushing experiments of either rocks or breccia despite its reported presence among gases released by acid hydrolysis of fines (BURLINGAME *et al.*, 1970) and heating of both fines and rocks (e.g., GIBSON and JOHNSON, 1971) (see also Table 4). WELLMAN (1970) has shown that CO should be a major constituent of the gas phase during crystallization of the Apollo 11 holocrystalline rocks. If CO were in molecular form, we should have observed at least trace amounts of it during crushing. Heating at low temperatures, then, must either release CO from a weakly-bound chemical state, as in carbonyls, or generate it through catalyzed oxidation-reduction reactions. Hydrolysis of an Apollo 12 rock chip indicated less than 2 ppm C as CO (EGLINTON *et al.*, 1971). Thus, it appears that a large part of the CO observed in the rocks results from the extraction procedure and that any trapped CO (with hydrogen) may have reverted to its low-temperature, thermodynamically stable form, CH_4.

Despite the poor gas retention of vesicles and vugs in the crystalline rocks, we believe that the gases we find are representative of the magmatic environment during crystallization (with possible thermodynamic modifications). Whether these gases were incorporated during accretion of the moon, generated by chemical reactions during earlier lunar history, or are brought to the moon by meteorites and comets, cannot be answered. Nevertheless, the overall composition of the gas phase is inert and reducing as expected from chemical and mineralogical observations. Additionally, no trace of H_2O has been found in any of the gases released by crushing lunar rocks.

Meteorites and terrestrial basalt

The gas composition in the limited number of meteorites crushed is generally similar to that found in the lunar rocks (Table 1)—reducing, with no CO detected, but with N_2 predominant. Abee, a relatively unmetamorphosed enstatite chondrite (E4), presents an interesting case. Crushing released up to 10^{-4} cc STP/g of H_2, N_2 and CH_4 and 10^{-6}–10^{-7} cc STP/g of H_2O, H_2S, CO_2, and C_2–C_3 hydrocarbons. BELSKY and KAPLAN (1970) have reported releasing comparable quantities of methane and C_2–C_3 hydrocarbons by crushing a sample of Abee at room temperature. As mentioned previously, the noble gases released from Abee were experimentally contaminated by solar gases. The data for Abee in Table 1 are not corrected. However, by assuming all the He found in Abee to be contamination from lunar breccia, then corresponding amounts of the other breccia gases are in all cases negligible. Thus the data are valid as listed except for He. The noble gases in the other meteorite samples show the same trends as found in the lunar rocks (Table 2).

Terrestrial basalts crystallize at a far greater oxygen fugacity than either lunar or meteoritic matter, and this is borne out by the more oxidizing nature of the trapped gases released—H_2, N_2, CH_4 are present, but H_2O and CO_2 predominate, as in terrestrial volcanos. This particular sample was a fresh, non-vesicular olivine basalt from Pechsteinkopf, Pfalz, Germany. The large differences in trapped gas composition may reflect the heterogeneous nature of the magmatic gas phase during extrusion and solidification (CHAIGNEAU, 1962; TAZIEFF and TONANI, 1963). This natural chemical fluctuation in the eruptive gas phase may also account to some degree for the compositional variations observed in the gases released from meteorites and lunar basalts.

Lunar fines and breccia

The lunar breccia contain 10^2–10^3 times the amount of trapped gas as do the crystalline rocks. Noteworthy is the 10–100 fold difference in gas content of a bubble-rich lithic rock fragment in 10046 and the matrix breccia material (Table 1). Solar wind helium constitutes more than 95% of the gases released from the breccia. The remaining gases are comprised of H_2S and C_2H_4, C_2H_6, C_3H_6, C_3H_8, all of which were not found in any of the crystalline rocks, and H_2, N_2, CH_4 and the other noble gases.

The void space in breccia 10046,1, estimated from the difference in its density and that of a typical crystalline rock (Table 3), is 6% of its total volume. Crushing the same sample opened up only one-quarter of the total internal volume occupied by the gases released. The total amount of gas trapped in the pore spaces of this breccia then would be about 3×10^{-3} cc STP/g with an internal gas pressure of 300 mm Hg (Table 3). This pressure could reflect the ambient atmosphere during the formation of this breccia, but probably is more related to a transient trapping of gases diffusing from grain surfaces.

Generally, the noble gas isotopic and elemental abundance ratios, including Kr and Xe, reflect those found in the total gas analysis of each sample (Table 2). He and Ne are released preferentially to Ar (and Kr and Xe). This creates a rather obvious disparity in the $^{20}Ne/^{36}Ar$ ratios which are consistently 4–5 times the ratio found in the total analyses. HEYMANN and YANIV (1971) have also reported such an enhancement in multiple crushing experiments on breccia 10065. They note a gradual decrease in the $^{20}Ne/^{36}Ar$ ratio as the breccia is further comminuted. Our incremental crushing steps are far less efficient, thus we would not expect to see such a trend. It appears that we are merely releasing solar-type gases that have diffused from grain boundaries into internal cavities. The elemental and isotopic ratios would then correspond to the trends observed during step-wise temperature extractions (e.g., PEPIN *et al.*, 1970). Helium, however, has further diffused from such collecting sites and has been mostly lost to the lunar atmosphere. Enhanced $^{20}Ne/^{36}Ar$ ratios are also evident in the gases released by crushing lunar crystalline rocks and the Bruderheim sample (Table 2). The same mechanism would apply here.

The release efficiency for the noble gases is surprisingly about the same as that in the crystalline rocks, despite the fact that the solar gases are contained in the outer few microns of grain surfaces (EBERHARDT *et al.*, 1970; KIRSTEN *et al.*, 1970), while in the crystalline rocks the noble gases are generated uniformly within the mineral grains. One must consider then that we are releasing gases from comparable sites in both types of samples, i.e., from grain boundaries and pore spaces. The hydrogen to helium ratios for lunar fines and breccias determined from temperature extractions range from 4–12 (EPSTEIN and TAYLOR, 1970, 1971; FRIEDMAN *et al.*, 1970; HANNEMAN, 1970; HINTENBERGER *et al.*, 1970, 1971a; ORÓ *et al.*, 1971; STOENNER *et al.*, 1970) yet H/He ratios in the gases released by crushing are about 0.01. As with the crystalline rocks much of the hydrogen is evidently chemically bound. If we assume a release efficiency of 10^{-3} for the other gases evolved during crushing, then the breccia contain up to 10^{-2} cc STP/g ($\sim$10 ppm) molecular CH_4 and N_2. This amount of N_2 is very close to the 15 ppm total N in lunar fines (HINTENBERGER *et al.*, 1971a). (MOORE *et al.* (1970, 1971) find values about 10 times higher). There is also some indication of the presence of either nitridic or ammonium nitrogen (CHANG *et al.*, 1971; HINTENBERGER *et al.*, 1970).

We infer on the order of 1–10 ppm trapped CH_4 in lunar breccia. This range agrees well with the amounts of indigenous CH_4 released by vacuum crushing and acid hydrolysis from other samples of fines and breccia (CHANG *et al.*, 1971; EGLINTON *et al.*, 1971). The total carbon content of typical breccia and fines from the Apollo 11 and Apollo 12 landing sites ranges from about 100–250 ppm (e.g., MOORE *et al.*, 1970, 1971). The presence of carbides has been confirmed by acid hydrolysis (e.g., EGLINTON *et al.*, 1971); however, the physical or chemical state of the bulk of the carbon is unknown. No CO or CO_2 was released during crushing, nor were any traces of water observed.

The most probable source of all the gases released from the breccias during crushing is, directly or indirectly, the solar wind. A meteoritic gas component may be present, but at levels below 2% in the soil (ANDERS *et al.*, 1971) and it is probably mostly lost during impact (MOORE *et al.*, 1970), thus such contributions seem negligible. Solar wind bombardment over a period of 10^5 years can easily provide sufficient reagents and energy to form such compounds as CH_4 and other low-molecular weight hydrocarbons, N_2 and H_2S. Indeed, EGLINTON *et al.* (1971) show that CH_4 is located near grain surfaces. The efficiency of these simple reactions is probably quite high (ZELLER *et al.*, 1966), yet it is difficult to understand why NH_3 was not detected.

RESULTS AND DISCUSSION FROM HEATING

The predominant gases extracted from lunar material by heating at 225° and 300°C are CO_2, CO, H_2 and CH_4 (Table 4). The quantities of gas released by heating, on a per gram sample basis, exceed by factors of 100 or more the amounts of gas released by crushing; and, similar to the crushing results, more gases evolve from the breccia (and fines) than the crystalline rocks. Hydrocarbons such as C_2H_6, C_2H_4 (or C_2H_2) and C_3H_6 were detected only in fines and breccias in the 300°C experiments. The CH_4/C_2H_6 ratios range from 6–17 and are about the same as those found by

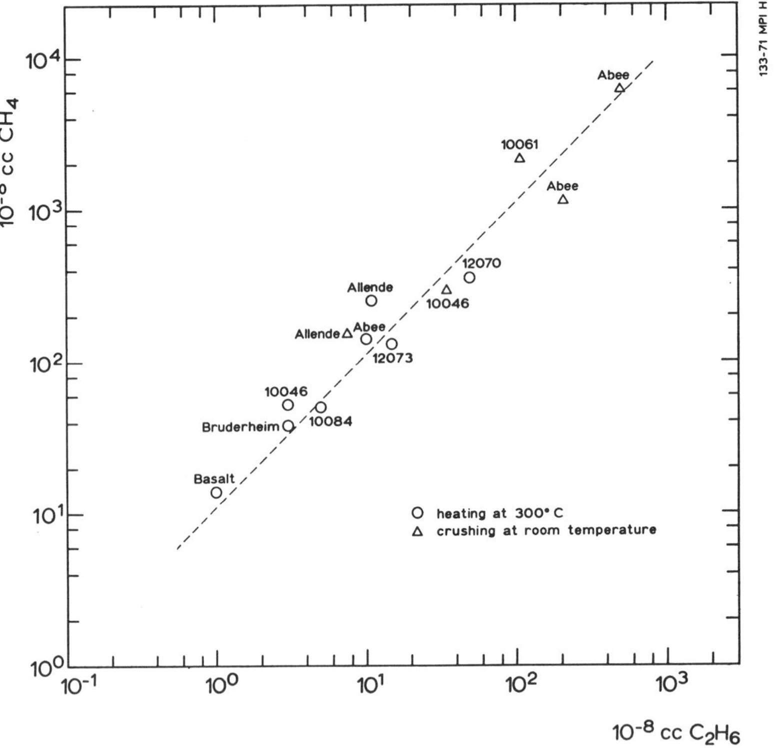

Fig. 4. CH_4 versus C_2H_6 gas amounts released at 300°C are plotted for lunar fines, breccia, some meteorites and a terrestrial basalt. Also our data on crushing at room temperature are included for comparison. Gas amounts for heating are in units of 10^{-8} cc STP/5 mg sample, for crushing in units of 10^{-8} cc STP/g sample. Dashed line represents the average CH_4/C_2H_6 ratio of 11 found by BELSKY and KAPLAN (1970) for C-, L- and E-group meteorites.

crushing breccia samples (Table 1) and lunar fines (ABELL *et al.*, 1970b) and C-, L-, and E-group meteorites (BELSKY and KAPLAN, 1970) at room temperature. Fig. 4 illustrates the relatively constant and comparable CH_4/C_2H_6 ratios found in gases released at room temperature and 300°C.

On the other hand, CH_4/C_2H_4 ratios are consistently lower in the heated material than in the gases released only by crushing. The same trend is also noted with C_3 hydrocarbons: CH_4/C_3H_8 ratios in gases released both by heating and crushing are about the same, whereas CH_4/C_3H_6 ratios are lower in the temperature extractions. Thus the alkanes appear little affected by heat release at low temperatures and are probably indigenous, whereas alkenes are, to some extent, chemically produced during gas extraction.

As already mentioned, CO and CO_2 were detected in all samples that were heated, yet these molecules were absent in the gases released by crushing at room temperature. The generation of the observed amounts of CO, CO_2 and H_2 starting from C and H_2O by the water gas reaction is rather improbable due to the apparent absence of water in the samples (even at 300°C) and the non-stoichiometric quantities of products evolved. The only exception is the case of Abee at 300°C. Other sources for CO and CO_2 could be the interaction of elemental carbon with oxygen of the silicates. However, it would be surprising that this reaction starts at the low temperatures applied.

It is interesting to note that despite the relatively high N_2 content of gases released by crushing, no N_2 was detected in any of the samples during the heating experiments, with the exception of Abee which had the most N_2 of any samples analyzed (Table 1). If the H_2, N_2, CH_4 and other gases are indeed indigenous, it is difficult to understand why N_2 behaves so uniquely. More data are necessary for a proper interpretation of the behavior of N_2.

Search for Carbonyls in Fines 10084

Because of the absence of CO in the crushing experiments and its easy and relative abundant release on heating, we felt that CO might be present in a chemically bound state, possibly as carbonyls.

Volatile carbonyls such as $Fe(CO)_5$ or $Ni(CO)_4$ are easily detectable in μg quantities by heat decomposition yielding a metallic mirror. We, therefore, slowly heated up to 240°C about 300 mg of untreated bulk fines in a small quartz apparatus in vacuum. The flask containing the fines was connected with a capillary tube which was heated continuously at about 350°C. No metal decomposition occurred indicating less than a few ppm of volatile carbonyls in this sample.

For testing the apparatus and estimating the sensitivity for carbonyl detection about 400 mg of a basalt powder were spiked with 20 μg of $Fe(CO)_5$. Shortly after the experiment was started, a distinct metallic mirror formed in the hot zone of the capillary tube. Subsequently, the mirror was dissolved with HCl and iron was quantitatively determined by a colorimetric method.

Conclusions

Lunar crystalline rocks contain ppb amounts of trapped gases that can be released by crushing at room temperature. These gases probably are representative of the magmatic gas phase. The presence (or absence) of CO is equivocal. Carbon monoxide has been considered a dominant constituent of trapped lunar volatiles, and is a reactant in many catalyzed low-temperature reactions. The absence of CO in the gases released by crushing and acid hydrolysis experiments, yet its ubiquitous presence in the gases evolved by heating at low temperatures suggests catalytic synthesis during extraction. The presence of N_2 in the gases released by crushing, yet its absence in the low-temperature fractions raises the question of its physical and chemical state in the rocks compared to the other indigenous gases evolved.

The breccia and fines, on the other hand, contain trapped gases which are derived either directly or indirectly from the solar wind. The presence of low-molecular weight hydrocarbons has important implications, for here is a mechanism, as Zeller and Ronca (1968) have suggested, by which low molecular weight organic molecules are synthesized in a primitive environment in substantial quantities (e.g., 10^{-2} cc STP/g of gas is equivalent to approximately 1 atom percent in the outer reactive 1000 Å of grain surface). Organic matter, then, need not have been created as a veneer on dust during condensation or accretion from the solar nebula, but made by continuing processes of interactions of solar wind atoms on surfaces of bodies and cosmic dust. Energy and reactant mobility provided by low grade metamorphism or even impact events could conceivably create local concentrations of higher molecular weight compounds. In this respect, the meteorite Abee resembles lunar breccias—both are brecciated, both contain the same sort of gases (except H_2O, and CO_2), even the methane to alkane/alkene ratios are similar.

Simple compaction and low temperature heating release relatively large quantities of inorganic and low molecular weight organic gases from samples of breccias and fines. Transient perturbations of the lunar regolith such as meteoritic impact, local volcanism or even large scale slumping or tidal strains would release pulses of such gases, which would be photoionized and accelerated by the electric field associated with the solar wind. For example, an impact that degasses a 3-meter thick regolith to the extent of 1% in a volume bounded by a circle of 100 meters diameter would produce about 10^{26} molecules or 10^{18} molecules/cm^3. Observations of such transient phenomena may well be occurring; e.g., impulses recorded by the ALSEP suprathermal ion detector and the probable detection of such molecules as C_2, CH, H_2, N_2 in luminescent spectra from local areas on the lunar surface (KOZYREV, 1963, 1969).

Acknowledgments—We thank the National Aeronautics and Space Administration for providing the lunar samples. The assistance of D. Kaether, W. Ehrhardt and H. Graf is gratefully acknowledged. We are thankful to Dr. H. E. Schmid, Research Laboratory of Brown-Boveri Company, Heidelberg, for performing the X-ray photography. Valuable discussions with Dr. H. O. Mulfinger, Jenaer Glaswerk Schott u. Gen., Mainz, and Prof. E. O. Fischer, Technische Universität München, are appreciated.

REFERENCES

ABELL P. I., DRAFFAN C. H., EGLINTON G., HAYES J. M., MAXWELL, J. R., and PILLINGER C. T. (1970a) Organic analysis of the returned Apollo 11 lunar sample. *Proc. Apollo 11 Lunar Sci. Conf., Geochim. Cosmochim. Acta* Suppl. 1, Vol. 2, pp. 1757–1773. Pergamon.

ABELL P. I., EGLINTON G., MAXWELL J. R., PILLINGER C. T., and HAYES J. M. (1970b) Indigenous lunar methane and ethane. *Nature* **226**, 251–252.

ANDERS E., LAUL J. C., KEAYS R. R., GANAPATHY R., and MORGAN J. W. (1971) Elements depleted on lunar surface: implications for origin of moon and meteorite influx rate. Second Lunar Science Conference (unpublished proceedings).

BELSKY T. and KAPLAN I. R. (1970) Light hydrocarbon gases, C^{13}, and origin of organic matter in carbonaceous chondrites. *Geochim. Cosmochim. Acta* **34**, 257–278.

BOGARD D. D., FUNKHOUSER J. G., SCHAEFFER O. A., and ZÄHRINGER J. (1970) Noble gas abundances in lunar material—II. Cosmic ray spallation products and radiation ages from Mare Tranquillitatis and Oceanus Procellarum. Submitted to *J. Geophys. Res.*

BURLINGAME A. L., CALVIN M., HAN J., HENDERSON W., REED W., and SIMONEIT B. R. (1970) Study of carbon compounds in Apollo 11 lunar samples. *Proc. Apollo 11 Lunar Sci. Conf., Geochim. Cosmochim. Acta* Suppl. 1, Vol. 2, pp. 1779–1791. Pergamon.

CHAIGNEAU M. (1962) Sur les gaz volcaniques de l'Etna (Sicile). *Compt. Rend.* **254**, 4060–4062.

CHANG S., KVENVOLDEN K., LAWLESS J., PONNAMPERUMA C., and KAPLAN I. R. (1971) Carbon, carbides, and methane in an Apollo 12 sample. *Science* **171**, 474–477.

D'AMICO J., DEFELICE J., and FIREMAN E. L. (1970) The cosmic-ray and solar-flare bombardment of the moon. *Proc. Apollo 11 Lunar Sci. Conf., Geochim. Cosmochim. Acta* Suppl. 1, Vol. 2, pp. 1029–1036. Pergamon.

EBERHARDT P., GEISS J., GRAF H., GRÖGLER N., KRÄHENBÜHL U., SCHWALLER H., SCHWARZMÜLLER J., and STETTLER A. (1970) Trapped solar wind noble gases, exposure age and K/Ar-age in Apollo 11 lunar fine material. *Proc. Apollo 11 Lunar Sci. Conf., Geochim. Cosmochim. Acta* Suppl. 1, Vol. 2, pp. 1037–1070. Pergamon.

EGLINTON G., ABELL P. I., CADOGAN P. H., MAXWELL J. R., and PILLINGER C. T. (1971) Survey of lunar carbon compounds. Second Lunar Science Conference (unpublished proceedings).

EPSTEIN S. and TAYLOR H. P., JR. (1970) The concentration and isotopic composition of hydrogen, carbon and silicon in Apollo 11 lunar rocks and minerals. *Proc. Apollo 11 Lunar Sci. Conf., Geochim. Cosmochim. Acta* Suppl. 1, Vol. 2, pp. 1085–1096. Pergamon.

EPSTEIN S. and TAYLOR H. P., JR. (1971) O^{18}/O^{16}, Si^{30}/Si^{28}, D/H, and C^{13}/C^{12} ratios in lunar samples. Second Lunar Science Conference (unpublished proceedings).

FIREMAN E. L. and GOEBEL R. (1970) Argon 37 and argon 39 in recently fallen meteorites and cosmic-ray variations. *J. Geophys. Res.* **75**, 2115–2124.

FRIEDMAN I., GLEASON J. D., and HARDCASTLE K. G. (1970) Water, hydrogen, deuterium, carbon and C^{13} content of selected lunar material. *Proc. Apollo 11 Lunar Sci. Conf., Geochim. Cosmochim. Acta* Suppl. 1, Vol. 2, pp. 1103–1109. Pergamon.

FRIEDMAN I., O'NEIL J. R., GLEASON J. D., and HARDCASTLE K. (1971) The carbon, hydrogen content and isotopic composition of some Apollo 12 materials. Second Lunar Science Conference (unpublished proceedings).

GIBSON E. K., JR. and JOHNSON S. M. (1971) Thermal analysis—inorganic gas release studies of lunar samples. Second Lunar Science Conference (unpublished proceedings).

HANNEMAN R. E. (1970) Thermal and gas evolution behavior of Apollo 11 samples. *Proc. Apollo 11 Lunar Sci. Conf., Geochim. Cosmochim. Acta* Suppl. 1, Vol. 2, pp. 1207–1211. Pergamon.

HEYMANN D. and YANIV A. (1971) Inert gases in breccia 10065: Release by vacuum crushing. Second Lunar Science Conference (unpublished proceedings).

HINTENBERGER H., WEBER H. W., VOSHAGE H., WÄNKE H., BEGEMANN F., and WLOTZKA F. (1970) Concentrations and isotopic abundances of the rare gases, hydrogen and nitrogen in Apollo 11 lunar matter. *Proc. Apollo 11 Lunar Sci. Conf., Geochim. Cosmochim. Acta* Suppl. 1, Vol. 2, pp. 1269–1282. Pergamon.

HINTENBERGER H., VOSHAGE H., and SPECHT S. (1971a) Heat extraction and mass spectrometric analysis of gases from lunar samples. Second Lunar Science Conference (unpublished proceedings).

HINTENBERGER H., WEBER H., and TAKAOKA N. (1971b) Concentrations and isotopic abundances of the rare gases in lunar matter. Second Lunar Science Conference (unpublished proceedings).

JESSBERGER E. and GENTNER W. (1971) Mass spectrometric analysis of gas inclusions in tektites and impactites. In preparation.

KAPLAN I. R. and PETROWSKI C. (1971) Carbon and sulfur concentration and isotopic composition in Apollo 12 lunar samples. Second Lunar Science Conference (unpublished proceedings).

KIRSTEN T., KRANKOWSKY D., and ZÄHRINGER J. (1963) Edelgas- und Kalium-Bestimmungen an einer größeren Zahl von Steinmeteoriten. *Geochim. Cosmochim. Acta* **27**, 13–42.

KIRSTEN T., MÜLLER O., STEINBRUNN F., and ZÄHRINGER J. (1970) Study of distribution and variations of rare gases in lunar material by a microprobe technique. *Proc. Apollo 11 Lunar Sci. Conf., Geochim. Cosmochim. Acta* Suppl. 1, Vol. 2, pp. 1331–1343. Pergamon.

KOZYREV N. A. (1963) Volcanic phenomena on the moon. *Nature* **198**, 979–980.

KOZYREV N. A. (1969) A red spot inside the lunar crater Aristarchus observed on April 1, 1969. *Astron. Zh.* **47**, 179–181.

MOORE C. B., GIBSON E. K., LARIMER J. W., LEWIS C. F., and NICHIPORUK W. (1970) Total carbon and nitrogen abundances in Apollo 11 lunar samples and selected achondrites and basalts. *Proc. Apollo 11 Lunar Sci. Conf., Geochim. Cosmochim. Acta* Suppl. 1, Vol. 2, pp. 1375–1382. Pergamon.

MOORE C. B., LEWIS C. F., DELLES F. M., GOOLEY R. C. and, GIBSON E. K., JR. (1971) Total carbon and nitrogen abundances in Apollo 12 lunar samples. Second Lunar Science Conference (unpublished proceedings).

Müller O. and Gentner W. (1968) Gas content in bubbles of tektites and other natural glasses. *Earth Planet. Sci. Lett.* **4**, 406–410.

Müller O. and Zähringer J. (1966) Chemische Unterschiede bei uredelgashaltigen Steinmeteoriten. *Earth Planet. Sci. Lett.* **1**, 25–29.

Mulfinger H. O. (1966) Physical and chemical solubility of nitrogen in glass melts. *J. Amer. Ceram. Soc.* **49**, 462–467.

Oró J., Flory D. A., Gibert J. M., McReynolds J., Lichtenstein H. A., Wikstrom S., and Gibson E. K., Jr. (1971) Abundances and distribution of organogenic elements and compounds in Apollo 12 lunar samples. Second Lunar Science Conference (unpublished proceedings).

Pepin R. O., Nyquist L. E., Phinney D., and Black D. C. (1970) Rare gases in Apollo 11 lunar material. *Proc. Apollo 11 Lunar Sci. Conf., Geochim. Cosmochim. Acta* Suppl. 1, Vol. 2, pp. 1435–1454. Pergamon.

Roedder E. (1965) Liquid CO_2 inclusions in olivine bearing nodules and phenocrysts from basalts. *Amer. Mineral.* **50**, 1746–1782.

Stoenner R. W., Lyman W. J., and Davis R., Jr. (1970) Cosmic-ray production of rare-gas radioactivities and tritium in lunar material. *Proc. Apollo 11 Lunar Sci. Conf., Geochim. Cosmochim. Acta* Suppl. 1, Vol. 2, pp. 1583–1594. Pergamon.

Tazieff H. and Tonani F. (1963) Fluctuations rapides et importantes de la phase gazeuse éruptive. *Compt. Rend.* **257**, 3985–3987.

Wellman T. R. (1970) Gaseous species in equilibrium with the Apollo 11 holocrystalline rocks during their crystallization. *Nature* **225**, 716–717.

Zeller E. J., Ronca L. B. and Levy P. W. (1966) Proton-induced hydroxyl formation on the lunar surface. *J. Geophys. Res.* **71**, 4855–4860.

Zeller E. J. and Ronca L. B. (1968) The surface geochemistry of solid bodies in space. In Origin and Distribution of the Elements (edited L. H. Ahrens), pp. 509–519. Pergamon Press.

Proceedings of the Second Lunar Science Conference, Volume 2, pp. 1397–1406.
The M.I.T. Press, 1971.

Carbon and sulfur isotope studies on Apollo 12 lunar samples

I. R. KAPLAN and CHARI PETROWSKI
Institute of Geophysics and Planetary Physics, University of California,
Los Angeles, California 90024

(*Received 26 February 1971; accepted in revised form 30 March 1971*)

Abstract—The concentration and isotopic content of carbon and sulfur in three lunar fines ranged from 37 to 119 ppm ($\delta C^{13}_{PDB} = -1.6$ to $+12.4‰$) and 620 to 714 ppm ($\delta S^{34} = +8.8$ to $+4.9‰$), respectively. An inside and an outside fragment of a basalt (No. 12022) were also analyzed, yielding a carbon and sulfur content of 14 ppm ($\delta C^{13} = -25.4‰$) and 754 ppm ($\delta S^{34} = -0.2‰$) respectively for the interior and 21 ppm ($\delta C^{13} = -24.9‰$) and 805 ppm ($\delta S^{34} = -0.1‰$) for the exterior fragment. The two fines containing the highest carbon content were most enriched in C^{13}. The basalt samples showed values for carbon content and isotopic abundance similar to a Hawaiian basalt freshly collected from a recent flow. Analysis of carbon-containing gases released by acid hydrolysis of lunar material, generally reflected the isotopic content of the intact sample. Stepwise pyrolysis of samples over a temperature range of 150°–1100°C yielded a gas mixture with a spread in isotopic values, confirming the presence of carbon in different forms.

The fines presently studied are less enriched in C^{13} than in samples from Apollo 11 mission. The isotopic composition (for both C and S) in the basalt is typical of terrestrial basalts and no significant difference could be detected between an inside, and an outside fragment of the same rock exposed to solar radiation. The data argue against a simple origin of the fines from the basalt and also suggest that carbon-rich and sulfur-poor material is being added to the regolith.

INTRODUCTION

STUDIES ON THE chemistry of carbon from Apollo 11 lunar samples indicated that this element can be distributed in a variety of forms either as organic or inorganic molecules. The strongest evidence suggests that the carbon was mainly present as inorganic carbon (ABELL *et al.*, 1970; CHANG *et al.*, 1970; KAPLAN *et al.*, 1970; MURPHY *et al.*, 1970) metal carbides, oxides of carbon (carbon dioxide and carbon monoxide) and probably elemental carbon. There is little reliable evidence for carbon–hydrogen compounds, more complex than methane, being indigenous to the original samples in concentration greater than 1 ppm. As there was evidence for terrestrial contamination of the returned samples from measurements at the Lunar Receiving Laboratory (LSPET, 1969; FLORY *et al.*, 1969), as well as from measurements by principal investigators after sample distribution (KAPLAN *et al.*, 1970; MURPHY *et al.*, 1970), studies were undertaken by us on Apollo 12 material to evaluate the major forms of carbon present in the uncontaminated material.

Measurements were performed on five samples from Apollo 12 of approximately 5 g each sent directly to UCLA. These include three bulk fines (12023,10, 12032,12, and 12042,11) and an olivine-rich porphyritic basaltic rock, from which an internal fragment (12022,77) and external fragment (12022,71) were supplied. In addition, studies were also carried out in collaboration with a research group at NASA, Ames

---

* Publication No. 897 Institute of Geophysics and Planetary Physics.

Research Center on sample 12023 supplied directly to Dr. C. Ponnamperuma (CHANG *et al.*, 1971) and on a fragment of breccia from the Apollo 11 collection (sample 10002,54).

RESULTS

Carbon

Being aware of the ease for introduction of terrestrial contamination, samples were only opened immediately prior to analyses. Blank measurements were always made before and after a sample was analyzed. For total carbon combustion, following the method described by KAPLAN *et al.* (1970), levels of CO_2 obtained in a clean system were <0.01 cc. Sand blanks generally yielded slightly higher volumes of CO_2, amounting to $\simeq 5$ ppm C or less during analysis of 1 or 2 g samples. Two sand monitors prepared at the Lunar Receiving Laboratory yielded 5.4 and 6.7 ppm C, respectively (Table 1), indicating that this is the probable level of contamination which may be expected in the handling, packaging, and analytical procedures.

The concentration of carbon in bulk fines appears to be lower in the Apollo 12 samples than in the Apollo 11 samples. In the latter samples the values ranged between 116 to 170 ppm. Values above this range were thought by KAPLAN and SMITH (1970) to have resulted from contamination. In Apollo 12, two fines (12023 and 12042) yielded similar total carbon values in the range 109 to 119 ppm (Table 2). One measurement of 140 ppm may have resulted from slight contamination. Sample 12032, which was much coarser and more poorly sorted than the other two bulk fines analyzed, contained significantly less carbon. The samples also showed less enrichment in C^{13} than Apollo 11 bulk fines (Table 2). The range of the most reliable measurements for $\delta^{13}C$ is $+7.0$ to $+12.4‰$, and one value of $+3.4‰$ (in the sample which yielded 140 ppm carbon content). It is interesting to note that the two samples of 12023 which were most enriched in C^{13} ($\delta^{13}C = +11.3$ and $+12.4‰$) were measured on the bulk sample sent to Ames Research Center, whereas the two lower values ($\delta^{13}C = +7.0$ and $+7.1‰$) were measured on a sample sent to UCLA. The isotopic values for sample 12032 ($\delta^{13}C = -14.6$ and $-1.6‰$) are the most enriched in C^{12} of any of the fines

Table 1. Isotopic composition of blanks, reagents, and containers from LRL used in the Apollo 12 mission.

Sample description	C %	$\delta^{13}C$* ‰	S %	$\delta^{34}S$† ‰
1. Sand monitor from organic cabinet in curators lab. Exposure time Mar. 11–30, 1970, no. 115,003.	6.7 (ppm)	—		
2. Sand monitor from F-201 vacuum chamber, no. 1296,003.	5.4 (ppm)	—		
3. Dashpot fluid (new bottle).	20	−52.3		
4. Teflon bag container (4D), no. 1071,003.	24	−58.1		
5. Retro engine exhaust fuel from trap A.	—	−34.7		
6. Sulfide sealant used in dry nitrogen box.			5.8	−5.0
7. Molybdenum disulfide lubricant for sample return box, no. 1069,003.			34.6	+3.4

* Relative to PDB standard.
† Relative to Canyon Diablo meteorite (troilite) standard.

Table 2. Concentration and isotopic composition of carbon and sulfur.

Sample	Carbon		Sulfur		
	Conc ppm	$\delta^{13}C$ ‰	Conc ppm	FeS %	$\delta^{34}S$ ‰
Bulk fines					
12023	112*	+11.3*	620	0.17	+8.3
	109*	+12.4*	509†	0.14†	+8.8
	140*	+3.4*			
	116	+7.0			
	119	+7.1			
12032	86	−14.6	714	0.20	+4.9
	37	−1.6			
12042	112	+9.9	—	—	—
	111	+10.3			
Olivine-rich basalt					
12022,77 Interior	19	−31.6‡	754	0.21	−0.2
	14	−25.4			
12022,71 Exterior	21	−24.9	805	0.22	−0.1
			603§	0.17§	−0.3§

* Results on samples submitted to Ames Research Center.
† Minimum value due to suspected loss.
‡ This sample was combusted at 1100° without crushing and yielded 10.3 ppm C, but a reliable $\delta^{13}C$ could not be made on the CO_2. The sample was then removed from the combustion boat, crushed, and reignited, yielding a value of 8.8 ppm for C. The mass spectrometer measurement was made on the gas at a reduced pressure, and is therefore only accurate to within ±5‰.
§ This sample treated with phosphoric acid.

measured by us to date. It is possible that the value of $\delta^{13}C = -14.6$ may be due to contamination, as once again it coincides with the greatest carbon content.

Accepting possible contaminations, it is still apparent from these results, and from measurements undertaken by MOORE *et al.* (1971) that the carbon content of the regolith material is far from homogeneous. Inconsistencies are obvious in both the present data and those of Moore. Although such inconsistencies were earlier explained by EPSTEIN and TAYLOR (1970) and by KAPLAN *et al.* (1970) as due to contamination, there may also be some inhomogeneity due to use of unsorted samples.

Two samples of an olivine-rich basalt yielded values of 14–21 ppm C and $\delta^{13}C$ values of −24.9 to −31.6‰ (Table 2). It is considered that there is no significant difference between the internal and external fragment. Sample 12022,71 has been described by Dr. M. Duke at LRL to be an external fragment of the rock facing upward, with a surface area of approximately 3 to 4 cm^2. The isotopic ratios ($\delta^{13}C \simeq -25$‰) are approximately equal to those measured in Apollo 11 basaltic rocks. One may therefore conclude that the isotopic composition of the volcanic rocks is (i) similar to volcanic basaltic rocks on earth (ii) significantly different from the bulk fines at the lunar surface and that (iii) no marked effect has occurred at the surface of the rock to either preferentially remove C^{12} or supply C^{13}.

Previous studies on Apollo 11 by several investigators indicated that CO, CO_2, and CH_4 could be obtained during pyrolysis of a lunar sample. CHANG *et al.* (1971) showed that for sample 12023, the three gases could be removed in the approximate ratio 5:3:2 (CO_2:CO:CH_4) at temperatures up to 500°C. However, pyrolysis at elevated temperatures yielded largely CO. Isotopic measurement of pyrolytic gas mixtures yielded values similar to those measured by KAPLAN *et al.* (1970) for Apollo 11 bulk fines; greatest enrichment in C^{12} at temperatures below 500°C and greatest enrichment in C^{13} at temperatures of pyrolysis between 500° and 750°C. A further attempt was made in the present investigation to separate the gases by separating them into liquid nitrogen condensable and noncondensable gases. It was assumed that the former would contain essentially only CO_2, whereas the latter fraction would be composed of carbon monoxide and methane. These gases were then recombusted to CO_2, and C^{13}/C^{12} isotopic ratios measured by mass spectroscopy. Attempts were first made to measure the small volume of gas directly on the mass spectrometer, but this often proved impossible, especially if the sample contained slight impurities. Samples which were not lost in this attempt, were measured by making a gas dilution with a known volume of a standard CO_2 sample. Values obtained from such measurements are noted in Table 3.

On the assumption that most of the CO_2 is liberated below 500°C and the amount of CO liberated at 500°–750°C is several times greater than that liberated at lower temperatures, pyrolysis was carried out by slowly elevating the furnace temperature between the intervals 150°–500°C, 500°–750°C, and 750°–1100°C. For the breccia, which contained nearly twice the carbon content of the Apollo 12 fines, the temperature intervals were smaller.

The data in Table 3 confirms previous results indicating that the carbon compounds do not represent a single homogenized source. For example, the difference in $\delta^{13}C$ for uncondensable gases obtained at 150°–750°C from sample 12023 and from sample 12042, may be due to the relative amounts of methane. By comparison with the data

Table 3. $\delta^{31}C$ of liquid nitrogen condensable and noncondensable gases produced during stepwise pyrolysis of Apollo 12 bulk fines and Apollo 11 breccia.

Sample No.	T °C	LN_2 condensable C (ppm)	LN_2 condensable $\delta^{13}C$ (‰)	LN_2 noncondensable C (ppm)	LN_2 noncondensable $\delta^{13}C$ (‰)
Bulk fines					
12023	150–750	4.7	—†	25.5	+8.6
	750–1000	6.1	—†	50.7	+11.4
		Σ 10.8		Σ 76.2	
	1100 Residue‡	Not run			
12032	150–750	2.2	—†	8.1	−3.8
	750–1100	3.4	−5.5*	13.9	−10.4
		Σ 5.6		Σ 22.0	
	1100 Residue‡	19.1	−7.7		
12042	150–750	1.6	—†	21.8	+20.4
	750–1100	0.4	+2.2*	48.6	+1.9
		Σ 2.0		Σ 70.4	
	1100 Residue‡	7.9	+10.8*		
Breccia					
10002	200–500	21.2	−9.9	12.4	−28.4
	500–650	2.4 }	−8.8	12.0	−8.7
	650–800	4.2 }		43.2	+4.7
	800–1100	11.7	+0.9	33.7	+6.1
		Σ 39.5		Σ 101.3	
	1100 Residue‡	89.1	−7.2		

* Sample analyzed by gas dilution with a substandard.
† Insufficient sample for isotope analysis.
‡ Residue from pyrolysis combusted at 1100°C in oxygen.

Table 4. Carbon containing gases produced during stepwise pyrolysis of Apollo 12 fines: relative abundance, total carbon, and carbon isotopic composition (from CHANG *et al.*, 1971).

Trial	T (°C)	CO	CO_2 (%)	CH_4	Total C ppm	$\delta^{13}C$ ‰
1	500	37	46	18	25	−10.0
	750	>98	<0.08	<0.7	49	±13.6
	1100	94	6	0	61	+12.6
	1100 (O_2)*				—	—
2	500	15	68	17	16	−17.7
	750	>98	<1.1	<0.4	27	+22.0
	1100	86	14	0	73	+8.0
	1100 (O_2)*				20	+7.0
3	500	32	38	30	16	−16.5
	750	95	4	1	24	+26.4
	1100	93	7	0	69	+7.4
	1100 (O_2)*				20	−7.0
Fe_3C	500	—	—	—	5†	—
	750	73	27	0	18†	−26.2
	1100	70	30	0	77†	−19.6
	1100 (O_2)*				0	

* Sample combusted in oxygen at 1100°C.
— Not measured.
† Percentage of total carbon originally in synthetic Fe_3C.

presented by CHANG *et al.* (1971, Table 4) sample 12023 would contain a greater proportion of an isotopically light component than sample 12042. It is seen from analysis of the breccia (10002; Table 3) that this component is not the CO_2. It is most probably the methane, which, by terrestrial analogy could attain values of $\delta^{13}C = -50$ to -60‰ in equilibrium with the CO_2 and CO.

Presently, the evidence is insufficient to determine whether the gases are present as such within the lunar material or whether they form by oxidation of reduced carbon during heating. Experiments by CHANG *et al.* (1971) suggests that CH_4 and possibly some CO_2 may be present within trapping sites and are released on crushing of the sample. No confirmation could be obtained for a previous claim on Apollo 11 samples (BURLINGAME *et al.*, 1970), that CO as a gas was present within vesicles in the glass spheres. This gas probably arises from oxidation of reduced carbon by oxides and possibly silicates within the sample.

In an attempt to differentiate carbide–carbon from residual carbon, samples were treated either with hydrochloric or with phosphoric acids and the hydrolysis products captured. The data presented in Table 5 indicate that in the basalts, the residual carbon, remaining after hydrofluoric acid degradation, contains most of the total carbon and has the same isotopic ratio as the intact sample. The hydrolysis gas of the bulk fine, 12032, is isotopically more enriched in C^{13} than the residual carbon. This was also found by experiments on 12023 bulk fines (CHANG *et al.*, 1971). Gases released from two hydrolysis experiments, primarily as methane, were oxidized to CO_2 and the $\delta^{13}C$ values measured. The more reliable value yielded 21 ppm C with $\delta^{13}C = +14$‰, the other result was 26 ppm C with $\delta^{13}C = +5$‰. These values are considered as minima, because hydrolysis of synthetic iron carbide yielded hydrocarbons depleted in C^{13} relative to the starting material.

Table 5. Carbon content and $\delta^{31}C$ of gases produced during acid hydrolysis and combustion of residues from Apollo 12 fines and basalt.

Sample No.	Hydrolysis gases C (ppm)	Hydrolysis gases $\delta^{13}C$ (‰)	Residues C (ppm)	Residues $\delta^{13}C$ (‰)
12032	5	−2.3*	38	−20†
12022,71 Ext	3	−25.6*	24	−24‡
	—	—	33	−24†
12022,77 Int	—	—	59	−22†

* Measurement made by gas dilution with known volume of substandard.

† Gas from combustion of residue purified by passage through 10% $Ba(OH)_2$.

‡ Sample residue from phosphoric acid hydrolysis was not treated with hydrofluoric acid.

Sulfur

The sulfur content and its isotopic distribution in the fines is similar to the data found for Apollo 11 samples. The S content in two bulk fines (12023 and 12032) varied between 620 and 714 ppm and the $\delta^{34}S$ varied between +8.8 and +4.9 mil.

The basalt samples showed a marked decrease in sulfur content, equaling only $\frac{1}{3}$ of the amount analyzed in two Apollo 11 samples. Moreover, there is a significant difference in the isotopic values. Whereas in Apollo 11 the values were in the range $\delta^{34}S = +1.2$ to $+1.3$‰, in the present study they fell in the range $\delta^{34}S = -0.1$ to -0.3‰. If this difference is shown by other basaltic rocks in Oceanus Procellarum, it may be important evidence in support of differentiation processes occurring within the upper layers of the moon. The value for the basaltic rocks in Apollo 12 fall close to the value for troilite in meteorites, and sulfide in tholeiitic terrestrial basalt, whereas the value of $\delta^{34}S$ in Apollo 11 basalts are close to the average for terrestrial olivine alkali basalts (KAPLAN and HULSTON, 1966; SCHNEIDER, 1970).

DISCUSSION

The results of analyses in this laboratory, on a very limited number of samples from Apollo 11 and Apollo 12 missions, indicate that the carbon and sulfur contents of the materials are relatively low and that a wide range of $\delta^{13}C$ values (−25 to +20‰) and $\delta^{34}S$ values (−0.3 to +8.8‰) are represented. The sulfur content is greater than that found in freshly forming terrestrial basalts (e.g., a Hawaiian basalt freshly collected after ejection as a bomb, yielded <20 ppm S), however, the carbon content and isotopic value of the carbon in lunar basalts appear to be very similar to the Hawaiian basalt ($\Sigma C \approx 30$ ppm, $\delta^{13}C \approx -24$‰). Values for total carbon are similar to those given by MOORE *et al.* (1971); EPSTEIN and TAYLOR (1971); and by EGLINTON *et al.* (1971) for the same samples and values for sulfur contents are similar to those presented by MORRISON *et al.* (1971) and by WILLIS *et al.* (1971).

However, the data do show a spread in values, even within replicate measurements on a particular sample. It is probable that this in part represents contamination, especially in the case of high values for carbon, but it is also due to inhomogeneity. Sulfur appears to exist almost entirely as the iron sulfide troilite (CAMERON, 1971) although reports of trace amounts of chalcocite (EVANS *et al.*, 1971), mackinawite, and chalcopyrrhotite (RAMDOHR *et al.*, 1971) have also been reported. These sulphides exist as discrete particles often associated with metal grains (GOLDSTEIN and YAKOWITZ, 1971) and very probably not uniformly distributed. It is therefore surprising that the data from Apollo 11 and Apollo 12 analyses of fines fall within the narrow range of 600 to 800 ppm, whereas the igneous rocks and breccias are distributed over the range 600–2300 ppm.

No comparisons in the isotopic values for sulfur are presently available, but the data presented here for C^{13}/C^{12} on the carbon compounds are in general agreement with those of EPSTEIN and TAYLOR (1971) for the bulk fines. The $\delta^{13}C$ values for total carbon are isotopically lighter in the Apollo 12 samples than in those from Apollo 11. There is, however, an unresolved disagreement with the data of FRIEDMAN *et al.* (1971) who consistently obtained isotopically lighter values. As we analyzed a different set of fines from those of Friedman, and as inhomogeneity can occur, we will not attempt to discuss these differences. Our results for $\delta^{13}C$ values of carbon in basaltic rocks are very similar to those of FRIEDMAN *et al.* in the range of -24 to $-25‰$. These values are very similar to those obtained from analysis of carbon in terrestrial basaltic rocks, and if they can be reproduced in future measurements, they could be accepted as indicative of the values of indigenous lunar carbon. This would imply that carbon extruded in some basaltic flows on earth either contain juvenile carbon or else sample an average of terrestrial sedimentary carbon which isotopically averages juvenile carbon, and is not necessarily representative of organic carbon as is often considered. It would also indicate that the lunar carbon is significantly different in its C^{13}/C^{12} ratio from the average carbon in carbonaceous chondrites (types I and II $\delta^{13}C = -6$ to $-12‰$, type III $\delta^{13}C = -18‰$) and from carbides and graphite in iron meteorites ($\delta^{13}C = -4$ to $-8‰$; SMITH and KAPLAN, 1970).

The presence of isotopically heavy carbon (and sulfur) implies a contribution from an external source (i.e., solar wind) or else a mechanism for removal of the light isotopes. In considering the possible alternatives from studies on Apollo 11 material, KAPLAN *et al.* (1970) suggested "hydrogen-stripping" by protons as a possible process. Such a mechanism could not be confirmed in the present study from the comparative analysis of an internal and external fragment of basalt. However, it is probably that the surface area exposed to solar wind protons was too small to be effective. Support for the proposed mechanism appears to have come from the fluorination experiments of EPSTEIN and TAYLOR (1971) on bulk fines. During rate controlled release of oxygen and silicon fluoride (SiF_4) from the particles it was found that the gases first released (interpreted as coming from outer surfaces) were most enriched in $\delta^{18}O$ and $\delta^{30}Si$. Particle-track (COMSTOCK *et al.*, 1971) analysis suggest that the rate of surface erosion of grains can be as high as 2×10^{-8} cm/yr. There is some evidence from the data presented in the results that an isotopically light component resides in trapped methane. This methane could have formed by reaction of solar wind hydrogen with lunar carbon. The possibility that an anomalous enrichment in heavy isotopes may have arisen from contamination by materials used in containers and at the Lunar Receiving Laboratory is probably unfounded (see Table 1). The only compound analyzed which falls approximately in the range of lunar samples is the lubricant molybdenum disulfide. However, no Mo anomaly was apparently observed (LSPET, 1970). The possibility of the influence of diisopropyl disulfide recently detected as a minor contaminant (GILBERT *et al.*, 1971) is discounted.

As would be expected from its greater mass, the enrichment of S^{34} in the sulfur of the fines is less than that for C^{13}. That some common enriching process in heavy isotopes is effective, may be judged from the data on sample 12032 from the Bench Crater. This was a relatively coarse sample with significantly lower carbon content than the other bulk fines analyzed in this study, and less enrichment in C^{13} as well as S^{34}. It is of interest to note that the basalt samples yielded an average value for $\delta^{34}S$ ($-0.17‰$) within the range of troilite from iron meteorites (KAPLAN and HULSTON, 1966) and significantly different from the average value for basaltic sulphur from Apollo 11 samples (ave. $\delta^{34}S = +1.2‰$). This would suggest that compositional differences are present (as found by many other workers) and supports the model involving differentiation of the upper crust.

Carbon in the samples is present in numerous forms and probably originates from a number of sources. Data presented by GIBSON and JOHNSON (1971) and by ORÓ *et al.* (1971) indicate that methane, carbon dioxide, and carbon monoxide can be released at different temperatures and are held within the lunar material by different binding energies. The studies of CHANG *et al.* (1971) and of EGLINTON *et al.* (1971) indicate that a significant proportion of the carbon is in a nonvolatile form. This is supported by the studies of BLOCH *et al.* (1971), who were unable to detect any CO or CO_2 in crushed unheated samples, whereas detectable quantities of methane were released from all the crushed samples. Furthermore, preferential release of CO_2 at lower temperatures of pyrolysis, added to the isotopically light values for the released CO_2 (in the range of atmospheric CO_2 which is $\delta^{13}C = -7$ to $-8‰$, see Table 3), suggests that some proportion of the carbon dioxide identified is of terrestrial origin, strongly adsorbed onto grain boundaries. From the above arguments, and from the hydrolysis experiments of CHANG *et al.* (1971) and EGLINTON *et al.* (1971), we may conclude that the light hydrocarbon gases (predominantly methane) are the only significant (>1 ppm) volatile carbon compounds trapped within the lunar material so far studied. Such a conclusion, of course, agrees with the model of their continuous formation by solar-wind proton reaction with non volatile carbon. Although a significant proportion of the total carbon (10–20%) can be accounted for by hydrolysis experiments of carbides, the residue has not yet been characterized. KAPLAN *et al.* (1970) suggested that it may be elemental carbon, because of its unreactive nature. One may also consider that during acid hydrolysis, iron-nickel carbide disproportionates into hydrocarbons and a nonsoluble carbon-rich phase which remains as a residue.

The spread in isotopic values from pyrolysis measurements (Table 3) also indicate that carbon compounds from multiple sources are present within the sample. The differences appear to be too large to be explained by experimental temperature fractionation effects. Three sources of carbon must be considered—lunar, meteoritic, and solar wind. If, according to ANDERS *et al.* (1971), $\approx 2\%$ type 1 carbonaceous chondrite material is evident on the lunar surface, this could account for 800 ppm of carbon in the bulk fines. Some of this material is sufficiently volatile to disappear on impact, but some proportion of the carbon may have been converted into a non-volatile form. At the same time, one cannot eliminate the possibility that carbon has been enriched in certain lunar facies during differentiation and is being added to the regolith. Unfortunately, criteria to separate these various possible sources are presently not available due to an incomplete understanding of the carbon chemistry and due to limited sampling. However, arguments for dominantly a lunar origin of the carbon in the fines are strong when related to the distribution of sulfur. The regolith material is depleted in sulfur when compared with the rocks. Carbonaceous chondrites would add sulfur in approximately the same amounts as carbon. Solar wind activity would, however, add a significantly smaller quantity of sulfur. These two processes act as additives and not diluents to the original sulfur in the rocks. Therefore, either sulfur is preferentially removed from the fines, or as an alternative mechanism which we prefer, it is diluted by a "carbon-rich, sulfur-poor" component of the moon.

Acknowledgments—We wish to thank George Claypool, Ed Ruth, and Dr Sherwood Chang for their assistance. This study was supported by NASA contract NAS 9–8843.

References

Abell P. I., Eglinton G., Maxwell J. R., and Pillinger C. T. (1970) Indigenous lunar methane and ethane. *Nature*, **226**, 251–252.

Anders E., Laul J. C., Keays R. R., Ganapathy R., and Morgan J. W. (1971) Elements depleted on lunar surface: Implication for origin of moon and meteorite influx rate. Second Lunar Science Conference (unpublished proceedings).

Bloch M., Fechtig H., Funkhauser J., Gentner W., Jessberger E., Kirsten T., Muller O., Neukum G., Schneider E., Steinbrunn F., and Zahringer J. (1971) Location and variation of rare gases in Apollo 12 lunar samples and active and noble gases, and total pressures in bubbles and internal cavities of lunar material. Second Lunar Science Conference (unpublished proceedings).

Burlingame A. L., Calvin M., Han J., Henderson W., Reed W., and Simoneit B. R. (1970) Study of carbon compounds in Apollo 11 lunar samples. *Proc. Apollo 11 Lunar Sci. Conf., Geochim. Cosmochim. Acta* Suppl. 1, Vol. 2, pp. 1779–1791. Pergamon.

Cameron E. N. (1971) Opaque minerals in certain lunar rocks from Apollo 12. Second Lunar Science Conference (unpublished proceedings).

Chang S., Smith J. W., Kaplan I. R., Lawless J., Kvenvolden K. A., and Ponnamperuma C. A. (1970) Carbon compounds in lunar fines from Mare Tranquillitatis IV. Evidence for oxides and carbides. *Proc. Apollo 11 Lunar Sci. Conf., Geochim. Cosmochim. Acta* Suppl. 1, Vol. 2, pp. 1857–1868. Pergamon.

Chang S., Kvenvolden K. A., Lawless J., Ponnamperuma C. A., and Kaplan I. R. (1971) Carbon in an Apollo 12 sample: Concentration, isotopic composition, pyrolysis products and evidence for indigenous carbides and methane. *Science* **171**, 474–477.

Comstock G. M., Evwaraye A. O., Fleischer R. L., and Hart H. R. (1971) The particle track record of lunar soil. Second Lunar Science Conference (unpublished proceedings).

Eglinton G., Abell P. I., Cadogan P. H., Maxwell J. R., and Pillinger C. T. (1971) Survey of lunar carbon compounds. 1. The presence of indigenous gases and hydrolysable carbon compounds in Apollo 11 and Apollo 12 samples. Second Lunar Science Conference (unpublished proceedings).

Epstein S., and Taylor H. P. (1970) The concentration and isotopic composition of hydrogen, carbon, and silicon in Apollo 11 lunar rocks and minerals. *Proc. Apollo 11 Sci. Conf., Geochim. Cosmochim. Acta* Suppl. 1, Vol. 2, pp. 1085–1096. Pergamon.

Epstein S., and Taylor H. P. (1971) O^{18}/O^{16}, Si^{30}/Si^{28}, D/H, and C^{13}/C^{12} ratios in lunar samples. Second Lunar Science Conference (unpublished proceedings).

EVANS H. T., TOULMIN P., BARTON P., and ROSEBOOM E. (1971) Second Lunar Science Conference (unpublished proceedings).

FLORY D. A., SIMONEIT B. R., and SMITH D. H. (1969) Apollo 11 organic contamination history. NASA Manned Spacecraft Center, Report.

FRIEDMAN I., O'NEIL J. R., GLEASON J. D., and HARDCASTLE K. (1971) The carbon and hydrogen content and isotopic composition of some Apollo 12 materials. Second Lunar Science Conference (unpublished proceedings).

GIBERT J., FLORY D., and ORÓ J. (1971) Identity of Apollo 11 lunar fines and Apollo 12 York meshes. *Nature* **229**, 33–34.

GIBSON E. K., and JOHNSON D. M. (1971) Thermal analysis-inorganic gas release studies of lunar samples. Second Lunar Science Conference (unpublished proceedings).

GOLDSTEIN J. I., and YAKOWITZ H. (1971) Metallic inclusions and metal particles in the Apollo 12 lunar soil. Second Lunar Science Conference (unpublished proceedings).

KAPLAN I. R., and HULSTON J. R. (1966) The isotopic abundance and content of sulfur in meteorites. *Geochim. Cosmochim. Acta* **30**, 479–496.

KAPLAN I. R., and SMITH J. W. (1970) Concentration and isotopic composition of carbon and sulfur in Apollo 11 lunar samples. *Science* **167**, 541–543.

KAPLAN I. R., SMITH J. W., and RUTH E. (1970) Carbon and sulfur concentration and isotopic composition of Apollo 11 lunar samples. *Proc. Apollo 11 Lunar Sci. Conf., Geochim. Cosmochim. Acta* Suppl. 1, Vol. 2, pp. 1317–1330. Pergamon.

LSPET (Lunar Sample Preliminary Examination Team) (1969) Preliminary examination of the lunar samples from Apollo 11. *Science* **165**, 1211–1227.

LSPET (Lunar Sample Preliminary Examination Team) (1970) Preliminary examination of the lunar samples from Apollo 12. *Science* **167**, 1325–1339.

MOORE C. B., LEWIS C. F., DELLES F. M., GOOLEY R. C., and GIBSON E. K. (1971) Total carbon and nitrogen in Apollo 12 samples. Second Lunar Science Conference (unpublished proceedings).

MORRISON G. H., GERARD J. T., POTTER N. M., GANGADHARAM E. V., ROTHENBERG A. M., and BURDO R. A. (1971) Elemental abundances of lunar soil and rocks from Apollo 12. Second Lunar Science Conference (unpublished proceedings).

MURPHY R. C., PRETI G., NAFISSI M. M., and BIEMANN K. (1970) Search for organic material in lunar fines by mass spectrometry. *Science* **167**, 755–757.

ORÓ J., FLORY D. A., GIBERT J. M., MCREYNOLDS J., LICHTENSTEIN H. A., WIKSTROM S., and GIBSON E. K. (1971) Abundances and distribution of organogenic elements and compounds in Apollo 12 samples. Second Lunar Science Conference (unpublished proceedings).

RAMDOHR P., GORESY A. E., and TAYLOR L. A. (1971) The opaque minerals in the lunar rocks from Oceanus Procellarum. Second Lunar Science Conference (unpublished proceedings).

SCHNEIDER A. (1970) The sulfur isotope composition of basaltic rocks. *Contrib. Mineral. Petrol.* **25**, 95–124.

SMITH J. W., and KAPLAN I. R. (1970) Endogenous carbon in carbonaceous meteorites. *Science* **167**, 1367–1370.

WILLIS J. P., AHRENS L. H., DANCHIN R. V., ERLANK A. J., GURNEY J. J., HOFMEYR P. K., MCCARTHY T. S., and ORREN M. J. (1971) Some interelement relationships between lunar rocks and fines, and stony meteorites. Second Lunar Science Conference (unpublished proceedings).

Proceedings of the Second Lunar Science Conference, Volume 2, pp. 1407–1415.
The M.I.T. Press, 1971.

The carbon and hydrogen content and isotopic composition of some Apollo 12 materials*

IRVING FRIEDMAN, JAMES R. O'NEIL, JIM D. GLEASON, and KENNETH HARDCASTLE
U.S. Geological Survey, Denver, Colorado 80225

(*Received 24 February 1971; accepted in revised form 5 April 1971*)

Abstract—Analysis of the total carbon in lunar dust 12070, 12030, and 12001, and rocks 12051, 12004, and 12021 by combustion in oxygen at 900°C after outgassing in vacuo at 150°C gave values of approximately 140 ppm with a δC^{13} of -10 to $-15‰$ PDB for the dust, and approximately 20 to 60 ppm of carbon with a δC^{13} of -22 to $-25‰$ for the crystalline rocks. On stepwise heating of the dust to 1000°C in vacuo there is evidence of heavier carbon of δC^{13} of approximately $+2‰$, which is released as CO and perhaps as CO_2. However, combustion of the pyrolysis residue yields additional carbon with a δC^{13} of $-25‰$ and giving a total carbon of approximately -10 to $-15‰$. These results are in agreement with our previous data on Apollo 11 materials and in disagreement with the very heavy values for lunar carbon in the Apollo 11 samples as quoted by EPSTEIN and TAYLOR (1970) and by KAPLAN *et al.* (1970). The deuterium contents of lunar dust, hydrogen and water appear to be somewhat heavier than for the same kinds of materials from that obtained on Apollo 11. The deuterium content of the hydrogen and water in the interior of a sample of Apollo 12 basalt is higher than that of any terrestrial water. The δD of $+307‰$ may be the result of deuterium production by spallation.

INTRODUCTION

THE AMOUNT AND ISOTOPIC compositions of hydrogen and carbon were determined on samples of Apollo 12 dust and rocks by techniques that were similar to those used by us on Apollo 11 samples. Two basic techniques were used to extract the water, hydrogen, and carbon compounds from the sample and to convert them to hydrogen gas and carbon dioxide for isotopic analysis.

The first procedure was a pyrolysis: heating the sample contained in a Vycor tube in vacuum stepwise to $\sim 1000°C$, collecting the water, hydrogen, CO_2, and CO, and measuring and analyzing each species separately. Upon completion of the vacuum heating, the sample was removed from the pyrolysis equipment, crushed if not already dust, mixed with previously combusted CuO, and placed in the combustion apparatus where any additional unreacted carbon species were combusted in oxygen at 900°C. The second procedure was a direct combustion without prior pyrolysis.

Since there was apparent disagreement among the groups (EPSTEIN and TAYLOR, 1970; FRIEDMAN *et al.*, 1970; KAPLAN *et al.*, 1970) that analyzed the Apollo 11 materials for the stable isotopes of carbon (Fig. 1), particular emphasis was placed on the analysis of carbon. We have always believed that the apparent disagreement was due to sample inhomogeneity, and to the fact that different carbon-containing species—CO_2, CO, carbides, elemental carbon—probably have different δC^{13}. Our experience with Apollo 12 samples acts to reinforce this opinion. We believe that lunar materials,

---

* Publication authorized by the Director, U.S. Geological Survey.

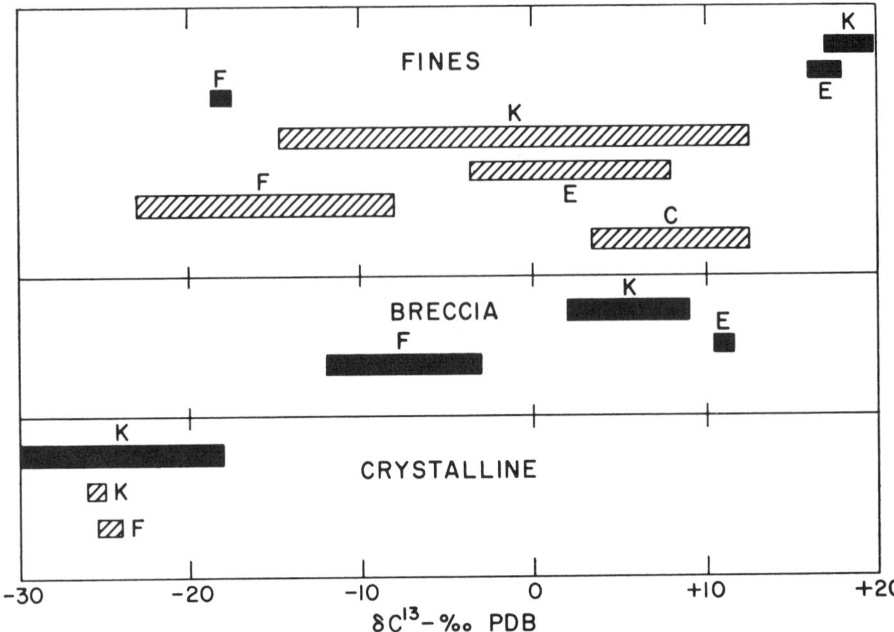

Fig. 1. Relative abundances of the stable isotopes of carbon in lunar materials. The analyses are given in ‰ deviation from the PDB standard. Solid bars = Apollo 12; diagonal lined bars = Apollo 11. C = CHANG *et al.* (1971); E = EPSTEIN and TAYLOR (1970, 1971); F = FRIEDMAN *et al.* (1970); K = KAPLAN *et al.* (1970) and KAPLAN and PETROWSKI (1971).

particularly lunar dust, contain both CO and CO_2, and that this "volatile" carbon has a δC^{13} that is heavy—up to +3‰ in our measurements (significantly heavier in those of EPSTEIN and TAYLOR, 1970, and KAPLAN *et al.*, 1970). However, the remaining non-"volatile" carbon in lunar materials has a δC^{13} of about −25‰.

Previous analysis of hydrogen on Apollo 11 dust and breccia gave values that were depleted in deuterium compared to most earth materials. For example FRIEDMAN *et al.* (1970) reported values for δD that range from −830 to −960‰; EPSTEIN and TAYLOR (1970) found a somewhat smaller range from −830 to −873‰; while HINTENBERGER *et al.* (1970) obtained a value for deuterium depleted by at least a factor of 3 over earth hydrogen (∼ −660‰). Analysis of water extracted from Apollo 11 samples gave somewhat heavier values, −580 to −870 (FRIEDMAN *et al.*, 1970), −261 to −421 (EPSTEIN and TAYLOR, 1970). The problem of contamination by terrestrial water is a serious one, but all authors agree that much if not all of the hydrogen in the dust and breccia is of solar wind origin and that this hydrogen is depleted in deuterium as compared to terrestrial materials.

Our present analyses give a δD of −292 to −338 for Apollo 12 dust, and values ranging from −155 to +307 for Apollo 12 rocks.

Pyrolysis Procedure

The samples were handled with cleaned forceps, and care was taken to avoid organic contamination. For the pyrolysis, the samples were not crushed, but they were heated as received, usually in one or, at most, four pieces. The samples were placed in a previously outgassed Vycor tube which was connected to a vacuum line. The Vycor tube and sample were then flamed several times to aid outgassing. A furnace was then placed in proximity to the sample and the sample heated to 60°C. Evacuation was continued for several hours, until the thermocouple vacuum gauge showed no pressure drop (pressure $<10^{-3}$ mm) when the system, whose total volume was 30 cc, was isolated from the vacuum for 10–15 minutes. The temperature, as measured by a thermocouple, was then slowly increased to 150°C to 200°C and kept at that temperature for several hours. The water given off at the end of this time was continuously condensed in a liquid N_2-cooled trap. At the completion of this phase of the experiment, the trap was raised to dry-ice temperature. The vacuum gauge showed that no CO_2, CO, H_2, or He was evolved at this temperature. The water, evolved below ~200°C, was considered to be contamination, but was nonetheless converted to H_2 and analyzed isotopically. The minute amounts of hydrogen gas were transferred by means of a small mercury diffusion pump into the leak of the specially constructed isotope ratio 6-inch 60 degree Nier-type mass spectrometer. The mass 2 beam intensity had previously been calibrated in terms of μg H_2. We had no difficulty in reproducing our δH values to $\pm 10‰$ on samples of 6 μg H_2 (54 μg H_2O).

The second and third phases were carried out by raising the temperature in two steps to a higher value, maintaining this higher temperature for several hours, while freezing out H_2O and CO_2 in a liquid N_2 trap. At the completion of each phase, the noncondensables were allowed to react with CuO at 550°C; the H_2O and CO_2 so formed were kept isolated from the CO_2 and H_2O evolved from the sample as such. Any remaining noncondensable gases (e.g., He) were then pumped into the mass spectrometer and the mass 4 and mass 3 peaks measured. The various lots of H_2O (the H_2O evolved as such and the H_2O formed by reaction of H_2 with the CuO) were separated from their companion CO_2 (CO_2 evolved as such and the CO_2 formed by CuO oxidizing CO) at dry-ice temperature. The CO_2 samples were transferred to another isotope ratio 6-inch 60 degree Nier-type mass spectrometer for δC^{13} determination. The intensity of the mass 44 beam was used to measure the amount of carbon for small samples. Manometric methods were used for samples of carbon larger than 10 μg. The two water samples were then individually processed over uranium and pumped into the spectrometer. For all samples some CO and CO_2 (in varying proportions) were evolved below ~600°. At higher temperatures only CO was evolved. Both H_2 and H_2O were evolved below 600° in the rocks and below 435° in the dust. We do not know whether this "high-temperature" (~400°C) water was present in the sample as H_2O, or was formed by reaction of H_2 with iron oxides present in the sample.

Dust

For the dust, the total H_2, both as H_2 and H_2O released from 160°C to 430°C, is 14 ppm with a δD of -260. This δ value is not as light as that found in the Apollo 11 materials. Epstein and Taylor (1971) also found that the hydrogen in the Apollo 12 dust appears to be isotopically heavier than that in Apollo 11 dust. The 16 ppm H_2 released from 430°C to 1025°C is lighter, $-347‰$. (It is interesting to note that the equilibrium constant [(D/H) water]/[(D/H) gas] for the 160°C to 430°C samples of water and hydrogen correspond to isotopic equilibrium at 200°C, whereas the 430°C to 1025°C samples correspond to equilibrium at 650°C; Suess, 1949; Bottinga, 1969). If we exclude the "contamination" water evolved from 50°C to 160°C, we get 30 ppm H_2 with a δD of -325.

Crystalline rock

The pyrolysis of 6.7 g of the interior of rock 12051 would be expected to yield carbon and hydrogen unaffected by solar wind. The first phase of the experiment was carried out by heating the rock from 60°C to 200°C. Only water (0.4 ppm H_2; $\delta D = -177$) was evolved, together with a trace (<0.1 ppm C) of CO_2. The second phase of gas collection was carried out between 200°C and 600°C. Hydrogen, 0.1 ppm, and some water (0.5 ppm H_2; $\delta D = -37$), CO (0.4 ppm C), and CO_2 (0.9 ppm C) were evolved. The CO_2 and CO were combined for δC^{13} analysis, and the combination gave a value of $\sim -34‰$ ($\pm 5‰$). Upon final heating of the sample to 1025°C, an additional 2.3 ppm of carbon ($\delta C^{13} = -21.8$) and 6 ppm H_2 ($\delta D = +341$) were obtained. We believe this high-temperature hydrogen to be lunar material and not terrestrial contamination. Its high deuterium content relative to that of the hydrogen of the earth ($\delta D = \sim 0$) may be due to entrapment of deuterium produced by spallation. Combustion of the pyrolyzed samples of both dust and rock gave significant amounts of carbon of about $\delta C^{13} = -25‰$. In fact, the major part of the rock carbon was evolved by combustion, rather than by pyrolysis.

Combustion Procedure

Direct combustion of lunar samples converts all carbon present to CO_2 and hydrogen present to H_2O. The simplicity and reproducibility of direct combustion is the method of choice for determining total amounts and isotopic composition of carbon in the lunar materials.

The combustion procedure is similar to that described by Craig (1953). The sample and an excess of treated CuO are weighed into a porcelain boat and degassed at 150°C until no further evolution of gas is detectable with a Hastings thermocouple gauge (0.5 to 1 hour). Inasmuch as the lunar surface is exposed to this temperature during the lunar day, these gases are considered terrestrial contaminants and are discarded. Some contaminant water is likely to be retained after this degassing procedure, but higher degassing temperatures may result in loss of lunar carbon. After degassing, oxygen is circulated through the system and pumped away. The boat is then pushed into the combustion zone of the apparatus which is at 900°C. Oxygen, purified by passing over hot CuO and through a trap cooled with liquid N_2, is added and the gases are circulated over traps by means of an automatic Toepler pump. Water is first collected in a trap held at dry ice temperature and then a second trap is cooled to liquid nitrogen temperature to collect the CO_2. A $PbCrO_4$ plug held at about 150°C removes oxides of sulfur. After 40 minutes to an hour the excess oxygen is pumped away and the CO_2 and H_2O are transferred and processed for mass spectrometric analysis.

Inasmuch as the amounts of carbon and hydrogen in lunar samples are small, great care was taken to avoid contamination. Blanks were run on the combustion apparatus and yielded an amount of CO_2 that produced a mass-44 signal of 135 to 380 mV. The smallest samples we analyzed gave a mass-44 signal of 5–6 V, and so for these samples the blank would amount to 2–6%. It is not possible to determine C^{13}/C^{12} ratios on the blanks, but presumably they are isotopically light. For most of the analyzed samples the carbon blank is insignificant.

In order to demonstrate the precision of carbon isotope analyses of minute samples, we made combustions of spectrographic carbon (NBS–21) and filter paper. The filter paper was used, because very small amounts can be easily and precisely weighed without worry of transfer losses—a major problem with powders. A 16.8-mg sample of filter paper was combusted and yielded 540 μ moles of CO_2 with a δ value of -24.7. A 0.5-mg sample of the same material yielded 15 μ moles of CO_2 with a δ value of -24.8. A value of -27.8 was obtained for NBS–21 spectrographic carbon; this value is identical to Craig's published value (Craig, 1957). These data are convincing evidence that we can determine precise carbon isotope ratios on the tiny amounts of carbon present in lunar samples. This is particularly important considering the gross disagreement in reported carbon isotope ratios in the Apollo 11 samples (Friedman *et al.*, 1970; Epstein and Taylor, 1970; Kaplan *et al.*, 1970).

The water blank on the combustion line was about 7 μg (as H_2) and had a δ value of -207, a value close to that obtained for our "contamination" water collected at 150°C ($\delta = -180$ to -207). The source of the carbon and water blank in the combustion line was probably the oxidation of the stopcock grease, plus water picked up by the previously combusted CuO during the time that it was weighed and mixed with the sample. Contamination water in addition to that picked up by the CuO would have been adsorbed by the lunar sample. If we subtract the deuterium blank from the combusted samples, we derive the concentrations of hydrogen and δD as given in

Table 1. In all probability these results represent an upper limit for hydrogen and deuterium. The comparison of the deuterium and carbon data on "similar" samples run by pyrolysis and combustion and by combustion alone is given in Table 2.

Discussion

The hydrogen in the dust as determined by the two techniques agrees both in amount and in δ value. The amounts of hydrogen (as water plus H_2 gas) of Apollo 12 samples are about half those found for the dust and breccia samples from Apollo 11, and the δD is heavier than that for Apollo 11 samples. The amount of carbon present in the dust is less than that found in Apollo 11 samples, again by a factor of about 2. These results are in agreement with the analyses of the preliminary examination team (LSPET, 1970). Again, allowing for sample inhomogeneity, the two methods gave similar results—116 to 139 ppm carbon, $\delta C^{13} = -8$ to -24. The pyrolysis, as in our Apollo 11 analysis, showed the presence of "heavy" CO_2 and CO, $\sim +2‰$, as well as much lighter carbon (as carbide or elemental carbon), which together gave a total δC^{13} of -8 to -24.

The most remarkable result is the heavy hydrogen found in the interior or rock 12051. The hydrogen from this sample had a δD of $+307‰$ and was present to the amount of 7 ppm. This large rock was described as blocky ejecta from a fresh 2-meter crater. A 6.7-g sample of the interior of this rock was heated in vacuum, and 3.6 ppm of water (0.4 ppm H_2), with a δD of -177, was given off between 60°C and 200°C. We believe that this water was geoatmospheric contamination. Between 200°C and 600°C another increment of water, 4.5 ppm (0.5 ppm H_2), $\delta D = -37$, was evolved, together with 0.1 ppm of hydrogen gas (δD = ?). It is possible that this water was also contamination and that it was relatively heavy, inasmuch as it could have equilibrated isotopically with the hydrogen gas (0.1 ppm), whose δD could not be measured. The very heavy hydrogen that evolved between 600° and 1025°($\delta D = +341$) cannot easily be accounted for as contamination or as an artifact of our experimental technique; and we conclude that this "heavy" hydrogen is truly lunar and may have been in part caused by spallation. Most, if not all, of this hydrogen was present in the rock as hydrogen, rather than as water, because we observed that upon heating the rock above 600°, the pressure first rose as hydrogen and CO were evolved, then fell as the H_2 and CO reacted with the iron oxide in the rock to give CO_2 and H_2O, which condensed in the liquid nitrogen-cooled trap.

To substantiate the conclusion that the "heavy" deuterium value is not an artifact or memory effect of our equipment, we have the fact that the sample analyzed just prior to the analysis of the heavy sample was light, $\delta D = -177$, as was the following sample, $\delta D = -340$. In addition, the δD of small samples (6 μg H_2) of water of known δD were measured several times and gave results in good agreement ($\pm 10‰$) with the known values determined on 10,000-μg samples. The analysis of this heavy hydrogen was carried out on a relatively large sample of 39 μg H_2. Another sample of 12051 also gave "heavy" ($\delta D = +408$) values for hydrogen that was liberated between 300° and 960°. Unfortunately, some of the gases were accidentally lost and the partial results are not reported.

Several authors presented data on carbon and hydrogen contents of Apollo 12 samples at the Second Lunar Science Conference. Table 3 lists the results of the carbon

Table 1. Results of heating and com-

Sample			Vacuum Heating								
			Temp.	H_2			H_2O			CO_2	
No.	Type	Wt. (g)	°C	ppm	D/H × 10^6	δD	ppm	D/H × 10^6	δD	ppm	δC^{13}
12030,30	Dust	1.00	50–160				8	131	−169	2	—
			160–430	4	94	−407	10	126	−200	12	—
			430–1025	9	84	−438	7	109	−313	9	—
12070,49	Dust	0.91	25–125							Tr	—
			175–435							3	—
			435–1008							5	—
12051,66	Basalt	6.80	60–200	—			0.4	130	−177	Tr	—
			200–600	0.1	—		0.5	152	−37	0.9	combined →†
			600–1025	6.0	212	+341					
12001,37	Dust	1.08									
12004,38	Olivine basalt	1.98									
12021,32	Gabbro	2.00									
12051,9	Basalt	2.87									

* Blank of 7.2 μg H_2 (δD = −207‰ has been subtracted from the raw results.
† This total does not include water evolved at or below 200°C. Water collected on combustion of vacuum-heated samples not reported, because it is considered to be mostly contamination.
‡ 0.9 ppm CO_2 combined with 0.4 ppm (CO) to get value of −34 for δC^{13}.
Concentrations = ±20% of amount present.
D/H × 10^6 = ±20%; δC^{13} = ±0.2‰; δO^{18} = ±0.2‰; δD = ±10‰

Table 2. Comparison of results of analysis by two techniques.

Sample	Pyrolysis and Combustion ppm*	Pyrolysis and Combustion δ‰	Combustion ppm	Combustion δ‰
Hydrogen				
Dust 12001,37			46	−338
Dust 12030,30	30	−325		
Rock 12051,9, exterior, some sawed faces			9	−155
Rock 12051,66, interior	6.5	+307		
Carbon				
Dust 12001,37			139	−15.0
Dust 12030,30	116	−24		
Dust 12070,49	133	−8.0		
Rock 12051,9, exterior, some sawed faces			42	−24.2
Rock 12051,66, interior	27	−23		

* The amounts of carbon reported here do not agree with the figures reported at the Second Lunar Science Conference. An error in calculations was discovered after the paper was presented in Houston.

analysis carried out by several different techniques on samples of Apollo 12 dust. It is evident that either the samples are very inhomogeneous or some analyses are in error. The lunar crystalline rocks have far less carbon than the dust. The problem of contamination of the rocks by handling and sawing with a diamond-impregnated wire or saw blade is serious. Diamonds have a δC^{13} of −3 to −6‰. Both the present authors and KAPLAN and PETROWSKI (1971) get about 8 ppm C with a δC^{13} of −25 ± 1‰ for carbon in the lunar basalts analyzed. Again, the amounts of carbon are variable from one part of the sample to another, and this may be due either to variable contamina-

bustion of selected Apollo 12 material.

CO		Combustion					Total				
		H*			C		H†			C	
ppm	δC^{13}	ppm	D/H × 10^6	δD	ppm	δC^{13}	ppm	D/H × 10^6	δD	ppm	δC^{13}
Tr	—										
44	−23.4				49	−26.0	30	107	−325	116	∼−24
5.7	—										
71	+2.1				48	−24.7				133	−8.0
—	—										
0.4 =	−34										
2.3	−21.8				23	−23.8	6.6	207	+307	27	−23
		46	105	−338	139	−15.0	46	105	−338	139	−15.0
		3	144	−86	61	−25.2	3	144	−86	61	−25.2
		3	147	−68	21	−25.4	3	147	−68	21	−25.4
		9	134	−155	42	−24.2	9	134	−155	42	−24.2

sC^{13} values are per mil deviations from PDB.
δO^{18} values are per mil deviations from SMOW.
δD values are per mil deviations from SMOW.
The D/H × 10^6 was calculated from the δD SMOW values, using the relation D/H × 10^6 = 0.158 (1000 + δD‰).

Table 3. Total carbon content of some Apollo 12 samples (ppm).

Sample	CHANG *et al.* (1971)	EPSTEIN and TAYLOR (1971)	FRIEDMAN *et al.* (this report)	KAPLAN and PETROWSKI (1971)	and MOORE *et al.* (1971)	NAGY *et al.* (1971)	ORÓ *et al.* (1971)
12070		145	133				
12001			139		100 to 130	115	
12033		30			23 to 60	108	
12042		125		111	100 to 165	141	
12023	109 to 140				140 to 155	109	
12037					30 to 80		
12032				37 to 86	25 to 60		
12028					70 to 200	96	230
12025					150 ± 30	111	
12051			27 to 42				
12004			61				
12021			21				
12030			116				
12022				14 to 21	24 to 70	90	

tion or to inhomogeneous distribution of a carbonaceous (graphitic?) phase in the rock.

The only hydrogen data reported by other investigators were determined from Apollo 12 dust. The problem of contamination of the dust by rocket exhaust, astronaut space-suit leakage, and geoatmospheric contamination is serious, especially in view of the small particle size and enormous surface of the dust. The problem is compounded by the fact that water and hydrogen gas can equilibrate at the temperature of extraction (200° to 1000°C), especially when catalyzed by the active surfaces present in the lunar dust. Water can be generated by the reaction of hydrogen with iron oxides; and, especially at temperatures above 900°C, it can be reduced to hydrogen by carbon, if any carbon is present in the sample. The reaction with iron oxides undoubtedly occurs, particularly above about 500°C; and if any contamination water is present at these high temperatures, it can form hydrogen, which will then be mixed with the hydrogen "initially" present in the sample. EPSTEIN and TAYLOR (1971) demonstrated that some such exchange between adsorbed "contamination" water and hydrogen can take place under the conditions of their experiment; namely, treatment of the sample with deuterium-depleted water vapor at 300°C before heating the sample to a higher temperature to liberate hydrogen gas. The hydrogen gas had a lower deuterium content than hydrogen that has evolved from an untreated sample.

CONCLUSIONS

(1) Lunar materials commonly contain CO and CO_2 with a δC^{13} that is enriched by approximately +3‰ relative to the PDB standard. In addition to this "volatile" carbon, both dust and rocks contain nonvolatile carbon with a δC^{13} of −25‰. (2) Lunar dust contains a large amount of hydrogen (30 to 50 ppm) depleted in deuterium by a factor of approximately two to five as compared to present earth hydrogen. The interior of a large basalt cobble contains a maximum of 7 ppm of hydrogen that is enriched in deuterium by approximately 30% as compared to present earth hydrogen. This enrichment in deuterium may be due to spallation-produced deuterium. The small amount of hydrogen places an upper limit on the amount of water present in the rock ($\sim$63 ppm H_2O).

Acknowledgments—This study was carried out under NASA contract T–75445.

REFERENCES

BOTTINGA Y. (1969) Calculated fractionation factors for carbon and hydrogen isotope exchange in the system calcite–carbon dioxide–graphite–methane–hydrogen–water vapor. *Geochim. Cosmochim. Acta* **33**, 49–64.

CHANG S., KVENVOLDEN K., LAWLESS J., PONNAMPERUMA C., and KAPLAN I. R. (1971) Carbon in an Apollo 12 sample: Concentration, isotopic composition, pyrolysis products, and evidence for indigenous carbides and methane. Second Lunar Science Conference (unpublished proceedings).

CRAIG H. (1953) The geochemistry of the stable carbon isotopes. *Geochim. Cosmochim. Acta* **3**, 53–92.

CRAIG H. (1957) Isotopic standards for carbon and oxygen and correction factors for mass-spectrometric analysis of carbon dioxide. *Geochim. Cosmochim. Acta* **3**, 133–149.

EGLINTON G., ABELL P. I., CADOGEN P. H., MAXWELL J. R., and PILLINGER C. T. (1971) Survey of lunar carbon compounds. Second Lunar Science Conference (unpublished proceedings).

EPSTEIN S. and TAYLOR H. P. (1970) The concentration and isotopic composition of hydrogen, carbon, and silicon. *Proc. Apollo 11 Lunar Sci. Conf., Geochim. Cosmochim. Acta* Suppl. 1, Vol. 2, pp. 1085–1096. Pergamon.

EPSTEIN S. and TAYLOR H. P., JR. (1971) O^{18}/O^{16}, Si^{30}/Si^{28}, D/H, and C^{13}/C^{12} ratios in lunar samples. Second Lunar Science Conference (unpublished proceedings).

FRIEDMAN I., GLEASON J. D., and HARDCASTLE K. G. (1970) Water, hydrogen, deuterium, carbon, and C^{13} content of selected lunar material. *Proc. Apollo 11 Lunar Sci. Conf., Geochim. Cosmochim. Acta* Suppl. 1, Vol. 2, pp. 1103–1109. Pergamon.

Hintenberger H., Weber H. W., Voshage H., Wänke H., Begemann F., and Wlotzka F. (1970) Concentrations and isotopic abundances of the rare gases, hydrogen, and nitrogen in Apollo 11 lunar matter. *Proc. Apollo 11 Lunar Sci. Conf., Geochim. Cosmochim. Acta* Suppl. 1, Vol. 2, pp. 1269–1282. Pergamon.

Kaplan I. R., Smith J. W., and Ruth E. (1970) Carbon and sulfur concentration and isotopic composition in Apollo 11 lunar samples. *Proc. Apollo 11 Lunar Sci. Conf., Geochim. Cosmochim. Acta* Suppl. 1, Vol. 2, pp. 1316–1329. Pergamon.

Kaplan I. R. and Petrowski C. (1971) Carbon and sulfur isotope studies on Apollo 12. Second Lunar Science Conference (unpublished proceedings).

LSPET (Lunar Sample Preliminary Examination Team) (1970) Preliminary examination of the lunar samples from Apollo 12. *Science* **167**, 1325–1339.

Moore C. B., Lewis C. F., Delles F. M., and Gooley R. C. (1971) Total carbon and nitrogen in Apollo 12 lunar samples. Second Lunar Science Conference (unpublished proceedings).

Nagy B., Drew C. M., Hamilton P. B., Modzeleksi J. E., Modzeleski V. E., Nagy L. A., Scott W. M., Thomas J. E., Urey H. C., and Ward R. (1971) CO, CO_2, CH_4, and lesser quantities of aromatic hydrocarbons on a sulfur compound were found in Apollo 12 fines and in an interior chip of a lunar rock. Second Lunar Science Conference (unpublished proceedings).

Oró J., Flory D. A., Gibert J. M., McReynolds J., Lichtenstein H. A., Wikstrom S., and Gibson E. K., Jr. (1971) Abundances and distribution of organogenic elements and compounds in Apollo 12 lunar samples. Second Lunar Science Conference (unpublished proceedings).

Suess H. E. (1949) Das Gleichgewicht $H_2 + HD \leftrightarrows HD + H_2O$ und die weitern Austauschgleichgewichte im System H_2, D_2, und H_2O. *Z. Naturforschung* **4a**, 328–332.

Proceedings of the Second Lunar Science Conference, Volume 2, pp. 1417–1420.
The M.I.T. Press, 1971.

Oxygen isotope fractionation in Apollo 12 rocks and soils

ROBERT N. CLAYTON,* NAOKI ONUMA,† and TOSHIKO K. MAYEDA
The Enrico Fermi Institute, University of Chicago, Chicago, Illinois 60637

(*Received 23 February 1971; accepted 22 March 1971*)

Abstract—Oxygen isotopic compositions of individual minerals and isotopic fractionations among coexisting minerals are remarkably similar for Apollo 11 and Apollo 12 crystalline rocks. Apollo 12 rocks show a somewhat lower plagioclase-ilmenite "isotopic temperature" of 1070°C, related to the late crystallization of ilmenite in these rocks.

OXYGEN ISOTOPE analyses have been carried out on separated minerals from seven crystalline rocks (12018, 12021, 12038, 12040, 12052, 12063, and 12064), on the olivine and matrix from two olivine vitrophyres (12008 and 12009), and on four soil samples (12033, 12037, 12057, 12070).

ANALYTICAL PROCEDURES

The procedures for mineral separation, oxygen extraction, and mass spectrometric analysis were the same as those used for the Apollo 11 samples (ONUMA *et al.*, 1970). Results of isotopic analyses are given in Table 1. Standard error in the mean of the duplicate measurements on each sample is estimated to be ± 0.07 per mil.

RESULTS AND DISCUSSION

As was the case with the Apollo 11 crystalline rocks, the seven Apollo 12 crystalline rocks are indistinguishable from one another in terms of the isotopic compositions (δ-values) of individual minerals (Table 1), and in terms of the mineral-pair isotopic fractionations (Δ-values) for mineral pairs (Table 2).

Comparison of isotopic fractionations in Apollo 11 rocks with those in Apollo 12 rocks shows a significant difference in mineral-pair fractionations involving olivine. Olivine was analyzed from only one Apollo 11 sample (10020), giving a δ-value of 5.14, which coincides with the values for Apollo 12 olivines. Texturally, the olivine in 10020 appears to be in reaction relation to clinopyroxene (DENCE *et al.*, 1970), and may not be in isotopic equilibrium with the other major phases.

There is also a small difference between Apollo 11 rocks and Apollo 12 rocks with respect to oxygen isotope fractionations involving ilmenite, the latter Δ-values being about 0.15 to 0.20 greater. This is consistent with the relative position of ilmenite in the crystallization sequences of the two sets of rocks, being early for Apollo 11 rocks (O'HARA *et al.*, 1970), and late for Apollo 12 rocks (BIGGAR *et al.*, 1971). Although there is obvious difficulty in attaching precise thermometric significance to isotopic fractionations between two minerals which crystallized over noncoincident

* Also Departments of Chemistry and Geophysical Sciences.
† Present Address: Department of Chemistry, University of Tokyo.

Table 1. Oxygen isotopic compositions* of Apollo 12 rocks and minerals.

<table>
<tr><th></th><th>Tr†</th><th>Pc</th><th>Pg</th><th>Cpx 3.36–3.47</th><th>Cpx 3.47–3.52</th><th>Pxf</th><th>Ol</th><th>Il</th><th>Mx</th><th>R</th></tr>
<tr><td>Rocks</td><td></td><td></td><td></td><td></td><td></td><td></td><td></td><td></td><td></td><td></td></tr>
<tr><td>12008,8</td><td></td><td></td><td></td><td></td><td></td><td></td><td>5.05</td><td></td><td>5.67</td><td></td></tr>
<tr><td>12009,26</td><td></td><td></td><td></td><td></td><td></td><td></td><td>5.20</td><td></td><td>5.81</td><td></td></tr>
<tr><td>12018,31</td><td></td><td>5.92</td><td></td><td>5.63</td><td>5.63</td><td></td><td>5.16</td><td>4.15</td><td></td><td></td></tr>
<tr><td>12021,28</td><td></td><td>5.96</td><td>5.76</td><td colspan="2">5.58</td><td></td><td></td><td>4.06</td><td></td><td></td></tr>
<tr><td>12038,26</td><td></td><td>6.19</td><td></td><td>5.68</td><td>5.66</td><td></td><td></td><td>4.08</td><td></td><td></td></tr>
<tr><td>12040,27</td><td></td><td>5.93</td><td>5.87</td><td colspan="2">5.61</td><td></td><td>5.09</td><td>4.10</td><td></td><td></td></tr>
<tr><td>12052,53</td><td></td><td>5.89</td><td></td><td>5.48</td><td>5.33</td><td></td><td></td><td>4.00</td><td></td><td></td></tr>
<tr><td>12063,56</td><td></td><td>5.98</td><td></td><td>5.64</td><td></td><td></td><td>5.24</td><td>4.22‡</td><td></td><td></td></tr>
<tr><td>12064,25</td><td>7.15</td><td>5.97</td><td></td><td>5.54</td><td>5.58</td><td>5.45</td><td></td><td>3.90</td><td></td><td></td></tr>
<tr><td>Soils</td><td></td><td></td><td></td><td></td><td></td><td></td><td></td><td></td><td></td><td></td></tr>
<tr><td>12033,55</td><td></td><td></td><td></td><td></td><td></td><td></td><td></td><td></td><td></td><td>5.90</td></tr>
<tr><td>12037,22</td><td></td><td></td><td></td><td></td><td></td><td></td><td></td><td></td><td></td><td>5.75</td></tr>
<tr><td>12057,74</td><td></td><td></td><td></td><td></td><td></td><td></td><td></td><td></td><td></td><td>5.72</td></tr>
<tr><td>12070,46</td><td></td><td></td><td></td><td></td><td></td><td></td><td></td><td></td><td></td><td>5.87</td></tr>
<tr><td>12070,46 (<0.3μ)</td><td></td><td></td><td></td><td></td><td></td><td></td><td></td><td></td><td></td><td>7.34</td></tr>
</table>

* δO^{18} defined as $\left(\frac{(O^{18}/O^{16})\text{ sample}}{(O^{18}/O^{16})\text{ standard}} - 1\right)$ 1000, where the standard is SMOW (Standard Mean Ocean Water)

† Abbreviations: Tr—tridymite; Pc—plagioclase; Pg—pigeonite; Cpx—calcium clinopyroxene; Pxf—pyroxferroite; Ol—olivine; Il—ilmenite; Mx—matrix; R—whole rock.

‡ Corrected by 0.21‰ for olivine and pyroxene contamination.

Table 2. Oxygen isotope fractionations* in Apollo 11 and Apollo 12 rocks.

	Apollo 11	Apollo 12
Pc–Cpx	0.45 ± 0.05	0.39 ± 0.07
Pc–Ol	1.04	0.78 ± 0.04
Pc–Il	1.76 ± 0.06	1.89 ± 0.10
Cpx–Ol	0.59	0.46 ± 0.04
Cpx–Il	1.31 ± 0.08	1.50 ± 0.07
Ol–Il	0.72	0.99 ± 0.01

* Fractionations are given as $\Delta = 1000 \ln \alpha$, where $\alpha_{AB} = \frac{(O^{18}/O^{16})_A}{(O^{18}/O^{16})_B}$ for two minerals A and B.

temperature ranges, straightforward application of the plagioclase-ilmenite thermometer (ONUMA *et al.*, 1970) gives mean isotopic temperatures of 1120°C for Apollo 11 rocks (both A and B types), and mean temperatures of 1070° for Apollo 12 rocks, in good agreement with the phase equilibrium data of BIGGAR *et al.* (1971).

BIGGAR *et al.* (1971) concluded that rocks 12038, 12018, and 12040 were derived from a common magma by progressive addition of cumulate olivine, and that 12064, 12021, and 12052 are similarly related by addition of cumulate pigeonite. The oxygen isotope data show no trends with respect to these groups of rocks.

The olivine vitrophyres, 12008 and 12009, appear to be simply rapidly crystallized equivalents of other more coarsely crystalline rocks. The olivine phenocrysts have the same isotopic compositions as the cumulate olivine in 12018 and 12040, and the matrix has the same composition as the whole-rock values estimated for other rocks from their mineral analyses.

In four cases the clinopyroxene fraction was separated on the basis of density into iron-rich and iron-poor portions; these were found to be isotopically indistinguishable. In rocks 12021 and 12040, pigeonite was separated by hand-picking, and in both cases was found to be enriched in O^{18} by about 0.2‰ relative to the Ca-clinopyroxene. Because of the separation procedure, the pigeonite analyzed is derived predominantly from the cores of large zoned crystals. The significance of the 0.2‰ difference is not clear. It may be the result of different isotopic fractionation factors for pyroxenes of different chemical composition, or may represent a departure from isotopic equilibrium resulting from either different positions in the crystallization sequence or to actual nonequilibrium crystallization. Rock 12064 also yielded an analyzable amount of pyroxferroite, which was found to have almost the same oxygen isotopic composition as the coexisting clinopyroxene. Tridymite from the same rock also appears to be in isotopic equilibrium, and has the same isotopic composition as cristobalite in Apollo 11 rock 10058 (ONUMA *et al.*, 1970).

In comparing δ-values of separated minerals and of soil samples, there appears to be a systematic difference between Apollo 11 and Apollo 12 rocks, the latter being lower in O^{18} by about 0.2‰. This difference, if real, is certainly very small, and reflects the small oxygen isotope effects associated with fractional melting or crystallization. This suggests that the range of about two permil found for terrestrial basalts from diverse localities (TAYLOR, 1968; GARLICK, 1966; ANDERSON *et al.*, 1971), previously considered a small range, is probably larger than can be accounted for only by processes at igneous temperatures, and reflects exchange or contamination with materials in the crust or upper mantle which have undergone low-temperature processes. Such low-temperature chemical processes are probably absent on the moon. The uniformity of oxygen isotopic composition of rocks and soil at the two sites analyzed and of rocks of different ages (TAYLOR and EPSTEIN, 1970) strengthens the assumption that this composition is typical of the lunar surface, and thus further strengthens the case against derivation of tektites or eucrites from the moon (ONUMA *et al.*, 1970; EPSTEIN *et al.*, 1970; FRIEDMAN *et al.*, 1970).

The four soil samples analyzed are almost identical in isotopic composition and are enriched in O^{18} by about 0.2‰ relative to the average whole-rock compositions of the crystalline rocks, as was also the case for the Apollo 11 soils. A fine-grained fraction of soil 12070 was separated by flotation in acetone (grain-size estimated to be <0.3 microns) and was found to have $\delta O^{18} = 7.34$‰. A sample similarly separated from Apollo 11 soil 10084 has a $\delta O^{18} = 7.60$‰. A similar enrichment in the very fine fraction of 10084 soil was noted by EPSTEIN and TAYLOR (1971), who attribute the the effect to a depletion of the light isotope during vaporization resulting from bombardment by high-energy particles. Other reasonable possibilities are volatilization resulting from heating in impact events, or addition of a fine-grained extra-lunar component enriched in O^{18} relative to the lunar rocks.

Our analysis of sample 12033 is identical with that reported for this sample by EPSTEIN and TAYLOR (1971). However, they report a $\delta\, O^{18}$ of 6.26 for soil sample 12070, whereas, our value for 12070,46 is the same as our value for 12033 ($\delta\, O^{18}$ =

5.87). On the basis of experiences to date with lunar soil samples, it would be surprising if this difference were due to sample heterogeneity unless the samples analyzed were of different grain size ranges.

CONCLUSIONS

Oxygen isotope variations in lunar igneous rocks of the mare regions are exceedingly small indicating the absence of large-scale, low-temperature processes. Isotopic fractionations within rocks correspond to equilibrium at temperatures near the solidus, independent of grain size or texture. This implies quenching of the isotopic compositions on solidification as a result of rapid cooling and the absence of water.

The main components of the regolith are derived from the crystalline rocks without change in oxygen isotopic composition, but some process has produced O^{18}-enriched material in the finest size-fractions.

Acknowledgments—This work was supported in part by NASA contract NAS–9–7888.

REFERENCES

ANDERSON A. T., CLAYTON R. N., and MAYEDA T. K. (1971) Oxygen isotope thermometry of mafic igneous rocks. *J. Geol.* (In press).

BIGGAR G. M., O'HARA M. J., and PECKETT A. (1971) Origin, eruption, and crystallization of protohypersthene basalts from Ocean of Storms. Second Lunar Science Conference (unpublished proceedings).

DENCE M. R., DOUGLAS J. A. V., PLANT A. G., and TRAILL R. J. (1970) Petrology, mineralogy, and deformation of Apollo 11 samples. *Proc. Apollo 11 Lunar Sci. Conf., Geochim. Cosmochim. Acta* Suppl. 1, Vol. 1, pp. 315–340. Pergamon.

EPSTEIN S. and TAYLOR H. P., JR. (1970) $^{18}O/^{16}O$, $^{30}Si/^{28}Si$, D/H and $^{13}C/^{12}C$ studies of lunar rocks and minerals. *Science* **167**, 533–535.

EPSTEIN S. and TAYLOR H. P., JR. (1971) O^{18}/O^{16}, Si^{30}/Si^{28}, D/H and C^{13}/C^{12} ratios in lunar samples. Second Lunar Science Conference (unpublished proceedings).

FRIEDMAN I., O'NEIL J. R., ADAMI L. H., GLEASON J. D., and HARDCASTLE K. (1970) Water, hydrogen, deuterium, carbon, carbon-13, and oxygen-18 content of selected lunar materials. *Science* **167**, 538–540.

GARLICK G. D. (1966) Oxygen isotope fractionation in igneous rocks. *Earth Planet. Sci. Lett.* **1**, 361–368.

O'HARA M. J., BIGGAR G. M., RICHARDSON S. W., JAMIESON B. G., and FORD C. E. (1970) The nature of seas, mascons, and the lunar interior in the light of experimental studies. *Proc. Apollo 11 Lunar Sci. Conf. Geochim. Cosmochim. Acta*, Suppl. 1, Vol. 1, pp. 695–710. Pergamon.

ONUMA N., CLAYTON R. N., and MAYEDA T. K. (1970) Oxygen isotope fractionation between minerals and an estimate of the temperature of formation. *Science* **167**, 536–538.

TAYLOR H. P., JR. (1968) The oxygen isotope geochemistry of igneous rocks. *Contrib. Mineral. Petrol.* **19**, 1–71.

TAYLOR H. P., JR. and EPSTEIN S. (1970) Oxygen and silicon isotope ratios of lunar rock 12013. *Earth Planet. Sci. Lett.* **9**, 208–210.

Proceedings of the Second Lunar Science Conference, Volume 2, pp. 1421–1441.
The M.I.T. Press, 1971.

O^{18}/O^{16}, Si^{30}/Si^{28}, D/H, and C^{13}/C^{12} ratios in lunar samples*

SAMUEL EPSTEIN and HUGH P. TAYLOR, JR.
Division of Geological and Planetary Sciences,
California Institute of Technology, Pasadena, California 91109

(*Received 27 February 1971; accepted in revised form 5 April 1971*)

Abstract—The total range in O^{18}/O^{16} in all analyzed whole-rock Apollo 11 and Apollo 12 samples is from $\delta O^{18} = 5.53$ to 6.60; the Si^{30}/Si^{28} ratio varies from $\delta Si^{30} = -0.49$ to $+0.31$. The lunar igneous rocks (ilmenite basalts and microgabbros) show a smaller variation, $\delta O^{18} = 5.53$ to 5.90 and $\delta Si^{30} = -0.28$ to 0.0, with lunar microbreccias and fines being enriched in both δO^{18} and δSi^{30} relative to igneous rocks. The whole-rock samples of the moon thus have an exceedingly homogeneous oxygen and silicon isotopic composition. The surfaces of grains of the lunar microbreccias and soils are, however, highly enriched in Si^{30} and O^{18}. The maximum measured δ-values of the surface oxygen and silicon are about +50 and +25 per mil, respectively, heavier than the bulk soils and rocks from the moon. These grain surfaces also appear to be depleted in total oxygen relative to total silicon. These O^{18} and Si^{30} enrichments fit a simple square-root-of-the-mass fractional vaporization model, and they also correlate very well with the concentrations of solar wind hydrogen in the lunar samples. The isotopic enrichments are therefore probably (1) the result of particle bombardment of the lunar grain surfaces by micrometeorites, heavy ions, and/or solar protons, or (2) the result of condensation of a Si^{30} and O^{18} enriched residue onto the grain surfaces. During vaporization due to meteorite bombardment the lighter isotopes may have preferentially escaped from the moon.

A maximum value for the deuterium concentration in lunar hydrogen gas (almost wholly of solar wind origin) has been estimated to be about 5 ppm. Taking into account the contribution of deuterium formed by cosmic-ray spallation processes the D/H ratio of the solar wind therefore probably is no larger than 3×10^{-6}. The δC^{13} values in the Apollo 12 breccia and soil samples are not as high as in the Apollo 11 samples, but are still higher than in terrestrial reduced carbon. This C^{13}-rich carbon is strongly concentrated in the lunar fines and breccias. Thus, the heavy isotopes of carbon, silicon, and oxygen are *all* enriched in the lunar fines, apparently most strongly in the surface coatings of the grains and/or the finest size-fraction.

OXYGEN AND SILICON ISOTOPE RELATIONSHIPS

THE NEW δO^{18} AND δSi^{30} data we have obtained on various Apollo 11 and Apollo 12 lunar samples are given in Table 1 and Figs. 1 and 2. These data supplement those already published on Apollo 11 rocks (TAYLOR and EPSTEIN, 1970a; EPSTEIN and TAYLOR, 1970b; O'NEIL and ADAMI, 1970; ONUMA *et al.*, 1970) and on the unusual, high-SiO_2 lunar rock 12013 (TAYLOR and EPSTEIN, 1970b). The most striking feature of all these data is that the Apollo 12 samples are essentially isotopically indistinguishable from the Apollo 11 rocks. The total range of O^{18}/O^{16} in *all* whole-rock lunar samples is from $\delta O^{18} = 5.53$ to 6.60, and for δSi^{30}, from -0.49 to $+0.31$. The lunar *igneous* rocks show much smaller variations, $\delta O^{18} = 5.53$ to 5.90, and δSi^{30} -0.28 to 0.0. The lunar microbreccias and fines are distinctly enriched in both δO^{18}

* Contribution number 1985 of the Division of Geological and Planetary Sciences, California Institute of Technology, Pasadena, California 91109.

Table 1. O^{18}/O^{16} and Si^{30}/Si^{28} ratios of some Apollo 11 and Apollo 12 rocks and minerals* (given as δO^{18} relative to SMOW and δSi^{30} relative to a standard quartz from the Rose Quartz pegmatite, Pala, California).

Sample		Whole rock	Plag	Px	Oliv	Crist	Ilm
10087	coarse fines	6.05, *+0.15*					
10084	(partially fluorinated)	5.88					
10084	extreme fines (acetone floats)	6.79, *+0.39*					
10084	fused fines	6.29					
10061A	fused breccia	6.60, *+0.29*					
10061B	breccia	6.39, *+0.30*					
10072	fused basalt	5.58, *−0.21*					
10068	fused breccia	6.35, *+0.26*					
12013	"granitic" breccia, #46A†	5.97, *−0.04*					
12070	fines	6.26, *+0.11*					
12033	trench fines	5.96, *+0.01*					
12034	trench breccia (interior)	5.98, *−0.25*					
12033	breccia	6.12, *+0.07*					
12035	troctolite	5.53‡	6.08, *+0.02*	5.69, *−0.23*	5.35		3.91
12064	basalt	5.82‡	6.19, *+0.11*	5.74, *−0.10*		6.73, *+0.31*	
12053	basalt	5.90, *−0.18*	6.07, *−0.01*	5.72, *−0.30*			
12021	basalt		6.05, *−0.05*	5.70, *−0.25*			
12002	basalt	5.85, *−0.23*					

* The δSi^{30} values are given in italics. All samples were analyzed at least twice; typically, the average deviation of each analysis from the mean is ±0.08 per mil for O^{18}/O^{16} and ±0.05 per mil for Si^{30}/Si^{28}. The notation "fused" indicates that the sample analyzed was the quenched glass remaining after melting in vacuum at 1200°–1300°C and removal of the H_2 and CO. Plag = plagioclase; Px = clinopyroxene; Oliv = olivine; Crist = cristobalite + tridymite; Ilm = ilmenite.

† See TAYLOR and EPSTEIN (1970b) for other analyses of rock 12013.

‡ Whole rock δO^{18} value calculated by material balance from δO^{18} of minerals.

and δSi^{30} relative to the igneous rocks, but this effect is more pronounced for the Apollo 11 samples than for Apollo 12.

Stable isotope measurements have thus now been made on lunar rocks having at least 3 distinctly different Rb–Sr ages, namely Apollo 12 basalts (~3.3 b.y.), Apollo 11 basalts (~3.7 b.y.), and rock 12013 (~4.0 b.y.) (ALBEE *et al.*, 1970). Inasmuch as these are geographically widely separated on the moon (12013 representing ejecta from Copernicus?), and as no appreciable δO^{18} differences have been observed in any of the rocks, it seems logical to conclude that as fas as O^{18}/O^{16} and Si^{30}/Si^{28} variations are concerned, the crust of the moon is isotopically quite homogeneous. This would be expected, for example, if it initially went through a largely molten stage and if all magmatic differentiation occurred at very high temperatures under low P_{H_2O} conditions.

Basalts and microgabbros

Some of the Apollo 12 basalts are slightly richer in O^{18} than the average Apollo 11 basalts (Table 1, 12053, 12002) but this is apparently due to their somewhat lower ilmenite contents, because the pyroxenes and plagioclases from both the Ocean of Storms and the Sea of Tranquillity are practically identical in δO^{18} within experimental error. In fact, if anything the δO^{18} values of the individual Apollo 12 minerals tend to lie at the low-O^{18} end of the isotopic spectrum shown by the Apollo 11 basalts, and they form a more coherent grouping than do the Apollo 11 samples. They correspond most closely in O^{18} to the Apollo 11 samples that contain the most anorthite-rich plagioclase; this is interesting in view of the fact that apparently none of the Apollo

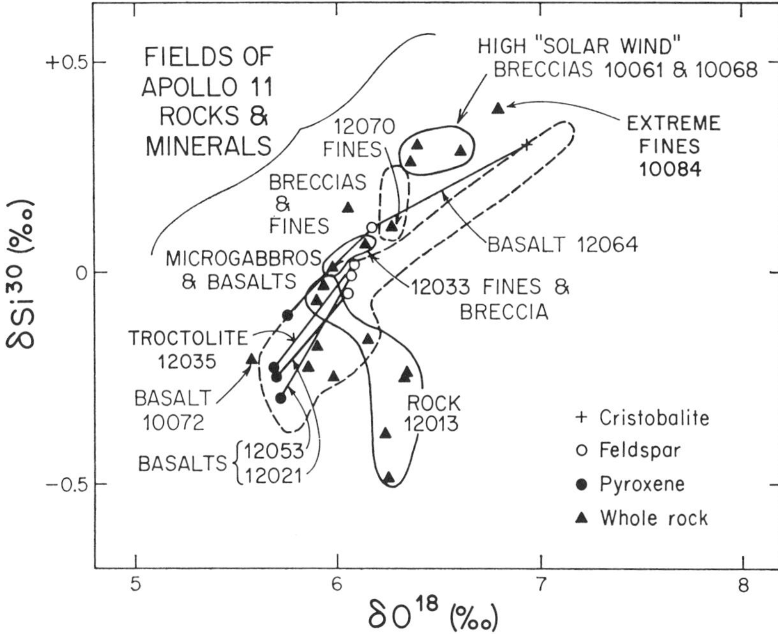

Fig. 1. Plot of δSi^{30} *vs.* δO^{18} for lunar samples analyzed in the present work and by TAYLOR and EPSTEIN (1970 b). Also shown are the 2 fields for lunar igneous rocks and for lunar breccias and fines (outlined by dashed lines) that encompass the data-points obtained previously on Apollo 11 samples (EPSTEIN and TAYLOR, 1970 b). Note that the long extension of the Apollo 11 field toward the upper right-hand corner of the diagram is due solely to a single analysis of cristobalite from 10044.

12 basalts contain as sodic a plagioclase as 2 high-O^{18} Apollo 11 plagioclases (An_{78}) analyzed by TAYLOR and EPSTEIN (1970a). Our measured plagioclase-pyroxene fractionations range from 0.23 to 0.47 (Apollo 11) and 0.35 to 0.41 (Apollo 12); this is excellent agreement considering the experimental error of ± 0.15.

The coarsest-grained Apollo 12 sample available to us was a troctolitic microgabbro, 12035. This olivine-rich rock has an average grain size slightly less than 1 mm and contains An_{90-93} plagioclase. It was of interest to see whether this rock (a cumulate?) might perhaps have undergone some measurable post-crystallization O^{18} exchange during cooling, analogous to the situation observed in terrestrial gabbros (see TAYLOR, 1968). Such an effect, if present, would best be seen in the plagioclase–ilmenite fractionation. However, as shown in Table 1, the ilmenite from this rock ($\delta = 3.91$) is isotopically identical to the Apollo 11 ilmenites (TAYLOR and EPSTEIN, 1970a). Utilizing the same geothermometric curve applied to the Apollo 11 samples, 12035 gives a calculated "temperature" of formation of 1220°C. As previously pointed out for the Apollo 11 samples, this temperature may be in error by as much as 100°C because of uncertainties in constructing the hypothetical plagioclase–ilmenite geothermometric curve. In view of the similar isotopic "temperatures" obtained on both the Apollo 11 and Apollo 12 igneous rocks, and because of the massive amount of experimental evidence obtained by various workers that these basaltic liquids crystallized over the

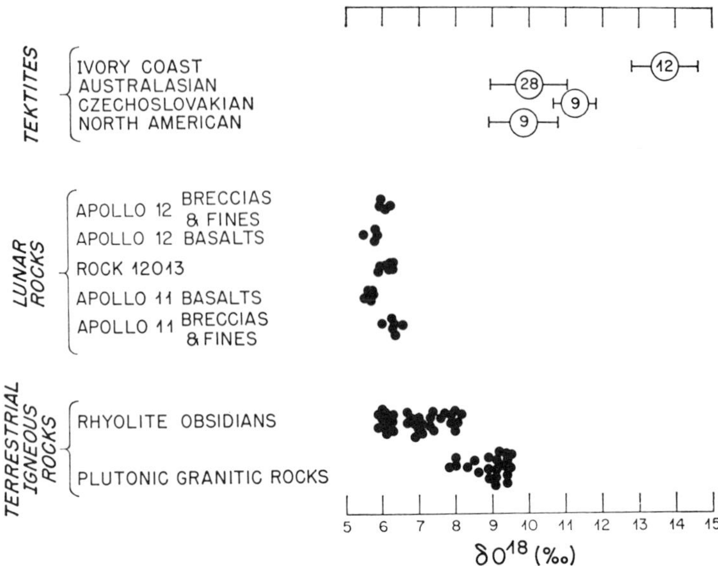

Fig. 2. Comparison of δO^{18} values of samples from rock 12013,10 (TAYLOR and EPSTEIN, 1970 b) and other Apollo 12 rocks analyzed in the present study with δO^{18} values previously obtained for Apollo 11 basalts (and microgabbros), breccias, and fines (TAYLOR and EPSTEIN, 1970a), terrestrial rhyolite obsidians (TAYLOR, 1968), terrestrial plutonic granitic rocks (TAYLOR, 1968; TAYLOR and EPSTEIN, 1962), and tektites (TAYLOR and EPSTEIN, 1966; TAYLOR and EPSTEIN, 1969). The plutonic granitic rocks include tonalites, granodiorites, quartz monzonites, and granites from batholiths and large stocks, excluding those that have exchanged with heated ground waters. The rhyolite obsidians with low δO^{18} values ($< +7.0$) are all from oceanic islands or the west coast of North America. For the tektites, the encircled number represents the number of different specimens analyzed.

temperature interval 1100°–1200°C, it is probably best to assume that our measured plagioclase–ilmenite fractionations correspond to equilibrium at such temperatures, and to use this as a fixed reference point in constructing a more accurate plagioclase–ilmenite geothermometer. This is the course we shall hereafter take, and *for such purposes* the following geothermometer will be used:

$$1000 \ln \alpha_{\text{plagioclase}(\text{An}_{90})-\text{ilmenite}} = 0.62 + 3.22\,(10^6 \text{T}^{-2})$$

Utilization of the above equation is equivalent to lowering our previously calculated plagioclase-ilmenite temperatures of the lunar basalts by about 50° to 70°C; it represents only a very minor change in the intercept value from 0.72 to 0.62, leaving the slope unchanged. Sample 12035 would then exhibit a temperature of 1170°C.

All the inferences made previously about the Apollo 11 basalts by us (TAYLOR and EPSTEIN, 1970a) therefore also apply with equal force to the Apollo 12 basalts, namely, (1) Crystallization and "freezing-in" of the oxygen isotope distribution patterns in the various coexisting minerals took place at very high temperatures (>1000°C); (2) The lunar ilmenite basalts are essentially identical in O^{18}/O^{16} to ordinary chondritic meteorites and to fresh, uncontaminated, unaltered basalts on earth; (3) The lunar basalts are distinctly different in δO^{18} from most other types of meteorites. Their pyroxenes and plagioclases are about 1.5 per mil higher in O^{18} than the corresponding minerals in the basaltic achondrites, hypersthene achondrites, and mesosiderites. They are also markedly richer in O^{18} than most type III carbonaceous chondrites, and lower in O^{18} than all ureilites and types I and II carbonaceous chondrites (TAYLOR *et al.*, 1965).

Given the geographic sampling coverage now available on the moon, it is probably safe to conclude that basaltic achondrites almost certainly cannot come from the moon; they must have come from a parent body in the solar system that underwent a volcanic history somewhat similar to that of the moon but one whose overall oxygen isotopic composition is at least 1 per mil lower than the moon. In view of the appreciable O^{18} variations found among various classes of meteorites, it can hardly be a coincidence that the moon, the ordinary chondritic meteorites, and the upper mantle of the earth (as judged by analyses of ultramafic rocks, basalts, and gabbros) all are (a) very homogeneous in O^{18}/O^{16}, and (b) essentially indistinguishable from one another in O^{18}/O^{16}. This strongly implies a genetic connection, probably one which goes back to the time of formation of the solar system.

Rock 12013 (*granitic microbreccia*)

Rock 12013, as the first lunar rock known to have a high SiO_2 content and to contain appreciable K feldspar, is of critical importance in interpreting the petrologic history of the Moon. The rock appears to have been a microbreccia that was injected and/or metasomatized by some granitic or rhyolitic material, at least in part in the form of a silicate melt. The isotopic data obtained on rock 12013 have already been reported on (TAYLOR and EPSTEIN, 1970b), but we include here a new analysis of another fragment of this rock ($\delta O^{18} = 5.97$ and $\delta Si^{30} = -0.04$, see Table 1). All of the presently available data are plotted on Fig. 1. These data again emphasize the oxygen isotopic homogeneity of the moon, particularly when it is noted that sample 12013 (as well as 12033 and 12034) may in part represent Fra Mauro formation (Imbrian ejecta) thrown out to the Apollo 12 site by the Copernicus impact event.

Even though rock 12013 is highly differentiated chemically, and portions of it approach a terrestrial rhyolite or granite in composition, it shows no appreciable difference in δO^{18} relative to the other Apollo 11 and Apollo 12 rocks (see TAYLOR and EPSTEIN, 1970b, and Figs. 1 and 2). In this respect it differs strongly from terrestrial plutonic granitic igneous rocks, which are typically 2 to 4 per mil richer in O^{18}. The δO^{18} values of rock 12013 do, however, closely resemble those of terrestrial rhyolite obsidians from areas where contamination with high-O^{18} sialic crust is known to be negligible (such as on oceanic islands, TAYLOR, 1968). The data are thus compatible with original formation of the high-SiO_2 portion of 12013 by magmatic differentiation in a volcanic or sub-volcanic environment from basaltic magma having a δO^{18} similar to that of the Apollo 11 and Apollo 12 basalts.

Inasmuch as portions of rock 12013 have SiO_2 contents in the same range as some tektites, it is natural to entertain the possibility that analogous material on the moon might represent the parent materials of tektites. This has been suggested, for example, by O'KEEFE (1970). However, the δO^{18} values of 12013 are totally outside the range of values for tektites so this hypothesis must be rejected (Fig. 2). Thus, even though high-SiO_2 rocks superficially resembling some tektites in chemical composition are now known to exist on the moon, they cannot be the parent materials of tektites.

If more granitic samples are obtained from the moon and if they have δO^{18} values similar to 12013, this could have very important implications regarding the early development of the granitic or sialic crust of the earth. It might suggest that the early terrestrial sialic crust had a δO^{18} of +6 to +7, whereas at present it is at least +8 to +11. This might imply that plutonic granitic rocks on earth ultimately have all been largely derived from or have incorporated considerable O^{18}-rich sedimentary material; they may be high in O^{18} *only* because there have been processes operating here such as chemical weathering and biologic and authigenic precipitation of silicates and carbonates. Inasmuch as these processes occur at low temperatures where O^{18} fractionations are very large, they can produce marked O^{18} enrichments.

The δSi^{30} data on rock 12013 are more difficult to interpret. Some portions of 12013 are isotopically "normal" as compared with most lunar rocks, but other parts have apparent δSi^{30} values somewhat lower than any yet obtained from the moon. Unfortunately, portions of sample 12013 are known to have high phosphorus contents, and P can cause anomalously low δSi^{30} values because PF_3^+ and POF_2^+ fall in the same part of the mass spectrum as SiF_3^+. We have demonstrated this isotopic effect on terrestrial mineral separates contaminated with apatite. Because of the small sizes of the individual samples analyzed, and to minimize contamination, it was not feasible to utilize an aqueous HCl treatment to dissolve any phosphates present prior to fluorination. Given these analytical problems, all we can conclude at present is that rock 12013 is essentially indistinguishable from the other lunar rocks in δSi^{30} (i.e. to within $\pm 0.3‰$).

Breccias and fines

The lunar soils and breccias at both the Apollo 11 and Apollo 12 sites are clearly enriched in δO^{18} and δSi^{30} relative to the local basalts from which they are presumably largely derived (see Table 1 and Figs. 1 and 2). The slight enrichment in δO^{18} shown by rock 12013 could, for example, have arisen because this rock is in part made up of a modified "breccia" component. There is a good correlation between the solar wind hydrogen contents of the soils, breccias, and basalts and the degree of enrichment in δO^{18} and δSi^{30}. The breccia with the highest H_2 concentration (10061, 49.8 μmole H_2/g, Table 4) also has the highest δO^{18} (+6.60). The soil sample with the lowest H_2 concentration (12033, 0.95 μmole H_2/g, Table 2) has the lowest δO^{18} (+5.96). The soil samples 12070 and 10084 (see Table 2 and EPSTEIN and TAYLOR, 1970a) have intermediate H_2 contents and also have intermediate δO^{18} and δSi^{30} values. Lowest of all in both H_2 content, δO^{18}, and δSi^{30} are, of course, the lunar basalts themselves. With respect just to δSi^{30}, the latter statement does not apply to rock 12013, possibly for the reasons outlined above. Similarly, the apparently anomalously low δSi^{30} of the 12034 breccia may be due to the same effect; this sample also has a very high phosphorus content.

These correlations all strongly suggest that the observed enrichment of O^{18} and Si^{30} in the soils and breccias is a result of the process of solar wind or other

Table 2. Isotopic and concentration data for hydrogen, carbon, and rare gases extracted from samples 12070, 12033, and 12042 from the Ocean of Storms.

		Hydrogen			Carbon		Water			Rare gas	
Sample	Fraction	Conc. μmole/g	δD ‰	D ppm	Conc. ppm	$\delta C^{13}/C^{12}$ ‰	Conc. μmole/g	δD ‰	D ppm	Conc. cc/g	H_2/rare gas
12070	A	—	—	—			7.8	−207	126		
fines	B	5.6	−841	25						0.027	4.6
1.00 g	C_1	9.6	−835	26	145	+4.7				0.055	3.9
	C_2	3.7	−567	68			4.0	−292	111	0.030	2.7
	D	—	—	—							
	Total	18.9	−784	34	145	+4.7	11.8	−236	120	0.112	3.7
12033	A										
trench	B	0.95	−523	75	30	−3.6	7.5	−97	142	0.007	~3
fines	C										
1.698 g	D	—									
	Total	0.95	−523	75	30	−3.6	7.5	−97	142	0.007	~3
12042	A	—	—	—			13.5	−146	134	0.062	3.1
fines	B	8.5	−815	29	29	−2.6				0.05	~1.3
0.972 g	C_1	3.0	−735	42							
	C_2	8.7*	−572*	67*							
	D^{2}	—	—	—	96	+11.2					
	Total	20.2	−698	47	125	+8.0	13.5	−146	134	0.112	2.3

* Probably contaminated by hydrogen from water fraction.

particle bombardment. A reasonable mechanism whereby this might be accomplished is the fractional vaporization of oxygen and silicon from the surfaces of the glass and mineral grains in the soils as they are bombarded by solar protons, or probably even more important, by heavier ions and micrometeorites. Any such fractional distillation would be expected to favor removal of Si^{28} and O^{16} relative to Si^{30} and O^{18}. We may note also that the δC^{13} values of carbon and δS^{34} values of sulfur in the breccias and fines are also abnormally high compared to the corresponding values in the lunar basalts (KAPLAN *et al.*, 1970; EPSTEIN and TAYLOR, 1970b). Inasmuch as fractional vaporization would always favor preferential removal of the light isotope, the isotopic data on carbon and sulfur also tend to support the proposed mechanism.

Particle bombardment should be more pronounced in the finest size-fraction of the soil, because this material has the highest surface area per unit weight and also has the highest solar wind rare gas content. Therefore, we analyzed that portion of the 10084 fines that remained in suspension after several minutes settling in acetone. As can be seen in Fig. 1, these "extreme fines" are very stongly enriched in both O^{18} and Si^{30}, in agreement with the proposed mechanism.

Large O^{18} and Si^{30} enrichments at the grain surfaces of the lunar fines and breccias

To investigate the proposed mechanism of O^{18} and Si^{30} enrichment in more detail we have undertaken a number of experiments. An extensive series of partial oxygen and SiF_4 extractions were performed on 4 different aliquots of the 10084 fines, as well as on the 10068 breccia, the 10061 breccia, and a breccia fragment from the 12033 trench fines. The results are shown in Fig. 3. In each experiment, about 0.5–1.0 g of sample was reacted with a large excess of fluorine for periods of 0.3 to 10 hours at various temperatures from 80° to 350°C. It was hoped that the initial O_2 and SiF_4

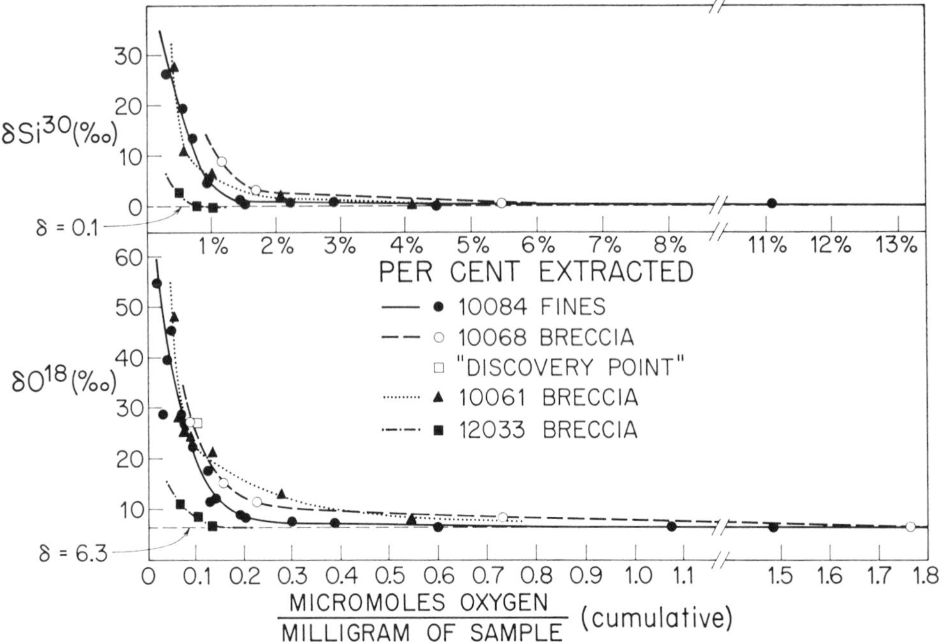

Fig. 3. Graph showing the variation of δO^{18} and δSi^{30} in the gases extracted from 10084 fines and 10068 breccia, 10061 breccia, and 12033 breccia by the various partial fluorination experiments described in the text. The δSi^{30} and δO^{18} values are *both* plotted against the cumulative μmoles/g of oxygen extracted during the fluorination experiments. The "per cent extracted" refers to the per cent of oxygen extracted relative to that present in the bulk sample. The δSi^{30} values of the initial 2 or 3 fractions extracted from 10084 and 10068 probably should be lowered by about 4 to 9 per mil (see text).

gas fractions released during these fluorination "stripping" experiments would come largely from the surfaces of the grains in the lunar fines and/or from the very finest-sized particles.

The experiments on the 10061 and 12033 breccias were done after all the other data plotted in Fig. 3 had been accumulated. These later experiments were specifically designed (1) to improve the accuracy of the δSi^{30} analytical data (see below) and (2) to see if any isotopic differences exist between a breccia with a very high solar wind H_2 content (10061) and one with very low solar wind H_2 content (12033). As shown in Fig. 3, the earliest fractions removed during fluorination of 10084, 10068, and 10061 are *enormously* enriched in O^{18} and Si^{30}, by amounts of about 50 and 25 per mil, respectively. There is an exponential decrease in the δO^{18} and δSi^{30} values with each successive cut, however, and after approximately 2% of the oxygen in the breccias and fines has been removed the isotopic compositions approach the constant values appropriate for the bulk samples. Note that the data points for the 10061 and 10068 breccias, which contain almost twice as much solar wind H_2 as does the 10084 soil, also plot consistently at higher values of δO^{18} and δSi^{30} than the data points obtained for the soil. Also note that 12033, which contains a factor of 40 to 50 less solar wind H_2 than either 10068 or 10061 also shows only a very tiny δO^{18} and δSi^{30} effect.

We knew from previous experience that no significant O^{18} fractionations accompany partial fluorinations of the type described above. However, no previous data were available on δSi^{30} effects in such a situation, so a series of identical experiments to those shown in Fig. 3 were undertaken on fresh, powdered basalt glass from the 1970 Kilauea eruption on Hawaii (see Table 3). Inasmuch as this glass is almost certainly homogeneous in δSi^{30} and δO^{18}, any isotope effects observed must be due to the analytical procedure. As expected, no significant oxygen isotope effects were observed in the initial cuts, which in fact tended to be 1 to 2 per mil *lower* in δO^{18} than the bulk glass, an effect in the opposite direction to that found on the lunar samples. However, significant Si^{30} enrichments were observed in the early sets of experiments done utilizing *exactly* the same procedure applied to the 10068 and 10084 samples shown in Fig. 3; the observed Si^{30} enrichments in the basalt glass are, however, *much* smaller than those found in the lunar samples. This suggests that a correction should be applied to the δSi^{30} values for 10068 and 10084 shown in Fig. 3, in order to subtract

Table 3. δO^{18} and δSi^{30} data obtained by successive partial fluorinations of lunar samples 10061B and 12033, and a Hawaiian basalt glass.*

Sample	Cumulative μmoles O_2 / Sample weight (mg)	δO^{18}	μmoles O_2	δSi^{30}	μmoles SiF_4	Ratio O_2/SiF_4
10061B breccia (0.950 g)						
a. 30 min./80°C	0.055	+47.9	52	+27.3	53	0.98
b. 30 min./95°C	0.063	+27.9	8 }	+11.3	23	0.83
c. 40 min./120°C	0.075	+25.0	11 }			
d. 30 min./165°C	0.089	+24.4	14 }	+6.5	57	1.04
e. 30 min./200°C	0.137	+21.1	45 }			
f. 30 min./240°C	0.279	+13.00	135	+2.16	108	1.25
g. 24 min./285°C	0.548	+8.06	256	+0.33	186	1.38
Whole rock		+6.39	412	+0.30	232	1.78
12033 breccia (0.775 g)						
a. 30 min./85°C	0.068	+11.1	52	+3.13	39	1.33
b. 60 min./130°C	0.105	+8.6	29	+0.19	26	1.12
c. 20 min./170°C	0.136	+6.9	17	+0.02	17	1.41
Whole rock		+6.12	601	+0.07	318	1.89
Hawaiian basalt glass, December 8, 1970 Kilauea eruption						
1–a. 60 min./150°C	0.067	+4.20	33.5	+9.64	6	5.6
2–a. 60 min./130°C	0.028	+7.23	13.5	+9.10	4.5	3.0
3–a. 30 min./105°C	0.158	+4.41	457	+0.15	278	1.64
4–a. 30 min./105°C	0.162	+4.64	431	+0.14	—	—
5–a. 20 min./70°C	0.076	+4.32	214	+0.08	126	1.70
6–a. 30 min./110°C	0.066	+4.15	135	+0.87	88	1.53
b. 30 min./135°C	0.093	+7.04	56	+0.33	36	1.56
c. 30 min./160°C	0.108	+6.08	29	+0.94	21	1.38

* Partial fluorinations were all done in sequence, (a) first, (b) second, (c) third, etc. Experiments (1) and (2) on the Hawaiian basalt were carried out using the low-temperature transfer of SiF_4 (see text); also, the samples in these 2 experiments were considerably coarser-grained than in (3, 4, 5), and (6). Experiments (3, 4, 5), and (6) on the Hawaiian basalt, and all experiments on 10061 B and 12033 were done using temperatures of 400°–450°C to transfer the SiF_4. Note that in the Hawaiian basalt experiments 6 separate samples (1–3 g each) were analyzed; generally, only the initial fluorination cut (a) was made, for direct comparison with the maximum isotopic effects observed in the lunar sample.

out that portion of the silicon isotope effect which is due to fractionation during the experimental procedure. The heaviest δSi^{30} values for 10068 and 10084 shown in Figs. 3 and 4 probably should be reduced by 8 to 9 per mil and the intermediate valùes by about 3 to 4 per mil.

Before carrying out any further partial fluorinations of lunar samples, we attempted to isolate and better understand this δSi^{30} effect observed in the basalt glass. It soon became clear that the analytical problem was simply that we were not getting 100% transfer of the SiF_4 out of the Ni reaction vessels at the low temperatures (85° to 150°C) used in the partial fluorination experiments. This can be seen by comparing the very high O_2/SiF_4 ratios obtained in these early experiments with the "normal" values obtained when SiF_4 transfer was carried out at temperatures of 400°–450°C (see Table 3). Apparently, isotopically light SiF_4 (due to a collision frequency fractionation?) was being preferentially adsorbed in the reaction vessel. In normal-sized samples this isotopic effect is negligible, but in the small samples obtained during the partial fluorinations it can be important. In any event, in all subsequent experiments of this type, after the partial fluorination reaction was complete and the O_2 had been collected, the temperature of the reaction vessel was raised to 400°–450°C and under these conditions we apparently get (1) complete transfer of SiF_4 and (2) a negligible δSi^{30} isotope effect (see Table 3).

For the above reasons, we believe that our most analytically precise δSi^{30} data are those obtained on 10061 and 12033, shown on Figs. 3 and 4. However, except for possible atmospheric H_2O contamination on the first fluorination cuts obtained on 10084 and 10068 (see below) the δO^{18} data for all the samples should be equally good. Also, again possibly because of incomplete SiF_4 transfer, the only precise data on the relative amounts of SiF_4 and O_2 produced in each "stripping" experiment are those obtained on 10061, 12033, and the later analyses of the Hawaiian basalt glass; therefore, only these lunar sample data are given in Table 3. We should point out that in all these later experiments the samples were *never* exposed to the laboratory atmosphere. The samples were opened, crushed, and loaded into the Ni reaction vessels in an atmosphere of dry nitrogen. Thus, whereas the first partial fluorination fractions of O_2 obtained in the earlier experiments on 10084 and 10068 may have suffered from a slight contamination by adsorbed atmospheric moisture, this is not a problem in the case of 10061 and 12033. This effect would cause the δO^{18} values of the former samples to be too low; these particular data points are therefore somewhat suspect, as indicated by the arrows on Fig. 4.

Even though several different samples of 10084 fines were analyzed and in spite of the fact that no attempt was made to duplicate the reaction times and temperatures on each successive cut, *all* the data for 10084 form a relatively smooth curve when plotted against per cent oxygen extracted (Fig. 3). In one case, an aliquot of the 10084 residue from a partial fluorination experiment was itself analyzed for δO^{18}, and as would be expected by material balance, its δO^{18} value (+5.88) is lower than that of the original 10084 sample (Table 1) and much more similar to the δO^{18} values of the local Apollo 11 basalts. Note that just as would be predicted on a micrometeorite or solar wind bombardment model, the samples with the highest solar wind H_2 content (10068 and 10061B) exhibit consistently higher δO^{18} and δSi^{30} ratios than 10084,

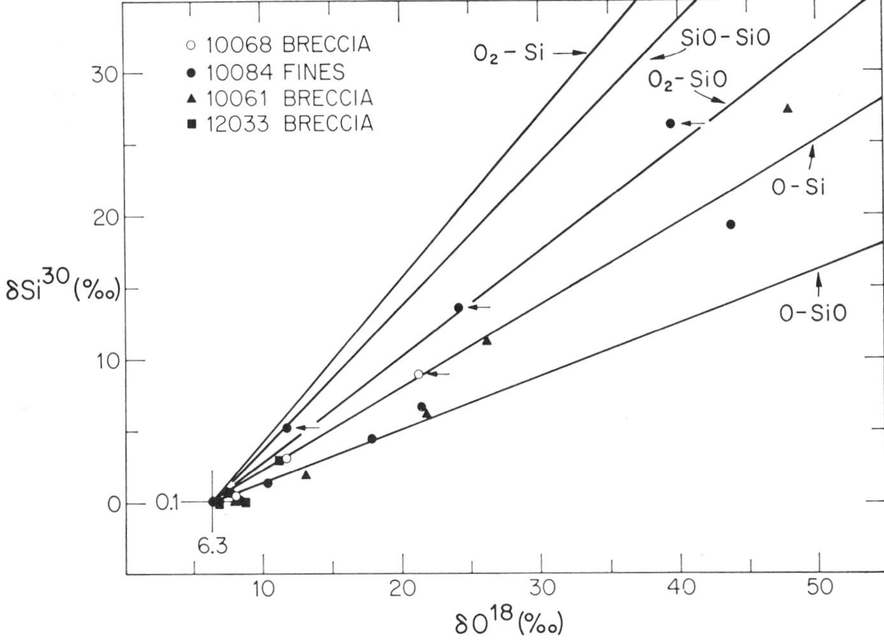

Fig. 4. Plot of δSi^{30} vs. δO^{18} for the gases extracted during the partial fluorination experiments shown in Fig. 3. The horizontal arrows show the directions in which the data-points have probably been shifted in several initial fluorination cuts because of contamination by small amounts of adsorbed low-δO^{18} water. Also, in order to take into account the δSi^{30} fractionations that accompany the chemical extraction procedure, the two highest δSi^{30} values for 10068 and 10084 should probably be lowered by about 6 to 9‰, and the four intermediate δSi^{30} values by about 3 to 4‰. The δSi^{30} data for 10061 and 12033 need no correction for either of the above effects, because they were not exposed to the laboratory atmosphere and the experiments were done in such a way as to essentially eliminate the δSi^{30} fractionation effects (see text). The curves shown on the diagram represent the isotopic relationships that would be observed if the fractional vaporization processes causing the enrichment in Si^{30} and O^{18} are dependent upon the square roots of the masses of the various chemical species of silicon and oxygen. Each curve is arbitrarily calculated on the basis of 1:1 loss of the designated species. The isotopic fractionations are then given by: O_2–Si ($\sqrt{(34/32)}$ and $\sqrt{(30/28)}$); SiO–SiO ($\sqrt{(46/44)}$ and $\sqrt{(46/44)}$, i.e., 45° line); O_2–SiO ($\sqrt{(34/32)}$ and $\sqrt{(46/44)}$); O–Si ($\sqrt{(18/16)}$ and $\sqrt{(30/28)}$); and O–SiO ($\sqrt{(18/16)}$ and $\sqrt{(46/44)}$). As an example of the effect of varying the proportions of the species, note that if instead of a 1:1 ratio, we assume continuous vaporization of 2 parts of O_2 and 1 part of SiO, the O_2–SiO curve would be shifted downward and would lie about halfway between the O–Si and O–SiO curves shown on the diagram. Also note that no account is taken of the possible O^{18} fractionations that might accompany vaporization of SiO species, except in the case of the SiO–SiO curve.

and the data points for 10084 in turn lie well above the data points obtained on 12033. This is particularly emphasized when its is remembered that the δSi^{30} values in the first 2 or 3 partial fluorination fractions obtained for 10084 and 10068 probably should be lowered by 4 to 9 per mil.

The data for experiments 3 to 6 on the Hawaiian basalt are not plotted on the figures, but within the errors of plotting on Figs. 3 and 4 no noticeable δO^{18} or δSi^{30} effect would be observed. If we make this simple assumption that fractional vaporization of the grain surfaces during particle bombardment results principally in the loss of O_2, O, Si, and/or SiO, then it is reasonable that the kinetic isotopic fractionations (e.g., in a diffusion-controlled process) might be proportional to the mean velocities of motion of the various molecular species, and thus to the square roots of the masses of these molecular species. The experimental data are compared with several such simple theoretical model curves in Fig. 4. The various diagonal lines shown on Fig. 4 represent different pairs of evaporating species of oxygen and silicon. Each fractionation line was constructed by arbitrarily assuming that the molecular species vaporize in a 1:1 ratio. For example, O and Si are assumed to vaporize in equal amounts along the O–Si curve, and thus the relative fractionations of the two species are given by $\sqrt{(18/16)}$ and $\sqrt{(30/28)}$. If, however, O is preferentially lost relative to Si, oxygen suffers a greater degree of isotopic fractionation and the curve would have a smaller slope. In the limit, obviously, if only O is lost the fractionation curve would be horizontal because there would be no Si^{30} enrichment at all.

The curves on Fig. 4 clearly cannot precisely describe the isotopic effects accompanying fractional vaporization because we cannot specify in detail the proportions, sequence, or types of species being vaporized. However, some data on the vaporization processes in lunar fines at high temperatures are available (DE MARIA *et al.*, 1971). They demonstrated the presence of gaseous species O, O_2, SiO, and SiO_2, with O_2 first being observed at 1000°–1150°C, SiO at 1300°C, and SiO_2 at 1600°–2000°C. Thermodynamically, the gaseous species "Si" is not stable, but it conceivably could be formed during nonequilibrium fractional vaporization.

Our own data on the O_2/SiF_4 ratio obtained in various fluorination cuts (Table 3) also provide some information regarding the evaporation of the oxygen and silicon species. The outermost surfaces of the grains, which presumably have suffered the greatest amount of vaporization, are markedly depleted in total oxygen relative to total silicon (by as much as a factor of 2 compared to the bulk fines). The O_2/SiF_4 ratio is as low as 0.98 in the first cut obtained from sample 10061, and it progressively increases with each successive cut; the value for the bulk sample is 1.78. Also note that this ratio is consistently larger in sample 12033 which contains lower amounts of solar wind H_2. If the grain surfaces are in fact depleted in oxygen, this implies that *none* of the 1:1 curves shown in Fig. 4 can be directly applicable because none of them produce sufficient loss of oxygen relative to silicon. It follows from this that free oxygen (O or O_2) *must* be lost relative to SiO or Si in a greater than 1:1 ratio, and therefore that each of the curves shown only represent upper limits; the curves describing the actual isotopic effects must have flatter slopes and are undoubtedly much more complicated than shown.

The data-points shown in Fig. 4 can be qualitatively explained from the results of DE MARIA *et al.* (1971) and the vaporization model discussed above. Giving greatest weight to the more precisely determined data-points on 10061 and 12033, it appears that initially there is strong O^{18} enrichment with little Si^{30} fractionation, and that subsequently greater vaporization of SiO occurs. Ultimately, the maximum isotopic fractionations perhaps result from simultaneous loss of O_2 and SiO with consequent movement along an O_2–SiO curve similar to the one shown in Fig. 4. Our isotopic data best fit the concept that O_2 and SiO are the principal species lost during vaporization, and that the overall O_2/SiO ratio in the evolved vapor phase was greater than 1.

Although we have so far only considered a fractional vaporization model to explain the large O^{18} and Si^{30} enrichments, the possibility exists that we are dealing with the condensation of Si^{30}- and O^{18}-enriched material *onto* the grain-surfaces of the lunar fines. This might come about, for example, during large meteorite or comet impacts on the lunar surface; the gases vaporized by such impacts could in part be lost from the moon and the lighter isotopic species in these gases would be preferentially lost. The residual heavier isotopic species would be preferentially retained and then would condense as fine dust and coatings on the grain-surfaces of the lunar fines. This type of fractional vaporization process could equally well explain the data-points in Fig. 4. If in the future some material with a high content of solar wind hydrogen that *lacks* an accompanying Si^{30} and O^{18} enrichment can be found on the moon, the condensation mechanism would then appear to be the most likely.

In principle, the relationships shown on Fig. 4 could also be explained by simple addition of Si^{30} and O^{18} from an extra-lunar source, such as the solar wind or cosmic rays. Arguing against this is the fact that the Si^{30} enrichment is accompanied by a Si^{29} enrichment that is approximately $\frac{1}{2}$ the magnitude of the former, as would be predicted by a fractional vaporization model.

The extreme δSi^{30} enrichments shown in Figs. 3 and 4 are *far* greater than anything ever observed before on natural samples (see TILLES, 1961; EPSTEIN and TAYLOR, 1970b). The extreme δO^{18} enrichments shown are also somewhat greater than any yet found on any minerals, either from terrestrial rocks or from meteorites. The closest approach to the $\delta O^{18} = +45$ to $+55$ shown in Fig. 3 is found in sedimentary silica deposits (chert) and in atmospheric CO_2.

ISOTOPIC COMPOSITION AND CONCENTRATION OF HYDROGEN

Analyses of the Apollo 11 samples (EPSTEIN and TAYLOR, 1970a) have shown that the element hydrogen is relatively abundant *only* in the lunar soils and breccias and is present in a form(s) which comes off both as hydrogen gas and water upon heating. The concentration of the hydrogen gas, which is dominantly or perhaps wholly of solar wind origin, was found to be greater than the concentration of the extractable water by factors of two to five. The deuterium concentrations in these two sources of hydrogen were markedly different. Some of the extracted fractions of the hydrogen gas contain as little as 18 ppm of deuterium as compared to the 157 ppm value assigned to mean ocean water (SMOW). On the other hand, the deuterium concentration in the H_2O extracted from the soils and breccias is as high as 133 ppm, a value similar to that for rain and water vapor in many temperate climatic zones. The similarity in the D/H ratios of the lunar "water" and terrestrial water together with the fact that the water found in lunar samples is even more readily extracted during heating of the lunar materials than the hydrogen gas itself, suggests that terrestrial water may have been introduced as a contaminant either on the moon or during the handling of the samples prior to our receipt of them. We further concluded that the measured deuterium concentration in the lunar hydrogen gas represent only an *upper* limit because our hydrogen gas extraction procedure involves some contamination by deuterium-rich hydrogen gas formed by reduction of the water in the lunar sample (by reactions with free iron, ilmenite etc.). Figure 5 shows this cross-contamination effect very clearly.

Apollo 12 fines and breccias

Our present report deals with further efforts to define the upper limit in deuterium concentration of lunar hydrogen gas as well as the determination of the concentration and isotopic composition of hydrogen in the Apollo 12 fines and breccias. The extraction procedure involves the heating of the lunar samples in a RF induction furnace and the trapping and analysis of the gases released in the *approximate* temperature intervals: 400°–500°C, designated as fraction A; 500°–550°C, fraction B; 550°–800°C, fraction C; and 800°C to melting as fraction D. The amounts and the δD values of hydrogen gas and water emitted at the various temperature ranges are shown

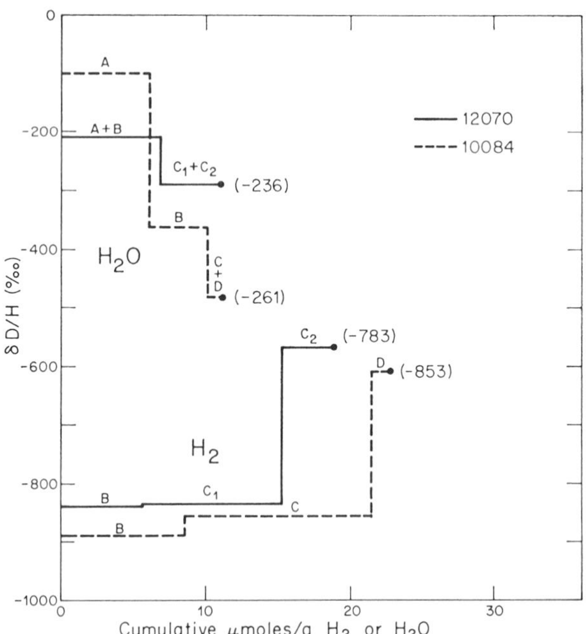

Fig. 5. Data for typical hydrogen (lower two curves) and water (upper curves) extractions from Apollo 11 and Apollo 12 soils, plotted as δD/H vs. cumulative μmoles/g. The letters A, B, C, C_1, C_2, and D represent the successive fractions extracted during heating (see text). The numbers in parentheses are the weighted average δD/H values of the various fractions from each sample.

in Fig. 5 and Table 2. Fraction A contains only water and represents more than half of the total water present in the lunar sample. Fractions B and C contain nearly all the hydrogen and fraction D contains most of the carbon as a mixture of CO and CO_2. The D/H ratios of the various fractions follow a pattern indicating cross contamination between the hydrogen and the water. It is also possible that the hydrogen in fractions D and C_2 is enriched in deuterium because some spallogenic deuterium is present in the lunar samples (see below). Note that the δD value (−783) of the hydrogen gas in sample 12070 is higher than the δD value (−853) of 10084, and also that the water in sample 12070 has a higher δD value (−236) than does the water from 10084 (−261). The difference in the D/H values for the H_2 from the two samples therefore can in large part be accounted for by different degrees of contamination with hydrogen derived from their respective coexisting waters.

The concentrations of H_2 gas in 12042 and 12070 are similar to those obtained on Apollo 11 soils, and are isotopically only moderately more enriched in deuterium. Similarly the D/H ratio of the water extracted from the Apollo 12 samples is slightly higher than in the Apollo 11 samples. Sample 12033 (trench fines) is very unusual in that its hydrogen gas concentration is lower by a factor of 20 compared to the hydrogen concentration in surface soil samples. This is probably due in large part to the less protracted solar wind bombardment history suffered by this buried soil sample or to the fact that it is coarser grained and thus has an overall lower surface area exposed to the solar wind. The high δD value of the hydrogen gas in sample 12033 is probably due to two factors. The H_2O content of this sample is about the same as for the other soils. The normal contamination of the solar wind hydrogen due to reduction of the deuterium-rich water will therefore be relatively larger in this case because there is so little solar wind hydrogen gas present. The other reason for a high D/H ratio in 12033 is the possible increased importance of spallation deuterium in this sample, since spallation deuterium should be independent of the surface to mass ratio and also should not be drastically affected by the limited amount of overburden that covered this sample on the moon. The δD value of H_2O in 12033 is the largest so far observed by us for "lunar" water, probably because of the much smaller contribution of oxidized solar wind hydrogen.

Exchange experiments with deuterium-free water

Although it seemed reasonable to conclude from our data that the hydrogen gas extracted from the lunar soils and breccias was contaminated by hydrogen formed from deuterium-rich water during the extraction procedure, it was desirable to show this experimentally. At the same time it was our aim to establish as closely as possible the actual D/H ratio in the lunar hydrogen gas. To accomplish these two aims several samples (1 soil [10084] and 2 breccias [10068 and 10061 A]) were allowed to exchange with 10–15 cm^3 (STP) of essentially deuterium-free water vapor ($\delta = -998$) at temperatures of about 300°C for a time ranging between one and three hours, prior to the extraction of the lunar hydrogen. An additional sample of breccia (10061B) was exchanged for 13 hours at about the same temperature. The exchange procedure involved the condensation of the deuterium-free water into the extraction vessel

(EPSTEIN and TAYLOR, 1970b) and the setting of the induction heating temperature to about 300°C. This temperature was only estimated from our RF generator setting and could be in error by perhaps ±50°C. It was anticipated that the pretreatment would eliminate a large fraction of the deuterium from the water present in the lunar samples and would thus decrease the amount of deuterium that the lunar hydrogen gas received from this water during the extraction procedure. At no time was any measurable hydrogen gas formed during the isotopic exchange experiment.

As the results in Table 4 and Figs. 6 and 7 show, a very noticeable decrease in the deuterium concentration in the lunar hydrogen gas did occur as a result of the pretreatment. The measured deuterium concentrations of several of the extracted hydrogen gas fractions range between 7–9 ppm. This is about a factor of three less deuterium than in the hydrogen gas extracted without pretreatment. Table 4 shows that the δD values for the water in the various extracted fractions are much lower than in the unexchanged samples, and that the water coming off at the higher temperatures (fractions B and C) is more difficult to exchange isotopically than the water that is extracted at lower temperatures (fraction A). Figure 6 shows the effect of time of exchange on the different samples. Breccia 10061 B was exchanged for the longest period of time (13 hours) and its average δ value of −922 is the lowest yet determined on any lunar sample.

Table 4. Isotopic and concentration data for hydrogen, carbon, and rare gases extracted from samples 10068, 10087, and 10061 from the Sea of Tranquillity. H_2O in lunar sample exchanged with D-free water at ~300°C.

		Hydrogen			Carbon		Water			Rare gas	
Sample	Fraction	Conc. μmole/g	δD ‰	D ppm	Conc. ppm	$\delta C^{13}/C^{12}$ ‰	Conc. μmole/g	δD ‰	D ppm	Conc. cc/g	H_2/ rare gas
10068	A	—	—	—			13.8	−559	69	—	
breccia	B	21.8	−903	15			2.2	−461	85	0.095	5.1
1.526 g	C	14.9	−858	23	84	7.4	3.1	−343	103	0.151	2.2
	D_1	2.6	−666	52			—	—		lost	
	D_2	—	—	—	81	14.2	—	—		—	
	Total	39.3	−870	20	165	10.7	19.1	−513	76	0.246	3.3
10087	A	—	—	—			25.1	−764	37		
coarse	B	7.5	−906	15						0.068	2.5
soil 2	C	10.3	−837	26			trace			0.149	1.5
chips	D	—	—	—	133	6.7	—	—			
0.444 g	Total	17.8	−866	21	133	6.7	25.1	−764	37	0.217	1.8
10061A	A	—	—	—			3.3	−613	61		
breccia	B_1	8.4	−941	9						0.034	5.5
vacuum	B_2	15.4	−956	7			1.3	−574	67	0.075	4.6
sealed	C_1	18.8	−908	14						0.113	3.7
1.332 g	C_2	6.9	−776	35			<0.3			0.221	0.7
	D	<0.3	—	—	181	15.5				0.045	0.2
	Total	49.8	−910	14	181	15.5	4.6	−601	63	0.488	2.3
10061 B	A	—	—	—			5.5	−877	19	—	—
breccia	B_1	16.2	−945	9			6.2	−724	43	—	—
sealed	B_2	17.0	−942	9	216	12.8	3.3	−543	72	—	—
in N_2	C	16.0	−876	19							
gas	D										
1.075 g	Total	49.2	−922	12	216	12.8	15.0	741	41	—	—

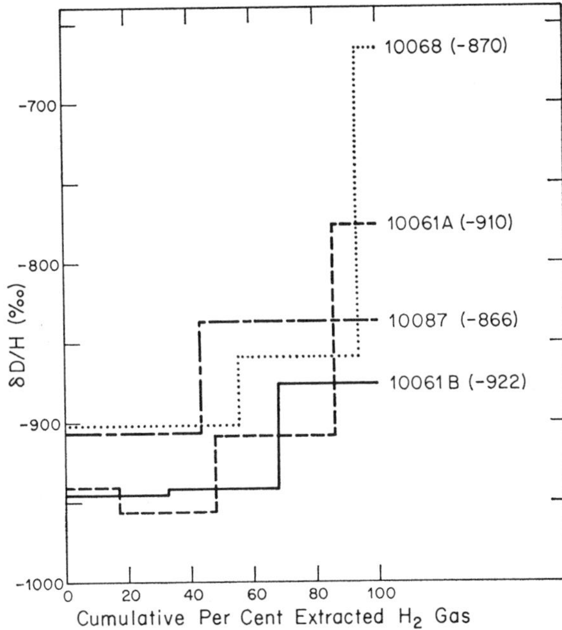

Fig. 6. Plot of δD/H vs. cumulative per cent of extracted hydrogen gas for all the data given in Table 4. These samples were all subjected to exchange with deuterium-free water prior to extraction of the hydrogen (see text).

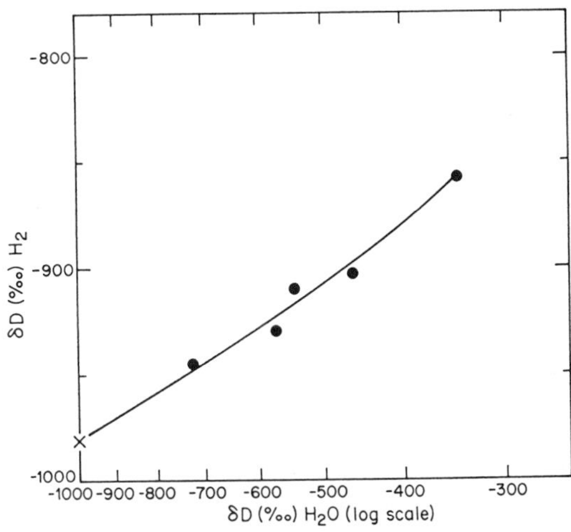

Fig. 7. Plot of δD for H_2 vs. δD for H_2O (log scale) in all the simultaneously extracted gas fractions obtained from samples that had previously been exchanged with D-free water (see Table 4). The solid curve represents a least-squares quadratic extrapolation to the situation that would presumably exist if *all* the H_2O had undergone complete isotopic exchange. The cross on the ordinate indicates the calculated concentration (3 ppm) of spallation deuterium in a lunar sample (see text). Note that FRIEDMAN *et al.* (1971) have observed D-rich hydrogen in an Apollo 12 basalt (δD $\approx$ +300 to +350), attributing this effect to the addition of spallation deuterium.

The maximum amount of deuterium in lunar hydrogen gas

We feel that we have as yet not experimentally obtained the lowest D/H value for pristine, uncontaminated lunar hydrogen gas. However, some estimate of this quantity can be obtained if we plot the δD values of various fractions of the H_2 gas against the δD value of the co-extracted waters (Table 4). Figure 7 shows this relationship. The various points on the curve in Fig. 7 were obtained by utilizing, for example, the δD values of water and H_2 gas in fraction B, of sample 10061A. Another point represents the δD values of H_2 gas and the water extracted from combined fractions B_2 and C in sample 10061B. The plotted curve in Fig. 7 simply shows that the higher the D/H ratio in the water the larger is the cross-contamination of deuterium in the coexisting hydrogen gas. Presumably, the best value of the D/H ratio of the H_2 gas could be obtained if all the deuterium could be exchanged away from the water by extrapolating to a $\delta D = -1000‰$ value (0 ppm D) for the water. Such an extrapolation gives an approximate δD of about $-980‰$ or a concentration of deuterium of about 3 ppm. It is possible to make a rough calculation as to how much deuterium is contributed to the breccia (10061) by spallogenic deuterium, by knowing the ratio of production rates of Xe^{126}/He^3 as given by HUNEKE *et al.* (1971) for lunar rocks, the Xe^{126} concentration in 10061 (PEPIN *et al.*, 1970), and the relative production rates of D/He^3 as can be inferred from the data of BADHWAR (1964). This calculation shows that the spallation deuterium contribution to breccia 10061 is about 3×10^{-10} moles D/g as compared to the 10^{-4} moles of H/g present in this rock (EPSTEIN and TAYLOR, 1970b). This would represent about 3.0 ppm of deuterium, and this value is shown by a cross on the ordinate of Fig. 7. This simple calculation compares remarkably well with the 3 ppm value obtained from our extrapolation of the curve in Fig. 7. Considering the difficulty in the extrapolation and the crudeness of the calculation these values for the D/H ratio in lunar H_2 gas agree rather well, and it now seems clear that the concentration of deuterium in the solar wind is exceedingly low and must definitely be less than 3 or 4 parts per million of the total solar wind hydrogen.

Better controlled hydrogen isotope exchange experiments of the lunar soils and breccias, or perhaps even better, utilization of samples *completely* free of terrestrial water contamination, may allow us to make even better estimates of the total D concentration in lunar hydrogen. At present, however, it appears unlikely that the total amount of deuterium from solar wind together with that derived from spallation processes can be appreciably greater than 5×10^{-6} that of the total hydrogen.

Partial fluorination of the lunar dust

An attempt was made to eliminate the water from the lunar soils by exposing a lunar soil to an atmosphere of fluorine at about 50°C for 15 min. Although a slight fluorination of the sample took place as indicated by the formation of tiny amounts of SiF_4 and O_2, the water was not eliminated. The fluorinated sample gave about the same amounts of H_2 and H_2O as the untreated samples.

Hydrogen in lunar basalt

An attempt was made to extract hydrogen from a sample of vacuum sealed basalt (10072) weighing about 0.7 g. Less than 0.3 μmoles/g of hydrogen gas and about the same amount of H_2O was extracted from this sample. No meaningful isotopic measurements can be made at this time with our present apparatus on such small samples.

Isotopic Composition and Concentration of Carbon

Our previous and present data on Apollo 11 soils and breccias show that carbon is present in concentrations ranging from about 133 to 262 ppm, and is higher in the breccias than the soils. The δC^{13} values relative to the PDB standard range from +6.7 for a sample of coarse soil chips (10087) to +18.6 for an Apollo 11 soil. Analyses of 3 different samples of 10061 give concentrations of 181, 216, and 262 ppm and respective δC^{13} values of +15.5, +12.8. and +10.8‰. The two samples having lower carbon concentrations and higher δC^{13} values were the ones subjected to prior heating at about 300°C with low-deuterium water; the samples were also outgassed to high vacuum at this elevated temperature. It is possible that some of the readily removable low-δC^{13} carbon (see Kaplan *et al.*, 1970) was lost in this manner. This particular carbon conceivably could represent low C^{13} carbon compounds from the exhaust of the lunar module during landing. Thus the more correct δC^{13} value for lunar carbon in the breccia 10061 might be the +15.5 value.

The Apollo 12 soil samples (fines) contain about the same concentrations of carbon as do the Apollo 11 soil samples. The δC^{13} values of the Apollo 12 soils are lower, however (+4.7 and +8.0 as compared to +16.3 and +18.6). Nonetheless all of our previous statements concerning the origin of lunar carbon in Apollo 11 samples also apply equally well to the Apollo 12 samples. As was the case for Apollo 11 samples, for comparable samples (12042) our results agree well with those of Kaplan and Petrowski (1971). Even the small difference between our δC^{13} values is consistent with the relationship observed by Kaplan *et al.* (1970); namely, for a specific class of materials like the surface lunar soil the lower the yield the lower the δC^{13} value. Our yield for 12042 is 125 ppm giving a δ value of +8.0 whereas Kaplan and Petrowski's yield for the same sample is 112 ppm giving a δC^{13} value of +10.8.

Our trench sample (12033) gives a low carbon concentration and a low δC^{13} value (30 ppm and $\delta C^{13} = -3.6$), a result also consistent with the data of Kaplan and Petrowski (1971). The low δC^{13} of the carbon in sample 12033 may be due to the increased importance of a low δC^{13} terrestrial contaminant (see Kaplan and Petrowski for δC^{13} values of possible earth contaminants) or to the fact that the sample was physically shielded from outside particle bombardment. As noted above, the surface coatings of grains from soils and breccias are highly enriched in Si^{30} and O^{18}. The low solar wind H_2 sample 12033 was more deeply buried and also is not as fine grained as the other soils. If the C^{13}-enriched lunar carbon is associated with solar wind bombardment it is reasonable that sample 12033 would have a low δC^{13} value. The qualitative correlation, previously pointed out (Epstein and Taylor, 1970a) between the concentration of carbon and solar wind hydrogen gas is therefore further substantiated in the Apollo 12 samples.

Note that our lunar samples analyzed for carbon do not come in contact with the laboratory atmosphere, nor with any container or material not previously outgassed at high temperatures. They are not handled except in transferring by clean forceps from the L.R.L. container into a platinum crucible. This is all carried out in a dry box filled with specially dried N_2 gas. All terrestrial forms of carbon that can contribute an impurity have low δC^{13} values compared to our measured δC^{13} values of lunar carbon.

The δC^{13} results of FRIEDMAN *et al.* (1971) on Apollo 12 samples disagree with those of Kaplan and Petrowski and with our results. It is difficult to assign a source of error to our analytical procedure. However, if there is something inherently wrong with the pyrolysis method, then the combustion method used by KAPLAN *et al.* (1970) should give results similar to those of FRIEDMAN *et al.*, who also used the combustion method for their carbon isotope determinations. The analytical agreement between ourselves and KAPLAN and PETROWSKI (1971) and KAPLAN *et al.* (1970), utilizing two different methods of analysis in which the sources of error should be different, suggests that our results probably cannot be in serious error.

Acknowledgments—We wish to thank L. T. Silver and D. S. Burnett for contributing certain samples analyzed in this study, as well as for some stimulating scientific discussions. Also, we wish to thank R. N. Clayton and J. R. O'Neil for their helpful criticism of the manuscript. Much of the laboratory work was done by P. Yanagisawa and J. Young. Financial support was provided by NASA Contract no. NAS 9–7944.

REFERENCES

ALBEE A. L., BURNETT D. S., CHODOS A. A., HAINES E. L., HUNEKE J. C., PAPANASTASSIOU D. A., PODOSEK F. A., RUSS G. P., III, and WASSERBURG G. J. (1970) Mineralogic and isotopic investigations on lunar rock 12013. *Earth Planet. Sci. Lett.* **9**, 137–163.

BADHWAR G. D. (1964) Isotopic composition of low energy hydrogen nuclei in the primary cosmic radiation. *J. Geophys. Res.* **69**, 4435–4439.

DE MARIA G., BALDUCCI G., GUIDO M., and PIACENTE V. (1971) Mass spectrometric investigation of the vaporization process of lunar samples. Second Lunar Science Conference (unpublished proceedings).

EPSTEIN S. and TAYLOR H. P., JR. (1970a) $^{18}O/^{16}O$, $^{30}Si/^{28}Si$, D/H and $^{13}C/^{12}C$ studies of lunar rocks and minerals. *Science* **167**, 533–535.

EPSTEIN S. and TAYLOR H. P., JR. (1970b) The concentration and isotopic composition of hydrogen, carbon, and silicon in Apollo 11 lunar rocks and minerals. *Proc. Apollo 11 Lunar Sci. Conf., Geochim. Cosmochim. Acta* Suppl. 1, Vol. 2, pp. 1085–1096. Pergamon.

FRIEDMAN I., O'NEIL J. R., GLEASON J. D., and HARDCASTLE K. (1971) The carbon, hydrogen content and isotopic composition of some Apollo 12 materials. Second Lunar Science Conference (unpublished proceedings).

HUNEKE J. C., PODOSEK F. A., BURNETT D. S., and WASSERBURG G. J. (1971) Rare gas studies of the galactic cosmic ray irradiation history of lunar rocks. *Geochim. Cosmochim. Acta.* (Submitted for publication).

KAPLAN I. R., SMITH J. W., and RUTH E. (1970) Carbon and sulfur concentration and isotopic composition in Apollo 11 lunar samples. *Proc. Apollo 11 Lunar Sci. Conf., Geochim. Cosmochim. Acta* Suppl. 1, Vol. 2, pp. 1317–1329. Pergamon.

KAPLAN I. and PETROWSKI C. (1971) Carbon and sulfur isotope studies on Apollo 12 lunar samples. Second Lunar Science Conference (unpublished proceedings).

O'KEEFE J. A. (1970) Tektite glass in Apollo 12 sample. *Science* **168**, 1209–1210.

O'NEIL J. R. and ADAMI L. H. (1970) Oxygen isotope analyses of selected Apollo 11 materials. *Proc. Apollo 11 Lunar Sci. Conf., Geochim. Cosmochim. Acta* Suppl. 1, Vol. 2, pp. 1425–1427. Pergamon.

ONUMA N., CLAYTON R. N., and MAYEDA T. (1970) Oxygen isotope fractionation between minerals and an estimate of the temperature of formation. *Science* **167**, 536–538.

PEPIN R. O., NYQUIST L. E., PHINNEY D., and BLACK D. C. (1970) Rare gases in Apollo 11 lunar material. *Proc. Apollo 11 Lunar Sci. Conf., Geochim. Cosmochim. Acta* Suppl. 1, Vol. 2, pp. 1435–1454. Pergamon.

TAYLOR H. P., JR. (1968) The oxygen isotope geochemistry of igneous rocks. *Contrib. Mineral. Petrol.* **19**, 1–71.

TAYLOR H. P., JR., and EPSTEIN S. (1962) Relationship between O^{18}/O^{16} ratios in coexisting minerals of igneous and metamorphic rocks. *Bull. Geol. Soc. Amer.* **73**, 461–480.

TAYLOR H. P., JR., DUKE M. B., SILVER L. T., and EPSTEIN S. (1965) Oxygen isotope studies of minerals in stony meteorites. *Geochim. Cosmochim. Acta* **29**, 489–512.

TAYLOR H. P., JR., and EPSTEIN S. (1966) Oxygen isotope studies of Ivory Coast tektites and impactite glass from the Bosumtwi Crater, Ghana. *Science* **153**, 173–176.

TAYLOR H. P., JR., and EPSTEIN S. (1969) Correlations between O^{18}/O^{16} ratios and chemical compositions of tektites. *J. Geophys. Res.* **74**, 6834–6844.

TAYLOR H. P., JR., and EPSTEIN S. (1970a) O^{18}/O^{16} ratios of Apollo 11 lunar rocks and minerals. *Proc. Apollo 11 Lunar Sci. Conf., Geochim. Cosmochim. Acta* Suppl. 1, Vol. 2, pp. 1613–1626. Pergamon.

TAYLOR H. P., JR. and EPSTEIN S. (1970b) Oxygen and silicon isotope ratios of lunar rock 12013. *Earth Planet. Sci. Lett.* **9**, 208–210.

TILLES D. (1961) Natural variations in isotopic abundances of silicon. *J. Geophys. Res.* **66**, 3003–3013.

Proceedings of the Second Lunar Science Conference, Volume 2, pp. 1443–1450.
The M.I.T. Press, 1971.

Vanadium isotopic composition and contents in lunar rocks and dust from the Ocean of Storms

M. E. LIPSCHUTZ
Departments of Chemistry and Geosciences, Purdue University, Lafayette, Indiana 47907
H. BALSIGER
Physikalisches Institut der Universität, Bern, Switzerland
and
I. Z. PELLY*
Department of Chemistry, Purdue University, Lafayette, Indiana 47907

(*Received 9 February 1971; accepted in revised form 25 March 1971*)

Abstract—The isotopic composition of vanadium from three Apollo 12 lunar crystalline rocks and fines is the same as that of terrestrial and meteoritic material to within 1%. The vanadium concentrations in these samples, as determined by isotopic dilution, range from 126–196 ppm for the rocks and is 116 ppm in the fines. As on Earth, vanadium apparently proxies for iron during magmatic evolution of lunar crystalline rocks.

INTRODUCTION

A NUMBER of noble gas nuclides constitute very sensitive monitors for the irradiation of solid matter by energetic charged particles and, based upon study of such nuclides (and less sensitive radionuclides and particle tracks), it is clear that lunar surface material has been thus irradiated during the last several hundred million years. It is also possible that much earlier in its history, lunar material as part of the solar system was exposed to a strong charged particle irradiation (FOWLER *et al.*, 1962; BERNAS *et al.*, 1967) although studies of extraterrestrial material have failed to uncover any unambiguous evidence for such a process. It could be that if an early irradiation occurred, it took place under conditions such that gaseous elements and other radiation monitors would not be retained or would subsequently be lost. Thus an effective radiation monitor must be sought among the nongaseous elements.

It has been noted previously (SHIMA and HONDA, 1963; BURNETT *et al.*, 1966; BALSIGER *et al.*, 1969) that one of the most sensitive monitors among the nongaseous elements is the vanadium isotopic composition. Thus, effects arising from an early charged particle irradiation would be expected to be detectable by comparative study of the $^{50}V/^{51}V$ ratios in terrestrial, meteoritic, and lunar samples unless either (a) all of the matter which now comprises these different objects was thoroughly mixed or (b) the integrated particle fluxes and relative proportions of irradiated to shielded material happened to be virtually identical in those parts of the solar system where these bodies were formed (cf. BURNETT *et al.*, 1965; 1966). In previous studies (BALSIGER *et al.*, 1969; PELLY *et al.*, 1970) we established that the vanadium isotopic composition in terrestrial samples was the same as that in eight chondrites (principally primitive

* Present Address: Department of Geology, Negev University, Beer Sheva, Israel.

ones of chemical-petrologic grades 1–4) to within 1%. In another study ALBEE *et al.* (1970) noted that the vanadium isotopic composition in a sample of fines and a rock from Apollo 11 was the same as that in terrestrial samples to within 2% and 3%. Since these limits could well conceal substantial effects arising from charged particle irradiation, it is still worthwhile to conduct a thorough investigation of the vanadium composition in lunar material and, in concert with this, to determine accurately the vanadium concentration in these same samples.

EXPERIMENTAL TECHNIQUES

The lunar samples studied and their type as described by LSPET (1970) are listed in Table 1. All our crystalline rock samples were obtained from the interior of the hand specimens. For mass-spectrometric analysis it was necessary that the final vanadium precipitate include as little Ti and Cr as possible since these elements also contain stable mass-50 isotopes. As will be seen, the concentrations of these interfering elements in the pristine lunar samples are 10 to 100 times higher than that of vanadium—the ranges being 1.6–2.0% Ti and 0.20–0.62% Cr (ANNEL *et al.*, 1971; BRUNFELT *et al.*, 1971; ROSE *et al.*, 1971; WILLIS *et al.*, 1971). Thus a particularly good chemical separation of vanadium from these troublesome elements is required. Our technique yielded decontamination factors of ~ 500 for both Ti and Cr and allowed vanadium to be recovered with chemical yields of 20–40% as monitored by ^{49}V carrier-free tracer. The reagent blank for the complete process was determined by isotope dilution as 0.15 micrograms of vanadium and by X-ray fluorescence as < 0.3 micrograms.

Only quartz and plastic lab-ware were used for processing the samples and the reagents were purified as described by PELLY *et al.* (1970). Prior to dissolving them, two of the lunar rocks—12021 and 12038—were etched briefly with 1:1 HNO_3 to remove any surface contamination. The samples (1–2 grams) were dissolved (PELLY *et al.*, 1970) and $\sim \frac{1}{3}$ of the solution was reserved for determination of the vanadium concentration by isotope dilution. The vanadium in the remaining $\frac{2}{3}$ of the solution was separated chemically by the technique described by PELLY *et al.* (1970) and its isotopic composition was determined mass-spectrometrically as described by BALSIGER *et al.* (1969).

We determined the vanadium concentration in our samples by isotope dilution after spiking the original aliquot reserved for this purpose with vanadium containing 35% ^{50}V. The chemical separation steps used for the spiked samples were identical to those described (PELLY *et al.*, 1970) and were carried out in a separate laboratory. In order to prevent contamination with spike, stringent measures were taken to prevent transfer of reagents and lab-ware from one laboratory to the other.

Table 1. Vanadium isotopic ratios and concentrations in Apollo 12 lunar samples.

Sample	Type	$^{50}V/^{51}V$*	V (ppm) This work	V (ppm) Other investigations†
12002,128	B (olivine dolorite)	$(2.455 \pm 0.014) \times 10^{-3}$	196	175(a); 190(b); 212, 223, 227(c)
12021,49	B (pigeonite dolorite)	$(2.470 \pm 0.027) \times 10^{-3}$	160	130(d); 147(e); 190(f); 192, 217(c)
12038,43	A (basalt)	$(2.464 \pm 0.012) \times 10^{-3}$	126	70(g); 120(b); 126(d); 144, 154(c)
12070,52	D (contingency fines)	$(2.469 \pm 0.033) \times 10^{-3}$	116	49(h); 64(g); 72(f); 110(b); 110(i); 110, 114(e); 112(a); 121(d); 130(j); 142, 143(c)

* The uncertainty listed for each sample includes a statistical error (three estimated standard errors) as well as those errors arising from correction of the mass-50 peak (see text).

† References: (a) WILLIS *et al.* (1971); (b) TAYLOR *et al.* (1971); (c) BRUNFELT *et al.* (1971); (d) ANNELL *et al.*, (1971); (e) ROSE *et al.* (1971); (f) MORRISON *et al.* (1971); (g) LSPET (1970); (h) BOUCHET *et al.* (1971); (i) SMALES *et al.* (1971); (j) WAKITA *et al.* (1971).

Results and Discussion

Table 1 includes the vanadium isotopic ratios and concentrations in the lunar samples studied. The uncertainty listed for each sample includes a statistical error (three estimated standard errors, $3\sqrt{(\sum_i (\Delta_i)^2/n(n-1))}$) as well as those errors arising from correction of the mass-50 peak for contributions of ^{50}Ti and ^{50}Cr. In order to make these corrections we assumed that the isotopic compositions of Ti and Cr in our lunar samples were identical to those we determined for terrestrial standards. During each run we repeatedly scanned the mass range in the neighborhood of mass-50 and computed the amounts of ^{50}Ti and ^{50}Cr from measurement of the peaks corresponding to the masses of the other stable Ti and Cr isotopes. As in our previous studies, the differing volatilities of the Ti, vanadium, and Cr species with time resulted in a variation in the mass-50 correction during each run. Changes in our chemical separation technique, primarily in the design of our ion exchange columns, resulted in an improved separation of vanadium from Ti and Cr hence in a reduction of the mass-50 correction to 3–10%. The ^{50}V/^{51}V ratios were not corrected for source or multiplier discrimination. The errors listed do not include the error due to a possible variation in mass discrimination from one run to another.

If the ^{50}V/^{51}V ratios in vanadium from different sources are to be compared for clues to a possible irradiation early in the history of the solar system, in principle a correction should be applied for ^{50}V produced in the recent past by cosmic rays and energetic solar particles. In meteorites this effect is negligible because of the relatively short exposure times involved (Balsiger *et al.*, 1969). At the lunar surface cosmic ray bombardment over a period of 500 million years may be estimated to produce a change of 0.1% in the ^{50}V/^{51}V ratio. The production rate of ^{50}V by energetic solar particles could be as high as or even higher than that by cosmic rays (cf. Shedlovsky *et al.*, 1970). However track data (cf. Comstock *et al.*, 1971) generally indicate that lunar material has been close to the surface (thus permitting irradiation by solar particles) for periods short compared with their cosmic ray exposure ages. Thus energetic solar particles probably changed the ^{50}V/^{51}V ratio at the Apollo 12 site by less than 0.1%.

The vanadium isotope ratios listed in Table 1 are plotted in Fig. 1 together with our previous results for chondrites and terrestrial samples. (One datum, the ^{50}V/^{51}V ratio for Plainview, was listed incorrectly in Pelly *et al.* (1970) and, instead, should be $(2.428 \pm 0.032) \times 10^{-3}$.) Clearly none of our isotopic measurements of the individual lunar samples (Table 1, Fig. 1) or their mean (Table 2) differ significantly from our previous results for terrestrial or meteoritic samples. The tendency for meteoritic ^{50}V/^{51}V to be slightly lower than data obtained from terrestrial and lunar samples is somewhat puzzling. As yet we cannot determine whether there were any systematic changes (i.e., a variation in mass discrimination) between the times the meteoritic and terrestrial samples and the lunar samples were run. Since the trend of the data might be an artifact arising from such a systematic change we are currently conducting a series of measurements on additional terrestrial and meteoritic samples. The group means listed in Table 2 were obtained by inversely weighting each analysis by the square of its experimental error (cf. Table 1). The estimated standard errors of

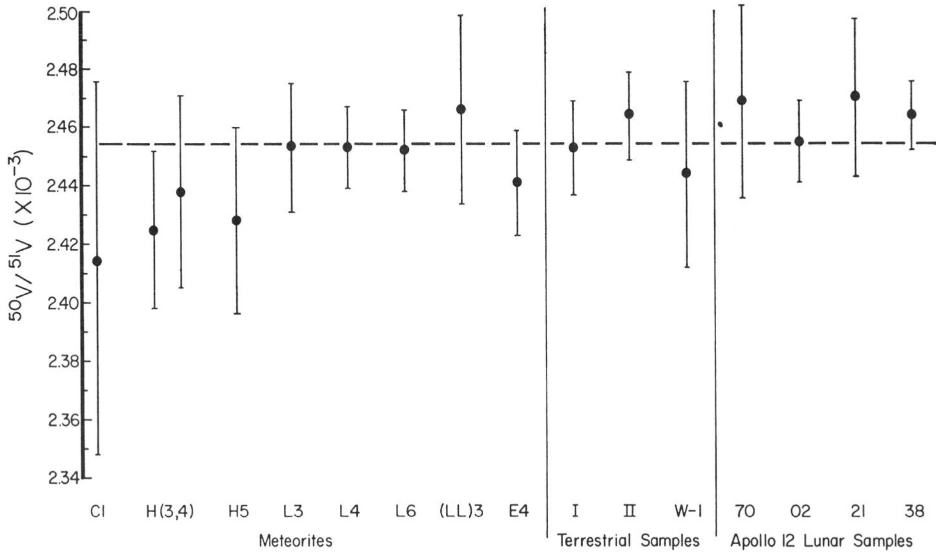

Fig. 1. $^{50}V/^{51}V$ isotopic ratios in lunar dust, lunar rocks, terrestrial diabase W-1, and reagent vanadium (I, II), and chondrites (uncertain chondrite classifications are placed in parentheses). The weighted mean ratio of all samples is indicated by the dashed line.

Table 2. Vanadium isotopic ratios in terrestrial, meteoritic, and lunar samples.

Material	No. investigated	Weighted $^{50}V/^{51}V$*
Lunar	4	$(2.462 \pm 0.003) \times 10^{-3}$
Terrestrial	3	$(2.457 \pm 0.005) \times 10^{-3}$
Meteoritic	9	$(2.447 \pm 0.004) \times 10^{-3}$
All	16	$(2.454 \pm 0.003) \times 10^{-3}$

* The uncertainty listed for each group is the estimated standard error of the mean calculated from the dispersions of the measurements of the individual samples.

these means were calculated from the dispersion of the measurements of the samples belonging to the respective group.

From our previous results on terrestrial and meteoritic samples we had calculated upper limits for possible differences of integrated fluxes ($\Delta\phi$) for two types of proton spectra (BALSIGER *et al.*, 1969). Our present data could be interpreted according to several possible models, one of which is implicitly assumed in our calculation of the means listed in Table 2, i.e., that the samples constituting each group have the same vanadium isotopic composition. We have calculated $\Delta\phi$ limits based upon a maximum difference of the group means of 1% which corresponds to the lunar-meteoritic difference plus two estimated standard errors, $(0.015 \pm 2 \times 0.005) \times 10^{-3}$. The resulting limits are 1.8×10^{18} "hard" protons/cm^2 (cosmic-ray spectrum, $E > 30$ MeV) and 9×10^{19} "soft" protons/cm^2 (solar flare spectrum, $E > 10$ MeV), respectively. Despite their poorer experimental precision of 2–3% ALBEE *et al.* (1970) calculated a $\Delta\phi$ value of 4×10^{17} particles/cm^2 ($E \geq 50$ MeV), which is considerably lower than ours. This seeming paradox appears due to their use of a production cross-section of 50 mb (G. P. RUSS, personal communication) which is about 4 times higher than our mean cross-section for the cosmic-ray spectrum and LSPET's (1969) possibly low value of 42 ppm for the total vanadium concentration in Apollo 11 fines.

In order to obtain the limits listed above one important parameter is, of course, the vanadium content of the samples. Table 1 includes both our results and those obtained by other investigations of the same samples. We conservatively estimate the errors associated with our concentration values as $\leq 5\%$, which includes the errors associated with spike calibration, isotopic ratio measurements, and that associated with possible variations in mass discrimination. Most of the other analyses tend to cluster about our values to within about 25% although the extreme results differ by factors of 2–3, a difference which probably is beyond that due to sample variability alone. Some discrepancies seem systematic, the results of BRUNFELT *et al.* (1971) being somewhat higher, those of LSPET (1970) being about a factor of two lower. The latter also appeared true for Apollo 11 as well since in 5 cases the results of LSPET (1969) were consistently lower by factors of 2–3 than those obtained by ANNELL and HELZ (1970) and MORRISON *et al.* (1970). The remaining small differences (cf. Table 1) may be ascribed either to sample variability or to experimental difficulties (NICHIPORUK, 1971). The elaborate reagent purification and long time required by our isotope dilution technique make it more tedious than other analytical methods for vanadium. Nevertheless, the inherent accuracy of the isotope dilution method prompts us to favor our results for the remaining discussion.

Our results indicate that the vanadium contents of crystalline rocks are somewhat higher than that of fines and a factor of 2–4 higher than those of chondrites (PELLY *et al.*, 1970; cf. NICHIPORUK, 1971). The range for the crystalline rocks falls within that for terrestrial igneous rocks such as diorites, andesites, gabbros, and basalts (GOLDSCHMIDT, 1954) in which vanadium follows chromium, titanium or, most often, trivalent iron. The ranges in concentration reported for the ferromagnesian elements chromium, iron, magnesium, and titanium in rocks 12002, 12021, 12038, and fines 12070 ordinarily vary in the extreme by less than 5% (ANNELL *et al.*, 1971; BRUNFELT *et al.*, 1971; ROSE *et al.*, 1971; WILLIS *et al.*, 1971) and we may therefore use the mean values of these data to study the geochemical relationships between these elements and vanadium in lunar crystalline rocks and fines.

These mean concentrations are plotted in the upper portion of Fig. 2 and illustrate that the crystalline rocks listed in Table 1 are arranged in order of decreasing ferromagnesian element content. Thus the general decrease in vanadium concentrations (which are also plotted in Fig. 2) is just that expected by analogy with terrestrial magmatic processes. Clearly, in crystalline rocks the trend of the vanadium data is most closely matched by that of iron, suggesting that as in terrestrial magmatic material lunar vanadium is proxying for iron in its minerals rather than being concentrated in ilmenite as suggested by ALBEE *et al.* (1970).

Since in terrestrial minerals vanadium proxies for trivalent rather than divalent iron, the relationship between these elements in lunar material must be a consequence

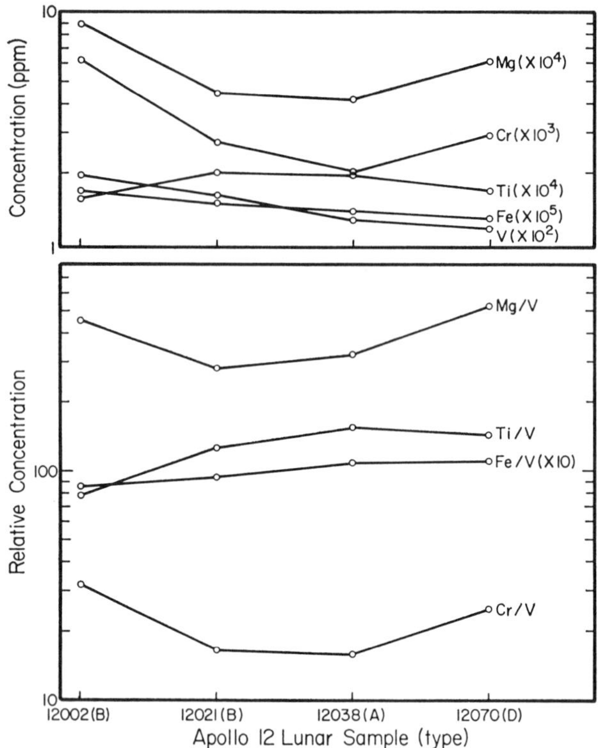

Fig. 2. Concentrations of some major and minor ferromagnesian elements and vanadium in lunar crystalline rocks and fines which, except for vanadium, are mean values obtained by other investigations of these materials. The relationships suggest that vanadium is proxying for iron.

of the extremely low lunar oxygen partial pressure. The ferromagnesian element contents relative to that of vanadium are plotted in the bottom portion of Fig. 2. On the basis of these data we cannot state with certainty that the apparent increase from 850–1100 in the iron/vanadium ratio with decreasing iron content is a real effect reflecting an iron-vanadium fractionation. Clearly, however, the process of conversion of crystalline rocks to fines is not accompanied by substantial fractionation between these two elements.

Acknowledgments—We thank Professor J. Geiss (Bern) for his aid and encouragement and Professor H. Rubin (Purdue) for stimulating discussions. This research was supported by the U.S. National Aeronautics and Space Administration (Contract NAS 9–10325) and the Swiss National Science Foundation (Grant 2.213.69).

References

Albee A. L., Burnett D. S., Chodos A. A., Eugster O. J., Huneke J. C., Papanastassiou D. A., Podosek F. A., Russ G. P. II, Sanz H. G., Tera F., and Wasserburg G. J. (1970) Ages, irradiation history, and chemical composition of lunar rocks from the Sea of Tranquillity. *Science* **167**, 463–446.

ANNELL C., and HELZ A. (1970) Emission spectrographic determination of trace elements in lunar samples from Apollo 11. *Proc. Apollo 11 Lunar Sci. Conf., Geochim. Cosmochim. Acta* Suppl. 1, Vol. 2, pp. 991–994. Pergamon.

ANNELL C. S., CARRON M. K., CHRISTIAN R. P., CUTTITTA F., DWORNIK E. J., HELZ A. W., LIGON D. T., JR., and ROSE H. J. JR. (1971) Chemical and spectrographic analyses of lunar samples from Apollo 12 mission. Second Lunar Science Conference (unpublished proceedings).

BALSIGER H., GEISS J., and LIPSCHUTZ M. E. (1969) Vanadium isotopic composition in meteoritic and terrestrial matter. *Earth Planet. Sci. Lett.* **6**, 117–122.

BERNAS R., GRADSZTAJN E., REEVES H., and SCHATZMANN E. (1967) On the nucleosynthesis of lithium, beryllium, and boron. *Ann. Phys.* **44**, 426–478.

BRUNFELT A. O., HEIER K. S., and STEINNES E. (1971) Determination of 40 elements in Apollo 12 materials by neutron activation analysis. Second Lunar Science Conference (unpublished proceedings).

BOUCHET M., KAPLAN G., VOUDON A., and BERTOLETTI N. J. (1971) Spark mass spectrometric analysis of major and minor elements in six lunar samples and age determination of four rocks. Second Lunar Science Conference (unpublished proceedings).

BURNETT D. S., FOWLER W. A., and HOYLE F. (1965) Nucleosynthesis in the early history of the solar system. *Geochim. Cosmochim. Acta* **29**, 1209–1241.

BURNETT D. S., LIPPOLT H. J., and WASSERBURG G. J. (1966) The relative isotopic abundance of K^{40} in terrestrial and meteoritic samples. *J. Geophys. Res* **71**, 1249–1269, 3609.

COMSTOCK G. M., EVWARAYE A. O., FLEISCHER R. L., and HART H. R., JR. (1971) The particle track record from the Ocean of Storms. Second Lunar Science Conference (unpublished proceedings).

FOWLER W. A., GREENSTEIN J. L., and HOYLE F. (1962) Nucleosynthesis during the early history of the solar system. *Geophys. J.* **6**, 148–220.

GOLDSCHMIDT V. M. (1954) *Geochemistry*, Oxford University Press.

LSPET (Lunar Sample Preliminary Examination Team) (1969) Preliminary examination of lunar samples from Apollo 11. *Science* **165**, 1211–1227.

LSPET (1970) Preliminary examination of the lunar samples from Apollo 12. *Science* **167**, 1325–1339.

MORRISON G. H., GERARD J. T., KASHUBA A. T., GANGADHARAM E. V., ROTHENBERG A. M., POTTER N. M., and MILLER G. B. (1970) Elemental abundances of lunar soil and rocks. *Proc. Apollo 11 Lunar Sci. Conf., Geochim. Cosmochim. Acta* Suppl. 1, Vol. 2, pp. 1383–1392. Pergamon.

MORRISON G. H., GERARD J. T., POTTER N. M., GANGADHARAM E. V., ROTHENBERG A. M., and BURDO R. A. (1971) Elemental abundances of lunar soil and rocks from Apollo 12. Second Lunar Science Conference (unpublished proceedings).

NICHIPORUK W. (1971) Vanadium (23). To be published in *Elemental Abundances in Meteoritic Matter* (editor B. Mason). Gordon and Breach.

PELLY I. Z., LIPSCHUTZ M. E., and BALSIGER H. (1970) Vanadium isotopic composition and contents in chondrites. *Geochim. Cosmochim. Acta* **34**, 1033–1036.

ROSE H. J., JR., CUTTITTA F., ANNELL C. S., CARRON M. K., CHRISTIAN R. P., DWORNIK E.J., HELZ A. W., and LIGON D. T., JR. (1971) Semimicroanalysis of Apollo 12 samples. Second Lunar Science Conference (unpublished proceedings).

SHEDLOVSKY J. P., HONDA M., REEDY R. C., EVANS J. C., JR., LAL D., LINDSTROM R. M., DELANEY J. C., ARNOLD J. R., LOOSLI H. H., FRUCHTER J. S., and FINKEL R. C. (1970) Pattern of bombardment-produced radionuclides in rock 10017 and in lunar soil. *Proc. Apollo 11 Lunar Sci. Conf., Geochim. Cosmochim. Acta* Suppl. 1, Vol. 2, pp. 1503–1532. Pergamon.

SHIMA M., and HONDA M. (1963) Isotopic abundance of meteoritic lithium, *J. Geophys. Res.* **68**. 2844–2854.

SMALES A. A., MAPPER D., WEBB M. S. W., WEBSTER R. K., WILSON J. D., and HISLOP J. S. (1971) Elemental composition of lunar surface material (part 2). Second Lunar Science Conference (unpublished proceedings).

TAYLOR S. R., KAYE M., GRAHAM A., RUDOWSKI R., and MUIR P. (1971) Trace element chemistry of lunar samples from the Ocean of Storms. Second Lunar Science Conference (unpublished proceedings).

WAKITA H., REY P., and SCHMITT R. A. (1971) Abundances of the 14 rare earth elements plus 22 major, minor, and trace elements in ten Apollo 12 rock and soil samples. Second Lunar Science Conference (unpublished proceedings).

WILLIS J. P., AHRENS L. H., DANCHIN R. V., ERLANK A. J., GURNEY J. J., HOFMER P. J., McCARTHY T. J., and ORREN M. J. (1971) Some inter-element relationships between lunar rocks and fines, and stony meteorites. Second Lunar Science Conference (unpublished proceedings).

Proceedings of the Second Lunar Science Conference, Volume 2, pp. 1451–1459.
The M.I.T. Press, 1971.

Variations in beryllium and chromium contents in lunar fines compared with crystalline rocks

R. E. Sievers, K. J. Eisentraut, D. J. Griest,
M. F. Richardson, and W. R. Wolf
Aerospace Research Laboratories, ARL/LJ,
Wright-Patterson Air Force Base, Ohio 45433

W. D. Ross
Monsanto Research Corporation,
Dayton, Ohio 45407

and

N. M. Frew and T. L. Isenhour
Department of Chemistry, University of North Carolina,
Chapel Hill, North Carolina 27514

(*Received 22 February 1971; accepted in revised form 31 March 1971*)

Abstract—Beryllium and chromium have been determined by gas chromatography in Apollo 11 and Apollo 12 lunar dust and rock. The technique was also used to determine the amounts of these metals present in some U.S.G.S. standard rocks. In addition, beryllium was determined in a sample of the titanium-rich Buddington's Brown Point gabbro and in samples of the Allende and Murchison meteorites. Gas chromatographic analysis of beryllium in lunar dust and rock from Apollo 11 and Apollo 12 sites has revealed an interesting phenomenon termed the beryllium anomaly. The beryllium content of the lunar dust is much higher than that of any of the crystalline rocks analyzed. The content of beryllium in the crystalline rocks is less than 1 ppm, while lunar dust contains as much as 3.2 ppm. It appears unlikely that appreciable amounts of the extra beryllium can be attributed to either meteorites or the solar wind; however, a cryptic component is a potential source of the extra beryllium. There is no apparent correlation of the beryllium contents with silica contents. Mass spectrometry has been used to determine chromium isotope ratios in the lunar samples. These ratios are essentially identical with those of Earth and meteorite samples, indicating that either the lunar material was not subjected to intense proton irradiation in the early solar system or it has undergone homogenization as has been demonstrated for Earth and meteoritic matter.

Introduction

Ultra-trace analysis for beryllium and chromium has been conducted on Apollo 11 and Apollo 12 lunar samples using electron capture gas chromatography. The applicability of gas chromatography to the analysis of metal chelates has been demonstrated previously (Moshier and Sievers, 1965; Ross and Sievers, 1970); and with this work the technique has been extended to geological samples. For some metals this technique is more sensitive than any other analytical method known. The method consists of converting the metal to a volatile β-diketonate chelate with subsequent separation and analysis by electron capture gas chromatography. This technique is particularly well suited to the analysis of lunar and extraterrestrial samples since only very small amounts (typically 10 mg) are required for analysis. Analyses of lunar samples have revealed some interesting phenomena, relating in particular to an anomaly in beryllium content. Mass spectrometric measurements were made in order to determine whether differences in chromium isotope ratios exist. MURTHY and SANDOVAL (1965) have pointed out that ^{54}Cr is an excellent spallation indicator. They have calculated that one might expect a several-fold increase in the relative abundance of ^{54}Cr as against ^{52}Cr if intense proton irradiation occurred in the early solar system.

ANALYTICAL METHODS

Beryllium analysis

Samples in the form of rock chips are powdered in a "diamond" mortar. The rock powder or lunar dust sample is then mixed with approximately five times its weight of sodium carbonate and the mixture is fused in a platinum crucible. After dissolving the fusion mixture in dilute hydrochloric acid, the solution is quantitatively transferred to a 30 ml polyethylene bottle. The pH of the solution is adjusted with dilute sodium hydroxide to approximately 4. A sodium acetate-acetic acid buffer (pH 5.5) is added, followed by 1 ml of a 0.1 *N* solution of Na_2EDTA. (Addition of Na_2EDTA is necessary to mask interfering metals.) The bottle is then shaken for five minutes on a shaker and then heated in a water bath at 95°C for five minutes. After the solution cools, 10.0 ml of 0.0824 *N* solution of trifluoroacetylacetone, H(tfa), in "Nanograde" benzene is added using a pipet, and the polyethylene bottle shaken on the shaker for fifteen minutes. In separate experiments conducted in this study it was found that fifteen minutes was an optimum shaking time since shaking for longer periods of time resulted in the extraction of unacceptable amounts of $Al(tfa)_3$ which would interfere with the detection of the $Be(tfa)_2$. It was also found that essentially all the beryllium was recovered in that time. After shaking, the mixture is placed in a separatory funnel, the aqueous layer drained, and the benzene layer washed for five seconds with 0.1 *N* sodium hydroxide to remove any excess ligand (which if not removed would interfere with the analysis). The aqueous sodium hydroxide layer is immediately discarded and the benzene layer is ready for analysis by gas chromatography. Typically one microliter samples of the benzene layer are injected.

A Hewlett-Packard F & M Division Model #810 Research Gas Chromatograph equipped with a tritium electron capture detector was used. A thick-walled Teflon column (3 mm i.d., length 56 cm) packed with 10% SE-30 (a polydimethylsiloxane) coated on 70/80 mesh Gas Chrom Z was used for the beryllium analysis. The column was operated isothermally at 120°C, with the electron capture detector at 176°C, and the injection port at 161°C. The dried carrier gas was 10% methane/90% argon at a measured flow rate of 60 ml/min. The peak for $Be(tfa)_2$ appeared 1.7 minutes after injection. Measurement of peak heights was used for the analysis since the $Be(tfa)_2$ peak was extremely sharp without noticeable tailing. Blanks on the reagents were run periodically, and it was found that there was no contribution to the beryllium peak from the reagents. Measurement of beryllium was typically at the picogram level, the lower detection limit being about 4×10^{-14} gram. Calibration curves of concentration versus peak height were linear throughout the ranges studied. The efficacy of the method was established through separate analysis of aqueous beryllium nitrate solutions of various known concentrations. The average of 34 determinations showed 99.8% recovery with a relative standard deviation of 13%.

Standard solutions prepared from pure sublimed $Be(tfa)_2$ were intermittently injected during the analysis of unknowns. Fresh standards were prepared each week and a new calibration curve was generated each day. Results were obtained on data from one to four separate fusions of each sample usually with 4–5 aliquots of each distinct fused sample analyzed separately by injecting the organic layer at least five times.

Chromium analysis

The chromium content of the lunar samples has been determined by the following procedure: The sample is put into solution by fusion and dissolution as described above for beryllium. The pH is adjusted to *ca.* 4 and the solution is diluted with distilled, demineralized water to a fixed value. Final concentration of the sample in the aqueous solution is on the order of 2.5 mg/ml.

A 250 microliter aliquot (containing 0.6 mg of the original rock sample) is pipetted into a one-ml glass vial and 100 microliters of distilled, pure trifluoroacetylacetone, H(tfa), is added. The vial is cooled to dry ice temperature and sealed. The sealed vial is heated for two hours at 105°C, allowed to cool to room temperature and to stand overnight to ensure that the reaction has proceeded to completion. The vial is cooled in dry ice, opened and the entire vial and contents placed in a 30 ml polyethylene bottle containing 10.0 ml of "Nanograde" benzene which extracts and dissolves the $Cr(tfa)_3$. Ten ml of 1.0 *N* NaOH is then added to the bottle and the bottle and its contents are shaken mechanically for 30 minutes to wash from the benzene layer any excess ligand and the major fraction of aluminum chelate present. A 1.0 ml aliquot of the benzene layer is then diluted volumetrically to a final volume so that the chromium content is in the range of $2\text{–}4 \times 10^{-11}$ g Cr per microliter. One-microliter samples of this solution are then injected into the gas chromatograph for analysis.

A Hewlett-Packard Model 402 gas chromatograph equipped with a ^{63}Ni electron capture detector was used for the chromium analysis. The column was 122 cm by 3 mm i.d. Teflon, packed with 10% SE-30 on Gas Chrom Z (70/80 mesh), column temperature 170°C. The carrier gas was 10% methane/90% argon at flow rate of 60 ml/min.

Standard solutions of known concentration of $Cr(tfa)_3$ were made by dissolving a weighed amount of the pure chelate in "Nanograde" benzene. Analyses were performed by alternate injections of unknown and standard solutions, and the amount of Cr in the unknown sample was then obtained by a ratio of the peak heights of the unknown and standard solutions. The method was first shown to be valid for geological samples by studying samples of U.S.G.S. standard rock, BCR-1 (basalt), to which known amounts of chromium were added. Since the level of chromium naturally occurring in BCR-1 (28 ppm reported by THOMPSON *et al.*, 1969) is much lower than that in the lunar samples (> 2000 ppm) this sample could effectively be used as a blank to determine what, if any, interferences or chromium contamination may be introduced into the analysis. Taking the BCR-1 sample through the identical procedure used in analysis of the lunar samples showed neither interfering peaks in the chromatograms nor significant contamination of the sample with chromium from the reagents used.

The basalt samples were then spiked with a known amount of chromium, as the dichromate, in order to bring their chromium content up to the levels of the lunar samples. The recovery of this amount of added chromium was found to be $88.8 \pm 1.9\%$ for seven individual determinations. All subsequent chromium analyses were corrected for a recovery of 88.8%.

Measurement of chromium isotope ratios

Measurements were made by electron impact mass spectrometry after conversion of chromium into a volatile chelate. This method is especially suitable due to the limited amount of lunar sample available for study. Volatile chelates may be produced quantitatively from very small samples and show extreme sensitivity in the mass spectrometer. In this work, $Cr(tfa)_3$ was synthesized directly after dissolution of the lunar materials. A full discussion of microanalysis of geological samples for trace metals by mass spectrometry of the volatile chelates is given in FREW and ISENHOUR (1971). A brief description of the techniques used in this work follows.

Portions of the lunar samples, weighing from one to two milligrams, were dissolved by heating to 80°C with 20 microliters of concentrated HF in a sealed polyethylene tube for two hours. The solution products were oxidized during a second heating for two hours at 180°C with 50 microliters of concentrated $HClO_4$ in a sealed Pyrex tube. Direct reaction with H(tfa) in a sodium acetate buffer occurred in a third heating period of two hours at 110°C. $Cr(tfa)_3$, $Fe(tfa)_3$, and $Al(tfa)_3$ were all produced in large quantities and were taken up in one to two ml of hexane. In order to remove excess ligand and $Fe(tfa)_3$, which acts as an interference to the mass spectrometric measurement of ^{54}Cr, the hexane solution was given a quick washing with 1 *N* NaOH, shaking vigorously for 2–3 minutes. The organic layer was then drawn off and aliquots were introduced to the mass spectrometer. Standard fragments of $Cr(tfa)_3^+$, $Cr(tfa)_2^+$, $Cr(tfa)_2^+ - CF_3$, and others were readily identifiable, proving the existence of the chromium compound in the mass spectrometer.

The instrument used was an Associated Electronics Industries MS-902 high resolution mass spectrometer equipped with a direct insertion probe. Because of various effects leading to signal irreproducibility, e.g., sample volatilization rate, instrument noise and drift, etc., absolute abundance measurements were not attempted; instead, isotope ratios were measured using the peak switching circuitry of the MS-902, which alternately scans two different masses. The mass spectrometer is coupled to a Digital Equipment Corporation PDP-8 computer and Analog-to-Digital converter through an Applied Data Research interface. Slight modification of this system, using the peak switching circuitry of the mass spectrometer and some simple data collection and integration programs produces direct measurement of peak ratios to a relative standard deviation of 0.25% under ideal conditions. All measurements were made on the $Cr(tfa)_2^+$ fragment because this peak is the most intense in the spectrum and also appears in a mass region (m/e 356–360) which is virtually free of independent organic interferences. In order to calculate the exact isotope distributions, corrections must be applied for the contributions of ^{13}C, ^{2}H, and ^{18}O. However, direct comparison of the measured ratios between terrestrial and lunar samples may be used since the corrections are identical. Standard terrestrial samples were prepared using portions of U.S.G.S. BCR-1 which were spiked with 3.5 micrograms of chromium prepared from NBS Standard Isotope Reference Material 979. One to two milligram samples of lunar fines 12070,8 and lunar rock 12002,145 were measured and compared to the Earth standards.

DISCUSSION OF RESULTS OF BERYLLIUM ANALYSES

The concentrations of beryllium in the U.S.G.S. standard rocks, BCR-1 and W-1, as determined in this study agreed well with those obtained by spectrochemical analyses (FLANAGAN, 1967) and with those using fluorimetry (MEEHAN, 1969; SILL and WILLIS, 1962).

The beryllium concentrations determined for the various lunar samples are shown in Table 1, while the results for the terrestrial and meteoritic samples are shown in Table 2. In all cases except the Allende meteorite, results from different fusions indicated a fairly homogeneous distribution of beryllium throughout the sample, and average values are given in these tables. However, since the Allende results indicated inhomogeneity, the individual results are presented to illustrate the range of beryllium content in the 1.7 g fragment which was analyzed.

A homogeneity study of three separate 30 to 60 mg fragments from rock 12002,141 resulted in beryllium concentrations of 0.89 ± 0.06 ppm, 0.62 ± 0.05 ppm and 0.57 ± 0.05 ppm. The fragments were taken from locations approximately 0.5 cm apart in the rock. Consequently it was concluded that with respect to bulk composition the beryllium was distributed reasonably homogeneously throughout the available sample.

Table 1. Beryllium concentrations in lunar samples

Sample	Type	Beryllium Concentration (ppm)
12052,46	Apollo 12 crystalline rock	0.74 ± 0.07
12063,68	Apollo 12 crystalline rock	0.89 ± 0.07
12038,60	Apollo 12 crystalline rock	0.90 ± 0.13
12002,141	Apollo 12 crystalline rock	0.65 ± 0.06
12070,8	Apollo 12 fines	3.21 ± 0.12
10084,145	Apollo 11 fines	1.60 ± 0.16
10069,2	Apollo 11 breccia	2.20 ± 0.13
10070,2	Apollo 11 crystalline rock	0.90 ± 0.06

Table 2. Beryllium concentrations in terrestrial and meteoritic samples

Sample	Beryllium Concentration (ppm)		
	This Work	Spectrochemical Analyses	Fluorimetry
U.S.G.S. Standard Basalt, BCR-1	1.53 ± 0.18	<2‡	1.59 ± 0.11§
U.S.G.S. Standard Diabase, W-1	0.63 ± 0.05	—	0.623 ± 0.006‖
Buddington's Brown Point Gabbro	0.34 ± 0.05	—	—
Allende Meteorite NMNH 3529*	0.12 ± 0.01		
	0.030 ± 0.004	—	—
	0.047 ± 0.002		
Murchison Meteorite NMNH 5376†	0.029 ± 0.003	—	—

* Type III carbonaceous chondrite-interior fragment. Results are for three separate 80 to 100 mg fusions of the fragment.

† Type II carbonaceous chondrite-fragment about 0.5 cm away from the fusion crust.

‡ Flannagan (1967).

§ Meehan (1969).

‖ Sill and Willis (1962).

This study has revealed an interesting phenomenon regarding differences in the concentration of beryllium in the lunar samples. The beryllium contents of the lunar fines and breccia are much greater than those of any of the crystalline rock samples analyzed. This phenomenon will be referred to as the "beryllium anomaly" in subsequent discussion. It is clear from the work of the NASA Preliminary Examination Team and others that the chemical composition of the lunar dust is quite different in many respects from that of the crystalline rocks. Certainly the lunar fines cannot have been derived directly from any of the crystalline rock types so far examined. Either the dust has a different origin or it has been subjected to selenochemical modification. The beryllium concentrations of lunar breccias would be expected to lie closer to those of the dust samples rather than the crystalline rocks since breccias are considered to be essentially compacted regolith.

For some elements meteoritic enrichment of the dust has been postulated. However, in all cases where reliable methods have been used for meteorite analyses, the beryllium content of meteorites was much lower than in any of the lunar rocks. (For example, see the values for the Allende and Murchison meteorites in Table 1 and the values reported for chondritic, achondritic, and iron meteorites by Sill and Willis (1962).) This fact would appear to eliminate meteorites as a source of the extra beryllium.

The solar wind is another cause of elemental enrichment in the lunar dust, although it appears doubtful that much extra beryllium could be attributed to the solar wind. Assuming an average flux of 2.5×10^8 ions/cm^2 sec over the last 4.6×10^9 years (a reasonable upper value for the age of the Oceanus Procellarum region) and a beryllium ion concentration of 0.22 ppb in the solar wind based on the solar photospheric abundances of Müller (1968), a rough value of 7.9×10^{15} Be ions/cm^2 can be derived for the beryllium influx from the solar wind. With a lunar soil density of 1.8 g/cm^3 (Holt and Rennilson, 1970) and assuming complete vertical mixing throughout the regolith to an average depth of 2.25 m (Warner, 1970 and Shoemaker *et al.*, 1970), the beryllium enrichment in the soil would be only about 0.0003 ppm. Even if the unlikely assumption were made that no significant regolith mixing occurred and that the contingency sample were all from the top cm of soil, the solar wind contribution would be only about 0.07 ppm, which is still orders of magnitude less than the beryllium anomaly observed. Additionally, it has been reported (WARNER, 1970) that the concentration of the main solar wind constituents is greater in Apollo 11 fines than in the fines from Apollo 12. If the solar wind were a major source of the extra beryllium in the soil, a logical conclusion would be that the Apollo 11 soil would contain more beryllium than the Apollo 12 soil. Yet our analyses show that just the opposite is true, indicating that the solar wind cannot be a major contributor. Considerable attention has been given by various investigators to the presence of a cryptic component, an uncommon rock type which may account for various anomalies in the lunar soil composition (e.g., GOLES *et al.*, 1970; WOOD *et al.*, 1970a, b; GOLES, 1971). If the cryptic component hypothesis is valid, the cryptic material might be expected to contain a considerably higher beryllium concentration than the crystalline rocks.

Beryllium is known to be concentrated in the last stages of magmatic crystallization in silica-rich fractions. In the Apollo 12 samples, however, there is no correlation between silica and beryllium content. The fines 12070,8 have almost identical silica content (LSPET, 1970) to that of the crystalline rock 12052,46, and yet the beryllium content is several times greater in the fines. The silica content of rock 12038,60 is higher yet; in spite of this the beryllium concentration is low. Clearly, other factors must be responsible for the observed abundances.

DISCUSSION OF RESULTS OF CHROMIUM ANALYSES

Samples of U.S.G.S. standards DTS-1 (dunite) and PCC-1 (peridotite) were analyzed; these samples have chromium levels similar to the lunar samples. The results of these analyses are listed in Table 3. The excellent agreement with the accepted values for these extensively analyzed U.S.G.S. standards allowed us to apply this technique with confidence to the analysis of the lunar samples.

A minimum of five individual determinations was made on each of the lunar samples allotted to us by NASA and the results are presented in Table 3. Also listed in this table are previously reported values for the chromium content of several of these samples. The value of MORRISON *et al.* (1970) is taken as a representative value of

Table 3. Determination of chromium content in several lunar samples and U.S.G.S. Standards

Sample	Concentration of Cr Found (percent)	Values Reported in Literature (percent)
DTS-1	0.394 ± 0.017	0.397*
PCC-1	0.273 ± 0.008	0.285*
10084,145	0.209 ± 0.009	0.20†
12070,8	0.270 ± 0.016	0.28‡
12038,60	0.208 ± 0.019	0.22‡
12052,46	0.374 ± 0.011	0.37‡
12063,68	0.328 ± 0.004	—
12002,141	0.621 ± 0.031	—

* THOMPSON *et al.* (1969).
† MORRISON *et al.* (1970).
‡ LSPET (1970).

several reported for sample 10084. The variety of sample types represented in this data is of such limited range that no significant geological conclusions can be drawn from this data. However, it is interesting to note that sample number 12002,141 has a chromium content significantly higher than any of the other samples and indeed seems to have a higher chromium content than any other rocks examined by the NASA Preliminary Examination Team in their initial investigations. Almost all of the Apollo 11 and Apollo 12 rocks have chromium concentrations at much lower levels. The two examples of values reported in a higher range (12004, 0.58% and 12009, 0.52%) (LSPET, 1970) were for rocks gathered from the same small area on the lunar surface as sample 12002,141, which might indicate that they all came from a single source. GOLES *et al.* (1971) and BRUNFELT *et al.* (1971) report values of 0.56% and 0.54 to 0.67%, respectively, for the chromium content of rock 12002.

CHROMIUM ISOTOPE RATIO IMPLICATIONS

Effects of cosmic-ray induced spallation reactions in extraterrestrial bodies have received considerable attention because of the possible information they provide on exposure ages and on variations in the cosmic radiation flux over time. Numerous studies on iron meteorites and on Apollo 11 samples have demonstrated cosmogenic enrichments of stable and radioactive isotopes normally present in low abundances. Chromium isotope distributions are of interest in that the stable isotope ^{54}Cr is a sensitive indicator of spallation processes, being unshielded with respect to decay of the short-lived isobars ^{54}V and ^{54}Mn. In addition, the production rate of ^{54}Cr is sensitive to shielding effects and to the low energy particle flux. The abundances of chromium typically encountered in extra-terrestrial samples, however, are on the order of 3500 ppm (LSPET, 1970 and SCHMITT *et al.* 1965). Thus, given an experimental uncertainty of about 1% in chromium isotope determinations, the estimated ^{54}Cr production of a few ppb for present flux intensities would not be measurable.

A wide survey by SVEC *et al.* (1962) and FLESCH *et al.* (1960) of Cr isotopic abundances in major terrestrial deposits of chromite and in secondary minerals has revealed no significant variations. MURTHY and SANDOVAL (1965) have measured chromium isotopic distributions of various meteoritic samples, including both stony and iron meteorites and individual chondrules. They report that no variations exist, either among different meteorites or between meteoritic and terrestrial samples, and they conclude either that there has been no significant high energy proton irradiation in the early solar system, or that all of the meteorites studied are secondary objects derived from a primordial condensate after complete isotopic homogenization. Evidence and opposition to a primordial irradiation by neutrons has been presented in studies of ^{157}Gd/^{158}Gd ratios in Apollo 11 samples by MURTHY *et al.* (1970). In the present work, the measurement of chromium isotopic distributions in lunar samples was undertaken to determine whether this material was perhaps of different origin from the Earth or whether the same mixing process had taken place.

DISCUSSION OF RESULTS OF CHROMIUM ISOTOPE MEASUREMENTS

The results of multiplicate chromium isotope ratio measurements on 12070,8 and 12002,145 are given, along with a summary of terrestrial and meteoritic ratios measured

Table 4. Summary—chromium isotope ratios

	54/52	53/52	50/52
NBS 979*	0.02822	0.11339	0.05186
Terrestrial†	0.0284	0.1135	0.0520
Meteoritic‡ (stone)	0.0282	0.1135	0.0519
(iron)	0.0280	0.1138	0.0522
(chondrules)	0.0282	0.1136	0.0517
12070,8 (fines)§	0.0283 ± 0.0003	0.1135 ± 0.0007	0.0523 ± 0.0012
12002,145 (rock)§	0.0283 ± 0.0002	0.1133 ± 0.0006	0.0521 ± 0.0014

* SHIELDS et al. (1966).
† SVEC et al. (1967).
‡ MURTHY and SANDOVAL (1965).
§ This work.

to date, in Table 4. The standard deviations of the lunar ratios are generally one percent or less, with the exception of the $^{50}Cr/^{52}Cr$ ratio, where a spurious interference peak raised the noise level. The lunar distributions are seen to be in agreement with terrestrial values to better than one percent for all three ratios (measured as $^{50}Cr/^{52}Cr$, $^{53}Cr/^{52}Cr$, and $^{54}Cr/^{52}Cr$); further, no significant differences are observed between the crystalline rock and the fine materials.

Following considerations analogous to those of Murthy and Sandoval (1965), our order-of-magnitude estimate of ^{54}Cr production for the high proton flux postulated by Fowler et al. (1962) is roughly 2000–3000 ppm. The expected value for the $^{54}Cr/^{52}Cr$ ratio would thus be in the neighborhood of unity, measurably different from terrestrial values.

The observed uniformity of the results leads us to conclude that either (1) the samples of the lunar regolith studied do not constitute primordial matter and have undergone the same isotopic homogenization as that theorized for terrestrial and meteoritic samples; or (2) the high energy particle flux intensity during the early formative stages of the solar system was much lower than has been estimated.

Acknowledgements—The authors would like to thank Mr. Charles W. Harris and Drs. Ashley S. Hilton, William G. Scribner, and James L. Booker for assistance. We would also like to thank Dr. Gordon Goles for the sample of Buddington's Brown Point Gabbro, Drs. Brian H. Mason and Roy S. Clarke, Jr. for the Allende and Murchison meteorite samples, and Dr. Francis J. Flanagan for U.S.G.S. standard rock samples. We thank Dr. D. Rosenthal for his cooperation in our use of the mass spectrometer facilities of the Center for Mass Spectrometry, Research Triangle Institute, Durham, N.C. We also wish to acknowledge support of the mass spectrometric measurements from the Biotechnology Resources Branch of the Division of Research Resources, NIH, under grant PR-330. We are grateful for partial financial support of this project by NASA.

REFERENCES

BRANDT J. C. (1970) *Introduction to the Solar Wind*, p. 150, W. H. Freeman and Co.

BRUNFELT A. O., HEIER K. S., and STEINNES E. (1971) Determination of 40 elements in Apollo 12 materials by neutron activation analysis. Second Lunar Science Conference (unpublished proceedings).

FLANAGAN F. J. (1967) U.S. Geological Survey silicate rock standards. *Geochim. Cosmochim. Acta* **31**, 289–308.

FLESCH G. D., SVEC H. J., and STALEY H. G. (1960) The absolute abundance of the chromium isotopes in chromite. *Geochim. Cosmochim. Acta* **20**, 300–309.

FOWLER W. A., GREENSTEIN J., and HOYLE F. (1962) Nucleosynthesis during the early history of the solar system. *Geophys. J.* **6**, 148–220.

FREW N. M. and ISENHOUR T. L. (1971) Microanalysis of geological samples for trace metals by mass spectrometry of the volatile chelates. Submitted to *Anal. Chem.*

GOLES G. G. (1971) Comments on the genesis and evolution of Apollo XI "soil". *Lithos* **4**, 71–81.

GOLES G. G., DUNCAN A. R., OSAWA M., MARTIN M. R., BEYER R. L., LINDSTROM D. J., and RANDLE K. (1971) Analysis of Apollo 12 specimens and a mixing model for Apollo 12 "soils". Second Lunar Science Conference (unpublished proceedings).

GOLES G. G., RANDLE K., OSAWA M., LINDSTROM D. J., JEROME D. Y., STEINBORN T. L., BEYER R. L., MARTIN M. R., and MCKAY S. M. (1970) Interpretations and speculations on elemental abundances in lunar samples. *Proc. Apollo 11 Lunar Sci. Conf., Geochim. Cosmochim. Acta* Suppl. 1, Vol 2, pp. 1177–1194. Pergamon.

HOLT H. E. and RENNILSON J. J. (1970) Geologic investigation of the landing site. Part B. In Apollo 12 preliminary science report. NASA SP-235, p. 182.

LSPET (Lunar Sample Preliminary Examination Team) (1970) Preliminary examination of lunar samples from Apollo 12. *Science* **167**, 1325–1339.

MEEHAN W. R. (1969) Beryllium in six U.S.G.S. standard rocks. *Earth Planet. Sci. Lett.* **7**, 1–2.

MORRISON G. H., GERARD J. T., KASHUBA A. T., GANGADHARAM E. V., ROTHENBERG A. M., POTTER N. M., and MILLER G. B. (1970) Multielement analysis of lunar soil and rocks. *Science* **167**, 505–507.

MOSHIER R. W. and SIEVERS R. E. (1965) *Gas Chromatography of Metal Chelates*, Pergamon.

MÜLLER E. A. (1968) The solar abundances. In *Origin and Distribution of the Elements* (editor L. H. Ahrens), pp. 155–176, Pergamon.

MURTHY V. R. and SANDOVAL P. (1965) Chromium isotopes in meteorites. *J. Geophys. Res.* **70**, 4379–4382.

MURTHY V. R., SCHMITT R. A., and REY P. (1970) Rubidium-strontium age and elemental and isotopic abundances of some trace elements in lunar samples. *Science* **167**, 476–479.

ROSS W. D. and SIEVERS R. E. (1970) Trace metal analysis by gas chromatography. In *Developments in Applied Spectroscopy*, Vol. 8, (editor E. L. Grove), pp. 181–192, Plenum.

SCHMITT R. A., SMITH R. H., and GOLES G. G. (1965). Abundances of Na, Sc, Cr, Mn, Fe, Co, and Cu in 218 individual meteoritic chondrules via activation analysis. *J. Geophys. Res.* **70**, 2419–2444.

SHIELDS W. R., MURPHY T. J., CATANZARO E. J., and GARNER E. L. (1966). Absolute isotopic abundance ratios and the atomic weight of a reference sample of chromium. *Nat. Bur. Standards 70A* (*U.S.*) (Phys. and Chem.), 193–197.

SHOEMAKER E. M., HAIT M. H., SWANN G. A., SCHLEICHER D. L., DAHLEM D. H., SCHABER G. G., and SUTTON R. L. (1970). Lunar regolith at Tranquillity Base. *Science* **167**, 452–455.

SILL C. W. and WILLIS C. P. (1962). The beryllium content of some meteorites. *Geochim. Cosmochim. Acta* **26**, 1209–1214.

SVEC H. J., FLESCH G. D., and CAPELLEN J. (1962). The absolute abundance of the chromium isotopes in some secondary minerals. *Geochim. Cosmochim. Acta* **26**, 1351–1353.

THOMPSON G., BANKSTON D. C., and PASLEY S. M. (1969). Trace element data for U.S.G.S. reference silicate rocks. *Chem. Geol.* **5**, 215–221.

WARNER J. (1970). Apollo 12 lunar-sample information. NASA TR R–353.

WOOD J. A., DICKEY J. S. JR., MARVIN U. B., and POWELL B. N. (1970a). Lunar anorthosites and a geophysical model of the moon. *Proc. Apollo 11 Lunar Sci. Conf., Geochim. Cosmochim. Acta* Suppl. 1, Vol 1, pp. 965–988. Pergamon.

WOOD J. A., MARVIN U. B., POWELL B. N., and DICKEY J. S. Jr. (1970b). Mineraology and petrology of the Apollo 11 lunar sample. Smithsonian Astrophysical Observatory, Spec. Rep. No. 307.

Proceedings of the Second Lunar Science Conference, Volume 2, pp. 1461–1469.
The M.I.T. Press, 1971.

Li, B, Mg, and Ti isotopic abundances and search for trapped solar wind Li in Apollo 11 and Apollo 12 material

O. Eugster and R. Bernas
Centre de Spectrométrie Nucléaire et de Spectrométrie de Masse du C.N.R.S., 91 Orsay, France.

(*Received 22 February 1971; accepted in revised form 29 March 1971*)

Abstract—Isotopic abundances of Li, B, Mg, and Ti were measured by means of a sputtering ion source mass spectrometer in breccias 10046 and 10074, in rocks 12022 and 12065, in a coarse fine fragment from 10084 and in the fines 12070. No differences were found between lunar and terrestrial isotopic abundances within experimental uncertainties. The average isotopic ratio for all samples analyzed differ from terrestrial values by half a percent or less. In the breccias and in the coarse fine fragment, the Li isotopic composition was studied as a function of depth and at sites where lunar Li is strongly depleted but no isotopic variations due to a possible admixture of trapped solar wind Li or cosmic ray produced spallation Li were detected. Leaching experiments for the extraction of Li from the surface layers of lunar dust grains yielded terrestrial Li isotopic ratios, indicating that the proportion of solar Li is too low to be detected. It is shown that trapped solar wind Li should be looked for in a target material which contains 0.1 ppb Li or less and in which spallation Li production is relatively low.

Introduction

The purposes of this work were (1) to compare lunar and terrestrial isotopic compositions of Li, B, Mg, and Ti; and (2) to search for trapped solar wind Li, which is expected to be isotopically different from lunar Li, in breccias and in the outer layers of grains from the fines. All analyses were performed by means of a mass spectrometer equipped with a sputtering ion source (Yiou *et al.*, 1968). This technique has been successfully applied to the study of the distribution of major and trace elements in lunar material (Andersen *et al.*, 1970). In the present work we show that a sputtering ion source mass spectrometer is well suited for the isotopic study of Li, Mg, and Ti and that it is sensitive enough for the direct isotopic analysis of trace elements such as boron, which is technically difficult to analyze using other mass spectrometric techniques. For the direct analysis of a rock surface it is, of course, required that the isotopes of the element under study do not seriously interfere with the isotopes of the neighboring elements.

In previous studies no isotopic variations of B and Ti due to radioactive decay or nuclear reactions have been found in comparing terrestrial and extraterrestrial material. This is not surprising since no long lived radioactivities are known which could be responsible for anomalous abundances of B or Ti isotopes, and spallation effects are much too small. In a careful study of the meteoritical and lunar Mg isotopic abundances Schramm *et al.* (1970) applied chemical processing of the samples and used the surface ionization technique. They did not find any isotopic anomalies caused by the decay of extinct Al^{26} that were larger than their detection limit of a few parts in 10^4 and discussed the significance of this uniformity.

Table 1. Isotopic composition

Sample	Lithium $\Delta_{7,6} \pm 2\sigma_\Delta$	Boron $\Delta_{11,10} \pm 2\sigma_\Delta$	Magnesium $\Delta_{25,24} \pm 2\sigma_\Delta$	Magnesium $\Delta_{26,24} \pm 2\sigma_\Delta$
10046	$+0.1 \pm 1.4$	-1.9 ± 2.1	$+0.3 \pm 1.8$	$+1.6 \pm 1.8$
10084	$+0.7 \pm 1.9$	-2.7 ± 6.0	—	—
12022	-1.1 ± 2.0	$+1.1 \pm 1.7$	-1.3 ± 2.0	-0.5 ± 1.6
12065	$+2.5 \pm 2.8$	$+0.9 \pm 3.7$	-1.2 ± 3.0	-1.3 ± 3.0
12070	-1.1 ± 2.2	0 ± 5.0	$+0.1 \pm 1.7$	-0.3 ± 1.6
Average for all samples	$+0.2$	-0.5	-0.5	-0.1

For definition of $\Delta_{i,k}$, see text.

Analyses of lunar material by several authors have already revealed the presence of large amounts of implanted solar wind particles in the grain surface of lunar fines and breccias (cf. Eberhardt *et al.*, 1970). On the basis of theoretical considerations (Bernas *et al.*, 1967) the isotopes Li^6 and Li^7 are partly depleted by proton capture reactions at the bottom of the solar convective zone, the depletion being two orders of magnitude stronger for Li^6 than for Li^7. In fact, measurements of sun spot spectra yield a minimum value for Li^7/Li^6 of 20 (Schmal and Schröter, 1965) which is much higher than the terrestrial ratio of 12.5. The determination of the solar wind Li isotopic composition would yield valuable information on nuclear reactions having occurred below the surface of the sun.

Experimental Procedure

Rock samples were analyzed in the following way: a fragment of about 20 mg was bombarded by a 9 keV Cs^+ ion beam of 10^{-6} to 10^{-9} Amp intensity causing a fraction of the atoms at the rock surface to be sputtered away as positive ions which were separated in the mass spectrometer and recorded using an ion counting system. Sample consumption amounted to about 10^{-8} g for the isotopic analysis of a major element and up to 10^{-6} g for the analysis of a trace element for which a more intense primary Cs beam had to be used. Preparation of the rock samples involved no chemical processing, thus minimizing laboratory contamination. Lunar fines and, for comparison, samples from the USGS standard diabase W–1 were dissolved using concentrated HF and $HClO_4$ in a closed Teflon pot (Sanz and Wasserburg, 1969) under a stream of argon. An aliquot of the solution corresponding to about 10^{-4} g of solid sample was deposited on a previously cleaned and checked gold disc and the sample analyzed in the same way as the surface of the rock fragments.

Each isotopic analysis of a lunar sample was immediately preceded or followed by an analysis of a terrestrial standard in order to minimize errors due to instrumental isotopic fractionation effects. For each pair of lunar and terrestrial analysis $\Delta_{i,k}$ values were calculated using the following relation:

$$\Delta_{i,k} = \frac{(i/k)_{\text{lunar}} - (i/k)_{\text{terrestrial}}}{(i/k)_{\text{terrestrial}}} \times 100 \tag{1}$$

where i and k are the mass numbers of the isotopes analyzed.

Isotopic Composition of Li, B, Mg, and Ti

Table 1 summarizes the results we have obtained for the isotopic analyses of Li, B, Mg and Ti. The $\Delta_{i,k}$ values for each isotopic ratio were calculated according to relation 1. The following terrestrial standards were used: for Li a muscovite fragment

of Li, B, Mg, and Ti.

Titanium			
$\Delta_{46,48} \pm 2\sigma_\Delta$	$\Delta_{47,48} \pm 2\sigma_\Delta$	$\Delta_{49,48} \pm 2\sigma_\Delta$	$\Delta_{50,48} \pm 2\sigma_\Delta$
-0.9 ± 1.8	-0.6 ± 1.9	-0.9 ± 1.7	$+0.4 \pm 1.8$
—	—	—	—
-0.8 ± 1.9	-0.3 ± 2.5	$+1.5 \pm 2.5$	$+0.3 \pm 2.5$
$+1.9 \pm 2.0$	$+0.5 \pm 1.7$	$+0.1 \pm 2.0$	-0.1 ± 2.0
$+1.9 \pm 2.4$	-1.3 ± 2.4	-0.5 ± 1.6	-0.5 ± 1.6
$+0.5$	-0.4	$+0.05$	$+0.03$

$2\sigma_\Delta = 2(\sigma_{sample} + \sigma_{standard})$.

separated from a granite rock; for Mg the same muscovite and an olivine sample; for B a tourmaline; and for Ti a rutile sample. Typical analytical errors corresponding to a 95% confidence level were 2% for Li and B and 1% for Mg and Ti. For almost every pair of lunar and terrestrial measurements the error limits overlap, indicating no differences in the isotopic composition of these four elements. The errors given for the Δ values in Table 1 are the sum of the 2σ errors of the lunar and the corresponding terrestrial standard analysis. The average isotopic ratios for all samples analyzed differ from terrestrial values by half a percent or less. Hence an upper limit for a natural isotopic fractionation for these four elements in lunar material with respect to the terrestrial values can be set: for Li and B such an isotopic fractionation is less than 1% and for Mg and Ti less than half a percent per mass unit.

At present, the precision of our instrument is not sufficient for the detection of small scale isotopic anomalies. However, its high sensitivity for the isotopic analysis of many elements otherwise difficult to ionize, the minimal laboratory contamination and sample consumption, the possibility of choosing sample locations with dimensions of a few hundred microns and the rapidity and simplicity of the analytical procedure are important advantages.

Search for Trapped Solar Wind Li

In lunar surface material four Li components should be present: lunar Li, spallation-produced Li, trapped cosmic ray Li, and trapped solar wind Li. In order to compare the expected concentration of trapped solar wind Li with that of the three other Li components we estimated for the fines the amounts of the four Li components present in a surface layer of 0.2 μm which corresponds to the penetration depth of the solar wind ions in lunar material (Eberhardt *et al.*, 1970).

(1) Lunar Li: an average Li concentration of 12.5 ppm (Tera *et al.*, 1970) corresponds to 5×10^{13} atoms Li/cm^2 in a layer of 0.2 μm. (2) Li produced by spallation reaction within lunar material: assuming an exposure age of 500 m.y., an average production cross-section for Li of 20 mb, a total galactic cosmic ray flux of 1 proton/cm^2 sec and a differential kinetic energy spectrum proportional to $E^{-2.5}$, we calculate a concentration of 4×10^8 Li atoms/cm^2. The ratio Li^7/Li^6 after the decay of Be^7 should be about 2 (Bernas *et al.*, 1967). (3) Trapped galactic cosmic ray Li: in the

interstellar space Li is produced by spallation reactions of the galactic cosmic ray particles with interstellar matter. The cosmic ray abundance ratio Li/H is 5×10^{-4}. By assuming a mixing or accumulation rate for the lunar regolith of one meter per 500 m.y. the residence time of the grains in a galactic Li implanted surface layer of about 10 cm is 50 m.y. For a cosmic ray flux of one proton/cm^2 sec we calculate 10^6 atoms/cm^2 of cosmic ray Li trapped in a 0.2 μm layer. The ratio Li^7/Li^6 for this component should also be 2. (4) Trapped solar wind Li: optical measurements of photospheric and sun spot spectra by GREVESSE (1968) yielded a Li/H abundance ratio of 2.6×10^{-12}. Assuming this abundance ratio for the solar wind the amount of Li in lunar surface material can be estimated on the basis of the amounts of the surface correlated rare gases which are mainly trapped solar wind particles. This procedure takes into account a possible admixture of particles from the lunar atmosphere implanted on the grain surfaces by ion pumping processes (HEYMANN *et al.*, 1970; MANKA and MICHEL, 1970). The most abundant solar wind trapped rare gas element is He. However, because He may partly be lost by diffusion we based our estimate on the amount of trapped solar wind Xe, for which diffusion losses should be much lower and whose solar abundance should be comparable to that of Li. Since the solar abundance of Xe is not well known we use the cosmic abundance ratio Xe/H $= 3 \times 10^{-11}$ (ALLER, 1961) and find that Li is about ten times less abundant in the solar wind than Xe. The Xe surface concentration in solar wind irradiated lunar material has been found to be of the order of 10^9 atoms/cm^2 (cf. EBERHARDT *et al.*, 1970) so that the solar wind Li surface concentration would be 10^8 atoms/cm^2.

These estimates show that a complex mixture of Li atoms from different origins and presumably of varying isotopic composition could be found in a target material in which lunar Li is strongly depleted. Having available solar wind-irradiated lunar material as well as a mass spectrometer which samples just the surface layer corresponding to the solar wind implantation depth and which allows the investigation of single spots of a few hundred microns diameter, we decided to analyze the isotopic composition of Li at the surface of breccia fragments and in the lunar fines in which solar wind implanted particles are known to be extremely abundant. The following experiments were performed in order to study the sputtering characteristics of our instrument and its sensitivity for the detection of surface-implanted Li ions.

Erosion rate of the probe beam

The erosion rate due to the sputtering effect of our probe beam was measured as follows. A gold foil whose surface was enriched by Li^6 ions ($Li^6/Li^7 = 20$) was covered by a muscovite foil of 2 μm thickness containing Li of normal isotopic composition. Thereupon the muscovite foil surface was bombarded by the Cs^+ ion beam of 9 keV with an intensity of 4×10^{-7} Å on a surface of about 2×10^{-3} cm^2, while the isotopic ratio of the sputtered Li ions was monitored constantly. During 6 hours and 40 min the Li^7/Li^6 ratio stayed the same as for a standard Li sample. After that time the relative abundance of Li^6 started to increase rapidly indicating that the muscovite foil had been sputtered away at a rate of about one angstrom per sec at the spot where the probe beam hit the muscovite surface.

Sensitivity for detecting surface implanted Li ions

The sensitivity of our technique for the detection of surface-implanted ions was investigated carefully. By means of a specially designed ion implanter three previously cleaned and checked gold foils were bombarded by fluxes of 3×10^9 to 6×10^{11} Li^6 ions/cm^2 of 6 keV. This energy corresponds to that of the Li ions in the solar wind. After the implantations we measured the Li isotopic ratio at the surface of the gold foils as a function of depth. In Fig. 1 the depth dependent variation of the ratio Li^6/Li^7 are plotted as $\Delta_{6,7}$ values for the three implantation experiments. Note that for better clarity of the presentation of the Li^6 enrichments $\Delta_{6,7}$ values are used instead of the $\Delta_{7,6}$ values. At an irradiation dosage of 3×10^9 atoms Li^6/cm^2 we were not able to detect a Li^6 enrichment. The gold foils irradiated by 6×10^{10} and 6×10^{11} Li^6 atoms/cm^2 showed enhancements of up to a factor of 2 and 6, respectively, for the Li^6 abundance within a layer of about 0.1 μm. Thus we estimate an irradiation dosage of about 10^{10} atoms Li^6/cm^2 to be the minimum dosage necessary for the detection of a Li^6 enrichment. Since the surface area sampled by the sputtering probe beam is about 2×10^{-3} cm^2 the amounts of Li^6 for these two analyses correspond to 10^{-15} g and 10^{-14} g, respectively. By scanning a large surface area the detection limit could probably be lowered. However, two factors reduce the possibility of detecting solar wind Li in lunar material: (1) it seems very difficult to find material which is as Li poor as our gold foils; and (2) the sensitivity for the detection of a Li^7 excess is an order of magnitude lower than that for Li^6 because of the high natural ratio Li^7/Li^6.

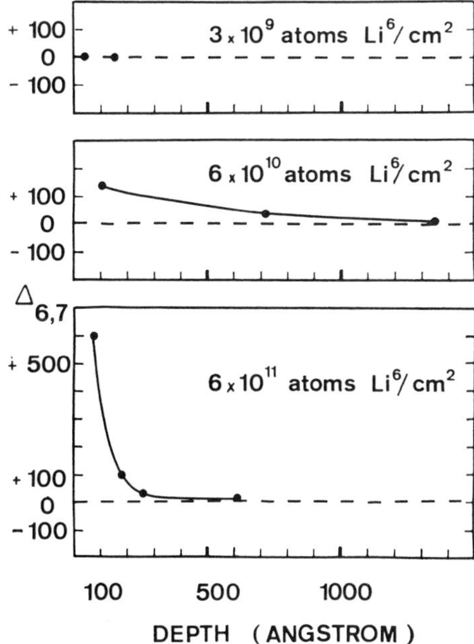

Fig. 1. Li isotopic ratios measured in the surface of three gold foils after irradiation by Li^6 ions. The $\Delta_{6,7}$ values are plotted as a function of depth below the surface. For definition of $\Delta_{6,7}$ values, see text.

Investigation of sample surfaces

Despite these expected difficulties we have looked for variations of the Li isotopic composition in breccias 10046 and 10074 and at the surface of a grain (diameter ~0.15 cm) of the coarse fines 10084. Figure 2 shows the $\Delta_{7,6}$ values obtained as a function of depth for the surface of a fragment of breccia 10074. Here we assume that the sputtering rate of the probe beam for this sample is the same as determined for the muscovite foil (1 angstrom/sec). Within experimental errors all Li^7/Li^6 ratios agree with the terrestrial one. For the breccia 10046 a systematic investigation of the surface of a fragment of 24 mg was performed. Using a probe beam of 0.04 cm diameter the Li isotopic composition and the intensity of the Li beam were measured at 35 different sites of the sample surface (Fig. 3, closed circles) and the $\Delta_{7,6}$ values were plotted versus the Li beam intensity for the spot analyzed. This experiment was repeated for a grain of the coarse fines 10084, whose surface was analyzed at seven different sites (Fig. 3, open circles). It is immediately apparent from Figs. 2 and 3 that our results do not indicate any systematic variations of the Li isotopic composition. For a constant probe beam the counting rate for Li ions is proportional to the number of Li ions sputtered from the sample surface. This number of Li ions is a function of the Li concentration and of the emission characteristics of the particular site. Thus the large variations of the Li counting rate are only partly due to the differences in the Li concentration of the spots analyzed. Nevertheless, it is evident that even at strongly Li depleted sites solar wind implanted or spallation produced Li are overshadowed by lunar Li.

The lunar fines have been studied differently. In order to check the possibility of extraction of trapped solar wind Li from an irradiated surface we performed a one

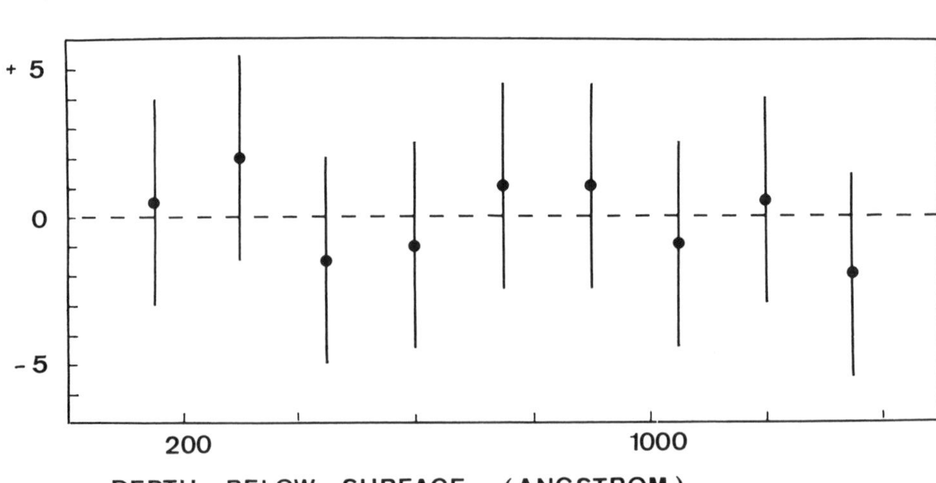

Fig. 2. Li isotopic ratios measured in the surface of a fragment of breccia 10074. The $\Delta_{7,6}$ values are plotted as a function of depth. For definition of $\Delta_{7,6}$ values, see text.

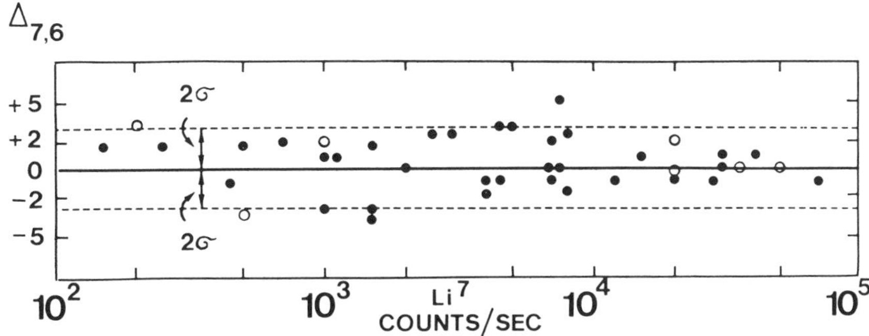

Fig. 3. Li isotopic ratios measured on the surface of breccia 10046 (closed circles) and of a fragment of the coarse fines 10084 (open circles). The $\Delta_{7,6}$ values are plotted as a function of the Li counting rate of the spot analyzed. For definition of $\Delta_{7,6}$ values, see text.

Table 2. Isotopic composition and amount of Li leached by H_2O from lunar fines 12070

	$\Delta_{7,6}$	$Li^7(10^{-12}$ g)
7.3 mg of unseparated fines		
Li dissolved after a treatment for 1 hour (treatment 1)	-2.5 ± 3	175
Same sample, after treatment 1:		
Li dissolved after a treatment for 3 days (treatment 2)	0 ± 3	77
Same sample, after treatment 2:		
Li dissolved after a treatment for 1 hour (treatment 3)	-1.5 ± 3	10
13.7 mg of 50 μm grain size fraction		
Li dissolved after a treatment for 1 hour	0 ± 3	105

For definition of $\Delta_{7,6}$, see text.

hour H_2O leaching experiment at room temperature for a gold foil which had been irradiated in the SIDONIE isotope separator of our laboratory by 2×10^{15} atoms Li^6/cm^2 at 6 keV. We measured a Li^7/Li^6 ratio of 0.075 and a Li^6 concentration of 2×10^{14} atoms/cm^2 of irradiated and leached gold surface. This corresponds to ten percent of the implanted Li^6. We are aware of the fact that the penetration and extraction characteristics for gold and lunar material are different, and damage to the crystal structures of a rock is more serious than that of pure gold. It is possible, too, that part of the Li may have been deposited and not implanted at the gold surface. Therefore, an additional irradiation of a rock surface similar to that of lunar rocks is planned.

In an attempt to extract trapped solar wind Li preferentially from lunar soil, a 7 mg sample of fines 12070 was treated three times successively with 0.5 cc H_2O. For aliquots of the three solutions an isotopic dilution experiment with a Li^6 tracer was carried out. The concentration and isotopic composition of the Li dissolved were determined and are summarized in Table 2. This experiment was repeated with a fraction of separated grains with diameters of about 50 μm. Repeated checks of the total blank corresponding to the Li content of H_2O including the possible Li contamination introduced by the handling procedure yielded 3×10^{-12} g Li/cc H_2O.

The concentrations of Li dissolved from the fines were between 2×10^{-10} g and 10^{-11} g and decreased with successive dissolution experiments. These small amounts of Li are of the order of magnitude of those typically analyzed in this laboratory in the course of the measurements of the isotopic cross sections for the production of Li, Be, and B in spallation reactions (YIOU *et al.*, 1968). The fact that even after a second and a third treatment of the same soil sample the amount of Li dissolved did not decrease drastically suggests that the Li measured is not a terrestrial surface contamination. Again we found that within experimental errors the Li isotopic abundances are the same as the terrestrial ones. The amount of Li dissolved far exceeds the expected amount of solar wind-implanted Li. Therefore, we did not perform an etching experiment with dilute acid which certainly would yield an even larger quantity of Li.

These experiments show that the proportion of lunar Li is too high for the detection of the other Li components. As discussed above the spallation component in a basaltic target material should exceed the solar wind component by a factor of four. Therefore, solar wind Li should be looked for in special materials such as the Fe-Ni pellets found in lunar dust (MASON *et al.*, 1970) for which the spallation cross section for the Li production is much smaller than for basaltic material. Assuming a solar wind Li/H ratio of 2.6×10^{-12}, a solar Li^7/Li^6 ratio of 20, a sensitivity for the detection of Li isotopic variations of 10% the Fe–Ni target material should contain 0.1 ppb Li or less to allow the detection of isotopic variations due to the admixture of solar wind Li.

Acknowledgments—We thank E. Gradsztajn for discussions and for writing the proposal for the solar wind Li experiment for NASA. We are very grateful to Christiane Carle for her competent help in operating the mass spectrometer. We thank M. Salomé and Mme Ligonnière for technical assistance, Dr. J. Chaumont and M. Baran for performing the Li implantations, and Drs. J. C. Allègre and Mme Christophe for providing us with the terrestrial samples used as references.

REFERENCES

ALLER L. H. (1961) *The Abundance of the Elements.* Interscience.

ANDERSEN C. A., HINTHORNE J. R., and FREDRIKSSON K. (1970) Ion microprobe analysis of lunar material from Apollo 11. *Proc. Apollo 11 Lunar Sci. Conf., Geochim. Cosmochim. Acta* Suppl. 1, Vol. 2, pp. 159–167. Pergamon.

BERNAS R., GRADSZTAJN E., REEVES H., and SCHATZMAN E. (1967) On the nucleosynthesis of lithium, beryllium and boron. *Ann. Phys.* **44**, 426–478.

EBERHARDT P., GEISS J., GRAF H., GRÖGLER N., KRÄHENBÜHL U., SCHWALLER H., SCHWARZMÜLLER J., and STETTLER A. (1970) Trapped solar wind noble gases, exposure age and K/Ar-age in Apollo 11 lunar fine material. *Proc. Apollo 11 Lunar Sci. Conf., Geochim. Cosmochim. Acta* Suppl. 1, Vol. 2, pp. 1037–1070. Pergamon.

GREVESSE N. (1968) Solar abundances of lithium, beryllium and boron. *Solar Physics* **5**, 159–180.

HEYMANN D., YANIV A., ADAMS J. A. S., and FRYER G. E. (1970) Inert gases in lunar samples. *Science* **167**, 555–558.

MANKA R. H. and MICHEL F. C. (1970) Lunar atmosphere as a source of argon-40 and other lunar surface elements. *Science* **169**, 278–280.

MASON B., FREDRIKSSON K., HENDERSON E. P., JAROSEWICH E., MELSON W. G., TOWE K. M., and WHITE J. S., JR. (1970) Mineralogy and petrography of lunar samples. *Science* **167**, 656–659.

SANZ H. G. and WASSERBURG G. J. (1969) Determination of an internal ^{87}Rb–^{87}Sr isochron for the Olivenza chondrite. *Earth Planet. Sci. Lett.* **6**, 335–345.

SCHMAL G. and SCHRÖTER E. H. (1965) Die Li-Häufigkeit und das Isotopenverhältnis Li^6/Li^7 in Sonnenflecken. *Z. Astrophys.* **62**, 143–153.

SCHRAMM D. N., TERA F., and WASSERBURG G. J. (1970) The isotopic abundance of ^{26}Mg and limits on ^{26}Al in the early solar system. *Earth Planet. Sci. Lett.* **10**, 44–59.

TERA F., EUGSTER O., BURNETT D. S., and WASSERBURG G. J. (1970) Comparative study of Li, Na, K, Rb, Cs, Ca, Sr, and Ba abundances in achondrites and in Apollo 11 lunar samples. *Proc. Apollo 11 Lunar Sci. Conf., Geochim. Cosmochim. Acta* Suppl. 1, Vol. 2, pp. 1637–1657. Pergamon.

YIOU F., BARIL M., DUFAURE DE CITRES J., FONTES P., GRADSZTAJN E., and BERNAS R. (1968). Mass-spectrometric measurement of lithium, beryllium and boron isotopes produced in O^{16} by high-energy protons, and some astrophysical implications. *Phys. Rev.* **166**, 968–974.

Proceedings of the Second Lunar Science Conference, Volume 2, pp. 1471–1485.
The M.I.T. Press, 1971.

Rubidium–strontium chronology and chemistry of lunar material from the Ocean of Storms

W. COMPSTON, H. BERRY, and M. J. VERNON
Department of Geophysics and Geochemistry, Australian National University, Canberra, Australia

and

B. W. CHAPPELL and M. J. KAYE
Department of Geology, Australian National University, Canberra, Australia

(Received 22 February 1971; accepted in revised form 29 March 1971)

Abstract—Chemical data suggest that the Apollo 12 igneous rocks are derivatives of at least six episodes of partial melting in the lunar mantle. Five mineral isochrons give ages for these events grouped around 3.3 b.y. but one with the apparently younger age of 2.94 ± 0.11 b.y. The initial $^{87}Sr/^{86}Sr$ for the igneous rocks ranges from 0.69938 ± 0.00006 to 0.69988 ± 0.00006 and at least five groups of igneous rocks can be distinguished on the basis of strontium isotopic composition. The fines contain a transferred component rich in many trace elements that have been added in greater amount to the light-colored fines. The igneous rocks, fines, and Apollo 11 group 2 igneous rocks and fines align on a model 4.5 b.y. isochron. This implies that there has been a lack of equilibrium between radiogenic and common strontium in the lunar mantle prior to magma generation and that the transferred component in the fines had retained a 4.5 b.y. model age prior to incorporation in the regolith.

INTRODUCTION

WE HAVE MADE a combined dating and chemical study of thirteen igneous rocks, one breccia, and three fines from the Apollo 12 material. As in our previous study of Apollo 11 samples (COMPSTON *et al.*, 1970a, 1970b) the prime purpose of the study of the major and trace element compositions was to test whether any specimens could be cogenetic and thus might validly be grouped on a ^{87}Sr-evolution diagram. As with the Apollo 11 samples, we have found that Rb–Sr ages on mineral separates are necessary to obtain satisfactory age-determinations.

ANALYTICAL METHODS

Trace and major element analyses (Table 1) were made by X-ray fluorescence spectrometry using the same methods as for the Apollo 11 samples. For specimens 12004,37 and 12051,39 the Apollo 11 chemical procedures were used for the isotope dilution analyses also, and precision for Rb in the pigeonite and ilmenite of these specimens (Table 2) was limited by the variability of the Rb blank (±1 ng). For all other specimens, variability of Rb blank was reduced to ±0.1 ng (σ) by minimizing exposure to laboratory air. Sr contamination was negligible for all analyses.

Comparative isotope-dilution (ID) and X-ray fluorescence (XRF) analyses (Table 3) for the total-rock samples show satisfactory agreement for Rb and for Sr except for 12009 and 12010, in which duplicate ID Sr analyses (not shown) failed to agree either with each other or with the XRF results. Analyses of 12009 by PAPANASTASSIOU and WASSERBURG (personal communication) are identical in Rb/Sr to our XRF result. Several other specimens also gave higher ID results but duplicates agreed with XRF. In retrospect, these difficulties with ID only seem explicable in such terms as loss of spike relative to sample by spraying during dissolution.

Table 1. X-ray fluorescence analyses

Rock type	Igneous rocks						
Sample	12004	12009	12018	12020	12035	12038	12040
SiO_2	44.59	45.03	44.21	44.66	43.17	46.56	44.08
TiO_2	2.88	2.90	2.62	2.73	2.28	3.31	2.41
Al_2O_3	8.02	8.59	7.42	7.31	8.03	12.53	7.18
FeO	22.03	21.03	21.73	21.58	22.20	17.99	21.27
MnO	0.29	0.28	0.28	0.28	0.29	0.27	0.28
MgO	12.66	11.55	15.05	13.91	15.49	6.71	16.21
CaO	9.05	9.42	8.25	8.73	8.08	11.62	8.10
Na_2O	0.20	0.23	0.20	0.21	0.21	0.66	0.19
K_2O	0.068	0.064	0.059	0.064	0.054	0.073	0.040
P_2O_5	0.08	0.07	0.07	0.08	0.06	0.14	0.06
S	0.07	0.06	0.05	0.06	0.05	0.06	0.04
Cr_2O_3	0.55	0.55	0.52	0.55	0.49	0.27	0.55
	100.49	99.77	100.46	100.16	100.40	100.19	100.41
$O \equiv S$	0.03	0.03	0.02	0.03	0.02	0.03	0.02
	100.46	99.74	100.44	100.13	100.38	100.16	100.39
Trace elements (ppm)							
Ba	55	60	55	60	45	120	40
Rb	1.13	1.03	0.94	1.03	0.83	0.48	0.49
Sr	96.0	95.6	86.1	91.4	100.4	185.8	80.6
Th	<1	<1	<1	<1	<1	<1	<1
U	<1	<1	<1	<1	<1	<1	<1
Zr	110	107	89	97	81	160	57
Nb	7	6	5	5	4	7	2
Y	36	34	30	32	29	46	22
La	5	4	3	4	3	10	2
Ce		10	6	11	6	25	2
V	145	153	140	146	130	104	153
Cr	3750	3790	3540	3780	3360	1840	3760
Co	52	49	54	50	52	25	52
Ni	52	52	55	50	33	2	40
Cu	9	(41)*	8	13	2	8	13
Zn	3	(30)*	<2	4	<2	3	8
Ga	1.2	2.0	1.5	1.8	1.8	3.8	1.6

* ? contamination.

Table 2. Isotope dilution analyses

	Sample Wt. (mg)	Rb* (ppm)	Sr† (ppm)	$^{87}Rb/^{86}Sr$*	$^{87}Sr/^{86}Sr$‡
12004,37					
Plagioclase	17.4	1.42 ± 0.08	328.3	0.0125 ± 0.0007	0.70039
Pigeonite	12.2	1.32 ± 0.13	231.5	0.0164 ± 0.0016	0.70055
Augite (1)	105.5	1.155 ± 0.018	128.2	0.0260 ± 0.0005	0.70094
Augite (2)	51.9	1.35 ± 0.035	156.5	0.0249 ± 0.008	0.70098
Total (1)	100.4	1.34 ± 0.02	100.0	(0.0377 ± 0.0008)	0.70133
Total (2)	107.7	1.123 ± 0.010	94.3	0.0344 ± 0.0007	0.70133
Ilmenite	8.2	1.86 ± 0.20	45.6	0.123 ± 0.013	0.70521
12004,59					
Ilmenite	11.7	1.749 ± 0.017	48.6	0.1036 ± 0.0014	0.70421
12051,39					
Plagioclase (1)	47.6	0.42 ± 0.04	394.1	0.0030 ± 0.0003	0.69969
Plagioclase (2)	51.5	0.72 ± 0.03	406.3	0.0051 ± 0.0003	0.69984
Total	102.1	0.96 ± 0.01	146.3	0.0189 ± 0.0005	0.70048
Ilmenite	14.3	1.30 ± 0.12	54.2	0.069 ± 0.006	0.70325
12038,45					
Plagioclase (1)	23.2	0.131 ± 0.008	391.0	0.00097 ± 0.00007	0.69948
Plagioclase (2)	28.3	0.172 ± 0.007	397.2	0.00131 ± 0.00006	0.69940

* Uncertainties shown are 95% confidence limits of precision.
† Coefficient of variation for precision is taken as 0.2% for all Sr analyses.

of Apollo 12 samples.

12051	12052	12064	12065	Breccia 12010	Fines 12033	12057	12070
45.07	46.13			46.27	46.96	45.74	45.83
4.62	3.35			3.30	2.50	2.91	2.81
9.96	9.95			10.04	13.99	12.13	12.48
20.25	20.70			20.33	14.65	17.43	16.81
0.28	0.28			0.27	0.19	0.23	0.23
7.21	8.07			8.29	8.96	9.90	10.18
11.45	10.89			10.81	10.68	10.44	10.45
0.28	0.26			0.29	0.66	0.45	0.43
0.080	0.071			0.124	0.41	0.25	0.27
0.09	0.08			0.13	0.43	0.26	0.31
0.09	0.07			0.07	0.07	0.07	0.12
0.26	0.46			0.43	0.29	0.35	0.30
99.64	100.31			100.35	99.69	100.16	100.22
0.04	0.03			0.03	0.03	0.03	0.06
99.60	100.28			100.32	99.76	100.13	100.16
	70	70		125	585	370	350
1.02	1.22	1.00	1.21	2.15	10.10	6.36	6.33
147.9	113.7	134.8	114.2	116.0	171.5	142.8	143.3
<1	<1	<1		1.9	10.3	5.8	6.6
<1	<1	<1		<1	2.7	1.2	1.6
128	121	114		175	762	497	512
7	7	7		11	44	29	30
48	40	41	40	51	162	114	111
5	5	5		9	48	33	29
	12	13		23	131	90	62
102	149	119		134	75	97	91
1780	3140	2020		2940	1990	2410	2080
33	34	25		34	32	40	45
6	6	7		32	108	158	186
6	8	7		11	19	7	6
<2	9	<2		6	14	8	6
2.9	2.1	3.1	2.4	2.4	3.1	2.4	2.5

of minerals concentrates.

	Sample Wt. (mg.)	Rb* (ppm)	Sr† (ppm)	$^{87}Rb/^{86}Sr$*	$^{87}Sr/^{86}Sr$‡
12308,45 (cont.)					
Total	99.8	0.60 ± 0.01	186.0	(0.0094 ± 0.0002)	0.69972
Pyroxene	29.5	0.345 ± 0.007	56.4	0.0176 ± 0.0007	0.70018
Ilmenite (1)	11.3	0.90 ± 0.02	64.0	0.0406 ± 0.0008	0.70132
Ilmenite (2)	8.3	0.65 ± 0.03	68.0	0.02745 ± 0.0014	0.70062
12040,13					
Total	108.3	0.52 ± 0.01	80.9	0.0185 — 0.0004	0.70047
Plagioclase	14.9	0.450 ± 0.013	364.5	0.00356 ± 0.00014	0.69975
Ilmenite	30.2	1.47 ± 0.02	33.8	0.1256 ± 0.0025	0.70534
12052,52					
Plagioclase (1)	36.9	2.14 ± 0.03	293.6	0.0211 ± 0.0004	0.70075
Plagioclase (2)	35.3	2.20 ± 0.03	301.2	0.0210 ± 0.0004	0.70066
Total	104.2	1.26 ± 0.02	113.8	0.0320 ± 0.0006	0.70123
Ilmenite	33.8	2.46 ± 0.04	98.7	0.0721 ± 0.014	0.70309
12010,24					
Feldspar – 100 ≠	13.4	3.73 ± 0.05	306.5	0.0352 ± 0.0007	0.70185
Feldspar – 350 ≠	24.4	3.51 ± 0.05	200.6	0.0505 ± 0.0010	0.70252

‡ 95% confidence limits of precision are taken as ±0.00010 for all ratios.

Table 3. Isotope dilution and X-ray fluorescence analyses for Rb, Sr, and $^{87}Sr/^{86}Sr$ for total samples of Apollo 12 rocks and fines.

Sample No.		Rb (ppm)*	Sr (ppm)*	$^{87}Rb/^{86}Sr$*	$^{87}Sr/^{86}Sr$†
12009,36	ID				0.70127
	XRF	1.03 ± 0.05	95.6 ± 0.2	0.0310 ± 0.0016	
12010,24	ID				0.70243
	XRF	2.15 ± 0.05	116.0 ± 0.3	0.0535 ± 0.0014	
12018,59	ID	0.915	85.1	0.0310	0.70114
	XRF	0.94 ± 0.06	86.1 ± 0.3	0.0313 ± 0.0021	
12020,33	ID	1.16	91.7	(0.0351)	0.70112
	XRF	1.033 ± 0.05	91.4 ± 0.2	0.0335 ± 0.0017	
12021,88	ID	1.15	122.9	0.02705	0.70090
12022,100	ID	0.77	149.9	0.0148	0.70011
12035,14	ID	0.93	106.5	0.0252	0.70073
	XRF	0.826 ± 0.05	100.4 ± 0.4	0.0237 ± 0.0015	
12038,45	ID	0.60	186.0	(0.0094)	0.69972
	XRF	0.475 ± 0.05	185.8 ± 0.3	0.00737 ± 0.00089	
12040,13	ID	0.52	80.9	0.0185	0.70047
	XRF	0.49 ± 0.06	80.6 ± 0.2	0.0175 ± 0.0021	
12052,52	ID	1.26	113.8	0.0320	0.70123
	XRF	1.22 ± 0.05	113.7 ± 0.2	0.0310 ± 0.0013	
12051,39	ID	0.96	146.3	0.0189	0.70048
	XRF	1.02 ± 0.05	147.9 ± 0.2	0.0200 ± 0.0010	
12064,33	ID	1.046		0.0226	0.70056
	XRF	1.00 ± 0.06	134.8 ± 0.3	0.0214 ± 0.0013	
12065,81	ID	1.185	113.2	0.0302	0.70102
	XRF	1.21 ± 0.10	114.2 ± 0.4	0.0305 ± 0.0028	
12004,37	ID	1.12	94.3	0.0344	0.70133
	XRF	1.13 ± 0.05	96.0 ± 0.3	0.0341 ± 0.0014	
12033,92	ID	10.03	173.3	0.1670	0.71008
	XRF	10.10 ± 0.09	171.5 ± 0.3	0.1699 ± 0.0017	
12057,75	ID	6.52	143.5	0.1311	0.70771
	XRF	6.36 ± 0.06	142.8 ± 0.4	0.1285 ± 0.0016	
12070,47	ID	6.387	144.6	0.1274	0.70753
		6.415	145.3		
	XRF	6.33 ± 0.06	143.3 ± 0.2	0.1275 ± 0.0014	

* The uncertainties shown for XRF represent 95% confidence limits of counting precision. The coefficient of variation for isotope dilution measurements of $^{87}Rb/^{86}Sr$ is conservatively estimated as 1.0%.

† 95% confidence limits of precision are taken as 0.00010 for all ratios.

The mass-spectrometric procedures are identical to those described previously, except that nearly all ratios have been measured using a $2 \times 10^{10}\ \Omega$ input resistor with ^{88}Sr beam intensity set at 1.0×10^{-10} amp as a standard operating condition. Our value for $^{87}Sr/^{86}Sr$ for the E. & A. standard under this condition is 0.70818 ± 0.00002 (2σ), as calculated from mixtures of the standard with ^{84}Sr spike, and our value for the C.I.T. seawater standard is 0.70931 ± 0.00003.

Chemistry

Chemical data on fifteen samples are given in Table 1. Samples 12021 and 12022, measured by mass-spectrometry, were not analyzed for major and trace elements and the data on 12064 and 12065 are incomplete. As LSPET (1970) have pointed out, the significant variation in the Apollo 12 igneous rocks is among the elements that favor ferromagnesian minerals. This variation suggests that the different rocks might have been derived by fractional crystallization from a common parent liquid. The operation of such a process would be critical to our interpretation of the Sr-isotopic data and hence we have considered it in some detail. Such a treatment has been possible only because we have had available information on the petrography of the igneous rocks and the nature and composition of the possible liquidus phases (GREEN *et al.*, 1971).

Liquidus phases in the Apollo 12 rocks in the earlier stages of crystallization are olivine and spinel, together with pigeonite in the quartz-normative types such as 12051 and 12052. As a hypothesis, one can suppose that all Mg-rich varieties, such as 12018 and 12040, which contain 24.3% and 27.5% normative olivine respectively, are cumulates from the same liquid that produced Mg-poor varieties such as 12051 and 12052 upon removal of olivine and spinel. Our calculations have been made on the basis of using 12009 composition as the primary liquid, although this precise composition is not essential to the establishment, or otherwise, of a simple fractionation relationship. Petrographically, 12009 is a quenched rock, and it can be assumed that it was either entirely liquid or contained at the most 7% of olivine before quenching on the lunar surface. Using the 12009 composition, we have calculated the effect of addition or removal of olivine; specifically we have sought to match the MgO and FeO contents of all the other rocks and thus to establish uniquely both the amount and composition of the olivine required to achieve these compositions. Certain constraints must be placed on the composition of the olivine (generally in the range Fo_{65} to Fo_{75}), and this, together with the calculated concentrations of all the other elements, provides a check on the feasibility of the process. If a marked discrepancy remains in Cr content after olivine addition or removal, the other possible liquidus phase, a chrome-spinel, can also be added or subtracted; this will affect the V content of the recalculated analysis which serves as a check on spinel movement. The olivine and spinel compositions used were those observed in the Apollo 12 rocks. Calculations have been made for all the igneous rocks analyzed.

The addition of 17.1 wt. % olivine with composition $Fo_{70.2}$ to 12009 gives good agreement with the observed 12018 composition for both major and trace elements. Similarly, if 11.2 wt. % of $Fo_{69.4}$ is added to 12009 the composition of 12020 is closely approximated. In the other direction, the subtraction of 11.8% of $Fo_{73.2}$ olivine from 12009 gives a match for MgO and FeO in 12052, and the additional removal of 0.24% spinel balances the Cr content with a remarkably good fit for other elements. The composition of the olivine on the liquidus of 12009 is Fo_{75} and in 12065, a similar rock to 12052, it is Fo_{67}. Thus in the above cases, involving addition or removal of olivine in the compositional range $Fo_{69.4}$ to $Fo_{73.2}$, the hypothesis of relating compositions by simple fractionation would seem to hold. It would not affect the validity of the calculations if a more Mg-rich composition than 12009 were chosen as the starting point; the rocks would still form a coherent and possibly cogenetic group.

Similar calculations show that all other rocks examined by us not only cannot be related simply to the 12009 group but are themselves each distinctive. Thus the addition of 7.4% olivine to 12009 gives a good agreement with 12004 except for slightly low Rb, Sr, Zr, and Y, but the composition of the olivine required, $Fo_{57.5}$, is much too Fe-rich to occur alone in the crystallization of 12009, and hence 12004 cannot have been fractionated from it or any liquid in the 12009 group. 12051, a quartz-normative rock, is superficially similar to 12052 and is presumably also a differentiate from an olivine-rich variety. Its derivation from 12009 would give a good fit for major elements, except Ti, but this is not true for trace elements (high Rb, low Sr, and Y).

Addition of 21.3% of $Fo_{67.8}$ to 12009 gives a match for MgO and FeO in 12035 but the Al_2O_3 content is 1 wt. % too high and, more seriously, the calculated Sr content is 22 ppm too low. A good major element fit can be obtained with 12040 but the Rb, Zr, and Y contents of that rock are too low for it to be simply related to any of the others. 12038 is clearly different; it is comparatively high in Na, more than two times any other rock, high in P, Ba, Zr, Y, REE, and low in Rb. For the two other rocks on which partial analyses are available, the data suggest that 12064 might be grouped with 12051 and 12065 with 12052.

Thus in terms of a simple olivine fractionation model we would be forced to place the 11 igneous rocks into 6 groups. In order of decreasing Rb/Sr ratios, these are

	Rb/Sr
12009 12018 12020 12052 (12065)	0.019 (average)
12004	0.0085
12035	0.0083
12051 (12064)	0.0069
12040	0.0061
12038	0.0026

We suggest that each of these groups represent a different episode of partial melting in the lunar mantle. Within each group, such as that of 12009, the variation can be ascribed to fractional crystallization of olivine and spinel either in a magma chamber, during flow, or during crystallization in a lava lake. There is some geographical correlation here. Thus 12009, 12018, and 12020 (with 12004) were all collected northwest of LM during EVA-1; all other samples came from the south during EVA-2.

The most useful trace elements in drawing the above distinctions are Rb, Sr, Y, and Zr. The abundances of Rb and Sr are somewhat erratic but Y and Zr, together with the REE, are well correlated. Thus relative to the 12009-18-20-52 series, considered on an olivine-free basis, 12004 is high in Y, Zr, and REE, 12035 is comparable, 12038 is high, 12040 is low, and 12051 high. The differing abundances of these elements could readily be explained by varying degrees of partial melting of a source region or regions in which on a large scale they were homogeneously distributed. This implies appreciable heterogeneity in Rb, Sr, and Rb/Sr in the source regions. Thus, for example, 12038 and 12051 are relatively high in Zr, Y, and REE, implying a lesser degree of partial melting. Rb is, on the other hand, relatively low in these rocks, suggesting that the source regions were appreciably lower in Rb. Such heterogeneity in Rb and Sr has important implications in the Rb–Sr dating and Sr-isotopic studies to be considered later.

We have analyzed three fines samples, the contingency fines 12070 collected 10 m northwest of LM during EVA-1, 12057, the fines from the bottom of ALSRC collected during EVA-2, and 12033, the light-colored fines taken from a trench near Head Crater (Table 1). Major and trace element compositions of the first two are very similar and as these two samples are from separate areas of the regolith this might suggest that the surface fines are of fairly uniform composition. The different composition implied by the lighter color of 12033 is borne out by the analysis. All fines show enrichment in certain trace elements, these having been added in greatest amount to 12033. The elements that show the greatest increase in the fines over their mean abundance in the igneous rocks, relative to their variability in the igneous rocks, are in decreasing order: Th, Rb, K, Zr, Nb, Ba, La, Y, and P (utilizing the LSPET data on Th in the igneous rocks). For these elements, the ratios of the increase in the light-colored fines over the other fines, above the mean igneous rock compositions, are closely grouped around 1.69. This suggests that the light-colored fines have been produced by the addition of a transferred component containing some 69% more of the elements listed above. This could have occurred through addition either of more of the same material or of a component in which these elements were more abundant.

The differences between the surface fines and 12033 show that the transferred component contains significant amounts of Si, Al, Na, K, P, Ba, Rb, Sr, Th, U, Zr, Nb, Y, REE, Cu, Zn, and Ga. An estimate of the relative amounts of some of these components can be made. Thus the transferred component would have a much higher Rb/Sr ratio than the igneous rocks. Again, the presence of P implies a phosphate component; Albee and Chodos (1970) have shown that the "whitlockite" in the Apollo 11 fines contains appreciable Y and REE. Thus these elements could have been added to the Apollo 11 fines in the phosphate phase. Th and U are also probably related to that phase; Burnett *et al.* (1970) have shown that U concentrations in rock 12013 are frequently associated with phosphates. Another important component could be alkali feldspar which would account for the increases in Si, Al, Na, K, Ba, Rb, Sr, and Ga.

We have analyzed 12010, a rock originally identified as a breccia by LSPET (1970) on the basis of its noble gas content. Table 1 shows that 12010 is remarkably similar to 12052 in its basic composition but it contains systematically higher amounts of those transferred components that characterize the fines. With the exception of Sr which is not significantly different from 12052, the other elements have been added consistently by 25–30% of the amount added to the surface fines.

Ages of Crystallization

Figure 1 summarizes our results for five internal Rb–Sr isochrons, which are detailed in Figs. 2–6. The value $1.39 \times 10^{-11}\ y^{-1}$ is taken for the ^{87}Rb decay constant. All points fit their respective isochrons to within the assigned error limits. The ages for three specimens—12038, 12052, and 12040—agree at 3.3 b.y. For another, 12051, the age is less precise at 3.58 ± 0.30 b.y. because of a wider variation in Rb blank during analysis of the particular mineral separates. It overlaps the error limits for 12038 and 12052, but may be slightly older than 12040. The age of 12004 is precisely indicated as 2.94 ± 0.11 b.y., substantially younger than the others. All specimens (except 12051 on these data) are significantly younger than the Apollo 11 lavas. Our age for 12051 agrees to within its wider uncertainty with the determinations of Papanastassiou and Wasserburg (1970) and of Turner (1971) who respectively give 3.26 ± 0.10 b.y. and 3.27 ± 0.05 b.y. for the same rock.

The younger age for 12004 was first indicated by the data for 12004,37 mineral separates. Subsequently, under conditions of lower variability in the Rb blank, another ilmenite (from chip 12004,59) was analyzed specifically to check this first

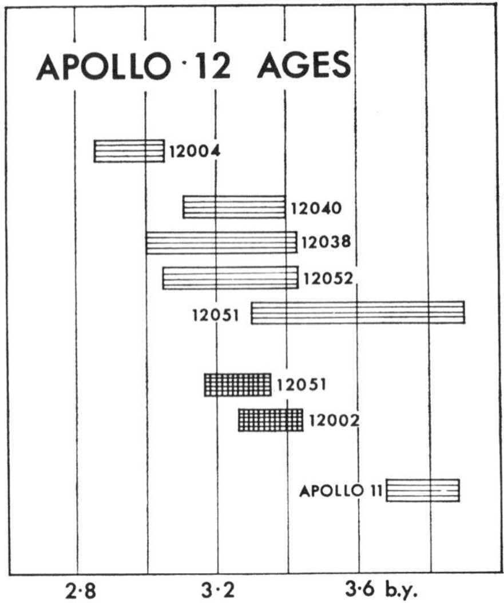

Fig. 1. Summary of Rb–Sr crystallization ages as 95% confidence intervals for five Apollo 12 rocks. Results of PAPANASTASSIOU and WASSERBURG (1970) for 12002 and 12051 (darker shading), and of COMPSTON *et al.* (1970a) for their group 1 Apollo 11 rocks are shown also for comparison.

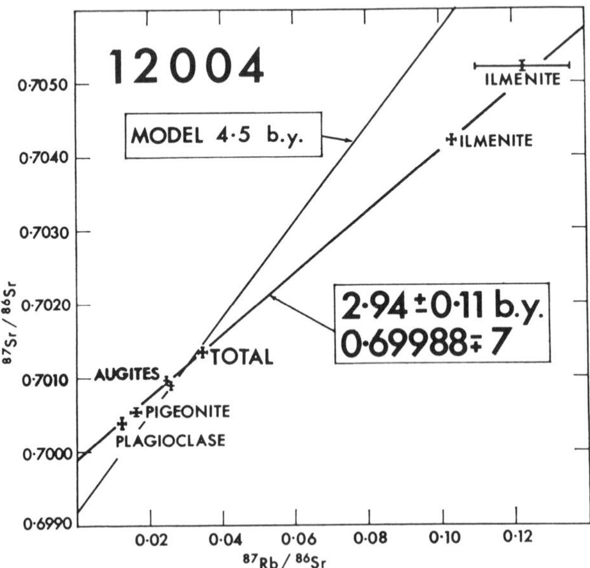

Fig. 2. Isochron diagram for 12004, a basalt which appears significantly younger than the others. A reference isochron of 4.5 b.y. through the basaltic achondrite initial $^{87}Sr/^{86}Sr$ is shown in this and in subsequent figures.

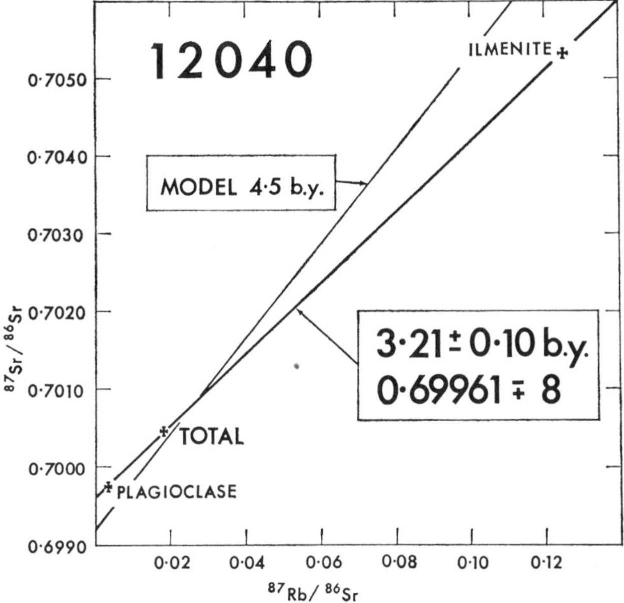

Fig. 3. Isochron diagram for 12040.

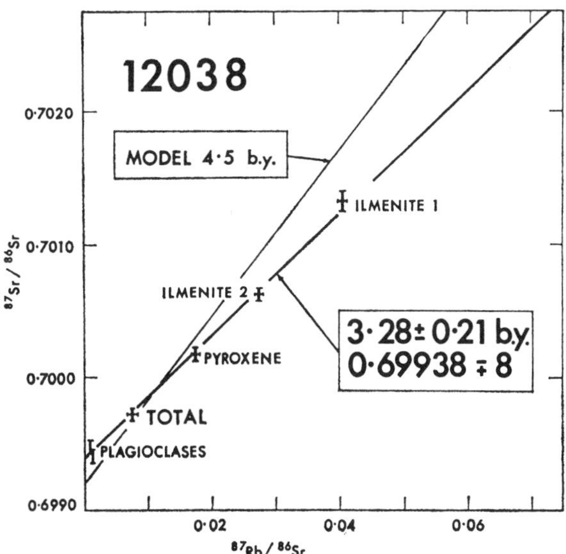

Fig. 4. Isochron diagram for 12038.

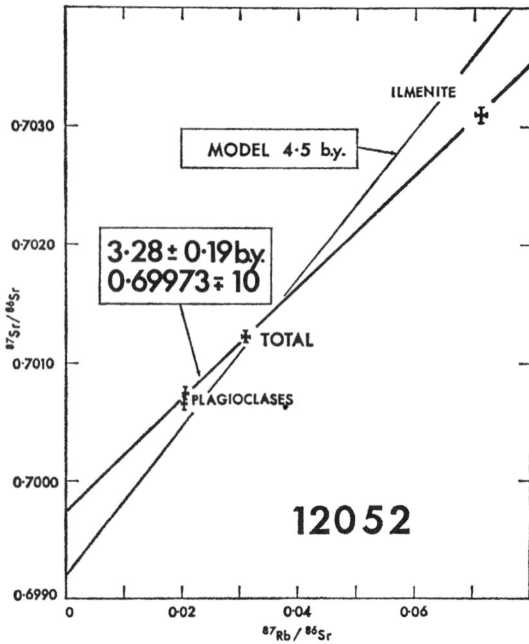

Fig. 5. Isochron diagram for 12052. Less dense material was not separated from the plagioclases, and evidently is contributing to their comparatively high Rb/Sr.

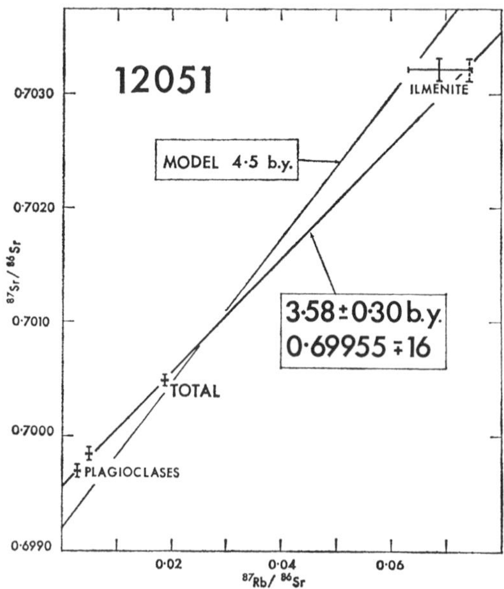

Fig. 6. Isochron diagram for 12051. The wide error limits for the ilmenite arise from Rb blank variability.

indication. The second analysis confirms it. A younger age for 12004 was also found by MURTHY *et al.* (1971).

Our age for 12040 is in good agreement with that of LUNATIC ASYLUM (1971) for the same rock. However, we cannot easily explain the disagreement for both slope and intercept between our results for 12052 and those of MURTHY *et al.* (1971).

VARIATIONS IN INITIAL $^{87}Sr/^{86}Sr$

The isochron data in Figs. 2–6 show that different Apollo 12 rocks had different $^{87}Sr/^{86}Sr$ when they crystallized at the lunar surface. 12038 is clearly lower than any of the others, and 12052 probably higher than 12040 and 12051. 12004 is certainly higher than 12040 and 12051 even if its age is taken as 3.2 b.y. PAPANASTASSIOU and WASSERBURG (1970) detect a difference in initial $^{87}Sr/^{86}Sr$ between 12002 and 12051. (Their value for 12051 is approximately 0.0002 lower than ours, as was also observed with Apollo 11 data and with the seawater standard, which is clearly due to a constant

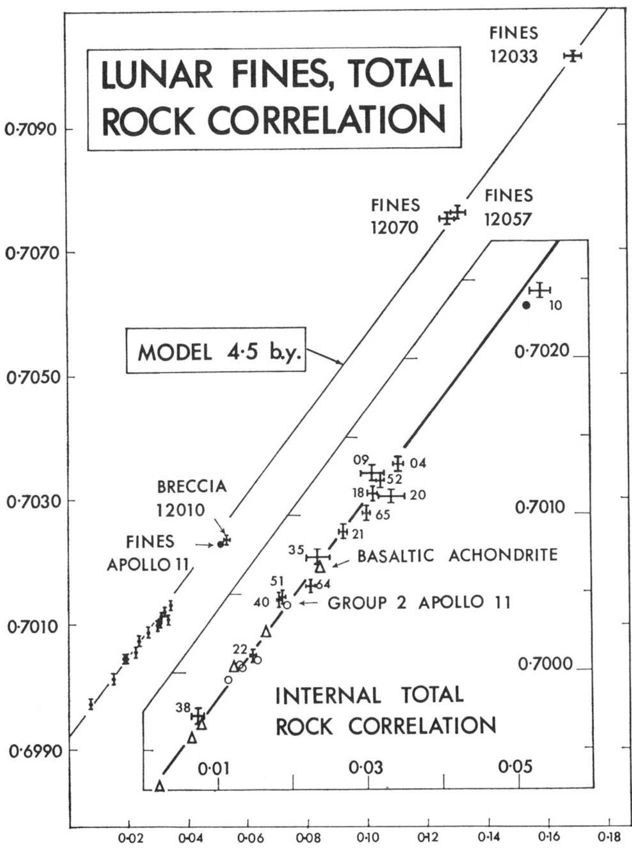

Fig. 7. Isochron diagram showing alignment of Apollo 12 fines and total rocks along the 4.5 b.y. reference isochron.

interlaboratory bias.) Such differences in $^{87}Sr/^{86}Sr$ signify that only certain of these rocks can be related to each other by any single process such as crystal fractionation from an originally uniform, primary melt. More than one melt or more than one process are required.

The situation for our data on Apollo 12 rocks is displayed in Fig. 7. The differences in initial $^{87}Sr/^{86}Sr$ are clearly associated with a good, though not perfect, alignment of the total rock points equivalent to a 4.50 b.y. isochron. The Apollo 11 group 2 rocks participate in this alignment, as do the basaltic achondrites, and also the fines from Apollo 11 and Apollo 12. If the age of the lunar mantle is taken as 4.50 b.y. and its initial $^{87}Sr/^{86}Sr$ as 0.69920 (PAPANASTASSIOU and WASSERBURG, 1969; BABI adjusted to our scale), the alignment of Fig. 7 can be interpreted as an array of single-stage Rb–Sr systems. It implies that little or no exchange of radiogenic and common Sr occurred within the lunar mantle during the period from 4.50 b.y. to the time of magma generation, at least over the dimensional scale at which the mantle has been sampled by the lavas. The Apollo 11 group 1 lavas would be one clear exception. A second interpretation of Fig. 7, that the basalts have been contaminated to varying degrees

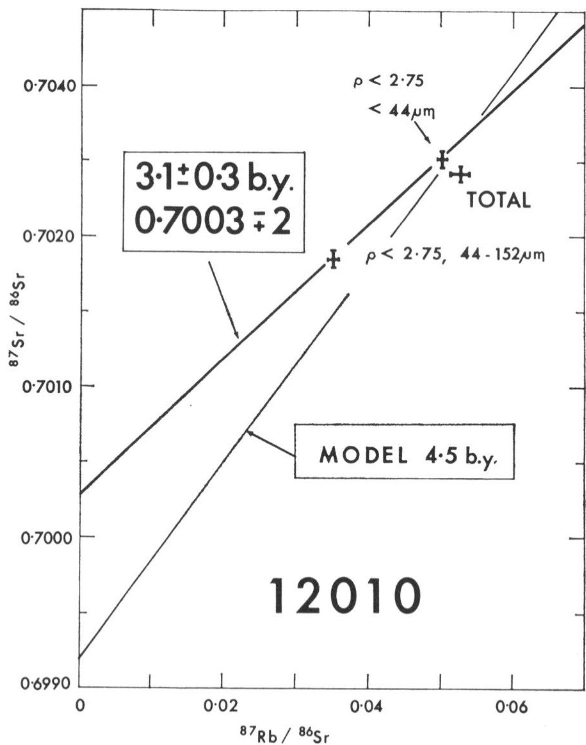

Fig. 8. Isochron diagram for breccia 12010. The two mineral concentrates, $\rho < 2.75$, are principally plagioclases but evidently contain traces of Rb-bearing phases. The two-point "isochron" through the minerals is unlikely to have direct time significance, in view of the composite origin of the rock.

by the fines during eruption, is not supported by the distribution of trace elements. Contamination should produce a strong correlation of Rb with elements enriched in the fines such as Zr and Y, and no such correlation is present in the igneous rocks. However, it is present in the breccia 12010, as already discussed (Fig. 8).

Consider a "model" magma defined as having a uniform initial chemistry and as maintaining a uniform $^{87}Sr/^{86}Sr$ during later crystal fractionation. Figure 7 shows that rock 12038 has a uniquely low initial $^{87}Sr/^{86}Sr$ among these Apollo 12 specimens, and would thus represent a particular model magma. Rock 12022 also seems lower in $^{87}Sr/^{86}Sr$ than the others, and may represent another magma. Rocks 12040, 12051, and 12064 could fit a single 3.2 b.y. isochron and may be representatives of a third magma, while 12052, 12009, 12018, 12002 could represent a fourth. Rock 12020 could fit either with the fourth group or with 12035, 12021, and 12065 to make a fifth group. Rock 12004 strongly resembles the 12009 group in isochron coordinates but cannot belong if its age is 2.94 b.y. This subdivision using initial $^{87}Sr/^{86}Sr$ is consistent with the grouping of specimens already made using trace and major elements. It successfully resolves the 12009 olivine-fractionated group (highest Rb/Sr), the high-Na rock 12038 (lowest Rb/Sr), and the intermediate 12051 and 12064.

The fines at the Apollo 12 site clearly contain a transferred component having a high Rb/Sr which, as shown by Fig. 7, has retained a 4.50 b.y. model age. It may be material which formed on the lunar surface at 4.50 b.y., or it may be material younger than 4.50 b.y. which, like the Apollo 12 basalt, bears no Rb–Sr record of its extrusion. The basalt data show that no separation of Rb and radiogenic Sr need occur during partial melting in the lunar mantle, so that $^{87}Sr/^{86}Sr$ in the total rock at the time of surface crystallization will correlate with Rb/Sr. It follows that $^{87}Sr/^{86}Sr$ in the Rb-rich transferred component may have been abnormally high when the latter first formed on the lunar surface, and that no younger limit to its age can be set without data from its mineral components.

Acknowledgments—We thank Dr. D. H. Green for helpful discussion and M. Cowan, Z. Wasik, G. Lea, Mrs. Agnes Brown, and Mrs. K. Phillips for their assistance with the experimental work.

References

Albee A. L., and Chodos A. A. (1970) Microprobe investigations on Apollo 11 samples. *Proc. Apollo 11 Lunar Sci. Conf., Geochim. Cosmochim. Acta* Suppl. 1, Vol. 1, pp. 135–157. Pergamon.

Burnett D. S., Monnin M., Seitz M., Walker R., Woolum D., and Yuhas D. (1970) Charged particle track studies in lunar rock 12013. *Earth Planet. Sci. Lett.* **9**, 127–136.

Compston W., Arriens P. A., Vernon M. J., and Chappell B. W. (1970a) Rubidium–strontium chronology and chemistry of lunar material. *Science* **167**, 474–476.

Compston W., Chappell B. W., Arriens P. A., and Vernon M. J. (1970b) The chemistry and age of Apollo 11 lunar material. *Proc. Apollo 11 Lunar Sci. Conf., Geochim. Cosmochim. Acta* Suppl. 1, Vol. 2, pp. 1007–1027. Pergamon.

Green D. H., Ringwood A. E., Ware N. G., Hibberson W. O., Major A., and Kiss E. (1971) Experimental petrology and petrogenesis of Apollo 12 basalts. Second Lunar Science Conference (unpublished proceedings).

LSPET (Lunar Sample Preliminary Examination Team) (1970) Preliminary examination of lunar samples from Apollo 12, *Science* **167**, 1325–1339.

LUNATIC ASYLUM (1971) Rb–Sr ages, chemical abundance patterns and history of lunar rocks. Second Lunar Science Conference (unpublished proceedings).

MCINTYRE G. A., BROOKS C., COMPSTON W., and TUREK A. (1966) The statistical assessment of Rb–Sr isochrons. *J. Geophys. Res.* **71**, 5459–5468.

MURTHY V. R., EVENSEN N. M., JAHN B., and COSCIO M. R., JR. (1971) Rb–Sr ages and elemental abundances of K, Rb, Sr, and Ba in samples from the Ocean of Storms. Second Lunar Science Conference (unpublished proceedings).

PAPANASTASSIOU D. A., and WASSERBURG G. J. (1969) Initial strontium isotopic abundances and the resolution of small time differences in the formation of planetary objects. *Earth Planet. Sci. Lett.* **5**, 361–376.

PAPANASTASSIOU D. A., and WASSERBURG G. J. (1970) Rb–Sr ages from the Ocean of Storms. *Earth Planet. Sci. Letters* **8**, 269–278.

TURNER G. (1971) ^{40}Ar–^{39}Ar ages from the Lunar Maria. Second Lunar Science Conference (unpublished proceedings).

APPENDIX. DESCRIPTION OF MINERAL CONCENTRATES

12004,37

Plagioclase 0.017 gm (−100 mesh). Freed from cristobalite but containing traces of olivine and ilmenite.

Pigeonite 0.0122 gm (−150/200 mesh fraction). Containing traces only of plagioclase olivine and ilmenite in composite grains.

Augite 1 0.1055 gm (−150/200 mesh fraction). Containing pyroxene/ilmenite and plagioclase/ ilmenite composite grains in small amounts.

Augite 2 0.0519 gm (−150/200 mesh fraction). Grain purity about 98%. No discernable plagioclase but traces of discrete ilmenite.

Ilmenite 0.0082 gm (−350 mesh). Contains about 10% of ilmenite, largely in discrete grains but not separable from the associated pyroxene.

12004,59

Ilmenite 0.0117 gm (−200 mesh). A complex of the heavy minerals clinopyroxene, ilmenite, and olivine containing approximately 50% ilmenite as discrete grains. Very few composites and no discernible plagioclase.

12010,24

Plagioclase 0.0134 gm (−100/350 mesh fraction). About 95% grain purity. Trace of olivine but negligible ilmenite. No visible pyroxene. Cristobalite not separated.

Plagioclase 0.0244 gm (−350 mesh fraction). Containing a very small percentage of opaque material, chiefly ilmenite. Not less than 90% grain purity.

12038,45

Plagioclase 1 0.0232 gm (−100 mesh). Very pure plagioclase containing very minor inclusions of ilmenite. Free from pyroxene and olivine. Cristobalite not separated.

Plagioclase 2 0.0283 gm (−100 mesh). Duplicate of Plagioclase 1.

Pyroxene 0.0295 gm (−100/350 fraction). Handpicked pyroxene consisting of augite and pigeonite, the former predominating.

Ilmenite 1 0.0113 gm (−100/200 fraction). A concentrate of heavy minerals containing approximately 10% of ilmenite in discrete form and the remainder pyroxene with a few pyroxene/ilmenite composite grains.

Ilmenite 2 0.0083 gm (100/350 fraction). A concentrate similar to Ilmenite 1 in composition but containing approximately 30% of free ilmenite grains.

12040,13

Plagioclase 0.0149 gm (−200/350 mesh fraction). Approaching 100% pure mineral. Very slight trace of ilmenite in composites but no pyroxene or olivine. Cristobalite not separated.

Ilmenite 0.0302 gm (−100/350 mesh fraction). From 40 to 50% free ilmenite with olivine as main impurity. Occasional grains of pyroxene/ilmenite and olivine/ilmenite composites. Plagioclase not visible but may be present in trace amounts as composite grains.

12051,39

Plagioclase 1 0.0476 gm (−100/200 mesh fraction). Freed from cristobalite. A high purity plagioclase containing traces of ilmenite and pyroxene but only in composite grains.

Plagioclase 2 0.0515 gm (−200 mesh). Approximately 100% grain purity and freed from cristobalite. Contains a trace of ilmenite in composite grains.

Ilmenite 0.0143 gm (−100 mesh). A concentrate containing approximately 70% of discrete mineral, about 20% of dark pyroxene (augite) and 10% of pyroxene/ilmenite composite grains. No free plagioclase but possibly a trace in composite grains.

12052,52

Plagioclase 1 0.0369 mg (−350 mesh). Containing very minor inclusions of ilmenite and traces of pyroxene/plagioclase composites. Cristobalite not separated.

Plagioclase 2 0.0353 gm (−350 mesh). Possibly slightly better grain purity than Plagioclase 1. Lighter in color due to lower percentage of included matter. Cristobalite not separated.

Ilmenite 0.0338 gm (−350 mesh). Chiefly clinopyroxene with a small proportion in composite grains with ilmenite. Negligible plagioclase or olivine but containing from 10 to 15% of discrete ilmenite.

Proceedings of the Second Lunar Science Conference, Volume 2, pp. 1487–1491.
The M.I.T. Press, 1971.

Sr isotopic measurements in Apollo 12 samples

MICHAEL L. BOTTINO
Marshall University, Huntington, West Virginia 25701 and Planetology Branch, Goddard Space Flight Center, Greenbelt, Maryland 20771

PAUL D. FULLAGAR
University of North Carolina, Chapel Hill, North Carolina 27514

and

CHARLES C. SCHNETZLER and JOHN A. PHILPOTTS
Planetology Branch, Goddard Space Flight Center, Greenbelt, Maryland 20771

(*Received 22 February 1971; accepted in revised form 31 March 1971*)

Abstract—Rb–Sr isotopic composition studies of eight Apollo 12 regolith samples yield model ages that range from 4.2 to 5.1 b.y. and average 4.66 b.y. The ages and variation in ages can be explained in terms of a simple mixing model involving a material similar to the dark-colored portion of 12013,10 and the normal Apollo 12 igneous rocks.

The isochron age of 12013,10 based on "whole-rock" chips and powders is 4.01 ± 0.15 b.y. with an initial ratio of 0.7047 ± 0.0015. This represents a minimum age for the formation of rock 12013. The high initial ratio and high average model age (4.5 b.y.) suggest this rock formed 4.5 b.y. ago and underwent a thermal or shock event at 4.0 b.y. which was sufficient to redistribute Sr.

INTRODUCTION

THIS PAPER summarizes our Rb, Sr, and Sr isotopic analyses of 27 samples of igneous rocks, soils, core, breccia, and mineral separates from Apollo 12 material. For a description of the samples and other trace element results, see SCHNETZLER *et al.* (1970) and SCHNETZLER and PHILPOTTS (1971). All data were obtained by mass spectrometric stable isotope dilution. The techniques employed in this study were the same as previously reported by this laboratory (SCHNETZLER *et al.*, 1970). All Sr isotopic and concentration measurements were made using a ^{84}Sr spike. Sr and Rb blanks were 0.07 and 0.003 μg, respectively. The ^{87}Sr/^{86}Sr value obtained for six analyses of the Eimer and Amend $SrCO_3$ standard, determined during the period of this investigation, was 0.7075 ± 0.0002 (2 $\bar{\sigma}$). All ^{87}Sr/^{86}Sr measured values were normalized to ^{86}Sr/^{88}Sr = 0.1194. The error of a Sr isotopic measurement is estimated to be less than 0.1%. The error for ^{87}Rb/^{86}Sr is estimated to be ±3%.

RESULTS AND DISCUSSION

The data are given in Table 1. The last column lists model ages calculated using an initial ^{87}Sr/^{86}Sr ratio of 0.6990.

The igneous rocks have very low ^{87}Rb/^{86}Sr ratios, similar to the Apollo 11 low alkali type, and consequently have little enrichment of radiogenic strontium. This is consistent with the results reported by other laboratories (PAPANASTASSIOU and WASSERBURG, 1970; COMPSTON *et al.*, 1971; MURTHY *et al.*, 1971; HUBBARD *et al.*, 1971; CLIFF *et al.*, 1971). Because of the small enrichment of radiogenic Sr, we have not calculated model ages for these rocks.

Table 1. Rb–Sr analytical results for Apollo 12 materials

		wt. (mg)	Rb (ppm)	Sr (ppm)	$^{87}Rb/^{86}Sr$	$^{87}Sr/^{86}Sr$ (a)	Model age (b)
Soils, etc.							
12001	Soil, fine	241	6.48	145.5	0.130	0.7074	4.50
12032	Soil, fine	195	9.24	161.4	0.166	0.7112	5.10
12070	Soil, fine	197	6.44	144.0	0.1295	0.7079	4.78
		198	6.50	144.4	0.1304	0.7078	4.70
12073	Breccia	131	10.2	164.8	0.179	0.7109	4.63
12025,71	Core	110	6.84	144.4	0.135	0.7074	4.35
12028,65	Core	96	0.613	80.9	0.0219	0.7004	—
12028,89	Core	92	7.84	155.2	0.147	0.7078	4.18
12028,120	Core	103	7.93	152.7	0.151	0.7100	5.06
12028,144	Core	113	8.96	154.9	0.168	0.7105	4.75
12032	Coarse-fine frag.	23	0.498	124.7	0.00399	0.6996	—
12013,10							
10 dark-colored portion		10.0	25.8	200	0.373	0.7260	5.03
15 dark-colored portion		5.2	18.7	157	0.346	0.7251	5.23
8 feldspar from dark portion		3.5	38.4	329	0.339	0.7210	4.52
23 light and dark chip		68.1	47.4	120	1.15	0.7727	4.46
37 + 24 light-colored portion		33.4	56.7	131	1.26	0.7784	4.40
8 light-colored portion		18.4	57.2	161	1.03	0.7583	4.02
15 light-colored portion		14.5	127	126	2.98	0.8777	4.19
15A $\rho < 2.62$		1.3	197	151	3.85	0.920	4.02
15B $2.62 < \rho < 2.93$		7.7	112	159	2.07	0.8256	4.28
15C $2.93 < \rho < 3.21$		8.3	48.8	96.8	1.47	0.7857	4.12
15D $3.21 < \rho < 3.50$		7.1	15.4	50.3	0.891	0.7608	4.82
15E $\rho > 3.50$		1.8	4.69	8.3	1.66	(0.795)	(4.07)
Igneous Rocks							
12018	Whole rock	175	1.04	89.3	0.0336	0.7006	—
12021	Martix	178	1.15	128.5	0.0259	0.7015	—
12038	Whole rock	124	0.604	190	0.00926	0.699	—
12040	Whole rock	148	1.00	85.5	0.0340	0.7002	—
12052	Whole rock	204	1.26	116	0.0309	0.699	—

(a) Normalized to $^{86}Sr/^{88}Sr = 0.1194$. (b) In 10^9 years, calculated, assuming initial $^{87}Sr/^{86}Sr = 0.6990$ and $\lambda_{Rb}87 = 1.39 \times 10^{-11}$ yr − 1.

Our data for subsamples of 12013,10 are given in Table 1 and plotted in Fig. 1, a standard Rb–Sr isochron plot. These data include new determinations, not previously reported in our preliminary paper on this sample (SCHNETZLER *et al.*, 1970). Calculated model ages for the eleven subsamples range from 4.02 to 5.23 b.y. and average 4.46 b.y. (sample No. 15E was excluded from all calculations because of high blank corrections). Excluding the density separates, the remaining seven samples yield an average model age of 4.55 b.y. This "age" is shown as a dashed-line in Fig. 1. This average model age is in excellent agreement with the average model age of 4.52 ± 0.06 b.y. obtained on a number of fragments by LUNATIC ASYLUM (1970). They also found a range in model ages—from 4.17 to 5.1 b.y. Thus, 4.5 b.y. might be considered a "whole-rock" model age, as it is the average of a large number of subsamples (approximately 17).

However, the best-fit isochron, calculated using the YORK (1966) least-squares method for eleven samples, yields an age of 4.08 + 0.12 (all errors quoted as one σ) and an initial ratio of 0.7036 + 0.0014. This isochron is shown on Fig. 1 by the solid line. Using only data for the six "whole-rock" chips and powders (which,

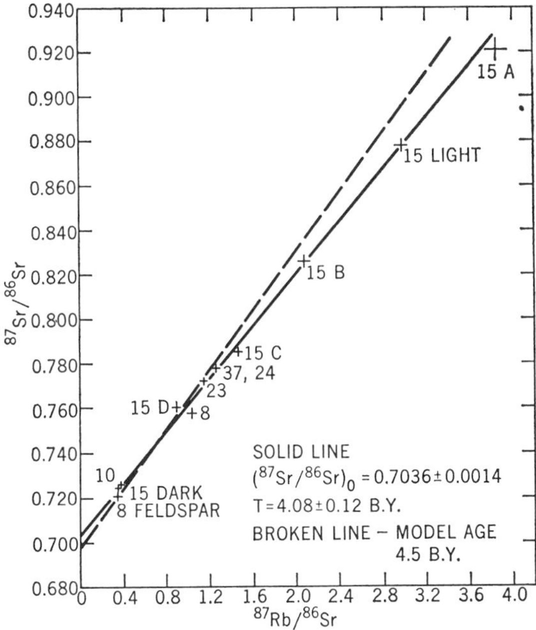

Fig. 1. Sr evolution diagram for 12013,10 subsamples.

because of larger sample size had generally smaller blank corrections), we obtain an age of 4.01 ± 0.15 and initial ratio of 0.7047 ± 0.0015, statistically indistinguishable from the results for all the data. Subsample No. 15 light was the lightest colored portion of the fragments of 12013 we received. In addition to a "whole-chip" portion, we analyzed five density fractions of this light-colored material. Again excluding 15E, we obtain an isochron age, for subsample No. 15 light and density fractions, of 3.84 ± 0.16 b.y. with initial ratio of 0.7104 ± 0.0037.

The above results are consistent with the Rb–Sr isochron age of 4.00 ± 0.05 b.y. obtained by Lunatic Asylum (1970) and the ^{40}Ar–^{39}Ar age of 4.03 ± 0.07 b.y. obtained by Turner (1971). The scatter in our data suggest that the portions of 12013 were not equilibrated. This is also seen in the results of Lunatic Asylum (1970), where different initial ratios were obtained from isochrons for different subsamples. Previous interpretations of the history of 12013 still seem valid (Lunatic Asylum, 1970; Schnetzler *et al.*, 1970). The high initial ratios and old model age indicate this rock formed at about 4.5 b.y. and underwent a thermal event at 4.0 b.y. sufficient to release all the argon and partially redistribute Sr. Alternately, the light-colored portion formed at 4.5 b.y. and at 4.0 b.y. intruded the black material at a temperature sufficient to redistribute Sr.

The data for 3 soils, 1 breccia, 5 core samples, and 1 coarse fine layer at a depth (uncorrected) of 13.2–14.4 cm. (LSPET, 1970) is quite different in Rb, Sr, and Sr isotopic compositions from the other core and soil samples, but similar to the normal igneous rocks. This is consistent with other trace element data (Schnetzler and Philpotts, 1971; Ganapathy *et al.*, 1970). The light-colored fragment from 12032

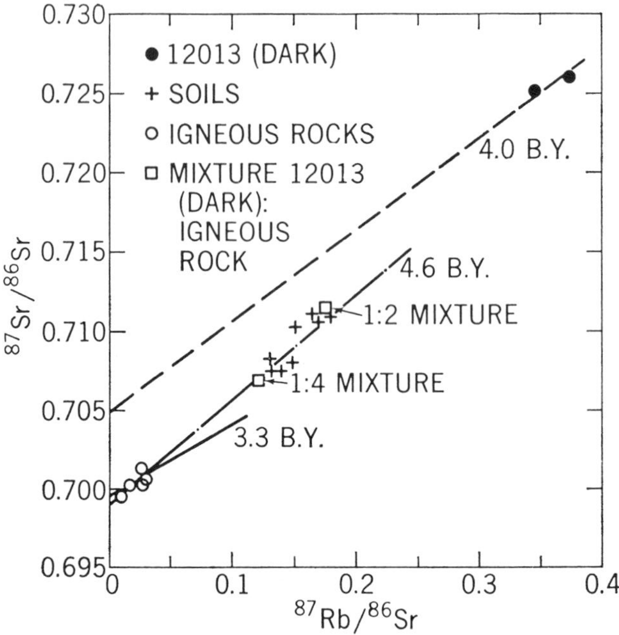

Fig. 2. Sr evolution diagram for soils, igneous rocks, dark-colored portions of 12013,10, and soil mixture calculations.

consists of plagioclase and low-calcium pyroxene. The other regolith samples are similar in Rb–Sr systematics and quite different from the igneous rocks. Calculated model ages for eight soil, breccia, and core samples range from 4.18 to 5.10 b.y. and average 4.66 b.y. The range of individual model ages ($\pm 10\%$ of the average) is twice as high as our anticipated precision of a single-age analysis. The soils therefore show model ages older than the isochron ages of the Apollo 12 rocks (approximately 3.3 b.y.—e.g., PAPANASTASSIOU and WASSERBURG, 1970; COMPSTON *et al.*, 1971) and, in addition, seem to have variable model ages (BOTTINO *et al.*, 1970).

In addition to the old and apparently variable model ages, the soils are characterized by high concentrations of certain trace elements. For example, Li, K, Rb, Ba, and rare-earth elements are approximately ten times more abundant in the soils than the igneous rock (BOTTINO *et al.*, 1970; SCHNETZLER and PHILPOTTS, 1971; HUBBARD *et al.*, 1971; MURTHY *et al.*, 1971; COMPSTON *et al.*, 1971). We have previously suggested, on the basis of trace element concentrations, that the dark-colored portion of 12013 is similar to the "magic component" which must be added to the normal rocks to produce the soils (SCHNETZLER *et al.*, 1970). Figure 2 shows the data we have obtained on 12013 dark-colored portion, soils, breccia, cores, and normal igneous rocks. Also shown are the results of mixing 12013 dark-colored portion and normal igneous rocks in a 1:4 and 1:2 proportion. From this figure, it is apparent that mixing of various amounts of 12013 dark-colored portion could account not only for the high trace element content of the Apollo 12 soils, but also their high and variable model ages.

Acknowledgments—We thank Dr. G. J. Wasserburg for providing the 12013,10 samples; C. W. Kouns for the phase separations; S. Schumann, P. Shadid, A. Sos and M. Robinson for analytical assistance. M. L. Bottino and P. D. Fullagar were provided support by National Science Foundation Grants G.A. 16501 and G.A. 18448, respectively.

REFERENCES

BOTTINO M. L., SCHNETZLER C. C., FULLAGAR P. D., and PHILPOTTS J. A. (1970) On the cryptic component of the Apollo 12 soil. *Trans. Amer. Geophys. Union* **51**, 772.

CLIFF R. A., LEE-HU C., and WETHERILL G. W. (1971) Rb–Sr and U, Th–Pb measurements on Apollo 12 material. Second Lunar Science Conference (unpublished proceedings).

COMPSTON W., BERRY H., VERNON M. J., CHAPPELL B. W., and KAYE M. J. (1971) Rubidium-strontium chronology and chemistry of lunar material from the ocean of storms. Second Lunar Science Conference (unpublished proceedings).

GANAPATHY R., KEAYS R. R., and ANDERS E. (1970) Apollo 12 lunar samples: Trace element analysis of a core and uniformity of the regolith. *Science* **170**, 533–535.

HUBBARD N. J., GAST P. W., and MEYER C. (1971) The origin of the lunar soil based on REE, K, Rb, Sr, P, and $^{87}Sr/^{86}Sr$ data. Second Lunar Science Conference (unpublished proceedings).

LSPET (Lunar Sample Preliminary Examination Team) Preliminary examination of the lunar samples from Apollo 12. *Science* **167**, 1325–1339.

Lunatic Asylum (1970) Mineralogic and isotopic investigations on lunar rock 12013. *Earth Planet. Sci. Lett.* **9**, 137–163.

MURTHY V. R., EVENSEN N. M., JAHN B., and COSCIO M. R., JR. (1971) Rb–Sr isotopic relations and elemental abundances of K, Rb, Sr, and Ba in Apollo 11 and Apollo 12 samples. Second Lunar Science Conference (unpublished proceedings).

PAPANASTASSIOU D. A. and WASSERBURG G. J. (1970) Rb–Sr Ages from the Ocean of Storms. *Earth Planet. Sci. Lett.* **8**, 269–278.

SCHNETZLER C. C., PHILPOTTS J. A., and BOTTINO M. L. (1970) Li, K, Rb, Sr, Ba, and rare-earth concentrations, and Rb–Sr age of lunar rock 12013. *Earth Planet. Sci. Lett.* **9**, 185–192.

SCHNETZLER C. C. and PHILPOTTS J. A. (1971) Alkali, alkaline-earth, and rare-earth concentrations in some Apollo 12 soils, rocks, and separated phases. Second Lunar Science Conference (unpublished proceedings.

TURNER G. (1971) ^{40}Ar–^{39}Ar ages from the lunar maria. Second Lunar Science Conference (unpublished proceedings).

YORK D. (1966) Least square fitting of a straight line. *Can. J. Phys.* **44**, 1079–1086.

Proceedings of the Second Lunar Science Conference, Volume 2, pp. 1493–1502.
The M.I.T. Press, 1971.

Rb–Sr and U, Th–Pb measurements on Apollo 12 material

R. A. CLIFF, C. LEE-HU, and G. W. WETHERILL
Department of Planetary and Space Science, University of California, Los Angeles, California 90024

(*Received 22 February 1971; accepted in revised form 25 March 1971*)

Abstract—Apollo 12 basalts 12002, 12021, and 12063 have low K, Rb, U, Th, and Pb concentrations relative to terrestrial basalt. Sr isotopic composition is consistent with a 4.6 b.y. model age. An internal isochron for 12021 yields an age of 3.30 ± 0.10 b.y., with 0.6993 initial ratio. The U/Pb age of 12063 is consistent with a two-stage evolution in a 4.65 b.y. old system which underwent U/Pb fractionation at 3.3 b.y. ago; thus it does not contradict the Rb/Sr result.

Soil samples 12070, 12032, 12033 are greatly enriched in K, Rb, U, Th, and Pb relative to other lunar material, except rock 12013. They have Rb/Sr model ages close to 4.6 b.y. Fractions of microbreccia and glass from 12070 and 12032 are even more enriched in K and Rb and have slightly younger model ages. U/Pb ages on 12070 and 12032 are strongly discordant; they are compatible with evolution in a 4.65 b.y. old system but must have suffered fairly recent lead loss, about 1.5 b.y. ago. A volatilization experiment shows extensive loss of K, Rb, and Pb at the melting point in vacuo. Absence of Rb/Sr fractionation between lunar basalts and their source places restrictions on the importance of this process on the moon, however.

INTRODUCTION

Rb/Sr and U, Th/Pb measurements on samples of both rocks and fines are described. These provide information on the ages of material present at the Apollo 12 site in Oceanus Procellarum, as well as placing some constraints on geochemical fractionation processes that may have operated on the moon. Coupled with a preliminary volatilization experiment on comparable terrestrial material the analyses allow the role of volatile transfer in the evolution of the moon to be evaluated.

The Apollo 12 fines are remarkable for the presence of large amounts of material not represented among the larger, presumably locally-derived, rocks; hereafter this material is referred to as exotic. By hand picking petrographically distinct fractions it has been possible to characterize isotopically some of the exotic material.

RUBIDIUM–STRONTIUM MEASUREMENTS

Rubidium, strontium, and potassium concentrations and strontium isotopic compositions have been measured by isotope dilution in 12002, 12063 whole rocks and in 12021 whole rock and separated mineral fractions. Fines samples 12070, 12032, and 12033 and hand picked fractions from 12070, 12032, and 10084 were also analyzed. Analytical techniques were as described previously (GOPALAN *et al.*, 1970). Blank corrections were insignificant except for 12021 tridymite which required a small strontium correction. The analytical data are presented in Table 1. The data for whole rocks are plotted on a strontium evolution diagram in Fig. 1. The Rb^{87}/Sr^{86} ratios of the basalts are very low, in the range 0.01 to 0.03, comparable to the low-Rb group

Table 1. Rb/Sr analytical data.

Sample	K μg/g	Rb μg/g	Sr^{86} μg/g	Rb^{87}/Sr^{86}	Sr^{87}/Sr^{86}
Rocks					
12002,148	475	0.977	8.501	0.03199	0.7009
12063,50	556	0.837	14.53	0.01622	0.6999
12021,51 (a)	518	1.139	11.19	0.02833	0.7002
(b)	510	1.126	11.16	0.02799	0.7008
12021 tridymite	3855	7.99	8.196	0.2716	0.7121
12021 plagioclase	885	1.127	38.27	0.00820	0.6998
12021 pyroxene	86	0.297.	1.284	0.0644	0.7023
Fines					
12070,60	2056	6.59	13.86	0.1324	0.7081
12070 (breccia)	3825	11.87	16.56	0.1974	0.7117
12032,31	3155	9.18	16.36	0.1521	0.7094
12032 (breccia)	4448	14.16	18.91	0.2085	0.7124
	4471	14.95	18.4	—	—
12032 (< 37 μ)	3302	9.25	17.83	0.1436	0.7089
12033	3338	10.08	16.71	0.1681	0.7100
10084 (lithic frags)	1771	3.34	17.32	0.0536	0.7027

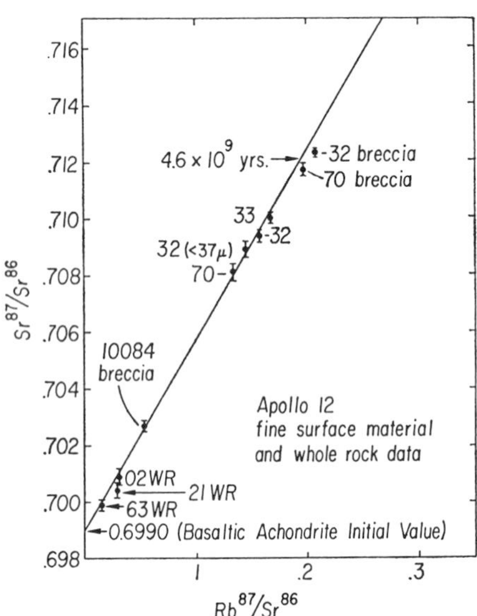

Fig. 1. Strontium evolution diagram for Apollo 12 whole rocks and fines.

of Apollo 11 samples and considerably less than typical values for terrestrial basalts (0.05 to 0.15). No samples corresponding to the high-Rb group of Apollo 11 rocks have been found. The Sr^{87}/Sr^{86} ratios are also low; the radiogenic enrichment for 12002 and 12063 is consistent with evolution in a 4.6 b.y. old system having the Rb/Sr ratio measured in the rocks now; 12021 whole rock appears to have a younger model age. The Sr^{87}/Sr^{86} value used in calculation is the mean of four runs on one dissolution and two on a second. The data on the first dissolution scatter more than would be predicted from the within-run statistics; the two runs on the second dissolution differ by less than the within-run error, and possibly yield the more reliable result. The data from the second dissolution give a 4.6 b.y. model age, similar to the other basalts analyzed. Consequently there is no evidence for major Rb/Sr fractionation in the basalts or their source since very early in the evolution of the moon.

In order to determine the age of crystallization of one rock, 12021, mineral separates were analyzed. The results for tridymite ($\rho < 2.37$), pyroxene (hand picked), and plagioclase ($2.70 < \rho < 2.80$), are presented in Table 1 and plotted on a strontium evolution diagram on Fig. 2.

The analyzed minerals show considerable variation in Rb/Sr ratio, with the tridymite enriched by a factor of ten relative to the whole rock. The data lie within analytical error on an isochron giving an age of 3.30 ± 0.10 b.y. with an initial Sr^{87}/Sr^{86} ratio of 0.6993. The error quoted takes into account the uncertainty in the whole rock data discussed above.

This result is in good agreement with another determination of this sample (LUNATIC ASYLUM, 1971) and with other internal isochrons on Apollo 12 rocks (PAPANASTASSIOU and WASSERBURG, 1970; LUNATIC ASYLUM, 1971; COMPSTON *et al.*, 1971). These results are distinctly younger than the 3.65 b.y. old basalts from the Apollo 11 site in Mare Tranquillitatis (PAPANASTASSIOU *et al.*, 1970, COMPSTON *et al.*, 1970, GOPALAN *et al.*, 1970), and they indicate that conditions for mare volcanism obtained at widely separated times in the evolution of the moon. The age difference between Apollo 11 and Apollo 12 rocks is consistent with relative ages based on crater densities (GAULT, 1970) but its magnitude is smaller than that predicted on the basis of a constant flux of crater-producing objects through time. Ages based on the two methods can be reconciled if a flux which decreases with time is assumed.

Results for the fines are included in the strontium evolution diagram, Fig. 1. Enrichment in K, Rb, and Sr^{87} of 12070 relative to the basalts and also the Apollo 11

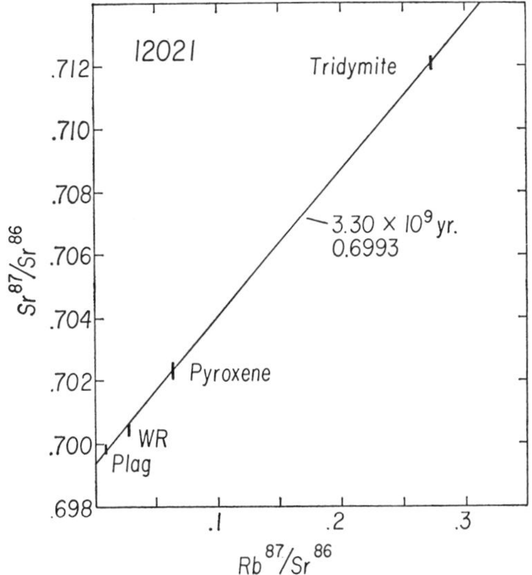

Fig. 2. Strontium evolution diagram for separated minerals from rock 12021.

fines has already been reported (PAPANASTASSIOU *et al.*, 1970, HUBBARD *et al.*, 1970). Our data confirm this result and show even more enrichment in 12032 and 12033. At the Apollo 11 site the soil could be approximately described in terms of a two component mixture of the two basalt types, although addition of minor amounts of a third component was indicated (LUNATIC ASYLUM, 1970a, PAPANASTASSIOU *et al.*, 1970, COMPSTON *et al.*, 1970). In the case of Apollo 12 the fines clearly contain, in addition to the local basalts, material with higher alkali concentrations. Within experimental error, samples 12070, 12032, and 12033 lie on a 4.6 b.y. isochron with the basaltic achondrite initial ratio.

In an attempt to relate the high alkali contents to petrographically identifiable components of the fines, fractions were obtained, by sieving and hand picking, from 12070 and 12032. Results for these fractions are also plotted on Fig. 1. The very fine fraction ($< 37\ \mu$) of 12032 also lies on the 4.6 b.y. isochron but it is slightly less radiogenic than the bulk sample. Two samples of glassy microbreccia fragments from the 190–1000 μ fraction of 12032 were also analyzed. These show approximately 50% enrichment in rubidium relative to the bulk sample. Good isotopic composition data were only obtained for one of these samples: it plots close to, but slightly below, the 4.6 b.y. model isochron. The microbreccia from 12070 has lower alkali concentrations than that from 12032 but the Rb/Sr ratio is similar. It also plots slightly below the 4.6 b.y. model isochron. Considering all the analyses on the fines together, it is apparent that while both bulk samples and fractions have model ages close to 4.6 b.y., they are more closely aligned along a 4.05 b.y. isochron with a higher initial ratio of 0.7009. For such a complex mixture of many components this good alignment is surprising. Possibly it may indicate a real time of isotopic homogenization, associated with a large impact, which resulted in metamorphic products ranging from rock fragments to microbreccia to glass. Both the age and initial ratio are comparable to results on part of 12013 (LUNATIC ASYLUM, 1970b). Given the large radiogenic enrichment of the microbreccia, it was interesting to see if similar material in the Apollo 11 fines could account for the slight deviation from the two-basalt mixing line. An analysis of hand-picked lithic fragments from 10084 does show higher alkali concentrations and slight radiogenic enrichment but much smaller than in the Apollo 12 fines.

URANIUM, THORIUM–LEAD MEASUREMENTS

A complete U, Th, Pb analysis on 12063 and U, Th concentration analysis of 12002 has been made. Two fines samples, 12070 and 12032, have also been analyzed. Analytical techniques were essentially those of TATSUMOTO (1966). For concentration analyses samples were spiked with U^{235}, Th^{230}, and Pb^{208} or Pb^{206} before dissolution.

Early dissolutions were made in closed teflon beakers, later ones in sealed teflon bombs at 140°C. Uranium and thorium were separated on Dowex 1 columns and run as nitrates or single rhenium filaments; lead was extracted by dithizone and loaded using a modified silica gel-phosphoric acid technique (CAMERON *et al.*, 1969). Electron multiplier collection was used for all mass spectrometry and in the case of lead runs, an empirical mass discrimination correction based on analyses of NBS 982 was made to all raw ratios; the correction amounts to 0.67% per mass unit. Lead blanks were run with each sample and were generally about 90 ng but fluctuated between 40 and 150 ng Pb; the blank correction was consequently large and constitutes the major source of uncertainty in the lead analysis; especially in the case of the basalts. Uranium and thorium blanks were < 0.006 and 0.002 μ g respectively and necessitated correction of less than 1% except for 12063 where the correction was 5% and 3% for uranium determinations (a) and (b) respectively. The analytical data are given in Table 2 and are plotted on a concordia diagram in Fig. 3 together with some results for Apollo 11 samples (TATSUMOTO, 1970a, SILVER, 1970, GOPALAN *et al.*, 1970) and Apollo 12 fines data presented at this meeting (TATSUMOTO, 1971, SILVER, 1971).

The analysis of 12063 is subject to larger errors because concentrations are an order of magnitude lower than in the fines. Because of the low concentration the lead isotopic composition was determined from two spiked runs. The Pb^{207} concentrations obtained from the two runs differed appreciably but could be brought into agreement by assuming the blank corrections might be anywhere within the range of measured

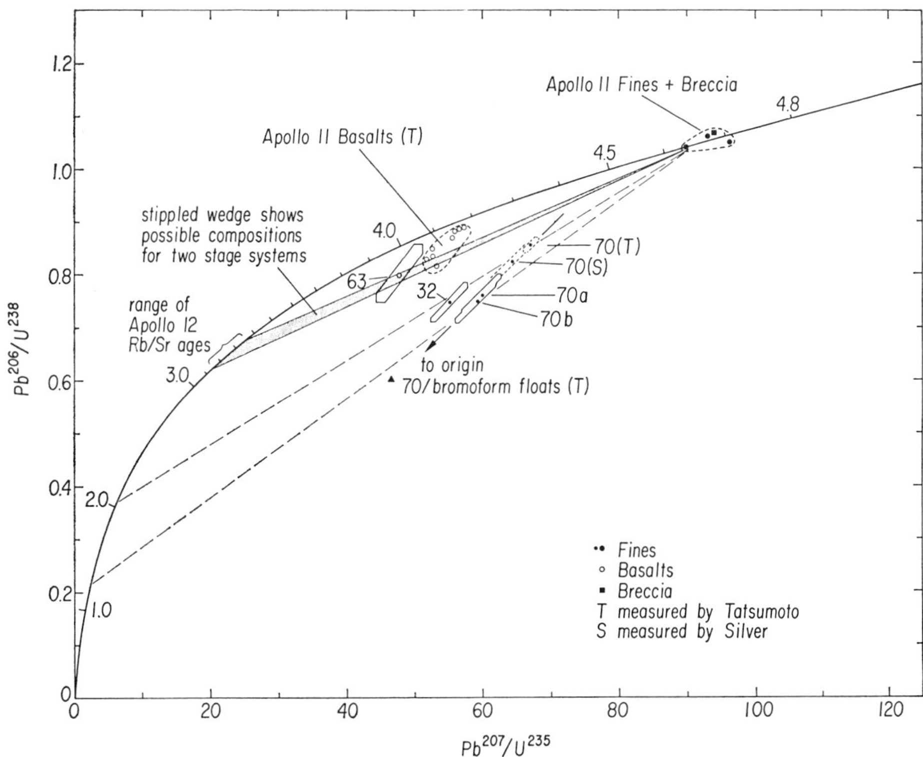

Fig. 3. Concordia diagram illustrating U/Pb systematics of Apollo 12 samples. Apollo 11 results of TATSUMOTO (1970), SILVER (1970), and GOPALAN *et al.* (1970) shown for reference. Points marked (T) and (S) are from TATSUMOTO (1971) and SILVER (1971) respectively. The dashed lines indicate the range of ages for a young lead loss event to satisfy a two stage evolution of the fines.

Table 2. U, Th, and Pb data for

Sample	Concentrations (ppm)*						Lead isotopic composition (before blank correction)		
	U	Th	Pb^{208}	Pb^{207}	Pb^{206}	Pb^{204} $\times 10^3$	$\frac{208}{204}$	$\frac{207}{204}$	$\frac{206}{204}$
12070,60 (a)	1.85	—	—	0.709	1.209	2.16	—	247	421
(b)	1.63	5.99	—	0.635	1.089	2.53	—	129	218
(c) volatilization extraction							101.1	53.7	83.8
12032,31 (a)	2.35	8.72	—	0.814	1.508	0.47	—	245	448
(b) composition by acid dissolution							278	154	278
12063,50 (a)	0.188	0.708	—	0.0635	0.1357	0.72	102	41.6	—
(b)	0.189	0.687	0.1554	0.0635	—	0.97	—	34.4	63.5
12002,148	0.219	0.747	—	—	—	—			

* Corrected for blank.

blank values rather than equal to that specifically relating to each analysis. The composition of 12063 was thus calculated after varying the blank correction to make the Pb^{207} concentrations consistent; the exact agreement of the Pb^{207} concentrations in Table 2 is a trivial result of this procedure. The error envelope in Fig. 3 includes the extreme possibility that all Pb^{204} measured was laboratory contamination. 12063 appears somewhat discordant, with a younger apparent age than the Apollo 11 basalts. The calculated ages are given in Table 2 but are of doubtful significance. The data have been corrected solely for a primitive, meteorite initial lead; no correction has been applied for any other lead present at the time of crystallization. Data on young terrestrial basalts (TATSUMOTO, 1966) show considerable initial lead concentrations, indeed many samples show a depletion of the U/Pb ratio of the basalt relative to its source at the time of formation. The isotopic composition of initial leads will correspond to that in the source of the magmas. Since, as reported for samples from both Apollo 11 and Apollo 12 (TATSUMOTO, 1970, 1971; SILVER, 1970, 1971; GOPALAN *et al.*, 1970) measured U^{238}/Pb^{204} ratios are much greater than terrestrial values, it is likely that lunar initial leads are distinctly radiogenic, and a correction based on meteorite, or a terrestrial model, lead would result in serious overestimation of ages. There is insufficient data on the lead isotopic evolution of the moon to allow a meaningful correction; all that can be said is that if the Rb/Sr ages of 3.3 b.y. correctly give the age of crystallization, and the lunar basalts derive from a source with uniform U/Pb ratio between the formation of the solar system, 4.65 b.y. ago, and magma generation, then Apollo 12 basalts should plot along a chord connecting 4.65 b.y. with the time of crystallization of the basalts on the concordia diagram (this follows from the theory given by WETHERILL (1956)). This is indicated by the stippled wedge on Fig. 3 and it is clear that our data is consistent with this model and thus does not conflict with the Rb/Sr results. Provided attempts in progress to drastically reduce laboratory contamination levels are successful, further lead isotopic studies may provide an indication whether prolonged mare volcanism repeatedly tapped the same source, which then would not have had an unfractionated U/Pb ratio from the time of formation of the solar system, or whether each phase of mare volcanism resulted from melting of a previously unaffected part of the moon, in which case the closed system model proposed above would still hold.

12070,60, 12032,31, and 12063,50.

Lead isotopic composition (corrected for blank)					Apparent ages			
$\frac{208}{206}$	$\frac{207}{206}$	$\frac{208}{204}$	$\frac{207}{204}$	$\frac{206}{204}$	$\frac{207}{206}$	$\frac{206}{238}$	$\frac{207}{235}$	$\frac{208}{232}$
—	0.5838	—	323	554	4.511	3.633	4.216	—
—	0.5805	—	247	426	4.495	3.683	4.223	3.402
1.0146	0.6001	—	—	—				
—	0.5367	—	1699	3164	4.404	3.632	4.145	3.385
0.9538	0.5423	720	410	755				
1.1387	0.4658	156	75	186	4.086	3.812	3.992	3.746

The results for fines samples 12070 and 12032 are similar to Rb/Sr results in that they contain higher concentrations of uranium, thorium, and radiogenic lead than both the basalts and the Apollo 11 fines. The results are plotted on Fig. 3, and it is remarkable that the Apollo 12 fines also contrast with those from Apollo 11 in being strongly discordant. The Apollo 11 fines were consistent with a 4.65 b.y. old closed system; the Apollo 12 fines resemble a system of similar age with a variable deficiency in lead relative to uranium and thorium. At least some of this apparent lead loss occurred since the crystallization of the mare basalt and the data can be reconciled with a simple two-stage model in which lead loss occurred about 1.5 b.y. ago. Data presented by TATSUMOTO (1971) indicate a more complex model is required, but it is interesting that fission track dates on rock 12013 are consistent with an event of this age (BURNETT *et al.*, 1971). Fig. 3 also shows two other results for the 12070 fines. The results of SILVER (1971) and TATSUMOTO (1971) differ somewhat from one another and more seriously from our result. All four analyses are aligned along a line through the origin, and it is important to consider whether the differences might result from systematic errors in the concentration analyses, such as incorrect spike calibrations. There is good evidence of consistency between uranium analyses presented here and by TATSUMOTO (1971) on rock 12063. Direct comparison of the lead analyses is less easy in view of the large blank corrections but there is no reason to suppose spike calibration errors greater than 2% in this case also. Both our concentration results and those of TATSUMOTO show strong fluctuations in uranium content between replicates on the same sample, and it is thus tempting to propose real fluctuations in the proportion of a soil component or components which have high uranium and thorium concentrations but are greatly deficient in lead. TATSUMOTO's analyses of bromoform floats from 12070 indicate that such a component is present in the soil and has a composition which would be consistent with this explanation.

EXOTIC COMPONENTS IN THE FINES

The analyses reported above further support the result, already reported by others (LSPET, 1970, PAPANASTASSIOU and WASSERBURG, 1970; HUBBARD *et al.*, 1970; SCHNETZLER *et al.*, 1970) that exotic material is important in the fines at the Apollo 12 site. This material has high concentrations of alkalis, rare earths, barium, uranium, thorium, lead, and phosphorus, a model Rb/Sr age close to 4.6 b.y. and discordant U/Pb ages indicating recent lead loss.

The question arises whether all the chemical and isotopic characteristics result from a single exotic component or whether several are involved. Secondly, how are variations in the proportion of exotic material, indicated by the chemical data, reflected in the modal composition of the fines examined. 12033 has the greatest proportion of exotic component but our sample (50 mg) was too small for detailed analysis. 12032 is also strongly enriched in exotic material relative to 12070. Based on petrographic observation, 12032 has more than 50% exotic material; this is mainly dark, compact microbreccia, frequently glass coated, but also includes "ropy and fluted" glass fragments with well developed flow texture and trails of incorporated crystal fragments. There appears to be a gradation between these two types, and together they constitute our 12032 "breccia" sample. In addition light-colored polycrystalline fragments composed of orthopyroxene and plagioclase make up a few percent of 12032. By contrast, 12070 contains more basalt and locally derived crystal fragments and the "ropy and fluted" glass is almost totally lacking. However, dark microbreccias, often glass coated, are an important constituent.

The potassium and rubidium concentrations in 12070 "breccia" are 20% lower than in 12032 "breccia"; this may indicate real variation in the microbreccias, but it may also be related to the lower proportion of glass. The large fluctuations in uranium, thorium, and lead concentrations between aliquots as well as large differences in alkali concentrations between laboratories would be more readily explained if these were concentrated in a less abundant component such as the glass. FRICK *et al.* (1971) report a uranium analysis of glass coatings on fragments in breccia 12073 which would support this idea. The variable discordance of U/Pb ages of the two soil samples might also relate to variable proportions of this glass component; conceivably the lead loss responsible for the discordance may relate to the time of production of the glass.

To a first approximation the chemical composition and uranium/lead isotope data are not inconsistent with a mixture of glass and microbreccia and local basalts. Such a model of the soil has been proposed (HUBBARD *et al.*, 1971). The strontium isotope data, however, suggest that a simple two component mixture model is inadequate since the best fit line to the data does not pass through the data points for the basalts. The data could be explained by Rb/Sr ratio variations in a suite of genetically related rocks. The best fit line to our fines Rb/Sr data suggest a further similarity between the exotic component and sample 12013 which has yielded 4.0 b.y. internal isochrons (LUNATIC ASYLUM, 1970b). However, petrographic observation indicates that exotic material in the soil could not be produced simply by fragmentation of 12013. On the other hand, impact metamorphism of rocks like 12013 or simultaneous production of glass, microbreccia, and 12013 under differing conditions in a large impact would be compatible with the data.

THE ROLE OF VOLATILIZATION IN GEOCHEMICAL FRACTIONATION ON THE MOON

Several authors have suggested that volatilization may have been important in geochemical fractionation on the moon, particularly of lead and the heavy alkalis

Table 3. Volatile element loss from terrestrial basalt (Heating: 2.5 hours from 20° C to 1200°C, 3 hours at 1200°C; Vacuum $<10^{-5}$ mm Hg, $fO_2 < 10^{-14}$ atm.)

	Starting material	Residue	% Decrease
K/Sr	9.03	1.64	81.9
Rb/Sr	0.0183	0.00324	82.3
Pb/Sr	0.0123	0.00202	83.6
K/Rb	494	506	−2.4

(COMPSTON *et al.*, 1970, O' HARA *et al.*, 1970). Leaving aside possible large-scale volatile fractionation at the formation of the moon, which is suggested by the low Pb^{204} and alkali concentration, the process may have operated on a smaller scale during eruption of the basaltic magmas and again during formation of the soil by impacts. The enrichment of the Apollo 11 soil in old radiogenic lead (SILVER, 1970) might result from either of these mechanisms.

As a preliminary experiment in evaluating the effect of volatilization, a sample of terrestrial basalt was heated at its melting point in vacuo in a stainless steel crucible. The concentration of strontium in the residue was essentially unchanged; however, 80% of the lead, potassium, and rubidium were lost. The changes in concentration ratios of elements analyzed are given in Table 3. Assuming that this result is applicable to the lunar basalts, it seems possible that exposure of magma to the lunar vacuum could result in severe fractionation of K, Rb, and Pb. However, since the observed Rb and Pb fractionations are similar and none of the lunar basalts give any indication of depletion of Rb/Sr ratio relative to their source, it seems very unlikely that this process was an important agent of fractionation in lunar magmas. Clearly more extensive experiments will be necessary before definitive conclusions can be drawn. In particular the relation between alkali loss and lead loss at lower temperature needs study.

SUMMARY OF RESULTS

(1) Sample 12021 has a Rb–Sr "mineral" age of 3.30 b.y. with an initial Sr^{87}/Sr^{86} ratio of 0.6993. (2) Sample 12063 has a slightly discordant whole rock U–Pb age consistent with a two stage evolution in a 4.65 b.y. old system and with a crystallization age consistent with the Rb–Sr results. (3) Data from a preliminary volatilization experiment suggest alkali and lead fractionation by volatile transfer has not been an important process in lunar magmatism. (4) Three Apollo 12 fines samples lie on a 4.6 b.y. model isochron with a basaltic achondrite initial ratio. The alkali enrichment of the fines is concentrated in glass and microbreccia fragments. Concentrates of this material from 12032 and 12070, together with the analyses of bulk fines show a better fit to a 4.0 b.y. isochron with elevated initial ratio. (5) Uranium, thorium-lead data in two fines samples show strong discordance consistent with a 4.65 b.y. old system which suffered lead loss about 1.5 b.y. ago.

REFERENCES

BURNETT D., MONNIN M., SEITZ M., WALKER R., and YUHAS D. (1971) Lunar astrology: U–Th distributions and fission track dating of lunar samples. Second Lunar Science Conference (unpublished proceedings).

CAMERON A. E., SMITH D. H., and WALKER R. L. (1969) Mass spectrometry of nanogram-size samples, of lead. *Anal. Chem.* **41**, 525–526.

COMPSTON W. A., ARRIENS P. A., VERNON M. J., and CHAPPELL B. W. (1970) The chemistry and age of Apollo 11 material. *Proc. Apollo 11 Lunar Sci. Conf., Geochim. Cosmochim. Acta* Suppl. 1, Vol. 2, pp. 1007–1027. Pergamon.

FRICK C., HUGHES T. C., LOVERING J. F., REID A. F., WARE N. G., and WARK D. A. (1971) Electron probe, fission track and activation analysis of lunar samples. Second Lunar Science Conference (unpublished proceedings).

GAULT D. E. (1970) Saturation and equilibrium conditions for impact cratering on the lunar surface: Criteria and implications. *Radio Science* **5/2**, 273–291.

GOPALAN K., KAUSHAL S., LEE-HU C., and WETHERILL G. W. (1970) Rb–Sr and U–Th–Pb Ages of lunar materials. *Proc. Apollo 11 Lunar Sci. Conf., Geochim. Cosmochim. Acta* Suppl. 1, Vol. 2, pp. 1195–1205. Pergamon.

HUBBARD N. J., GAST P. W., and WIESMANN H. (1970) Rare earth, alkaline earth, and alkali metal and Sr^{87}/Sr^{86} data for subsamples of lunar sample 12013. *Earth Planet. Sci. Lett.* **9**, 181–184.

HUBBARD N. J., MEYER C., JR., GAST P. W., and WIESMANN H. (1971) The composition and derivation of Apollo 12 soils. Second Lunar Science Conference (unpublished proceedings).

LSPET (Lunar Sample Preliminary Examination Team) (1970) Preliminary examination of the lunar samples from Apollo 12. *Science* **167**, 1325–1339.

LUNATIC ASYLUM (1970a) Ages, irradiation history, and chemical composition of lunar rocks from the Sea of Tranquillity. *Science* **167**, 463–466.

LUNATIC ASYLUM (1970b) Mineralogic and isotopic investigations on lunar rock 12013. *Earth Planet. Sci. Lett.* **9**, 137–163.

LUNATIC ASYLUM (1971) Rb–Sr ages, chemical abundance patterns and history of lunar rocks. Second Lunar Science Conference (unpublished proceedings).

PAPANASTASSIOU D. A., and WASSERBURG G. J. (1970) Rb–Sr ages from the Ocean of Storms. *Earth Planet. Sci. Lett.* **8**, 269–278.

PAPANASTASSIOU D. A., WASSERBURG G. J., and BURNETT D. S. (1970) Rb–Sr ages of lunar rocks from the Sea of Tranquillity. *Earth Planet. Sci. Lett.* **8**, 1–19.

SCHNETZLER C. C., PHILPOTTS J. A., and BOTTINO M. L. (1970) Li, K, Rb, Sr, Ba, and rare earth concentrations and Rb–Sr age of lunar rock 12013. *Earth Planet. Sci. Lett.* **9**, 185–192.

SILVER L. T. (1970) Uranium-thorium-lead isotopes in some Tranquillity Base samples and their implications for lunar history. *Proc. Apollo 11 Lunar Sci. Conf., Geochim. Cosmochim. Acta* Suppl. 1, Vol. 2, pp. 1533–1574. Pergamon.

SILVER L. T. (1971) U–Th–Pb isotope relations in Apollo 11 and Apollo 12 lunar samples. Second Lunar Science Conference (unpublished proceedings).

TATSUMOTO M. (1966) Isotopic composition of lead in volcanic rocks from Hawaii, Iwo Jima, and Japan. *J. Geophys. Res.* **71**, 1721–1733.

TATSUMOTO M. (1970a) Age of the moon: An isotopic study of U–Th–Pb systematics fo Apollo 11 lunar samples, II. *Proc. Apollo 11 Lunar Sci. Conf., Geochim. Cosmochim. Acta* Suppl. 1, Vol. 2, pp. 1595–1612. Pergamon.

TATSUMOTO M. (1970b) U–Th–Pb age of Apollo 12 rock 12013. *Earth Planet. Sci. Lett.* **9**, 193–200.

TATSUMOTO M. (1971) U–Th–Pb systematics of Apollo 12 lunar samples. Second Lunar Science Conference (unpublished proceedings).

WETHERILL G. W. (1956) Discordant uranium-lead ages I. *Trans. Amer. Geophys. Union* **37/3**, 320–326.

Proceedings of the Second Lunar Science Conference, Volume 2, pp. 1503–1519.
The M.I.T. Press, 1971.

Lunar astrology—U–Th distributions and fission-track dating of lunar samples

D. BURNETT,* M. MONNIN,† M. SEITZ, R. WALKER, and D. YUHAS
Laboratory for Space Physics, Washington University, St. Louis, Missouri 63130

(*Received 19 February 1971; accepted in revised form 30 March 1971*)

Abstract—Uranium distributions in minerals of lunar rocks 12040, 12013, and lunar anorthosites were measured using induced fission track maps of the rock sections. The uranium is concentrated in zirconium and phosphorus phases, with the major uranium bearing phases being baddeleyite, whitlockite, and Zr–Ti phases in 12040 and apatite, whitlockite, zircon, and a Zr–Ti phase in 12013. Having average concentrations of 7 ppm U, the potassium-rich residual magmas of 12040 contain most of the uranium bearing grains, thus the well-defined K–U correlation in whole lunar rocks is valid for 12040 on a microscopic scale as well. In rocks 12040 and 12013 there is a distinct inverse relation of U concentration to grain size indicating, at least for the simple case, that the U increases rapidly in the melt even in the final stages of crystallization. The new Zr–Ti phases have the highest U concentration although their total U content is minor in 12040 and 12013 but not negligible. In rock 12013 apatite, whitlockite, and zircon are of equal importance in total U content. 78% of the U in 12040 is in whitlockite, representing a more equilibrated distribution than in 12013. Anorthosites, having $\sim$0.04 ppm, show concentrations of U in phosphates and a Zr–Ti phase.

Thorium–uranium ratios measured in minerals of 12013 range from 0.2 to 20. Zircons have the lowest Th/U ratios, and whitlockites the highest. Calculated Pb^{208}/Pb^{206} ratios in apatite and the Zr–Ti phase are compatible with total rock ratios measured in components of 12013. If plutonium acts chemically like Th, some geochemical separation of Pu and U will occur and the previously measured Pu/U value for the early solar nebula will be lower.

Fossil fission tracks in apatites of 12013 give an apparent age of 1.3 b.y. This low age, originally attributed to thermal fading, may date a distinct thermal or shock event in the rock's history. Whitlockites of 12040 have an apparent age of 6.5 b.y. The track excess causing the anomalously old age must be attributed to causes other than the spontaneous fission of Pu^{244}. Etched apatites in one anorthosite show no evidence for a Pu anomaly. Dates from fission tracks registered in feldspars that cap U inclusions have been measured in 12013 and 12040. Although the data are preliminary, the results indicate that this technique is practical.

INTRODUCTION

IN THIS PAPER we report measurements of the distribution of U among various phases in lunar rocks 12040 and 12013, Th/U ratios for phases in 12013, and fission-track ages.

The distribution of U and Th in lunar rocks is of interest, first, because it provides a microscopic view of the chemical fractionation processes which occur within a given

* Permanent address: Division of Geological Science, California Institute of Technology, Pasadena, California 91109.

† Permanent address: University of Clermont-Ferrand, Clermont-Ferrand, France.

sample during the cooling and crystallization of a lunar lava. Comparatively little data of this kind is available for trace elements. Second, information on U and Th distribution is necessary for a complete understanding of Pb isotopic data on lunar samples. This is of particular importance in view of the differences in the model Pb–U–Th ages (see, for example, TATSUMOTO *et al.*, 1971; SILVER, 1971) and the Rb–Sr internal isochron ages (see, for example, PAPANASTASSIOU and WASSERBURG, 1970) for lunar rocks.

Previous studies on Apollo 11 samples showed that U was localized in minor phases (CROZAZ *et al.*, 1970; FLEISCHER *et al.*, 1970; and LOVERING and KLEEMAN, 1970); however these phases were not well characterized. From the well-defined correlation of U with K it could be inferred (TERA *et al.*, 1970) that U in Apollo 11 rocks was localized in interstitial phases of complex mineralogy but characterized by the presence of K-rich glasses (ROEDDER and WEIBLEN, 1970). Preliminary reports of U distribution in rock 12013 (BURNETT *et al.*, 1970; and LUNATIC ASYLUM, 1970) also showed U to be highly localized and associated with phosphates. Zircon was also present (LUNATIC ASYLUM, 1970) but did not appear to be a major U-bearing phase.

There are several reasons why it is interesting to measure fission track ages in lunar samples. Perhaps the most important is the possibility of identifying small fragments that were formed near the beginning of the solar system. Previous studies in meteorites (CANTELAUBE *et al.*, 1967; WASSERBURG *et al.*, 1969; FLEISCHER *et al.*, 1965, 1968; and SCHIRK *et al.*, 1968) have shown the existence of excess fission tracks that have been attributed to the spontaneous fission of extinct Pu^{244}. In some cases the number of fission tracks exceeds the number expected from uranium spontaneous fission by a factor of ~ 60. The half-life of Pu^{244} of $\sim 8 \times 10^7$ years should give rise to lunar fission track anomalies in the interesting time interval of 4.5 to $\sim 4 \times 10^9$ years.

Fission-track dating of the major minerals is also important for its implications for other fossil-track studies in lunar materials. Surface residence times and erosion rates calculated from the study of solar flare and galactic cosmic ray tracks in lunar rocks have assumed that no track fading occurred while the rocks were heated on the surface during the lunar day (CROZAZ *et al.*, 1970; FLEISCHER *et al.*, 1970; LAL *et al.*, 1970; and PRICE and O'SULLIVAN, 1970). One way to prove that no appreciable fading has taken place is to measure old fission-track ages for the minerals (pyroxene or feldspar) that are used for the cosmic-ray work.

It is likely that fission-track ages will be much more susceptible to environmental effects such as shock (AHRENS *et al.*, 1970) than ages from other methods. Thus fission tracks may be able to measure discrete events in ths history of a rock that might be missed by methods such as Rb–Sr dating.

Fission-track dating of lunar samples is feasible because uranium is found in small local concentrations; otherwise, it would be difficult to separate out the fission tracks from the large background of cosmic-ray tracks.

This paper represents an expansion of our manuscript in the unpublished proceedings of the Second Lunar Science Conference. U distribution data for lunar rocks are also contained in papers by FRICK *et al.* (1971); COMSTOCK *et al.* (1971); and RICE and BOWIE (1971) from this conference. Fission-track age determinations are reported by BHANDARI *et al.* (1971) and by COMSTOCK *et al.* (1971).

Experimental Techniques

Polished sections were photographed in reflected light at either 50× or 100× magnification to produce a "section map." The rock's texture shows up in a very similar way in secondary electron image photographs from the SEM (scanning electron microscope). It is thus easy to locate precisely a particular region of the rock section in the SEM.

The rock sections were then covered with preannealed muscovite mica or another suitable detector and irradiated with thermal neutrons. Pieces of standard uranium glass with mica detectors were placed in close proximity to the rock sections to monitor the neutron flux. We have avoided the use of Lexan detectors, because, as shown by Fleischer *et al.* (1970) and from our own experience, serious track fading can occur during the reactor irradiation. If identical fading has not occurred for the U standard and the sample, erroneous U concentrations and ages can result, although qualitative or relative conclusions concerning the distribution of U will be unaffected. To provide a general orientation on the mica, the outline of the rock section was scratched on the outer surface of the mica under the microscope. For the larger rock sections (~ 10 mm^2) a grid of about 0.4 mm squares was scratched on the mica to act as fiducial markings. Tiny grains, fractures, and veins in the rock and in the mica also served as fiducial markings. A series of photographs of the rock and detector recorded their relative position before demounting. After demounting the mica from the rock, points on the mica and rock which were adjacent during the neutron irradiation could be located to about 15 microns with these photographs.

The positioning of locations on the rock section also automatically becomes more precise as the mapping proceeds. Thus, if we consistently find the location of a reasonable U-bearing mineral displaced by a fixed (small) distance from the predicated locations of uranium-rich regions, we ascribe the difference to small errors in positioning and correct future position predictions accordingly.

The error of 15 μ in location of uranium-rich regions means that a photograph taken at a magnification of 2000 is certain to contain the source of uranium fission. At this magnification, SEM microprobe pictures taken in the P–Zr channel [1.9–2.1 keV] easily show the location of even very small ($\gtrsim 1$ μ) grains.

After the irradiation the mica was etched to reveal fission fragment tracks. In all samples we found the tracks to occur in clusters or "stars." The number of induced fission tracks from specific grains can be obtained from the mica star maps. Locating particular stars on the mica with the SEM is simple: the heavily etched regions tend to charge up and give a vivid contrast in the SEM. Thus, even at low magnifications, the pattern of stars is easily visible.

U Distributions

In many cases the stars are associated into "constellations" or are closely grouped in localized regions giving more or less uniform and high track densities. Comparatively few tracks were found not associated with stars. In other areas the density is comparable to the fission-track background from the mica itself (typically 200 t/cm^2 for 3×10^{16} n/cm^2). Comparing the fission-track distribution with the rock section, we find that the large pyroxene, ilmenite, feldspar, and olivine grains are noticeably devoid of tracks (≤ 2 ppb U).

The fission star maps are analogous in many ways to true star maps of the heavens. At the 100× magnification used for the main reference map, only the brighter stars are clearly visible. These group into characteristic constellations that soon become familiar. As the magnification is increased, and particularly when one passes to the SEM, fainter stars forming new constellations appear. Most of our work has been done on the first magnitude stars.

We systematically examined locations on the section map corresponding to observed stars in the mica. Some grains having irregular shapes were identified as the source of the fission tracks by the shape of the stars in the mica. In other areas, grains occur in groups having patterns that are recognizable as constellations on the mica. In cases where an identification can be made, the uranium occurs in either phosphorus or zirconium phases. Some sources cannot be identified because of the complexity of the regions in which they occur. However, the sources are usually found to be buried beneath the surface as indicated by the short track lengths in the mica when an identification cannot be made.

In rock 12013, 76 star locations from 2 sections comprising 3.0 mm^2 were examined. The majority of the identified stars were phosphate (20 stars) and zircon (16 stars) grains. These phosphate grains fall into two chemical groups. The first group, which invariably contains a measurable amount of rare earths but no chlorine or other major elements, we call whitlockite (7 stars). The second group, containing no detectable rare earths but containing chlorine, we call chloroapatite (13 stars). Our data suggest this nomenclature but do not positively establish the mineral identity of these phases. However the existence of two distinct phosphate phases is unequivocal. A third mineral phase, containing primarily Zr and Ti but perhaps containing Fe, Ca, and Si, was found to be the source of 5 large stars. This Zr–Ti phase may be identical to the phase B reported by FRICK *et al.* (1971); however, our 12013 grains were too small to permit a reliable chemical analysis.

The dark regions of 12013 (DRAKE *et al.*, 1970; LUNATIC ASYLUM, 1970) contain uniformly distributed U (i.e., uniform track density for a total neutron dose of $\sim 1 \times 10^{17}$ n/cm^2) in addition to stars. In one dark area examined, the distributed U corresponds to about 6 ppm. This distributed U is probably related to the distributed P reported in the dark portions of 12013 by DRAKE *et al.* (1970). We estimate that roughly 50–75% of U in the dark portions of 12013 is distributed. In our sections about 10–20% of the area was composed of dark material.

We mapped the uranium in four sections of rock 12040 ($\sim$10 mm^2 each). The uranium distribution in each section is qualitatively similar to that described here. From one section (11.6 mm^2) we found 9885 tracks in the mica giving an average U concentration of 0.24 ppm. As in rock 12013 we found the tracks to occur in stars. These stars are usually clustered in regions of more or less uniform and high-track density. The majority (80%) of the fission tracks in this section occur in 11 distinct regions having well-defined boundaries. The largest of these regions containing 5100 tracks has an average concentration of 6.9 ppm U. These regions correspond to areas in the rock section consisting of finely textured transparent material but containing some fine grained opaque minerals as well. We found these areas to be rich in potassium compared to the major minerals and to contain grains of troilite, apatite, and baddeleyite of 1 to 50 microns in size in a matrix rich in K, Al, and Si. Larger ilmenite and plagioclase grains often form a border of these regions. In some parts of these regions it is difficult to establish exactly in what mineral phase the uranium is concentrated because of the complicated mineralogy on a 10 micron scale. Where an identification can be made, the uranium occurs in either calcium-phosphate or baddeleyite grains. Furthermore, some areas containing only ilmenite or troilite grains do not show any uranium.

In rock 12040 we will call the calcium-phosphates whitlockite owing to their detectable rare earths and lack of a chlorine peak. However, we have not unequivocally identified this phase. Although no chemical differences are detected by the x-ray system, we have evidence for two groups of phosphates from chemical etching. After a 20 sec treatment with 0.12% HNO_3 (22°C) one group shows nicely formed particle tracks while the other group is hardly affected. Furthermore, the etched whitlockites often show unetched filaments which contain a trace of chlorine (see Fig. 1).

Twenty-two other fission-track stars containing more than 10 tracks each were found in areas outside the main track-rich regions described above. The source of these stars was identifiable in 15 cases. Eight of these stars originate from whitlockite grains of which seven grains occur in a potassium-rich matrix. Only one whitlockite grain occurs in a matrix with no detectable potassium. Three of the stars originate from baddeleyite grains occurring in K-rich areas and one star originates from a Fe–Ti–Zr–Y–Si phase in a K-rich matrix. This Zr–Ti phase may be identical to the Y-rich phase reported by FRICK *et al.* (1971). Since we are looking at one plane of a 3-dimensional rock, these small K-rich areas are probably only a part of a larger area located above or below the plane in which we are viewing and are similar to the large

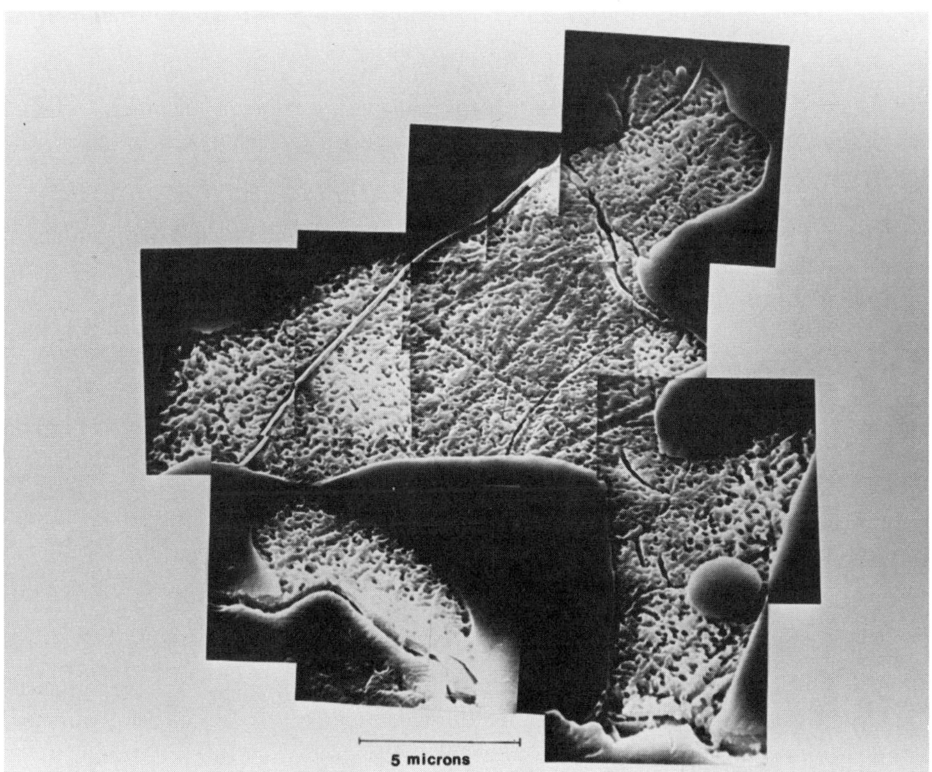

Fig. 1. A montage of SEM photographs showing a whitlockite grain in rock 12040. The crystal has been etched in 0.12% HNO_3 for 20 sec at 22°C and the nicely formed fission tracks give an apparent age of 6.4 billion years. A filament showing a trace of chlorine runs down from right to left in the crystal.

areas described above. Thus these areas contain virtually all the uranium of the rock and are probably quenched residual magmatic liquids as described by ROEDDER and WEIBLEN (1970).

Our results for 12040 show that the correlation between K and U, which is well-defined for most whole-rock terrestrial (WASSERBURG *et al.*, 1964) as well as lunar samples, holds up well, even on an ultramicroscopic scale, for lunar samples. Our results demonstrate what was previously presumed, namely that K and U have a pronounced tendency to remain in the fluid phase until the last moments of crystallization of a magma. Only then do these two elements segregate into discrete phases. This explains why K and U are highly correlated, although they are not found in the same minerals. The lunar data also suggest that the correlation of K and U in terrestrial samples is not significantly influenced by the presence of high concentrations of volatiles (H_2O or CO_2) in terrestrial magmas.

Three grains bearing uranium occurred in a pyroxene crystal several hundred microns away from any ilmenite, troilite, or K-rich areas and so appear distinct from the residual liquid phase. The largest grain is 8 μ long and 2.6 μ wide. The other two grains are approximately 2.5 μ in diameter. Investigations of these grains with the microprobe show them to be Zr–Ti–Ca–Fe–Si and are possibly the unidentified mineral phase seen in 12013.

For both 12013 and 12040 we searched for P- and Zr-rich phases not associated with stars by taking 400 sec x-ray photographs using a single channel 2.05–2.55 analyzer at 500× magnification. In these photographs phosphorus, zirconium, and sulfur-rich grains greater than 1 micron in size appear as bright spots. For 12040 twenty photographs taken at intervals to cover the areas not showing fission tracks revealed no phosphate or zirconium minerals. Troilite as well as the major minerals were present in these photographs. No detectable potassium was found in regions which did not contain uranium. This search included areas in the rock most similar in appearance to the uranium-bearing regions described above, as well as other interstitial regions. For 12013 some large whitlockite and apatite grains were found to be associated with only small stars even for an irradiation of 10^{17} n/cm^2 whereas, as indicated above, these same minerals also give rise to first magnitude stars. This indicated that there are wide variations in U content, even within the same mineral, in 12013.

Tables 1 and 2 present U contents measured for individual grains of U-bearing minerals in 12040 and 12013 respectively. U contents were calculated

$$U = U_s \cdot \frac{\rho}{\rho_s} \cdot \frac{R_s}{R}, \tag{1}$$

where U_s is U content of the standard glass (SCHREURS *et al.*, 1970), ρ is the measured track density, and R is the fission fragment range in mg/cm^2. R_s/R has been estimated assuming that the range of fission fragments stopping in element Z is proportional to $Z^{1/2}$ following MORY *et al.* (1970). This correction is less than 30% for all cases here and is not an important source of error. Track densities for stars were usually obtained by dividing the total number of tracks in the star by the area of the grain. This is the best means of calculating ρ, particularly for small grains, because in some

Table 1. U distribution in lunar rock 12040 (ppm).

Whitlockite	Whitlockite *cont.*	Baddeleyite
800†	55‡	1500†
640†	55	640†
550†	55	580†
410†	50‡	
210†	45	Zr–Ti phase
200†	40*	1800†
200	40*	1600†
160	10†	1400†
65	10*	1250†
60		980†
		Zr–Ti–Fe–Ca–Y phase
		100*

* U concentration calculated from track density at center of star.

† Grain area $\leq 10^{-6}$ cm^2; U concentration probably a lower limit.

‡ U concentration calculated from fossil track density in the grain assuming the same "age" as for the dated whitlockites.

Table 2. U and Th distribution in lunar rock 12013.

U (ppm)	Th/U (weight)	U (ppm)	Th/U (weight)	U (ppm)	Th/U (weight)	U (ppm)	Th/U (weight)
Whitlockite		Apatite		Zircon		Zr–Ti phase	
190	6.8 ± 0.6‖	200	≤5.0 ± 1.0§	1400†	0.2 ± 0.06	2000†	2.9
880*†	—	≤110‡		1500†	0.49 ± 0.12		
680†	9.7 ± 0.7	430		140	3.6 ± 0.4		
830†	—	94*		—	3.0 ± 0.6		
69	20 ± 2	80*		530†	0.35 ± 0.09		
85	10 ± 1	160*		310†	—		
—	13 ± 1.5	190†		400†	—		
		120†		690†	—		
		150†		360†	—		
		110		300†	—		
		270		155†	—		
				360†	—		
				80†	—		

* U concentration calculated from track density at center of star.
† Grain area $\leq 10^{-6}$ cm^2; U concentrations probably are lower limits.
‡ Upper limit, because it was not possible to unambiguously associate a star with this grain.
§ Identification of Th star uncertain.
‖ Quoted errors are two standard deviations from track statistics.

areas the mica was slightly separated from the section during the irradiation. For some larger stars where the track density appeared uniform at the center of the star, ρ was calculated as the number of tracks per unit area on the mica. U contents and the procedure used for each star are indicated in Tables 1 and 2. Equation 1 is only valid for grains whose dimensions are large compared to the mean range of fission fragments ($\sim 10\ \mu$), whereas many of the grains investigated are smaller than 10^{-6} cm^2 in area (see Figs. 2 and 3). In general U contents calculated by equation 1 for small grains can either be high or low depending on the variation of the grain area with

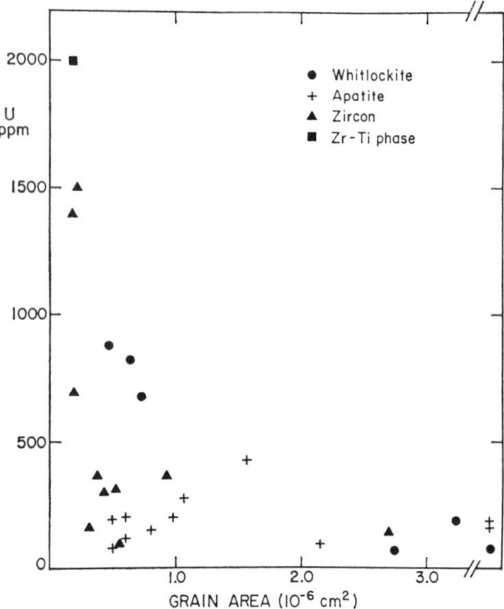

Fig. 2. U concentration vs. grain size in minerals of rock 12013. This graph shows the inverse relationship between the two quantities. The constant uranium concentration in apatite as compared to whitlockite or zircon is apparent. The three points on the right side of the graph extend to 3.7×10^{-6} cm^2.

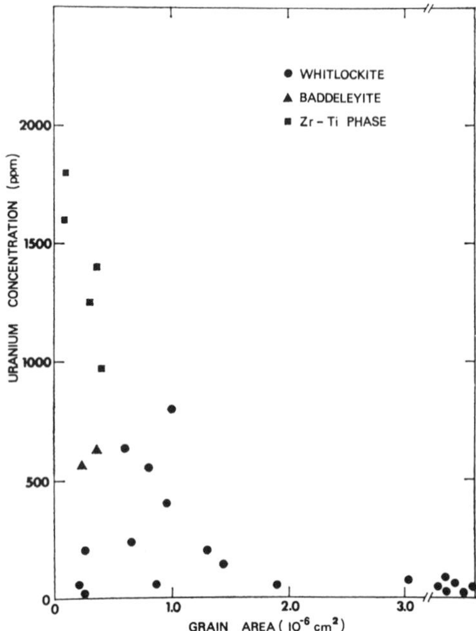

Fig. 3. U concentration vs. grain size in minerals of rock 12040. As in rock 12013, the uranium concentration shows an inverse relationship with grain size. The six points on the right side extend to 9×10^{-6} cm^2.

depth; however, for very small grains the calculated U contents will tend to be lower limits.

As shown in Figs. 2 and 3 there is a distinct inverse dependence of U concentration with grain size for both 12013 and 12040 although there is considerable scatter. Such an inverse dependence is frequently observed in suites of terrestrial zircons (SILVER, 1963). In 12013 the apatites have comparatively well-defined U contents of 100–200 ppm, whereas the whitlockite and zircon show a wider spread. In both rocks the Zr–Ti phases contain the highest U concentrations (>0.1% U).

U distribution in 12040 can be qualitatively understood if we assume that, even in the final moments of crystallization, the U concentration were increasing rapidly in the residual melt and that the smaller grains nucleated later and thus equilibrated with a melt of higher U content. It is not clear that this simple interpretation can be applied to 12013 because 12013 does not represent material which crystallized directly from a melt.

Because of the complexity of U distribution in these samples, it does not appear possible to draw unambiguous conclusions from our data concerning the partitioning of U between co-existing phases in lunar rocks. Conceivably, partition coefficients could be derived from binary grains of U bearing phases; however, no suitable cases were observed in our studies.

It is possible to make crude estimates of U material balance from our data; however because of the complexity of U distribution and because of the limited sampling—particularly for an extremely inhomogeneous sample such as 12013, the results of this calculation should be used with extreme caution. We have weighted the observed U concentrations of each grain by its area and calculated the average concentration and fraction of U contained for each phase (Table 3). For 12013 we have included the distributed U in the dark regions and *assumed* that this represents 20% of the U in 12013 corresponding to the estimated fraction of dark material in 12013. Taking the calculation at face value, we would conclude that apatite, whitlockite, and zircon are of comparable importance and that the Zr–Ti phases are not necessarily a negligible source of U in 12013.

Rock 12040 shows a different trend. Most of the uranium (78%) is contained in the whitlockite phase. Baddeleyite and the Zr–Ti phases are low in total uranium and

Table 3. U material balance for lunar rocks 12013 and 12040.

Phase	12013 U ppm	12013 % of total U	12040 U ppm	12040 % of total U
Apatite	170	25	—*	—
Whitlockite	190	25	104	78
Zircon	330	17	—*	—
Baddeleyite	—*	—	840	10
Zr–Ti phases	2000	10	1300	5
Distributed	6†	20		7‡

* Phase not present.
† Dark regions.
‡ From tracks in major minerals both from buried sources and distributed.

of comparable importance. This rock may represent a more equilibrated distribution than does rock 12013.

We also studied several anorthositic fragments from the Apollo 11 and Apollo 12 lunar soil. These were chosen from the lunar gravel on the basis of their light color and distinctive fine texture. In three anorthosites from the Apollo 11 gravel we found the uranium concentration to be about 0.04 ppm. This is consistent with the anorthosites being from the highlands, where from the absence of Po^{210}, the uranium is thought to be low (TURKEVICH *et al.*, 1970). As in other rocks, the tracks in the anorthosites occur in stars. Eleven stars were found in 3 fragments totalling 6.2 mm^2. These stars are usually associated with phosphates; however, one of these stars was associated with a Zr–Ti–Si phase similar to the ones found in 12013 and 12040.

TH–U RATIOS IN ROCK 12013

The Th/U ratios in 12 crystals of 12013 were measured using the α irradiation technique (HAIR *et al.*, 1971). The α particles fission both Th and U, whereas the neutrons fission only U^{235}. Using the relative number of tracks (N) from a given grain produced by α particles and by thermal neutrons, we calculate

$$\mathrm{Th/U} = \frac{N_\alpha}{N_n} \cdot \frac{\rho_n}{\rho_\alpha} - 1, \quad (2)$$

where ρ_n and ρ_α are the track densities measured in the standard glass for the neutron and α irradiations. This equation is valid for grains of all sizes. In principle accurate Th/U ratios can be obtained by this technique; however, measurements of N_α are hindered by the radiation damage effects produced in the mica and the section by the α irradiation (10 μahr.). Th data were obtained only for one small area on one section of 12013. This area did not contain a good apatite star; however, Th/U ratios for the other phases were obtained. These are presented in Table 2 and Fig. 4.

The Th/U ratios vary considerably from one mineral to the next ranging from 0.2

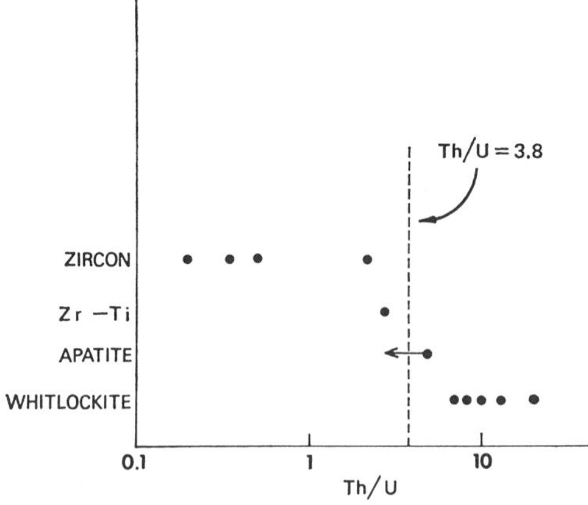

Fig. 4. Plot of the thorium–uranium ratios for different minerals of 12013. The Th/U values range over two orders of magnitude and show a strong dependence on mineral phase.

to 20. This result is not surprising considering the leach experiments of SILVER (1970) which have shown that Th and Pb^{208} are not uniformly distributed with U and its daughters throughout rock 10017. In rock 12013 the effect is so striking that it shows up in a simple visual examination of the stars; some are "bright" on the Th + U maps and "dim" on the U only map, while for others the reverse is true. The low (Th/U) for 12013 zircon compared to the total rock is consistent with data on terrestrial granites (compare, for example, BROWN and SILVER, 1956).

The apparent enrichment of Th in whitlockite is particularly interesting in view of the key role this mineral has played in studies of extinct Pu^{244} in meteorites. Assuming no geochemical segregation WASSERBURG *et al.* (1969) obtained an initial ratio of Pu/U of 0.03 for the St. Severin chondrite at the time the mineral had cooled to the gas retention temperature. Based on the amounts of spallation Xe^{126} in the whitlockite, the (rare earth)/U in St. Severin whitlockite is about five times larger than that for the meteorite as a whole (RUSS, unpublished calculations). Similarly, rare earths are enriched relative to U in 12013 whitlockite by about a factor of 8 (WAKITA and SCHMITT, 1970; LUNATIC ASYLUM, 1970). This similarity suggests that Th should also be enriched relative to U in St. Severin whitlockite. In view of evidence for fractionation of transuranic elements in St. Severin, the critical question is whether Pu behaves geochemically more like Th (or rare earths) or like U. The answer to this question is not known, but if, as suggested by FIELDS *et al.* (1966), Pu follows the rare earths, then the Pu/U ratio for the early solar nebula is considerably lower than 0.03, and the use of the 0.03 Pu/U ratio in calculations of the time scale for galactic nucleosynthesis (WASSERBURG *et al.*, 1969b; HOHENBERG, 1969) is not justified.

The average Th/U for the measured grains is 3.9, in good agreement with the total rock values (TATSUMOTO, 1970). However an area-weighted total rock Th/U ratio of about 6 is obtained using the average mineral Th/U ratios from Table 2 and U distribution given in Table 3. This suggests that our U distribution gives too much weight to whitlockite.

Volatilization and leaching studies have given important information concerning the existence of different Pb isotopic components in the lunar soil (SILVER, 1970). Given more extensive data on Th/U ratios in different lunar minerals and rocks, it may be possible to set limits on possible sources for the various isotopic components from the measured Pb^{208}/Pb^{206} ratios.

Assuming either iron meteorite or terrestrial common lead corrections, TATSUMOTO (1970) obtained discordant Pb–U–Th ages for 2 of 3 samples from 12013. Because the Th–Pb ages were discordant (either with respect to the Pb–Pb ages or with respect to the accepted 4.0×10^9 yr. age for 12013; LUNATIC ASYLUM, 1970; TURNER, 1970) and because our data show that zircon is not an important source of Th in 12013, the discordant ages cannot be attributed entirely to loss of radiogenic Pb from zircon or to incomplete dissolution of zircon during the Pb–U–Th analyses.

The radiogenic Pb^{208}/Pb^{206} ratios expected for the different radioactive mineral phases of 12013 are given in Table 4 and compared with the total rock ratios obtained by TATSUMOTO (1970). The two sets of values are compatible, but in view of the uncertainties in our U material balance and the lack of a reliable apatite Th/U value, any further interpretation does not appear profitable.

Table 4. Comparison of calculated radiogenic (Pb^{208}/Pb^{206}) for mineral phases in rock 12013 with measured total rock values.

	Calculated*	Total rock values†	
Zircon	0.07		
Apatite	1.1	Dark	1.0
Zr–Ti phase	0.7	Light	0.87
Whitlockite	2.8		

* Assuming average Th/U ratios for phases from Table 2 and an age of 4.0×10^9 yrs.
† TATSUMOTO (1970), based on two stage growth model.

FISSION TRACK DATING

There are two general methods of dating lunar rocks using fission tracks. The first method measures the fossil-track density in the U-bearing phase itself and is analogous to the standard fission-track technique. The second method relies on the fact that U-rich inclusions can act as fission-track sources producing tracks that register in adjacent major minerals such as feldspar or pyroxene.

We have applied both methods to rocks 12013, 12040, and one anorthosite fragment. The results to date are not completely satisfactory, and the results presented here should be considered as a preliminary report on this problem.

In one polished section of rock 12013 we first located phosphate grains in the microprobe, and then attacked these crystals for increasing times in a dilute solution of HNO_3 until etch pits appeared in apatites. We were not able to etch pits in whitlockite. The pits, one picture of which was published by us in an earlier report on rock 12013 (BURNETT *et al.*, 1970) are rather poorly developed in this rock. To obtain a fossil-track density due to spontaneous fission we subtracted the cosmic ray track background measured in adjacent feldspar grains. Following the initial attack, the sample was covered by mica and given a thermal neutron dose. The number of induced tracks was measured in the mica and an age was calculated from the following equation:

$$T = \frac{1}{\lambda_D} \ln \left[\frac{\sigma I \phi \lambda_D \rho_f}{2\lambda_{SF}\rho_i} + 1\right], \tag{3}$$

where σ is the fission cross section, I the U^{235}/U^{238} ratio, ϕ the neutron flux, λ_D and λ_{SF} the total and spontaneous fission decay constants, and ρ_f and ρ_i the fossil and induced fission-track densities. This equation makes the following assumptions: (1) The observed pits are due to spontaneous fission; (2) the registration efficiencies for fission tracks in mica and apatite are equal; (3) the apatite grain has a thickness $> 10\ \mu$ (the range of fission fragments); (4) an equivalent thickness of crystal was removed in polishing the section; (5) the cosmic-ray background in apatite and feldspar are the same; and (6) U concentration was not changed when the etch pits were developed.

In spite of the fact that the grains were quite small, the results shown in Table 5 show a consistent age of $\sim 1.4 \times 10^9$ years.

Extrapolation of the laboratory annealing results of NAESER and FAUL (1969) on terrestrial apatites indicated that between 50% and 100% of the tracks would be erased

Table 5. Apatite ages in rock 12013, ρ_{fa} is the fossil track density in the apatite; ρ_{im} is the induced track density in the mica.

ρ_{fa}* (10^8 cm^{-2})	ρ_{im} (10^8 cm^{-2})	ρ_{fa}/ρ_{im}§	Age (10^9 yrs.)
0.80	1.38†	0.57 ± 0.13	1.3 ± 0.2
0.63	1.39‡	0.45 ± 0.14	1.1 ± 0.2
1.16	2.74†	0.42 ± 0.14	1.0 ± 0.2
1.16	0.71‡	1.63 ± 0.56	3.2 ± 0.6
0.63	1.34‡	0.47 ± 0.11	1.1 ± 0.15
0.85	1.11‡	0.77 ± 0.26	1.7 ± 0.4
1.10	1.75†	0.63 ± 0.22	1.5 ± 0.3

* Corrected for cosmic ray track density.

† Calculated from total number of tracks divided by the area of grain.

‡ Calculated from track density in mica at center of star.

§ Quoted errors are two standard deviations calculated from track statistics.

during the surface exposure time of $\sim 10^7$ years (as measured by galactic cosmic-ray tracks) at the lunar daytime surface temperature of ~ 130°C. We therefore originally attributed this low age to simple thermal fading. Although this remains the most reasonable interpretation, the results on 12040 described below are inconsistent with fading, and the age may represent the time since a distinct heating or shock event occurred.

In rock 12040 we irradiated the section with a relatively low neutron dose prior to etching (20 sec in 0.12% HNO_3). The fossil track density was obtained by subtracting both the induced track density (measured in a mica external detector) and the cosmic-ray background. The subtraction amounted to only 10% of the total track density.

In contrast to rock 12013, the etch pits in these crystals were well formed and regular (Fig. 1). Five large whitlockite crystals containing 500 to 3000 tracks were found. As shown in Table 6 the crystals gave a consistent—but surprising—apparent age of $6.5 \pm 0.7 \times 10^9$ years. The quoted error is the 2σ limit.

Reimer *et al.* (1970) have shown that the factor of two that appears in the denominator of the logarithmic term in equation 3 is not strictly correct and will vary for different external detectors, etching conditions, and modes of observation. The use of short etching times and observations in the SEM tends to minimize this correction, and we estimate that at most the age could be lowered by 15% by including this effect.

There are a number of possible explanations of the high age—some trivial and some interesting. For example, it may be that spallation events produce etchable tracks with far more efficiency in whitlockite than in feldspar (and pyroxene). The cosmic-ray background correction that we applied would then be incorrect. A more interesting interpretation is to attribute the excess tracks to Pu^{244} or to a super-heavy element. However, the age of the rock would have to be 3.9×10^9 years if we assume an initial Pu/U ratio of 0.03. In view of the measured Rb/Sr age of 3.3×10^9 years (Papanastassiou and Wasserburg, 1971), this is a most unlikely possibility. A typical integrated thermal neutron flux for lunar material is 2×10^{16} n/cm^2 (Lunatic

Table 6. Whitlockite ages in rock 12040, ρ_{fw} is the fossil track density in the whitlockite; ρ_{im} is the induced track density in the mica.

ρ_{fw}* (10^8 cm^{-2})	ρ_{im} (10^8 cm^{-2})	ρ_{fw}/ρ_{im}†	Age (10^9 yrs.)
3.85	0.239	16.0 ± 1.7	6.6 ± 0.7
3.68	0.207	17.8 ± 1.3	7.0 ± 0.5
2.38	0.160	14.9 ± 1.6	6.3 ± 0.6
2.44	0.153	15.9 ± 1.3	6.6 ± 0.8
3.00	0.231	13.0 ± 2.2	5.8 ± 0.9

* Corrected for cosmic-ray track density (1.8×10^7 t/cm^2 as found in feldspar).

† Quoted errors are two standard deviations calculated from track statistics.

Asylum, 1970). If the irradiation occurred during the rock's recent residence on the lunar surface, this flux would account for about 15% of the excess tracks we see. However, a neutron irradiation at an earlier date produces more induced fission tracks owing to a building up of U^{235} as we go back in time. A neutron irradiation of the same magnitude occurring 1.8 billion years ago would account for all of the excess fission tracks. These or other explanations are possible. We are continuing our work to find the correct explanation for the anomalous age.

Two apatite grains in one fragment of anorthosite have been etched. The pits are somewhat irregular as in rock 12013. Although the uranium concentration is similar to the dated stars in 12040, the fossil-track density is only $\sim\frac{1}{3}$ as high. Thus in this one anorthositic fragment there is *no* evidence for Pu anomaly.

In rock 12013 we also attempted to measure ages by dating buried inclusions. Most buried uranium sources occurred along grain boundaries or in cracks and only 5–10% of the total were suitable for dating. We found four U stars in regions that showed a pure feldspar composition. Subsequent etching of the feldspar showed a spray of tracks that we interpreted as arising from the spontaneous fission in a buried inclusion. Optical examination of the star as registered in the mica confirmed that the tracks were produced by a buried source.

The use of mica as the external detector created a special problem in comparing fossil and induced track densities. The process of etched track production depends on the rate of ionization of the fission fragments. Below a threshold value, different for mica and feldspar, etchable tracks are not formed. Mica has a lower threshold than feldspar, and it is possible for a fission fragment to emerge from the feldspar with an energy such that it registers in the mica but *not* in the feldspar. The proportion of such particles increases as the depth of the inclusion increases. The net effect is that the effective efficiencies for track registration in feldspar and mica are different for each inclusion and depend critically on the burial depth and on the geometry (point source, thick slab, etc.) of the inclusion.

We took this effect into account by estimating the depth and geometry of the inclusions from optical measurements of track lengths in the mica and by constructing correction curves based on measurements of the registration efficiencies in mica and feldspar covered with different thicknesses of absorber. In three stars which appeared to emanate from thick sources (because of the presence of many short tracks) we found ages of $4.9 \pm 1.4 \times 10^9$, $4.1 \pm 1.4 \times 10^9$, and $3.6 \pm 1.3 \times 10^9$ years. The fourth star gave ages of 0.5 ± 1.9 and $3.2 \pm 1.6 \times 10^9$ years depending on whether or not we assumed the inclusion was a point source or a thick source.

The efficiency corrections are complicated, and the measurements of the fossil star track densities are difficult. Thus, we are not completely happy with the above results. Before drawing conclusions on track stabilities, etc., we prefer to wait for additional data. To minimize the efficiency correction problem we are now using polished plates of terrestrial feldspar as our external detectors. Using these detectors one age of $3.9 \pm 0.6 \times 10^9$ years has been obtained from a buried inclusion in rock 12040. Although more data from the same rock are needed, this result indicates that the feldspar detector technique is practical.

Acknowledgments—It is a pleasure to acknowledge the assistance of P. Swan for his contributions to all phases of the experimental work. We are also deeply grateful to J. Wood of the Smithsonian Astrophysical Observatory who kindly guided us in the selection of the anorthosites during a visit to his laboratory. This work was supported by NASA Contract NAS9–8165.

References

Ahrens T. J., Fleischer R. L., Price P. B., and Woods R. T. (1970) Erasure of fission tracks in glasses and silicates by shock waves. *Earth Planet. Sci. Lett.* **8**, 420–426.

Bhandari N., Bhat S., Lal D., Rajagopalan G., Tamhane A. S., and Venkatavaradan V. S. (1971) Fossil track studies in lunar materials—III: The spontaneous fission record of uranium and extinct transuranic elements in meteorites and Apollo rocks and fines. Second Lunar Science Conference (unpublished proceedings).

Brown H., and Silver L. T. (1956) The possibilities of securing long range supplies of uranium and thorium and other substances from igneous rocks. *Proceedings of the International Conference on the Peaceful Uses of Atomic Energy* (Geneva), Vol. 8, pp. 129–132, United Nations Publication.

Burnett D. S., Monnin M., Seitz M., Walker R., Woolum D., and Yuhas D. (1970) Charged particle track studies in lunar rock 12013. *Earth Planet. Sci. Lett.* **9**, 127–136.

Cantelaube Y., Maurette M., and Pellas P. (1967) Heavy ion tracks in the minerals of the Saint Severin chondrite. In *Radioactive Dating and Methods of Low Level Counting* (IAEA, Vienna, Austria), pp. 215–229.

Comstock G. M., Evwaraye A. O., Fleischer R. L., and Hart H. R., Jr. (1971) The particle track record of the Ocean of Storms. Second Lunar Science Conference (unpublished proceedings).

Crozaz G., Haack U., Hair M., Maurette M., Walker R. M., and Woolum D. (1970) Nuclear track studies of ancient solar radiations and dynamic lunar surface processes. *Proc. Apollo 11 Lunar Sci. Conf., Geochim. Cosmochim. Acta* Suppl. 1, Vol. 3, pp. 2051–2080. Pergamon.

Drake M. J., McCallum I. S., McKay G. A., and Weill D. F. (1970) Mineralogical studies of lunar rock 12013,10. *Earth Planet. Sci. Lett.* **9**, 103–123.

Fields P. R., Friedman A. M., Milsted J., Lerner J., Stevens C. M., Metta D., and Sabine W. K. (1966) Decay properties of plutonium-244 and comments on its existence in nature. *Science* **212**, 131–134.

Fleischer R. L., Price P. B., and Walker R. M. (1965) Spontaneous fission track from extinct Pu^{244} in meteorites and the early history of the solar system. *J. Geophys. Res.* **70**, 2703–2707.

Fleischer R. L., Price P. B., and Walker R. M. (1968) Identification of Pu^{244} fission tracks and the cooling of the parent body of the Toluca meteorite. *Geochim. Cosmochim. Acta* **32**, 21–31.

Fleischer R. L., Haines E. L., Hart H. R., Jr., Woods R. T., and Comstock G. M. (1970) The particle track record of the Sea of Tranquillity. *Proc. Apollo 11 Lunar Sci. Conf., Geochim. Cosmochim. Acta* Suppl. 1, Vol. 3, pp. 2103–2120. Pergamon.

FRICK C., HUGHES T. C., LOVERING J. R., REID A. F., WARE N. G., and WARK D. A. (1971) Electron probe, fission track, and activation analysis of lunar samples. Second Lunar Science Conference (unpublished proceedings).

HAIR M. W., KAUFHOLD J., MAURETTE M., and WALKER R. M. (1971) The microanalysis using fission tracks. In *Radiation Effects* (to be published).

HOHENBERG C. M. (1969) Radioisotopes and the history of nucleosynthesis in the galaxy. *Science* **166**, 212–215.

LAL D., MACDOUGALL D., WILKENING L., and ARRHENIUS G. (1970) Mixing of the lunar regolith and cosmic ray spectra: Evidence from particle-track studies. *Proc. Apollo 11 Lunar Sci. Conf., Geochim. Cosmochim. Acta.* Suppl. 1, Vol. 3, pp. 2295–2303. Pergamon.

LOVERING J. F., and KLEEMAN J. D. (1970) Fission track uranium distribution studies on Apollo 11 lunar samples. *Proc. Apollo 11 Lunar Sci. Conf., Geochim. Cosmochim. Acta* Suppl. 1, Vol. 1, pp. 627–631. Pergamon.

LUNATIC ASYLUM (1970) Mineralogy and isotopic investigations on lunar rock 12013. *Earth Planet. Sci. Lett.* **9**, 137–163.

MORY J., DEGUILLEBON D., and DELSARTE G. (1970) Measurement of the mean range of fission fragments using mica as a detector—Influence of the crystalline structure. *Radiation Effects* **5**, 37–40.

NAESER C. W., and FAUL H. (1969) Fission track annealing in apatite and sphene. *J. Geophys. Res.* **74**, 705–710.

PAPANASTASSIOU D. A., and WASSERBURG G. J. (1970) Rb–Sr ages from the Ocean of Storms. *Earth Planet. Sci. Lett.* **8**, 269–278.

PAPANASTASSIOU D. A., and WASSERBURG G. J. (1971) (to be published).

PRICE P. B., and O'SULLIVAN D. (1970) Lunar erosion rate and solar flare paleontology. *Proc. Apollo 11 Lunar Sci. Conf., Geochim. Cosmochim. Acta* Suppl. 1, Vol. 3, pp. 2351–2359. Pergamon.

REIMER G. M., STORZER D., and WAGNER G. A. (1970) Geometry factor in fission track counting. *Earth Planet. Sci. Lett.* **9**, 401–404.

RICE C. M., and BOWIE S. H. U. (1971) Distribution of uranium in Apollo 11 type B samples. Second Lunar Science Conference (unpublished proceedings).

ROEDDER E., and WEIBLEN P. W. (1970) Silicate liquid immiscibility in lunar magmas, evidenced by melt inclusions in lunar rocks. *Science* **167**, 641–643.

SCHIRK J., HOPPE M., MAURETTE M., and WALKER R. M. (1968) Recent fossil track studies bearing on extinct Pu^{244} in meteorites. *International Symposium in Meteorite Research* Vienna, Reidel Publishing Co.

SCHREURS J. W. H., FRIEDMAN A. M., ROKOP D. J., HAIR M. W., and WALKER R. M. (1970) Calibrated U–Th glasses for neutron dosimetry and determination of uranium and thorium concentrations by the fission track method. In *Radiation Effects* (to be published).

SILVER L. T. (1963) The use of cogenetic uranium–lead isotope systems in zircon in geochronology. In *Radioactive Dating*, pp. 279–287, IAEA, Vienna.

SILVER L. T. (1970) Uranium–thorium–lead isotopes in some Tranquillity Base samples and their implications for lunar history. *Proc. Apollo 11 Lunar Sci. Conf., Geochim. Cosmochim. Acta* Suppl. 1, Vol. 2, pp. 1533–1574. Pergamon.

SILVER L. T. (1971) U–Th–Pb isotope relations in Apollo 11 and Apollo 12 lunar samples. Second Lunar Science Conference (unpublished proceedings).

TATSUMOTO M. (1970) U–Th–Pb age of Apollo 12 rock 12013. *Earth Planet. Sci. Lett.* **9**, 193–200.

TATSUMOTO M., KNIGHT R. J., and DOE B. R. (1971) U–Th–Pb systematics of Apollo 12 lunar samples. Second Lunar Science Conference (unpublished proceedings).

TERA F., EUGSTER O., BURNETT D. S., and WASSERBURG G. J. (1970) Comparative study of Li, Na, K, Rb, Cs, Ca, Sr, and Ba abundances in achondrites and in Apollo 11 lunar samples. *Proc. Apollo 11 Lunar Sci. Conf., Geochim. Cosmochim. Acta* Suppl. 1, Vol. 2, pp. 1637–1657. Pergamon.

TURKEVICH A. L., PATTERSON J. H., FRANZGROTE E. J., SOWINSKI K. P., and ECONOMOW T. E. (1970) Alpha radioactivity of the lunar surface at the landing sites of Surveyors 5, 6, and 7. *Science* **167**, 1722–1724.

TURNER G. (1970) ^{40}Ar–^{39}Ar age determination of lunar rock 12013. *Earth Planet. Sci. Lett.* **9**, 177–180.

WAKITA H., and SCHMITT R. A. (1970) Elemental abundances in seven fragments from lunar rock 12013. *Earth Planet. Sci. Lett.* **9**, 169–176.

WASSERBURG G. J., MACDONALD G. J. F., HOYLE F., and FOWLER W. A. (1964) Relative contribution of uranium, thorium, and potassium to heat production in the earth. *Science* **143**, 465–467.

WASSERBURG G. J., HUNEKE J. C., and BURNETT D. S. (1969) Correlation between fission tracks and fission-type xenon from an extinct radioactivity. *Phys. Rev. Lett.* **22**, 212–215.

WASSERBURG G. J., SCHRAMM D. N., and HUNEKE J. C. (1969b) Nuclear chronologies for the galaxy. *Astrophys. J.* **157**, L91–L96.

Proceedings of the Second Lunar Science Conference, Volume 2, pp. 1521–1546.
The M.I.T. Press, 1971.

U–Th–Pb systematics of Apollo 12 lunar samples

MITSUNOBU TATSUMOTO, ROY J. KNIGHT, and BRUCE R. DOE
U.S. Geological Survey, Denver, Colorado 80225

(*Received 22 February 1971; accepted in revised form 31 March 1971*)

Abstract—The data determined in the U–Th–Pb system on rocks collected by the Apollo 12 mission confirm the great antiquity and extreme depletion of common lead in the surficial lunar rocks in maria. Unlike the rocks analyzed from the Apollo 11 mission, those from Apollo 12 do not define a good Pb–Pb isochron, although six of the eight rocks lie close to a 3900 to 4000 m.y. isochron. Apollo 11 rocks have a well-defined 4200 m.y. isochron, and the unusual Apollo 12 rock 12013 has a well-defined 4370 m.y. isochron. Although these Pb–Pb isochron ages for the Apollo rocks are about 10% older than Rb–Sr mineral isochron ages and $^{40}Ar/^{39}Ar$ whole-rock ages as determined by other workers, they are in the same relative sequence of ages. Concordia relations on whole rocks and U–Pb and Th–Pb isochrons on density fractions, even though poorly defined, are compatible with the ages determined by other methods if the values of $^{238}U/^{204}Pb$ observed for the Apollo 11 rocks and fragments from rock 12013 are proportional to the $^{238}U/^{204}Pb$ in the source materials of these rocks and if the initial lead was not homogenized in the magma-forming events. This proportionality constraint does not exist for the Apollo 12 rocks.

An attempt was made to resolve the conflict in the U–Th–Pb whole-rock data with the Rb–Sr mineral isochron and $^{40}Ar/^{39}Ar$ whole-rock age data by a concordia diagram treatment of U–Pb data on density fractions. In general, the U–Pb data as plotted by using this technique indicate an event younger than those events determined for the same rocks by other lead isotope or dating methods. This difference could be explained if the unknown amount of lead contamination due to the sample processing decreases with increase in the amount of sample processing, in spite of the fact that all liquids in the sample processing were distilled before use.

The agreement between apparent ages determined by U–Th–Pb methods and those ages determined by other dating techniques is much better on the soils and breccias, which are of complex origin, than it is on the rocks. In general, the U–Th–Pb data suggest that the soils were derived from source material about 4650 m.y. old that had undergone differentiation of U relative to lead no more than 4000 m.y. ago. U–Th–Pb data on Apollo 12 soils and breccia also indicate that alteration in U/Pb also has taken place at times much younger than 3000 m.y. ago by "third events" (perhaps alteration due to impact events?) which are not reflected in the Rb–Sr mineral isochron or in the $^{40}Ar/^{39}Ar$ whole-rock ages but the occurrence of which may be supported by partial alteration of whole-rock K–Ar ages. In addition, we emphasize that the occurrence of "third events" is supported by the data of all lead isotope investigators on soils and breccias, and we view this information as a most important discovery on Apollo 12 material, possibly second only in importance to the great antiquity of rock 12013.

INTRODUCTION

PRESENTED HEREIN are the lead isotopic compositions and, as determined by isotope dilution, the uranium, thorium, and lead concentrations in some samples collected by the Apollo 12 mission (volcanic rocks 12009, 12021, 12022, 12035, 12038, 12052, 12063, 12064; igneous breccia 12013; fines 12033 and 12070; and impact breccia 12034). The new data presented here on samples collected by the Apollo 12 mission

* Publication authorized by the Director, U.S. Geological Survey.

are compared primarily with data for samples returned by the Apollo 11 mission that has been previously discussed by TATSUMOTO and ROSHOLT (1970) and TATSUMOTO (1970a) to facilitate uniformity of the data used in the comparison. Certain aspects of the data on the igneous breccia 12013 are included here that were not previously included (i.e., analysis of sawdust from 12013) or that were not thoroughly discussed (i.e., data on fragment 09B) by TATSUMOTO (1970b) owing to the short time available for writing that manuscript.

DATA

Analytical procedures and accuracy

Certain facets of the U–Th–Pb dating of lunar samples conflict with the dating by other techniques. Some investigators have been tempted to attribute the conflict entirely to the U–Th–Pb system and due to analytical problems at least in part. Accordingly, a rather complete description of the aspects of the analytical procedures in question and the analytical accuracy seems warranted.

Sample preparation

All samples except the igneous breccia 12013 and the soils were first briefly washed with acetone by use of an ultrasonic vibrator. Fragments 42 and 09 from rock 12013 and the soil samples were analyzed without any precleaning. Sawed surfaces on fragment 45 of rock 12013 were abraded with a dental drill and washed briefly with double-distilled acetone using an ultrasonic vibrator. For most samples, a small chunk was crushed in a boron carbide mortar; one portion of the powder was used for determination of the lead isotopic composition and the other for determination of the uranium, thorium, and lead concentrations. Fragment 45 of rock 12013 was treated directly without crushing. Rock 12064 was processed by use of a tungsten carbide mortar rather than a boron carbide mortar. The whole-rock analysis of this sample was made on about 4.4 g of the sample, and the density separations were made on the remaining 13 g. The portion used for density separations was analyzed by the same procedures given previously by TATSUMOTO (1970a, p. 1596) except that the isodynamic separator used on sample 12064 was that in the clean mineral separation room of L. T. Silver at California Institute of Technology. The sample weights of the density fractions are 10084, fraction 0 (acetone suspension, Apollo 11 samples only)—0.2196 g, hand magnet fraction weight on this sample was not recorded; sample 12064, hand magnet fraction—0.008 g, fraction 1 (acetone suspension)—0.1216 g, fraction 2 (bromoform floats)—2.4261 g, fraction 3 (methylene iodide floats)—1.9635 g, fraction 4 (methylene iodide sinks, nonmagnetic)—5.8065 g, fraction 5 (methylene iodide sinks, magnetic)—1.6460 g; sample 12033, hand magnet fraction—a few grains, fraction 1—0.1226 g, fraction 2—0.1029 g, fraction 3—1.0351 g, fraction 4 (total methylene iodide sinks)—0.2510 g; sample 12070, hand magnet fraction—a few grains, fraction 1—0.1816 g; fraction 2—0.2154 g; fraction 3—0.9758 g; fraction 4 (methylene iodide sinks)—0.2510 g.

Density fractions of samples of rocks 10017 and 12064 and soils 10084, 12033, and 12070 were made by the same techniques as those previously described, which utilized immediately vacuum-predistilled heavy-density reagents and doubly-distilled acetone (TATSUMOTO, 1970a). There are still some reasons, nevertheless, to suspect lead contamination from the heavy liquids. For Apollo 12 materials, fraction 1 refers to material that remained in suspension in acetone after 2 minutes of ultrasonic agitation. On Apollo 11 (TATSUMOTO, 1970a, p. 1600), fraction 1 refers to bromoform floats, which is equivalent to fraction 2 on Apollo 12 materials; however, the acetone suspension was made on Apollo 11 samples. Analysis of this fraction on soil 10084 will be reported here.

Analysis

The lead contents of the lunar rocks are so little that we have had to develop new analytical techniques that have lowered the level of lead contamination to 4 to 7 nanograms (10^{-9} g) per analysis (TATSUMOTO, 1970b). Even so, this level of contamination is great enough that it still limits the size of the samples which can be analyzed. Dissolution of samples was made by a $HClO_4$–HNO_3–HF mixture in a Teflon digestion bomb followed by treatment of the residue with HNO_3 and HF in a Teflon beaker in a Teflon tank under nitrogen atmosphere. In most concentration determinations by isotope dilution, the ^{204}Pb, ^{207}Pb (and, for density fraction 1 of 12064 and 12033 and fragment 09B of rock 12013, the ^{208}Pb), and the ^{235}U, and ^{230}Th "spikes" were added to the bomb along with the powdered samples. Fragments 09A, 42, and 45 of rock 12013 were taken into solution first, then aliquoted, and spiked. Tests have been made on spiking before and after dissolution on oceanic tholeiites (which have lead contents comparable to those of lunar samples) and the reproducibility of the U/Pb is about 0.3% of the ratio. Therefore, there should be no problem of equilibration of the sample with the "spike," but there always is a small amount of white precipitate left in the lunar sample dissolution (metatitanic acid?) so that some problems in determinations of concentrations cannot be completely ruled out, nor can it be predicted at this time whether the effect would result in values of U/Pb or Th/Pb that are too great or too small. The bombs were also tested for memory by making sequential runs of ^{204}Pb and ^{207}Pb spikes usually used for the blank determination followed by ^{208}Pb spike, but no memory was observed.

The lead, which is coprecipitated from concentrated HNO_3 by $Ba(NO_3)_2$ and dissolved in water, is electrodeposited onto a platinum anode, converted to the phosphate, and is then analyzed by the silica gel-phosphate method (TATSUMOTO, 1970b). Early in the use of this technique the electrodeposition was sometimes not too efficient and the lead blank from the silica gel-phosphate method (0.03 ng) was more significant than it was later (where this blank was significant will be noted in the text where appropriate). The standard deviations of the mass spectrometric measurements of the lead isotope ratios are less than 1% for $^{206}Pb/^{204}Pb$ and less than 0.1% for $^{207}Pb/^{206}Pb$ and $^{208}Pb/^{206}Pb$. The overall lead blank is about 5 ng; this amount leads to an uncertainty in the lead isotope ratios for most samples of less than 3%.

Trace elements

The lead, uranium, and thorium concentrations in the Apollo 12 samples are shown in Table 1. The samples are basaltic and doleritic rocks whose concentrations range from 0.270 to 0.653 ppm Pb, 0.157 to 0.404 ppm U, and 0.615 to 1.41 ppm Th. In addition, we analyzed two fines and two breccia samples, of which rock 12013 is called an igneous breccia and 12034 is an impact breccia (WARNER, 1970). The lead, uranium, and thorium concentrations in the impact breccia and the fines of Apollo 12 are about 10 or more times greater than those of the igneous rocks.

The concentrations of lead, uranium, and thorium in the igneous breccia 12013 are exceedingly great—20 to 40 times greater than in the lunar igneous rocks—and, in addition, they reflect the extremely heterogeneous nature of the rock as previously stated (TATSUMOTO, 1970b). The values of Th/U reported by WAKITA and SCHMITT (1970) are low as compared to our values, their highest ratio of 3.0 is 10% less than our lowest ratio of 3.3 The difference in Th/U seems to be due to their systematically low thorium contents which appear to be low on the average by 10 to 20% as compared to our values; however, the rock does seem to be heterogeneous in Th/U. The LSPET (1970) value on the whole rock is 3.2, and the possibility remains that we analyzed fragments greater in Th/U than the average and that WAKITA and SCHMITT analyzed fragments less than average in Th/U. Use of the low Th/U values with our lead values

Table 1. Lead, uranium, and thorium concentration in Apollo 12 samples.

Sample No.	Type*	Description	Pb (ppm)	U (ppm)	Th (ppm)	Th^{232}/U^{238}	Approx. initial Pb (ppb) (after blank)
12009,22	A	Porphyritic olivine basalt	0.404	0.243	0.881	3.74	27
12021,122	B	Pigenite dolerite, pegmatite	0.419	0.261	0.932	3.69	11
12022,37	B	Olivine dolerite	0.309	0.198	0.710	3.70	23
12035,10,a	B	Troctolite	0.315	0.240	0.801	3.45	26
12035,10,b	B	Troctolite	0.270	0.199	0.682	3.54	—
12038,42	A	Basalt	0.283	0.157	0.615	4.05	19
12052,66,a	A	Olivine basalt	0.653	0.365	1.282	3.63	20
12052,66,b	A	Olivine basalt	0.585	0.347	1.231	3.67	—
12052,66,c	A	Olivine basalt	0.635	0.404	1.411	3.61	—
12063,49,a	A	Olivine basalt	0.332	0.191	0.679	3.67	13
12063,49,b	A	Olivine basalt	0.302	0.191	0.637	3.45	—
12064,21	B	Dolerite with crystobalite	0.532	0.278	0.977	3.64	30
12034,16,a	C	Crystal impact breccia with glass	4.16	3.576	13.00	3.76	82
12034,16,b	C	Crystal impact breccia with glass	3.82	3.497	13.29	3.93	—
12033,53,a	D	Fines	4.00	2.670	9.700	3.75	96
12033,53,b	D	Fines	4.43	3.269	12.14	3.84	—
12070,56	D	Fines (contingency)	3.16	1.641	6.020	3.79	190
12070,56	D	Fines (contingency)	3.86	2.103	7.700	3.78	—
12013,10	A	Igneous breccia					
09A			9.28	5.675	20.73	3.78	200
09B			12.30	5.871	22.94	4.04	760
42			16.31	10.80	34.29	3.28	200
45			11.43	5.752	19.05	3.42	220
Sawdust			25.34 (10.78)†	5.158	16.88	3.38	—

* Type: A, fine-grained igneous rock; B, medium-grained igneous rock; C, breccia; D, soil.
† Corrected for lead contamination in order to get a reasonable U–Th–Pb system as compared to that for other chips.

will, however, always give ages by Th–Pb methods lower than U–Pb ages. The newly analyzed sawdust must be contaminated with lead from some source apparently different from the copper wire used in the wire saw inasmuch as the sawdust contains 2.65% copper and the saw wire only 0.6 ppm lead, values which would result in a negligible lead addition. More likely the contamination comes from the vacuum cleaner bags involved in vacuuming up the sawdust. If all the ^{204}Pb in the sawdust comes from contamination, the contaminant in the sample is 14.8 ppm (versus about 0.2 ppm common-type lead in the three largest fragments analyzed from rock 12013) of the 25.3 ppm lead in the total sawdust sample. The isotopic composition of the contaminant lead in this calculation was assumed to be similar to that found in dust gathered from the filters of our clean laboratory ($^{206}Pb/^{204}Pb$: 18.48; $^{207}Pb/^{204}Pb$: 15.73; $^{208}Pb/^{204}Pb$: 38.38; T. J. Chow, private communication, 1970). On this basis, the corrected lead content is found to be comparable to that in the fragments. The sawdust was thought to be a representative, homogenized sample of this igneous breccia; however, the uranium and thorium concentrations in the sawdust (two elements that are probably not contaminated) are 5 and 17 ppm, respectively, contents that are about half the reported values obtained by gamma spectrometry as reported

Table 2. Lead, uranium, and thorium concentrations in Apollo 11 and Apollo 12 samples.

Samples	Pb (ppm)	U (ppm)	Th (ppm)	Th^{232}/U^{238}
		Apollo 11		
Rocks				
Group II (3)*	0.29 ~ 0.51	0.16 ~ 0.27	0.53 ~ 1.02	3.53 ~ 3.99
Group I (3)*	1.56 ~ 1.74	0.85 ~ 0.87	3.30 ~ 3.43	4.03 ~ 4.08
Breccia (1)*	1.7	0.67	2.6	3.94
Fines (1)*	1.4	0.54	2.1	3.97
		Apollo 12		
Rocks (8)*	0.28 ~ 0.64	0.16 ~ 0.40	0.61 ~ 1.41	3.61 ~ 3.92
Breccia (1)*	4.2	3.6	13.0	3.76
Fines (2)*	3.2 ~ 4.00	1.6 ~ 2.7	6.0 ~ 9.7	3.75 ~ 3.79
Igneous breccia [12013]** (4)***	9.2 ~ 16.3	5.7 ~ 10.8	19.1 ~ 34.3	3.28 ~ 4.04

* Number of samples analyzed; ** sample number; ***number of chips analyzed.

by LSPET (1970) on the whole rock and about 10 to 20% lower than the concentrations reported in other chips by TATSUMOTO (1970b). Rather than attributing the low results to dilution from extreme contamination (to avoid this, considerable time was spent in handpicking fibers from the sawdust which were probably introduced by the vacuum cleaner bags), we feel that uranium-rich and thorium-rich small particles (such as phosphates) were probably not recovered in the sawdust. At any rate, as analyzed, the sawdust is not a representative sample of the igneous breccia, and data on it must be interpreted with care.

A comparison of the lead, uranium, and thorium concentrations between samples returned by the Apollo 11 and Apollo 12 missions is given in Table 2. The contents of the three elements and the value of Th/U for the Apollo 12 mission rocks are similar to those of the group II rocks from the Apollo 11 mission (potassium poor) but distinctly lower than those of the group I rocks returned by the Apollo 11 mission (which are highly potassic). The group I and group II classification used here is adopted from COMPSTON *et al.* (1970). The lead, uranium, and thorium contents of impact breccia and fines from Apollo 12 are two to five times greater than in the equivalent rock types from Apollo 11, but are between those contents of the igneous rocks and the igneous breccia 12013 suggesting that the fines and impact breccia are a mixture of rocks and the igneous breccia. SCHNETZLER *et al.* (1970) and SCHNETZLER and PHILPOTTS (1970) have pointed out that the dark phase of rock 12013 is a good candidate for the nonmeteoritic "KREEP-like component" in the lunar soils. WOOD *et al.* (1971) has also identified crystallized material of similar composition in the coarse fines of the Apollo 12 mission and calls them norite, but some of this norite has as much as 2% K_2O. Potassium-rich, rare earth element-rich, and phosphorus-rich glass and rock fragments (KREEP material) which were also reported at the Conference (particularly by MEYER *et al.*, 1971; HUBBARD *et al.*, 1971), are rather similar in composition to the dark phase of rock 12013. Most investigators have compared KREEP-like material with common terrestrial basalts; however, the major element compositions of these materials are rather equivalent to those of more unusual terrestrial basalts called shoshonites (basalts with coexisting plagioclase and sanidine in the groundmass). Analyses of two samples of Cenozoic shoshonites for the U–Th–Pb system have been given by PETERMAN *et al.* (1970) [2.56 and 2.63 ppm uranium, 9.54 and 11.01 ppm thorium, and 22.6 and 20.0 ppm lead]. Comparison of the shoshonite data with those on the dark phase of rock 12013 indicates that uranium and thorium in rock 12013 is enriched relative to the shoshonite data available by a factor of about 2, whilst the common lead content still follows the lunar depletion pattern found for the other igneous rock types and is depleted in rock 12013 by about a factor of 100 relative to the amount in the available shoshonite data (the lead in lunar samples is mainly radiogenic and the common lead component in fragment 12013 is only about 0.2 ppm). This lead depletion factor is actually larger than that found for other lunar rock types as compared to their terrestrial equivalents which, is about a factor of 10 to 20.

The lead content of lunar samples is sufficiently small that some workers have attributed all the common lead found to contamination at some stage of sample handling on the moon, in the Lunar Receiving Laboratory, or in the lead isotope analyses procedures. Thallium, for example, which like lead is a "volatile element" that is less abundant on the moon than on the earth, is found on the moon (0.3 to 1.1 ppb; GANAPATHY *et al.*, 1970) at about $\frac{1}{10}$ to $\frac{1}{100}$ that in terrestrial basalts (about 20 ppb; GANAPATHY *et al.*, 1970). Indeed, our computed common lead contents of lunar rocks are just about the values expected from the thallium depletion factor. We have reason to believe, therefore, that a common or initial lead correction of some sort is required for the lunar samples.

Isotope ratios

The isotopic compositions of Apollo 12 samples (Tables 3 and 4) are extremely radiogenic and the observed $^{238}U/^{204}Pb$ values—so-called μ values—range from 470 to 2700. These great μ values indicate that the volatile element lead was greatly depleted relative to uranium and thorium on the lunar surface as compared to that on the surface of the earth, as has been shown in the Apollo 11 papers (TATSUMOTO, 1970a; SILVER, 1970; GOPALAN *et al.*, 1970). The isotopic compositions of the rocks fall into two groups at least: one group (12009, 12022, 12035, 12038, and 12064) that has a value of about 300 to 450 for $^{206}Pb/^{204}Pb$ and $^{208}Pb/^{204}Pb$ and another group (12021, 12052, and 12063) that has values of about 700 for the $^{206}Pb/^{204}Pb$ ratio and 650 to 800 for $^{208}Pb/^{204}Pb$. Sample 12070 ("contingency fines") has a value of $^{206}Pb/^{204}Pb$ intermediate between the values for the two groupings of igneous rocks, but fines sample 12033, the impact breccia 12034, and the igneous breccia 12013 have exceedingly great values for the ratio.

A comparison of the isotopic compositions of the Apollo 12 samples with those of Apollo 11 is given in Table 5. The two groupings of Apollo 12 rocks by $^{206}Pb/^{204}Pb$ and $^{208}Pb/^{204}Pb$ are rather similar to those in rocks from the Apollo 11 mission where the group I (high-potassium) rocks have values of $^{206}Pb/^{204}Pb$ and $^{208}Pb/^{204}Pb$ that are greater than 400 and the group II (potassium-poor) rocks have values for $^{206}Pb/^{204}Pb$ of about 300 to 400 and for $^{208}Pb/^{204}Pb$ of 300 to 450. No parallelism is found, however, in the lead, uranium and thorium contents of the Apollo 12 rocks. WARNER and ANDERSON (1971) have further subdivided the group II rocks into

Table 3. Isotopic composition of lead of Apollo 12 samples.

		Atomic Ratio						
		Raw data			Corrected for blank			
Sample	Type*	Pb^{206}/Pb^{204}	Pb^{207}/Pb^{204}	Pb^{208}/Pb^{204}	Pb^{206}/Pb^{204}	Pb^{207}/Pb^{204}	Pb^{208}/Pb^{204}	U^{238}/Pb^{204}
12009,22	A	357.7	160.4	383.1	396.6	168.9	403.3	505.6
12021,122	B	606.3	244.6	608.0	672.2	270.3	671.9	871.9
12022,37	B	377.0	162.1	378.6	410.7	172.3	402.3	487.5
12035,10	B	247.0	83.96	259.4	293.4	97.85	304.3	447.0
12038,42,a	A	338.9	139.6	376.4	391.2	158.8	428.5	473.0
12038,42,b	A	207.8	89.6	239.8	225.5	96.5	258.6	—
12052,66	A	714.9	298.1	710.3	796.7	331.3	789.2	985.6
12063,49,a	A	645.1	269.4	638.6	720.2	299.9	710.5	863.3
12063,49,b	A	551.6	230.9	545.6	739.9	306.9	724.8	972.9
12064,21	B	397.8	208.5	395.7	449.6	234.9	444.6	510.6
12034,16,a	C	926.4	459.7	850.1	1388	660.3	1217	2485
12034,16,b	C	555.7	280.05	524.7	924.2	462.9	863.0	1765
12033,53	D	985.5	502.8	896.2	1134	577.6	1028	1581
12070,56	D	400.9	233.2	387.7	433.4	251.7	415.6	495.5
12013,10	A							
09A		898.8	481.4	904.8	1201.7	641.6	1203	1612
09B					(415)†	(230)†	(423)†	(435)†
42		1892	965.6	1631	2059	1050	1773	2773
45		1202	643.2	1062	1438	768.1	1266	1510
Sawdust		39.63	26.77	56.4	(1329)‡	(707)‡	(1149)‡	(1398)‡

* Type: A, fine-grained igneous rock; B, medium-grained igneous rock; C, breccia; D, soil.
† Corrected for Pb^{208} spike. The lead isotopic composition was calculated using Pb^{208}/Pb^{206} ratio in 09A.
‡ A reasonable value was calculated assuming lead contamination was 14.56 ppm for sawing.

Tablc 4. Isotopic composition of lead and concentration of lead, uranium, and thorium in mineral concentrates.

Sample	Fraction	Pb (ppm)	U (ppm)	Th (ppm)	Th^{232}/U^{238}	Corrected for blank Pb^{206}/Pb^{204}	Pb^{207}/Pb^{204}	Pb^{208}/Pb^{204}	U^{238}/Pb^{204}
12033	Whole rock	4.00	2.67	9.70	3.75	1134	577.6	1028	1581
	1. Very fine (float in acetone)	4.65	2.95	11.69	4.10	237.5	124.3	231.8	317.1
	2. $\rho < 2.9$ plag. (~98%)	4.24	2.60	8.91	3.54	224.4	128.0	218.8	293.4
	3. $2.9 < \rho < 3.3$ mixture (glass ~70%; compound grain 30%)	3.70	3.13	11.06	3.66	1391	711.6	1271	2453
	4. $3.3 < \rho$ (Py. ~85%; Ol. 10% others)	1.05	0.700	1.38	2.04	79.44	42.77	75.52	108.9
	Σ Sum of fractions	3.09	2.68	9.37	3.42				
12070	Whole rock	3.16	1.64	6.02	3.79	433.4	251.7	415.6	495.5
	1. Very fine (float in acetone)	6.39	2.44	9.35	3.97	449.2	285.4	437.8	394.2
	2. $\rho < 2.9$ plag. (> 95%)	12.62	8.58	31.42	3.79	158.3	93.97	165.2	247.3
	3. $2.9 < \rho < 3.3$ mixture (glass ~60%; compound grains ~40%)	1.26	0.588	2.16	3.79	484.6	268.5	469.0	496.1
	4. $3.3 < \rho$ ilm. (Py. 40%; Ol. ~40%)	0.606	0.343	1.10	3.32	249.9	126.3	231.4	286.4
	Σ Sum of fractions	3.17	1.79	6.59	3.73				
12064	Whole rock	0.532	0.278	0.977	3.64	449.6	234.9	444.6	510.6
	1. Very fine (float in acetone)	1.68	0.451	1.58	3.61	(42.79)*	(30.76)*	(41.76)*	(28.56)*
	2. $\rho < 2.9$ (plag. > 95%)	0.307	0.144	0.462	3.31	124.7	64.47	142.8	136.6
	3. $2.9 < \rho < 33$ mixture (compound grains)	0.579	0.353	1.14	3.35	152.7	68.78	163.2	204.7
	4. $3.3 < \rho$ non-magnetic (Py. ~60%, ilm. ~40%)	0.216	0.147	0.520	3.65	187.4	79.39	203.6	279.8
	5. $3.3 < \rho$ magnetic (ilm. ~80%)	0.759	0.497	1.77	3.69	354.3	135.8	364.3	487.7
	Σ Sum of fractions	0.384	0.231	0.793	3.54				
10084	0. acetone suspension	3.68	0.850	3.26	3.96	83.1	66.5	98.8	50.0

* Corrected for Pb^{208} spike ($^{206}Pb/^{208}Pb$ is assumed to be the same as for the whole rock).

Table 5. Lead isotopic composition of Apollo 11 and Apollo 12 samples

Samples	Pb^{206}/Pb^{204}	Pb^{207}/Pb^{204}	Pb^{208}/Pb^{204}
	Apollo 11		
Rocks			
Group II (3)*	280 ~ 420	140 ~ 200	290 ~ 450
Group I (3)*	410 ~ 1240	190 ~ 590	440 ~ 1280
Breccia (1)*	260	170	270
Fines (1)*	260	170	270
	Apollo 12		
Rocks (5)*	390 ~ 410	160 ~ 170	400 ~ 430
(3)*	670 ~ 720	260 ~ 300	640 ~ 710
Breccia (1)*	1390	660	1220
Fines [12070]† (1)*	430	252	416
[12033] (1)*	1130	580	1030
Igneous breccia [12013] (3)‡	1200 ~ 2060	640 ~ 1050	1200 ~ 1770

* Number of samples analyzed; † sample number; ‡ number of chips analyzed.

porphyritic basalts (also olivine-rich) [of which 12009, 12021, 12022, and 12052 are included in this study] and granular or ophitic basalts (also olivine-poor) [of which 12035, 12038, 12063, and 12064 are included in this study]; however, no correlation of $^{206}Pb/^{204}Pb$ with this subgrouping is found, either. Two of four rocks from the Apollo 12 mission classified as granular and ophitic (group II for Apollo 11 material) are in the great $^{206}Pb/^{204}Pb$ category, whereas one of four porphyritic basalts is in the low $^{206}Pb/^{204}Pb$ category. Apparently the subdivision of group II is not of significance in the U–Th–Pb system.

DISCUSSION

Isochron age methods

There are four isochron techniques of dating in the U–Th–Pb system, three of which are independent; two of three for the U–Pb system ($^{207}Pb/^{204}Pb$–$^{206}Pb/^{204}Pb$, $^{207}Pb/^{204}Pb$–$^{235}U/^{204}Pb$, $^{206}Pb/^{204}Pb$–$^{238}U/^{204}Pb$) and one for the Th–Pb system ($^{208}Pb/^{204}Pb$–$^{232}Th/^{204}Pb$). The Pb–Pb isochron, which was the first dating technique to give precise ages of meteorites, will be discussed first because of the simplicity of the method, its proven reliability, and the insensitivity of the method to a variety of analytical uncertainties. This discussion is followed by discussions of the $^{206}Pb/^{204}Pb$–$^{238}U/^{204}Pb$, $^{207}Pb/^{204}Pb$–$^{235}U/^{204}Pb$, and $^{208}Pb/^{204}Pb$–$^{232}Th/^{204}Pb$ isochrons which are analogous to the $^{87}Sr/^{86}Sr$–$^{87}Rb/^{86}Sr$ or BPI (Bernard Price Institute) isochron dating techniques.

Pb–Pb isochron

Owing to the fact that $^{238}U/^{235}U$ in lunar and terrestrial bodies is, for all practical purposes, a physical constant of 137.8 (see ROSHOLT and TATSUMOTO, 1970, for measurements on lunar samples), the Pb–Pb isochron has several advantages over other isochron methods. For example, this dating technique is not sensitive to analytical uncertainties in the determination of U/Pb and Th/Pb, and, in fact, the uranium, thorium, and lead contents need not be determined at all! As in all isochron treatments, no initial lead composition need be determined or estimated if several coeval samples with the same initial lead isotopic compositions are available that had a spread in their ratios of radioactive parent to the stable isotope of the daughter element, ^{204}Pb, for long periods of time. Even better, isochron lines that pass near the origin as do those for many lunar samples are insensitive to blank correction because the lunar samples are very old and highly radiogenic, whereas the isotopic composition of the blank lead is always near the origin (however, our blanks on whole-rock and soil analyses are less than 1% the lead in the sample analyzed). The only effect of the blank is to determine the position of the sample on the isochron, but it does not affect the slope (i.e., age) of the isochron. Lastly, the lunar samples are so old that errors in ^{204}Pb measurement (the least abundant lead isotope) also have little effect.

Naturally, the technique has some constraints. These are that all samples must be of the same age, have started with the same initial lead isotopic composition and have a spread in the lead isotope ratios. The linear relationship should not be the result of mixing. If there is no spread in the isotope ratios—for example, if one has only one sample—an initial lead must be assumed (this statement was oversimplified in TATSUMOTO, 1970a, p. 1598). The samples also should not have evolved in greater than two-stage systems if the third and higher order stages occurred at times significantly greater than $t = 0$. If these constraints are not met, irregularities in the isochron development will probably occur. Other aspects of the Pb–Pb isochron will be discussed where appropriate.

In comparison of rock-lead dating with Rb-Sr dating on terrestrial materials, good agreement has generally been obtained between the two methods. ROSHOLT *et al.* (1970) get 1820 ± 110 m.y. for the Pb–Pb isochron and 1810 ± 50 m.y. for the Rb–Sr isochron on a sample of granite from Saskatchewan, Canada. The U–Pb and Th–Pb isochrons are also in good agreement with the ages determined by the other techniques. FARQUHARSON and RICHARDS (1970) report ages of 1553 ± 29 m.y. for the U–Th–Pb isochrons and 1565 ± 52 m.y. for the Rb–Sr isochrons for a microgranite and pegmatite from Mount Isa, Australia. PETERMAN *et al.* (1971) give ages of 2950 ± 110 m.y. for the Pb–Pb isochron and 2925 ± 80 m.y. for the Rb–Sr isochron on a paragneiss from the Granite Mountains, Wyoming, and their age of 2610 ± 70 m.y. by the Rb–Sr isochron is within analytical uncertainties of the 2820-m.y. age obtained by the Pb–Pb and Th–Pb isochrons (ROSHOLT and BARTEL, 1969) on a recently altered granite that has also suffered metamorphism and intrusion at about 1600 m.y. (ROSHOLT and PETERMAN, 1969; PETERMAN *et al.*, 1971).

Rocks. All the samples analyzed from the Apollo 12 mission are plotted in Fig. 1, and an enlarged diagram of part of Fig. 1 is given on Fig. 2. The data on Fig. 1 are dominated by those for the igneous breccia 12013 and discussion of this sample first will help to illustrate the controversy existing between the different kinds of dating techniques. All samples of 12013, including the sawdust, lie very close to an isochron of 4370 m.y. that passes near the common lead field *no matter whether the analyses are or are not corrected for blank lead*, as if the lead initially present in the rock developed in a source with a small value of ^{238}U/^{204}Pb. Even though fragment 42

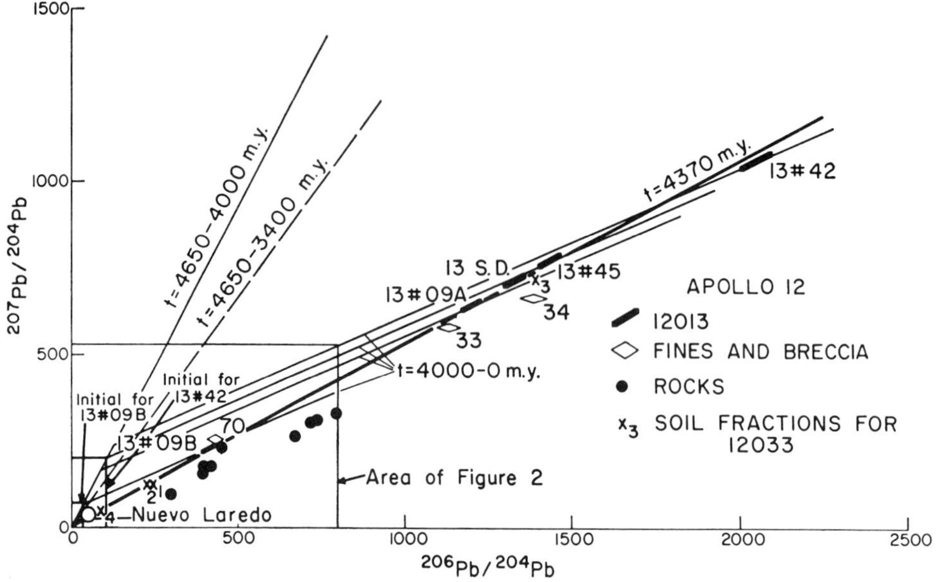

Fig. 1. The $^{206}Pb/^{204}Pb$—$^{207}Pb/^{204}Pb$ evolution for Apollo 12 whole rock and soil samples, density fractions (×) of soil 12033, and a meteorite (Nuevo Laredo). The heavy isochron line is drawn through the data on igneous breccia 12013. Data for rock 12013 would lie on the 4650–4000 m.y. isochron if it evolved in a U–Pb system between 4650 and 4000 m.y. with a $^{238}U/^{204}Pb$ of zero since 4000 m.y. ago. Isochron labeled 4650–3400 m.y. is included for comparison. Isochrons labeled 4000–0 m.y. define the locus of points of the $^{238}U/^{204}Pb > 0$ since 4000 m.y. ago. The intersection of the 4000–0 m.y. isochron with the 4650–4000 m.y. isochron would represent the lead isotopic composition initially in the rock at 4000 m.y. ago. See text for discussion. Size of boxes indicates the experimental error.

departs from the isochron beyond analytical uncertainties, the departure is not great and could be due to a minor process. This age of 4370 m.y. is in conflict with the two ages given by the Rb–Sr dating system (4500 m.y. for a whole-rock isochron composed of dark fragments, and 4000 m.y. for a mineral isochron; LUNATIC ASYLUM, 1970) and the age given by the $^{40}Ar/^{39}Ar$ method (3800 m.y.; TURNER, 1970b). In any controversy about the dating, however, we should not forget that all data indicate the sample is old, probably not younger than 4000 m.y. nor much older than 4600 m.y. Although the 4370-m.y. age by the Pb–Pb isochron is a valid age estimate, the intermediate age shown by the method relative to Rb–Sr dating could be explained if the $^{238}U/^{204}Pb$ that developed in each fragment at 4000 m.y. and existing to the present (second stage) is exactly proportional (enriched in ^{238}U over ^{204}Pb by a factor of about 2) to the $^{238}U/^{204}Pb$ existing in all fragments except 42 between 4600 m.y. and 4000 m.y. (first stage). For fragment 42, the $^{238}U/^{204}Pb$ in the second stage is enriched in ^{238}U over ^{204}Pb by a factor of about 2.4 and therefore departs only a little from that proportionality constant that governs the other samples.

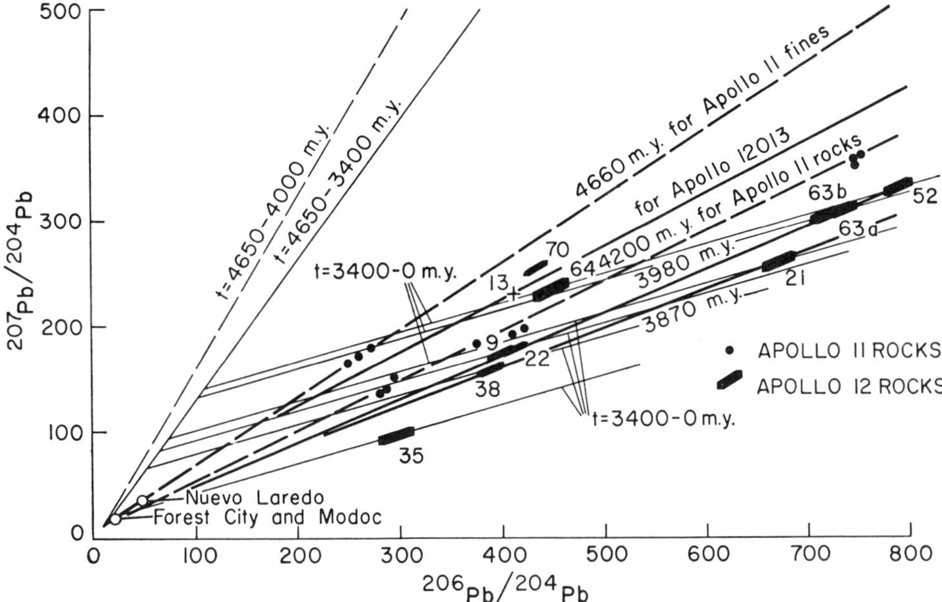

Fig. 2. Enlarged $^{206}Pb/^{204}Pb$—$^{207}Pb/^{204}Pb$ evolution diagrams showing data on Apollo 12 rocks in greater detail. The heavy isochron lines are drawn from primordial lead for several different ages. Data for the rocks would lie on the 4650–3400 m.y. isochron if they evolved in a U–Pb system during that time period and subsequently had a value of $^{238}U/^{204}Pb$ of zero. Isochron labeled 4650–4000 m.y. is included for comparison. Isochrons labeled 3400–0 m.y. define the locus of points if the $^{238}U/^{204}Pb > 0$ since 3400 m.y. ago. The intersection of the 3400–0 m.y. isochron with the 4650–3400 m.y. isochron would represent the isotopic composition of the lead initially in the rock. See Fig. 1 caption for additional explanation of symbols.

In this approach to resolving the age discrepancy, the values of $^{206}Pb/^{204}Pb$ and $^{207}Pb/^{204}Pb$ in the source of the magmas also cannot have been homogenized 4000 m.y. ago. If such systematics are the answer to the age discrepancies, then we feel that the most logical interpretation of the history of rock 12013, of the two interpretations given from Rb-Sr dating (LUNATIC ASYLUM, 1970), is that the entire rock was formed at one time (perhaps 4500 m.y.) and internally reconstituted by some thermal event (impact?) at a later time (probably 4000 m.y.) rather than that an old gabbro was intruded by a younger granite. This kind of interpretation is also supported by the petrographic study of JAMES (1971). If such regularities as described above are considered in the U–Pb system, they should also be considered in the Rb–Sr system. For example, if some lead was volatilized from 12013 in a proportional manner relative to uranium by a heating event, volatilization of some rubidium relative to strontium in a proportional manner is also reasonable (CLIFF *et al.*, 1971, report volatilization of both lead and rubidium). Such a process would result in an age of rock formation that is too young on the Pb–Pb isochron and too old on a Rb–Sr isochron (see discussion in SILVER, 1970, p. 1571). If the age of rock 12013 is older than that given by the Pb–Pb isochron (4370 m.y.), it is also probably younger than that given by the Rb–Sr "whole-rock" isochron (4500 m.y.). The best estimate of the age of formation of rock 12013 seems therefore to be in the range 4370 to 4500 m.y.

In the rocks collected by the Apollo 11 mission, the lead isotope data plot into two groups. Those with the greater value of $^{206}Pb/^{204}Pb$ are the group I, high-potassium basalts and those with the lesser values of $^{206}Pb/^{204}Pb$ are the group II, low-potassium basalts. As was found for rock 12013, an isochron line connecting the two groups of samples also passes through the common lead field with an indicated age of close to 4200 m.y. (Fig. 2), again as if the initial lead developed in a source with a small value of $^{238}U/^{204}Pb$. (The data on Fig. 2 of this paper has not been corrected for initial lead. TATSUMOTO (1970a) had corrected for a radiogenic initial lead on his Fig. 3 where he obtained a 4000-m.y. age.) The 4000-m.y. age given by TATSUMOTO (1970a) is a lower limit on the calculated age. Note that if one of the two groups of basalt is on the 3980-m.y. isochron (by using its observed $^{206}Pb/^{204}Pb$ but changing the $^{207}Pb/^{204}Pb$ until the point falls on that isochron) and the other remains on the 4200-m.y. isochron, a line connecting the two groups would not pass close to the common lead field. We therefore hardly appear to be mixing rocks of significantly different ages. Just as in rock 12013, because of the arrangement of the data, we would quite confidently interpret the rocks as being 4200 m.y. old in the absence of other information, and that they started with a nonradiogenic initial lead isotopic composition relative to terrestrial common lead. Again, however, the lead isotope data are in conflict with the age of the rocks determined by other means, strongly so for Group I basalts because the value of Rb/Sr does not appear to be close to equal to that in its source (the whole-rock Rb–Sr data for the group I basalts do not lie near the 4500-m.y. whole-rock isochron).

As for rock 12013, the Pb–Pb isochron age is older than the Rb–Sr mineral isochron age (PAPANASTASSIOU *et al.*, 1970) and the $^{40}Ar/^{39}Ar$ whole-rock age (TURNER, 1970a). The lead isotope data also could be interpreted to agree with the ages determined by other methods through assuming that the values of $^{238}U/^{204}Pb$ in the rocks are proportional to those values in the source material for the rocks, as for rock 12013. In this interpretation, the enrichment factor for $^{238}U/^{204}Pb$ of Apollo 11 rocks would be about 2.7 for all rocks (using an age of formation of 3800 m.y.). This is the interpretation preferred by COMPSTON *et al.* (1970) and considered by SILVER (1970) and by TATSUMOTO (1970a, p. 1606) in his concordia discussion for these rocks. The disturbing thing about this interpretation is the precise enrichment factor required. We have to choose, then, between two rather unsatisfying hypotheses: one, that the rocks are about 4200 m.y. old and that they had a nonradiogenic initial lead; and the other, that the rocks are of some age (perhaps between 3600 and 3800 m.y. old), and that they had a very precise enrichment factor for $^{238}U/^{204}Pb$. Of the two hypotheses, we still tend to favor the first (4200-m.y. age), because of the high quantity of the isochron and because several lines of information suggest that the initial lead in the rocks may not have been very radiogenic. We realize that this interpretation leaves a conflict with those arrived at by other dating techniques.

Consideration of one greater aspect in complexity of the Apollo 11 rocks is instructive. Suppose that the group I rocks are about 200 m.y. younger than the group II rocks (for our calculations, 3600 m.y. rather than 3800 m.y.). A Pb–Pb isochron for all rocks that goes through the common lead field could still be obtained by judicious regulation of the values of $^{238}U/^{204}Pb$ between the time of rock formation and the present. For this case, the enrichment factor in $^{238}U/^{204}Pb$ for the group II rocks would still appear to be about 2.7 as before, but for the group I rocks it would be only about 2.4 Amazingly, the enrichment factors would compensate for the age difference just the right amount to allow the isochron to still pass through the common lead field rather than above it or below it. This aspect is better illustrated in Apollo 12 rocks.

The lead isotopic data for rocks returned from the Apollo 12 mission (other than rock 12013) contrast markedly to those for rock 12013 and to those for rocks collected by the Apollo 11 mission because of the irregularity of the data on a Pb–Pb isochron plot. Data on six rocks lie near an isochron of about 3900 to 4000 m.y. that passes through the common lead field, but data on rock 12064 lie well above that isochron and those on rock 12035 lie well below it. These last two rocks clearly could not have had the same initial lead as the others, and they could be of a different age. If the samples are of about the same age (as suggested by the Rb–Sr mineral isochron from the LUNATIC ASYLUM (1971) and $^{40}Ar/^{39}Ar$ whole-rock data from TURNER (1971)), the value of $^{238}U/^{204}Pb$ initially in the rock was also not proportional, at least for 12064 and 12035, to that in its source material. Note that a line passed either through 12064 or 12035 and through any other sample does not pass through the common lead field. Thus, in the analyzed rock assemblage from Apollo 12, there is clear failure in the assumptions necessary for using the Pb–Pb isochron. The data do establish, however, that proportionality of $^{238}U/^{204}Pb$ in a magma relative to the source of the magma is not necessarily the rule. It may be worth noting that the isochron that passes through the bulk of the data and the common lead field indicates a younger age than that found for the rocks collected by the Apollo 11 mission which in turn are younger than rock 12013. The order is the same as that established by either of the other dating methods and there is some similarity to the intervals between apparent ages. We know of no obvious criteria to cause us to discard either rock 12064 and 12035 from the Pb–Pb isochron treatment, and conclusions regarding Apollo 12 rocks achieved solely by this technique are suspect. Sufficient sample was obtained for rock 12064 to permit some investigation of it for a possible internal isochron.

Density fractions on rocks 10017 and 12064

The isotopic compositions of the density fractions separated from rocks 10017 (TATSUMOTO, 1970a) and 12064 (Table 4) are plotted on Fig. 3. Both sets of density-fraction data have a considerable spread in their isotopic ratios that permit construction of Pb–Pb isochrons. In both rocks the slope of the isochron is markedly less than the slopes for the whole-rock specimens, and the resultant apparent ages are 3950 m.y. for 10017 (versus 4200 m.y. for the whole-rock isochron of Apollo 11) and 3650 m.y. for 12064 (versus 4300 m.y. for an isochron through primordial lead and the whole rock). A similar line drawn through the data on a concentrated HNO_3 leach of 10017 and its residue indicates an apparent age of about 3950 m.y., similar to the

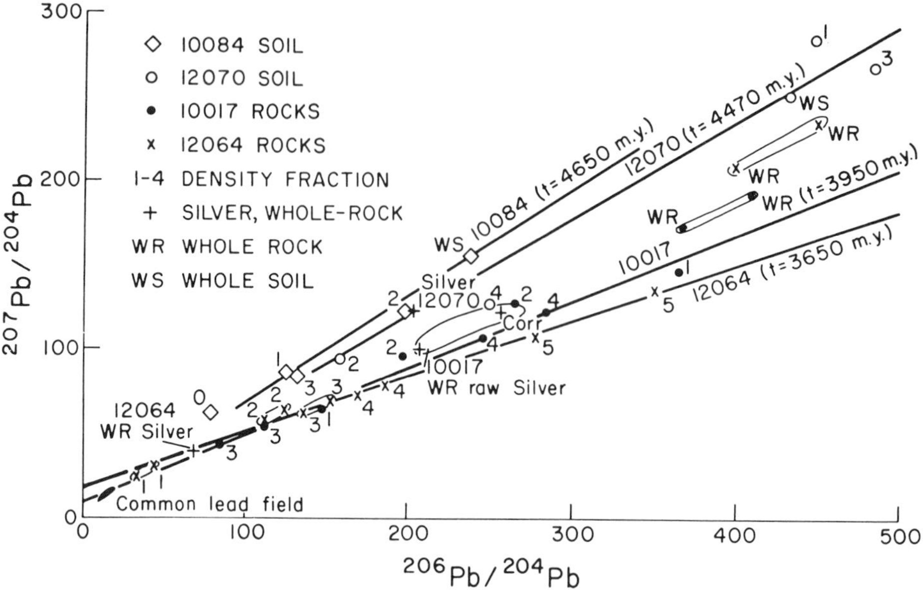

Fig. 3. The $^{206}Pb/^{204}Pb$—^{207}Pb–^{204}Pb evolution for density fractions from selected Apollo 12 samples. For each density fraction on the rocks, 10017 and 12064, the left symbol represents the raw data and the symbol more to the right is the data corrected for chemistry blank. The data represented by + are for whole-rock 10017, SILVER (1970) and for whole-rock 12064, SILVER (1971).

age on the density fractions on the same sample obtained by TATSUMOTO (1970a). While these ages are closer to the Rb–Sr mineral isochron ages (3400 m.y. for Apollo 12 rocks by PAPANASTASSIOU and WASSERBURG, 1970) and $^{40}Ar/^{39}Ar$ whole-rock ages (TURNER, 1971), they are still high by 5 to 10 percent. A disturbing feature of both Pb–Pb isochrons arising from density fractions is that neither isochron includes its respective whole-rock analysis. Such a relationship could occur in nature if the whole rock contained a relict phase rich in lead that either was lost in the unanalyzed hand-magnet fraction, lost as very fine particles that remained in suspension (a candidate might be phosphate minerals), or dissolved and lost in the heavy liquids (the sum of the leads in the density fractions does not equal the lead content of the whole rock). Unfortunately, all heavy liquids were combined after use, so this explanation cannot be tested. Even though the liquids were distilled before use, the heavy liquids more likely have contributed a greater lead blank to the density fractions than we have estimated. Fraction 2 of rock 10017, for example, lies above the density-fraction isochron that is described by the rest of the fractions, and it lies along a line that connects the whole rock with the common lead field. This relationship suggests difficulties in blank correction. The fact that all density fractions are less radiogenic in $^{206}Pb/^{204}Pb$ than their whole rocks is also compatible with a contamination problem. The greater age for the Pb–Pb isochron relative to those of other dating methods is suggestive that the amount of lead contamination is inversely proportional to the observed $^{206}Pb/^{204}Pb$ values in the fractions.

The density-fraction data would have agreed with the Rb–Sr data had they fallen on a line of 3600-m.y. slope for 10017 that passes through the whole-rock value, and on a line of 3400-m.y. slope for rock 12064 that passes through the whole-rock value. Considering the uncertain state of the blank, we cannot now say that such is not true. The approximate values of $^{206}Pb/^{204}Pb$ for which lines through the common lead field and the observed density-fraction data intersect the isochron of approximate age that includes the whole-rock value are: for 10017, fraction 1—930; 2—400 (same as the whole rock), 3—540; and 4—670; and for 12064, fraction 1—300; 2—600; 3—900; 4—1100; and 5—1400. The data do support the concept that the mineral-fraction data lie on an isochron of less slope, and therefore of younger age, than the isochron for the whole rocks. Otherwise, all the density-fraction data would lie along lines connecting the whole rock to the common lead field.

The contamination explanation does seem to require two baffling coincidences, however, that continuously seem to be needed to obtain ages compatible with the Rb–Sr mineral isochron and $^{40}Ar/^{39}Ar$ whole-rock ages. First, the contamination during our mineral separation must have occurred in a rather regular manner for the density-fraction data to retain linearity. Both random scatter of data and even negative slopes of isochrons are possible. Second, the data of SILVER (1970) on the whole-rock 10017 lie near our density-fraction isochron for 10017 and his data on whole-rock 12064 (SILVER, 1971) plot very near our mineral isochron for rock 12064. Both of Silver's whole rocks do lie near lines connecting our respective analyses and the common lead field so that coincidental amounts of contamination could account for the differences—but this could hardly have been predicted. Therefore, the presence of a randomly distributed relict component cannot yet be completely abandoned. Also, we are not yet prepared to say whether the lead-isotope data on density fractions either agree or disagree with the ages determined by other methods.

Soils

The lead-isotope data on soils 12033 and 12070 and breccia 12034 returned by the Apollo 12 mission are shown on Figs. 1 and 2. All these soil and breccia data clearly lie below the isochron drawn between the Apollo 11 soils and breccia and the common lead field. There is perhaps some irony in the fact that the model whole-soil age reported by PAPANASTASSIOU *et al.* (1970) of 4670 m.y. is in excellent agreement with the Pb–Pb isochron age (4660 m.y.) on the Apollo 11 soils by TATSUMOTO (1970a) inasmuch as the soil and breccia are clearly complex material that contain components from local rocks, rock detritus thrown in by impacts at some distance (including a "KREEP-like component"), and meteorites. If such agreement is surprising for these complex materials, the irony is only compounded by the excellent agreement between the model "whole-soil" Rb–Sr age on soil sample 12070 of 4440 m.y. (PAPANASTASSIOU and WASSERBURG, 1970) and the Pb–Pb isochron model age determined by us of 4470 m.y. The close agreement is probably more representative of excellent mixing of components in the soil rather than of the age of soil formation. (There is some conflict in the model Rb–Sr ages reported on these samples in the unpublished proceedings of the Apollo 12 Lunar Science Conference. Both MURTHY *et al.* (1971) and CLIFF *et al.* (1971) state that 12070 has a model age of 4600 m.y. within their analytical uncertainties.) Other soil (12033) and breccia (12034) samples from the Apollo 12 mission are younger in their model Pb–Pb isochron ages, down to 4200 m.y. For 12033, MURTHY *et al.* give a model age of 4200 m.y., which is satisfyingly close to the Pb–Pb model age of 4320 m.y.

Density fractions on 10084, 12070, and 12033

Density separations were carried out on three soil samples—10084 from the Apollo 11 mission (TATSUMOTO, 1970a) and 12070 and 12033 from the Apollo 12 mission. Soil data are plotted for 10084 and 12070 on Fig. 3 and for 12033 on Fig. 1. The data on the Apollo 12 soils are included in Table 3 and those for 10084 may be obtained from the original report (TATSUMOTO, 1970a). The data on 10084 lie reasonably close to the 4650-m.y.-old isochron drawn through primordial lead and the whole-soil. The new analysis of the acetone suspension fraction of 10084 (Table 4) is less radiogenic than the whole soil and is in good agreement with the most easily leachable lead reported by SILVER (1970). Soil 12033 gives a Pb–Pb isochron in agreement with the primordial lead, whole-soil Pb–Pb isochron. Sample 12070 contains two fractions that express the heterogeneity of this soil sample. Fraction one (acetone suspension) could represent a mobilized lead component such as that postulated by SILVER (1970) and TATSUMOTO (1970a, p. 1606) and fraction 3 could represent some of the KREEP-like component. These components apparently balance each other in the whole soil.

Parent-Daughter Isochron Relationships

Unlike the Pb–Pb isochron, whole-rock lead isotope data on U–Pb and Th–Pb diagrams (Fig. 4) do not define any definite isochrons, although the ages involved are clearly very old. The reasons for the scatter are not entirely understood. We feel at this time that the scatter is more likely to be due to volatilization of some lead from the rocks caused by the relatively recent impact event responsible for the rock now being on the surface rather than to exceedingly great analytical uncertainties. However, the impact may cause redistribution of the lead within the rock rather than volatilization of lead from the rock. If this surmise is correct, some parts of the rock may have become enriched in lead, and others may have been depleted of lead. Conceivably, we might learn something by passing "best-fit" lines through the scatter of the data. For rock 12013, the ages are then indicated to be about 3900 m.y. for the ^{238}U–^{206}Pb isochron, 4200 m.y. for the ^{235}U–^{207}Pb isochron, and 4000 m.y. for the ^{208}Pb–^{232}Th isochron, the limiting uncertainties are drawn on Fig. 4. The average of these crude ages (4030 m.y.) is surprisingly close to the ages determined by the Rb–Sr mineral isochron (LUNATIC ASYLUM, 1970) and $^{40}Ar/^{39}Ar$ methods (TURNER, 1970b). When the same best-fit lines are drawn through the scatter of the data on the density fractions for rock 10017 (TATSUMOTO, 1970a), the ages are about 3660 m.y., 3950 m.y., and roughly about 2390 m.y. (little variation in the $^{208}Pb/^{204}Pb$), respectively, in the same order as given for 12013. The average of the U–Pb isochrons (3800 m.y.) is in general agreement with the Rb–Sr mineral isochron ages (PAPANASTASSIOU *et al.*,

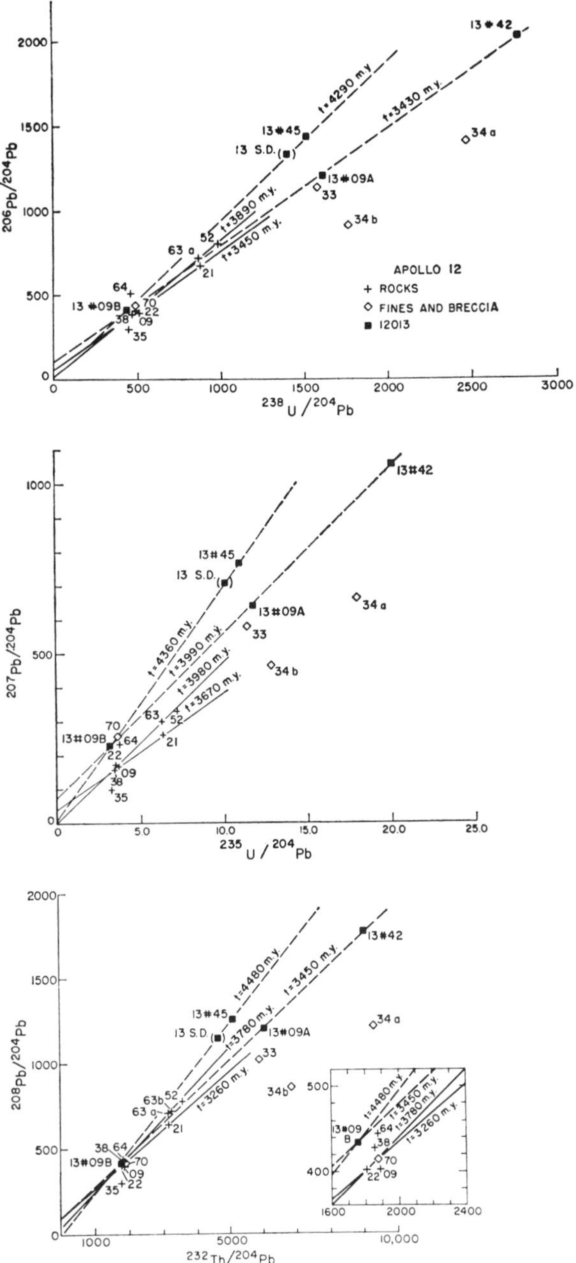

Fig. 4. The ^{206}Pb–^{238}U, ^{207}Pb–^{235}U, and ^{208}Pb–^{232}Th isochrons for whole rocks of Apollo 12 samples. Numbers on data points are final digits of sample numbers, and appended a's and b's indicate duplicate determinations. The size of the symbols approximate the analytical error (1 σ).

1970) and the $^{40}Ar/^{39}Ar$ whole-rock ages (TURNER, 1970a). For rock 12064 density fractions, these ages are 3320 m.y., 3550 m.y., and 3570 m.y., respectively. Again, the average isochron age (3480 m.y.) is in remarkable agreement, considering the scatter, with the Rb–Sr mineral isochron generally observed in Apollo 12 rocks (COMPSTON *et al.*, 1971; MURTHY *et al.*, 1971; CLIFF *et al.*, 1971; LUNATIC ASYLUM, 1971) and $^{40}Ar/^{39}Ar$ whole-rock ages (TURNER, 1971).

The density-fraction isochron data on the Apollo 11 rocks, the rock 12064 and igneous breccia 12013, although not of as good quality as the whole-rock data, do appear to support the contention that we may be dealing with an acute initial lead problem with some coincidental proportionalities of the $^{238}U/^{204}Pb$ values in the magma and its source rocks, but we feel that the data are not of sufficient quality to compel this conjecture.

CONCORDIA AGE RELATIONSHIPS

Rocks

The U–Pb system is plotted on a U–Pb concordia diagram (Fig. 5). Using this diagram requires knowledge of the uranium and lead contents as well as of the lead isotopic composition. An initial lead must also be determined or assumed. This last requirement is not as difficult to meet for lunar samples as might be imagined, inasmuch as the initial lead in the magmas is primordial lead plus some radiogenic addition. The systematics of the U–Pb system are such that correction of the observed ratios for primordial lead will give a data point that still lies on the straight line drawn between the age of the source material and the time of formation of the magma but to the right of the position where it would lie if corrected for the true initial lead. This line, called a discordia line, is drawn for times representing the two limiting conditions of the data, i.e., between 4650 and 3400–3900 m.y. The Apollo 11 rocks lie in one patch and the Apollo 12 rocks (except for 12035 and 12064) in another patch. Though there is not a straightforward relationship between the ^{206}Pb–^{207}Pb isochron and the concordia diagram, the two patches reflect the younger age shown by the ^{206}Pb–^{207}Pb isochron for the Apollo 12 rocks relative to that for the Apollo 11 rocks.

For the concordia ages to agree with the $^{40}Ar/^{39}Ar$ and internal Rb/Sr isochron ages, data for rock 12013 should lie on a line between about 4650 m.y. and 4000 m.y., rocks of Apollo 11 near a line between 4650 m.y. and about 3700 m.y., and rocks of Apollo 12 near a line from 4650 m.y. and about 3400 m.y., assuming that the age of all the source rocks for these materials is a primordial source and that only two stages have been involved. The data do not clearly reflect what was expected because rock 12013 data are on a distinctly different discordia line and the data for Apollo 11 rocks lie on or below a 3900 or 4650-m.y. discordia line. Note that the sawdust from 12013 lies very near the concordia curve when corrected for modern lead ($^{206}Pb/^{204}Pb$—18.8; $^{207}Pb/^{204}Pb$—15.8; $^{208}Pb/^{204}Pb$—38.8) as the initial lead in spite of the abundant lead contamination. There seem to be two groups in the Apollo 12 igneous rocks (except 12035 and 12064) as shown by the dashed lines in Fig. 5. The groupings also shown in the parent-daughter diagram (U–Pb, Th–Pb), may indicate that at least two parent magmas were involved. The data for Apollo 12 rocks scatter con-

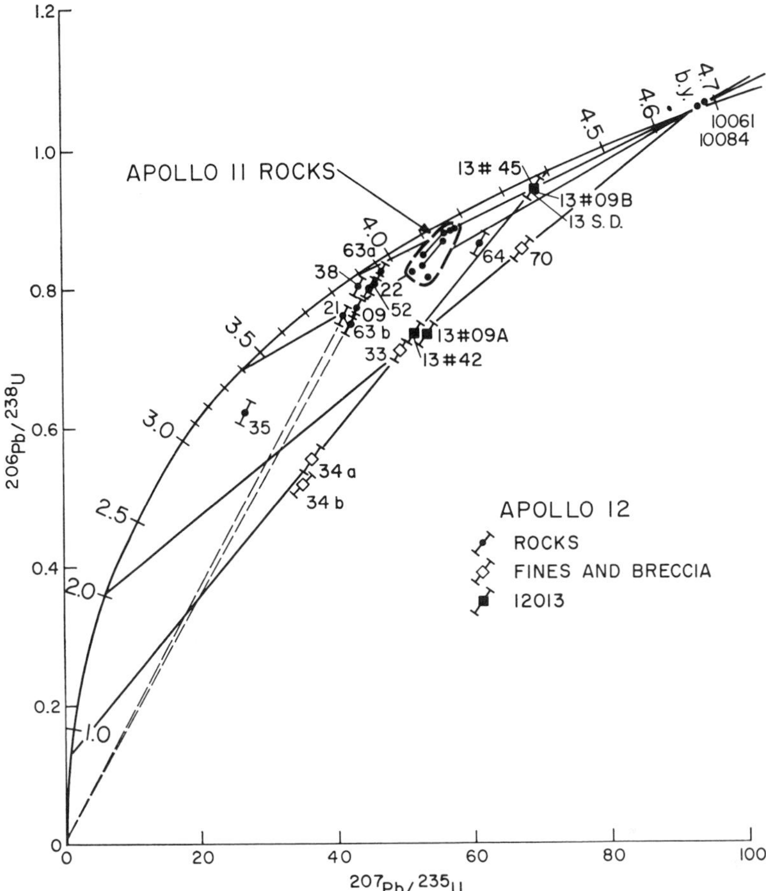

Fig. 5. The U–Pb evolution diagram. Plotted points are ($^{207}Pb_{observed}$–$^{207}Pb_{primordial}$)/^{235}U against ($^{206}Pb_{observed}$–$^{206}Pb_{primordial}$)/^{238}U. Numbers on data points are final digits of sample numbers, and numbers on concordia indicate 10^9 years. Discordia lines connected limiting values for Apollo 11 (4660–3900 m.y.) and Apollo 12 (4660–3400 m.y.) rocks according to Rb–Sr mineral isochrons (PAPANASTASSIOU *et al.*, 1970); all soil samples (10061, 10084, 12033, and 12070); and fragments of igneous breccia 12013 that also includes impact breccia 12034. Dashed discordia lines give the limiting $^{207}Pb/^{206}Pb$ slopes of all Apollo 12 rocks except 12035 and 12064.

siderably about the 4650 to 3400-m.y. discordia line but they may lie within all the uncertainties as estimated from the multiple analyses of 12063 and 12052 (sample inhomogeneity plus laboratory treatment) that affect the data of the 4650 to 3400-m.y. discordia line. Rock 12035 lies drastically below such a discordia line, but we are somewhat concerned about this sample. It is very friable and perhaps some key material in the U–Th–Pb system was lost by the third events (impact?) or in the collection and processing of the material. Such loss would be an episodic event at time zero and would move the data point towards zero. The loss cannot account for the lateral offset of 12035 from the grouping of six analyses.

If it is assumed that all rocks are concordant and that the problem is one of not knowing what the initial lead really was in the rock, the initial lead and $^{238}U/^{204}Pb$ in the source rocks may be calculated (Table 6). If the uranium content in each of the magmas and source rocks is assumed to be constant, then the Apollo 12 rocks have lost about five-sixths of their lead relative to that in the source rocks. For comparison, Apollo 11 rocks have lost about two-thirds of their lead relative to that in source rocks.

As pointed out previously (TATSUMOTO, 1970b), the data on rock 12013 clearly do not fit the 4650 to 3400 or 3900-m.y. discordia lines but seem to reflect some event much younger than 3000 m.y. and are evidence of third event systems on the moon (it should be kept in mind that chiplet 09B was a separate piece from chip 09A, both of which came from fragment 09).

Density fractions on rock 12064

The data on rock 12064 are given in Fig. 6. The density fraction data for this rock appear to lie reasonably well along a 4400 to 2500-m.y. trend that would also include rock 12035 reasonably well. The 4400-m.y. intersection is also common to that for the igneous breccia 12013 and the impact breccia 12034 and the soil 12033. Though we expected that the data would lie on a 4650 to 3400-m.y. discordia line, the data cannot easily be made to fit such a line. Fraction 1 (acetone suspension) lies well above it and fractions 3, 4, and 5 lie below it. Even though the observed arrangement of the data could be due to the possible contamination, previously discussed, if the degree of contamination is the reciprocal of the amount of sample handling, again we are surprised at the regularity of the data if contamination is the cause. The 2500-m.y. intercept, however, does not seem to be reflected in the results determined by any of the other treatments of the U–Th–Pb system or in those determined by the Rb–Sr mineral isochron or $^{40}Ar/^{39}Ar$ methods. The only other information suggesting an approximate age of 2500 m.y. is the K–Ar ages in the Apollo 12 preliminary report (LSPET, 1970), we, however, tend to interpret those apparent ages as intermediate ages between the time of volcanism and the younger age of impact throwout.

Table 6. Calculated first stage parameters for concordant age (t_1 in million years) by two-stage model*

Sample	Condordant t_1	$(U^{238}/Pb^{204})_{t_0 \sim t_1}$	$(Pb^{206}/Pb^{204})_{t_1}$	$(Pb^{207}/Pb^{204})_{t_1}$
12009	3214	161.3	73.63	89.10
12021	3401	183.2	73.57	93.79
12022	3770	74.05	28.02	37.66
12035	2908	67.89	41.51	46.29
12038	3725	61.90	25.72	33.90
12052	3511	256.0	91.67	121.3
12063,a	3742	153.0	49.14	68.04
12064	3087	302.9	139.6	164.1

* Assuming $t_0 = 4.63$. For $t_0 = 4.60$ and $t_0 = 4.65$, the calculated ages change less than 30 m.y.

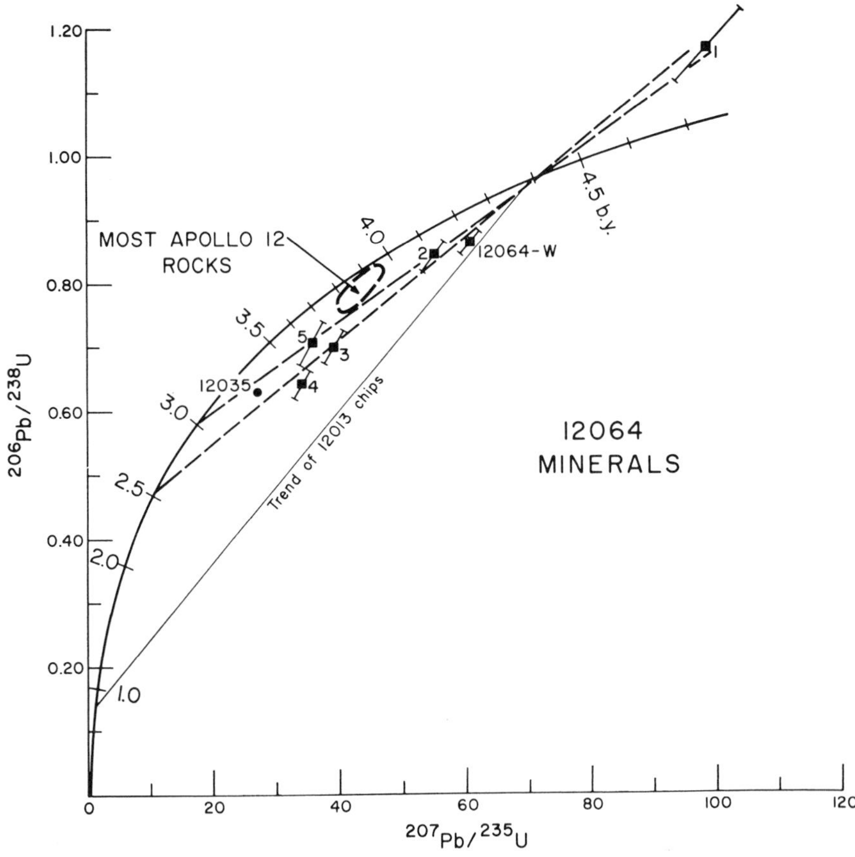

Fig. 6. The U–Pb evolution diagram for rock 12064 and its separated density fractions. Dashed lines are drawn to fit the data. For comparison, rock 12035, the field encompassing most Apollo 12 rocks, and the trend of the fragments of igneous breccia 12013 are given. The definitions of the density fractions 1–5 are given in Table 4. The bars express the estimated analytical uncertainties.

Soils

In contrast to the rocks of Apollo 11 and Apollo 12 and soils and breccias of Apollo 11 which seem to approximate single- (no alteration of $^{238}U/^{204}Pb$ since the moon formed) or two-stage development (line t_1–t_0 in Fig. 7), the soils and breccias of Apollo 12 show distinct development of a third much younger stage (Fig. 5) and are more accurately represented by such lines as t_2–Q_1 in Fig. 7. All other lead isotope data reported in the unpublished proceedings of the Second Lunar Science Conference on Apollo 12 are in agreement on this point (Cliff *et al.*, 1971; Silver, 1971; and Huey *et al.*, 1971). The only alternative explanation is that some new phase, not dissolved or leached of its uranium and lead, is present in Apollo 12 soil and breccias but is not present in any other rock of either Apollo 11 or Apollo 12 or in Apollo 11 soil and breccia. No such phase was reported at the Second Lunar Sample Conference,

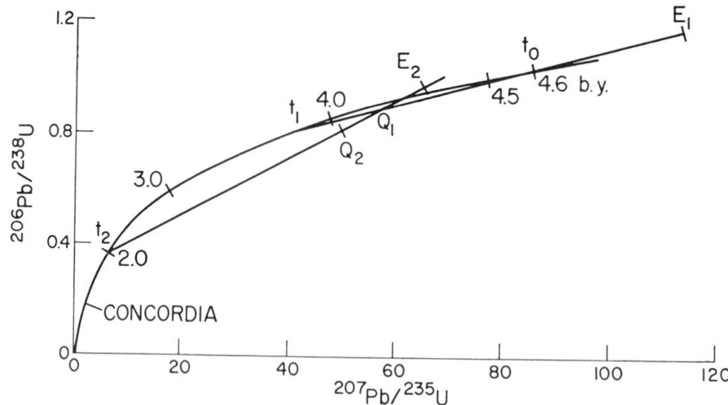

Fig. 7. U–Pb evolution diagram illustrating processes involved in second and third stage events (for detailed explanation see Tatsumoto, 1970a).

and in fact, one possibly troublesome phase, ilmenite, is less in the Apollo 12 samples. Some unusual occurrences did happen in rock 12013 that are suggestive of a third event, as already discussed. In these events, the KREEP-like component either may be young, much less than 3000 m.y., or may have been altered at a young age (at about 800 m.y. ?). Third events are the most likely explanation for the configuration of the 12013 data and almost certainly is the explanation for the soil and impact breccia data from the Apollo 12 mission. We do not know the precise age or ages of the young, third-stage event or events; conceivably, they might be related to formation of impact glasses and alteration of the U/Pb ratio in some rocks, such as 12013, by some relatively young impact such as Copernicus, a ray from which the Apollo 12 samples may in fact have been collected.

Density fractions of soils 12033 and 12070

Some insight into the complexity of Apollo 12 soils and breccias may be given by the density-fraction data for soils 12070 and 12033 (Fig. 8). The data point of the extremely fine fraction of 12070 (fraction 1) falls nicely on a t_1–t_0 line. The data represented by this line could be for soil that might have been derived from 4550-m.y.-old source material at about 3200 m.y. ago at which time fraction 1 was enriched somewhat in lead over uranium. Fractions 3 (MI float) and 4 (MI sink) also fall this line. Fraction 2 (BF float) falls on a line drawn through it and one other bit of fraction data that is much younger. The oldest intersection with concordia for this younger event would be a line connecting fractions 1 and 2 which would intersect concordia at about 1.0 b.y. Fraction 2 primarily contains feldspar and scoriaceous glass which may be a younger (or relatively recently altered) KREEP-like material (note the great lead, uranium and thorium concentrations for fraction 2 in Table 4). The behavior of the fractions from 12033 differs markedly from the behavior of the density-fractions of 12070. It is almost as if 12033 was largely reconstituted in terms of the U–Pb system at some fairly recent time rather than just containing some added

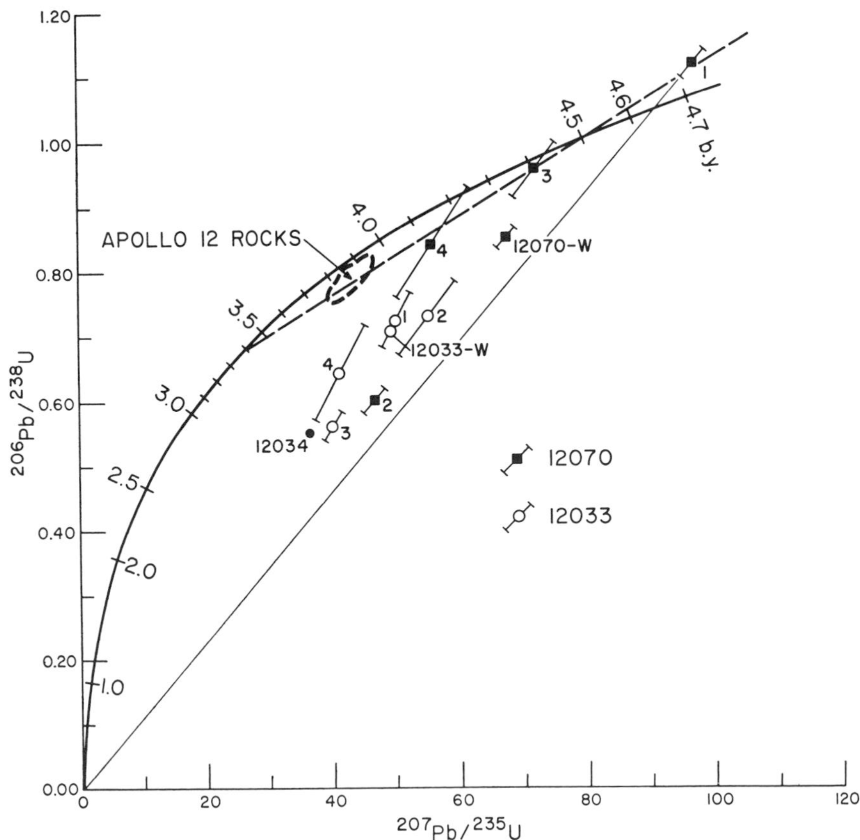

Fig. 8. U–Pb evolution diagram for soils 12033 and 12070 and their density fractions. The definitions of density fractions 1–4 are given in Table 4. The bars express the estimated analytical uncertainties. Breccia sample 12034 is shown for comparison.

definable component. Even though the soil and breccia data of Apollo 12 are interpreted only with great difficulty to any degree of precision, evidence clearly indicates a component greater than 4 b.y. in age and another component much younger than 3 b.y. in age. Neither the $^{40}Ar/^{39}Ar$ method nor the internal Rb/Sr isochron methods is reflecting this younger third event; however, these methods are applied to the soil only with great difficulty.

Conclusions

In lunar samples, differences in the ages obtained by the Rb–Sr and $^{40}Ar/^{39}Ar$ methods and those obtained by U–Th–Pb dating seem to be attributed, by some investigators, mainly to analytical problems in the U–Th–Pb system and complexities in the interpretation of that system. The Pb–Pb isochron is, however, the dating method least sensitive to analytical uncertainties. In addition, if the U/Pb and Th/Pb values in the volcanic rocks have been enriched over those in the isotopically inhomogeneous source rocks by volatilization of lead from the magma, depletion of the magma in rubidium relative to strontium is also likely. The general effects of such volatilization would be Pb–Pb whole-rock isochron ages that are younger than the age of the source rocks for the magma and Rb–Sr whole-rock isochron ages that are older than the age of the source rocks. If the high quality Pb–Pb isochron for the Apollo 11 rocks and rock 12013 is due to a coincidental constant enrichment factor for U/Pb in the volcanic rocks relative to the source rocks, the approximate agreement of the low-potassium whole-rock Rb-Sr isochron ages (which are interpreted as ages of the magma source rocks) with the "age of the moon" also may be a coincidence.

We feel that the Pb–Pb isochrons on the Apollo 11 rocks and rock 12013 are of sufficiently high quality that abandonment of a 4200-m.y.-old age for Apollo 11 rocks and 4370-m.y.-old age for rock 12013 is premature. The lead isotope data on Apollo 12 rocks, other than 12013, does not form a simple Pb–Pb isochron so that ages must be derived from concordia diagrams. In the absence of other kinds of data, we would interpret the concordia relations as suggesting that the Apollo 12 rocks were derived about 3000 m.y. ago from a 4650-m.y.-old source. There is sufficient scatter in these data, however, to permit a derivation of the rocks at 3400 m.y. from a 4650-m.y.-old source, a derivation age that is in essential agreement with the Rb–Sr mineral isochron and $^{40}Ar/^{39}Ar$ whole-rock methods. Although the lead isotope data on the lunar soils are subject to alternative interpretations, the 4650 $\pm$ 50 m.y. age determined by U–Th–Pb methods for the Apollo 11 soils is probably a consequence of the Apollo 11 soils being a well-mixed composite of lead enriched and depleted materials, and the age still furnishes the best working hypothesis for the age of the moon. The most significant feature of the lead isotope data on Apollo 12 soils is evidence of "third events" in these soils that are much less than 3000 m.y. in age. These "third events" are probably attributable to formation of or volatilization of lead from components in the soil due to young impacts. These "third events" are not clearly shown in the $^{40}Ar/^{39}Ar$ or Rb–Sr dating methods.

A speculation on the origin of the moon is not discussed in this paper because of page limitations and because of the rapid advance of physical astronomy; however, our concept of the lunar origin is not much different from those of other investigators (for example, GANAPATHY *et al.*, 1970; ANDERS, 1970–1971) and the symposium on the evolution of the solar system (SHIMAZU, 1967). This concept is that the moon originated from the beginning as a satellite of the earth in the process of solar evolution.

The radiation effect of the proto-sun at the Hayashi phase (high luminosity stage; HAYASHI *et al.*, 1962) could account for the depletion of lead and other volatile elements when the particle-size in the proto-earth-moon system reached millimeter to meter size (ONO and FUJIMOTO, 1967). The moon, which was formed from the outer particles, became more depleted in volatile elements than the earth, which was accreted from interior particles of the proto-earth-moon system.

Acknowledgments—We thank George Reed of the Argonne National Laboratory for his suggestion that we try anodic electrodeposition of lead in the purification procedure rather than cathodic electrodeposition, and T. J. Chow of the University of California for the lead isotopic analysis of dust lead on filters. We also thank G. A. Izett and R. E. Wilcox of the U.S. Geological

Survey, for petrographic examination of mineral fractions. We are indebted to D. M. Unruh for laboratory assistance. We wish to thank S. Matsuo of the Tokyo University of Education for discussions on the origin and evolution of the solar system, planets, satellites, and meteorites. We benefited also by our discussions with our colleagues in the U.S. Geological Survey, particularly Zell E. Peterman, Robert E. Zartman, and John N. Rosholt. This study was supported in part by NASA Contract T–75445.

References

Anders E. (1970) Water on the moon? *Science* **169**, 1309–1310.

Anders E. (1971) Meteorites and the early solar system. *Ann. Rev. Astron. Astrophys.* **9**, 1–33.

Cliff R. A., Lee-Hu C., and Wetherill G. W. (1971) Rb–Sr and U, Th–Pb measurements on Apollo 12 material. Second Lunar Science Conference (unpublished proceedings).

Compston W., Chappell B. W., Arriens P. A., and Vernon M. J. (1970) The chemistry and age of Apollo 11 lunar material. *Proc. Apollo 11 Lunar Sci. Conf., Geochim. Cosmochim. Acta* Suppl. 1, Vol. 2, pp. 1007–1027. Pergamon.

Compston W., Berry H., Vernon M. J., Chappell B. W., and Kaye M. J. (1971) Rubidium-strontium chronology and chemistry of lunar material from the Ocean of Storms. Second Lunar Science Conference (unpublished proceedings).

Farquharson R. B. and Richards J. R. (1970) Whole-rock U–Th–Pb and Rb–Sr ages of the Sybella microgranite and pegmatite, Mount Isa, Queensland. *J. Geol. Soc. Aust.* **17**, 53–58.

Ganapathy R., Keays R. R., Laul J. C., and Anders E. (1970) Trace elements in Apollo 11 lunar rocks: Implications for meteorite influx and origin of moon. *Proc. Apollo 11 Lunar Sci. Conf., Geochim. Cosmochim. Acta* Suppl. 1, Vol. 2, pp 1117–1142. Pergamon.

Gopalan K., Kaushal S., Lee-Hu C., and Wetherill G. W. (1970) Rb–Sr and U, Th–Pb ages of lunar materials. *Proc. Apollo 11 Lunar Sci. Conf., Geochim. Cosmochim. Acta* Suppl. 1, Vol. 2, pp. 1195–1205. Pergamon.

Hayashi C., Hoshi R., and Sugimoto D. (1962) Evolution of the stars. *Suppl. Prog. Ther. Phys.* **22**, 183 pp.

Hubbard N. J., Gast P. W., and Meyer C. (1971) The origin of the lunar soil based on REE, K, Rb, Ba, Sr, P, and Sr^{87}/Sr^{86} data. Second Lunar Science Conference (unpublished proceedings).

Huey J. M., Ihochi H., Black L. P., Ostic R. G., and Kohman T. P. (1971) Lead isotopes and volatile transfer in the lunar soil. Second Lunar Science Conference (unpublished proceedings).

James O. B. (1971) Petrology of lunar microbreccia 12013,6. U.S. Geol. Survey–NASA Inter-agency Rpt., *Astrogeol.* **23**, 39 pp.

LSPET (Lunar Sample Preliminary Examination Team) (1970) Preliminary examination of the lunar samples from Apollo 12. *Science* **167**, 1325–1339.

Lunatic Asylum (1970) Mineralogic and isotopic investigations on lunar rock 12013. *Earth Planet. Sci. Lett.* **9**, 137–163.

Lunatic Asylum (1971) Rb–Sr ages, chemical abundance patterns and history of lunar rocks. Second Lunar Science Conference (unpublished proceedings).

Meyer C., Jr., Aitken F. K., Brett R., McKay D., and Morrison D. (1971) Rock fragments and glasses rich in K, REE, P in Apollo 12 soils: Their mineralogy and origin. Second Lunar Science Conference (unpublished proceedings).

Murthy V. R., Evenson N. M., Jahn B.-M., and Coscio M. R., Jr. (1971) Rb–Sr isotopic relations and elemental abundances of K, Rb, Sr, and Ba in Apollo 11 and Apollo 12 samples. Second Lunar Science Conference (unpublished proceedings).

Ono S. and Fujimoto Y. (1967) Origin of the solar system (in Japanese). *Kagaku* **39**, 547–551.

Papanastassiou D. A., Wasserburg G. J., and Burnett D. S. (1970) Rb–Sr ages of lunar rocks from the Sea of Tranquillity. *Earth Planet. Sci. Lett.* **8**, 1–19.

Papanastassiou D. A. and Wasserburg G. J. (1970) Rb–Sr ages from the Ocean of Storms. *Earth Planet. Sci. Lett.* **8**, 269–278.

PETERMAN Z. E., DOE B. R., and PROSTKA H. J. (1970) Lead and strontium isotopes in rocks of the Absaroka volcanic field, Wyoming. *Contrib. Mineral. Petrol.* **27**, 121–130.

PETERMAN Z. E., HILDRETH R. A., and NKOMO I. (1971) Precambrian geology and geochronology of the Granite Mountains, central Wyoming. *Geol. Soc. Amer. Abstracts with Programs* **3**, no. 5.

ROSHOLT J. N. and BARTEL A. J. (1969) Uranium, thorium, and lead systematics in Granite Mountains, Wyoming. *Earth. Planet. Sci. Lett.* **7**, 141–147.

ROSHOLT J. N. and PETERMAN Z. E. (1969) Uranium, thorium, and lead systematics in the Granite Mountains, Wyoming. *Geol. Soc. Amer. Abstracts with Programs*, Part 5, 70.

ROSHOLT J. N., PETERMAN Z. E., and BARTEL A. J. (1970) U–Th–Pb and Rb–Sr ages in granite reference sample from southwestern Saskatchewan. *Can. J. Earth Sci.* **7**, 184–187.

ROSHOLT J. N. and TATSUMOTO M. (1970) Isotopic composition of uranium and thorium in Apollo 11 samples. *Proc. Apollo 11 Lunar Sci. Conf., Geochim. Cosmochim. Acta* Suppl. 1, Vol. 2, pp. 1499–1502. Pergamon.

SCHNETZLER C. C. and PHILPOTTS J. A. (1970) Trace element abundances in Apollo 12 samples. *Geol. Soc. Amer. Abstracts with Programs* **2**, no. 7, 676.

SCHNETZLER C. C., PHILPOTTS J. A., and BOTTINO M. L. (1970) Li, K, Rb, Sr, Ba, and rare-earth concentrations, and Rb–Sr age of lunar rock 12013. *Earth Planet. Sci. Lett.* **9**, 185–192.

SHIMAZU Y. (editor) (1967) Symposium of solar system (in Japanese): *Kagaku* **37**, 514–571.

SILVER L. T. (1970) Uranium–thorium–lead isotopes in some Tranquillity Base samples and their implications for lunar history. *Proc. Apollo 11 Lunar Sci. Conf., Geochim. Cosmochim. Acta* Suppl. 1, Vol. 2, pp. 1533–1574. Pergamon.

SILVER L. T. (1971) U–Th–Pb relations in Apollo 11 and Apollo 12 lunar samples. Second Lunar Science Conference (unpublished proceedings).

TATSUMOTO M. (1970a) Age of the moon: An isotopic study of U–Th–Pb systematics of Apollo 11 lunar samples, II. *Proc. Apollo 11 Lunar Sci. Conf., Geochim. Cosmochim. Acta* Suppl. 1, Vol. 2, pp. 1595–1612. Pergamon.

TATSUMOTO M. (1970b) U–Th–Pb age of Apollo 12 rock 12013. *Earth Planet. Sci. Lett.* **9**, 193–200.

TATSUMOTO M. and ROSHOLT J. N. (1970) Age of the moon: An isotopic study of uranium–thorium–lead systematics of lunar samples. *Science* **167**, 461–463.

TURNER G. (1970a) Argon-40/argon-39 dating of lunar rock samples. In *Proc. Apollo 11 Lunar Sci. Conf., Geochim. Cosmochim. Acta* Suppl. 1, Vol. 2, pp. 1665–1684. Pergamon.

TURNER G. (1970b) ^{40}Ar–^{39}Ar age determination of lunar rock 12013. *Earth Planet. Sci. Lett.* **9**, 177–180.

TURNER G. (1971) ^{40}Ar–^{39}Ar ages from the lunar maria. Second Lunar Science Conference (unpublished proceedings).

WAKITA H. and SCHMITT R. A. (1970) Elemental abundances in seven fragments from lunar rock 12013. *Earth Planet. Sci. Lett.* **9**, 169–176.

WARNER J. (compiler) (1970) Apollo 12 lunar-sample information. NASA TR R–353.

WARNER J. L. and ANDERSON D. H. (1971) Lunar crystalline rocks—Petrography, geology, and origin. Second Lunar Science Conference (unpublished proceedings).

WOOD J. A., DICKEY J. S., Jr., MARVIN U. B., and POWELL B. N. (1970) Lunar anorthosites and a geophysical model of the moon. *Proc. Apollo 11 Lunar Sci. Conf., Geochim. Cosmochim. Acta* Suppl. 1, Vol. 1, pp. 965–988. Pergamon.

WOOD J. A., MARVIN U., REID J. B., TAYLOR G. J., BOWER J. F., POWELL B. N., and DICKEY J. S., Jr. (1971) Relative proportions of rock types, and nature of the light-colored lithic fragments in Apollo 12 soil samples. Second Lunar Science Conference (unpublished proceedings).

Proceedings of the Second Lunar Science Conference, Volume 2, pp. 1547–1564.
The M.I.T. Press, 1971.

Lead isotopes and volatile transfer in the lunar soil

JAMES M. HUEY, HARUHIKO IHOCHI,* LANCE P. BLACK,† RONALD G. OSTIC,‡ and TRUMAN P. KOHMAN
Department of Chemistry, Carnegie-Mellon University, Pittsburgh, Pennsylvania 15213

(*Received 8 January 1971; accepted in revised form 9 April 1971*)

Abstract—The high-temperature volatility of lead isotopes in the lunar fines has been investigated by stepwise heating and mass spectrometry of lead in the volatile condensates. Well-defined "labile" and "refractory" components are present, with relatively "old" and "young" character, respectively. Single-stage-model Pb^{207}/Pb^{206} apparent ages of 4.67 Gy (Gy = 10^9 years) for Apollo 11 and 4.50 Gy for Apollo 12 result from differing proportions of the two components. The labile component is believed to consist of exogenous solar-type lead and lunar lead enriched in the regolith by volatile transfer associated with early volcanic activity, as suggested by SILVER (1970), TATSUMOTO (1970), and COMPSTON *et al.* (1970), and the refractory component probably resides mainly in crystalline rock fragments. Fit to a two-stage three-component model indicates a mean age of ~4.1 Gy for the fractionation of the two lunar-derived components. Alternate interpretations of Silver's Apollo 11 fines data do not require the assumption of a lunar age greater than those of the meteorites and the earth.

INTRODUCTION

ELEMENT FRACTIONATIONS based on relative volatility have played important roles in the early solar system. The distributions of trace elements in meteorites suggest progressive depletion of volatile elements in the sequence carbonaceous chondrites types I, II, III, and "ordinary" chondrites (LARIMER, 1967; LARIMER and ANDERS, 1967, 1970; ANDERS, 1968). In the latter, elements such as Bi and Tl correlate negatively with degree of metamorphism or equilibration (LAUL *et al.*, 1970a, b). The varying densities of the planets can be accounted for, at least in part, by different degrees of incorporation or retention of volatile elements and compounds (UREY 1952, 1966, and many other papers).

Analyses of lunar surface materials indicate an extreme general depletion of volatile elements, including the rare gases, H, C, N, S, and the halogens, as well as less volatile metallic elements (KEAYS, 1970; GANAPATHY *et al.*, 1970a, b; REED *et al.*, 1970a, b; WASSON and BAEDEČKER, 1970), relative to primordial solar-system matter as sampled by the type I carbonaceous chondrites ("C1"). The regolith, however, contains greater amounts of these elements, apparently as a result of admixture of exogenous solar-type matter from the solar wind, meteorites, and cometary matter (KEAYS, 1970; GANAPATHY *et al.*, 1970a, b; WASSON and BAEDECKER, 1970).

---

* Present address: Chemical Abstracts Service, Ohio State University Branch, Columbus, Ohio 43210.

† Present address: Department of Geology and Mineralogy, Oxford Univerity, Oxford, England.

‡ Present address: Lindsay Vocational and Technical School, Lindsay, Ontario, Canada.

Among the depleted elements is primordial Pb, as indicated by Pb^{204} (Gopalan *et al.*, 1970a, b; Kohman *et al.*, 1970a, b; Silver, 1970a, b; Tatsumoto and Rosholt, 1970; Tatsumoto, 1970a; Wanless *et al.*, 1970a, b). U and Th are enhanced relative to primordial solar-system matter (LSPET, 1969, 1970), resulting in extreme radiogenic character for all lunar Pb so far examined. The Pb of the Apollo 11 soil is less radiogenic than that of most of the crystalline rocks as a result of admixture of the exogenous Pb as well as, apparently, older Pb-enriched materials from which the Pb-depleted basalts were fractionated, so that the mixed fines yield apparent single-stage-model ages approximately equal to the ages of many meteorites and the earth, taken here to be 4.56 Gy. (The value 4.55 $\pm$ .07 Gy derived by Patterson (1956) from data on 5 meteorites has been refined most recently by Kanasewich (1968, Table 2), using a least-squares treatment of 53 meteorite analyses, to 4.56 $\pm$ .03 Gy.) Leaching experiments by Silver (1970b) have shown that the more labile Pb is older and the acid-resistant Pb younger than the average, and he suggested that volatile transfer of parentless Pb was responsible for this distribution. Tatsumoto (1970a) favored lead loss by volatilization from magma or by impact shock as an explanation of the relative ages of the crystalline rocks and the dust and breccia. Compston *et al.* (1970) suggested that volatile transfer of Pb occurred as a consequence of volcanic activity in the maria, and incorporated the idea into a semiquantitative interpretation of the Apollo 11 Pb-isotope data.

We have continued studies of lunar fines Pb utilizing vacuum volatilization as an isolation technique (Kohman *et al.*, 1970a, b), and have employed differential thermal release as a means of fractionating the several Pb components of various origins.

Samples

A 10.35-g sample of Apollo 11 fines from Mare Tranquillitatis (NASA–LRL 10084,45 = CMU No. L–1) and a 5.00-g sample of Apollo 12 fines from Oceanus Procellarum (12070,51 = L–2) were utilized. The former was ground in a motor-shaken tungsten-carbide capsule-and-pestle and probably was contaminated by Pb from chipping of the pestle. Portions of the latter sample were ground separately in an agate mortar-and-pestle, and are not necessarily identical in composition.

Techniques

Chemistry

The procedures previously described (Kohman *et al.*, 1970a, b) were used with little modification. Lead and thallium were separated from the sample matrix by vacuum volatilization (Fig. 1) over a temperature range of 700 to 1100°C. In most cases, after heating at a given temperature the air-cooled exit tube containing the sublimate was cut off and a new tube was sealed on. The heating schedules are given in Table 1. The volatile sublimate inside the exit tube was dissolved in an 8-*M* HCl + 16-*M* HNO_3 mixture. An aliquot of this solution was spiked with enriched Pb^{206} and Tl^{203} for the determination of abolute abundances. Another aliquot was used for determination of isotopic composition. Lead was separated from thallium using a Dowex–1–X8 anion-exchange column and further purified by electrodeposition on platinum.

Chemical blanks were determined by replicates of the chemical procedure including all but the heating. Separate heating blanks were determined and added to the chemical blanks to give total blanks for each analysis. Values of chemical, heating, and total blanks for each analysis are pre-

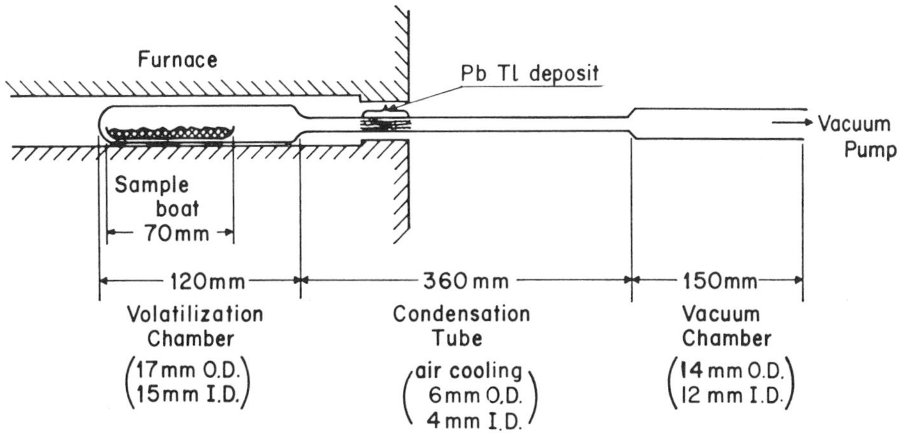

Fig. 1. Vacuum volatilization system for isolation of lead and thallium from solids. Tubing and boat constructed of high-purity fused silica (Amersil Quartz Division, Englehard Industries).

Table 1. Heating schedules of lunar fines samples for lead volatilization.

Designation NASA–LRL	Designation CMU	Mass Heated (g)	Temperature (°C)	Time (hr)	Designation NASA–LRL	Designation CMU	Mass Heated (g)	Temperature (°C)	Time (hr)
10084,45	L–1–A	2.99949	{800	20	12070,51	L–2–A	0.93013	{800	48
			850	20				900	48
			950}	20				950	48
								1100}	48
	L–1–B	3.00350	{900	42					
			975	28		L–2–B_1	3.47905	700	24
			1050}	27		B_2		800	24
	L–1–C_1	4.25085	400	23		B_3		900	48
	C_2		800	23		B_4		950	48
	C_3		{900	48		B_5		1100	48
			950}	47					
	C_4		1100	47					

sented with lead concentration data in Table 2. Contamination of the L–1 samples by Pb in the WC and Co binder of the capsule-and-pestle was estimated by Pb analyses of high-purity fused silica ground under similar conditions and W analyses (see Acknowledgments) of the SiO_2 and L–1 to be ~0.23 ppm Pb and ~3.7 ppb Pb^{204} in L–1–A, L–1–B, and L–1–C_1, under the assumption that the contamination would be "labile" and would go into the first volatility fraction in the differential-heating experiment. Since this contamination is poorly determined and probably variable, no correction for this effect was applied, but the Pb^{204} contents (and to a less significant extent those of Pb^{206}, Pb^{207}, and Pb^{208}) of the three samples mentioned are undoubtedly somewhat high as a result.

Mass spectrometry

A technique based on the work of Cameron *et al.* (1969) was used for lead analysis. 10 μl of a water suspension of fine-grained silica-gel particles, the lead sample in 20 μl of 5% HNO_3, and 1 μl of 0.25-*M* H_3PO_4 are successively loaded, without interim drying, onto a single rhenium mass-spectrometer filament (zone-refined, high-purity rhenium measuring 0.030″ × 0.0012″; supplied

Table 2. Lead isotopic composition of lunar-fines samples.

Sample	Observed Ratios					Blank-corrected Ratios				
	$\frac{206}{204}$	$\frac{207}{204}$	$\frac{208}{204}$	$\frac{207}{206}$	$\frac{208}{206}$	$\frac{206}{204}$	$\frac{207}{204}$	$\frac{208}{204}$	$\frac{207}{206}$	$\frac{208}{206}$
L–1–A	90.2	62.8	108.0	0.696	1.197	99.6	69.0	117.1	0.693	1.176
L–1–B	101.10	68.84	116.77	0.681	1.155	109.60	74.32	124.77	0.678	1.138
L–1–C_1	65.04	50.22	86.59	0.772	1.331	67.98	52.42	89.61	0.771	1.318
C_2	117.70	72.14	124.43	0.613	1.057	147.49	89.11	150.08	0.604	1.018
C_3	131.46	75.47	132.68	0.574	1.009	161.87	91.58	157.89	0.566	0.975
C_4										
C_Σ						92.74	63.49	108.4	0.685	1.170
L–2–A	130.76	79.48	144.36	0.608	1.104	172.10	103.00	183.14	0.5985	1.064
L–2–B_1	133.11	89.06	168.65	0.669	1.267	149.58	99.62	187.23	0.6660	1.252
B_2	75.81	48.03	89.61	0.621	1.182	97.86	59.19	109.08	0.6048	1.115
B_3	584.80	318.25	523.63	0.544	0.895	771.96	418.27	683.79	0.5418	0.886
B_4	78.92	48.23	80.97	0.611	1.026	97.84	58.44	94.13	0.5973	0.962
B_5										
B_Σ						238.8	140.3	242.8	0.588	1.018

Sample	Mass heated (g)	Chemistry Blank (μg)			Lead Content (ppm)				
		Total	Prespike	Chemical	204	206	207	208	Total
L–1–A	2.99949	0.07	0.02	0.05	0.00582	0.5857	0.4077	0.6951	1.6943
L–1–B	3.00350	0.12	0.07	0.05	0.00561	0.6205	0.4242	0.7152	1.7655
L–1–C_1	4.25085	0.08	0.03	0.05	0.00445	0.3081	0.2388	0.4104	0.9618
C_2		0.07	0.02	0.05	0.00085	0.1262	0.0766	0.1296	0.3332
C_3		0.065	0.015	0.05	0.00088	0.1451	0.0825	0.1430	0.3716
C_4		0.055	0.005	0.05	(0.00066)	(0.0131)	(0.0120)	(0.0276)	(0.0534)
C_Σ		0.27	0.07	0.20	0.00684	0.5925	0.4100	0.7106	1.7199
L–2–A	0.93013	0.12	0.07	0.05	0.00507	0.8815	0.5300	0.9472	2.3637
L–2–B_1	3.47905	0.08	0.03	0.05	0.00242	0.3664	0.2451	0.4633	1.0772
B_2		0.07	0.02	0.05	0.00081	0.0796	0.0485	0.0897	0.2186
B_3		0.06	0.01	0.05	0.00083	0.6455	0.3520	0.5776	1.5759
B_4		0.055	0.005	0.05	0.00081	0.0802	0.0480	0.0778	0.2068
B_5		0.055	0.005	0.05	(0.00034)	(0.0066)	(0.0061)	(0.0141)	0.0271
B_Σ		0.32	0.07	0.25	0.00521	1.1783	0.6997	1.2225	3.1057

by the Rembar Co.). The sample is then dried by resistance heating in air at 1.8 A for 5 minutes. For ionization the filament temperature is raised to 800°C in about 10 minutes, then increased by approximately 75°C every five minutes to 1170–1190°C. Data are collected after a stabilization period of 10 minutes. If satisfactory stability is not achieved after this time, the filament current is taken to approximately twice its usual reading for about a second, then returned to its normal value (suggested by MOORBATH, 1970).

Analysis of NBS standards (Table 3) suggest that this treatment does not alter isotopic composition. The data suggest that neither isotopic fractionation nor contamination (e.g., by lead or hydrocarbons) is responsible for the small inter-laboratory differences.

Calculations

All calculations in this paper were made using

$$\lambda_5 = \text{disintegration constant of } U^{235} = 0.9722 \text{ Gy}^{-1},$$
$$\lambda_8 = \text{disintegration constant of } U^{238} = 0.1537 \text{ Gy}^{-1},$$
$$(U^{238}/U^{235})_{now} = 137.8,$$

the values preferred by RUSSELL and FARQUHAR (1960) and KANASEWICH (1968). Gy = 10^9 year. Calculations were facilitated by a set of tables of separately computed functions (KOHMAN 1970).

Table 3. CMU mass spectrometry of NBS lead standards.

NBS No.	Determined by	Pb^{204}/Pb^{206}	Pb^{207}/Pb^{206}	Pb^{208}/Pb^{206}
SRM–981	NBS*	0.059042 ± 0.000037	0.91464 ± 0.00033	2.1681 ± 0.0008
	CMU†	0.05911 ± 0.00002	$0.9146 \pm 0.0002_5$	2.1659 ± 0.0008
SRM–983	NBS*	0.000371 ± 0.000020	0.071201 ± 0.000040	0.013619 ± 0.000024
	CMU†	0.000368 ± 0.000004	0.07124 ± 0.00007	$0.01369 \pm 0.00001_3$

* Catanzaro *et al.* (1968). Uncertainty is "overall limit of error", approximately 2σ
† This work. Uncertainty is standard deviation of mean of measured ratios.

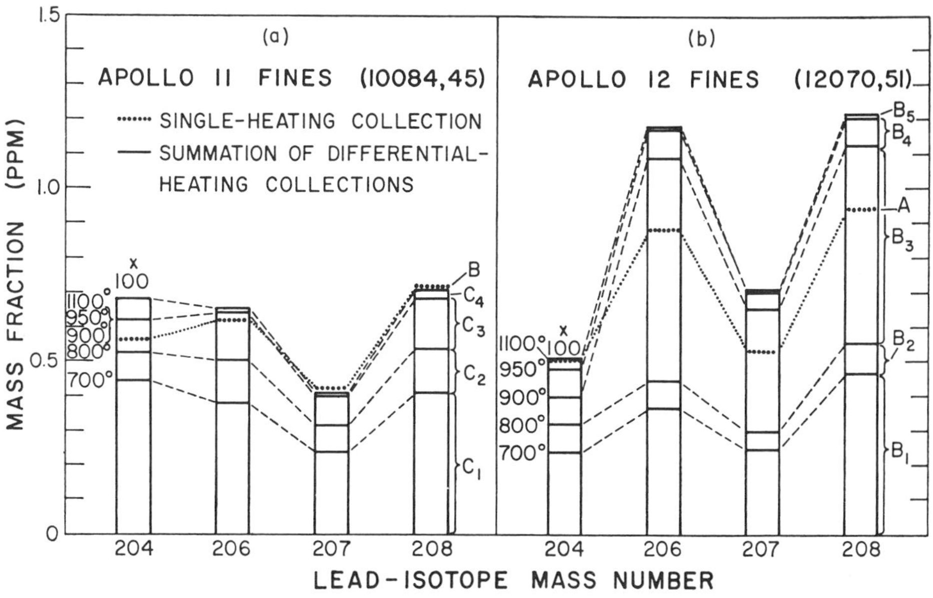

Fig. 2. Differential thermal release of lead isotopes from lunar fines.

Results

Table 2 gives the raw mass-spectrometric data, the chemistry blanks applied, the blank-corrected isotope ratios, and the corrected content of each isotope as calculated from the isotope-dilution analyses. Figure 2 displays the amounts of each isotope released during each stage of the differential heatings L–1–C and L–2–B, with the values in total heatings L–1–B and L–2–A for comparison. The very small amounts of Pb in the final heating fractions suggest that volatilization of this element was essentially complete.

Discussion

Differential-volatility fractionation of lunar-fines Pb

Figures 3 and 4 are plots of Pb^{207} content versus Pb^{206} content for Apollo 11 and Apollo 12 fines, respectively. These plots are relatively insensitive to the effects of contamination and resulting variations in Pb^{204}. Apollo 11 data of several other

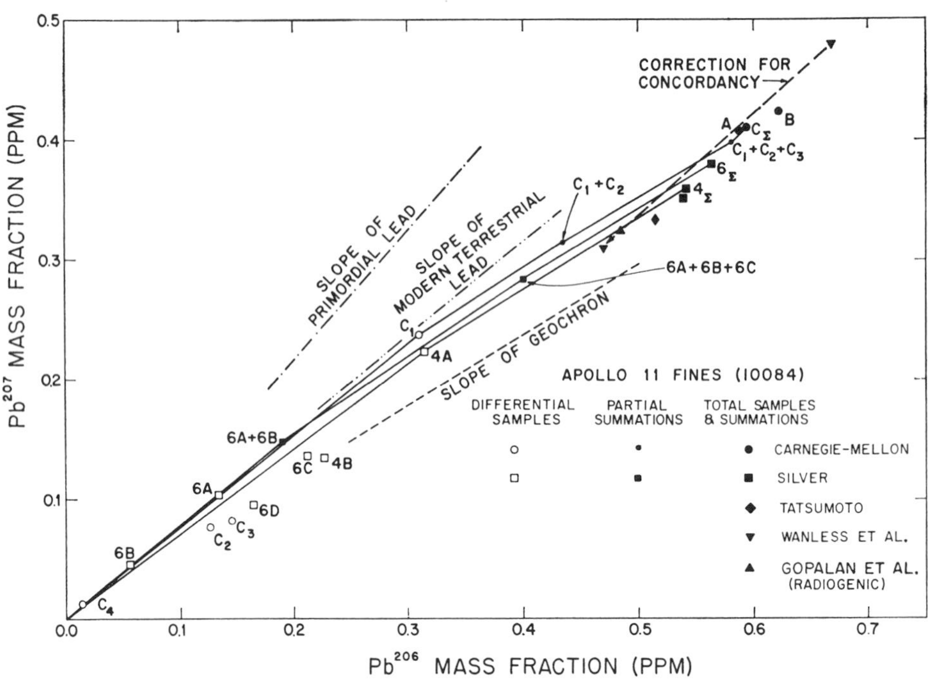

Fig. 3. Pb^{206} and Pb^{207} contents of Mare Tranquillitatis fines (Apollo 11).

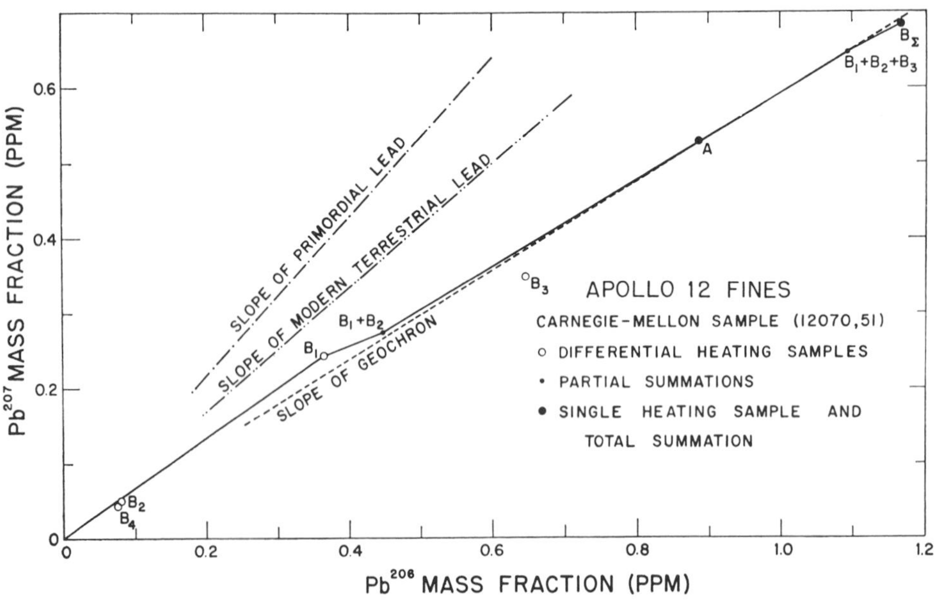

Fig. 4. Pb^{206} and Pb^{207} contents of Oceanus Procellarum fines (Apollo 12).

investigators are shown for comparison. Agreement of the L–1–A, L–1–B, and L–1–C_Σ points corresponds to the fact that the sample was homogenized after grinding. The spread of values for various investigators is probably due partly to heterogeneity of the 10084 soil and partly to variable amounts of terrestrial Pb contamination. The substantial difference between the L–2–A and L–2–B_Σ points indicates significant variability with respect to minor ("accessory") U- and Th-rich components.

The Apollo 12 fines show clearly a distinction between labile Pb, released nearly completely at 700°C, and refractory Pb, liberated rapidly only at ~900°C. The Apollo 11 fines have less of the refractory component and less total Pb, but otherwise the release pattern is similar. The labile component has relatively high Pb^{207}/Pb^{206}, indicating an old character in the sense of a decrease of $\mu = (U^{238}/Pb^{204})_{now}$ during its evolution, whereas the opposite is the case for the refractory component. The differential-volatility pattern thus resembles the differential-leaching pattern found by SILVER (1970b). The older component, which also contains most of the primordial Pb, has apparently been separated from some of the U and enriched in the regolith in a chemically labile form rather early in the moon's history. Volatile transfer (SILVER, 1970b; TATSUMOTO, 1970a; COMPSTON *et al.*, 1970) is a reasonable mechanism though not the only possibility.

Additional information can be obtained from isotope-ratio plots, especially of $y = Pb^{207}/Pb^{204}$ versus $x = Pb^{206}/Pb^{204}$. Because of the extremely small amounts of Pb^{204} in the lunar materials, the position of a sample on such a plot is quite sensitive to contamination. Comparison of the absolute levels of Pb^{204} in our isolated samples (Table 2) with those observed in Apollo 11 soils by TATSUMOTO (TATSUMOTO and ROSHOLT, 1970; TATSUMOTO, 1970a) (1.95 ppb), GOPALAN *et al.* (1970b) (2.16 ppb), and SILVER (1970a, b) (3.10, 3.32, 4.85 ppb), and deduced by WANLESS *et al.* (1970b) (1.93 ppb), and with the levels of other volatile trace elements in Apollo 11 and Apollo 12 soils (KEAYS *et al.*, 1970; GANAPATHY *et al.*, 1970a, b) (see next section), suggests that about two-thirds of the Pb^{204} in our Apollo 11 sample and about one-half of that in the Apollo 12 sample result from terrestrial contamination not accounted for in our chemical blanks (presumably acquired in handling and grinding prior to heating). In the following we assume arbitrarily that exactly half of the Pb^{204} in L–2–A and in each L–2–B fraction is from terrestrial common Pb of mean 300–My–old composition ($Pb^{204}/Pb^{206}/Pb^{207}/Pb^{208}$ = 1/18.24/15.77/38.52: KANASEWICH, 1968, Appendix 2), which corresponds closely to that observed in gasolines, atmospheric aerosols, and snow (TATSUMOTO and PATTERSON, 1963; CHOW and JOHNSTONE, 1965; CHOW, 1968). The resulting corrected ratios are given in Table 4. This moves all points in an x, y plot (Fig. 5), away from the origin while preserving their relative positions. The corrections to the absolute amounts of Pb^{206}, Pb^{207}, and Pb^{208} are relatively small.

Since the points for the several fractions of the differentiation experiment L–2–B do not lie along a single mixing line in Fig. 5, at least three components must be present.

Two-stage three-component model for lunar-soil lead

To account for these observations we adopt a model suggested by COMPSTON *et al.*, (1970). Assume that at a time $-T_P$ primordial Pb (point P in Fig. 5 with coordinates

Table 4. Isotopic composition of Apollo 12 fines lead after correction for assumed pre-chemistry contamination (50% of Pb^{204}).

Sample	Pb^{206}/Pb^{204}	Pb^{207}/Pb^{204}	Pb^{208}/Pb^{204}
L–2–A	326	190	328
L–2–B_1	281	183	336
B_2	177	103	180
B_3	1526	821	1329
B_4	177	101	150
B_Σ	459	265	447

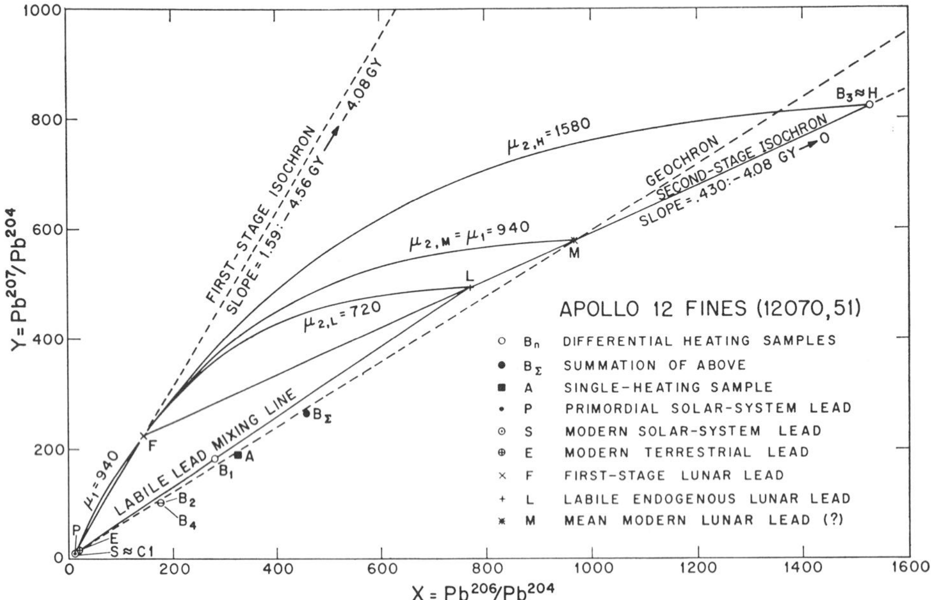

Fig. 5. Pb^{207}/Pb^{204} versus Pb^{206}/Pb^{204} in Apollo 12 fines, showing hypothetical evolution in two-stage three-component model. (Gy = 10^9 years).

x_P, y_P) and U were incorporated into the outer lunar regions with U^{238}/Pb^{204} at some large value μ_1 (decay-corrected to the present time). At a subsequent time $-T_F$ a large-scale fractionation occurred, yielding two major fractions, with lower and higher U^{238}/Pb^{204}, $\mu_{2,L}$, and $\mu_{2,H}$, respectively. We assume for the present that some kind of volatile transfer was involved. The low-μ fraction was enriched in the regolith, with Pb in a chemically labile form, where it subsequently became admixed with extralunar Pb from meteoritic and cometary accretions, also in a chemically labile form, while fragments of the underlying crystalline rocks with higher μ were added to the regolith by cratering. The first L–2 heating fraction (B_1) contains the endogenous low-μ Pb (L) and the solar-system Pb (S), while the refractory component (B_3) consists essentially of endogenous high-μ Pb (H).

The parameters of the model are the primordial isotope ratios x_P and y_P; the ages T_P and T_F; the chemical ratios μ_1, $\mu_{2,L}$, and $\mu_{2,H}$; and the mass fractions of exogenous, labile endogenous, and refractory Pb^{204}: $f_{204,S}$, $f_{204,L}$, and $f_{204,H}$, respectively. It is assumed that $T_P = 4.56$ Gy, the age of the earth-meteorite system as derived from lead-isotope studies (KANASEWICH, 1968), with $x_P = 9.346$ and $y_P = 10.218$ as observed in iron meteorites (OVERSBY, 1970).

From a consideration of all apparently reliable data on the *nonvolatile* (siderophile) trace elements Co, Ni, Ir, and Au in Apollo 11 and Apollo 12 fines and crystalline rocks and in C1 meteorites available in 1970 (Table 5) and following the reasoning of KEAYS *et al.* (1970) and GANAPATHY *et al.* (1970a, b), we deduce that the mean fractional content of C1-like matter in the soil is $f_S = 1.72\%$. From the Pb^{204} content of the C1 Orgueil, 45.3 ppb (REED *et al.*, 1960), we obtain $f_{204,S} = 0.78$ ppb.

The remaining parameters can be evaluated from the Apollo 12 differential-heating data. From the Pb^{204} content of fraction B_3, $f_{204,H} = 0.42$ ppb. Subtracting $f_{204,S}$ from $f_{204,B_1} = 1.21$ ppb yields $f_{204,L} = 0.43$ ppb. Assuming that S (= C1) Pb has close to primordial isotopic composition (MARSHALL, 1962), subtraction of the corresponding amounts of Pb^{206} and Pb^{207} from B_1 Pb gives $x_L = 770$, $y_L = 495$ (point L). Points L and H (= B_3) define an isochron for the second-stage growth from $-T_F$ to the present. Even though L and B_3 may be mixtures of several components of different μ, the assumption that these components are all cogenetic means that the mixing lines and the isochron coincide. The slope of the isochron is 0.430,

Table 5. Comparison of siderophile trace elements in lunar materials and type I carbonaceous chondrites (C1).

	Ni		Co		Ir		Au	
	f(ppm)	(refs)	f(ppm)	(refs)	f(ppb)	(refs)	f(ppb)	(refs)
Carb. chondrites, type I (C1)	10330	(cgnrs)	511	(gnrs)	438	(de)	136	(de)
Apollo 11:								
Cryst. rocks (AB)	6	(a)	20.6	(abfhijkmqtuv)	.14	(fw)	.24	(fw)
Breccias (C)	191	(abjmopqu)	31.7	(abfhijkmoquv)	8.49	(fw)	2.41	(fw)
Fines (D)	203	(abjlmopqtu)	28.9	(abfjkmoqtuv)	8.03	(fw)	2.48	(fw)
Excess (D–AB)	197		8.3		7.89		2.24	
$f_s = \frac{\text{D–AB}}{\text{C1}}$ (%)	1.91		1.62		1.80		1.65	
					1.75	(f)	1.60	(f)
					1.4	(w)	.9	(w)
Apollo 12:								
Cryst. rocks (AB)					.333	(x)	.013	(x)
Fines (D)					8.63	(x)	2.03	(x)
Excess (D–AB)					8.30		2.02	
$f_s = \frac{\text{D–AB}}{\text{C1}}$ (%)					1.89		1.48	
					1.88	(x)	1.35	(x)

a. ANNELL and HELZ (1970g)
b. COMPSTON *et al.* (1970g)
c. CHRISTIE (1914)
d. CROCKET *et al.* (1967)
e. EHMANN *et al.* (1970)
f. GANAPATHY *et al.* (1970g)
g. GREENLAND and LOVERING (1965)
h. GOLES *et al.* (1970g)
i. GOLES *et al.* (1970g)
j. HASKIN *et al.* (1970g)
k. LSPET (1969)
l. MAXWELL *et al.* (1970s)
m. MORRISON *et al.* (1970g)
n. NICHIPORUK *et al.* (1967)
o. SMALES *et al.* (1970g)
p. TAYLOR *et al.* (1970g)
q. TUREKIAN and KHARKAR (1970g)
r. WIIK (1956)
s. WIIK (1963)
t. WIIK and OJANPERA (1970s)
u. WÄNKE *et al.* (1970g)
v. WAKITA *et al.* (1970g)
w. WASSON and BAEDECKER (1970g)
x. GANAPATHY *et al.* (1970b)

(1970s): The Moon Issue of *Science*, **167**, 415–784 (1970 January 30). (1970g): *Proc. Apollo 11 Lunar Sci. Conf., Geochim. Cosmochim. Acta* Suppl. 1, Vol. 2, 991–1936. Pergamon.

whence $T_F = 4.08$ Gy. The slope of the isochron beginning at P for the first stage growth from $-T_P$ to $-T_F$ can now be calculated to be 1.59. Its intersection with the extension of the $-T_F \rightarrow 0$ isochron gives point F representing the uniform lead at $-T_F$, with $x_F = 145$, $y_F = 225$.

From the lengths of the isochron segments $P - F$, $F - L$, and $F - H$, one can calculate $\mu_1 = 940$, $\mu_{2,L} = 720$, and $\mu_{2,H} = 1580$. For a hypothetical fraction with $\mu_{2,M} = \mu_1$ during the second growth stage, the present lead composition would be represented by the point M with $x_M = 970$, $y_M = 580$. Of course, it lies on the "geochron" (MURTHY and PATTERSON, 1962), the isochron for the single-stage lead evolution from point P in the assumed age of the solar system, 4.56 Gy (KANASEWICH, 1968). Such lead may exist in the moon below the depth of the maria basalts or in the lunar highlands. On the other hand the prefractionation μ-value of ~ 940 may have purely local significance, since it is much higher than the prefractionation values of ~ 100 to ~ 500 calculated by TATSUMOTO (1970a) from two-stage-model fits to the Apollo 11 crystalline rocks.

These quantitative results are sensitive to the values used for f_S and to the terrestrial contamination corrections, but from the radiogenic Pb^{206} and Pb^{207} contents alone one can derive a less sensitive functional relationship between $Q = (Pb^{207}/Pb^{206})_{\text{radiogenic}}$, $E = \mu_2/\mu_1 =$ enrichment factor for U relative to Pb, and the age T_F of the fractionation, given the total age T_P of the system:

$$Q = \left(\frac{Pb_1^{207} + Pb_2^{207}}{Pb_1^{206} + Pb_2^{206}}\right)_{\text{radiogenic}} = \alpha \frac{(e^{\lambda_5 T_P} - e^{\lambda_5 T_F}) + E(e^{\lambda_5 T_F} - 1)}{(e^{\lambda_8 T_P} - e^{\lambda_8 T_F}) + E(e^{\lambda_8 T_F} - 1)},$$

where $\alpha = (U^{235}/U^{238})_{\text{now}} = 1/137.8$. Figure 6 shows the computed relationships for $T_P = 4.56$ Gy and selected values of Q, including those for the low- and high-temperature fractions of both Apollo 11 and Apollo 12 fines. The former Q values were calculated assuming that two-thirds of the blank-corrected Pb^{204} in L–1–C_1 and L–1–C_3 represents prechemistry contamination. The ratios and the positions of the curves are relatively independent of the various subtractions made. The pair of points for Apollo 12 soil for the less certain T_F value derived through use of the Pb^{204} data is shown:

$$E_L = \mu_{2,L}/\mu_1 = 0.76; \; E_H = \mu_{2,H}/\mu_1 = 1.68.$$

The deduced age T_F is greater than the ages of the basalts of the Apollo 12 site and even of most Apollo 11 rocks. The single age, ~ 4.1 Gy, characterizing the simplified model is presumably actually a weighted mean of the ages of many fractionations which contributed to the material now in the soil, some of which occurred elsewhere on the moon (WOOD *et al.*, 1970a, b). The surviving rocks, however, are only the *youngest* of the igneous products in the immediate region, earlier-formed rocks which reached the surface having been largely comminuted by impacts. If the fractionation were related to the volcanism which flooded the maria, evidently a prolonged process (BALDWIN, 1963, 1970; MUTCH, 1970), its weighted mean age could be considerably greater than that of the topmost basalt layers, particularly those of the relatively young O. Procellarum (SHOEMAKER and HACKMAN, 1962; OFFIELD and POHN, 1970). If, on the other hand, the fractionation was impact-related, the concentration of

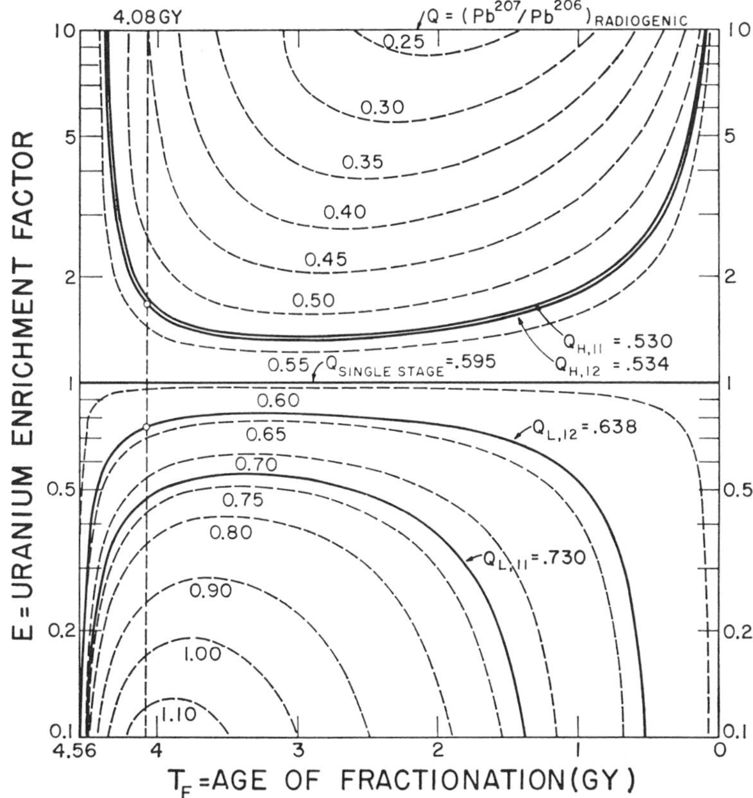

Fig. 6. Functional relationship between age of Pb-U fractionation (T_F), U-enrichment factor (E), and present radiogenic Pb^{207}/Pb^{206} ratio (Q) in two-stage evolution, for total age of 4.56 Gy. L = labile lead; H = refractory lead; 11 = Apollo 11 fines; 12 = Apollo 12 fines. (Gy = 10^9 years).

impact events in the earliest part of the moon's history (Hartmann, 1965, 1966, 1970) leads to a similar conclusion.

Apparent age of the lunar soil

Single-stage-model calculations of the apparent age of the lunar soil from various measurements of the Pb^{204}, Pb^{206}, and Pb^{207} relative abundances in Apollo 11 and Apollo 12 fines are summarized in Table 6. These and Rb^{87}–Sr^{87} model-age calculations yield values close to the accepted age of the earth-meteorite system, indicating that the moon has about the same age and that its soil is a fairly representative mixture of lunar matter (Albee *et al.*, 1970; Gopalan *et al.*, 1970a; Kohman *et al.*, 1970a, b; Tatsumoto, 1970a).

However, the agreement is not exact, the Pb^{207}/Pb^{206} ages of Apollo 11 soil being significantly greater than the best value for the earth and meteorites from the same radioactive systems, 4.56 Gy (Kanasewich, 1968). The opposite appears to be the

Table 6. Single-stage-model apparent ages of lunar fines and breccias.

Lunar Site, Sample Type, Sample No.	Observers	Nature of Sample or Experiment	Pb^{207}/Pb^{206} Total	Pb^{207}/Pb^{206} Radiogenic	Model Age (Gy)
Apollo 11, Fines, 10084	GOPALAN *et al.* (1970a, b)	Whole sample	0.697	0.680	4.76
	KOHMAN *et al.* (1970)	Sample A, total heating	0.694	0.651	4.69
	This work	Sample B, total heating	0.678	0.639	4.67
		Sample C, diff. heating fract. 1	0.771	0.720	4.84
		diff. heating fract. 2	0.604	0.571	4.50
		diff. heating fract. 3	0.566	0.533	4.40
		summation of fractions	0.685	0.639	4.67
	SILVER (1970a)	Whole sample, Expt. 2	0.649	0.623	4.63
	SILVER (1970b)	Expt. 4, Leachate	0.736	0.719	4.84
		. Residue	0.591	0.551	4.45
		Expt. 6, Leachate A	0.774	0.736	4.87
		Leachate B	0.796	0.736	4.87
		Leachate C	0.645	0.629	4.64
		Residue	0.581	0.527	4.38
	TATSUMOTO and ROSHOLT (1970)	Whole Sample	0.653	0.637	4.66
	TATSUMOTO (1970)	Density fraction $\rho < 2.9$	0.687	0.654	4.70
		Density fraction $2.9 < \rho < 3.3$	0.614	0.591	4.55
		Density fraction $3.3 < \rho$	0.633	0.598	4.57
	WANLESS *et al.* (1970a)	Whole sample, corr. for reag. blank	0.724	0.642	4.67
	WANLESS *et al.* (1970b)	Same, subtraction of mod. terr. Pb to produce concordancy	0.669	0.653	4.70
Apollo 11, Breccia, 10060	SILVER (1970a, b)	Whole sample	0.664	0.609	4.59
Apollo 11, Breccia, 10061	TATSUMOTO and ROSHOLT (1970)	Whole sample a	0.655	0.638	4.66
	TATSUMOTO (1970)	Whole sample b	0.653	0.638	4.66
Apollo 12, Fines, 12070	This work	Sample A, total heating	0.598	0.570	4.50
		Sample B, diff. heating fract. 1	0.666	0.638	4.66
		diff. heating fract. 2	0.605	0.553	4.45
		diff. heating fract. 3	0.542	0.535	4.40
		diff. heating fract. 4	0.597	0.545	4.43
		summation of fractions	0.588	0.567	4.49

case at the Apollo 12 site. Our best single-stage-model ages for Apollo 11 and Apollo 12 soils are 4.67 and 4.50 Gy, respectively. From the differential-volatility results it is evident that these apparent ages reflect the relatively low amount of "young" refractory Pb in the M. Tranquillitatis fines and the relatively large amount in those from O. Procellarum, and neither is a true age of the moon or its regolith.

SILVER (1970b) has given two interpretations of his Apollo 11 U–Th–Pb observations which require an age of part of the Pb of the fines to be considerably older than 4.6 Gy, and the moon to be at least 4.95 ± 0.1 Gy old. We feel that alternate interpretations are possible, and that none of the lunar U–Th–Pb data are inconsistent with a lunar age close to those of the earth and the bulk of the meteorites.

The first argument is based on the total Pb-isotope and U contents of the fines and and assumption that the Pb^{204} is predominantly derived from terrestrial common Pb contamination. Subtracting corresponding amounts of Pb^{206} and Pb^{207} and radiogenic derivatives of the U^{238} and U^{235} now present during 4.18 Gy, the age of the regolith as deduced from the ages of the underlying basalts, an "excess" component remained with Pb^{207}/Pb^{206} averaging 1.96 ± 0.14. For the two-stage model implied, the first stage (with higher μ) ending 4.18 Gy ago must have begun 4.94 ± 0.15 Gy ago. However, some endogenous primordial Pb^{204} is probably present, and assuming that all of the Pb^{204} is from that would give for Silver's sample 10084,35 (total-sample analysis) an excess $Pb^{207}/Pb^{206} = 1.60$. Evolution at constant μ by 4.18 Gy ago need only have begun 4.49 Gy ago. Similar calculations for leach-experiment composite data yield excess ratios of 1.42 and 1.37, whose generation would require starting times no earlier than 4.20 and 4.15 Gy ago, respectively. An appropriate mixture of primordial and contamination Pb would be consistent with a first-stage evolution beginning about 4.56 Gy ago.

Silver's second argument is based on the assumption that his second soil-leach fraction (No. 6B) approximates a mixture of coherent U–Pb systems, the Pb^{207}/Pb^{206} ratio (terrestrial-contamination-corrected) of 0.767 indicating an apparent age of 4.93 Gy. However, here the age calculation assumes constant μ (a single stage ending at the present). A two-stage model with decreasing μ could easily account for this lead in a total time less than 4.6 Gy. Moreover, assumption that some of the Pb^{204} represents primordial Pb yields a lower radiogenic Pb^{207}/Pb^{206} ratio.

Volatile transfer as a lunar process

The fit to the observations of the two-stage three-component model based tentatively on volatile transfer, with reasonable values of the parameters, suggests that this is a satisfactory hypothesis, provided appropriate mechanisms exist. However, alternate hypotheses should be examined.

The magnitudes of the U–Pb fractionations for any reasonable T_F (Fig. 6) could easily be accounted for by igneous processes such as partial melting or magmatic crystallization. But this could hardly explain the lability of the Pb in the low-μ fractions. It seems more likely that the strong difference in volatility of the labile and refractory components of lunar fines reflects a difference in physical state.

If the labilization of the Pb were due to radiation damage of U- and Th-rich accessory components and grain boundaries in the rocks (CROZAZ *et al.*, 1970a, b; LOVERING and KLEEMAN, 1970; RICHARDSON *et al.*, 1970a, b), the most labile Pb would appear younger than the average rather than older in terms of Pb^{207}/Pb^{206} ratios.

Labilization of Pb through shock damage would require that the materials most heavily exposed to impact also have older lead in the sense of an enrichment of Pb *prior* to the shocking. Since impact is most severe in the uppermost layers, this would require that the chemical fractionation be accompanied by extreme stratification, with thin low-μ matter everywhere capping higher-μ rocks. Volatile transfer seems the only way to achieve such stratification. Shock damage may well maintain or enhance the lability of the topmost matter.

Volatile transfer thus seems the most likely explanation of the early enrichment of Pb relative to U in the fraction of the lunar soil from which the Pb is most easily removed by volatilization or acid leaching. Two mechanisms for producing vapor phases are obvious: impact and volcanism. Impact volatilization of a given portion of matter would probably be either complete or insignificant, so that no element fractionation would occur at this stage. The less volatile vapor constituents would probably condense and precipitate first, but would be re-mixed with the later condensates on the surface, especially when large numbers of impacts are involved. Impact melting probably involves relatively small amounts of matter, and any consequent liquid-vapor separation would probably be of minor extent and likewise subject to nullification by re-mixing. Thus, impact effects can hardly provide the large-scale chemical fractionation needed.

Lunar volcanism, on the other hand, appears to be quite adequate. Lead is a frequent constituent of terrestrial volcanic emanations (ZIES, 1938a, b; WHITE and WARING, 1963; HOUTERMANS *et al.*, 1964; KRAUSKOPF, 1967). It often occurs in deposits of sublimates from volcanic vents and fumaroles as cotunnite, $PbCl_2$, indicating an enhancement of lead volatility in the presence of HCl or other chlorides. PbF_2, PbO, and PbS are also moderately volatile, and in the strongly reducing environment of lunar magmas formation of elementary Pb vapor might be significant. The lower gravity might allow Pb-bearing vapors to originate at greater depths in the moon than in the earth. The lack of an atmosphere might facilitate the escape of vapors from the lunar crust. The residual magmas and rocks derived therefrom would be even more depleted in lead and other volatiles than the average lunar matter.

This transfer could be vertical or lateral, or both. Vertical volatile transfer refers to processes in which the lead is brought from depth to the surface, and any sublimate effused into the atmosphere condenses and falls back onto the surface in the same general locality from which it was derived. It would also include the revolatilization of labile Pb from the near-surface materials by a subsequent lava flow, so that some of the older Pb was redeposited on the new surface. Lateral volatile transfer refers to situations in which the vapors and their condensates spread over large areas before settling back to the surface. For readily condensable volatiles like lead, lateral volatile transport from sites of effusion while still in the vapor phase would probably be of minor importance.

Lateral distribution of vapor-enriched matter could, however, occur through subsequent impacts which ejected solid surface matter in ballistic trajectories or transported smaller particles in an accompanying wind of expanding gas. The distribution patterns of lunar-crater ejecta and the presence in the mare fines of particulate matter presumably of highlands origin (WOOD *et al.*, 1970a, b) indicate that this kind of post-fractionation distribution of labile lead must have been important. Since earlier volcanism may have characterized the development of the highlands, this may contribute to the great mean age of fractionation indicated by our analysis of the mare fines.

In summary, volatilization of old lead accompanying lunar volcanism, as suggested particularly by COMPSTON *et al.* (1970), seems to have been responsible for the vertical transfer of lead from depths in the lunar body, its enrichment in the surface materials in a chemically labile form, and its depletion in the maria basalt rocks. Any lateral transport which is implied by the observations could be accounted for mainly by post-volcanic impacts. Shock damage from such impacts may contribute to the lability of the older lead exposed at the surface.

Acknowledgments—Mr. Mark W. Haramic provided technical assistance with the mass spectrometer. Drs. Stanley E. Church and Michael Bikerman made helpful suggestions concerning the geochemical interpretation of the results. We thank Mr. John D. Johnston and the Westinghouse Electric Corporation, Astronautics Division, Large, Pennsylvania, for X-ray-fluorescence W analyses. James M. Huey expresses appreciation for a NASA Graduate Traineeship. This work was supported principally by the National Aeronautics and Space Administration under Contract NAS–9–8073, with assistance from the United States Atomic Energy Commission under Contract No. AT(30–1)–844.

References

Albee A. L., Burnett D. S., Chodos A. A., Eugster O. J., Huneke, J. C., Papanastassiou D. A., Podosek F. A., Russ G. P., II, Sanz H. G., Tera F., and Wasserburg G. J. (1970) Ages, irradiation history, and chemical composition of lunar rocks from the Sea of Tranquillity. *Science* **167**, 463–466.

Anders E. (1968) Chemical processes in the early solar system, as inferred from meteorites. *Accts. Chem. Res.* **1**, 289–298.

Baldwin R. B. (1963) Analyses of earlier theories of the moon's history. In *The Measure of the Moon*, pp. 305–307. University of Chicago Press.

Baldwin R. B. (1970) Summary of arguments for a hot moon. *Science* **170**, 1297–1300.

Cameron A. E., Smith D. H., and Walker R. L. (1969) Mass spectrometry of nanogram-size samples of lead. *Anal. Chem.* **41**, 525–526.

Catanzaro E. J., Murphy T. J., Shields W. R., and Garner E. L. (1968) Absolute isotopic abundance ratios of common, equal-atom, and radiogenic lead isotope standards. *J. Res. Nat. Bur. Stand.* **72A**, 261–267.

Chow T. J. (1968) Isotope analysis of seawater by mass spectrometry. *J. Water Pollution Control Federation* **40**, 399–411.

Chow T. J. and Johnstone M. S. (1965) Lead isotopes in gasoline and aerosols of Los Angeles Basin, California. *Science* **147**, 502–503.

Christie W. A. K. (1914) A carbonaceous aerolite from Rajputana. *Rec. Geol. Soc. India* **44**, pt. 1, 41–51.

Compston W., Chappell B. W., Arriens P. A., and Vernon M. J. (1970) The chemistry and age of Apollo 11 lunar material. *Proc. Apollo 11 Lunar Sci. Conf., Geochim. Cosmochim. Acta* Suppl. 1, Vol. 2, pp. 1007–1027. Pergamon.

Crocket J. H., Keays R. R., and Hsieh S. (1967) Precious metal abundances in some carbonaceous and enstatite chondrites. *Geochim. Cosmochim. Acta.* **31**, 1615–1623.

Crozaz G., Haack U., Hair M., Hoyt H., Kardos J., Maurette M., Miyajima M., Seitz M., Sun S., Walker R., Wittels M., and Woolum D. (1970a) Solid state studies of the radiation history of lunar samples. *Science* **167**, 563–566.

Crozaz G., Haack U., Hair M., Maurette M., Walker R., and Woolum D. (1970b) Nuclear track studies of ancient solar radiations and dynamic lunar surface processes. *Proc. Apollo 11 Lunar Sci. Conf., Geochim. Cosmochim. Acta* Suppl. 1, Vol. 3, pp. 2051–2080. Pergamon.

Ehmann W. D., Baedecker P. A., and McKown D. M. (1970) Gold and iridium in meteorites and some selected rocks. *Geochim. Cosmochim. Acta* **34**, 493–507.

Ganapathy R., Keays R. R., Laul J. C., and Anders E. (1970a) Trace elements in Apollo 11 lunar rocks: Implications for meteorite influx and origin of the moon. *Proc. Apollo 11 Lunar Sci. Conf., Geochim. Cosmochim. Acta* Suppl. 1, Vol. 2, pp. 1117–1142. Pergamon.

GANAPATHY R., KEAYS R. R., and ANDERS E. (1970b) Apollo 12 lunar samples: Trace element analysis of a core and the uniformity of the regolith. *Science* **170**, 533–535.

GOPALAN K., KAUSHAL S., LEE-HU C., and WETHERILL G. W. (1970a) Rubidium-strontium, uranium, and thorium–lead dating of lunar material. *Science* **167**, 471–473.

GOPALAN K., KAUSHAL S., LEE-HU C., and WETHERILL G. W. (1970b) Rb–Sr and U, Th–Pb ages of lunar materials. *Proc. Apollo 11 Lunar Sci. Conf., Geochim. Cosmochim. Acta* Suppl. 1, Vol. 2, pp. 1195–1205. Pergamon.

GREENLAND L. and LOVERING J. F. (1965) Minor and trace element abundances in chondritic meteorites. *Geochim. Cosmochim. Acta* **29**, 821–858.

HARTMANN W. K. (1965) Secular changes in meteoritic flux through the history of the solar system. *Icarus* **4**, 207–213.

HARTMANN W. K. (1966) Early lunar cratering. *Icarus* **5**, 406–418.

HARTMANN W. K. (1970) Preliminary note on lunar cratering rates and absolute time scale. *Icarus* **12**, 131–133.

HOUTERMANS F. G., EBERHARDT A., and FERRARA G. (1964) Lead of volcanic origin. In *Isotopic and Cosmic Chemistry* (editors Craig, Miller, and Wasserburg), Chap. 18. North Holland.

KANASEWICH E. R. (1968) The interpretation of lead isotopes and their geological significance. In *Radiometric Dating for Geologists* (editors Hamilton and Farquhar), pp. 147–223. Interscience-Wiley.

KEAYS R. R., GANAPATHY R., LAUL J. C., ANDERS E., HERZOG G. F., and JEFFERY P. M. (1970) Trace elements and radioactivity in lunar rocks: Implications for meteorite infall, solar-wind flux, and formation conditions on the moon. *Science* **167**, 490–493.

KOHMAN T. P. (1970) Tables of uranium–thorium–lead decay-growth functions. United States Atomic Energy Commission Report NYO–844–79.

KOHMAN T. P., BLACK L. P., IHOCHI H., and HUEY J. M. (1970a) Lead and thallium isotopes in Mare Tranquillitatis surface material. *Science* **167**, 481–483.

KOHMAN T. P., BLACK L. P., IHOCHI H., and HUEY J. M. (1970b) Lead and thallium isotopes in Mare Tranquillitatis surface material. *Proc. Apollo 11 Lunar Sci. Conf., Geochim. Cosmochim. Acta* Suppl. 1, Vol. 2, pp. 1345–1350. Pergamon.

KRAUSKOPF K. B. (1967) Volcanic Gases. In *Introduction to Geochemistry*, Chap. 16. McGraw-Hill.

LARIMER J. W. (1967) Chemical fractionations in meteorites, I. Condensation of the elements. *Geochim. Cosmochim. Acta* **31**, 1215–1238.

LARIMER J. W. and ANDERS E. (1967) Chemical fractionations in meteorites, II. Abundance patterns and their interpretation. *Geochim. Cosmochim. Acta* **31**, 1239–1270.

LARIMER J. W. and ANDERS E. (1970) Chemical fractionations in meteorites, III. Major element fractionations in chondrites. *Geochim. Cosmochim. Acta* **34**, 367–388.

LAUL J. C., CASE D. R., SCHMIDT-BLEEK F., and LIPSCHUTZ M. E. (1970a) Bismuth contents of chondrites. *Geochim. Cosmochim. Acta* **34**, 89–103.

LAUL J. C., PELLY I., and LIPSCHUTZ M. E. (1970b) Thallium contents of chondrites. *Geochim. Cosmochim. Acta* **34**, 909–920.

LOVERING J. F. and KLEEMAN J. D. (1970) Fission track uranium distribution studies on Apollo 11 lunar samples. *Proc. Apollo 11 Lunar Sci. Conf., Geochim. Cosmochim. Acta* Suppl. 1, Vol. 1, pp. 627–631. Pergamon.

LSPET (Lunar Sample Preliminary Examination Team) (1969) Preliminary examination of lunar samples from Apollo 11. *Science* **165**, 1211–1227.

LSPET (Lunar Sample Preliminary Examination Team) (1970) Preliminary examination of lunar samples from Apollo 12. *Science* **167**, 1325–1339.

MARSHALL R. R. (1962) Mass spectrometric study of the lead in carbonaceous chondrites. *J. Geophys. Res.* **67**, 2001–2015.

MOORBATH S. Private communication to Lance Black (1970).

MURTHY V. R. and PATTERSON C. C. (1962) Primary isochron of zero age for meteorites and the earth. *J. Geophys. Res.* **67**, 1161–1167.

MUTCH T. A. (1970) *Geology of the Moon: A Stratigraphic View*. Princeton University Press.

NICHIPORUK W., CHODOS A., HELIN E., and BROWN H. (1967) Determination of iron, nickel, cobalt, calcium, chromium and manganese in stony meteorites by X-ray fluorescence. *Geochim. Cosmochim. Acta* **31**, 1911–1930.

OFFIELD T. W. and POHN H. A. (1970) Lunar crater morphology and relative-age determination of lunar geologic units, Part 2. Applications. In *Geological Survey Research 1970:* U.S. Geol. Surv. Prof. Paper 700-C, 163–169.

OVERSBY V. M. (1970) The isotopic composition of lead in iron meteorites. *Geochim. Cosmochim. Acta* **34**, 65–75.

PATTERSON C. (1956) Age of meteorites and the earth. *Geochim. Cosmochim. Acta* **10**, 230–237.

REED G. W., JR., JOVANOVIC S., and FUCHS L. H. (1970a) Trace elements and accessory minerals in lunar samples. *Science* **167**, 501–503.

REED G. W., JR., and JOVANOVIC S. (1970b) Halogens, mercury, lithium, and osmium in Apollo 11 samples. *Proc. Apollo 11 Lunar Sci. Conf., Geochim. Cosmochim. Acta* Suppl. 1, Vol. 2, pp. 1487–1492. Pergamon.

REED G. W., JR., KIGOSHI K., and TURKEVICH A. (1960) Determinations of concentrations of heavy elements in meteorites by activation analysis. *Geochim. Cosmochim. Acta* **20**, 122–140.

RICHARDSON K. A., MCKAY D. S., GREENWOOD W. R., and FOSS T. H. (1970a) Alpha-particle activity of Apollo 11 samples. *Science* **167**, 516–517.

RICHARDSON K. A., MCKAY D. S., GREENWOOD W. R., and FOSS T. H. (1970b) Alpha-particle activity of Apollo 11 samples. *Proc. Apollo 11 Lunar Sci. Conf., Geochim. Cosmochim. Acta* Suppl. 1, Vol. 1, pp. 763–767. Pergamon.

RUSSELL R. D. and FARQUHAR R. M. (1960) *Lead Isotopes in Geology*. Interscience.

SHOEMAKER E. M. and HACKMAN R. J. (1962) Stratigraphic basis for a lunar time scale. In *The Moon* (editors Kopal and Mikhailov), pp. 289–300. Academic.

SILVER L. T. (1970a) Uranium-thorium-lead isotope relations in lunar materials. *Science* **167**, 468–471.

SILVER L. T. (1970b) Uranium–thorium–lead isotopes in some Tranquillity Base samples and their implications for lunar history. *Proc. Apollo 11 Lunar Sci. Conf., Geochim. Cosmochim. Acta* Suppl. 1, Vol. 2, pp. 1533–1574. Pergamon.

TATSUMOTO M. (1970a) Age of the moon: An isotopic study of U–Th–Pb systematics of Apollo 11 lunar samples, II. *Proc. Apollo 11 Lunar Sci. Conf., Geochim. Cosmochim. Acta* Suppl. 1, Vol. 2, pp. 1595–1612. Pergamon.

TATSUMOTO M. and PATTERSON C. C. (1963) The concentration of common lead in sea water. In *Earth Science and Meteorics*, pp. 74–89. North-Holland.

TATSUMOTO M. and ROSHOLT J. N. (1970) Age of the moon: An isotopic study of uranium–thorium–lead systematics of lunar samples. *Science* **167**, 461–463.

UREY H. C. (1952) Chemical processes during the formation of the planets. In *The Planets: Their Origin and Development*, Chap. 4. Yale University Press.

UREY H. C. (1966) Chemical evidence relative to the origin of the solar system. *Mon. Not. R. Astr. Soc. (London)* **131**, 199–223.

WANLESS R. K., LOVERIDGE W. D., and STEVENS R. D. (1970a) Age determinations and isotopic abundance measurements on lunar samples. *Science* **167**, 479–480.

WANLESS R. K., LOVERIDGE W. D., and STEVENS R. D. (1970b) Age determinations and isotopic abundance measurements on lunar samples (Apollo 11). *Proc. Apollo 11 Lunar Sci. Conf., Geochim. Cosmochim. Acta* Suppl. 1, Vol. 2, pp. 1729–1739. Pergamon.

WASSON J. T. and BAEDECKER P. A. (1970) Ga, Ge, In, Ir, and Au in lunar, terrestrial and meteoritic basalts. *Proc. Apollo 11 Lunar Sci. Conf., Geochim. Cosmochim. Acta* Suppl. 1, Vol. 2, pp. 1741–1750. Pergamon.

WHITE D. E. and WARING G. A. (1963) Volcanic emanations (Data of Geochemistry, Sixth edition, Chapter K). U.S. Geological Survey Prof. Paper *440-K*, 1–29.

WIIK H. B. (1956) The chemical composition of some stony meteorites. *Geochim. Cosmochim. Acta* **9**, 279–289.

WIIK H. B. (1963) Private communication to B. Mason, *Space Sci. Rev.* **1**, 621–646.

WOOD J. A., DICKEY J. S., JR., MARVIN U. B., and POWELL B. N. (1970a) Lunar anorthosites. *Science* **167**, 602–604.

WOOD J. A., DICKEY J. S., JR., MARVIN U. V., and POWELL B. N. (1970b) Lunar anorthosites and a geophysical model of the moon. *Proc. Apollo 11 Lunar Sci. Conf., Geochim. Cosmochim. Acta* Suppl. 1, Vol. 1, pp. 965–988. Pergamon.

ZIES E. G. (1938a) The concentration of the less familiar elements through igneous and related activity. *Amer. J. Sci.* **35a**, 385–404.

ZIES E. G. (1938b) The concentration of the less familiar elements through igneous and related activity. *Chem. Rev.* **23**, 47–64.

Proceedings of the Second Lunar Science Conference, Volume 2, pp. 1565–1570.
The M.I.T. Press, 1971.

Activation analysis determination of uranium and ^{204}Pb in Apollo 11 lunar fines

ANTHONY TURKEVICH
Enrico Fermi Institute and Department of Chemistry,
University of Chicago, Chicago, Illinois 60637

G. W. REED, JR.
Argonne National Laboratory, Argonne, Illinois 60439 and
Enrico Fermi Institute, University of Chicago, Chicago, Illinois 60637

H. R. HEYDEGGER
Department of Chemistry, Purdue University, Hammond, Indiana 46323 and
Enrico Fermi Institute, University of Chicago, Chicago, Illinois 60637

and

J. COLLISTER*
Enrico Fermi Institute, University of Chicago, Chicago, Illinois 60637

(Received 22 February 1971; accepted in revised form 13 March 1971)

Abstract—The ^{238}U abundance in the Apollo 11 type D material (10084,58) has been determined via the $^{238}U(n, 2n)^{237}U$ reaction with a resultant mean value from four replicates of 0.54 ± 0.02 ppm.

Neutron activation analysis has also been applied to the determination of the ^{204}Pb in the same sample via the $^{204}Pb(n, 2n)^{203}Pb$ reaction. Three samples gave results varying from 2.2 to 7.1 ppb. In addition, some data were obtained on the bismuth and thallium contents of this sample.

INTRODUCTION

THIS REPORT summarizes the results obtained on the abundances of some of the heaviest elements in returned lunar material. These elements, thallium, lead, bismuth, thorium, and uranium, although present in only trace amounts, are important in understanding the geochemical history of lunar material, as well as the heat production in the moon. There have been several reports on the abundances of these elements in Apollo 11 material. The data for uranium and thorium appear to be well established (e.g., FIELDS *et al.*, 1970; GOPALAN *et al.*, 1970; SILVER, 1970; TATSUMOTO, 1970). Likewise, the lead has been shown to be very radiogenic (e.g., GOPALAN *et al.*, 1970; SILVER, 1970; TATSUMOTO, 1970), and the absolute amounts found by various workers appear to be in reasonable agreement.

The situation in the case of the other two elements of this group is less satisfactory. KOHMAN *et al.* (1970) and GANAPATHY *et al.* (1970) find an overall thallium abundance in Apollo 11 fines of up to 3 ppb, but the latter, as well as ANDERS *et al.* (1971) get results as low as 0.3 ppb, with an indicated inverse correlation with particle size. The values for bismuth in lunar fines are likewise from GANAPATHY *et al.* (1970) and ANDERS *et al.* (1971), who find values up to 1.5 ppb with, again, evidence for a dependence of the results on particle size. The low amounts of these elements make

* Present Address: York University, Toronto, Canada.

their determination difficult. There is, however, great interest in their abundances since, in meteorites, they have been found to be sensitive indicators of the geochemical history of these objects.

The determination of the amount of ^{204}Pb in lunar material is in somewhat less satisfactory condition. Lunar lead is so radiogenic that the corrections for laboratory contamination in isotope-dilution measurements make the published values for the amount of this nuclide (again in the range of ppb) particularly uncertain. There has even been some question raised (e.g., SILVER, 1970) as to whether ^{204}Pb indigenous to lunar material has actually been detected in some of the samples. At the same time, there is considerable interest in the amount of primordial lead in such material since, as in the cases of thallium and bismuth, it is likely to be an indicator of the possible role of the volatility of an element in the geochemical history of the material.

Because of this interest in the heavy elements, it was considered worthwhile to pursue their determination by neutron activation techniques. Although these are almost always less accurate than isotope-dilution methods, results obtained thereby have the advantage of being less likely to be invalidated by contamination of the sample. In addition, in the case of ^{204}Pb, the technique used determines specifically this isotope even in the presence of hundreds of times the abundance of the heavier, radiogenic, isotopes. A major effort was, therefore, devoted to the lead measurement, but concurrent bismuth and thallium determinations were made when possible, not only to establish their abundances, but also as internal checks on the lead results. It has been shown (REED, KIGOSHI, and TURKEVICH, 1960) that, in meteorites at least, these three elements exhibit parallel abundance trends.

The nuclear reactions that have been used in the present work and the radiations of the products that were measured are summarized in Table 1. It is seen that both in the case of ^{238}U and ^{204}Pb use has been made of the $(n, 2n)$ reaction to supplement the more usual slow neutron induced reactions. Attempts to measure the ^{208}Pb via the (n, γ) production of the 3.3 hr ^{209}Pb isotope were unsuccessful. Other work, cited above, has adequately established the amount of this radiogenic isotope.

EXPERIMENTAL

Except for one irradiation of a breccia, the material investigated was the powdered lunar "fines" from the Apollo 11 mission (sample 10084). The samples for our experiments were taken from the shipment containers and transferred inside a N_2 dry box to Super-Sil fused silica tubes, and were sealed immediately after removal from the box. Monitor samples, between 1 μg and 1 mg

Table 1. Nuclear processes utilized in neutron activation determination of heavy elements

Element	Nuclear reaction	Half-life	Radiations
Uranium	$^{238}U(n, 2n)^{237}U$	6.75 d	β^-, γ
	$^{235}U(n, f)^{140}Ba$	12.8 d	β^-, γ
	↓		
	^{140}La	38.0 hr	β^-, γ
Bismuth	$^{209}Bi(n, \gamma)^{210}Bi$	5.0 d	β^-
	↓		
	^{210}Po	138 d	α
Lead	$^{204}Pb(n, 2n)^{203}Pb$	52 hr	X, γ
Thallium	$^{203}Tl(n, \gamma)^{204}Tl$	3.8 y	β^-

of the element, were evaporated in similar fused silica tubes. Inert MgO was often added to help retain the recoiling fission products of uranium. Samples and monitors were placed side by side in irradiation cans. The irradiations were carried out in the core of the High Flux Beam Reactor at the Brookhaven National Laboratory. The high energy (> 1 MeV) neutron flux in this reactor is of the order of 10^{13} $cm^{-2}sec^{-1}$, while the thermal flux is $\sim 10^{14}$ $cm^{-2}sec^{-1}$. In order to try to isolate the 3.3 hr ^{209}Pb, the samples were worked up within a few hours of removal from the reactor in the Hot Laboratory facilities at the Brookhaven National Laboratory.

After irradiation, the samples were quantitatively transferred to a zirconium crucible containing appropriate carriers and tracers and were fused with Na_2O_2, Na_2CO_3, and NaOH. The HNO_3 solution of the melt was first passed through a silica-gel column, and then the Pb and Bi were separated by electrolytic deposition. Additional recovery of the Pb was accomplished by precipitation of the nitrates of Sr, Ba, and Pb. The Pb samples were purified by repetition of cycles including sulfide, sulfate, and nitrate precipitations and chloroform extractions with dithizone.

The uranium was separated from most of the solution of the fusion melt by an anion-exchange column and purified by ether extractions and removal of contaminating radioactivities by scavenging precipitations and solvent extractions. It was finally electroplated onto platinum, with the recovery being measured by the recovery of the ^{234}U tracer added at the start. In the case of lead, barium, and thallium, the final samples were in the form of chromates. The final bismuth sample was in the form of the oxychloride.

The samples from the monitors were usually subjected to less extensive chemical purification, but were finally prepared for radioactivity measurements in the same fashion as were the samples isolated from the irradiated lunar material.

For each nuclide of interest, the measurements of the radioactivity were carried out in such a manner as to enhance the sensitivity of the technique, while at the same time making the measurements as specific as possible for that nuclide. In the case of ^{237}U, this involved measuring the 60-keV γ rays in coincidence with the β^- decay of the nuclide. In the case of the electron-capturing ^{203}Pb, the Tl X-rays were measured with a NaI(Tl) crystal operated in coincidence with the 0.28-MeV γ rays as detected by a 42 cm^3 Ge(Li) crystal. The counting rates of the samples were very low (of the order of events per 1000 min). However, the low backgrounds, the specificity, and the possibility of detecting and identifying any contaminating species made these detection systems very suitable for this program.

Results

The results of our measurements on the uranium and ^{204}Pb contents of Apollo 11 samples (principally soil) are shown in Table 2. The errors quoted are the statistical errors (1 σ) from the radioactivity measurements. It is evident that the uranium contents of the soil as determined by the $^{238}U(n, 2n)$ reaction in different irradiations agree within the errors. The standard deviation of the mean for sample 10084,58, i.e., 0.54 ppm, is ± 0.02 ppm.

Table 2. Uranium and ^{204}Pb contents of Apollo 11 samples

Irradiation	Sample Mass (mg)	Sample Number	Uranium content (in ppm) via $^{238}U(n, 2n)^{237}U$	Uranium content (in ppm) via $^{235}U(n, f)^{140}Ba$	^{204}Pb content (in ppb)
L-2	206	10062,24	—	0.25 ± 0.02	—
	192	10084,58	—	0.43 ± 0.01	—
L-4	103	10084,2	0.49 ± 0.06	0.48 ± 0.05	—
	70	10084,58	0.50 ± 0.07	0.47 ± 0.05	—
L-6	817	10084,58	0.51 ± 0.04	—	2.2 ± 0.8
L-7	797	10084,58	0.62 ± 0.07	—	7.1 ± 1.5
L-8	730	10084,58	0.55 ± 0.02	—	4.7 ± 0.4
Weighted average		10084,58	0.54 ± 0.02		

There are only two measurements of uranium via the $^{235}U(n, f)$ reaction to compare directly with those via the $^{238}U(n, 2n)$ reaction. Although in both cases they give slightly lower answers, these are well within even the statistical errors. In view of the larger number of $^{238}U(n, 2n)$ determinations, the average of these is taken to be the best estimate of the uranium content of sample 10084,58.

There are three determinations of the ^{204}Pb in lunar soil presented in Table 2. In view of the possible uncertainties in the determinations of this nuclide by other techniques, it is worth presenting some details of the present results. Figures 1 and 2 give the evidence that the lead radioactivity isolated in irradiation L-8 was actually ^{203}Pb ($t_{1/2}$ = 52 hr). Figure 1 indicates that, in spite of the low counting rates for the X-ray/γ-coincidence system, an acceptable 52 hr decay was followed. Figure 2 shows that the X-ray spectrum in coincidence with the γ-ray was that expected in the decay of ^{203}Pb. Although this was the best example, the other two samples behaved adequately in regard to these two tests of radiochemical purity.

On the other hand, the three values for ^{204}Pb in Table 2 scatter appreciably more than can be accounted for by the statistical uncertainties. This may be a reflection of other, less recognizable, errors in the experiment; it may, however, indicate a variability in the ^{204}Pb content even within a given sample.

In addition to the results presented in Table 1, a value of 7.7 ± 1.4 ppb for the thallium content and an upper limit of 5.4 ppb for the bismuth content were obtained on sample 10084,58 (run L-8). A value of about 1.7 ppb was found for the bismuth content of the same sample in run L-7.

DISCUSSION

The average values obtained in this work for the uranium content and the ^{204}Pb content of lunar soil are compared with the more precise of the published experi-

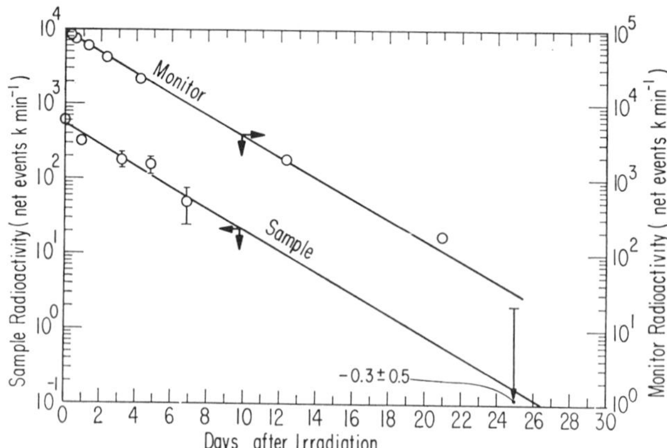

Fig. 1. Decay of ^{203}Pb ($t_{1/2}$ = 52 hr) radioactivity in lead separated from neutron irradiated lunar fines (run L-8). The radiation measured is the (73 ± 15)-keV X-ray in coincidence with (278 ± 20)-keV γ ray. A background rate of (2.3 ± 0.3) events kmin^{-1} has been subtracted. The lunar sample (lower curve, left hand ordinate) is seen to decay with the same halflife as the monitor lead (upper curve, right hand ordinate).

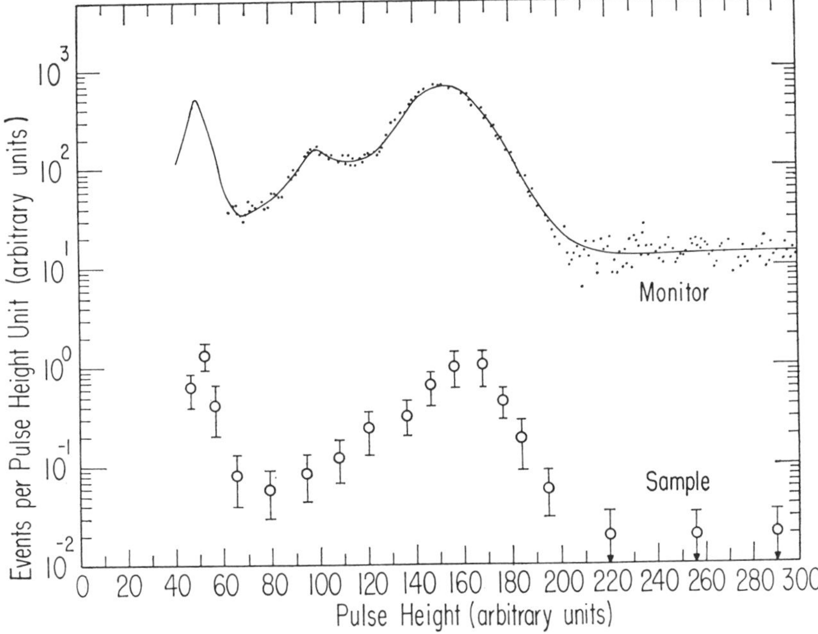

Fig. 2. ^{203}Pb in lead separated from neutron irradiated lunar fines (run L-8). The spectrum is that of X-rays in coincidence with (278 ± 20)-keV γ rays. The abscissae are proportional to the energy of the X-rays. The top spectrum is that of the lead monitor; the lower spectrum is that of the sample.

Table 3. Uranium and ^{204}Pb contents of Apollo 11 soil—comparison of different determinations

Sample	Uranium (ppm)	^{204}Pb (ppb)	Reference
10084,58	0.54 ± 0.02	2.2, 4.7, 7.1	This work
10084,75	0.591 ± 0.018		Fields *et al.* (1970)
10084,25	0.549 ± 0.005	2.16*	Gopalan *et al.* (1970)
10084	0.544 ± 0.005	1.95*	Tatsumoto (1970)
10084,35	0.562 ± 0.005	3.10*	Silver (1970)
10084,45		5.6,* 5.8,* 6.8*	Huey *et al.* (1971)

* These values are after correction for laboratory blanks.

mental values for these abundances in Table 3. The weighted average value found in the present work, 0.54 ± 0.02 ppm, for the uranium content of sample 10084 is seen to be in agreement with the other values, whose unweighted average is 0.56 ppm. The meager data obtained in this study on the isotopic composition of the uranium are consistent with the terrestrial value as established by other workers (Silver, 1970; Tatsumoto, 1970).

The ^{204}Pb values obtained in this work are also compared in Table 3 with those deduced by other workers, usually after making considerable blank corrections. It is seen that the present investigation tends to confirm the existence in lunar soil of ^{204} Pb at the levels previously reported. Those values lie within the range of the present more direct determinations. Likewise, the bismuth and thallium results presented here agree with the other published work (GANAPATHY *et al.*, 1970; KOHMAN *et al.*, 1970; ANDERS *et al.*, 1971).

This confirmation of ^{204}Pb in Apollo 11 lunar soil must still be taken with some caution. There is always the possibility that the extremely small amounts detected are the result of contamination of the samples with terrestrial lead before analysis by the various workers. The observed variation of the ^{204}Pb content within the same sample tends to support this possibility.

Acknowledgment—The authors gratefully acknowledge the cooperation of the Brookhaven National Laboratory in this work in providing the irradiation facilities of the High Flux Beam Reactor, the Hot Laboratory facilities of the Department of Applied Sciences and the laboratory and radioactivity measurement equipment of the Department of Chemistry. This work was supported mostly by NASA contract NAS–9–7883; most of the radioactivity detection equipment employed in this work was purchased under AEC contract AEC AT(11–1)–1167.

REFERENCES

ANDERS E., LAUL J. C., KEAYS R. R., GANAPATHY R., and MORGAN J. W. (1971) Elements depleted on lunar surface: Implications for origin of moon and meteorite influx rate. Second Lunar Science Conference (unpublished proceedings).

FIELDS P. R., DIAMOND H., METTA D. N., STEVENS C. M., ROKOP D. J., and MORELAND P. E. (1970) Isotopic abundances of actinide elements in lunar material. *Proc. Apollo 11 Lunar Sci. Conf., Geochim. Cosmochim. Acta* Suppl. 1, Vol. 2, pp. 1097–1102. Pergamon.

GANAPATHY R., KEAYS R. R., LAUL J. C., and ANDERS E. (1970) Trace elements in Apollo 11 lunar rocks: Implications for meteorite influx and origin of moon. *Proc. Apollo 11 Lunar Sci. Conf., Geochim. Cosmochim. Acta* Suppl. 1, Vol. 2, pp. 1117–1142. Pergamon.

GOPALAN K., KAUSHAL C., LEE-HU C., and WETHERILL G. W. (1970) Rb–Sr and U, Th–Pb ages of lunar materials. *Proc. Apollo 11 Lunar Sci. Conf., Geochim. Cosmochim. Acta* Suppl. 1, Vol. 2, pp. 1195–1205. Pergamon.

HUEY J. M., IHOCHI H., BLACK L. P., OSTIC R. G., and KOHMAN T. P. (1971) Lead isotopes and volatile transfer in the lunar soil. Second Lunar Science Conference (unpublished proceedings).

KOHMAN T. P., BLACK L. P., IHOCHI H., and HUEY J. M. (1970) Lead and thallium isotopes in Mare Tranquillitatis surface material. *Proc. Apollo 11 Lunar Sci. Conf., Geochim. Cosmochim. Acta* Suppl. 1, Vol. 2, pp. 1345–1350. Pergamon.

REED G. W., KIGOSHI K., and TURKEVICH A. (1960) Determinations of concentration of heavy elements in meteorites by activation analysis. *Geochim. Cosmochim. Acta* **20**, 122–140.

SILVER L. T. (1970) Uranium–thorium–lead isotopes in some Tranquillity Base samples and their implications for lunar history. *Proc. Apollo 11 Lunar Sci. Conf., Geochim. Cosmochim. Acta* Suppl. 1, Vol. 2, pp. 1533–1574. Pergamon.

TATSUMOTO M. (1970) Age of the moon: An isotopic study of U–Th–Pb systematics of Apollo 11 lunar samples. II. *Proc. Apollo 11 Lunar Sci. Conf., Geochim. Cosmochim. Acta* Suppl. 1, Vol. 2, pp. 1595–1612. Pergamon.

WANLESS R. K., LOVERIDGE W. D., and STEVENS R. D. (1970) Age determinations and isotopic abundance measurements on lunar samples (Apollo 11). *Proc. Apollo 11 Lunar Sci. Conf., Geochim. Cosmochim. Acta* Suppl. 1, Vol. 2, pp. 1729–1739. Pergamon.

Proceedings of the Second Lunar Science Conference, Volume 2, pp. 1571–1576.
The M.I.T. Press, 1971.

Isotopic abundances of actinide elements in Apollo 12 samples

P. R. Fields, H. Diamond, D. N. Metta, C. M. Stevens, and D. J. Rokop
Chemistry Division, Argonne National Laboratory, Argonne, Illinois 60439

(*Received 22 February 1971; accepted in revised form 31 March 1971*)

Abstract—The abundances of uranium and thorium were measured mass spectrometrically as 1.70 ± 0.05 ppm and 6.52 ± 0.2 ppm, respectively, in the soil and 2.32 ± 0.07 ppm and 9.06 ± 0.3 ppm, respectively, in a breccia from Apollo 12. The mol ratio $^{235}U:^{238}U$ was 0.007253 ± 0.000020 in the soil and 0.00739 ± 0.0002 in the breccia, compared to 0.007257_7 for terrestrial uranium. The $^{235}U/^{238}U$ ratio data for Apollo 11 fines has been re-evaluated to be 0.007233 ± 0.000015. The following upper limits were set for the soil: $^{239}Pu \leq 1 \times 10^{-9}$ ppm and $^{244}Pu \leq 6 \times 10^{-11}$ ppm. No unusual alpha particles or spontaneous fissions were observed. ^{236}U was observed in the soil and in the breccia. This surprising result is under further investigation.

Introduction

In previous publications (Fields *et al.*, 1970) we reported the abundances and isotopic composition of uranium and thorium in Apollo 11 soil and the results of a search for the long lived nuclides, ^{236}U, ^{244}Pu, ^{247}Cm, and other transuranium isotopes. Two techniques were utilized: isotopic dilution followed by mass spectrometry, and alpha particle energy analysis. In this paper we report the results of a similar series of investigations of Apollo 12 soil and breccia samples. The analytical results are then compared with the earlier Apollo 11 values.

Experimental Procedure and Results

Two samples were received from the Apollo 12 collection, 5 g of soil (sample 12070,91) and 5 g of breccia (sample 12073,34). The breccia sample was a one cubic centimeter section from a portion of the upper end of a larger rock. Figure 1 shows the breccia sample and a white inclusion which appeared to be a homogeneous mineral phase identified as an anorthositic plagioclase feldspar. The white feldspar section was removed before we processed the breccia and will be used for other studies by another group of investigators.

A 1 g and a 3 g sample of the soil and a 1 g sample of breccia were dissolved for analysis. ^{230}Th, ^{233}U, ^{242}Cm, and ^{236}Pu tracers were added to both 1 g samples, but only ^{236}Pu was added to the 3 g sample. The tracers were required for the isotopic dilution analysis, but the large sample was used for high precision and high sensitivity mass analysis of the uranium and plutonium and for alpha particle analysis of thorium, uranium, and transplutonium element fractions.

The bulk of Apollo 12 soil dissolved more readily in HF and $HClO_4$ than the corresponding Apollo 11 soil, but approximately 5% resisted further attack by the acid solution. The residue was heated with 15 *M* NaOH and then the resulting solid

Fig. 1. Breccia sample from Apollo 12 rock 12073,34 showing white inclusion (anorthositic plagioclase feldspar).

dissolved completely in dilute HNO_3. In contrast to this behavior, the Apollo 11 soil dissolved completely in the initial acid solution.

The chemical procedure employed in separating the various actinide elements from the solution was described in (FIELDS *et al.*, 1970b). The uranium, thorium, and plutonium fractions contained more inert impurities than had been encountered in the processing of Apollo 11 material and it was necessary to add another chemical purification step for each of these elements to produce a satisfactory sample for mass spectrometric analysis. Very faint traces of $HClO_4$ reduce the sensitivity of the mass spectrometric analyses. Hence, it was necessary to heat the dry sample for 24 hours in a sand bath at 150–200°C in order to remove the last traces of $HClO_4$ used earlier in the procedure to destroy any organic residues. Additional purification steps were also added to yield a purer transplutonium element fraction.

The isotopic composition of the separated uranium and thorium fractions from the 1 g samples of lunar dust and breccia were measured in the same 30-cm mass spectrometer that was used to analyze the Apollo 11 samples. The uranium and thorium abundances, determined by averaging many mass spectrometric measurements, are given in Table 1. The reported errors, based upon standard deviations, came from estimates of uncertainties in sampling, counting geometry, alpha counting, and mass spectrometry. No allowance was made for the uncertainties in the half-lives of the ^{233}U (1.62×10^5 yrs) and ^{230}Th (8.0×10^4 yrs) tracers. The reagent blank for uranium was 0.2% of the sample and the thorium blank was 0.4%.

In order to observe the alpha particles and possible spontaneous fission decay of thorium, uranium and the transplutonium elements, a 3 g sample of soil was processed with only ^{236}Pu added as a tracer. A portion of the uranium fraction from this sample

Table 1. Micrograms (metal wt) of actinide element per gram of lunar sample.

	Apollo 11	Apollo 12	
	soil	soil	breccia
Uranium	0.591 ± 0.018	1.70 ± 0.05	2.41 ± 0.07
Thorium	2.24 ± 0.06	6.52 ± 0.2	9.06 ± 0.3
^{244}Pu	$\leq 9 \times 10^{-11}$	$\leq 6 \times 10^{-11}$	$\leq 1 \times 10^{-9}$
^{239}Pu	$\leq 5 \times 10^{-10}$	$\leq 1 \times 10^{-9}$	$\leq 1 \times 10^{-9}$
^{240}Pu	$\leq 3 \times 10^{-10}$	$\leq 1.7 \times 10^{-10}$	—
^{247}Cm	$\leq 1.25 \times 10^{-10}$	—	—

Table 2. Isotopic composition of uranium in lunar samples.

	Apollo 11	Apollo 12	
	soil	soil	breccia
U^{238}	1	1	1
U^{236}	$\leq 3 \times 10^{-9}$	200×10^{-9}	44×10^{-9}
U^{235}	0.007233 ± 0.000015*	0.007253 ± 0.000020†	0.00739 ± 0.0002‡

* 4 sample loadings, 15 sets of data, 250 $^{235}U/^{238}U$ ratios; 6 samples of NBS normal uranium, 17 sets of data 200 $^{235}U/^{238}U$ ratios.

† 2 sample loadings, 3 sets of data, 51 $^{235}U/^{238}U$ ratios; 5 samples of NBS U500 and U150 standard samples, 5 sets of data, 60 $^{235}U/^{238}U$ ratios.

‡ This result is a by-product of the uranium determination in Table 1.

was also analyzed for its isotopic composition in the 30-cm machine (results in Table 2) and the bulk of the uranium was used to search for ^{236}U in a special tandem mass spectrometer capable of determining neighboring isotopes of widely disparate abundances. The isotopic composition of the 1 and 3 g samples together with the corresponding Apollo 11 results are summarized in Table 2. The rather high ^{236}U content in Apollo 12 soil cannot be explained at the present time. Contamination in the spectrometer appeared to be ruled out by a series of natural uranium runs that showed the ^{236}U to be less than 0.7×10^{-9} relative to ^{238}U, and the possibility of hydride formation was ruled out by the use of a high pass energy filter that usually reduces the lower energy uranium hydride by several orders of magnitude (MORELAND *et al.*, 1970) but no change was observed in the $^{236}U/^{235}U$ ratio. A third source of mass 236, $^{187}Re^{16}O_2{}^{17}O$ whose mass is 60 millimass units lower than ^{236}U, would have appeared about one-half a peak-width away from the observed peak. Additional determinations of ^{236}U will be carried out as samples become available.

The results of measurements of the $^{235}U/^{238}U$ ratio in Apollo 11 (10084) fines were re-evaluated (Table 2). The ratio was shown to be possibly lower than the ratio for terrestrial uranium (0.007257_7) by 3.4 ± 2.0 parts per thousand. The significance of this difference depends upon the similarity of the conditions for the analyses of the lunar and standard samples. The analyses of the NBS natural uranium standards were done on samples of 0.5 to 2 μg, the same size as the lunar uranium sample. However, the terrestrial samples were high purity uranium, whereas the lunar material contained visible amounts of impurities resulting from the chemical separation, which may result in a different mass discrimination factor. The yield of the uranium separation chemistry was estimated at 60%, so isotopic fractionation cannot be ruled out.

The $^{235}U/^{238}U$ ratio for Apollo 12 (120070) fines, using fewer data, is -0.7 ± 2.7 parts per thousand compared to the terrestrial value.

A 10^7 yr difference in the time of nucleosynthesis of the uranium isotopes would be reflected in a 1% difference in the $^{235}U/^{238}U$ ratios of lunar and terrestrial uranium. Thus, the similarity between terrestrial and lunar uranium shows that the uranium in the two bodies had a common origin (within 5 million years).

Alpha pulse analysis of the uranium fraction from Apollo 12 soil is shown in Fig. 2. The radioactivity ratio $^{238}U:^{234}U$ is 1.06 ± 0.02 (standard deviation). Mass analysis showed no change in the $^{235}U/^{238}U$ ratio, eliminating the possibility that some ^{238}U had been introduced accidently into the sample. The alpha pulse analyzer was carefully calibrated with a sample of terrestrial uranium that gave a radioactivity ratio, $^{238}U:^{234}$ of 1.02 ± 0.02.

The rather surprising ratio of 1.06, differing from our Apollo 11 measurements and the extensive measurements of ROSHOLT and TATSUMOTO (1970) led us to consider the possibility that the members of the ^{238}U decay chain could escape the lunar gravitational field due to the recoil energy following alpha or beta decay. While calculations confirmed the recoil escape possibility, the enrichment of ^{238}U required a model in

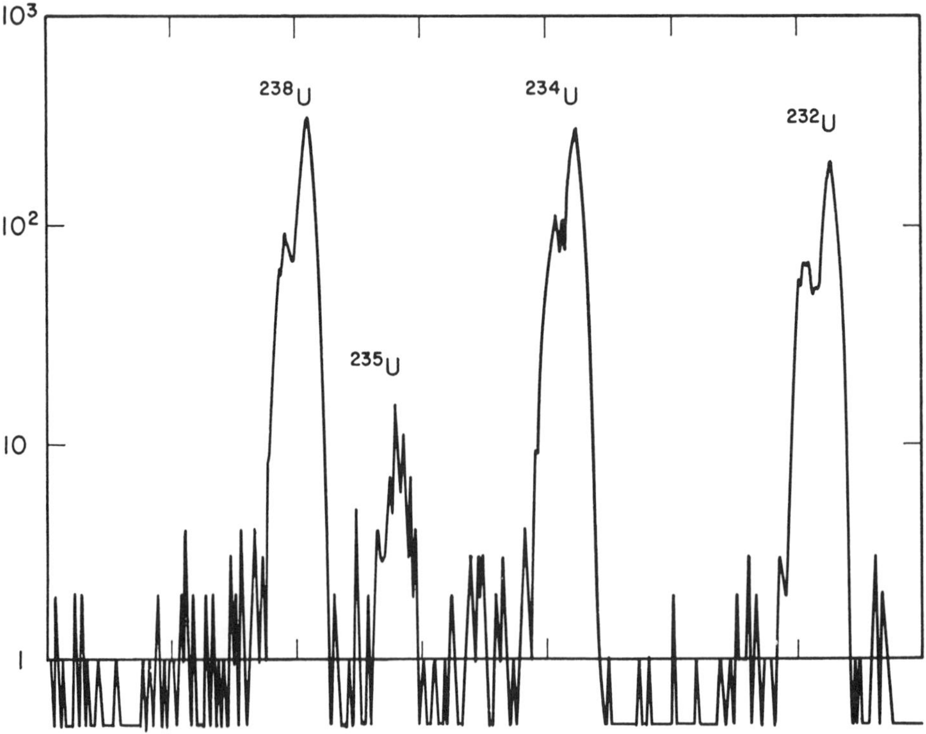

Fig. 2. Alpha pulse analysis spectrum of uranium fraction from Apollo 12 soil.

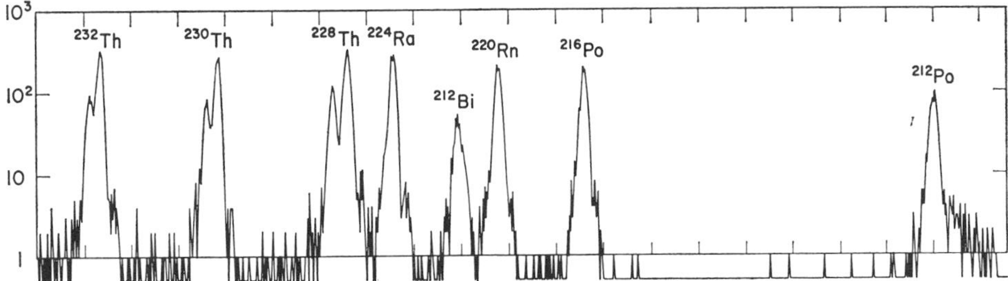

Fig. 3. Alpha pulse analysis spectrum of thorium fraction from Apollo 12 soil.

which the uranium is deposited at grain boundaries of crystals and is subsequently freed as a very fine dust on the surface of the moon as the crystalline rocks disintegrated. Under these circumstances the ^{234}Th in the top 200 Å of uranium dust could have escaped from the lunar surface. This hypothesis was checked experimentally by allowing the coarser grains from a 1 g sample of Apollo 12 fines to settle through water at greater than 0.17 cm/min. The very fine particles ($\sim$0.1 g) thus collected and the water used for the wet-sieving were processed and the isolated uranium was pulse analyzed for 49 days. The resulting ^{238}U:^{234}U radioactivity ratio of 1.05 $\pm$ 0.06 was inconclusive.

The purified thorium fraction from the three grams of soil was also alpha pulse analyzed (Fig. 3). The ratio of the activities, ^{232}Th:^{230}Th was 1.240 $\pm$ 0.034. This result is not in agreement with a similar measurement by ROSHOLT and TATSUMOTO (1971). If the half-lives of ^{232}Th and ^{238}U are 1.41 $\times$ 10^{10} yrs and 4.51 $\times$ 10^9 yrs, respectively, then the resultant mass ratio, ^{232}Th:^{238}U is 3.88 $\pm$ 0.11 in good agreement with 3.84 $\pm$ 0.09 obtained from isotope dilution measurements on other samples (Table 1) of the soil. This agreement implies that the last separation of uranium and thorium could not have occurred within the last 6 $\times$ 10^5 yrs. Earlier thorium pulse analyses (FIELDS *et al.*, 1970) of the Apollo 11 soil indicated an excess of ^{227}Th in the thorium fraction. This anomaly was traced to terrestrial actinium contamination in our lanthanum carrier used in the chemical separations, but no such contamination was detected in the Apollo 12 analysis.

The plutonium fractions from all the samples were analyzed in a 250-cm mass spectrometer. The average results of these analyses, corrected for background, and a summary of the previous Apollo 11 analyses are presented in Table 1. The maximum sensitivity of the mass spectrometer was utilized in setting the upper limit of 6 $\times$ 10^{-17} grams of ^{244}Pu/gram of sample for the Apollo 12 soil.

About 44 mg of fine soil was spread over an area of approximately 1.5 cm^2, covered with 0.25 mg/cm^2 of aluminum foil, and counted in a 20 day interval for spontaneous fissions. No fissions were observed, giving a limit of 0.016 spontaneous fissions per mg of lunar fines per day. If we take this sample as characteristic of the lunar surface we expect less than 0.03 spontaneous fissions per day per cm^2 of lunar surface. The transplutonium element fraction from the three grams of soil (12073) was also counted for spontaneous fissions. None were observed in a 28 day interval. To achieve this low background it has been necessary to convert the fission counter to battery operation. An examination of the transplutonium actinide fraction by alpha pulse analyses for 8 days revealed no unusual activity.

DISCUSSION AND CONCLUSIONS

The ratio of thorium to uranium remains approximately 3.8 despite a threefold variation between their concentrations in the fines from Apollo 11 and from Apollo 12. As in the case of Apollo 11, the mass ratio $^{235}U:^{238}U$ in Apollo 12 samples is very close to terrestrial uranium. The absence of ^{244}Pu confirms the conclusion reached in our Apollo 11 analysis, namely, there were no nearby supernova explosions within the last one or two billion years. The limit of the ^{239}Pu in Apollo 12 soil sets an upper limit of ~ 80 neutrons cm^{-2} sec^{-1} for the thermal neutron flux over the last $\sim 10^5$ yrs.

The observation of ^{236}U in variable abundances that imply neutron or proton irradiations several orders of magnitude higher than seen in other irradiation products in the soil, is surprising. Further measurements to try to duplicate the data are in progress. The unique 2.4×10^7 yr half-life of U^{236} makes it much more sensitive to an increase in irradiation that might have occurred 10^7–10^8 years ago, than would other irradiation products. The apparent observation of ^{236}U in variable amounts might imply a meteoritic infusion from material that had been irradiated more extensively (and recently) than most of the lunar material, and which has been imperfectly mixed with the lunar surface material. Such an effect has been mentioned by BARBER *et al.* (1971) in connection with the density of high atomic number tracks in some lunar silicates.

A search for 2.2×10^6 yr ^{237}Np in the neptunium fraction of the 3 gram sample 12070 is planned. A search for spontaneous fissions in a volatile fraction of Apollo 11 is in progress.

Acknowledgment—We wish to thank L. Fuchs of this Laboratory for identifying the inclusion in the breccia sample and Doris Huff for her aid in many of the mass spectrometer analyses. The work was supported by the USAEC and NASA Contract T–76536.

REFERENCES

BARBER D. J., HUTCHEON I., and PRICE P. B. (1971) Extra lunar dust in Apollo cores? *Science* **171**, 372–374.

FIELDS P. R., DIAMOND H., METTA D. N., STEVENS C. M., ROKOP D. J., and MORELAND P. E. (1970) Isotopic abundances of actinide elements in lunar material. *Science* **167**, 499–450.

FIELDS P. R., DIAMOND H., METTA D. N., STEVENS C. M., ROKOP D. J., and MORELAND P. E. (1970b) Isotopic abundances of actinide elements in lunar material. *Proc. Apollo 11 Lunar Science Conf., Geochim. Cosmochim. Acta* Suppl. 1, Vol. 2, pp. 1097–1102. Pergamon.

MORELAND P. E., JR., ROKOP D. J., and STEVENS C. M. (1970) Observations of uranium and plutonium hydrides formed by ion-molecule reactions. *Intl. J. Mass Spectrometry and Ion Physics* **5**, 127–136.

ROSHOLT J. N., and TATSUMOTO M. (1970) Isotopic composition of uranium and thorium in Apollo 11 samples. *Proc. Apollo 11 Lunar Sci. Conf., Geochim. Cosmochim Acta* Suppl. 1, Vol. 2, pp. 1499–1502. Pergamon.

ROSHOLT J. N., and TATSUMOTO M. (1971) Isotopic composition of uranium and thorium in Apollo 12 samples. Second Lunar Science Conference (unpublished proceedings).

Proceedings of the Second Lunar Science Conference, Volume 2, pp. 1577–1584.
The M.I.T. Press, 1971.

Isotopic composition of thorium and uranium in Apollo 12 samples*

J. N. ROSHOLT and M. TATSUMOTO
U.S. Geological Survey, Denver, Colorado 80225

(*Received 22 February 1971; accepted in revised form 30 March 1971*)

Abstract—The isotopic composition of uranium and thorium was determined by mass spectrometry and alpha spectrometry in 14 lunar samples from Apollo 12. The U^{238}/U^{235} ratio in samples measured is the same as that for terrestrial uranium within experimental error (137.8 ± 0.3). The U^{234} daughter is in radioactive equilibrium with parent U^{238} in the samples; however, it could not be demonstrated that Th^{230} is in equilibrium with U^{238} in some rock samples as measured by alpha spectrometry.

One interpretation of the α-particle spectral data of thorium would suggest the existence of a significant amount of a shape isomer of Th^{232} (Th^{232m}) accompanied by a decay series collateral to the regular decay series for Th^{232}. Different modes are postulated for the production of Th^{232m} as a fraction of primordial thorium and for the production of radiogenic Th^{232} by decay of Pu^{244} progenitor.

THE ISOTOPIC composition of uranium and thorium was determined in 14 samples from Apollo 12. The isotopic ratios of uranium (U^{238}, U^{235}, U^{234}) were identical, within experimental error, with those of terrestrial reference samples in radioactive equilibrium. However, unexpected variations in the isotopic ratios of thorium (Th^{232}, Th^{230}, Th^{228}) may provide clues to the variable cosmogenic history of thorium. The description of chemical separations and the measurements for uranium and thorium were given in ROSHOLT and TATSUMOTO (1970) and TATSUMOTO (1970a). After the separation of lead for the determination of the isotopic abundances of lead isotopes (TATSUMOTO, 1970b), uranium and thorium were separated from the lunar samples. Both mass spectrometry and alpha spectrometry were used in determination of the isotopic composition of uranium. Radioactivity ratios of U^{234} to U^{238} were determined first; then uranium was removed from the platinum counting disk and U^{238}/U^{235} ratios were determined by mass spectrometry on 0.4-microgram quantities of uranium. The results (Table 1) indicate that U^{234} is in radioactive equilibrium with U^{238}, and the U^{238}/U^{235} ratio in samples measured is the same as terrestrial uranium (137.8 ± 0.3). Uncertainties in radioactivity ratios are standard deviations based on counting statistics.

The radioactivity ratios of Th^{232}/Th^{230} in lunar samples (Table 2) were determined by α-particle spectrometry on aliquots of the same solution of the rock sample used in determining the concentrations of uranium and thorium (TATSUMOTO, 1970b); however, separate splits of the fines were used for determination of the concentrations of uranium and thorium. Th^{228} data is not included in this paper because, to evaluate properly its relation to Th^{232}/Th^{230} variations, more time is required to measure Th^{228} decay and Ra^{228} growth from Th^{232}. Th^{232} has been used as a natural tracer for the

* Publication authorized by the Director, U.S. Geological Survey.

Table 1. Concentrations and isotopic composition of uranium.

Sample	Rock type	U (ppm)	U^{234}/U^{238} (activity ratio)	U^{238}/U^{235} (atom ratio)
12013,10,09	Breccia	5.67	1.01	137.7
12013,10,42	Breccia	10.80	0.99	138.0
12013,10,45	Breccia	5.75	1.02	137.6
12034,16	Breccia	3.58	1.00	137.7
12033,53	Fines	2.67	0.99	137.7
12070,56	Fines (contingency)	1.64	1.01	137.8
12009,22	Crystalline rock	0.243	1.00	138.0
12021,122	Crystalline rock	0.261	1.00	137.8
12022,37	Crystalline rock	0.198	0.99	137.7
12035,10	Crystalline rock	0.199	0.99	137.6
12038,42	Crystalline rock	0.157	1.01	137.6
12052,66	Crystalline rock	0.365	1.00	137.8
12063,49	Crystalline rock	0.191	0.99	137.7
12064,21	Crystalline rock	0.278	1.01	137.9
Error range (%)			2	0.25

Table 2. Radioactivity ratios of thorium isotopes.

Sample	Th^{232}/U^{238} (atom ratio)	$[Th^{232}/U^{238}]_{atom} \times [\lambda_{232}/\lambda_{238}]$* (expected Th^{232}/Th^{230} activity ratio)	Th^{232}/Th^{230} (measured activity ratio)	Expected ratio / Measured ratio
			Breccias and fines	
12013,10,09	3.78	1.20	1.30	0.92
12013,10,42	3.28	1.04	1.03	1.01
12013,10,45	3.42	1.09	1.09	1.00
12034,16	3.75	1.19	1.23	0.97
12033,53(a)	3.75	1.19	1.20	0.99
12033,53(b)	3.84	1.22	1.18	1.03
12070,56(a)	3.79	1.20	1.18	1.02
12070,56(b)	3.78	1.20	0.99	1.21
			Crystalline rocks	
12009,22	3.74	1.18	0.73	1.63
12021,122	3.69	1.17	0.69	1.69
12022,37	3.70	1.17	0.95	1.24
12035,10	3.54	1.12	0.95	1.19
12038,42(a)	3.92	1.24	1.03	1.24
12038,42(b)	3.97	1.26	1.05	1.21
12062,66	3.63	1.15	1.02	1.13
12063,49(a)	3.67	1.17	0.73	1.60
12063,49(b)	3.45	1.09	0.61	1.80
12064,21	3.64	1.16	1.02	1.13
Error range (%)		2	2–3	

* $\lambda_{232} = 4.88 \times 10^{-11}\ yr^{-1}$; $\lambda_{238} = 1.537 \times 10^{-10}\ yr^{-1}$.
(a) and (b) indicate different portions of sample.

determination of the state of radioactive equilibrium between Th^{230} and parent U^{238} in terrestrial rocks (Rosholt *et al.*, 1967). The 4.0 MeV α-particle emitted from Th^{232} was measured to determine its radioactivity. A typical α-particle spectra of lunar crystalline rock is shown in Fig. 1. If Th^{230} is in radioactive equilibrium with U^{238}, the activity ratio expected for thorium isotopes can be calculated, using the atomic ratio of Th^{232}/U^{238} as determined by isotope dilution (Tatsumoto, 1970b) and using

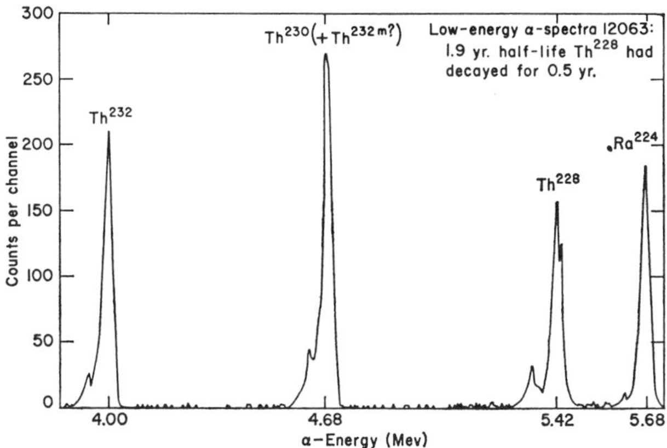

Fig. 1. Low-energy α-particle spectra of sample 12063.

the decay constants for Th^{232} and U^{238}. The expected activity ratio is

$$(Th^{232}/Th^{230})_{expected} = (Th^{232}/U^{238})_{atom} \times (\lambda_{232}/\lambda_{238}),$$

where λ_{232} and λ_{238} are the decay constants for Th^{232} and U^{238}. Values for the measured activity ratios of thorium isotopes and the expected activity ratios of Th^{232}/Th^{230} are shown in Table 2.

Comparison of the expected ratio/measured ratio (Table 2) indicates that fragments (12013,10,42 and 12013,10,45), breccia (12034), and fines (12033) are similar in thorium isotopic composition to terrestrial reference samples; however, the sample of the dark portion of rock 12013 (12013,10,09) has a low value for this comparison. A split of contingency fines (12070) has an excess of radioactivity over that expected at the α-particle energy of Th^{230} (4.66 ± 0.06 MeV peak). All the crystalline rocks analyzed have an excess of radioactivity at this α-particle energy. Unusually large excesses are indicated for three of the samples (12009, 12021, 12063) where the expected ratio/measured ratio (Table 2) exceeds unity by 60 to 80%. A distribution diagram of the comparison of the expected ratio/measured ratio (Fig. 2) shows the variations of thorium isotopic composition in different rock types. The pattern of variations suggests agreement with the KREEP hypothesis of HUBBARD *et al.* (1971) because the dark portion of rock 12013 has a thorium isotopic composition at one extreme, the basalts have a composition at the other extreme, and the fines have an intermediate composition. We do not interpret the variations in the Th^{232}/Th^{230} ratios as due to disequilibrium in the U^{238}–U^{234}–Th^{230} decay sequence because of (a) the equilibrium conditions demonstrated between U^{234} and U^{238} in all samples measured (Table 1), (b) the Th^{230} equilibrium demonstrated in samples 12013, 12033, 12034, and in some of the Apollo 11 samples (ROSHOLT and TATSUMOTO, 1970), and (c) the concordancy between Pb^{206}/U^{238} ages and Pb^{208}/Th^{232} ages in the two-stage lead evolution model by TATSUMOTO (1970a, b).

Continuing measurements of the α-particle spectra of thorium separates of samples from both Apollo 11 and Apollo 12 indicate that samples with deviations in the Th^{232}/Th^{230} ratios also show deviations in the radioactivity ratios of Th^{228} and its

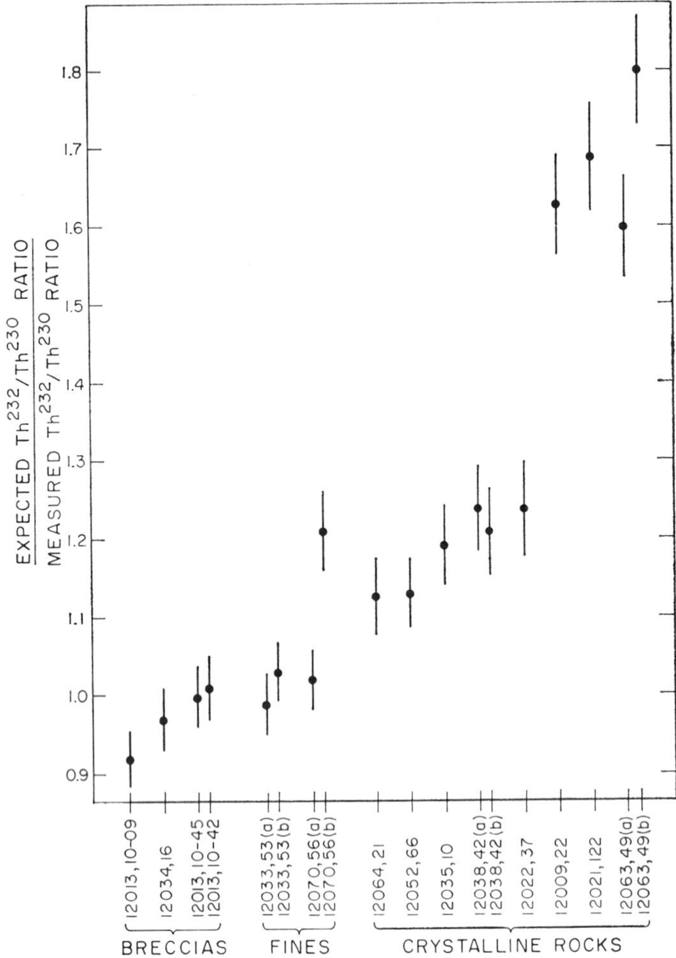

Fig. 2. Comparison of ratios of expected Th^{232}/Th^{230} to measured Th^{232}/Th^{230} showing variations of thorium isotopic composition in different rock types. Vertical bars represent uncertainties in the expected ratio/measured ratio values calculated from standard deviations based on counting statistics and a 2% uncertainty in the Th^{232}/U^{238} ratio.

immediate daughter product Ra^{224}. Ra^{224} has a half-life of 3.64 days, and the radioactivity ratio of Ra^{224}/Th^{228} should be unity after approximately 30 days following the electrodeposition of thorium. An example is illustrated in Fig. 1 for sample 12063, measured 6 months after thorium was electrodeposited on the platinum counting disk. The Ra^{224}/Th^{228} α-radioactivity ratio in this sample at this time of measurement is 1.08 ± 0.03; Th^{228}, with a 1.9 year halflife, has decayed to 83% of its original activity. The amount of recoil atom loss in production of Ra^{224} from electrodeposited thorium has not been evaluated completely; thus, a quantitative comparison of Ra^{224} excess to 4.6 MeV α-particle excess cannot be made at this time.

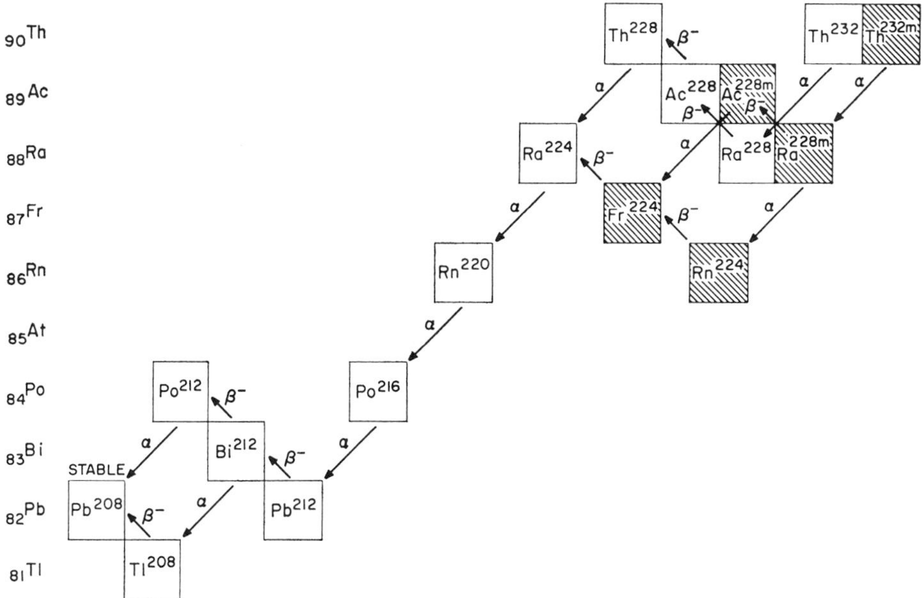

Fig. 3. Block diagram of suggested decay series (cross-hatched) of Th^{232m} collateral to regular decay series of Th^{232}.

A possible interpretation of the α-spectral data, in lunar samples with excess 4.6 MeV α-activity, is that some of the Ra^{224} is produced by a mode of radioactive decay other than direct α-particle emission of Th^{228}. A collateral radioactive series, with the long-lived parent being an isomer of Th^{232} (Th^{232m}), could account for the α-particle data observed. A block diagram of this suggested collateral series (indicated cross-hatched blocks) and its relation to the Th^{232} decay series is shown in Fig. 3. Such a collateral series would require an isomer of Ra^{228} (Ra^{228m} as daughter of Th^{232m}) and possibly an isomer of Ac^{228}, both of which have significant components of α-particle emission in their decay modes to bypass the Th^{228} branch of the series. The possibility for rare α-particle emission in Ra^{228} has been considered from the time of classical investigations of natural radioactivity; however, FEATHER *et al.* (1957) have presented evidence for only very slight amounts of α-emission in Ra^{228} with an upper limit of the α/β branching ratio of about 1.5×10^{-8}. The radioactive properties of the parent isomer, Th^{232m}, as indicated by the lead data in lunar samples, suggest that its halflife would be similar to that of Th^{232} ($\sim 1.4 \times 10^{10}$ years) and the energy of its α-particle emission, as suggested by α-spectrometry, would be about 4.7 MeV. Positive identification of its presence is apparently masked by the 4.68 MeV α-particle emitted by Th^{230}. Our best spectral resolution, using 300 mm^2 silicon surface barrier detectors, has been 30 keV, FWHM, and there is only slight evidence of the existence of an α-particle with an energy of 5–10 keV greater than that emitted by Th^{230}. The suggested decay scheme and α-particle energy for Th^{232m} in relation to the decay properties of Th^{232}, Th^{230}, and Th^{228} are shown in Fig. 4.

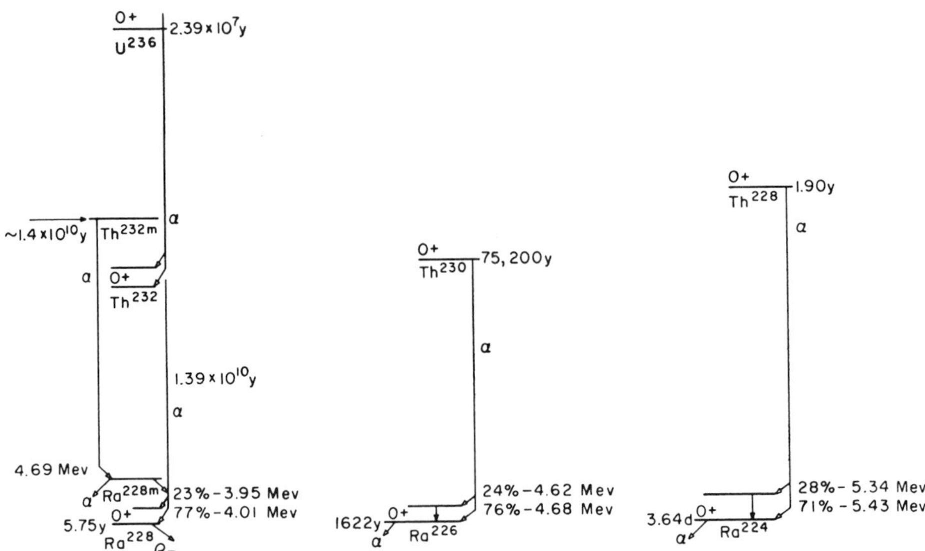

Fig. 4. Suggested decay scheme and α-particle energy of Th^{232m} and its relations to Th^{232}, Th^{230}, and Th^{228}.

Table 3. Concentrations and alpha radioactivity ratios of thorium isotopes.

Sample	Th (ppm)	4.6 MeV alpha counts of Th / 4.0 MeV alpha counts of Th (expected)*	(measured)	$\frac{Th^{232m}}{Th^{232} + Th^{232m}}$ (calculated activity ratio)
			Breccias and fines	
12013,10,09	20.73	0.834	0.772	-0.035 ± 0.030
12013,10,42	34.29	0.960	0.968	0.004 ± 0.030
12013,10,45	19.05	0.920	0.916	-0.002 ± 0.035
12034,16	13.00	0.838	0.814	-0.010 ± 0.030
12033,53(a)	9.70	0.839	0.835	-0.003 ± 0.030
12033,53(b)	12.14	0.821	0.847	0.014 ± 0.035
12070,56(a)	6.02	0.831	0.847	0.009 ± 0.035
12070,56(b)	7.70	0.833	1.01	0.090 ± 0.030
			Crystalline rocks	
12009,22	0.881	0.842	1.37	0.22 ± 0.03
12021,122	0.932	0.853	1.44	0.24 ± 0.03
12022,37	0.710	0.851	1.06	0.10 ± 0.05
12035,10	0.682	0.890	1.06	0.08 ± 0.04
12038,42(a)	0.615	0.778	0.969	0.10 ± 0.04
12038,42(b)	0.816	0.794	0.964	0.09 ± 0.04
12052,66	1.28	0.868	0.981	0.06 ± 0.03
12063,49(a)	0.679	0.858	1.37	0.22 ± 0.02
12063,49(b)	0.637	0.913	1.65	0.28 ± 0.03
12064,21	0.977	0.865	0.980	0.06 ± 0.04

* $\lambda_{232} = 4.88 \times 10^{-11}\ yr^{-1}$; $\lambda_{238} = 1.537 \times 10^{-10}\ yr^{-1}$.
(a) and (b) indicate different portions of sample.

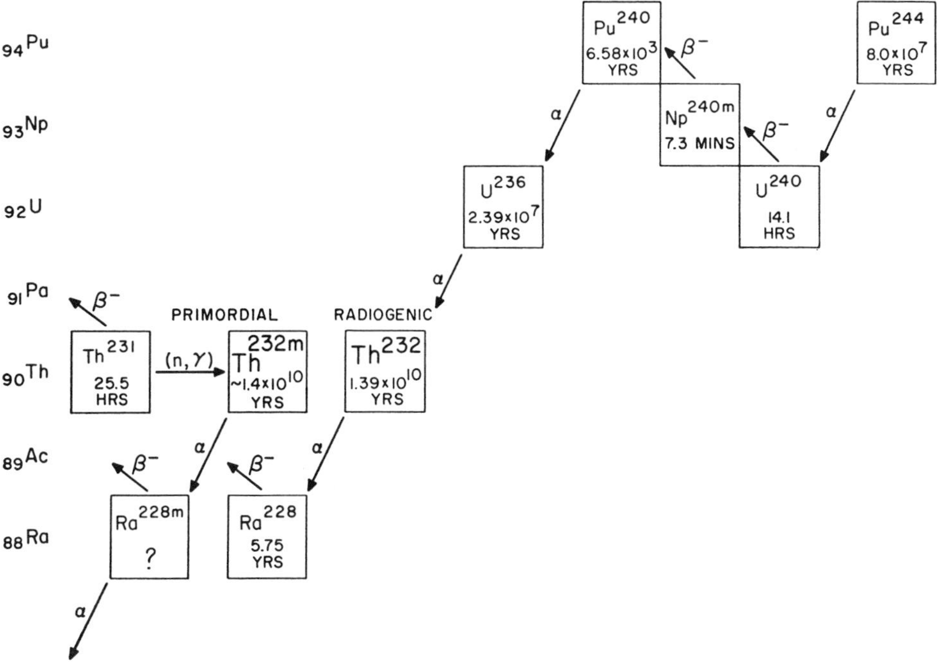

Fig. 5. Block diagram of different modes suggested for the production of primordial Th^{232m} component of thorium and radiogenic Th^{232}.

If such an isomer exists in lunar samples and accounts for the excess α-radioactivity in the 4.6 MeV spectra of thorium, then a calculation of its activity abundance in individual samples can be made using the following equation:

$$\frac{Th^{232m}}{Th^{232} + Th^{232m}} = \frac{R_{measured} - R_{expected}}{R_{measured} + 1},$$

where R = (4.6 MeV α-counts of Th)/(4.0 MeV α-counts of Th). The concentrations of thorium, the expected and measured ratios, R, and the calculated activity ratios of $Th^{232m}/(Th^{232} + Th^{232m})$ are given in Table 3. Values of about 0.25 for this ratio are indicated for the most anomalous samples of crystalline rock. Significant variations exist among the 8 crystalline rocks analyzed, suggesting fractionation between Th^{232} and Th^{232m}. Preliminary measurements (ROSHOLT, unpublished data) indicate that such an isomer may exist in some terrestrial rocks in the amount of about 3% of the terrestrial thorium. It should be noted that no long-lived isomers of even proton-even neutron nuclei of heavy elements have been reported in the literature.

Primordial thorium, in r-process nucleosynthesis, would have been produced by beta decay of neutron-rich nuclides and through neutron capture by Th^{231}. If a significant fraction of this thorium was produced as the isomer Th^{232m}, a variable cosmogenic history for Th^{232} and Th^{232m} may exist (ROSHOLT and TATSUMOTO, 1970).

Wasserburg *et al.* (1969) have given evidence for "last-minute" synthesis of heavy elements at about 4.8 billion years ago in their model for nuclear chronology in the galaxy. Thus, radiogenic Th^{232} would have been produced up to 400 million years after the "last-minute" synthesis and after the formation of the solar system, by decay of Pu^{244} progenitor. Different modes for the production of primordial Th^{232m} and radiogenic Th^{232} are suggested in the block diagram shown in Fig. 5.

Acknowledgments—We thank R. J. Knight and D. M. Unruh for laboratory assistance. This study was carried out under NASA Contract T-75445.

References

Feather N., Miller N., and Peat S. W. (1957) Search for a rare α-emission in ^{228}Ra (MsTh 1). *Proc. Phys. Soc.* (*London*) **70A**, 478–480.

Hubbard N. J., Meyer C., Jr., Gast P. W., and Wiesmann H. (1971) The composition and derivation of Apollo 12 soils. *Earth Planet. Sci. Lett.* **10**, 341–350.

Rosholt J. N., Jr., Peterman Z. E., and Bartel A. J. (1967) Reference sample for determining the isotopic composition of thorium in crustal rocks. *U.S. Geol. Surv. Prof. Paper* **575-B**, B133–B136.

Rosholt J. N., and Tatsumoto M. (1970) Isotopic composition of uranium and thorium in Apollo 11 samples. *Proc. Apollo 11 Lunar Sci. Conf., Geochim. Cosmochim. Acta* Suppl. 1, Vol. 2, pp. 1499–1502. Pergamon.

Tatsumoto M. (1970a) Age of the moon: An isotopic study of U–Th–Pb systematics of Apollo 11 lunar samples, II. *Proc. Apollo 11 Lunar Sci. Conf., Geochim. Cosmochim. Acta* Suppl. 1, Vol. 2, pp. 1595–1612. Pergamon.

Tatsumoto M. (1970b) U–Th–Pb age of Apollo 12 rock 12013. *Earth Planet. Sci. Lett.* **9**, 193–200.

Wasserburg G. J., Schramm D. N., and Huneke J. C. (1969) Nuclear chronologies for the galaxy. *Astrophys. J.* **157**, L91–L96.

Proceedings of the Second Lunar Science Conference, Volume 2, pp. 1585–1589.
The M.I.T. Press, 1971.

A search for superheavy elements in lunar material

J. J. WESOLOWSKI, W. JOHN, and D. NEASE*
Lawrence Radiation Laboratory, University of California, Livermore, California 94550

(*Received 12 February 1971; accepted in revised form 5 April 1971*)

Abstract—A search has been made in lunar material for the superheavy elements whose existence in nature was recently suggested by theoretical calculations. The technique involved exposing the sample to thermal neutrons and measuring the kinetic energies of the resulting fission fragments. The upper limits on the abundance of superheavy elements was determined to be 2×10^{-12} parts by weight for elements with atomic number 110 and 6×10^{-13} for elements with greater atomic number.

INTRODUCTION

CALCULATIONS by theorists from various countries, using a combination of the liquid drop model and Nilsson orbitals have predicted the existence of relatively stable nuclei much heavier than any known nuclei (TSANG and NILSSON, 1970). Some of the estimated half lives are long enough to raise the possibility that some isotopes of these superheavy elements exist in nature. For example, the isotope $Z = 110$, $A = 294$ is estimated to be beta-stable and to have a half life of 10^{10} y for spontaneous fission and 10^8 y for alpha decay (TSANG and NILSSON, 1970). Hartree–Fock calculations are consistent with the general prediction of a relatively stable superheavy region, but preliminary results indicate that isotopes around $Z = 120$ would be the most stable (BASSICHIS and KERMAN, 1970).

Predictions of the chemical properties of the superheavy elements are rather uncertain. Hartree–Dirac–Slater calculations show rather small energy differences between alternative electronic configurations (WABER *et al.*, 1969; TUCKER *et al.*, 1968). Additional uncertainties are associated with the possible oxidation states. A simple extrapolation of the periodic table implies that element 110 is a homologue of Pt, 114 a homologue of Pb, and 120 a homologue of Ra. This extrapolation, although uncertain, is useful as a starting place in the search.

There are a number of scientific groups, particularly in the Soviet Union and the United States, actively engaged in the search for superheavy elements in nature (SEABORG, 1968). It is useful to employ various methods in the search for a new element since each method involves different assumptions about the nuclear and chemical properties. Our searches have been based on the assumption that some isotopes of the superheavy elements would be fissionable with thermal neutrons. The technique used is to expose a thin sample of material to a high flux of thermal neutrons and to measure the kinetic energies of the resulting coincident fission fragments. This method is highly sensitive and is capable of distinguishing the superheavy elements from

* Present address, Department of Physics, Cornell University, Ithaca, New York.

lighter fissionable elements. In fission, the kinetic energies of the fragments are produced by the Coulomb repulsion between the nuclear charges. A plot of the average total kinetic energy, E_k, of the fragments versus $Z^2/A^{1/3}$ for known isotopes is fitted well by a straight line (HYDE, 1964). By extrapolation, we expect E_k for $^{294}110$ to be 215 MeV compared to 172 MeV for ^{235}U. (A correction must be made because the fragments are detected after neutron emission. The postneutron energies are 208 MeV for $^{294}110$ and 170 MeV for ^{235}U.) The heavier superheavy elements will have correspondingly larger values for E_k (see Fig. 2). Recent calculations by Nix are in agreement with these extrapolations (NIX, 1969).

Previously we have searched for superheavy elements in Pt, Pb, and Au ores using this technique, assuming the possible presence of superheavy homologues (WESOLOWSKI *et al.*, 1969). However, the uncertainties associated with the chemical properties of the superheavy elements are magnified by the environmental changes over the surface of the earth which have concentrated the elements in certain minerals in various locations. In this respect a search in lunar material may offer an advantage owing to the less violent changes of the surface of the moon. It is also conceivable that the cosmological history of the moon is different from that of the earth. Furthermore, the surface of the moon is an excellent target for heavy elements which may be present in the cosmic wind.

PREPARATION OF THE LUNAR SAMPLE

The lunar material (sample 12070,96) was obtained from the contingency sample collected by the astronauts of Apollo 12. It was composed of fine grains which had sufficient cohesion to make pouring from the containers difficult. Because of the short range of fission fragments it was necessary to make a thin target of the lunar material. A sputtering technique was used in an attempt to minimize differential concentration of the elements. About 1.3 g of the lunar material was first compressed into a 1-inch diameter wafer under 114,500 psi pressure for six minutes, using water as a binder. After being dried under a heat lamp the wafer was glued with epoxy to a copper disc.

The sample was then placed in the sputtering chamber facing a nickel substrate which had approximately 25 ug/cm^2 of aluminum previously evaporated onto it. Approximately 3000 Å of NaCl separated the aluminum from the nickel. Lunar material of 87 ug/cm^2 thickness was sputtered onto the aluminum-nickel substrate. The substrate was then removed from the sputtering chamber and placed in a vacuum evaporation chamber where approximately 25 ug/cm^2 of aluminum was evaporated onto it. The foil was sprayed with acrylic for reinforcement while mounting. The Al—lunar material—Al sandwich foil was placed on an inclined plane and carefully slid into distilled water. The NaCl dissolved, releasing the foil from the Ni substrate. The floating foil was picked up from the water with an aluminum mounting ring. A few small spots of adhesive were applied at the edge of the foil. The acrylic was then washed off the foil with acetone. Prior attempts to prepare lunar foils without the aluminum layers were unsuccessful.

EXPERIMENTAL PROCEDURE

A high thermal neutron flux was needed for this experiment in order to obtain reasonable sensitivity. Suitable neutron fluxes exist inside the thermal columns of reactors. Unfortunately, they are accompanied by a flux of fast neutrons and gamma rays. The response of a conventional heavy ion detection system in such a hostile environment is poor. We have developed a system using cooled phosphorus-diffused Si detectors and fast electronics which maintains good resolution for several weeks inside the thermal column of a reactor in a slow neutron flux of 2×10^{11} n/cm^2·sec (cadmium ratio = 600).

Two phosphorus-diffused silicon detectors of 300 ohm-cm resistivity with a surface area of 300 mm^2 were placed on opposite sides of a target wheel which could be rotated to any one of four positions. The four wheel sectors contained two lunar samples, one aluminum blank for background measurements, and a calibration source consisting of 1.7×10^{-9} g of ^{235}U on a thin Au foil. A Au cover foil was used to insure that no uranium transferred to the detectors. The faces of the detectors were only about 2 mm from the target foils in order to achieve maximum solid angle. The assembly was mounted inside a small evacuated aluminum chamber which was then placed inside the thermal column of the reactor. A 2 meter-long evacuated tube carried the target shaft, detector leads, cooling lines, and thermocouple leads outside the thermal column. The detectors were cooled to about -23°C by circulating refrigerated alcohol.

A detailed description of the electronics system can be found in an earlier paper (Wesolowski *et al.*, 1970). A multi-parameter analyzer was used to record the energies measured by the two detectors for each coincident event. Each event was also tagged with clock time. Because of the importance attached to observing even one high energy event an additional precaution was taken. The authenticity of any interesting coincident fragment event could be checked by examining a picture of the fast analog signals on 35 mm film.

Results

Figure 1 is a typical single parameter lunar sample spectrum from one of the fission detectors. It represents 44 hours of data collection taken after the system had been in the thermal column of the reactor for about nine days. The counts are attributed to the fission of uranium present in the target material since the energy distribution is essentially the same as that from the ^{235}U calibration source. About half the uranium is from the lunar material itself and about half from the aluminum backing material. The resolution is reasonable considering the large solid angle subtended by the detectors. The amount of uranium present in the lunar material was estimated to be

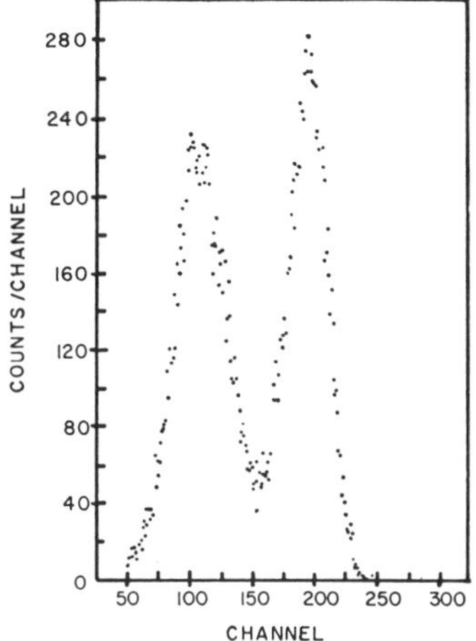

Fig. 1. Energy spectrum of fission fragments from ^{235}U in lunar sample.

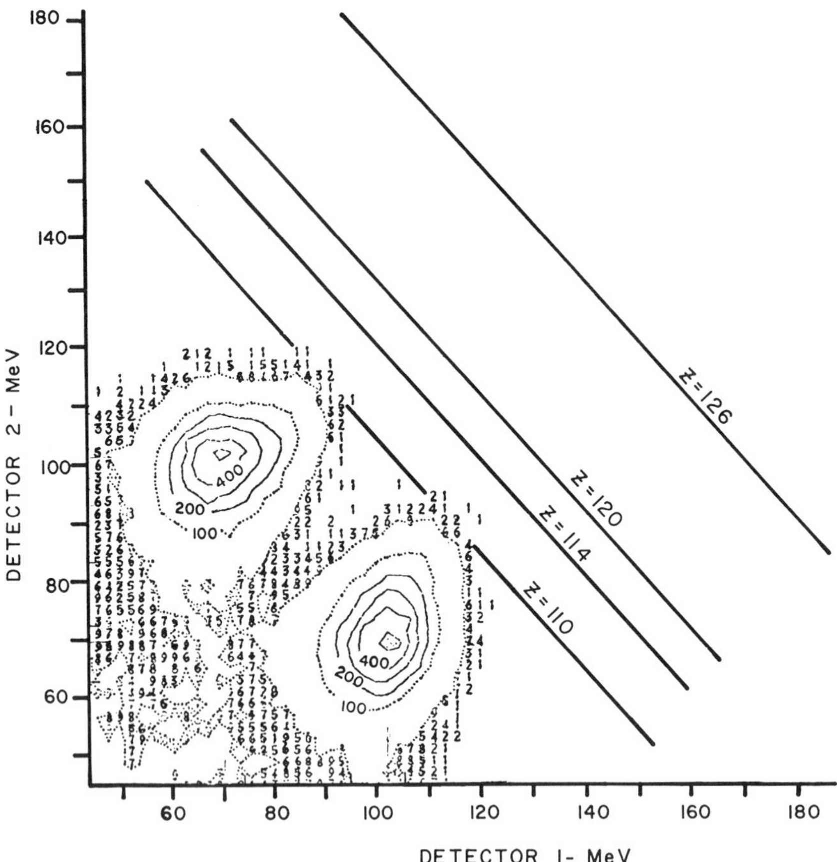

Fig. 2. Energy data array of coincident fission fragments from lunar sample. The contours and the numbers outside the contours give the number of events per 2 MeV squared. The observed counts are attributed to an impurity of 10^{-12} g ^{235}U in the target. The diagonal lines are located at the average energies expected for the fission of superheavy elements.

about 0.2 ppm by weight. This must be considered a lower limit for the concentration in the original sample since some preferential fractionation could have occurred during the sputtering process. Gamma-ray analysis of sample 12070 at the Radiation Counting Laboratory in Houston yielded a value of 1.5 ppm (LSPET, 1970). This difference need not be attributed entirely to preferential fractionation since the total weight of the target material used in the fission experiment was very small (0.2 mg) compared to the total sample size and there could be significant concentration variations.

The two-parameter energy array for the coincident fission fragments for the entire 141 hour run on the lunar material is shown in Fig. 2. The contours encircle fission peaks attributed to the uranium in the target. Numbers <10 are shown directly. Also shown are the predicted total kinetic energy lines corresponding to superheavy elements 110, 114, 120, and 126. The lines have been corrected for energy loss through the target, pulse-height defect in the detectors, and loss of kinetic energy due to neutron emission. It is estimated that $\bar{\nu}$, the average number of neutrons emitted per fission, for superheavy elements is 10–11 (NIX, 1969). Using the line shown in Fig. 2 we conclude that the upper limit for Z = 110 in lunar sample 12070 is 2×10^{-12} parts by weight, assuming the same thermal neutron fission cross section as for ^{235}U. It may be remarked that the fission of Z = 110 would be expected to be more nearly symmetric than for ^{235}U (HYDE, 1964). Taking this into account would improve the sensitivity somewhat.

Figure 2 shows that there are no counts in the vicinity of the Z = 114, 120, and 126 kinetic energy lines. If one makes the somewhat arbitrary assumption that at least ten events would have to be observed to demonstrate the fission of a superheavy element, the upper limit of detection for Z = 114, 120, and 126 would be about 6×10^{-13} parts by weight. From the present work there is therefore no evidence for the presence of superheavy elements in lunar material.

Acknowledgments—We wish to thank W. Brunner and R. L. Smith for their assistance in making the lunar target. We were greatly assisted by the electronics and mechanical support groups of the Livermore reactor. We thank Dr. Eugene Goldberg for his support of this experimental program. We also wish to thank the staff of NASA for expediting the handling of the lunar samples. Work performed under the auspices of the U.S. Atomic Energy Commission.

REFERENCES

BASSICHIS W. H., and KERMAN A. K. (1970) Self-consistent calculations of shell effects including the proposed island of stability. *Phys. Rev.* **C2**, 1769–1776.

HYDE E. K. (1964) *The Nuclear Properties of the Heavy Elements*. Prentice-Hall.

LSPET (Lunar Sample Preliminary Examination Team) (1970) Preliminary examination of lunar samples from Apollo 12. *Science* **167**, 1325–1339.

NIX J. R. (1969) Predicted properties of the fission of super-heavy nuclei. *Phys. Lett*, **30B**, 1–4.

SEABORG G. (1968) *Ann. Rev. Nucl. Sci.* **18**, 53–152.

TSANG C. F., and NILSSON S. G. (1970) Further theoretical results on the stability of superheavy nuclei, and earlier work referenced herein. *Nucl. Phys.* **A140**, 289–304.

TUCKER T. C., ROBERTS L. D., NESTOR C. W., JR., CARLSON T. A., and MALIK F. B. (1968) Calculation of the electron binding energies and X-ray energies for the superheavy elements 114, 126, and 140 using relativistic self-consistent field atomic wave functions. *Phys. Rev.* **174**, 118–124.

WABER J. T., CROMER D. T., and LIBERMAN D. (1969) SCF Dirac-Slater calculations of the translawrencian elements. *J. Chem. Phys.* **51**, 664–681.

WESOLOWSKI J. J., JOHN W., and HELD J. (1970) The use of silicon heavy ion detectors in high radiation fields. *Nucl. Instr. and Methods* **83**, 208–212.

WESOLOWSKI J. J., JOHN W., JEWELL R., and GUY F. (1969) A search for element 110 in platinum ore via neutron-induced fission. *Phys. Lett.* **28B**, 544–545.

Proceedings of the Second Lunar Science Conference, Volume 2, pp. 1591–1605.
The M.I.T. Press, 1971.

Kr^{81}–Kr and K–Ar^{40} ages, cosmic-ray spallation products, and neutron effects in lunar samples from Oceanus Procellarum

K. MARTI and G. W. LUGMAIR
Chemistry Department, University of California at San Diego, La Jolla, California 92037

(*Received 20 February 1971; accepted in revised form 30 March 1971*)

Abstract—Cosmic-ray exposure ages of six lunar rocks from Oceanus Procellarum ranging from 94 to 303 million years are reported. Depth-dependent rare-gas concentration profiles were obtained in rock 12002. Solar particle effects are apparent in the radioisotope Kr^{81}, but only very small effects are found in the integrating stable isotopes. The K–Ar^{40} ages obtained from our Apollo 12 bulk rock samples range from 1.34 to 2.98 billion years. Integrated thermal neutron fluxes in Apollo 12 materials as obtained from the gadolinium isotopic anomalies range from 4.6 to 10.5 × 10^{15} n/cm^2. The correlation of the very high relative yield of Xe^{131} with spallation and neutron capture effects is discussed. Xe^{131}/Xe^{126} ratios and Kr^{78}/Kr^{83} ratios which reflect a varying degree of shielding are found to be correlated. A model calculation of average effective burial depth of the rocks is presented. First results on rare-gas concentration profiles in the double core samples are reported.

INTRODUCTION

INVESTIGATIONS OF Apollo 11 samples revealed that noble gases in lunar materials have different origins. This is again found in our Apollo 12 samples. The soil, to a depth of at least 40 cm, contains large amounts of trapped solar-wind gases which almost completely mask other components. In the crystalline rocks, on the other hand, cosmic ray-produced spallation and radiogenic gases are predominant. In some rocks, there is no evidence for trapped gases, but fission xenon from U is a minor component. We have studied these gases in detail and derived also concentration profiles and mass yield spectra. Cosmic-ray ages were obtained by using the Kr^{81}–Kr dating method. Ar^{40} results are combined with K data to give bulk rock K–Ar ages which may then be compared to ages obtained by other methods. We have determined accurate integrated thermal neutron fluxes in Apollo 12 materials making use of the anomalies in the gadolinium isotopic composition which have already proved useful in determining the irradiation history of Apollo 11 samples (ALBEE *et al.*, 1970; EUGSTER *et al.*, 1970; MARTI *et al.*, 1970). The neutron slowing down densities, combined with spallation data, make it possible to estimate the average burial depth of a sample below the lunar surface.

EXPERIMENTAL TECHNIQUES

Measurements of rare gases, of gadolinium, and of potassium were made on split fractions of about 1-gram rock and 50-mg core samples. Rock samples (except aliquots from rock 12002, Arnold slice) were first cleaned ultrasonically to remove adhering dust particles. Rare gases were extracted at about 1700°C after previous removal of adsorbed atmospheric gases (at 80°C for several days). Experimental techniques have already been reported (MARTI *et al.*, 1969, MARTI *et al.*, 1970). Vacuum baking of our double walled quartz extraction systems at ~800°C lowered the ^{84}Kr and ^{132}Xe blanks to $\sim 0.2 \times 10^{-12}$ and $< 0.1 \times 10^{-12}$ cc STP, respectively. Potassium analyses were performed using an atomic absorption spectrometer. Samples of 20 to 60 mg were dissolved in teflon beakers using HF and HCl and were diluted to volume. All solutions were in a matrix of 1:1 dilution of constant boiling HCl. Ultrapure KCl was used as the standard. Standards were also used which contained in proper percentages all the major elements of the rocks studied, but no potassium. Techniques used and the accuracy of the results obtained by two different methods will be discussed in detail elsewhere (MARTI and MEYERS, 1971). For the isotopic analyses of Gd ~100 mg and ~30 mg aliquots of rock and core samples respectively were used. Ion exchange techniques were used for the chemical separation of Gd which are similar to those reported by EUGSTER *et al.* (1969). Gd then was loaded as Gd $(NO_3)_3$ on a single Re filament, oxidized, and then analyzed in a programmable mass spectrometer as GdO^+ in the current integration mode with digital data output. The techniques of the chemical separation and mass spectrometric analyses of Gd are discussed in detail by LUGMAIR and MARTI (1971).

RESULTS

Kr^{81}–Kr ages

The time intervals of exposure to cosmic rays of Apollo 12 rocks has been determined by the Kr^{81}–Kr method (MARTI, 1967). This method is based on the assumption that the relative production rates of the krypton isotopes remained constant throughout the irradiation interval. This assumption may not be valid if the irradiation history of a rock is complex, as will be discussed later. Results are given in Table 1. Exposure ages range from 94 to 303 m.y., with pairs of rocks at 95 and 130 m.y. Additional Kr^{81}–Kr ages of four rocks are reported by BOCHSLER *et al.* (1971) covering a range of from 59 to 225 m.y. Our Kr^{81}–Kr ages are systematically higher than those obtained from our He^3 concentrations and a production rate of $P(He^3) = 1 \times 10^{-8}$ cc STP/g per m.y. This may indicate either diffusion losses of He^3 or an overestimation of the He^3 production rate because of shielding. Similar results were reported by HINTENBERGER *et al.* (1971). All analyzed samples were chips from the interior of the rocks. The depth dependent production of spallation Kr has been studied in rock 12002. Aliquots of samples from various depths of the Arnold slice were analyzed, and some results are plotted in Fig. 1. While the He^3 and Ne^{21} concentration profiles, as well as the Ne^{22}/Ne^{21} spallation ratio, indicate only small variations, the Kr^{81}/Kr^{83} ratio decreases sharply in the topmost one cm and parallels the Al^{26} and Na^{22} profiles found in aliquot samples (FINKEL *et al.*, 1971). The Al^{26} and Na^{22} profiles in rock

Table 1. Kr^{81}–Kr^{83} ages of Apollo 12 rocks.

Rock	Kr^{83}/Kr^{81}	Kr^{83}_{sp}/Kr^{81}	P_{81}/P_{83}	Kr^{81}–Kr^{83} age in m.y.	$Kr^{81} \times 10^{-12}$ cc STP/g
12002,A 9/20	535 ± 30	526	0.591	94 ± 6	0.125 ± 0.007
12009,34	790 ± 140	780	0.575	136 ± 24	0.047 ± 0.008
12018,39	1090 ± 80	1090	0.589	195 ± 16	0.073 ± 0.005
12021,61	1660 ± 100	1660	0.603	303 ± 18	0.133 ± 0.008
12052,85	745 ± 40	740	0.575	129 ± 7	0.078 ± 0.004
12063,64	562 ± 30	550	0.567	95 ± 5	0.142 ± 0.008

$P_{81}/P_{83} = 0.95\left(\frac{Kr^{80} + Kr^{82}}{2Kr^{83}}\right)_{sp}$. Errors in the ages do not include uncertainties in the Kr^{81} halflife and in the P_{81}/P_{83} production ratios; Kr^{81} concentrations include relative errors only.

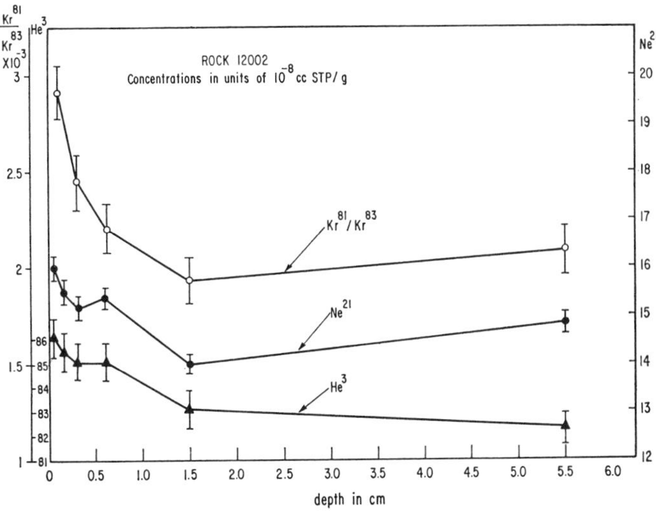

Fig. 1. Concentration profiles in rock 12002. Errors given do not include uncertainties of He and Ne concentrations in the standards.

Table 2. K–Ar ages of Apollo 12 rocks.

Rock	$Ar^{40} \times 10^{-8}$ cc STP/g	K (ppm)	Age*
12002,D	1,010 ± 60	460 ± 40	2.51 ± 0.14
12002,A	990 ± 70	440 ± 30	2.54 ± 0.13
12009,34	890 ± 70	440 ± 25	2.40 ± 0.13
12018,39	300 ± 20	370 ± 25	1.34 ± 0.09
12021,61	1,690 ± 90	550 ± 35	2.98 ± 0.12
12052,85	990 ± 60	500 ± 30	2.37 ± 0.11
12063,64	1,270 ± 80	520 ± 30	2.65 ± 0.12

* $\frac{\lambda_K}{\lambda_\beta} = 0.124$, $\lambda = 5.46 \times 10^{-10}\ y^{-1}$ and $K^{40} = 0.0118\%$ were used. Errors are quadratically added.

12002 result from the superposition of solar flare proton activation onto the galactic cosmic-ray produced effects and are discussed in detail by Finkel *et al.* (1971). The increase of the Kr^{81}/Kr^{83} ratio close to the surface is also ascribed to solar particle effects, and this increased production of Kr^{81} during the last few million years on the top surface of rock 12002 invalidates the Kr^{81}–Kr method. It is necessary, therefore, that rock samples from at least one cm depth be analyzed to obtain reliable Kr^{81}–Kr ages.

Potassium-argon ages

Ages obtained from the K and Ar^{40} data in bulk rock samples of our six igneous rocks are reported in Table 2. With the exception of rock 12018, the ages are similar to those obtained for Apollo 11 rocks. Rock 12018 has retained extremely little radiogenic Ar^{40} which suggests a complex thermal history. Rock 12018 has also the

Table 3. Krypton contents and isotopic com-

Rock	$Kr^{83} \times 10^{-12}$ cc STP/g	Kr^{78}	Kr^{80}	Kr^{81}
12002,A 9/20	67 ± 10	16.53 ± 0.22	48.4 ± 0.4	0.187 ± 0.010
12009,34	37 ± 6	15.24 ± 0.28	46.4 ± 0.6	0.127 ± 0.023
12018,39	79 ± 12	16.44 ± 0.20	48.8 ± 0.4	0.092 ± 0.007
12021,61	220 ± 33	17.47 ± 0.15	50.7 ± 0.4	0.060 ± 0.004
12052,85	58 ± 9	15.23 ± 0.22	47.3 ± 0.5	0.134 ± 0.007
12063,64	80 ± 12	14.66 ± 0.20	45.5 ± 0.4	0.183 ± 0.010

Table 4. Xenon contents and isotopic com-

Rock	$Xe^{131} \times 10^{-12}$ cc STP/g	Xe^{124}	Xe^{126}	Xe^{128}	Xe^{129}
12002,A 9/20	30 ± 4	9.39 ± 0.14	15.95 ± 0.25	25.34 ± 0.25	33.33 ± 0.30
12009,34	6.4 ± 0.8	8.25 ± 0.22	14.00 ± 0.30	22.99 ± 0.45	31.87 ± 0.45
12018,39	28 ± 4	8.72 ± 0.11	15.57 ± 0.18	24.78 ± 0.20	30.42 ± 0.30
12021,61	120 ± 15	11.06 ± 0.12	18.94 ± 0.20	29.43 ± 0.25	36.54 ± 0.25
12052,85	20 ± 3	7.28 ± 0.11	12.73 ± 0.12	20.74 ± 0.20	25.39 ± 0.30
12063,64	38 ± 5	6.82 ± 0.10	12.62 ± 0.20	20.56 ± 0.25	27.31 ± 0.30

lowest K content of our rocks. A short K–Ar age for this rock has already been reported by SCHAEFFER *et al.* (1970). It is interesting to compare the K–Ar ages with the U–$Xe^{136}_{fission}$ ages. Upper limits for these ages are obtained from the Xe^{136} contents by assuming that all Xe^{136} of the rocks as reported in Table 4 is due to U fission. In the case of rock 12018, this upper limit is 2.3 b.y. if a U content of 0.248 ppm (SILVER, 1971) is used. With the exception of rock 12009, upper limits of the U–Xe^{136} ages are compatible with reported Rb–Sr isochron ages and Ar^{39}–Ar^{40} ages of Apollo 12 rocks (PAPANASTASSIOU and WASSERBURG, 1970; TURNER, 1971). Using a U content of 0.243 ppm for rock 12009 (TATSUMOTO *et al.*, 1971), we obtain an upper limit for this U–$Xe^{136}_{fission}$ age of 1.8 b.y. which is smaller than the K–Ar^{40} age given in Table 2. This result will be discussed later in a discussion of spallation target element composition.

Spallation gases

The light noble gas isotopes He^3, He^{20-22}, $Ar^{36,38}$ in our rocks are almost pure spallation gases, similar in isotopic composition to those found in meteorites. These results will be reported and discussed in a separate paper. In Tables 3 and 4, we report the Kr and Xe data obtained from our rock samples. The Kr and Xe isotopic compositions indicate nearly pure cosmic-ray spallation products. However, small fission components due to U, and in at least some rocks, small trapped gas components are also found. Judging from the isotopic composition of all the rare gases, there is no evidence for a trapped component in rock 12021. We have subtracted from the mea-

position in Apollo 12 rocks ($Xr^{83} \equiv 100$).

Kr^{82}	Kr^{83}	Kr^{84}	Kr^{86}	$\left(\frac{Kr^{78}}{Kr^{83}}\right)_{sp}$
75.9 ± 0.4	100	55.6 ± 0.8	3.98 ± 0.12	0.167 + 0.002
74.7 ± 0.6	100	52.7 ± 1.5	3.22 ± 0.30	0.154 ± 0.003
75.3 ± 0.5	100	52.4 ± 0.5	2.01 ± 0.12	0.165 ± 0.002
76.2 ± 0.4	100	50.0 ± 0.5	1.55 ± 0.08	0.175 ± 0.002
73.7 ± 0.5	100	53.6 ± 0.8	3.19 ± 0.15	0.153 ± 0.002
73.8 ± 0.5	100	53.5 ± 0.8	2.00 ± 0.15	0.147 ± 0.002

position in Apollo 12 rocks ($Xe^{131} \equiv 100$).

Xe^{130}	Xe^{131}	Xe^{132}	Xe^{134}	Xe^{136}	$\left(\frac{Xe^{131}}{Xe^{126}}\right)_{sp}$
17.11 ± 0.15	100	21.55 ± 0.30	6.69 ± 0.10	4.99 ± 0.10	6.10 + 0.26 − 0.34
15.92 ± 0.12	100	18.86 ± 0.35	5.06 ± 0.20	3.81 ± 0.20	7.12 + 0.18 − 0.15
17.25 ± 0.15	100	18.22 ± 0.25	2.32 ± 0.10	1.17 ± 0.10	6.42 ± 0.07
19.62 ± 0.10	100	20.67 ± 0.30	2.66 ± 0.08	1.09 ± 0.04	5.28 ± 0.06
14.21 ± 0.15	100	17.00 ± 0.25	5.42 ± 0.12	3.80 ± 0.10	7.83 + 0.10 − 0.07
14.22 ± 0.18	100	17.23 ± 0.20	3.00 ± 0.08	1.85 ± 0.05	7.88 + 0.16 − 0.25

sured mass spectra fission components due to U using a U content of 0.261 ppm (TATSUMOTO *et al.*, 1971) and a U–$Xe^{136}_{fission}$ age of 3.25 b.y. which is an average of Apollo 12 Rb–Sr and Ar^{39}–Ar^{40} ages (PAPANASTASSIOU and WASSERBURG, 1970; TURNER, 1971). This correction amounts to 47% for Xe^{136} and 2.2% for Kr^{86}. WETHERILL's (1953) data are used for spontaneous fission of U. The remaining mass yield spectra allow us to give new upper limits for the spallation yields for Xe^{134} and Xe^{136}. The upper and lower limits for spallation yields of Kr and Xe in rock 12021 are shown in Figs. 2 and 3, respectively. Although the Xe^{136} yield ($Xe^{136}/Xe^{126} = 0.031$) in rock 12021 is considerably lower than that reported for rock 10017 (MARTI *et al.*, 1970), the Kr^{86} yield ($Kr^{86}/Kr^{83} = 0.0152$) is quite similar in these two rocks. There is, therefore, an indication that although the Xe^{136} spallation yield may be very close to zero, the Kr^{86} yield may not be zero but around 1.5% of the Kr^{83} yield. In the calculations of spallation ratios of Kr and Xe in our rocks, we have adopted Kr^{86} and Xe^{136} spallation yields as obtained from rock 12021 and discussed above.

Most striking again are the very high but variable relative yields of Xe^{131} in all our rocks. The highest values of 7.9 for Xe^{131}/Xe^{126} ratios are similar to the highest values of this ratio found in Apollo 11 type A rocks. ALEXANDER *et al.* (1971) report a slightly smaller spallation ratio Xe^{131}/Xe^{126} in rock 12002, which was obtained from a stepwise release of the Xe. The relative yield of Xe^{131}, however, is not the only one to vary in our samples. Varying relative yields are observed also for some other Xe isotopes and also for Kr isotopes, especially Kr^{78}. Because it was one of our aims in the Apollo 12 investigations to try to explain the high Xe^{131} yields, further discussion of this problem will be given in later sections.

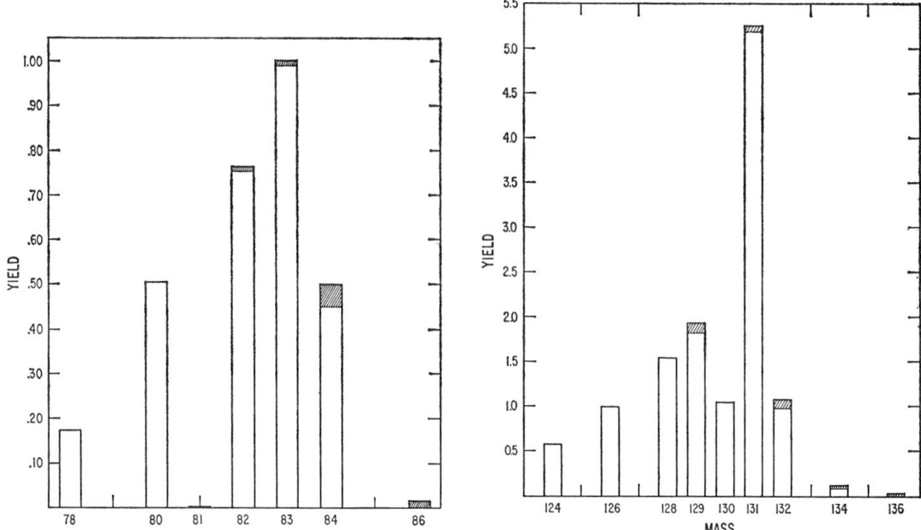

Fig. 2. Mass yields of Kr in rock 12021, normalized to $Kr^{83} = 1.00$. A small amount of fission krypton due to U (see text) has been subtracted. If all Kr^{86} is assumed to belong to the spallation component, then the upper limits of the bars represent the relative Kr spallation mass yields. If all the Kr^{86} is assumed to belong to a trapped component, the relative spallation yields are obtained by subtracting this trapped component (hatched area).

Fig. 3. Mass yields of Xe in rock 12021 normalized to $Xe^{126} = 1.00$. Fission xenon due to U has been subtracted (see text). If all Xe^{136} is assumed to belong to the spallation component, then the upper limits of the bars represent the relative Xe spallation mass yields. If all the Xe^{136} is assumed to belong to a trapped component, the relative spallation yields are obtained by subtracting this trapped component (hatched area).

Rare-gas concentration profiles in the double core 12025–12028

In Fig. 4, the concentrations of He^4, Ne^{20}, Ar^{36}, and Kr^{86}, which represent almost pure solar wind-implanted gases, and of Ar^{40} are plotted versus depth for 5 samples from various depths in the double core. With the exception of the sample from a 3-cm depth, there is a general decrease in gas concentration with depth. If a constant grain size distribution in these samples is assumed, the profile may indicate decreasing mixing rates with depth. The integrated thermal neutron flux (Table 5) decreases slightly from the deepest (40-cm) core sample towards that measured in our topmost 3-cm sample, which tends to support this contention. Similar isotopic anomalies of Gd were found by ALBEE *et al.* (1971) in their samples of the double core. It is interesting to note that there is very little variation in the relative noble-gas elemental abundances. The Ar^{40} content does not correlate with the solar wind-implanted gases, and increases smoothly with depth, although K does not seem to vary much in our samples. If all the Ar^{40} is combined with measured K contents, K–Ar^{40} ages of 3.4 and 4.8 b.y. are calculated for the topmost and deepest samples, respectively. Excess Ar^{40}, therefore, is clearly indicated. The anti-correlation of Ar^{40} contents with rare gases of solar

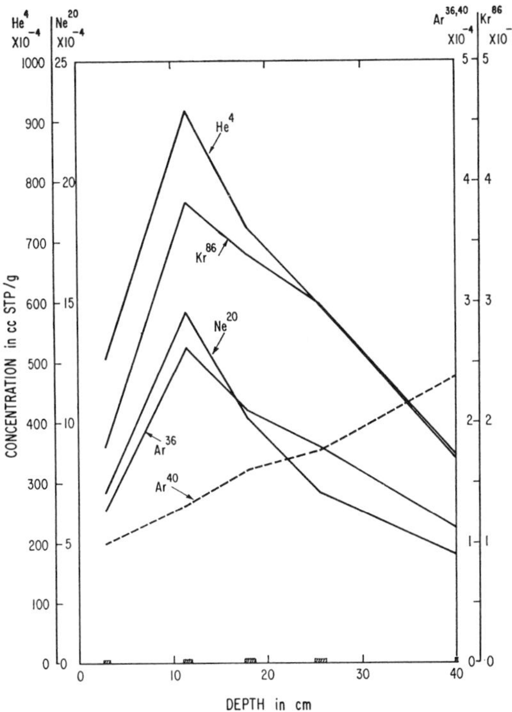

Fig. 4. Rare gas concentration profiles in double core 12025–12028. The bars indicate the depth ranges from which the samples were taken.

origin puts some restrictions on proposed models to account for the origin of excess Ar^{40} (Heymann and Yaniv, 1970; Manka and Michel, 1970).

Thermal neutron effects

Because of the very large capture cross-sections of Gd^{155} and Gd^{157}, gadolinium is a sensitive indicator for thermal neutrons. We have studied the Gd isotopic anomalies due to neutron capture in Apollo 12 samples. In Table 5, we report the measured isotopic ratios of $^{158}GdO/^{157}GdO$ and the resulting integrated neutron fluxes in six rock and three double-core samples. The calculated fluxes cover a range of from 4.6 to 10.5 × 10^{15} n/cm^2. The capture cross-sections used to calculate the theoretical correlation (Fig. 5). are $\sigma_{157} = 2.54 \times 10^5$ barn and $\sigma_{155} = 6.1 \times 10^4$ barn for ^{157}G and ^{155}Gd, respectively, recommended by Goldberg *et al.* (1966). The good correlation of the isotopic ratios $^{156}GdO/^{155}GdO$ and $^{158}GdO/^{157}GdO$ shows that the variation of these two ratios can be explained by the capture of thermal neutrons in Gd of terrestrial composition. Eugster *et al.* (1970) reported the Gd isotopic composition in a large number of Apollo 11 samples and arrived at the same conclusion. It is interesting to note that in their correlation plot, as well as in ours, an effective ratio $\sigma_{155}/\sigma_{157}$ slightly larger than that used for calculating the theoretical correlation line

Table 5. Integrated thermal neutron fluxes in Apollo 12 samples.

Sample	$^{158}GdO/^{157}GdO$	10^{15} n/cm^2*
12002 bottom	1.587 79 ± 9	4.57 ± 0.18
12009,34	1.588 76 ± 10	6.05 ± 0.19
12018,39	1.589 28 ± 26	6.84 ± 0.40
12021,61	1.588 92 ± 9	6.29 ± 0.18
12052,85	1.588 89 ± 10	6.25 ± 0.19
12063,64	1.587 86 ± 8	4.68 ± 0.17
Core		
12025,34 (2.5–3.3 cm)	1.590 25 ± 18	8.32 ± 0.30
12028,200 (25.4–26.1 cm)	1.591 24 ± 20	9.83 ± 0.32
12028,162 (39.8–40.0 cm)	1.591 68 ± 21	10.48 ± 0.34
TERR. GdO	1.584 79 ± 8	≡ 0

* $\sigma^{157}Gd$ used is 2.54×10^5 barn; errors given for the isotopic ratios are 2σ mean; errors in neutron fluxes are obtained by quadratically adding uncertainties in lunar and terrestrial ratios.

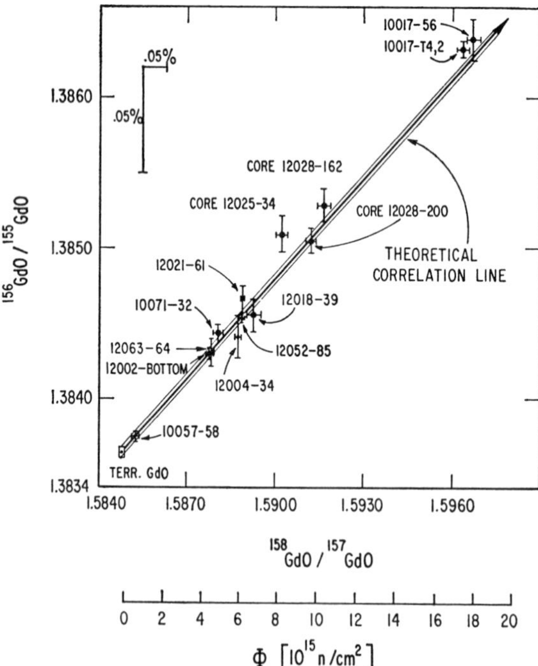

Fig. 5. Correlation diagram of $^{156}GdO/^{155}GdO$ versus $^{158}GdO/^{157}GdO$ in Apollo 11 and Apollo 12 samples. The theoretical correlation line represents the variation of the isotopic ratios of terrestrial GdO due to integrated neutron fluxes Φ. Errors given are $2\sigma_{mean}$.

is indicated. This may be due to a minor contribution from epithermal resonance neutrons (≥ 0.4 eV) (LUGMAIR and MARTI, 1971).

DISCUSSION

Isotopic composition of spallation krypton

The isotopic composition of spallation Kr in meteorites was found to vary with hardness of irradiation (EUGSTER *et al.*, 1969; MARTI *et al.*, 1969). A variation in the Kr^{78}/Kr^{83} spallation ratio was also found in Apollo 11 rocks and ascribed to varying shielding of the rocks (MARTI *et al.*, 1970; PEPIN *et al.*, 1970). In our Apollo 12 rocks we find systematic variations of all Kr isotopic ratios which clearly can be attributed to a varying degree of shielding of the rock samples. Similar results were reported by BOCHSLER *et al.* (1971). Figure 6 shows the correlation of P_{81}/P_{83} with the Kr^{78}/Kr^{83} spallation ratio. The P_{81}/P_{83} spallation production ratio is used in the calculation of Kr^{81}–Kr^{83} ages (see Table 1). The observed correlation is very good, partly because small uncertainties in the spallation Kr spectra will least affect these ratios. However, all the Kr spallation ratios are found to correlate with each other. The fact that Kr isotopic ratios show a smooth variation with the average energy of cosmic-ray particles enables us to use them to evaluate the average burial depth of a sample. The linear correlation seen in Fig. 6 rules out any significant effects in Kr^{80} and Kr^{82} from neutron capture in Br. In any samples where Kr^{80} and Kr^{82} would show anomalies from neutron capture in Br, the Kr^{78}/Kr^{83} spallation ratio may be used to obtain P_{81}/P_{83} for Kr dating. For the chemical composition of Apollo 12a rocks (see Table 7) the relation is

$$\frac{P_{81}}{P_{83}} = 1.262\,\frac{Kr^{78}}{Kr^{83}} + 0.381 \quad \left(\text{for } 0.145 \leq \frac{Kr^{78}}{Kr^{83}} \leq 0.175\right).$$

Rock 12063 which belongs to the 12b subgroup also falls on this line, while Apollo 11 rocks do not because of very different Zr/Sr ratios.

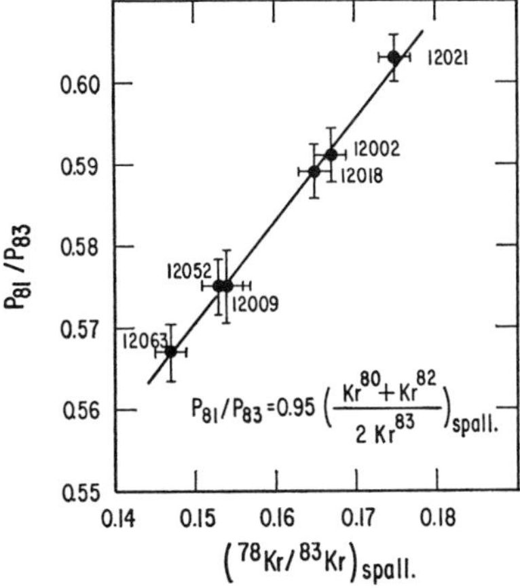

Fig. 6. Correlation of the Kr production ratio P_{81}/P_{83} with Kr^{78}/Kr^{83}. Correlation line is a least squares fit to the data.

A model for the estimation of the average effective irradiation depths of the rock samples

Table 6 gives the integrated thermal neutron fluxes Φ of three Apollo 11 and six Apollo 12 rock specimens along with their Kr^{81}–Kr ages (T_r), taken from Table 1 and MARTI *et al.* (1970). In column 4 the average neutron fluxes Φ/T_r (in n/cm² sec) are calculated and may be compared to the Kr spallation ratios Kr^{78}/Kr^{83} given in column 5. In Fig. 7, the shape of the depth dependent thermal neutron flux below the lunar surface is shown such as calculated by LINGENFELTER *et al.* (1970) for a chemical

Table 6. Thermal neutron fluxes and $^{78}Kr/^{83}Kr$ spallation ratios in Apollo 11 and Apollo 12 samples.

Sample	Φ 10^{15} n/cm²	T_r m.y.	Φ/T_r n/cm² sec	$\frac{Kr^{78}}{Kr^{83}}$	T_n m.y.
12021,61	6.29 ± 0.13	303 ± 18	0.66 ± 0.04	0.175	≡ 303
12002 bottom	4.57 ± 0.18	94 ± 6	1.54 ± 0.12	0.167	145
12018,39	6.84 ± 0.40	185 ± 16	1.11 ± 0.11	0.165	≡ 195
12009,34	6.05 ± 0.19	136 ± 24	1.41 ± 0.25	0.154	125
12052,85	6.25 ± 0.19	129 ± 7	1.54 ± 0.10	0.153	≡ 129
12063,64	4.68 ± 0.17	95 ± 5	1.56 ± 0.10	0.147	
10071 (A)	4.92 ± 0.17	372 ± 22	0.42 ± 0.03	0.191	
10057 (A)	0.78 ± 0.11	47 ± 2	0.52 ± 0.08	0.167	
10017 (A) (av.)	17.74 ± 0.36	509 ± 29	1.10 ± 0.06	0.172	

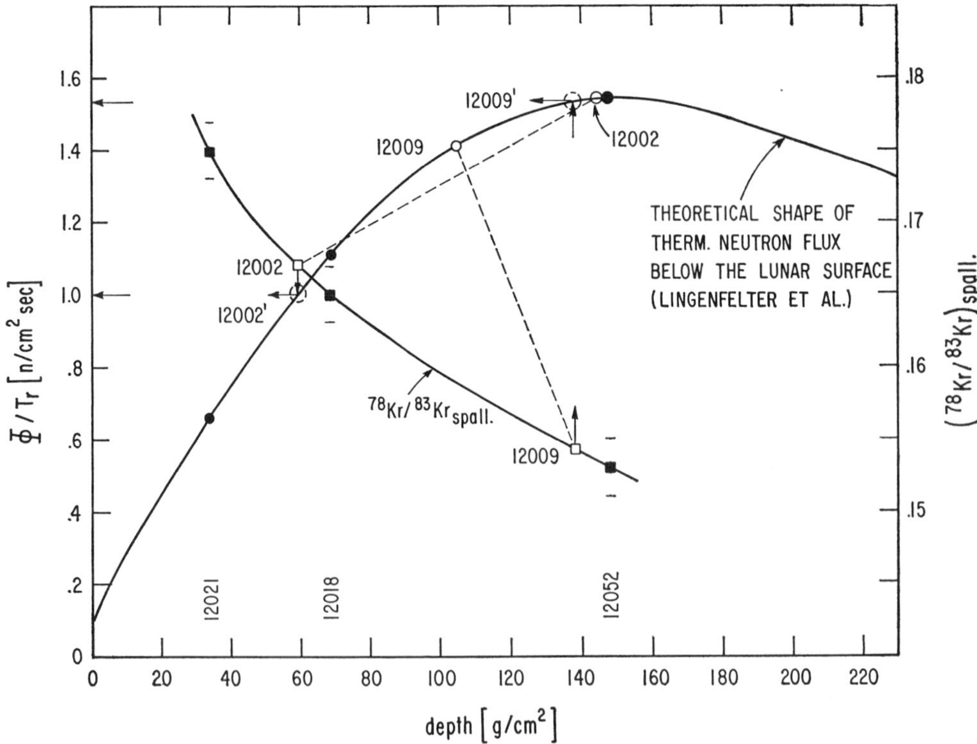

Fig. 7. Model calculation of average effective irradiation depth of Apollo 12 rocks and corresponding effective neutron irradiation ages (T_n).

composition of Apollo 11 soil. In order to get a (relative) depth scale, we fix the maximum of the flux (at 147 g/cm^2) to a neutron flux value of 1.55 n/cm^2 sec which is the largest value found so far (Table 6, rocks 12052 and 12063). The Kr^{81} contents of rocks 12021 and 12052 normalized to the concentrations of their major target elements indicate light and heavy shielded locations respectively. Such locations are also suggested by their Kr^{78}/Kr^{83} spallation ratios. This fact, therefore, suggests that the irradiation histories of rocks 12021 and 12052 were relatively simple ones. This is probably also true for rock 12018, but for the remaining rocks the irradiation histories may be more complex. In Fig. 7 we obtain the effective irradiation depth for rocks 12018, 12021, and 12052 from their calculated neutron fluxes (Table 6). These three rocks (all of group 12a) show approximately the variation with depth of the spallation ratio Kr^{78}/Kr^{83} which is also shown in Fig. 7. From a comparison of the burial depths obtained from the neutron and spallation Kr curves respectively, it appears that the history of rock 12002 must in fact have been complex. In the last column of Tab!e 6, we have calculated effective neutron irradiation ages (T_n) which are required to obtain agreement between the neutron and spallation Kr depth scales. Disagreement between T_r and T_n therefore indicates a complex history. The T_r of rock 12009 has large errors which overlap with T_n. No good case can be made for rock 12063 because its chemical composition (group 12b) is somewhat different. If the flux maximum for thermal neutrons should turn out to be larger than 1.55 n/cm^2 sec, the obtained burial depths would decrease accordingly while T_n would remain practically unchanged.

Effects of secondary neutrons on the relative Xe^{131} yield

It was found (Marti *et al.*, 1970; Pepin *et al.*, 1970) that the varying Xe^{131}/Xe^{126} spallation ratios measured in Apollo 11 rocks correlate to some degree with Kr^{78}/Kr^{83} ratios. Eberhardt *et al.* (1970) have plotted all the available data and confirmed such a correlation. They concluded that all Apollo 11 data can be fitted by one correlation line which would indicate that varying chemical composition has little effect on the position in a Xe^{131}/Xe^{126} versus Kr^{78}/Kr^{83} diagram. These Xe and Kr ratios found in our six Apollo 12 rocks are plotted in Fig. 8. The results of Bochsler *et al.* (1971) were also included in this plot, as are some of our Apollo 11 data for reference. The same assumptions for spallation as discussed above ($Kr^{86}/Kr^{83} = 0.0152$) were used. Because there are reasons to believe that the chemical composition of a rock will affect its position in this diagram, we have subdivided the Apollo 12 rocks in groups 12a and 12b (Table 7) according to differences in the abundance ratios of their major target elements Sr and Zr. Rock 12009 is not included in the best fit of group 12a rocks because the abundances of the target elements Sr, Y, Zr, Ba, REE in our specimen appear to be different from reported values (Annell *et al.*, 1971; Compston *et al.*, 1971; Haskin *et al.*, 1971; Murthy *et al.*, 1971). This is concluded from the very low Kr^{81} content, from the small Xe^{131} ratio and because of a covariation of U and Ba or REE from the U^{238}–Xe^{136} Kr^{83} age, which is smaller than its K–Ar age. For the same reason, rock 12009 is also excluded from the fit shown in Fig. 9 (Rock 12004 is also excluded because of the possibility of sample mislabeling: discrepancy in the Xe

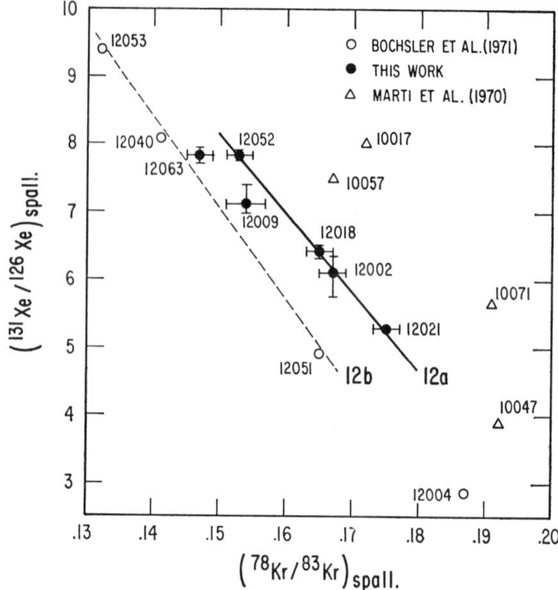

Fig. 8. Isotopic correlation of spallation Xe and Kr. The straight line is a least squares fit to the data (rocks 12004 and 12009 excluded, see text). All points were given the same weight.

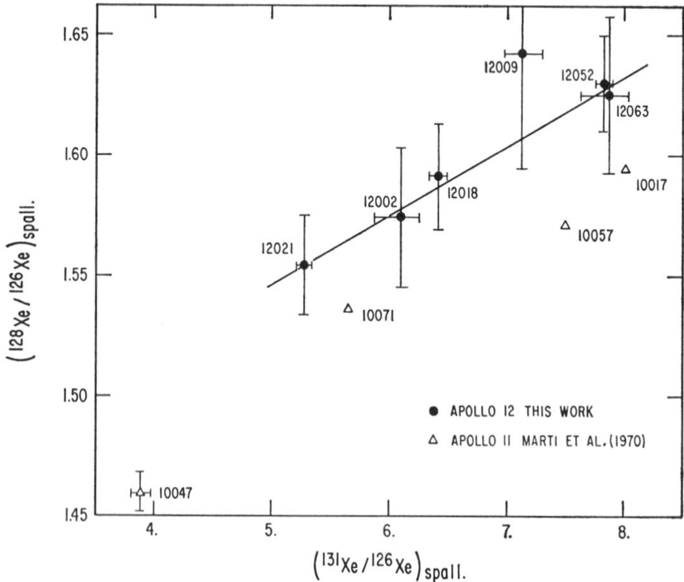

Fig. 9. Isotopic correlation of Xe^{126}, Xe^{128}, and Xe^{131}. The straight line is a least squares fit to the data (rock 12009 excluded, see text).

Table 7. Abundance ratios of major target elements for spallation Kr and Xe.

Rock Type	$\frac{Zr}{Sr}$	$\frac{Y}{Sr}$	$\frac{Y + Zr}{Sr}$	$\frac{La + Ce + Nd}{Ba}$
Apollo 11, A	3.63	1.12	4.86	0.51
Apollo 11, B	1.69	0.68	2.37	1.09
Apollo 12, a	1.18	0.42	1.61	0.58
Apollo 12, b	0.99	0.39	1.38	0.60

Apollo 11, A is average for rocks 10017, 10057, 10069, 10071, 10072.

Apollo 11, B is average for rocks 10044, 10047.

Apollo 12, a is average for rocks 12002, 12004, 12009, 12018, 12021, 12052.

Apollo 12, b is average for rocks 12040, 12051, 12053, 12063.

results reported by Bochsler *et al.*, 1971 and Bogard *et al.*, 1970). In Fig. 9, it is seen that the Xe^{128}/Xe^{126} spallation ratio correlates with Xe^{131}/Xe^{126}. Other Xe isotope ratios also appear to correlate with Xe^{131}/Xe^{126}, but larger uncertainties prevent any firm conclusions. The good linear correlation of spallation Xe and spallation Kr ratios strongly suggests that fast neutrons are responsible for a varying but high relative Xe^{131} yield. However, the present data do not rule out conclusively some contribution to the Xe^{131} yield from resonance neutron capture in Ba^{130}. It was shown (Albee *et al.*, 1970; Eugster *et al.*, 1970; Marti *et al.*, 1970) that thermal neutron capture of Ba^{130} accounts for only a few percent of the Xe^{131} production. The Apollo 11 rock data shown in Figs. 8 and 9 do not fit onto the Apollo 12 rock correlation lines. This is most likely due to differences in the target element composition.

Acknowledgments—The capable assistance of K. R. Goldman, B. Lightner, M. Meyers, and A. Schimmel in this work is acknowledged. We have benefited from discussions with J. R. Arnold, T. W. Armstrong, G. Arrhenius, D. S. Burnett, P. Eberhardt, H. Graf, G. J. Wasserburg, and R. C. Reedy. We thank J. R. Arnold for supplying us with aliquot samples of rock 12002 and for his collaboration. Support for this research from NASA contract NAS 9–8107 is acknowledged.

References

Albee A. L., Burnett D. S., Chodos A. A., Eugster O. J., Huneke J. C., Papanastassiou D. A., Podosek F. A., Price G. R., Sanz H. G., Tera F., and Wasserburg G. J. (1970) Ages, irradiation history and chemical composition of lunar rocks from the Sea of Tranquillity. *Science* **167**, 463–466.

Albee A. L., Burnett D. S., Chodos A. A., Haines E. L., Huneke J. C., Podosek F. A., Papanastassiou D. A., Price G. R., Tera F., and Wasserburg G. J. (1971) The irradiation history of lunar samples. Second Lunar Science Conference (unpublished proceedings).

Alexander E. C., Jr., Davis P. K., Kaiser W. A., Lewis R. S., and Reynolds J. H. (1971) Some specific rare gas studies in Apollo samples. Second Lunar Science Conference (unpublished proceedings).

Annell C. S., Carron M. K., Christian R. P., Cuttitta F., Dwornik E. J., Helz A. W., Ligon D. T., Jr., and Rose H. J., Jr. (1971) Chemical and spectrographic analyses of lunar samples from Apollo 12 mission. Second Lunar Science Conference (unpublished proceedings)

BOCHSLER P., EBERHARDT P., GEISS J., GRAFF H., GRÖGLER N., KRÄHENBÜHL V., MÖRGELI M., SCHWALLER H., and STETTLER A. (1971) Potassium-argon ages, exposure ages and radiation history of lunar rocks. Second Lunar Science Conference (unpublished proceedings).

BOGARD D. D., FUNKHOUSER J. G., SCHAEFFER O. A., and ZÄHRINGER J. (1970) Noble gas abundances in lunar material, II. Cosmic ray spallation products and radiation ages from Mare Tranqillitatis and Oceanus Procellarum. Preprint.

COMPSTON W., BERRY H., VERNON M. J., CHAPPELL B. W., and KAYE M. J. (1971) Rubidium–strontium chronology and chemistry of lunar material from the Ocean of Storms. Second Lunar Science Conference (unpublished proceedings).

EBERHARDT P., GEISS J., GRAF H., GRÖGLER N., KRÄHENBÜHL V., SCHWALLER H., SCHWARZMÜLLER J., and STETTLER A. (1970) Correlation between rock type and irradiation history of Apollo 11 igneous rocks. *Earth Planet. Sci. Lett.* **10**, 67–72.

EUGSTER O., EBERHARDT P., and GEISS J. (1969) Isotopic analyses of krypton and xenon in 14 stone meteorites. *J. Geophys. Res.* **74**, 3874–3896.

EUGSTER O., TERA F., BURNETT D. S., and WASSERBURG G. J. (1970) The isotopic composition of Gd and neutron capture effects in some meteorites. *J. Geophys. Res.* **75**, 2753–2768.

EUGSTER O., TERA F., BURNETT D. S., and WASSERBURG G. J. (1970) The isotopic composition of Gd and the neutron capture effects in samples from Apollo 11. *Earth Planet. Sci. Lett.* **8**, 20–30.

FINKEL R. C., ARNOLD J. R., IMAMURA M., REEDY R. C., FRUCHTER J. S., LOOSLI H. H., EVANS J. C., and DELANY A. C. (1971) Depth variation of cosmogenic nuclides in a lunar surface rock and lunar soil. Second Lunar Science Conference (unpublished proceedings).

GOLDBERG M. D., MUGHABGHAB S. F., PUROHIT S. N., MAGURNO B. A., and MAY V. M. (1966) Neutron cross sections. BNL 325, 2nd ed., Suppl. 2, 64–155–1 and 64–157–1.

HASKIN L. A., ALLEN R. O., HELMKE P. A., ANDERSON M. R., KOROTEV R. L., and ZWEIFEL K. A. (1971) Rare earths and other trace elements in Apollo 12 lunar materials. Second Lunar Science Conference (unpublished proceedings).

HEYMANN D., and YANIV A. (1970) Ar^{40} anomaly in lunar samples from Apollo 11. *Proc. Apollo 11 Lunar Sci. Conf., Geochim. Cosmochim. Acta* Suppl. 1, Vol. 2, 1261–1267. Pergamon.

HINTENBERGER H., WEBER H., and TAKAOKA N. (1971) Concentrations and isotopic abundances of the rare gases in lunar matter. Second Lunar Science Conference (unpublished proceedings).

LINGENFELTER R. E., CANFIELD E. H., and HAMPLE (1970) Private communication.

LUGMAIR G. W., and MARTI K. (1971) Neutron effects in lunar samples. To be submitted to *Earth Planet. Sci. Lett.*

MANKA R. H., and MICHEL F. C. (1970) Lunar atmosphere as a source of argon-40 and other lunar surface elements. *Science* **169**, 278–280.

MARTI K. (1967) Mass-spectrometric detection of cosmic-ray-produced Kr^{81} in meteorites and the possibility of Kr–Kr dating. *Phys. Rev. Lett.* **18**, 264–266.

MARTI K., SHEDLOVSKY J. P., LINDSTROM R. M., ARNOLD J. R., and BHANDARI N. G. (1969) Cosmic-ray-produced radionuclides and rare gases near the surface of Saint-Severin meteorite. In *Meteorite Research* (editor P. M. Millmann), pp. 246–266. Reidel.

MARTI K., LUGMAIR G. W., and UREY H. C. (1970) Solar wind gases, cosmic-ray spallation products and the irradiation history of Apollo 11 samples. *Proc. Apollo 11 Lunar Sci. Conf., Geochim. Cosmochim. Acta* Suppl. 1, Vol. 2, 1357–1367. Pergamon.

MARTI K., and MEYERS M. E. (1971) K and Ar in lunar materials. In preparation.

MURTHY V. R., EVENSEN N. M., JAHN B., and COSCIO M. R., JR. (1971) Rb–Sr isotopic relations and elemental abundances of K, Rb, Sr, and Ba in Apollo 11 and Apollo 12 samples. Second Lunar Science Conference (unpublished proceedings).

PAPANASTASSIOU D. A., and WASSERBURG G. J. (1970) Rb–Sr ages from the Ocean of Storms. *Earth Planet. Sci. Lett.* **8**, 269–278.

PEPIN R. O., NYQUIST L. E., PHINNEY D., and BLACK D. C. (1970) Rare gases in Apollo 11 lunar material. *Proc. Apollo 11 Lunar Sci. Conf., Geochim. Cosmochim. Acta* Suppl. 1, Vol. 2, 1435–1454. Pergamon.

SCHAEFFER O. A., FUNKHOUSER J. G., BOGARD D. D., and ZÄHRINGER J. (1970) Potassium-argon ages of lunar rocks from Mare Tranquillitatis and Oceanus Procellarum. *Science* **170**, 161–162.

SILVER L. T. (1971) U–Th–Pb isotope relations in Apollo 11 and Apollo 12 lunar samples. Second Lunar Science Conference (unpublished proceedings).

TATSUMOTO M., KNIGHT R. J., and DOE B. R. (1971) U–Th–Pb systematics of Apollo 12 lunar samples. Second Lunar Science Conference (unpublished proceedings).

TURNER G. (1971) ^{40}Ar–^{39}Ar ages from the lunar maria. Second Lunar Science Conference (unpublished proceedings).

WETHERILL G. W. (1953) Spontaneous fission yields from uranium and thorium. *Phys. Rev.* **92**, 907.

Proceedings of the Second Lunar Science Conference, Volume 2, pp. 1607–1625.
The M.I.T. Press, 1971.

Concentrations and isotopic abundances of the rare gases in lunar matter

H. HINTENBERGER, H. W. WEBER, and N. TAKAOKA
Max-Planck-Institut für Chemie
(Otto-Hahn-Institut)
Mainz, Germany

(*Received 22 February 1971; accepted in revised form 30 March 1971*)

Abstract—Rare gas concentrations and isotopic abundances have been measured in bulk material of fines (10087,8; 12001,43; 12044,11; 12070,62), in bulk material of breccias (10021,20; 10046,16; 10061,11); in grain size fractions of fines and breccias (10087,8; 12070,62; 10021,20; 10061,11) and in crystalline rocks (12002,120; 12004,33; 12018,41; 12020,37; 12052,94; 12053,20; 12063,69; 12064,15; 12065,29). Characteristic differences have been found between fines and breccias, not only in the absolute concentrations but more striking in the relative abundance distributions of the rare gas nuclides. The relative abundances of neon and usually also of xenon are distinctly higher in Apollo 11 breccias than in fines. Temporal variations and processes of repeatedly fractionated gas implantation and gas release are discussed as possible causes for the observed differences in the abundance distribution of the rare gases between breccias and fines.

The ratios of the cosmogenic nuclides $(^{3}He/^{21}Ne)_{sp}$ and $(^{3}He/^{38}Ar)_{sp}$ are always considerably lower for fines and breccias than for crystalline rocks, whereas the ratio $(^{21}Ne/^{38}Ar)_{sp}$ is in the same order of magnitude. This can be explained by a diffusion loss of ^{3}He or (and) by an additional contribution to the ^{21}Ne and ^{38}Ar content by solar flares.

Exposure ages and gas retention ages have been determined from 5 Apollo 11 and 9 Apollo 12 crystalline rocks. Groupings are observed in the $(^{38}Ar/^{21}Ne)_{sp}$ ratios.

INTRODUCTION

THE EARLY examination of Apollo 11 samples has already revealed the presence of rare gas nuclei of three different origins in lunar material: (1) rare gases which were introduced into the material by the solar wind and which are trapped within the surface layers of the mineral grains; (2) rare gas nuclides which were mainly produced by nuclear interactions of cosmic rays with lunar matter and which are distributed throughout the whole volume of the bombarded mineral grains; and (3) radiogenic rare gas nuclides from the decay of long-lived radio nuclides (LSPET, 1969 and 1970; EBERHARDT *et al.*, 1970; FUNKHOUSER *et al.*, 1970a and b; HEYMANN *et al.*, 1970; HEYMANN and YANIV, 1970; HINTENBERGER *et al.*, 1970a and b; HOHENBERG *et al.*, 1970; KIRSTEN *et al.*, 1970; MARTI *et al.*, 1970; PEPIN *et al.*, 1970).

If the rare gas extraction is performed on grain size fractions of lunar material, it is possible to measure which portions of the gases are mainly surface correlated, and which ones volume correlated, so that the solar wind, the spallation, and the radiogenic components can be determined separately. Cosmic-ray exposure ages and radiogenic ages can be calculated therefrom, and information can be obtained on the composition of the solar wind gases and on the history and trapping properties of the mineral grains.

Formerly, we reported on investigations of the rare gases in the bulk fines 10084,18 and bulk breccia 10021,20, in grain size fractions of 10084,18, and in a few crystalline Apollo 11 rocks (HINTENBERGER *et al.*, 1970a and b). In the present paper, these investigations are extended to bulk material of the fines 10087,8; 12001,43; 12044,17; and 12070,62; to bulk material of the breccias 10046,16 and 10061,11; and to grain size fractions of 10087,8; 12070,62; 10021,20; and 10061,11. In addition, nine Apollo 12 crystalline rocks were investigated and dated, and a lithic fragment from a breccia.

FINES AND BRECCIAS: BULK SAMPLES

Table 1 shows the results on bulk samples of fines and breccias. The data are mean values calculated from the results obtained on a number of different portions of the same sample. A considerable variation far outside of the analytical errors is observed for the concentrations of all rare gas nuclides in a given sample, but particularly in breccias. The number of aliquots measured from the same sample and the variation intervals observed for the ^{4}He contents are also listed in Table 1. In spite of the rather

Table 1. Rare gas concentrations in cm^3 STP/g in bulk material of lunar fines and breccias. Due to the heterogeneity of the samples, the absolute concentrations measured on different positions of the same sample vary far outside the analytical errors (see Table 2). The concentrations given in this table are mean values measured of aliquots (0.2–0.8 mg each) of the same sample. The number of aliquots measured and the variation intervals observed for ^{4}He are given below.

Nuclides and Ratios	10084,18	10087,8	10021-20	10046,16	10061,11	12070,62	12044,11	12001,43
^{3}He	$8\cdot27\cdot10^{-5}$	$6\cdot89\cdot10^{-5}$	$1\cdot27\cdot10^{-4}$	$6\cdot82\cdot10^{-5}$	$8\cdot57\cdot10^{-5}$	$3\cdot01\cdot10^{-5}$	$3\cdot27\cdot10^{-5}$	$4\cdot11\cdot10^{-5}$
^{4}He	$2\cdot19\cdot10^{-1}$	$1\cdot91\cdot10^{-1}$	$3\cdot73\cdot10^{-1}$	$2\cdot00\cdot10^{-1}$	$2\cdot55\cdot10^{-1}$	$7\cdot82\cdot10^{-2}$	$8\cdot33\cdot10^{-2}$	$1\cdot10\cdot10^{-1}$
^{20}Ne	$2\cdot28\cdot19^{-3}$	$2\cdot39\cdot10^{-3}$	$5\cdot62\cdot10^{-3}$	$3\cdot06\cdot10^{-3}$	$5\cdot02\cdot10^{-3}$	$1\cdot30\cdot10^{-3}$	$1\cdot23\cdot10^{-3}$	$1\cdot89\cdot10^{-3}$
^{21}Ne	$6\cdot02\cdot10^{-6}$	$6\cdot29\cdot10^{-6}$	$1\cdot43\cdot10^{-5}$	$8\cdot04\cdot10^{-6}$	$1\cdot34\cdot10^{-5}$	$3\cdot55\cdot10^{-6}$	$3\cdot37\cdot10^{-6}$	$5\cdot07\cdot10^{-6}$
^{22}Ne	$1\cdot83\cdot10^{-4}$	$1\cdot91\cdot10^{-4}$	$4\cdot40\cdot10^{-4}$	$2\cdot41\cdot10^{-4}$	$4\cdot03\cdot10^{-4}$	$1\cdot03\cdot10^{-4}$	$9\cdot74\cdot10^{-5}$	$1\cdot50\cdot10^{-4}$
^{36}Ar	$4\cdot36\cdot10^{-4}$	$3\cdot80\cdot10^{-4}$	$7\cdot45\cdot10^{-4}$	$4\cdot34\cdot10^{-4}$	$8\cdot57\cdot10^{-4}$	$2\cdot66\cdot10^{-4}$	$2\cdot46\cdot10^{-4}$	$2\cdot85\cdot10^{-4}$
^{38}Ar	$8\cdot36\cdot10^{-5}$	$7\cdot31\cdot10^{-5}$	$1\cdot53\cdot10^{-4}$	$8\cdot59\cdot10^{-5}$	$1\cdot67\cdot10^{-4}$	$5\cdot21\cdot10^{-5}$	$4\cdot87\cdot10^{-5}$	$5\cdot68\cdot10^{-5}$
^{40}Ar	$4\cdot73\cdot10^{-4}$	$6\cdot02\cdot10^{-4}$	$1\cdot89\cdot10^{-3}$	$1\cdot21\cdot10^{-3}$	$1\cdot70\cdot10^{-3}$	$1\cdot54\cdot10^{-4}$	$1\cdot54\cdot10^{-4}$	$1\cdot69\cdot10^{-4}$
^{84}Kr	$2\cdot4\cdot10^{-7}$	$2\cdot2\cdot10^{-7}$	$4\cdot0\cdot10^{-7}$	$2\cdot1\cdot10^{-7}$	$4\cdot1\cdot10^{-7}$	$1\cdot6\cdot10^{-7}$	$1\cdot6\cdot10^{-7}$	$2\cdot0\cdot10^{-7}$
^{132}Xe	$3\cdot1\cdot10^{-8}$	$3\cdot0\cdot10^{-8}$	$9\cdot3\cdot10^{-8}$	$4\cdot4\cdot10^{-8}$	$5\cdot1\cdot10^{-8}$	$2\cdot0\cdot10^{-8}$	$2\cdot1\cdot10^{-8}$	$2\cdot4\cdot10^{-8}$
$^{4}He/^{3}He$	2650	2770	2940	2930	2980	2600	2550	2680
$^{20}Ne/^{22}Ne$	12.5	12.5	12.8	12.7	12.5	12.6	12.6	12.6
$^{22}Ne/^{21}Ne$	30.4	30.4	30.8	30.0	30.1	29.0	28.9	29.6
$^{36}Ar/^{38}Ar$	5.22	5.20	4.87	5.05	5.13	5.11	5.05	5.02
$^{40}Ar/^{36}Ar$	1.08	1.58	2.54	2.79	1.98	0.579	0.626	0.593
$^{4}He/^{20}Ne$	96.1	79.9	66.4	65.4	50.8	60.2	67.7	58.2
$^{4}He/^{36}Ar$	502	503	501	461	298	294	339	386
$^{20}Ne/^{36}Ar$	5.23	6.29	7.54	7.05	5.86	4.89	5.00	6.63
$^{84}Kr/^{36}Ar$	$5\cdot6\cdot10^{-4}$	$5\cdot7\cdot10^{-4}$	$5\cdot4\cdot10^{-4}$	$4\cdot8\cdot10^{-4}$	$4\cdot8\cdot10^{-4}$	$6\cdot0\cdot10^{-4}$	$6\cdot7\cdot10^{-4}$	$7\cdot2\cdot10^{-4}$
$^{132}Xe/^{36}Ar$	$7\cdot1\cdot10^{-5}$	$8\cdot0\cdot10^{-5}$	$1\cdot2\cdot10^{-4}$	$1\cdot0\cdot10^{-4}$	$5\cdot9\cdot10^{-5}$	$7\cdot6\cdot10^{-5}$	$8\cdot6\cdot10^{-5}$	$8\cdot5\cdot10^{-5}$
number of aliquots	9	2	3	3	2	6	3	3
Variation interval for helium-4	±23%	±5%	±10%	±30%	±12%	±7%	±7%	±5%

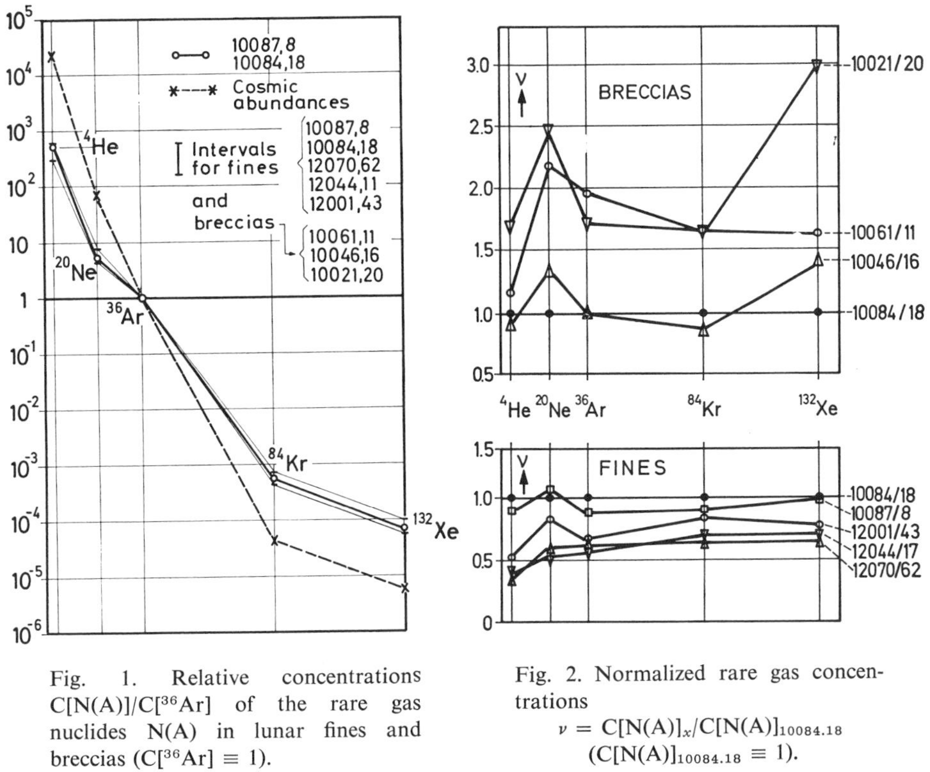

Fig. 1. Relative concentrations $C[N(A)]/C[^{36}Ar]$ of the rare gas nuclides N(A) in lunar fines and breccias ($C[^{36}Ar] \equiv 1$).

Fig. 2. Normalized rare gas concentrations

$$\nu = C[N(A)]_x/C[N(A)]_{10084.18}$$
$$(C[N(A)]_{10084.18} \equiv 1).$$

large variability of the concentrations it is possible to recognize differences in the gas concentrations and abundance distribution patterns which appear to be correlated with the sample number, and therefore which reflect true differences in the histories of the samples.

The relative concentrations of ^{4}He, ^{20}Ne, ^{36}Ar, ^{84}Kr, and ^{132}Xe, normalized to ^{36}Ar = 1, are shown in Fig. 1. Although the ^{4}He and ^{132}Xe concentrations differ from each other by 7 orders of magnitude, an agreement within a factor of two is always observed for the relative concentration values of single nuclides in different samples (see bars in Fig. 1). The relative cosmic abundances (Aller, 1961; Suess and Urey, 1958), however, are clearly outside of these intervals.

The characteristic differences between fines from Apollo 11 and fines from Apollo 12, and between fines and breccias from Apollo 11, are evident from Fig. 2. This figure shows the concentrations of ^{4}He, ^{20}Ne, ^{36}Ar, ^{84}Kr, and ^{132}Xe in different breccias and fines relative to their concentrations in the fines 10084,18. The concentrations of the rare gases in the fines 10084,18 and 10087,8 agree within the concentration ranges of different aliquots of these samples. Nevertheless small but significant differences in their abundance distributions can be observed. The Apollo 11 breccias show much more characteristic differences with respect to the relative abundance distributions and, to a lesser degree, with respect to the absolute gas concentrations than do the fines 10084 and 10087. The concentrations in the breccia 10021,20 are higher by a factor of about 2 to 3 depending on the nuclide considered. In particular, the neon and xenon contents in the breccias show remarkable excesses as compared to the fines 10084 and 10087.

Fines of Apollo 12 show lower concentrations as compared to the Apollo 11 fines, especially in the light nuclides.

FINES AND BRECCIAS: GRAIN SIZE FRACTIONS

Grain size fractions of the fines 10087,8 and 12070,62; and of the breccias 10021,20 and 10061,11 were investigated. Nylon sieves of 5 μm and of 10 μm mesh width and stainless steel sieves of 20 μm, 25 μm, 35.5 μm, 54 μm, 75 μm, 120 μm, and 200 μm mesh width were used. Five to nine grain-size fractions were made from each sample. The largest grains investigated had diameters between 120 and 200 μm, the finest smaller than 5 μm. The gas concentrations are compiled in Tables 2 to 5.

It is well known from investigations by EBERHARDT *et al.*, 1970; HEYMANN *et al.*, 1970; HEYMANN and YANIV, 1970; and HINTENBERGER *et al.*, 1970a and b that the concentrations of all rare gases decrease systematically with increasing grain size. Our results on the rare gases in grain size fractions of the fines 10084,18 have already been published (HINTENBERGER *et al.*, 1970b) and are used, in the present paper, for comparison. Our new data show roughly the same grain size dependence of gas concentrations.

For grain sizes between 10 μm and 200 μm the rare gas concentrations in the fines 10087,8 and in the breccia 10021,20 are almost inversely proportional to the particle size. The fines 12070,62, however, do not show this anticorrelation in such a distinct manner as the Apollo 11 samples (see Fig. 3 a, b, and c). Besides, it may be mentioned that the finest fractions always contain much less gas than would be expected from

Table 2. Rare gas concentrations in cm^3 STP/g in grain size fractions of fines from Apollo 11 sample 10087,8. Errors in isotope ratios 2%, errors in concentrations of He, Ne, and Ar about 3%, of Kr and Xe about 10%. Sample weights between 0.2–4 mg.

Nuclides and Ratios	< 5 μm	5–10 μm	10–20 μm	20–25 μm	25–35.5 μm	35.5–54 μm	54–75 μm	75–120 μm	120–200 μm
^{3}He	$1.57\cdot10^{-4}$	$7.00\cdot10^{-5}$	$7.89\cdot10^{-5}$	$4.17\cdot10^{-5}$	$3.11\cdot10^{-5}$	$2.23\cdot10^{-5}$	$1.23\cdot10^{-5}$	$1.10\cdot10^{-5}$	$1.11\cdot10^{-5}$
^{4}He	$4.26\cdot10^{-1}$	$1.95\cdot10^{-1}$	$2.10\cdot10^{-1}$	$1.11\cdot10^{-1}$	$8.51\cdot10^{-2}$	$5.96\cdot10^{-2}$	$2.95\cdot10^{-2}$	$2.60\cdot10^{-2}$	$2.69\cdot10^{-2}$
^{20}Ne	$5.15\cdot10^{-3}$	$2.53\cdot10^{-3}$	$2.55\cdot10^{-3}$	$1.56\cdot10^{-3}$	$1.25\cdot10^{-3}$	$8.99\cdot10^{-4}$	$3.96\cdot10^{-4}$	$3.71\cdot10^{-4}$	$4.07\cdot10^{-4}$
^{21}Ne	$1.29\cdot10^{-5}$	$6.86\cdot10^{-6}$	$6.90\cdot10^{-6}$	$4.33\cdot10^{-6}$	$3.66\cdot10^{-6}$	$2.79\cdot10^{-6}$	$1.59\cdot10^{-6}$	$1.42\cdot10^{-6}$	$4.58\cdot10^{-6}$
^{22}Ne	$4.09\cdot10^{-4}$	$2.03\cdot10^{-4}$	$2.04\cdot10^{-4}$	$1.24\cdot10^{-4}$	$9.75\cdot10^{-5}$	$7.30\cdot10^{-5}$	$3.23\cdot10^{-5}$	$3.00\cdot10^{-5}$	$3.34\cdot10^{-5}$
^{36}Ar	$8.20\cdot10^{-4}$	$3.88\cdot10^{-4}$	$3.97\cdot10^{-4}$	$2.50\cdot10^{-4}$	$2.07\cdot10^{-4}$	$1.31\cdot10^{-4}$	$4.99\cdot10^{-5}$	$7.30\cdot10^{-5}$	$7.42\cdot10^{-5}$
^{38}Ar	$1.61\cdot10^{-4}$	$7.65\cdot10^{-5}$	$7.71\cdot10^{-5}$	$4.97\cdot10^{-5}$	$4.20\cdot10^{-5}$	$2.68\cdot10^{-5}$	$1.05\cdot10^{-5}$	$1.49\cdot10^{-5}$	$1.52\cdot10^{-5}$
^{40}Ar	$1.30\cdot10^{-3}$	$6.23\cdot10^{-4}$	$6.29\cdot10^{-4}$	$3.91\cdot10^{-4}$	$3.40\cdot10^{-4}$	$2.39\cdot10^{-4}$	$1.03\cdot10^{-4}$	$1.50\cdot10^{-4}$	$1.78\cdot10^{-4}$
^{84}Kr	$4.0\cdot10^{-7}$	$1.9\cdot10^{-7}$	$2.2\cdot10^{-7}$	$1.3\cdot10^{-7}$	$1.0\cdot10^{-7}$	$9.0\cdot10^{-8}$	$2.6\cdot10^{-8}$	$3.7\cdot10^{-8}$	$4.2\cdot10^{-8}$
^{132}Xe	$5.4\cdot10^{-8}$	$3.0\cdot10^{-8}$	$3.1\cdot10^{-8}$	$1.9\cdot10^{-8}$	$1.5\cdot10^{-8}$	$1.2\cdot10^{-8}$	$4.5\cdot10^{-9}$	$7.1\cdot10^{-9}$	$7.8\cdot10^{-9}$
^{4}He/^{3}He	2710	2790	2660	2660	2740	2670	2400	2360	2420
^{20}Ne/^{22}Ne	12.6	12.5	12.5	12.6	12.8	12.3	12.3	12.4	12.2
^{22}Ne/^{21}Ne	31.7	29.6	29.6	28.6	26.6	26.2	20.3	21.1	21.1
^{36}Ar/^{38}Ar	5.09	5.07	5.15	5.03	4.93	4.89	4.75	4.90	4.88
^{40}Ar/^{36}Ar	1.59	1.61	1.58	1.56	1.64	1.82	2.06	2.05	2.40
^{4}He/^{20}Ne	82.7	77.1	82.4	71.2	68.1	66.3	74.5	70.1	66.1
^{4}He/^{36}Ar	520	503	529	444	411	455	591	356	363
^{20}Ne/^{36}Ar	6.28	6.52	6.42	6.24	6.04	6.86	7.94	5.08	5.49
^{84}Kr/^{36}Ar	$4.9\cdot10^{-4}$	$4.9\cdot10^{-4}$	$5.4\cdot10^{-4}$	$5.1\cdot10^{-4}$	$4.8\cdot10^{-4}$	$6.8\cdot10^{-4}$	$5.3\cdot10^{-4}$	$5.1\cdot10^{-4}$	$5.7\cdot10^{-4}$
^{132}Xe/^{36}Ar	$6.6\cdot10^{-5}$	$7.7\cdot10^{-5}$	$7.8\cdot10^{-5}$	$7.6\cdot10^{-5}$	$7.2\cdot10^{-5}$	$9.2\cdot10^{-5}$	$9.0\cdot10^{-5}$	$9.7\cdot10^{-5}$	$1.1\cdot10^{-4}$

Table 3. Rare gas concentrations in cm^3 STP/g in grain size fractions of fines from Apollo 12 sample 12070,62. Errors in isotope ratios 2%, errors in concentrations of He, Ne, and Argon about 3%, of Kr and Xe about 10%. Sample weights between 0.2 and 4 mg.

Nuclides and Ratios	< 25 μm	25–35.5 μm	35.5–53 μm	35.5–75 μm	53–75 μm	75–120 μm	120–200 μm	200–300 μm	300–500 μm
^{3}He	$5.88 \cdot 10^{-5}$	$5.46 \cdot 10^{-5}$		$1.89 \cdot 10^{-5}$		$1.40 \cdot 10^{-5}$	$9.55 \cdot 10^{-6}$	$7.87 \cdot 10^{-6}$	$8.54 \cdot 10^{-6}$
	$5.27 \cdot 10^{-5}$	$4.96 \cdot 10^{-5}$	$2.03 \cdot 10^{-5}$	$1.95 \cdot 10^{-5}$	$1.51 \cdot 10^{-5}$		$1.02 \cdot 10^{-5}$	$8.14 \cdot 10^{-6}$	
^{4}He	$1.59 \cdot 10^{-1}$	$1.47 \cdot 10^{-1}$		$4.84 \cdot 10^{-2}$		$3.37 \cdot 10^{-2}$	$2.20 \cdot 10^{-2}$	$1.72 \cdot 10^{-2}$	$2.09 \cdot 10^{-2}$
	$1.42 \cdot 10^{-1}$	$1.33 \cdot 10^{-1}$	$5.17 \cdot 10^{-2}$	$5.02 \cdot 10^{-2}$	$3.79 \cdot 10^{-2}$		$2.32 \cdot 10^{-2}$	$1.90 \cdot 10^{-2}$	
^{20}Ne	$2.27 \cdot 10^{-3}$	$2.43 \cdot 10^{-3}$		$9.32 \cdot 10^{-4}$		$6.67 \cdot 10^{-4}$	$4.11 \cdot 10^{-4}$	$3.34 \cdot 10^{-4}$	$4.06 \cdot 10^{-4}$
							$4.52 \cdot 10^{-4}$		
	$2.12 \cdot 10^{-3}$	$2.22 \cdot 10^{-3}$	$9.60 \cdot 10^{-4}$	$9.01 \cdot 10^{-4}$	$7.43 \cdot 10^{-4}$			$3.54 \cdot 10^{-4}$	
^{21}Ne	$5.98 \cdot 10^{-6}$	$6.26 \cdot 10^{-6}$		$2.76 \cdot 10^{-6}$		$2.16 \cdot 10^{-6}$	$1.48 \cdot 10^{-6}$	$1.25 \cdot 10^{-6}$	$1.37 \cdot 10^{-6}$
							$1.57 \cdot 10^{-6}$		
	$5.59 \cdot 10^{-6}$	$5.75 \cdot 10^{-6}$	$2.89 \cdot 10^{-6}$	$2.75 \cdot 10^{-6}$	$2.26 \cdot 10^{-6}$			$1.26 \cdot 10^{-6}$	
^{22}Ne	$1.80 \cdot 10^{-4}$	$1.93 \cdot 10^{-4}$		$7.43 \cdot 10^{-5}$		$5.35 \cdot 10^{-5}$	$3.35 \cdot 10^{-5}$	$2.65 \cdot 10^{-5}$	$3.24 \cdot 10^{-5}$
							$3.65 \cdot 10^{-5}$		
	$1.69 \cdot 10^{-4}$	$1.77 \cdot 10^{-4}$	$7.77 \cdot 10^{-5}$	$7.37 \cdot 10^{-5}$	$5.98 \cdot 10^{-5}$			$2.86 \cdot 10^{-5}$	
^{36}Ar	$3.68 \cdot 10^{-4}$	$4.27 \cdot 10^{-4}$		$1.72 \cdot 10^{-4}$		$1.22 \cdot 10^{-4}$	$8.79 \cdot 10^{-5}$	$6.69 \cdot 10^{-5}$	$1.06 \cdot 10^{-4}$
							$9.29 \cdot 10^{-5}$		
	$3.32 \cdot 10^{-4}$	$4.21 \cdot 10^{-4}$	$1.65 \cdot 10^{-4}$	$1.77 \cdot 10^{-4}$	$1.51 \cdot 10^{-4}$			$8.96 \cdot 10^{-5}$	
^{38}Ar	$7.20 \cdot 10^{-5}$	$8.24 \cdot 10^{-5}$		$3.35 \cdot 10^{-5}$		$2.39 \cdot 10^{-5}$	$1.73 \cdot 10^{-5}$	$1.31 \cdot 10^{-5}$	$2.10 \cdot 10^{-5}$
							$1.83 \cdot 10^{-5}$		
	$6.42 \cdot 10^{-5}$	$8.17 \cdot 10^{-5}$	$3.26 \cdot 10^{-5}$	$3.44 \cdot 10^{-5}$	$2.96 \cdot 10^{-5}$			$1.78 \cdot 10^{-5}$	
^{40}Ar	$1.89 \cdot 10^{-4}$	$2.13 \cdot 10^{-4}$		$1.03 \cdot 10^{-4}$		$7.91 \cdot 10^{-5}$	$7.14 \cdot 10^{-5}$	$7.70 \cdot 10^{-5}$	$7.73 \cdot 10^{-5}$
							$7.01 \cdot 10^{-5}$		
	$1.69 \cdot 10^{-4}$	$2.11 \cdot 10^{-4}$	$9.70 \cdot 10^{-5}$	$1.07 \cdot 10^{-4}$	$9.36 \cdot 10^{-5}$			$7.65 \cdot 10^{-5}$	
^{84}Kr	$2.3 \cdot 10^{-7}$	$2.6 \cdot 10^{-7}$		$1.1 \cdot 10^{-7}$		$8.3 \cdot 10^{-8}$	$6.2 \cdot 10^{-8}$	$4.2 \cdot 10^{-8}$	$7.4 \cdot 10^{-8}$
							$6.4 \cdot 10^{-8}$		
	$2.1 \cdot 10^{-7}$	$2.7 \cdot 10^{-7}$	$9.9 \cdot 10^{-8}$	$1.1 \cdot 10^{-7}$	$9.3 \cdot 10^{-8}$			$6.1 \cdot 10^{-8}$	
^{132}Xe	$2.7 \cdot 10^{-8}$	$3.0 \cdot 10^{-8}$		$1.4 \cdot 10^{-8}$		$9.8 \cdot 10^{-9}$	$7.7 \cdot 10^{-9}$	$7.0 \cdot 10^{-9}$	$1.1 \cdot 10^{-8}$
							$8.0 \cdot 10^{-9}$		
	$2.9 \cdot 10^{-8}$	$3.4 \cdot 10^{-8}$	$1.2 \cdot 10^{-8}$	$1.3 \cdot 10^{-8}$	$1.1 \cdot 10^{-8}$			$9.4 \cdot 10^{-9}$	
^{4}He/^{3}He	2700	2690		2560		2410	2300	2190	2450
							2270		
	2690	2680	2550	2570	2510			2330	
^{20}Ne/^{22}Ne	12.6	12.6		12.5		12.5	12.3	12.6	12.5
							12.4		
	12.5	12.5	12.4	12.2	12.4			12.4	
^{22}Ne/^{21}Ne	30.1	30.8		26.9		24.8	22.6	21.2	23.6
							23.2		
	30.2	30.8	26.9	26.8	26.5			22.7	
^{36}Ar/^{38}Ar	5.11	5.18		5.13		5.10	5.08	5.11	5.05
							5.08		
	5.17	5.15	5.06	5.15	5.10			5.03	
^{40}Ar/^{36}Ar	0.514	0.499		0.599		0.648	0.812	1.15	0.729
							0.755		
	0.509	0.501	0.588	0.605	0.620			0.854	
^{4}He/^{20}Ne	70.0	60.5		51.9		50.5	53.5	51.5	51.5
							51.3		
	67.0	59.9	53.9	55.7	51.0			53.7	
^{4}He/^{36}Ar	432	344		281		276	250	257	197
							250		
	428	316	313	284	251			212	
^{20}Ne/^{36}Ar	6.17	5.69		5.42		5.47	4.68	4.99	3.83
							4.87		
	6.39	5.27	5.82	5.09	4.92			3.95	
^{84}Kr/^{36}Ar	$6.3 \cdot 10^{-4}$	$6.1 \cdot 10^{-4}$		$6.2 \cdot 10^{-4}$		$6.8 \cdot 10^{-4}$	$7.1 \cdot 10^{-4}$	$6.3 \cdot 10^{-4}$	$6.9 \cdot 10^{-4}$
							$6.8 \cdot 10^{-4}$		
	$6.4 \cdot 10^{-4}$	$6.5 \cdot 10^{-4}$	$6.0 \cdot 10^{-4}$	$6.3 \cdot 10^{-4}$	$6.1 \cdot 10^{-4}$			$6.8 \cdot 10^{-4}$	
^{132}Xe/^{36}Ar	$7.4 \cdot 10^{-5}$	$7.1 \cdot 10^{-5}$		$7.9 \cdot 10^{-5}$		$8.1 \cdot 10^{-5}$	$8.8 \cdot 10^{-5}$	$1.0 \cdot 10^{-4}$	$1.0 \cdot 10^{-4}$
							$8.6 \cdot 10^{-5}$		
	$8.8 \cdot 10^{-5}$	$8.0 \cdot 10^{-5}$	$7.1 \cdot 10^{-5}$	$7.5 \cdot 10^{-5}$	$7.3 \cdot 10^{-5}$			$1.0 \cdot 10^{-4}$	

Table 4. Rare gas concentrations in cm^3 STP/g in grain-size fractions from the Apollo 11 breccia 10021,20. Errors in isotope ratios 2%, errors in concentrations of He, Ne, and Ar about 3%, of Kr and Xe about 10%. Sample weights between 0.2–4 mg.

Nuclides and Ratios	< 5 μm	10–20 μm		25–35.5 μm		35.5–75 μm	75–120 μm	120–200 μm
^{3}He	1·71·10^{-4}	1·19·10^{-4}	1·31·10^{-4}	5·74·10^{-5}	6·62·10^{-5}	3·00·10^{-5}	1·75·10^{-5}	1·07·10^{-5}
^{4}He	5·19·10^{-1}	3·54·10^{-1}	3·96·10^{-1}	1·68·10^{-1}	1·99·10^{-1}	8·38·10^{-2}	4·50·10^{-2}	2·54·10^{-2}
^{20}Ne	7·76·10^{-3}	4·95·10^{-3}	5·45·10^{-3}	2·40·10^{-3}	2·91·10^{-3}	1·28·10^{-3}	6·12·10^{-4}	3·63·10^{-4}
^{21}Ne	1·90·10^{-5}	1·28·10^{-5}	1·41·10^{-5}	6·73·10^{-6}	7·97·10^{-6}	3·99·10^{-6}	2·30·10^{-6}	1·64·10^{-6}
^{22}Ne	6·05·10^{-4}	4·00·10^{-4}	4·29·10^{-4}	1·92·10^{-4}	2·30·10^{-4}	1·04·10^{-4}	4·95·10^{-5}	2·97·10^{-5}
^{36}Ar	8·10·10^{-4}	6·26·10^{-4}	6·24·10^{-4}	2·90·10^{-4}	3·46·10^{-4}	1·57·10^{-4}	7·88·10^{-5}	4·98·10^{-5}
^{38}Ar	1·63·10^{-4}	1·21·10^{-4}	1·27·10^{-4}	5·81·10^{-5}	6·82·10^{-5}	3·18·10^{-5}	1·63·10^{-5}	1·06·10^{-5}
^{40}Ar	2·05·10^{-3}	1·43·10^{-3}	1·57·10^{-3}	6·99·10^{-4}	8·27·10^{-4}	4·03·10^{-4}	2·35·10^{-4}	1·70·10^{-4}
^{84}Kr	4·7·10^{-7}	3·4·10^{-7}	3·6·10^{-7}	1·6·10^{-7}	1·7·10^{-7}	7·5·10^{-8}	3·8·10^{-8}	2·4·10^{-8}
^{132}Xe	1·0·10^{-7}	7·5·10^{-8}	8·8·10^{-8}	4·1·10^{-8}	3·9·10^{-8}	1·7·10^{-8}	8·4·10^{-9}	5·9·10^{-9}
^{4}He/^{3}He	3040	2970	3020	2930	3010	2790	2570	2370
^{20}Ne/^{22}Ne	12.8	12.4	12.7	12.5	12.7	12.3	12.4	12.2
^{22}Ne/^{21}Ne	31.8	31.3	30.4	28.5	28.9	26.1	21.5	18.1
^{36}Ar/^{38}Ar	4.97	5.17	4.91	4.99	5.07	4.94	4.83	4.70
^{40}Ar/^{36}Ar	2.53	2.28	2.52	2.41	2.39	2.57	2.98	3.41
^{4}He/^{20}Ne	66.9	71.5	72.7	70.0	68.4	65.5	73.5	70.0
^{4}He/^{36}Ar	641	565	635	579	575	534	571	510
^{20}Ne/^{36}Ar	9.58	7.91	8.73	8.28	8.41	8.15	7.77	7.29
^{84}Kr/^{36}Ar	5·9·10^{-4}	5·4·10^{-4}	5·8·10^{-4}	5·5·10^{-4}	4·8·10^{-4}	4·8·10^{-4}	4·9·10^{-4}	4·8·10^{-4}
^{132}Xe/^{36}Ar	1·2·10^{-4}	1·2·10^{-4}	1·4·10^{-4}	1·4·10^{-4}	1·1·10^{-4}	1·1·10^{-4}	1·1·10^{-4}	1·2·10^{-4}

Table 5. Rare gas concentrations in cm^3 STP/g in grain size fractions from the Apollo 11 breccia 10061,11. Errors in isotope ratios 2% errors in concentrations of He, Ne, and Ar about 3%, of Kr and Xe about 10%. Sample weights between 0.5 and 5 mg.

Nuclides and Ratios	< 25 μm	25–35.5 μm	35.5–75 μm	75–120 μm	> 120 μm
^{3}He	1·96·10^{-4}	7·40·10^{-5}	4·26·10^{-5}	2·32·10^{-5}	4·46·10^{-6}
^{4}He	5·68·10^{-1}	2·13·10^{-1}	1·16·10^{-1}	6·21·10^{-2}	8·75·10^{-3}
^{20}Ne	7·82·10^{-3}	2·84·10^{-3}	1·85·10^{-3}	9·58·10^{-4}	1·28·10^{-4}
^{21}Ne	1·93·10^{-5}	7·88·10^{-6}	5·75·10^{-6}	3·45·10^{-6}	9·48·10^{-7}
^{22}Ne	6·24·10^{-4}	2·30·10^{-4}	1·49·10^{-4}	7·89·10^{-5}	1·10·10^{-5}
^{36}Ar	1·21·10^{-3}	4·66·10^{-4}	2·93·10^{-4}	1·72·10^{-4}	2·06·10^{-5}
^{38}Ar	2·27·10^{-4}	8·88·10^{-5}	5·57·10^{-5}	3·23·10^{-5}	4·61·10^{-6}
^{40}Ar	2·23·10^{-3}	9·13·10^{-4}	5·94·10^{-4}	3·64·10^{-4}	7·20·10^{-5}
^{84}Kr	5·3·10^{-7}	2·5·10^{-7}	1·7·10^{-7}	1·1·10^{-7}	6·10^{-9}
^{132}Xe	1·3·10^{-7}	4·2·10^{-8}	4·5·10^{-8}	2·0·10^{-8}	3·10^{-9}
^{4}He/^{3}He	2910	2880	2720	2673	1960
^{20}Ne/^{22}Ne	12.5	12.3	12.4	12.1	11.6
^{22}Ne/^{21}Ne	32.3	29.2	25.9	22.9	11.6
^{36}Ar/^{38}Ar	5.33	5.25	5.26	5.32	4.5
^{40}Ar/^{36}Ar	1.83	1.96	2.03	2.12	3.5
^{4}He/^{20}Ne	72.6	75.0	62.7	64.8	68.4
^{4}He/^{36}Ar	469	457	396	361	425
^{20}Ne/^{36}Ar	6.46	6.09	6.31	5.57	6.21
^{84}Kr/^{36}Ar	4·4·10^{-4}	5·3·10^{-4}	5·9·10^{-4}	6·3·10^{-4}	3·10^{-4}
^{132}Xe/^{36}Ar	1·1·10^{-4}	0·91·10^{-4}	1·5·10^{-4}	1·1·10^{-4}	1·10^{-4}

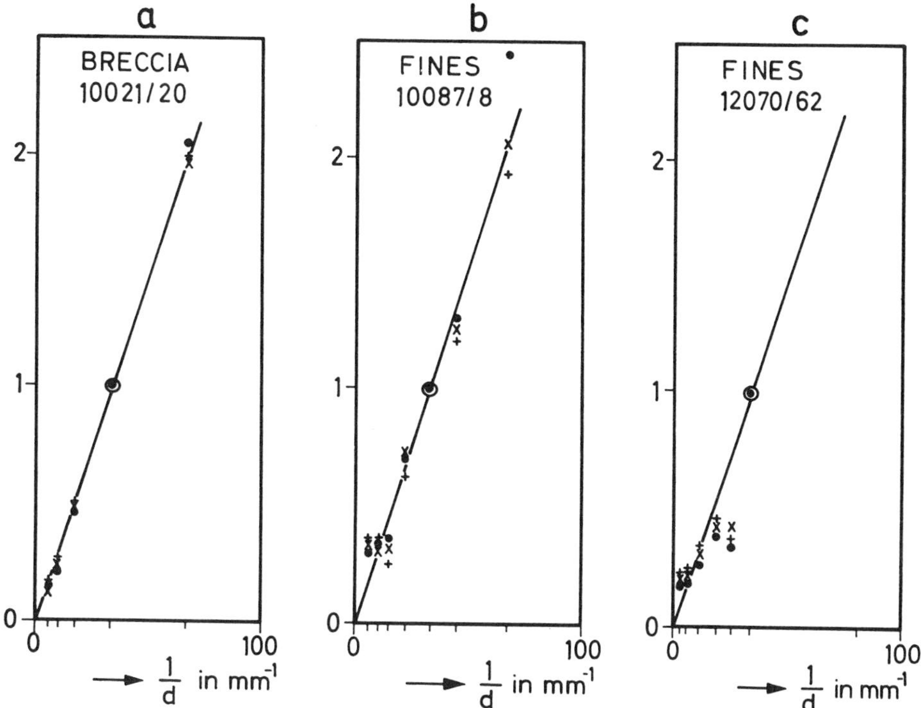

Fig. 3. Normalized rare gas concentrations vs. inverse mean grain size which is approximated by the arithmetic mean of the limiting mesh widths. ● = ^{4}He, × = ^{20}Ne, + = ^{36}Ar, ⊙ = normalization point ($\bar{d}$ = 30.2 μm, concentration ≡ 1. The 1/d values marked at the abscissa correspond to the following mean grain size fractions: fig.a: 15 μm, 30.2 μm, 55.2 μm, 97.5 μm, and 160 μm; fig. b: 15 μm, 22.5 μm, 30.2 μm, 44.7 μm, 64.5 μm, 97.5 μm, and 160 μm; fig. c: 12.5 μm, 30.2 μm, 44.7 μm, 55 μm, 97.5 μm, and 250 μm.

extrapolations of the linear relationships shown in Fig. 3a–3b (for the grain size fractions < 5 μm, often only 20 to 30% of the expected values, compare Tables 2 to 5).

One observes not only a grain size dependence of the gas concentrations but also of the ratios ^{4}He/^{3}He, ^{22}Ne/^{21}Ne, ^{36}Ar/^{38}Ar, and ^{40}Ar/^{36}Ar. This allows us to calculate separately the contents of cosmic ray components (e.g., ^{21}Ne) or of radiogenic ^{40}Ar and the contents of solar wind components.

If the ^{21}Ne values are plotted versus the ^{20}Ne values for different grain size fractions (Fig. 4a) a linear correlation is obtained so that the ordinate intercept represents the ^{21}Ne spallation component and the slope the ^{21}Ne/^{20}Ne-ratio of the solar wind component of neon (HEYMANN *et al.*, 1970). This is true for $(^{20}\text{Ne})_m \gg (^{20}\text{Ne})_{sp}$, the exact relation being given by

$$(^{21}\text{Ne})_m = (^{21}\text{Ne})_{sp} + \{(^{20}\text{Ne})_m - (^{20}\text{Ne})_{sp}\}\left(\frac{^{21}\text{Ne}}{^{20}\text{Ne}}\right)_{tr}$$

(m = measured; sp = spallation; tr = trapped).

According to another method developed by EBERHARDT *et al.* (1970), from a

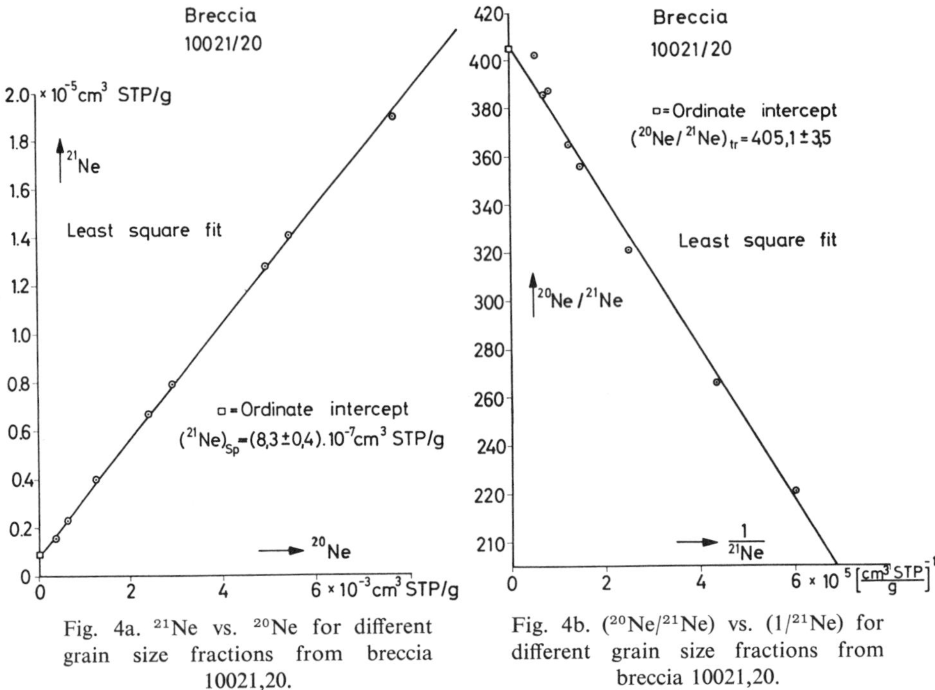

Fig. 4a. ^{21}Ne vs. ^{20}Ne for different grain size fractions from breccia 10021,20.

Fig. 4b. (^{20}Ne/^{21}Ne) vs. (1/^{21}Ne) for different grain size fractions from breccia 10021,20.

(^{20}Ne/^{21}Ne) versus (1/^{21}Ne) plot, the ratio (^{20}Ne/^{21}Ne)$_{tr}$ of the solar wind component can be determined from the ordinate intercept, whereas the slope is given by the expression

$$({}^{21}\mathrm{Ne})_{sp} \cdot \left\{ \left(\frac{{}^{20}\mathrm{Ne}}{{}^{21}\mathrm{Ne}} \right)_{sp} - \left(\frac{{}^{20}\mathrm{Ne}}{{}^{21}\mathrm{Ne}} \right)_{tr} \right\}.$$

Fig. 4b shows such a plot for the breccia 10021,20 as an example for the second method. Both methods were used to determine the concentrations and isotope abundance ratios of the spallation and solar wind components and to determine radiogenic argon. The results are presented in Table 6.

Table 6. Isotope ratios of trapped rare gas concentrations (cm^3 STP/g) of spallogenic and radiogenic rare gas nuclides and tentative figures for "exposure ages" (10^6 years) of fines and breccias. Ratios and concentrations are mean values of the results obtained by the different methods explained in the text.

	10084,18	10087,8	10021,20	10061,11	12070,62
(^{4}He/^{3}He)$_{tr}$	2620 ± 10	2780 ± 40	3070 ± 20	2930 ± 20	2740 ± 10
(^{20}Ne/^{21}Ne)$_{tr}$	408 ± 5	409 ± 6	408 ± 4	420 ± 20	430 ± 30
(^{22}Ne/^{21}Ne)$_{tr}$	32.6 ± 0.4	32.5 ± 0.4	32.7 ± 0.4	34 ± 1	32.6 ± 0.7
(^{36}Ar/^{38}Ar)$_{tr}$	5.21 ± 0.05	5.14 ± 0.04	5.06 ± 0.09	5.35 ± 0.03	5.18 ± 0.03
(^{40}Ar/^{36}Ar)$_{tr}$	0.95 ± 0.02	1.51 ± 0.03	2.18 ± 0.01	1.81 ± 0.02	0.44 ± 0.01
(^{3}He)$_{sp}$	(8 ± 3)·10^{-7}	(1.3 ± 0.5)·10^{-6}	(2.6 ± 0.5)·10^{-6}	(1.5 ± 0.1)·10^{-6}	(1.5 ± 0.3)·10^{-6}
(^{21}Ne)$_{sp}$	(4.5 ± 0.5)·10^{-7}	(0.58 ± 0.05)·10^{-6}	(0.8 ± 0.1)·10^{-6}	(1.0 ± 0.2)·10^{-6}	(0.5 ± 0.1)·10^{-6}
(^{38}Ar)$_{sp}$	(9 ± 6)·10^{-7}	(1.1 ± 0.4)·10^{-6}	(0.8 ± 1.3)·10^{-6}	(0.5 ± 0.5)·10^{-6}	(0.5 ± 0.2)·10^{-6}
(^{40}Ar)$_{rad}$	(7.5 ± 0.6)·10^{-5}	(3.5 ± 0.7)·10^{-5}	(7.0 ± 0.6)·10^{-5}	(6 ± 1)·10^{-5}	(2.7 ± 0.3)·10^{-5}
T(^{3}He)	80 ± 30	130 ± 50	260 ± 50	150 ± 10	150 ± 30
T(^{21}Ne)	330 ± 40	430 ± 40	600 ± 70	800 ± 200	360 ± 80
T(^{38}Ar)	600 ± 400	700 ± 200	600 ± 900	400 ± 300	300 ± 200

The grain size fractions were produced by pressing the fines or the pulverized breccias through stainless steel or nylon sieves with a stainless steel pistil. Breccias were pulverized by soft grinding in a stainless steel mortar. No liquids were used. It is assumed that the grain size of a given fraction can be characterized well enough by the arithmetic mean of the mesh widths of the two sieves used for that particular fraction. But admixtures of smaller grains adhering to larger ones or to each other can, of course, not be excluded. Linear relationships which exist between concentrations of surface-correlated components and the reciprocal grain diameter $1/d$ are thereby certainly falsified (see, e.g., Fig. 3c).

Such admixtures do, however, not affect the method described above for determining separately the concentrations of surface-correlated and volume-correlated gases. The postulates for this method are only: (1) that the thickness Δr of the surface layer considered is small compared to the grain diameter d; (2) that the surface concentrations of the surface correlated component are equal for all grain sizes; and (3) that the volume concentrations of the volume correlated component are equal for all grain sizes. The postulates are fulfilled if the grain size fractions are not too much different in their mineralogical and chemical composition, if the irradiation history of all grains is similar, and if the grains used are not too small. It appears that the method proposed may well be applied as long as the measured data points fall on a straight line in diagrams of the kind shown in Figs. 4a and 4b.

It is interesting to compare the relative concentrations in different grain size fractions of the same sample. Fig. 5 shows as an example the concentrations of the

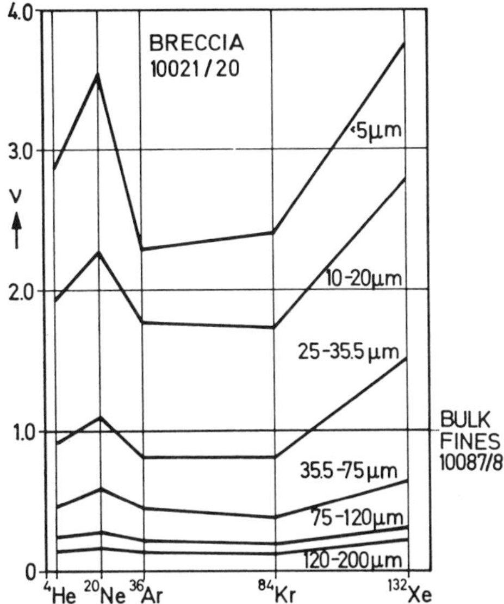

Fig. 5. Normalized rare gas concentrations

$$\nu = C[N(A)]_{10021,20}/C[N(A)]_{10087,8}$$

for different grain size fractions of the breccia 10021,20. (Concentration in bulk fines 10087,8 $\equiv$ 1).

main rare gas nuclides in different grain size fractions of the breccia 10021,20 relative to these concentrations in the bulk fines 10087,8. The excess of Ne and Xe already established from the bulk sample data of the breccia 10021,20 is manifested again in each single grain size fraction. The smaller grain size show higher excesses than the larger grain size fractions.

The comparison of the concentrations in the different grain size fractions of the breccia 10021,20 and the concentrations in the fractions of corresponding grain size of the fines 10087,8 shows again the characteristic relative increase of the neon and xenon concentrations in the breccia. This is especially accentuated in the small grain size fractions.

In spite of many possible objections one may tentatively calculate "exposure ages" of fines and breccias from their spallogenic components. For this calculation we used the following figures as production rates: $P_3 = P\,(^3He) = 1.10^{-8}\ cm^3/g.10^6$ years; $P_{21} = P\,(^{21}Ne) = 1.35.10^{-9}\ cm^3/g.10^6$ years; $P_{38} = P(^{38}Ar) = 1.43.10^{-9}\ cm^3/g.10^6$ years. According to BOGARD *et al.* (1970), these figures for P_{21} and P_{38} are valid for lunar samples with a mean (Si + Mg) content of 24% and a mean Ca content of 8.4%.

FINES AND BRECCIAS: DISCUSSION

The results obtained on the various grain size fractions of all breccias and fines investigated show that the rare gas concentrations to a large extent are surface correlated. These gas contents originate predominantly from the irradiation of lunar surface soil by solar wind particles. The abundance distribution of the trapped solar wind gases, however, does not correspond to the distribution in the actual solar wind. This is evident from the observation that, for example for the fines 10084, the abundance distribution in the bulk sample is distinctly different from the distributions in certain mineral fractions as, for example, in ilmenite fractions (EBERHARDT *et al.*, 1970) or in iron grain fractions (HINTENBERGER *et al.*, 1970a and b). A thorough study of the gas abundance distributions in various lunar samples may, however, lead to information with respect to the indigenous abundance distribution of these gases in the solar wind and with respect to modifications of this distribution caused by secondary processes acting on the lunar material.

Figs. 2 and 5 show clearly that the relative elemental abundances of neon, and usually also of xenon, are distinctly higher in the Apollo 11 breccias than in the fines. It is not possible to explain these excesses by a different exposure of the lunar material to a solar wind of a constant abundance pattern and a single fractionation process during or after implantation. Simple fractionation processes which might have occurred during trapping or by diffusion processes can only result in a monotonically mass dependent modification of the original rare gas abundance distribution. Simple fractionation processes might, at most, be made responsible for the difference in the gas abundance distributions of the Apollo 12 fines 12044 or 12070, and that of the fines 10084. The differences observed in the abundance patterns in Apollo 11 breccias and fines require, however, much more complicated models for their explanation. One might think of the following possibilities:

(1) The abundance distribution of the rare gases in the solar wind might have been variable with the time so that the Apollo 11 breccia material "saw" another abundance spectrum of the solar wind than did the Apollo 11 fines during their respective integrated exposure times. The exposure of the fines to solar wind might have lasted to the present time, whereas the exposure of the breccias ended at the latest with the formation of the breccia. It can therefore be expected that temporal variations in the solar wind composition might be most easily detected by comparing the composition of trapped gases in fines and breccias.

(2) Although the differences in the composition of the gases trapped in fines and breccias cannot be explained by a single fractionation mechanism, one certainly can invent complicated sequences of mass discriminating fractionation processes by which the gases partially and repeatedly are implanted and released. One possibly can construct suitable models for any observed gas composition. Mechanisms for gas implantation which might be considered in this context are, for example, (a) the direct implantation of the solar wind gases in the surface layers of mineral grains; (b) the implantation of indigenous lunar gases according to mechanisms which have been discussed in connection with the ^{40}Ar problem (see, e.g., HEYMANN and YANIV, 1970; EBERHARDT *et al.*, 1970), and (c) the implantation by shock effects (see, e.g., EBERHARDT *et al.*, 1970).

Mass discrimination and elemental fractionation may still have been enhanced by saturation effects. Repeatedly occurring partial gas losses as a result of solar or impact heating may, in addition, have played an important role. Such possibilities were already mentioned earlier, for example, by EBERHARDT *et al.* (1970), by HEYMANN and YANIV (1971), and by REYNOLDS (private communication). Considering all the various processes effective on the lunar surface as, for example, the periodic temperature changes due to solar irradiation or as the impact gardening of the lunar soil or as the processes which have caused the brecciation, one must expect that rather complicated models are just appropriate for the explanation of the various gas abundance distributions found in different lunar samples.

(3) Impacts might have lead to the formation of transient lunar atmospheres. These atmospheres might have contained released solar wind gases, radiogenic rare gases, and even rare gases of extralunar (e.g., meteoritic or cometary) origin. The implantation of gases from such atmospheres into lunar soil may have lead to the formation of a trapped gas component which is completely different from a trapped gas component directly implanted by the solar wind.

(4) A further mechanism is described by YANIV and HEYMANN (1971). They found in the breccia 10065, in particular in pores of that sample, rare gases with a considerably modified abundance distribution. This gas should not be surface correlated.

Considering the body of processes which possibly might have played a role it is difficult to select those which really are responsible for the observed peculiarities in the abundance patterns. But it appears certain that at least the processes mentioned above under (2) were partially effective. The question, to which extent also temporal variations in the primary solar wind composition or to which extent additionally implanted gases which did not originate from the solar wind played a role, will be discussed in another publication. For such a discussion further results of our investigation on 2 other breccias have to be taken into account which are not described in the present paper.

If, in spite of all doubts, "exposure ages" are calculated from the $(^3He)_{sp}$, $(^{21}Ne)_{sp}$ and $(^{38}Ar)_{sp}$ concentrations and from production rates as they usually are calculated, taking into account the (Mg + Si) and Ca content of rocks, then ^3He "ages" are obtained from 80 to 260 million years and ^{21}Ne ages from 330 to 800 million years. Moreover, the high ^{21}Ne "ages" and the ^{38}Ar "ages" agree within their limits of error. The ^{38}Ar ages are, however, less certain due to the large errors of the $(^{38}Ar)_{sp}$ values. But all ^3He ages are considerably lower than the ^{21}Ne "ages."

In order to avoid the term "age" one may say that the ratios $(^3He/^{21}Ne)_{sp}$ and $(^3He/^{38}Ar)_{sp}$ are always lower for fines and breccias than for crystalline rocks. According to the concentrations given in the Tables 6 and 9, the $(^3He/^{21}Ne)_{sp}$-ratios are <4.5 for fines and breccias and 5 to 10 for crystalline rocks, whereas the respective $(^3He/^{38}Ar)_{sp}$ ratios are <7.5 and 5 to 12.6. The $(^{21}Ne/^{38}Ar)_{sp}$ ratios, however, range from 0.5 to 2 for fines, breccias, and crystalline rocks.

The low $^3He/^{21}Ne$ ratios and the low ^3He ages might be interpreted as being caused by diffusion losses which preferentially would be effective for ^3He. Another possible explanation is that solar flare particles in contrast to cosmic ray particles would enhance the spallation components of ^{21}Ne and ^{38}Ar relatively more than that of ^3He. Rare gas nuclides produced by solar flare particles have been identified by YANIV *et al.* (1971) in single grains of the fines 10084 and 12070.

In spite of all uncertainties with the interpretation, the spallation isotope concentrations and, hence, the "exposure age" values are positively correlated with the sample number and, thus, are believed to reflect true differences in the histories of different samples.

Crystalline rocks

The concentrations and isotope ratios of rare gases extracted from crystalline rocks are compiled in Tables 7 and 8. Table 9 shows the spallogenic components of ^3He, ^{21}Ne, and ^{38}Ar, as well as the radiogenic ^4He and ^{40}Ar components. From the spallogenic components exposure ages were determined (Fig. 6). The production rates were calculated with the expressions given by BOGARD *et al.* (1970).

$P_3 = 1.0 \cdot 10^{-8}$ cm^3 STP/g sample $\cdot 10^6$ years,

$P_{21} = 0.56 \cdot 10^{-8}$ cm^3 STP/g (Mg + Si) $\cdot 10^6$ years,

$P_{38} = 1.7 \cdot 10^{-8}$ cm^3 STP/g Ca $\cdot 10^6$ years,

and the Mg, Si, and Ca concentrations compiled in Table 10. The exposure ages thus obtained are presented in columns 7, 8, and 9 of Table 9. Columns 10 and 11 of this table show the gas retention ages T_4 and T_{40} calculated from the radiogenic ^4He and ^{40}Ar of Table 9, and the K, U, and Th content of Table 10 (see Fig. 7). In most cases the rare gases and the Mg, Si, and K content, and in some cases the U and the Th content, have been measured on the same samples. Samples 10017, 10044, 10057, 12004, 12018, 12052, 12053, and 12063 are exchange samples from H. WÄNKE.

In most cases the ^3He and ^{21}Ne exposure ages agree to within better than 10%.

Our ^{3}He and ^{21}Ne ages agree also very well with the ages given by BOGARD *et al.* (1970).

It is interesting to compare our ^{3}He, ^{21}Ne, and ^{38}Ar ages with ^{81}Kr–Kr ages of other authors, because these ages are based on the ^{81}Kr halflife (see Table 11). In general the ^{81}Kr–Kr ages agree better with the ^{38}Ar ages than with the ^{21}Ne and ^{3}He ages. In cases where our ages show an increase $T_3 < T_{21} < T_{38}$, as for rocks 10017,33 and 12063,69, the ^{81}Kr–Kr age is higher than our ^{38}Ar age. Gas loss would be the simplest explanation. A comparison of the (U–He) and the (K–Ar) gas retention ages, however, show gas loss only for rock 12063,69.

The rocks can be grouped according to their (^{38}Ar/^{21}Ne)-ratios. This ratio is for the high Mg rocks 12002, 12004, 12018, and 12020 between 0.49 and 0.62, whereas for

Table 7. Rare gas concentrations in cm^3 STP/g in lunar rocks from Apollo 11 and Apollo 12. For the magnitude of errors see Table 2. Sample weights about 100 mg.

Nuclides and Ratios	10003,11	10049,20	12002,120	10202,120	12020,37	12064,15	12065,29
^{3}He	$1{\cdot}13\cdot10^{-6}$	$2{\cdot}84\cdot10^{-7}$	$8{\cdot}93\cdot10^{-7}$	$8{\cdot}94\cdot10^{-7}$	$7{\cdot}71\cdot10^{-7}$	$2{\cdot}05\cdot10^{-6}$	$1{\cdot}82\cdot10^{-6}$
^{4}He	$2{\cdot}24\cdot10^{-4}$	$8{\cdot}89\cdot10^{-4}$	$1{\cdot}51\cdot10^{-4}$	$1{\cdot}43\cdot10^{-4}$	$1{\cdot}42\cdot10^{-4}$	$1{\cdot}81\cdot10^{-4}$	$1{\cdot}51\cdot10^{-4}$
^{20}Ne	$1{\cdot}05\cdot10^{-6}$	$1{\cdot}64\cdot10^{-6}$	$2{\cdot}24\cdot10^{-7}$	$1{\cdot}80\cdot10^{-7}$	$2{\cdot}38\cdot10^{-7}$	$2{\cdot}95\cdot10^{-7}$	$3{\cdot}78\cdot10^{-7}$
^{21}Ne	$1{\cdot}29\cdot10^{-7}$	$3{\cdot}15\cdot10^{-8}$	$1{\cdot}59\cdot10^{-7}$	$1{\cdot}63\cdot10^{-7}$	$1{\cdot}19\cdot10^{-7}$	$3{\cdot}13\cdot10^{-7}$	$2{\cdot}94\cdot10^{-7}$
^{22}Ne	$2{\cdot}13\cdot10^{-7}$	$1{\cdot}61\cdot10^{-7}$	$1{\cdot}81\cdot10^{-7}$	$1{\cdot}83\cdot10^{-7}$	$1{\cdot}48\cdot10^{-7}$	$3{\cdot}62\cdot10^{-7}$	$3{\cdot}44\cdot10^{-7}$
^{36}Ar	$2{\cdot}86\cdot10^{-7}$	$3{\cdot}03\cdot10^{-7}$	$6{\cdot}88\cdot10^{-8}$	$5{\cdot}90\cdot10^{-8}$	$6{\cdot}06\cdot10^{-8}$	$1{\cdot}91\cdot10^{-7}$	$1{\cdot}74\cdot10^{-7}$
^{38}Ar	$2{\cdot}20\cdot10^{-7}$	$1{\cdot}02\cdot10^{-7}$	$8{\cdot}33\cdot10^{-8}$	$8{\cdot}12\cdot10^{-8}$	$6{\cdot}33\cdot10^{-8}$	$2{\cdot}94\cdot10^{-7}$	$2{\cdot}61\cdot10^{-7}$
^{40}Ar	$2{\cdot}63\cdot10^{-5}$	$6{\cdot}47\cdot10^{-5}$	$9{\cdot}28\cdot10^{-6}$	$9{\cdot}71\cdot10^{-6}$	$8{\cdot}99\cdot10^{-6}$	$1{\cdot}65\cdot10^{-5}$	$1{\cdot}32\cdot10^{-5}$
^{84}Kr	$2{\cdot}8\cdot10^{-10}$	$5{\cdot}4\cdot10^{-10}$	$7{\cdot}8\cdot10^{-11}$	$6{\cdot}4\cdot10^{-11}$	$1{\cdot}3\cdot10^{-10}$	$2{\cdot}2\cdot10^{-10}$	$1{\cdot}6\cdot10^{-10}$
^{132}Xe	$9{\cdot}9\cdot10^{-11}$	$3{\cdot}1\cdot10^{-10}$	$3{\cdot}3\cdot10^{-11}$	$2{\cdot}6\cdot10^{-11}$	$4{\cdot}8\cdot10^{-11}$	$6{\cdot}5\cdot10^{-11}$	$5{\cdot}2\cdot10^{-11}$
^{4}He/^{3}He	198	3130	169	160	184	88.3	83.0
^{20}Ne/^{22}Ne	4.93	10.2	1.24	0.984	1.61	0.815	1.10
^{22}Ne/^{21}Ne	1.65	5.11	1.14	1.12	1.24	1.16	1.17
^{36}Ar/^{38}Ar	1.30	2.97	0.826	0.727	0.957	0.650	0.667
^{40}Ar/^{36}Ar	92	214	135	165	148	86.4	75.9

Table 8. Rare gas concentrations in cm^3 STP/g in lunar rocks from Apollo 11 and Apollo 12. For the magnitude of errors, see Table 2. Sample weights about 100 mg.

Nuclides and Ratios	10017,33	10044,33	10057,40	12004,33	12018,41	12053,20	12063,69	12052,94
^{3}He	$2{\cdot}92\cdot10^{-6}$	$6{\cdot}37\cdot10^{-7}$	$3{\cdot}82\cdot10^{-7}$	$6{\cdot}32\cdot10^{-7}$	$1{\cdot}83\cdot10^{-6}$	$7{\cdot}90\cdot10^{-7}$	$6{\cdot}54\cdot10^{-7}$	$1{\cdot}18\cdot10^{-6}$
^{4}He	$4{\cdot}91\cdot10^{-4}$	$1{\cdot}49\cdot10^{-4}$	$5{\cdot}56\cdot10^{-4}$	$2{\cdot}32\cdot10^{-4}$	$4{\cdot}99\cdot10^{-5}$	$2{\cdot}09\cdot10^{-4}$	$1{\cdot}69\cdot10^{-4}$	$1{\cdot}58\cdot10^{-4}$
^{20}Ne	$7{\cdot}92\cdot10^{-7}$	$1{\cdot}18\cdot10^{-7}$	$6{\cdot}26\cdot10^{-8}$	$1{\cdot}78\cdot10^{-6}$	$4{\cdot}93\cdot10^{-7}$	$1{\cdot}01\cdot10^{-6}$	$2{\cdot}91\cdot10^{-7}$	$2{\cdot}76\cdot10^{-7}$
^{21}Ne	$4{\cdot}66\cdot10^{-7}$	$6{\cdot}71\cdot10^{-8}$	$5{\cdot}34\cdot10^{-8}$	$8{\cdot}99\cdot10^{-8}$	$3{\cdot}57\cdot10^{-7}$	$1{\cdot}33\cdot10^{-7}$	$1{\cdot}00\cdot10^{-7}$	$2{\cdot}15\cdot10^{-7}$
^{22}Ne	$5{\cdot}59\cdot10^{-7}$	$8{\cdot}38\cdot10^{-8}$	$6{\cdot}40\cdot10^{-8}$	$2{\cdot}36\cdot10^{-7}$	$4{\cdot}11\cdot10^{-7}$	$2{\cdot}20\cdot10^{-7}$	$1{\cdot}29\cdot10^{-7}$	$2{\cdot}50\cdot10^{-7}$
^{36}Ar	$4{\cdot}65\cdot10^{-7}$	$9{\cdot}54\cdot10^{-8}$	$4{\cdot}92\cdot10^{-8}$	$3{\cdot}59\cdot10^{-7}$	$1{\cdot}56\cdot10^{-7}$	$1{\cdot}92\cdot10^{-7}$	$9{\cdot}96\cdot10^{-8}$	$1{\cdot}28\cdot10^{-7}$
^{38}Ar	$6{\cdot}83\cdot10^{-7}$	$1{\cdot}24\cdot10^{-7}$	$8{\cdot}10\cdot10^{-8}$	$1{\cdot}21\cdot10^{-7}$	$2{\cdot}02\cdot10^{-7}$	$1{\cdot}36\cdot10^{-7}$	$1{\cdot}08\cdot10^{-7}$	$1{\cdot}82\cdot10^{-7}$
^{40}Ar	$4{\cdot}95\cdot10^{-5}$	$4{\cdot}47\cdot10^{-5}$	$5{\cdot}63\cdot10^{-5}$	$1{\cdot}33\cdot10^{-5}$	$6{\cdot}07\cdot10^{-6}$	$1{\cdot}33\cdot10^{-5}$	$1{\cdot}99\cdot10^{-5}$	$1{\cdot}41\cdot10^{-5}$
^{84}Kr	$7{\cdot}2\cdot10^{-10}$	$2{\cdot}0\cdot10^{-10}$	$1{\cdot}2\cdot10^{-10}$	$6{\cdot}0\cdot10^{-10}$	$5{\cdot}6\cdot10^{-10}$	$4{\cdot}7\cdot10^{-10}$	$5{\cdot}4\cdot10^{-10}$	$4{\cdot}9\cdot10^{-10}$
^{132}Xe	$3{\cdot}5\cdot10^{-10}$	$1{\cdot}1\cdot10^{-10}$	$8{\cdot}3\cdot10^{-11}$	$1{\cdot}5\cdot10^{-10}$	$1{\cdot}7\cdot10^{-10}$	$1{\cdot}7\cdot10^{-10}$	$1{\cdot}5\cdot10^{-10}$	$1{\cdot}4\cdot10^{-10}$
^{4}He/^{3}He	168	234	1460	367	27.3	265	258	134
^{20}Ne/^{22}Ne	1.42	1.41	0.978	7.54	1.20	4.59	2.26	1.10
^{22}Ne/^{21}Ne	1.20	1.25	1.20	2.63	1.15	1.65	1.29	1.16
^{36}Ar/^{38}Ar	0.681	0.769	0.607	2.97	0.772	1.41	0.922	0.703
^{40}Ar/^{36}Ar	106	469	1140	37.0	38.9	69.3	200	110

Table 9. Cosmogenic and radiogenic components of rare gases in lunar crystalline rocks from the Apollo 11 and Apollo 12 missions.

Sample Number	Spallogenic components in 10^{-7} cm^3 STP/g			Radiogenic components in 10^{-6} cm^3 STP/g		Exposure ages in 10^6 years			U/Th–He and K–Ar ages 10^9 years	
	$^3He_{sp}$	$^{21}Ne_{sp}$	$^{38}Ar_{sp}$	^4He_r	$^{40}Ar_r$	T_3	T_{21}	T_{38}	T_4	T_{40}
10003,11	11.0	1.27	1.86	134‡	26.0	110	100	140	1.9	4.0
10017,33	29.1	4.65	6.70	449	49.0	290	340	510	2.0	2.7
10021,20*	20.8†	2.96	5.4§	—	30.9	210	—	—	—	—
10044,32	6.35	0.669	1.22	142	44.6	64	50	93	2.0	3.9
10049,20	2.28†	0.275	0.44‡	737	64.4	23	21	36	2.8	2.9
10057,40	3.82	0.534	0.806	554	56.3	38	41	58	2.3	2.9
12002,120	8.93	1.63	0.802	138	9.65	89	96	89	1.8	2.5
12004,33	5.95	0.857	0.53§	128‡	13.1	60	53	45	2.1	3.1
12018,41	18.3	3.57	1.96	35.1	5.94	180	210	200	0.62	2.0
12020,37	7.68	1.19	0.583	132	8.94	77	71	56	2.2	2.7
12052,94	11.8	2.15	1.79	150	14.0	120	140	130	1.7	2.8
12053,20	7.70	1.31	1.01†	154†	13.2	77	87	83	2.4	2.8
12063,69	6.50	0.995	1.00	155	19.8	65	69	72	2.5	3.1
12064,15	20.5	3.13	2.94	175	16.3	210	220	190	2.6	2.7
12065,29	18.2	2.94	2.57	140	13.0	180	200	200	2.0	2.4

* Lithic fragment from breccia 10021,20. Because of the unknown chemistry, T_3 only has been determined. † = error between 5 and 10%, ‡ = error between 10 and 20%, § = error between 20 and 30%; otherwise errors of concentrations smaller than 5%. For errors of ages see text.

Table 10. Concentrations of Mg, Si, and Ca and of K, U, and Th used for the determination of exposure ages and of radiogenic ages.

Sample	Mg (wt. %)	Si (wt. %)	Ca mean (wt. %)	K (ppm)	U (ppm)	Th (ppm)
10003,11	4.9[e]	17.9[e]	8.0[e,f,d]	485[c,d]	0.26[b]	1.01[b]
10017,33	4.8[a]	19.6[a]	7.7[a,d,g,h,j]	2060[a]	0.69[a]	3.94[a]
10044,32	3.9[a]	20.1[a]	7.7[a,k,g,j]	860[a]	0.28[a]	0.98[a]
10049,20	4.4[a]	20.0[a]	7.1[a,f]	2280[a]	0.81[a]	4.03[a]
10057,40	4.2[a]	18.9[a]	8.2[a,l,k,m,i]	2010[a]	0.80[a]	3.94[a]
12002,120	8.90[n]	20.8[n]	5.3[n,t,u,v]	450[n]	0.30[o]	1.1[o,t]
12004,33	7.49[n]	21.5[n]	6.9[n,w,z,α,β,γ]	410[n]	0.238[n]	0.82[n]
12018,41	8.69[n]	21.1[n]	5.8[n,y,x,t,z]	410[n]	0.252[n]	0.85[n]
12020,37	9.44[n]	20.5[n]	6.1[n,ξ,y,w]	380[n]	0.242[p]	0.71[n]
12053,20	4.86[n]	22.1[n]	7.2[n,α,v]	500[n]	0.242[n]	0.87[n]
12063,69	5.05[n]	20.9[n]	8.2[n,v,z,w]	630[n]	0.236[n]	0.82[n]
12064,15	3.85[r]	21.7[r]	9.3[n,v,η]	670[n]	0.25[q]	0.91[q]
12065,29	5.12[n]	21.9[n]	7.5[n,z,β,γ,t]	660[n]	0.282[s]	0.991[s]
12052,94	5.10[n]	22.1[n]	8.0[n,β,y]	560[n]	0.356[n]	1.28[n]

[a] WÄNKE *et al.* (1970). [b] O'KELLEY *et al.* (1970). [c] GAST *et al.* 1(70). [d] COMPSTON *et al.* (1970). [e] GOLES *et al.* (1970). [f] ROSE *et al.* (1970). [g] TERA *et al.* (1970). [h] MAXWELL *et al.* (1970). [i] MORRISON *et al.* (1970). [j] WAKITA *et al.* (1970). [k] ENGEL and ENGEL (1970). [l] BEGEMANN *et al.* (1970). [m] GANAPATHY *et al.* (1970). [n] WÄNKE *et al.* (1971). [o] TAYLOR *et al.* (1971). [p] Mean of rocks 04,18, 53, and 63. [q] Mean of SILVER (1971) and TATSUMOTO *et al.* (1971). [r] BIGGAR *et al.* (1971). [s] RANCITELLI *et al.* (1971). [t] GOLES *et al.* (1971). [u] FINKEL *et al.* (1971). [v] WILLIS *et al.* (1971). [w] WAKITA *et al.* (1971). [x] ROSE *et al.* (1971). [y] COMPSTON *et al.* (1971). [z] BOUCHET *et al.* (1971). [α] MORRISON *et al.* (1971). [β] MAXWELL and WIIK (1971). [γ] SMALES (1971). [η] SCOON (1971). [ξ] ANNELL *et al.* (1971).

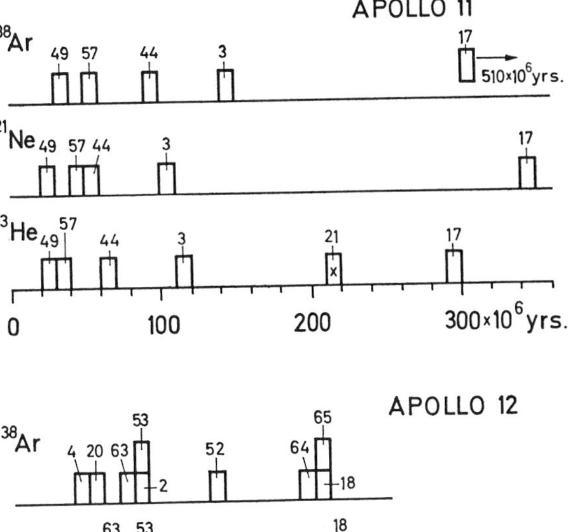

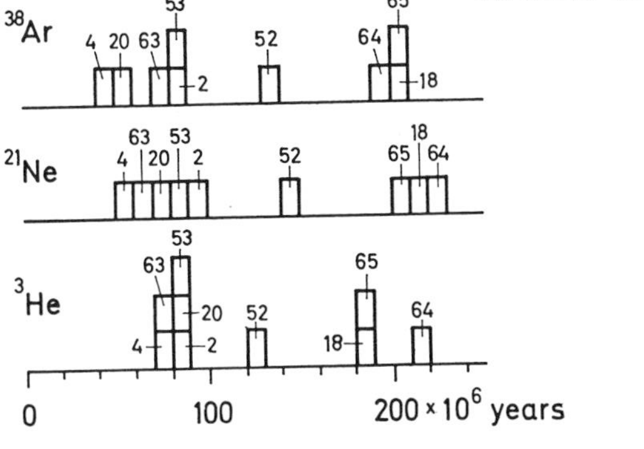

Fig. 6. Exposure ages of lunar crystalline rocks.

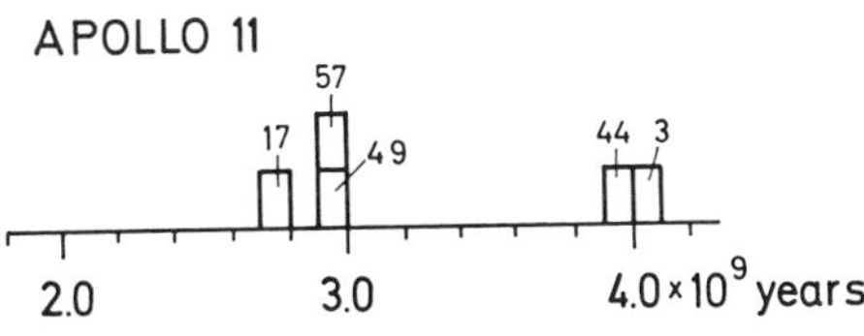

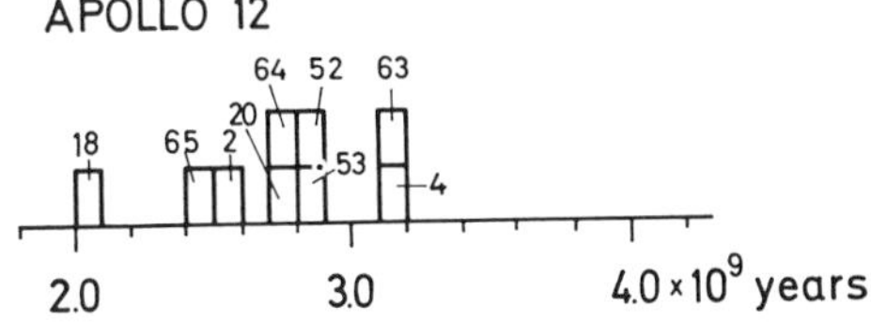

Fig. 7. K–Ar ages of lunar crystalline rocks.

Table 11. Comparison of the ^{3}He, ^{21}Ne, and ^{38}Ar exposure ages of the present paper, with the ^{81}Kr–Kr exposure ages determined by MARTI *et al.* and by BOCHSLER *et al.* Ages in millions of years.

1	2[a]	3[a]	4[a]	5[b,c]	6[d]
Sample	T_3	T_{21}	T_{38}	T_{81}	T_{81}
10003	110	100	140	—	129
10017	290	340	510	{509 {449	480
10057	38	41	58	47	52.5
12002	89	96	89	94	—
12004	60	53	45	—	58
12018	180	210	200	195	—
12053	77	87	83	—	99
12063	65	69	72	95	—

[a] Present paper; [b] MARTI *et al.* (1970); [c] MARTI and LUGMAIR (1971); [d] BOCHSLER *et al.* (1971).

the low Mg rocks 12053, 12063, 12064, and 12054, ratios between 0.88 and 1.02 were found. The Apollo 11 rocks range between 1.45 and 1.81 (see tables 7 and 8).

Rare gas concentrations were also determined on a crystalline rock inclusion from the breccia 10021,20 (see Table 9). Since the chemical composition is not perfectly enough known, only the ^{3}He ages was determined for this sample. In addition, magnetically separated mineral fractions of the crystalline rock 12064,15 were investigated. A high ^{131}Xe content was found but the details will be presented elsewhere. In the present paper it may be mentioned only that the ^{3}He contents in the nonmagnetic, in the strong-magnetic, and in the bulk material are $6.4 \cdot 10^{-7}$, $2.9 \cdot 10^{-6}$, and $2.8 \cdot 10^{-6}$ cm^3 STP/g, respectively, and that obviously more than 75% helium was lost from the nonmagnetic plagioclase fraction.

Acknowledgments—The technical assistance of Christa Müller and of Ingrid Raczek is gratefully acknowledged. We thank NASA for the generous supply of lunar material and to the Bundesministerium für Bildung und Wissenschaft for giving support for the NASA contract.

REFERENCES

ALLER L. H. (1961) *The Abundance of the Elements.* Interscience.

ANNELL C. S., CARRON M. K., CHRISTIAN R. P., CUTTITTA F., DWORNIK E. J., HELZ A. W., LIGON D. T., and ROSE H. J. (1971) Chemical and spectrographic analyses of lunar samples from Apollo 12 mission. Second Lunar Science Conference (unpublished proceedings).

BEGEMANN F., VILCSEK E., RIEDER R., BORN W., and WÄNKE H. (1970) Cosmic-ray produced radioisotopes in lunar samples from the Sea of Tranquillity (Apollo 11). *Proc. Apollo 11 Lunar Sci. Conf., Geochim. Cosmochim. Acta* Suppl. 1, Vol. 2, pp. 995–1005. Pergamon.

BIGGAR G. M., O'HARA M. J., and PECKETT A. (1971) Origin, eruption and crystallization of protohypersthene basalts from the Ocean of Storms. Second Lunar Science Conference (unpublished proceedings).

BOCHSLER P., EBERHARDT P., GEISS J., GRAF H., GRÖGLER N., KRÄHENBÜHL U., MÖRGELI M., SCHWALLER H., and STETTLER A. (1971) Potassium-argon ages, exposure ages, and radiation history of lunar rocks. Second Lunar Science Conference (unpublished proceedings).

BOGARD D. D., FUNKHOUSER J. G., SCHAEFFER O. A., and ZÄHRINGER J. (1970) Noble gas abundances in lunar material, II. Cosmic ray spallation products and radiation ages from Mare Tranquillitatis and Oceanus Procellarum. *J. Geophys. Res.* (submitted).

Bouchet M., Kaplan G., Voudon A., and Betroletti M.-J. (1971) Spark mass spectrometric analysis of major and minor elements in six lunar samples. Second Lunar Science Conference (unpublished proceedings).

Compston W., Chappell B. W., Arriens P. A., and Vernon M. J. (1970) The chemistry and age of Apollo 11 lunar material. *Proc. Apollo 11 Lunar Sci. Conf., Geochim. Cosmochim. Acta* Suppl. 1, Vol 2, pp. 1007–1027. Pergamon.

Compston W., Berry H., Vernon M. J., Chappell B. W., and Kaye M. J. (1971) Rubidium–strontium chronology and chemistry of lunar material from the Ocean of Storms. Second Lunar Science Conference (unpublished proceedings).

Eberhardt P., Eugster O., Geiss J., and Marti K. (1966) Rare gas measurements in 30 stone meteorites. *Z. Naturforsch.* **21a**, 414–426.

Eberhardt P., Geiss J., Graf H., Grögler N., Krähenbühl U., Schwaller H., Schwarzmüller J., and Stettler A. (1970) Trapped solar wind noble gases, exposure age and K/Ar-age in Apollo 11 lunar fine material. *Proc. Apollo 11 Lunar Sci. Conf., Geochim. Cosmochim. Acta* Suppl. 1, Vol. 2, pp. 1037–1070. Pergamon.

Engel A. E. J. and Engel C. G. (1970) Lunar rock compositions and some interpretations. *Proc. Apollo 11 Lunar Sci. Conf., Geochim. Cosmochim. Acta* Suppl. 1, Vol. 2, pp. 1081–1084. Pergamon.

Finkel R. C., Arnold J. R., Reedy R. C., Fluchter J. S., Loosli H. H., Evans J. C., Shedlovsky J. P., Imamura M., and Delany A. C. (1971) Depth variations of cosmogenic nuclides in a lunar surface rock. Second Lunar Science Conference (unpublished proceedings).

Funkhouser J. G., Schaeffer O. A., Bogard D. D., and Zähringer J. (1970a) Gas analysis of the lunar surface. *Science* **167**, 561–563.

Funkhouser J. G., Schaeffer O. A., Bogard D. D., and Zähringer J. (1970b) Noble gas abundances in lunar material I. Solar wind implanted gases in Mare Tranquillitatis and Oceanus Procellarum. *J. Astrophys.* (submitted).

Ganapathy R., Keays R. R., Laul J. C., and Anders E. (1970) Trace elements in Apollo 11 lunar rocks: Implications for meteorite influx and origin of moon. *Proc. Apollo 11 Lunar Sci. Conf., Geochim. Cosmochim. Acta* Suppl. 1, Vol. 2, pp. 1117–1142. Pergamon.

Gast P. W., Hubbard N. J., and Wiesmann H. (1970) Chemical compositions and petrogenesis of basalts from Tranquillity Base. *Proc. Apollo 11 Lunar Sci. Conf., Geochim. Cosmochim. Acta* Suppl. 1, Vol. 2, pp. 1143–1163. Pergamon.

Goles G. G., Randle K., Osawa M., Schmitt R. A., Wakita H., Ehmann W. D., and Morgan J. W. (1970) Elemental abundances by instrumental activation analyses in chips from 27 lunar rocks. *Proc. Apollo 11 Lunar Sci. Conf., Geochim. Cosmochim. Acta* Suppl. 1, Vol. 2, pp. 1165–1175. Pergamon.

Goles G. G., Duncan A. R., Osawa M., Martin M. R., Beyer R. L., Lindstrom D. J., and Randle K. (1971) Analyses of Apollo 12 specimens and mixing model for Apollo XII "soils." Second Lunar Science Conference (unpublished proceedings).

Heymann D., Yaniv A., Adams J. A. S., and Fryer G. E. (1970) Inert gases in lunar samples. *Science* **167**, 555–558.

Heymann D., and Yaniv A. (1970) Inert gases in the fines from the Sea of Tranquillity. *Proc. Apollo 11 Lunar Sci. Conf., Geochim. Cosmochim. Acta* Suppl. 1, Vol. 2, pp. 1247–1259. Pergamon.

Heymann D., and Yaniv A. (1971) Inert gases in breccia 10065: Release by vacuum crushing. Second Lunar Science Conference (unpublished proceedings).

Hintenberger H., Weber H. W., Voshage H., Wänke H., Begemann F., Vilcsek E., and Wlotzka F. (1970a) Rare gases, hydrogen, and nitrogen: Concentrations and isotopic composition in lunar material. *Science* **167**, 543–545.

Hintenberger H., Weber H. W., Voshage H., Wänke H., Begemann F., and Wlotzka F. (1970b) Concentrations and isotopic abundances of the rare gases, hydrogen and nitrogen in Apollo 11 lunar matter. *Proc. Apollo 11 Lunar Sci. Conf., Geochim. Cosmochim. Acta* Suppl 1, Vol. 2, pp. 1269–1282. Pergamon.

Hohenberg C. M., Davis P. K., Kaiser W. A., Lewis R. S., and Reynolds J. H. (1970) Trapped and cosmogenic rare gases from stepwise heating of Apollo 11 samples. *Proc. Apollo 11 Lunar Sci. Conf., Geochim. Cosmochim. Acta* Suppl. 1, Vol. 2, pp. 1283–1309. Pergamon.

KIRSTEN T., MÜLLER O., STEINBRUNN F., and ZÄHRINGER J. (1970) Study of distribution and variations of rare gases in lunar material by a microprobe technique. *Proc. Apollo 11 Lunar Sci. Conf., Geochim. Cosmochim. Acta* Suppl. 1, Vol. 2, pp. 1331–1343. Pergamon.

LSPET (Lunar Sample Preliminary Examination Team) (1969) Preliminary examination of lunar samples from Apollo 11. *Science* **165**, 1211–1227.

LSPET (Lunar Sample Preliminary Examination Team) (1970) Preliminary examination of lunar samples from Apollo 12. *Science* **167**, 1325–1339.

MARTI K., LUGMAIR G. W., and UREY H. C. (1970) Solar wind gases, cosmic-ray spallation products, and the irradiation history of Apollo 11 samples. *Proc. Apollo 11 Lunar Sci. Conf., Geochim. Cosmochim. Acta* Suppl. 1, Vol. 2, pp. 1357–1367. Pergamon.

MARTI K., and LUGMAIR G. W. (1971) ^{81}Kr–Kr and K–^{40}Ar ages, cosmic-ray spallation products, and neutron effects in Apollo 11 and Apollo 12 lunar samples. Second Lunar Science Conference (unpublished proceedings).

MAXWELL J. A., PECK L. C., and WIIK H. B. (1970) Chemical composition of Apollo 11 lunar samples 10017, 10020, 10072, and 10084. *Proc. Apollo 11 Lunar Sci. Conf., Geochim. Cosmochim. Acta* Suppl. 1, Vol. 2, pp. 1369–1374. Pergamon.

MAXWELL J. A., and WIIK H. B. (1971) Chemical composition of Apollo 12 lunar samples 12004, 12033, 12051, 12052, and 12065. Second Lunar Science Conference (unpublished proceedings).

MORRISON G. H., GERARD J. T., KASHUBA A. T., GANGADHARAM E. V., ROTHENBERG A. M., POTTER N. M., and MILLER G. B. (1970) Elemental abundances of lunar soil and rocks. *Proc. Apollo 11 Lunar Sci. Conf., Geochim. Cosmochim. Acta* Suppl. 1, Vol. 2, pp. 1383–1392. Pergamon.

MORRISON G. H., GERARD J. T., POTTER N. M., GANGADHARAM E. V., ROTHENBERG A. M., and BURDO R. A. (1971) Elemental abundances of lunar soil and rocks from Apollo 12. Second Lunar Science Conference (unpublished proceedings).

NYQUIST L. E. and PEPIN R. O. (1971) Rare gases in Apollo 12 surface and subsurface fine materials. Second Lunar Science Conference (unpublished proceedings).

O'KELLEY G. D., ELDRIDGE J. S., SCHONFELD E., and BELL P. R. (1970) Primordial radionuclide abundances, solar proton and cosmic ray effects and ages of Apollo 11 lunar samples by nondestructive gamma-ray spectrometry. *Proc. Apollo 11 Lunar Sci. Conf., Geochim. Cosmochim. Acta* Suppl. 1, Vol. 2, pp. 1407–1423. Pergamon.

PEPIN R. O., NYQUIST L. E., PHINNEY D., and BLACK D. C. (1970) Rare gases in Apollo 11 lunar material. *Proc. Apollo 11 Lunar Sci. Conf., Geochim. Cosmochim. Acta* Suppl. 1, Vol. 2, pp. 1435–1454. Pergamon.

RANCITELLI L. A., PERKINS R. W., FELIX W. D., and WOGMAN N. A. (1971) Cosmogenic and primordial radionuclide measurements in Apollo 12 lunar samples by nondestructive analysis. Second Lunar Science Conference (unpublished proceedings).

ROSE H. J., CUTTITTA F., DWORNIK E. J., CARRON M. K., CHRISTIAN R. P., LINDSAY J. R., LIGON D. T., and LARSON R. R. (1970) Semimicro X-ray fluorescence analysis of lunar samples. *Proc. Apollo 11 Lunar Sci. Conf., Geochim. Cosmochim. Acta* Suppl. 1, Vol. 2, pp. 1493–1497. Pergamon.

ROSE H. J., CUTTITTA F., ANNELL C. S., CARRON M. K., CHRISTIAN R. P., DWORNIK E. J., HELZ A. W., and LIGON D. T. (1971) Semimicroanalysis of Apollo 12 samples. Second Lunar Science Conference (unpublished proceedings).

SCOON J. H. (1971) Quantitative chemical analyses of lunar samples 12040,36 and 12064,38. Second Lunar Science Conference (unpublished proceedings).

SILVER L. T. (1971) U–Th–Pb isotope relations in Apollo 11 and Apollo 12 lunar samples. Second Lunar Science Conference (unpublished proceedings).

SMALES A. A. (1971) Elemental compositions of lunar surface material. Part 2. Second Lunar Science Conference (unpublished proceedings).

SUESS H. E., and UREY H. C. (1958) Häufigkeit der Elemente in den Planeten und Meteoriten. In *Handbuch der Physik* (editor S. Flügge), Vol. 51, pp. 296–323. Springer.

TATSUMOTO M., KNIGHT R. J., and DOE B. R. (1971) U–Th–Pb systematics of Apollo 12 lunar samples. Second Lunar Science Conference (unpublished proceedings).

TAYLOR S. R., KAYE M., GRAHAM A., RUDOWSKI R., and MUIR P. (1971) Trace element chemistry of lunar samples from the Ocean of Storms. Second Lunar Science Conference (unpublished proceedings).

TERA F., EUGSTER O., BURNETT D. S., and WASSERBURG G. J. (1970) Comparative study of Li, Na, K, Rb, Cs, Ca, Sr, and Ba abundances in achondrites and in Apollo 11 lunar samples. *Proc. Apollo 11 Lunar Sci. Conf., Geochim. Cosmochim. Acta* Suppl. 1, Vol. 2, pp. 1637–1657. Pergamon.

TURNER G. (1970) Argon-40/argon-39 dating of lunar rock samples. *Proc. Apollo 11 Lunar Sci. Conf., Geochim. Cosmochim. Acta* Suppl. 1, Vol. 2, pp. 1665–1684. Pergamon.

WÄNKE H., RIEDER R., BADDENHAUSEN H., SPETTEL B., TESCHKE F., QUIJANO-RICO M., and BALACESCU A. (1970) Major and trace elements in lunar material. *Proc. Apollo 11 Lunar Sci. Conf., Geochim. Cosmochim. Acta*, Suppl. 1, Vol. 2, pp. 1719–1727. Pergamon.

WÄNKE H., WLOTZKA F., TESCHKE F., BADDENHAUSEN H., SPETTEL B., BALACESCU A., QUIJANO-RICO M., JAGOUTZ E., and RIEDER R. (1971) Major and trace elements in Apollo 12 samples and studies on lunar metallic iron particles. Second Lunar Science Conference (unpublished proceedings).

WAKITA H., SCHMITT R. A., and REY P. (1970) Elemental abundances of major, minor, and trace elements in Apollo 11 lunar rocks, soil and core samples. *Proc. Apollo 11 Lunar Sci. Conf., Geochim. Cosmochim. Acta* Suppl. 1, Vol. 2, pp. 1685–1717. Pergamon.

WAKITA H., REY P., and SCHMITT R. A. (1971) Abundances of the 14 rare earth elements plus 22 major, minor, and trace elements in the Apollo 12 rock and soil samples. Second Lunar Science Conference (unpublished proceedings).

WILLIS J. P., AHRENS L. H., DANCHIN R. V., ERLANK A. J., GURNEY J. J., HOFMEYR P. K., McCARTHY T. S., and ORREN M. J. (1971) Some interelement relationships between lunar rocks and fines, and stony meteorites. Second Lunar Science Conference (unpublished proceedings).

YANIV A., and HEYMANN D. (1971) Inert gases from Apollo 11 and Apollo 12 fines: Reversals in the trends of relative element abundances. Second Lunar Science Conference (unpublished proceedings).

YANIV A., TAYLOR G. J., ALLEN S., and HEYMANN D. (1971) Stable rare gas isotopes produced by solar flares in single particles of Apollo 11 and Apollo 12 fines. Second Lunar Science Conference (unpublished proceedings).

Proceedings of the Second Lunar Science Conference, Volume 2, pp. 1627–1641.
The M.I.T. Press, 1971.

Rare gas measurements in three mineral separates of rock 12013,10,31

W. A. KAISER
Department of Physics, University of California, Berkeley, California 94720

(*Received 24 February 1971; accepted in revised form 20 April 1971*)

Abstract—He, Ne, Ar, Kr, and Xe were measured in 3 different mineral separates of 12013,10,31 selected by handpicking and identified by microprobe analysis (WALTER, 1970). About 10 mg of each fraction were available for the rare gas analyses. The rare gas measurements showed clear differences due to different concentrations of K, Ba, rare earth elements (REE) etc. The K–Ar ages were in good agreement with TURNER's (1970) results. The Kr and Xe amounts were predominantly produced by cosmic ray bombardment. The ^{126}Xe-exposure ages varied from 40×10^6 y to 47×10^6 y). In one sample cosmogenic ^{126}Xe was, by calculation, 98.5% from Ba as a target; in another, this percentage was 76. Thus we could make a good measurement of the cosmogenic Xe spectrum for Ba and a rough estimate of the spectrum for REE. The spectra were generally in agreement with published target data for 730 MeV protons (FUNK *et al.*, 1967; HOHENBERG and ROWE, 1970). No trapped Xe and Kr was found in the separates. No ^{129}Xe coming from in situ decay of ^{129}I was measured. No evidence for or against the ^{244}Pu hypothesis was seen, confirming the results of ALEXANDER (1970) and LUNATIC ASYLUM (1970).

INTRODUCTION

THIS WORK WAS STARTED under the impression that Apollo rock 12013 had a crystallization age of 4.5×10^9 y (WASSERBURG, 1970), in which case it should be highly possible to detect traces of extinct radioactivities in lunar material. The extraordinarily high abundances of Ba, REE, U, and nearly all other trace elements reported by the LSPET (1970), pointed to this rock as a unique source for studies. We should have been able to see radiogenic ^{129}Xe coming from decay in situ of ^{129}I and a fissiogenic Xe contribution from ^{244}Pu.

Thus, the situation appeared unusually favorable in 12013. Since ALEXANDER (1970) was investigating a homogenized sample 24 + 37, mostly light material, and the LUNATIC ASYLUM (1970) had measured several bulk pieces, our approach was to make separations, in hopes of achieving in one or the other of the fractions enrichments of the trace elements, especially U. We were successful in this respect as judged by microprobe investigations on representative material, which provided us with mineralogical data we could use to identify our samples with samples measured elsewhere for trace element abundances. Meanwhile several investigations (TURNER, 1970; LUNATIC ASYLUM, 1970; TATSUMOTO, 1970) showed that the actual crystallization age of this rock was closer to 4.0×10^9 y, making it doubtful that evidences of extinct radioactivity could be detected.

EXPERIMENTAL TECHNIQUES

Sample 12013,10,31 was an interior chip consisting of a light and a dark phase. We crushed the sample in an agate mortar and separated by sieving particles of a size between 50 and 150 μ, in order to acquire as many pure crystals (without intergrowths) as possible. Then, we put the whole sample in reagent grade acetone and cleaned it in an ultrasonic-cleaner for 5 min, separating the grains from each other in the process. Afterwards, we sieved the sample again using a special filtering funnel. This consisted of 3 quartz sections, which enabled us to sieve and filter simultaneously. We used a "membrane" filter with a pore size of 0.04 μ to remove the fine particles. The crystals larger than 50 μ were collected on a stainless steel screen. We made separates from this material handpicking under a stereomicroscope according to color. We were able to separate 4 different mineral agglomerates: milky white, clear, black vesicular, and greenish, but we present results for only three of the fractions in this paper in the interests of brevity. (Additional measurements on sample 12013,10,31 including a detailed stepwise temperature analysis of the residual fine material, will be described elsewhere (KAISER, 1971).

The samples were wrapped in Reynolds Al-foil (heavy duty), previously cleaned by polishing with a tissue wipe and ultrasonic agitation in acetone. The Al-foil was always handled with gloves after cleaning. The samples were measured in BMS 4, previously described as system 1 in HOHENBERG *et al.* (1970), a Pyrex glass-spectrometer with an all-metal-sample system. Separate analyses were made of the gases released by one-hour heating at 700°C and 1550°C (well above the melting point for all samples). The temperature was measured by two thermocouples, which were consistently in good agreement with each other. The temperatures specified (referring to the base of crucible) are correct to within ± 15°C. We have made numerous measurements of the blanks. Since the samples weighed only about 10 mg each and were wrapped in about 200 mg of Al-foil, it was important to include the Al-foil in our blanks. Therefore, we ran pure Al-foil of the same kind and amounts as used to wrap our samples under the same temperature and measurement conditions as the samples. The results for the different gases were in good agreement with each other except for Kr. The isotopic compositions found were those of air. Three Al-samples were run (Table 1).

All results are corrected for spectrometer background and for discrimination except for He, where no discrimination-value was available. For He we applied a discrimination factor of 1. (Results of He measurements on an interlaboratory standard sample of the Bruderheim meteorite indicated that this discrimination factor is essentially unity.) The spectrometer background corrections were essentially negligible except in instances to be noted below. The errors listed in the tables include all known sources and are based upon observed reproducibilities of peak-ratios, air-pipette analyses, or blank runs as the case may be. A few of the errors had to be increased further for causes we now note: During this series of runs we had some problems with the ^{20}Ne background, caused by a saturation of the Vac-ion pump of the spectrometer and requiring a somewhat large background correction. The error assigned for this correction was the full value of the background and is certainly too high. The ^{20}Ne values were also corrected for doubly charged ^{40}Ar but this correction was small. We had a high "dirt" background for ^{78}Kr. The error listed in this case includes the full value of this correction.

TRACE ELEMENT ABUNDANCES

Mineralogical data were obtained for each of the three fractions by electron microprobe analysis of a few grains (WALTER, 1970). Mineral agglomerate 31A, a

Table 1. Rare gases in the Al-foil [$\times 10^{-8}$ cc STP/g Al]. The foil is Reynolds Wrap, heavy duty, 0.001″ thickness. Except for Kr, we have averaged these separate runs.

T	^{4}He	^{22}Ne	^{36}Ar	^{83}Kr Weight			^{130}Xe
				132 mg	167 mg	186 mg	
700°C	31.9 ± 3.8	0.013 ± 0.001	0.029 ± 0.001	0.00006	0.00026	0.00027	0.000022 ± 0.000002
1550°C	2.7 ± 0.3	0.005 ± 0.001	0.056 ± 0.003	0.00020	0.00058	0.00023	0.000041 ± 0.000008
Total	34.6 ± 4	0.018 ± 0.002	0.085 ± 0.004	0.00026	0.00084	0.00050	0.000063 ± 0.000010

Table 2. Estimates of trace element abundances for the separates.

	31A Milky white	31B Clear	31C Black vesicular		Remarks
K (%)	4.32 ± 0.78	1.67 ± ?	$\frac{0.57 + 0.48^g}{2} = 0.53 \pm 0.07^i$		[a] $Y(31A, 31B) = \frac{\overline{Y(15, 18, 41, 44, 37 + 24)}}{Y(\text{chondrites})} \times Y(\text{chondrites})$ Y(chondrites = 1.9 ppm) (WAKITA and SCHMITT, 1970, p. 174, Fig. 3)
Ba (ppm)	6360 ± 970	3450 ± ?	$\frac{895 + 1330^g}{2} = 1112 \pm 217^i$		[b] Average value for 15, 18, 31, 44, 37 + 24; WAKITA and SCHMITT, 1970, p. 171, Table 1)
Ce (ppm)	96 ± 20	169 ± ?	374 ± ?		[c] Zr(31A, 31B) = Zr/Hf × Hf; $(Zr/Hf)_{\text{Apollo 12 soil}} = 27$; (GOLES *et al.*, 1971)
Eu (ppm)		2.8 ± ?	3.48[g] ± ?		[d] Zr value measured by optical spectroscopy (LSPET, 1970)
Sr (ppm)	145 ± 10	180[f] ± ?	$\frac{157 + 219^f}{2} = 190^i \pm 31$		[e] $U = \frac{K}{K/U}$; K/U [%/ppm] = 0.19 ± 0.02 (average value of 15, 18, 41, 44, 37 + 24; WAKITA and SCHMITT, 1970, p. 171, Table 1)
Y (ppm)	247[a] ± ?	247[a] ± ?	589[h] ± ?		[f] Sr(31B) = (Sr/Eu) × Eu; $(Sr/Eu)_{\text{light colored phase}} = 64.3$ (HUBBARD *et al.*, 1970, p. 182, Table 1)
Hf (ppm)	20.3[b] ± ?	20.3[b] ± ?	36[g] ± ?		[g] Chemical abundances for 06 (WAKITA and SCHMITT, 1970, p. 171, Table 1, a sample similar to 15—dark (SCHNETZLER *et al.*, 1970))
Zr (ppm)	548[c] ± ?	548[c] ± ?	972[c] ± ?		[h] $Y(31C) = \frac{Y(06)}{Y(\text{chondrites})} \times Y$ (chondrites) (WAKITA and SCHMITT, 1970, p. 174, Fig. 3)
	(2200)[d] ± ?	(2200)[d] ± ?	(2200)[d] ± ?		[i] Average values of 06 (WAKITA and SCHMITT, 1970) and 15—dark (SCHNETZLER *et al.*, 1970)
U (ppm)	22.7[e] ± 4.8	8.8 ± ?	6[g] ± ?	6.6[j] ± 0.9	[j] $U = \frac{K}{K/U}$; K/U [%/ppm] = 0.08 (06) (WAKITA and SCHMITT, 1970, p. 171, Table 1)

Sample 31A corresponds to an average of samples 15, 15A, and 15B (SCHNETZLER *et al.*, 1970). Sample 31B corresponds to sample 8—light (SCHNETZLER *et al.*, 1970). Sample 31C corresponds to sample 15—dark (SCHNETZLER *et al.*, 1970). The errors assigned are $\Delta\bar{x} = \sqrt{\frac{\Sigma(\bar{x} - x_i)^2}{n(n-1)}}$ type.

milky white phase, contained mainly plagioclase (An_{50}) and pyroxene; sample 31B, a clear phase, was mostly plagioclase (An_{80}); sample 31C, a black phase, was mostly low Ca pyroxene with some feldspar and a high abundance of opaque minerals e.g., ilmenite. In a private communication SCHNETZLER (1970) compared our samples with those of SCHNETZLER *et al.* (1970) using both the mineralogical data by WALTER and direct microscopic observation. On this basis he supplied us with the correspondences specified in Table 2. Trace element abundances were then estimated for the three fractions using those correspondences and data of SCHNETZLER *et al.* (1970); WAKITA

Table 3. Rare gases from one-hour heating of separates

	12013,10,31A (8.95) mg Al foil (199.1 mg)			
	700°C	1550°C	Total	700°C
$^{3}He/^{4}He$	0.000089 ± 0.0000006[a]	0.000128 ± 0.000003[a]	0.00009 ± 0.0000008[a]	0.00015 ± 0.000004[a]
$^{4}He[10^{-3}$ cc STP/g]	**3.14 ± 0.06**	**0.15 ± 0.003**	**3.29 ± 0.06**	**1.26 ± 0.03**
$^{20}Ne/^{22}Ne$	2.64 ± 1.22	0.84 ± 2.22	1.72 ± 1.42	5.28 ± 0.09
$^{21}Ne/^{22}Ne$	0.67 ± 0.01	0.852 ± 0.02	0.77 ± 0.01	0.53 ± 0.01
$^{22}Ne[10^{-8}$ cc STP/g]	**2.73 ± 0.10**	**3.45 ± 0.14**	**6.18 ± 0.20**	**3.43 ± 0.13**
$^{38}Ar/^{36}Ar$	0.91 ± 0.03	1.15 ± 0.02	1.05 ± 0.02	0.68 ± 0.02
$^{40}Ar/^{36}Ar$	21630 ± 780	22090 ± 440	21890 ± 420	7530 ± 170
$^{36}Ar[10^{-8}$ cc STP/g]	**4.31 ± 0.07**	**5.81 ± 0.1**	**10.12 ± 0.13**	**5.24 ± 0.09**
$^{40}Ar_r[10^{-4}$ cc STP/g]	**9.27 ± 0.37**	**12.80 ± 0.34**	**22.07 ± 0.50**	**3.85 ± 0.07**
$^{78}Kr/^{83}Kr$	0.59 ± 1.2	0.10 ± 0.26	0.19 ± 0.25	0.03 ± 0.95
$^{80}Kr/^{83}Kr$	0.33 ± 0.03	0.44 ± 0.02	0.42 ± 0.02	0.38 ± 0.03
$^{82}Kr/^{83}Kr$	0.78 ± 0.09	0.78 ± 0.03	0.78 ± 0.03	0.96 ± 0.07
$^{84}Kr/^{83}Kr$	2.73 ± 0.19	1.06 ± 0.05	1.35 ± 0.05	3.46 ± 0.16
$^{86}Kr/^{83}Kr$	0.72 ± 0.03	0.29 ± 0.01	0.37 ± 0.02	1.07 ± 0.04
$^{83}Kr[10^{-12}$ cc STP/g]	**75 ± 3**	**350 ± 10**	**425 ± 10**	**40 ± 2**
$^{124}Xe/^{130}Xe$	0.38 ± 0.02	0.41 ± 0.01	0.41 ± 0.01	0.29 ± 0.01
$^{126}Xe/^{130}Xe$	0.67 ± 0.03	0.78 ± 0.02	0.78 ± 0.02	0.50 ± 0.02
$^{128}Xe/^{130}Xe$	1.17 ± 0.05	1.23 ± 0.03	1.23 ± 0.03	0.98 ± 0.04
$^{129}Xe/^{130}Xe$	2.76 ± 0.13	1.62 ± 0.05	1.66 ± 0.05	3.07 ± 0.09
$^{131}Xe/^{130}Xe$	4.81 ± 0.18	4.31 ± 0.12	4.32 ± 0.12	4.22 ± 0.13
$^{132}Xe/^{130}Xe$	2.47 ± 0.09	1.19 ± 0.03	1.24 ± 0.03	3.01 ± 0.13
$^{134}Xe/^{130}Xe$	0.79 ± 0.03	0.24 ± 0.006	0.26 ± 0.01	0.93 ± 0.04
$^{136}Xe/^{130}Xe$	0.68 ± 0.03	0.17 ± 0.004	0.19 ± 0.004	0.83 ± 0.04
$^{130}Xe[10^{-12}$ cc STP/g]	**13 ± 0.5**	**368 ± 10**	**381 ± 10**	**10 ± 0.5**

and SCHMITT (1970); HUBBARD *et al.* (1970); GOLES *et al.* (1971) (see Table 2 for details).

RESULTS AND DISCUSSION

The results for He, Ne, Ar, Kr, and Xe for separates 31A, 31B, and 31C are presented in Tables 3 to 11.

Helium

All samples investigated had very high amounts of He with low $^{3}He/^{4}He$ ratios (Table 3). Sample 31A contained the lowest $^{3}He/^{4}He$ ratio observed. The total amount of ^{3}He was low compared with 31C, the sample with the highest ^{3}He concentration. Assuming a similar cosmic ray exposure history for all the minerals, this indicates a severe He loss by diffusion in 31A. The correction for trapped ^{4}He and for cosmogenic ^{4}He (using a cosmogenic $^{3}He/^{4}He$ ratio of 0.0266; KAISER, 1971) could be neglected. Sample 31B showed a $^{3}He/^{4}He$ ratio comparable to 31C. However, the total ^{3}He and ^{4}He concentrations were the lowest observed. The total ^{3}He-content is comparable to 31A, but the ^{4}He content is nearly a factor of 3 lower, understandable because of the low U concentration in 31B. The total ^{3}He and ^{4}He contents in 31C were the highest observed. This sample, mostly pyroxene, showed higher retentivity against diffusive loss of He than the two plagioclase fractions 31A and 31B.

Neon

Most of the Ne measured in the separates was cosmogenic (Table 4). Two kinds

from rock 12013. Errors given are total errors (1σ).

12013,10,31B (9.99 mg) Al-foil (245.7 mg) 1550°C	Total	700°C	12013,10,31C (9.26 mg) Al-foil (243.8 mg) 1550°C	Total
0.00036 ± 0.000007[a]	0.000161 ± 0.0000007[a]	0.00014 ± 0.000001[a]	0.00017 ± 0.000012[a]	0.000146 ± 0.000001[a]
0.05 ± 0.001	**1.31 ± 0.03**	**3.37 ± 0.06**	**0.33 ± 0.006**	**3.70 ± 0.06**
1.23 ± 0.24	3.83 ± 0.26	3.52 ± 0.11	1.07 ± 0.05	2.15 ± 0.06
0.865 ± 0.023	0.65 ± 0.01	0.66 ± 0.02	0.855 ± 0.04	0.77 ± 0.02
1.91 ± 0.07	**5.34 ± 0.15**	**3.70 ± 0.15**	**4.66 ± 0.20**	**8.36 ± 0.25**
1.08 ± 0.02	0.89 ± 0.02	0.65 ± 0.02	1.05 ± 0.03	0.88 ± 0.02
12000 ± 220	9830 ± 140	5100 ± 160	3210 ± 90	4030 ± 80
5.56 ± 0.10	**10.80 ± 0.13**	**3.70 ± 0.06**	**4.77 ± 0.08**	**8.47 ± 0.01**
6.63 ± 0.12	**10.48 ± 0.14**	**1.82 ± 0.07**	**1.49 ± 0.05**	**3.31 ± 0.09**
0.11 ± 0.28	0.11 ± 0.28	1.00 ± 2.2	0.14 ± 0.29	0.24 ± 0.30
0.44 ± 0.02	0.44 ± 0.02	0.38 ± 0.06	0.46 ± 0.03	0.45 ± 0.02
0.78 ± 0.03	0.80 ± 0.03	1.32 ± 0.11	0.82 ± 0.04	0.88 ± 0.04
1.23 ± 0.03	1.50 ± 0.03	3.07 ± 0.24	1.10 ± 0.05	1.34 ± 0.05
0.29 ± 0.008	0.39 ± 0.01	0.96 ± 0.08	0.26 ± 0.01	0.34 ± 0.02
293 ± 8	**333 ± 8**	**42 ± 3**	**300 ± 12**	**342 ± 13**
0.39 ± 0.02	0.39 ± 0.02	0.17 ± 0.02	0.51 ± 0.03	0.49 ± 0.03
0.71 ± 0.05	0.70 ± 0.05	0.33 ± 0.03	0.83 ± 0.04	0.80 ± 0.04
1.23 ± 0.09	1.22 ± 0.08	0.89 ± 0.08	1.39 ± 0.08	1.36 ± 0.07
1.65 ± 0.12	1.72 ± 0.12	5.02 ± 0.25	2.17 ± 0.13	2.36 ± 0.13
4.39 ± 0.32	4.39 ± 0.32	5.05 ± 0.26	4.36 ± 0.25	4.41 ± 0.24
1.46 ± 0.13	1.53 ± 0.13	4.88 ± 0.28	1.72 ± 0.09	1.93 ± 0.09
0.23 ± 0.02	0.26 ± 0.02	1.89 ± 0.09	0.59 ± 0.03	0.68 ± 0.03
0.13 ± 0.01	0.16 ± 0.01	1.63 ± 0.08	0.53 ± 0.03	0.61 ± 0.03
236 ± 13	**246 ± 13**	**7 ± 0.3**	**93 ± 4**	**100 ± 4**

[a] The $^3He/^4He$ ratios are not discrimination corrected.
We applied the discrimination factor 1.

of trapped Neon were present: Ne coming from the Al-foil and an additional component, probably a surface correlated air contamination. The separation of trapped and cosmogenic components was based only on the $^{21}Ne/^{22}Ne$ ratio. The cosmogenic $^{21}Ne/^{22}Ne$ value used was 0.865, the highest value observed in rock 12013,10,31 (31B) which is in good agreement with values found in other lunar samples and in meteorites. Cosmogenic ^{21}Ne measured in 31A and 31B was low compared to 31C, further supporting the observation that these samples have suffered diffusive loss of the light rare gases.

Argon

The Ar was dominated by large amounts of radiogenic ^{40}Ar due to the high K-concentration in this rock. The ^{36}Ar and ^{38}Ar were mainly cosmogenic (Table 5). The correction for cosmogenic ^{40}Ar was small because according to LÄMMERZAHL and ZÄHRINGER (1966) the cosmogenic $^{40}Ar/^{38}Ar$ ratio is less than 0.2. A "pure" cosmogenic ratio ($^{38}Ar/^{36}Ar = 1.52$) was taken from the 1200°C fraction in rock 10044 (HOHENBERG *et al.*, 1970). The trapped $^{38}Ar/^{36}Ar$ ratio chosen was that for air. Sample 31A had the highest $^{40}Ar/^{36}Ar$ ratio measured (21890); 31C, the lowest (4030). The total ^{36}Ar amount was comparable in all the separates. The fraction of ^{36}Ar which is cosmogenic (X) was calculated by means of the formula

$$(^{38}Ar/^{36}Ar)_m = (^{38}Ar/^{36}Ar)_{cosm}X + (1 - X)(^{38}Ar/^{36}Ar)_{air}.$$

Table 4. Trapped and cosmogenic ^{22}Ne in the separates. Units are 10^{-8} cc STP/g.

Sample	700°C cosm.	700°C trapped	1550°C cosm.	1550°C trapped	Total cosm.	Total trapped
31A	2.09 ± 0.10	0.64 ± 0.05	3.4 ± 0.15	0.05 ± 0.09	5.49 ± 0.17	0.69 ± 0.1 (0.40)[a]
31B	2.06 ± 0.10	1.37 ± 0.07	1.91 ± 0.07	≡0	3.97 ± 0.12	1.37 ± 0.07 (0.45)[a]
31C	2.78 ± 0.13	0.92 ± 0.10	4.61 ± 0.27	0.05 ± 0.21	7.39 ± 0.30	0.97 ± 0.23 (0.47)[a]

[a] Total concentrations of $\frac{^{22}\text{Ne}}{\text{gram of sample}}$ attributed to Al-foil.

Table 5. Trapped and cosmogenic ^{36}Ar in the separates. Units are 10^{-8} cc STP/g.

Sample	700°C cosm.	700°C trapped	1550°C cosm.	1550°C trapped	Total cosm.	Total trapped
31A	2.52 ± 0.12	1.79 ± 0.12	4.52 ± 0.13	1.29 ± 0.11	7.04 ± 0.18	3.08 ± 0.16 (1.89)[a]
31B	2.10 ± 0.08	3.14 ± 0.09	4.03 ± 0.12	1.53 ± 0.10	6.13 ± 0.14	4.67 ± 0.14 (2.10)[a]
31C	1.40 ± 0.07	2.30 ± 0.08	3.34 ± 0.13	1.43 ± 0.12	4.74 ± 0.15	3.73 ± 0.14 (2.24)[a]

[a] Total concentrations of $\frac{^{36}\text{Ar}}{\text{gram of sample}}$ attributed to Al-foil.

The Al-foil did not account for all the trapped Ar suggesting again that there was some adsorbed air.

Xenon

Xe results will be discussed first because the treatment of the Kr results are dependent upon them. Xe was predominantly cosmogenic in all the samples. Trapped Xe and Xe produced in decay by spontaneous fission of the U–Pu system were present. *Trapped* Xe. ALEXANDER (1970) and LUNATIC ASYLUM (1970) reported a trapped Xe component the composition of which could not be established because of the dominant cosmogenic Xe in rock 12013. The isotopic composition of this trapped component can make a large difference in determining the cosmogenic yields for the higher masses and the yield of Xe produced in fission. Because of the small weight of sample involved, it was important to have good data for blanks, including Al-foil. We assume for the moment that the trapped Xe component is atmospheric in composition; we shall see if this leads to consistent results. The ^{130}Xe concentrations were quite different in the three separates (Table 3) but the ^{130}Xe amounts in the Al-foil per gram of sample were nearly the same because the samples had similar weights and the amount of Al-foil used was roughly the same. Thus, the percentage of Al-trapped Xe in the total ^{130}Xe was quite different from one sample to another. For the 1550°C gas fractions these percentages, based on the foil measurements, were 2.5, 4.3, and 11.6 in fractions A, B, and C. The *maximum* possible (terminal) percentages of trapped ^{130}Xe—assuming that all ^{136}Xe was trapped, i.e., neglecting the fission contribution for this isotope—were 8, 5, and 25.

The comparison between the percentage value of trapped ^{130}Xe due to the Al-foil and the terminal percentage value is especially interesting in the case of 31B. The predicted ^{130}Xe-percentage value of the Al-foil experiment almost agreed with the terminal percentage value, showing that little trapped Xe beyond that contributed by the Al-foil was present. This led us to the conclusion that no trapped Xe component other than the Al-trapped Xe was in the separates. The trapped Xe in the 700°C fractions of the separates, obtained from a two component plot (^{124}Xe/^{130}Xe vs. ^{126}Xe/^{130}Xe—see HOHENBERG *et al.*, 1970, p. 1298; the pure cosmogenic point was computed as described below) was in fair agreement with the Xe-amounts measured in the 700°C fraction of the Al-foil supporting this conclusion. Since an assumption of atmospheric composition for the trapped Xe component has thus led to consistent results, we see no need to invoke another trapped component of different isotopic composition.

Cosmogenic Xe. The measured xenon was primarily cosmogenic in all the samples of 12013,10,31 investigated (Tables 3 and 6). This is in agreement with ALEXANDER (1970) and LUNATIC ASYLUM (1970). The cosmogenic Xe amounts were obtained by subtracting the trapped and fission contributions: First we subtracted the trapped Xe due to the Al-foil; then we subtracted a component from U-fission assuming all the remaining ^{136}Xe to be from such a source. Sample 31A contained an extremely high abundance of Ba (6360 ppm). The REE concentration (Ce = 96 ppm) was low, and, therefore, the cosmogenic ^{126}Xe present was 98.5% from Ba isotopes, based on spallation systematics (RUDSTAM, 1966, as used by HOHENBERG *et al.*, 1967). The cosmogenic Xe in 31B is also mostly from Ba (95.6%). Thus, these samples give an essentially pure cosmogenic Xe spectrum from Ba in rock 12013 (Fig. 1, Tables 7a and 7b). The cosmogenic yields are based on the 1550°C fractions only. The cosmogenic Xe

Table 6. Cosmogenic and trapped ^{130}Xe [$\times 10^{-12}$ cc STP/g]. See text for the method used to calculate these components.

Sample	700°C		1550°C		Total	
	cosm.	trapped	cosm.	trapped	cosm.	trapped
31A	11 ± 1	2 ± 0.5	359 ± 10	9 ± 2	370 ± 10	11 ± 2
31B	7 ± 0.5	3 ± 3	226 ± 13	10 ± 2	233 ± 13	13 ± 4
31C	2 ± 0.5	5 ± 1.5	82 ± 5	11 ± 2	84 ± 5	16 ± 5

Table 7a. Isotopic composition of cosmogenic Xe in the separates.

Sample	124	126	128	129	130	131	132	134
31A	0.42 ± 0.01	0.80 ± 0.02	1.25 ± 0.03	1.49 ± 0.06	1.00 ± 0.00	4.28 ± 0.13	0.99 ± 0.05	0.08 ± 0.02
31B	0.41 ± 0.03	0.74 ± 0.06	1.26 ± 0.09	1.44 ± 0.13	1.00 ± 0.00	4.36 ± 0.33	1.21 ± 0.14	0.09 ± 0.03
31C	0.58 ± 0.04	0.95 ± 0.06	1.52 ± 0.10	1.60 ± 0.22	1.00 ± 0.00	4.22 ± 0.33	0.88 ± 0.19	0.06 ± 0.07

Table 7b. Isotopic composition of cosmogenic Xe from Ba and REE.

Ba:	124	126	128	129	130	131	132	134
Ba[a]	0.54 ± 0.05	1.00 ± 0.00	1.63 ± 0.15	1.91 ± 0.18	1.30 ± 0.09	5.62 ± 0.54	1.44 ± 0.26	0.11 ± 0.03
$BaCl_2$[b]	0.72 ± 0.02	1.00 ± 0.00	1.14 ± 0.04	1.12 ± 0.04	0.72 ± 0.03	1.31 ± 0.06	0.55 ± 0.04	0.05 ± 0.10
REE[c]	0.83 ± 0.29	1.00 ± 0.00	1.50 ± 0.79	0.96 ± 1.17	0.25 ± 0.38	0.71 ± 2.50	≤1.2	≤0.29
$CeCl_3$[d]	0.83 ± 0.02	1.00 ± 0.00	1.08 ± 0.04	1.08 ± 0.09	0.24 ± 0.10	1.15 ± 0.10	0.22 ± 0.17	0.0015 ± 0.0015

[a] Inferred from samples 31A and 31B.
[b] FUNK *et al.* (1967).
[c] Inferred from samples 31A, 31B, and 31C.
[d] HOHENBERG and ROWE (1970).

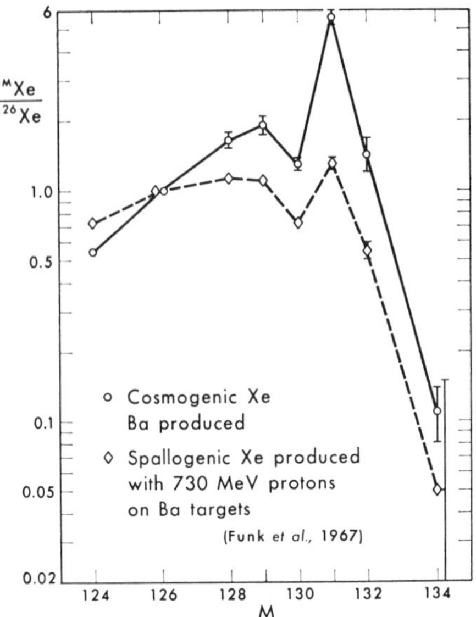

Fig. 1. Spectrum of cosmogenic Xe from lunar Ba inferred from separates 31A and 31B. It is compared to the spectrum of Xe produced by 730 MeV protons in a Ba target (FUNK *et al.*, 1967).

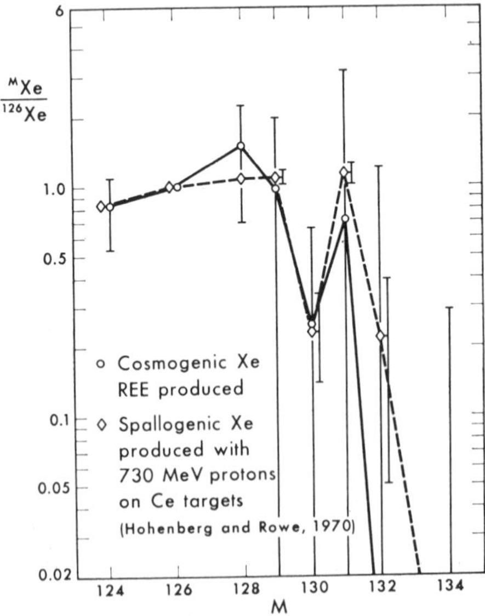

Fig. 2. Spectrum of cosmogenic Xe from lunar REE obtained by subtraction of Xe from separates in rock 12013. It is compared to the spectrum of Xe produced by 730 MeV protons in a Ce target (HOHENBERG and ROWE, 1970).

spectrum obtained in 31C is different from 31A and 31B. The $^{129}Xe/^{126}Xe$ and $^{131}Xe/^{126}Xe$ yields are lower. The ^{126}Xe produced in this sample contained, unlike the other separates, a significant portion (24%) of cosmogenic ^{126}Xe from REE targets according to RUDSTAM calculations. This made it possible to estimate the cosmogenic Xe spectrum from REE targets in this rock (Fig. 2, Table 7b). The spectrum is marked by a high $^{124}Xe/^{126}Xe$, and by low $^{129}Xe/^{126}Xe$, $^{130}Xe/^{126}Xe$, and $^{131}Xe/^{126}Xe$ ratios. The $^{132}Xe/^{126}Xe$ and $^{134}Xe/^{126}Xe$ ratios seem to be very low. The errors listed are based on the Xe measurements only. No errors for the chemical abundances were taken into account. We also plotted the spallation spectra obtained by bombardment of Ba targets (FUNK *et al.*, 1967) and Ce targets (HOHENBERG and ROWE, 1970) with 730 MeV protons. The spallation spectrum for Ce agrees well with the cosmogenic spectrum for REE.

The spallation spectrum for Ba disagrees with the cosmogenic Ba spectrum. The spectra (except for mass 131) could be brought into fair agreement by rotating the spallation spectrum counterclockwise around the intersection point at mass 126. But this is generally what would be expected for particles of lower energy than the somewhat high value of 730 MeV used in the target experiment, since high energy protons accentuate the spallation yields at masses far removed from the target. The discrepancy at mass 131 is of course augmented by the now famous ^{131}Xe anomaly in lunar rocks.

Fissiogenic Xe. All the sample of 12013,10,31 contained Xe produced by spontaneous fission decay of the U–Pu system. However, because of the dominance of and the uncertainties in the cosmogenic yields for Xe it was impossible to determine the isotopic composition of fission Xe in any of the samples studied. Thus, we used the Xe spectrum for spontaneous fission decay of ^{238}U (WETHERILL, 1953). The amounts of fission ^{136}Xe were obtained by subtracting the ^{136}Xe due to the Al-foil from the total ^{136}Xe and assuming all the remaining ^{136}Xe to be fissiogenic (Table 8).

Krypton

Trapped Kr. The Kr measurements for the Al-foil were not as consistent as the results obtained for the other rare gases. Three experiments with Al-foil (two temperatures each, namely 700°C and 1550°C) led to discordant results for this gas (Table 1). We have no explanation for the nonreproducibility in Al-trapped Krypton. The procedure for separating trapped, cosmogenic, and fission Kr was necessarily changed from that used for Xe. The fission corrections, which were small for the 1550°C and negligible otherwise, were based on the fission Xenon found. We used the fission yields for spontaneous fission of ^{238}U measured by WETHERILL (1953). The remaining ^{86}Kr was

Table 8. Fissiogenic ^{86}Kr and ^{136}Xe [$\times 10^{-12}$ cc STP/g].

Sample	Xenon 700°C	Xenon 1550°C	Xenon Total	Krypton Total
31A	5 ± 1	44 ± 4	49 ± 4	6 ± 4.5
31B	1 ± 1	8 ± 2	9 ± 2	1 ± 2
31C	1 ± 1	26 ± 5	27 ± 5	4 ± 4

assumed to be trapped (i.e., $^{86}Kr_{cosm} = 0$). Results of the calculations (Tables 8–10) were cosmogenic Kr spectra and concentrations for trapped and cosmogenic Krypton. The Kr is mostly cosmogenic as illustrated by sample 31C where at most 17% of the 300×10^{-12} cc STP/g ^{83}Kr of the 1550°C fraction could be trapped, based on the ^{86}Kr content. This 17% trapped ^{83}Kr corresponds to a ^{83}Kr value of 1.9×10^{-12} cc STP/g Al-foil for the 1550°C fraction, a value lower than measured in any of the foils. The trapped ^{83}Kr percentage values for the other separates were 31B = 19% (2.3×10^{-12} cc STP/g Al-foil in 1550°C) 31A = 19% (3.0×10^{-12} cc STP ^{83}Kr/g Al-foil in 1550°C). The amounts of trapped ^{83}Kr in the 700°C fractions of the separates were, if attributed to the Al-foil in the range of concentrations found at 700°C in the experiments with the foil (Tables 1 and 9).

Cosmogenic Kr. Cosmogenic Kr, as obtained by subtracting trapped and fission contributions, was more abundant in the 1550°C fractions and we therefore adopt the cosmogenic spectra obtained for those fractions (Table 10). Using these spectra, we were able to deduce the concentrations of cosmogenic Kr in the 700°C fractions by means of the ratio $^{84}Kr/^{83}Kr$.

AGES

Radiogenic ages

The K–Ar ages of 31A, 31B, and 31C were nominally 3.8×10^9 y, 4.1×10^9 y, and 4.1×10^9 y, respectively (Table 11). These values are in very good agreement with each other within the errors assigned, and with the results obtained by TURNER (1970), LUNATIC ASYLUM (1970), and TATSUMOTO (1970). We also confirm TURNER's (1970) result that neither the light phase nor the dark phase had suffered diffusive loss of ^{40}Ar. We note that the K–Ar ages obtained on 31A, 31B, and 31C confirm the estimated chemical compositions. The ^{4}He–U–Th ages are lower in the plagioclases 31A and 31B. The ages, both of which were 0.7×10^9 y, indicate an equally severe diffusive loss of He in both samples. This leads to similar ^{4}He–U–Th ages despite different U concentrations. A ^{4}He–U–Th age of 2.3×10^9 y was found in 31C. This particular ^{4}He–U–Th age was in good agreement with values reported by ALEXANDER (1970) and LUNATIC ASYLUM (1970).

Table 9. Cosmogenic and trapped ^{83}Kr [$\times 10^{-12}$ cc STP/g].

	700°C		1550°C		Total	
Sample	cosm.	trapped	cosm.	trapped	cosm.	trapped
31A	34 ± 8	41 ± 7	287 ± 8	63 ± 3	321 ± 11	104 ± 8
31B	23 ± 4	17 ± 4	237 ± 7	56 ± 2	260 ± 8	73 ± 5
31C	24 ± 8	18 ± 8	251 ± 13	49 ± 2	275 ± 15	67 ± 8

Table 10. Isotopic composition of cosmogenic Kr in the separates.

Sample	78	80	82	83	84
31A	0.12 ± 0.24	0.49 ± 0.02	0.73 ± 0.04	1.00 ± 0.00	0.20 ± 0.05
31B	0.13 ± 0.28	0.50 ± 0.02	0.73 ± 0.03	1.00 ± 0.00	0.35 ± 0.03
31C	0.15 ± 0.29	0.51 ± 0.03	0.79 ± 0.05	1.00 ± 0.00	0.34 ± 0.04

Table 11. Ages.

	31A Milky-white	31B Clear	31C Black
Measured rare gas components			
$^{3}He_{cosm}$($\times$ 10^{-8} cc STP/g)	29.6 $\pm$ 0.5	21.1 $\pm$ 0.5	54.2 $\pm$ 0.9
$^{4}He_{rad}$($\times$ 10^{-3} cc STP/g)	3.29 $\pm$ 0.06	1.31 $\pm$ 0.03	3.71 $\pm$ 0.06
$^{21}Ne_{cosm}$($\times$ 10^{-8} cc STP/g)	4.75 $\pm$ 0.15	3.43 $\pm$ 0.10	6.39 $\pm$ 0.26
$^{40}Ar_{rad}$($\times$ 10^{-4} cc STP/g)	22.1 $\pm$ 0.5	10.48 $\pm$ 0.14	3.31 $\pm$ 0.09
$^{83}Kr_{cosm}$($\times$ 10^{-12} cc STP/g)	320 $\pm$ 11	260 $\pm$ 8	275 $\pm$ 15
$^{126}Xe_{cosm}$($\times$ 10^{-12} cc STP/g)	296 $\pm$ 11	173 $\pm$ 17	80 $\pm$ 5
$^{136}Xe_{sf}$($\times$ 10^{-12} cc STP/g)	49 $\pm$ 4	9 $\pm$ 2	27 $\pm$ 5
Inferred ages			
^{4}He–U–Th age ($\times$ 10^{9} y)	0.7 $\pm$ 0.1	0.7 $\pm$ 0.1	2.3 $\pm$ 0.4
^{40}Ar–^{40}K age ($\times$ 10^{9} y)	3.8 $\pm$ 0.4	4.1 $\pm$?	4.1 $\pm$ 0.3
$^{136}Xe_{sf}$–^{244}Pu–^{238}U age ($\times$ 10^{9} y)	4.0 $+$ 0.2 $-$ 0.8	2.2 $\pm$?	4.3 $+$ 0.1 $-$ 0.15
^{3}He-exposure age ($\times$ 10^{6} y)	30 $\pm$ 0.5	21 $\pm$ 0.5	54 $\pm$ 1
^{21}Ne-exposure age ($\times$ 10^{6} y)	32 $\pm$ 1.5	23 $\pm$ 1	43 $\pm$ 2
^{83}Kr-exposure age ($\times$ 10^{6} y)	215 $\pm$ 66(102 $\pm$ 32)[a]	180 $\pm$ 55(85 $\pm$ 26)[a]	107 $\pm$ 34(73 $\pm$ 23)[a]
^{126}Xe-exposure age ($\times$ 10^{6} y)	40 $\pm$ 13	41 $\pm$ 13	47 $\pm$ 16

[a] This value is based on spectroscopic data for Zr (LSPET, 1970).

Exposure ages

The plagioclases (31A and 31B) had lower ^{3}He concentrations than the pyroxene (31C). Assuming a production rate of 1 $\times$ 10^{-8} cc STP ^{3}He/g 10^{6} y (KIRSTEN *et al.*, 1970) these amounts lead to ^{3}He exposure ages of 30 $\times$ 10^{6}, 21 $\times$ 10^{6}, and 54 $\times$ 10^{6} y, respectively. Together with the low ^{4}He–Th–U ages for 31A and 31B, this indicates a diffusive loss of He in the plagioclases (Table 11). The ^{21}Ne amounts show essentially the same pattern. The plagioclases (31A and 31B) have lower cosmogenic ^{21}Ne concentrations than 31C. Using 0.15 $\times$ 10^{-8} cc STP ^{21}Ne/g 10^{6} y (KIRSTEN *et al.*, 1970) as the production rate, we arrive at exposure ages of 32 $\times$ 10^{6} y, 23 $\times$ 10^{6} y, and 43 $\times$ 10^{6} y (31C), respectively. We want to note here, that the assumption of a universal production rate for Ne is certainly an oversimplification (KAISER, 1971). But even taking chemical effects into account would not change the point we intend to express here: 31A and 31B both suffered diffusive loss of Ne.

The cosmogenic ^{126}Xe concentrations were significantly different in the separates. Despite this, the exposure ages based on ^{126}Xe, are in good agreement. They range from 40 $\times$ 10^{6} y in 31A to 47 $\times$ 10^{6} y in 31C. The samples are strongly enriched in Ba and REE, the target elements for cosmogenic-produced Xe. The production rates used (assuming a 2π geometry) were

$$1.07 \times 10^{-15} \text{ cc STP } ^{126}\text{Xe}/10^{-6} \text{ g Ba } 10^{6} \text{ y},$$

and

$$1.01 \times 10^{-15} \text{ cc STP } ^{126}\text{Xe}/10^{-6} \text{ g Ce } 10^{6} \text{ y (REE)}$$

These values were obtained from the achondrite Nuevo Laredo (MUNK, 1967) and RUDSTAM calculations for Pasamonte (HOHENBERG *et al.*, 1967) assuming an exposure age 16 $\times$ 10^{6} y for Nuevo Laredo (HEYMANN *et al.*, 1968). These exposure ages are the most trustworthy ones we obtained for rock 12013. This fact is not evident in view of the large errors assigned, but they are due almost entirely to the 30% error in $^{126}Xe_{cosm}$ for the Nuevo Laredo meteorite.

The ^{83}Kr exposure ages are questionable. The essential elements are Sr, Y, and Zr. The production rates used were

$$3.56 \times 10^{-15} \text{ cc STP } ^{83}Kr/10^{-6} \text{ g Sr } 10^6 \text{ y,}$$
$$1.76 \times 10^{-15} \text{ cc STP } ^{83}Kr/10^{-6} \text{ Y } 10^6 \text{ y}$$

(Y value for Nuevo Laredo was estimated by SCHMITT (1970) and GOLES (1970) to be 23 ppm.)

$$0.98 \times 10^{-15} \text{ cc STP } ^{83}Kr/10^{-6} \text{ g Zr } 10^6 \text{ y}$$

and were based on Nuevo Laredo (MUNK, 1967) and Pasamonte calculations (HOHENBERG *et al.*, 1967) as before. The exposure ages obtained were much higher than the ^{126}Xe exposure ages and ranged from 107×10^6 y in 31C to 215×10^6 y in 31A.

There are two ways of explaining the high Kr exposure ages: either the production rates used were wrong or the chemical abundances were underestimated. Therefore, we checked our production rates on lunar rock 10017, studied by MARTI *et al.* (1970). This rock has a ^{81}Kr–^{83}Kr exposure age of 509×10^6 y. The ^{83}Kr exposure age obtained using MARTI's ^{83}Kr concentration value, the chemical values for Sr, Y, and Zr (WAKITA *et al.*, 1970) and the production rates previously stated was 473×10^6 y. Therefore, we believe that the chemical abundances—especially the Zr values—used for the calculations are underestimated. The Zr values are very uncertain. None of the investigators working with samples of 12013 remeasured Zr, although one of them inferred a value substantially lower than the 2200 ppm reported by LSPET (1970). However, using the LSPET Zr value in the calculations, the ^{83}Kr exposure ages become lower and more nearly agree with the ^{126}Xe exposure ages (listed in Table 11 in brackets). Thus, our results support the Zr value reported by LSPET (1970).

Unfortunately, the samples studied were too small and the background for ^{81}Kr too high to perform ^{81}Kr–^{83}Kr exposure age determinations.

^{136}Xe–^{244}Pu–^{238}U *ages*

Sample 31A showed the highest fissiogenic Xe concentration and highest U concentration. The resulting ^{136}Xe–Pu–U age was 4.0×10^9 y and was in agreement with the K–Ar age of 3.8×10^9 y (Table 11). A ^{136}Xe spontaneous fission yield of 6% was used for ^{238}U as well as ^{244}Pu. The spontaneous fission half lives applied were taken from FLEISCHER and PRICE (1964) for ^{238}U, and from FIELDS *et al.* (1966) for ^{244}Pu. A $^{244}Pu/^{238}U$ ratio of 0.015 at the time of the formation of the solar system (4.6×10^9 y ago) was assumed (PODOSEK, 1970; LUNATIC ASYLUM, 1970). But, one gets the same age (to three significant figures) for assumed $^{244}Pu/^{238}U$ ratios of 0.0 and 0.047 (HOHENBERG, 1970). If the U and fissiogenic ^{136}Xe data are taken at face value for 31C unreasonably old ages and contributions to fission Xe from ^{244}Pu are required.

But a much more likely explanation for the data is that the U-value was underestimated. Use of the fissiogenic ^{136}Xe data and the crystallization age of TURNER

(1970) leads to a K–U ratio = 0.06 (%/ppm) for the dark colored phase. There would be negligible ^{136}Xe from ^{244}Pu decay for these values.

Conclusions

We confirm, in essence, the results reported by Alexander (1970) and Lunatic Asylum (1970), that no ^{129}Xe coming from in situ decay of ^{129}I is present. Also, no real evidence for fission Xe originating from ^{244}Pu was established. The K–Ar ages are in good agreement with results obtained by Turner (1970) and with the Rb–Sr ages (Lunatic Asylum, 1970), for distinct mineral phases. Because of pronounced chemical differences among the different fractions, we were able to derive cosmogenic Xe spectra from Ba targets and REE targets in this moon sample. The ^{126}Xe exposure ages confirm the data reported by Alexander (1970) and the Lunatic Asylum (1970). We did not find any indications of trapped Kr and Xe components in the separates. Instead, we could attribute all the trapped Kr and Xe to the Al-foil in which the samples were wrapped.

The ^{83}Kr exposure ages were very high and in disagreement with each other, suggesting much higher Zr concentrations than inferred by Wakita and Schmitt (1970) and supporting the value by LSPET (1970).

Acknowledgments—The author is particularly pleased to acknowledge the inspiring advice and helpful criticism of John H. Reynolds. I also benefited from Paul K. Davis for his assistance in writing the computer programs used. I wish to thank E. C. Alexander, R. S. Lewis, G. McCrory, and D. Overskei for many encouraging discussions and M. C. Malin for taking some of the Al data. This work was supported in part by NASA and by the U.S. Atomic Energy Commission and bears AEC Code UCB–34P32–77.

References

Alexander E. C. (1970) Rare gases from stepwise heating of lunar rock 12013. *Earth Planet. Sci. Lett.* **9**, 201–208.

Bochsler P., Eberhardt P., Geiss J., Graf H., Grögler N., Krähenbühl U., Mörgeli M., Schwaller H., and Stettler A. (1971) Potassium–argon ages, exposure ages and radiation history of lunar rocks. Second Lunar Science Conference (unpublished proceedings).

Drake M. J., McCallum I. S., McKay G. A., and Weill D. F. (1970) Mineralogy and petrology of Apollo 12 sample no. 12013: a progress report. *Earth Planet. Sci. Lett.* **9**, 103–124.

Eberhardt P., Geiss J., Graf H., Grögler N., Krähenbühl U., Schwaller H., Schwarzmüller J., and Stettler A. (1970) Correlation between rock type and irradiation history of Apollo 11 igneous rocks. *Earth Planet. Sci. Lett.* **10**, 67–72.

Eugster O., Eberhardt P., and Geiss J. (1967) The isotopic composition of krypton in unequilibrated and gas rich chondrites. *Earth Planet. Sci. Lett.* **2**, 385–393.

Fields P. R., Friedman A. M., Milsted J., Lerner J., Stevens C. M., Metta D., and Sabine W. K. (1966) The decay properties of plutonium 244 and comments on its existence in nature. *Nature* **212**, 131–134.

Fleischer R. L., and Price P. B. (1964) Decay constant for spontaneous fission of ^{238}U. *Phys. Rev.* **B63**.

Funk H., Podosek F., and Rowe M. W. (1967) Spallation yields of krypton and xenon from irradiation of strontium and barium with 730 MeV protons. *Earth Planet. Sci. Lett.* **3**, 193–196.

Gay P., Brown M. G., and Rickson K. O. (1970) Mineralogical studies of lunar rock 12013,10. *Earth Planet. Sci. Lett.* **9**, 124–127.

GOLES G. (1970) Private communication.

GOLES G., DUNCAN A. R., OSAWA M., MARTIN M. R., BEYER R. L., LINDSTROM D. J., and RANDLE K. (1971) Analyses of Apollo 12 specimens and a mixing model for Apollo 12 "soils." Second Lunar Science Conference (unpublished proceedings).

HEYMANN D., MAZOR E., and ANDERS E. (1968) Ages of calcium-rich achondrites, I. Eucrites. *Geochim. Cosmochim. Acta* **32**, 1241–1268.

HOHENBERG C. M., MUNK M. N., and REYNOLDS J. H. (1967) Spallation and fissiogenic xenon and krypton from stepwise heating of the Pasamonte achondrite; the case of extinct plutonium 244 in meteorites; relative ages of chondrites and achondrites. *J. Geophys. Res.* **72**, 3139–3176.

HOHENBERG C. M., and ROWE M. W. (1970) Spallation yields of xenon from irradiation of Cs, Ce, Nd, Dy, and a rare earth mixture with 730 MeV protons. *J. Geophys. Res.* **75**, 4205–4209.

HOHENBERG C. M., DAVIS P. K., KAISER W. A., LEWIS R. S., and REYNOLDS J. H. (1970) Trapped and cosmogenic rare gases from stepwise heating of Apollo 11 samples. *Proc. Apollo 11 Lunar Sci. Conf., Geochim. Cosmochim. Acta* Suppl. 1, Vol. 2, pp. 1283–1309. Pergamon.

HOHENBERG C. M. (1970) Radioisotopes and the history of nucleosynthesis in the galaxy. *Science* **166**, 212–215.

HUBBARD N. J., GAST P. W., and WIESMANN H. (1970) Rare earth, alkaline and alkali metal and $^{87/86}$Sr data for subsamples of lunar sample 12013. *Earth Planet. Sci. Lett.* **9**, 181–185.

KAISER W. A. (1971) Unpublished data.

KIRSTEN T., MÜLLER O., STEINBRUNN F., and ZÄHRINGER J. (1970) Study of distribution and variation of rare gases in lunar material by a microprobe technique. *Proc. Apollo 11 Lunar Sci. Conf., Geochim. Cosmochim. Acta* Suppl. 1, Vol. 2, pp. 1331–1343. Pergamon.

LÄMMERZAHL P., and ZÄHRINGER J. (1966) K–Ar–Altersbestimmungen an Eisenmeteoriten II: Spallogenes ^{40}Ar and ^{40}Ar–^{38}Ar Bestrahlungsalter. *Geochim. Cosmochim. Acta* **30**, 1059–1075.

LSPET (Lunar Sample Preliminary Examination Team) (1970) Preliminary examination of the lunar samples from Apollo 12. *Science* **167**, 1325–1339.

LUNATIC ASYLUM (1970) Mineralogic studies of lunar rock 12013. *Earth Planet. Sci. Lett.* **9**, 137–164.

MARTI K., LUGMAIR G. W., and UREY H. C. (1970) Solar wind gases, cosmic ray spallation products, and irradiation history of Apollo 11 samples. *Proc. Apollo 11 Lunar Sci. Conf., Geochim. Cosmochim. Acta* Suppl. 1, Vol. 2, pp. 1357–1367. Pergamon.

MARTI K., and LUGMAIR G. W. (1971) Kr^{81}–Kr and K–Ar^{40} ages, cosmic-ray spallation products, and neutron effects in Apollo 11 and Apollo 12 lunar samples. Second Lunar Science Conference (unpublished proceedings).

MORRISON G. H., GERARD J. T., KASHUBA A. T., GANGADHARAM E. V., ROTHENBERG A. M., POTTER N. M., and MILLER G. B. (1970) Multielement analysis of lunar soil and rocks. *Science* **167**, 505–507.

MUNK M. N. (1967) Argon, krypton, and xenon in Angra dos Reis, Nuevo Laredo, and Norton County achondrites. *Earth Planet. Sci. Lett.* **3**, 457–465.

NIER O. A. (1950) A redetermination of the relative abundance of the isotopes of neon, krypton, rubidium, xenon, and mercury. *Phys. Rev.* **79**, 450–454.

PODOSEK F. A. (1970) The abundance of ^{244}Pu in the early solar system. *Earth Planet. Sci. Lett.* **8**, 183–187.

RUDSTAM G. (1966) Systematics of spallation yields. *Z. Naturforsch.* **21a**, 1027–1041.

SCHMITT R. A. (1970) Private communication.

SCHNETZLER C. C., PHILPOTTS J. A., and BOTTINO M. L. (1970) Li, K, Rb, Sr, Ba, and rare earth concentrations, and Rb–Sr age of lunar rock 12013. *Earth Planet. Sci. Lett.* **9**, 185–193.

SCHNETZLER C. C. (1970) Private communication.

TATSUMOTO M. (1970) U–Th–Pb age of Apollo rock 12013. *Earth Planet. Sci. Lett.* **9**, 193–201.

TURNER G. (1970) ^{40}Ar–^{39}Ar age determination of lunar rock 12013. *Earth Planet. Sci. Lett.* **9**, 177–181.

WAKITA H., and SCHMITT R. A. (1970) Elemental abundances in seven fragments from lunar rock 12013. *Earth Planet. Sci. Lett.* **9**, 164–169.

WAKITA H., SCHMITT R. A., and REY P. (1970) Elemental abundances of major, minor, and trace elements in Apollo 11 lunar rocks, soil, and core samples. *Proc. Apollo 11 Lunar Sci. Conf., Geochim. Cosmochim. Acta* Suppl. 1, Vol. 2, pp. 1685–1717. Pergamon.

WALTER L. S. (1970) Unpublished data.

WASSERBURG G. J. (1970) COSPAR, Leningrad Russia.

WETHERILL G. W. (1953) Spontaneous fission yields from uranium and thorium. *Phys. Rev.* **92**, 907–912.

YORK D. (1966) Least squares fitting of a straight line. *Can. J. Phys.* **44**, 1079–1084.

Proceedings of the Second Lunar Science Conference, Volume 2, pp. 1643–1650.
The M.I.T. Press, 1971.

Spallogenic Ne, Kr, and Xe from a depth study of 12002

E. C. ALEXANDER, JR.
Department of Physics, University of California, Berkeley, California 94720

(*Received 22 February 1971; accepted in revised form 30 March 1971*)

Abstract—Isotopic abundances and concentrations of neon, krypton, and xenon were measured in the top, middle, and bottom of rock 12002. The rare gases are mainly spallation produced and *no* evidence of a depth variation was detected in the relative isotopic yields. Apparent depth variations do exist in the concentration of xenon and krypton. The rare gas data along with the results of other workers are used to construct a three stage model of the irradiation history of 12002. The model is as follows: Stage 1. A long ($\sim$100 m.y.) irradiation at greater than one meter burial. Stage 2. A shorter (20 to 50 m.y.) irradiation at a shallow burial of 0 to 10 cm. Stage 3. A very short (2 to 5 m.y.) surface residence period. Stages 2 and 3 may not be separate stages but may reflect a noncatastrophic continuous movement of 12002 to the surface.

INTRODUCTION

THE SPALLATION YIELDS of stable rare gas isotopes in lunar rocks were some of the more enigmatic results reported at the Apollo 11 Lunar Science Conference. The high and variable yield of ^{131c}Xe was particularly puzzling. A number of authors (ALBEE *et al.*, 1970; FUNKHOUSER *et al.*, 1970; HINTENBERGER *et al.*, 1970; HOHENBERG *et al.*, 1970; MARTI *et al.*, 1970; PEPIN *et al.*, 1970; BOGARD *et al.*, 1971) have offered possible explanations which fall into two broad categories. The two-fold variation in the ^{131c}Xe spallation yield could be due to: (1) differences in chemical abundances of target elements and/or (2) differences in the effective irradiation energy spectrum integrated by each rock due to varying shielding histories. A companion paper (ALEXANDER *et al.*, 1971) discusses the effects of chemical differences between the minerals of rock 12013 on the spallation yields. In this paper we present the results of a study of self-shielding effects on the spallation yields within rock 12002.

EXPERIMENTAL TECHNIQUES

Rare gas mass spectroscopy was used to measure both the concentration and isotopic abundance of the rare gas isotopes. The rare gases were extracted from $\sim$250 mg samples in three temperature steps. The glass extraction system has been previously described as System 2 (HOHENBERG *et al.*, 1970). Experimental procedures and data reduction were essentially unchanged from those described for System 2.

SAMPLE DESCRIPTION

We received three pieces of Apollo rock 12002. The samples were 12002,80, 12002,83, and 12002,88—the top, "middle," and bottom of pieces 12002,28 respectively. Piece 12002,28 was column B3, an approximately 1.5 × 1.5 × 6.5 cm bar extending through rock 12002 (see the Apollo 12 Lunar Receiving Laboratory Orientation Drawings of Lunar Rock No. 12002). The top 1 to 2 mm of sample 12002,

80 and the bottom 3 to 4 mm of sample 12002,88 were removed to eliminate the surface solar wind component; i.e., the bottom of the top piece and the top of the bottom piece were analyzed. Sample 12002,83 was completely interior.

RESULTS

After the analysis of 12002 was completed, it was discovered that the He and Ar data were obtained in a region of nonlinearity in the mass spectrometer and are erroneous. Therefore, the discussion will be confined to the Ne, Kr, and Xe data which were not affected by the nonlinearity. Neon and krypton data are given in Table 1 and xenon data are given in Table 2. The errors listed for the isotope ratios are one standard deviation (1 σ) in the fit of the observed ratios to the drift line. The ^{40}Ar peak was monitored during the Ne analyses and was used to correct the ^{20}Ne data for $^{40}Ar^{+2}$ contamination. The correction was greater than 1% only in the ~700°C

Table 1. Neon and krypton from lunar rock 12002

	12002,80 (210.8 mg)			
Temp. °C	~700	1000	1800	~700
21/20	0.755 ± 0.017	0.997 ± 0.004	1.009 ± 0.003	0.799 ± 0.036
22/20	0.959 ± 0.020	1.145 ± 0.005	1.148 ± 0.005	0.968 ± 0.043
$[^{20}Ne]^* \times 10^{-8}$ cc/gm	0.52	8.53	9.31	0.59
Blank ^{20}Ne $\times 10^{-8}$ cc/gm			0.47	
78/84	0.00 ± 0.02	0.15 ± 0.02	0.18 ± 0.01	0.01 ± 0.02
80/84	0.076 ± 0.006	0.49 ± 0.02	0.58 ± 0.02	0.10 ± 0.01
82/84	0.24 ± 0.01	0.84 ± 0.02	0.96 ± 0.03	0.26 ± 0.02
83/84	0.26 ± 0.01	1.08 ± 0.04	1.30 ± 0.05	0.27 ± 0.02
86/84	0.307 ± 0.006	0.188 ± 0.009	0.131 ± 0.005	0.307 ± 0.010
$[^{84}Kr]^* \times 10^{-12}$ cc/gm	15.2	24.5	41.5	11.1†
Blank ^{84}Kr $\times 10^{-12}$ cc/gm			20.3	

	12002,83 (292.6 mg)		12002,88 (251.8 mg)		
Temp. °C	1000	1800	~700	1050	1800
21/20	1.028 ± 0.003	1.048 ± 0.002	0.790 ± 0.033	1.031 ± 0.003	1.048 ± 0.002
22/20	1.167 ± 0.002	1.167 ± 0.004	0.981 ± 0.038	1.138 ± 0.004	1.144 ± 0.002
$[^{20}Ne]^* \times 10^{-8}$ cc/gm	8.75	10.2	0.79	9.27	10.1
Blank ^{20}Ne $\times 10^{-8}$ cc/gm		0.15			0.22
78/84	0.13 ± 0.01	0.17 ± 0.01	0.02 ± 0.01	0.14 ± 0.01	0.13 ± 0.02
80/84	0.48 ± 0.02	0.50 ± 0.01	0.061 ± 0.006	0.45 ± 0.01	0.39 ± 0.01
82/84	0.82 ± 0.04	0.86 ± 0.02	0.21 ± 0.01	0.76 ± 0.03	0.73 ± 0.02
83/84	1.00 ± 0.05	1.15 ± 0.02	0.23 ± 0.01	0.97 ± 0.04	0.91 ± 0.02
86/84	0.199 ± 0.008	0.173 ± 0.007	0.285 ± 0.015	0.193 ± 0.005	0.171 ± 0.005
$[^{84}Kr]^* \times 10^{-12}$ cc/gm	19.9†	37.2†	19.1	27.9	30.7
Blank ^{84}Kr $\times 10^{-12}$ cc/gm		8.3†			15.3

* The concentrations of ^{20}Ne and ^{84}Kr were determined using the "peak height" method and are reproducible to about ±10%.

† The concentrations of krypton in sample 12002,83 may be in error by ±30% due to erratic sensitivities in the associated calibration runs. The krypton data have been corrected for background as discussed in the text.

Table 2. Xenon from lunar rock 12002

Sample	12002,80			
Temp. °C	~700	1000	1800	~700
124/132	0.0268 ± 0.0020	0.396 ± 0.016	0.370 ± 0.009	0.0307 ± 0.0017
126/132	0.0237 ± 0.0046	0.729 ± 0.019	0.642 ± 0.022	0.0397 ± 0.0016
128/132	0.122 ± 0.010	1.154 ± 0.033	1.045 ± 0.030	0.119 ± 0.009
129/132	1.028 ± 0.042	1.550 ± 0.039	1.410 ± 0.034	0.995 ± 0.022
130/132	0.165 ± 0.011	0.768 ± 0.014	0.713 ± 0.012	0.167 ± 0.005
131/132	0.905 ± 0.027	4.591 ± 0.107	4.222 ± 0.089	0.904 ± 0.015
134/132	0.385 ± 0.020	0.284 ± 0.007	0.366 ± 0.007	0.428 ± 0.018
136/132	0.346 ± 0.008	0.214 ± 0.003	0.270 ± 0.005	0.363 ± 0.009
[^{132}Xe]* × 10^{-12} cc/gm	1.4	6.4	5.3	0.9
Blank ^{132}Xe × 10^{-12} cc/gm			0.8	

12002,83		12002,88		
1000	1800	~700	1050	1800
0.408 ± 0.010	0.358 ± 0.007	0.0251 ± 0.0027	0.435 ± 0.011	0.317 ± 0.010
0.729 ± 0.016	0.623 ± 0.008	0.0279 ± 0.0017	0.744 ± 0.013	0.570 ± 0.006
1.126 ± 0.023	0.976 ± 0.016	0.117 ± 0.007	1.167 ± 0.036	0.886 ± 0.020
1.547 ± 0.027	1.414 ± 0.027	1.063 ± 0.026	1.549 ± 0.019	1.361 ± 0.011
0.752 ± 0.018	0.685 ± 0.017	0.178 ± 0.005	0.775 ± 0.018	0.632 ± 0.014
4.470 ± 0.098	3.934 ± 0.072	0.927 ± 0.022	4.652 ± 0.120	3.823 ± 0.049
0.291 ± 0.003	0.352 ± 0.014	0.382 ± 0.011	0.310 ± 0.006	0.360 ± 0.016
0.208 ± 0.010	0.270 ± 0.005	0.332 ± 0.011	0.213 ± 0.004	0.262 ± 0.012
4.9	4.5	1.2	5.6	3.1
	0.4			0.8

* The concentrations of ^{132}Xe were determined using the "peak height" method and are reproducible to ±10%.

fractions. The maximum correction was 38% for the ~700°C fraction of 12002,80. The light isotopes of krypton have been corrected for background (as measured in the spectrometer immediately before the krypton was admitted), and the error shown includes the added uncertainty due to the background correction. None of the data were corrected for blanks. Data for an 1800°C blank run before each sample are included in the tables.

The data in Tables 1 and 2 are typical of those previously reported for lunar crystalline rocks. The rare gases are almost completely due to spallation, and there is no evidence of a trapped or "solar wind" component. The small amount of non-spallation gas is probably due to blank contamination.

A search for a depth effect in the relative ^{131c}Xe yield was the main purpose of our analysis of 12002. Figure 1 is a plot of ^{131}Xe/^{136}Xe vs. ^{126}Xe/^{136}Xe in the manner described by ALEXANDER (1970). A small fission correction was applied to the data (see the previous reference for details). In a plot such as Fig. 1 the data will form a linear array if the heating experiment samples different mixtures of uniform spallation and trapped components. The slope of the plot is then the relative spallation yield ^{131c}Xe/^{126c}Xe. The data in Fig. 1 define a line well within the error limits. If there were a depth variation in the ^{131c}Xe yield, the data would form two or more lines or scatter badly. Therefore, to a depth of 6.5 cm in rock 12002 and to the error limits of the

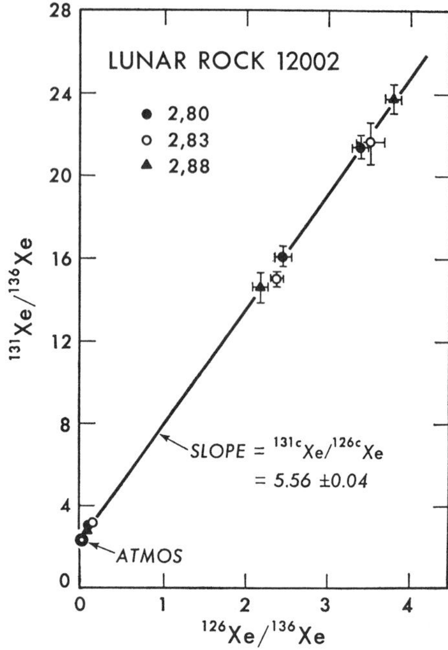

Fig. 1. Correlation of $^{131}Xe/^{136}Xe$ vs. $^{126}Xe/^{136}Xe$ for lunar rock 12002. Data have been corrected for spontaneous fission of ^{238}U in the manner described by ALEXANDER (1970). The line is the least-squares fit of the data using the method of YORK (1966). The regression line was forced through the atmospheric value. The well-defined line demonstrates the absence of a depth variation in the spallation yield of ^{131c}Xe relative to ^{126c}Xe.

Table 3. Relative spallation yields of xenon and krypton.*

Xenon isotopes	124	126	128	129	130	131	132	134	136
	56.3 ± 0.6	≡100	150.4 ± 0.9	121 ± 0.5	91.1 ± 0.6	556 ± 0.4	46 ± 4	7.0 ± 0.7	≡0.0
Krypton isotopes		78	80	82	83	84	86		
		21 ± 1	65 ± 2	≡100	134 ± 2	61 ± 3	≡0.0		

* The relative yields listed in this table are the slopes of the least-squares fits to data in Fig. 1 and in similar figures constructed for the other isotopes.

experiment (a few percent) there is *no* depth variation in the yield of ^{131c}Xe relative to the other spallation xenon isotopes.

Plots similar to Fig. 1 were constructed for all of the xenon and krypton (normalized to ^{86}Kr) isotopes. No evidence of depth variations were found in the relative yields of any xenon or krypton isotopes. Table 3 contains the relative spallation yields for xenon and krypton isotopes as calculated from the slopes of Fig. 1 and similar plots. Although there are no depth effects in the relative isotopic yields, apparent depth variations do exist in the concentration of spallation gases. Figure 2 is a plot of the relative concentrations of ^{130c}Xe, ^{83c}Kr, and ^{21c}Ne as a function of depth. Both

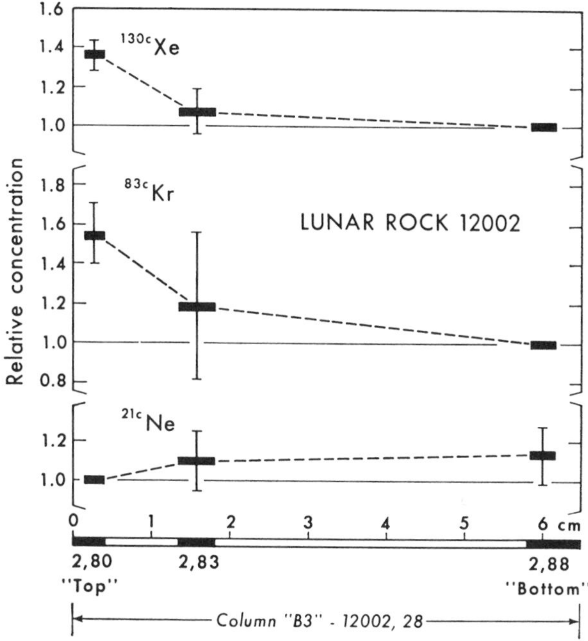

Fig. 2. Relative concentration of ^{130c}Xe, ^{83c}Kr, and ^{21c}Ne in column "B3" of lunar rock 12002. Shaded areas in the horizontal cm scale give the positions of the samples within column B3.

the ^{130c}Xe and the ^{83c}Kr decrease with depth. In contrast, the ^{21c}Ne appears to increase slightly with depth although it could be uniform within the experimental error.

Using the method described by HOHENBERG *et al.* (1970), which is based on the spallation xenon content of the Nuevo Laredo meteorite, and the Ba and REE concentrations of HUBBARD *et al.* (1971), we calculate a production rate of 7.0×10^{-14} cc ^{130c}Xe/gm/m.y. $\pm$ 30% in rock 12002. As the concentration of ^{130c}Xe varies with depth, the apparent "^{130c}Xe exposure age" also varies from 105 $\pm$ 32 m.y. at the top to 77 $\pm$ 25 m.y. at the bottom. Using a production rate of 0.12×10^{-8} cc ^{21c}Ne/gm/m.y. we calculate a "^{21c}Ne exposure age" of 161 $\pm$ 20 m.y.

DISCUSSION

The lack of a depth variation in the yield of ^{131c}Xe relative to the other xenon isotopes appears to argue against the specific suggestion of HOHENBERG *et al.* (1970) that the high ^{131c}Xe yield might be a surface effect. Mechanisms involving solar flare protons (ALBEE *et al.*, 1970; BOGARD *et al.*, 1971) also predict a strong surface effect in the ^{131c}Xe yield (as the range of solar flare protons is only a few mm) and are not supported by our results. Unfortunately, the irradiation history of 12002 is particularly unfavorable for observing surface effects.

The irradiation history of 12002 is complicated, and at least a three stage model is necessary to accommodate observations of other workers to the results of this work. The model can be summarized as follows:

Stage 1. A long irradiation under at least one meter (~350 to 400 gm/cm^2) of regolith. This stage lasted at least on the order of 100 m.y. but could have been much longer if 12002 was buried deeper. The Gd isotope observation of ALBEE *et al.* (1971) that 12002 accumulated most of its neutron flux in a very shielded location necessitates this stage.

Stage 2. A shorter irradiation at a slightly shielded (0 to ~10 cm) depth for from 20 to 50 m.y. The tracks of the heavy cosmic rays (BHANDARI *et al.*, 1971 and ALBEE *et al.*, 1971) and the ^{3c}He/2^{3}H exposure ages of D'AMICO *et al.* (1971) document this stage. (The first and second stages could easily have occurred in the opposite order.)

Stage 3. A very short irradiation (compared to the total irradiation) on the actual lunar surface. BHANDARI *et al.* (1971) calculate ~2 m.y. for the length of Stage 3. D'AMICO *et al.* (1971) calculate 8.8 ± 6.5 m.y. and by combining the ^{3}H profile of D'AMICO *et al.* (1971) with the ^{3}He profile of MARTI and LUGMAIR (1971) we calculate ~2.6% of the total irradiation, or ~3 m.y., as the length of the surface irradiation. A short surface residence time is compatible with the observation of FINKEL *et al.* (1971) and others that 12002 did not tumble once it reached the surface.

Possible variations in the ^{131c}Xe yield due to surface irradiations may simply be masked by the longer irradiations at depth. The larger question of whether differences of shielding from galactic cosmic rays cause the variation in the ^{131c}Xe yield (EBERHARDT *et al.*, 1970 and MARTI and LUGMAIR, 1971) is not answered by the results from 12002. The scale of any energy effects from galactic cosmic rays is expected to be decimeters to meters and rock 12002 is too small to monitor such variations.

The ^{21c}Ne and ^{130c}Xe exposure ages of 161 ± 20 m.y. and 77 ± 25 to 105 ± 32 m.y. respectively can be compared to MARTI and LUGMAIR's (1971) Kr^{81}–Kr age of 94 ± 6 m.y. and to the ^{3c}He/2^{3}H ages of from 63 ± 5 to 144 ± 10 m.y. and the ^{38c}Ar/(^{37}Ar + ^{39}Ar) ages of from 125 ± 15 to 155 ± 15 m.y. of D'AMICO *et al.* (1971). In view of the complex history of rock 12002, however, the "exposure ages" based on meteoritic production rates are of doubtful physical meaning.

The variations in the ^{130c}Xe and ^{83c}Kr concentrations in Fig. 2 might reflect variations in the abundances of the target trace elements within 12002 or they might be an artifact due to variations in the sensitivity of the spectrometer. However, it is possible that the concentration gradient is real and nontrivial, in which case it reflects the flux gradient in Stage 2 of the above model. Stage 3 is simply not long enough to establish concentration gradients as large as were observed. However, if the gradient were established in Stage 2, then rock 12002 had the same orientation in Stages 2 and 3. It is unlikely that a catastrophic event would have moved the rock without changing its orientation. Therefore, the model might be modified to a two stage model. Namely, a deeply buried Stage 1 followed by a catastrophic movement to a shallow burial under fine regolith in Stage 2. The covering regolith was then slowly, gently, and continuously removed over a span of tens of millions of years by the mechanism of GOLD (1971) or by some other method. Finally rock 12002 reached the surface some two to five m.y. ago.

It is unfortunate for the ^{131c}Xe anomaly question that rock 12002 has such a short surface residence time and complex history. Depth studies on future large lunar rocks will hopefully yield much less ambiguous results.

Acknowledgments—I wish to acknowledge Professor J. H. Reynolds for his many helpful suggestions and critical evaluation; P. K. Davis for his assistance in programming the data reduction, and Dr. W. A. Kaiser, R. L. Lewis, G. McCrory, M. Malin, and D. Overskei for their aid and suggestions. This work was supported in part by NASA and the AEC, it bears AEC Code No. UCB–34P32–78.

References

ALBEE A. L., BURNETT D. S., CHODOS A. A., EUGSTER O. J., HUNEKE J. C., PAPANASTASSIOU D. A., PODOSEK F. A., RUSS G. P., II, SANTZ H. G., TERA F., and WASSERBURG G. J. (1970) Ages, irradiation history, and chemical composition of lunar rocks from the Sea of Tranquillity. *Science* **167**, 463–466.

ALBEE A. L., BURNETT D. S., CHODOS A. A., HAINES E. L., HUNEKE J. C., PAPANASTASSIOU D. A., PODOSEK F. A., RUSS G. P., TERA F., and WASSERBURG G. J. (1971) The irradiation history of lunar samples. Second Lunar Science Conference (unpublished proceedings).

ALEXANDER E. C., JR. (1970) Rare gases from a stepwise heating of lunar rock 12013. *Earth Planet. Sci. Lett.* **9**, 201–207.

ALEXANDER E. C., JR., DAVIS P. K., KAISER W. A., LEWIS R. S., and REYNOLDS J. H. (1971) Further rare gas studies in rock 12013. Second Lunar Science Conference (unpublished proceedings).

BHANDARI N., BHAT S., LAL D., RAJAGOPALAN G., TAMHANE A. S., and VENKATAVARADAN V. S. (1971) Fossil track studies in lunar materials, I: High resolution time averaged (millions of years) data on chemical composition and energy spectrum of cosmic ray nuclei of Z = 22–28 at 1 A.U. Second Lunar Science Conference (unpublished proceedings).

BOGARD D. D., FUNKHOUSER J. G., SCHAEFFER O. A., and ZÄHRINGER J. (1971) Noble gas abundances in lunar material. Cosmic ray spallation products and radiation ages from the Sea of Tranquillity and the Ocean of Storms. *J. Geophys. Res.* **76**, 2757–2779.

D'AMICO J., DEFELICE J., FIREMAN E. L., JONES C., and SPANNAGEL G. (1971) Tritium and argon radioactivities and their depth variations in Apollo 12 samples. Second Lunar Science Conference (unpublished proceedings).

EBERHARDT P., GEISS J., GRAF H., GRÖGLER N., KRÄHENBÜHL U., SCHWALLER H., SCHWARZMÜLLER J., and STETTLER A. (1970) Correlation between rock type and irradiation history of Apollo 11 igneous rocks. *Earth Planet. Sci. Lett.* **10**, 67–72.

FINKEL R. C., ARNOLD J. R., REEDY R. C., FRUCHTER J. S., LOOSLI H. H., EVANS J. C., SHEDLOVSKY J. P., IMAMURA M., and DELANY A. C. (1971) Depth variation of cosmogenic nuclides in a lunar surface rock. Second Lunar Science Conference (unpublished proceedings).

FUNKHOUSER J. G., SHAEFFER O. A., BOGARD D. D., and ZÄHRINGER J. (1970) Gas analysis of the lunar surface. *Science* **167**, 561–563.

GOLD T. (1971) Evolution of mare surface. Second Lunar Science Conference (unpublished proceedings).

HINTENBERGER H., WEBER H. W., VOSHAGE H., WÄNKE H., BEGEMAN F., and WLOTZKA F. (1970) Concentrations and isotopic abundances of the rare gases, hydrogen and nitrogen in Apollo 11 lunar matter. *Proc. Apollo 11 Lunar Sci. Conf., Geochim. Cosmochim. Acta* Suppl. 1, Vol. 2, pp. 1269–1282. Pergamon.

HOHENBERG C. M., DAVIS P. K., KAISER W. A., LEWIS R. S., and REYNOLDS J. H. (1970) Trapped and cosmogenic rare gases from stepwise heating of Apollo 11 samples. *Proc. Apollo 11 Lunar Sci. Conf., Geochim. Cosmochim. Acta* Suppl. 1, Vol. 2, pp. 1283–1309. Pergamon.

HUBBARD N. J., GAST P. W., and MEYER C. (1971) The origin of the lunar soil based on REE, K, Rb, Ba, Sr, P, and $Sr^{87/86}$ data. Second Lunar Science Conference (unpublished proceedings).

MARTI K., LUGMAIR G. W., and UREY H. C. (1970) Solar wind gases, cosmic-ray spallation products and the irradiation history of Apollo 11 samples. *Proc. Apollo 11 Lunar Sci. Conf., Geochim. Cosmochim. Acta* Suppl. 1, Vol. 2, pp. 1357–1367. Pergamon.

MARTI K., and LUGMAIR G. W. (1971) Kr^{81}–Kr and K–Ar^{40} ages, cosmic-ray spallation products and neutron effects in Apollo 11 and Apollo 12 lunar samples. Second Lunar Science Conference (unpublished proceedings).

PEPIN R. O., NYQUIST L. E., PHINNEY D., and BLACK D. C. (1970) Rare gases in Apollo 11 Lunar Material. *Proc. Apollo 11 Lunar Sci. Conf., Geochim. Cosmochim. Acta* Suppl. 1, Vol. 2, pp. 1435–1454. Pergamon.

YORK D. (1966) Least-square fitting of a straight line. *Can. J. Phys.* **44**, 1079.

Proceedings of the Second Lunar Science Conference, Volume 2, pp. 1651–1669.
The M.I.T. Press, 1971.

Location and variation of trapped rare gases in Apollo 12 lunar samples

T. Kirsten, F. Steinbrunn, and J. Zähringer
Max-Planck-Institut für Kernphysik,
69 Heidelberg, Germany

(*Received 23 February 1971; accepted in revised form 31 March 1971*)

Abstract—Local variations of solar wind helium up to a factor of 20 within 10 μ have been observed by in situ analysis of breccia 12073, whereas on a mm-scale, gas concentrations are homogeneous and correspond to those of 12070 soil. Highest enrichments of up to 1.5 cc He/g occur in fine grained, Ti- and Cr-rich areas. The light dark structured breccia 12010 contains trapped gases only in its dark portion. Solar wind He in the surface of crystalline rock 12063 is more homogeneously distributed than in breccias. The orientation of this rock on the lunar surface is inferred from the He distribution. Direct evidence for turnover of rock 12063 is presented. Trapped He contents of 500 single grains of 12070 soil differ by orders of magnitude. The highest concentrations occur in pure ilmenites and polymineralic fragments, the lowest in pure plagioclases and glasses. The redistribution of solar wind gases after implantation is influenced by crystal imperfections and impurities. He, Ne and Ar isotopes in 162 single grains of 12070 soil show large variations in concentrations and elemental ratios, even for specimens of the same type and size. The $^{20}Ne/^{36}Ar$ ratio in a particular olivine is as high as 38. A two-stage origin of trapped gases is suggested. A new minimum value for the $^{40}Ar/^{36}Ar$ ratio of the trapped component is 0.15 ± 0.03. Trapped gases in 12070 depend on grain size according to $r^{-0.6}$. Mean penetration depths for bulk soil are 3, 2, and 1 μ for He, Ne and Ar, respectively. For pure minerals, penetration depths are smaller. Some percent of "trapped" gases are volume-related. Large enrichments of radiogenic gases are found in some anorthosites and glasses of 12070 soil and in a lithic fragment of breccia 12010. In some grains of 12070, cosmogenic gases correspond to exposure ages around 1 b.y., three times higher than the bulk soil exposure age, which is 350 m.y. The K–Ar gas retention age of 12070 soil is 1.7 b.y.

Introduction

Rare gas analyses of Apollo 11 lunar samples revealed the presence of a trapped component in lunar fines, breccias, and rock surfaces (LSPET, 1969, Eberhardt *et al.*, 1970, Funkhouser *et al.*, 1970, Heymann and Yaniv 1970, Kirsten *et al.*, 1970 and others). It is mainly attributed to the solar wind bombardment of the lunar surface. Large variations of absolute concentrations as well as isotope and elemental ratios made it clear that secondary alterations prevent a direct interpretation in terms of solar wind abundances and fluxes. A detailed knowledge of the distribution patterns on a microscale can serve as a tool to study these alterations (Kirsten *et al.*, 1970). This may then yield information regarding the processes at the lunar surface as well as ancient radiation conditions including solar abundances. In situ analysis with high local resolution (Kirsten *et al.*, 1970), single grain analysis (Heymann and Yaniv, 1970) and etching experiments (Eberhardt *et al.*, 1970, Hintenberger *et al.*, 1970) of Apollo 11 samples gave, among others, the following results: (1) Trapped gases are surface related as expected for a low energy radiation like the solar wind; (2) There is a continuous turnover of the lunar soil; (3) Gases are redistributed, mainly by solid state diffusion; (4) Chemical and structural differences influence the individual behavior of each grain; and (5) ^{40}Ar abundances can not be explained by solar wind.

In this paper we extend these studies with improved techniques to Apollo 12 samples. It was known from LSPET (1970) that these samples are also loaded with trapped gases, though to a lesser extent than comparable Apollo 11 materials. (Data on radiogenic and cosmogenic isotopes of Apollo 12 fines, presented at the Houston Conference, will be published in a separate paper).

EXPERIMENTS AND RESULTS

Investigations with the Micro-He-Probe

With this method, gases are released by the electron beam of a microprobe and simultaneously analyzed in a mass spectrometer (Zähringer 1966, KIRSTEN *et al.* 1970). Three modes of operation have been used: (1) Scans across polished sections. (2) Scans on original rough surfaces with continuous focus adjustment. (3) One-shot melting of singly mounted grains. At the same time, chemical analyses may be obtained as in a regular microprobe. With sensitivity 5×10^{-10} cc ^4He/sec, electron beam 1 μA, beam diameter 10 μ, scanning speed 16 μ/sec, and volume degassed $(3 \pm 1) \times 10^{-9}$ cc/sec, gas concentrations have to be in the order of 5×10^{-2} cc/g or 10^{-4} cc/cm^2 to become detectable. This restricts the method to the analysis of solar wind helium. Further experimental details are described in KIRSTEN *et al.* (1970).

In situ analysis. Helium has been detected within breccias and on top of exposed rock surfaces. The ^4He distribution within polished sections of breccias 12073,32 and 12010,26 (40 scans each, 100 μ distant, 4 mm length) is very inhomogeneous, with fluctuations up to a factor of 20 within a 10 μ range. Maximum concentrations are 1.5 cc ^4He/g ($\sim 10^{-2}$ cc/cm^2 cross section) for 12073,32 (40 times above average) and 0.8 cc ^4He/g ($\sim 5 \times 10^{-3}$ cc/cm^2 cross section) for 12010,26 (10 times above average "dark"). Absolute errors are about $\pm 50\%$ because of the inaccurate knowledge of the degassed volume.

Large enrichments are always restricted to small spots of $\sim$10–20 μ. He is concentrated in the fine grained matrix, especially in spots with high Ti-concentrations, while crystalline inclusions are free of trapped He (Fig. 1).

The degree of correlation between Ti and He was derived from all registered scans by a computer analysis, applying the general principles of statistics. For each breccia, 240 40 μ-intervals of microprobe scans were evaluated in all regions with ^4He above detection limit. The mathematically well defined "sample correlation coefficient" r allowed to reject the assumption "Ti and He not correlated" on confidence levels of $>99.5\%$ for breccias 12073,32 and 10019,13, $>90\%$ for breccia 12010,26. This results in the following intervals for the "population correlation coefficients" ρ, which determine the strength of a correlation ($\rho = 0$: no correlation, $0 < \rho \leq 1$: correlation, $0 > \rho \geq -1$: anticorrelation): $0.4 < \rho < 0.6$ for Apollo 11 breccia 10019,13, $0.15 < \rho < 0.4$ for breccia 12073,32 and $0 < \rho \leq 0.25$ for breccia 12010,26.

Scans across larger crystals gave two distinct He peaks at the grain surfaces. It was also noticed that zones of very high Cr-content ($>1\%$) are especially rich in He (Fig. 2).

Breccia 12010,26 displays a light-dark structure, similar to light-dark structured meteorites. It was recognized as a breccia only after gas analysis by LSPET. Solar wind He is strongly restricted to the dark veins (>1 cc ^4He/g), He-scans exactly identify the actual border between crystalline areas and fine grained matrix, which were probably joined during an impact process.

Samples 12063,106 (Top) and 12063,104 (Bottom) are documented with respect to external surfaces of this fine grained crystalline rock. While the interior was always free of solar wind

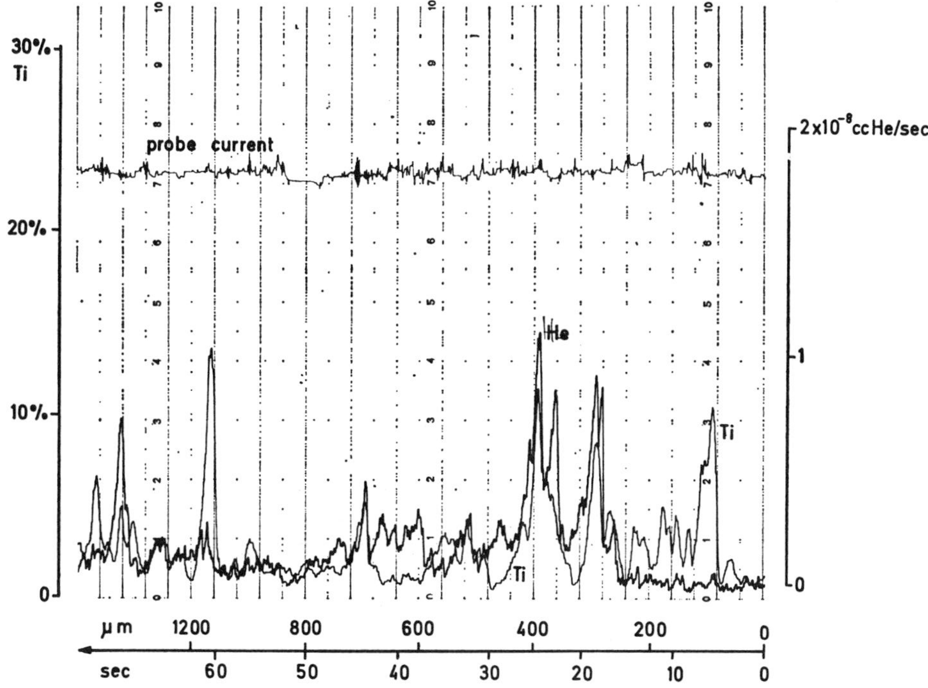

Fig. 1. Ti–He correlation in breccia 12073,32.

He, both (unpolished) original surfaces were found to be loaded with He, up to 10^{-2} cc He/cm^2 (80 scans of 1 cm length). Although some enrichments up to a factor of 10 occur within 10 μ distance, the distribution in general is more uniform than across polished sections of breccias. Areas up to 500 × 500 μ display homogeneous He loading within a factor of 2.

Three scans went across a 250 μ glass coated microcrater on the 12063,106 surface. In the crater bottom, the He concentration was less than 10^{-4} cc/cm^2, right next to the crater it was between 10 and 50 × 10^{-4} cc/cm^2.

The distribution of He on the surface of crystalline rock 12063,104 is shown in Fig. 3. For parallel scans, 100 μ apart, vertical bars proportional to the maximum He concentrations occurring in areas of 15 × 40 μ are plotted. Ti-concentrations were also registered. However in the case of rough surfaces, X-ray absorption causes larger errors, in the order of 20%. Nevertheless, a strong correlation between Ti-rich areas and He-concentrations holds for crystalline surfaces as well. The gas free area on the right hand side of Fig. 3 must be caused by mechanical erosion, probably during sample transfer. Figure 4 shows how one can infer the original orientation of a crystalline rock at the lunar surface for 12063,106 (Top). Microphotographs were taken under different angles of light incidence. There is only one direction of incidence for which the upper parts of shadowed areas correspond to He-free areas in the He-distribution map. Results similar to those of 12063,106 were obtained for an exterior surface of crystalline rock 12021,16. With the He-microprobe, the face directly exposed to the sun was easily recognized.

Single grain analysis. We have analyzed He in 500 grains of 8 types, handpicked from Apollo 12 soil 12070,68: Pyroxene, Olivine, Plagioclase light and dark, Ilmenite, Glass, Glass spherules, and polymineralic fragments. Grain sizes ranged between 100 and 400 μ. With the micro helium probe, a large number of grains can be analyzed in a relatively short time.

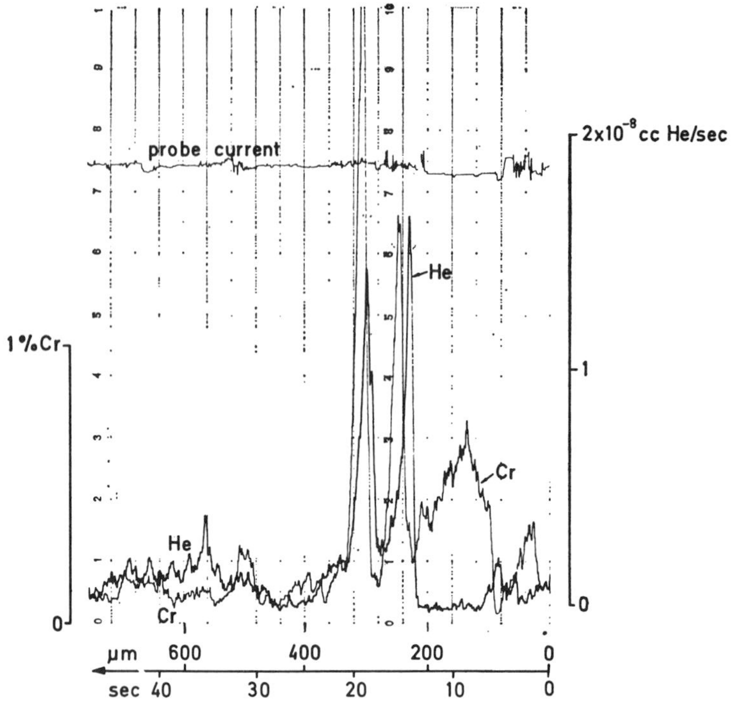

Fig. 2. Cr–He correlation in breccia 12073,32.

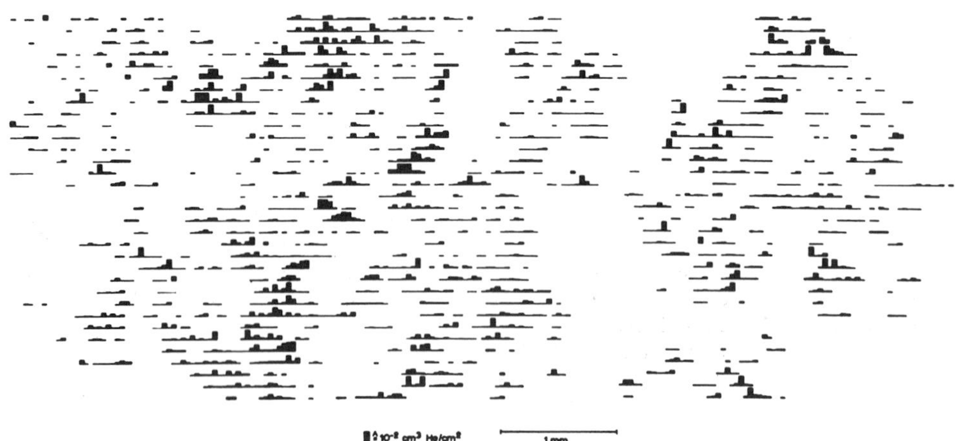

Fig. 3. Distribution of solar helium in rock surface 12063,104.

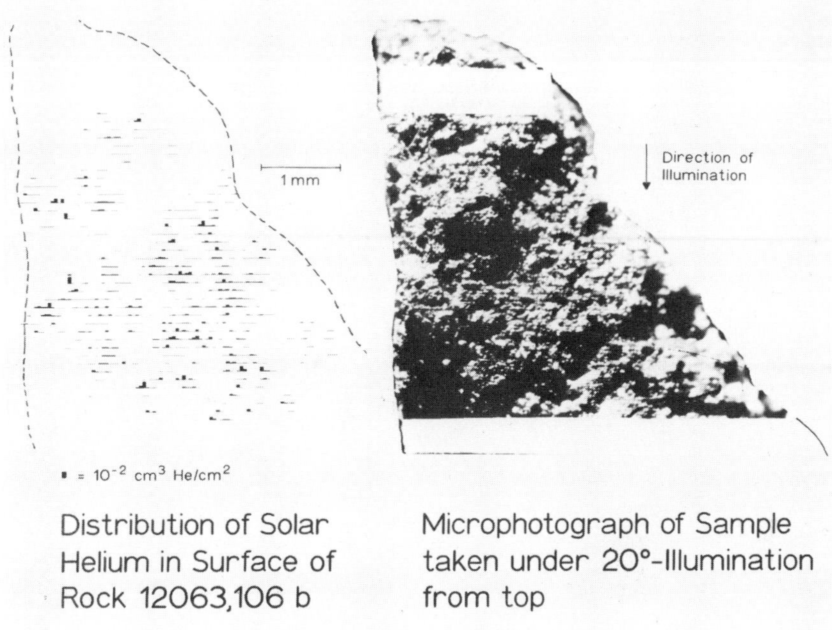

Fig. 4. Distribution of solar helium in rock surface 12063,106. Microphotograph of sample taken under 20° illumination from top.

A non-destructive and unequivocal mineral identification on small single grains is not possible. As a compromise, minerals were selected under the binocular and evaluated according to their external features and their behavior in polarized light. We can therefore not exclude the possibility of misjudgment in some cases. It is also evident that lunar minerals may not be homogeneous crystallographically, even if they appear to be in the binocular. In discussing the results, the possibility of unclean or mislabeled samples should be kept in mind. This applies especially for "ilmenites," where we did not succeed in selecting 100% pure specimens.

Grains were washed in acetone, and ultrasonically stirred to remove adherent dust particles after which they were weighed on a microbalance. They were then glued together with spodumene standard grains on a glass slide, carbon coated, mounted in the He-microprobe, and melted by 10 μA electron pulses. With a response of 0.1 sec, He peak heights are proportional to the amount of He released, within 10% accuracy. This was checked with irradiated spodumene standards of different grain sizes. The results are given in Fig. 5. It shows the distribution of He-concentrations per gram for grains of all analyzed sizes.

Only 15 of 55 glass spherules had smooth and shiny surfaces. They all had He contents $< 80 \times 10^{-4}$ cc/g. Other spherules contain much more He, due to rough surfaces and adherent microcrystals. A similar difference is found between pure colorless plagioclase and grey feldspathic particles. For olivines it was noticed that those which melt into light green spherules have much lower He concentrations than those which are dark after melting. Generally, the degree of such impurity increases the He concentration.

This is different, however, for ilmenites. Clean minerals melting into deep black spherules have He concentrations up to 1 cc/g, one order of magnitude higher than those which melt into more brownish spherules.

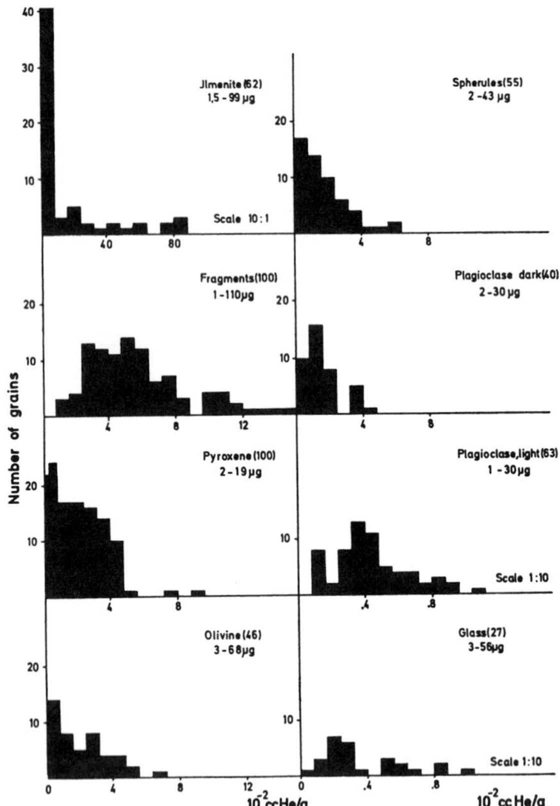

Fig. 5. ^{4}He concentrations in single grains of 12070,68 fines. Micro–He–probe analysis.

Investigations by static Mass Spectrometry

A Nier-type mass spectrometer was used for the experiments described in this section. For normal operation, blank values were (in 10^{-10} cc): ^{3}He $\sim$ 1, ^{4}He $\sim$ 60, ^{20}Ne $\sim$ 3, ^{21}Ne $\sim$ 0.05, ^{22}Ne $\sim$ 2, ^{36}Ar $\sim$ 0.2, ^{38}Ar $\sim$ 0.1, ^{40}Ar $\sim$ 100. If not mentioned otherwise, standard procedures were applied. Samples were predegassed for 6 hours at 180°C.

Single grain analysis. To extend the He-microprobe single grain analysis to the isotopes of He, Ne, and Ar, static mass spectrometry had to be applied at the price of a much more elaborate procedure. Since it was our intention to analyze single crystals weighing between 10 and 100 micrograms (μg), common analytical techniques had to be modified. The most serious requirement arose from the desire to measure accurate ^{40}Ar/^{36}Ar ratios in such small grains, in order to contribute information to the "trapped ^{40}Ar-problem" (HEYMANN and YANIV, 1970a). Common ^{40}Ar-blank values are in the order of 10^{-8} cc, while we wanted to detect 10^{-11} cc ^{40}Ar. To do so, we dropped the grains into an extremely small Mo-crucible (merely a hot plate) while the mass spectrometer, continuously registering mass peak 40, was directly connected to the extraction part in which a Ti-getter was kept at 400°C. We managed to reduce the increase on the ^{40}Ar peak in the static system to a value corresponding to 2×10^{-11} cc ^{40}Ar/min, which remained unchanged when the crucible was heated to 2000°C. Under these conditions, 1×10^{-11} cc ^{40}Ar, released when the sample was dropped, could easily be recognized as a clear step in the mass number 40 recording. However, under these extreme circumstances, we found that sample wrapping materials such as Al, Au, or Pt foils do essentially contribute to the blank values. Even if materials themselves are extremely pure (as e.g., a certain gold foil), hot metal vapor condensing at the cooled furnace walls leads to appreciable ^{40}Ar release. This required unwrapped samples. Handling of such small unwrapped particles under vacuum is extremely difficult. We developed a special technique to do so. However, in performing multiple sample loading, it was not possible to differentiate individual samples loaded at one time. We therefore decided to use for each loading about 20 grains of the same mineral of approximately equal weights (determined with a microbalance, spread 20%). We feel that this method is extremely powerful also for other applications, e.g., detection of ^{3}He in cosmic spherules.

Mass peaks other than ^{40}Ar were relatively constant under the applied conditions. However, it should be noted that the procedure does not allow separation of Ar from Ne. Therefore, mass number 20 has to be corrected for doubly ionized ^{40}Ar. In our instrument, this contribution was always 4.8%. The corresponding correction was mostly below 10%, except for a few specimens with particular ^{40}Ar enrichments. Time extrapolation was applied to take care of the controlled ^{40}Ar-increase between the recording of mass numbers 20 and 40.

In general, mineral selection was as described above. However, for these experiments only relatively large grains (180–350 μ diameter) could be used. It is not a simple task to pick about 20 pure specimens of this size from each component, when the weight difference within one type has to be less than 20%. It is even more difficult when only 300 mg bulk fines are assigned to this purpose.

The results obtained on 162 individual samples are plotted in Figs. 6 and 7 and in part listed in Table 1. This concerns ^{4}He, ^{20}Ne and ^{36}Ar, the major isotopes of the trapped rare gas component. Absolute errors are mainly determined by the weight

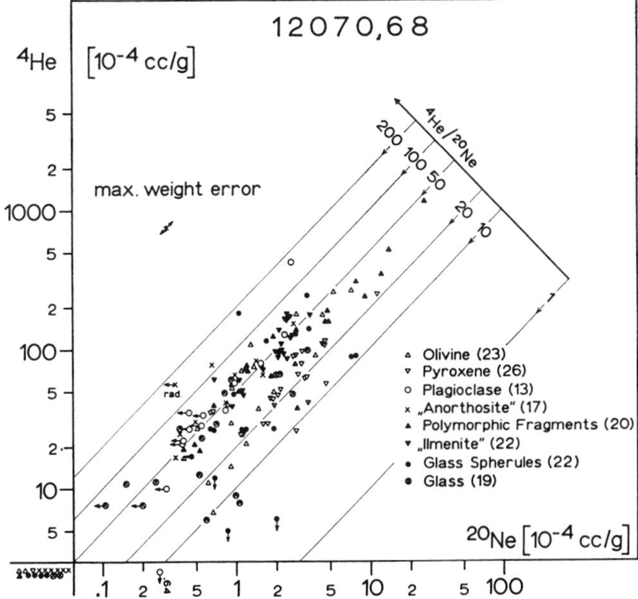

Fig. 6. ^{4}He, ^{20}Ne, and $^{4}He/^{20}Ne$ in single grains of 12070,68 fines. Static mass spectrometry analysis.

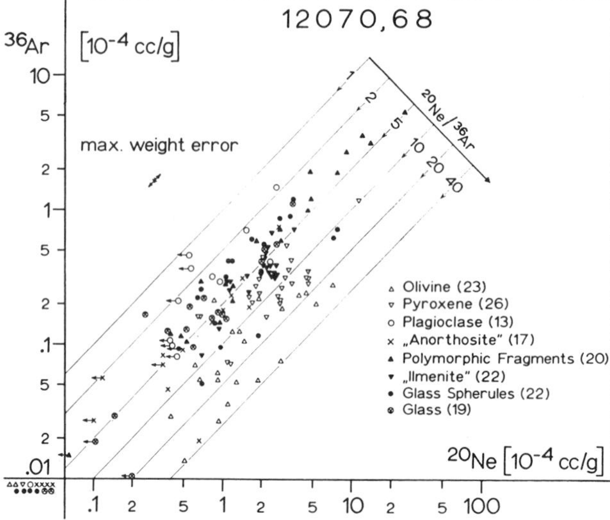

Fig. 7. ^{20}Ne, ^{36}Ar, and ^{20}Ne/^{36}Ar in single grains of 12070,68 fines. Static mass spectrometry analysis.

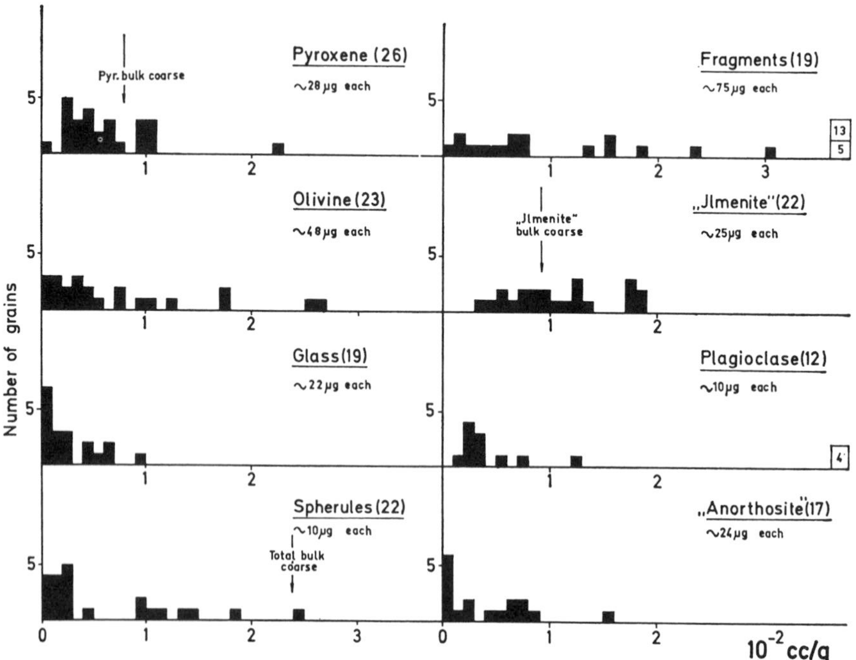

Fig. 8. ^{4}He concentrations in single grains of 12070,68 fines. Static mass spectrometry analysis.

Table 1. Variation of rare gases in individual grains (> 200 μ) of 12070,68 fines

	Approximate weight of grain	Concentration in 10^{-4} cc/g										Elemental ratios							
		^{4}He			^{20}Ne			^{36}Ar			$^{40}Ar_{rad}$	^{4}He/^{20}Ne			^{20}Ne/^{36}Ar			^{40}Ar/^{36}Ar	
	μg	max.	min.	med.	max.	min.	med.	max.	min.	med.	max.	max.	min.	med.	max.	min.	med.	max.	min.
Pyroxene	28	229	< 1	43	11.4	< 0.1	2.8	1.2	< 0.01	0.28	0.4	54	9.3	22	16	4.3	10	23	0.24
Olivine	48	264	< 1	74	7.2	< 0.05	1.9	0.28	< 0.005	0.11	~ 0.1	84	10	40	38	4.1	17	1.4	0.21
Glass	30	98	< 1	28	3.4	< 0.1	0.93	1.1	< 0.01	0.23	3.3	72	7.4	31	7	1.5	4.0	47	0.27
Spherules	10	244	< 2	73	7.8	< 0.3	2.1	1.2	< 0.02	0.44	~ 0.1	170	< 3	35	16	2.8	4.8	1.0	0.15
Ti-rich fragments	77	1280	≤ 1	192	26	< 0.1	4.8	5.3	0.015	1.2	2.8	65	26	40	6.5	2.4	4.0	184	0.41
"Ilmenite"	25	181	39	106	3.6	0.7	2.0	0.72	0.07	0.36	1.1	88	36	53	8.4	3.6	5.6	3.2	0.85
Feldspars	10	427	0.64	75	2.6	< 0.3	0.94	1.5	< 0.02	0.37	0.4	160	2.4	80	5.6	< 1.2	2.5	1.6	0.41
"Anorthosite"	24	155	< 1	41	2.7	< 0.1	0.66	0.75	< 0.01	0.13	15	170 rad.	41	62	35	< 2.1	5.1	220	0.57
Total		1280	< 1	—	26	< 0.05	—	5.3	< 0.005	—	15	170	2.4	—	38	< 1.2	—	220	0.15
For comparison																			
Bulk dust	0–225 μ			945			17.5			3.2				54			5.5	—	0.50 —
Bulk dust	150–225 μ			246			5.2			1.23	~ 0.3			47			4.2	—	0.82 —
Pyroxene-bulk	100–225 μ			79			2.4			0.25	0.15			32			9.7	—	1.25 —
"Ilmenite"-bulk	100–225 μ			91			1.8			0.31	0.4			51			5.8	—	1.39 —

differences, $\pm 10\%$, as marked in Figs. 6 and 7. Spallogenic contributions have also been calculated from the other He-, Ne-, and Ar-isotopes. Due to the extremely small sample sizes, average absolute spallogenic contributions per grain are about 10^{-10} cc for ^{3}He and 10^{-11} cc for ^{21}Ne and ^{38}Ar. This causes relatively large errors and we will not further discuss this component except to mention that the highest apparent exposure age (in a glass particle) is $\gtrsim 1$ b.y. Corrected values are given in Figs. 6 and 7 for the few cases where a cosmogenic correction affects either ^{4}He, ^{20}Ne, or ^{36}Ar by more than 1%. A thorough discussion of all these data, together with similar data obtained on single grains of Apollo 11 soil and breccias is rather voluminous and will be published in a separate paper. The same applies also for ratios such as $(^4He/^3He)_{tr}$, $(^{20}Ne/^{22}Ne)_{tr}$, or $(^{38}Ar/^{36}Ar)_{tr}$.

Figure 8 shows the distribution of ^{4}He concentrations for the different separates. Similarly, ^{20}Ne/^{36}Ar and ^{40}Ar/^{36}Ar ratios are shown in Figs. 9 and 10. Table 1 compiles extreme and mean values for concentrations and ratios within each group. For comparison, bulk data are also given in Table 1.

Trapped rare gases in soil and breccias. Bulk samples as well as six grain size fractions of 12070,68 fines were analyzed for He-, Ne-, and Ar- isotopes. The results for ^{4}He, ^{20}Ne and ^{36}Ar are given in Fig. 11 and Table 1. By etching with a HF–H_2SO_4 mixture (average radius reduction 3 μ for the 25–50 μ fraction, 6 μ for the 50–100 μ fraction, results see Fig. 11), gas concentrations are lowered by factors of 3.5 and 8 for He, 8 and 15 for Ne, and 18 and 21 for Ar, respectively.

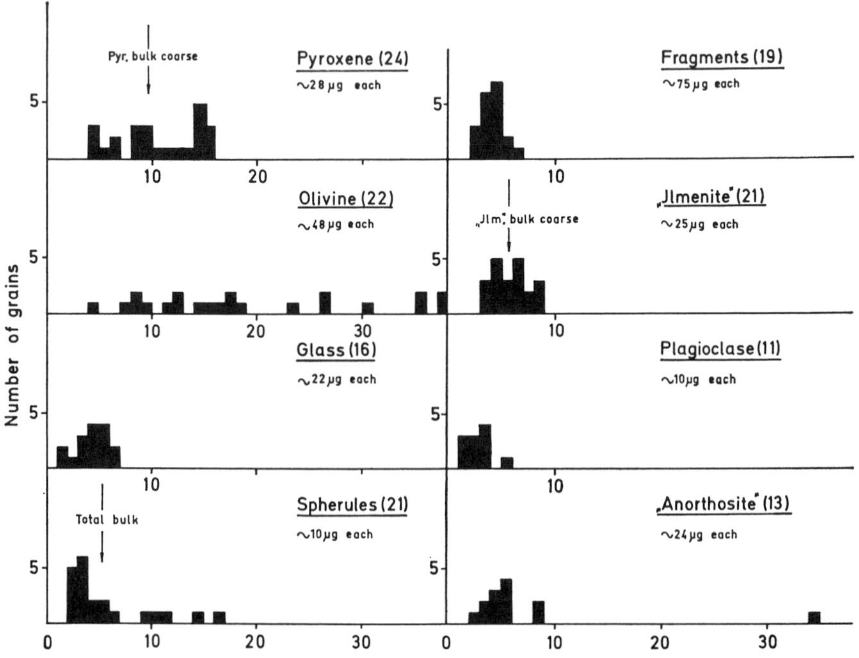

Fig. 9. ^{20}Ne/^{36}Ar ratios in single grains of 12070,68 fines.

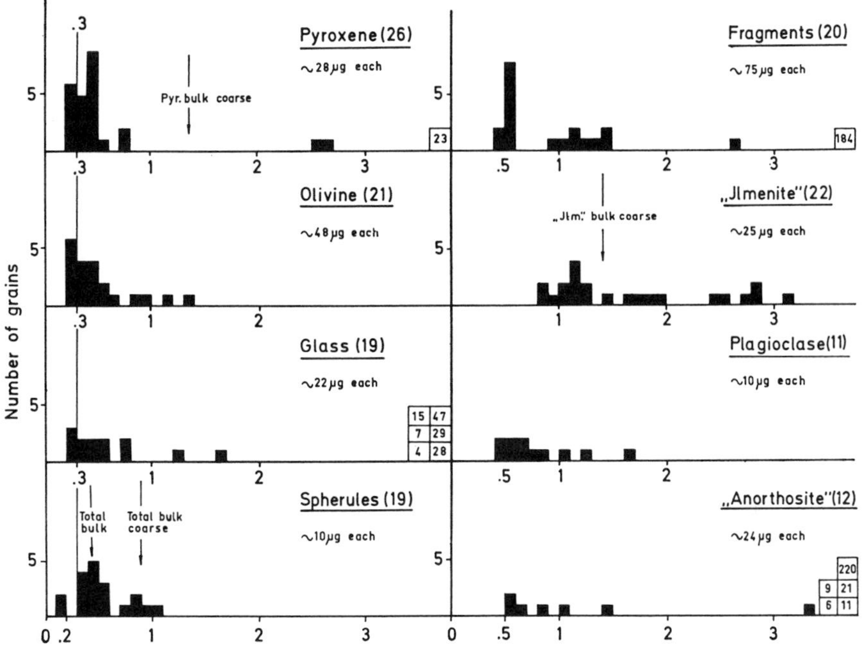

Fig. 10. $^{40}Ar/^{36}Ar$ ratios in single grains of 12070,68 fines.

Further etching experiments have been carried out on pyroxene and "ilmenite" separates (grain size 100–225 μ) for radius reductions of 0, 1, 3, 6, and 10 μ. The results are given in Fig. 12 and Table 1.

The distribution of rare gases in breccias has also been investigated in selected chips as well as in homogeneous aliquots. The results are presented in Table 2. Data for all light isotopes of 12070 bulk soil are also included in Table 2 for comparison. Also for these samples, radiogenic and spallogenic components and isotopic ratios of trapped gases are discussed in a separate paper. Main results are K–Ar ages of 1.7 b.y. and 1.13 b.y. for 12070 soil and 12073 breccia, respectively. From the exposure age of 12070 soil (350 m.y.) it is clear that spallogenic contributions do not affect the discussion of ^{4}He, ^{20}Ne and ^{36}Ar in this paper.

All data given in this section are accurate within $\pm 3\%$.

Discussion

Bulk Data

Absolute gas concentrations of Apollo 12 bulk soil 12070 (Table 1) are lower than in Apollo 11 bulk soil 10084 by about a factor of 2 for ^{4}He and by about 20% for ^{20}Ne and ^{36}Ar. Our data are 20% higher than those reported by Hintenberger *et al.* (1971) and 50% higher than the values given by LSPET (1970). Possibly our sample contained relatively more fine grained material. Even larger inhomogeneities are

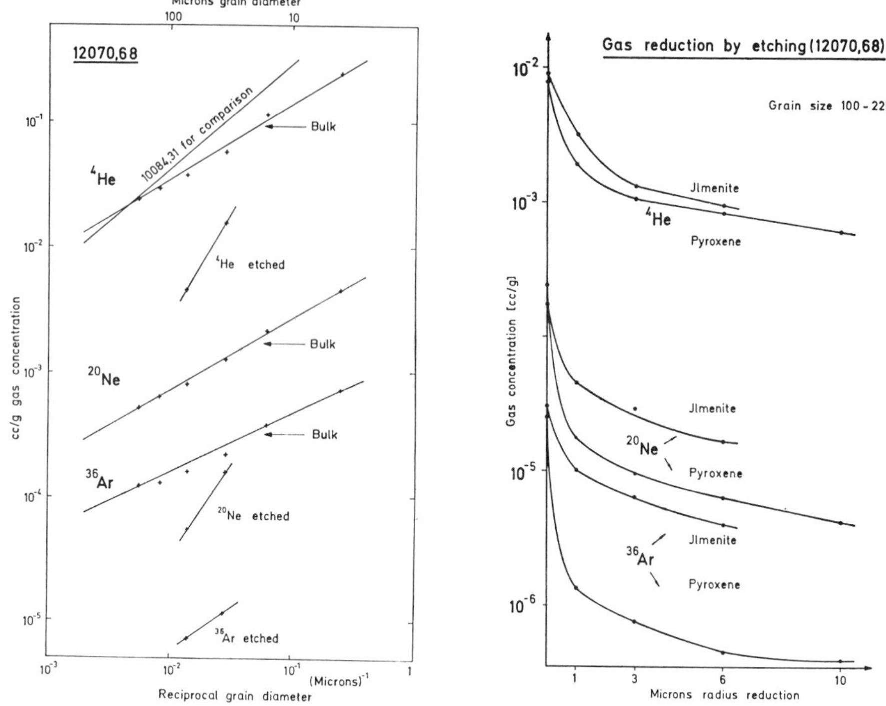

Fig. 11. He, Ne, and Ar in grain size fractions of 12070,68 fines. Data for etched samples (average radius reduction 3 μ for 25–50 μ fraction, 6 μ for 50–100 μ fraction) are also given.

Fig. 12. Reduction of trapped gas contents by etching of ilmenite and pyroxene from 12070,68 fines.

Table 2. Variation of rare gas concentrations in breccias (selected data)

					10^{-4} cc/g				
		^{3}He	^{4}He	^{20}Ne	^{21}Ne	^{22}Ne	^{36}Ar	^{38}Ar	^{40}Ar
12010,26	Light feldspathic chip	0.0036	74.2	0.0806	0.00069	0.00693	0.0195	0.0052	25.9
	Lithic rock fragment	0.0030	36.4	—	0.00086	0.00184	0.0027	0.0009	2.23
	Gabbroic (shocked) fragment	0.0128	2.68	0.029	0.00242	0.00521	0.0038	0.0025	0.265
	Homogeneous dark chip	0.270	775	23.6	0.0637	1.99	2.42	0.46	7.07
12073,032	Homogeneous fragment	0.171	440	8.24	0.0268	0.652	1.51	0.289	1.300
	same, adjacent	0.166	427	8.09	0.0270	0.632	1.40	0.272	1.075
	same, 1 cm apart	0.165	407	7.86	0.0270	0.609	1.37	0.264	1.025
	Aliquot, 50–100 μ	0.160	384	7.63	0.0245	0.590	1.462	0.280	1.05
	same, etched	0.044	79	0.95	0.0089	0.083	0.089	0.021	0.242
10046,19	Light rock fragment	0.038	11.3	0.019	0.0065	0.0075	0.0103	0.0081	0.72
	Homogeneous dark chip	0.74	2170	48	0.1255	3.67	4.48	0.835	10.0
	Aliquot, 50–100 μ	0.96	3120	52.5	0.1266	3.95	7.02	1.31	14.8
	same, etched	0.114	305	1.135	0.0111	0.100	0.074	0.0226	0.344
12070,68	Bulk soil, 0–225 μ	0.348	945	17.5	0.0447	1.34	3.21	0.596	1.60
	Bulk soil, 50–100 μ	0.147	368	7.72	0.0215	0.60	1.55	0.298	0.986

indicated from MEGRUES (1971) results. He observes variations up to a factor of 2 within three mg-size samples, with our values falling in this range.

To a first approximation, the inverse relationship between grain size and gas concentrations of 12070 (Fig. 11) indicates a surface related distribution. The same has previously been observed for Apollo 11 soil (EBERHARDT *et al.*, 1970, KIRSTEN *et al.*, 1970, among others). More accurately, gas concentrations per gram are proportional to $r^{-0.6}$ rather than r^{-1} (Fig. 11), which means that concentrations per surface area are less for smaller grains. This effect is more pronounced in Apollo 12 soil than in Apollo 11 soil. The ^{4}He slope difference between soils 10084 and 12070 in Fig. 11 indicates a relative enhancement of a volume related component in 12070. Note that for large grain sizes, 12070 contains more ^{4}He than 10084. One could imagine that radiogenic ^{4}He is responsible for this reversal, especially since 12070 bulk contains less total gas and four times more U and Th than 10084 (SILVER, 1971). If gas retention would have been complete during a maximum of 4.5 b.y., radiogenic ^{4}He could contribute up to 30×10^{-4} cc/g. This is certainly overvalued, since the K–Ar gas retention age of 12070 was found to be 1.7 b.y. only. Even if radiogenic ^{4}He accounts for a part of the volume component, it can not be the main source since similar trends are observed for ^{20}Ne and ^{36}Ar. Other possible reasons could be composite grain surfaces or more penetrating solar flare particles.

The mean penetration depth b of trapped gases can be determined from the results of the etching experiments. A simplified model is assumed in which the gas concentration G decreases exponentially from the surface of a grain with radius R towards the interior according to $G(r) = G_{\text{surf.}} \exp \{r - R)/b\}$ (see KIRSTEN *et al.*, 1970). For 12070,68 bulk (Fig. 11), average penetration depths of 3 μ for He, 2 μ for Ne and 1 μ for Ar result. They exceed the expected penetration depths of solar wind particles, which are in the order of 100 Å.

There again exists, as observed in similar studies of Apollo 11 fines (KIRSTEN *et al.*, 1970), a grain size dependency such that apparent mean penetration depths increase with grain size. This is especially pronounced for Ar.

In other words, gases in smaller grains are less equilibrated than in larger ones. After being saturated with solar wind in a relatively short time at the lunar surface, smaller grains are more readily covered and therefore suffer less heating by direct radiation during lunar day. This could possibly explain the reduced migration of gases in smaller grains.

From the ^{4}He content of 150–225 μ bulk 12070, a surface concentration of 2.2×10^{-4} cc/cm^2 can be calculated. With the solar wind ^{4}He flux directly measured by GEISS *et al.* (1970), it would take only 30 years to build up this level. The actual residence time of a 200 μ grain at the top surface is much larger. It can be estimated to be $\sim 10^5$ y. from the turnover rate based on the cosmic ray exposure age, which is $\sim$80 cm/350 m.y. for the Oceanus Procellarum. Since smaller grains have lower residence times and lower gas concentrations per surface area, it is suggested that by some mechanism gas concentrations continue to increase after the fast initial build up ($\sim$10 y.) during the whole residence time at the very surface, though at a much lower rate. This is supported by the results from the in situ analysis of exposed crystalline rock surfaces ("grain size" centimeters). 12063,106 (Top) contains up to

40 times more He per unit area than 150–225 μ bulk soil 12070-68. Apart from inward diffusion, it is suggested that all types of energetic solar flare particles with ranges of some 10 μ drive solar wind already implanted in the grains deeper into the interior. This would occur during the total surface residence time of the soil particles as governed by their grain size. For heavier gases, the effect would be less effective than for lighter gases. The mechanism would explain: (1) The larger concentrations per surface area for larger grains. (2) The larger mean penetration depth for larger soil particles. (3) The enhanced surface concentration of heavier gases for a given grain size.

It seems that about 90% of the gas contents are surface concentrated, while a remainder of 2–10% penetrates far into the interior. This can also be seen from the results of the etching experiments of mineral separates (Fig. 12). For a depth of etching of 1 μ, the mean penetration depths for pyroxene are 1.4 μ for He and about 0.5 μ for Ne and Ar. Again, going to larger depths, the volume component becomes important and results in larger apparent penetration depths. It may well be that even the 1 μ values are not the end of this trend, since initial implantation depths are even lower. This is indicated from the results of Eberhardt *et al.* (1970) for Apollo 11 soil. By etching more gently, these authors obtained a mean penetration of 0.2 μ in ilmenite. The more surface concentrated profiles for separated minerals as compared to bulk soil can be explained by the fact that undisturbed crystal lattices allow less diffusion (and probably impingement) than the abundant polymineralic fragments, which determine the average behavior of the bulk soil. Gas concentrations per surface area in larger bulk grain size fractions are overestimated since one disregards the internal surfaces of these polymorphic aggregates.

Turning to breccias, sample 12010 shows a marked light-dark structure similar to light-dark structured meteorites. Two lithic fragments contain only very little solar wind gases, but extremely large ^{4}He concentrations (Table 2). ^{4}He/^{3}He ratios are 20600 and 12200. This must be due to radiogenic He from high U-concentrations in the order of 10 ppm. Similar enrichments have been found in sample 12013 (Wakita and Schmitt, 1970). The analyzed chip of the dark portion of 12010 is very rich in solar wind gases (Table 2), we found about three times more gas than reported by LSPET (1970). The whole pattern may be explained by an impact process which combined large light crystalline fragments with solar wind loaded fine grained material.

Contrary to 12010, breccia 12073 is very homogeneous for different mg-samples, both in appearance and in gas concentrations. There are no distinct differences between 3 chips taken at different locations and an aliquot prepared from the total sample. If equal grain size fractions are compared, this breccia contains about the same amounts of trapped gases as 12070,68 soil (Table 2).

For comparison, similar studies have been carried out for breccia 10046,19. In this case, gas concentrations of average chips are comparable to the aliquot, while light rock fragments again show much less solar wind implanted gases (Table 2).

From these results and from results of Eberhardt *et al.* (1970) and Yaniv and Heymann (1971) it is evident that solar wind gases are redistributed at the lunar surface and that the physical and chemical properties of the different minerals and fragments affect the actual gas concentrations. This leads to the discussion of the results obtained from single grains.

Single Grain Data

One might consider presenting the results given in Figs. 6, 7, and 8 in cc/cm^2 rather than cc/g. However, due to the irregular shape of the soil components and the existence of internal surfaces, the error of such numbers would be at least a factor of 2. This corresponds to the largest occurring diameter difference of all grains analyzed by static mass spectrometry. Therefore, similar grain size is assumed in the discussion of Figs. 6, 7, and 8 (but not Fig. 5). For an approximate transformation, divide concentrations per gram by 100 to obtain concentrations per cm^2.

The most obvious result from the analysis of individual soil constituents is the extent of noble gas variations for individuals of the same type (Table 1), even apart from the known grain size dependency. Furthermore, elemental ratios differ by more than an order of magnitude. Apparently bulk dust data yield rather arbitrary averages which can only be understood if one studies individual variations. Possible reasons for such variability are: (1) Different surface exposure times (including zero); (2) Differing degrees of saturation; (3) Time variable particle flux and composition; (4) Differing ratios of irradiated/unirradiated surfaces and geometrical shielding; (5) Differing capture probabilities; (6) Differing residence times in surface and subsurface layers, that is, differing mixing rates; (7) Continuous and episodic diffusion, inwards and outwards, depending on differing temperatures; (8) Change of retentivity by impurities; (9) Ignorance of internal surfaces of aggregates; (10) Gas loss due to ion sputtering; (11) Multiple irradiation periods; and (12) Multiple sources, not necessarily surface irradiation, (e.g., primordial gas inclusions, radiogenic enrichments).

The distribution of ^{4}He concentrations for the different fractions measured by static mass spectrometry is shown in Fig. 6. In general, mean concentrations are lower than those obtained from microprobe grain analysis (Fig. 5). This is because the latter data include many much smaller grains (for the sake of statistics). If microprobe data are normalized to the same grain size, patterns agree with one exception: Very pure ilmenites, which melt into deep black spherules, give very high surface concentrations, of up to 60×10^{-4} cc ^{4}He/cm^2. Ilmenites are pure only for small grain sizes and are therefore missing in Fig. 6. The mean values of all analyzed grains fit, within statistics, the bulk values of comparable samples (Table 1). Polymineralic fragments have the largest spread, the highest absolute concentrations, and the highest mean value. They bear most of the total trapped gases of lunar soil. This is explained by their large internal surfaces. They may themselves be considered as "microbreccias," resulting in a quasi-volume proportional gas concentration, modulated by different admixtures of gas rich ilmenite. Ilmenites have the highest He-concentrations of all true individual grains, especially if they are very pure. This explains the Ti–He correlation observed by in situ microprobe analysis (Fig. 1). The minimum He-concentration for single ilmenites (220 μ grain size) is 39×10^{-4} cc/g. There are no "empty" ilmenites in contrast to all other components. This excludes the possibility that "empty" grains were never exposed to the solar wind, otherwise there should be "empty" ilmenites as well. We conclude that grains without solar wind suffered strong outgassing followed by immediate shielding. This could occur during a larger impact. The most retentive ilmenites undergo only partial outgassing at these occasions. By the same argument, glassy particles suffer strong diffusion losses and thus have the lowest concentrations. Differences within each group are best explained by differing surface residence times and the influence of imperfections, which make minerals "leaky" (e.g., see Plagioclase "light" and "dark" in Fig. 5). The time needed for initial surface saturation is too short to cause the observed differences.

Elemental ratios of trapped gases in related bulk samples show little variation (MARTI and LUGMAIR, 1971). Nevertheless, large individual variations exist for the components of a given sample. $^{4}He/^{20}Ne$ ratios range from 2.4 to 170 (Table 1). Original $^{4}He/^{20}Ne$ ratios tend to be lowered by diffusion. Even the maximum value of 170 observed in a glass spherule is much below the directly observed ratios of 430 and 620 (GEISS *et al.*, 1970). The lowest individual ratios occur for plagioclase and glass spherules. This could be explained by their low gas retentivity. However, the largest ratio occurs also in a glass spherule. This can be explained in terms of a model proposed by HEYMANN and YANIV (1971), according to which trapped gases are composed of: (1) remainders of an early loading, fractionated diffusively by a following impact heating which either relatively enriches heavier gases or leads to complete outgassing; and (2) a more recent reloading followed by slight diffusion loss at moderate lunar surface temperatures. Our results favor this model, with high $^{4}He/^{20}Ne$-glasses being recently reloaded and low $^{4}He/^{20}Ne$-glasses covered immediately after strong, but not complete outgassing. "Empty" grains are completely outgassed, while for retentive ilmenites temperatures were not sufficient to release all the gases. The $^{20}Ne/^{36}Ar$ ratios range from 1.2 to 38. The largest spread as well as the largest mean value is observed in retentive olivines, while polymineralic fragments tend to average individual differences and thus have the smallest spread (Figure 9). The lowest $^{20}Ne/^{36}Ar$ ratios occur again for less retentive glasses and feldspars, down to <1.2. Different from $^{4}He/^{20}Ne$ ratios however, the largest values are observed for the more retentive olivines and pyroxenes, which suffered less diffusion loss. The absence of high $^{20}Ne/^{36}Ar$ ratios in *less* retentive sites is then explained by "preimpact"-Ar, which was, contrary to Ne, not completely outgassed.

No correlation exists between $^{4}He/^{20}Ne$ ratios and total ^{4}He concentrations. The same is true for $^{20}Ne/^{36}Ar$ vs. ^{20}Ne (Fig. 7, 8). A simple one step fractionation which simultaneously lowers elemental ratios and total gas concentrations can therefore not apply. This is also prohibited by the observation that $^{4}He/^{20}Ne$ and $^{20}Ne/^{36}Ar$ ratios do not follow the same trend for olivines and pyroxenes (with apparent Ne-excess) and for plagioclases (with apparent Ne-deficit). $^{4}He/^{20}Ne$ and $^{20}Ne/^{36}Ar$-ratios are, if at all, inversely related, with the possible exception of olivine. It is interesting to note that a similar trend is observed for elemental ratios of various breccias (Table 2). HEYMANN and YANIV (1970) have analyzed Apollo 11 constituents and, also in this case, there is no positive relation between $^{4}He/^{20}Ne$ and $^{20}Ne/^{36}Ar$ ratios. Predegassed grains with "preimpact"-Ar causing low $^{20}Ne/^{36}Ar$ ratios have still the highest $^{4}He/^{20}Ne$ ratios, resulting from the recent "postimpact" irradiation.

After correction for spallogenic gases, there are no large variations in the isotopic ratios of light trapped rare gases of 12070 single grains, with the exception of the $^{40}Ar/^{36}Ar$ ratio. It has been shown that lunar fines contain surface related trapped ^{40}Ar which correlates with solar wind gas concentrations but can itself not be of solar wind origin (HEYMANN and YANIV (1970a). The most likely source is reimplantation of ^{40}Ar from a latent lunar atmosphere, transported by the solar wind (MANKA and MICHEL, 1971). The different $^{40}Ar/^{36}Ar$ ratios of 12070 (0.5, see Table 2) and 10084 as well as MARTI and LUGMAIR'S (1971) core tube analysis have stressed the different origins of implanted ^{36}Ar and ^{40}Ar. By analyzing single grains of Apollo 12 soil, it was our intention to establish a low upper limit for the $^{40}Ar/^{36}Ar$ ratio in the solar wind itself. Seven grains have ratios definitely below 0.25, with a minimum of 0.15 ± 0.03 for a glass spherule (Fig. 10). Since this value is three times below the bulk ratio, one may conclude that the latent lunar atmosphere is rather variable with time. To compensate for the low values, at least part of ratios above 0.5 would then also reflect implanted gas ratios. Alternatively, differences could also be caused by differing implantation depths for ^{40}Ar and ^{36}Ar and subsequent diffusion. In general, we can not differentiate between trapped ^{40}Ar and in situ radiogenic Ar for grains with higher $^{40}Ar/^{36}Ar$-ratios. The latter ratios are not correlated with absolute ^{36}Ar-concentrations. It may be noticed however that the eight grains with the largest amounts of in situ ^{40}Ar are all anorthosites and glasses (up to 15×10^{-4} cc $^{40}Ar/g$, Table 1). It is probable that these particular specimens are "KREEP" (HUBBARD *et al.*, 1971), highly enriched in potassium.

In situ Analysis

The outlined features of the distribution of trapped rare gases are completed and confirmed by the results of the in situ analysis of breccias and crystalline rock surfaces. Direct evidence for the surface concentration arises from the gas enrichment in fine grained material and grain boundaries (Fig. 1). Similar results, although with lower local resolution, were reported by MEGRUE (1971). The inhomogeneity of the gas content indicates that breccias were formed from constituents which differed distinctly in their initial gas concentrations, with particular enrichments in Ti-rich components. The different strengths of the Ti–He correlation for different breccias must be due to redistribution of trapped gases from their initial sites by diffusion and by shock effects during formation of the breccias.

The Cr–He correlation (Fig. 2) is explained by the coexistence of ilmenite and chromium spinel (SIMPSON and BOWIE, 1971). On top of crystalline rock surfaces, trapped gases are more uniformly distributed than in breccias because of the common radiation conditions for the whole rock as compared to the multicomponent origin of breccias. Local differences could be due to differing capture probabilities and geometrical shielding of regions within the rough surface. Another factor may be solar wind induced surface erosion by ion sputtering.

By order of magnitude, measured surface concentrations agree with saturation levels expected from the data given by LORD (1968), and with MEGRUE'S (1971) results from in situ analysis of trapped gases by Laser beam extraction. Absolute concentrations obtained for breccias fit the data obtained from bulk analysis (Table 2).

Crystalline rock surfaces contain the highest gas concentrations per unit surface area since they were still undergoing bombardment at the time of sample collection, with no time for diffusion loss while shielded from reloading. The detection of trapped

He on Top and Bottom of sample 12063 provides direct evidence for the turnover and the orientation of this rock.

Further information on temperatures and surface residence times of particular rocks and breccias could be obtained from concentration profiles measured with high local resolution (Å) by ion sputtering mass spectrometry combined with secondary ionization of rare gases and by diffusion experiments.

Acknowledgements—We are grateful to NASA for providing us with lunar samples. The valuable assistance of H. Richter, R. Schwan and H. Weber is appreciated.

References

Eberhardt P., Geiss J., Graf H., Grögler N., Krähenbühl U., Schwaller H., Schwarzmüller J., and Stettler A. (1970) Trapped solar wind noble gases, exposure age and K/Ar-age in Apollo 11 lunar fine material. *Proc. Apollo 11 Lunar Sci. Conf., Geochim. Cosmochim. Acta* Suppl. 1, Vol. 2, pp. 1037–1070. Pergamon.

Funkhouser J. G., Schaeffer O. A., Bogard D. D., and Zähringer J. (1970) Gas analysis of the lunar surface. *Proc. Apollo 11 Lunar Sci. Conf., Geochim. Cosmochim. Acta* Suppl. 1, Vol. 2, pp. 1111–1116. Pergamon.

Geiss J., Eberhardt P., Bühler F., Meister J., and Signer P. (1970) Apollo 11 and Apollo 12 solar wind composition experiments: Fluxes of He and Ne isotopes. *J. Geophys. Res.* **75**, 5972–5979.

Heymann D. and Yaniv A. (1970) Inert gases in the fines from the Sea of Tranquillity. *Proc. Apollo 11 Lunar Sci. Conf., Geochim. Cosmochim. Acta* Suppl. 1, Vol. 2, pp. 1247–1259. Pergamon.

Heymann D. and Yaniv A. (1970a) Ar^{40} anomaly in lunar samples from Apollo 11. *Proc. Apollo 11 Lunar Sci. Conf., Geochim. Cosmochim. Acta* Suppl. 1, Vol. 2, pp. 1261–1267. Pergamon.

Hintenberger H., Weber H. W., Voshage H., Wänke H., Begemann F., and Wlotzka F. (1970) Concentrations and isotopic abundances of the rare gases, hydrogen and nitrogen in Apollo 11 lunar matter. *Proc. Apollo 11 Lunar Sci. Conf., Geochim. Cosmochim. Acta* Suppl. 1, Vol. 2, pp. 1269–1282. Pergamon.

Hintenberger H., Weber H., and Takaoka N. (1971) Concentrations and isotopic abundances of the rare gases in lunar matter. Second Lunar Science Conference (unpublished proceedings).

Hubbard N. J., Gast P. W., and Meyer C. (1971) The origin of the lunar soil based on REE, K, Rb, Ba, Sr, P, and $^{87}Sr/^{86}Sr$ data. Second Lunar Science Conference (unpublished proceedings).

Kirsten T., Müller O., Steinbrunn F., and Zähringer J. (1970) Study of distribution and variations of rare gases in lunar material by a microprobe technique. *Proc. Apollo 11 Lunar Sci. Conf., Geochim. Cosmochim. Acta* Suppl. 1, Vol. 2, pp. 1331–1343. Pergamon.

Lord H. C. (1968) Hydrogen and helium ion implantation into olivine and enstatite: retention coefficients, saturation concentrations and temperature release profiles. *J. Geophys. Res.* **73**, 5371–5280.

LSPET (Lunar Sample Preliminary Examination Team) (1969) Preliminary examination of lunar samples from Apollo 11. *Science* **165**, 1211–1227.

LSPET (Lunar Sample Preliminary Examination Team) (1970) Preliminary examination of the lunar samples from Apollo 12. *Science* **167**, 1325–1339.

Manka R. H. and Michel F. C. (1971) Lunar atmosphere as a source of lunar surface elements. Second Lunar Science Conference (unpublished proceedings).

Marti K. and Lugmair G. W. (1971) Kr^{81}-Kr and K-Ar^{40} ages, cosmic ray spallation products and neutron effects in Apollo 11 and Apollo 12 lunar samples. Second Lunar Science Conference (unpublished proceedings).

Megrue G. H. (1971) Distribution and origin of helium, neon, and argon isotopes in Apollo 12 samples by In Situ Analysis with a Laser-Probe mass spectrometer. Second Lunar Science Conference (unpublished proceedings).

SILVER L. T. (1971) U–Th–Pb isotope relations in Apollo 11 and Apollo 12 lunar samples. Second Lunar Science Conference (unpublished proceedings).

SIMPSON P. R. and BOWIE S. H. U. (1971) Opaque phases in Apollo 12 samples. Second Lunar Science Conference (unpublished proceedings).

WAKITA H. and SCHMITT R. A. (1970) Elemental abundances in seven fragments from lunar rock 12013. *Earth and Planet. Sci. Lett.* **9**, 169–176.

YANIV A. and HEYMANN D. (1971) Inert gases from Apollo 11 and Apollo 12 fines: Reversals in the trends of relative element abundances. Second Lunar Science Conference (unpublished proceedings).

ZÄHRINGER J. (1966) Primordial helium detection by microprobe technique. *Earth Planet. Sci. Lett.* **1**, 20–22.

Proceedings of the Second Lunar Science Conference, Volume 2, pp. 1671–1679.
The M.I.T. Press, 1971.

The irradiation history of lunar samples

D. S. BURNETT, J. C. HUNEKE, F. A. PODOSEK, G. PRICE RUSS III, and G. J. WASSERBURG
Lunatic Asylum of The Charles Arms Laboratory, Division of Geological and Planetary Sciences,* California Institute of Technology, Pasadena, California 91109

(*Received 10 March, 1971; accepted 3 May, 1971*)

Abstract—From new data on Apollo 12 samples illustrated in this paper and data available in the literature, the galactic cosmic ray and solar wind exposure of lunar samples has been investigated by means of rare gas and Gd isotopic measurements. Neutron exposures obtained from Gd isotopic measurements on Apollo 11 and Apollo 12 soil samples are the same to within 10% and Xe/(Ba + 1.65 Ce) to within 40%. These data permit regolith mixing depths of 10–20 meters to be calculated for the Apollo 11 and Apollo 12 sites. The Apollo 12 double core shows only a small, but significant, gradient in neutron exposure ($\sim 10\%$ larger at the bottom). Consequently, the core could not have been stratified in its present configuration for more than 50×10^6 yr. Apollo 12 rocks have neutron exposures which are similar to the Apollo 11 rocks. No variations in neutron exposure are observed for samples from different depths in 12002.

Strong evidence for a range in irradiation depths for lunar rocks is obtained from the relative yields of rare gas spallation products and neutron dosages calculated from the measured Gd isotopic variations. However, it appears that the average irradiation depths were not large (probably ≤ 60 cm). Almost all rocks must have resided at depths greater than ~ 4 meters most of the time since their original crystallization.

An internally consistent set of spallation product exposure ages has been calculated for 14 Tranquillity Base rocks. Six of seven "low-K" rocks appear to be grouped with an exposure age of around 100 m.y.

Thermal release data show large variations ($\sim \Delta M/M$) in the composition of SUCOR (surface correlated) Kr and Xe indicating that SUCOR may not be well defined. SUCOR Xe appears to be complex and may contain, in addition to solar wind, additional components from the lunar atmosphere (for example, fission Xe) which have been implanted by solar wind action.

INTRODUCTION

THE LUNAR SURFACE is under constant bombardment by cosmic rays and solar wind ions which penetrate only the outermost layers of the moon. The concentration of cosmic ray spallation products and trapped solar wind ions in lunar material are functions of both the time of exposure and the average irradiation depth of the sample and the thermal cycling of the lunar soil.

The measurement of these cosmic ray and solar wind effects will yield information about lunar surface processes. The galactic cosmic ray and solar wind exposure of lunar samples has been investigated by the measurement of spallogenic and trapped rare gases and of Gd^{158} excesses from the capture of spallation produced neutrons by Gd^{157}. Of these, Gd is particularly important, since it is not easily susceptible to loss by diffusion (e.g., H, He, etc.).

* Contribution Number 2025.

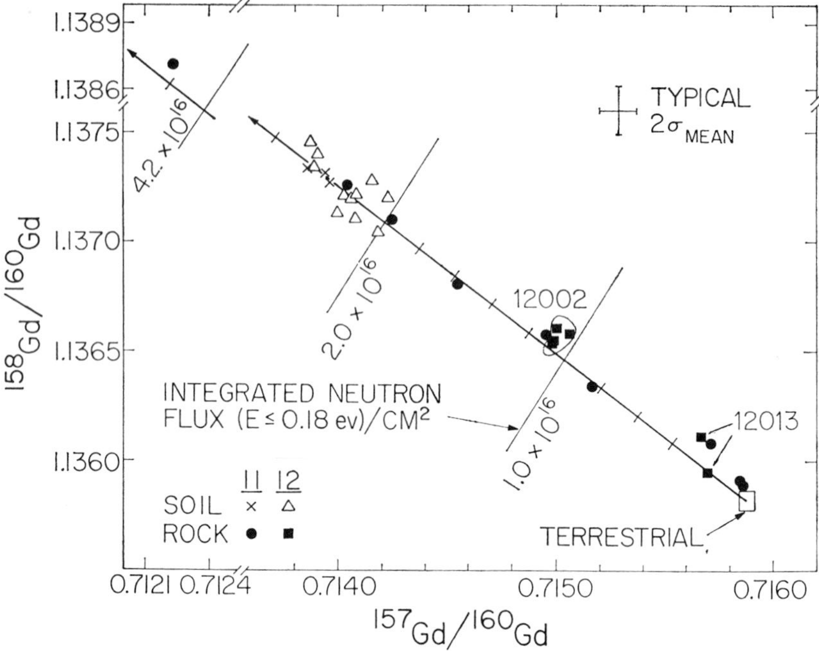

Fig. 1. Gd^{158}/Gd^{160} vs. Gd^{157}/Gd^{160} correlation diagram. The position of a point along the predicted correlation line for samples which have been exposed to a neutron flux is a measure of the dosage received. A dosage scale (n/cm^2) is given. The soils from Apollo 11 and Apollo 12 have rather similar dosages and are higher than almost all of the rocks. Samples from different positions within the rock 12002 separated by up to 5 cm are encircled.

NEUTRON DOSAGE

The Gd isotopic data (relative precision is typically $2\sigma = \pm 2$ parts in 10^4) are displayed on a Gd^{158}/Gd^{160}–Gd^{157}/Gd^{160} correlation diagram (Fig. 1). The observed Gd isotopic variations result from the capture by Gd^{157} of low energy (< 0.2 eV) neutrons (produced by the interactions of cosmic rays with lunar materials). That these isotopic effects are the result of neutron capture is demonstrated by the inversely correlated variations in Gd^{157}/Gd^{160}–Gd^{158}/Gd^{160} and Gd^{155}/Gd^{160}–Gd^{156}/Gd^{160} along the lines predicted for these reactions.

The measured isotopic shifts can be used to calculate low energy neutron dosages (n/cm^2), utilizing an average Gd^{157} capture cross section calculated from the theoretical lunar neutron energy spectrum of LINGENFELTER, CANFIELD, and HAMPLE (unpublished). The neutron dosages calculated from this spectrum are about a factor of 2.5 higher than those we estimated previously (EUGSTER *et al.*, 1970) based on an assumed Maxwell-Boltzmann energy distribution. We estimate that 95% of the Gd^{157} capture occurs below 0.18 eV and our neutron dosages refer only to neutrons in the 0–0.18 eV region. The calculated neutron dosages are not sensitive to fluxes at higher energies.

The position of a point along the correlation line in Fig. 1 is a measure of the neutron dosage. The neutron dosage scale shows that lunar rocks exhibit a wide range in dosage from 0 to 4×10^{16} n/cm². In contrast, the dosages for Apollo 11 and a variety of Apollo 12 (12070, 12042, 12025, and 12033) soil samples are very similar. The measured dosages for soils are the average of the dosages of the individual soil particles. Consequently, the systematically higher dosages for soil samples as compared to the rocks means that the average soil particle has resided in the outer few meters of the lunar surface for a longer period of time than a typical rock.

Samples from known positions in rock 12002 separated by up to 5 cm show no measurable variations in neutron dosage (see Fig. 1). The calculations of LINGENFELTER *et al.* indicate that the neutron flux rises very sharply with depth to a maximum at around 120 g/cm² beneath the lunar surface. For a separation of 5 cm with this gradient, it would require 230×10^6 yr to build up an observable difference in the Gd isotopic composition.

AVERAGE IRRADIATION DEPTH

In Fig. 2 we plot the measured neutron dosages versus the concentration of Xe^{126} produced from spallation reactions on Ba utilizing both our data and that from other laboratories. The Xe^{126} (Ba) concentrations include the effects of REE spallation:

$$Xe^{126} \equiv \frac{[Xe^{126}]_{\text{spallation}}}{[Ba] + 1.65\,[Ce]}$$

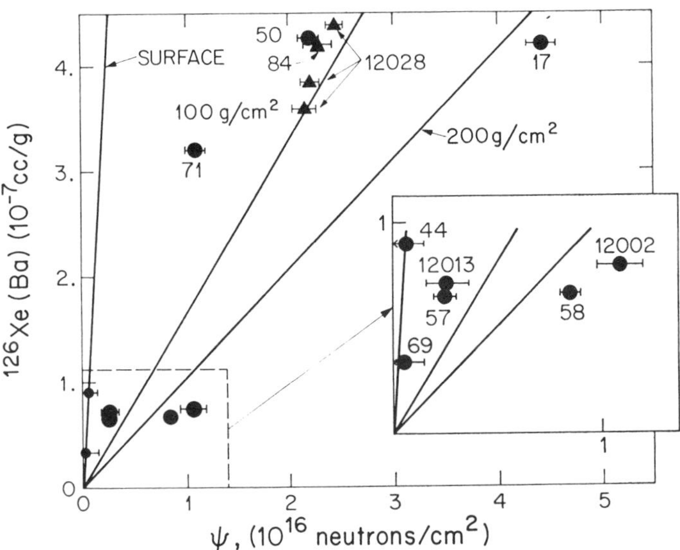

Fig. 2. Correlation plot of the measured neutron dosage with the amount of Xe^{126} produced by Ba spallation. Abbreviated numbers refer to Apollo 11 rocks. The spread indicates that lunar rocks have been irradiated at a variety of irradiation depths. The Gd and some Xe and Ba data are from our laboratory. Additional Ba data are obtained from data presented by various authors in the Proceedings of the Apollo 11 Lunar Science Conference. Other Xe data are from BOGARD *et al.* (1971), PEPIN *et al.* (1971), and J. H. REYNOLDS (personal communication).

The relative Xe^{126} yields from rare earth and Ba spallation were obtained from our analyses of samples of differing Ba/REE from rock 12013 (LUNATIC ASYLUM, 1970). Samples which have been irradiated at a single irradiation depth for various lengths of time will plot along a line on this diagram. A sample buried at 100 g/cm^2, for example, will lie at the origin for zero irradiation time and will move outward along the 100 gm/cm^2 line proportional to the irradiation time. The spread of data points indicates that these rocks have been irradiated at a wide variety of irradiation depths. Only the data for rocks 10044 and 10069 are compatible with irradiation on the lunar surface. The other rocks have greater average irradiation depths. We have made crude estimates of the positions of lines of constant irradiation depth on Fig. 2 using the calculated neutron flux gradient of LINGENFELTER *et al.* and an estimate of the Xe^{126} production rate gradient based on the iron meteorite spallation product depth studies of SIGNER and NIER (1960) and the thick-target irradiation data of HONDA (1962). The positions of these lines are quite sensitive to the parameters in the production rate equations, particularly for large depths. The sense of the error will probably be to overestimate irradiation depths. From the Xe–Gd systematics, it appears that the majority of the rocks have comparatively shallow (~ 100 g/cm^2) irradiation depths, with the exception of 10017, 10058, and 12002.

A shallow irradiation depth for many lunar rocks is also indicated by the comparatively good agreement in the cosmic-ray exposure ages calculated for different rare gas spallation products using production rates obtained from measurements of *radioactive* spallation products. These production rates are appropriate for a rock sitting on the lunar surface, whereas the relative production rates of rare gas nuclei at large depths are expected to be different from their relative values at the surface.

EXPOSURE AGES

Based on the above arguments, calculation of exposure ages using surface production rates appears to be at least provisionally justified. We have calculated an internally consistent set of exposure ages for Apollo 11 rocks utilizing Xe^{126} data whenever possible, and Ne^{21} data if Xe^{126} measurements were not available. These ages are plotted in Fig. 3 versus the spallogenic Ar^{38}/Ne^{21} ratio as a shielding indicator. High values of Ar^{38}/Ne^{21} correspond to less shielding. We take a surface production rate of 1×10^{-7} cc STP Xe^{126}/gm Ba/10^8 yr. There is a good correlation between the exposure ages and chemical groups in that 6 of 7 low-K Tranquillity Base rocks (47, 48, 20, 3, 58, and 44) have ages around 100×10^6 yr and may represent debris from a single major impact event at that time. The ages of the low-K rocks do not show any pronounced correlation with the degree of shielding which a rock has experienced.

The high-K rocks fall into two exposure age groups at $\sim 40 \times 10^6$ yr (49, 57, 69) and at $\sim 400 \times 10^6$ yr (17, 22, 71, 72), differing in average shielding. The youngest group has larger Ar^{38}/Ne^{21} ratios, indicating less average shielding than for the other high-K group. Significant amounts of Xe^{126} in the high-K samples with lower Ar^{38}/Ne^{21} may have been produced at larger depths prior to exposure on the surface, in which case the exposure ages are not well defined.

It is evident from the concentration data that almost all of the rocks must have been below a depth of ~ 600 gm/cm^2 (≈ 3 to 4 meters) for most of the 3.3–3.6 $\times 10^9$ yr

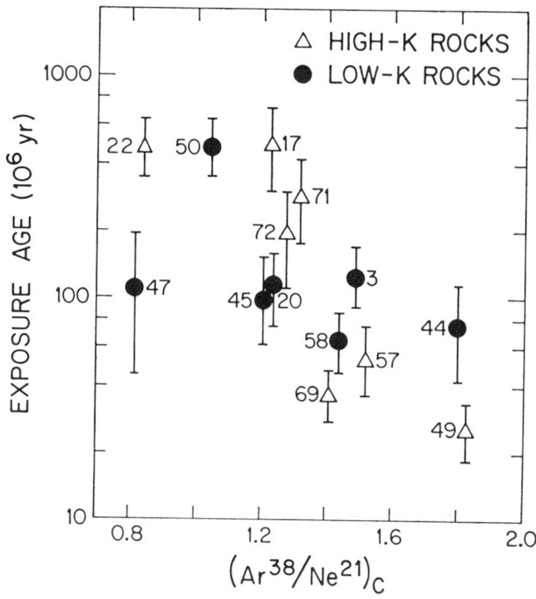

Fig. 3. Exposure ages of lunar samples calculated from spallation Xe^{126} or Ne^{21} amounts using lunar surface production rates. Higher spallogenic Ar^{38}/Ne^{21} indicates less shielding. Low-K rocks, with the exception of 50, appear to form a well-defined group at $\sim 100 \times 10^6$ yr. The exposure ages and Ar^{38}/Ne^{21} ratios are calculated from data obtained in our laboratory and from data presented in the *Proceedings of the Apollo 11 Lunar Science Conference* by Funkhouser *et al.* (1970), Hintenberger *et al.* (1970), Hohenberg *et al.* (1970), Marti *et al.* (1970), and Pepin *et al.* (1970).

since their crystallization. These rocks were then relatively recently ($\sim 10^8$ yr) excavated from either bedrock at this depth or a less mature regolith containing abundant large lithic fragments.

Core and Soil Samples

Fig. 4 summarizes our Gd and Xe data for the Apollo 12 double core. This figure shows the depth in the double core tube and the corrected depth for the undisturbed soil (Carrier *et al.*, 1971). The hachured region at 20 cm undisturbed soil depth corresponds to the coarse layer (VI) (see LSPET, 1970). The neutron dosage Ψ, spallation Xe^{126}(Ba), and Xe^{132}_{sw} concentration from the solar wind are shown for different depths. To a first approximation, the core appears to be completely mixed with respect to both nuclear reaction products and solar wind gases. The lower two samples have about a 10% higher neutron dose. If this change is interpreted to be the result of long-term stratification, a stratification time of 20–50 $\times 10^6$ yr is required. The measurements could equally well be interpreted by assuming that the bottom layers of the core formed from a soil with a slightly higher average dosage. This would imply that there is a fundamental change in the properties of the soil at either the top or bottom of layer IV.

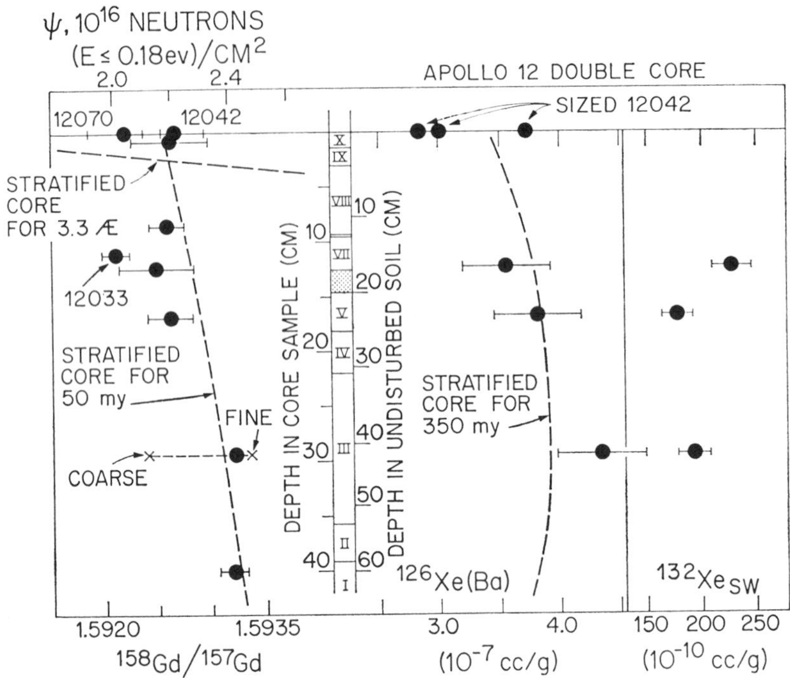

Fig. 4. Neutron dosage, spallation Xe^{126}, and solar wind Xe^{132} as a function of depth in the Apollo 12 double core. From the relatively uniform concentration of spallation and solar wind products, it appears that the core is well mixed; there is no clear-cut correlation in our measured quantities with the pronounced visual stratification in the core at layer VI. The time for deposition ofthe core is less than about 10^8 yr.

The Xe^{126} (Ba) data appear to show a regular increase with depth, although the data points are all the same within the limits of error. Because of the comparatively slow variation in the estimated Xe^{126} production rate as a function of depth, the Xe^{126} data do not set very strong limits on the stratification time. This may be seen for the dashed curve shown in Fig. 4, which corresponds to irradiation for 3.5×10^8 yr of a stratified soil which was not preirradiated prior to deposition. A 50×10^6 yr irradiation of previously irradiated and mixed material would also fit the Xe^{126} data. The solar wind Xe^{132} contents appear to be constant within 30%. This indicates that all portions of the core contain similar amounts of material which have been on the surface for about the same period of time, assuming the Xe is not saturated.

The positions of the calculated gradients in neutron dosage for 3.3×10^9 yr and Xe_c^{126} concentrations for 350×10^6 yr indicate that the amounts of nuclear reaction products in the lunar soil are significantly diluted with unirradiated material compared to what one would expect for irradiation of material near (≤ 1 meter) the lunar surface for 3.3×10^9 yr. This is also true for the solar wind Xe^{132}.

A simple model to explain these results is to assume that the soil samples studied represent the irradiation of material accumulated on the lunar surface over a time T

and mixed recently (within the last 10^8 yr) through a depth L. If R is the total number of atoms of a nuclear reaction product or solar wind ion produced or implanted on 1 cm^2 of lunar surface in one second, then the observed concentration C is just given by $C = TR/L$. For Xe^{126}, for example, $R = Ba \cdot \int_0^L P_{126}(r)\,dr$ where $P_{126}(r)$ is the depth dependence of the Xe^{126} production rate from spallation of Ba. If we take T to be 3.3×10^9 yr, then values of the mixing length L of 15, 11, and 3 meters can be calculated from the neutron dosage, Xe_c^{126} and Xe_{SW}^{132}, respectively. The Xe_{SW}^{132} flux was calculated from the estimated solar abundances ($Xe^{132}/He^4 = 1 \times 10^{-9}$) and a solar wind He^4 flux of 7×10^6 α/cm^2 sec. The mixing length of 3 meters is in basic agreement with the other calculations, considering the uncertainty in solar Xe/He. Smaller values of T yield proportionally smaller values of L. If (L/T) is interpreted as an accumulation rate for the lunar soil, a value of ~ 0.4 cm/10^6 yr is obtained from the neutron dosage or Xe_c^{126}.

The similar neutron dosages for the Apollo 11 and Apollo 12 soils indicate that the irradiation and mixing history of the soil at these two sites has been very similar. From this observation, the maria appear to be uniformly well-mixed seas. The Xe^{126} (Ba) from two fractions of 12042 are somewhat lower than measured in the Apollo 12 core samples which may, in contrast to the dosage values, indicate a somewhat different irradiation history; however, spallation Xe^{126} may not be strictly proportional to Ba in a soil sample (e.g., a recent addition to 12042 of an unirradiated component such as rock 12013 rich in Ba and with a high Ba/Gd ratio could explain the differences). Consequently, the inference of uniform mixing based on the neutron dosage appears more reliable at this time, particularly since Xe could be lost by outgassing. The ratio of neutron dosage to Xe_c^{126} (Ba) for the soil (Fig. 2) requires mixing to *at least* a depth of 0.5–1 meter.

Xe AND Kr IN LUNAR SOIL

High precision (better than 0.1% for the major isotopic ratios) Xe and Kr analyses have been made on grain size fractions from 10084 and 12042 soil samples, and on samples from both the Apollo 11 and Apollo 12 cores. We have interpreted our data in terms of a surface correlated (SUCOR; primarily solar wind) component and a volume correlated (primarily spallation) component. In a manner similar to that of EBERHARDT *et al.* (1970), we assume that

$$\left(\frac{Xe^i}{Xe^{136}}\right)_m = \left(\frac{Xe^i}{Xe^{136}}\right)_{\text{SUCOR}} + P_i \cdot T \cdot \frac{Ba}{Xe_m^{136}}$$

where the subscript m refers to the measured quantity. $P_i \cdot Ba$ is the production rate of isotope i by spallation, assumed to be proportional to the measured Ba concentration; and T is an effective exposure time. An analogous equation can be written for $(Kr^i/Kr^{86})_m$ with Sr replacing Ba, and with Kr^{86} replacing Xe^{136}. Extrapolation of plots of $(Xe^i/Xe^{136})_m$ vs. Ba/Xe_m^{136} can be used to estimate the isotopic composition of SUCOR Xe and Kr. However, these plots do not yield particularly good linear arrays, indicating that the assumptions inherent in the above equation are not completely satisfied. Nevertheless, for the more abundant isotopes where the spallation contributions are less important, comparatively precise values for the isotopic composition of SUCOR Xe and Kr are obtained. The SUCOR Xe isotopic composition from 12042 agrees well with that obtained by us previously (PODOSEK *et al.*, 1971) for Apollo 11 soils. However, the 12042 SUCOR Kr composition is identical to that of atmospheric Kr, whereas that obtained from 10084 had a systematic mass fractionation of 0.7% per mass unit favoring the lighter isotopes in the atmosphere. These differences are not understood, but may indicate that a unique SUCOR composition does not exist.

A stepwise degassing experiment of a very fine ("acetone floats") fraction of 12042 has provided evidence that the surface correlated Xe (SUCOR) consists of a mixture of several components. In Fig. 5 we present a correlation diagram of the measured (Xe^{130}/Xe^{132}) vs. (Xe^{136}/Xe^{132}) for our Apollo 12 samples. If these samples contained only varying mixtures of single SUCOR and spallation components, then all of the data points would plot along a straight line on this diagram. The total sample points are consistent with a mixture of just two well-defined components; however the points corresponding to the various temperature fractions of the acetone floats (ACE) sample spread roughly perpendicular to the total sample line, indicating that SUCOR is not a single component. Correcting the temperature release points for spallation produces a displacement indicated by the arrows on Fig. 5. The corrected points define a trend which is consistent with mass fractionation or with the addition of a

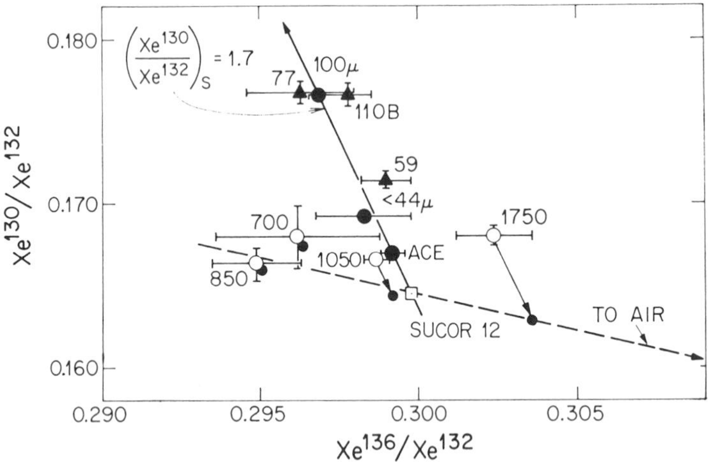

Fig. 5. Three-isotope correlation diagram for analyses of grain size fractions from 12042 soil (large solid circles) and 12028 core samples (triangles). The open circles refer to temperature fractions of a very fine fraction (less than 10 μm) of 12042. The total point for this same fraction is plotted as ACE. A mixing line for mixtures of a surface correlated component (SUCOR 12) with spallation Xe having $(Xe^{130}/Xe^{132})_s = 1.7$ has been drawn. The arrows indicate the effect of correcting the temperature release points for spallation. The corrected points are indicated by the small, filled circles. It is evident that the total soil samples *appear* to be a mixture of SUCOR and spallation Xe. However, the temperature fractions indicate that SUCOR is a mixture of several components related by fractionation.

fission component. The data for the Kr and for the other Xe isotopes are compatible with this result.

We do not believe that the isotopic compositions of SUCOR Xe and Kr can be unambiguously identified with that of the solar wind. Our conclusion is based on the evidence from the temperature release studies and because there appears to be no simple relation between SUCOR, atmospheric, and meteoritic (carbonaceous chondrite) Xe and Kr. SUCOR Xe and Kr may contain, in addition to solar wind, additional implanted components from the lunar atmosphere. For example, highly fractionated SUCOR components released at different temperatures can be generated by the release of trapped solar wind gases by impacts and reimplantation of these gases at lower energies from the lunar atmosphere.

Acknowledgments—We gratefully acknowledge technical support from T. Wen, F. Tera, P. Young, and A. Massey. This research was supported by NASA Grant 9–8074.

References

Bogard D. D., Funkhouser J. G., Shaeffer O. A., and Zähringer J. (1971) Noble gas abundances in lunar material—Cosmic ray spallation products and radiation ages from the Sea of Tranquillity and the Ocean of Storms. *J. Geophys. Res.* **76**, 2757–2779.

Carrier W. D. III, Johnson S. W., Werner R. A., and Schmidt R. (1971) Disturbance in samples recovered with the Apollo core tubes. Second Lunar Science Conference (unpublished proceedings).

Eberhardt P., Geiss J., Graf H., Grögler N., Krähenbühl U., Schwaller U., Schwarzmüller J., and Stettler A. (1970) Trapped solar wind noble gases, exposure age and K/Ar ages in Apollo 11 lunar fine material. *Proc. Apollo 11 Lunar Sci. Conf., Geochim. Cosmochim. Acta* Suppl. 1, Vol. 2, pp. 1037–1070. Pergamon.

Eugster O., Tera F., Burnett D. S., and Wasserburg G. J. (1970) The isotopic composition of Gd and neutron capture effects in Apollo 11 samples. *Earth Planet. Sci. Lett.* **8**, 20–30.

Funkhouser J. G., Schaefer O. A., Bogard D. D., and Zähringer J. (1970) Gas analysis of the lunar surface. *Science* **167**, 561–563.

Hintenberger H., Weber H. W., Voshage H., Wänke H., Begemann F., and Wlotzka F. (1970) Concentrations and isotopic abundances of the rare gases, hydrogen, and nitrogen in lunar matter. *Proc. Apollo 11 Lunar Sci. Conf., Geochim. Cosmochim. Acta* Suppl. 1, Vol. 2, pp. 1269–1282. Pergamon.

Hohenberg C. M., Davis P. K., Kaiser W. A., Lewis R. S., and Reynolds J. H. (1970) Trapped and cosmogenic rare gases from stepwise heating of Apollo 11 samples. *Proceedings of the Apollo 11 Lunar Sci. Conf., Geochim. Cosmochim. Acta* Suppl. 1, Vol. 2, pp. 1283–1310. Pergamon.

Honda M. (1962) Spallation products distributed in a thick iron target bombarded by 3-BeV protons. *J. Geophys. Res.* **67**, 4847–4858.

Marti K., Lugmair G. W., and Urey H. C. (1970) Solar wind gases, cosmic-ray spallation products and the irradiation history of Apollo 11 samples. *Proc. Apollo 11 Lunar Sci. Conf., Geochim. Cosmochim. Acta* Suppl. 1, Vol. 2, pp. 1357–1369. Pergamon.

LSPET (Lunar Sample Preliminary Examination Team) (1970) Preliminary examination of lunar samples from Apollo 12. *Science* **167**, 1325–1339.

Lunatic Asylum (1970) Mineralogic and isotopic investigations on lunar rock 12013. *Earth Planet. Sci. Lett.* **9**, 137–163.

Pepin R. O., Nyquist L. E., Phinney D., and Black D. C. (1970) Rare gases in Apollo 11 lunar material. *Proc. Apollo 11 Lunar Sci. Conf., Geochim. Cosmochim. Acta* Suppl. 1, Vol. 2, pp. 1435–1454. Pergamon.

Podosek F. A., Huneke J. C., Burnett D. S., and Wasserburg G. J. (1971) Isotopic composition of xenon and krypton in the lunar soil and in the solar wind. *Earth Planet. Sci. Lett.* **10**, 199–216.

Signer P. and Nier A. O. C. (1960) The distribution of cosmic-ray-produced rare gases in iron meteorites. *J. Geophys. Res.* **65**, 2947–2964.

Proceedings of the Second Lunar Science Conference, Volume 2, pp. 1681–1692.
The M.I.T. Press, 1971.

Breccia 10065: Release of inert gases by vacuum crushing at room temperature

DIETER HEYMANN and AKIVA YANIV*
Departments of Geology and Space Science, Rice University, Houston, Texas 77001

(*Received 19 February 1971; accepted in revised form 31 March 1971*)

Abstract—The release of inert gases from breccia 10065 was studied by vacuum crushing of a small sample (117 mg) of the rock at room temperature. At the end of the crushing, approximately 60% of He; 30% of Ne; 12% of Ar; 30% of Kr; and 40% of Xe was released. The evidence indicates that the gas is contained in pores (bubbles, vugs, voids, grain boundaries), not in the solids proper. The variation of He^4/Ne^{20} (which increases) and Ne^{20}/Ar^{36} (which decreases) during the crushing strongly suggests that the breccia was formed from gas-rich parent materials in the presence of a gasphase which became physically trapped in the pores. Subsequent diffusion of inert gases from the solid phases into the pores must have occurred. The inferred elemental and isotopic composition of the poregas is grossly similar to that of the trapped solar-wind gas in fines from the Sea of Tranquillity, which suggests that the poregas was produced by the heating of regolith materials during an impact that produced the breccia. However, the uncommonly high Ne^{20}/Ar^{36} ratio of at least 25 (as against 5–6 in the fines) and the presence of substantial amounts of the heavy inert gases Kr and Xe implies that the gasphase cannot have been formed by simple, quantitative outgassing of regolith materials. Our preferred model calls for the formation of a hot gascloud near the point of impact, with the lighter inert gases He, Ne, and Ar rapidly diffusing out of the cloud into the surrounding vacuum. Fractional release of inert gases from regolith material (in part from fines carried along by the expanding gascloud, in part from moderately heated regolith away from the point of impact) at lower temperatures continuously replenishes the light inert gases in the cloud. We conclude that the breccia was not formed at the instant of impact, but in a base surge type event as proposed by MCKAY *et al.* An alternative possibility is that the poregas came from the impacting projectile, which might have been a carbonaceous chondrite like object. In that case the gas now present in the breccia could have been trapped instantaneously, i.e., in the primary impact event.

INTRODUCTION

APOLLO 11 BRECCIAS contain very large amounts of all the inert gases; in face, these rocks are usually more gas-rich than the fines (FUNKHOUSER *et al.*, 1970). A number of investigators have suggested that breccias represent "lithified" soil (FREDRIKSSON *et al.*, 1970; FRONDEL *et al.*, 1970; MCKAY *et al.*, 1970). This would seem to imply the following sequence: solar wind ions became implanted into individual particles of the regolith (cf. EBERHARDT *et al.*, 1970), and the breccia were subsequently formed from this gas-rich material. The gases now found in the interior of centimeter-sized breccias cannot have been trapped directly from the solar wind *after* the formation of the rock.

However, this simple scheme raises a number of questions. The breccia of Apollo 11 are richer in gas than the fines, and the average Ar^{40}/Ar^{36} ratio of 2.2 is greater

* On leave of absence from the Department of Physics, Tel-Aviv University, Ramat Aviv, Israel.

than that of the Apollo 11 fines, 1.1 (FUNKHOUSER *et al.*, 1970). FUNKHOUSER *et al.* (1970) have suggested that the breccia were not locally derived but come from fines of another site than Tranquillity Base. A more fundamental objection is that the He^4/Ne^{20} and Ne^{20}/Ar^{36} ratios (diagnostic features of gas loss by diffusion) in bulk breccias are nearly identical to those in the fines (FUNKHOUSER *et al.*, 1970). It is difficult to understand how these rocks could have become shock-lithified without substantial fractionation of the light inert gases.

MCKAY *et al.* (1970) have suggested that breccias were formed during a base-surge type event, in the presence of a substantial gasphase. These authors have pointed out that most breccias, in contrast with crystalline rocks are quite porous, having porosities up to about 35%. Their theory raises the possibility that gas became physically trapped in porespaces during breccia formation. In order to test this possibility we have analyzed breccia 10065 by vacuum crushing at room temperature instead of by melting or stepwise heating, the two most widely used methods for gas extraction. (This suggestion was made to us by Dr. D. S. MCKAY of NASA–MSC, Houston, Texas.) Additional information was obtained by the melt-extraction of three other samples: one aliquot of the original breccia, and two samples of the crushed material.

PROCEDURE

The procedures for melt-extraction as well as the mass-spectrometry have been described elsewhere (HEYMANN and YANIV, 1970a). The vacuum crusher is shown diagrammatically in Fig. 1. The sample was placed on the bottom plate of the stainless steel cylinder beneath the steel plunger. The plunger was lifted with external magnets (hand operated) released and dropped on the sample. Each drop was called a "hit." The cylinder was connected to the Pyres-glass extraction line via a

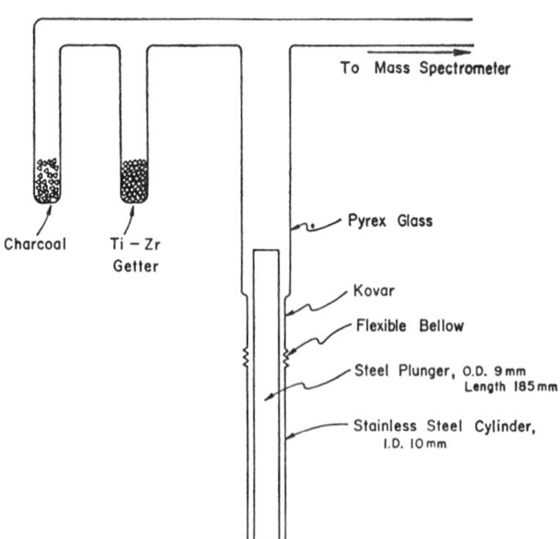

Fig. 1. Schematic representation of the vacuum crusher. The samples were placed on the bottom of the stainless steel cylinder. The steel plunger could be lifted with external magnets, and be released to "hit" the sample. The gases given off came in contact only with the Ti–Zr getter at room temperature. The charcoal served to adsorb Ar, Kr, and Xe in order to allow He and Ne to be measured separately.

flexible bellow and a Kovar-Pyrex joint. Although the cylinder was amply supported from below, the flexible bellow was used to minimize strain on the glass during the crushing. The bellow also permitted us to tap the cylinder vigorously from the side, with the plunger in the raised position, to promote mixing of the crushed material. This was usually done after every 100 hits.

The gas-extraction was done in a vacuum of better than 10^{-7} mm Hg and at about 25°C. The evolved gas came in contact with a Ti–Zr getter at room temperature and charcoal at liquid nitrogen temperature. Apparently only small quantities of gases such as N_2, H_2O, CH_4, CO, CO_2, etc., which normally require gettering temperatures of several hundred °C, were given off, because we never observed any interference from these molecules in the mass-spectrometric procedure. H_2 was apparently absorbed quite effectively on the Ti–Zr getter at room temperature, because we observed no significant increase of the H_2 interference. The temperature of the sample during the crushing was not monitored; however, the outside of the steel cylinder never became noticeably warmer than the surrounding air, even after 2000 hits.

The amounts of inert gas given off by the materials of the crusher were negligible, as determined in a "blank" run, hence no corrections were necessary except for the usual background peaks in the mass spectrometer itself.

Results

Inert gas data are given in Table 1; elemental and isotopic ratios in Table 2. Sample I is an 821 μg chip from the original allotment of 10065. This sample is very gas-rich as shown by its He^4 content of 0.445 cm^3 STP/g, greater than 6 out of 7 Apollo 11 breccias reported by Funkhouser *et al.* (1970). The gas in this sample is unusual because of its high Ne^{20}/Ar^{36} ratio of 43 and its high Ar^{36}/Ar^{38} ratio of 5.65. Relatively high Ne^{20}/Ar^{36} ratios of 24–33 are known to occur in ilmenite (Eberhardt *et al.*, 1970), but in bulk fines and bulk breccias the ratio is always less than 10, and is typically in the range 5–7. All Ar^{36}/Ar^{38} ratios in bulk fines and breccias are below 5.32 ± 0.05. Ratios greater than 5.4; however, have been observed in the low temperature fraction of Ar released in the stepwise heating of fines (Hohenberg *et al.*, 1970; Pepin *et al.*, 1970).

Sample II is a 117 mg portion of the breccia which was crushed in three steps as shown in Table 1. Large quantities of He^4 were given off; in fact, so large that we removed the material from the crusher and continued with a 13 mg portion, called sample III. Another portion of 3.309 mg, sample IV, was used for gas extraction by melting. Judging from the results of samples II and IV, the crushing of sample II by 41 hits had removed at most about 1% of the gas contained in the original sample II.

Sample III was crushed in seven steps as shown in Table 1. The amounts of gas given off per hit decreased regularly, and were roughly proportional to the amount of gas still present in the sample (Fig. 2). In the course of the crushing the He^4/Ne^{20} ratio increased from a low of about 60 to a high of about 130; then seemed to level off. At the same time the Ne^{20}/Ar^{36} ratio decreased from about 25 to about 10–15 (Fig. 3).

After 3601 hits the experiment was terminated. Fig. 4 shows the release of inert gases from sample III. After 3601 hits the following fractions had been given off: $He^4 \sim 60\%$; $Ne^{20} \sim 30\%$; $Ar^{36} \sim 12\%$; $Kr^{84} \sim 30\%$; and $Xe^{132} \sim 40\%$. Sample V, an aliquot of the powdered breccia after 3601 hits still contained He^4, Ne^{20}, $Ar^{36} \sim 50\%$; Kr^{84}, $Xe^{132} \sim 40\%$. The material balance is reasonably good, except for Ar^{36} which shows an apparent recovery of only 62%. We do not believe that any Ar^{36} was lost during the experiment, or has gone undetected. We suspect that either sample

Table 1. Inert gases

Sample	He^3	He^4	Ne^{20}	Ne^{21}	Ne^{22}	Ar^{36}
			(Units: cc STP/g)			
I	1.22×10^{-4}	4.45×10^{-1}	4.66×10^{-3}	1.23×10^{-5}	3.51×10^{-4}	1.09×10^{-4}
			(Units: cc STP)			
II, 1	2.80×10^{-9}	8.32×10^{-6}	1.46×10^{-7}	3.60×10^{-10}	1.09×10^{-8}	5.86×10^{-9}
2	6.90×10^{-8}	1.99×10^{-4}	3.27×10^{-6}	8.18×10^{-9}	2.48×10^{-7}	1.38×10^{-7}
3	5.81×10^{-8}	1.60×10^{-4}	2.29×10^{-6}	5.64×10^{-9}	1.71×10^{-7}	8.67×10^{-8}
4	1.30×10^{-7}	3.67×10^{-4}	5.71×10^{-6}	1.42×10^{-8}	4.30×10^{-7}	2.31×10^{-7}
III, 1	3.69×10^{-8}	9.87×10^{-5}	1.25×10^{-6}	3.10×10^{-9}	9.40×10^{-8}	5.50×10^{-8}
2	2.33×10^{-7}	6.27×10^{-4}	5.45×10^{-6}	1.41×10^{-8}	4.27×10^{-7}	2.52×10^{-7}
3	1.75×10^{-8}	4.72×10^{-5}	3.56×10^{-7}	8.79×10^{-10}	2.65×10^{-8}	1.82×10^{-8}
4	2.90×10^{-7}	7.96×10^{-4}	6.13×10^{-6}	1.52×10^{-8}	4.60×10^{-7}	3.92×10^{-7}
5	9.39×10^{-9}	2.47×10^{-5}	1.86×10^{-7}	4.60×10^{-10}	1.40×10^{-8}	1.35×10^{-8}
6	2.91×10^{-7}	8.01×10^{-4}	6.82×10^{-6}	1.70×10^{-8}	5.14×10^{-7}	5.39×10^{-7}
7	4.55×10^{-9}	1.33×10^{-5}	1.09×10^{-7}	2.66×10^{-10}	8.21×10^{-9}	1.07×10^{-8}
8	8.83×10^{-7}	2.41×10^{-3}	2.03×10^{-5}	5.11×10^{-8}	1.54×10^{-6}	1.28×10^{-6}
IV, 1	3.70×10^{-7}	1.03×10^{-3}	1.53×10^{-5}	4.13×10^{-8}	1.18×10^{-6}	2.80×10^{-6}
2	1.31×10^{-5}	3.64×10^{-2}	5.41×10^{-4}	1.46×10^{-6}	4.17×10^{-5}	9.90×10^{-5}
3	1.12×10^{-4}	3.11×10^{-1}	4.62×10^{-3}	1.25×10^{-5}	3.56×10^{-4}	8.46×10^{-4}
V, 1	8.12×10^{-8}	3.37×10^{-4}	5.22×10^{-6}	1.42×10^{-6}	4.05×10^{-7}	9.80×10^{-7}
2	4.18×10^{-6}	1.73×10^{-2}	2.68×10^{-4}	7.30×10^{-5}	2.08×10^{-5}	5.04×10^{-5}

* Original 117 mg sample contains approximately 1% more gas.
† Amounts of gas too small for accurate measurements.

Table 2. Elemental and isotopic ratios.

Sample	He^4/Ne^{20}	Ne^{20}/Ar^{36}	Ar^{36}/Kr^{84}	Kr^{84}/Xe^{132}	He^3/He^4 ($\times 10^4$)	Ne^{20}/Ne^{22}	Ne^{21}/Ne^{22}	Ar^{36}/Ar^{38}	Ar^{40}/Ar^{36}
I	82	43	1200	3.2	3.21	13.3	0.035	5.65	4.94
II, 1	57	24.9	—	—	3.37	13.4	0.033	5.43	2.48
2	61	23.8	1236	2.6	3.47	13.2	0.033	5.42	2.37
3	70	26.4	924	3.5	3.64	13.4	0.033	5.46	2.41
4	64	24.7	1125	2.9	3.54	13.3	0.033	5.45	2.38
III, 1	79	22.8	874	2.1	3.74	13.3	0.033	5.41	2.86
2	115	21.6	533	1.7	3.72	13.2	0.033	5.44	3.20
3	133	19.6	682	1.6	3.70	13.4	0.033	5.39	2.90
4	130	15.6	498	3.2	3.65	13.3	0.033	5.38	2.86
5	133	13.8	317	1.6	3.80	13.3	0.033	5.37	2.82
6	117	12.7	972	2.8	3.63	13.3	0.033	5.44	2.31
7	122	10.2	428	1.6	3.42	13.3	0.032	5.41	2.65
8	126	15.5	642	2.4	3.66	13.3	0.033	5.42	2.71
IV, 1	68	5.5	1600	3.5	3.58	13.0	0.035	5.38	2.05
V, 1	45	5.3	1360	3.4	3.40	12.9	0.035	5.34	2.02

IV or V, or both were not wholly representative for the material from which they were taken. This conclusion seems justified when one compares samples I and IV: the former contains 1.09×10^{-4} cm^3 STP/g of Ar^{36}, the latter 8.46×10^{-4} cm^3 STP/g; yet both are portions of the same breccia. Apparently the Ar-content does vary appreciably for random samples at the 1 mg level. At the end of the crushing the mean particle size was about 1 mμ.

from breccia 10065.

Ar^{38}	Ar^{40}	Kr^{84}	Xe^{132}	Comments
1.94×10^{-5}	5.41×10^{-4}	9.1×10^{-8}	2.8×10^{-8}	821 μg chip; melt
1.08×10^{-9}	1.45×10^{-8}	†	†	117 mg chip; 1 hit
2.54×10^{-8}	3.26×10^{-7}	1.1×10^{-10}	4.3×10^{-11}	20 hits
1.59×10^{-8}	2.09×10^{-7}	9.4×10^{-11}	2.7×10^{-11}	20 hits
4.24×10^{-8}	5.50×10^{-7}	2.0×10^{-10}	7.0×10^{-11}	$\sum$ 41 hits
1.02×10^{-8}	1.57×10^{-7}	6.3×10^{-11}	3.0×10^{-11}	13 mg from 117 mg; 40 hits
4.63×10^{-8}	8.31×10^{-7}	4.7×10^{-10}	2.8×10^{-10}	400 hits
3.37×10^{-9}	5.26×10^{-8}	2.7×10^{-11}	1.7×10^{-11}	40 hits
7.28×10^{-8}	1.12×10^{-6}	7.9×10^{-10}	2.5×10^{-10}	1000 hits
2.50×10^{-9}	3.79×10^{-8}	4.2×10^{-11}	2.7×10^{-11}	40 hits
9.9×10^{-8}	1.24×10^{-6}	5.5×10^{-10}	2.0×10^{-10}	2000 hits
1.97×10^{-9}	2.83×10^{-8}	2.5×10^{-11}	1.5×10^{-11}	40 hits
2.36×10^{-7}	3.47×10^{-6}	2.0×10^{-9}	8.2×10^{-10}	$\sum$ 3560 hits
5.20×10^{-7}	5.74×10^{-6}	1.75×10^{-9}	5.0×10^{-10}	3.309 mg from 117 mg; melt
1.84×10^{-5}	2.03×10^{-4}	6.19×10^{-8}	1.8×10^{-8}	Recalculated to 117 mg*
1.57×10^{-4}	1.74×10^{-3}	5.29×10^{-7}	1.5×10^{-7}	Recalculated to cc STP/g*
1.84×10^{-7}	1.98×10^{-6}	7.2×10^{-10}	2.1×10^{-10}	2.275 mg after 3601 hits; melt
9.46×10^{-6}	1.02×10^{-4}	3.70×10^{-8}	1.1×10^{-8}	Recalculated to 117 mg

Errors in gas content are generally less than ±5%.

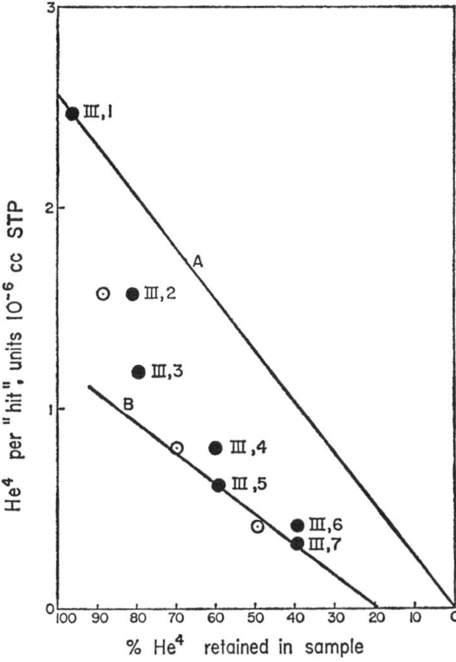

Fig. 2. He^4 given off per "hit" as a function of percentage of He^4 still in the sample for runs III, 1–7. Closed circles represent percentage of He^4 retained at the end of each run. Open circles represent the mean of percentage of He^4 retained during each run.

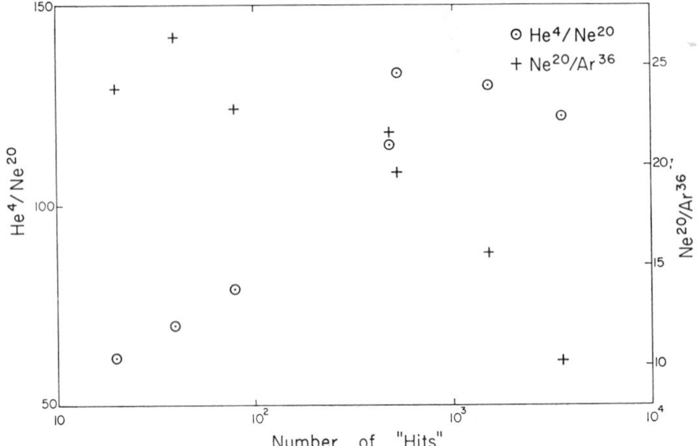

Fig. 3. He^4/Ne^{20} and Ne^{20}/Ar^{36} trends during the crushing. Note that He^4/Ne^{20} increases from a low of about 60 to a high of about 130, whereas Ne^{20}/Ar^{36} decreases from a high of about 25 to a low of about 10. These trends cannot have been caused by variable proportions or radiogenic He^4 or cosmogenic He, Ne, or Ar. The trends cannot be explained either by diffusion of inert gases from the solid phase of the breccia into *initially empty pores*. From this we conclude that gas was physically trapped in the pores when the breccia formed. This *poregas* had a relatively low He^4/Ne^{20} ratio of less than 60, and a relatively high Ne^{20}/Ar^{36} ratio of greater than 25.

DISCUSSION

Components

The inert gases in lunar materials and meteorites consist of several components: radiogenic components (He^4, Ar^{40} from in situ decay of U, Th, and K); cosmogenic components (inert gases produced in situ by galactic and solar cosmic rays); and trapped components (in lunar samples the only trapped component that has been firmly identified represents trapped solar wind ions). Each component has its own characteristic elemental and isotopic composition. Any variation in the relative proportions of the components shows up as variation in the raw mass-spectrometric data (Table 2). We must therefore examine whether the variations in He^4/Ne^{20} and Ne^{20}/Ar^{36} (Fig. 3) could be due to variations in radiogenic, cosmogenic, and trapped gas.

The He^4/Ne^{20} ratio can increase because of two reasons: an increased proportion of radiogenic He^4, or an increased proportion of cosmogenic He and Ne (for material of breccia composition one predicts cosmogenic He^4/Ne^{20} production ratio in the range of 20–30). Both possibilities can be dismissed. If one assumes that run II, 1 with $He^4/Ne^{20} = 57$ contains *no* radiogenic He^4, then the first explanation requires that about $\frac{1}{2}$ of all He^4 in runs III 3–8 is radiogenic He^4. The amount of He^4 observed could not have been produced in the breccia within 4.5 billion years from typical U($\sim$0.5 ppm) and Th($\sim$2 ppm) contents reported for such rocks (O'KELLEY *et al.*, 1970). Furthermore, one would predict that the He^3/He^4 ratio in runs III 3–8 should be about $\frac{1}{2}$ of the value in run II, 1; this is not the case as seen from Table 2: the ratio, if anything increases slightly during the crushing. The (He^4/Ne^{20}) cosmogenic production ratio of 20–30 implies that roughly $\frac{1}{2}$ of the He^4 in runs III 3–8 must be cogmogenic if the second explanation is correct. But if as much as $\frac{1}{2}$ of He^4 is cosmogenic in these runs, then the He^3/He^4 ratio should have increased to about 0.1 since the $(He^3/He^4)_c$ production ratio in silicates is about 0.2 (HEYMANN, 1967). This is not the case.

The decrease of Ne^{20}/Ar^{36} during the crushing could only be due to a cosmogenic component, since no radiogenic components are involved. We have estimated that the $(Ne^{20}/Ar^{36})_c$ production ratio in material of breccia composition is approximately 1.5. If we assume again that run II, 1 contains no cosmogenic Ne or Ar, then the reduction of the Ne^{20}/Ar^{36} ratio from about 25 to about 10 in run III, 7 requires that about 10% of the Ne^{20}, and about 64% of the Ar^{36} in run III, 7 is of cosmogenic origin. But this much cosmogenic Ar^{36} should reduce the Ar^{36}/Ar^{38} ratio appreciably. With $(Ar^{36}/Ar^{38})_c \sim 0.6$, one calculates that the Ar^{36}/Ar^{38} ratio in III, 7 should be about 1.0. This is not the case; the observed ratio is 5.41 (Table 2).

Hence the elemental and isotopic trends seen in the course of the crushing cannot be explained by the presence in variable proportions of radiogenic and cosmogenic components. If any are present, they are masked by "trapped" components whose origin we shall discuss in a later section.

Siting of the gas. We must now consider the question: Where does the gas, released in the crushing come from? There are three possibilities. The gas is either contained in the porespaces of the breccia, or dissolved in the solids proper, or both. Release from the solid phase requires heating; or, if the gas is located near surfaces of grains, the "grinding up" of these surfaces at the submicron level to the extent that the dissolved gas can escape. Release from pores (bubbles, voids, vugs, grain boundaries) requires merely that the pore is cracked open such that the gas escapes into the vacuum of the extraction apparatus. That the large amounts of inert gases released (Fig. 4) could be due to heating of the sample is very unlikely because of the high abundance of the heavy elements Xe and Kr in the early stages of the crushing (I, 1–3). Judging from the work of PEPIN *et al.* (1970) the temperatures required to explain the observed Kr and Xe release (800–900°C) would imply severe shock conditions during the crushing. This cannot have been the case.

What seems to speak against destruction of grain surfaces at the submicron level are the very large fractions of inert gases given off: it is difficult to believe that as much as 60% of the He^4 contained in the sample could have been liberated in this manner. It is also difficult to understand how this mechanism can account for the opposing He^4/Ne^{20} vs. Ne^{20}/Ar^{36} trends. We conclude therefore that most, if not all of the gas liberated during the crushing has come out of the porespaces.

He^4 release. It is instructive to consider Fig. 2 once again. We have plotted He^4 released per hit vs. percentage of He^4 still remaining in sample III. The points for steps 2, 4, and 6, in which percentage of He^4 retained changes appreciably, have been placed at the midpoint of the range. The amount of He^4 given off in step III, 1 is about 2.5×10^{-6} cm^3 STP/hit. If all the gas is contained in porespaces of equal size (and if there is uniform He^4 partial pressure in all of the pores), one expects the points to fall

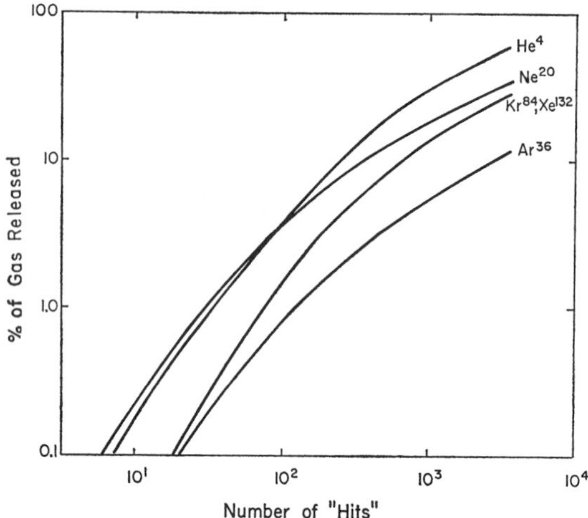

Fig. 4. Fractions of He, Ne, Ar, Kr, and Xe given off from sample II as a function of the number of "hits." Note that at the end of the crushing the release is: about 60% of He; 30% of Ne; 12% of Ar; 30% of Kr; and 40% of Xe. We have concluded that the gas given off is contained in pores, not in the solids proper of the breccia.

near the line A. If only a fraction of the He^4, say 50%, is contained in the porespaces, the remainder being present in the solids proper, one expects variation along a straight line intersecting the abscissa at 50% He^4 retained. The results cannot be fitted to any such simple model; rather, the data seem to suggest that the release of He^4 in the initial steps of the crushing comes from relatively large, or interconnected pores, whereas the release from smaller or nonconnected pores becomes the dominant mechanism in the later steps. It is interesting to note that points 4–7 lie fairly closely to a straight line B, which seems to imply that gas from large pores became exhausted at about the 4th or 5th step.

The bulk density of breccia 10065 is reported at 2.34 g/cm^3 (KANAMORI *et al.*, 1970) which implies a porosity of about 25%. The He^4 content is $\sim$0.3 cm^3 STP/g (IV, 3) or about 1.2 cm^3 STP/cm^3. Since at least 60% of the He^4 is probably contained in the porespaces, this implies a partial He^4 pressure of at least 3 atm.

Model. Let us now consider a model that can, at least qualitatively, explain the major features of the gas release. *We shall assume that the breccia was formed from gas-rich fines in the presence of an ambient gasphase which became physically trapped in the pores.* Any gas present in the solids proper, immediately after the formation of the breccia shall be called *dissolved gas*; the gas present in the pores shall be called *poregas.*

The He^4/Ne^{20} fractionation (Fig. 3) suggests that some of the dissolved gas diffused out of the solids into the pores after the breccia had formed, probably immediately after the breccia was consolidated and the rock was still warm. The He^4/Ne^{20} ratio of the bulk fines 10084 is about 90–100 (HEYMANN and YANIV, 1970a; HINTENBERGER *et al.*, 1970; MARTI *et al.*, 1970; PEPIN *et al.*, 1970). Let us assume that the He^4/No^{20} ratio of the dissolved gas was not much below this value, say in the range 50–100. Let us further assume that the He^4/Ne^{20} ratio of the poregas could not have been higher than the lowest value seen in the crushing, i.e., 57 (II, 1). The diffusion should have been attended by a strong fractionation such that the gas entering the pores was enriched in He^4 relative to Ne^{20}. (PEPIN *et al.* (1970) report a fourfold enrichment of He^4 in the gas released from fines at 300°C; at 400°C the enrichment is still twofold.) The diffusing gas, richer in He^4 than the dissolved gas, becomes mixed with the poregas already present, increasing the He^4/Ne^{20} ratio in the pores. However, the small pores have a larger surface-to-volume ratio than the large pores, hence the increase should have been greater in the small than in the large pores. We have already suggested that the large, or interconnected pores contribute most of the released gas during the earlier steps of the crushing. Hence one expects He^4/Ne^{20} to increase during the crushing. This explanation is further supported by the observation that the material remaining after the crushing (V, 1) has the lowest He^4/Ne^{20} ratio of 45. The trend of increasing He^3/He^4 ratios (Table 2) is also consistent with the model: gas released from fines at relatively low temperatures shows a significant enrichment of He^3 relative to He^4 (HOHENBERG *et al.*, 1970; PEPIN *et al.*, 1970).

The paradox that the Ne^{20}/Ar^{36} ratio decreases during the crushing is only apparent. Let us assume that the ratio in the poregas was fairly high, at least 25 (run II, 1); perhaps as high as 43 (run I). The ratio in the dissolved gas on the other hand was relatively low, probably in the range 5–7, a typical value for Apollo 11 fines (HEYMANN and YANIV, 1970a). Although the Ne–Ar fractionation in the low temperature release from fines may be appreciable (PEPIN *et al.*, 1970), the diffusing gas apparently never had a Ne^{20}/Ar^{36} ratio much greater than 10. This would have caused the ratio in the pores to decrease, and the decrease would be the greatest in the smallest pores.

Origin of the poregas. Let us now speculate on the origin of the poregas. The composition of this gas cannot be firmly established; however, judging from the preceding section one can safely conclude that the poregas had an uncommonly high Ne^{20}/Ar^{36} ratio of at least 25, perhaps as high as 43. The He^4/Ne^{20} ratio was probably less than 60. The poregas contained Kr and Xe with abundances (relative to Ar) comparable to fines ($Ar^{36}/Kr^{84} \sim 1200$; $Kr^{84}/Xe^{132} \sim 3$). Isotopic ratios in the poregas were similar to those now seen in bulk fines, with the exception of Ar^{36}/Ar^{38} which was at least 5.4, perhaps as high as 5.65; and Ar^{40}/Ar^{36} which was at least 2.5, perhaps as high as 5 (Table 2).

The gross similarity between the composition of the poregas and that of Apollo 11 bulk fines 10084 suggests that the poregas was obtained by the outgassing of fines. However, simple, quantitative outgassing of fines from Tranquillity Base cannot account for the composition of the poregas, mainly because the Ne^{20}/Ar^{36} ratio would then be much too low, ~6. FUNKHOUSER *et al.* (1970) who were the first to report the systematic difference of the Ar^{40}/Ar^{36} ratios in fines and breccias from Apollo 11 have suggested that the breccia came from another location on the Moon. This seems a reasonable hypothesis to explain all the differences between the poregas and the gas in fines from Apollo 11. However, this would require that the poregas was produced by heating of regolith material at an unknown site X, that the gas was then transported to another unknown site Y where the breccia was formed and the poregas was trapped (fines in Y had $Ar^{40}/Ar^{36} \sim 2$), and finally the breccia was transported from Y to Tranquillity Base.

This is not a very attractive theory. Let us therefore attempt to explain the poregas composition by fractional outgassing of known fines, such as 10084. We shall assume that the breccia is impact produced, and that this impact also caused heating of the regolith material, which in turn produced the gasphase from which the poregas was trapped. First we note that the relative abundances of Kr and Xe in the poregas are at least as high, if not greater than in 10084. This would require near-quantitative outgassing of fines at high temperatures (800–900°C; PEPIN *et al.*, 1970). Regolith material near the point of impact could easily account for the high-temperature outgassing. However, the near-quantitative outgassing produces a gasphase with $Ne^{20}/Ar^{36} \sim 5$; too low for the observed ratio in the poregas. We are therefore forced to conclude that the gasphase produced near the point of impact was not immediately trapped in the breccia but expanded radially away from the point of impact, with the light inert gases He, Ne, and Ar diffusing out of the hot gascloud, producing a gasphase enriched in Xe and Kr.

The hot gasphase moving along the surface continued to become depleted in the light gases, but at the same time fresh gas was added from two sources. Regolith material away from the point of impact, less strongly heated than the material near the point of impact, would release inert gases strongly enriched in the light elements (PEPIN *et al.*, 1970; HOHENBERG *et al.*, 1970), in particular He and Ne. The gas cloud picked up fines which it carried along; and again, moderate heating of these fines would have resulted in release of gas strongly enriched in He and Ne. Apparently He and Ne release continued to be near-quantitative, because the He^4/Ne^{20} ratio in the poregas does not indicate any strong fractionation. On the other hand, Ar release into the gasphase, near-quantitative at the point of impact, became increasingly less quantitative away from the point of impact. Thus, we suspect that He^4/Ne^{20}; Ne^{20}/Ar^{36}; Ar^{36}/Kr^{84} in the gasphase were initially very similar to the values now seen in the fines, decreased rapidly in the beginning due to the escape of the lighter elements; then rose once again due to the addition of gas released from regolith material at lower temperatures. The fractional release of Ar at lower temperatures may then account for $Ar^{36}/Ar^{38} > 5.4$ in the poregas, because low-temperature release from fines is known to produce such an increased ratio (PEPIN *et al.*, 1970; HOHENBERG *et al.*, 1970). With near-quantitative release of He and Ne, the H^3/He^4 and Ne^{20}/Ne^{22} ratios in the poregas would not be significantly increased above the values now seen in the fines.

The extralunar object. An alternative explanation for the poregas is that it was in part derived from the impacting projectile. Carbonaceous chondrites are suitable candidates because of their relatively high trapped Ne^{20}/Ar^{36} ratios (MAZOR *et al.*, 1970). Quantitative outgassing of a carbonaceous chondrite such as Mokoia would produce a gasphase with $He^4/Ne^{20} \sim 300$; $Ne^{20}/Ar^{36} \sim 20$; $Ar^{36}/Kr^{84} \sim 100$; and $Kr^{84}/Xe^{132} \sim 1$. Thus, the extra-lunar (meteoritic) hypothesis can also account for the abundant presence of the heavy inert gases in the poregas; in fact, dilution with light inert gases from heated regolith material is required to raise the Ar^{36}/Kr^{84} ratio from ~ 100 to the observed values of $\sim 500 \simeq 1000$ (Table 2). A crucial test for the meteoritic hypothesis would be an accurate measurement of the isotopic composition of Xe released in the crushing, because of the known, substantial difference between trapped Xe in carbonaceous chondrites and lunar fines. Regrettably, the amounts released in our experiment were not sufficient for an accurate isotopic measurement, hence this question must remain unsettled.

The Ar^{40} problem. FUNKHOUSER *et al.* (1970) were the first to point out that both the fines as well as the breccia from Tranquillity Base contain Ar^{40} not supported by K, and that the average Ar^{40}/Ar^{36} ratio of the breccia is about twice that of the fines. HEYMANN and YANIV (1970b) have proposed that the so-called "excess" Ar^{40} in the fines represents Ar^{40} given off by the Moon, reimplanted into surfaces of regolith grains by an acceleration mechanism in the lunar atmosphere. FUNKHOUSER *et al.* (1970) have suggested that the breccias were not locally derived, but came from another site in which the fines have a higher Ar^{40}/Ar^{36} ratio than that observed in Apollo 11 fines. There are, however, alternative explanations. The model presented here allows the outgassing of large volumes of crystalline rock. Crystalline rocks contain substantial amounts of radiogenic Ar^{40}; their Ar^{40}/Ar^{36} ratios are usually greater than 100 (FUNKHOUSER *et al.*, 1970). Another possibility comes from the work of MARTI and LUGMAIR (1971) on the double core 12025 in which the Ar^{40}/Ar^{36} increases with depth, and at 40 cm has a value of about 2.5. This profile seems to suggest that the outgassing of regolith fines down to several tens of centimeters, perhaps a few meters, may produce a gasphase with Ar^{40}/Ar^{36} significantly greater than from fines such as 10084, collected from the top of the regolith.

CONCLUSIONS

MCKAY *et al.* (1970) have proposed a base surge type origin of lunar breccia. "They [base surges] provide a mechanism which could transport hot gas and ejecta (including both preexisting regolith and bedrock) radially outward from larger lunar impacts, possibly to considerable distances. They may also erode and heat large quantities of regolith in the vicinity of the crater and deposit this material with the ejecta." We conclude that our first model, in which all of the poregas is derived from heated regolith material has much in common with the base surge hypothesis of MCKAY *et al.* We too require a hot gas cloud produced by a large impact; a cloud that spreads laterally along the surface; picking up and carrying with it regolith materials that become welded together. We conclude that the gasphase that became incorporated in the breccia pores cannot have been shock implanted, because we require the mixing of gas given off by regolith materials at high temperatures near the point of impact with gas that was given off at lower temperatures outward from the point of impact. It is in this sense that our results support the base surge hypothesis of MCKAY *et al.* (1970).

If on the other hand the poregas was derived in part from a carbonaceous chondrite like projectile, the poregas probably could have been shock implanted in an "instant" formation of the breccia.

Acknowledgments—We wish to thank Dr. D. S. McKay, who was the first to suggest this experiment to us. The work was performed with support from the National Aeronautics and Space Administration, Contract NAS 9–7899.

REFERENCES

EBERHARDT P., GEISS J., GRAF H., GRÖGLER N., KRÄHENBÜHL U., SCHWALLER H., SCHWARZ-MÜLLER J., and STETTLER A. (1970) Trapped solar wind noble gases, exposure age and K/Ar age in Apollo 11 lunar fine material. *Proc. Apollo 11 Lunar Sci. Conf., Geochim. Cosmochim Acta* Suppl. 1, Vol. 2, pp. 1037–1070. Pergamon.

FREDRIKSSON K., NELEN J., and MELSON W. G. (1970) Petrography and origin of lunar breccias and glasses. *Proc. Apollo 11 Lunar Sci. Conf., Geochim. Cosmochim. Acta* Suppl. 1, Vol. 1, pp. 419–432. Pergamon.

FRONDEL C., KLEIN C., ITO J., and DRAKE J. C. (1970) Mineralogical and chemical studies of Apollo 11 lunar fines and selected rocks. *Proc. Apollo 11 Lunar Sci. Conf., Geochim. Cosmochim. Acta* Suppl. 1, Vol. 1. pp. 445–474. Pergamon.

FUNKHOUSER J. G., SCHAEFFER O. A., BOGARD D. D., and ZÄHRINGER J. (1970) Gas analysis of the lunar surface. *Proc. Apollo 11 Lunar Sci. Conf., Geochim. Cosmochim. Acta* Suppl. 1, Vol. 2, pp. 1111–1116. Pergamon.

HEYMANN D. (1967) On the origin of hypersthene chondrites: Ages and shock-effects of black chondrites. *Icarus* **6**, 189–221.

HEYMANN D., and YANIV A. (1970a) Inert gases in the fines from the Sea of Tranquillity. *Proc. Apollo 11 Lunar Sci. Conf., Geochim. Cosmochim. Acta* Suppl. 1, Vol. 2, pp. 1247–1260. Pergamon.

HEYMANN D., and YANIV A. (1970b) Ar^{40} anomaly in lunar samples from Apollo 11. *Proc. Apollo 11 Lunar Sci. Conf., Geochim. Cosmochim. Acta* Suppl. 1, Vol. 2, pp. 1261–1268. Pergamon.

HEYMANN D., and YANIV A. (1971) Inert gases from Apollo 11 and Apollo 12 fines: Reversals in the trends of relative element abundances. Second Lunar Science Conference (unpublished proceedings).

HINTENBERGER H., WEBER H. W., VOSHAGE H., WÄNKE H., BEGEMANN F., and WLOTZKA F. (1970) Concentrations and isotopic abundances of the rare gases, hydrogen and nitrogen in lunar matter. *Proc. Apollo 11 Lunar Sci. Conf., Geochim. Cosmochim. Acta* Suppl. 1, Vol. 2, pp. 1269–1282. Pergamon.

HOHENBERG C. M., DAVIS P. K., KAISER W. A., LEWIS R. S., and REYNOLDS J. H. (1970) Trapped and cosmogenic rare gases from stepwise heating of Apollo 11 samples. *Proc. Apollo 11 Lunar Sci. Conf., Geochim. Cosmochim. Acta* Suppl. 1, Vol. 2, pp. 1283–1310. Pergamon.

KANAMORI H., NUR A., CHUNG D. H., and SIMMONS G. (1970) Elastic wave velocities of lunar samples at high pressures and their geophysical implications. *Proc. Apollo 11 Lunar Sci. Conf., Geochim. Cosmochim. Acta* Suppl. 1, Vol. 3, pp. 2289–2294. Pergamon.

MARTI K., LUGMAIR G. W., and UREY H. C. (1970) Solar wind gases, cosmic-ray spallation products and the irradiation history of Apollo 11 samples. *Proc. Apollo 11 Lunar Sci. Conf., Geochim. Cosmochim. Acta* Suppl. 1, Vol. 2, pp. 1357–1368. Pergamon.

MARTI K., and LUGMAIR G. W. (1971) Kr^{81}–Kr and K–Ar^{40} ages, cosmic-ray spallation products and neutron effects in Apollo 11 and Apollo 12 lunar samples. Second Lunar Science Conference (unpublished proceedings).

MAZOR E., HEYMANN D., and ANDERS E. (1970) Noble gases in carbonaceous chondrites. *Geochim. Cosmochim. Acta* **34**, 781–824.

MCKAY D. S., GREENWOOD W. R., and MORRISON D. A. (1970) Origin of small lunar particles and breccia from the Apollo 11 site. *Proc. Apollo 11 Lunar Sci. Conf., Geochim. Cosmochim. Acta* Suppl. 1, Vol. 1, pp. 673–694. Pergamon.

O'KELLEY G. D., ELDRIDGE J. S., SCHONFELD E., and BELL P. R. (1970) Primordial radionuclide abundances, solar proton and cosmic ray effects and ages of Apollo 11 lunar samples by non-destructive gamma-ray spectrometry. *Proc. Apollo 11 Lunar Sci. Conf., Geochim. Cosmochim. Acta* Suppl. 1, Vol. 2, pp. 1407–1424. Pergamon.

PEPIN R. O., NYQUIST L. E., PHINNEY D., and BLACK D. C. (1970) Rare gases in Apollo 11 lunar material. *Proc. Apollo 11 Lunar Sci. Conf., Geochim. Cosmochim. Acta* Suppl. 1, Vol. 2, pp. 1435–1454. Pergamon.

VINOGRADOV A. P. (1971) Preliminary data on lunar ground brought to Earth by automatic probe "Luna-16." Second Lunar Science Conference (unpublished proceedings).

Proceedings of the Second Lunar Science Conference, Volume 2, pp. 1693–1703.
The M.I.T. Press, 1971.

Stepwise heating analyses of rare gases from pile-irradiated rocks 10044 and 10057

P. K. DAVIS, R. S. LEWIS, and J. H. REYNOLDS
Physics Department, University of California,
Berkeley, California 94720

(*Received 23 February 1971; accepted in revised form 5 April 1971*)

Abstract—Argon, krypton, and xenon from stepwise heating of two lunar rocks which had been irradiated in a pile (integrated flux 1.35×10^{19} slow neutrons/cm^2) were examined mass spectrometrically. The argon data are the basis for the so-called $^{40}Ar/^{39}Ar$ method of K–Ar dating, which gives an age of $(4.00 \pm 0.07) \times 10^9$ years for rock 10044 but is not useful for rock 10057, where the argon is very lightly bound. The krypton and xenon results led to release curves for "trapped" and cosmogenic gases plus the gases produced by neutron capture (or fission) in Ba, Br, I, and U. Inferred concentrations for rocks 10044 and 10057, respectively, are: Ba: 84 and 202 ppm; Br: 12.4 and 47 ppb; I: 0.9 and 0.7 ppb; U from krypton: 0.15 and 1.37 ppm; U from xenon (preferred): 0.19 and 0.59 ppm. Despite the fact that we have not recovered the rare gases from sites where they were lightly bound (in later experiments we shall), our inferred trace element compositions are in acceptable agreement with other workers, except for iodine where there are not yet definitive analyses for lunar samples.

INTRODUCTION

AS ONE MEANS at our disposal to increase the scientific return from lunar samples, we have undertaken a rather broad study of rare gases in specimens which have been strongly neutron irradiated in a reactor. Initial objectives were to obtain K–Ar ages by the $^{40}Ar/^{39}Ar$ technique (MERRIHUE and TURNER, 1966), I–Xe ages from release of ^{128}Xe correlated with excess ^{129}Xe from extinct ^{129}I (JEFFERY and REYNOLDS, 1961), and Pu–U–Xe ages from release of xenon from pile-induced fission of ^{235}U correlated with fission xenon from extinct ^{244}Pu (PODOSEK, 1970). We also expected to obtain concentrations of certain trace elements which produce rare gas nuclides by slow neutron absorption (MERRIHUE, 1966). This paper is more nearly a progress report than a definitive study; the program is ongoing with present results indicating rather well what can be obtained and what cannot, as follows. The K–Ar ages are often obtainable with high precision as TURNER (1970a) has already so beautifully demonstrated. The I–Xe and U–Pu–Xe dating techniques are not yet applicable to lunar rocks since we have yet to find a lunar rock which is old enough to exhibit fossil xenon from extinct radioactivities. The trace elements Ba, Br, I, and U are readily measured in the lunar rocks by our technique, even in the presence of complicated preexisting krypton and xenon spectra. Indeed, temperature release curves for the gases from these trace elements can be calculated, as a possible means of distinguishing surficial and other lightly bound fractions of an element from more retentively sited components.

EXPERIMENTAL PRELIMINARIES

The work was focused on the two lunar rocks for which HOHENBERG *et al.* (1970) have obtained complete data for stepwise temperature release of rare gases, rock 10057, which is a fine-grained and high rubidium (type A) basalt, and rock 10044, which is a coarse-grained and low-rubidium (type B) basalt (PAPANASTASSIOU, *et al.*, 1970). Because of the variety of effects involved, we chose a neutron spectrum for the irradiation with a smaller ratio of fast to thermal neutrons than TURNER (1970a) has used. The irradiation was carried out in the pool of the General Electric Test Reactor, Vallecitos Nuclear Center, Pleasanton, California. The nominal integrated fluxes, as supplied by the Vallecitos Center were

thermal (E < 0.17 ev) 1.35×10^{19} neutrons/cm^2
epithermal (0.17 ev < E < 0.18 Mev) 0.411×10^{19} neutrons/cm^2
fast (E > 0.18 Mev) 0.0869×10^{19} neutrons/cm^2.

We have not yet run the monitors which can confirm those fluxes. Tentatively we use the nominal values with an assigned error of $\pm 20\%$. The capsule was designed to rotate continuously about a vertical axis parallel to the core of the reactor, but unfortunately the mechanism for rotation failed soon after the irradiation began. As a substitute for continuous rotation under these circumstances, the capsule was rotated 180° midway through the irradiation. Counting of cobalt-doped flux wires packed with the samples indicates that the spatial variations in the total neutron exposure for the samples were no more than 2 to 3%.

The samples were packed in evacuated quartz tubes, but it was not possible to analyze the rare gases which escaped into the gas phase during the neutron irradiation or during the time the samples were stored in the extraction system (and mildly heated, maximum temperature 110° C) prior to the actual run. In subsequent work these "lost" gases are also being analyzed. But in the present report, we must emphasize, the sites in the rocks of very low retentivity for argon, krypton, and xenon have not been sampled.

^{40}Ar/^{39}Ar AGE STUDIES

In favorable rocks the argon data permit reliable K–Ar ages to be determined (MERRIHUE and TURNER, 1966). If during the last part of the release there is a temperature regime where the argon is coming almost entirely from highly retentive sites, the ratio ^{40}Ar/^{39}Ar will exhibit a plateau. Both isotopes are coming from potassium sites: The ^{40}Ar was formed there by natural radioactive decay of ^{40}K; the ^{39}Ar was formed there during the pile irradiation by fast neutrons via the (n, p) reaction on ^{39}K. The ratio of the isotopes seen at the plateau is proportional to the ^{40}Ar/^{40}K ratio in retentive minerals, a quantity which is a well known function of the age of the minerals. Figure 1 shows ^{40}Ar/^{39}Ar ratios from Apollo 11 rocks 10044 and 10057 determined at a progression of temperatures (half hour heatings) and plotted as a function of the cumulative ^{39}Ar release at the end of the heatings. The data for the plot are set out in Table 1. In determining the ordinate we have subtracted small amounts of ^{40}Ar due to combined contributions of solar wind and cosmic-ray spallation (the ratio ^{40}Ar/^{36}Ar assumed to be unity in this component); similarly we have subtracted small amounts of ^{39}Ar from the (n, α) reaction on ^{42}Ca, using the monitoring reaction ^{40}Ca (n, α) ^{37}Ar to make this correction. One can see from Table 1 that both corrections are usually small.

For rock 10044, one can obtain a valid date by the method. It exhibits a well-defined plateau, which is almost identical to that found by TURNER (1970a). Note that Turner's points are for the upper midpoints of the rectangles in the implied histogram,

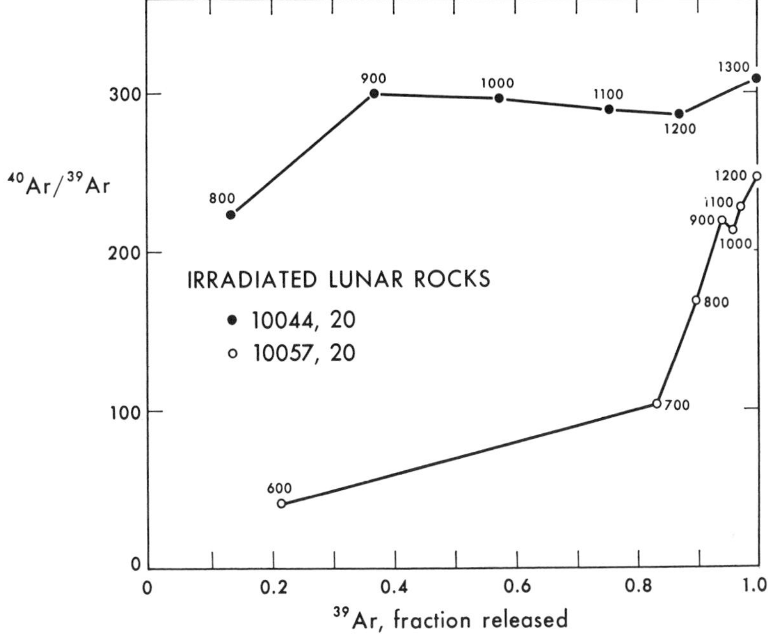

Fig. 1. Comparative release of ^{40}Ar and ^{39}Ar from pile-irradiated lunar rocks. The numbers on the points are release temperatures in degrees Centigrade (progressive half hour heatings). Rock 10044, with a good high temperature plateau, can be precisely dated by the technique. Rock 10057 cannot.

Table 1. Data for $^{40}Ar/^{39}Ar$ dating of lunar rocks.

Heating Temperature °C	Cumulative fractional release of ^{39}Ar after heating	$\left(\frac{^{40}Ar}{^{39}Ar}\right)_{measured}$	$\left(\frac{^{40}Ar}{^{39}Ar}\right)_{corrected}$
	Rock 10044		
800	0.133*	221.3†	223.3‡
900	0.372	293.0	300.9
1000	0.577	288.0	297.3
1100	0.758	281.5	289.7
1200	0.872	254.7	286.5
1300	1.000	203.2	309.4
	Rock 10057		
600	0.214*	41.7†	40.5‡
700	0.830	109.9	103.8
800	0.898	176.5	168.9
900	0.942	217.9	218.9
1000	0.959	214.6	213.7
1100	0.973	213.5	228.9
1200	1.000	200.4	247.2

* Errors in ^{39}Ar increments are 10% of the increments.

† Errors in the measured $^{40}Ar/^{39}Ar$ ratios are 1.5%.

‡ The additional error in the corrected $^{40}Ar/^{39}Ar$ ratios is 14% of the correction. ^{40}Ar has been corrected for cosmogenic and trapped argon. ^{39}Ar has been corrected for production by the (n, α) reaction on ^{42}Ca.

whereas we plot the upper right-hand corner. The difference in method of plotting makes it appear that our sample was less retentive of ^{40}Ar than his, which is not the case (our sample showed 3.2% loss of ^{40}Ar; Turner's showed 7% loss). The maximum fractional variation in the ^{40}Ar/^{39}Ar ratios above 800° C in this rock corresponds to a fractional age variation of only 3%, a respect in which our results are identical with Turner's. The age we calculate for rock 10044 is $(4.00 \pm 0.07) \times 10^9$ years, based on a meteorite standard, Karoonda. Karoonda is one of a group of meteorites found to have a common age when dated by the ^{40}Ar/^{39}Ar method (PODOSEK 1971). We assume, with PODOSEK, that the most precisely dated meteorite of this group (St. Severin) is 4.6×10^9 years old; in that case, the age of Karoonda inferred from PODOSEK's data is $(4.58 \pm 0.05) \times 10^9$ years. The error in this last quantity is the largest component in the error computed for rock 10044. In addition, any shift in the age of St. Severin will shift the age of rock 10044 essentially the same amount. Our result does not agree well with TURNER's (1970a) value of $(3.74 \pm 0.08) \times 10^9$ years for this rock, but it coincides, within the errors, with the average age of TURNER's other type B rocks from Apollo 11, namely $(3.88 \pm 0.08) \times 10^9$ years. TURNER (1970b) has noted that "plateau ages" obtained for type B rocks from Apollo 11 are systematically older than those for type A. Our datum supports this observation.

Rock 10057, on the other hand, exhibits no plateau whatsoever and cannot be dated by the method. From agreement between the Rb–Sr ages on these two rocks (PAPANASTASSIOU *et al.*, 1970) we would expect ^{40}Ar/^{39}Ar ratios in retentive minerals in the two rocks to have the same values within 5%. But note how lightly the ^{39}Ar is bound in rock 10057: 90% of the isotope is released even before the 900° C heating was started. It is clear from our work that the isotope ^{40}Ar is also lightly bound and poorly retained.

The difference between rocks 10044 and 10057 for argon dating could well have been anticipated from the release patterns obtained for unirradiated samples of these rocks and tabulated by HOHENBERG *et al.* (1970). Both rocks show double peaks in their ^{40}Ar release, at ~800° C and ~1200° C. But most of the release in rock 57 is associated with the low temperature peak, whereas virtually all of the release in rock 44 is associated with the high temperature peak. For both rocks the release curves for ^{40}Ar after irradiation (this work and TURNER, 1970a) are different from before. For rock 44 the release curve is shifted down in temperature by about 200 centigrade degrees, without change in shape. For rock 10057 the broad lower temperature release peak is sharpened and shifted down in temperature by 100 to 200 centigrade degrees. The higher temperature peak, of less importance than the other in both irradiated and unirradiated samples, is shifted down in temperature by about 300 centigrade degrees and broadened. Clearly ^{40}Ar release patterns are altered by the neutron irradiation. This does not seem to affect the validity of the method, which, of course, depends only upon effects for the two isotopes, 39 and 40, being the same.

Kr AND Xe RELEASE

Krypton and xenon spectra from irradiated lunar rocks are very complex because of the presence of so many gas components. Fortunately the isotopic compositions of the various components are known and the number of isotopes exceeds the number of components. In this circumstance, the problem is amenable to analysis. For krypton we have 7 isotopes: the usual stable isotopes plus 10.7 year ^{85}Kr produced in the neutron-induced fission of ^{235}U. There are five possible gas components: spallogenic krypton from cosmic-rays, "trapped" krypton, fissiogenic krypton from n-fission of ^{235}U, monoisotopic ^{83}Kr from *n*-capture in ^{82}Se, and a mixture of ^{80}Kr and ^{82}Kr from *n*-capture in bromine. We use quotation marks in referring to the "trapped" component because this component is probably a mixture of atmospheric contamination and of solar-wind krypton in dust particles adhering to the rock sample; in neither case is the gas trapped in the rock in the usual sense. For xenon we have 9 isotopes and 6 components of known isotopic composition: spallogenic xenon from cosmic rays, "trapped" xenon, fission xenon from *n*-fission of ^{235}U (completely dominating the traces of xenon present from spontaneous fission of ^{238}U), an extra monoisotopic component, ^{136}Xe, from pile-neutron capture in short-lived fissiogenic ^{135}Xe, monoisotopic ^{128}Xe from *n*-capture in iodine, and monoisotopic ^{131}Xe from *n*-capture in barium (in lunar rocks completely dominating ^{131}Xe from *n*-capture in tellurium).

Most of these components are known and constant in composition throughout the release. Where there were uncertainties in composition we adopted a single isotopic composition taken from HOHENBERG *et al.* (1970), but assigned large enough errors to include the possible variations. We refer here to "trapped" xenon where the isotopic composition may be like the atmosphere, or like the solar wind, or in between; and to cosmogenic ^{129}Xe, ^{131}Xe, and ^{132}Xe which showed occasional, sporadic variations in relative abundance during the experiments with the unirradiated samples.

METHOD OF CALCULATION

How best to unravel these spectra and determine the amounts (and errors in the amounts) of each component present in each temperature fraction posed an interesting mathematical problem. We solved the problem to our satisfaction with a least squares method involving matrices which permitted us to use all the data in a simple computational format. We describe the process briefly here as it applies to krypton. Although straightforward, the method appears not to be published and we plan to publish a more adequate account elsewhere. Each of the samples is a mixture of seven isotopes which can be represented by a vector $\vec{A}$ in seven dimensional space. Each of the samples is also a mixture of five components which can be represented by a vector $\vec{C}$ in five dimensional space. The fractional isotopic abundances of the various pure components, written as an array, define a 7×5 matrix, **F**, which relates the vectors $\vec{A}$ and $\vec{C}$:

$$\vec{A} = \mathbf{F}\vec{C}.$$

We measure $\vec{A}$ for a sample and seek $\vec{C}$. The best value of $\vec{C}$ is one for which the sum of squares:

$$S = \sum_{j=1}^{7} \left[a_j - \sum_{i=1}^{5} f_{ji} c_i \right]^2$$

is a minimum. The solution to this least squares problem is $\vec{C} = (\mathbf{F}'\mathbf{F})^{-1}\mathbf{F}'\vec{A}$ where $\mathbf{F}'$ is the transpose of $\mathbf{F}$ and $(\mathbf{M})^{-1}$ is the inverse of $\mathbf{M}$. The error estimates in $\vec{C}$ are also obtained by a matrix method which utilizes the errors in both the components of $\vec{A}$ and the elements of $\mathbf{F}$. The basic matrix relation used in solving the error problem was

$$(\mathbf{P} + \varepsilon\mathbf{Q})^{-1} \to \mathbf{P}^{-1} - \varepsilon\mathbf{P}^{-1}\mathbf{Q}\mathbf{P}^{-1} \qquad \text{as } \varepsilon \to 0.$$

by means of which the errors in the matrix $(\mathbf{F}'\mathbf{F})^{-1}\mathbf{F}'$ were obtained from the errors in $\mathbf{F}$.

Results and Discussions

The totaled results from all temperature fractions are set out in Table 2. Release curves for the various components are shown as Figs. 2 through 5. The errors plotted in the figures are 1 σ errors with contributions included, as they would propagate through our matrix solutions, from all known sources, *except* the uncertainties in mass spectrometer sensitivity. These last errors affect equally all components of a gas at each release temperature. For krypton this additional source of error is judged to be 25%; for xenon, 20%. The errors in Table 2 are total errors, including the errors in spectrometer sensitivity. The final errors are sometimes quite large, in part because we were generous in assigning errors to the isotopic abundances of those components where a range of values would be possible (see above).

Table 2. Totaled gas components from pile-irradiated lunar rocks.

Component	Amount	Inferred Quantity	Comparison	Reference
		Kr Components		
		Rock 10044		
Kr "trapped"	400 ± 103		1450 ± 450 unirradiated	(1)
$Kr^{80,82}$ *n*-capture Br	292 ± 75	12.4 ± 3.2 ppb Br	190 ppb Br	(2)
Kr cosmogenic	258 ± 72		523 ± 156 unirradiated	(1)
Kr *n*-fission U^{235}	30 ± 14	0.15 ± 0.07 ppm U	0.28 ppm U	(3)
Kr^{83} *n*-capture Se	2 ± 12	134 ± 770 ppb Se	~800 ppb Se (rock 10045)	(4)
		Rock 10057		
Kr "trapped"	1907 ± 486		2060 ± 370 unirradiated	(1)
$Kr^{80,82}$ *n*-capture Br	1105 ± 287	47 ± 12 ppb Br	25.2 ppb Br	(5)
Kr cosmogenic	350 ± 153		326 ± 59 unirradiated	(1)
Kr *n*-fission U^{235}	280 ± 76	1.37 ± 0.37 ppm U	0.80 ppm U	(3)
Kr^{83} *n*-capture Se	16 ± 64	990 ± 3900 ppb Se	~700 ppb Se (rock 10022)	(4)
		Xe Components		
		Rock 10044		
Xe^{131} *n*-capture Ba	2070 ± 415	84 ± 17 ppm Ba	95 ppm Ba	(7)
Xe *n*-fission U^{235}	224 ± 55	0.19 ± 0.05 ppm U	0.28 ppm U	(3)
Xe "trapped"	89 ± 62		52 ± 7 unirradiated	(1)
Xe cosmogenic	85 ± 90		128 ± 9 unirradiated	(1)
Xe^{128} *n*-capture I	13 ± 12	0.93 ± 0.85 ppb I	≥10 ppb I	(2)
		Rock 10057		
Xe^{131} *n*-capture Ba	4939 ± 989	202 ± 40 ppm Ba	208 ppm Ba	(3)
Xe *n*-fission U^{235}	686 ± 139	0.59 ± 0.12 ppm U	0.80 ppm U	(3)
Xe "trapped"	370 ± 83		669 ± 44 unirradiated	(1)
Xe cosmogenic	112 ± 64		296 ± 21 unirradiated	(1)
Xe^{128} *n*-capture I	10.3 ± 7.9	0.74 ± 0.55 ppb I	≥4.7 ppb I (rock 10017)	(6)

Units are 10^{-12} cc STP/g unless otherwise specified. Errors in this table are total errors (~1 σ). References: (1) Hohenberg *et al.* (1970); (2) Reed and Jovanovic (1970); (3) Wänke *et al.* (1970); (4) Haskin *et al.* (1970); (5) Ganapathy *et al.* (1970); (6) Reed and Jovanovic (1971); (7) Tera *et al.* (1970).

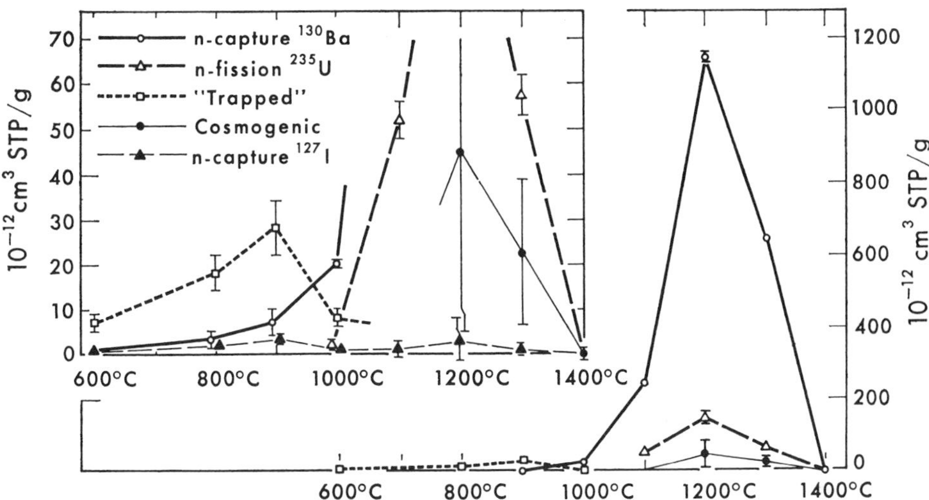

Fig. 2. Release curves for xenon components from pile-irradiated lunar rock 10044. The spectra are dominated by ^{131}Xe from *n*-capture in ^{130}Ba. Errors (1 σ) include all known sources except for variation from one temperature to another in xenon sensitivity ($\pm 20\%$).

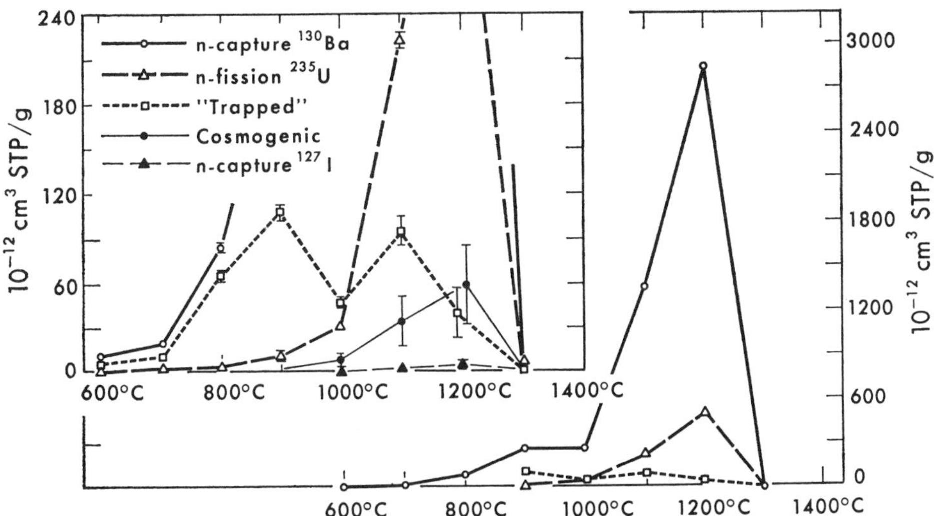

Fig. 3. Release curves for xenon components from pile-irradiated lunar rock 10057. The spectra are dominated by ^{131}Xe from *n*-capture in ^{130}Ba. For comment on errors, see caption for Fig. 2.

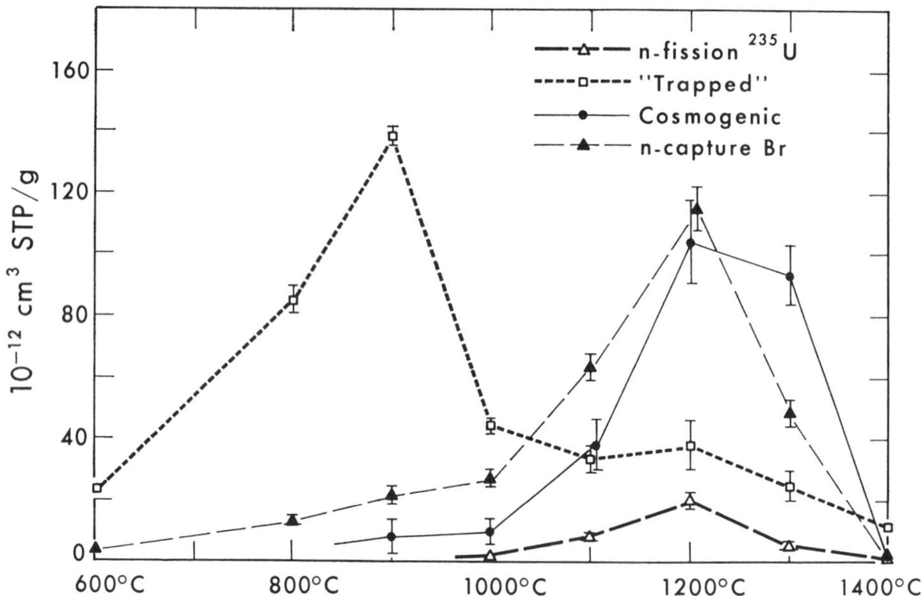

Fig. 4. Release curves for krypton components from pile-irradiated lunar rock 10044. Errors (1 σ) include all known sources except for variation from one temperature to another in krypton sensitivity ($\pm 25\%$).

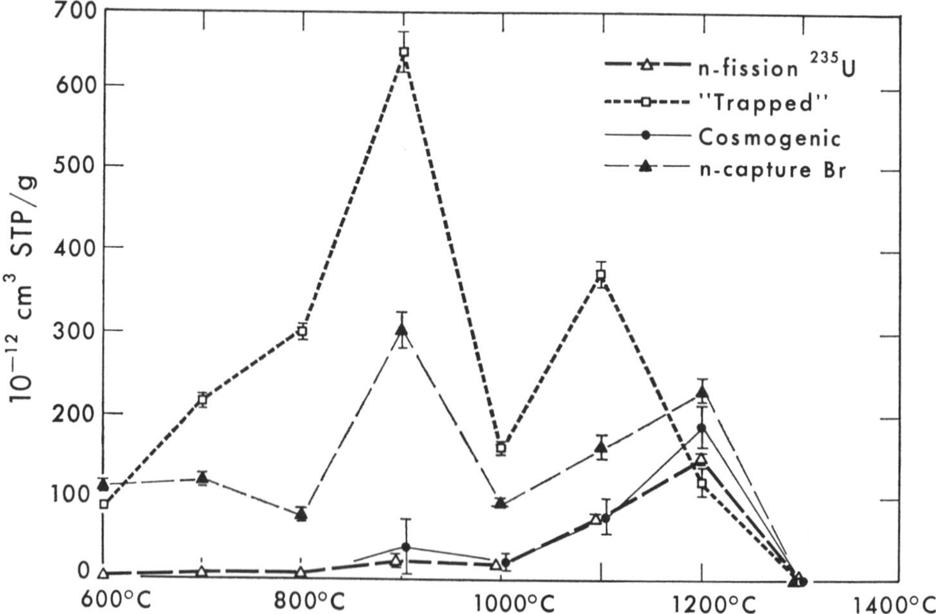

Fig. 5. Release curves for krypton components from pile-irradiated lunar rock 10057. For comment on errors, see caption for Fig. 4.

No serious discrepancies between results in irradiated and unirradiated samples can be noted. There appears to be less "trapped" krypton in 10044 and less "trapped" xenon in 10057 after the irradiation than before, but since the "trapped" gases are probably a mixture of atmospheric contamination and of solar wind gas dissolved in dust particles adhering to the grains, their concentrations would not be expected to be highly reproducible. The substantially larger amounts of "trapped" xenon and krypton in our sample of 10057, relative to the amounts we saw in 10044, was again seen in the irradiated samples and proved that the effect is not instrumental. The release patterns for the "trapped" gases in irradiated rock 10044 differ from those obtained with the unirradiated samples (see Fig. 4, HOHENBERG *et al.*, 1970) but the latter were of such shape as to suggest that most of the "trapped" gases in 10044 are system blanks.

One would not expect the amounts or release patterns for the cosmogenic gases to be altered by the irradiation, and no such effects were seen. An exception was an apparent loss of cosmogenic xenon in irradiated rock 10057. But the effect is statistically of low significance. There is a large 1 σ error (aside from the 20% uncertainty in xenon sensitivity for the mass spectrometer) for cosmogenic xenon in irradiated 10057. At the 2 σ level, the errors would almost overlap.

The release curves in Figs. 2–5 are useful for getting an overall picture of the krypton and xenon in the irradiated rocks. Most striking is the dominance of the xenon by ^{131}Xe from n-capture in ^{130}Ba. The effect is so large as to suggest that neutrons are in some way responsible for the large and variable ^{131}Xe anomaly in the Apollo rocks. Unfortunately the ^{131}Xe anomaly cannot be attributed to capture of natural slow neutrons in the rocks, a process which can be independently monitored in the rocks by the measurement of n-capture in ^{157}Gd (ALBEE *et al.*, 1970; MARTI *et al.*, 1970).

The xenon components generated in the pile have single-peaked release patterns rather similar to the release of the cosmogenic component, except for the component from iodine, which is more diffuse in its release.

For krypton there is no dominant component corresponding to Xe^{131}; the various components tend to be comparable in size. Most of the components generated in the pile have release patterns similar to the cosmogenic release pattern, but again the component from the halogen, now bromine, is released more diffusely. Also we note that the krypton components in 10057 tend to be generally more diffusely released than in 10044. The krypton from n-capture in bromine in 10057 is especially diffusely released, and is almost four times more abundant than its counterpart in 10044.

TRACE ELEMENT ESTIMATES

The concentrations of Ba, U, Br, I, and Se inferred from our rare gas concentrations (see Table 2) depend upon the nominal slow-neutron flux for the irradiation, which we have not yet checked, and upon values for slow neutron cross sections taken from the literature. Some confidence in these parameters is generated by the generally good agreement with radiochemical data for the barium and uranium concentrations we have measured. These elements are lithophile and nonvolatile.

They can thus be expected to be sited away from grain boundaries where the rare gases produced will be reasonably well retained. Such appears to be the case, judging from both our concentrations and release curves.

Our values for selenium are too imprecise to be very useful. It is doubtful if our method will ever give much better selenium values, except in particularly favorable samples.

There are relatively few data for bromine and iodine in the lunar rocks. Bromine has been measured by MORRISON *et al.* (1970, 1971) but in their Apollo 12 report they include this element in a group for which there are analytical problems associated with the determinations. Bromine has been measured in rock 10057 by GANAPATHY *et al.* (1970) at 25.2 ppb and in rock 10044 by REED and JOVANOVIC (1970) at 190 ppb. Our results for the bromine (at reasonably gas-retentive sites) in these rocks are 47 and 12.4 ppb respectively, so that we agree reasonably well with GANAPATHY *et al.* (1970) but poorly with REED and JOVANOVIC. In other comparisons (see GANAPATHY *et al.*, 1970) the bromine values of REED and JOVANOVIC (1970) tend to be high. While rock 10044 is a type B rock and conceivably could differ greatly in its bromine content from Type A rocks, ANDERS *et al.* (1971) recently have reported a low value (29 ppb) from bromine in rock 10047, of Type B. We thus are inclined to question the high value reported by REED and JOVANOVIC (1970). But we really have to reserve judgment until we have studied the "lost" gases from irradiated lunar rocks and have gained some idea of the concentrations of surficial bromine.

The situation with respect to iodine analyses is presently unclear. Here we can compare only our values for "interior" iodine with values of "leachable" iodine given by REED and JOVANOVIC (1970) as lower limits. Obviously we are comparing what may be mutually exclusive results. For rock 10044 their value is ≥ 10 ppb; ours is 0.9 ppb. For type A rocks we had no samples in common, but their value of ≥ 4.7 ppb for rock 10017 can perhaps be compared with our value of 0.7 ppb for rock 10057. The question of the abundance of iodine in lunar samples is clearly still open.

Acknowledgments—This work was supported in part by NASA and the U.S. Atomic Energy Commission. It bears AEC Code No. UCB-34P32-76.

REFERENCES

ALBEE A. L., BURNETT D. S., CHODOS A. A., EUGSTER O. J., HUNEKE J. C., PAPANASTASSIOU D. A., PODOSEK F. A., RUSS G. P., II, SANZ H. G., TERA F., and WASSERBURG G. J. (1970) Ages, irradiation history, and chemical composition of lunar rocks from the Sea of Tranquillity. *Science* **167**, 463–466.

ANDERS E., LAUL J. C., KEAYS R. R., GANAPATHY R., and MORGAN J. W. (1971) Elements depleted on lunar surface: Implications for origin of moon and meteorite influx rate. Second Lunar Science Conference (unpublished proceedings).

GANAPATHY R., KEAYS R. R., LAUL J. C., and ANDERS E. (1970) Trace elements in Apollo 11 lunar rocks: Implications for meteorite influx and origin of moon. *Proc. Apollo 11 Lunar Sci. Conf., Geochim. Cosmochim. Acta* Suppl. 1, Vol. 2, pp. 1117–1142. Pergamon.

HASKIN L. A., ALLEN R. O., HELMKE P. A., PASTER T. P., ANDERSON M. R., KOROTEV R. L., and ZWEIFEL K. A. (1970) Rare earths and other trace elements in Apollo 11 lunar samples. *Proc. Apollo 11 Lunar Sci. Conf., Geochim. Cosmochim. Acta* Suppl. 1, Vol. 2, pp. 1213–1231. Pergamon.

HOHENBERG C. M., DAVIS P. K., KAISER W. A., LEWIS R. S., and REYNOLDS J. H. (1970) Trapped and cosmogenic rare gases from stepwise heating of Apollo 11 samples. *Proc. Apollo 11 Lunar Sci. Conf., Geochim. Cosmochim. Acta* Suppl. 1, Vol. 2, pp. 1283–1309. Pergamon.

JEFFERY P. M. and REYNOLDS J. H. (1961) Origin of excess Xe^{129} in stone meteorites. *J. Geophys. Res.* **66**, 3582–3583.

MARTI K., LUGMAIR G. W., and UREY H. C. (1970) Solar wind gases, cosmic ray spallation products, and the irradiation history. *Science* **167**, 548–550.

MERRIHUE C. M. (1966) Xenon and krypton in the Bruderheim meteorite. *J. Geophys. Res.* **71**, 263–313.

MERRIHUE C. M. and TURNER G. (1966) Potassium–argon dating by activation with fast neutrons. *J. Geophys. Res.* **71**, 2352–2857.

MORRISON G. H., GERARD J. T., KASHUBA A. T., GANGADHARAM E. V., ROTHENBERG A. M., POTTER N. M., and MILLER G. B. (1970) Elemental abundances of lunar soil and rocks. *Proc. Apollo 11 Lunar Sci. Conf., Geochim. Cosmochim. Acta* Suppl. 1, Vol. 2, pp. 1383–1392. Pergamon.

MORRISON G. H., GERARD J. T., POTTER N. M., GANGADHARAM E. V., ROTHENBERG A. M., and BURDO R. A. (1971) Elemental abundances of lunar soil and rocks from Apollo 12. Second Lunar Science Conference (unpublished proceedings).

PAPANASTASSIOU D. A., WASSERBURG G. J., and BURNETT D. S. (1970) Rb–Sr ages of lunar rocks from the Sea of Tranquillity. *Earth Planet. Sci. Lett.* **8**, 1–19.

PODOSEK F. A. (1971) Neutron-activation potassium-argon dating of meteorites. *Geochim. Cosmochim. Acta* **35**, 157–173.

REED G. W., Jr. and JOVANOVIC S. (1970) Halogens, mercury, lithium, and osmium in Apollo 11 samples. *Proc. Apollo 11 Lunar Sci. Conf., Geochim. Cosmochim. Acta* Suppl. 1, Vol. 2, pp. 1487–1492. Pergamon.

REED G. W. and JOVANOVIC S. (1971) The halogens and other trace elements in Apollo 12 soil and rocks; Halides, platinum metals and mercury on surfaces. Second Lunar Science Conference (unpublished proceedings).

TERA F., EUGSTER O., BURNETT D. S. and WASSERBURG G. J. (1970) Comparative study of Li, Na, K, Rb, Cs, Ca, Sr and Ba abundances in achondrites and in Apollo 11 lunar samples. *Proc. Apollo 11 Lunar Sci. Conf., Geochim. Cosmochim. Acta* Suppl. 1, Vol. 2, pp. 1637–1657. Pergamon.

TURNER G. (1970a) Argon–40/argon–39 dating of lunar rock samples. *Science* **167**, 466–468.

TURNER G. (1970b) Argon–40/argon–39 dating of lunar rock samples. *Proc. Apollo 11 Lunar Sci. Conf., Geochim. Cosmochim. Acta* Suppl. 1, Vol. 2, pp. 1665–1684. Pergamon.

WÄNKE H., RIEDER R., BADDENHAUSEN H., SPETTEL B., TESCHKE F., QUIJANO-RICO M., and BALACESCU A. (1970) Major and trace elements in lunar material. *Proc. Apollo 11 Lunar Sci. Conf., Geochim. Cosmochim. Acta* Suppl. 1, Vol. 2, pp. 1719–1727. Pergamon.

Proceedings of the Second Lunar Science Conference, Volume 2, pp. 1705–1715.
The M.I.T. Press, 1971.

Stable rare gas isotopes produced by solar flares in single particles of Apollo 11 and Apollo 12 fines

A. YANIV,* G. J. TAYLOR,† S. ALLEN, and D. HEYMANN
Departments of Geology and Space Science, Rice University,
Houston, Texas 77001

(*Received 22 February 1971; accepted in revised form 31 March 1971*)

Abstract—Mass spectrometric measurements of the inert gases in single particles from Apollo 11 and Apollo 12 fines show that these particles contain not only trapped components but also detectable, often significant amounts of cosmogenic He, Ne, and Ar. In order to study the origin of the cosmogenic components the composition of the particles was determined by electron microprobe analysis prior to the inert gas measurement. After correcting the cosmogenic components to a common composition, i.e., that of the bulk fines, we compare the *observed* $(He^3/Ne^{21})_c$ and $(Ne^{21}/Ar^{38})_c$ ratios to those *predicted* for bombardment of fines by galactic cosmic rays: 7.42 and 1.48, respectively. There are substantial differences between observed and predicted values. The difference could be in part due to He^3_c diffusion losses; however, the evidence strongly suggests that the particles contain significant amounts of Ne^{21}_c and Ar^{38}_c produced in them by solar cosmic rays. Particularly interesting is the fact that Ne^{21}_c excess occurs mainly in Mg-rich, Ca-poor; Ar^{38}_c excess in Ca-rich, Mg-poor particles. Production of Ne by solar flares apparently leads to variable $(Ne^{20}/Ne^{21})_c$ ratios with Al-rich, Mg-poor particles generally showing higher ratios than Mg-rich, Al-poor particles. The presence of solar-flare produced inert gases leads to "apparent galactic radiation ages" of the particles from 3×10^6 to 3.1×10^9 years, the latter being much higher than radiation ages of crystalline rocks.

INTRODUCTION

IN OUR PREVIOUS report on Apollo 11 samples (HEYMANN and YANIV, 1970), we have shown that stable inert gases produced by high energy protons can be detected in single particles of the fines, notwithstanding the presence of trapped (solar wind) gas. The question arises whether any of these stable nuclides were produced by solar flare protons, or by galactic protons alone. In order to answer this question, let us make the following rough estimates. We assume that, on the average, the flux of solar protons (>25 MeV) is 100 times the flux of galactic protons. We restrict ourselves to the case of Ne^{21}. We shall assume an effective range of 0.3 cm for solar flare protons and of 30 cm for galactic protons. Because the production of Ne^{21} is chemistry dependent we adopt for this calculation, somewhat arbitrarily, 10% (by weight) Mg, 5% Al, and 20% Si; typical soil values. We adopt 20 mb as the effective cross section for Ne^{21} production from Mg, Al, and Si by cosmic rays; and 40 mb for the effective production from Mg by solar flares. The latter number is based on our preliminary data for Ne^{21} production by 15–45 MeV protons on Mg targets.

---

* On leave of absence from the Department of Physics, Tel Aviv University, Ramat Aviv, Israel.

† Present address: Smithsonian Astrophysical Observatory, Cambridge, Massachusetts.

If the particle during its exposure to galactic cosmic rays had never been within the range of solar flare protons the only Ne^{21} production is obviously due to galactic cosmic rays. On the other hand, if the particle had spent all of the above time within the effective range of solar flare protons, the ratio of Ne^{21} produced by solar as compared to galactic protons is estimated as about 50. If the turnover rate in the top 30 cm is much shorter than the exposure time to galactic protons of this layer, the ratio decreases to about 0.5. Since we are dealing with *individual* particles, whose location in time in the regolith may correspond to any of these cases, solar flare produced Ne^{21} is expected to be detectable.

Since the rare gas production by high energy protons is chemistry dependent it is imperative to know the chemical composition of the particles which are analyzed for rare gases. To this end the major elements were determined by electron microprobe analysis prior to the rare gas analysis.

Experimental Techniques

Samples of <1 mm fines were examined with a binocular microscope. The objective was to select homogeneous particles for the probe measurements, because these measurements could only be done on a few selected spots of the particle. Since glassy particles and mineral fragments are most likely to be chemically homogeneous, we tried to restrict ourselves to these types. Glass particles were chosen on the basis of optical clarity and contained no visible inclusions of crystal fragments or incompletely fused soil. Samples that had an uncommon amount of material coating their surfaces were rejected. The samples were then broken to expose fresh, uncontaminated interior surfaces. The largest fragments were pressed into lengths of indium wire, which in turn were glued to glass slides for microprobe analysis. The samples could not be glued directly to the glass slides because of the strong possibility of the glue affecting the mass spectrometric results by introducing hydrocarbons into the vacuum system. The analyses were made on two different ARL–EMX microprobes. Particles for which only Si, Al, Ca, Fe, Mg, and Ti results are given were analyzed at the Manned Spacecraft Center, Houston, Texas. Those for which Cr, K, Na, and Mn data are given in addition to the previous six elements were analyzed at the Department of Earth and Space Sciences, SUNY, Stony Brook, N.Y. As many as 10 points were analyzed on a given sample but usually only three to five were measured because of small sample size and irregular surfaces. Measurements were made on the most horizontal and smooth surfaces available. In two cases, sample numbers 7 and 8 (brown glasses from 10084), no surface was of reasonable quality, so the results for these are not as accurate as for the others. The analyses were corrected for drift, background, mass absorption, atomic number, and fluorescence using the data reduction program at NASA–MSC, which is a modified version of Rucklidge's program. Judging from the variability from point to point on a given sample, the analyses for major elements are accurate to within $\pm 5\%$ *of the amount reported.* For values of <1 wt %, the accuracy is lower, about ± 0.1 wt %. The values for K and Na are accurate to within ± 0.05 wt %, as special care was taken in making these measurements (i.e., long counting times, background measurements).

The rare gases were measured by mass-spectrometry as already described elsewhere (Heymann and Yaniv, 1970).

Results and Discussion

Description of the particles and their chemical composition is given in Table 1. The composition of the colorless glass was taken from Wood *et al.* (1970).

As a first step we have subtracted from the raw data given in Table 2 the estimated amounts of trapped He^3_t, Ne^{21}_t, and Ar^{38}_t as follows. We adopted $(He^3/He^4)_t = 3.7 \times 10^{-4}$, based on our previous measurements (Heymann and Yaniv, 1970). This

Table 1. Description and chemical composition of the samples.

Sample No., Color	Weight µg.	Size micron.	SiO_2	Al_2O_3	CaO	FeO	MgO	TiO_2	Cr_2O_3	K_2O	Na_2O	MnO	Total
1 Green	113	500	41.5	30.4	14.5	5.0	6.0	0.5		<0.1(a)			98.0
2 Green	93	450	41.1	31.1	15.0	3.7	8.2	0.5		<0.1(a)			99.6
3 Green	261	800	45.5	26.9	15.7	6.2	7.5	0.6		<0.1(a)			102.4
4 Green	97	450	44.8	27.4	16.3	4.8	9.8	0.6		<0.1(a)			103.7
5 Yellow-Green	129	500	46.0	18.3	8.9	9.2	8.3	2.4					93.1
6 Brown	179	500	39.6	15.9	10.1	12.0	7.9	9.0		0.1(a)			94.5
7 Brown	71	350	41.6	13.1	9.0	7.9	9.1	9.1		0.1(a)			89.8
8 Brown	1000	1500	37.9	8.1	10.7	10.2	8.8	13.1		0.1(a)			88.7
9 Brown	135	500	44.6	13.0	9.8	13.1	9.4	5.7		0.1(a)			95.6
10 Brown	646		41.9	9.5	9.6	14.6	8.4	10.5		0.1(a)			94.4
11 Brown	51												
13 Brown	171												
16 Colorless (a)	519		21.4 (a)	18.7 (a)	13.4 (a)	—	0.1 (a)	—		<0.1(a)			
17 Yellow (b)	420												
18 Green (b)	467												
24 Colorless	4422	74–88											
25 Colorless	774	105–250											
27 Colorless	210	700											
29 Green (b)	627	700											
41 Amber (c)	145		35.8	0.7	0.3	29.6	32.8	0.2	0.3	0.05	0.14	0.2	100.1
42 Amber (c)	143		36.5	0.8	0.5	27.2	35.3	0.2	0.3	0.05	0.12	0.2	101.3
43 Amber (c)	113		36.5	0.7	0.4	30.3	34.2	0.1	0.3	0.05	0.13	0.2	103.0
44 Amber (c)	115		36.5	0.7	0.3	26.4	36.7	0.1	0.4	0.04	0.11	0.2	101.5

(a) Numbers refer to elements, not oxides. Data from WOOD *et al.* (1970). These numbers have been adopted for all of our colorless particles. (b) No microprobe data available. Adopted average composition of particles in same color group. (c) Amber particles 41–44 are from 12070, all others are from 10084.

Table 2. Inert gas contents (units 10^{-8} cm^3 STP/g).

Sample	He^4	He^3	Ne^{20}	Ne^{21}	Ne^{22}	Ar^{36}	Ar^{38}	Ar^{40}
1	609,000 ± 31,000	254 ± 24	10,500 ± 520	59.3 ± 4.3	809 ± 58	3,800 ± 200	732 ± 69	—
2	77,900 ± 14,000	76 ± 17	1,360 ± 110	94 ± 12	158 ± 22	442 ± 36	275 ± 24	641 ± 59
3	296,000 ± 11,000	202 ± 8.5	3,610 ± 220	31.2 ± 4.8	293 ± 30	933 ± 81	190 ± 16	9,510 ± 470
4	179,000 ± 9,600	73 ± 14	2,170 ± 140	33 ± 10	207 ± 16	409 ± 24	66.8 ± 9.3	2,940 ± 240
5	350,000 ± 14,000	152 ± 16	2,420 ± 100	32.2 ± 4.1	208 ± 17	820 ± 50	155 ± 18	626 ± 50
6	—	17.7 ± 8.8	2,390 ± 81	14.0 ± 4.3	190 ± 15	251 ± 12	43.9 ± 3.7	853 ± 27
7	437,000 ± 31,000	187 ± 25	9,020 ± 650	57 ± 21	329 ± 29	737 ± 59	192 ± 17	283 ± 20
8	77,400 ± 2,100	32.1 ± 2.1	709 ± 14	4.89 ± 0.81	58 ± 3	94.6 ± 2.5	26.9 ± 0.9	267 ± 33
9	227,000 ± 12,000	104 ± 10	4,250 ± 180	41 ± 6	357 ± 24	900 ± 39	181 ± 10	2,300 ± 98
10	156,000 ± 4,100	66.2 ± 3.3	4,040 ± 59	15.1 ± 1.6	316 ± 9	750 ± 22	150 ± 3.6	1,400 ± 35
11	90,000 ± 45,000	78 ± 12	3,340 ± 180	34 ± 17	285 ± 25	587 ± 32	120 ± 8	19,900 ± 2,600
13	359,000 ± 16,000	159 ± 9	13,200 ± 290	66 ± 7	1,040 ± 30	1,440 ± 33	287 ± 8	37,600 ± 1,100
16	203,000 ± 3,900	80 ± 4	2,160 ± 41	34 ± 2	190 ± 8	445 ± 20	126 ± 3	7,180 ± 580
17	285,000 ± 7,400	235 ± 63	3,330 ± 630	46 ± 5	302 ± 14	202 ± 6	38 ± 1	3,150 ± 200
18	244,000 ± 6,300	88 ± 5	7,590 ± 130	69 ± 4	646 ± 17	920 ± 16	190 ± 5	6,050 ± 210
24	1,820,000 ± 80,000	778 ± 10	22,800 ± 290	76 ± 5	1,740 ± 22	6,430 ± 82	1,240 ± 16	15,500 ± 200
25	521,000 ± 7,200	226 ± 10	5,040 ± 76	31 ± 2	406 ± 9	1,870 ± 26	404 ± 8	17,800 ± 330

value is the same as the one derived by EBERHARDT *et al.* (1970) from ilmenite grain size fractions. We have assumed that all He^4 is of the trapped variety. The uncertainty in the results due to the possible presence of radiogenic He_r^4 is difficult to assess because we do not know the U and Th contents of our particles. However, the U and Th contents can be estimated in an indirect way, either from K contents measured in this work or from data by WOOD *et al.* (1970), given in Table 1. Using $K/U = 2000$, $Th/U = 4$, and an accumulation time of 4×10^9 yr for radiogenic He_r^4, we have estimated that the He_c^3 values given in Table 4 could be too low by 20% on the average. Only He_c^3 in particles 8 and 11 could be too low by about 50%. We have used $(Ne^{21}/Ne^{20})_t = 2.7 \times 10^{-3}$, the lowest value from our 10084 size fractions (HEYMANN and YANIV, 1970). This value is about 10% higher than that given by EBERHARDT *et al.* (1970) for instance, but our calculated Ne_c^{21} data are generally not very sensitive to the uncertainty in this number, because $(Ne^{21}/Ne^{20})_{measured}$–$(Ne^{21}/Ne^{20})_t$ is nearly always much greater than the uncertainty. All of Ne^{20} was assumed to be trapped. We have used $(Ar^{36}/Ar^{38})_t = 5.35$ and $(Ar^{36}/Ar^{38})_c = 0.6$.

In order to make the results of cosmogenic gases in the particles more useful, we decided to reduce all the data to a common chemical composition, namely, that of the bulk soil, as follows:

In the first step we calculated the $(He_c^3/Ne_c^{21})_{bulk}$ and $(Ne_c^{21}/Ar_c^{38})_{bulk}$ expected from galactic cosmic rays only. This was done by taking the corresponding, known ratios in hypersthene chondrites, adjusting these to the chemistry of bulk fines using the first three relations of Table 3, and the composition of hypersthene chondrites from MASON (1965). The composition of bulk fines was taken from ENGEL and ENGEL (1970), MAXWELL *et al.* (1970), and PECK and SMITH (1970). The relations were taken from HEYMANN *et al.*, (1968) but we added a small correction for Ti. The calculated ratios are given in Table 3 and are plotted as straight lines in Figs. 1, 2, 3, and 4. In the second step we have adjusted the calculated Ne_c^{21} and Ar_c^{38} of the single particles, using the same equations again to bulk composition. These normalized values are the ones plotted in Figs. 1, 2, 3, and 4, and are given in Table 4.

Since there is some question whether the meteoritic production ratios by galactic cosmic rays can be used as standards for the exposure of lunar regolith material, we have applied the relations of Table 3 to results available on interior samples of crystalline rocks (data from FUNKHOUSER *et al.*, 1970) and have obtained excellent agreement (within less than $\pm 15\%$) between calculated and observed $(He^3/Ne^{21})_c$ ratios. There is, however, a discrepancy for $(Ne^{21}/Ar^{38})_c$ ratio, which has been noted in a recent paper by BOGARD *et al.* (1970). These authors find that a 50% greater coefficient for Ca in the Ar_c^{38} production formula of Table 3 gives a better fit to the

Table 3. Production rates of cosmogenic nuclides as a function of composition.

$He_c^3 = 1.00\ [0.01179\ 0 + 0.572] \times 10^{-8}$ cc STP g^{-1} m.y.$^{-1}$ (2π geometry)
$Ne_c^{21} = 0.00347\ [2.2\ Mg + 1.35\ Al + Si + 0.17\ Ca + 0.017\ (Fe + Ni + Ti)] \times 10^{-8}$ cc STP g^{-1} m.y.$^{-1}$ (2π geometry)
$Ar_c^{38} = 0.000597\ [16.5\ Ca + Fe + Ni + 2.5\ Ti] \times 10^{-8}$ cc STP g^{-1} m.y.$^{-1}$ (2π geometry)
$(He_c^3/Ne_c^{21})_{bulk} = 7.42$
$(Ne_c^{21}/Ar_c^{38})_{bulk} = 1.48$

Elements in weight percent.

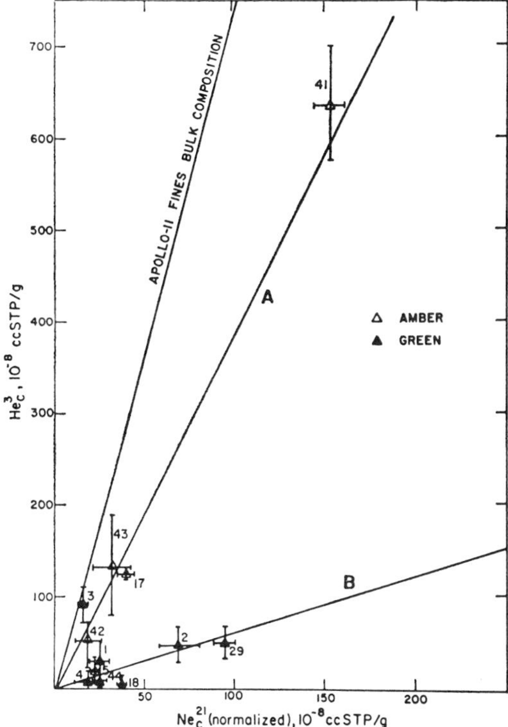

Fig. 1. Cosmogenic He^3 plotted vs. cosmogenic Ne^{21} for green and amber particles. The straight line marked Apollo 11 fines bulk composition represents the locus of all points produced in material of bulk chemistry by galactic cosmic rays. The data points for the particles have been normalized to bulk composition also. Points can be displaced downward or to the right from the bulk line either by diffusion loss (He^3_c is lost preferentially) or by solar flare production of Ne^{21} and He^3 (Ne^{21} is produced preferentially). Lines A and B represent the trends for the amber and green particles, respectively.

rock data. If we adopt the increased Ca-coefficient from BOGARD *et al.* (1970), then the slope of the $(Ne^{21}/Ar^{38})_c$ bulk lines in Figs. 2 and 4 decreases by about a factor of 1.5. The position of the data points for colorless and brown particles will change only little because their normalization factors remain virtually unchanged. The points for green particles, which are richer in Ca than the bulk fines will be displaced horizontally to the left, but generally by less than 15% of their normalized Ar^{38}_c values. The amber particles will be displaced to the right, by less than 25%. If the cosmogenic isotopes had been produced in the particles by galactic cosmic rays only, all the points should fall on the bulk line, assuming no diffusion loss. Any deviation of the points from the line is either due to diffusion losses or to solar flare production, provided that production ratios such as $(He^3/Ne^{21})_c$ are substantially different for solar flares in a given particle with its given chemistry, than for galactic cosmic rays. We surmise that the observed deviations from the bulk line are at least partly, if not mainly, due to solar flare produced He^3, Ne^{21}, and Ar^{38}.

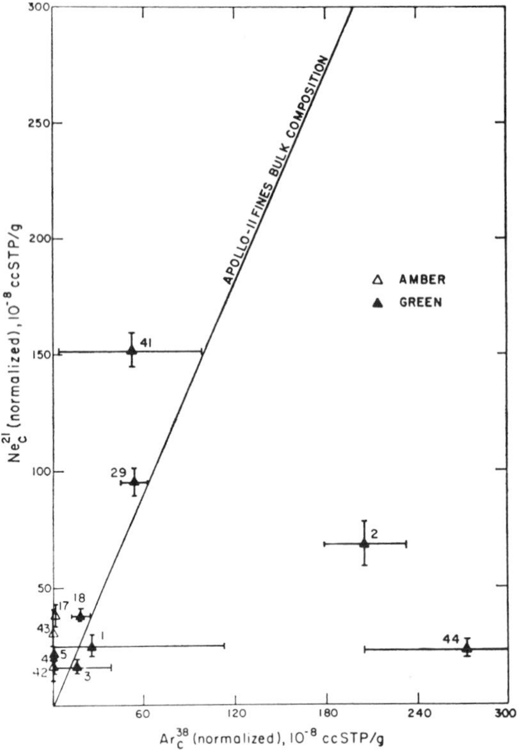

Fig. 2. Plot of Ne_c^{21} vs. Ar_c^{38} both normalized to bulk composition. Green and amber particles.

In Figs. 1 and 2 the results obtained on two groups of particles are presented—green glasses and amber particles. The amber particles have the composition of olivine, and were selected from the Apollo 12 fines; the green glasses are from the Apollo 11 fines.

We see two main reasons why the points in Fig. 1 fall below or to the right of the bulk line. Diffusion losses will decrease He_c^3/Ne_c^{21} and will therefore displace the points below the line whereas solar flare production of these isotopes will displace points to the right. Unpublished results by our laboratory on He^3 and Ne^{21} cross sections in Mg, Al, and Si targets bombarded with proton of energies below 50 MeV show that Ne^{21} produced from Mg exceeds He^3 production from any of the targets by as much as a factor 100. In Fig. 2 on the other hand one can see that most of the points lie above the bulk line which speaks against diffusion losses alone. If the Ar_c^{38} normalization had been made with the increased production from Ca according to BOGARD *et al.* (1970) this conclusion would not change. The fact that all but one of the Mg-rich and Ca-poor amber particles are displaced more towards the Ne_c^{21}-rich, Ar_c^{38}-poor portion of the diagram than the green ones which are poorer in Mg and rich in Ca is, we believe, strong support of our contention that the displacements in

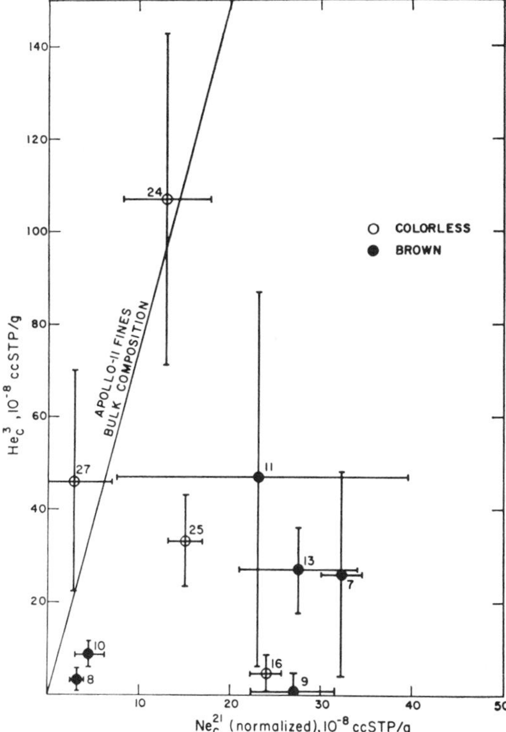

Fig. 3. Plot of He_c^3 vs. Ne_c^{21} (normalized) for colorless and brown particles.

Figs. 1 and 2 are primarily caused by solar flare production of Ne_c^{21} and Ar_c^{38}, mainly in Mg and Ca, rather than diffusion loss of He_c^3 and Ne_c^{21}.

Figure 1 contains an apparent paradox: the Mg-rich amber particles are displaced less to the Ne_c^{21} rich portion than are the green particles. This difference could be due to substantial differences in He_c^3 retention of the two groups, such that the green particles lose He_c^3 much more readily than do the amber ones. However, there is another possible explanation. Any point below and to the right of the bulk line can be reached by different ratios of exposure time to galactic and solar cosmic rays. Thus it is possible that the green particles (Apollo 11) have on the average experienced a relatively longer exposure to solar flares than the amber particles (Apollo 12). Recent, unpublished results on green particles from the Apollo 12 site from our laboratory are similar in this respect to the ones from the Apollo 11 site. We believe therefore that the location on the Moon from which the particles were collected is not a major factor contributing to this apparent difference in exposure history. We think that the green particles both in the Apollo 11 and Apollo 12 sites received most of their solar flare exposure elsewhere on the Moon, perhaps in the highlands, and were transported from there to their present location by impacts.

In Figs. 3 and 4 the point corresponding to brown and colorless particles are plotted. The brown particles, like the green ones, are substantially displaced below

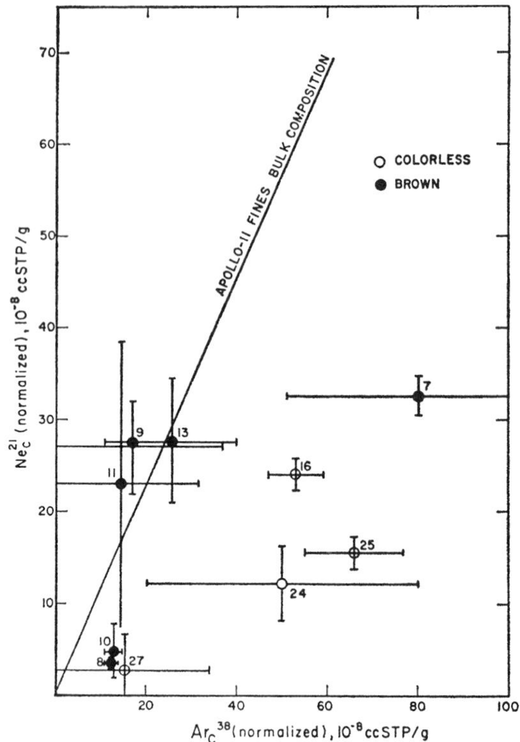

Fig. 4. Plot of Ne_c^{21} vs. Ar_c^{38} (normalized) for colorless and brown particles.

the bulk line, but on the average to a lesser extent. Also the points for the brown particles scatter more than those for the green particles. Both the brown and green particles come from the Apollo 11 landing site. The distinct difference in the trends between the two groups may mean either that the brown particles have better He_c^3 retention than the green ones, or since the two groups have very similar Mg contents, different radiation histories.

The colorless glasses which are very poor in Mg but rich in Ca show a trend expected in terms of solar flare production. Very little He_c^3 or Ne_c^{21} production by solar flares is expected, hence one predicts that the points should fall on the bulk line. Two of the four points indeed do, indicating that diffusion losses are sometimes negligible. The clear glasses are interesting in this respect because they are compositionally very similar to the feldspar in eucritic meteorites, which is known to have very poor He retentivity. It is also interesting that the very two particles which show no He_c^3 loss are displaced to the right in Fig. 4. Surely this displacement cannot be due to Ne_c^{21} diffusion loss, but must be assigned to Ar_c^{38} production by solar flares in this Ca-rich material. In fact, all of the clear glasses in Fig. 4 lie far to the right of the bulk line as one would expect for this Ca-rich, Mg-poor glass.

In Fig. 5 we have plotted the raw Ne data for a number of samples from Apollo 11 and Apollo 12, including basaltic lithic fragments, in a three-isotope representation.

Table 4. Cosmogenic gases, radiation ages.

Sample	Gas contents, 10^{-8} cm^3 STP/g			Normalization factors		Normalized gas contents 10^{-8} cm^3 STP/gr		Apparent ages, m.y.		
	He_c^3	Ne_c^{21}	Ar_c^{38}	Ne_c^{21}	Ar_c^{38}	Ne_c^{21}	Ar_c^{38}	He^3	Ne^{21}	Ar^{38}
1	31 ± 24	30.8 ± 5.1	27 ± 88	1.21	1.06	25.5	26	29	170	300
2	48 ± 21	90 ± 14	216 ± 28	1.31	1.06	69	205	45	467	2300
3	91.6 ± 8.9	19.2 ± 2.0	18 ± 25	1.25	1.08	15.4	17	86	100	190
4	7 ± 12	27 ± 11	—	1.33	1.20	20	—	7	130	—
5	21 ± 14	25.6 ± 4.3	—	1.16	0.696	22	—	20	150	—
6	—	7.6 ± 2.4	—	1.00	0.877	7.6	—	—	50	—
7	26 ± 22	33 ± 22	61 ± 23	1.04	0.766	32	80	25	220	910
8	3.1 ± 2.3	3.0 ± 0.8	10.3 ± 1.1	0.914	0.828	3.3	12	3	22	140
9	1.1 ± 8.3	29.5 ± 6.3	14 ± 16	1.09	0.820	27.1	17	—	180	190
10	7.8 ± 3.1	4.2 ± 1.6	10.8 ± 6.1	0.912	0.837	4.6	12.9	7	30	150
11	47 ± 41	25 ± 17	12 ± 12	1.09	0.824	23	15	44	150	170
13	27 ± 9	30.2 ± 7.4	21 ± 12	1.09	0.824	27.6	26	25	180	290
16	4.5 ± 3.9	28 ± 2	49.0 ± 5.6	1.18	0.926	24	53	4	160	600
17	124 ± 6	37.0 ± 4.6	0.4 ± 1.6	1.16	0.696	38.2	0.6	120	250	6
18	—	48.3 ± 3.3	20.9 ± 6.5	1.27	1.10	37.9	19	—	250	220
24	107 ± 36	14.6 ± 5.5	47 ± 29	1.18	0.926	12.4	51	100	83	580
25	32.8 ± 9.9	17.7 ± 2.0	61 ± 10	1.18	0.926	15.1	66	31	100	750
27	46 ± 25	3 ± 5	14 ± 17	1.18	0.926	2.7	15	43	18	170
29	50 ± 28	121 ± 8	59.2 ± 9.7	1.27	1.10	95	53.9	47	630	610
41	636 ± 64	215 ± 12	8 ± 7	1.43	0.165	151	47	600	1000	540
42	52 ± 31	25 ± 10	—	1.47	0.167	17	—	49	110	—
43	133 ± 53	43 ± 12	—	1.45	0.167	30	—	120	200	—
44	8.3 ± 5.6	44.8 ± 5.3	41 ± 10	1.50	0.148	22.1	270	8	150	3100

Radiogenic He^4 corrections could increase the He^3 exposure age by as much as 25% in the particles of very low He^4 content when K/U = 2000 and Th/U = 4 are assumed.

The errors on the apparent ages are about equal, percentagewise to those on He_c^3, Ne_c^{21} and Ar_c^{38}, respectively.

The solid line was drawn through the Ne-composition produced by galactic cosmic rays, and through the basaltic lithic fragments. Most of the points are seen to fall on or close to this line, but there are several interesting exceptions. The colorless glasses appear to be displaced systematically above the line; in fact, the highest Ne^{20}/Ne^{22} ratios were observed in these Al-rich, Mg-poor particles. On the other hand the two extreme cases below the line are Al-poor, Mg-rich amber particles. The correlation, however is not perfect; some of the clear glasses are close to the line, while two amber particles lie above it. We believe that these trends are due again to the exposure of the particles to solar flares such that the isotopic composition of the solar wind trapped gas was modified by the addition of low-energy proton produced Ne. Our unpublished results, mentioned before, support the general Al-Mg trend. In Al bombardments there is an energy range below 20 MeV, where the cross sections are substantial, and where the three Ne isotopes are produced in proportions such that any point representing an Al-rich, Mg-poor particle will be strongly displaced upward. For Mg bombardments on the other hand there is an energy range, mainly between 30 and 50 MeV where a Mg-rich, Al-poor particle would be displaced downward. The scatter of the points may be due to the fact that the excitation functions of the Ne-isotopes in production from Al and Mg, and also their relative production rates vary greatly over the entire energy range from 50 MeV down

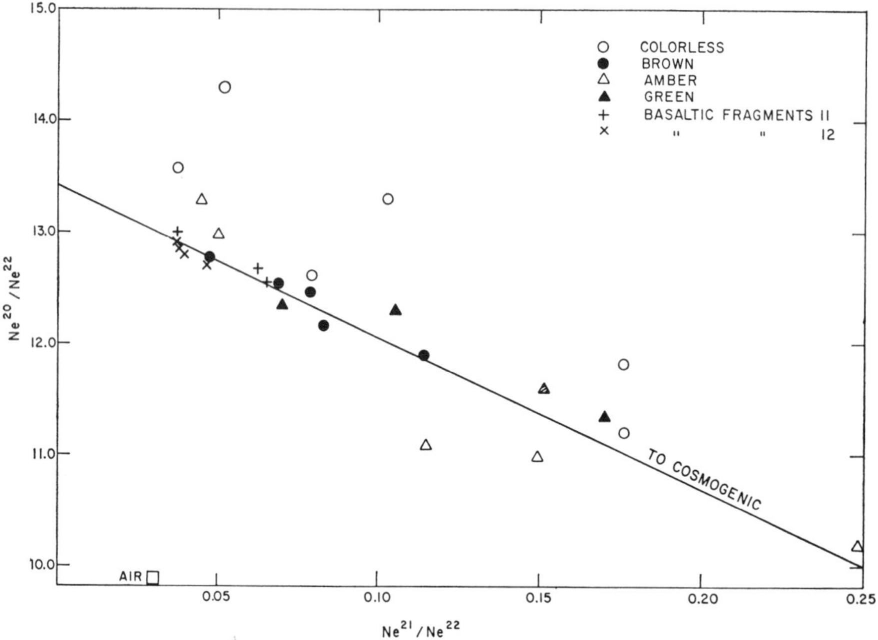

Fig. 5. Three-isotope plot for raw Ne-data of basaltic lithic fragments from Apollo 11 and Apollo 12 fines and particles from this work. The solid line is drawn through the "cosmogenic" point (Ne production by galactic cosmic rays) and the points of basaltic lithic fragments. The extreme points above the line are Al-rich, Mg-poor colorless glasses, whereas the extreme points below are Al-poor, Mg-rich amber particles of olivine composition.

Thus the general trend can be reversed depending on the degree of shielding, hence the energy spectra of solar protons which the individual particles were exposed to.

The evidence presented here strongly suggests that solar flare produced He, Ne, and Ar is present in the lunar fines. Mg-rich, Ca-poor particles often show an "overbundance" of Ne_c^{21}, whereas Mg poor, Ca rich material often shows an "overabundance" of Ar_c^{38} relative to the expected galactic proportions. In this respect it is especially interesting to note that the apparent "cosmic ray exposure ages" of three of the particles, when calculated with production rates for galactic cosmic rays are greater than 1, 2, and 3 billion years, much greater than the exposure age of the crystalline rocks (BOGARD *et al.*, 1970). We see no obvious reasons why the exposure ages of small particles should be that much greater than those of "rocks" a few centimeter in dimension other than solar flare production in submillimeter particles.

Acknowledgments—We would like to thank two laboratories for the use of their microprobe facilities, and microprobe standards: NASA Manned Spacecraft Center, Houston, Texas, (Drs. D. S. McKay and Arch M. Reid) and Department of Earth and Space Sciences, SUNY, Stony Brook, N.Y. (Dr. A. E. Bence). We would also like to thank Dr. John Wainwright of NASA–MSC for his assistance with the data reduction. The work was performed with support from the National Aeronautics and Space Administration, Contract NAS–9–7899.

References

BOGARD D. D., FUNKHOUSER J. G., and SCHAEFFER O. A. (1970) Noble gas abundances in lunar material, II. Cosmic ray spallation products and radiation ages from Mare Tranquillitatis and Oceanus Procellarum. *J. Geophys. Res.* (in press).

ENGEL A. E. J. and ENGEL C. G. (1970) Lunar rock compositions and some Interpretations. *Proc. Apollo 11 Lunar Sci. Conf., Geochim. Cosmochim. Acta* Suppl. 1, Vol. 2, pp. 1081–1084. Pergamon.

FUNKHOUSER J. G., SCHAEFFER O. A., BOGARD D. D., and ZÄHRINGER J. (1970) Gas analysis of the lunar surface. *Proc. Apollo 11 Lunar Sci. Conf., Geochim. Cosmochim. Acta* Suppl. 1, Vol. 2, pp. 1111–1116. Pergamon.

HEYMANN D., MAZOR E., and ANDERS E. (1968) Ages of calcium-rich achondrites, I. Eucrites. *Geochim. Cosmochim. Acta* **32,** 1241–1268.

HEYMANN D. and YANIV A. (1970) Inert gases in the fines from the Sea of Tranquillity. *Proc. Apollo 11 Lunar Sci. Conf., Geochim. Cosmochim. Acta* Suppl. 1, Vol. 2, pp. 1247–1253. Pergamon.

MASON B. (1965) The chemical composition of olivine–bronzite and olivine–hypersthene chondrites. *Amer. Museum Novitates,* **2223,** 1–38.

MAXWELL J. A., PECK L. C., and WIIK H. B. (1970) Chemical composition of Apollo 11 lunar samples 10017, 10020, 10072, and 10084. *Proc. Apollo 11 Lunar Sci. Conf., Geochim. Cosmochim. Acta* Suppl. 1, Vol. 2, pp. 1369–1374. Pergamon.

PECK L. C. and SMITH V. C. (1970) Quantitative chemical analysis of lunar samples, *Science* **167,** p. 532.

WOOD J. A., DICKEY J. S., MARVIN U. B., and POWELL B. N. (1970) Lunar anorthosites and a geophysical model of the moon. *Proc. Apollo 11 Lunar Sci. Conf., Geochim. Cosmochim. Acta* Suppl. 1, Vol. 1, pp. 965–988. Pergamon.

Proceedings of the Second Lunar Science Conference, Volume 2, pp. 1717–1728.
The M.I.T. Press, 1971.

Lunar atmosphere as a source of lunar surface elements

R. H. MANKA and F. C. MICHEL
Space Science Department, Rice University, Houston, Texas 77001

(*Received 22 February 1971; accepted in revised form 30 March 1971*)

Abstract—The characteristics of the lunar atmosphere are reviewed which are relevant to analysis of lunar samples and interpretation of ion detector data, and a number of general properties of the atmosphere are calculated. The mechanism is presented by which atmospheric ions are accelerated by fields in the solar wind, impact the moon, and are trapped in the lunar surface. The variation of lunar ion trajectory and impact energy is calculated as a function of ion mass and interplanetary magnetic field strength. Fluctuation in the direction of the interplanetary magnetic field as a cause of widely varying ion trajectories, and smearing of the selenographic distribution of trapped atmosphere, is illustrated. Ar^{40} in the lunar atmosphere is examined in more detail and, using the results of previous trapping calculations, an order of magnitude estimate for the neutral density of Ar^{40} in the lunar atmosphere of 10^2 to 10^3 cm^{-3} is obtained. The isotopic mass fractionation is studied for solar wind gases which are released into the atmosphere and retrapped; this theory predicts an enhancement of a lighter over a heavier retrapped isotope, and comparisons are made to data obtained for initial gas releases during stepwise heating of lunar samples.

INTRODUCTION

THE ANALYSIS OF lunar samples continues to exhibit unexpected compositions in some gaseous elements. One of the most pronounced anomalies is the excess of surface correlated Ar^{40}, compared to solar wind implanted Ar^{36}, which was discussed by several researchers, including HEYMANN *et al.* (1970), who proposed a lunar source for the Ar^{40}, being produced by K^{40} decay in the moon. In a paper showing that the lunar atmosphere is the likely source of the surface Ar^{40}, MANKA and MICHEL (1970) suggested that the atmosphere would also be the source of other unexpected surface elements, or unexpected concentrations of elements, including solar wind elements which have been cycled through the atmosphere and reimplanted in a manner characteristic of atmospheric ions. However, the atmospheric contribution to the elements is not as obvious as in the case of Ar^{40}; the nonradiogenic noble gas isotopes are not as readily attributable to lunar origin as in Ar^{40}, and the gases directly implanted from the solar wind tend to mask those cycled through the atmosphere. In addition, gases other than the noble gases can react chemically when they impact the surface, and are then difficult to distinguish from trace elements in the sample minerals.

The lunar atmosphere, moon, and solar wind form a dynamically interacting system, and while the lunar atmosphere is not dense, it can affect critical isotope ratios in the surface. The lunar atmosphere is a good example of a prototype planetary atmosphere and has been reviewed by BERNSTEIN *et al.* (1963), MICHEL (1964), and HINTON and TAEUSCH (1964), who discuss sources such as solar wind accretion and volcanism, and losses such as ionization and interaction with the solar wind, chemical reactions, and gravitational escape; mean lifetimes against ionization are 10^6 to 10^7

seconds. Using the ion trajectories obtained by MANKA and MICHEL (1970), it will be possible to calculate more accurately the largest loss mechanism, interaction with fields in the solar wind. Some species generally expected in the lunar atmosphere are solar wind and radiogenic gases such as H, He, Ne, Ar, Kr, and Xe, and volcanized gases such as H_2O, CO_2, N_2, O_2, and their dissociative products. Some less volatile gases which are easily cold-trapped (for example Hg, REED and JOVANOVIC, 1970) could have densities in the lunar atmosphere determined by the amount stored on the dark side.

Atmospheric species except hydrogen and helium are gravitationally bound (have lifetimes against gravitational escape much greater than against ionization) and thus form part of the equilibrium lunar atmosphere. The density of each species decreases approximately exponentially with height

$$n(r) \simeq n_0 e^{-r/h} \tag{1}$$

where h is the scale height for the species given by

$$h = \frac{kT}{mg} \tag{2}$$

and n_0 is the density per cm^3 at the surface, k is the Boltzmann constant, T and m are the species temperature and mass, and g is the lunar gravitational acceleration. The Ar^{40} scale height is about 50 km on the sunlit hemisphere.

The properties of ions which are reviewed in this paper should also apply to ions measured by detectors in the ALSEP package. The Suprathermal Ion Detector (FREEMAN *et al.*, 1971) detects "cloudlike" sporadic bursts of ions and also ions associated with Apollo vehicle impacts. The Solar Wind Spectrometer (SNYDER *et al.*, 1971) has detected ions associated with vehicle impact events and obtained the directional spectrum of the ions. The relationship of these measurements to the ion trajectories reviewed in this paper will be discussed briefly later.

The purpose of this paper is to present a number of properties of the lunar atmosphere, and to outline the calculations of some general results relevant to the study of the lunar surface and its particle environment.

TRAJECTORIES OF ATMOSPHERIC IONS

When an ion is formed in the lunar atmosphere, whether it escapes the moon or is accreted to the surface will depend on the interplanetary electric field, and in some cases on electric and magnetic fields at the lunar surface. The trajectories of ions formed in the lunar atmosphere and accelerated by fields in the solar wind have been discussed by MANKA and MICHEL (1970). In a frame of reference at rest with respect to the moon the interplanetary electric field is given by

$$\mathbf{E}_{SW} = -\mathbf{V}_{SW} \times \mathbf{B}_{SW} \tag{3}$$

where $\mathbf{V}_{SW}$, $\mathbf{B}_{SW}$, and $\mathbf{E}_{SW}$ are, respectively, the solar wind velocity, magnetic field, and electric field. The general features of the trajectory are obtained if we assume that $\mathbf{V}_{SW}$, $\mathbf{B}_{SW}$, and $\mathbf{E}_{SW}$ are perpendicular. For an ion formed at rest at $x = y = 0$, then its orbit in the crossed electric and magnetic fields is an ordinary cycloid and is given by

$$x = \frac{-V_{\mathrm{sw}}}{\omega_{\mathrm{c}}} \sin \omega_{\mathrm{c}} t + V_{\mathrm{sw}} t \tag{4}$$

and

$$y = \frac{V_{\mathrm{sw}}}{\omega_{\mathrm{c}}} (1 - \cos \omega_{\mathrm{c}} t) \tag{5}$$

where ω_c is the angular cyclotron frequency. The electrostatic force far exceeds the gravitational force so the initial motion of an ion is along E_{sw} and as the ion gains energy the magnetic force curves the ion in the direction of the solar wind flow with a resulting cycloidal orbit. The height of the cycloid, twice the cyclotron radius for the ion in the lunar rest frame, is much greater than lunar dimensions. Thus the ion's trajectory from formation to impact is the initial part of a cycloid, and the motion is nearly parallel to E_{sw}, and most of the flux of lunar ions to the surface is in a direction perpendicular to the solar wind flow. In general, ions formed in the lower sunlit atmosphere are driven up (with respect to E_{sw}) into the moon while ions formed at the equator and in the upper hemisphere escape, as illustrated in Fig. 1. Depending upon the direction of B_{sw}, the interplanetary electric field is generally upward or downward out of the solar ecliptic plane; and, when the direction of B_{sw} reverses several times during each solar rotation (as is common due to the sector structure of the interplanetary magnetic field), the direction of E_{sw} and the ion flux also reverse.

The energy of the ion at impact is just the energy gain along the interplanetary electric field

$$\mathscr{E} = eE_{\mathrm{sw}} y_i \tag{6}$$

where e is the ion charge (we assume single ionization) and y_i is the y-coordinates at impact. The coordinates and other parameters of impact are found by solving the equations for the cycloid with the equation for the locus of the lunar surface. As shown in Fig. 1, an ion formed at some height, say one scale height, above the surface and at an "electric" latitude λ_1 impacts at λ_2 with an impact angle α and an energy $\mathscr{E}$. (The latitude is chosen with respect to the direction to $\bar{E}_{\mathrm{sw}}$; thus if $\bar{E}_{\mathrm{sw}}$ reverses direction, the $\lambda = +90°$ and $-90°$ poles also exchange positions with respect to the fixed selenographic coordinates.)

There is a critical starting latitude for which an ion just grazes the surface, and ions formed at latitudes nearer the equator, and in the upper hemisphere, escape. The sketch of Fig. 1 illustrates, with some exaggeration, the trajectories of three ions of quite different mass. Actually, hydrogen is not gravitationally bound, scale height does not have a meaning in the usual sense, and most hydrogen ions are formed at much greater heights than for bound species. Tables 1A and 1B show the impact angle, latitude, and energy for ions formed at altitude h and latitude λ_1. The average interplanetary magnetic field varies considerably about 6γ and is oriented at about 45 degrees to the flow direction. The characteristic features of the trajectory can be obtained by assuming $\bar{B}_{\mathrm{sw}}$ perpendicular to $\bar{V}_{\mathrm{sw}}$ and using this assumption, the orbit parameters are calculated numerically for both $B_{\mathrm{sw}} = 5\gamma$ and 10γ (which give $E_{\mathrm{sw}} = 2$ volts km^{-1} and 4 volts km^{-1}, respectively) where 5γ is probably closer to the actual average conditions.

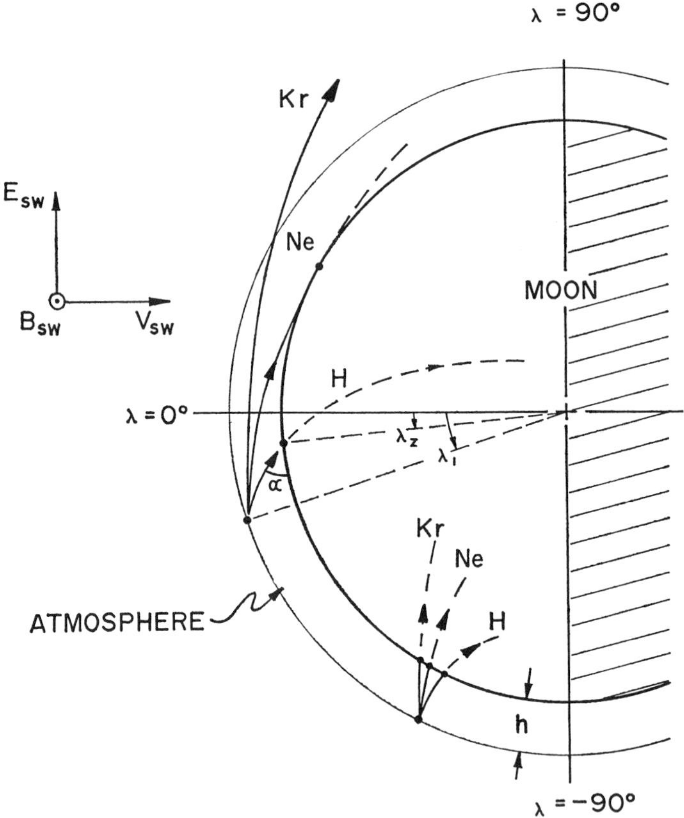

Fig. 1. Sketch of the orbits of ions formed a height h above the lunar surface. This "electric" latitude is chosen with respect to the direction of the interplanetary electric field. The trajectory for hydrogen is shown only to illustrate the greater curvature of a light ion's trajectory, though the curvatures shown for all three species are exaggerated. The scale height h differs for each specimen.

Examination of the impact coordinates shows relatively small differences for the two cases; however the impact energies for the 10γ case are about twice those for the 5γ case since the energies depend on E_{sw} as shown in equation (6). Physically we can see that since only the initial part of the cycloid is involved, the ion trajectory is primarily along $\bar{E}_{sw}$, and thus variations in the magnitude of $\mathbf{B}_{sw}$ do not greatly alter the impact coordinates; however, variations in the magnitude of $\mathbf{B}_{sw}$ do proportionally change the magnitude of $\mathbf{E}_{sw}$ and thus the value of $\mathscr{E}$. The impact energies are higher near the equator than at the poles so atmospheric ions should be more easily trapped near the equator. The impact energy is also approximately inversely proportional to the mass of the ion.

In the geometry of the trajectories discussed so far, the plane of the trajectory is determined by the direction of $\mathbf{E}_{sw}$, which in turn is orthogonal to $\mathbf{B}_{sw}$. While $\mathbf{B}_{sw}$ lies on the average in the vicinity of the plane of the ecliptic, it actually fluctuates

Table 1A. Orbit parameters for ions starting at rest one scale height from the lunar surface. Solar wind parameters are V_{SW} = 400 km/sec and B_{SW} = 5 gamma. Calculations are for the lunar noon-midnight meridian plane.

λ_1 (deg)	λ_2 (deg)	α (deg)	$\mathscr{E}$ gain (eV)
	Neon—20 (h = 100 km)		
−90	−90.1	93.3	200
−60	−58.3	61.6	228
−30	−24.5	28.9	397
−14.9	0	6.8	951
−13.7	8.2	0	1363
	Argon—40 (h = 50 km)		
−90	−90.0	91.6	100
−60	−59.1	60.8	115
−30	−27.2	29.4	199
−11.1	0	4.2	687
−10.4	4.7	0	935
	Krypton—84 (h = 23.8 km)		
−90	−90.0	90.8	48
−60	−59.6	60.4	55
−30	−28.7	29.7	95
−8.0	0	2.4	487
−7.6	2.6	0	627

Table 1B. Orbit parameters for ions starting at rest one scale height from the lunar surface. Solar wind conditions are V_{SW} = 400 km/sec and B_{SW} = 10 gamma. Calculations are for the noon-midnight meridian plane.

λ_1 (deg)	λ_2 (deg)	α (deg)	$\mathscr{E}$ gain (eV)
	Neon—20 (h = 100 km)		
−90	−90.2	94.6	400
−60	−58.3	63.1	451
−30	−24.8	30.9	763
−13.7	0	9.3	1747
−11.3	11.7	0	2857
	Argon—40 (h = 50 km)		
−90	−90.0	92.3	200
−60	−59.1	61.5	228
−30	−27.3	30.4	392
−10.3	0	5.6	1286
−9.1	6.8	0	1960
	Krypton—84 (h = 23.8 km)		
−90	−90.0	91.1	96
−60	−59.6	60.7	110
−30	−28.7	30.2	189
−7.5	0	3.4	912
−6.9	3.7	0	1289

widely in time and may even be at right angles to the ecliptic. In this latter case, the electric field lies in the ecliptic, as does the plane of the ion trajectory. Thus although the ion flux is generally in the sunlit portion of the upper or lower hemisphere, this distribution is smeared considerably owing to fluctuations of $\mathbf{B}_{SW}$. Figure 2 contains data of BURLAGA and NESS (1968) showing the distribution of magnetic field directions in the plane of, and perpendicular to, the ecliptic; also shown is an example of one

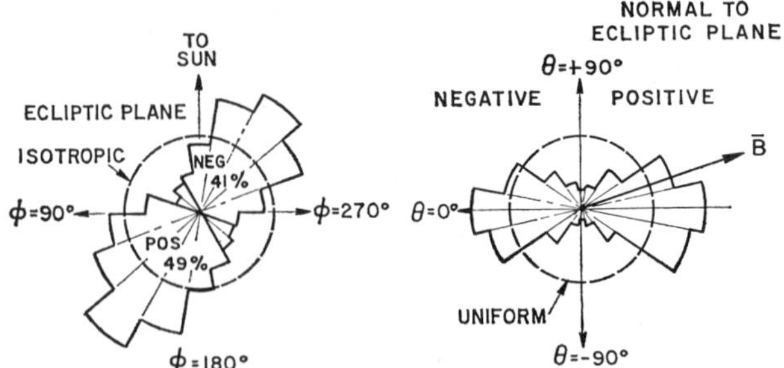

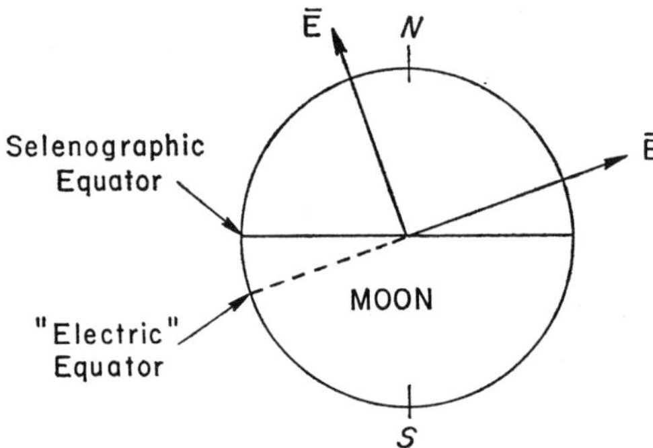

Fig. 2. Directional histograms of interplanetary magnetic fields parallel and perpendicular to the ecliptic as measured by Pioneers 6 and 7 (BURLAGA and NESS, 1968). Also shown is an example of a particular orientation of B_{sw}, and corresponding E_{sw}, projected onto the moon.

magnetic field and corresponding electric field orientation projected on the moon as seen looking along the solar wind flow direction.

Clearly the considerable fluctuations of B_{sw} out of the ecliptic plane will rotate the planes of lunar trajectories. While the interplanetary electric field may be the principal energy source of the ions, the lunar surface electric and magnetic fields could play important roles in changing an ion's direction and modifying its energy as it approaches the surface. The Suprathermal Ion Detector (FREEMAN *et al.*, 1971) is orientated so that it looks in the ecliptic plane and sees sporadic "clouds" of ions arriving at the lunar surface. The fluctuations in the direction of $\bar{E}_{sw}$, may combine with the surface fields to allow the detection of these bunches of ions. The measured energies of the ions, of tens to hundreds of electron volts, are reasonable for $\bar{E} \times \bar{B}$ acceleration. The Solar Wind Spectrometer (SNYDER *et al.*, 1971) looks both vertical and horizontal to the local surface and the majority of the ion flux detected after the Apollo 12 S–IVB impact event was directed horizontal to the local surface and from north to northeast. That is, the trajectories lie in a plane generally perpendicular to the ecliptic plane and which is qualitatively in agreement with the predictions of MANKA and MICHEL (1970) for the case when $\bar{B}_{sw}$ is generally in the ecliptic plane.

TRAPPING OF IONS

For an ion whose trajectory causes it to impact the moon, whether the ion is firmly implanted in the surface or is quickly released will depend on both the impact energy and the trapping probability $\eta_t(\mathscr{E})$ for the lunar material. Studies at Berne on ion implantation in aluminum foils with their associated oxide layer must presently be assumed to be similar to the characteristics for silicate grains. An example of the trapping data obtained by BÜHLER *et al.* (1966) is shown in Fig. 3. The trapping probability rises nearly linearly for energies to 1 keV and then levels off to unity at about 3 keV. This data was previously approximated by a three-part fit (MANKA and MICHEL, 1970),

$$\eta_t(\mathscr{E}) = \begin{cases} 0.6\mathscr{E}, & 0 \le \mathscr{E} < 1 \text{ keV} \\ 1 - 1.8 \exp(-1.5\mathscr{E}), & 1 \le \mathscr{E} < 3 \text{ keV} \\ 1, & \mathscr{E} \ge 3 \text{ keV}, \end{cases} \tag{7}$$

which is awkward to integrate but which is useful because it clearly indicates the dependence of trapping on energy: The approximately linear dependence below 1 keV and the total trapping above 3 keV.

For the calculations in this paper we use the equally good fit obtained with the single function

$$\eta_t(\mathscr{E}) = \tanh(0.75\mathscr{E}). \tag{8}$$

From the energies calculated in Table 1, we can see that it is this initial part of the trapping curve, up to a few keV, that will determine the implantation efficiency of the lunar ionosphere. An important feature of the trapping curve is whether there is a finite, and perhaps not very small, cut-off energy at which the trapping probability

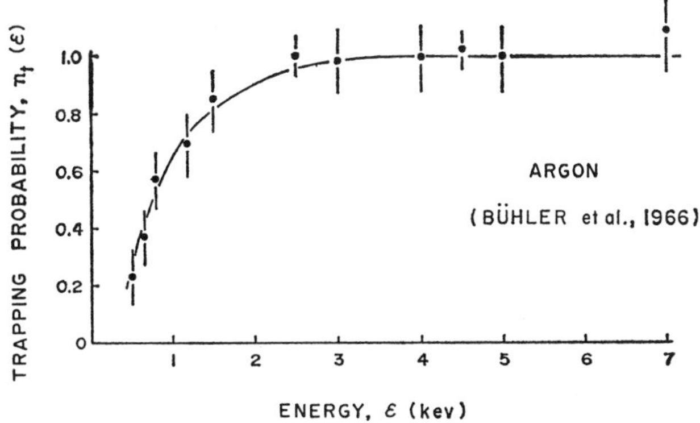

Fig. 3. Trapping probability for argon ions bombarding aluminum foil as a function of bombarding energy (from the measurements of BÜHLER *et al.*, 1966). The curve fit to the data is copied from Bühler's graph; however the appearance of a hyperbolic tangent fit is similar.

goes to zero. Physically this is expected and the data of KORNELSEN (1964) for the trapping of noble gas ions in tungsten indicate such a cut-off as do the detailed measurements of MEISTER (1969), for the trapping of nobel gases in aluminum, who has considered a function containing a cut-off energy to fit the data.

ARGON-40 IN THE LUNAR ATMOSPHERE

From analysis of the acceleration of atmospheric ions, some further properties of the lunar atmosphere can be deduced. The approximately equal concentrations of Ar^{40} and Ar^{36} in the lunar soil implies that the respective ion fluxes are related by the trapping efficiencies and geometries appropriate for the trapping of lunar atmosphere or solar wind ions and allows us to estimate the density of Ar^{40} in the lunar atmosphere.

First, we estimate the flux of Ar^{36} in the solar wind by relating it to known fluxes of helium and neon. It is difficult to obtain a true flux for an element which is not plentiful in the solar wind as the exposure time may not be sufficient to detect the element; also, the ratio of two elements may be strongly affected by diffusion in the trapping material. Table 2 gives ratios of elements in the solar wind as determined from several sources of information (some numbers are calculated from other data presented by the author).

The effects of diffusion can be seen, for example, in the ratio of He^4/Ne^{20} for lunar fines, ilmenite, and the Solar Wind Composition foil. If the true average solar wind ratio were the same in all three cases, then it appears that the ilmenite retains He^4 better than the fines but the He^4/Ne^{20} ratio is still below that of the foil where the time scale and temperature conditions should prevent significant diffusion loss of He^4. The Ne^{20}/Ar^{36} ratio was not detected by the foil but a value as large as 33 was obtained in the ilmenites. Since the ratio of these heavier elements will not be as strongly affected by diffusion, EBERHARDT *et al.* (1970) assumed this to be a lower limit to the true solar wind ratio. An average of the results of the two SWC measurements gives a He^4 flux of 7.2×10^6 cm^{-2} sec^{-1}, and a He^4/Ne^{20} ratio of 525 (GEISS *et al.*, 1970). Since this He^4 flux agrees well with an average value for the He^4 flux $=$ 8.4×10^6 cm^{-2} sec^{-1} which can be calculated from the data of ROBBINS *et al.* (1970) for average satellite measurements in the solar wind, we can have some confidence that solar wind conditions at the time of Apollo 11 and Apollo 12 approximated the average conditions. Using this He^4/Ne^{20} ratio and the Ne^{20}/Ar^{36} ratio for ilmenites, we calculate

$$\frac{He^4}{Ar^{36}} = \frac{He^4}{Ne^{20}} \times \frac{Ne^{20}}{Ar^{36}} \simeq 1.7 \times 10^4 \quad (9)$$

Table 2. Estimated and measured elemental ratios for the solar wind as deduced from various sources of information.

Source	He^4/Ne^{20}	Ne^{20}/Ar^{36}	He^4/Ar^{36}
Cosmic (CAMERON 1968)	980	11	1.1×10^4
Lunar Fines (HEYMANN *et al.*, 1970)	30–110	6–8	350–700
Lunar Ilmenite (EBERHARDT *et al.*, 1970)	(230)	≥33	≥7600
SWC Foil (GEISS *et al.*, 1970; Apollo 11 and Apollo 12 results have been averaged)	525		

(which is close to the value given by CAMERON (1968) for cosmic abundances—see Table 2). The corresponding Ar^{36} flux is

$$\Phi_{36} \cong 4.2 \times 10^2 \text{ cm}^{-2} \text{ sec}^{-1}. \tag{10}$$

We now relate this flux to that of Ar^{40} ions from the atmosphere. At height r above the sunlit lunar surface, the number of ions per cm^3 is given by the neutral number density divided by the ionization time τ_i. Using the cross sections for photoionization and charge exchange given by BERNSTEIN *et al.* (1963), and a hydrogen flux in the solar wind of 2×10^8 cm^{-2} sec^{-1}, we obtain $\tau_i = 1.5 \times 10^6$ seconds for argon. If we assume the simple case where the field drives the ions straight in and at right angles to the surface, then the flux per cm^2 second is approximately the number of ions formed per second in a 1 cm^2 column above the surface. The number of neutral argon-40 atoms in this column is

$$N \cong \int_0^\infty n_0 e^{-z/h}\,dz = n_0 h \tag{11}$$

where n_0 is the density of neutrals in the atmosphere near the lunar surface, z is the height above the surface, and h is the scale height. Thus the flux of argon-40, Φ_{40} is approximately given by $N/\tau_i \cong n_0 h/\tau_i$, or

$$n_0 \cong \frac{\Phi_{40}\tau_i}{h}. \tag{12}$$

From previous trapping calculations (MANKA and MICHEL, 1970), we estimate that about 8.5% of the Ar^{40} is trapped compared to $\sim$100% for the solar wind Ar^{36} so that for equal content in the lunar soil, the Ar^{40} flux must be about twelve times the Ar^{36} flux. Using a scale height of 50 km in equation (12) gives a number of 1.5×10^3 cm^{-3}; we thus obtain an order of magnitude estimate for n_0 (Ar^{40}) of 10^2 to 10^3 cm^{-3}. The lower range of values is included since a more detailed calculation must treat the ion trajectories and trapping at the equator more exactly. The ions impacting at the equator travel obliquely through the atmosphere and the proper trapping efficiency to use is one calculated specifically for the equator; these considerations could reduce the number density in the atmosphere necessary to provide the required flux to the surface.

It should be noted that this approach to calculating n_0 (Ar^{40}) is essentially an indirect measurement utilizing lunar samples as the sensor. The approximate values obtained here agree roughly with previous estimates by BERNSTEIN *et al.* (1964) and HINTON and TAEUSCH (1964), who get about 10^3 to 10^4 cm^{-3}. The variation could be due to different assumptions in the calculations, could imply that the outgassing of Ar^{40} from the moon is lower than assumed in the calculations, or the rate of removal of ions from the atmosphere by the acceleration mechanism described here could be more rapid than the mechanisms assumed in some calculations.

INITIAL RELEASE GASES IN STEPWISE HEATING—A TEST FOR POSSIBLE ATMOSPHERIC TRAPPING OF OTHER ELEMENTS

As mentioned earlier, it is likely that noble gases implanted by the solar wind are also being cycled through the lunar atmosphere, with the possibility that some isotope ratios are being altered by gases from the lunar interior. There is also the possibility of some mass fractionation in the solar wind isotopes which are retrapped from the atmosphere. The initial isotope ratios observed in stepwise heating of samples typically shows an enhancement in the lighter isotope compared with the ratio observed at higher temperatures (PEPIN *et al.*, 1970; HOHENBERG *et al.*, 1970). NYQUIST and PEPIN 1971) indicated at the Second Lunar Science Conference that later data shows the same enhancement found for Apollo 11 samples. One possible explanation is an atmospheric effect (J. H. REYNOLDS, personal communication); we would also expect the possibility of mass fractionation on the basis of atmospheric trapping. The possible mass effect can be qualitatively seen from the model: Two isotopes of a given element should have nearly identical trapping efficiencies, but the lighter isotope has a slightly greater scale height and thus the ion travels further in the interplanetary electric field, impacts with greater energy, and is trapped with slightly greater efficiency. However, to calculate this increased trapping efficiency, the trapping probability must be integrated over the incident ion energy spectrum for each isotope

$$N_t = \int_0^\infty \eta_t(\mathscr{E})n(\mathscr{E})\,\mathrm{d}\mathscr{E} \qquad (13)$$

where $n(\mathscr{E})$ is determined from the exponentially decreasing number density in the atmosphere and the resulting trajectory along the interplanetary electric field into the moon.

The results are presented in Table 3. The experimental lunar fine values are the enhancement of the given isotopic ratio measured at initial release temperatures over the ratio measured at peak release temperatures. The values for krypton were read from the straight line shown in the data of PEPIN *et al.* (1970). The values for argon and neon were taken from the table of HOHENBERG *et al.* (1970); the ratio at initial release was calculated from the data for the two lowest temperatures by taking an average weighted by the quantity of gas released at each temperature. Similarly the high temperature ratio was calculated from a weighted average of the ratio for the two temperatures giving peak gas release of the trapped component. The enhancement expected from atmospheric mass fractionization was calculated by numerically integrating the trapping function of equation (8) over an approximate energy spectrum

Table 3. Enhancement of the isotopic ratios, for gases released at the initial temperature compared to the peak release temperature, measured from lunar samples in a stepwise heating process. The experimental enhancement is compared to the inverse mass ratio and to the enhancement predicted by atmospheric trapping theory (Krypton data from PEPIN *et al.* (1970); argon and neon data from HOHENBERG *et al.* (1970).

Ratio	Ratio Enhancement at Low Temperature: Lunar Fines	$\frac{m_2}{m_1}$	Atmospheric Trapping
Kr^{83}/Kr^{84}	1.015	1.012	1.012
Kr^{83}/Kr^{86}	1.036	1.036	1.034
Kr^{84}/Kr^{86}	1.02	1.024	1.022
Ar^{36}/Ar^{38}	1.053	1.056	1.050
Ne^{20}/Ne^{22}	1.099	1.100	1.086

for the ions. As mentioned previously, the trapping function probably has a finite cut-off energy which will affect the above integration; however the ratio of the integrals should not be greatly changed. An approximation in the calculation is the use of the same trapping function for Ne, Ar, and Kr. Also, only the first ionization go-around is calculated; of the ions formed in the sunlit atmosphere, about 60% escape the moon without impacting and of the remaining 40% which hit the moon, part are trapped and the remainder are released into the atmosphere where they can be reaccelerated, etc., thus possibly adding an additional contribution to the fractionation. Finally we assume that the isotopic ratio of the solar wind gases initially released into the atmosphere is just the ratio observed at peak gas release temperatures. It was pointed out by Hohenberg *et al.* (1970) that the enhancement is proportional to the inverse mass ratio of the isotopes, which is included for comparison.

Inspection of the data of Hohenberg *et al.* (1970) for the amount of gas in the initial release compared to total gas content of the element shows a decrease with increasing element mass which is consistent with the idea of the initial release gas coming from solar wind which has diffused out of the surface and been weakly re-implanted. However the same trend would also appear if the whole effect were just a preferential diffusion of light isotopes out when the sample is heated in the laboratory, though it is not clear that the isotopic ratios would then vary linearly with the inverse mass. A question arises because of the tightly bound Ar^{40}, which is presumed to be atmospheric and yet has the same peak release temperatures as does Ar^{36}. Another complicated phenomenon involves the amount of He^3 in the initial releases which could be somewhat large for atmospheric trapping since helium in the lunar atmosphere is not entirely gravitationally bound.

However the obvious feature of the results in Table 3 is the close agreement with experimental observations. It is clear that solar wind gases will be leaving the lunar surface and once in the atmosphere they can be retrapped and the trapping mechanism can cause some mass fractionation. The question is one of amount—can atmospheric trapping provide the amounts of gas observed for the various elements, and does the trapping effect dominate over the many complicated surface effects.

Conclusions

A number of general properties of the lunar atmosphere have been calculated. It is possible that $\bar{E} \times \bar{B}$ acceleration of ions, with help from the surface electric and magnetic fields, can explain the observations of ALSEP ion detector experiments. An order of magnitude value for the neutral Ar^{40} density in the lunar atmosphere has been obtained and in a sense, analysis of lunar samples constitutes an indirect measurement of this species. More detailed calculations should make it possible to define this component of the lunar atmosphere and other components as well. Clearly a direct measurement of the atmospheric ion fluxes of certain critical isotopes would greatly enhance our understanding of the complete system.

Acknowledgments—We thank Dr. J. H. Reynolds for bringing to our attention the mass dependence seen in the enhanced isotopic ratios of initial release gases; we also thank Drs. J. Meister and J. Geiss for discussions and for much helpful information on the trapping of ions. We appreciate helpful comments by Drs. D. Heymann, A. Yaniv, and a number of other lunar sample and ALSEP Investigators.

This work was supported in part by the National Aeronautics and Space Administration under grant NAS 9–5911.

REFERENCES

BERNSTEIN W., FREDRICKS R. W., VOGL J. L., and FOWLER W. A. (1963) The lunar atmosphere and the solar wind. *Icarus* **2**, 233–248.

BÜHLER F., GEISS J., MEISTER J., EBERHARDT P., HUNECKE J. C., and SIGNER P. (1966) Trapping probability of the solar wind in solids, Part I: Trapping probability of low energy He, Ne, and Ar ions. *Earth Planet. Sci. Lett.* **1**, 249–255.

BURLAGA L. F., and NESS N. F. (1968) Macro- and micro-structure of the interplanetary magnetic field. *Can. J. Phys.* **46**, S962.

CAMERON A. G. W. (1968) A new table of abundances of the elements in the solar system. In *Origin and Distribution of the Elements* (editor L. H. Ahrens), pp. 125–143. Pergamon.

EBERHARDT P., GEISS J., GRAF H., GRÖGLER N., KRÄHENBÜHL U., SCHWALLER H., SCHWARZMÜLLER J., and STETTLER A. (1970) Trapped solar wind noble gases, exposure age and K/Ar-age in Apollo 11 lunar fine material. *Proc. Apollo 11 Lunar Sci. Conf., Geochim. Cosmochim. Acta* Suppl. 1, Vol. 2, pp. 1037–1070. Pergamon.

FREEMAN J. W., HILLS H. K., and FENNER M. A. (1971) Some results from the Apollo 12 suprathermal ion detector. Second Lunar Science Conference (unpublished proceedings).

GEISS J., EBERHARDT P., BÜHLER F., and MEISTER J. (1970) Apollo 11 and Apollo 12 solar wind composition experiments: Fluxes of He and Ne isotopes. *J. Geophys. Res.* **75**, 5972–5979.

HEYMANN D., YANIV A., ADAMS J. A. S., and FRYER G. E. (1970) Inert gases in lunar samples. *Science* **167**, 555–558.

HINTON F. L., and TAEUSCH D. R. (1964) Variation of the lunar atmosphere with the strength of the solar wind. *J. Geophys. Res.* **69**, 1341–1347.

HOHENBERG C. M., DAVIS P. K., KAISER W. A., LEWIS R. S., and REYNOLDS J. H. (1970) Trapped and cosmogenic rare gases from stepwise heating of Apollo 11 samples. *Proc. Apollo 11 Lunar Sci. Conf., Geochim. Cosmochim. Acta* Suppl. 1, Vol. 2, pp. 1283–1309. Pergamon.

KORNELSEN E. V. (1964) The ionic entrapment and thermal desorption of inert gases in tungsten for kinetic energies of 40 eV to 5 keV. *Can. J. Phys.* **42**, 364–381.

MANKA R. H., and MICHEL F. C. (1970) Lunar atmosphere as a source of argon-40 and other lunar surface elements. *Science* **169**, 278–280.

MEISTER J. (1969) Ein Experiment zur Bertimmung der Zusammenstezung und Isotopenverhaeltnisse des Sonnenwindes: Einfangverhalten von Aluminium für niederenergetische Edelgasionen. Ph.D. thesis, University of Berne.

MICHEL F. C. (1964) Interaction between the solar wind and the lunar atmosphere. *Planet. Space Sci.* **12**, 1075–1091.

NYQUIST L. E., and PEPIN R. O. (1971) Rare gases in Apollo 12 surface and subsurface fine materials. Second Lunar Science Conference (unpublished proceedings).

PEPIN R. O., NYQUIST L. E., PHINNEY D., and BLACK D. C. (1970) Isotopic composition of rare gases in lunar samples. *Science* **167**, 550–553.

REED G. W., JR., and JOVANOVIC S. (1970) Halogens, mercury, lithium, and osmium in Apollo 11 samples. *Proc. Apollo 11 Lunar Sci. Conf., Geochim. Cosmochim. Acta* Suppl. 1, Vol. 2, pp. 1487–1492. Pergamon.

REYNOLDS J. H. (1971) Personal communication.

ROBBINS D. E., HUNDHAUSEN A. J., and BAME S. J. (1970) Helium in the solar wind. *J. Geophys. Res.* **75**, 1178–1187.

SNYDER C. W., CLAY D. R., and NEUGEBAUER M. (1971) An impact-generated plasma cloud on the moon. Second Lunar Science Conference (unpublished proceedings).

Proceedings of the Second Lunar Science Conference, Volume 2, pp. 1729–1745.
The M.I.T. Press, 1971.

Calculation of cosmogenic radionuclides in the Moon and comparison with Apollo measurements

T. W. ARMSTRONG and R. G. ALSMILLER, JR.
Oak Ridge National Laboratory, Oak Ridge, Tennessee 37830

(*Received 19 February 1971; accepted in revised form 24 March 1971*)

Abstract—Calculations have been carried out to determine the time and spatial dependence of the production of various radionuclides in the moon from solar and galactic proton bombardment. The calculational method utilizes Monte Carlo techniques to obtain a detailed description of the induced nucleon-meson cascade. All required spallation cross sections are computed using the intranuclear-cascade-evaporation model. The calculated depth dependence of ^{26}Al and ^{22}Na is in good agreement with Apollo 11 and Apollo 12 measurements. The depth-dependent neutron spectra are also directly available from the calculations. The calculated thermal neutron flux is in good agreement with the flux obtained from measured ^{60}Co concentrations. Irradiation ages obtained from the calculated thermal flux and the thermal fluence based on measured gadolinium concentrations are consistent with the ages determined by other methods.

INTRODUCTION

CALCULATIONS HAVE been carried out to determine the time and spatial dependence of the production of various radionuclides in the moon from solar and galactic cosmic-ray bombardment. The primary purposes of the calculations are to obtain specific results with which direct comparisons with the Apollo measurements can be made in order to evaluate the applicability of the calculational method for present purposes and to obtain additional results that are not available experimentally which will aid in the interpretation of some of the Apollo measurements.

A description of the calculational method is given in the next section. The method of calculation used here takes into account the details of the physical processes involved in the development of the nucleon-meson cascade induced by cosmic-ray bombardment and differs markedly from previous approaches (ARNOLD *et al.*, 1961; HONDA, 1962; KOHMAN and BENDER, 1967). Monte Carlo methods are utilized to obtain a detailed description of the nucleon-meson cascade and to compute, via the intranuclear-cascade-evaporation model for energies $\gtrsim 3$ GeV (BERTINI, 1969; GUTHRIE, 1970) and an extrapolation method based on the intranuclear-cascade model at higher energies (GABRIEL *et al.*, 1970), all required nucleon-nucleus cross sections (above 15 MeV) and pion-nucleus cross sections. Thus, a unique feature of the calculational method is that cross sections for various spallation products, which are often not available experimentally, are not required as input since they are computed as needed using the intranuclear-cascade-evaporation model. Another advantage of the calculational method is that a complete description of the cascade is obtained so that results of interest other than radionuclide production, such as the depth dependence of the neutron spectrum, are directly available. Results obtained using this method of calculation for estimating the development of the nucleon-meson cascade and the production of residual nuclei in thick targets have been compared with experimental data for several simple source-geometry configurations (for example, ARMSTRONG and ALSMILLER, 1968; ARMSTRONG, 1969; ARMSTRONG and BISHOP, 1970), and, in general, good agreement has been found.

Results for the calculated induced activity and neutron spectra are given in the third section. The calculated depth distribution for ^{26}Al and ^{22}Na are compared with Apollo 11 and Apollo 12 measurements. The calculated thermal neutron flux and the thermal neutron fluence based upon observed gadolinium concentrations in various Apollo 11 samples are used to obtain a cosmic-ray exposure age for the samples, and these ages are compared with the ages determined by other methods. The calculated thermal neutron flux is also compared with the flux deduced from measurements of the ^{60}Co concentration in an Apollo 12 rock.

A discussion of the results and calculational method is given in the last section.

DESCRIPTION OF CALCULATION

Calculational method

The nucleon-meson cascade and residual nuclei production induced by the galactic and solar protons have been computed by utilizing Monte Carlo techniques in conjunction with the intranuclear-cascade-evaporation model for treating nonelastic nuclear interactions. The Monte Carlo simulation of the cascade involves selecting primary proton energies from either a galactic or solar proton spectrum and computing the trajectories of the primary protons and all secondary and higher order particles produced in nuclear collisions. The energy loss of charged particles between nuclear collisions due to the excitation and ionization of atomic electrons is treated using the continuous slowing-down approximation, with an approximate expression for the density-effect correction (ARMSTRONG and ALSMILLER, 1970). At each nonelastic nuclear collision, a calculation based on the intranuclear-cascade-evaporation model is performed to determine the particle types, multiplicity, energy, and direction of the collision products and the charge and mass of the residual nucleus. The particles produced in the collision may be protons, neutrons, charged pions, and neutral pions from the intranuclear cascade, and protons, neutrons, deuterons, tritons, ^{3}He's, and alpha particles from the evaporation. The neutral pions are assumed to decay immediately into two photons. These photons are not important for present purposes and are neglected. The deuterons, tritons, ^{3}He's, and alpha particles from nuclear interactions are assumed to slow down and come to rest at their point of origin since for the low energies involved the range of these particles is very short and the probability of undergoing nuclear interaction before coming to rest is very small. Each particle (nucleon and charged pion) in the cascade is followed until it eventually disappears by escaping from the lunar surface, undergoing nuclear absorption, coming to rest due to energy losses from ionization and excitation of atomic electrons, or, in the case of pions, decaying. Positively charged pions that come to rest are assumed to decay. Negatively charged pions that come to rest are assumed to undergo nuclear capture, and the capture products are included in the cascade. It has previously been shown that the intranuclear-cascade model describes the π^--capture process quite well (GUTHRIE *et al.*, 1968). An energy cutoff of 15 MeV is used for neutrons and protons. Protons that slow down to 15 MeV (or are produced below 15 MeV) are assumed to come to rest without undergoing nuclear interaction. The treatment of neutrons below 15 MeV is discussed later. A complete description of each "event" (nuclear interaction, stopped charged particle, etc.) that occurs during the transport calculation is stored on magnetic tape. These cascade-history tapes are then analyzed to obtain the results of interest. Thus, a single-transport calculation provides not only the depth dependence of the residual-nuclei production but also much useful ancillary information.

The calculations are carried out using the intranuclear-cascade program of BERTINI (1969) and the evaporation program of GUTHRIE (1970). The present version of Bertini's intranuclear-cascade program is limited to pion energies $\gtrsim$2.5 GeV and nucleon energies $\gtrsim$3 GeV. This upper-energy limitation is not restrictive in computing the cascade induced by solar protons but is not nearly high enough for treating the cascade induced by galactic protons. For the galactic case, an extrapolation method (GABRIEL *et al.*, 1970) was used to obtain the description of nonelastic-collision products induced by nucleons > 3 GeV and pions > 2.5 GeV. This extrapolation method uses the results from 3-GeV nucleon-nucleus collisions and 2.5-GeV pion-nucleus collisions as predicted by the intranuclear-cascade-evaporation model, together with energy, angle, and multiplicity scaling relations which are consistent with the sparse experimental data available for high-energy interactions, to estimate the particle production at the higher energies. It should be noted that the basic input data required for this method of treating nonelastic collisions are differential particle-particle cross sections $\gtrsim$3 GeV, which are rather well known. Particle-nucleus cross sections are then computed using the model. Thus, cross sections for various spallation products are not required as input since they are computed in the course of the transport calculation.

The calculational method discussed here has been used previously to estimate the development of the nucleon-meson cascade and the production of residual nuclei in thick targets for several simple source-geometry configurations for which experimental data exist (ARMSTRONG and ALSMILLER, 1968; ARMSTRONG, 1969; ARMSTRONG *et al.*, 1971). In general, good agreement between the calculations and measurements has been found.

A calculational method different from that described above was used for the low-energy neutrons. The transport of neutrons below 15 MeV was carried out using the discrete ordinates method (ENGLE, 1967) since for the problem of interest discrete ordinates is somewhat more efficient than Monte Carlo in this energy region. The required neutron cross sections below 15 MeV were taken from an experimental neutron cross-section library (JOANOU and DUDEK, 1963).

Lunar composition

The calculations have been made for the lunar composition given in Table 1. This composition is taken from Apollo 11 measurements for type D fine material (LSPET, 1969). The small amount of "other" elements indicated in Table 1 were not included in the calculations, with the exception that for one calculation Gd, Sm, and Dy were included to determine the effects of rare earths on the low-energy neutron flux. The following nominal concentrations based on Apollo 11 measurements (GAST and HUBBARD, 1970; HASKIN *et al.*, 1970; PHILPOTTS and SCHNETZLER, 1970; MORRISON *et al.*, 1970) were used for the rare earths: Gd, 17 ppm; Sm, 14 ppm; and Dy, 20 ppm.

A density of 1.5 g/cm^3, which corresponds to the density of Apollo 11 fine material, was used in all of the calculations. However, the results are essentially independent of density if distances are expressed in g/cm^2. Since the density enters the calculation only in determining the probability of charged-pion decay per unit distance, the results of interest here are not sensitive to the relatively small ($\sim$a factor of two) density variations found for lunar material.

Table 1. Lunar composition used.

Element	Weight %	Atom %
O	42.07	60.47
Si	20.00	16.37
Al	6.90	5.88
Fe	12.40	5.11
Ca	8.60	4.94
Mg	4.80	4.54
Ti	4.20	2.02
Na	0.40	0.40
Mn	0.18	0.07
K	0.10	0.06
"other"	0.35	0.14
	100.00	100.00

Proton source data

The calculations have been carried out for an isotropic flux of galactic or solar protons incident on a half-space of lunar material. Both solar minimum and solar maximum galactic proton spectra have been considered. The solar minimum spectrum was taken from the review by McDONALD (1969). This spectrum is based on several sets of balloon and satellite measurements made near 1965 and is rather well defined. The solar maximum spectrum is not as well known at present. For the calculations here, the 1959 spectrum predicted by the solar wind modulation theory of DURGAPRASAD *et al.* (1967) was used. The galactic proton spectra used are given elsewhere (ARMSTRONG and ALSMILLER, 1971a).

The solar proton-source data used are shown in Fig. 1. In this figure the proton intensity is given in terms of the omnidirectional proton flux above 30 MeV and represents the integral of the intensity over the life of the flare. The flare spectra have been divided into three classifications according to their characteristic rigidity, P_0. The data for solar cycle 19 (which here is taken to be January 1954 through December 1964) were taken from the work of MODISETTE *et al.* (1965) and are based primarily on the evaluation made by WEBBER (1963). The data from the beginning of solar cycle 20 through November 1969 were taken from various sources (HSIEH, 1970; YUCKER, 1970; RADKE, 1969; BAKER, 1969; ESSA, 1970). The data of HSIEH (1970) were used for the three flares that took place immediately before the Apollo 11 landing. The intensity for the November 3, 1969, flare, which took place immediately before the Apollo 12 flight, was taken from Explorer 41 measurements (ESSA, 1970). In many cases, no spectral information was available for flares occurring during solar cycle 20. In these cases, the rigidity was arbitrarily placed in the $75 < P_0 < 125$ MV interval.

To take into account solar proton bombardment prior to solar cycle 19, the flare data for solar cycle 19 were used for all previous solar cycles.

Time dependence of radionuclide production from solar flares

It is not practical to take into account explicitly the energy dependence of the incident protons for each individual flare because of the large number of flares and because the energy dependence

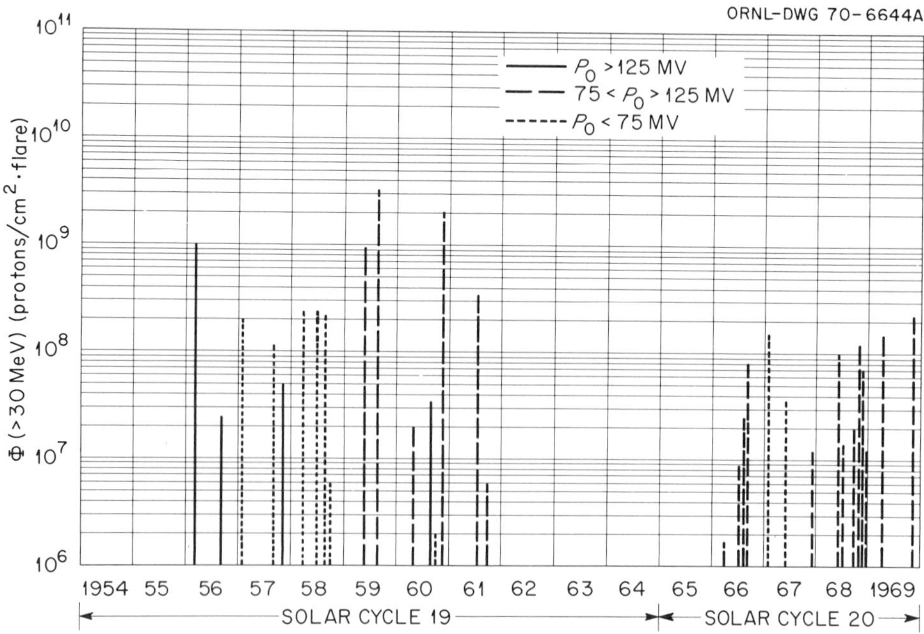

Fig. 1. Solar proton intensity during solar cycle 19 and the first part of solar cycle 20.

is not well known for most flares. Therefore, an approximate procedure was used in which transport calculations were carried out only for flares having characteristic rigidities of 50, 100, and 200 MV and unit omnidirectional fluxes above 30 MeV. The results from the transport calculations for $P_0 = 50$, 100, and 200 MV were used to represent the spectra of all flares having rigidities in the intervals $P_0 < 75$, $75 < P_0 < 125$, and $P_0 > 125$ MV, respectively. These results were then combined with the flare intensity and frequency data given in Fig. 1 to obtain the depth-dependent radionuclide production from flare bombardment as a function of time. The details of this procedure are given elsewhere (ARMSTRONG and ALSMILLER, 1971a).

Restrictions on the calculations

Several simplifying assumptions have been made in carrying out the calculations. Only the incident protons in the solar and galactic spectra have been considered. The isotopic changes due to nuclear interactions and stopping by incident multiply charged nuclei are confined to depths very near the surface (LAL *et al.*, 1967). The relative importance of multiply charged nuclei to the production of a particular isotope depends strongly on the isotope in question (LAL and VENKATAVARADAN, 1967). For the isotopes calculated here for which comparisons with Apollo measurements are made, the contribution of incident multiply charged nuclei is not expected to be large.

Incident protons below 30 MeV have been neglected. Because of this and the omission of nuclear interactions by protons below 15 MeV in the transport calculation, the production of isotopes with low proton-production thresholds, e.g., $^{56}Fe(p, n)^{56}Co$, is not completely accounted for.

There is one feature of the calculational method which can be a limitation in some instances. Because of the stochastic nature of the calculation, it is difficult to obtain statistically meaningful estimates of those residual nuclei which have very low relative production rates. In many cases this difficulty can be circumvented. For example, if thin-target production cross sections are available, then these cross sections can be used with the calculated depth and energy-dependent flux spectra to obtain the thick-target production. However, for this initial study this alternate procedure was not used.

RESULTS

Induced activity

To indicate the relative production rate of various residual nuclei, Fig. 2 shows the calculated production rate integrated over all depths due to galactic proton bombardment at solar minimum. The curves have been drawn to connect those nuclei having the same charge number. The error bars for the calculated results in this and subsequent figures correspond to statistical errors of one standard deviation. The production shown in Fig. 2 has been calculated from nuclear interactions induced by nucleons above 15 MeV and charged pions of all energies. The production from neutrons below 15 MeV, which is due primarily to thermal-neutron capture, can be obtained straightforwardly from the calculated neutron spectra and available neutron cross sections. The production shown for ^{2}H, ^{3}H, ^{3}He, and ^{4}He is due entirely to nuclear evaporation and represents a lower limit for the actual production since the production contributed by the stopping of incident multiply charged nuclei has been neglected. The isotopic production curves in Fig. 2 are approximately Gaussian in shape. This feature is in agreement with the work of Rudstam on the systematics of spallation yields in thin targets (RUDSTAM, 1966).

The depth dependence of the ^{26}Al production is now considered. Figure 3 shows the production from solar protons having energy spectra corresponding to characteristic rigidities of $P_0 = 50$, 100, and 200 MV and unit intensities. By combining these results

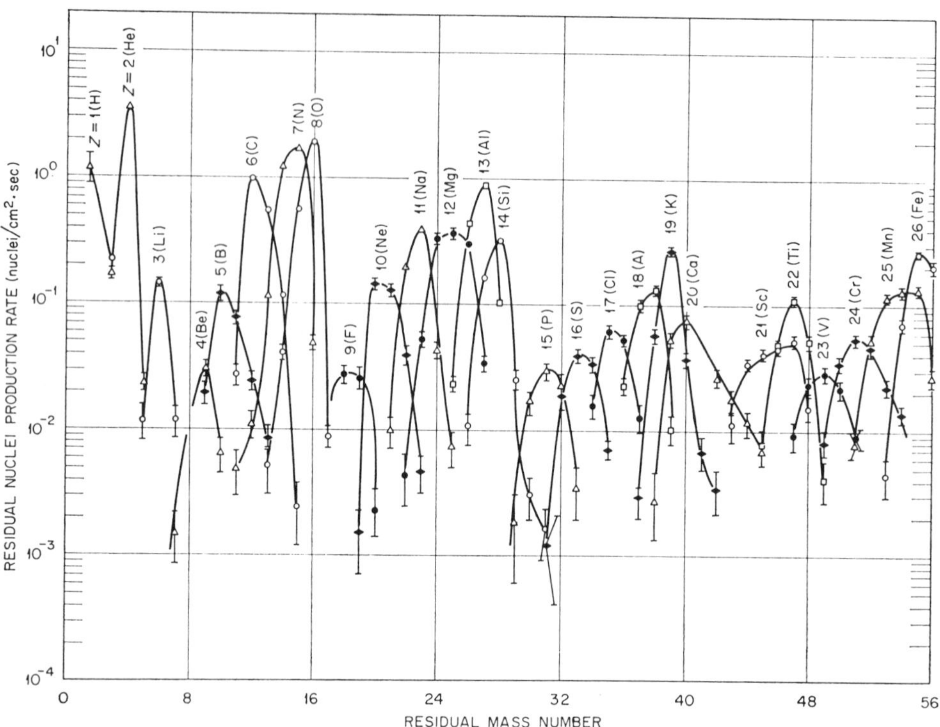

Fig. 2. Residual-nuclei production from galactic proton bombardment at solar minimum.

with the flare data for solar cycle 19 shown in Fig. 1 and using the flare data for solar cycle 19 for all previous solar cycles, the ^{26}Al activity due to solar protons is obtained. This activity from solar cosmic rays (SCR) is given in Fig. 4(a), together with the activity from galactic cosmic rays (GCR) at solar maximum and solar minimum. In Fig. 4(b) the total activity, i.e., the sum of the solar contributions and the average of the galactic contributions at solar minimum and solar maximum, is compared with the activity measured by SHEDLOVSKY *et al.* (1970) for rock 10017 and the data of FINKEL *et al.* (1971) for rock 12002. This comparison between the calculated and experimental data is absolute in the sense that there are no free parameters in the theory. The ^{26}Al activity for depths $\gtrsim 10$ g/cm^2 is due primarily to solar protons whereas the activity at greater depths is primarily due to galactic protons. Therefore, the good agreement between the calculated and experimental results in Fig. 4(b) confirms the validity of the calculated results for both the solar and galactic protons. It is significant that the good agreement between measured and calculated ^{26}Al activities has been obtained using recent estimates of the solar and galactic proton fluxes. This further supports the findings of others (for example, ARNOLD *et al.*, 1961; LAL and VENKATAVARADAN, 1966; FINKEL *et al.*, 1971) that the average solar and galactic proton intensities over a past time interval comparable to the halflife of ^{26}Al ($\sim 10^6$ y.) have been approximately the same as observed recently.

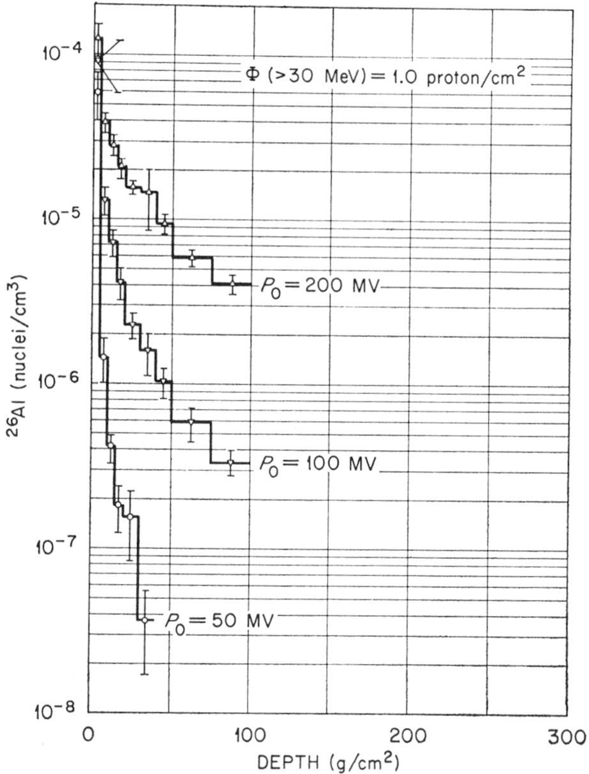

Fig. 3. Production of ^{26}Al vs. depth from individual solar flares having characteristic rigidities of $P_0 = 50$, 100, and 200 MV and unit omnidirectional proton fluxes above 30 MeV.

The depth dependence of the ^{22}Na activity is shown in Fig. 5(a). The procedure used to obtain the activity from solar protons is the same as that discussed above for ^{26}Al. However, whereas the ^{26}Al activity is constant over a time interval of one solar cycle because of the long ^{26}Al halflife (7.4×10^5 y.), the activity due to ^{22}Na (halflife = 2.62 y.) varies widely over a solar cycle and the time dependence following the production from individual flares must be considered. To illustrate, Fig. 6 shows the time dependence of the ^{22}Na activity from solar protons in the 5- to 10-g/cm² depth interval from the beginning of solar cycle 19 to the time of the Apollo 12 flight. Figure 5(b) shows a comparison of the depth dependence of the total (SCR plus average GCR) calculated ^{22}Na activity and the activity measured by SHEDLOVSKY *et al.* (1970) for Apollo 11 rock 10017. Taking into account the November 3, 1969, flare, the calculated depth distribution at the time of the Apollo 12 flight is very similar in shape and $\sim$15% higher than the distribution shown in Fig. 5(b) for Apollo 11. This is not in agreement with the measured distribution of FINKEL *et al.* (1971) for rock 12002, which has a steeper gradient and is $\sim$50% higher near the surface than the measured distribution for rock 10017. This discrepancy may be due to the fact that incident protons below

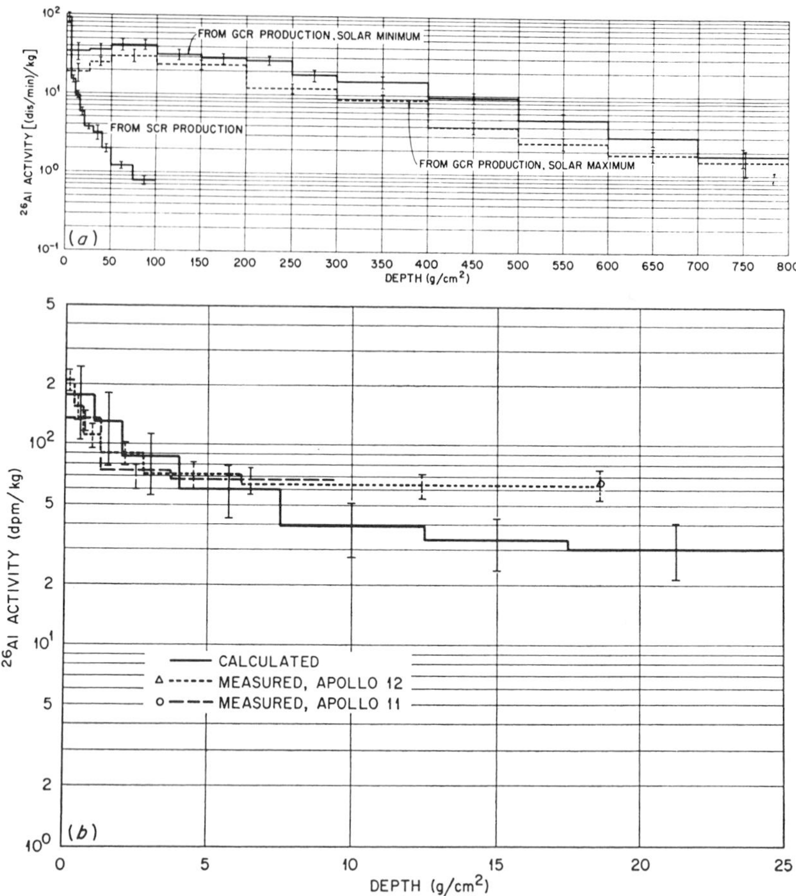

Fig. 4(a). ^{26}Al activity vs. depth for solar and galactic proton bombardment; (b). Comparison of calculated and measured depth dependence for ^{26}Al. The Apollo 11 data are from SHEDLOVSKY *et al.* (1970) for rock 10017; the Apollo 12 data are from FINKEL *et al.* (1971) for rock 12002.

30 MeV were neglected in the calculations. Since the November 3, 1969, flare was relatively rich in low-energy protons ($P_0 \sim 70$ MV), it is possible for the incident protons below 30 MeV to contribute appreciably to the ^{22}Na activity at small depths at the time of the Apollo 12 flight.

Some additional comparisons between calculated and measured activities are given in Table 2.

Neutron flux

Information on the depth dependence of the particle spectra is also available from the cascade calculations. The induced neutron flux from GCR bombardment will be discussed here; other particle flux results obtained from both GCR and SCR bombardment are given elsewhere (ARMSTRONG and ALSMILLER, 1971a).

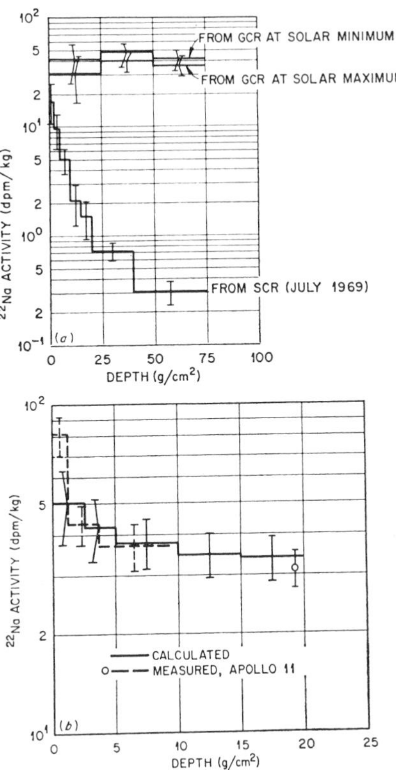

Fig. 5(a). Depth dependence of ^{22}Na activity from solar proton bombardment at the time of the Apollo 11 flight and from galactic proton bombardment; (b). Comparison of calculated and measured depth distribution for ^{22}Na activity. The Apollo 11 data are from SHEDLOVSKY *et al.* (1970).

The neutron leakage spectrum is of some interest since it has been proposed that this spectrum be measured by a lunar orbiting satellite to obtain information on the average lunar composition (LINGENFELTER *et al.*, 1961). The neutron leakage flux spectrum is shown in Fig. 7 for GCR bombardment at solar minimum and solar maximum without rare earths in the lunar composition, and for GCR bombardment at solar minimum with rare earths included. Also shown for comparison is the spectrum from an approximate calculation by LINGENFELTER *et al.* (1961) using a basaltic-type composition. The two calculations are in excellent agreement except in the thermal-energy region. This difference is most likely due to the relatively large amount of hydrogen contained in the composition used by LINGENFELTER *et al.* The dips in the spectra near 0.5 and 1.0 MeV are due to resonances in the cross section for oxygen. The statistical error for that portion of the spectrum calculated by the discrete ordinates method (<15 MeV) is not shown in Fig. 7. This portion of the spectra is, of course, subject to statistical fluctuation since the neutron source for the discrete ordinates calculation is obtained by Monte Carlo. The statistical error below 15 MeV has not been calculated but is estimated to be about 20% or less.

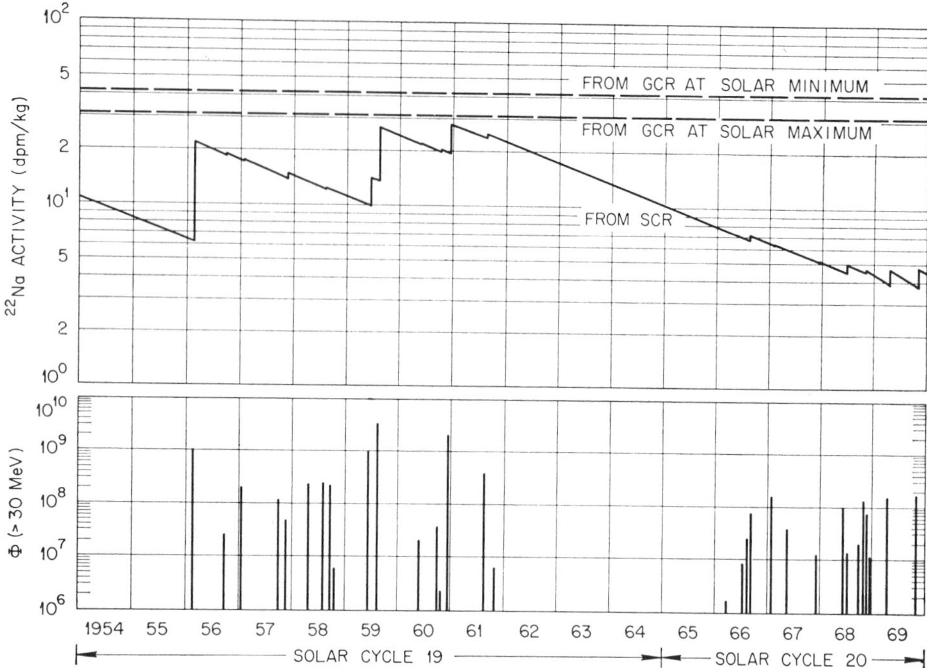

Fig. 6. ^{22}Na activity in the depth interval 5–10 g/cm^2 from solar and galactic proton bombardment. The vertical lines on the abscissa indicate the times at which flares occurred, and the magnitude of the lines correspond to the omnidirectional proton flux per flare above 30 MeV.

Table 2. Comparison of calculated and measured induced activity in Apollo 11 rock 10017 (depth: 12–30 mm)

Nuclide	Activity (dpm/kg) Calculated	Measured*
^{26}Al	55 ± 11	57 ± 9
^{22}Na	39 ± 10	37 ± 6
^{54}Mn	5 ± 2	10 ± 11
^{49}V	3 ± 2	7 ± 2†
^{36}Cl	16 ± 4	16 ± 2

* From Shedlovsky *et al.* (1970).
† Estimated by Shedlovsky *et al.* (1970) from bulk fines.

The neutron flux in various energy groups versus depth for GCR bombardment at solar minimum with rare earths included in the composition is shown in Fig. 8. This spectral information can be combined with known neutron cross sections to estimate the residual-nuclei production below 15 MeV which was omitted in Fig. 2. Figure 9 shows the depth dependence of the thermal neutron flux induced by GCR bombardment. The thermal flux here includes all neutrons with energies below 0.4 eV.

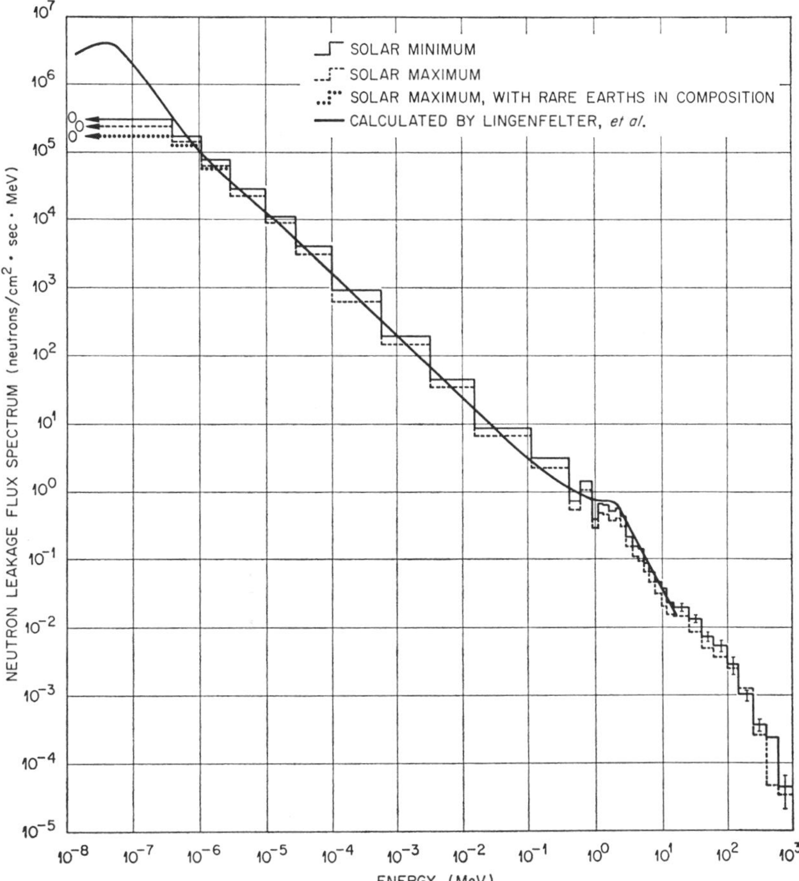

Fig. 7. Neutron albedo from galactic proton bombardment. Also shown for comparison is the albedo calculated by LINGENFELTER *et al.* (1961) for a basaltic composition.

The inclusion of the rare earths in the composition, using the nominal rare earth concentration given earlier, reduces the thermal flux by $\sim 30\%$ at all depths.

From measurements of the time-integrated thermal flux (i.e., fluence) based on Gd concentrations and the calculated thermal fluxes, an estimate of the cosmic-ray exposure age can be made. Table 3 shows the exposure age range obtained by using the thermal flux at the surface and the maximum thermal flux. An average of the solar minimum and solar maximum fluxes (with rare earths present in the composition), shown in Fig. 9, was used in computing the exposure age. As shown in Table 3, the exposure ages computed in this manner are consistent with the ages obtained by other methods. More specific irradiation ages, as opposed to the irradiation age ranges given in Table 3, cannot be deduced from the calculated results because the average irradiation depth of the measured samples is not known. However, the calculated

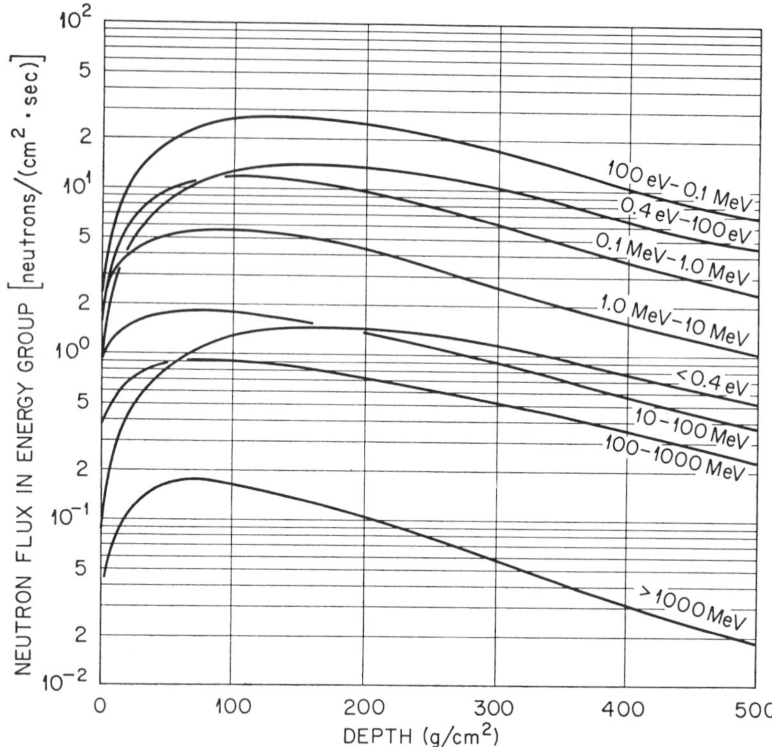

Fig. 8. Depth dependence of neutron flux in various energy groups induced by galactic protons at solar minimum with rare earths included in the composition.

Table 3. Exposure age of Apollo 11 samples based on ^{158}Gd concentrations.

Sample	Measured thermal fluence* (10^{15} n/cm^2)	Calculated exposure age (10^6 y.)	Exposure age from other methods (10^6 y.)
Soil	9.0 ± 0.5	200–3500	700†
Rock 17-32	17.4 ± 0.6	390–6900	510‡
Rock 50-24	8.7 ± 0.4	200–3500	410§
Rock 57-39	0.94 ± 0.23	21–372	47‡
Rock 71-118	4.3 ± 0.4	100–1700	340‡

* From EUGSTER *et al.* (1970).
† Based on ^{126}Xe concentration (ALBEE *et al.*, 1970).
‡ Based on ^{81}Kr–Kr dating (MARTI *et al.*, 1970; FUNKHOUSER *et al.*, 1970).
§ Based on ^{3}He concentration (EBERHARDT *et al.*, 1970).

results can be used in an alternate manner to estimate the average irradiation depth. The measured thermal fluence and measured exposure age yield an average thermal flux. The average irradiation depth which would yield this flux can be obtained from Fig. 9. Thus, the calculated depth dependence of the thermal (and higher energy) flux can be used in conjunction with measured residual-nuclei production to study the irradiation history of individual rocks and to develop soil-mixing models (EUGSTER *et al.*, 1970).

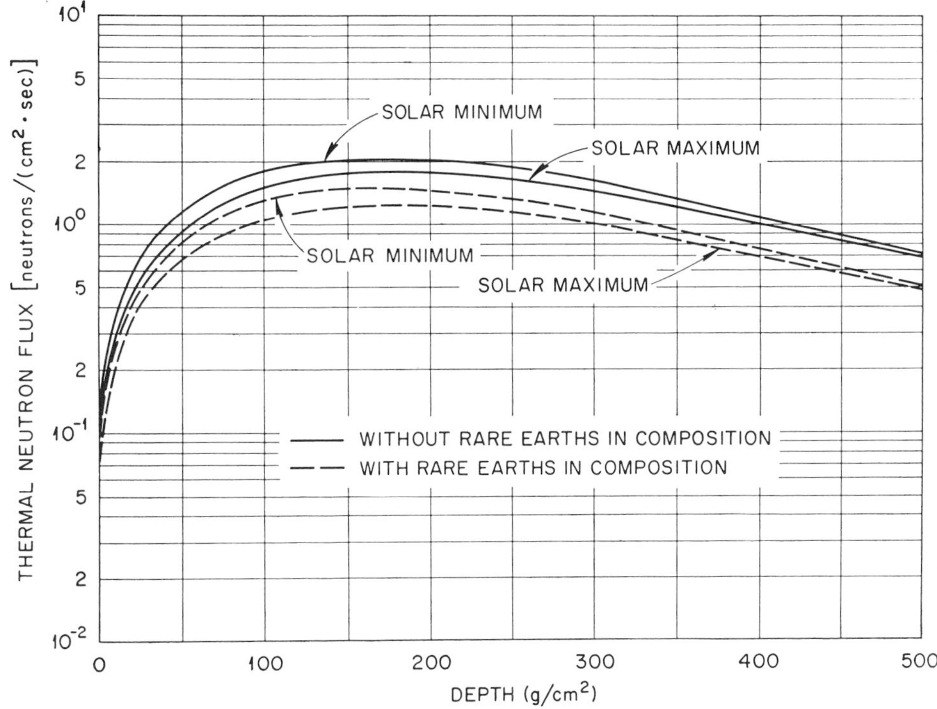

Fig. 9. Depth dependence of thermal-neutron flux induced by galactic proton bombardment.

The calculated thermal flux can also be compared with the thermal flux derived from measured ^{60}Co concentrations since there is evidence that the principal source of ^{60}Co is from neutron capture by ^{60}Co (O'KELLEY *et al.*, 1970). The ^{60}Co derived flux allows a more definite comparison with the calculated flux than does the Gd derived flux in the sense that the question of irradiation history is essentially eliminated in the ^{60}Co case. Since ^{60}Co has a halflife of only 5.3 years, the position of the sample on the lunar surface during the relevant irradiation period can be taken to be the same as when the sample was found.

Detailed depth-dependent ^{60}Co concentrations from Apollo samples are not available for comparison at present. However, the results of O'KELLEY *et al.* (1971) yield a thermal flux of 0.33 $\pm$ 0.17 neutrons/cm^2·sec based on the whole-rock ^{60}Co concentration for rock 12002. This rock is roughly 20 g/cm^2 thick. The average thermal flux for the first 20 g/cm^2 from Fig. 11 is 0.33 neutrons/cm^2·sec for the case of no rare earths in the composition and 0.23 neutrons/cm^2·sec for the case where nominal rare-earth concentrations based on Apollo 11 measurements are included. These calculated values represent an average of the solar minimum and solar maximum fluxes and have an estimated statistical standard deviation of $\sim$25%. More definitive comparisons between calculated and measured thermal fluxes will be possible as additional data on ^{60}Co and rare-earth concentrations become available.

DISCUSSION

The purpose of this work has been to investigate the applicability of a calculational approach based on Monte Carlo methods, which was developed primarily to aid in the shield design for high-energy particle accelerators, to the problem of predicting the induced activity and other quantities of interest due to cosmic-ray bombardment of the moon. Although exhaustive comparisons between calculated results and Apollo measurements have not been made, the good agreement for the few comparisons that have been made indicate the general validity of the calculated results. This good agreement provides some confidence in applying the calculational methods to obtain other results to aid in the interpretation of the Apollo measurements.

The advantages of the calculational method used here include the following: (1) essentially all details of the nucleon-meson cascade are available from a single-transport calculation; (2) all required nucleon-nucleus and pion-nucleus cross sections ($\gtrsim$15 MeV) are computed from rather well known particle-particle cross sections, so that the cross sections for various spallation products, which often are not available experimentally, are not required as input; and (3) the method is extremely general in terms of geometry, composition, and incident spectra which can be specified.

On the other hand, considerable computation time is required for this method of calculation. For example, the transport and analysis calculations for the case of incident galactic protons required approximately one hour on the IBM 360/91 computer. Therefore, the calculational method used here should be considered as complementary to, rather than a replacement for, the more conventional semiempirical approaches, such as that of ARNOLD *et al.* (1961), which are less general but require less computation time.

For this initial study, we have not attempted to incorporate in the calculations many of the refinements which could have been included. For example, all of the calculated results have been obtained for an isotropic proton flux incident on a half-space of lunar material (i.e., 2π incident flux), whereas the flux incident on the rocks for which the calculated induced activities are compared is over a solid angle somewhat greater than 2π. The computer code used for the calculations is sufficiently general that the exact geometry of the rocks can be taken into account. Also, all of the calculations have been carried out for the same lunar composition, and the effect of the different rock composition on the induced-activity results was not considered.

It is evident from the calculated results for the induced activity that considerable improvement of the statistical precision is desirable. In most cases, the statistics can be improved substantially by computing, separate from the transport calculation, the spallation cross sections for each of the target elements using the intranuclear-cascade model, and then folding these cross sections into the flux spectra from the transport calculation to obtain the induced activity. This alternate method is especially effective in improving the statistics for those cases where the target isotope is present in small amounts.

The calculations described here also yield the photon-source distribution needed for computing the photon albedo from the moon. This photon albedo has been calculated and is the subject of a separate paper (ARMSTRONG and ALSMILLER, 1971b).

Acknowledgments—We are very grateful to W. W. Engle, Jr., and coworkers of the Oak Ridge National Laboratory for carrying out the low-energy neutron transport calculations. The programming assistance provided by H. S. Moran, B. L. Bishop, and K. C. Chandler of the Oak Ridge National Laboratory is appreciated. We thank K. C. Hsieh of the University of Chicago, E. Schonfeld of the Manned Spacecraft Center, G. D. O'Kelley of the Oak Ridge National Laboratory, and J. R. Arnold and R. C. Reedy of the University of California, San Diego, for helpful discussions and making data available prior to publication.

This research is funded by the National Aeronautics and Space Administration, Order T–85613, under Union Carbide Corporation's contract with the U.S. Atomic Energy Commission.

References

Albee A. L., Burnett D. S., Chodos A. A., Eugster O. J., Huneke J. C., Papanastassiou D. A., Podosek F. A., Russ G. Price II, Sanz H. G., Tera F., and Wasserburg G. J. (1970) Ages irradiation history, and chemical composition of lunar rocks from the Sea of Tranquillity *Science* **167**, 463–466.

Armstrong T. W., and Alsmiller R. G., Jr. (1968) Monte Carlo calculations of the nucleon-meson cascade in iron initiated by 1- and 3-GeV protons and comparison with experiment. *Nucl. Sci. Eng.* **33**, 291–296.

Armstrong T. W. (1969) Monte Carlo calculations of residual nuclei production in thick iron targets bombarded by 1- and 3-GeV protons and comparison with experiment. *J. Geophys. Res.* **74**, 1361–1373.

Armstrong T. W., Alsmiller R. G., Jr., and Chandler K. C. (1971) Monte Carlo calculations of high-energy nucleon-meson cascades and applications to galactic cosmic-ray transport. Oak Ridge National Lab. Rep. ORNL-TM-3319; also to be published in *Proc. National Sym. on National and Manmade Radiation in Space*, Las Vegas, March 1–5, 1971.

Armstrong T. W., and Alsmiller R. G., Jr. (1970) An approximate density-effect correction for the ionization loss of charged particles. *Nucl. Instr. Meth.* **82**, 289–290.

Armstrong T. W., and Alsmiller R. G., Jr. (1971a) Calculation of cosmogenic radionuclides in the moon and comparison with Apollo measurements. Oak Ridge National Lab. Rep. ORNL–TM–3267 (in press).

Armstrong T. W., and Almiller R. G., Jr. (1971b) Calculation of the lunar photon albedo and induced dose rate from galactic and solar proton bombardment. Oak Ridge National Lab. Rep. ORNL–TM–3268 (in press).

Arnold J. R., Honda M., and Lal D. (1961) Record of cosmic ray intensity in the meteorites. *J. Geophys. Res.* **66**, 3519–3531.

Baker M. B., Santina R. E., and Masley A. J. (1969) Modeling of solar cosmic ray events based on recent observations. *A.I.A.A.J.* **7**, 2105–2109.

Bertini H. W. (1969) Intranuclear-cascade calculation of the secondary nucleon spectra from nucleon-nucleus interactions in the energy range 340 to 2900 MeV and comparisons with experiment. *Phys. Rev.* **188**, 1711–1730.

Durgaprasad N., Fichtel C. E., and Guss D. E. (1967) Solar modulation of cosmic rays and its relationship to proton and helium fluxes, interstellar travel, and interstellar secondary production. *J. Geophys. Res.* **72**, 2765–2782.

Eberhardt P., Geiss J., Graf H., Grögler N., Krähenbühl U., Schwaller H., Schwarzmüller J., and Stettler A. (1970) Trapped solar wind noble gases, Kr^{81}/Kr exposure ages and Kr/Ar ages in Apollo 11 lunar material. *Science* **167**, 558–560.

Engle W. W., Jr. (1967) A users manual for ANISN, a one-dimensional discrete ordinates transport code with anisotropic scattering. Computing Technology Center, Union Carbide Corp. Rep. K–1693.

ESSA (ESSA Research Laboratories) (1970) Solar-geophysical data. Environmental Research Laboratories, Boulder, Colo. Rep. Number 309–Part II.

Eugster O., Tera F., Burnett D. S., and Wasserburg G. J. (1970) The isotopic composition of Gd and the neutron capture effects in samples from Apollo 11. *Earth Planet. Sci. Lett.* **8**, 20–30.

Finkel R. C., Arnold J. A., Reedy R. C., Fruchter J. S., Loosli Heinz-Hugo, Evans J. C., Shedlovsky J. P., and Delany A. C. Depth variation of cosmogenic nuclides in a lunar surface rock. Second Lunar Science Conference (unpublished proceedings).

Funkhouser J. G., Schaeffer O. A., Bogard D. D., and Zähringer J. (1970) Gas analysis of the lunar surface. *Science* **167**, 561–563.

Gabriel T. A., Alsmiller R. G., Jr., and Guthrie M. P. (1970) An extrapolation method for predicting nucleon and pion differential production cross sections from high-energy (> 3 GeV) nucleon-nucleus collisions. Oak Ridge National Lab. Rep. ORNL–4542.

Gast P. W., and Hubbard N. J. (1970) Abundance of alkali metals, alkaline and rare earths, and strontium-87/strontium-86 ratios in lunar samples. *Science* **167**, 485–487.

Guthrie M. P., Alsmiller R. G., Jr., and Bertini H. W. (1968) Calculation of the capture of negative pions in light elements and comparison with experiments pertaining to cancer radiotherapy. *Nucl. Instr. Meth.* **66**, 29–36.

Guthrie Miriam P. (1970) EVAP-4: Another modification of a code to calculate particle evaporation from excited compound nuclei. Oak Ridge National Lab. Rep. ORNL–TM–3119.

Haskin L. A., Helmke P. A., and Allen R. O. (1970) Rare-earth elements in returned lunar samples. *Science* **167**, 487–490.

Honda M. (1962) Spallation products distributed in a thick iron target bombarded by 3-BeV protons. *J. Geophys. Res.* **67**, 4847–4858.

Hsieh K. C. (1970) Private communication.

Joanou C. D., and Dudek J. S. (1963) GAM II, a B3 code for the calculation of fast neutron spectra and associated multigroup constants. General Atomic Rep. GA-4265.

Kohman T. P., and Bender M. L. (1967) Nuclide production by cosmic rays in meteorites and in the moon. In *High Energy Nuclear Reactions in Astrophysics* (editor B. S. P. Shen), pp. 169–246, Benjamin.

Lal D., and Venkatavaradan V. S. (1966) Low-energy protons: average flux in interplanetary space during the last 100,000 years. *Science* **151**, 1381–1384.

Lal D., Rajan R. S., and Venkatavaradan V. S. (1967) Nuclear effects of "solar" and "galactic" cosmic-ray particles in near-surface regions of meteorites. *Geochim. Cosmochim. Acta* **31**, 1859–1869.

Lal D., and Venkatavaradan V. S. (1967) Activation of cosmic dust by cosmic-ray particles. *Earth Planet. Sci. Lett.* **3**, 299–310.

Lingenfelter R. E., Canfield E. H., and Hess W. N. (1961) The lunar neutron flux. *J. Geophys. Res.* **66**, 2665–2671.

LSPET (Lunar Sample Preliminary Examination Team) (1969) Preliminary examination of lunar samples from Apollo 11. *Science* **165**, 1211–1227.

Marti K., Lugmair G. W., and Urey H. C. (1970) Solar wind gases, cosmic ray spallation products and the irradiation history. *Science* **167**, 548–550.

McDonald F. B. (1969) IQSY observations of low-energy galactic and solar cosmic rays. In *Annals of the IQSY*, Vol. 4 (editor A. C. Strickland), p. 187, M.I.T. Press.

Modisette J. L., Vinson T. M., and Hardy A. C. (1965) Model solar proton environments for manned spacecraft design. NASA Rep. NASA TN D–2746.

Morrison G. H., Gerard J. T., Kashuba A. T., Gangadharam E. V., Rothenberg A. M., Potter N. M., and Miller G. B. (1970) Multielement analyses of lunar soil and rocks. *Science* **167**, 505–507.

O'Kelley G. D., Eldridge J. S., Schonfeld E., and Bell P. R. (1970) Primordial radionuclide abundances, solar proton and cosmic-ray effects and ages of Apollo 11 lunar samples by nondestructive gamma ray spectrometry. *Proc. Apollo 11 Lunar Sci. Conf., Geochim. Cosmochim. Acta* Suppl. 1, Vol. 2, pp. 1407–1423. Pergamon.

O'Kelley G. D., Eldridge J. S., Schonfeld E., and Bell P. R. (1971) Comparative radionuclide concentrations and ages of Apollo 11 and Apollo 12 samples from nondestructive gamma-ray spectrometry. Second Lunar Science Conference (unpublished proceedings).

Philpotts J. A., and Schnetzler C. C. (1970) Potassium, rubidium, strontium, barium, and rare-earth concentrations in lunar rocks and separated phases. *Science* **167**, 493–495.

RADKE G. (1969) Solar flare dose rates in a near earth polar orbit. *Aerospace Med.* **40**, 1495–1503.

RUDSTAM G. (1966) Systematics of spallation yields. *Z. Naturforsch.* **21a**, 1027–1041.

SHEDLOVSKY J. P., HONDA M., REEDY R. C., EVANS J. C., JR., LAL D., LINDSTROM R. M., DELANY A. C., ARNOLD J. R., LOOSLI H. H., FRUCHTER J. S., and FINKEL R. C. (1970) Pattern of bombardment-produced radionuclides in rock 10017 and in lunar soil. *Proc. Apollo 11 Lunar Sci. Conf., Geochim. Cosmochim. Acta* Suppl. 1, Vol. 2, pp. 1503–1532. Pergamon.

WEBBER W. R. (1963) An evaluation of the radiation hazard due to solar-particle events. Boeing Rep. D2–90469.

YUCKER W. R. (1970) Statistical analysis of solar cosmic ray proton fluence. McDonnell Douglas Astronautics Rep. MDAC Paper WD 1320.

Proceedings of the Second Lunar Science Conference, Volume 2, pp. 1747–1755.
The M.I.T. Press, 1971.

Cosmogenic radionuclide concentrations and exposure ages of lunar samples from Apollo 12

G. Davis O'Kelley and James S. Eldridge
Oak Ridge National Laboratory, Oak Ridge, Tennessee 37830
and
Ernest Schonfeld and P. R. Bell*
Manned Spacecraft Center, Houston, Texas 77058

(*Received 22 February 1971; accepted in revised form 29 March 1971*)

Abstract—Cosmogenic radionuclide abundances in a suite of samples from the Ocean of Storms were determined nondestructively by gamma-ray spectrometers of low background. Samples investigated were crystalline rocks 12002, 12004, 12039, 12052, 12053, 12054, 12062, and 12064; breccias 12013, 12034, and 12073; fines 12032 and 12070. The general concentration patterns of spallogenic radionuclides resemble those observed for Apollo 11 samples, but with some differences in detail. Cosmogenic radionuclides determined in this study were ^{22}Na, ^{26}Al, ^{46}Sc, ^{48}V, ^{52}Mn, ^{54}Mn, ^{56}Co, and ^{60}Co. Despite delays in obtaining samples during the preliminary examination, 5.7-day ^{52}Mn was determined in two rocks and 16-day ^{48}V was determined in four rocks.

Solar protons and galactic protons are both involved in the production of ^{22}Na, ^{26}Al, and ^{54}Mn in surface samples; however, several rocks show evidence of shielding. Concentrations of radionuclides in rock 12034 are consistent with production by galactic protons at depth, shielded from the effects of solar protons. Sample 12002,30 from the top of rock 12002 exhibited high concentrations of nuclides produced by solar flare protons, in confirmation of the orientation of 12002.

From the ^{60}Co concentration in rock 12002, a thermal neutron flux of 0.35 ± 0.18 neutrons cm^{-2} sec^{-1} was estimated. Estimates of cosmic-ray exposure ages were calculated by the ^{22}Na–^{22}Ne method. The results for seven samples are in good agreement with ^{3}He exposure ages by other investigators and range from 48 to 251 million years.

Introduction

The extensive studies of nuclides produced in the bombardment of meteorites by the solar and galactic cosmic rays have revealed much detailed information concerning the intensity and energy spectra of the incident radiations and their constancy with time. Lunar samples are even more suitable objects for such studies than meteorites, since the lunar samples have been irradiated in known orientations in space and are free from atmospheric ablation. A number of studies on radionuclide concentrations in Apollo 11 lunar samples (Begemann *et al.*, 1970; Herzog and Herman, 1970; O'Kelley *et al.*, 1970a, 1970b; Perkins *et al.*, 1970; Shedlovsky *et al.*, 1970; Wrigley and Quaide, 1970) clearly demonstrated the potential of such information for elucidating the bombardment history of the lunar material, the histories of the incident particle fluxes, the erosion rates of rocks, and the rate of turnover of the lunar surface due to impact.

The absence of atmospheric ablation makes possible the detailed study of the effects of recent solar flares and long-term solar particle bombardment. Because of

* Present address: Oak Ridge National Laboratory, Oak Ridge, Tennessee 37830.

the short range of the solar particles and the high yields of some of the nuclear reaction products, gamma-ray spectrometry has been used quite successfully to determine the most recent orientation of rocks on the lunar surface (O'KELLEY *et al.*, 1970a, 1970b; PERKINS *et al.*, 1970; SCHONFELD and O'KELLEY, 1971).

Several radionuclides of interest have short half-lives. For this reason, much of the data reported below were recorded at the Lunar Receiving Laboratory (LRL), Houston, Texas, during the preliminary examination of the Apollo 12 samples. An early account of the results on some of the samples was given in LSPET (1970). Since the publication of the preliminary examination report the data analyses have been refined and further samples have been analyzed.

EXPERIMENTAL PROCEDURES

Several gamma-ray spectrometers were used in the course of this study. A NaI(Tl) scintillation coincidence spectrometer with an associated on-line, data acquisition system described by O'KELLEY *et al.* (1970b), together with a Ge(Li) spectrometer permitted rapid analyses of samples at the LRL during the quarantine period, so that nuclides of short half-life could be determined. Some studies at later times were carried out at Oak Ridge National Laboratory on a NaI(Tl) spectrometer similar to the scintillation spectrometer at the LRL.

The first Apollo 12 sample for radioactivity determination (12002,0) was received from the LRL Sample Laboratory on November 28, 1969, about 8.4 days after liftoff from the moon. During quarantine, samples were mounted in stainless steel containers for gamma-ray analysis. After quarantine, samples were generally sealed inside thin teflon bags for measurement. Methods of data acquisition and data analysis were essentially the same as those we used to analyze Apollo 11 samples.

For analyses of data on samples measured during the preliminary study, calibration of the LRL coincidence spectrometer was established by recording a library of spectra from cylindrical radioactive standards prepared by dispersing known amounts of radioactivity in quantities of iron powder. When recording the library of standard spectra, the standard sources were placed inside the steel containers actually used.

Spectrum libraries used for analyzing samples 12002,0; 12002,20; 12013,11; 12032,16; 12034,0; 12070,0; and 12073,0 were obtained from replicas which accurately reproduced the electronic and bulk densities of the lunar samples. Procedures for preparation of the cylindrical standards and the replicas were described earlier by O'KELLEY *et al.* (1970a, 1970b). A more detailed description of the analytical procedures employed for the Apollo 12 studies was given by O'KELLEY *et al.* (1971).

RESULTS AND DISCUSSION

Cosmogenic radionuclide concentrations

Our results on spallogenic radionuclides are given in Table 1. The general concentration patterns resemble those we observed in the Apollo 11 samples (O'KELLEY *et al.*, 1970a, 1970b); however, a number of subtle differences were noted due to effects of chemical composition and shielding. The data of Table 1 were recorded on large samples, usually a rock or a large fragment of a rock. Sample weights are listed in a companion paper by O'KELLEY *et al.* (1971). As observed in all gamma-ray spectrometry studies of Apollo 11 samples, the high concentrations of Th and U in lunar material makes difficult the determination of weak gamma-ray components. Because of the short times available for some of the measurements, it was not possible

Table 1. Concentrations (dpm/kg) of spallogenic radionuclides in Apollo 12 samples.* Values for short-lived nuclides have been corrected to 1426 GMT, Nov. 20, 1969.

Sample†	Type‡	^{22}Na	^{26}Al	^{46}Sc	^{48}V	^{52}Mn	^{54}Mn	^{56}Co	^{60}Co
12002,0	B	42 ± 3	75 ± 6	3.5 ± 1.0	13 ± 3	31 ± 12	38 ± 3	33 ± 4	0.55 ± 0.30
12002,20	B	47 ± 3	67 ± 5						0.73 ± 0.65
12002,30	B	86 ± 3	126 ± 6				50 ± 5	148 ± 20	
12004,1	A	53 ± 5	90 ± 6	3.7 ± 1.5			35 ± 4	34 ± 8	< 2.6
12039,0	B	43 ± 5	95 ± 7	< 6.0			37 ± 6	40 ± 10	
12052,1	A	40 ± 6	75 ± 6				27 ± 7	26 ± 10	
12053,0	A	40 ± 6	81 ± 12	7.0 ± 2.0	20 ± 5		35 ± 5	32 ± 6	< 1.0
12054,0	B	39 ± 7	50 ± 10	5.0 ± 2.0			36 ± 5	40 ± 10	< 1.0
12062,0	AB	30 ± 5	57 ± 9	5.0 ± 2.0	9 ± 3		31 ± 6	7 ± 4	
12064,0	B	40 ± 5	51 ± 5	5.0 ± 2.0	22 ± 6	33 ± 18	35 ± 3	32 ± 6	< 1.0
12013,0	C	50 ± 10	115 ± 16	< 15			< 66	50 ± 30	< 8.0
12013,11	C	26 ± 10	90 ± 10						
12034,0	C	29 ± 5	45 ± 5	< 10	< 60		16 ± 8	< 16	< 4.0
12073,0	C	63 ± 7	110 ± 10	< 10			28 ± 7	47 ± 12	
12032,16	D	48 ± 6	100 ± 7	< 10			27 ± 7	< 30	< 2.0
12070,0	D	70 ± 8	146 ± 16				41 ± 10	55 ± 14	< 1.5

* Upper limits are 2 σ evaluated from least-squares analysis.
† A zero following the 5-digit sample number designates a whole rock or fines sample.
‡ Petrologic type according to LSPET (1970).

to determine all 8 nuclides listed in Table 1 for all of the samples. Rock 12002 was the most carefully studied of all our samples and a rather complete radionuclide pattern was obtained. Except for some exceptions noted below, agreement within experimental error was obtained in the few cases where other radionuclide measurements on the same samples could be compared (RANCITELLI *et al.*, 1971; FINKEL *et al.*, 1971).

As was noted previously, ^{22}Na and ^{26}Al are produced both by solar and galactic cosmic rays (SHEDLOVSKY *et al.*, 1970; PERKINS *et al.*, 1970; O'KELLEY *et al.*, 1970a,b). Because the chemical composition of lunar material favors production of these nuclides, because they can be measured nondestructively by gamma-gamma coincidence methods with high sensitivity, and because their different half-lives (2.6 and 7.4×10^5 years) probe different regions of geologic time, their yields are of great interest.

The concentrations of ^{22}Na and ^{26}Al from Table 1 may be compared with calculated concentrations produced by galactic protons alone. This permits an estimate of the solar proton component. The ^{26}Al production in a 2π geometry was estimated by the method of FUSE and ANDERS (1969) and the ^{22}Na production was estimated by the method of BEGEMANN *et al.* (1970). Chemical compositions were taken from the best values from the Apollo 12 Lunar Science Conference and from the Apollo 12 Lunar Sample Catalog (WARNER, 1970).

The comparison between measured and calculated values is shown in Table 2. As a test of the calculations we show in Table 2 data on a sample 10017, ARA which was taken from the bottom of a well-oriented rock, as discussed by O'KELLEY *et al.* (1970b). This bottom piece from 10017 was shielded by about 14 g/cm^2 of rock, which effectively absorbed the solar protons. Agreement between calculation and experiment is good. Rock 12034 is a breccia recovered from a trench on the north rim of Head Crater; its burial depth was estimated as 15 cm (SHOEMAKER *et al.*, 1970). The low values for the concentrations of ^{22}Na and ^{26}Al obtained experimentally

Table 2. Comparison between measured concentrations of ^{26}Al and ^{22}Na in lunar rocks and fines compared with concentrations calculated for galactic production only

Sample	^{26}Al (dpm/kg) Measured	^{26}Al (dpm/kg) Calculated*	^{22}Na (dpm/kg) Measured	^{22}Na (dpm/kg) Calculated*	Remarks
Crystalline rocks					
10017,ARA	50 ± 7	41	30 ± 5	33	Bottom piece
12002,0	75 ± 6	42	42 ± 3	41	
12002,30	126 ± 6	42	86 ± 3	41	top slice
12004,1	90 ± 6	43	53 ± 5	40	
12052,1	75 ± 6	47	40 ± 6	37	
12053,0	81 ± 12	46	40 ± 6	37	
12062,0	57 ± 9	45	30 ± 5	36	
12064,0	51 ± 5	49	40 ± 5	35	
Breccias					
12013,0	115 ± 16	58	50 ± 10	41	
12034,0	45 ± 5	52	29 ± 5	40	buried 15 cm
12073,0	110 ± 10	50	63 ± 7	38	
Fines					
12032,16	100 ± 7	51	48 ± 6	41	
12070,0	146 ± 16	50	70 ± 8	40	

* Production in 2π geometry by method of FUSE and ANDERS (1969).
† Production rates estimated by method of BEGEMANN *et al.* (1970).

show that the rock was shielded from recent solar-proton bombardment. Consideration of the solar-proton spectrum (EBEOGLU and WANIO, 1966; LAL *et al.*, 1967) and the available information on variations in ^{22}Na and ^{26}Al concentrations with depth in lunar materials (FINKEL *et al.*, 1971; RANCITELLI *et al.*, 1971; ELDRIDGE *et al.*, 1971) conservatively specify a burial depth of $\gtrsim 8$ cm. Agreement between measured and calculated nuclide concentrations shown in Table 2 for 12034 is also good. It appears that the calculation of BEGEMANN *et al.* (1970) overestimates the ^{22}Na yields slightly.

Samples 12002,30 was a 46-g piece cut from the top of oriented rock 12002,0 and was investigated to obtain depth variations of cosmogenic nuclides by FINKEL *et al.* (1971). Before 12002,30 was submitted to destructive analysis, the data in Table 2 were obtained. As expected, high concentrations of ^{26}Al and ^{22}Na were seen, in excess of the production by galactic protons. It will be noted from Table 1 that the concentration of ^{54}Mn has also been enhanced over the nominal value by the solar proton bombardment while ^{56}Co which is almost totally produced by solar flares manifests a large surface concentration gradient. In contrast, breccia 12034 was shielded from solar protons and shows a low concentration of ^{54}Mn and undetectable ^{56}Co.

For the other rocks of Table 2 large excesses of ^{26}Al over that produced by galactic protons is observed, with moderate excesses of ^{22}Na. These results, together with the ^{56}Co concentrations of Table 1 show that the rocks in question were at least partially exposed on the lunar surface.

Rocks 12054, 12062, and 12064 show evidence of recent low exposure. Within experimental errors it is not possible to decide whether the low values of ^{26}Al are due to partial shielding from solar protons or whether the ^{26}Al did not attain saturation. It will be shown below that galactic proton exposure ages suggest that 12062 and 12064 have been near but not on the lunar surface for the last 150–200 m.y., which may indicate that these rocks received a low exposure to solar protons. Another possible explanation would be a high surface erosion rate, but it is difficult to understand why certain rocks erode rapidly while others do not.

It may be noted that for samples whose radionuclide concentrations can be compared (12062, 12034, 12070) our ^{22}Na concentrations agree with those of RANCITELLI *et al.* (1971) within experimental errors, but our ^{26}Al concentrations appear to be consistently lower.

The two soil samples we examined appear to have been taken from quite different depths. The high concentrations of ^{22}Na and ^{26}Al, and especially the high ^{56}Co, are consistent with near surface sampling for 12070. The sample of 12032 apparently came from a deeper zone, about 5 cm below the surface.

The concentrations of ^{46}Sc in samples from Apollo 12 are about 2.4 times lower than those we found in Apollo 11 samples. This reduction reflects the lower concentration of Ti target nuclei in the samples from the Ocean of Storms.

Despite delays in obtaining samples during the preliminary examination, 5.7-day ^{52}Mn was determined in two rocks and 16-day ^{48}V was determined in four rocks. The ^{52}Mn yields are approximately as expected from the chemical composition and correlate well with the ^{54}Mn yields. Most of the ^{48}V is produced by solar protons via the reaction ^{48}Ti(p, n)^{48}V. To correlate the observed ^{48}V yields with chemical composition, it was necessary to estimate the production of ^{48}V from spallation of iron by high-energy protons. This estimate was derived from the ^{48}V and ^{56}Co concentrations of rock 12062, which showed low exposure to solar-flare protons. By assuming that all ^{56}Co in 12062 was produced by solar-flare protons, the corresponding concentration of ^{48}V was estimated by use of the chemical composition in WARNER (1970) and the (p, n) cross sections for producing ^{48}V and ^{56}Co as measured by TANAKA and FURUKAWA (1959). Of the 9 dpm/kg of ^{48}V shown in Table 1 for 12062, about 3 dpm/kg could be attributed to solar flare production. The 6 dpm/kg of ^{48}V produced by high-energy spallation is not expected to vary significantly among the crystalline rocks of Table 1 because of the nearly constant concentration of iron.

In Fig. 1 we show that the yields of ^{48}V corrected to November 20, 1970, correlate well with the average titanium concentrations reported in the literature for 12002, 12053, and 12064. The flare responsible for the solar ^{48}V occurred on November 3, 1970; if the solar contribution is corrected to that date, the difference between the solid and dashed lines of Fig. 1 will be doubled.

Thermal neutron flux

Cobalt-60 has a half-life of only 5.3 years and is produced with a high cross section (37 barns) by thermal-neutron capture in ^{59}Co. Production of ^{60}Co either by spallation or by the (n, p) reaction in Ni is very low because of the small abundance of the target isotopes and the low cross sections for the nuclear reactions concerned. The concentration of ^{60}Co in lunar material can be employed to calculate the neutron flux characteristic of recent, steady-state production on the lunar surface. Such

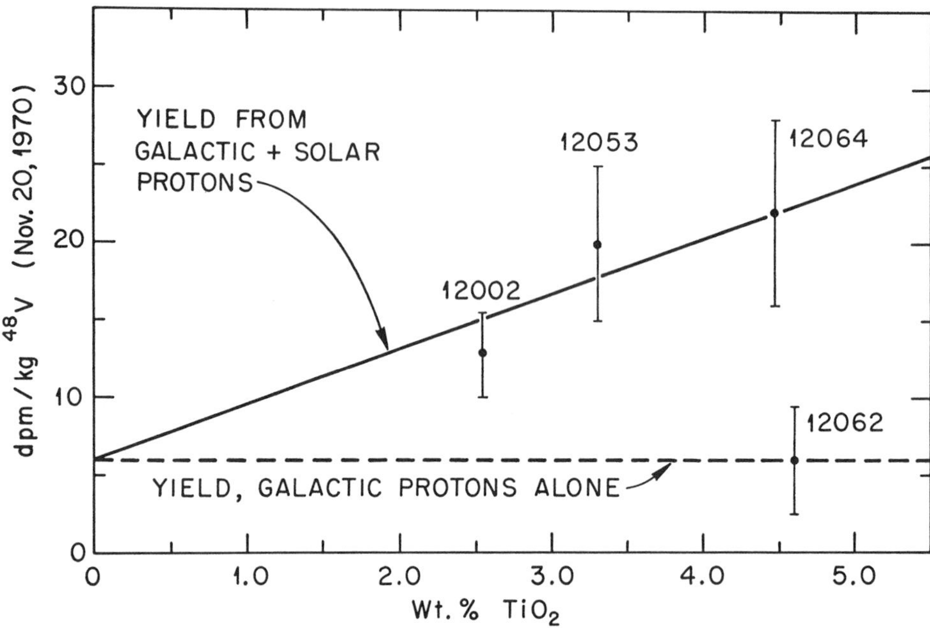

Fig. 1. Correlation between ^{48}V induced by solar-flare protons and titanium concentration. The solid line includes both the yield from solar protons calculated, as described in the text, and the yield from galactic protons.

information is a useful complement to fluxes deduced from mass spectrometric measurements of isotopic anomalies in Gd. The Gd isotope ratios yield an integrated thermal-neutron flux which requires a meaningful exposure age before an average flux can be obtained. Lunar rocks endure such a complex history that the average flux obtained mass spectrometrically may not represent the most recent flux to the accuracy desired.

Although ^{60}Co can be determined in lunar samples by gamma-ray spectrometry, rather large samples are required because the stable ^{59}Co target nuclide is present in such low concentration. Further, the intense interferences from abundant U and Th and cosmogenic radionuclides make difficult the resolution of small quantities of ^{60}Co.

In Table 1 we show that in the case of rock 12002,0 a value of 0.55 $\pm$ 0.30 dpm/kg was obtained for the ^{60}Co concentration. Based on an average Co concentration of 70 ppm in 12002, the thermal-neutron flux was found to be 0.35 $\pm$ 0.18 neutrons cm^{-2} sec^{-1}. The average mass density of rock 12002 was approximately 20 g/cm^2. Our result for a flux in a 20 g/cm^2 sample is in good agreement with the depth dependence of thermal neutron fluxes measured mass spectrometrically by MARTI and LUGMAIR (1971) in lunar material of about 18 to 150 g/cm^2. Our result for 12002 is also in agreement with the theoretical value of 0.23 $\pm$ 0.06 neutrons cm^{-2} sec^{-1} calculated by ARMSTRONG and ALSMILLER (1971), who averaged the solar maximum and minimum fluxes and included nominal Apollo 11 rare-earth concentrations in the lunar surface composition.

Exposure ages

Estimates of cosmic-ray exposure ages were made by the ^{22}Na–^{22}Ne method as discussed by O'KELLEY *et al.* (1970a, 1970b). It was assumed that the effective cross sections for production of ^{22}Na and ^{22}Ne were equal. Concentrations of Ne were obtained from the literature. The spallogenic ^{22}Ne was estimated to be 1.10 ^{21}Ne. Radioactive concentrations of ^{22}Na were taken from Table 1 and corrected for excess ^{22}Na of solar origin by a semiempirical factor.

In Table 3 we compare our exposure ages from the ^{22}Na–^{22}Ne method with ^{3}He exposure ages. The ^{3}He exposure ages were calculated from a production rate of 10^{-8} cm^3 STP ^{3}He/g per 10^6 years exposure. The agreement in Table 3 is gratifying and suggests that the ratio of production rates assumed for the ^{22}Na–^{22}Ne method is substantially correct.

The rocks of relatively shorter exposure age (12002, 12004, 12013, 12053) all were collected (SUTTON and SCHABER, 1971) in the Ocean of Storms north of a line connecting the north rim of Bench Crater and the center of Surveyor Crater. Rocks 12062, 12064, and 12065 have significantly longer exposure ages and were collected south of this line, which appears to be a boundary associated with Middle Crescent Crater. As shown by WARNER and ANDERSON (1971), most of the crystalline rocks north of this diffuse boundary are porphyritic basalts, while those to the south are granular and ophitic basalts. The model proposed by WARNER and ANDERSON to account for this distribution tentatively associated with Middle Crescent Crater suggests that the area north of the boundary would be strewn with ejecta of somewhat more recent exposure than the region to the south, which might be rich in older regolith material. Although this conclusion is speculative and is based on relatively few exposure ages, our data lend qualitative support to the WARNER and ANDERSON model.

Table 3. Estimation of exposure ages of Apollo 12 lunar samples.

Sample	Exposure age (10^6 y)		Ref. gas data
	^{3}He	^{22}Na–^{22}Ne	
12002	89	96	a
12004	61	58	b
12013	40	48	c
12053	79	99	b
12062	150	153	b
12064	205	251	b
12065	182	217	b

a. ^{3}He from HINTENBERGER *et al.* (1971); ^{21}Ne from MARTI and LUGMAIR (1971).
b. ^{3}He and ^{21}Ne concentrations from HINTENBERGER *et al.* (1971).
c. ^{3}He and ^{21}Ne concentrations from SCHAEFFER *et al.* (1970).

Acknowledgments—The authors gratefully acknowledge contributions to the work reported here by R. S. Clark, J. E. Keith, V. A. McKay, K. J. Northcutt, W. R. Portenier, M. K. Robbins, R. T. Roseberry, and R. E. Wintenberg. We thank J. R. Arnold and P. W. Gast for helpful discussions and the management and staff of the Lunar Receiving Laboratory for their hospitality. This research was carried out under Union Carbide's contract with the U.S. Atomic Energy Commission through interagency agreements with the National Aeronautics and Space Administration.

REFERENCES

ARMSTRONG T. W. and ALSMILLER R. G., JR. (1971) Calculation of cosmogenic radionuclides in the moon and comparison with Apollo measurements. Second Lunar Science Conference (unpublished proceedings).

BEGEMANN, F., VILCSEK E., RIEDER R., BORN W., and WÄNKE H. (1970) Cosmic-ray produced radioisotopes in lunar samples from the Sea of Tranquillity (Apollo 11). *Proc. Apollo 11 Lunar Sci. Conf., Geochim. Cosmochim. Acta* Suppl. 1, Vol. 2, pp. 995–1007. Pergamon.

EBEOGLU D. B. and WAINIO K. M. (1966) Solar proton activation of the lunar surface. *J. Geophys. Res.* **71**, 5863–5872.

ELDRIDGE J. S., O'KELLEY G. D., SCHONFELD E., and NORTHCUTT K. J. (1971) Unpublished data.

FINKEL R. C., ARNOLD J. R., REEDY R. C., FRUCTER J. S., LOOSLI H. H., EVANS J. C., SHEDLOVSKY J. P., IMAMURA M., and DELANY A. C. (1971) Depth variation of cosmogenic nuclides in a lunar surface rock. Second Lunar Science Conference (unpublished proceedings).

FUSE K. and ANDERS E. (1969) Aluminum-26 in meteorites. VI. Achondrites. *Geochim. Cosmochim. Acta* **33**, 653–670.

HERZOG G. F. and HERMAN G. F. (1970) Na^{22}, Al^{26}, Th and U in Apollo 11 lunar samples. *Proc. Apollo 11 Lunar Sci. Conf., Geochim. Cosmochim. Acta* Suppl. 1, Vol. 2, pp. 1239–1247. Pergamon.

HINTENBERGER H., WEBER H., and TAKAOKA N. (1971) Concentrations and isotopic abundances of the rare gases in lunar matter. Second Lunar Science Conference (unpublished proceedings).

LAL D., RAJAN R. S., and VENKATAVARADAN V. S. (1967) Nuclear effects of "solar" and "galactic" cosmic-ray particles in near-surface regions of meteorites. *Geochim. Cosmochim. Acta* **31**, 1859–1869.

LSPET (Lunar Sample Preliminary Examination Team) (1970) Preliminary examination of lunar samples from Apollo 12. *Science* **167**, 1325–1339.

MARTI K. and LUGMAIR G. W. (1971) Kr^{81}–Kr and K–Ar^{40} ages, cosmic-ray spallation products and neutron effects in Apollo 11 and Apollo 12 lunar samples. Second Lunar Science Conference (unpublished proceedings).

O'KELLEY G. D., ELDRIDGE J. S., SCHONFELD E., and BELL P. R. (1970a) Elemental compositions and ages of lunar samples by nondestructive gamma-ray spectrometry. *Science* **167**, 580–582.

O'KELLEY G. D., ELDRIDGE J. S., SCHONFELD E., and BELL P. R. (1970b) Primordial radionuclide abundances, solar-proton and cosmic-ray effects and ages of Apollo 11 lunar samples by nondestructive gamma-ray spectrometry. *Proc. Apollo 11 Lunar Sci. Conf., Geochim. Cosmochim. Acta* Suppl. 1, Vol. 2, pp. 1407–1423. Pergamon.

O'KELLEY G. D., ELDRIGE J. S., SCHONFELD E., and BELL P. R. (1971) Abundances of the primordial radionuclides K, Th and U in Apollo 12 lunar samples by nondestructive gamma-ray spectrometry: Implications for origins of lunar soils. Second Lunar Science Conference (unpublished proceedings).

PERKINS R. W., RANCITELLI L. A., COOPER J. A., KAYE J. H., and WOGMAN N. A. (1970) Cosmogenic and primordial radionuclide measurements in Apollo 11 lunar samples by nondestructive analysis. *Proc. Apollo 11 Lunar Sci. Conf., Geochim. Cosmochim. Acta* Suppl. 1, Vol. 2, pp. 1455–1471. Pergamon.

RANCITELLI L. A., PERKINS R. W., FELIX W. D., and WOGMAN N. A. (1971) Cosmogenic and primordial radionuclide measurements in Apollo 12 lunar samples by nondestructive analysis. Second Lunar Science Conference (unpublished proceedings).

SCHONFELD E. and O'KELLEY G. D. (1971) The selenographic orientation of Apollo 12 rocks determined by nondestructive gamma-ray spectroscopy. Second Lunar Science Conference (unpublished proceedings).

SCHAEFFER O. A., FUNKHOUSER J. G., BOGARD D. D., and ZÄRINGER J. (1970) Potassium-argon ages of lunar rocks from Mare Tranquillitatus and Oceanus Procellarum. *Science* **170**, 161–162.

SHEDLOVSKY J. P., HONDA M., REEDY R. C., EVANS J. C., JR., LAL D., LINDSTROM R. M., DELANY A. C., ARNOLD J. R., LOOSLI H. H., FRUCHTER J. S., and FINKEL R. C. (1970) Pattern of bombardment-produced radionuclides in rock 10017 and in lunar soil. *Proc. Apollo 11 Lunar Sci. Conf., Geochim. Cosmochim. Acta* Suppl. 1, Vol. 2, pp. 1503–1533. Pergamon.

SHOEMAKER E. M., BATSON R. M., BEAN A. L., CONRAD C. JR., DAHLEM D. H., GODDARD E. N., HAIT M. H., LARSON K. B., SCHABER G. G., SCHLEICHER D. L., SUTTON R. L., SWANN G. A., and WATERS A. C. (1970) Geology of the Apollo 12 landing site. In *Apollo 12 Preliminary Science Report, NASA SP–236*, NASA Manned Spacecraft Center, Houston, pp. 113–156.

SUTTON R. L. and SCHABER G. G. (1971) Lunar locations and orientations of rock samples from Apollo missions 11 and 12. Second Lunar Science Conference (unpublished proceedings).

TANAKA S. and FURUKAWA M. (1959) Excitation functions for (p, n) reactions with titanium, vanadium, chromium, iron and nickel up to $E_p = 14$ MeV. *J. Phys. Soc. Japan*, **14**, 1269–1275.

WARNER J. L., Compiler (1970) *Apollo 12 lunar sample information.* NASA Manned Spacecraft Center report S-243, Houston, Texas.

WARNER J. L. and ANDERSON D. H. (1971) Lunar crystalline rocks: Petrology, geology and origin. Second Lunar Science Conference (unpublished proceedings).

WRIGLEY R. C. and QUAIDE W. L. (1970) Al^{26} and Na^{22} in lunar surface materials: Implications for depth distribution studies. *Proc. Apollo 11 Lunar Sci. Conf., Geochim. Cosmochim. Acta* Suppl. 1, Vol. 2, pp. 1751–1757. Pergamon.

Proceedings of the Second Lunar Science Conference, Volume 2, pp. 1757–1772.
The M.I.T. Press, 1971.

Erosion and mixing of the lunar surface from cosmogenic and primordial radionuclide measurements in Apollo 12 lunar samples

L. A. RANCITELLI, R. W. PERKINS, W. D. FELIX, and N. A. WOGMAN
Battelle Memorial Institute, Pacific Northwest Laboratories
Richland, Washington 99352

(*Received 2 March 1971; accepted in revised form 31 March 1971*)

Abstract—Apollo 12 lunar samples have been analyzed for their cosmogenic and primordial radionuclide contents by nondestructive gamma-ray spectrometry. The radionuclides ^{22}Na, ^{26}Al, ^{40}K, ^{46}Sc, ^{48}V, ^{54}Mn, ^{56}Co, ^{60}Co, ^{232}Th, and ^{238}U were measured on lunar rock specimens 12002,93 through 97, 12005,0, 12016,0, 12034,9, 12051,1, 12051,3, 12053,38, 12053,41 through 44, 12062,0, 12063,0, and 12065,0; on lunar soil sample 12070,3; and on lunar core tube samples 12025,4 through 14 and 12028. Compared with Apollo 11 samples the Apollo 12 samples show a significantly wider range in cosmogenic radionuclide concentrations and a much wider range in the primordial radionuclides. The recent solar flare produced radionuclides ^{48}V (16.1d) and ^{56}Co (77.3d) were near the concentrations observed in Apollo 11 samples, while the ^{46}Sc concentrations were about one half, in agreement with the lower concentration of Ti. Concentration gradient measurements of ^{22}Na and ^{26}Al in the double core tube sample 12025, 12028, in vertical sections from rocks 12002 and 12053, and in whole rocks, indicated mixing of the top 3 cm of lunar soil in 10^5 years and erosion rates of lunar rocks of 1 to 5 mm per million years. Aluminum-26 to ^{22}Na ratios are consistent with constant solar activity during the past million years. The U and Th concentration in the 9 to 16 cm deep region of the double tube 12024, 12028 was two-fold lower than sections above or below this level. This indicates the deposition of a unique type of material during this period of the regolith formation and places a limit on the depth of lunar soil mixing.

INTRODUCTION

THE APOLLO 11 lunar studies established the fact that the radionuclide production in the first centimeter of the lunar surface was the result of an intense solar proton bombardment with a small contribution from higher energy galactic cosmic rays. This was in marked contrast with the results of meteorite studies which, over the years, have provided us with a history of the galactic cosmic ray spectrum but because of their surface ablation during atmospheric entry have yielded little information on the solar irradiation. Since radionuclide production in the top centimeter of the lunar surface results primarily from solar proton bombardment, spallation products can serve as unique indicators of the recent and past history of the sun. Moreover, their measurement also provides a basis for determining dynamic lunar processes such as the erosion of rocks, the mixing of soil, and changes in rock orientations on the lunar surface. While these latter possibilities were touched upon, the major emphasis in the Apollo 11 studies was directed toward establishing the solar contribution to the radionuclide production on the lunar surface (SHEDLOVSKY *et al.*, 1970a; PERKINS *et al.*, 1970a, b). Physical processes such as erosion and lunar rock orientation have since received more detailed attention in several independent studies based on track measurements (CROZAZ *et al.*, 1971; BARBER *et al.*, 1971; FLEISCHER *et al.*, 1971) and cosmogenic radionuclide measurements (FINKEL *et al.*, 1971). The information gained through Apollo 11 studies served as a basis for a much more detailed study of lunar surface radioactivity as observed in the Apollo 12 samples and for the interpretation of the radionuclide measurements in terms of a variety of dynamic lunar surface and solar processes. The measurement of cosmogenic and primordial radionuclides in Apollo 12 lunar samples has now added an important segment to our knowledge of the lunar surface.

A high degree of accuracy was maintained in radionuclide measurements which permitted the observation of subtle differences in radionuclide content which may be related to differences in specimen size, shape, degree of burial, cosmic irradiation history, erosion rate, and chemical composition. Particular emphasis was placed on determining near surface concentrations of cosmogenic radionuclides in the rocks, and the concentration gradients in both rocks and core tube samples. The concentration gradients in the rock samples provided a basis for a model which describes both radionuclide production and lunar rock erosion, while the gradients in the fines provide a basis for defining mixing rates.

PROCEDURE

Extreme care was taken to ensure the highest possible accuracy in all of the Apollo 12 radionuclide measurements. The general methods employed for sample mockup preparations which were described earlier (PERKINS *et al.*, 1970a, 1970b) were employed with only minor modifications. The mockups were prepared from a mixture of casting resin (Titan Casting Resin, Titan Corp., Seattle, Washington), iron powder and aluminum oxide which contain precisely known radionuclide additions. A mockup produced by this procedure could be made to precisely duplicate the lunar rock geometry, physical density, and electron density and did not degrade or change in any observable manner with time. For the ^{22}Na and ^{26}Al measurements on whole rocks, errors associated with counting statistics were usually on the order of 1–2%. The absolute errors for ^{22}Na and ^{26}Al measurements based on all analytical uncertainties including the error in the radioisotope standards ranged from 2 to 5%.

Multiple-gamma coincidence counting techniques (PERKINS *et al.*, 1970b) permitted the estimation of radionuclide concentration gradients of lunar rocks from which lunar surface orientation and any recent changes in this orientation could be recognized. In addition, newly developed beta-multiple gamma coincidence counting techniques were employed which permitted the surface concentrations of ^{26}Al in the rock samples to be determined.

The beta-multiple gamma detection system employed two 30 cm diameter by 20 cm thick NaI (Tl) crystals with their associated 30 cm by 10 cm thick NaI light pipes as the principal gamma ray detectors. A 12.7 cm diameter by 1 cm thick P-10 gas proportional beta detector with a 500 microgram per square centimeter gold-plated mylar window was located against the front face of each detector. The lunar specimen in its double teflon bags plus one polyethylene bag was positioned in a reproducible mode between the two beta detectors for counting. The entire beta-gamma-gamma spectrometer system as shown in Fig. 1 was shielded from natural and cosmic radiation by a 4 inch thick lead cave. Each beta detector acted as a gate for a 4096 channel memory which operated as a 64 × 64 channel array to record both the single and coincidence gamma-ray events observed by the two NaI (Tl) crystals. Beta-particles leaving the upper surface of the lunar sample could trigger the upper beta detector, while the lower beta-detector was triggered by beta particles from the lower surface of the sample. This beta-multiple gamma coincidence counting mode provided an extremely low background and allowed the surface concentrations of ^{26}Al to be estimated by comparison with known standard mockups. The range of the beta particles, about 0.2 and 0.05 mm (for twofold attenuation) for ^{26}Al and ^{22}Na, respectively, defines the maximum depth from which these radionuclides are detected by the coincidence system. Since the Apollo 12

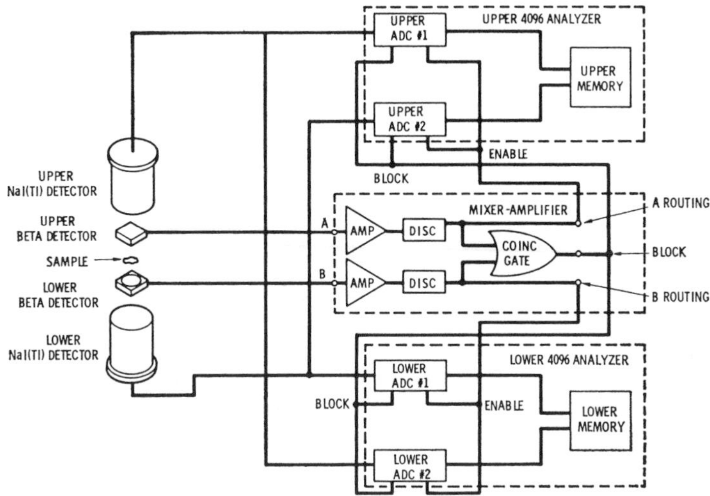

Fig. 1. Beta-multiple gamma-ray spectrometer.

rocks were counted in two Teflon bags plus a thin polyethylene bag, additional attenuation of beta particles occurred, which essentially stopped all of the beta particles associated with the decay of ^{22}Na.

Results and Discussion

Radionuclide concentration and production rates

A summary of radionuclide concentrations in the rock samples and bulk fines is presented in Table 1. The ^{26}Al and ^{22}Na in sections through two vertical rock slices are included in Table 2. The ^{26}Al, ^{22}Na, Th, and U in the double core tube soil and in a soil sample from the Surveyor III scoop are presented in Table 3. The cosmogenic radionuclide concentrations show general similarities to those in Apollo 11 samples, but there are some very striking differences. For example, the ^{26}Al:^{22}Na ratio of 2.56 for the 2360 gram crystalline rock 12063,0 is far higher than previously observed ratios. Conversely, the ^{26}Al:^{22}Na ratio is 1.13 in the 482 gram rock 12005,0 the lowest yet observed. This ratio is of particular interest because of the unusually high ^{22}Na concentration of 72 dpm/kg which is about twice that in the other rocks. To explain this high ^{22}Na content on the basis of chemical composition would not be reasonable unless the Mg or Na content was far higher in rock 12005,0 than seems likely based on the observed ranges reported by the preliminary examination team (LSPET, 1970) and at the Second Lunar Science Conference. As discussed later in this article, it appears more likely that the low ratio in rock 12005,0 and the high ratio in rock 12065,0 are due to their shape, their lunar surface orientation, and their erosion rates.

The concentrations of both ^{26}Al and ^{22}Na are about 20% higher in the bulk fines of Apollo 12 versus Apollo 11 samples. As is indicated later, this is easily accounted for by a slightly shallower sampling depth for Apollo 12 fines.

Table 1. Cosmogenic and primordial

Radionuclides	12005,0	12063,0	12016,0	12051,1
	(A)*	(A)	(AB)	(AB)
^{22}Na (dpm/kg)	72 ± 2	38 ± 2	44 ± 2	40 ± 2
^{26}Al	81 ± 2	78 ± 2	75 ± 2	93 ± 2
^{46}Sc	5.5 ± 0.8	6.0 ± 0.8	5.6 ± 1.3	7.0 ± 1.2
^{48}V	—	19 ± 12	—	—
^{54}Mn	37 ± 4	37 ± 4	36 ± 4	29 ± 4
^{56}Co	46 ± 6	30 ± 3	14 ± 4	—
^{60}Co	0.50 ± 0.29	0.81 ± 0.42	0.56 ± 0.34	0.36 ± 0.31
K (%)	0.026 ± 0.001	0.055 ± 0.002	0.044 ± 0.002	0.058 ± 0.002
Th (ppm)	0.403 ± 0.017	0.653 ± 0.030	0.570 ± 0.025	0.940 ± 0.050
U (ppm)	0.106 ± 0.010	0.178 ± 0.015	0.157 ± 0.015	0.234 ± 0.020
Sample weight (grams)	482	2360	2028	545
Rock weight (grams)	482	2426	2028	1660

Values prefaced with < represent 2σ of the gross counting rate in the photopeak area.
* Rock type.

Table 2. ^{22}Na and ^{26}Al concentrations in rock slices 12002,93–97 and 12053,38–44.

Sample	Average Depth† (g/cm^2)	^{22}Na (dpm/kg)	^{26}Al (dpm/kg)	^{26}Al/^{22}Na
12002,93	1.5	78 ± 9	95 ± 9	1.22 ± 0.18
12002,94	5.8	49 ± 8	58 ± 6	1.18 ± 0.23
12002,95	9.5	35 ± 4	52 ± 4	1.48 ± 0.20
12002,96	14.6	42 ± 5	44 ± 5	1.05 ± 0.17
12002,97	18.9	39 ± 8	52 ± 8	1.33 ± 0.34
12053,38	1.2	66 ± 10	152 ± 12	2.30 ± 0.39
12053,41	4.9	45 ± 13	75 ± 14	1.67 ± 0.57
12053,44	(8.1)*	(32 ± 4)	(60 ± 5)	(1.88 ± 0.28)
12053,43				
12053,42	14.7	38 ± 8	58 ± 8	1.53 ± 0.38

* Values in parentheses are average values of 12053,41, 44, 43.

Table 3. Summary of radionuclide content of double core tube 12025, 12028 and surveyor dust.

Sample	Average Depth† (g/cm^2)	^{22}Na (dpm/kg)	^{26}Al (dpm/kg)	Th(ppm)	U(ppm)
12025,14	0.4	197 ± 80	246 ± 69	6.0 ± 2.8	1.4 ± 0.7
12025,13	1.4	109 ± 36	241 ± 31	6.5 ± 1.1	2.2 ± 0.4
12025,13, 12	1.8	70 ± 35	202 ± 29	6.1 ± 1.6	3.1 ± 0.6
12025,12	2.7	62 ± 62	176 ± 51	6.4 ± 2.3	2.4 ± 0.6
12025,11	3.8	53 ± 29	169 ± 33	6.0 ± 1.1	1.9 ± 0.4
12025,10	5.2	56 ± 25	86 ± 21	7.0 ± 1.1	1.9 ± 0.4
12025,9, 8, 7	8.5	35 ± 8	54 ± 7	6.8 ± 0.4	2.1 ± 0.2
12025,6, 5, 4	13.5	34 ± 7	62 ± 8	6.3 ± 0.4	2.1 ± 0.2
12028,11, 13, 14, 16, 17, 18	23.8	44 ± 6	49 ± 4	3.2 ± 0.2	0.90 ± 0.09
12028,20, 21, 22, 23, 25, 26	35.5	37 ± 6	45 ± 5	7.5 ± 0.3	2.0 ± 0.2
12028,27, 28, 30, 31, 32, 36	49.2	36 ± 6	49 ± 5	8.4 ± 0.3	2.3 ± 0.2
12028,37, 38, 40, 41, 43, 44	64.3	41 ± 6	38 ± 5	6.6 ± 0.3	1.9 ± 0.2
Surveyor Soil	5*	44 ± 14	85 ± 14	5.9 ± 0.7	

* Estimated from ^{22}Na and ^{26} Al concentrations in 12025,10.
† In core tube.

radionuclides in lunar materials.

12051,3	12062,0	12065,0	12034,9	12070,13
(AB)	(AB)	(AB)	(C)	(D)
37 ± 2	33 ± 2	32 ± 2	34 ± 2	80 ± 2
81 ± 2	76 ± 2	82 ± 2	60 ± 3	165 ± 4
5.8 ± 1.1	4.4 ± 4.5	5.4 ± 1.1	—	5.9 ± 1.6
—	—	7 ± 11	—	—
32 ± 4	33 ± 4	31 ± 4	<40	21 ± 8
16 ± 4	—	22 ± 8	<56	57 ± 7
0.26 ± 0.26	0.42 ± 0.27	<1.2	1.5 ± 1.3	1.7
0.058 ± 0.002	0.059 ± 0.002	0.054 ± 0.002	0.432 ± 0.15	0.199 ± 0.006
0.864 ± 0.043	0.871 ± 0.044	0.991 ± 0.049	13.9 ± 0.5	6.73 ± 0.20
0.234 ± 0.020	0.241 ± 0.020	0.282 ± 0.025	3.53 ± 0.17	1.70 ± 0.09
765	727	2088	124	276
1660	739	1209	155	1102

The ^{46}Sc concentrations in the Apollo 12 samples are about half those observed in Apollo 11. This is in accord with the substantially lower concentrations of the major target element Ti (LSPET, 1970) in the Apollo 12 samples. The ^{56}Co (77.3d) and ^{48}V (16.1d), which are both produced by low energy (p, n) reactions on Fe and Ti, respectively, are at similar concentrations to those observed in Apollo 11 lunar samples. Their production is almost entirely due to solar flare proton bombardment and their concentrations provide an indication of recent integrated solar flux at the Apollo 12 site. The ratio of ^{48}V to ^{46}Sc, which are both produced from Ti, was three times higher in Apollo 12 samples than the Apollo 11 rocks. This is due to the fact that the time interval between the major solar flares which produced most of the ^{48}V and sample collection was much shorter for the Apollo 12 mission. This also explains the comparable amounts of ^{48}V in Apollo 11 and Apollo 12 samples even though the target element abundance for its production was much lower in Apollo 12 material. The cosmogenic radionuclide ^{54}Mn is produced by a (p, $2pn$) reaction on ^{56}Fe. The substantially higher ^{54}Mn concentration in the rock samples than in the fines is partially due to the higher stable iron content of the rocks, but is also apparently due to an increase in secondary production with depth in the rocks where significant secondary flux buildup may occur.

The very low ^{60}Co concentrations (see Table 1) in the lunar materials are to be expected from the low nickel and cobalt concentrations of the lunar surface. Based on integral neutron flux estimates of 10^{15}–10^{16} (ALBEE *et al.*, 1970), neutron capture by stable cobalt could have produced the observed ^{60}Co concentrations. Only upper limits were obtained for ^{7}Be (53d) and ^{51}Cr (28d), and these were similar to those observed in the Apollo 11 samples.

It has been shown that rather steeply declining concentration gradients for some cosmogenic radionuclides do exist in lunar rocks (SHEDLOVSKY *et al.*, 1970a, 1970b; O'KELLY *et al.*, 1970a, b; PERKINS *et al.*, 1970a, b), and of course similar gradients would exist in the lunar soil. Of the cosmogenic radionuclides, ^{22}Na and ^{26}Al concentrations are most easily measured with good accuracy by nondestructive techniques. In addition, they are by virtue of the differences in their half-lives, a convenient pair for studies of lunar surface orientation of lunar rocks, any recent changes in orientation, erosion rates of the lunar rocks, mixing rates and the mixing depth of lunar soil, and of major temporal changes in the cosmic-ray flux or energy spectrum. The fraction of the total ^{26}Al and ^{22}Na produced in Apollo 12 rocks by solar protons as estimated from galactic cosmic ray production in meteorites is comparable with the estimated production in Apollo 11 rocks (PERKINS *et al.*, 1970b). This supports our earlier conclusion that the solar activity has remained relatively constant for the past million years.

Radionuclide production rate calculations

To interpret ^{22}Na and ^{26}Al concentrations in terms of lunar surface processes, it is essential to know their expected production rates as a function of depth in the lunar surface. Since radionuclide production in the first centimeter of the lunar surface is mainly due to solar flare protons, a very rapidly descending production rate exists. The major target elements in lunar material for ^{22}Na production are Mg and Na below 20 MeV, and Mg, Si, and Al at higher energies. The major target elements for ^{26}Al production are Al, Si, and Mg below about 25 MeV and Si and Al at higher energies.

A calculation of the production rate depth gradients for ^{22}Na and ^{26}Al requires a knowledge of both the excitation functions for their production from the elemental composition of lunar materials and the energy spectrum of the incident solar protons.

Known excitation functions for ^{22}Na production were used. Direct measurements of excitation functions for ^{26}Al production were used where available (RANCITELLI and WOGMAN, 1971); otherwise they were estimated from analogous reactions.

For the calculations a representative shape factor for the solar cosmic ray energy spectrum was determined from satellite data collected during the November 1968, December 1968, and April 1969 flares (HSIEH and SIMPSON, 1970). Expressed in the kinetic power law form, the solar proton energy distribution can be stated as

$$\frac{dJ}{dE} = ke^{-\alpha}$$

where J is the proton flux (P/cm^2-sec-steradian-MeV), E is the particle energy (MeV), and k is a constant determined from the flare intensity. The shape function, α, was 3.1. for two of the flares and 3.5 for the third flare. For production rate calculation an α value of 3.1 was used.

The calculations assumed the lunar surface to consist of an infinite plane. The laminae thicknesses within the plane were set at 0.5 mm and given a cross section compatible with thin target calculations. Activation within the laminae due to second particles was assumed negligible. For a unit incident flux within a specific angular distribution, energy attenuation was calculated as a function of the depth of the laminae within the plane for 2 MeV increments from 10 to 400 MeV. Activation within each laminae was calculated by integrating the product of the flux and the excitation function over the depth corrected energy distribution.

Production rates based on both sun angle irradiation, which assumes that all of the bombarding particles arrive at the lunar surface along the lines of sunlight, and

2π isotropic irradiation were calculated. The actual exposure on the lunar surface is probably a combination of both sun angle and isotropic irradiation. However, an isotropic irradiation flux which produces a slightly steeper production rate gradient than a sun angle irradiation was assumed in the analysis of observed concentrations. A concentration gradient for zero erosion was calculated for the lunar fines composition and this curve was then normalized to the observed ^{22}Na and ^{26}Al in the first 5.4 g/cm^2 of depth (Fig. 2). A correction for the galactic contribution was made prior to this normalization. The normalization constant derived in this manner was subsequently employed to adjust the calculated production rates in rocks 12002 and 12053.

The ^{22}Na concentration in lunar material reflects only the solar flare activity during the past decade, and of course is very dependent on the most recent large solar flares. The calculated concentration gradients for ^{22}Na and ^{26}Al based on isotropic irradiation and the chemical composition (BRUNFELT *et al.*, 1971; GOLES *et al.*, 1971; WILLIS *et al.*, 1971) of rock 12002 are presented in Figs. 3a and 3b. The four gradients for ^{26}Al are those which would exist if the lunar surface were of density 3 and were eroding at rates of zero, 0.1, 1.0, and 10 mm per million years. The observed concentrations of ^{22}Na and ^{26}Al at depths of several cm are about 35 and 55 dpm per kg respectively, while the calculated surface concentrations indicate ^{22}Na and ^{26}Al

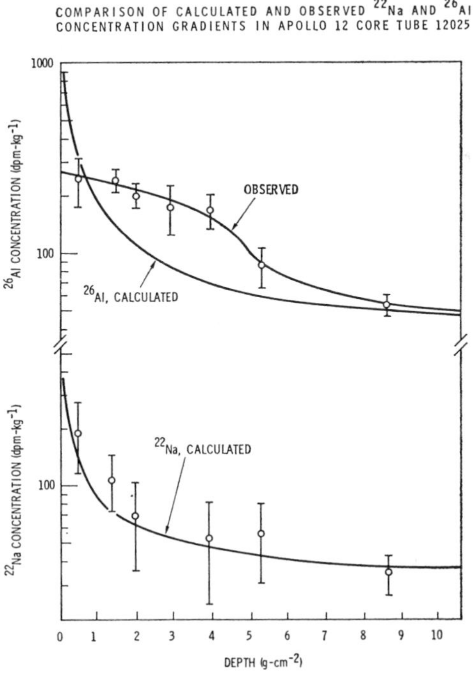

Fig. 2. Comparison of calculated and observed ^{22}Na and ^{26}Al concentration gradients in Apollo 12 double core tube 12025.

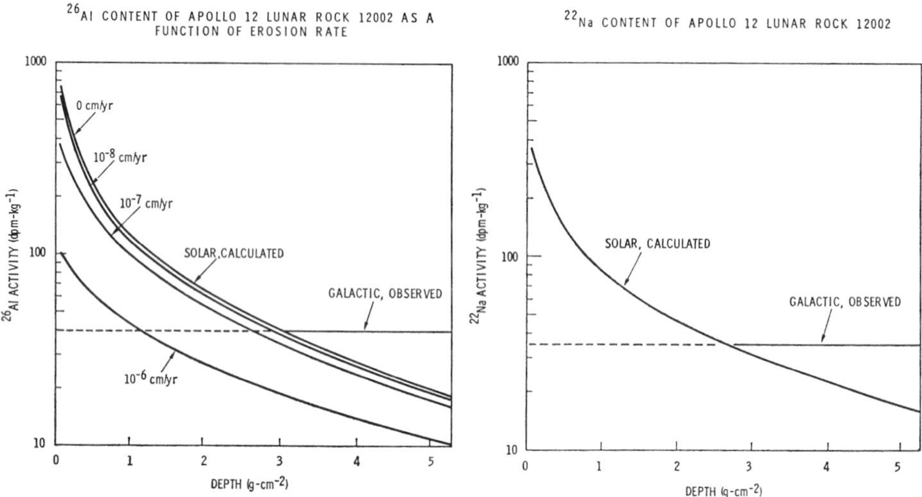

Fig. 3 (a) Calculated ^{26}Al Content of Lunar rock 12002 as a function of erosion rate (left); (b) calculated ^{22}Na content of Apollo 12 lunar rock 12002 (right).

values of about 400 and 800 dpm per kg respectively, in the first 0.5 mm. Thus, concentration changes of some fifteen-fold for ^{26}Al between the top and bottom of a few cm thick rock could exist if no erosion occurred. Erosion rates of 10 mm per million years and 1 mm per million years respectively, would lower the concentration in the first 0.5 mm by about eightfold and twofold, respectively. Since erosion would not affect the ^{22}Na concentration gradient, a tenfold change in concentration between the top and bottom surface could exist.

Orientations and surface radionuclide concentrations

Two methods were employed for lunar surface orientation studies of the lunar rocks. The first involves the analysis of gamma-ray spectrometric data accumulated in the normal coincidence counting of the samples. Coincidence counting rates in the two large NaI (Tl) crystals which view the samples are particularly sensitive to radionuclide location within the sample, especially where three gamma-rays are emitted per disintegration, as is the case for ^{22}Na and ^{26}Al. Thus the surface(s) with the highest ^{22}Na and ^{26}Al can be identified. A second and more sophisticated technique involved the beta-multiple gamma detection system described above. Table 4 summarizes the results on lunar rock orientation and ^{26}Al surface concentrations as determined by normal gamma coincidence counting and the beta-multiple gamma coincidence techniques.

From normal gamma coincidence of lunar rocks 12005,0, 12016, 12051,1, 12051,3 and 12063,0, the highest ^{22}Na concentrations were found to be at the top for the designated orientations as designated in Table 4 and Fig. 4. The highest ^{26}Al concentrations were also on the top for all except 12063,0. These observations indicate that the designated orientations in Table 4 were similar to the most recent lunar

Table 4. Lunar rock orientations and surface concentrations of ^{26}Al.

Rock	^{26}Al Surface Concentrations Top	Bottom	Most Recent Rock Orientation* (γ-Coincidence)
12005,0	164 ± 17	91 ± 12	A
12016,0	135 ± 16	119 ± 12	B
12051,1	160 ± 20	159 ± 18	C
12051,3	242 ± 18	120 ± 10	D
12063,0	—	—	E

* A Approximately as shown in Fig. 4; B Inverted from orientation shown in Fig. 4; C, D Approximately as shown in Fig. 4 but with small end partially submerged; E Inverted from orientation shown in Fig. 4.

surface orientation. However, since rock 12063 showed the highest ^{22}Na concentration on the designated top and the highest ^{26}Al on the bottom, then its most recent orientation must have been opposite to the orientation indicated in Fig. 4. Also, to have the highest ^{26}Al on the bottom of the rock's orientation on the lunar surface would require that the rock be turned over in the past few hundred thousand years. This recent change in the orientation of rock 12063,0 has also been suggested from track measurements (Crozoz *et al.*, 1971). The measured ^{26}Al surface concentrations by beta-multiple gamma coincidence counting of the top and bottom halves of lunar rocks 12005,0, 12016,0; 12051,1; and 12051,3 for their designated orientation are also shown in Table 4. These surface concentrations confirm the orientation observed by the normal counting procedures for all except rock 12051,1, where statistical uncertainty precludes this judgment. Of much greater interest, however, are the relative and actual surface concentrations of ^{26}Al. If the designated bottom surfaces were not exposed to solar protons, within the last few million years, their concentrations should be on the order of 55 dpm per kg. There may be some bias in our measured values due to differences in surface roughness of the samples and the mockup standards and plastic bag folding; however, at present it appears that this bias could be no greater than 10 to 20%. However, it is apparent that the designated rock bottoms are significantly higher in ^{26}Al than the expected 55 dpm/kg had they been submerged in the lunar surface. Of particular interest is the fact that the top surfaces show maximum concentrations of less than 250 dpm per kg in the top 0.2 mm as measured by the beta-multiple gamma coincidence counting. If they were completely exposed to the solar proton flux, they would have surface ^{26}Al concentrations on the order of 800 dpm ker kg, assuming zero erosion. This estimated value of 800 dpm per kg versus a calculated value of about 1000 dpm per kg for the first 0.2 mm of depth would be expected because of curved surfaces of the lunar rocks. If these lunar rocks were precisely half buried in the lunar soil, then one could safely estimate the ^{26}Al concentration of the exposured surface from our knowledge that the buried surface must have a concentration of about 55 dpm per kg. The difference between our measured bottom surface concentration and the true bottom concentration would then be due to the fact that part of our designated bottom was actually exposed on the lunar surface. From these considerations we estimate that the average surface concentrations (in the top 0.2 mm) could have been as high as 400 dpm per kg, or

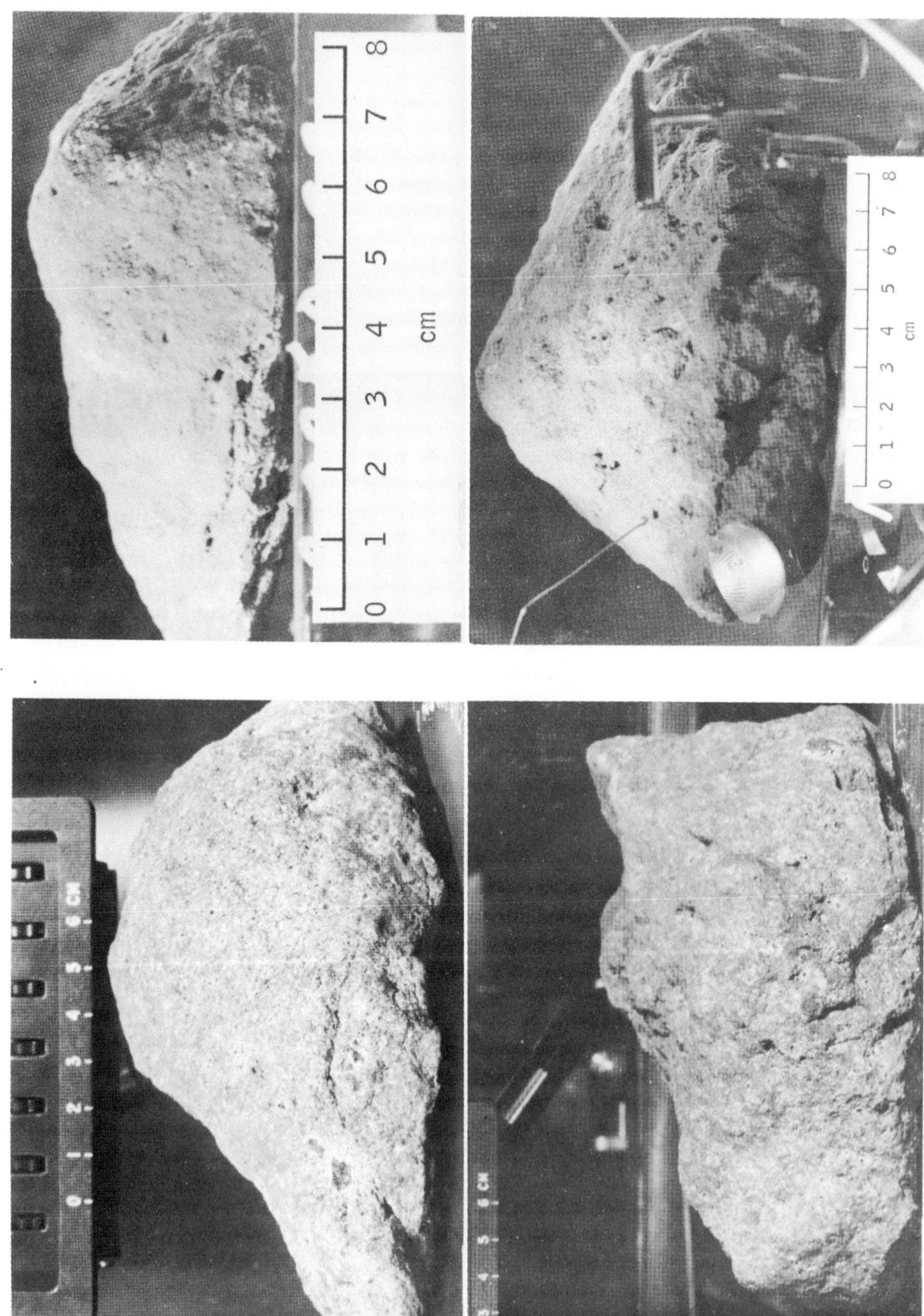

Fig. 4. Apollo 12 samples for which the lunar surface orientation has been determined. Rocks 12005 (upper left); 12016 (upper right); 12051 (lower left); 12063 (lower right).

about one-half of the equilibrium concentration. This rather low concentration requires an erosion rate of about 1 mm per million years, but must be considered with caution since the actual burial depth of these samples is not known.

Erosion and mixing rates

A much more precise estimate of lunar rock erosion rates can be obtained from examination of the ^{26}Al concentrations in "vertical sections" through rocks 12053 and 12002. The concentrations in various depth sections from the top to bottom surfaces of these rocks are included in Table 2. The calculated ^{26}Al contents as a function of erosion rate in the top centimeter of lunar rocks with the chemical compositions of 12002 and 12053 are plotted in Fig. 5. The highest ^{26}Al concentration of 152 dpm per kg in the top 0.8 cm section of rock 12053 is consistent with a surface concentration of about 400 dpm per kg and an erosion rate of about 1 mm per million years. On comparing the ^{26}Al concentration in this top specimen (12053,38) with that in an equivalent weight thickness of lunar fines (Table 3 and Fig. 2), and correcting for chemical composition, it appears that the ^{26}Al is equal to about one-half the saturation value and this would also be achieved with an erosion rate of about 1 mm per million years.

The ^{26}Al concentration in the surface specimen of rock 12002 is about two-thirds that of 12053,38. Based on chemical composition, rock 12002 should have a 5% lower ^{26}Al content and a 50% higher ^{22}Na content than 12053 for equal erosion rates. That rock 12002 is at saturation seems clear from a comparison of the ^{26}Al concentrations in deep sections of this rock with similar depths in rock 12053. The relatively low ^{26}Al concentration in specimen 12002,93 might possibly have resulted if it occupied a position of low vertical angle in the lunar orientation of the rock. Its relatively low ^{22}Na content supports the possibility that it occupied a low vertical angle on the moon. The alternate explanation that the erosion rate was substantially higher, on

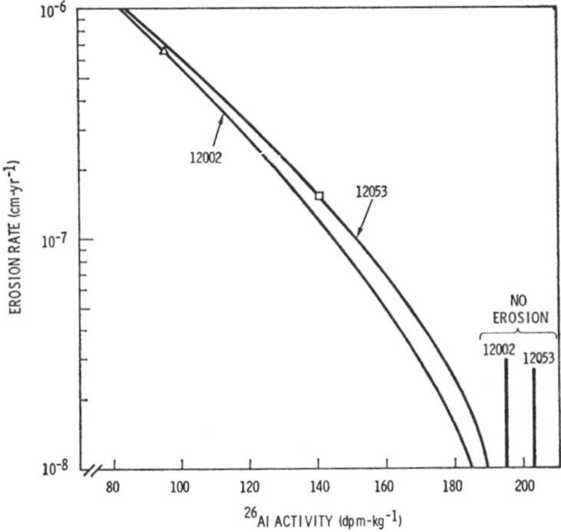

Fig. 5. Calculated ^{26}Al content in the top centimeter of lunar rocks as a function of erosion rate.

the order of 4 or 5 mm per million years cannot be precluded. However, this erosion rate is substantially higher than estimates based on cosmic ray track measurements on other rocks (Crozaz *et al.*, 1971; Barber *et al.*, 1971; Fleisher *et al.*, 1971).

Plotted in Fig. 2 are the vertical concentration profiles of ^{22}Na and ^{26}Al in the first 6 cm of depth of the double core tube 12025. The depth profile in units of g/cm^2 for ^{26}Al in the soil (Table 3) is very similar to that in the lunar rocks (Table 2) beyond the first few centimeters. For this comparison a density of 1.8 g/cm^3 (Apollo 12 Preliminary Science Report, 1970) was assumed. The very striking observation, however, is that the ^{26}Al concentration gradient in the first 2 cm is nearly flat. That this is not due to mixing of this area of the core either during or subsequent to sampling is attested to by the fact that a steeply descending gradient is observed for the 2.6 yr ^{22}Na (See Figs. 3a and 3b). To obtain the observed concentration profile for the ^{26}Al (7.6×10^5 yr) would require rather complete mixing of soil on the lunar surface to a depth of 2 or 3 cm every 10^5 years. The uniform stratigraphy of this section of the core tube supports mixing to this depth, and precludes the possibility that the flat ^{26}Al gradient has resulted from recent deposition of new material. The concentrations of both ^{22}Na and ^{26}Al in the double core tube 12025, 12028 decrease to a relatively constant value at depths greater than 10 cm. The average ^{26}Al content in the 10 cm to 39 cm depth section where galactic cosmic ray production predominates is 45 dpm/kg. This value agrees well with a value of 48 dpm/kg which was calculated by assuming a 2π exposure and using the method of Fuse and Anders (1969) which relates ^{26}Al content to chemical composition of meteorites. This rather good agreement would strongly suggest that the average galactic cosmic ray flux in meteorites is very similar to that at 1 A.U. The ratio of ^{26}Al to ^{22}Na in the deep portion of the core tube is about 1.0. This is substantially higher than the ratio of 0.78 in chondrites (Rancitelli *et al.*, 1969, Rancitelli *et al.*, 1971), but is properly accounted for by the lower magnesium content of lunar material and hence the lower ^{22}Na production in lunar soil.

The depth gradients for ^{22}Na and ^{26}Al are particularly significant because of the information they provide on soil mixing and radionuclide production rates, but they also provide an important base line for other purposes. From the ^{26}Al and ^{22}Na concentrations in the Apollo 12 bulk fines samples 12070,3, an average sampling depth of 3.5 cm is indicated. The soil from the Surveyor II Scoop (Table 3) indicates a burial depth of 3 cm, and assuming similar mixing rates at Tranquillity Base, an average sampling depth for the bulk fines sample, 10084, of 5 cm is indicated. The ^{22}Na and ^{26}Al concentrations in the trench rock 12034 (see Table 1) are consistent with a burial depth of greater than 5 cm. The ^{26}Al to ^{22}Na ratio of 1.13 with the high ^{22}Na concentration of 72 dpm per kg in rock 12005,0 is consistent with concentrations present in bulk fines sample 12070,3 and possible lunar rock erosion rates. Rock 12005,0 is a thin relatively flat rock of about 4 cm maximum thickness. Its ^{22}Na and ^{56}Co concentrations, although twice that of the other lunar rocks, were the same as those in fines sample 12070,3, while ^{26}Al was one-half that in the fines. Such a thin rock would have a minimum ^{26}Al production from galactic or higher energy solar protons, but would, of course, have a large surface area for production from low energy solar protons. Rapid erosion over the large surface of a rock with this geometry could explain the observed ^{22}Na and ^{26}Al concentrations. It is recognized, of course, that a very unusual chemical composition could also explain the unusual ^{26}Al:^{22}Na ratios and that our explanation must be regarded as tentative until chemical analyses are available.

Cosmic ray exposure ages

Estimates of the cosmic ray exposure age for rock 12062 were made by the ^{22}Na–^{21}Ne method based on reported noble gas measurements (LSPET, 1970). For this calculation it was assumed that the production rates of ^{22}Na and ^{22}Ne were equal and the ^{22}Ne–^{21}Ne ratio was 1.05. This calculation gives a ^{22}Na–^{21}Ne exposure age for rock 12062 of 164 million years, suggesting a long surface or near surface life.

Primordial radionuclides

The concentrations of the primordial radionuclides K, U, and Th in lunar materials allow a comparison of the moon's geochronology with other objects in the solar system and help establish the types of processes which have shaped the moon's surface.

The concentrations of potassium, uranium, and thorium in several crystalline rocks, a breccia (12034), and the bulk fines (12070), are included in Table 1. The potassium content of five rocks fall within the narrow range of 0.044 to 0.059% in agreement with the earlier findings of LSPET (1970), while rock 12005 contains 0.026% K, the lowest yet observed in lunar material. The Th and U content of the crystalline rocks varied over a wider range, 0.416 to 0.99 and 0.109 to 0.241 ppm, respectively, with rock 12005,0 containing the lowest concentrations observed in lunar materials, suggesting an origin independent of the other crystalline rocks reported on to date. The K, U, and Th content of the Apollo 12 crystalline rocks is one-third to one-fourth those observed on the Apollo 11 samples with one notable exception. Apollo 11 rock 10003, classified as a group 2 rock by COMPSTON *et al.*, 1970, contained Th, U, and K in concentrations similar to those of the Apollo 12 rocks. In marked contrast to the Apollo 11 samples, the K, U, and Th concentrations in Apollo 12 breccias and fines are very high compared with the crystalline rocks. The K, U, and Th of the breccia 12034 are seven to tenfold higher than in the crystalline rocks. The Apollo 12 bulk fines, 12070, were also enriched in their primordial radionuclide content relative to the crystalline rocks, but to a lesser degree than the breccia. It is therefore evident that the soil is not a simple erosion product of these crystalline rocks and the breccia are not simple impact products of surrounding soil. It is tempting to suggest that the soil is a mixture of material composed primarily of crystalline rocks with a 10 to 15% admixture of material with the high primordial radioisotope composition of the feldspathic differentiate rock 12013, however, this is unsatisfactory since the low K/U ratios indicate the soil and breccia are on the average more highly differentiated than any of the crystalline rocks. As suggested by several groups at the Apollo 12 lunar science conference, the soil contains a highly differentiated component similar to that reported in 12031 which could account for the apparent highly differentiated nature of the soil.

The relative atomic abundances of Th/U in the six Apollo 12 crystalline rocks were constant at 3.8 ± 0.1, in excellent agreement with our previous measurement of 3.8 ± 0.2 in three Apollo 11 rocks and the calculated value of 3.8 ± 0.3 for the present day solar system (FOWLER and HOYLE 1960). The ratio in the breccia 12034 and fines 12070 of 4.1 is close to the values in the rocks when the statistical uncertainty is considered. The K/U ratio of the rocks which ranged from 1900 to 3100 is larger than that reported by LSPET (1970) and comparable to the range reported in the Apollo 11 samples (O'KELLEY *et al.*, 1970a, PERKINS *et al.*, 1970a). We confirm the LSPET (1970) findings that the K/U ratio in the breccia and soil were one-half those of the crystalline rocks at the Apollo 12 site, a marked contrast to the Apollo 11 findings where the soil and rocks had about the same K, U, and Th content. These low K/U ratios in the soil and breccias, as stated above, offer a strong indication the material from which the soil was derived underwent extensive differentiation unique from the processes that formed the crystalline rocks at the Apollo 12 site.

It is interesting to note, although this may be an artifact of the small number of samples, that the K/U ratios of crystalline rocks measured here and by LSPET (1970) fall into several distinct groupings, 1800–2000, 2400–2500, and 2800 to 3100. The constant K content of the Apollo 12 rocks suggests that these rocks may represent samples from a separate basalt flow and the low K rock 12005 is a sample from an independent flow. If this proves to be the case, the K/U ratios may result from localized segregation from an initially homogeneous lava as suggested by COMPSTON *et al.*, 1970 for the group 2 Apollo 11 rocks.

The Th content of the double core tube 12025, 12028 (see Table 3) was constant within the uncertainty of the measurement to a depth of 9 cm and comparable to the Th content of the bulk fines 12070,13. A discontinuity in the Th content was noted in the 9.4 to 16.4 cm depth section of the core. The average Th content of the six samples 12028, 11, 13, 14, 16, 17, 18 was 3.2 ± 0.2 ppm, about one-half the value for overlying material. The uranium content of this section of 0.90 ± 0.09 ppm was also low compared with the bulk fines 12070, however, the Th/U ratio remained unchanged at 3.6 ± 0.4. The existence of this discontinuity in the primordial radionuclide content of the core tube in the region of 9 to 16 cm depths indicates the soil is not well mixed to these depths and that a unique type of material was deposited during this period of regolith formation. The Th and U content of the core tube 12028 at depths below the 16.4 cm depth are significantly higher than the top of the core tube and the bulk fines while the Th/U content remains relatively constant.

The rare gas content of the type AB rock 12062 was reported earlier by LSPET (1970). On the basis of their rare gas concentrations and the primordial radionuclide measurements of this work, K–Ar and U–Th–He gas retention ages were calculated. This crystalline rock showed respective ages of 2.61 and 1.50 billion years, indicating a substantial difference in the rate of rare gas loss.

Acknowledgments—We wish to thank R. M. Campbell, D. R. Edwards, J. G. Pratt and J. H. Reeves of this Laboratory for their aid in standards preparation and in data acquisition. The unique and sensitive instrumentation which made this work possible was developed during the past decade under sponsorship of the United States Atomic Energy Commission, Division of Biology and Medicine.

This paper is based on work supported by the National Aeronautics and Space Administration—Manned Spacecraft Center, Houston, Texas under Contract NAS 9–7881.

References

ALBEE A. L., BURNETT D. S., CHODOS A. A., EUGSTER O. J., HUNEKE J. C., PAPANSTASSIOU D. A. PODOSEK F. A., PRICE R. G., II, SANZ H. G., TERA F., and WASSERBURG G. J. (1970) Ages, irradiation history, and chemical composition of lunar rocks from the Sea of Tranquillity. *Science* **167**, 463–466.

Apollo 12 Preliminary Science Report, NASA SP–235, 1970.

BARBER D. J., HUTCHEON I., PRICE P. B., RAJAN R. S., and WENK H. R. (1971) Exotic particle tracks and lunar history. Second Lunar Science Conference (unpublished proceedings).

BRUNFELT A. O., HEIER K. S., and STEINNES E. (1971) Determination of 40 elements in Apollo 12 materials by neutron activation analysis. Second Lunar Science Conference (unpublished proceedings).

COMPSTON W., CHAPPELL B. W., ARRIENS P. A., and VERNON M. J. (1970) The chemistry and age of Apollo 11 lunar material. *Proc. Apollo 11 Lunar Sci. Conf., Geochim. Cosmochim. Acta* Suppl. 1, Vol. 2, pp. 1007–1027. Pergamon.

CROZAZ G., WALKER R., and WOOLUM D. (1971) Nuclear track studies of dynamic surface processes on the moon and the constancy of solar activity. Second Lunar Science Conference (unpublished proceedings).

FINKEL R. C., ARNOLD J. R., REEDY R. C., FRUCHTER J. S., LOOSLI H. H., EVANS J. C., SHEDLOVSKY J. P., IMAMURA M., and DELANY A. C. (1971) Depth variations of cosmogenic nuclides in a lunar surface rock. Second Lunar Science Conference (unpublished proceedings).

FLEISCHER R. L., HART H. R., Jr., and COMSTOCK G. M. (1971) Very heavy solar cosmic rays: Energy spectrum and implications for lunar erosion. Second Lunar Science Conference (unpublished proceedings); *Science* (in press).

FOWLER W. A., and HOYLE F. (1960) Nuclear cosmochronology. *Ann. Phys.* **10**, 280.

FUSE K. and ANDERS E. (1969) Aluminum-26 in meteorites—VI. Achondrites. *Geochim. Cosmochim. Acta* **33**, 653–670.

GOLES G. G., DUNCAN A. R., OSAWA M., MARTIN M. R., BEYER R. L., LINDSTROM D. J., and RANDLE K. (1971) Analyses of Apollo 12 specimens and a mixing model for Apollo 12 "Soils". Second Lunar Science Conference (unpublished proceedings).

HSIEH J. and SIMPSON J. (1970), Personal communication. University of Chicago, Chicago, Illinois.

LSPET (Lunar Sample Preliminary Examination Team) (1970) Preliminary examination of lunar samples from Apollo 12. *Science*, **167**, 1325–1339.

O'KELLEY G. D., ELDRIDGE J. S., SCHONFELD E., and BELL P. R. (1970a) Elemental compositions and ages of lunar samples by nondestructive gamma-ray spectrometry. *Science* **167**, 580–582.

O'KELLEY G. D., ELDRIDGE J. S., SCHONFELD E., and BELL P. R. (1970b) Primordial radionuclide abundances, solar proton and cosmic-ray effects and ages of Apollo lunar samples by nondestructive gamma ray spectrometry. *Proc. Apollo 11 Lunar Sci. Conf., Geochim. Cosmochim. Acta* Suppl 1, Vol. 2, pp. 1407–1423. Pergamon.

PERKINS R. W., RANCITELLI L. A., COOPER J. A., KAYE J. H., and WOGMAN N. A. (1970a) Cosmogenic and primordial radionuclides in lunar samples by nondestructive gamma-ray spectrometry. *Science* **167**, 577–580.

PERKINS R. W., RANCITELLI L. A., COOPER J. A., KAYE J. H., and WOGMAN N. A. (1970b) Cosmogenic and primordial radionuclide measurements in Apollo 11 lunar samples by nondestructive analysis. *Proc. Apollo 11 Lunar Sci. Conf., Geochim. Cosmochim. Acta* Suppl. 1, Vol. 2, pp. 1455–1469. Pergamon.

RANCITELLI L. A., PERKINS R. W., COOPER J. A., KAYE J. H., and WOGMAN N. A. (1969) Radionuclide composition of the Allende meteorite from nondestructive gamma-ray spectrometric analysis. *Science* **166**, 1269–1272.

RANCITELLI L. A., PERKINS R. W., and WOGMAN N. A. (1971) Unpublished data on the Denver, Dwaleni, Kiffa, Lost City, Malakal, Nejo and Ucera meteorites.

RANCITELLI L. A. and WOGMAN N. A. (1971) Unpublished data. Battelle Memorial Institute, Pacific Northwest Laboratories, Richland, Washington.

Shedlovsky J. P., Honda M., Reedy R. C., Evans J. C., Jr., Lal D., Lindstrom R. M., Delany A. C., Arnold J. R., Loosli H., Fruchter J. S., and Finkel R. C. (1970a) Pattern of bombardment-produced radionuclides in rock 10017 and in lunar soil. *Science* **167**, 574–576.

Shedlovsky J. P., Honda M., Reedy R. C., Evans J. C., Jr., Lal D., Lindstrom R. M., Delany A. C., Arnold J. R., Loosli H., Fruchter J. S., and Finkel R. C. (1970b) *Proc. Apollo 11 Lunar Sci. Conf., Geochim. Cosmochim. Acta*, Suppl. 1, Vol. 2, pp. 1503–1532. Pergamon.

Willis J. P., Ahrens L. H., Danchin R. V., Erlank A. J., Gurney J. J., Hofmeyr P. K., McCarthy T. S., and Orren M. J. Some interelement relationships between lunar rocks and fines, and stony meteorites. Second Lunar Science Conference (unpublished proceedings).

Proceedings of the Second Lunar Science Conference, Volume 2, pp. 1773–1789.
The M.I.T. Press, 1971.

Depth variation of cosmogenic nuclides in a lunar surface rock and lunar soil

R. C. FINKEL, J. R. ARNOLD, M. IMAMURA, R. C. REEDY, J. S. FRUCHTER, H. H. LOOSLI,* J. C. EVANS,† and A. C. DELANY‡
Department of Chemistry, University of California, San Diego, La Jolla, California 92037

and

J. P. SHEDLOVSKY
NCAR, Boulder, Colorado 80301

(*Received 22 February 1971; accepted in revised form 31 March 1971*)

Abstract—Two pieces of lunar rock 12002 were divided into depth fractions nominally 0–1, 1–2, 2–4, 4–9, 9–20, 20–60, and ~60 mm depth below the upper surface. Depth profiles of short- and long-lived radionuclides were obtained by analyzing these fractions. Effects of solar particle bombardment are visible down to the 9–20 mm fraction and may extend deeper. Very high values (731 dpm/kg for Fe^{55}, 533 dpm/kg for Co^{56}) are observed in the top layer.

The comparison of short- and long-lived activities can be made in much more detail with these data than was previously possible. The closely parallel profiles of such species as 2.6 yr Na^{22} and 7.4×10^5 yr Al^{26}, produced by similar reactions, permit a firm inference that the flux and spectral shape of the solar particle spectrum averaged over 10^6 years has been similar to that observed recently. In this and perhaps in other ways the sun's radiation has varied little. From the depth profiles of isotopes with different half-lives, the model developed by SHRELLDALFF (1970) and extended by REEDY and ARNOLD (1971) is used to calculate fluxes and spectra for various periods into the past.

The profile for 3.7×10^6 yr Mn^{53} suggests that some surface erosion has taken place, perhaps of the order of 0.5 mm/10^6 years.

We also measured Mn^{53} depth profiles in Apollo 11 core samples 10004 and 10005. Both core samples show gradients which suggest that there has been no appreciable gardening at the Apollo 11 site on a depth scale of centimeters for times on the order of several million years.

INTRODUCTION

LUNAR SAMPLES PERMIT us to study the history of solar and galactic cosmic rays in materials which come from a known location in the solar system, and which have never been ablated. In addition to the expected effects of galactic cosmic radiation, clear evidence of long term bombardment by solar particles in the region of tens of MeV per nucleon was obtained from studies of Apollo 11 samples. These studies included measurements of radioactivity (SHRELLDALFF, 1970; O'KELLEY *et al.*, 1970; PERKINS *et al.*, 1970), stable nuclides (MARTI *et al.*, 1970), and nuclear tracks (LAL *et al.*, 1970; FLEISCHER *et al.*, 1970; PRICE and O'SULLIVAN ,1970; CROZAZ *et al.*, 1970).

From our own work on radionuclides in Apollo 11 rock 10017 it was possible to define approximate mean solar proton fluxes during various periods of the past, but

* Present address: Univ. Bern, Institut Physik, Bern, Switzerland.
† Present address: Brookhaven National Labs., Upton, Long Island, N.Y.
‡ Present address: NCAR, Boulder, Colorado 80301.

the depth effects were not defined with sufficient precision to allow us to make quantitative statements about the spectral index (hardness) of the bombarding radiation. In addition the separation of the solar and galactic components of production was very difficult.

In the work on samples of Apollo 12 rock 12002 reported here, we have studied the activity of a number of long- and short-lived nuclides as a function of depth down to about 6 cm (the rock thickness). These profiles have been used to estimate the mean fluxes and rigidities for periods comparable with the mean life of each nuclide. In order to study lunar surface processes, we have also measured Mn^{53} in Apollo 11 core samples 10004 and 10005.

EXPERIMENTAL TECHNIQUES

We received two pieces of rock 12002, a vertical slice (slice A 12002,22, wt. 145.88 g), and a horizontal slice (slice D–1 12002, 34, wt. 51.18 g), and two sets of core samples from Apollo 11, 10004 (0, 3.3, 6.6, 9.9, 13.2 cm depth) and 10005 (0, 2.6, 5.2, 7.8, 10.4 cm depth). The location of the rock samples is shown in Fig. 1. As in the case of rock 10017 previously, we are indebted to E. Schonfeld of the Radiation Counting Laboratory, Houston, for determining the orientation of this rock on the lunar surface (SCHONFELD and O'KELLEY, 1971). This determination was confirmed by track data (ARRHENIUS *et al.*, 1971), by pit counts, and by the differing physical properties of the top and bottom surfaces of slice A as observed during grinding.

After removing surface grains for track analysis (ARRHENIUS *et al.*, 1971), the samples were divided into seven depth regions, as in Table 1, combining material from both pieces down to the ~8 mm thickness of slice D–1. The samples were removed (from the top down) by a grinding wheel attachment to a dental drill. Sample and wheel were kept wet with isopropanol, and the ground material washed off frequently into a bath of isopropanol. The depth removed was monitored frequently at 42 points on the surface with a micrometer caliper (slice D–1), and at 17 points with a vernier (slice A). The standard deviation for thickness removed was ± 0.2 mm for the large area slice D–1, and somewhat larger for slice A. Material was removed easily, with little pressure

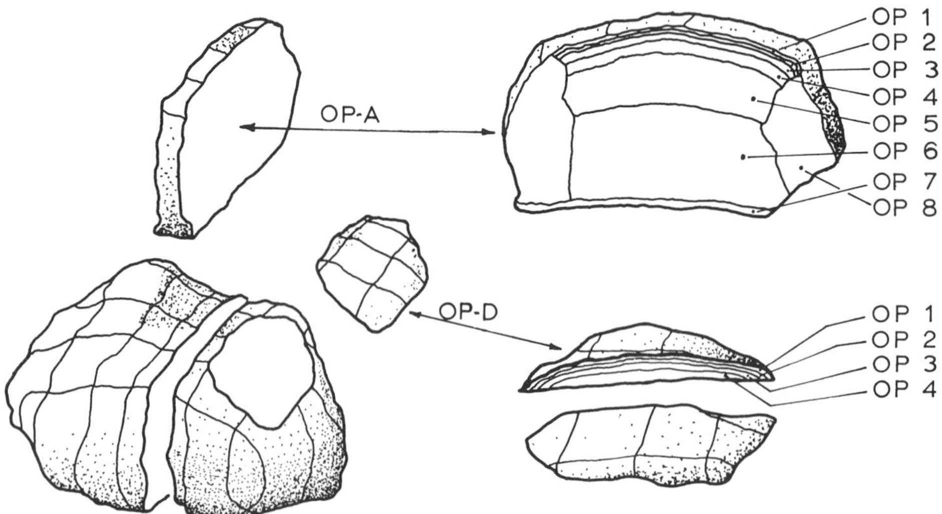

Fig. 1. Schematic of rock 12002, showing location of samples (on left) and subdivision of samples (on right).

Table 1. Activities in rock 12002 (all activities in dpm/kg)

Isotope	$T_{1/2}$	OP–1 9.9 g (0–1 mm)	OP–2 9.5 g (1–2 mm)	OP–3 17.7 g (2–4 mm)	OP–4 25.3 g (4–9 mm)	OP–5 18.2 g (9–20 mm)	OP–6 50.4 g (20–60 mm)	OP–7 4.4 g (60 mm)
Be^{10}	2.5×10^6 y					15.4 ± 3.0	11.5 ± 1.4	
Na^{22}	2.6 y	166 ± 18	122 ± 14	91 ± 11	71 ± 8	54 ± 7	39 ± 6	
Al^{26}	7.4×10^5 y	209 ± 26	154 ± 21	111 ± 16	91 ± 12	72 ± 11	64 ± 6	66 ± 11
Cl^{36}	3.1×10^5 y	13.4 ± 3.8		9.7 ± 2.5	14.7 ± 2.4	8.4 ± 2.1	9.4 ± 1.4	
Mn^{53}	3.7×10^6 y	98 ± 6	86 ± 5	73 ± 5	67 ± 4	52 ± 3	47 ± 3	51 ± 3
Mn^{54}	303 d	98 ± 17	77 ± 12	53 ± 11	50 ± 9	39 ± 10	31 ± 5	
Fe^{55}	2.6 y	731 ± 68	454 ± 46	285 ± 30	171 ± 23	101 ± 17	56 ± 12	
Co^{56}	77 d	533 ± 70	205 ± 30	80 ± 15	32 ± 9			
Co^{57}	270 d	16 ± 6	<6					

Table 2. Chemical Composition of Rock 12002

	This Work	Adopted		This Work	Adopted
Be (ppm)	1.6	1.0	Ti (%)	2.9	1.5
Na (ppm)	1600	1500	V (ppm)	230	204
Mg (%)	8.8	8.8	Cr (ppm)		5460
Al (%)	3.9	4.3	Mn (ppm)	2100	2100
Si (%)		20	Fe (%)	16.8	16.8
K (ppm)	450	450	Co (ppm)	85	68
Ca (%)	5.1	5.4	Ni (ppm)	140	101
Sc (ppm)	35	41			

on the tool, from the top surface. It became gradually more difficult down to a depth of 3–4 mm, where the grinding behavior was similar to that of an oceanic basalt. The presumed cause of this gradient is micro-meteorite impact. The bottom sample (OP–7) showed no trace of this surface friability, confirming the presumption of little or no surface exposure.

After separation, samples were removed for rare gas studies (MARTI and LUGMAIR, 1971) and track analysis. The material was ground to <120 μm size under isopropanol, where necessary. The suspensions were then dried and placed in the teflon still.

The still chemistry was virtually the same as SHRELLDALFF (1970) except for the inclusion of a C^{14} collection system. The C^{14} system, which consisted of a CuO furnace at 550°C followed by two bubblers containing ammoniacal $SrCl_2$, was placed after the main H_2O traps. The exit gases were then collected in an evacuated tank as before. At this stage carriers were added for Be, Cl, V, Co, Ni, and Pb. Be, Na, Al, Cl, Mn, Fe, and Co were then separated and purified following SHRELLDALFF (1970).

The chemical composition of rock 12002, knowledge of which was necessary for determining the percent chemical yield of relevant elements and the concentrations of target nuclei, was measured by atomic absorption spectrometry. The values we measured and the values adopted after considering the analyses of other workers (BRUNFELT *et al.*, 1971; GAST and HUBBARD, 1971; GOLES *et al.*, 1971; LIPSCHUTZ, 1971; SIEVERS *et al.*, 1971; TAYLOR *et al.*, 1971; WILLIS *et al.*, 1971) are given in Table 2.

In the core samples, each of which was several tens of milligrams of fines, only Mn^{53} was determined. We adopted a Mn chemistry similar to that used for the rock. The 2.6 cm and 5.2 cm samples of 10004 were combined, as were the 7.8 cm and 10.4 cm samples, because individually these samples were not large enough to allow an accurate determination of Mn^{53}. Content of Fe and Mn were determined in each sample by atomic absorption spectrometry. The contents of these elements were quite uniform through all samples, the ratio of Fe to Mn being 75. The details of the chemistry will be described elsewhere (FRUCHTER *et al.*, 1971).

Most of the counting was done in detectors which have been previously described (SHRELLDALFF, 1970). However, two new detectors were employed. In the case of Mn^{54} an X–γ coincidence system utilizing a liquid scintillation cell as a coincidence gate for a 3″ NaI(Tl) well detector was used. The Mn sample was dissolved as $MnCl_2$ in a commercially available xylene-based scintillation solution marketed under the trade name Aquasol (New England Nuclear Corp.). A 10 ml quartz scintillation vial containing the sample was placed in the well of a NaI(T1) detector, and the entire assembly was contained in a light-tight box. The efficiency of the liquid scintillation counter, with a single PM tube, for the 5.4 keV Cr X-ray emitted by Mn^{54} was 55%. Since the 3″ NaI(Tl) well crystal had an 11.5% efficiency for the 835 keV γ-ray emitted by Mn^{54}, the total efficiency was 6.3% in the X-γ coincidence system. The background of the counter in a Geiger umbrella was 0.022 cpm in the full-width at the half maximum region of the 835 keV gamma. A detailed description of this system will be published elsewhere (FRUCHTER, 1971).

In the case of Fe^{55} the existing X-ray counting system, briefly described previously (SHRELL-DALFF, 1970), was modified by placing the 153 cm² detector inside a pressure can and operating it at 2 atmospheres absolute pressure, with a flow of P–10 gas. The pressure can had a 10mg/cm² Al window to permit external calibration as described previously. Operation at higher pressure resulted in a shift to a higher energy of the minimum ionization peak due to absorption of residual gammas, thus producing far less structure in the background for the Fe^{55} peak region and consequently facilitating the calculation of the background by the baseline subtraction method. Furthermore, the background was reduced from 0.09 to 0.065 cpm, while achieving an increase in efficiency from 6.9% to 9% because of the increased X-ray absorption.

Another modification of our Apollo 11 procedures was the use of an activation technique for Mn^{53}. The concentration of Mn^{53} was determined in aliquots of OP1–7 by the thermal neutron activation technique ($Mn^{53}[n, \gamma]Mn^{54}$) first proposed by MILLARD (1965). The long half-life of this nuclide coupled with its large thermal neutron capture cross section (σ_{th} = 80 barns for $T_{1/2}$ = 3.7×10^6 yrs [M. HONDA, private communication]) enables us to determine it very precisely in relatively small samples (200 mg) of lunar rock. The increased precision results not only because of the high count rates, but because the small sample size and short counting times allow the measurement of replicate samples.

For each sample triplicate aliquots as well as various required flux monitors were irradiated with a total dose of approximately 2×10^{18} neutrons/cm² in the JRR-3 reactor at Ibaragi, Japan. This reactor was selected because of its very high thermal to fast neutron ratio which prevented large interferences from the $Mn^{55}(n, 2n)Mn^{54}$ and $Fe^{54}(n, p)Mn^{54}$ reactions. The relatively low flux of this reactor gave an amplification factor, (activity 54)/(activity 53), of only 600, but the resulting activity was sufficient for the experiment. This amplification factor was determined by including in the irradiation X-ray standardized samples of Mn^{53} extracted from the Grant meteorite. Corrections from stable $Mn^{55}(n, 2n)Mn^{54}$ were of the order of 15% of the total count rate. After the samples were received in La Jolla, they were subjected to a purification chemistry con-consisting of a mixed-solvent anion column, a TTA extraction, and a cation column. They were then checked for radiochemical purity on a 40 cc Ge(Li) detector and finally counted in a 3″ NaI(Tl) well detector.

The close agreement between our Mn^{53} values and Mn^{53} values determined independently by HONDA *et al.* (private communication) at the University of Tokyo increases our confidence in the accuracy of our numbers.

Manganese 53 in the core samples was determined in a similar way, but the Brookhaven High Flux Beam Reactor was used for their radiation in order to get a higher integrated thermal neutron flux (2×10^{19} n_{th}/cm²). Full details of these procedures will appear in a later paper (FRUCHTER *et al.*, 1971).

RESULTS

Our results for rock 12002 are given in Table 1. Activities are corrected to November 21, 1969. The errors quoted include all known sources of error: counting statistics, calibration, chemical yield, and self-adsorption. The upper limit given for Co^{57} corresponds to a 2σ error added to a possible small signal.

Background counts were taken where possible, using as counting samples for each isotope blanks which had been carried through the same chemical procedures as the actual lunar samples. Efficiencies were determined by means of standard sources prepared from calibrated solutions in forms and thicknesses comparable to the actual counting samples. Self-absorption corrections were calculated by the method of LIBBY (1956) for β emitters. For X-ray emitters the Gold integral for extended sheet sources was used (ROSSI, 1952).

Our results can be compared, for some isotopes, with the nondestructive measurements of rock 12002 by O'KELLEY *et al.* (1971) and RANCITELLI *et al.* (1971). The sample 12002,30 of O'KELLEY *et al.* is the same as our slice OPD (Fig. 1). Their results for Al^{26}, Na^{22}, and Mn^{54} correspond to a suitably weighted average of OP1, OP2, OP3, and OP4. Their measured values and our weighted averages are respectively, Al^{26} (126 $\pm$ 6 dpm/kg, 127 $\pm$ 9 dpm/kg), Na^{22} (86 $\pm$ 3 dpm/kg, 101 $\pm$ 6 dpm/kg), and Mn^{54} (50 $\pm$ 5 dpm/kg, 63 $\pm$ 5 dpm/kg). The errors in our weighted averages are the propagated errors of the various depth fractions. There is reasonable agreement between the two determinations.

RANCITELLI *et al.* measured Na^{22} and Al^{26} in a slab of rock 12002 parallel to our slice A (Fig. 1). Their profile of activity versus depth agrees well with ours for both isotopes, although their Al^{26} values for the deep samples are somewhat smaller than ours. The Mn^{53} data for the core samples are given in Table 3 with the contents of Fe and Mn. The errors are estimated in the same way as in the case of rock 12002.

INTERPRETATION

In order to understand the data summarized in Table 1 and Table 3, we must be able to compare the measured activities with the production profiles expected from various possible sources. The methods of calculating the expected production by galactic and solar cosmic rays are detailed below.

Galactic cosmic-ray production

The expected galactic cosmic-ray (GCR) production of radionuclides was calculated in a manner similar to that used in SHRELLDALFF (1970) and ARNOLD *et al.* (1961). The details for this model are discussed in REEDY and ARNOLD (1971).

Table 3. Mn^{53} in the Apollo 11 core samples.

Depth in the Core (cm)	Estimated Undisturbed Depth (cm)	Fe(%)	Mn(ppm)	Mn^{53} (dpm/120 g Fe)
		Sample 10004		
0	0	12.5	1680	58 $\pm$ 6
3.3	1.5–2	11.9	1580	34 $\pm$ 4
6.6	4 –5	12.4	1650	42 $\pm$ 4
9.9	7.5–11	11.6	1590	29 $\pm$ 6
13.2	14 –20	11.8	1590	26 $\pm$ 4
		Sample 10005		
0	0	12.3	1650	60 $\pm$ 6
2.6, 5.2	1–3.5	12.5	1670	51 $\pm$ 11
7.8, 10.4	5–12	12.3	1660	27 $\pm$ 5

The galactic cosmic-rays are high-energy particles, mostly protons (about 87%) and alpha particles (about 12%), which are stopped in the moon largely by nuclear interactions. The secondaries produced in these interactions can cause further nuclear reactions. Secondary neutrons are the dominant particles for low energy reactions below 100 MeV, and both primary and high-energy secondary particles are important at energies above 100 MeV.

The model for these GCR calculations uses one flux spectrum to represent the energy distribution of all strong-interacting particles, both primaries and secondaries. The energy spectrum used for all particles with energies above 100 MeV was approximated by

$$\frac{\mathrm{d}J}{\mathrm{d}E} = K(\alpha + E)^{-2.5}, \tag{1}$$

where both α, which determines the shape of the spectrum, and K, which is a normalization constant determined from the integral flux of particles with energies above 1 GeV/nucleon, vary with depth.

The flux spectrum used for energies below 100 MeV is different from that used in SHRELLDALFF (1970), and was derived from more recent measurements of the neutron spectrum in the Earth's atmosphere (TAJIMA *et al.*, 1967) and from theoretical calculations for the intensity and shape of the neutron leakage spectrum from the moon (ARMSTRONG and ALSMILLER, 1971). The low-energy portion of this spectrum, especially in the region of 2 to 20 MeV, was also checked by comparison with the observed production of Ar^{37} using experimental cross sections for the $Ca^{40}(n, \alpha)Ar^{37}$ reaction. The calculated production rate of Ar^{37} agrees very well with the measured activities of D'AMICO *et al.* (1970) given below. The details for this low energy portion of the GCR flux spectra are given in REEDY and ARNOLD (1971). The flux spectra used for the depths of 0 and 11 g cm^{-2} (OP–1 and OP–6) are shown in Fig. 2.

The production rate for a given radionuclide at a depth d, $P(d)$, was obtained by using the flux spectrum at that depth, $(dJ/dE)(E, d)$, in the expression

$$P(d) = \sum_i N_i \int \frac{\mathrm{d}J}{\mathrm{d}E}(E, d)\sigma_i(E)\, dE, \tag{2}$$

where $\sigma_i(E)$ is the excitation function for the production of this radionuclide from target element i, and N_i is the number of target atoms per kilogram of rock. The cross sections used for energies below 100 MeV were for neutrons as the incident particle. For energies above 100 MeV, proton spallation cross sections were used. Excitation functions were obtained from experimental results, analogous reactions, or spallation formulae, and were similar to those used in SHRELLDALFF (1970).

Because of the high solar cosmic-ray production of radionuclides in the top layers of the rock, the values for OP–6, the deep sample, were used to compare with the theoretical values obtained using this model. The results are given in Table 4. The calculated values for the surface (OP–1) range from 74% of the OP–6 value (Fe^{55}) to 99% (Be^{10}). The production rates calculated for the radionuclides Ar^{37}, Ar^{39}, and H^3 in rock 12002 at a depth of 11 g cm^{-2} were 28, 6, and 194 atoms min^{-1} kg^{-1}, respectively, which compare to the values obtained for a depth of 4.9–6.4 cm in this rock by D'AMICO *et al.* (1971) of 27.5, 8.0, and 155 dpm/kg, respectively.

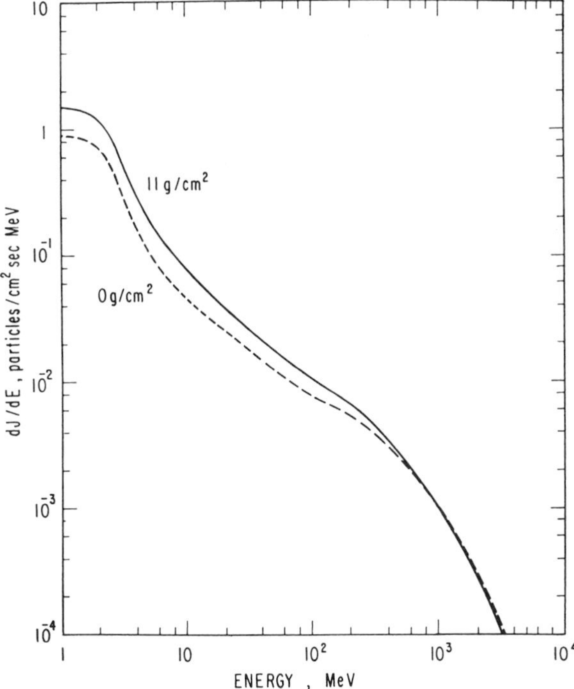

Fig. 2. Galactic flux spectrum in the moon for depths of 0 g/cm² (OP1) and 11 g/cm² (OP6).

Table 4. Production by galactic cosmic rays at 6 cm depth in rock 12002

	Measured dpm/kg	Calculated dpm/kg
Mn^{53}	47	23
Be^{10}	11.5	17
Al^{26}	64	43
Cl^{36}	9.4	7.8
Fe^{55}	56	91
Na^{22}	39	40
Mn^{54}	31	27

The average agreement of the calculated and measured activities for these ten radionuclides is relatively good, with scatter in both directions. Much of the disagreement is probably due to uncertainties in the cross sections for the important reactions producing these radionuclides. This is especially true for Be^{10} and Mn^{53}. There are no systematic differences which could be attributed to errors in the spectral shape used in these calculations.

There is a slight correlation between the disagreement of the calculated and observed production rates and the half-lives of the radionuclides (see Table 4). For the long-lived radionuclides the experimental activities are larger than those calculated, although Be^{10} does not show this trend. There is thus perhaps a suggestion that the GCR flux in the rock was higher several million years ago than it is at present. This conclusion, however, is very tentative.

Solar cosmic-ray production calculations

The solar cosmic-ray (SCR) production of radionuclides was calculated in a manner slightly different from that used in SHRELLDALFF (1970). The calculated values of dpm/cm² in SHRELLDALFF for the short-lived isotopes were too high by a factor of 2; there are also numerical errors in the calculated long-term flux. However, the discussion and general conclusions in SHRELLDALFF are correct. Details of the present model and calculations are given in REEDY and ARNOLD (1971).

The solar cosmic rays are relatively low energy particles, mainly protons, with few particles having $E > 100$ MeV. These particles are emitted from the sun in bursts during certain solar flares. Almost all are emitted in the few years around the period of maximum of the 11-year solar cycle.

The energy distribution of the solar cosmic-ray particles varies from flare to flare and is usually described by an exponential rigidity distribution or a power law distribution. Here an exponential rigidity distribution is used,

$$\frac{dJ}{dR} = k \exp(-R/R_0), \tag{3}$$

where R is the rigidity of the particle, momentum per unit charge, $R = pc/ze$ in units of megavolts (MV) (a 10 MeV proton has a rigidity of 137 MV). k is an integration constant for that flare. Typical values for the shape parameter, R_0, are in the range of 50 to 200 MV for solar flare particles in the energy range of 10 to 100 MeV. An alternative is the kinetic energy power law,

$$\frac{dJ}{dE} = k'E^{-\gamma} \tag{4}$$

where γ typically has values in the range 2.5–4.

Because of the relatively low energy of the solar protons, most of them are stopped by ionization-energy losses before undergoing a nuclear reaction. A 50 MeV proton has a range in lunar material of only 3 g cm^{-2} while its nuclear interaction length is about 100 g cm^{-2}. In the model used to calculate the production of radionuclides by solar protons, the number of secondary particles is assumed to be negligible. The flux at a given point is determined solely by the primary flux and the rate of energy loss by ionization.

The flux of solar protons changes very rapidly with depth, as can be seen from Fig. 3, which shows the differential flux of the solar protons at various depths for a 2π bombardment of an infinite plane. The free space energy distribution used in this example has an exponential rigidity shape with $R_0 = 100$ MV, and a time-averaged flux above 10 MeV of 50 protons/cm² sec. The top two curves show the rapid change in flux between the surface of the rock and a depth of about 1 millimeter. The bottom curve is the calculated flux at the bottom of rock 12002.

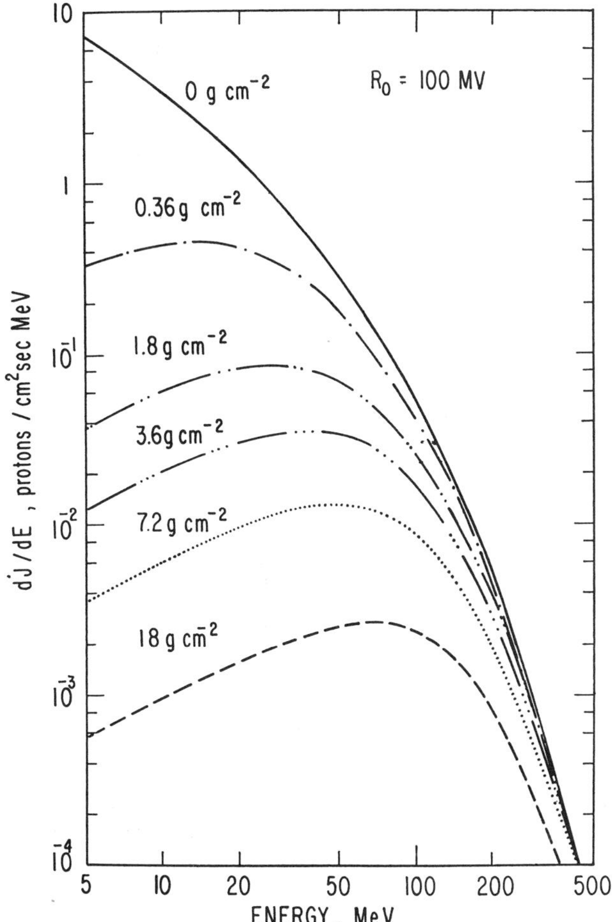

Fig. 3. Differential flux of solar protons at various depths in the moon for a 2π bombardment of an infinite plane. The free space energy distribution has an exponential rigidity shape with $R_0 = 100$ MV and an integral flux above 10 MeV of 50 protons/cm^{2}sec.

The production rates of various radionuclides were obtained for various depths by integrating the product of the differential flux at that depth and the excitation function for the reaction using eq. (2). The excitation functions used for Co^{56}, Fe^{55}, Mn^{54}, Na^{22}, and Al^{26} were taken from thin-target proton cross-section data. The excitation function for the production of Mn^{53} from iron was derived for low energies from measured (p, α) reactions on iron. A further discussion of excitation functions is given in Reedy and Arnold (1971).

The production rate in a layer of finite thickness must be obtained as an average of the production rates at all depths in that layer. The production rate can vary by as much as an order of magnitude over the first millimeter of rock. The solar proton

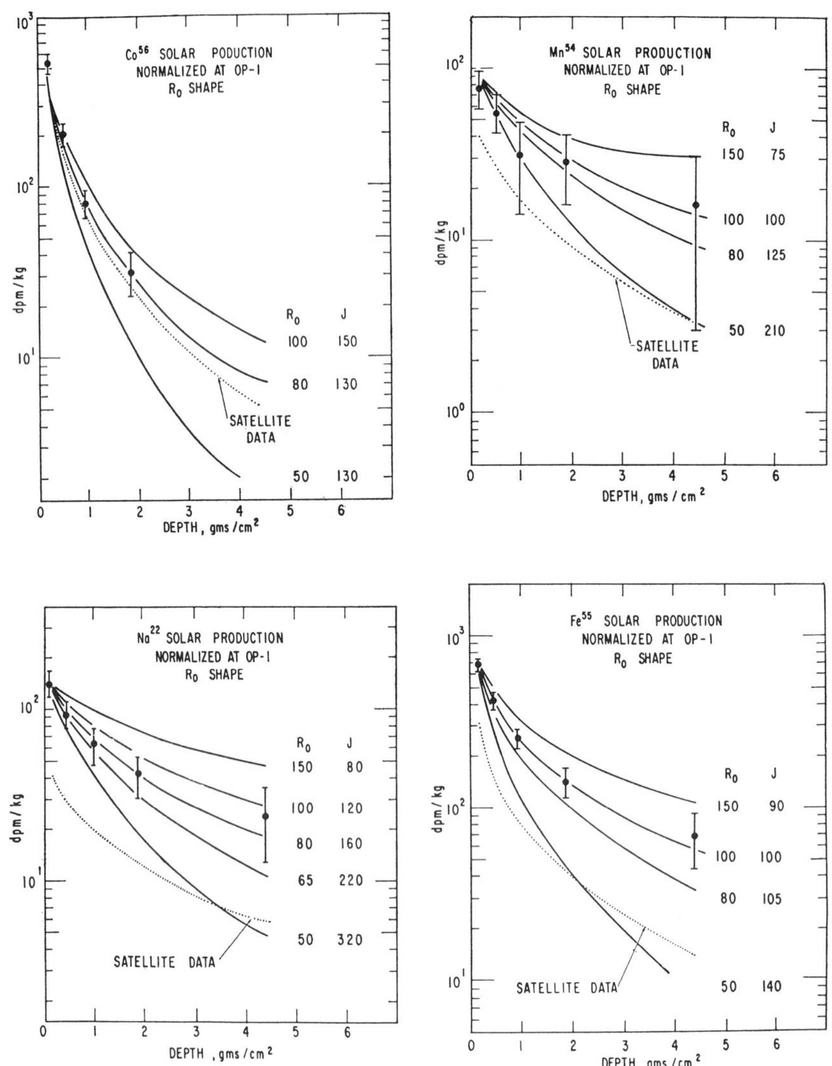

Fig. 4. Observed and calculated solar production for short-lived nuclides. Experimental points are given after galactic production has been subtracted. Solid lines are theoretical curves which have been normalized to the experimental points at OP–1. Dotted lines are production calculations based on absolute determinations of solar flare spectra by satellites and ground-based instrumentation (see text). J is the 4π integral flux above 10 MeV in units of protons/cm^{2}sec. R_0 is the mean rigidity in units of MV.

production profile versus depth are shown for several isotopes and for several theoretical spectral shapes by the solid lines in Figs. 4 and 5. The details for these calculations are given in REEDY and ARNOLD (1971).

In order to compare the above calculations with the measured activities, it is necessary to correct the measurements for the galactic contribution. For the radionuclides Fe^{55}, Mn^{54}, Mn^{53}, Al^{26}, and Na^{22}, the net solar proton-produced activity

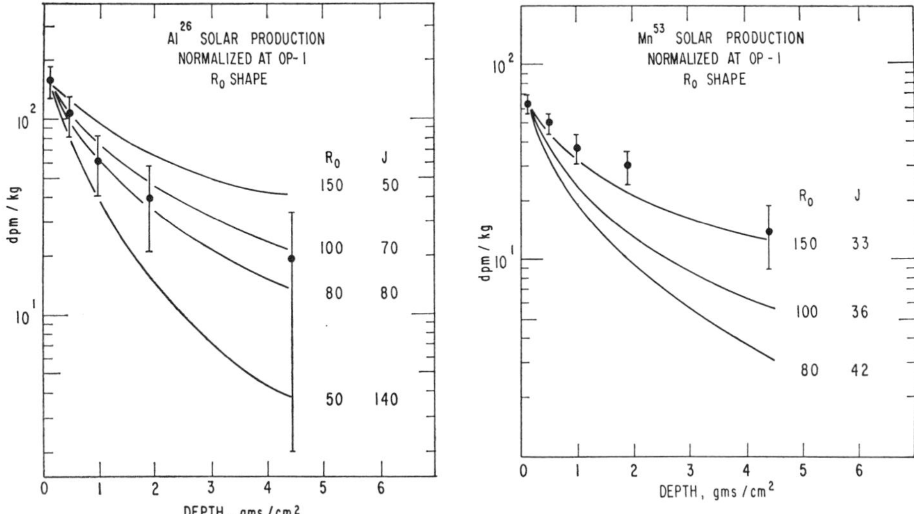

Fig. 5. Observed and calculated solar production for long-lived nuclides. Experimental points are given after galactic production has been subtracted. Solid lines are theoretical curves which have been normalized to the experimental points at OP–1. J is the 4π integral flux above 10 MeV in units of protons/cm^{2}sec. R_0 is the mean rigidity in units of MV.

was calculated for each layer by subtracting the galactic contribution in each layer from the observed activities. No galactic correction was made for Co^{56} as its galactic production rate in the rock is very low, of the order of a few dpm kg^{-1}. For other nuclides a two step procedure was used to determine the galactic contribution to each layer. First, the measured activity in OP–6 was assumed to be of pure galactic origin and was used to normalize the theoretically calculated galactic depth profile (described above) in the rock. Using the calculated galactic profile and the normalization to OP–6, the solar contributions to layers OP–1 to OP–5 were obtained by subtraction. The solar proton spectrum deduced from these values was then used to estimate the solar contribution to OP–6 (usually a few dpm kg^{-1}). This solar contribution was then subtracted from the OP–6 activity to get the pure galactic activity in OP–6 and the galactic activities in the other layers were recalculated using this normalization. The errors ascribed to the net solar activities in each layer were obtained from the experimental error for that layer and the experimental error for the OP–6 layer. These net solar proton-produced activities are the values plotted for these isotopes as points in Figs. 4 and 5.

For short-lived radionuclides it is possible to check these calculations by comparing our measured solar produced activity profiles with profiles calculated by using directly observed solar flares. Data for solar cycles 19 and 20 were required to calculate the production expected for the short-lived radionuclides Co^{56}, Mn^{54}, Na^{22}, and Fe^{55}.

The data for cycle 19 (1954–1964) were taken from WEBBER (1963). The solar proton fluxes during the major flares of cycle 19 were determined by several experiments, mainly ground based, and usually were given for energies above 30 MeV.

The integral fluxes of protons above 10 MeV were estimated for cycle 19 flares by extrapolations from higher energies assuming exponential rigidity shapes. The average shape of the solar proton spectrum during cycle 19 corresponded to a value of R_0 of about 100 MeV. The average omnidirectional integral flux of protons above 10 MeV over the whole cycle was 66 protons/cm^{2}sec.

The experimental flux and spectral data for solar protons emitted during cycle 20 (1965–present) have been taken from BOSTROM *et al.* (1967–70). [The preliminary data for cycle 20 kindly made available by HSIEH and SIMPSON (private communication) and by KINSEY and MCDONALD (private communication), used by SHRELLDALFF (1970) and in our preliminary paper FINKEL *et al.* (1971), are not in final form at this time. We hope to include a full comparison with these experiments in a later publication.] These data were measured by their experiment on IMP–IV (1967–69) and IMP–V (1969–present) for energies above 10 MeV. Data of KINSEY (1969) from IMP–III (1965–67) shows that there was little solar proton emission prior to May, 1967. During the first five years of cycle 20 (1965–67), the average omnidirectional integral flux of protons above 10 MeV was determined to be about 39 protons/cm^2 sec and the average shape corresponded to a value of R_0 of about 60 MV.

The disintegration rates expected for each isotope and each layer from these flares were determined from the observed solar proton fluxes for cycle 20 and the WEBBER data for cycle 19. The flux was assumed to be isotopic and the production calculated for each flare was corrected for decay from the time of the flare to the time of Apollo 12 (November 21, 1969). The calculated disintegration rates for the short-lived isotopes from the observed flares are shown as dotted lines for the isotopes Co^{56}, Mn^{54}, Fe^{55}, and Na^{22} in Fig. 4.

The agreement for 77 day Co^{56} is good. For this isotope two events, those of April 12 and November 3, 1969, contribute 94% of the production. For Mn^{54} the calculated production is too low by about a factor of 2. However, incorporation in our calculations of recent experimental excitation functions (brought to our attention too late to be incorporated here) (JENKINS and WAIN, 1970) for the production of Mn^{54} from Fe^{56} would reduce this discrepancy. For Na^{22} and Fe^{55} the deviations are larger. Since the less accurately measured flares of cycle 19 contribute at least 50% of the production of these nuclides, the deviations observed do not seem alarming. Although there seems to be a systematic tendency for observed fluxes to be smaller than those required to produce the observed activities, we consider the agreement at present to be about as good as may be expected. It is reassuring that Co^{56}, for which the observations are the best, and for which there is no significant galactic production to be subtracted, shows the best agreement.

DISCUSSION

The depth profiles observed in rock 12002 for all the measured nuclides are consistent with the interpretation proposed by SHRELLDALFF (1970) which explained the observed activity gradients in rock 10017 by the rapid decrease of production due to solar cosmic rays with increasing depth below the lunar surface. The gradients for different isotopes correlate with the characteristic energy necessary for their production in a lunar rock: Co^{56}, made by low energy protons on iron, $Fe^{56}(p, n)Co^{56}$, has the lowest threshold and the steepest measured production profile, while Cl^{36} and Be^{10}, made by high energy reactions, have the highest thresholds and the shallowest production profiles.

Similar gradients have been observed by other workers and reported at the Second Lunar Science Conference. Of particular interest are the rare gas measurements of MARTI and LUGMAIR (1971), especially those made on fragments of our sample. The radioactive Kr^{81} ($T_{1/2} = 2.1 \times 10^5$ yrs) shows a gradient similar to ours, while the stable nuclides show a smaller gradient. This and other evidence demonstrates that a long bombardment took place below the surface. D'AMICO *et al.* (1971) reported measurements of Ar^{37} (35 days), H^3 (12.2 yrs), and Ar^{39} (270 yrs) on samples of rock 12002 adjacent to ours. O'KELLEY *et al.* (1971) and RANCITELLI *et al.* (1971) measured data on Al^{26} and other species in various samples of rock 12002. Nuclear track studies of both solar and galactic Fe group nuclei in lunar material are reported by BARBER *et al.* (1971) as are track data for the Surveyor III window.

The agreement between calculated and observed profiles for the short-lived nuclides, discussed above, gives us some confidence in drawing conclusions based on profiles of the longer-lived nuclides. These profiles provide the first detailed look at the mean solar particle flux in the earth-moon region over long periods of time. Of the possible long-lived nuclides in rock 12002, C^{14} will be reported later by BOECKL and SUESS (private communication). We report Cl^{36}, Al^{26}, Mn^{53}, and Be^{10}.

The low absolute production of Cl^{36} and Be^{10}, which is a consequence of their being high energy products, coupled with the small expected gradient, makes any solar effect too small to be observed. Figure 5 shows the observed depth profiles for Al^{26} and Mn^{53} and those calculated for various assumed mean fluxes and rigidities. These calculations assume a negligible erosion rate. Erosion, if significant, would have the effect of lowering the apparent mean flux J and increasing the mean rigidity R_0. The fact that Mn^{53} does require a lower J and higher R_0 compared to Al^{26} suggests the presence of erosion. An erosion rate was calculated on the assumption that J and R_0 have been the same over the mean lives of Al^{26} and Mn^{53}. The best fit to both the Al^{26} and Mn^{53} data was found for an erosion rate of the order of 0.5 mm/10^6 yrs and a constant flux over the past several million years of $J = 100$ protons/cm^2 sec and $R_0 = 80$ MV.

The uncertainty of such a conclusion is great, however. The mean solar proton flux or rigidity, or the erosion rate, or both, may have varied over this period. It is also possible that Mn^{53} is not saturated in rock 12002 because of recent exhumation to the surface. However, our Mn^{53} data for Apollo 11 core samples 10004 and 10005, discussed below, indicate that rock 12002 is, in fact, substantially in equilibrium with surface bombardment conditions. We would prefer therefore to rely on estimates of erosion rates taken from observations of phenomena such as nuclear tracks which vary more rapidly with depth than do radioisotope concentrations.

The striking qualitative fact is that the profile and estimated flux and rigidity for Al^{26} is similar to that observed for the short-lived species. This can be seen most directly by looking at the pair Na^{22}–Al^{26}, produced by similar reactions but with different half lives. The detailed analysis confirms the immediate impression of similarity.

The solar flare activity, as observed at the moon, averaged over various periods of the past, then appears similar in intensity and spectral shape to that observed today. This result is for us somewhat unexpected. Given that a few major events near sunspot maximum are responsible for most of the energetic proton flux observed in each cycle, the near absence of such particles in the years around solar minimum, and finally the striking differences between the two most recent cycles, we might well have expected much larger differences.

In terms of sunspot activity, the present cycle 20 is in the middle range of the cycles observed over the last two hundred years. The mean proton numbers observed for Al^{26} and less clearly for Mn^{53} are much less than those averaged over the years 1958–61 around solar maximum, and very much greater than those averaged over the period 1963–66 around solar minimum (essentially zero).

The million-year mean life of Al^{26} corresponds more or less with the Pleistocene age of terrestrial history, one of the rare periods of ice advance and major climatic fluctuation. One possible source of this fluctuation is the sun itself. Changes on this time scale in the total thermal flux from the sun of the order of 1–2% (comparable to the MILANKOVICH (1938) changes in flux at certain regions of the earth due to changes in the earth's motion) might be sufficient to cause or trigger these events. Such changes have been postulated by climatic theorists. The question of the relation of solar high energy particle flux to the total solar thermal emission is worth exploring.

DRUMMOND (1970), KONDRATYEV and NIKOLSKY (1970), and JERZYKIEWICZ and SERKOWSKI (1968) discuss the available data on the observed variation of the total solar thermal flux (solar constant) over the last few decades. Interpretations conflict but the weight of evidence seems to be that the flux varied less than 0.5%, and perhaps very much less.

If there is no great difference in the dependence of solar thermal flux on solar activity (as measured by proton flux) on a time scale of 10^4 to 10^6 years, and the dependence (or rather independence) observed on a scale of years, then we may conclude that the solar thermal flux has varied little (less than 0.1 times the difference between 1959 and 1964) on the average over the Pleistocene. Of course the stated assumption, while plausible, is unproved. It is also possible that short intervals of fluctuation might have triggered events on the earth without changing the rock record very much. We believe, however, that long-term variations in solar thermal flux, on a scale of more than a percent, must now be considered unlikely.

In order to investigate lunar surface processes in more detail, we have measured the Mn^{53} gradients in two Apollo 11 cores (Table 3). Although there was some distortion during the sampling of the Apollo 11 cores, thermoluminescence properties (DALRYMPLE and DOELL, 1970; HOYT *et al.*, 1970) indicate that the relative position of samples in the core remained intact. The undisturbed depth (Table 3, column 2) is estimated from data in CARRIER *et al.* (1971). The bulk density of the undisturbed soil, although not well known, is estimated to be 1.6 g/cm^2.

Although the discussion of the core data is limited by ambiguities in the sample depth, we can get some information about mixing and deposition at the very surface (≤ 10 cm) of the Apollo 11 site during the mean life of Mn^{53}. The observed activities near the top of the core are consistently higher than those at the bottom. The bottom samples, when adjusted for differences in target abundance (Fe), correspond to 38 ± 6 dpm/kg in rock 12002, in agreement with the OP–6 and OP–7 values. The solar cosmic ray production profile is thus relatively well preserved, although there are some indications of irregularities. However, any mixing or deposition would have to be slight since the very steep production profile of Mn^{53} by solar cosmic rays is easily altered. The higher values at the top lead us to conclude that on the time scale of several million years the site of the Apollo 11 cores has been relatively undisturbed. Using 100 protons/cm^2 sec and $R_0 = 80$ MV for solar cosmic rays we calculate an upper limit of continuous deposition of 2 mm/10^6 yr. If we assume continuous mixing, the upper limit of the mixing rate would be such that an average particle in the soil is displaced vertically 3 cm in 10^7 yr.

Our Apollo 11 data might be compared with the Al^{26} profile of an Apollo 12 core measured by RANCITELLI *et al.* (1971), or the track profiles, the spallation rare gases, and the neutron capture product profiles which have been measured by many workers. However, any detailed discussion should be made from data on the same core. We hope to give a more quantitative discussion after further work on the Apollo 14 samples.

The possibilities for further work on solar particle history by these methods include: measurement of the profile of 8×10^4 yr Ni^{59} produced mainly in Apollo 11 and Apollo 12 rocks by solar alpha particles, and the study of profiles in rocks recently exposed on the lunar surface (12063 and 12064 may be examples). A combination of measurements on selected rocks and cores can now give us information on gardening and other lunar surface processes.

Acknowledgments—We are indebted as before to E. Schonfeld for the correct orientation of rock 12002 and for valuable discussions. Edward Stone, Carl Bostrom, James Kinsey, Frank McDonald, John Hsieh, and John Simpson furnished much useful information on solar flare particles. We have profited by discussions with many other colleagues. Norman Fong, Florence Kirchner, Jack Hollon, Donald Sullivan, and Lawrence Finnin furnished essential support to the work. This research was supported by NASA Contract NAS 9–7891.

REFERENCES

ARMSTRONG T. W. and ALSMILLER R. G., JR. (1971) Calculation of cosmogenic radionuclides in the moon and comparison with Apollo measurements. Second Lunar Science Conference (unpublished proceedings).

ARNOLD J. R., HONDA M., and LAL D. (1961) Record of cosmic-ray intensity in the meteorites. *J. Geophys. Res.* **66**, pp. 3519–3531.

ARRHENIUS G., ASUNMAA S. K., LIANG S., MACDOUGALL D., and WILKENING L. (1971) Irradiation and impact in Apollo rock and soil samples. Second Lunar Science Conference (unpublished proceedings).

BARBER D. J., HUTCHEON I. D., PRICE P. B., RAJAN R. S., and WENK R. (1971) Exotic particle tracks and lunar history. Second Lunar Science Conference (unpublished proceedings).

BOSTROM C. O., WILLIAMS D. J., and ARENS J. F. (1967–1970) Solar proton monitor experiment. Solar Geophysical Data, ESSA, Boulder, Colorado, Vols. 282–309.

BRUNFELT A. O., HEIER K. S., and STEINNES E. (1971) Determination of 40 elements in Apollo 12 materials by neutron activation analysis. Second Lunar Science Conference (unpublished proceedings).

CARRIER W. D., III, STEWART W. J., WERNER R. A., and SCHMIDT R. (1971) Disturbance in samples recovered with the Apollo core tubes. Second Lunar Science Conference (unpublished proceedings).

CROZAZ G., HAACK U., HAIR M., MAURETTE M., WALKER R., and WOOLUM D. (1970) Nuclear track studies of ancient solar radiations and dynamic lunar surface processes. *Proc. Apollo 11 Lunar Sci. Conf., Geochim. Cosmochim. Acta* Suppl. 1, Vol. 3, pp. 2051–2080. Pergamon.

DALRYMPLE G. B. and DOELL R. R. (1970) Thermoluminescence of lunar samples from Apollo 11. *Proc. Apollo 11 Lunar Sci. Conf., Geochim. Cosmochim. Acta* Suppl. 1, Vol. 3, pp. 2081–2092. Pergamon.

D'AMICO J., DEFELICE J., FIREMAN E. L., JONES C., and SPANNAGEL G. (1971) Tritium and argon radioactivities and their depth variations in Apollo 12 samples. Second Lunar Science Conference (unpublished proceedings).

DRUMMOND A. J. (1970) Recent measurements of the solar radiation incident on the atmosphere. COSPAR XIII, Leningrad, Paper A.2.7.

FINKEL R. C., ARNOLD J. R., IMAMURA M., REEDY R. C., FRUCHTER J. S., LOOSLI H. H., EVANS J. C., JR., SHEDLOVSKY J. P., and DELANY A. C. (1971) Depth variation of cosmogenic nuclides in a lunar surface rock. Second Lunar Science Conference (unpublished proceedings).

FLEISCHER R. L., HAINES E. L., HART H. R., JR., WOODS R. T., and COMSTOCK G. M. (1970) The particle track record of the Sea of Tranquillity. *Apollo 11 Lunar Sci. Conf., Geochim. Cosmochim. Acta* Suppl. 1, Vol. 3, pp. 2103–2120. Pergamon.

FRUCHTER J. S. (1971) Development of a high sensitivity liquid scintillation-NaI(Tl) coincidence detector for X–γ emitters in lunar samples. In preparation.

FRUCHTER J. S., IMAMURA M., SHIMOMURA K., and HONDA M. (1971) Neutron activation analysis of Mn^{53} in lunar samples. In preparation.

GAST P. W. and HUBBARD N. J. (1971) Rare earth abundances in soil and rock from the Ocean of Storms. Second Lunar Science Conference (unpublished proceedings).

GOLES G. G., DUNCAN A. R., OSAWA M., MARTIN M. R., BEYER R. L., LINDSTROM D. J., and RANDLE K. (1971) Analyses of Apollo 12 specimens and a mixing model for Apollo 12 "soils." Second Lunar Science Conference (unpublished proceedings).

HOYT H. P., JR. KARDOS J. L., MIYAJIMA M., SEITZ M. G., SUN S. S., WALKER R. M., and WITTELS M. C. (1970) Thermoluminescence, X-ray and stored energy measurements of Apollo 11 samples. *Proc. Apollo 11 Lunar Sci. Conf., Geochim. Cosmochim. Acta* Suppl. 1, Vol. 3, pp. 2269–2287. Pergamon.

JENKINS I. L. and WAIN A. G. (1970) Excitation functions for the bombardment of Fe^{56} with protons. *J. Inorg. Nucl. Chem.* **32**, 1419–1425.

JERZYKIEWICZ M. and SERKOWSKI K. (1968) Causes of climatic change. *Meterol. Mon.* **8**, 142–143.

KINSEY J. H. (1969) A study of low energy cosmic rays at 1 A.U. Ph.D. thesis, University of Maryland and Goddard Space Flight Center report X-611-69-396.

KONDRATYEV K. YA and NIKOLSKY G. A. (1970) Solar radiation and solar activity. *Quart. J. R. Met. Soc.* **96**, 509–522.

LAL D., MACDOUGALL D., WILKENING L., and ARRHENIUS G. (1970) Mixing of the lunar regolith and cosmic ray spectra: Evidence from particle track studies. *Proc. Apollo 11 Lunar Sci. Conf., Geochim. Cosmochim. Acta* Suppl. 1, Vol. 3, pp. 2295–2303. Pergamon.

LIBBY W. F. (1956) Relations between energy and half-thickness for absorption of beta radiation. *Phys. Rev.* **103**, 1900–1901.

LIPSCHUTZ M. E. (1971) Vanadium isotopic composition and contents in lunar rocks and dust from the Ocean of Storms. Second Lunar Science Conference (unpublished proceedings).

MARTI K., LUGMAIR G. W., and UREY H. C. (1970) Solar wind gases, cosmic-ray spallation products, and the irradiation history of Apollo 11 samples. *Proc. Apollo 11 Lunar Sci. Conf., Geochim. Cosmochim. Acta* Suppl. 1, Vol. 2, pp. 1357–1367. Pergamon.

MARTI K. and LUGMAIR G. W. (1971) Kr^{81}–Kr and K–Ar^{40} age, cosmic-ray spallation products and neutron effects in Apollo 11 and Apollo 12 lunar samples. Second Lunar Science Conference (unpublished proceedings).

MILANKOVITCH M. (1938) Astronomische Mittel zur Erforschung der ergeschichtlichen Klimate. *Handl. I. Geophysik* **9**, pp. 593–698.

MILLARD H. T. (1965) Thermal neutron activation: Measurement of cross section for manganese–53. *Science* **147**, pp. 503–504.

O'KELLEY G. D., ELDRIDGE J. S., SCHONFELD E., and BELL P. R. (1970) Primordial radionuclide abundances, solar proton and cosmic-ray effects and ages of Apollo 11 lunar samples by nondestructive gamma-ray spectrometry. *Proc. Apollo 11 Lunar Sci. Conf., Geochim. Cosmochim. Acta* Suppl. 1, Vol. 2, pp. 1407–1423. Pergamon.

O'KELLEY G. D., ELDRIDGE J. S., SCHONFELD E., and BELL P. R. (1971) Comparative radionuclide concentrations and ages of Apollo 11 and Apollo 12 samples from nondestructive gamma-ray spectrometry. Second Lunar Science Conference (unpublished proceedings).

PERKINS R. W., RANCITELLI L. A., COOPER J. A., KAY J. H., and WOGMAN N. A. (1970) Cosmogenic and primordial radionuclide measurements in Apollo 11 lunar samples by nondestructive analysis. *Proc. Apollo 11 Lunar Sci. Conf., Geochim. Cosmochim. Acta* Suppl. 1, Vol. 3, pp. 1455–1470.

PRICE P. B., and O'SULLIVAN D. (1970) Lunar erosion rate and solar flare paleontology. *Proc. Apollo 11 Lunar Sci. Conf., Geochim. Cosmochim. Acta* Suppl. 1, Vol. 3, pp. 2351–2359. Pergamon.

RANCITELLI L. A., PERKINS R. W., FELIX W. D., and WOGMAN N. A. (1971) Cosmogenic and primordial radionuclide measurements in Apollo 12 lunar samples by nondestructive analysis. Second Lunar Science Conference (unpublished proceedings).

REEDY R. C., and ARNOLD J. R. (1971) Interaction of solar and galactic cosmic-ray particles with the moon. In preparation.

ROSSI B. (1952) *High Energy Particles*, Appendix V, Prentice-Hall, New York.

SCHONFELD E. and O'KELLEY G. D. (1971) The selenographic orientation of the Apollo 12 rocks determined by nondestructive gamma-ray spectroscopy. Second Lunar Science Conference (unpublished proceedings).

SHRELLDALFF: SHEDLOVSKY J. P., HONDA M., REEDY R. C., EVANS J. C., LAL D., LINDSTROM R. M., DELANY A. C., ARNOLD J. R., LOOSLI H. H., FRUCHTER J. S., and FINKEL R. C. (1970) Pattern of bombardment-produced radionuclides in rock 10017 and in lunar soil. *Proc. Apollo 11 Lunar Sci. Conf., Geochim. Cosmochim. Acta* Suppl. 1, Vol. 2, pp. 1503–1532. Pergamon.

SIEVERS R. E., EISENTRAUT K. J., JOHNSON D. G., RICHARDSON M. F., and WOLF N. R. (1971) Variations in beryllium and chromium concentrations in lunar fines compared with crystalline rocks. Second Lunar Science Conference (unpublished proceedings).

TAJIMA E., ADACHI M., DOKE T., KUBOTA S., and TSUKUDA M. (1967) Spectrum of cosmic-ray produced neutrons. *J. Phys. Soc. Japan* **22**, pp. 355–360.

TAYLOR S. R., KAYE M., GRAHAM A., RUDOWSKI R., and MUIR P. (1971) Trace element chemistry of lunar samples from the Ocean of Storms. Second Lunar Science Conference (unpublished proceedings).

WEBBER W. R. (1963) A summary of solar cosmic ray events. In *Solar Proton Manual* (editor Frank B. McDonald). NASA Report, NASA Tr R–169, pp. 1–17.

WILLIS J. P., AHRENS L. H., DANCHIN R. V., ERLANK A. J., GURNEY J. J., HOFMEYER P. K., MCCARTHY T. S., and ORREN M. J. (1971) Some interelement relationships between lunar rocks and fines, and stony meteorites. Second Lunar Science Conference (unpublished proceedings).

Proceedings of the Second Lunar Science Conference, Volume 2, pp. 1791–1796.
The M.I.T. Press, 1971.

Some cosmogenic and primordial radionuclides in Apollo 12 lunar surface materials

ROBERT C. WRIGLEY
Space Science Division, NASA-Ames Research Center, Moffett Field, California 94035

(*Received 22 February 1971; accepted in revised form 29 March 1971*)

Abstract—Samples from the Apollo 12 site were analyzed by nondestructive γ-ray spectrometry for cosmogenic radionuclides Na^{22} and Al^{26} and for primordial radionuclides thorium and uranium. Na^{22} and Al^{26} concentrations in Apollo 12 fines are $\sim 15\%$ higher than in Apollo 11 fines but have the same Na^{22}:Al^{26} ratio. Rock samples 12004,1, 12040,30, and 12075,5 have Na^{22}:Al^{26} ratios of 0.63, 1.09, and 0.69. Rock 12075, 5 has a Na^{22}:Al^{26} ratio comparable to Apollo 11 rocks but has very high Na^{22} and Al^{26} concentrations apparently due to a favorable shape. The exceptionally high Na^{22}:Al^{26} ratio for rock 12040,30 may be explained partly by chemical differences but may also indicate excavation from the regolith within the last million years or so. Thorium and uranium concentrations in the rocks cover a much lower range than Apollo 11 rocks; Apollo 12 fines have thorium and uranium concentrations about eight times those of the Apollo 12 igneous rocks and could not have been produced directly by comminution from them. Whatever additional process operated, if large areas of the moon are invoked, large amounts of radioactivity may be involved and a separate explanation for that fact may be necessary.

INTRODUCTION

SAMPLES OF THE lunar surface at the Apollo 12 site in Oceanus Procellarum were analyzed for cosmogenic and primordial radioactivity by nondestructive γ-ray spectrometry. Na^{22}, Al^{26}, thorium, and uranium were measured in two aliquots of sample 12070 (sieved fines, particle size < 1 mm), and in three igneous rocks, samples 12004,1, 12040,30, and 12075,5 (petrographic types A, B, and A, respectively). Cosmogenic radionuclides provide information about the history of the top few centimeters, or perhaps meters, of the lunar surface since they are created by solar and galactic cosmic rays which enter the surface and are attenuated rapidly (SHEDLOVSKY *et al.*, 1970). Primordial radionuclides provide information on more global processes of differentiation within the moon and its regolith.

EXPERIMENTAL TECHNIQUES

The equipment used to measure nondestructively the radionuclides of interest is identical to that used for the measurements of Apollo 11 samples (WRIGLEY and QUAIDE, 1970). The anti-coincidence shielded two-parameter γ-ray spectrometer is shown schematically in Fig. 1. Events detected in the anti-coincidence mantle cause rejection of any events in the NaI(Tl) crystals. Accepted events in the crystals are recorded during the same period of analysis in either singles or coincident modes by a 4096 channel, two-parameter pulse height analyzer (Northern Scientific Model 650) set up in a 32×128 channel array with equal energy ranges, 0–2.8 MeV. Such an asymmetric array provides 128 channel resolution for one detector for γ-ray identification and single γ-ray analysis without affecting the two-parameter analysis.

Apollo 12 fines samples were counted in two stainless steel containers sealed in dry nitrogen with Viton O-rings. The container for the 276 gm sample was identical to those used for Apollo 11

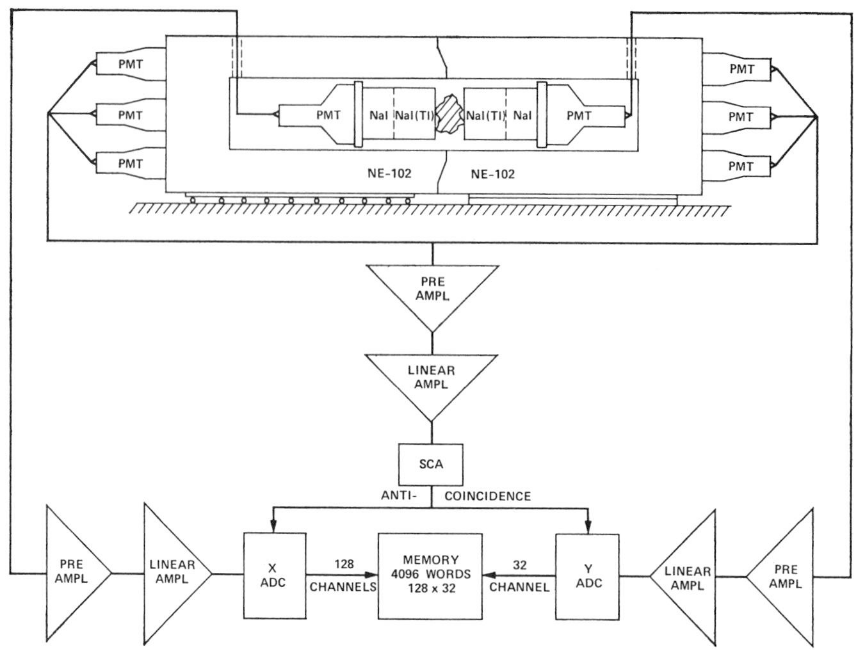

Fig. 1. Schematic diagram of anti-coincidence shielded two-parameter γ-ray spectrometer.

samples: 10 cm diameter by 2.5 cm thick with 0.4 mm walls. The other container was 11 cm diameter by 1.0 cm thick with 0.4 mm walls but a bolt flange restricted the sample to 7 cm diameter. Apollo 12 rocks were counted in a double layer of thin teflon bags inside a polyethylene bag.

Mock-ups of each Apollo 12 sample for each radionuclide of interest were prepared in a manner very similar to that of Apollo 11 samples (WRIGLEY and QUAIDE, 1970) except that a commercial solution of Al^{26} replaced the Al^{26} sealed pellets. Of the models used for making replicas of the rock samples, the one of sample 12004,1 was an epoxy model provided by the Lunar Sample Curator's office but the others were beeswax models made during this investigation by careful comparative measurements with a ruler. Reagent grade iron powder was used to fill replicas. Iron powder covers a density range of 2.8 to 3.4 gm/cc depending on the packing. Since the replica shell accurately represents the volume of the sample, the iron powder packs to the correct density. At this point the mock-up also matches the rock's electron density within 5%. Al^{26} mock-ups for the rock samples were also made by the method described by PERKINS *et al.* (1970): iron powder (65%) and gypsum are mixed with radioactive water and allowed to harden in a plaster mold of the appropriate rock. The two types of mock-ups agreed well for samples 12040,30 and 12075,5; compared to two gypsum mock-ups of 12004,1, the iron powder mock-up of 12004,1 was $\sim$15 standard deviations low and was not used. Thorium and uranium standards are analyzed ores from the New Brunswick Laboratory of the AEC, numbers 79 and 42 respectively, and are known to $\pm 1\%$. The former contains 1.01% thorium and 0.036% uranium; this necessitates a uranium correction to the thorium standard spectra. The plastic bags sealing the thorium and uranium pellets allowed no apparent loss of radon. The Na^{22} standard is an NBS solution, number 4922-E, known to $\pm 2\%$. The Al^{26} standard is a commercial solution with an error quoted at $\pm 6\%$. During the course of the investigation we noted this solution was 10% low in comparison to the Al^{26} pellets used previously and which had given good interlaboratory agreement. Consequently, the activity of the Al^{26} solution was adjusted upward 10% and the error increased from $\pm 6\%$ to $\pm 10\%$ to cover the possibility the previous interlaboratory agreement was incorrect.

Data were reduced by subtracting background and interfering interactions caused by nuclides previously determined and then comparing full-energy peaks with those of the appropriate mock-up (RANCITELLI, 1970). The full energy peaks used were 0.58 × 2.62 MeV (thorium), 0.61 × 1.12 MeV (uranium), 0.51 × 2.32 MeV (Al^{26}), and 0.51 × 1.79 MeV (Na^{22}).

RESULTS

The radionuclide concentration results for the Apollo 12 fines samples are shown in Table 1 as series of 3000 or 4000 minute analyses repeated at various times and as a set of means. Recalculated mean results for the Apollo 11 fines sample 10084,113–2 based on the present Al^{26} mock-up are shown for comparison. Standard deviations for individual results represent all counting statistics whereas those for the means also include possible mock-up errors and accuracy of radioisotope standards. The numbers in parentheses after the standard deviations of Al^{26} means represent the standard deviations due to all causes except the large standard deviation of the Al^{26} standard. Thus, Al^{26} means can be compared more meaningfully.

The Al^{26}, Na^{22}, and U means for sample 12070,13 agree very well with those given by RANCITELLI *et al.* (1971) for the same sample and with those given by O'KELLEY *et al.* (1971) for another sieved aliquot of sample 12070. The thorium mean for sample 12070,13 agrees within experimental error with the thorium concentration of 6.25 ± 0.50 ppm reported by O'KELLEY *et al.* (1971) although their quoted error is quite large. However, the thorium concentration of 6.73 ± 0.20 ppm reported for sample 12070,13 by RANCITELLI *et al.* (1971) is more than 10% higher than reported here; the discrepancy is unresolved. Comparison of the means for samples 12070,13 and 12070,12 shows excellent agreement and that the precision of the results for the 47 gm sample is virtually indistinguishable from the precision for the larger sample. Higher counting efficiencies of the smaller sample account for this. It is doubtful that further reductions in sample size would affect counting efficiencies significantly.

Table 1. Na^{22}, Al^{26}, thorium, and uranium in lunar surface fine material (<1 mm).

Sample		12070,13			10084,113–2
Mass		276 GM			327 GM
Time at Counting	146 Days	176 Days	380 Days	Mean	Mean
Na^{22}, dpm/kg	85 ± 9	74 ± 8	67 ± 9	75 ± 7	68 ± 6
Al^{26}, dpm/kg	171 ± 8	168 ± 7	175 ± 7	171 ± 18(4)	147 ± 16(3)
Thorium, ppm	6.11 ± 0.15	5.84 ± 0.12	5.90 ± 0.12	5.95 ± 0.12	2.19 ± 0.05
Uranium, ppm	1.70 ± 0.05	1.69 ± 0.04	1.75 ± 0.04	1.71 ± 0.03	0.62 ± 0.02
Sample		12070,12			
Mass		47 GM			
Time at Counting	101 Days	104 Days	114 Days	143 Days	Mean
Na^{22}, dpm/kg	72 ± 8	87 ± 8	86 ± 8	76 ± 9	80 ± 6
Al^{26}, dpm/kg	186 ± 9	165 ± 9	170 ± 9	173 ± 11	174 ± 19(8)
Thorium, ppm	5.94 ± 0.20	6.19 ± 0.21	6.19 ± 0.21	6.19 ± 0.26	6.13 ± 0.13
Uranium, ppm	1.69 ± 0.06	1.77 ± 0.06	1.60 ± 0.06	1.61 ± 0.08	1.67 ± 0.07

(1) All Na^{22} concentrations are corrected for decay to Nov. 20, 1969.

(2) Thorium and uranium concentrations assume terrestrial isotopic abundances and radioactive equilibrium in their decay chains.

(3) Numbers in parentheses represent standard deviations due to all causes except the uncertainty of the Al^{26} standards.

Our results for Na^{22} concentrations in Apollo 11 materials appeared to be high (WRIGLEY and QUAIDE, 1970) in comparison to other workers. Two postulates were put forth: (a) the Al^{26} pellet standards did not adequately represent a uniform distribution, and (b) Co^{56} contributed significantly to the Na^{22} counting region. The former postulate now seems to be the explanation since the uniformly distributed Al^{26} standard provides Na^{22} concentrations in close agreement with those of RANCITELLI *et al.* (1971) and PERKINS *et al.* (1970) for samples 12070,13 and 10084,113–2, respectively. Moreover, no effect of the 77-day Co^{56} can be seen in the Na^{22} results in sample 12070,12. An apparent decrease in Na^{22} concentration with time is noticeable for sample 12070,13 but this cannot be due to Co^{56}; at 146 days after collection only 15 ± 2 dpm/kg of Co^{56} would remain (RANCITELLI *et al.*, 1971) and this would contribute less than 2% of the Na^{22} counting rate. The apparent decrease in Na^{22} concentration seems to be simply a large statistical fluctuation. Similar large Na^{22} fluctuations are also noted for sample 12070,12 but they do not form a monotonic decrease.

Results for Apollo 12 rock samples 12004,1, 12040,30, and 12075,5 are given in Table 2 as a set of means only. Standard deviations are as given for the means of the samples of fine material. Results for 12004,1, the only rock analyzed by another group (O'KELLEY *et al.*, 1971) agree quite well. Their thorium concentration of 0.92 ± 0.09 ppm coupled with the thorium discrepancy noted earlier in the fines may indicate our thorium standard is low.

DISCUSSION

As with the Apollo 11 material, Na^{22} and Al^{26} concentrations are much higher than can be accounted for by galactic cosmic ray production, and lower energy solar radiation must be considered (i.e., SHEDLOVSKY *et al.*, 1970). Since the lower energy radiation creates a steep gradient in the top few centimeters, high Na^{22} and Al^{26} concentrations in Apollo 11 fine material were interpreted as due to a shallow sampling depth (PERKINS *et al.*, 1970). Apollo 12 Na^{22} and Al^{26} concentrations are higher still ($\sim 15\%$) and would seem to indicate even shallower sampling since chemical differences in the major target elements are small. The Na^{22}:Al^{26} ratios are virtually identical in Apollo 11 and Apollo 12 fines: 0.46 ± 0.04 and 0.44 ± 0.04, respectively. The situation is different for the rocks. Whereas the compilation of Na^{22}:Al^{26} ratios by

Table 2. Na^{22}, Al^{26}, thorium and uranium in Apollo 12 igneous rocks.

Sample Sample Mass/Rock Mass, GM	12004,1 506/585	12040,30 206/319	12075,5 173/233
Na^{22}, dpm/kg	56 ± 4	61 ± 7	74 ± 7
Al^{26}, dpm/kg	$89 \pm 11(7)$	$56 \pm 6(3)$	$107 \pm 12(5)$
Thorium, ppm	0.78 ± 0.03	0.47 ± 0.04	0.62 ± 0.05
Uranium, ppm	0.23 ± 0.02	0.16 ± 0.02	0.19 ± 0.02

(1) Na^{22} concentrations have been corrected for decay to Nov. 20, 1969.

(2) Thorium and uranium concentrations assume terrestrial isotopic abundances and equilibrium in their decay chains.

(3) Numbers in parentheses represent standard deviations due to all causes except the uncertainty of the Al^{26} standard.

HERZOG and HERMAN (1970) for Apollo 11 rocks spans the range from 0.41 to 0.76, samples 12004,1, 12040,30, and 12075,5 have ratios of 0.63 ± 0.07, 1.09 ± 0.13, and 0.69 ± 0.07, respectively. Although the latter falls in the previous range, it has the highest Na^{22} and Al^{26} concentrations for a major portion of an igneous rock to date. Such high concentrations of Na^{22} and Al^{26} for 12075,5 could be explained by the rock's peaked shape; its top surface (see below) presents a large surface area to the solar flux.

Rock 12040,30 has an excessively high $Na^{22}:Al^{26}$ ratio (1.09). Chemically, it has ~20% more magnesium and ~15% less aluminum than a more typical rock, 12004,1 (SCOON, 1971; MAXWELL and WINK, 1971). According to the tentative results of BEGEMANN *et al.* (1970), such differences in major target elements for Na^{22} and Al^{26} could account for ~15% of the $Na^{22}:Al^{26}$ ratio for 12040,30 thus leaving ~30% unaccounted for. The erosional mechanism proposed by RANCITELLI *et al.* (1971) as an explanation for the high $Na^{22}:Al^{26}$ ratio of 0.89 they measured for rock 12005 would not seem to apply to rock 12040,30 since it has neither a flat shape nor an obvious exposure to the meteoroid or secondary particle flux (surface pitting is unobservable). The explanation of the remaining discrepancy in the $Na^{22}:Al^{26}$ ratio for 12040,30 may lie in the possibility that the rock was excavated from the regolith within the last million years so that Al^{26} did not reach an equilibrium concentration.

Although no quantitative measurement of concentration gradients was made, very obvious differences in Al^{26} concentration (as determined by a mock-up with a uniform distribution of Al^{26}) were noted between the "X" and "Y" detectors for 12004,1 and 12075,5. Both rocks have top surfaces defined by numerous small impact pits and few, if any, pits on their reverse sides or bottoms. These differences were of the order of 20% for 12004,1 and 50% for 12075,5. In both cases, the high Al^{26} coincided with the pitted surface. Also, the large difference for 12075,5 confirms the above interpretation of sizable solar flux contributions. No noticeable difference existed for 12040,30. Differences between "X" and "Y" detectors for Na^{22} were apparent but much smaller, perhaps due to poorer statistics. The pitted sides again showed higher activity.

Concentrations of naturally occurring thorium and uranium are much higher in the fines at the Apollo 12 site than the fines at Apollo 11. Simultaneously, the igneous rocks at Apollo 12 cover a much lower range than Apollo 12 fines. Hence, while thorium and uranium concentrations of Apollo 11 fines could have resulted from a simple comminution of Apollo 11 igneous rocks, that cannot be true at the Apollo 12 site; the fines (particle size <1 mm) have thorium and uranium concentrations approximately eight times larger than igneous surface rocks of the regolith. Since thorium and uranium concentrate in late-stage differentiation products, i.e., the volatile minerals, the enhancement of thorium, uranium, and volatile elements in the fine material may be due to volatile transfer (HUEY *et al.*, 1971) or to mixing with an exotic component (HUBBARD *et al.*, 1971). However, if large portions of the moon are to be involved, such as debris from Imbrium, that involvement may require very high thorium and uranium concentrations for those large portions of the moon. If so, a separate explanation for the existence of such areas would be required. For instance, if the Fra Mauro Formation is to be the exotic component and if the part of the Fra Mauro Formation in question is considered representative of the whole formation, then it all must have high thorium and uranium concentrations. Since the Fra Mauro Formation is believed to be debris from the Imbrium impact, then the thickness of the material having high thorium and uranium must be of the order of tens of kilometers. Such an immense thickness of highly radioactive material would require a separate explanation as to its possible existence and its place in lunar differentiation.

REFERENCES

BEGEMANN F., VILCSEK E., RIEDER R., BORN W., and WÄNKE H. (1970) Cosmic-ray produced radioisotopes in lunar samples from the Sea of Tranquillity (Apollo 11). *Proc. Apollo 11 Lunar Sci. Conf., Geochim. Cosmochim. Acta* Suppl. 1, Vol. 2, pp. 995–1005. Pergamon.

HERZOG G. F. and HERMAN G. F. (1970) Na^{22}, Al^{26}, Th and U in Apollo 11 lunar samples. *Proc. Apollo 11 Lunar Sci. Conf., Geochim. Cosmochim. Acta* Suppl. 1, Vol. 2, pp. 1239–1245. Pergamon.

HUBBARD N. J., GAST P. W., and MEYER C. (1971) The origin of the lunar soil based on REE, K, Rb, Ba, Sr, P and $Sr^{87/86}$ data. Second Lunar Science Conference (unpublished proceedings).

HUEY J. M., IHOCHI H., BLACK L. P., OSTIC R. G., and KOHMAN T. P. (1971) Lead isotopes and volatile transfer in the lunar soil. Second Lunar Science Conference (unpublished proceedings).

MAXWELL J. A. and WIIK H. B. (1971) Chemical composition of Apollo 12 lunar samples 12004, 12033, 12051, 12052 and 12065. Second Lunar Science Conference (unpublished proceedings).

O'KELLEY G. D., ELDRIDGE J. S., SCHONFELD E., and BELL P. R. (1971) Comparative radionuclide concentrations and ages of Apollo 11 and Apollo 12 samples from nondestructive gamma-ray spectrometry. Second Lunar Science Conference (unpublished proceedings).

PERKINS R. W., RANCITELLI L. A., COOPER J. A., KAYE J. H., and WOGMAN, N. A. (1970) Cosmogenic and primordial radionuclide measurements in Apollo 11 lunar samples by nondestructive analysis. *Proc. Apollo 11 Lunar Sci. Conf., Geochim. Cosmochim. Acta* Suppl. 1, Vol. 2, pp. 1455–1469. Pergamon.

RANCITELLI L. A. (1970) Personal communication.

RANCITELLI L. A., PERKINS R. W., FELIX W. D., and WOGMAN N. A. (1971) Cosmogenic and primordial radionuclide measurements in Apollo 12 lunar samples by nondestructive analysis. Second Lunar Science Conference (unpublished proceedings).

SCOON J. H. (1971) Quantitative chemical analyses of lunar samples. Second Lunar Science Conference (unpublished proceedings).

SHEDLOVSKY J. P., HONDA M., REEDY R. C., EVANS J. C., JR., LAL D., LINDSTROM R. M., DELANY A. C., ARNOLD J. R., LOOSLI H.-H., FRUCHTER J. S., and FINKEL R. C. (1970) Pattern of bombardment-produced radionuclides in rock 10017 and in lunar soil. *Proc. Apollo 11 Lunar Sci. Conf., Geochim. Cosmochim. Acta* Suppl. 1, Vol. 2, pp. 1503–1532. Pergamon.

WRIGLEY R. C. and QUAIDE W. L. (1970) Al^{26} and Na^{22} in lunar surface materials: Implications for depth distribution studies. *Proc. Apollo 11 Lunar Sci. Conf., Geochim. Cosmochim. Acta* Suppl. 1, Vol. 2, pp. 1751–1755. Pergamon.

Proceedings of the Second Lunar Science Conference, Volume 2, pp. 1797–1802.
The M.I.T. Press, 1971.

Spallogenic ^{53}Mn ($T \sim 2 \times 10^6\, y$) in lunar surface material by neutron activation

W. HERR, U. HERPERS
Institut fuer Kernchemie der Universität Koeln
and
R. WOELFLE
Institut fuer Radiochemie der KFA Juelich, Germany.

(*Received 22 February 1971; accepted in revised form 31 March 1971*)

Abstract—Because of their usefulness for the evaluation of exposure ages, activity profiles, and long-term flux variations the long-lived spallation nuclides are of interest. Low-level techniques and large amounts of material, often exceeding 100 g, are necessary for quantitative determinations of ^{53}Mn by X-ray counting. A sensitive method for ^{53}Mn determination by neutron activation was used, increasing the activity rate by a factor of more than 10^4, so that ^{53}Mn could be measured in samples as small as 0.5 g. Three Apollo 12 lunar rock samples, two soil samples, and manganese extracted from layers of rock 10017 by the LaJolla group (J. R. Arnold and coworkers) were analyzed. The interfering (n, 2n) reaction contributed $\leq 50\%$ to the total ^{54}Mn activity. Corrections were made with activated Fe, Mn, and internal Zn standards. Simultaneously, the ^{53}Mn content of the Peace River meteorite was measured for comparison. A depth effect was noticed indicating that solar protons contribute to the production of ^{53}Mn.

INTRODUCTION

THE EFFECT OF cosmic-ray particles which are recorded in exposed matter has been studied in the past three decades mainly in meteorites (HONDA and ARNOLD, 1967; KIRSTEN and SCHAEFFER, 1970). Except for the Lost City and Příbram meteorites the orbits of meteoritic bodies are unknown. Furthermore, most of these meteorites have lost considerable mass in penetrating the atmosphere. Since samples from the lunar surface were brought to earth, one has targets which recorded cosmic-ray and also the lower-energy solar flare events. This may shed new light on the history of cosmic rays and solar flares. The long-lived radionuclides, such as ^{10}Be, ^{26}Al, ^{36}Cl, and especially ^{53}Mn, are of great value with regard to the measurement of the long-term constancy of cosmic and solar flare radiations. Also, informations on the erosion rate of rocks and the mixing rate of regolith should be accessible by radioactivity profiles.

METHODS

Cosmic ray produced ^{53}Mn, originating mainly from iron, has a relatively high production cross-section and a long halflife. Neither is well known. The direct counting of ^{53}Mn is intricate because of the soft K-radiation; also rather large amounts of precious material would be needed.

MILLARD (1965) pointed out the possibility of converting ^{53}Mn into the γ-emitting ^{54}Mn by a neutron capture. By developing this technique, and by assuming that the halflife of ^{53}Mn is 2×10^6 yr, we determined the ^{53}Mn-contents of about 30 iron meteorites (HERPERS *et al.*, 1967, 1969; HERR *et al.*, 1969). KAYE and CRESSY (1965) gave a ^{53}Mn halflife of $1.9 \pm 0.5 \times 10^6$ y and MILLARD (1965) deduced the product of the (n, γ) activation cross-section times halflife to be

$350 \pm 100 \cdot 10^6$ barns·y. The large uncertainty in this product, however, does permit only a relative determination of the ^{53}Mn content via the neutron activation method. On the other hand, the most sensitive activation technique is (apart from T and σ) very precise, and the error in the procedure itself can be minimized down to $\pm 3\%$. We endeavored to relate our ^{53}Mn activation data to recent direct X-ray counting results of ^{53}Mn, which Arnold and coworkers achieved for the Peace River chondrite (private communication 1970). They measured a ^{53}Mn decay rate of 72 ± 7 dpm/kg for this piece of meteorite. We have performed an activation experiment on the same Peace River chondrite and have obtained a ^{53}Mn product of $T \times \sigma = 260 \pm 55 \cdot 10^6$ barns·y. This product has been termed the *Millard factor*.

The first set of our experiments deals with the analysis of ^{53}Mn in extracted Mn samples from rock 10017 and from Apollo 11 soil. The former rock was investigated by the LaJolla group (SHEDLOVSKY *et al.*, 1970) for induced radioactivities and also for ^{53}Mn. They gave us their Mn concentrates ($MnCl_2$ in dilute HCl) for independent ^{53}Mn measurements. It was highly desirable to gather more ^{53}Mn data of "deeper" rock samples. In the second set we analyzed Apollo 12 rock and soil samples. The Mn and Fe content of the respective Apollo 12 samples are listed in Table 1.

The mineral samples were fused with $NaOH + Na_2O_2$ in a Ni crucible. The hydroxides, dissolved in 10 N HCl, were transferred to a Dowex 1×8 ion-exchange column. The Mn^{2+} solutions of both groups of lunar samples were purified further as $HMnO_4$ by a reduced pressure distillation from 10 M $H_2SO_4 + NaIO_4$. The chemical yield of manganese was checked by a carrier-free ^{52}Mn ($T = 6\ d$) tracer and independently determined by a reactivation analysis via ^{56}Mn. The manganese was freed expecially from traces of iron and then in the form of MnO_2 (mixed with MgO and ZnO) was sealed in 20 mm-long quartz tubes and irradiated for 6 weeks in the research reactor (FRJ–2) at Juelich. In the same irradiation Al capsule, a set of "normal" MnO_2 and one sample of MnO_2 extracted from the Sichote Alin iron meteorite were activated. The latter served as a monitor for different reactor irradiations. Also spec. pure Fe_2O_3 and ZnO samples were monitors. The flux was 8×10^{12} n $\times$ cm^{-2} $\times$ s^{-1} and the flux ratio of Φ_{fast}/Φ_{therm} was only ≤ 0.001 as measured by the reactions $^{31}P(n, p)^{31}Si$ and $^{59}Co(n, \gamma)^{60}Co$.

The radiometric measurements were done with a well-type $2'' \times 2''$ NaI(Tl) crystal and a 400-channel analyzer. The samples were independently checked with a 20 cc Ge(Li)-detector and ^{54}Mn halflife was checked. No other interfering radioactivities were observed near and beyond the 0.84 MeV photopeak of ^{54}Mn (see Fig. 1). Absolute neutron flux calibration was done by comparing the neutron induced activity in the internal Zn standard (^{65}Zn-activity, $\sigma_{therm} = 0.44$ b, BAUMGÄRTNER, 1961) with a standardized ^{22}Na sample on the basis of their annihilation radiation as measured in a γ,γ-coincidence circuit (WOELFLE *et al.*, 1968).

RESULTS and DISCUSSION

Our ^{53}Mn data are presented in Tables 2 and 3. It can be seen from the last two lines of Table 2 that a possible interference of the $^{54}Fe(n, p)^{54}Mn$ nuclear reaction is negligible (only 1 to 2%), and can be suppressed by chemical removal of iron traces prior to irradiation. However, the second nuclear reaction, the $^{55}Mn(n, 2n)^{54}Mn$, is more serious and may contribute up to 50% of the total ^{54}Mn activity. Both reactions are evidently dependent on the respective reactor positions.

Table 1. Manganese and iron abundances in Apollo 12 samples (^{55}Mn determined by γ,γ-coincidence spectroscopy).

Lunar sample	Mn content [ppm]	Fe content [%]
12021,57	1959 ± 98	14.22
12022,35	1994 ± 96	16.40
12053,39	1954 ± 94	15.29
12070,87 (soil)	1556 ± 78	13.17

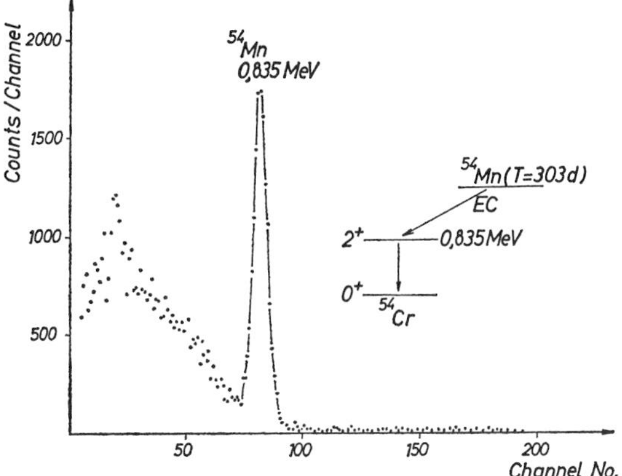

Fig. 1. ^{54}Mn γ-spectrum of lunar rock 12022,35; irrad. time: 1008 h; decay time: 44 d; Φ_{therm}: $8 \times 10^{12}[n \times cm^{-2} \times s^{-1}]$; count. time: 10 h.

The ^{53}Mn results on rock 10017 samples, taken from different layers, have been compared with those from the LaJolla group (Shedlovsky *et al.*, 1970) in Table 2, columns 8 and 10. The latter authors used "low-level" X-ray counting techniques, and more recently also the neutron activation method. We have been permitted to compare the LaJolla group's (J. R. Arnold and coworkers) unpublished activation results with our own (Table 2, column 9). The ^{53}Mn activation results are in rather good agreement with those of the LaJolla group, with the exception of the two "deeper" lying samples T–4I (12–30 mm) and T–3I ($\sim$60 mm). In case that our data should be confirmed, the shielding effect would be pronounced.

The capability of the activation method is seen by comparing with the X-ray counting results (columns 8 and 10). The statistical error of counting in the activation method is very small and can be held below 1%. The still remaining uncertainty in the ^{53}Mn halflife and its neutron cross section is however not considered. Our ^{53}Mn activation data, given in column 8, are related to the ^{53}Mn X-ray determination of the Peace River chondrite, from which we calculated a new Millard factor by assuming that a ^{53}Mn halflife of $T = 2 \times 10^6$ y is valid. A ^{53}Mn depth effect can be recognized. The direction of the effect seems a bit surprising because from laboratory measurements with high-energy proton-bombarded targets the ^{53}Mn production rate due to secondary particles increases with depth (Honda, 1962).

Apparently, the solar p and α bombardment is responsible for this result and gives evidence of the existence of solar flares in the distant past, $\sim 2 \times 10^6$ y ago. Moreover, the ^{10}Be and ^{36}Cl, which are produced only by higher energy reactions of galactic cosmic rays, do not show similar depth dependence (Shedlovsky *et al.*, 1970). With respect to our earlier paper on the ^{53}Mn content in the lunar regolith 10084 (grain size $\leq 100\ \mu$m) we published a ^{53}Mn disintegration rate of 30 dpm/kg (Herr *et al.*, 1970). In the course of this study, however, we have concluded that this value is too low. The recent application of the newly determined Millard factor leads to a value of 41 ± 7 dpm/kg for the ^{53}Mn content. This value is in much better agreement with the

Table 2. Determination of ^{53}Mn by

1 Samples from J. P. Shedlovsky *et al.* (1970)	2 Mn amount (mg)	3 Chem. Yield before irrad. (%)	after irrad.	4 Total ^{54}Mn end of irrad. (cpm)	5 ^{54}Mn (dpm/mg Mn)
T–4 D (0–4 mm)	2.42 ± 0.02	90.2*	90.1*	51.4 ± 0.5†	423
T–4 F (0–4 mm)	2.84 ± 0.08	90.8	93.4	61.4 ± 0.5	412
T–4 U (4–12 mm)	4.00 ± 0.13	91.0	94.8	73.9 ± 0.6	343
T–4 I (12–30 mm)	3.68 ± 0.11	90.5	92.2	61.0 ± 0.5	322
T–3 I (~60 mm)	3.57 ± 0.10	88.3	92.1	58.5 ± 0.5	325
T–1 (soil)	5.05 ± 0.14	92.2	96.0	129.2 ± 0.8	467
Mn–Std.	10.80	—	93.8	69.4 ± 0.6	112
Fe–Std.	6.8(Fe)	—	84.8	132.2 ± 1.0	228

* Error ± 1.5%.
† Counting error only.

Table 3. ^{53}Mn content of Apollo 12 materials.

Rocks	Sample Weight (g)	Chem. Yield before irrad. (%)	after irrad.	Total ^{54}Mn (end irrad.) (cpm)	^{54}Mn (dpm/mg Mn)	^{54}Mn corr. for (n, p) and $(n, 2n)$ (dpm/mg Mn)	^{53}Mn $(\times 10^{-3})$ (dpm/mg Mn)	^{53}Mn content (dpm/kg)
12021,57	0.8422	87.2*	96.5*	25.4 ± 0.3†	304	186	17.4	33 ± 4‡‡
12022,35	1.0377	90.2	91.5	35.7 ± 0.4	320	203	18.9	38 ± 4
12053,39	1.0697	97.3	96.0	41.2 ± 0.4	343	226	21.0	41 ± 3
12070,87 (Fines)	0.5007	67.5	92.5	14.1 ± 0.2	461	342	31.8	50 ± 4
"Peace River"	2.020	89.2	98.2	138.3 ± 0.9	446	319	29.7	72 ± 4
Mn–Std	11.67 (mg. Mn)	—	77.5	63.5 ± 0.6	116.8			
Fe–Std	7.7 (mg Fe)	—	89.3	160.4 ± 0.9				

‡‡ Errors include all systematic and counting errors (1σ) in the procedure, not include the uncertainty in the Millard factor.

respective value of Shedlovsky *et al.* (1970), who find 46 ± 11 dpm/kg in their sample of Apollo 11 regolith. Surely, a main source for ^{53}Mn in the surface are medium energy particles (10 to 100 MeV) emitted by occasional sun bursts or flares, the durations of which are short compared with the halflifes of the observed radionuclides. It seems also probable that energetic neutrons can partially contribute to ^{53}Mn production.

It has been pointed out by Finkel *et al.* (1971) that there is a possibility of calculating the contribution of each solar flare to the production of "short nuclides." On the other hand, by considering the integral flux and energy spectrum of solar flares, the possibility exists of obtaining information on the intensity and rigidity of the incident

neutron capture in Apollo 11 material.

6 ^{54}Mn corr. for (n, $2n$) and (n, p) (dpm/mg Mn)	7 ^{53}Mn ($\times 10^{-3}$ dpm/mg Mn)	8 ^{53}Mn content (dpm/kg)	9 ^{53}Mn (by n-activ.) ARNOLD and FRUCHTER§ (dpm/kg)	10 ^{53}Mn (by X-ray count.) from SHEDLOVSKY *et al.* (1970) (dpm/kg)
311	29.5	56 ± 5‡	62 ± 5	94 ± 24 (DF)
300	28.5	54 ± 4	57 ± 5	
231	21.9	42 ± 3	45 ± 4	43 ± 16
210	20.0	38 ± 3	48 ± 4	50 ± 12
213	20.2	38 ± 3	48 ± 4	48 ± 21 (IS)
354	33.7	54 ± 4	55 ± 5	46 ± 11
—	Soil 10084 see earlier value	41 ± 7 calculated with		46 ± 11
—	30.3 ± 5.5 dpm/kg HERR *et al.* (1970)	new Millard factor		

‡ Errors include all systematic and counting (1 σ) errors in the procedure; not included the uncertainty in the Millard factor.

§ ARNOLD J. R., and FRUCHTER J. S. (private communication, 1971); unpublished results.

particles with respect to long-term changes (SHEDLOVSKY *et al.*, 1970). This underlines the necessity for precise measurements of such induced radionuclides.

In Table 3 the results of five Apollo 12 minerals are presented. The three rock samples originate, as far as we were able to find out, from the inner and lower parts. Their ^{53}Mn-content is in the range of 34 to 41 dpm/kg. Due to the lack of other ^{53}Mn data, there is no way at present to make comparisons or to draw conclusions. However, the ^{53}Mn-value of the Apollo 12 regolith (50 dpm/kg) is apparently larger than that of the respective Apollo 11 soil (41 ± 7 dpm/kg). This may depend on the conditions under which the soil was collected. During the Apollo 11 mission larger amounts of soil were taken and probably deeper layers scratched, pointing out the importance of sampling. Still open and of great interest is the problem of to what degree the ^{53}Mn, which is produced by solar flare activities in the outermost lunar rock surface only, is balanced by the erosion. A solution of these questions should be attained by investigating well defined rock and core layers.

Acknowledgments—We wish to thank J. R. Arnold for the extracted Mn samples of lunar rock 10017 and the material of the Peace River meteorite. We are also indebted to him and his coworkers for having placed at our disposal unpublished results of their respective ^{53}Mn values. Our thanks are also due to NASA for supplying lunar material, and to the Deutsche Forschungsgemeinschaft for its support.

REFERENCES

BAUMGAERTNER F. (1961) Tabelle zur Neutronenaktivierung. *Kerntechnik* **3**, 356–369.

FINKEL R. C., ARNOLD J. R., REEDY R. C., FRUCHTER J. S., LOOSLI H. H., EVANS J. C., SHEDLOVSKY J. P., IMAMURA M., and DELANY A. C. (1971) Depth variation of cosmogenic nuclides in a lunar surface rock. Second Lunar Science Conference (unpublished proceedings).

HERPERS U., HERR W., and WOELFLE R. (1967) Determination of cosmic ray produced nuclides ^{53}Mn, ^{45}Sc, and ^{26}Al in meteorites by neutron activation and gamma coincidence spectroscopy. In *Radioactive Dating and Methods of Low-Level Counting*, pp. 199–205, International Atomic Energy Agency, Vienna.

HERPERS U., HERR W., and WOELFLE R. (1969) Evaluation of ^{53}Mn by (n, γ) activation, ^{26}Al, and special trace elements in meteorites by γ-coincidence techniques. In *Meteorite Research*, pp. 387–396, International Atomic Energy Agency, Vienna.

HERR W., HERPERS U., and WOELFLE R. (1969) Die Bestimmung von ^{53}Mn, welches in meteoritischem Material durch kosmische Strahlung erzeugt wurde, mit Hilfe der Neutronenaktivierung. *J. Radioanal. Chem.* **2**, 197–203.

HERR W., HERPERS U., HESS B., SKERRA B., and WOELFLE R. (1970) Determination of manganese-53 by neutron activation and other miscellaneous studies on lunar dust. *Science* **167**, 747–749.

HONDA M. (1962) Spallation products distributed in a thick iron target bombarded by 3-BeV protons. *J. Geophys. Res.* **67**, 4847–4858.

HONDA M., and ARNOLD J. R. (1967) Effects of cosmic rays on meteorites. In *Handbuch der Physik* XLVI/2, pp. 613–632, Springer-Verlag.

KAYE J. H., and CRESSY P. J. (1965) Half-life of manganese-53 from meteorite observations. *J. Inorg. Nucl. Chem.* **27**, 1889–1892.

KIRSTEN T. A., and SCHAEFFER O. A. (1970) High energy interactions in space. In *Interactions of Elementary Particle Research in Science and Technology* (editor L. C. L. Yuan), Academic Press, to be published.

MILLARD H. T., JR. (1965) Thermal neutron activation: Measurement of cross section for manganese-53. *Science* **147**, 503–504.

SHEDLOVSKY J. P., HONDA M., REEDY R. C., EVANS J. C., JR., LAL D., LINDSTROM R. M., DELANY A. C., ARNOLD J. R., LOOSLI H. H., FRUCHTER J. S., and FINKEL R. C. (1970) Pattern of bombardment-produced radionuclides in rock 10017 and in lunar soil. *Science* **167**, 574–576.

SHEDLOVSKY J. P., HONDA M., REEDY R. C., EVANS J. C., JR., LAL D., LINDSTROM R. M., DELANY A. C., ARNOLD J. R., LOOSLI H. H., FRUCHTER J. S., and FINKEL R. C. (1970) Pattern of bombardment-produced radionuclides in rock 10017 and in lunar soil. *Proc. Apollo 11 Lunar Sci. Conf., Geochim. Cosmochim. Acta* Suppl. 1, Vol. 2, pp. 1503–1532. Pergamon.

WOELFLE R., HERPERS U., and HERR W. (1968) Eine verbesserte, selektive γ-γ-Koinzidenz-Anordnung zur zerstörungsfreien Bestimmung von Submikro-Kupferspuren in den hochreinen Metallen Be, Bi, Pb, Se, Sn, und Tl. *Z. anal. Chemie* **233**, 241–252.

Proceedings of the Second Lunar Science Conference, Volume 2, pp. 1803–1812.
The M.I.T. Press, 1971.

Tritium in lunar material

P. Bochsler, P. Eberhardt, J. Geiss, H. Loosli,
H. Oeschger, and M. Wahlen*
Physikalisches Institut, University of Bern, Sidlerstrasse 5, 3000 Bern, Switzerland

(*Received 24 February 1971; accepted in revised form 30 March 1971*)

Abstract—We have measured the H^3 activity in Apollo 11 lunar fines 10084,47, in three different samples of the Apollo 11 breccia 10046, and at different depths in the Apollo 12 crystalline rock 12053,37. Our samples of breccia 10046 were most likely contaminated with man-made H^3. A two step extraction at 300°C and 1700°C was tested and applied for recognizing such a contamination. No indication for contamination was found in lunar fines 10084 and in rock 12053. In lunar fines 10084 the H^3 activity, extrapolated to date of collection, was 306 ± 19 dpm/kg (average of three determinations). Three interior samples of rock 12053, from 1.1 to 2.5 cm depth, had the same H^3 activity within ±10% with an average value of 261 ± 21 dpm/kg. This activity is in good agreement with meteoritic values, if the difference in irradiation geometry is considered. The $He^3/2H^3$ exposure age of rock 12053 is $(67 \pm 7) \times 10^6$ y. A comparison with the Kr^{81}/Kr exposure age of $(99 \pm 6) \times 10^6$ y shows that the $He^3/2H^3$ exposure age is most likely too low because of He^3 diffusion loss.

Introduction

Tritium in lunar material is produced mainly by interactions of galactic and solar cosmic radiation with the lunar material. A direct influx of H^3 produced in solar flares might also contribute to the H^3 content of lunar material. To investigate these different sources and to compare H^3 activities with other cosmic ray-produced radioactive or stable isotopes, we measured H^3 in Apollo 11 lunar fines (10084,47), three different samples of an Apollo 11 breccia (10046,43, 10046,44, and 10046,45), and at different depths in an Apollo 12 crystalline rock (12053,37). Reliable and reproducible results were obtained on the Apollo 11 lunar fines and on rock 12053. The three samples which we received from rock 10046 were, however, contaminated with man-made H^3. Therefore, special care has been taken to distinguish between contamination and cosmic ray-produced H^3.

Extraction Procedure and Measuring Techniques

Extraction

The extraction of hydrogen is carried out in an all-bakeable UHV system (Fig. 1). The system is designed to allow the simultaneous extraction of H_2 and the noble gases He, Ne, Ar, Kr, and Xe. The samples (0.7–1.5 g) are placed in cups made of aluminum foil and dropped into a molybdenum crucible where they are melted by induction heating in the presence of a tritium-free H_2 carrier obtained from Messer-Griesheim GmbH, Duisburg, Germany. Prior to the extraction the carrier is cleaned from noble and other gases by absorption in a titanium sponge. The sponge is then heated until the pressure in the extraction vessel reaches 100 Torr, corresponding to approxi-

* Present Address: Department of Chemistry, University of California at San Diego, La Jolla, California 92037.

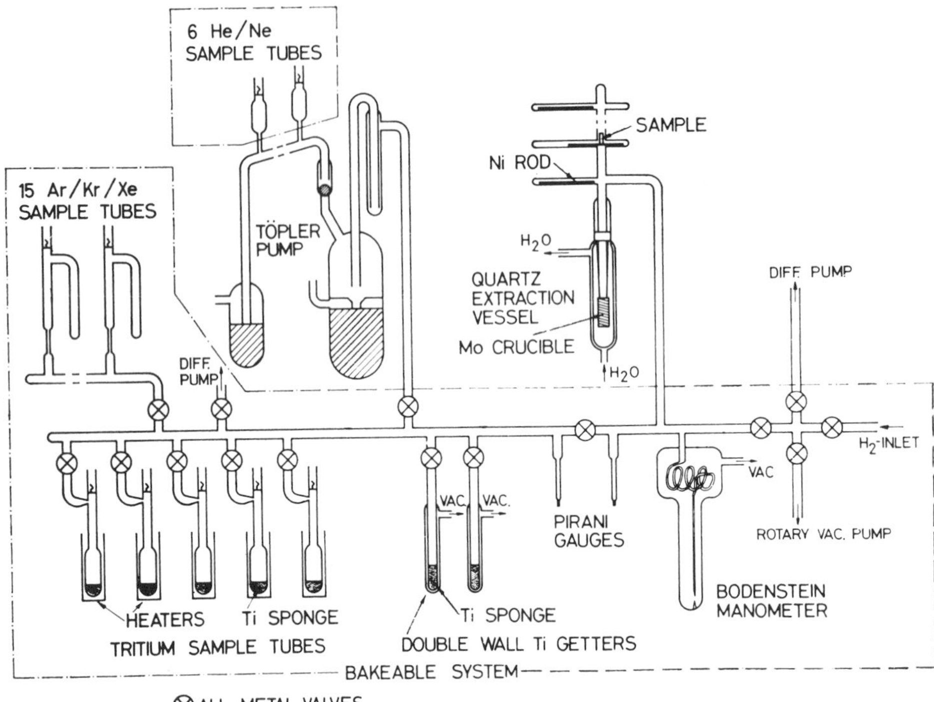

Fig. 1. System used for the simultaneous extraction of H^3 and noble gases.

mately 30 cm^3 STP H_2. A ratio of carrier to H_2 released from the sample of at least 30 can be expected. After the extraction the total amount of hydrogen is quantitatively absorbed in a hot titanium sponge (3 g) at appropriate temperatures. Then Xe/Kr and Ar are adsorbed on charcoal at $-120°C$ and $-180°C$ respectively, and He and Ne are collected with a Toepler pump. The whole extraction system, excluding the sample storage part, is kept at approximately 100°C to avoid H_2 and Xe losses.

Normally, samples were extracted by melting at 1700°C for 30 minutes. Most extractions were followed by a reextraction in order to guarantee a complete degassing and to avoid tritium losses. For the reextraction the molybdenum crucible is heated to 1750°C in the presence of fresh carrier for 30 minutes. Subsequently, the whole extraction vessel around the crucible is brought to 1100°C for 30 minutes by an external heater. In most cases little H^3 activity was present in this reextraction (typically 5%). The results given include the activities found in the reextraction, if any.

After we had some indications of a possible contamination with man-made tritium, the extraction was done in most cases in two steps: at 300°C and 1700°C. An external heater was used for heating the extraction vessel to 300°C for 40 minutes. We hoped that a distinction between cosmic-ray produced H^3 and contamination H^3 would be possible, expecting that below 300°C only a negligible part of cosmic-ray produced H^3 is released but that contamination H^3 is effectively removed at this temperature.

Several samples can be extracted in one series without opening the vacuum system. A typical series consists of blank of an empty Al-cup, sample heated at 300°C, sample heated at 1700°C, reheat of sample, Al-cup-blank, second sample heated at 300°C, sample heated at 1700°C, reheat of sample, and Al-cup-blank.

Gas conversion

After removal of the noble gases, the stored H_2 is transferred to a Ti-sponge in a quartz sample tube. In a gas conversion system the absorbed H_2 is reextracted at 900°C from the titanium by an automatic Toepler pump. The H_2, together with a sufficient amount of C_2H_4, is then circulated by means of a membrane pump over a palladium catalyst (220°C) and converted to C_2H_6, which is a very suitable proportional counting gas.

Counting technique

The ethane is measured in a low level proportional counter with a total volume of 12 cc (Tschudi, 1968) at about 5 atm pressure. No other gases are added, and tests of the counting characteristics show that the counter works in the range of limited proportionality. The counter is operated in anticoincidence with a cylindrical gas proportional guard counter, and the tritium decay spectrum is analyzed with a multi-channel analyzer. Energy calibration is done with a H^3-standard (NBS No. 4926) and for each sample with an external X-ray source (Am^{241}). This allows the same counting conditions to be obtained for every measurement.

The net spectra of the samples agree well with those of the H^3 standard (see Fig. 2). The measurement of the spectrum is not only important for identification of H^3 and exclusion of contamination by other radionuclides, but also for background reduction.

The efficiency of the whole procedure (extraction, conversion, and counting) is determined with tritiated hydrogen standards (prepared from NBS No. 4926) processed in the presence of basaltic samples. An efficiency of 0.55 with a reproducibility of ± 0.02 was obtained. The efficiency of the extraction alone was 0.95 or better. The errors quoted with our H^3 results include the counting errors of the various extraction fractions and blanks (1 σ) and the error of the efficiency. Our errors quoted are somewhat greater than those of other investigators (D'Amico *et al.*, 1971, Stoenner *et al.*, 1971). This is due to the smaller samples used, the analyses of several fractions (up to 4), and the inclusion of the uncertainties of the blanks (discussed later).

Background and procedural blank

The background of the counter (including the blank of the gas conversion system) was determined using H^3-free H_2 and C_2H_4 for conversion to C_2H_6. The counter background was typically between 0.06 and 0.08 cpm in the H^3 window (total background of counter 0.15 cpm). The H^3 blank originating from the H^3 extraction (procedural blank) was determined by processing empty cups made of aluminum foil. High purity Al-foil had to be used since ordinary foil showed some H^3 contamination. Net contribution in the tritium window of the procedural blanks was typically between 0.02 and 0.05 cpm. No increase was observed in the higher energy range, excluding contamination by other radionuclides. During an extraction series the procedural blank was constant within ± 0.01 cpm. The counter background and procedural blanks were frequently determined (typically before and after each lunar and other sample).

Meteorites and basalt processed in LRL

As a test of the extraction and measuring procedures and as a further blank check, we measured H^3 in two recently fallen meteorites (St. Séverin and Bruderheim), and a basalt sample (VB 2) processed at the Lunar Receiving Laboratory. The results are given in Table 1. The measurements on three aliquots of the St. Séverin chondrite gave very reproducible results. Our Bruderheim result agrees well with the results of Bainbridge *et al.* (1962) and Charalambus *et al.* (1969). Fireman and De Felice (1961) report a considerably lower H^3 activity for Bruderheim, and Fireman (1967) reports a similarly low H^3 activity for an olivine and troilite concentrate of the St. Séverin chondrite. Our values for St. Séverin and Bruderheim lie well within the range of tritium activities in chondrites reported earlier from our laboratory (Geiss *et al.*, 1960, 1960a). The basalt sample VB 2 showed a small, variable H^3 activity which indicates a possible contamination with man-made H^3.

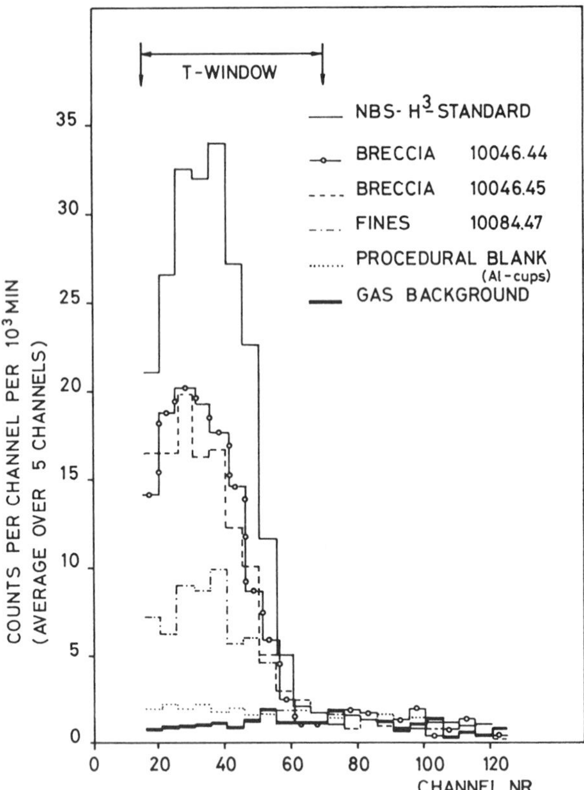

Fig. 2. Pulse height spectra of NBS H^3-standard, lunar samples, gas background, and procedural blank. The spectra 10046 are from a tritium contaminated breccia. The spectrum 10084 is from an uncontaminated sample of lunar fine material. Weights of lunar samples were between 1.1 g and 1.50 g. All lunar and meteoritic samples show similar spectra, in agreement with the spectra obtained from H^3 standards. From this we conclude that the net effects in the samples and also the contamination in the breccia 10046 are due to H^3.

Table 1. Results of H^3 determinations on two recently fallen meteorites and a basalt sample processed at the LRL

Sample	Total Activity of Sample dpm/kg	Activity Extrapolated to Date of Fall dpm/kg
Recently fallen meteorites		
St. Séverin (Be 347b)	363 ± 23	
(3 determinations)	377 ± 35	457 ± 21
Bruderheim (Be 102)	359 ± 33	
Basalt, processed at LRL	244 ± 24	442 ± 43
VB 2	30–90	
(range of 4 determinations)		

Sample sizes were between 1.1 g and 1.5 g.

Results on Lunar Samples

Apollo 11 Breccia 10046

We had received three samples of the Apollo 11 breccia 10046 in order to measure variations of H^3 with sample depth. The H^3 activity we found in the three samples of this rock was considerably higher than in the lunar fines (10084) (cf. Tables 2 and 3) Furthermore, measurements of the H^3 activity in three aliquots of the same sample (10046,45) gave a H^3 activity systematically decreasing with time (see Table 2), with an apparent "half-life" of 2 months. We concluded that the three samples of rock 10046 which we received must have been contaminated with man-made H^3.

Subsequent H^3 extractions on lunar samples were then carried out in two steps to check for possible H^3 contamination. A first H^3 fraction was obtained by heating the sample with H_2 carrier to 300°C for 40 minutes. Subsequently, the sample was heated to 1700°C, again with H_2 carrier and the remaining H^3 extracted. Little or no H^3 activity was found in any 300°C fraction of the lunar fines 10084 and the samples from rock 12053 (see Table 3). The 300°C fractions of rock 10046, however, contained approximately 30% of the total H^3 activity.

An artificially contaminated basalt sample (VB2) was used to further investigate the extraction behavior of contamination H^3. A watch with a H^3 luminous dial was

Table 2. Results of H^3 measurements on Apollo 11 breccia 10046

	Date of Measurement	Percentage of Total Activity in 300°C fraction	Total Activity of Sample* dpm/kg
Sample 10046,45			
1. Aliquot	2/24/70	n.e.	1120 ± 50
2. Aliquot	3/26/70	n.e.	920 ± 50
3. Aliquot	6/11/70	30%	430 ± 40
Sample 10046,44			
1. Aliquot	3/23/70	n.e.	740 ± 40
Sample 10046,43			
1. Aliquot	5/28/70	31%	400 ± 40

Sample size between 0.7 and 1.4 g. n.e.: Sample was not extracted at 300°C.
* Actually measured activities, not extrapolated to time of collection.

Table 3. Results of H^3 measurements on Apollo 11 fines 10084 and Apollo 12 rock 12053

	Sample Size	Sample Depth	Percentage of Total Activity in 300°C fraction	Total Activity at Time of Collection dpm/kg
Apollo 11 fines 10084,47				
1. Aliquot	1.22 g	—	n.e.	303 ± 33
2. Aliquot	1.21 g	—	n.e.	299 ± 27
3. Aliquot	1.51 g	—	<6%	316 ± 32
Apollo 12 rock 12053				
ca 13	1.10 g	0.3–0.7 cm	<6%	283 ± 35
ca 16	0.93 g	1.1–1.5 cm	n.e.	256 ± 34
ca 7	1.18 g	1.5–2.0 cm	<3%	261 ± 38
ca 3	1.22 g	2.0–2.5 cm	~8%	266 ± 35

n.e.: Sample was not extracted at 300°C.

placed face downwards on the powdered basalt sample for 22 h. A H^3 contamination of more than 40,000 dpm/kg resulted. A two step extraction was then made with the contaminated basalt. The 300°C fraction contained approximately 45% of the total H^3 contamination. A similar result was obtained on an old meteorite which showed some H^3 contamination (Mocs). In this case the 300°C fraction contained nearly 100% of the contamination H^3.

The presence of H^3 in the 300°C fraction is thus a good indication for contamination with man-made H^3. We may expect that between 50% and 100% of the total contamination is in the 300°C fraction. With this assumption we can correct our results on rock 10046 for the contamination. A H^3 activity between 160 dpm/kg and 300 dpm/kg would result.

Apollo 11 fines 10084

The results are given in Table 3. No H^3 activity was found in the 300°C fraction and we may assume that no contamination is present. Our results on the soil agree well with the measurements of D'Amico *et al.* (1970) on lunar fines 10084 and of Stoenner *et al.* (1970) on lunar fines 10002.

Apollo 12 rock 12053

We have received a slab from this rock (12053,37, or C-4) to measure the depth dependence of the H^3 activity and to investigate for a possible contribution from energetic solar particles. Figures 3 and 4 show the location of our slab in rock 12053. The orientation of this rock on the lunar surface is not quite certain (Shoemaker *et al.*, 1970). The results of Rancitelli *et al.* (1971) indicate that the orientation of rock 12053 with respect to top and bottom as shown in Fig. 3 must correspond roughly to the orientation on the lunar surface. As we were able to measure H^3 on 1 g samples, we could make a rather fine subdivision of the slab getting a detailed depth profile of the H^3 activity (see Fig. 5). Enough sample is available at each depth to allow at least two H^3 determinations. Up to now we have measured four different samples

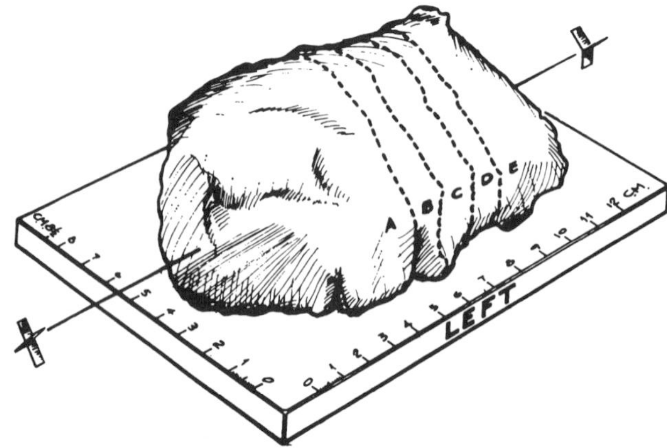

Fig. 3. Location of slice C in lunar rock 12053 (drawing from LRL, Houston).

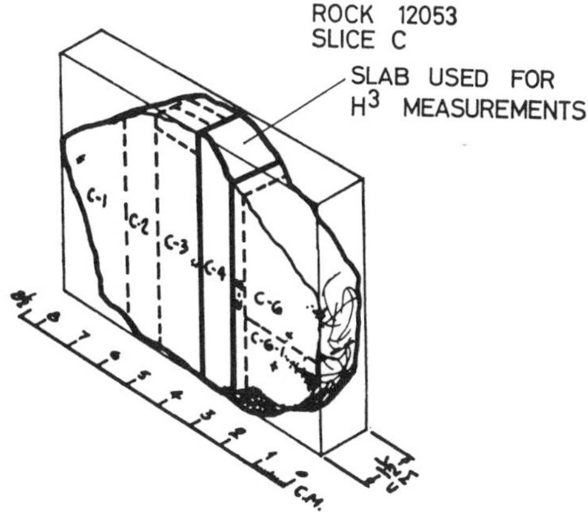

Fig. 4. Location of our slab (12053,37, or C-4) in slice C of lunar rock 12053 (adapted from LRL drawing).

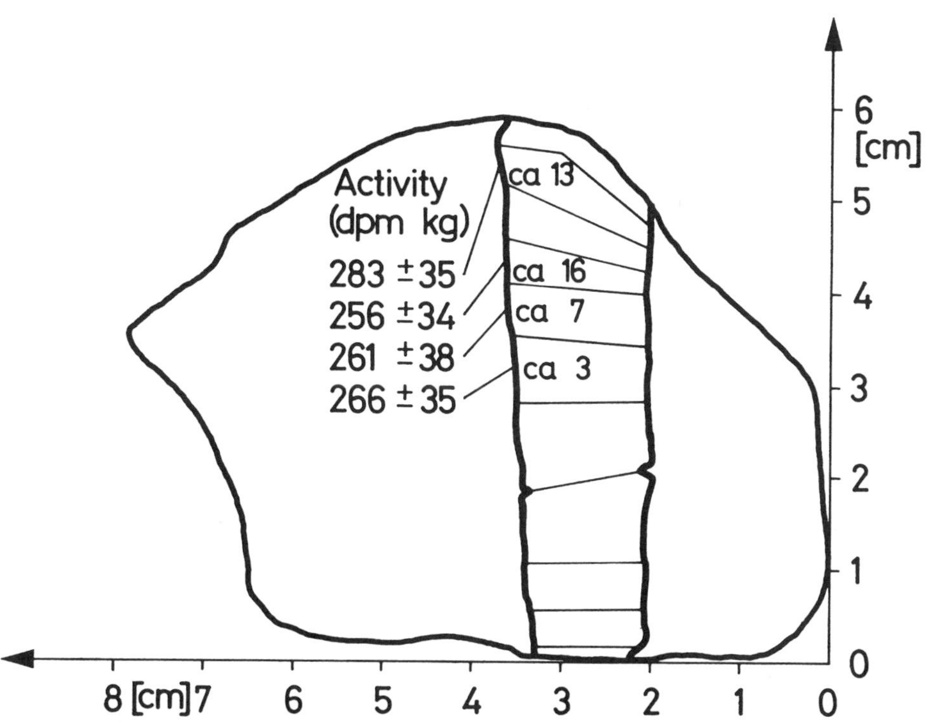

Fig. 5. Location and specific tritium activities of samples ca 13, ca 16, ca 7, and ca 3 in our slab of rock 12053.

(Table 3). The three interior samples ca 3, ca 7, and ca 16 have identical H^3 activities with an average of 261 $\pm$ 21 dpm/kg. This value agrees well with that obtained by D'Amico *et al.* (1971) on rock 12002 at similar depth. It is slightly higher than the H^3 activities found by D'Amico *et al.* (1970) and Stoenner *et al.* (1970) in Apollo 11 rocks. H^3 contamination of the 12053 samples is very unlikely as the 300°C fraction of ca 7 gave a very low H^3 limit.

Conclusions

Until now we do not have enough information on surface samples from rock 12053 to allow a discussion of the H^3 production by solar energetic particles and a possible direct influx of H^3 produced in solar flares. Therefore, this discussion is restricted to the H^3 activity in lunar material produced by the galactic cosmic radiation and to a comparison with the H^3 activities found in chondritic meteorites.

Based on surface H^3 data by D'Amico *et al.* (1971) and our preliminary value for a surface sample of rock 12053, as well as the observed depth dependence of other radioactive isotopes (Finkel *et al.*, 1971), we conclude that the contribution of solar energetic particles to the H^3 production for the samples ca 3 and ca 7 is less than 20%. Thus the H^3 production in rock 12053 at a depth of 2 cm by galactic cosmic rays is assumed to be 240 dpm/kg ($\pm$15%).

The H^3 data on chondrites obtained by Geiss *et al.* (1960, 1960a), Bainbridge *et al.* (1962), Charalambus *et al.* (1969), Begemann (1966), as well as our new results on the St. Séverin and Bruderheim chondrites lie very well in the range of H^3 production rates calculated by Trivedi and Goel (1969) from the results of a thick target irradiation with 3 GeV protons.

Of special importance for this comparison is the H^3 activity of (450 $\pm$ 30) dpm/kg measured in a sample of the meteorite Kiel which fell 1962 (Begemann, 1966). From the He^3/Ne^{21} ratio of 10.3 (Hintenberger *et al.*, 1964) it can be concluded that this meteorite had a very small preatmospheric size. Trivedi and Goel (1969) calculated a H^3 production rate between 400 and 480 dpm/kg for chondrites with a radius of 5 cm, in good agreement with the observed H^3 activity of the meteorite Kiel. The best way of comparing the H^3 activity in lunar rock 12053 with those determined in chondrites seems to us to assume that the irradiation by galactic cosmic radiation of this lunar rock corresponds to a 2π irradiation of a chondrite of comparable size, such as Kiel. This leads to an expected value of 225–275 dpm/kg for the center of lunar rock 12053. The range stems from the uncertainty in the location of the analyzed sample in the meteorite Kiel and was estimated from the depth effect curves given by Trivedi and Goel (1969).

We are aware that for such a comparison further corrections should be applied, taking into account the different chemical composition of chondrites and lunar samples, the contribution of energetic albedo neutrons, the not completely flat horizon on the moon, and the variation of the flux of the galactic cosmic radiation during the solar cycle. However, we estimate that these corrections are relatively small, approximately of the same magnitude as the errors of the H^3 measurements. The H^3 activities in lunar samples and in chondritic meteorites thus correspond to the same primary galactic cosmic ray flux. This makes any large radial cosmic-ray intensity variations between 1 and 3 A.U. unlikely.

The spallation He^3 concentration in sample ca 7 is $(68 \pm 5) \times 10^{-8}$ cm^3 STP/g. This leads for this interior sample to a $(He^3/2H^3)$ cosmic-ray exposure age of $(67 \pm 7) \times 10^6$ y. The diffusion loss of He^3 from lunar feldspar (EBERHARDT *et al.*, 1971) can easily explain the discrepancy between this age and the Kr^{81}/Kr exposure age of $(99 \pm 6) \times 10^6$ years measured in the same sample ca 7 (BOCHSLER *et al.*, 1971).

Acknowledgments—We should like to thank Dr. N. Grögler for discussions. We are indebted to K. Hänni for carrying out the gas conversions and E. Lenggenhager for technical help. This work is supported by the Swiss National Science Foundation (Grants NF 2.73.68, 2.213.69, 2.69.68, and 2.198.69).

REFERENCES

BAINBRIDGE A. E., SUESS H. E., and WÄNKE H. (1962) The tritium content of three stony meteorites and one iron meteorite. *Geochim. Cosmochim. Acta* **26**, 471–480.

BEGEMANN F. (1966) Tritium content of two chondrites. *Earth Planet. Sci. Lett.* **1**, 148–150.

BOCHSLER P., EBERHARDT P., GEISS J., GRAF H., GRÖGLER N., KRÄHENBÜHL U., MÖRGELI M., SCHWALLER H., and STETTLER A. (1971) Potassium-argon ages, exposure ages and radiation history of lunar rocks. Second Lunar Science Conference (unpublished proceedings).

CHARALAMBUS S., GOEBEL K., and STOETZEL-RIEZLER W. (1969) Tritium and argon-39 in stone and iron meteorites. *Z. Naturforsch.* **24a**, 234–244.

D'AMICO J., DE FELICE J., and FIREMAN E. L. (1970) The cosmic-ray and solar-flare bombardment of the moon. *Proc. Apollo 11 Lunar Sci. Conf., Geochim. Cosmochim. Acta* Suppl. 1, Vol. 2, pp. 1029–1036. Pergamon.

D'AMICO J., DE FELICE J., FIREMAN E. L., JONES C., and SPANNAGEL G. (1971) Tritium and argon radioactivities and their depth variations in Apollo 12 samples. Second Lunar Science Conference (unpublished proceedings).

EBERHARDT P., GEISS J., GRAF H., GRÖGLER N., KRÄHENBÜHL U., SCHWALLER H., and STETTLER A. (1971) To be published.

FINKEL R. C., ARNOLD J. R., REEDY R. C., FRUCHTER J. S., LOOSLI H. H., EVANS J. C., SHEDLOVSKY J. P., IMAMURA M., and DELANY A. C. (1971) Depth variation of cosmogenic nuclides in a lunar surface rock. Second Lunar Science Conference (unpublished proceedings).

FIREMAN E. L. and DE FELICE J. (1961) Tritium, argon-37 and argon-39 in the Bruderheim meteorite. *J. Geophys. Res.* **66**, 3547–3551.

FIREMAN E. L. (1967) Radioactivities in meteorites and cosmic-ray variations. *Geochim. Cosmochim. Acta* **31**, 1691–1700.

GEISS J., HIRT B., and OESCHGER H. (1960) Tritium- und Helium-Gehalte in Meteoriten. *Helv. Phys. Acta* **33**, 590–593.

GEISS J., OESCHGER H., and SIGNER P. (1960a) Radiation ages of chondrites. *Z. Naturforsch.* **15a**, 1016–1017.

HINTENBERGER H., KÖNIG H., SCHULTZ L., and WÄNKE H. (1964) Radiogene, spallogene und primordiale Edelgase in Steinmeteoriten. *Z. Naturforsch.* **19a**, 327–341.

RANCITELLI L. A., PERKINS R. W., FELIX W. D., and WOGMAN N. A. (1971) Cosmogenic and primordial radionuclide measurements in Apollo 12 lunar samples by nondestructive analysis. Second Lunar Science Conference (unpublished proceedings).

SHOEMAKER E. M., BATSON R. M., BEAN A. L., CONRAD C., DAHLEM D. M., GODDARD E. N., HAIT M. H., LARSON K. B., SCHABER G. G., SCHLEICHER D. L., SUTTON R. L., SWANN G. A., and WATERS A. C. (1970) Preliminary geologic investigation of the Apollo 12 landing site. Apollo 12 Preliminary Science Report, NASA SP–235, 113–156.

STOENNER R. W., LYMAN W. J., and DAVIS R., JR. (1970) Cosmic-ray production of rare-gas radioactivities and tritium in lunar material. *Proc. Apollo 11 Lunar Sci. Conf., Geochim. Cosmochim. Acta* Suppl. 1, Vol. 2, pp. 1583–1594. Pergamon.

Stoenner R. W., Lyman W. J., and Davis R., Jr. (1971) Radioactive rare gases in lunar rocks and and in the lunar atmosphere. Second Lunar Science Conference (unpublished proceedings).

Trivedi B. M. P. and Goel P. S. (1969) Production of Na^{22} and H^3 in a thick silicate target and its application to meteorites. *J. Geophys. Res.* **74**, 3909–3917.

Tschudi L. (1968) Datierung kleiner C^{14}-Proben. *Lizentiatsarbeit*, University of Bern (unpublished).

Proceedings of the Second Lunar Science Conference, Volume 2, pp. 1813–1823.
The M.I.T. Press, 1971.

Radioactive rare gases and tritium in lunar rocks and in the sample return container

R. W. Stoenner, Warren Lyman, and Raymond Davis, Jr.
Chemistry Department, Brookhaven National Laboratory, Upton, New York 11973

(*Received 23 February 1971; accepted in revised form 30 March 1971*)

Abstract—The radioactive isotopes ^{37}Ar (halflife 35.1 day) and ^{39}Ar (halflife 269 year) were measured on two lunar rocks from the tote-bag. The activities observed in dpm/kg were as follows: 12063, 27.8 ± 0.6 for ^{37}Ar, 7.8 ± 0.6 for ^{39}Ar; 12065, 49.5 ± 0.6 for ^{37}Ar, 7.8 ± 2 for ^{39}Ar. The ^{37}Ar activities were higher than observed on Apollo 11 rocks, and it is concluded that the increase may be attributed to the solar flare event of November 2, 1969. The observed ^{37}Ar/^{39}Ar ratios are compared to the production of these isotopes by 600-MeV protons and fast neutrons. The exposure ages of these two rocks were found to be rock 12063, 100 × 10^6 years and rock 12065, 280 × 10^6 years.

High amounts of tritium were observed diffusing from the surface of these rocks at room temperature, as high as 200 dpm from a 10-gram chip. The origin of the tritium activity could be attributed to a triton component of solar flare particles or to a tritium contamination within the space craft.

Radioactive rare gases were measured in the gas from the selected sample return container. The amounts observed were 0.040 ± 0.002 dpm ^{37}Ar, 0.0023 ± 0.0010 dpm ^{39}Ar, (8.6 ± 3.1) × 10^{-4} dpm ^{85}Kr or ^{81}Kr, and 4.7 ± 1.0 dpm ^{222}Rn. Laboratory experiments were performed to measure the recoil loss and diffusion of ^{37}Ar, ^{39}Ar, and tritium from simulated lunar fines bombarded with 600-MeV protons. It was concluded that the radioactive gases observed in the sample return container probably diffused out of the 2.8 kg of fine lunar material in the container.

Introduction

The radioactive ^{37}Ar (35-day halflife) and ^{39}Ar (269-year halflife) are produced in lunar surface rocks in high yield from the abundant elements Ca, Ti, and Fe by galactic cosmic rays and solar flare particles. The 35-day halflife of ^{37}Ar makes it a convenient isotope to observe irregular solar flare events. The longer lived ^{39}Ar can be used conveniently to measure the average particle fluxes over the past few hundred years. Tritium (12.3 years) is an important isotope to study in lunar materials. Tritium has been observed in solar flare particles (Fireman *et al.*, 1961) and the lunar surface would be an effective collector for these solar tritons. These three isotopes and other rare-gas radioactivities were measured in lunar rocks 12063 and 12065 at the Lunar Receiving Laboratory (LRL) during the biological quarantine period. Higher amounts of ^{37}Ar were observed in the Apollo 12 rocks than in rock samples from the Apollo 11 mission. The higher amounts of ^{37}Ar can be attributed to the solar flare event of November 2.

The sample return container (SRC) was closed on the surface of the moon and contains therefore an 18-liter sample of the lunar atmosphere. The rare gases were collected from the selected sample return container and a search was made for radioactive rare gases. A surprisingly large amount of ^{37}Ar was observed, 0.040 ± 0.002

dpm. In addition small amounts of ^{39}Ar, krypton activities, and ^{222}Rn were measured. To interpret the argon results, a simulated lunar fine material was bombarded with 600-MeV protons. A study was made of the recoil loss and diffusion of argon radioactivities and tritium from this fine material. These results were used to estimate the amount of these radioactivities expected in the lunar atmosphere by recoil processes, and in the gas in the SRC by diffusion from the fine material present. The amount of ^{37}Ar observed in the SRC can be accounted for by diffusion from the fine material in the container.

ARGON RADIOACTIVITIES FROM LUNAR ROCKS

Argon radioactivities and tritium were extracted from lunar rocks 12063 and 12065 by vacuum melting using the procedures described previously (STOENNER *et al.*, 1970). The samples used were chipped from the surface of the rocks during the biological quarantine period. At this time "before and after" photographs were taken, but it was difficult to establish the exact location of the chip on the drawings of the rock made later. The location of the chip used from rock 12063 was from the top at the south end, slice A in the LRL Curator's drawing (see NASA–MSC photos S–69–60600 before and S–69–61664 after). The chip from rock 12065 was taken from the center of the face of slice A as shown in the drawing (see NASA–MSC photos S–69–60581 before and S–69–61666 after). The chips used were exterior pieces approximately 1 cm thick containing micrometeorite craters on the surface.

The measured argon radioactivities are listed in Table 1 and the values are compared to those we measured in an interior and an exterior chip of an Apollo 11 rock (10057). The ^{39}Ar activities observed in the Apollo 12 rocks are considerably lower than those we observed in the Apollo 11 rocks and soil, and the ^{37}Ar activities are higher. A possible explanation of the low ^{39}Ar activities is that the chips which we measured were taken from the bottom of these relatively large rocks (12063—2.43 kg; 12065—approximately 1.8 kg; 10057—0.92 kg). However, FIREMAN and his associates (D'AMICO *et al.*, 1971) have measured the ^{39}Ar and ^{37}Ar from three depths in rock 12002 and found the ^{37}Ar and ^{39}Ar activities only decreased by 11 and 25% respectively (from the surface values) at an average depth of 5.6 cm. Hence it is unlikely that the low ^{39}Ar activities we observed compared to rock 10057 are explained by a depth effect. It will be shown later that the low ^{39}Ar contents of the Apollo 12 rocks can be explained by their low titanium content.

The argon radioactivities are formed as spallation products from high-energy cosmic ray and solar flare particles. In addition, these activities are produced by

Table 1. Argon radioactivities from lunar rocks.

Lunar rock	Activity (dpm/kg)			Composition in weight %			
	^{37}Ar	^{39}Ar	^{37}Ar/^{39}Ar	Fe	Ti	Ca	K
12063,—	27.8 ± 0.6	$7.80 + 0.6$	3.56 ± 0.15	16.7	3.1	7.9	0.065
12065,—	49.5 ± 0.6	7.76 ± 0.16	6.38 ± 0.14	17.1	2.3	9.0	0.06
10057,3 exterior	18.5 ± 0.9	11.2 ± 0.3	1.65 ± 0.09	15.5	7.5	7.1	0.15
10057,27 interior	23.7 ± 4.1	14.8 ± 0.4	1.60 ± 0.30				
10002,6 fines	18.7 ± 1.2	9.2 ± 0.4	2.03 ± 0.16	12.4	4.2	8.6	0.10

Table 2. $^{37}Ar/^{39}Ar$ ratios from lunar rocks.

Lunar rock	Observed ratio $^{37}Ar/^{39}Ar$	Calculated $^{37}Ar/^{39}Ar$ ratios* 600-MeV protons	14-MeV neutrons
12063,—	3.56 ± 0.15	3.5	4.8
12065,—	6.38 ± 0.14	4.0	4.8
10057	1.62 ± 0.10	2.5	4.3
10002,6 fines	2.03 ± 0.16	3.2	4.6

* Cross-sections used (see STOENNER *et al.*, 1970) 600 MeV protons: ^{37}Ar from Fe 5.6 mb, Ti 11.6 mb, and Ca 46.7 mb; ^{39}Ar from Fe 6.3 mb, Ti 15.7 mb, and Ca 2.0 mb. 14 MeV neutrons: $^{40}Ca(n, \alpha)^{37}Ar$ 180 mb, $^{40}Ca(n, 2p)^{39}Ar$ 35 mb, and $^{39}K(n, p)^{39}Ar$ 350 mb.

secondary neutrons by the reactions $^{40}Ca(n, \alpha)^{37}Ar$, $^{40}Ca(n, 2p)^{39}Ar$, and $^{39}K(n, p)^{39}Ar$. We have previously (STOENNER *et al.*, 1970) compared the observed $^{37}Ar/^{39}Ar$ ratios in lunar material with the ratio of the cross-sections for producing these isotopes by 600-MeV protons and 14-MeV neutrons on the elements iron, titanium, calcium, and potassium. Using the values of the cross-sections previously given and the elemental compositions given in Table 1, the corresponding $^{37}Ar/^{39}Ar$ ratios were calculated. The individual calculated ratios are given in Table 2 for 600-MeV protons and for 14-MeV neutrons. This method of estimating the relative production rates of ^{37}Ar and ^{39}Ar on the lunar surface by cosmic ray particles is relatively crude because of the broad energy range of the primary particles, and the secondary particles produced in the nuclear cascade processes. One notices though that there is a larger variation in the measured $^{37}Ar/^{39}Ar$ ratios than can be accounted for by the differences in the chemical composition of these three rocks. The high ^{37}Ar activities observed in the Apollo 12 rocks were produced by solar flare protons on November 2, 1969; this event will be discussed later.

As mentioned above, the ^{39}Ar activities are lower in the two Apollo 12 rocks, a result that may be explained by the chemical composition of these samples. It may be noticed from the chemical composition given in Table 1 that the calcium and iron compositions of these samples are all approximately the same, whereas the titanium composition varies considerably. Since ^{39}Ar is produced mainly from the elements calcium, titanium, and iron (potassium is very low) one would expect that the variations in the ^{39}Ar contents in these samples to be related to the widely differing titanium compositions. If the ^{39}Ar measurements presented here are combined with those of FIREMAN and his associates on rocks with similar calcium and iron contents (D'AMICO *et al.*, 1970; D'AMICO *et al.*, 1971) one finds the ^{39}Ar contents increase with the titanium composition. It is then clear that the range of ^{39}Ar activities observed in these lunar rocks can be accounted for mainly by the variations in the chemical composition.

The most striking comparison is that the Apollo 12 rocks have higher $^{37}Ar/^{39}Ar$ ratios than were observed in Apollo 11 material. D'AMICO *et al.* (1970) observed the following $^{37}Ar/^{39}Ar$ ratios: Apollo 11 samples, soil 2.25 ± 0.35; 10017,14 1.28 ± 0.20; 10072,11 1.62 ± 0.25; Apollo 12 rock 12002, top surface 0–0.8 cm, 2.95 ± 0.36; 0.8–3.1 cm depth, 3.05 ± 0.26; 4.9–6.4 cm depth, 3.43 ± 0.40. The higher ^{37}Ar activity observed in all Apollo 12 samples may be attributed to the solar flare of November 2, which occurred 17 days prior to the Apollo 12 landing. This flare produced the largest proton increase of 1969 in the vicinity of earth, as observed by Explorer 41 satellite and high latitude riometers (Solar-Geophysical Data, 1970). The spectrum was typically steep, and no increase was observed in ground-based neutron monitors. There was a total proton intensity of 10^8 cm^{-2} with energy above 30 MeV, which is approximately equivalent to the total proton intensity from nonflare cosmic rays in that energy interval in a year. The production of ^{37}Ar and ^{39}Ar from calcium is a low energy process and one would expect good yields of those isotopes from 30-MeV protons. We then attribute the high ^{37}Ar activities to this flare event. On the other hand the rocks from Apollo 11 were relatively free of flare-produced radioactivity, the only significant event preceding this mission occurred on April 12, 1969, 100 days prior to the mission. Because of the long halflife of ^{39}Ar, its value remains relatively constant and corresponds to the average value for the production of ^{39}Ar by galactic and solar cosmic rays over the last 500 years. The ^{37}Ar activity varies depending on solar-flare particle intensity and the solar modulation of the galactic cosmic ray intensity (FORMAN *et al.*, 1970; D'AMICO *et al.*, 1970). During the Apollo 11 mission the ^{37}Ar activity was below the average value and during the Apollo 12 mission it was above the average value.

EXPOSURE AGES

The ^{39}Ar activity and the measured spallation ^{38}Ar can be used to determine the exposure ages of the rocks. The rare gas contents of chips from the fragments used in this experiment were measured by the gas analysis laboratory (LRL). BOGARD *et al.* (1970) found the spallation ^{38}Ar contents to be 12063, 9.2×10^{-8} and 12065, 28×10^{-8} cm^3/g. Using these values, the 600-MeV proton cross-sections from STOENNER *et al.* (1970), and the cross section of 74 mb for the production of ^{38}Ar from calcium (BOGARD *et al.*, 1970) we calculate the exposure ages to be rock 12063, 100×10^6 years and rock 12065, 280×10^6 years.

It is difficult to evaluate the errors introduced in calculated exposure age by using the 600-MeV proton cross-sections to represent the relative production of ^{39}Ar and ^{38}Ar by galactic and solar cosmic rays. The approach used here does not account correctly for the production of these isotopes in lunar material by energetic neutrons and alpha particles. To obtain accurate ^{38}Ar–^{39}Ar exposure ages the relative production of these isotopes in thick target irradiations of lunar-like material is required. Measurements of this kind are in progress for the radioactive argon isotopes, but the measurement of stable ^{38}Ar by mass-spectroscopy is a difficult task that has not been attempted. In view of this, our exposure ages may be in error by perhaps as much as 50%.

It is interesting to compare our exposure ages with those of other workers. D'AMICO *et al.* (1971) estimate the exposure age of rock 12065 to be 160×10^6 years by the ^{3}He/2^{3}H method and 410×10^6 years by the ^{38}Ar/^{37}Ar + ^{39}Ar method. HINTENBERGER *et al.* (1971) estimate the exposure age of rock 12063 to be 84×10^6 years. They used the ^{38}Ar production rate of 1.43×10^{-9} cm^3/g$\cdot 10^6$ years, whereas we would estimate an ^{38}Ar production rate of 0.93×10^{-9} cm^3/g$\cdot 10^6$ years for rock 12063 based upon the measured ^{39}Ar activity and the 600-MeV cross-section ratio. MARTI and LUGMAIR (1971) obtained an exposure age of 95×10^6 years for rock 12063 using the ^{81}Kr–^{83}Kr method. This excellent method depends upon their estimate of the relative production rates of ^{81}Kr and ^{83}Kr. They obtain this ratio by averaging the ^{80}Kr and ^{82}Kr contents relative to the ^{83}Kr content in the sample. It would be well to examine the validity of this procedure by proton bombardment of Sr, Y, and Zr targets. To obtain accurate exposure ages it is essential to measure the relative production rates of the isotopes in lunar-like material as a function of the energy. Until these basic measurements are made, one can give only approximate exposure ages.

TRITIUM ANALYSES

The tritium contents of the samples were extremely high, too high to be explained by normal cosmic-ray particle bombardment. The samples were placed in the vacuum extraction system prior to melting (see STOENNER *et al.*, 1970 for details) and the crucible baked out. In the course of this operation the hydrogen recovered while heating the crucible was counted to check the tritium background in the system. This procedure was followed for each rock before it was melted. The tritium was isolated, purified, and counted; the results are given in Table 3. Normally the system background counting rates are less than 0.1 dpm of tritium. The observed counts were about 100-fold higher than expected, and the tritium appeared to be lightly adsorbed on the surface of the sample. To ensure that the activity measured was indeed tritium, the endpoint and beta spectra were determined with a proportional counter, and it was clear that the activity was caused by tritium.

Rocks 12063 and 12065 were collected during the second EVA and were transported in the tote-bag. During the return mission the tote-bag was kept in the command module and was exposed to the atmosphere in the spacecraft. On return to the earth the tote-bag and its contents were handled in the crew reception area of the LRL. It was then possible that these rocks became contaminated with tritium in the atmosphere of the spacecraft or in the LRL. A possible source of tritium in the atmosphere of the spacecraft arises from the lithium hydroxide absorber used to remove carbon dioxide. Cosmic ray produced neutrons will yield tritium by the high cross-section slow neutron reaction $^6Li(n, ^4He)T$. To examine this possibility, the tritium content of the lithium oxide was measured, and found to contain 80 dpm per gram. Since approximately 30 pounds of lithium hydroxide are used, the total amount of tritium present in the absorber was 10^6 dpm of tritium. One would expect the tritium produced in the lithium oxide to be retained as absorbed water and not be released to the atmosphere of the command module. However, because of the uncertainty in this

Table 3. Tritium activity from Apollo 12 rocks.

Sample	Activity (dpm)
Initial bake-out for 12063	1.62
Melted 9.53 g 12063	20.0
Initial bake-out for 12065	204
Melted 9.19 g 12065	24.5

question, and the presence of other possible sources of tritium in the spacecraft (watches and beta-lights), we cannot determine whether or not the tritium observed was contamination or was present on the surface of the moon. It is interesting to note that rock 12065 which was so very high in apparent tritium contamination also had a very high ^{37}Ar content, 49.5 dpm/kg.

We would like to point out again that tritium has been observed in solar flare particles by Fireman *et al.* (1961). It is entirely possible that the tritium observed here was deposited on the moon by the flare of November 2, 1969 and the tritons were lightly adsorbed on the lunar material. Because our measurements were made immediately after sample return during the biological quarantine we had the best opportunity to observe tritium. Other investigators did not observe the high levels of tritium reported here (D'Amico *et al.*, 1971). These investigators made a careful search for lightly adsorbed tritium using the identical procedure that we followed, and found less than 10^{-2} dpm of tritium adsorbed on samples from the sample return containers. They also studied a surface sample of rock 12065 and found less than 0.02 dpm tritium adsorbed on the surface of the rock. These studies were made six months after ours, and it is apparent that the tritium we observed had escaped. Small amounts of "man-made" tritium contamination was reported by Bochsler *et al.* (1971) in an Apollo 11 breccia sample. The questions raised by these observations will be pursued during the Apollo 14 mission if suitable samples are available.

Radioactive Rare Gases in the Sample Return Container

The sample return container (SRC) when closed on the surface of the moon includes an 18-liter sample of the lunar atmosphere. It is of great interest to search for radioactive gases present in the lunar atmosphere. There are several processes that would release radioactive gases from the fine lunar soil. These are the recoil of high-energy nuclear interaction products, the diffusion of gaseous natural radioactive decay products, and the release of fission products.

An attempt was made to observe ^{37}Ar, ^{39}Ar, ^{85}Kr, and ^{222}Rn in the selected sample return container. This container was returned with an internal pressure of 0.03 torr, and was processed in the F201 chamber. A device was attached to the surface of the container that pierced a thin portion of the wall with a $\frac{1}{4}$-inch diameter rod. The gas inside the container was then withdrawn through a 120-cm long flexible stainless steel hose to liquid helium-cooled and liquid nitrogen-cooled charcoal traps to collect the gases. The piercing rod was not withdrawn during this operation so that a very low pumping speed was achieved. The gas was collected over a period of approximately one hour. The final pressure in the SRC dropped to 0.003 torr, indicating 90% of the gas present was collected.

The rare gases collected were separated by gas phase chromatography on a charcoal column and the individual rare gases were counted in small proportional counters using the techniques already described by Stoenner *et al.* (1970). The observed activities are given in Table 4. The ^{37}Ar exhibited the correct pulse height spectrum and was fifteen times background. Figure 1 shows the pulse height spectrum obtained (see Stoenner *et al.*, 1970 for a description of the counting technique). The measure-

Table 4. Radioactive rare gases from the sample return container.

Rare gas	Activity (dpm)	Atoms per cm^3
^{37}Ar*	0.040 ± 0.002	0.160 ± 0.008
^{39}Ar	0.0023 ± 0.0010	24 ± 11
^{85}Kr	0.00086 ± 0.00031	0.39 ± 0.14
^{222}Rn†	~4.7 ± 1.0	2.1 ± 0.4

* Activity corrected to 20 Nov 0000 hours.
† Activity corrected to time of box puncture. The error quoted is the statistical counting error. The counter efficiency was estimated.

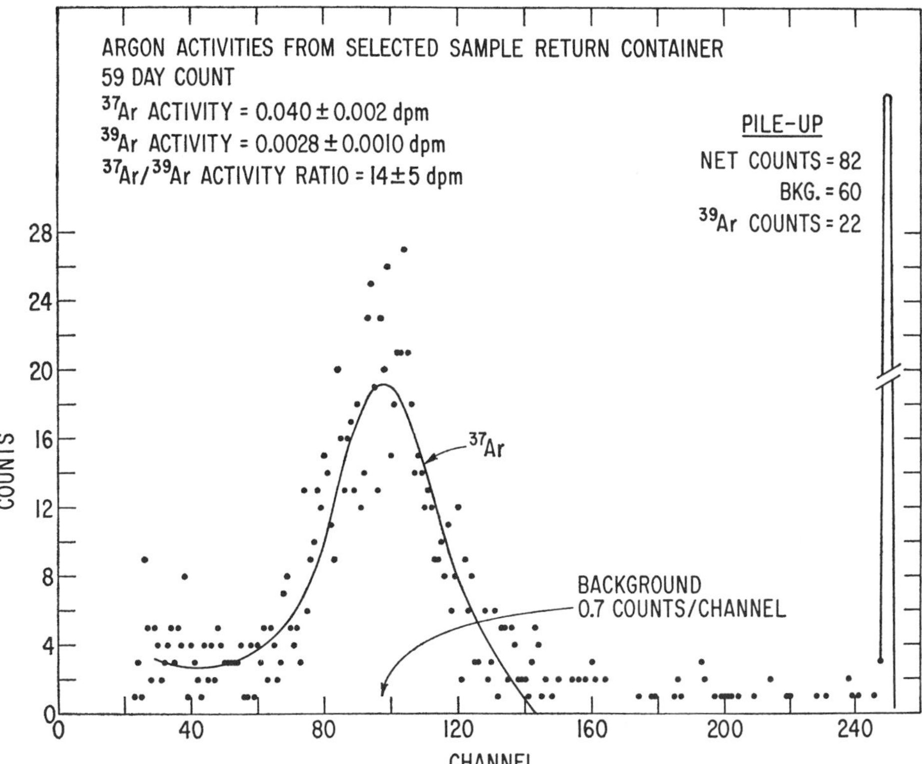

Fig. 1. Pulse height spectrum of argon radioactivities. The ^{37}Ar peak shown is at 2.8 keV. Pulses from ^{39}Ar greater than 7.8 keV are summed in channels 250–255.

ment of ^{39}Ar and ^{85}Kr was much less certain; the rates observed were only about 30% above counter background. The ^{222}Rn was characterized by its halflife, but there was a high ^{222}Rn background from the charcoal, and the counting efficiencies were only estimated.

The ^{37}Ar observed in the gas of the SRC could be present in the lunar atmosphere or it could arise by diffusion from the fine material in the container. To resolve this question a sample of lunar-like fines was prepared, and bombarded with 600-MeV

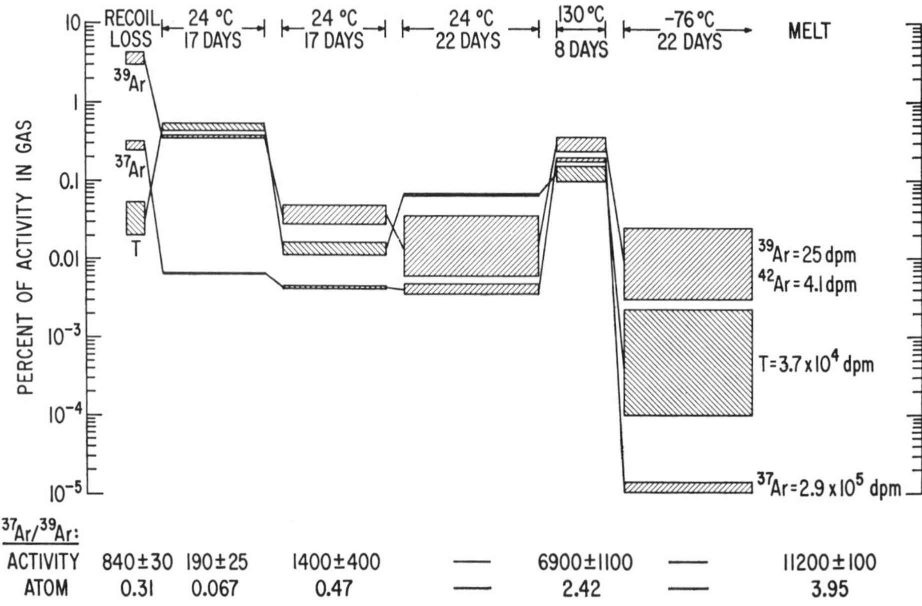

Fig. 2. The recoil loss and diffusion release of ^{37}Ar, ^{39}Ar, and tritium from samples of simulated lunar fines irradiated with 600-MeV protons.

protons. The simulated fines contained 40% pyroxene, 25% plagioclase, 10% olivine, 20% quartz, and 5% ilmenite (analysis 4.42% Ca, 1.90% Ti, 6.33% Fe, and 0.58% K). Each mineral component had a particle size distribution characteristic of lunar fines (25% 20–60 mesh, 10% 60–100 mesh, 10% 100–140 mesh, 25% 140–200 mesh, and 30% <200 mesh). The mineral samples were placed in a glass cell and evacuated and baked out for two days at 350°C. Two cells were prepared containing 25 grams of material, one was sealed following the evacuation described, and the other was filled with 600 torr of helium pressure and sealed. During the irradiation the evacuated cell was cooled to -196°C to ensure a high vacuum in the gas phase of the cell. After irradiation the cells were opened and carrier argon and hydrogen introduced. The carrier gas was removed, purified, and counted to measure ^{37}Ar, ^{39}Ar, and tritium activities. Following this, carrier gas was again introduced and the cells were stored at 24°C to measure the amount of those activities that diffused out of the grains. These room temperature diffusion measurements were repeated twice. Following these experiments the cells were heated to 130°C for seven days to simulate the surface temperature corresponding to a lunar day. A final extraction was made after a storage period at -76°C to simulate the conditions of a lunar night. Finally the bombarded fine material was vacuum melted to determine the total amount of ^{37}Ar, ^{39}Ar, ^{42}Ar, and tritium present. The results, expressed as percent of the activity obtained in the gas extractions, are given in Fig. 2. It may be observed that a higher percentage of the ^{39}Ar (3.6%) recoiled from the fine material than ^{37}Ar (0.28%). The ^{37}Ar diffused out at about the same rate for all three room temperature experiments, about 4×10^{-3}%, and the ^{39}Ar diffusion rate diminished during successive extractions. It is interesting to note that the tritium recoil yield was very low (0.03%), probably because the energy of the triton was sufficient to penetrate into neighboring grains. But, surprisingly, the first extraction at room temperature yielded 0.5% of tritium, and the bombarded fines adsorbed the carrier hydrogen so tenaciously that only a 60% yield was obtained. Apparently the loss of carrier results from a chemical reaction of the hydrogen gas with the dry and radiation-damaged fine material during the 17 day exposure. In subsequent extractions complete carrier recovery was achieved. The diffusion of all three isotopes increased greatly at 130°C, and dropped to a very low rate at dry-ice temperatures.

One may use the above data to estimate the amount of ^{37}Ar and ^{39}Ar that one might expect to diffuse out of the fine lunar material in the SRC. Let us presume there is 40 dpm/kg ^{37}Ar and 10 dpm/kg of ^{39}Ar in the fine material at the time the SRC was closed on the lunar surface. There was 2.8 kg of fine material in the container, 2.7 kg in a woven Teflon bag, and 100 grams in a capped core tube holder. Using ^{37}Ar and ^{39}Ar diffusion yields of 4×10^{-3} and 4×10^{-2}%, respectively, we estimate 0.0045 dpm of ^{37}Ar and 0.011 dpm of ^{39}Ar would be released. The ^{37}Ar observed in the SRC was about a factor of 9 above this value, and the ^{39}Ar was about a factor of 4 below this estimate. It is reasonable that ^{37}Ar might be released somewhat faster than ^{39}Ar in the actual retrieved lunar material from the Apollo 12 mission. This is because most of the ^{37}Ar in the fine material was produced by the November 2 flare, and this fine material was held at a low temperature on the dark side of the moon for two weeks prior to its recovery by the astronauts, whereas the ^{39}Ar in the samples was produced during the last few hundred years and has survived a few thousand lunar days. Hence the remaining ^{39}Ar resides in larger grains, and will diffuse out very slowly. We therefore conclude that the ^{37}Ar and ^{39}Ar observed in the gas in the sample return container diffused from the fine lunar dust in the container.

The recoil yields measured in this experiment can be used to estimate the amount of ^{37}Ar released to the lunar atmosphere by a sudden burst of flare particles. If a flare produces 100 dpm/kg of ^{37}Ar in the fines, and if the ^{37}Ar falls off with depth in the lunar surface with an exponential mean free path of 150 g/cm^2, then 3000 ^{37}Ar atoms per cm^2 would be released to the lunar atmosphere. Since the scale height L for argon in the lunar atmosphere is 50 kilometers, there should be 6×10^{-4} atoms of ^{37}Ar per cm^3 at the surface (^{37}Ar atoms cm^{-3} = $3000/L$). This amount of ^{37}Ar corresponds to 1.5×10^{-4} dis/min in an 18-liter volume, a factor of 200 below the rate we observed in the sample return container.

The radon loss from lunar soil and its behavior in the lunar atmosphere is an interesting question. Initially KRANER *et al.* (1966) estimated the amount of ^{222}Rn escaping from lunar soil and pointed out that a fraction of its decay products should be deposited on the lunar surface. YEH and VAN ALLEN (1969) attempted to observe the alpha particle emission from the moon and set an upper limit on the flux of 0.064 cm^{-2} sec^{-1} steradian^{-1}. During the surveyor missions TURKEVICH *et al.* (1970) observed a total surface alpha emission of 0.09 ± 0.03 cm^{-2} sec^{-1}. One might therefore expect approximately 0.06 ^{222}Rn decays min^{-1} in the 18-liter sample return container. In this calculation we have used the ^{222}Rn emission rate of TURKEVICH *et al.* (1970) of (58 ± 17) × 10^{-3} atom sec^{-1} cm^{-2}, and exponential scale height for ^{222} Rn of 10 km.

The amount observed in the SRC is over 100 times the amount that could be present in an 18 liter volume of the lunar atmosphere. Hence we conclude the ^{222}Rn must have emanated from the 2.8 kg of fine material in the container. Since the box was closed for 5.8 days prior to puncturing, the saturation ^{222}Rn emanation rate was 7.2 dpm. Noting that there was 2.8 kg of fine material in the container, we calculate the average ^{222}Rn emanation rate to be 2.6 × 10^{-3} ^{222}Rn atoms g^{-1} min^{-1}.

Using this radon release rate one can estimate the average depth of release of radon necessary to explain the surface alpha emission rate observed by TURKEVICH *et al.* (1970). Using a soil density of 1.6 g cm^{-3}, and assuming one-half of the decaying ^{222}Rn daughter atoms escape from the lunar atmosphere, ^{222}Rn produced to the very great depth of 7 meters in the soil would be required. To calculate the diffusion rate expected as a function of depth requires a knowledge of the diffusion coefficient and the temperature profile in the lunar soil. Even without the analysis one can say that the emanation rate we measure is too low to correlate with the observations from the Surveyor 5 alpha-scattering experiment.

It is further interesting to estimate the fraction of ^{222}Rn produced in lunar fines that escapes to the atmosphere. The total ^{222}Rn production rate in lunar fines containing 1.5 ppm uranium is 1.11 per gram per minute, and therefore the fraction of ^{222}Rn emanating is 2.3 × 10^{-3}. Although this fraction is small, the transport of daughter ^{206}Pb from depth to the surface may be an important consideration in the interpretation of lead ages of lunar soil.

Acknowledgments—We would like to acknowledge the generous cooperation of the staff of the Lunar Receiving Laboratory, and particularly the assistance of Don Bogard and his staff of the gas analyses laboratory (LRL). The staff of the Space Radiation Effects Laboratory (NASA) performed the 600-MeV proton irradiations. We would like to thank Miriam Forman for many helpful discussions. This work was supported by NASA and the AEC.

REFERENCES

BOCHSLER P., EBERHARDT P., GEISS J., LOOSLI H., OESCHGER H., and WAHLEN M. (1971) Tritium in lunar material. Second Lunar Science Conference (unpublished proceedings).

BOGARD D. D., FUNKHAUSER J. G., SCHAEFFER O. A., and ZÄHRINGER J. Noble gas abundances in lunar material, II. Cosmic ray spallation products and radiation ages from Mare Tranquillitatis and Ocean Procellarum. *J. Geophys. Res.* (in press).

D'AMICO J., DE FELICE J., and FIREMAN E. L. (1970) The cosmic-ray and solar-flare bombardment of the moon. *Proc. Apollo 11 Lunar Sci. Conf., Geochim. Cosmochim. Acta* Suppl. 1, Vol. 2, pp. 1029–1036. Pergamon.

D'AMICO J., DE FELICE J., FIREMAN E. L., JONES C., and SPANNAGEL G. (1971) Tritium and argon radioactivities and their depth variations in Apollo 12 samples. Second Lunar Science Conference (unpublished proceedings).

FIREMAN E. L., DE FELICE J., and TILLES D. (1961) Solar flare tritium in a recovered satellite. *Phys. Rev.* **123**, 1935–1936.

FORMAN M. A., STOENNER R. W., and DAVIS R., JR. Cosmic ray gradient measured by the argon-37/argon-39 ratio in the Lost City meteorite. *J. Geophys. Res.* (in press).

HINTENBERGER H., WEBER H., and TAKAOKA N. (1971) Concentrations and isotopic abundances of the rare gases in lunar matter. Second Lunar Science Conference (unpublished proceedings).

KRANER H. W., SCHROEDER G. L., DAVIDSON G., and CARPENTER J. W. (1966) Radioactivity of the lunar surface. *Science* **152**, 1235–1236.

MARTI K. and LUGMAIR G. W. (1971) Kr^{81}–Kr and K–Ar^{40} ages, cosmic-ray spallation products and neutron effects in Apollo 11 and Apollo 12 lunar samples. Second Lunar Science Conference (unpublished proceedings).

Solar-Geophysical Data (1970) No. 309, Part II, pp. 78–83.

STOENNER R. W., LYMAN W. J., and DAVIS R., JR. (1970) Cosmic-ray production of rare-gas radioactivities and tritium in lunar material. *Proc. Apollo 11 Lunar Sci. Conf., Geochim. Cosmochim. Acta* Suppl. 1, Vol. 2, pp. 1583–1594. Pergamon.

TURKEVICH A. L., PATTERSON J. H., FRANZGROTE E. J., SOWINSKY K. P., and ECONOMOU T. E. (1970) Alpha radioactivity of the lunar surface at the landing sites of Surveyors 5, 6, and 7. *Science* **167**, 1722–1724.

YEH R. S., and VAN ALLEN J. A. (1969) Alpha-particle emissivity of the moon: An observed upper limit. *Science* **166**, 370–372.

Proceedings of the Second Lunar Science Conference, Volume 2, pp. 1825–1839.
The M.I.T. Press, 1971.

Tritium and argon radioactivities and their depth variations in Apollo 12 samples

J. D'AMICO, J. DEFELICE, E. L. FIREMAN, C. JONES, and G. SPANNAGEL
Smithsonian Astrophysical Observatory, Cambridge, Massachusetts 02138

(*Received 22 February 1971; accepted 17 March 1971*)

Abstract—Tritium and argon radioactivities were measured in three sections of bar B_2 (6.4 cm length) of lunar rock 12002. The tritium was 392 $\pm$ 11 dpm/kg in the portion containing the top surface at 0–0.8 cm depth; in the 0.8–3.1 cm depth, 270 $\pm$ 5 dpm/kg; in the 4.9–6.4 cm portion containing the bottom surface, 155 $\pm$ 7 dpm/kg. Although the tritium decreased from the top to the bottom of the rock, the argon radioactivities were approximately constant. For the three portions, the Ar^{37} were 30 $\pm$ 4, 25.0 $\pm$ 1.5, and 27.5 $\pm$ 2.5 dpm/kg, and the Ar^{39} were 8.1 $\pm$ 0.7, 8.2 $\pm$ 0.5, and 8.0 $\pm$ 0.6 dpm/kg. The spallation He^3 and Ar^{38} contents were measured in the top and bottom samples. The ($He^3/2H^3$) exposure age increased with depth from 63 $\pm$ 5 m.y. for the top sample to 144 $\pm$ 10 m.y. for the bottom sample. The argon exposure ages were approximately constant with depth.

The depth variation of the ($He^3/2H^3$) exposure age is interpreted in terms of a solar-flare exposure age or a surface dwell time of 4.4 $\pm$ 2.2 m.y. and a cosmic-ray exposure age of 144 $\pm$ 10 m.y. The comparison with fossil-track cosmic-ray ages indicates that the solar-flare intensity averaged over the past 5 m.y. was not larger than its average over the past 30 years; in fact, it appears to have been smaller.

For the sections of bar B_2, the tritium was measured in temperature steps. The tritium contents released at room and higher temperatures were also measured in rock 12065 and soil samples 12001, 12032, 12033, 12042, and 12070. In no case was tritium released at room temperature during a 5-hour period with crucible bakeout; no tritium could be attributed to terrestrial contamination. The tritium contents are attributed to solar-flare and galactic cosmic-ray interactions for all samples. The bottom section of rock 12002 and soil sample 12033, which was taken from a trench approximately 15 cm deep, had the lowest total tritium contents. The tritium and argon radioactivities and their depth variations in lunar material were compared with those produced by 158-MeV protons acting upon a target of simulated lunar material and other target data, and an average solar-flare proton flux of approximately 4 $\times$ 10^9/cm^2 yr above 30 MeV with an differential energy spectrum proportional to between E^{-3} and $E^{-3.5}$ was obtained.

INTRODUCTION

TRITIUM, Ar^{37}, and Ar^{39} radioactivities in lunar material have two well-established origins: nuclear interactions by energetic solar-flare protons and nuclear interactions by galactic cosmic rays. The neighboring stable isotopes, He^3 and Ar^{38}, result from these two processes and from a third, solar-wind implantation. An implantation process for radioactivities in lunar material has not been established, although implanted tritium was observed in a recovered satellite (FIREMAN *et al.*, 1961). The absolute amounts of tritium and argon radioactivities in Apollo 11 samples (FIREMAN *et al.*, 1970; STOENNER *et al.*, 1970) indicated that approximately half these radioactivities arose from nuclear interactions by solar flares and half from nuclear interactions by cosmic rays and that very little, if any, arose from implantation. Results (STOENNER *et al.*, 1971) on the absolute amount of tritium and its temperature release indicated that the Apollo 12 samples were very different from the Apollo 11, and that there was either tritium implantation or tritium contamination of the former. To examine this possibility, we carried out a tritium survey of Apollo 12 samples taken from various lunar locations and returned in different sample containers.

We also received Apollo 12 samples taken from different depths in an oriented rock and fines from different depths in soil. These can provide more detailed informaton about solar flares than was obtainable from the Apollo 11 samples. In fact, the depth measurements can provide data for a new type of age, a solar-flare exposure age, in addition to the usual cosmic-ray exposure age. The solar-flare exposure age is related to the time that the rock had been resting uncovered on the lunar surface. This time has also been estimated from fossil tracks attributed to very heavy cosmic rays (CROZAZ *et al.*, 1970; FLEISCHER *et al.*, 1970). The comparison of these two ages should provide information about solar-flare intensities during approximately the past 10 m.y.

The depth variation of tritium and argon radioactivities, when compared to the production cross sections in targets bombarded by protons of different energies, should also provide information about the intensity and energy spectrum of solar-flare particles averaged over times approximately equal to the mean lives of the radioactivities.

TRITIUM AND ARGON MEASUREMENTS

The counting apparatus and counters used for the tritium and argon measurements were identical to those used for the Apollo 11 samples (FIREMAN *et al.*, 1970; D'AMICO *et al.*, 1970). One of the low-level shields was improved by additional shielding material and more anticoincidence guard counters, thus lowering the background of our 42-cm^3 tritium counter from 0.140 ± 0.006 count/min to 0.115 ± 0.004 count/min, and the background of our 14-cm^3 tritium counter from 0.065 ± 0.004 count/min to 0.049 ± 0.003 count/min. The backgrounds of our argon counters in the improved shield were reduced by 10%. The later measurements on Apollo 12 samples were done with the improved shield.

A sidearm in which the sample could be kept at room temperature and in contact with carrier gas while the crucible was heated was installed on the furnace of the extraction system. The sample was put into the sidearm and the pressure in the system reduced to 0.025 atm. Carrier hydrogen of several torr pressure was added to this air. A crucible bakeout step to check for any terrestrial tritium present in our system or for any very loosely bound tritium in the sample was then carried out. This step consisted of heating the empty crucible to 1600°C with the sample in the sidearm in the presence of the air and hydrogen and the argon carriers for approximately 2 hours. The gas was removed while the crucible was at 1600°C and then transferred to a section of the system with finely divided vanadium metal powder at 800°C, which removes the chemically active constituents. The vanadium powder was then slowly cooled to room temperature to absorb hydrogen as vanadium hydride. The noble gases were then removed and their volumes measured. The vanadium was reheated and the evolved hydrogen was collected. After its volume was measured, the hydrogen was added to a proportional counter that contained 400-torr pressure of P-10 gas (commercially available counting gas containing 90% argon and 10% methane). The counter was removed from the system. The resolution of the counter was then checked with an Fe^{55} source and the counting was done in the low-level shield. The results, which are given in Tables 1 and 2, show that no tritium was observed in the gas from the crucible bakeout step from seven samples. We conclude that there was no terrestrial tritium present in our system and that no tritium in the samples came off at room temperature.

The Apollo 12 samples were analyzed in our laboratory between 2 March and 18 November 1970; there was a delay of between 3.5 months and a year from the time samples were collected and our analysis. R. W. STOENNER and R. DAVIS analyzed a 10-g sample of rock 12065 shortly after Apollo 12 returned. They measured approximately 20×10^3 dpm/kg of tritium in the crucible bakeout step (private communication, 1970). On 24 June 1970, we analyzed a 1-g sample of 12065, which had a surface area of ~ 1 cm^2. We found less than 20 dpm/kg in the crucible bakeout step. The discrepancy of a factor of 1000 between the two measurements of the tritium content from the bakeout step for rock 12065 is surprising. There were no discrepancies between the measurements of STOENNER *et al.* (1970) and FIREMAN *et al.* (1970) for the Apollo 11 samples even though these measurements were done several months apart.

We received two sections of bar B_2 from rock 12002. Sample 12002 was a crystalline rock whose orientation was determined by pit counts and by nondestructive gamma-ray counting. The lack of pits on the bottom of the rock indicated that it had not been exposed on the surface of the moon. The section of the bar called 12002-57 extended from 0 to 3.1 cm in depth; it had 1.8 cm^2 of top surface, which had 64 glass-lined pits greater than 0.1 mm in diameter. We broke this section at a point between 0.7 and 0.8 cm from the top surface and analyzed the two pieces separately. The section of the bar called 12002,59 extended from 4.9–6.4 cm in depth and contained 1 cm^2 area of bottom surface. A small amount of material scraped from the pieces was used for the measurements of the stable rare gases.

After the crucible bakeout step, a sample was moved into the crucible from the sidearm with a magnet and heated to approximately 250°C for several hours. Although the amounts of tritium released from the central piece (0.8–3.1 cm) and the bottom piece (4.9–6.4 cm) were small, 11 ± 2 and 15 ± 2 dpm/kg, respectively, the tritium released from the top piece (0–0.8 cm) was 120 ± 5 dpm/kg. Though significant, this amount was only 30% of the total tritium released from this sample. After the 250°C heating, the samples of 12002 were melted; the amounts of tritium released from the melt (see Table 3) were 260 ± 9 dpm/kg for the top piece, 255 ± 3 dpm/kg for the central piece, and 128 ± 5 dpm/kg for the bottom sample. Very little tritium was obtained by remelting these samples: 12 ± 5, 4 ± 2, and 12 ± 3 dpm/kg, respectively. The total tritium contents were 392 ± 12 dpm/kg for depths of 0–0.8 cm, 270 ± 5 dpm/kg for 0.8–3.1 cm, and 155 ± 8 dpm/kg for 4.9–6.4 cm. The results are plotted in Fig. 1.

The noble gases from each temperature step were used as carriers for the next temperature step. After the final melt, the argon and heavier rare gases were purified over hot titanium and condensed on charcoal at liquid-nitrogen temperature. The argon was removed from the charcoal at dry-ice temperature and placed in the same small low-level proportional counters used for Apollo 11 samples (FIREMAN *et al.* 1970). The argon yields were above 95%. Methane (10%) was added to the argon. The counter was removed from the purification system and counted in the low-level system, where the counts between 0.4 and 8.0 kev were recorded on a 100-channel analyzer and those above 7.4 keV on scalers. Even though there was a delay of more than 3 months between sample recovery and the beginning of the 12002-sample counts, the Ar^{37} activity clearly appeared as a peak in the 2.8 ± 0.6 keV channels. Figure 2 is a plot of the argon data for a 12-day count from 4 to 16 March 1970 for a 12002 sample. The argon activities were followed for approximately 2 months; the Ar^{37} decayed with a 35-day half-life. The argon activities of rocks 12002 and 12065 are given in Table 1.

Tritium and Ar^{39} activities were measured in a number of Apollo 12 soil samples in order to search for ones that might have unusually high radioactivity contents and to check whether the soil samples had activities within the range observed for rock 12002. Table 2 lists the soil samples together with their activities. Most samples were ~ 1 g in size and were received too late for accurate Ar^{37} determinations. During these measurements, three samples of the nonmagnetic phase of the Lost City and Allende meteorites were analyzed to check consistency with previous measurements done on meteorite samples of larger size. The tritium in a 1.1-g whole-rock sample of Allende, 380 ± 35 dpm/kg agreed with previous determinations of 370 ± 30 and 410 ± 30 dpm/kg made with samples of 11 and 20 g (FIREMAN and GOEBEL, 1970). The tritium contents in the samples of the nonmagnetic phase of the Lost City meteorite ranged between 410 ± 30 and 425 ± 30 dpm/kg, which are 20% higher than the value 358 ± 5 dpm/kg obtained for a neighboring whole-rock sample. The tritium content of the nonmagnetic phase of a meteorite is usually higher than that in whole-rock samples, so these results are consistent with each other.

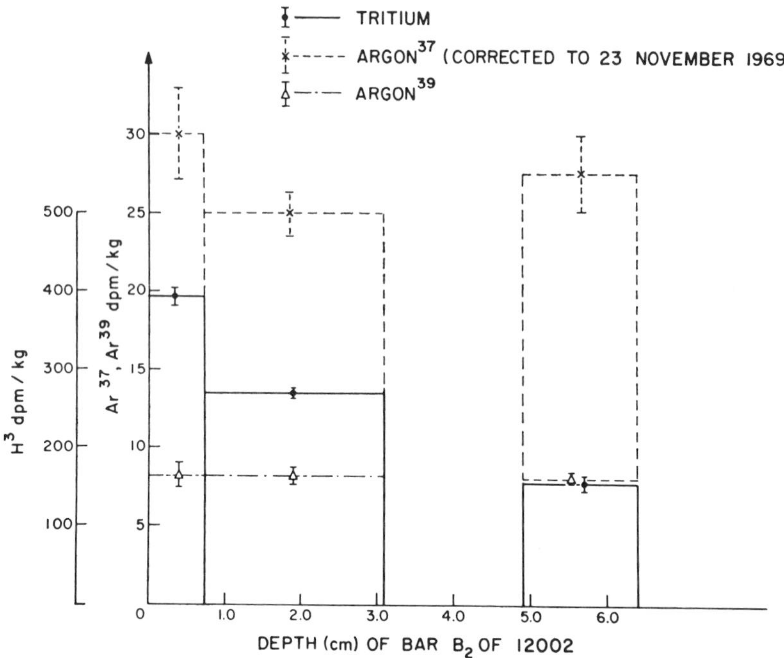

Fig. 1. Tritium, Ar^{37}, and Ar^{39} activities vs. depth in lunar rock 12002.

Apollo 12 lunar sample 12070, consisting of fines collected close to the Lunar Module, was returned in the contingency Teflon bag. Less than 2 dpm/kg of the tritium was observed in the crucible bakeout step. During the heating of this sample at 1600°C, some material was deposited on the furnace walls. Significant amounts of tritium were observed in reheats of the crucible until the wall deposits were severely treated with a torch. The total amount of tritium in the melt and the three reheats was 350 ± 40 dpm/kg.

Sample 12001 consisted of fines returned under vacuum conditions; its ALSRC container had 40–60 μ of pressure when received at the Lunar Receiving Laboratory. The sample was collected

Table 1. Tritium and argon radioactivities in Apollo 12 rock samples.

Sample	Wgt. (g)	Depth (cm)	Temperature steps	H^3 (dpm/kg)	Total H^3 (dpm/kg)	Ar^{37} 23 November 1969 (dpm/kg)	Ar^{39} (dpm/kg)	Dates of H^3 analysis (1970)
12002,57	5.4	0–0.8	Crucible bakeout	< 6				
		with top	20–250°C	120 ± 5	392 ± 11	30 ± 4	8.1 ± 0.7	13–22 March
		surface	250°C-melt	260 ± 8				
			Remelt	12 ± 5				
12002,57	16.0	0.8–3.1	Crucible bakeout					
			and 20–250°C	11 ± 2	270 ± 5	25.0 ± 1.5	8.2 ± 0.5	3–11 March
			250°C-melt	255 ± 3				
			Remelt	4 ± 2				
12002,59	8.4	4.9–6.4	Crucible bakeout	< 3				
		with bottom	20–250°C	15 ± 2	155 ± 7	27.5 ± 2.5	8.0 ± 0.6	6–14 March
		surface	250°C-melt	128 ± 5				
			Remelt	12 ± 3				
12065,110	1.013	Surface	Crucible bakeout	< 20				
			20–1000°C	294 ± 26	294 ± 26	—	8.5 ± 1.0	24 June–
			1000°C-melt	< 15				1 July
			Remelt	< 15				

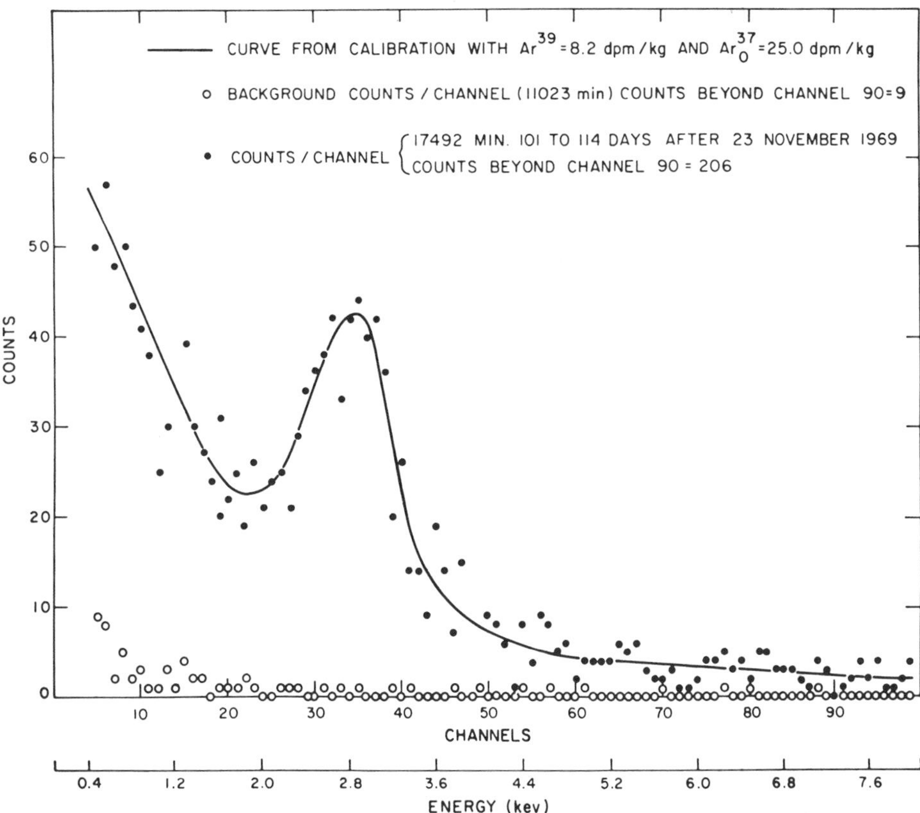

Fig. 2. Argon activities for lunar rock 12002 at 0.8–3.1 cm depth.

near the Lunar Module at the end of the first extravehicle activity. No tritium was observed in the crucible bakeout step. The amount of tritium observed when the sample was melted was 275 $\pm$ 15 dpm/kg; when the sample was remelted and a small deposit on the furnace walls heated, 60 $\pm$ 15 dpm/kg were observed. No additional tritium was obtained with further meltings and wall heatings. The total tritium content was 335 $\pm$ 25 dpm/kg.

Sample 12032 was fines returned to the Lunar Receiving Laboratory with $\sim$0.5 atm in its ALSRC container. It was obtained near the north rim of Bench Crater. The hydrogen from the crucible bakeout and that from the melting of the sample were combined and had 260 $\pm$ 20 dpm/kg; when the sample was remelted and the furnace walls heated, 30 $\pm$ 15 dpm/kg were obtained. No additional tritium was obtained from a second remelting and wall heating. The total tritium content was 290 $\pm$ 30 dpm/kg. The Ar^{39} activity in this sample was 11.9 $\pm$ 1.0 dpm/kg.

Sample 12033 consisted of fines also returned to the Lunar Receiving Laboratory in the ALSRC container with 0.5 atm. It was dug from the bottom of a 15-cm trench in the soil. The hydrogen from the crucible bakeout was combined with that from the melting of the sample; the tritium content was 143 $\pm$ 12 dpm/kg. The remelting of the sample and the heating of deposits on the furnace walls gave 72 $\pm$ 12 dpm/kg. No additional tritium was obtained from a second remelting of the sample and reheating of the furnace walls. The total tritium content was 215 $\pm$ 20 dpm/kg. The Ar^{39} in this sample was 11.5 $\pm$ 1.5 dpm/kg.

Sample 12042 was fines also returned in the container with 0.5 atm. It was taken on the outer flank of the Surveyor crater rim, approximately 20 m northwest of halo craters in an area of wrinkled texture. No tritium was observed in the crucible bakeout step. The amount of tritium

Table 2. Tritium and argon-39 in Apollo 12 soil samples and meteorite samples.

Sample	Wgt. (g)	Description	Temperature Steps	H^3 (dpm/kg)	Total H^3 (dpm/kg)	Ar^{39} (dpm/kg)	Dates of H^3 analysis (1970)
12070,86	(1.06)	Contingency,	Crucible bakeout	<2			
		near LM	Melt and 3 reheats	350 ± 30	350 ± 40	—	31 March–13 April
			Final remelt	<10			
12001,49	(0.97)	Near LM	Crucible bakeout	<10			
			Melt and 1 reheat	335 ± 20	335 ± 25	—	14–20 May
			2nd remelt	<15			
12032,37	(0.94)	Near north	Crucible bakeout				
		rim of Bench	and melt	260 ± 20			
		Crater	Remelt	30 ± 15	290 ± 30	11.9 ± 1.0	4–24 June
			2nd remelt	<10			
12033,35	(1.036)	Bottom of	Crucible bakeout				
		trench about	and melt	143 ± 12			
		15 cm deep	Remelt	72 ± 12	215 ± 20	11.1 ± 1.5	20 May–1 June
			2nd remelt	<12			
12033,95	(4.51)		Crucible bakeout	<5			
			$20 \simeq 550°C$	218 ± 5	239 ± 12	10.5 ± 1.3	12–22 December
			550°–Melt	21 ± 4			
			Remelt	<5			
12042,39	(0.906)	Outer flank of	Crucible bakeout	<15			
		Surveyor	Melt	367 ± 30	367 ± 30	9.4 ± 1.0	16–22 June
		Crater	Remelt	<5			
12042,66	(4.61)	Outer flank of	Crucible bakeout	<5			
		Surveyor	20–550°C	198 ± 5			
		Crater	550°–Melt	103 ± 5	311 ± 10	8.5 ± 0.7	10–20 November
			Remelt	10 ± 3			
Allende	(1.108)	—	Melt	300 ± 20			
			Remelt	80 ± 20	380 ± 35	—	11–16 April
			2nd remelt	<15			
	(1.057)	—	Melt	420 ± 30	420 ± 30	—	26–29 October
			Remelt	<10			
Nonmagnetic phase of Lost City	(1.025)	—	Melt	380 ± 15	410 ± 30	—	30 October–3 November
			Remelt	30 ± 15			
	(1.013)	—	Melt	425 ± 30	425 ± 30	—	3–6 November
			Remelt	<10			
Lost City whole rock	(3.38)	—	Melt	358 ± 5	358 ± 5	12.4 ± 0.9	23–29 March
			Remelt	<10			

from the melted sample was 367 ± 30 dpm/kg. No tritium was observed in the remelting of the sample and the heating of the furnace walls. The total tritium content was 367 ± 30 dpm/kg. The Ar^{39} activity in this sample was 9.4 ± 1.0 dpm/kg.

To obtain more information on the temperature release of tritium from the fines and to check room-temperature loss, we analyzed another sample of 12042 and 12033 approximately 6 months later at three temperatures. No tritium was observed from the crucible bakeouts. From several hours of heating with the crucible at a dull red color, approximately 550°C, 198 ± 5 dpm/kg of tritium were obtained from sample 12042 and 218 ± 5 dpm/kg from sample 12033. The samples were then melted and 101 ± 5 dpm/kg were obtained from 12042 and 21 ± 4 dpm/kg from 12033. Little or no tritium was obtained upon remelting the samples and heating the furnace walls. The total tritium contents of 311 ± 12 dpm/kg for sample 12042 and 239 ± 12 dpm/kg for sample 12033 indicated an insignificantly small amount of tritium loss from 12042 and none from 12033 during the 6 months. The bulk of the tritium was released from these fines at approximately 550°C. Since practically no deposits formed on the furnace walls during the stepwise heating, practically no tritium was observed upon remelting the samples and heating the furnace walls.

Discussion of the Tritium and Argon Radioactivities

The depth variation of tritium in rock 12002, which is given in Table 1, shows a clear decrease of the tritium contents with depth from 392 ± 11 dpm/kg to 155 ± 7

dpm/kg over 6.4 cm. This behavior is different from that expected from galactic cosmic-ray production and also different from that observed with thick targets bombarded with Gev protons (FIREMAN and ZÄHRINGER, 1957). On the other hand, a decrease with depth is expected for tritium produced by solar-flare particle interactions.

The tritium content of 294 ± 26 dpm/kg from a surface sample of rock 12065 is lower than that in the top surface sample of 12002 and indicates that the 12065 sample was taken from an exposed side rather than from the top of the rock. The tritium contents of four Apollo 11 rock samples range between 219 and 250 dpm/kg (FIREMAN *et al.*, 1970; STOENNER *et al.*, 1970). Differences in the amounts of shielding material could explain the tritium differences between Apollo 11 and Apollo 12 rock samples. STOENNER and DAVIS (private communication, 1970) observed 22,000 dpm/kg of tritium in rock 12065 shortly after the sample was recovered. This tritium was so loosely bound that most of it came off at room temperature. Our measurements on 12065, which gave no such loosely bound tritium, were done 6 months later. The Apollo 12 samples were collected toward the end of a lunar night period so that the samples were cold before being collected. It is possible to speculate that high contents of loosely bound tritium are located on surface regions saturated with solar wind.

The tritium contents of the Apollo 12 soil samples (see Table 2) are in the range observed for rock 12002. Soil samples 12070, 12001, 12042, and 12032 have amounts of tritium similar to those observed at less than 3.1-cm depth. The tritium contents are also close to those observed for two Apollo 11 soil samples, 325 ± 17 and 314 ± 13 dpm/kg (FIREMAN *et al.*, 1970; STOENNER *et al.*, 1970). Soil sample 12033, which was taken from a trench 15-cm deep, had 239 ± 12 dpm/kg of tritium, a value somewhat higher than that in the bottom of rock 12002 but lower than those in the other soil samples.

It is interesting to note that the temperature release of tritium from the top and bottom surface samples of 12002 is different. At 250°C, 120 ± 5 dpm/kg were released from the top and only 15 ± 2 dpm/kg from the bottom. The top sample contains a solar-wind component of stable rare gases, the bottom sample does not (see Table 4). If we ascribe the tritium difference 105 ± 7 dpm/kg at 250°C to surface-implanted tritium, we obtain a tritium flux of $5 \times 10^{-3}/cm^2$ sec. This is small compared to the flux of solar-wind hydrogen, $\sim 2 \times 10^7/cm^2$ sec. Regardless of whether this amount of tritium is attributed to implantation, there is tritium production by solar-flare particle interactions on the basis of the decrease of tritium from 0.8–6.4 cm depth.

The decrease of Ar^{37} and Ar^{39} with depth in 12002, which is plotted in Fig. 1, is slight. The Ar^{37} is 30 ± 4 dpm/kg at a depth of 0–0.8 cm; 25.0 ± 1.5 dpm/kg at 0.8–3.1 cm; and 27.5 ± 2.5 dpm/kg at 4.9–6.4 cm. The Ar^{39} activities at these depths are 8.1 ± 0.7, 8.2 ± 0.5, and 8.0 ± 0.6 dpm/kg, respectively. In an iron target bombarded by 6.2-GeV protons the Ar^{37} activity increased by 65% in 5-cm depth. The absence of an increase with depth indicates the presence of argon produced by solar flares. The Ar^{37}/Ar^{39} ratios in rock 12002 are approximately 3.0; the ratios in Apollo 11 rock samples ranged between 1.3 and 2.3 (FIREMAN *et al.*, 1970; STOENNER *et al.*, 1970). The higher Ar^{37}/Ar^{39} ratios in the Apollo 12 samples are caused in part by the higher Ca/Ti ratio in these samples. The 2 November 1969 flare also contributed to raising the Ar^{37}/Ar^{39} ratio. This flare had approximately 10^8 protons/cm² with energies above 30 Mev, and a differential energy spectrum proportional to E^{-3} (Lincoln, 1970). If this proton flux and energy spectrum are combined with the measured Ar^{37} cross sections given in Table 2, the flare contributed 10 dpm/kg of Ar^{37} activity in lunar rock 12002 on 23 November 1969. This amount of Ar^{37} activity is consistent with the observations.

The Solar-Flare Flux and Energy Spectrum

We attribute the tritium and argon activities in rock 12002 to solar-flare and galactic cosmic-ray particle interactions. The flare protons cause most of the nuclear interactions at small depths; the stripping of α particles can account for very little tritium.

An upper limit to the amount of tritium produced by α-particle stripping can be based on Co^{57} measurements. The α-particle stripping in Fe^{56} would produce Co^{57} and H^3 by the reaction

$$He^4 + Fe^{56} \longrightarrow Co^{57} + H^3.$$

Co^{57} is also produced by the (p, n) reaction on Fe^{57}. The amount of Co^{57} in rock 10017 was 5.8 ± 2.9 dpm/kg at a depth of 0 to 0.4 cm and less than 1.5 dpm/kg at 0.4–1.2 cm (Shedlovsky *et al.*, 1970). Similar amounts were obtained in rock 12002 (Finkel *et al.*, 1971). The stripping cross-section is not strongly dependent on the atomic number (Armstrong *et al.*, 1967; Gonzalez-Vidal and Wade, 1960; Priest and Vincent, 1969) so that the tritium produced by α-particle stripping at a depth between 0.4 and 1.2 cm is less than 20 dpm/kg, which is calculated by dividing the Co^{57} activity by the atomic percent of iron. This upper limit is small compared to the observed amounts of tritium, and we shall neglect the α-particle-stripping process.

Tritium production cross sections by protons on the principal lunar elements are necessary to estimate the solar-flare flux and energy spectrum. Cross-section measurements have been made on a number of elements at various energies. Most of these are summarized by Kirsten and Schaeffer (1970); the lowest proton energy was 32 MeV (Gonzalez-Vidal and Wade, 1960). The threshold energy for tritium production from oxygen in 20 MeV, and between 15 and 20 MeV for the other principal target elements. These cross-section measurements are well represented by the expression

$$\sigma = 40e^{-14/\sqrt{E-20}}\text{mb}, \qquad E > 20 \text{ MeV} \tag{1}$$

where E is the energy of the incident proton in MeV. This expression is a constant times a Coulomb factor for tritium escape from a reaction with a 20-MeV threshold. To examine cross sections for thick composite targets similar to lunar rocks, we bombarded two targets with 158-MeV protons in the Harvard cyclotron. The targets consisted of pellets of simulated lunar composition between iron absorbers. The tritium and Ar^{37} results for three depths are given in Table 3. The tritium cross section at zero and at 14.5 g/cm^2 depths are the same as those obtained for 150-MeV and 80-MeV protons on the Al targets. The decrease of tritium with increasing depth in the targets is not so rapid as in rock 12002; this indicates that more tritium was produced on the lunar surface by protons of less than 158 MeV than by protons of higher energies.

Table 3. Effective cross sections in thick target of simulated lunar material.

Sample depth (g/cm^2)	Protons/cm^2 ($\times 10^{13}$)	Proton energy (MeV)	H^3 Activity (dpm/g)	H^3 σ (mb)	Ar^{37} Activity (10^4 dpm/g)	Ar^{37} σ (mb)
0.0–0.2	5.3	158	1400 ± 100	10.0 ± 1.0	4.4 ± 0.2	2.4 ± 0.2
14.5–15.0	7.4	85*	1350 ± 100	6.9 ± 0.7	6.2 ± 0.3	2.5 ± 0.2
22.0–22.4	7.4	50*	1080 ± 80	5.5 ± 0.6	1.10 ± 0.05	0.43 ± 0.04

Incident proton energy: (158 ± 2) MeV.
Target composition in weight percent: O(43.0); Si(18.7); Fe(15.4); Mg(9.1); Ca(6.7); Al(5.3); Ti(1.8).
* Energy estimated from stopping power of target (iron absorbers between 0.5 and 14.5 g/cm^2 depth and between 15.0 and 22.0 g/cm^2 depth).

The flux and energy spectrum of the solar-flare protons can be calculated from the depth variation of tritum and expression (1). The tritium produced at depth t per kg min is

$$H(t) = A \int_{E_1}^{\infty} \int_{0}^{\theta} (\sigma F)\, \mathrm{d}E\, \mathrm{d}\Omega \tag{2}$$

where A is the number of atoms per kg (2.5×10^{25}), E_1 is the energy of protons with range t, Ω is the solid angle within the angle of incidence θ, and F is the flux of solar-flare protons per cm^2 min MeV ster. For a plane surface, protons of range R can reach the depth t within the angle

$$\int_0^{\theta} \mathrm{d}\Omega = 2\pi(1 - \cos\theta) = 2\pi\left(1 - \frac{t}{R}\right). \tag{3}$$

Relation (3) can be rewritten as

$$\int_0^{\theta} \mathrm{d}\Omega = 2\pi\left(1 - \frac{E_1^2}{E^2}\right). \tag{4}$$

because the range of protons is proportional to the square of its energy. The substitution of relations (4) and (1) into (2) gives

$$H(t) = 80\pi A \int_{E_1}^{\infty} e^{-14/\sqrt{E-20}}\left(1 - \frac{E_1^2}{E^2}\right) F\, \mathrm{d}E. \tag{5}$$

With the flux an inverse power of the energy

$$F = \frac{k}{E^n} \tag{6}$$

relation (5) becomes

$$H(t) = 80\pi \mathrm{A} k \int_{E_1}^{\infty} e^{-14/\sqrt{E-20}}\left(1 - \frac{E_1^2}{E^2}\right) \frac{\mathrm{d}E}{E^n}. \tag{7}$$

The integral was evaluated for $n = 2.5$, 3, 3.5, and 4, and the energy E_1 converted to depth t by the energy-range relation. The curves for $H(t)$ were converted into the histogram in Fig. 3 by integrating over the appropriate depth intervals. The tritium activities at a depth between 4.9 and 6.4 cm for all values of n were normalized to 50 dpm/kg.

The measured tritium activities are the sum of that produced by galactic cosmic rays and that by solar flares. The tritium produced by galactic cosmic rays with energies above 400 MeV has been calculated to be 85 ± 15 dpm/kg at depths of less than 25 g/cm² in lunar material (Fireman *et al.*, 1970). Trevidi and Goel (1970) calculated the tritium produced by galactic cosmic rays above 250 MeV on a plane lunar surface to be 110 dpm/kg at the surface of the moon and to increase with depth. The intensity of galactic cosmic rays between 250 MeV and 400 MeV is somewhat uncertain; nevertheless, the calculations agree at the lunar surface but differ by approximately 50 dpm/kg at 5-cm depth. The shape of rock 12002 causes significant departures from the plane-surface approximation at a depth of 4.9–6.4 cm but not at depths of less than 4.1 cm in bar B_2. With these uncertainties in mind, the tritium produced by

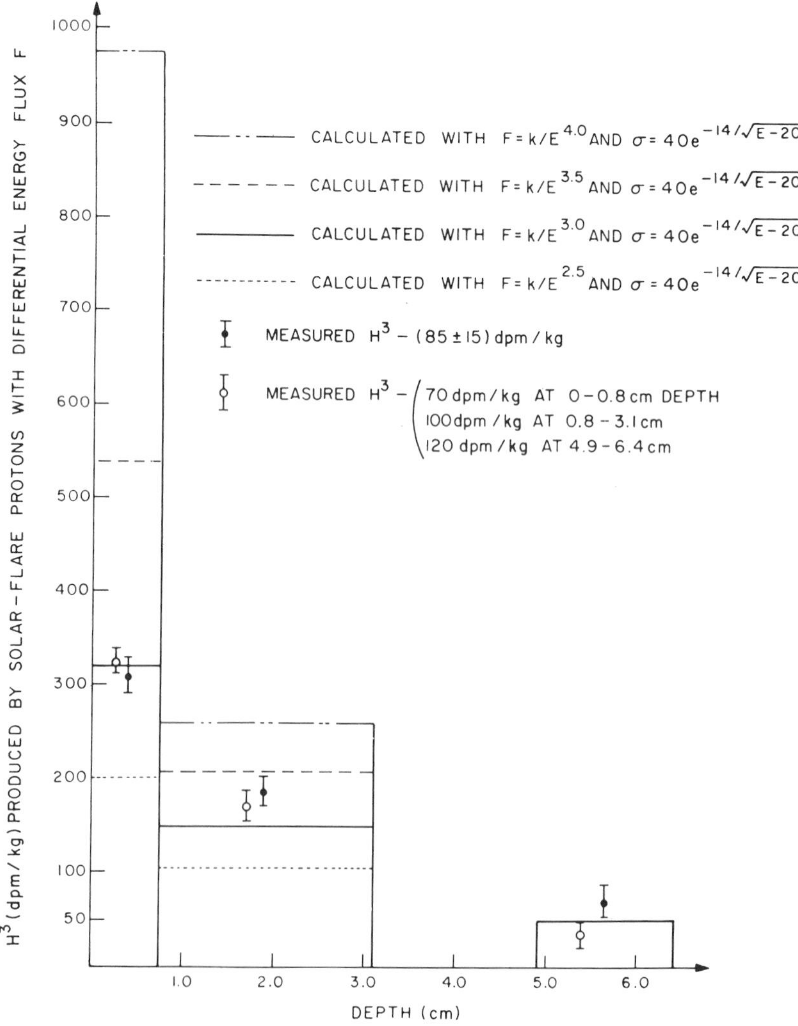

Fig. 3. Solar-flare-produced tritium vs. depth.

solar flares is given two ways in Fig. 3: (1) points representing the measured tritium activity in rock 12002 minus 85 ± 15 dpm/kg; and (2) points representing the measured activity minus 70 dpm/kg at 0–0.8 cm, 100 dpm/kg at 0.8–3.1 cm, and 120 dpm/kg at 4.9–6.4 cm.

The differential flux of $F = k/E^3$ fits the points at depths to 0.8 cm. A differential flux between k/E^3 and $k/E^{3.5}$ fits the points between 0.8 and 3.1 cm. The differential flux of k/E^4 gives far too much tritium at 0–0.8 cm to be consistent with the measurements. The $k/E^{2.5}$ histogram appears to be lower than the measurements, but not significantly. The differential spectrum for large solar flares in the range 50–400 MeV was $k/E^{4.5}$ (BISWAS and FICHTEL, 1965) so that the tritium gives a less steep energy spectrum than that of the large solar flares of the early 1960s. With 310 ± 20 dpm/kg of solar-flare-produced tritium at 0- to 0.8-cm depth and a differential flux of k/E^3, the number of solar-flare protons above 30-MeV energy is 1040 ± 100/cm^2 min ster or 3.4 × 10^9 cm^2 year. With a differential flux of $k/E^{3.5}$, the number of flare protons above 30 MeV is 4.9 × 10^9/cm^2/year. These fluxes are approximately the 30-year average before November 1969. KIRSTEN and SCHAEFFER (1970) estimate the flux of flare protons above 30-MeV energy averaged over the 3 high years 1956, 1957, and 1960 to be 10^{10}/cm^2 year. The flux obtained from the tritium activity is approximately one-third of the average of those years. These estimates are reasonably consistent with each other. The increase in Ar^{37} activity of rock 12002 is also consistent with the intensity of the 2 November 1969 solar flare.

THE HELIUM, NEON, AND ARGON MEASUREMENTS

A small amount of material, between 20 and 100 mg, was scraped from the lunar rock fragments and used for the analyses of stable rare gases. Since the tritium and the radioactive argon isotopes were measured on essentially the same samples, the ($He^3/2H^3$) and the [$Ar^{38}/(Ar^{37} + Ar^{39})$] exposure ages were determined on the same material.

The rare gases were measured in a fairly standard way. Samples ranging in size from 15 to 79 mg were wrapped in aluminum foil and sealed into the sidearm of a furnace. The samples were degassed at room temperature for several weeks. They were then dropped into a molybdenum crucible furnace surrounded by an evacuated envelope previously described (FIREMAN and DEFELICE, 1968). The samples were heated at 1600°C for 30 min. The evolved gases were purified over a copper oxide unit at 450°C, a titanium foil getter at 850°C, and a getterloy furnace at 850°C. The argon was frozen on charcoal at liquid-nitrogen temperature. The helium and neon were admitted in the mass spectrometer, a 6-in. metal Nuclide instrument operated statically. After the helium and neon had been measured, the argon was admitted to the spectrometer by raising the temperature of the charcoal. Before and after each measurement on a lunar sample, a meteorite sample with known rare gas was measured to check the overall sensitivity of the system. At the emission and multiplier settings used for helium, neon, and argon, the sensitivities were 10.2 ± 10^{-8} cc/v, 3.3 × 10^{-8} cc/v, and 3.8 × 10^{-9} cc/v, respectively. The sensitivities varied by less than 10% during the course of the measurements. Blanks determined by reheating the sample were negligible.

Table 4. Stable rare-gas isotopes.

Sample	Location	He^3 (10^{-8} cc/g)	He^4 (10^{-6} cc/g)	Ne^{21} (10^{-8} cc/g)	Ne^{20}/Ne^{22}	Ne^{21}/Ne^{22}
12002,57	0–0.8 cm	100 ± 5	208 ± 10	12.0 ± 0.5	2.28	0.56
12002,57	0–0.8 cm	101 ± 5	326 ± 10	11.5 ± 0.5	6.63	0.40
12002,59	4.9–6 cm	87 ± 4	118 ± 8	10.5 ± 0.5	0.86	0.90
12065,110	Surface	187 ± 10	222 ± 10	25.1 ± 0.7	3.10	0.69
		Ar^{38} (10^{-8} cc/g)	Ar^{36}/Ar^{38}	Ar^{40}/Ar^{36}	He^3_{sp} (10^{-8} cc/g)	Ar^{38}_{sp} (10^{-8} cc/g)
12002,57	0–0.8 cm	11.4 ± 0.7	1.32	64	97 ± 5	9.9 ± 0.7
12002,57	0–0.8 cm	15.8 ± 0.8	2.30	25	93 ± 5	10.3 ± 0.9
12002,59	4.9–6 cm	10.1 ± 0.6	0.71	127	87 ± 5	10.1 ± 0.6
12065,110	Surface	32.4 ± 1.8	1.11	30	187 ± 10	28.5 ± 2.2

The errors do not include the errors in absolute calibrations, which are 5% for He, 20% for Ne, and 10% for Ar.

The rare-gas data are summarized in Table 4. The results on rock 12065 agree with those obtained by Bogard *et al.* (1970). The 12002 samples from a depth of 0 to 0.8 cm contained solar-wind neon and argon in addition to spallation neon and argon on the basis of the Ne^{20}/Ne^{22} and Ar^{36}/Ar^{38} ratios. The sample from a depth of 4.9–6.4 cm contained pure spallation neon and argon. This is of interest since there was some bottom surface material in this sample. It appears that the bottom of rock 12002 had never faced the sun. The solar-wind He^4 contents in the top samples are the He^4 excesses over that in the bottom sample, which can be attributed to the decay of uranium and thorium. The solar-wind He^3 in the top samples are subtracted from the total He^3 contents to obtain the spallation He^3, denoted by He^3_{sp}. The average He^3_{sp} in the two top samples is $(95 \pm 4) \times 10^{-8}$ cc/g compared to $(87 \pm 5) \times 10^{-8}$ cc/g in the bottom. The He^3_{sp} difference from top to bottom of rock 12002 is small, only $(8 \pm 6) \times 10^{-8}$ cc/g. Marti and Lugmair (1971) measured the He^3 as a function of depth in rock 12002 and also found the variation to be small. Their more accurate determinations are: $(86 \pm 1) \times 10^{-8}$ cc/g near the top and $(82 \pm 1) \times 10^{-8}$ cc/g near the botton; the He^3_{sp} difference from top to bottom is $(4 \pm 2) \times 10^{-8}$ cc/g.

The spallation Ar^{38}, denoted by Ar^{38}_{sp}, is obtained from the Ar^{36} and Ar^{38} contents by using an Ar^{36}/Ar^{38} ratio of 5.28 for solar-wind argon and of 0.70 for spallation argon. The amounts of Ar^{38}_{sp} given in the last column of Table 4 show that there was no change of Ar^{38}_{sp} with depth.

The Ages of Rock 12002

The $(He^3_{sp}/2H^3)$ and the $[Ar^{38}_{sp}/(Ar^{37} + Ar^{39})]$ exposure ages calculated in the usual manner are given in Table 5. The $(He^3_{sp}/2H^3)$ age at a depth of 0–0.8 cm is 63 ± 5 m.y.; at 0.8–3.1 cm, 86 ± 5 m.y.; and at 4.9–6.4 cm, 144 ± 10 m.y. The simplest way to interpret this change in exposure age of rock 12002 with depth is that it was buried under sufficient material (~10 cm of soil) to protect it from solar flares at some time during its exposure to cosmic rays. The surface dwell time or solar-flare exposure age can be calculated from the change in spallation He^3, ΔHe^3_{sp}, from the top to the bottom of the rock. From Tables 1 and 4, the ΔH^3 is 237 ± 13

Table 5. The cosmic-ray and solar-flare exposure ages of rocks 12002 and 12065.

Sample	Depth (cm)	Cosmic-ray ages $\frac{He^3_{sp}}{2H^3}$ (10^6 years)	Cosmic-ray ages $\frac{Ar^{38}_{sp}}{Ar^{37}+Ar^{39}}$ (10^6 years)	Solar-flare age $\frac{\Delta He^3_{sp}}{2\Delta H^3}$ (10^6 years)	Gas-retention ages U, Th–He^4 (10^9 years)	Gas-retention ages K–Ar^{40} (10^9 years)
12002	0–0.8	63 ± 5	135 ± 15	8.8 ± 6.5	—	2.5 ± 0.3
12002*	0.8–3.1	86 ± 5	155 ± 15	9 ± 13	—	—
12002	4.9–6.4	144 ± 10	145 ± 15	—	1.8 ± 0.3	2.5 ± 0.3
12065,110	Surface	159 ± 15	410 ± 50	—	—	2.2 ± 0.3

* He^3_{sp} and Ar^{38}_{sp} are taken to be the average of the top and bottom samples.

dpm/kg, and the ΔHe^3_{sp} is $(8 \pm 6) \times 10^{-8}$ cc/g or, according to MARTI and LUGMAIR (1971), $(4 \pm 2) \times 10^{-8}$ cc/g. The solar-flare exposure age, $\Delta He^3_{sp}/2\Delta H^3$, is 8.8 ± 6.5 m.y. according to our measurements or 4.4 ± 2.2 m.y. if we use the more accurate He^3 measurements. This means that the rock 12002 had been sitting in its present position exposed to solar flares for 4.4 + 2.2 m.y., if the solar-flare intensity averaged over the past two cycles is representative of the past 4.4 m.y.

The fossil tracks due to heavy galactic cosmic rays ($Z > 20$), which are observed at >0.5-cm depth, gave surface dwell times for Apollo 11 rocks between 10 and 30 m.y. (CROZAZ *et al.*, 1970). Similar measurements on rock 12002 by COMSTOCK *et al.* (1971) gave a surface dwell time of 24 m.y. for its top surface and 0 m.y. for its bottom surface. Fossil-track measurements by BHANDARI *et al.* (1971) give two values, 5 and 50 m.y. with the 5 m.y. corresponding to the time the top surface is exposed and 50 m.y. corresponding to a burial at a depth of less than 10 cm. The solar-flare age is the same as or younger than the fossil-track cosmic-ray ages. This result leads to the conclusion that the solar-flare intensity averaged over the past 5 m.y. was either of the same intensity or, more likely, of smaller intensity than averaged over the past two solar cycles.

The $[Ar^{38}_{sp}/(Ar^{37} + Ar^{39})]$ exposure age at a depth of 0–0.8 cm is 135 ± 15 m.y.; at 0.8–3.1 cm, 155 ± 15 m.y.; and at 4.9–6.4 cm, 145 ± 15 m.y. As seen in Table 1, the argon radioactivities decrease only slightly over the 6.4-cm thickness of rock 12002. The decrease with depth of the solar-flare production of argon is approximately balanced by the increase with depth of its galactic cosmic-ray production. At a depth between 4.9 and 6.4 cm, the He^3_{sp}/H^3 and the $[Ar^{38}_{sp}/(Ar^{37} + Ar^{39})]$ ages are approximately equal, 144 ± 10 and 145 ± 15 m.y., respectively. The simplest way to interpret this result is to conclude that the He^3 diffusion loss was small for rock 12002 and that its galactic cosmic-ray exposure age is 144 ± 10 m.y.

We have taken the Ar^{38} production rate to be equal to the sum of the Ar^{37} and Ar^{39} activities. The basis for this assumption is the production cross sections. For iron and titanium, the Ar^{38} production cross-section equals the sum of the Ar^{37} and Ar^{39} cross sections (STOENNER *et al.*, 1970; KIRSTEN and SCHAEFFER, 1970). For calcium, the production of Ar^{37} occurs through the $(p, 3pn)$ reaction, and of Ar^{38} through the $(p, 3p)$ reaction, which according to nuclear systematics are similar, with the Ar^{37} cross section being slightly larger. Since Ar^{39} is not produced by protons on calcium, the $Ar^{37} + Ar^{39}$ activity would represent the Ar^{38} production quite well, even from calcium. The Ar^{37} activity is influenced by temporal changes in the cosmic-ray flux. The effect of the 2 November 1969 flare on the Ar^{37} + Ar^{39} activities was to increase the Ar^{37} activity by 25%. The Ar^{38} production rates estimated for the Apollo 11 and Apollo 12 rocks from the Ar^{37} + Ar^{39} activities are almost identical. Stoenner *et al.* (1970) and Bogard *et al.* (1970) estimated the Ar^{38} production rate from the Ar^{39} activity and the cross-section data. These estimates differed by a factor of 2 because the relative production of Ar^{38} and Ar^{39} from calcium depends strongly on target conditions. It therefore appears that the estimate of Ar^{38} production rate from the Ar^{37} + Ar^{39} activities has a smaller uncertainty than the estimate from the Ar^{39} activity alone.

Since there is no solar wind nor primordial neon or argon in sample 12002,59 (see Table 2), the He^4 content is the result of uranium and thorium decays. The amount of spallation He^4 estimated from the amount of He^3 is negligible. According to nondestructive gamma-ray measurements of O'Kelley and Schonfeld (private communication, 1970), the thorium content is 0.965 ppm; the uranium is 0.225 $\pm$ 0.03 ppm; and the potassium is 430 $\pm$ 40 ppm. The (U, Th–He^4) age is calculated to be $(1.83^{+0.30}_{-0.27}) \times 10^9$ years with the decay constants $\lambda_{238} = 1.54 \times 10^{-10}$/year, $\lambda_{235} = 9.8 \times 10^{-10}$/year, and $\lambda_{232} = 4.99 \times 10^{-11}$/year. The amounts of Ar^{40} in the three samples of 12002 were the same $(930 \pm 70) \times 10^{-8}$ cc/g. If $(900 \pm 70) \times 10^{-8}$ cc/g of Ar^{40} is attributed to the decay of K^{40}, then the K–Ar^{40} age is $(2.50 \pm 0.25) \times 10^9$ years. The K^{40}–Ar^{40} age is quite similar to that found for other Apollo 12 rocks (LSPET, 1970). The (U, Th–He^4) age appears to be slightly younger than the K–Ar^{40} age; however, the He^4 diffusion loss over 2.5×10^9 years was not very large.

The flux of micrometeorite particles of mass greater than 10^{-8} g is estimated to be $3 \times 10^{-9}/m^2$ sec from the solar-flare exposure age and the pit count on the top surface. There were 64 pits whose glass-lined diameter was greater than 0.1 mm on 1.8 cm^2 of top surface. With the solar-flare exposure age of 4.4 m.y. the production rate of the pits is $8 \times 10^{-6}/cm^2$ year or $3 \times 10^{-9}/m^2$ sec. If one assumes that the ratio of the pit diameter to the particle diameter is 5 and the density of the particles is 2.5, then the mass of the micrometeorites that produced the pits was greater than 10^{-8} g. This micrometeorite flux is 3 orders of magnitude smaller than that estimated from satellite data (Cour-Palais, 1969). If the top surface of rock 12002 were protected from particle impacts by a thin layer of dust, then our estimated flux would be a lower limit value.

Acknowledgments—This work was supported in part by contract NAS 9–8105 from the National Aeronautics and Space Administration. We would like to thank A. M. Koehler of the Harvard Cyclotron Laboratory for the cyclotron irradiation.

References

Armstrong D. D., Blair A. G., and Thomas H. C. (1967) (He^4, *t*) reaction on medium-weight nuclei. *Phys. Rev.* **155**, 1254–1260.

Bhandari N., Bhat S., Lal D., Rajagopalan G., Tamhane A. S., and Venkatavaradan V. S. (1971) Fossil track studies in lunar materials I. Second Lunar Science Conference (unpublished proceedings).

Biswas F. and Fichtel C. E. (1965) Composition of solar cosmic rays. *Space Sci. Rev.* **4**, 709–736.

BOGARD D. D., FUNKHOUSER J. G., SCHAEFFER O. A., and ZÄHRINGER J. (1970) Noble gas abundances in lunar material — 11. Cosmic ray spallation products and radiation ages from Mare Tranquillitatis and Oceanus Procellarum. Submitted for publication.

COMSTOCK G. M., EVWAPAYE A. O., FLEISCHER R. L., and HART H. R. JR. (1971) The particle track record of the Ocean of Storms. Second Lunar Science Conference (unpublished proceedings).

COUR-PALAIS B. G. (1969) Meteoroid environment model. NASA SP–8013 (March 1969).

CROZAZ G., HAAK U., HAIR M., MAURETTE M., WALKER R., and WOOLUM D. (1970) Nuclear track studies of ancient solar radiations and dynamic lunar surface processes. *Proc. Apollo 11 Lunar Sci. Conf., Geochim. Cosmochim. Acta* Suppl. 1, Vol. 3, pp. 2051–2080. Pergamon.

D'AMICO J., DEFELICE J., and FIREMAN E. L. (1970) The cosmic-ray and solar-flare bombardment of the moon. *Proc. Apollo 11 Lunar Sci. Conf., Geochim. Cosmochim. Acta* Suppl. 1, Vol. 2, pp. 1029–1036. Pergamon.

FINKEL R. C., ARNOLD J. R., REEDY R. C., FRUCHTER J. S., LOOSLI H. H., EVANS J. C., SHEDLOVSKY J. P., IMAMURA M., and DELANY A. C. (1971) Depth variation of cosmogenic nuclides in a lunar surface rock. Second Lunar Science Conference (unpublished proceedings).

FIREMAN E. L., DEFELICE J., and TILLES D. (1961) Solar flare tritium in a recovered satellite. *Phys. Rev.* **123**, 1935.

FIREMAN E. L., D'AMICO J., and DEFELICE J. (1970) Tritium and argon radioactivities in lunar material. *Science* **167**, 566–568.

FIREMAN E. L. and DEFELICE J. (1968) Rare gases in phases of the Deelfontein meteorite. *J. Geophys. Res.* **73**, 6115–6116.

FIREMAN E. L. and GOEBEL R. (1970) Ar^{37} and Ar^{39} in recently fallen meteorites and cosmic-ray variations. *J. Geophys. Res.* **75**, 215–224.

FIREMAN E. L. and ZÄHRINGER J. (1957) Depth variation of tritium and argon-37 produced by high-energy protons in iron. *Phys. Rev.* **107**, 1695–1698.

FLEISCHER R. L., HAINES E. L., HANNEMAN R. E., HART H. R. JR., KASPER J. S., LIFSHIN E., WOODS R. T., and PRICE P. B. (1970) Particle track, x-ray, thermal, and mass spectrometric studies of lunar material. *Science* **167**, 568–571.

GONZALEZ-VIDAL J. and WADE W. H. (1960) Survey of tritium-producing nuclear reactions. *Phys. Rev.* **120**, 1354–1359.

KIRSTEN T. A. and SCHAEFFER O. A. (1970) High energy interactions in space. Interactions of elementary particle research. In *Science and Technology* (editor L. C. L. Yuan), Academic Press (to be published).

LINCOLN J. V. (1970) Solar geophysical data. ESSA Rept. No. 309, part II, p. 81, May.

LSPET (Lunar Sample Preliminary Examination Team) (1970) Preliminary examination of lunar samples from Apollo 12. *Science* **167**, 1325–1339.

MARTI K. and LUGMAIR G. W. (1971) Kr^{81}–Kr and K–Ar^{40} ages, cosmic-ray spallation products and neutron effects in Apollo 11 and Apollo 12 lunar samples. Second Lunar Science Conference (unpublished proceedings).

PRIEST J. R. and VINCENT J. S. (1969) Study of the $Se^{45}(\alpha, t)\ Ti^{46}$ reaction at 41 MeV. *Phys. Rev.* **182**, 1121–1130.

SHEDLOVSKY J. P., HONDA M., REEDY R. C., EVANS J. C., LAL D., LINDSTROM R. M., DELANY A. C., ARNOLD J. R., LOOSLI H. H., FRUCHTER J. S., and FINKEL R. C. (1970) Pattern of bombardment-produced radionuclides in rock 10017 and in lunar soil. *Proc. Apollo 11 Lunar Sci. Conf., Geochim. Cosmochim. Acta.* Suppl. 1, Vol. 2, pp. 1503–1532. Pergamon.

STOENNER R. W., LYMAN W. J., and DAVIS R. JR. (1970) Cosmic ray production of rare gas radioactivities and tritium in lunar material. *Science* **167**, 553–555.

STOENNER R. W., LYMAN W. J., and DAVIS R. JR. (1971) Radioactive rare gases in lunar rocks and in the lunar atmosphere. Second Lunar Science Conference (unpublished proceedings).

TREVIDI B. M. P. and GOEL P. S. (1970) Nuclide production rates in stone meteorites and moon by galactic cosmic radiation. *J. Geophys. Res.* (in press).

ORGANIC GEOCHEMISTRY

Proceedings of the Second Lunar Science Conference, Volume 2, pp. 1843–1863.
The M.I.T. Press, 1971.

Survey of lunar carbon compounds. I. The presence of indigenous gases and hydrolysable carbon compounds in Apollo 11 and Apollo 12 samples

P. I. ABELL,* P. H. CADOGAN, G. EGLINTON, J. R. MAXWELL, and C. T. PILLINGER
School of Chemistry, University of Bristol, Bristol BS8, ITS, England.

(*Received 16 March 1971; accepted in revised form 31 March 1971*)

Abstract—Indigenous gases and chemical reaction products released by acid etch of Apollo 11 and Apollo 12 lunar samples have been examined by gas chromatography, mass spectrometry, and combined gas chromatography-mass spectrometry. Methane, ethane, ethylene, acetylene, and carbon monoxide are among the species identified and quantified. Hydrocarbons have been resolved into indigenous species and chemical reaction products by the use of deuterium-labelled reagents. The samples examined included size-differentiated fines, interior fragments of an igneous rock chip, and Apollo 12 fines from different locations. Possible sources of the indigenous hydrocarbons include solar wind implantation and a small primordial contribution. The hydrocarbons formed during acid etch probably arise from carbide or carbide-like materials contributed by meteoritic impact and solar wind implantation.

INTRODUCTION

ANALYSIS OF APOLLO 11 fines by a number of workers using a considerable variety of techniques has indicated that the quantities of organic compounds isolatable by solvent extraction were extremely low (ABELL *et al.*, 1970a; BURLINGAME *et al.*, 1970; HARE *et al.*, 1970; KVENVOLDEN *et al.*, 1970; HODGSON *et al.*, 1970; MEINSCHEIN *et al.*, 1970; SR. M. E. MURPHY *et al.*, 1970; R. C. MURPHY *et al.*, 1970; ORÓ *et al.*, 1970; RHO *et al.*, 1970). Preliminary mass spectrometric organic analysis of various Apollo 12 samples (LSPET, 1970) indicated that these might be expected to contain concentrations of solvent-extractable material even lower than those of Apollo 11 fines.

Other analytical methods applied at Bristol to the Apollo 11 fines included programmed heating, vacuum crushing and acid etching (ABELL *et al.*, 1970a). We have concentrated on the last two methods for the examination of indigenous gases and hydrolysable carbon compounds with particular reference to their possible origins. Preliminary investigations of the Apollo 11 fines showed that some of the carbon could be released as gaseous hydrocarbons on treatment with aqueous mineral acids. Mass spectrometric analysis of the gases released from hydrofluoric acid etch of the fines indicated the presence of CH_4 at parts per million concentrations (ABELL *et al.*, 1970a; BURLINGAME *et al.*, 1970). Analysis by gas chromatography (gc) (CHANG *et al.*, 1970) and gas chromatography–mass spectrometry (gc–ms) ORÓ *et al.*, 1970) of the

* Present address: Department of Chemistry, University of Rhode Island, Kingston, Rhode Island.

products of hydrochloric acid etch showed that CH_4, C_2H_2, C_2H_4, C_2H_6, C_3H_6, C_3H_8, and a number of unidentified C_3 and C_4 hydrocarbons were present. A less efficient etching reagent, phosphoric acid, released similar products but at reduced concentrations (CHANG *et al.*, 1970). The same reagent afforded hydrocarbons from breccias at similar concentrations to those observed for the fines, whereas the concentrations released from two crystalline rocks were lower by an order of magnitude (CHANG *et al.*, 1970). Release of carbon monoxide by hydrofluoric acid (BURLINGAME *et al.*, 1970) and carbon dioxide by sulphuric acid (ORÓ *et al.*, 1970) was also reported for two samples of the fines.

At the time of the Apollo 11 Conference it was uncertain whether the hydrocarbons released were formed by acid hydrolysis of carbides or were present as such. However, the presence of cohenite, $(Fe, Ni)_3C$, has been reported in the Apollo 11 fines (FRONDEL *et al.*, 1970) and unidentified carbide has been reported in breccia 10,046 (ADLER *et al.*, 1970). Hydrolysis of cohenite of meteoritic origin with aqueous mineral acid affords a mixture of hydrocarbon gases, including CH_4, C_2H_4, C_2H_6, C_3H_6, and C_3H_8 (CHANG *et al.*, 1970; ABELL *et al.*, 1970b).

Detailed studies of the gasous carbon compounds released from lunar samples are necessary to reveal the origin, location, and distribution of the carbon in question. The initial requirement of these studies is the distinction between indigenous entities and those formed as the result of the chemical treatment. We have devised an isotopic labelling method which distinguishes between indigenous hydrocarbon gases and those formed from other carbon compounds by hydrolysis during the etching procedures; The method has been previously outlined (ABELL *et al.*, 1970b); full experimental details are reported herein, together with a gas chromatographic procedure for the routine analysis of hydrocarbons, deuterocarbons, and carbon oxides. Application of these methods to a variety of samples provides a measure of the carbon present as indigenous hydrocarbons and hydrolysable species. The use of selective etching reagents should allow definition of the physical location and hence, the origin of these compounds. Since lunar carbon compounds are abiologically derived, their study in this way may afford an insight into (1) the nature of carbon compounds contributed to the primitive Earth and (2) the origin of carbon compounds and the fate of carbon in the solar system, studies until now based mainly on terrestrial experience.

EXPERIMENTAL TECHNIQUES

General

All Apollo 12 samples are referred to by sample number. Unless otherwise stated, Apollo 11 fines refers to aliquots of the 10,086 Bulk Fines D sample. Sample handling and storage was carried out in a clean area using the methods previously described (ABELL *et al.*, 1970a). Prior to etching or vacuum crushing, samples and reagents were outgassed in a vacuum system incorporating a liquid nitrogen trap, a mercury diffusion pump, and a rotary pump. The earliest experiments were carried out with a McLeod gauge in the system; this has now been replaced by a Penning gauge. In one experiment (see Results) a tungsten filament ion gauge was used.

Vacuum Crushing

Apollo 11 lunar fines (0.670 g) were pulverised by continuous shaking (16 hr) in an evacuated all-glass ball mill (50 ml. capacity and wall thickness 0.25 in.), containing two balls of diameter

0.5 in. and fitted with a break seal for gas analysis. This vessel was used in place of the metal capsule previously described (ABELL *et al.*, 1970a). Adsorbed terrestrial gases were removed as far as possible from the vessel and sample by baking at 150°C for one hour under vacuum (10^{-5} torr). The gases released during the crushing procedure were examined using the mass-spectrometer gas analysis system. Blank experiments were conducted by vibrating the empty ball mill.

Aliquots of Apollo 11 lunar fines (*ca.* 0.5 g) were also pulverised in the presence of deuterium oxide (isotopic purity 99.8%, 1 ml) in order to exchange as far as possible protonated species (particularly H_2O) with deuterated analogues. The D_2O (0.015g) was degassed using a freeze/pump/thaw cycle and sealed into a small glass ampoule which was crushed together with the sample. The interior chip of an olivine basaltic rock (12022,79; 1.54g) was broken into small fragments on pre-extracted aluminium foil with a stainless steel hammer and chisel. The fragments were then pulverised to powder in the evacuated ball mill and the gases released analysed. The fine powder was then used for acid etching experiments. Uncrushed fragments of rock remaining in the ball mill after 16 hr were re-crushed in vacuo.

Acid Etching

Acid etching experiments were performed using hydrofluoric acid, (40%; Hopkin and Williams "AnalaR" grade), deuterium chloride (38% in D_2O; Ciba, isotopic purity 99.5 atom percent D) and deuterium fluoride (20% in D_2O; Merck, Sharp and Dohme; isotopic purity >99.5%). The essential technical details for this experiment have already been reported (ABELL *et al.*, 1970a). Here we specify minor additions or modifications made to the established procedure. Immediately prior to mixing a sample and the acid, the actual acid to be used for the etch was examined for dissolved gaseous contaminants. This was achieved by means of the extra side arm and break seal incorporated into the reaction vessel (Fig. 1). To keep partially deuterated reaction products to a minimum when using deuterated acids, it was necessary to exchange as far as possible protonated species (particularly adsorbed H_2O) with deuterium. This was accomplished by treating aliquots of fines with an excess of D_2O during the degassing procedure (ABELL *et al.*, 1970b). The reaction time for acid etch was increased to 16 hr. In one experiment, using deuterium chloride as etching reagent, the reaction vessel was heated for 90 hr at 120°C. Meteoritic cohenite (4.4 mg; courtesy of Dr. R. Brett, Manned Spacecraft Center, Houston, Texas) was etched using the same procedure. Blank experiments were carried out with no sample in the reaction vessel. The gases released were examined by mass spectrometry, gc or gc–ms.

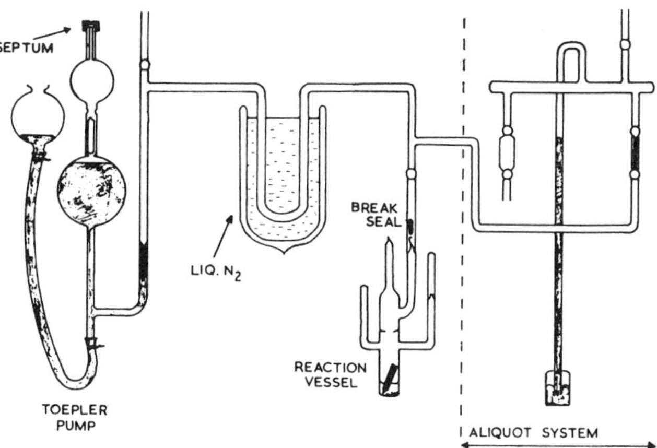

Fig. 1. Reaction and fractionation system for gas analysis.

Alkali etching

Alkali etching experiments were performed using sodium deuteroxide (40% in D_2O; B.D.H.; isotopic purity, 99%, 1 ml). The vessel used was identical to that used for acid etching, and the D_2O exchange and degas procedures were as outlined above. Unsieved (61.4 mg) and sieved Apollo 11 fines (200–240 mesh, 51.2 mg) were each etched for about 6 days. The gases released were analysed for CH_4 and CD_4 content at regular intervals by gas chromatography. In the second experiment, the total condensed fraction (see below) obtained after 6 days was analysed by gas chromatography for the presence of C_2 hydrocarbons.

Gas fractionation and concentration system

The apparatus shown schematically in Fig. 1 was used for the concentration of the gases released by etching prior to transfer to the gas chromatographic system for analysis. Greaseless stopcocks (Springham Valves, Ltd.) were used as valves throughout this system. Any gas released from the reaction vessel can be transferred to a point beneath the septum using the Toepler pump, withdrawn with a gas-tight syringe (100 or 250 μl, Scientific Glass Engineering) and analyzed by gc or gc-ms. The reaction vessel was attached to the system and the space above the break seal evacuated to *ca.* 10^{-3} torr with the Toepler pump. A sample of the gas removed was analyzed by gc as a contamination control. To avoid interference from either HF or DF gas and water vapour the reaction vessel was cooled to -65°C (chloroform slush bath) and the remaining volatile material was transferred to the analytical system by fracturing the break seal with a magnetic weight. The gaseous reaction products were fractionated separately into two volatility ranges by the liquid nitrogen trap. Components not condensed at -195°C (H_2, D_2, Ne, N_2, O_2, Ar, CO, CH_4, CD_4) were concentrated and removed for analysis. The trap was warmed to ambient and the less volatile fraction (CO_2, C_2H_4, C_2H_6, C_3H_6, C_3H_8, H_2S, etc.) treated in the same way. Prior to etching, the acid reagent was examined for the presence of gaseous contaminants using the same system.

To improve the recovery of the small amounts of gases released on NaOD etching, helium (50 μl) was injected into the system, via the Toepler pump septum, and allowed to equilibrate with the released gasses. Immediately prior to gas concentration, the reaction mixture was briefly warmed (50°C) to release gases trapped in the viscous reagent. JOHNSON *et al.* (1970) suggested that trace amounts of a number of species, in particular CO and CH_4, can be generated from a variety of hydrocarbons in a Toepler pump. Continuous recycling of ethane at 10 cm pressure for 60 minutes with a Toepler pump generated only submicrogram quantities of CH_4 and CO. It is unlikely that gas generated by this process would make a significant contribution to the gases under investigation. There was no evidence for the presence of CO when aliquots of standard gases, including methane and ethane, were processed through the complete analytical sequence.

Gas chromatography

Gas chromatography was performed using a Varian Aerograph 1200 series gas chromatograph fitted with a microkatharometer (Servomex Controls, Ltd.) in series with a flame ionisation detector (FID). The gases released from lunar samples were analysed using a 25′ × $\frac{1}{16}$″ o.d. stainless steel column packed with 60–80 mesh Graphon (DICORCIA and BRUNER, 1970) operated with 2–4 ml/min of helium as carrier gas (Fig. 2).

The non-condensed fraction was concentrated with the Toepler pump and analysed at -78°C (solid CO_2/acetone). Under these conditions an equimolar mixture of CD_4 (Stohler Isotopic Chemicals, Ltd.) and CH_4 gave a 50% valley (10,000 theoretical plates). CD_4 and CD_3H were not separated under these conditions. The CD_4 released from the lunar samples always exceed the CH_4 (in the ratio 2.5–7.6) and the effective separations between CD_4 and CH_4 were poorer (Fig. 3) than that measured for the standard equimolar mixture. The condensed fraction was analysed at 0°C to allow separation of C_2D_6 and C_2H_6. Retention times on Graphon under these conditions did not permit the measurement of C_3 or C_4 hydrocarbons or deuterocarbons. These species, released from two samples of Apollo 11 fines, were analysed at 100°C with a 20′ × $\frac{1}{16}$″ o.d.

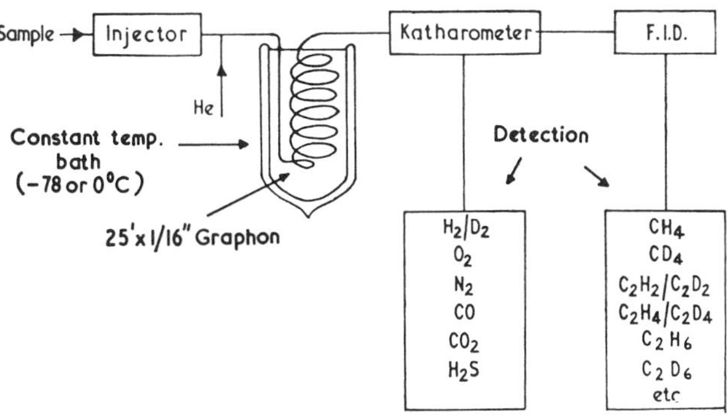

Fig. 2. Gas chromatographic system for analysis of low molecular weight gases.

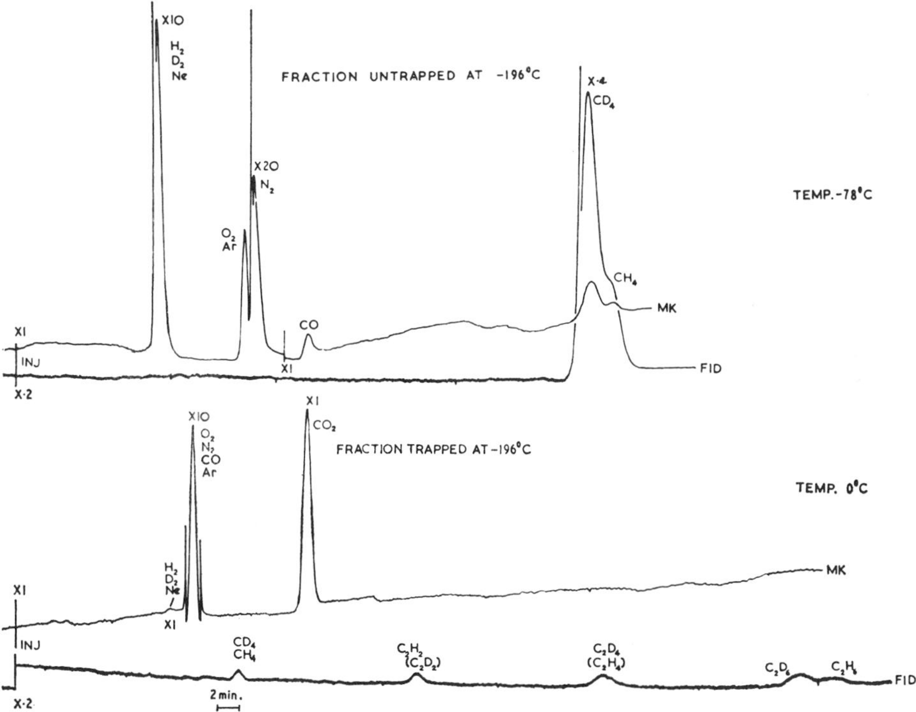

Fig. 3. Gas chromatographic analysis of gases released on DF treatment of lunar fines (12023, 47.4 mg). Column: 25′ × $\frac{1}{16}$″ od, 60–80 mesh Graphon, He 2 ml/min, MK = microkatharometer response, FID = Flame ionisation detector response. Attenuation factors are marked where appropriate.

stainless steel column packed with 100–120 mesh Porapak T with 6–8 ml/min of helium as carrier gas.

Calibration was carried out with known quantities of individual gases or mixtures (e.g., CH_4, C_2H_2, C_2H_4, C_2H_6, C_3H_8, n-C_4H_{10}, CO, CO_2) from the aliquot system (Fig. 1) and passing them through the complete analytical sequence. The gases were quantitated either by digital integration of the gas chromatograph output (Kent Chromalog 2) or by weighing peaks cut from the chart paper. An estimate of the overall error involved in the complete analytical sequence (fractionation and concentration system, syringe transfer, gas chromatographic peak integration) was made by processing aliquots through the complete sequence and by injecting standard aliquots of hydrocarbons directly into the chromatograph. The overall error in the sequence is $\pm 20\%$. Recoveries of the C_1 to C_3 hydrocarbons are greater than 90%.

Mass spectrometry

Mass spectrometry was used to analyse a limited number of samples of Apollo 11 fines (but was subsequently replaced by the gas chromatographic method). Gases released by in vacuo crushing and by DCl etching were analysed in the system described previously (ABELL *et al.*, 1970a). An isopentane/liquid N_2 slush bath ($-140°C$) was used to prevent DCl vapour from reaching the ion source.

Gas chromatography-mass spectrometry

Gas chromatography-mass spectrometry was performed using a Varian MAT CH-7 mass spectrometer coupled to a Varian Aerograph 1200 gas chromatograph, via an all-glass, single-stage, Watson-Biemann helium separator. Analyses were carried out with a $20' \times \frac{1}{16}''$ stainless steel column packed with 100–120 mesh Porapak T, operated with 4 ml/min of helium as carrier gas. The non-condensed and condensed fractions, released by DF etch from Apollo 11 fines (0.200g), were chromatographed isothermally at 20°C and 80°C, respectively. Mass spectra were recorded continuously by repetitive scanning (5 sec/decade) over the range m/e 10–80.

RESULTS

The use of deuterium-labelled reagents allows distinction between indigenous hydrocarbon gases and those formed by chemical reaction during isolation. Thus, for example, methane released as CH_4 represents an indigenous component of the samples. methane released as CD_4 represents a synthetic product. Quantitative data for all carbon compounds are expressed in μg/g as carbon in the samples.

Vacuum crushing

Crushing in vacuo was chosen as a possible non-chemical method for releasing indigenous gases (ABELL *et al.*, 1970a). Mass spectrometric analysis of the gases released by in vacuo crushing (Apollo 11 fines) revealed the presence of the rare gases He, Ne, and Ar. The quantities of these gases were substantially less than those obtained by the thermal extraction methods of other workers (e.g., HINTENBERGER *et al.*, 1970; HEYMANN and YANIV, 1970; EBERHARDT *et al.*, 1970); however, the relative abundances of the Ne and Ar isotopes were similar to those already observed. Methane (0.9 μg/g) and ethane (0.1 μg/g) were present in the released gases. A blank experiment performed on the empty ball mill showed only trace quantities of terrestrial atmospheric gases (O_2, N_2, ^{40}Ar, CO_2). Crushing two other samples in the presence of D_2O also liberated the rare gases; greater than 95% of the methane released could be accounted for as CH_4 (0.24 μg/g, 0.24 μg/g), CD_3H (0.16 μg/g, 0.06 μg/g) and CD_4 (0.36 μg/g, 0.43 μg/g). Similarly, about 75% of the ethane could be accounted for as C_2H_6 (0.06 μg/g, 0.03 μg/g), C_2D_6 (0.05 μg/g, 0.03 μg/g) and C_2D_5H (0.03 μg/g, 0.06 μg/g). The resolution of the mass spectrometer was not sufficient ($M/\Delta M$ = 800) to distinguish between N_2 and CO (m/e 28). The second most abundant ion in the mass spectrum of CO is m/e 12, but the observed abundance of this ion could be accounted for by the methane species present in the same fraction and it is unlikely that there was a major contribution by CO to the ion m/e 28. The only other species identified among the gases were O_2 and CO_2 at concentrations similar to those observed in the blank experiments. The crushing method indicated that indigenous methane and ethane are present as CH_4 and C_2H_6 in the fines. However, the method has certain disadvantages: (1) it is inefficient; less than 10% of the indigenous hydrocarbon gases released by acid etching are released by crushing and thus species present in low abundance are not observed; and (2) it also releases gases formed by the hydrolysis of carbides, presumably arising as a result of localised high temperatures and the presence of adsorbed terrestrial water.

Vacuum crushing has been used to prepare rock samples prior to acid etching. The gases released by crushing rock 12022,79 were analysed by gas chromatography. A small methane peak was detected (0.5 μg/g). No peak was detected for CO, and all of the CO_2 seen could be ascribed to terrestrial contamination.

Acid etching

The results of the acid etching experiments are described in the following three sections.

(1) *Mass Spectrometry.* Mass spectrometry was used to examine the hydrocarbons and deuterocarbons released by DCl treatment of two samples of Apollo 11 fines. The mass spectrum of the gases released from one of the samples (Fig. 4b) shows that 18%

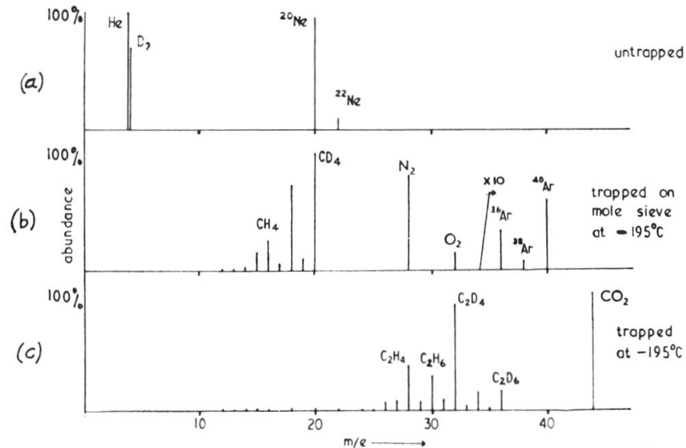

Fig. 4. Mass spectra of gases released by DCl treatment of Apollo 11 lunar fines and admitted to the mass spectrometer via gas inlet system. In (b) abundance of ions above m/e 34 are multiplied by a factor of 10.

of the carbon released as methane is in the form CH_4, almost all (90%) of the remainder being as CD_4. Likewise (Fig. 4c) 31% of the carbon released as ethane is in the form C_2H_6, 90% of the remainder is present as either C_2D_6 or C_2D_5H. The quantities of carbon released from the fines vary with the particle size. The CH_4 content of two samples of the unsieved bulk fines (1.3 μg/g, 1.3 μg/g) falls between the quantities of CH_4 liberated from 30–100 mesh fines (0.7 μg/g) and fines smaller than 200 mesh (1.8 μg/g). The CD_4 and CD_3H values follow a similar trend; the $CD_4 + CD_3H$ content of the unsieved bulk fines (5.8 μg/g, 5.8 μg/g) falls between the quantities liberated from 30–100 mesh fines (3.1 μg/g) and the fines smaller than 200 mesh (7.6 μg/g). It is noteworthy that the $(CD_4 + CD_3H)/CH_4$ ratio is approximately constant (4.2–4.5). Evidently, the quantities of the indigenous methane and the hydrolysis products (CD_4, CD_3H) vary with the particle size in a similar fashion.

The quantities of C_1 and C_2 hydrocarbons and deuterocarbons released from Apollo 11 bulk fines (by DCl etch) and measured by mass spectrometry, are smaller than those measured by the gas chromatographic method (using DCl and DF). It is likely that the quantitative data obtained by mass spectrometry are low because the response of the mass spectrometer does not have a linear relationship with the sample pressure. The lunar sample gases and calibrant gases were analysed in the same way but the source pressures were different. The mass spectrometric data for the DCl etch are self-consistent, as are the results similarly obtained in three crushing experiments. We believe that quantitation afforded by mass spectrometry alone is low, but that the trend observed for the DCl etch of size-differentiated fines is real.

Figure 4c shows the presence of much greater quantities of CO_2 than were observed in the crushing experiments. It was impossible to obtain quantitative data for this gas; a major portion of it was condensed in the $-140°C$ trap. However, CO_2 was present in sufficient quantities (compared to N_2 and O_2) to preclude terrestrial atmospheric background *in the system* as its source (see Discussion). The ion m/e 28 (Fig. 4b) is assigned as N_2 by an argument similar to that already presented for the crushing experiments. The rare gases (He, Ne and Ar) are observed in non-terrestrial isotopic abundances (Fig. 4a,b).

Examination of the DCl for hydrocarbon and deuterocarbon contaminants established that these were not present in quantities above the detection limits. The possibility of hydrocarbons arising from deuterocarbons by exchange was investigated by examining the reaction of DCl with cohenite; deuterocarbons, but not hydrocarbons, were observed.

(2) *Gas chromatography. Apollo 11 Samples* (Table 1). The detection limit for an individual hydrocarbon was equivalent to 0.1 μg/g as carbon for a 50 mg sample. Analysis of the gases released by etching Apollo 11 fines with undeuterated acids (HF and HF/HCl) indicates that 25–35 μg/g of carbon in the sample can be released as C_1–C_2 hydrocarbons. The use of deuterated acids (DCl and DF) indicates that the methane and ethane are mixtures of undeuterated and fully deuterated compounds. Deuterated and undeuterated compounds were not completely resolved. The small contributions of CD_3H and C_2D_5H identified by mass spectrometry and gc–ms (see below) were not separable from their fully deuterated counterparts; CD_3H and C_2D_5H

obviously arise from the same source as the fully deuterated analogues. Thus, the quantitative data expressed as CD_4 and C_2D_6 include a small contribution from these species. The ratio of carbon as CD_4 (19–22 μg/g) to CH_4 (4.8–5.2 μg/g) released on acid etch is reasonably consistent (3.6–4.3). The relative and absolute yields of C_2H_6 and C_2D_6 are more variable (0.16–0.9 and 1.2–3.6 μg/g, respectively). Analysis on Porapak T of the gases released by HF etch indicated the presence of species with retention times corresponding to propane and propene (cf. CHANG *et al.*, 1970; HENDERSON *et al.*, 1971). In the gases released from one sample, species with retention times of methylacetylene and allene were observed. Deuterated and undeuterated species are not resolved on Porapak and we are unable to decide at present whether the C_3 hydrocarbons occur as such, or represent hydrolysis products.

Only deuterocarbons were released (Table 1) by the action of DF on cohenite; acetylene was absent. The yield of carbon as gaseous deuterocarbons was low (*ca.* 1.0%, assuming Fe_3C); this could result, in part, from the relatively large size of the cohenite fragments (*ca.* 40–60 mesh). The ratio of CD_4/C_2D_6 from cohenite was similar to that observed for lunar fines (*ca.* 10:1).

Gas chromatography indicated that CO was released from the fines in amounts ranging from 1.2 to 3.0 μg/g. One value (8.6 μg/g) was obtained when an ion guage was incorporated in the vacuum system: CO is known to be synthesised on hot filaments in vacuum systems and this high value probably reflects a contribution from this source. In one experiment in which the sample was etched with DCl for 90 hours at 120°C, CO (*ca.* 65 μg/g as carbon) was detected. The detection limit for CO was 0.5 μg/g for a 50-mg sample.

CO_2 was found in the hydrolysis products at concentrations ranging from 4 to 76 μg/g. Initially, blank experiments were free from CO_2 but significant quantities appeared in later blanks.

(3) *Gas chromatography. Apollo 12 Samples.* (Table 2) Gas chromatographic analyses of gases released by DF etch of fines having high contents of total carbon (12001, 12023, 12042; *ca.* 130 μg/g; MOORE *et al.*, 1971) showed the presence of deuterated and undeuterated C_1 and C_2 hydrocarbons. The quantities of CD_4 and CH_4 released from these samples (8–15 μg/g and 2.1–2.2 μg/g) were somewhat smaller than those obtained from Apollo 11 fines and there appeared to be more variability in the CD_4/CH_4 ratios (3.8–7.2). The quantities of C_2 hydrocarbons released from these samples were similar to, or lower than, those obtained from Apollo 11 fines but with slightly smaller C_2D_6/C_2H_6 ratios. The CO released by DF at ambient was consistently less than 2 μg/g. Two samples (12023 and 12037) treated with DCl at 120°C (90 hr) released CO in quantities <1 μg/g.

DF etching of fines of lower carbon content (12037, total carbon 82 μg/g; MOORE *et al.*, 1971) released correspondingly smaller amounts of methane (1.2 μg/g CH_4; 3.0 μg/g CD_4); the quantities of C_2 hydrocarbons released barely exceeded the detection limit. Two samples of fines with very low total carbon content (12032, 12033; 25–60 μg/g; MOORE *et al.*, 1970) were also analysed and only very small quantities of methane were detected. Traces of C_2 hydrocarbons were observed. DF etching of an interior chip of an olivine basaltic rock (12022,79; total carbon 50 μg/g; MOORE *et al.*,

1970), previously crushed in vacuo, was carried out. In the two analyses, methane (0.78, 1.1 μg/g) and ethane (0.10, 0.14 μg/g) were undeuterated, indicating the absence of hydrolysable carbide. Acetylene and ethylene (*ca.* 0.1 μg/g) were observed but in neither case could the extent of deuteration be determined (see above). A trace of CO (*ca.* 0.1 μg/g) was detected in one analysis.

GC–MS

The application of gc–ms was limited because of the poor sensitivity arising from sample losses in the separator. However, we have used the method to analyse the gases released by DF etch of two samples of Apollo 11 fines. This experiment was designed to confirm the validity of the isotopic labelling/gc procedure and to determine the isotopic composition of species not separated by the gas chromatographic method. Continuous scanning of the column effluent confirmed the presence of CD_4 and CH_4, together with trace quantities of CD_3H. Similarly, only trace quantities of C_2D_5H accompanied C_2D_6 and C_2H_6. The mass spectrum indicated that ethylene comprised almost entirely C_2D_4. The isotopic composition of acetylene could not be determined because of the small quantities released. The chromatographic conditions allowed partial resolution of CO and N_2. There was no distinct peak corresponding to CO in the chromatogram. However, continuous scanning revealed the presence of an ion of m/e 12 in the tail of the nitrogen peak. If it is assumed that m/e 12 arose from CO, then, up to 3 μg/g of lunar carbon could have been released as CO from the lunar fines.

H_2S was identified but it is unlikely to be indigenous to the fines. The most likely origin is from isotopic exchange of D_2S synthesized by the reaction of DF on sulphide minerals. Isotopic exchange of D_2S occurs on the column used. Only H_2S was observed in the ion source of the mass spectrometer when D_2S was injected into the gc–ms system. Deuterium, presumably arising from the reaction of the acid on metals such as iron and nickel, accounted for the major part of the gas mixture.

Alkali etch

The gases evolved by etching one sample of Apollo 11 fines with sodium deuteroxide at ambient temperature were analysed by gas chromatography at regular intervals. In comparison with the acid etch, where gas evolution is vigorous, there was no visible evolution of gas during NaOD etch at ambient. The samples appeared to be unaltered. In one experiment in which unsieved fines were used, a large proportion (70% = 3.7 μg/g) of the CH_4 present (as determined by DF etching) was released after 120 hours. The CD_4 (3.4 μg/g) evolved amounted to $<20\%$ of that from etching with DF. Very little CH_4 and CD_4 were evolved after 120 hours. In a second experiment, etching of a 200–240 mesh aliquot also produced a high proportion (*ca.* 60%) of the CH_4 released by acid etch, whereas the quantity of CD_4 released was comparatively small (Fig. 5). At present little is known about the rate of hydrolysis of carbides under alkaline conditions. The rate of evolution of CH_4 was approximately constant throughout the reaction. The total quantity (1.1 μg/g) of C_2H_6 released after 139 hr. was comparable to, or greater than, that released by acid etch of unsieved fines. The

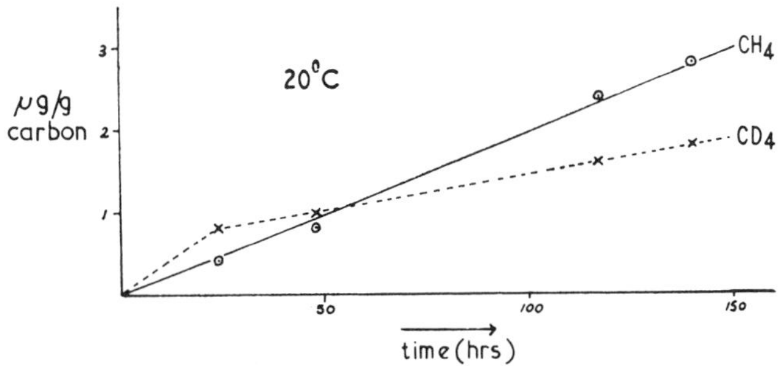

Fig. 5. Quantities of CD_4 and CH_4 released by NaOD etch of Apollo 11 fines. Total quantities of gas are plotted against duration of etch.

Table 1. C_1 and C_2 hydrocarbons and deuterocarbons released by acid etch of Apollo 11 fines (10,086 bulk fines D; *ca* 140 µg/g total C).

Sample wt. (mg.)	Reagent	Carbon (µg/g) released as						Ratio CD_4/CH_4	Carbon accounted for (to 1 µg)
		CH_4	CD_4	C_2H_6	C_2D_6	C_2H_4 and/ or C_2D_4	C_2H_2 and/ or C_2D_2		
50.9	HF (40%)	20		1.6		3.4	0.3	—	25
49.4	HF (40%)	32		1.4		1.0	0.2	—	35
101.0	HF/HCl (20%)	24		0.6		0.1	tr.	—	25
39.5	DF (20%)	5.2	19	0.9	3.6	2.2	1.3	3.6	32
60.6	DF (20%)	4.8	18	0.4	2.2	3.2	0.7	3.8	29
47.0	DCl (40%)	5.1	22	0.16	1.2	1.2	0.3	4.3	30
Cohenite (4.4 mg)	DF (20%)	*	500	*	58	16	*	—	574

tr. = trace.
* not detected.

C_2D_6 (0.26 µg/g), however, was considerably less abundant than that from acid etch of unsieved fines (*ca.* 2 µg/g).

Discussion

Table 1 lists the quantities of C_1 and C_2 hydrocarbons and deuterocarbons (measured by gas chromatography) released by acid etch of Apollo 11 fines. Several acidic reagents (HF, HF/HCl, DCl, and DF) have been used and the quantities of gases released are reasonably consistent although the samples were not size-differentiated, crushed or aliquotted. A significant fraction (25–35 µg/g) of the total carbon can be released as C_1 and C_2 hydrocarbon species. Henderson *et al.* (1971) and Chang *et al.* (1970) have reported values of 16 µg/g for the C_1 and C_2 hydrocarbon species released from 10,086 fines by HF and HCl etch, respectively.

The isotopic labelling method (Abell *et al.*, 1970b and the present paper), used for crushing experiments and acid-etching experiments, demonstrates that both indigenous hydrocarbons and hydrolysable carbon compounds are present in lunar fines. It is likely, as suggested by Chang *et al.* (1970), that the hydrolysable material is carbide, including cohenite [(Fe, Ni)$_3$C], as shown by the evolution of CD_4, C_2D_6, and almost completely deuterated hydrocarbons (eg., CD_3H). The quantities of CH_4, CD_4, C_2H_6, and C_2D_6 indicate that the indigenous hydrocarbons (CH_4 + C_2H_6)

contribute *ca.* 5 μg/g and carbides ($CD_4 + C_2D_6$) contribute *ca.* 22 μg/g to the carbon in the Apollo 11 fines.

Table 2 shows the quantities of C_1 and C_2 hydrocarbons and deuterocarbons released from a variety of Apollo 12 fines. The concentrations of indigenous hydrocarbons and carbides are less than those measured for the Apollo 11 fines. The CD_4/CH_4 ratios (Tables 1, 2) show that carbide and methane are also present in different proportions. One sample of the fines (12023) has been examined independently by Chang *et al.* (1971) using 6 *N* DCl: their values for CH_4 (1.6 μg/g) and carbide (6.7 μg/g) are similar to those shown in Table 2, although no C_2H_6 was reported by these authors. Henderson *et al.* (1971) have independently examined the hydrocarbon products of HF etch of all of the Apollo 12 fines listed in Table 2, with the exception of 12037 fines: their results for total C_1 and C_2 hydrocarbons are similar to those that can be derived from the values for hydrocarbons and deuterocarbons listed in Table 2.

The concentrations of carbon accounted for both as indigenous hydrocarbons and carbide probably represent minimum values. At present, the gas chromatographic method does not allow distinction between the indigenous and synthetic components of species other than methane and ethane. Thus, the isotopic composition of the ethylene, acetylene and higher hydrocarbons released, when determined, will provide more accurate values for the indigenous and synthetic components.

In addition, it is unlikely that all of the carbon present as hydrolysable carbides is evolved as hydrocarbons on acid treatment. A small sample of cohenite of meteoritic origin liberated only 1–2% of the carbon as C_1 and C_2 hydrocarbons or deuterocarbons when treated with HCl (Chang *et al.*, 1970) or DF, respectively (see above). However, the lunar carbide is likely to be much more finely divided than the fragments

Table 2. C_1 and C_2 species from DF etch of Apollo 12 samples.

		Carbon (μg/g) released as									
Sample	wt. (mg)	CH_4	CD_4	C_2H_6	C_2D_6	C_2D_2 and/ or C_2H_2	C_2D_4 and/ or C_2H_4	CO	Ratio CD_4CH_4	Carbon accounted for (to 1μg)	Total Carbon‡
12001 fines	55.4	2.2	12	0.38	0.76	*	1.0	†	5.0	16	130
12023 fines	47.4	2.1	8.0	0.08	0.16	0.08	0.12	1.8	3.8	12	135 166
12037 fines	53.5	1.2	3.0	†	0.06	0.10	0.12	1.3	2.5	6	82
12042 fines	49.9	2.1	15	†	*	0.36	1.1	†	7.2	19	130 140
12032 fines	101.9	0.28	1.7	†	†	0.24	0.40	0.3	6.1	3	60 25
12033 fines	105.0	0.11	0.84	†	tr	†	tr	†	7.6	1	50 23
12022 interior chip	218.0	1.1	†	0.10	†	0.12	0.07	<1	0.0	1	50
12022 interior chip	140.0	0.78	†	0.14	†	0.15	0.10	0.1	0.0	1	50

* not measured.
† not detected.
tr trace.
‡ Measured by combustion (Moore *et al.*, 1970).

of cohenite. The yield of deuterocarbons from lunar carbide is likely to be higher but not quantitative.

Application of the isotopic labelling procedure allows distinction between indigenous hydrocarbons and carbides. A closer examination of these two species can now be made.

Indigenous hydrocarbons in lunar fines

The quantities of carbon accounted for as CH_4 and C_2H_6 from acid etch (Table 2) approximately parallel the abundances of total carbon for the Apollo 12 fines as measured by MOORE *et al.* (1971). Three samples of the fines (12001, 12023, and 12042) are dark in colour and have high values for total carbon. These samples also contain correspondingly high values for CH_4 and C_2H_6. Three samples (12032, 12033, and 12037) of light-colored fines have been examined, CH_4 being the only indigenous species identified (Table 2). The 12032 and 12033 fines have low total carbon abundance and low CH_4 content, whereas 12037 has an intermediate total carbon abundance and an intermediate CH_4 content.

Apollo 11 bulk fines have the highest total carbon abundance (MOORE *et al.*, (1970), are very dark in colour, and have the highest indigenous hydrocarbon content. The relationship between CH_4 and total carbon abundance is also valid for size-differentiated fines. MOORE *et al.* (1970) observed that the fines were enriched in carbon compared to the igneous rocks and that, for the fines, the total carbon abundance increased with decreasing particle size. Our preliminary mass spectrometric data obtained by DCl etch of size-differentiated Apollo 11 fines show that CH_4 is also concentrated in the finest fines. These observations suggest that a surface-correlated phenomenon is the source of some of the carbon present in the lunar fines and that the indigenous hydrocarbons are derived from this implanted carbon.

Table 3 lists the quantities of CH_4 released by DF etch of Apollo 11 fines and various samples of Apollo 12 fines; also listed are a number of other available parameters for the same samples (^{36}Ar abundance, cosmic ray track density, cosmic ray exposure age and fraction of grains with an amorphous surface coating). The data in Table 3 indicate possible correlations between (1) the CH_4 content and the ^{36}Ar content. It is generally agreed that the high abundances of the rare gases in the fines could indicate a major contribution from solar wind implantation, although little is known about the respective diffusion losses of ^{36}Ar and CH_4; (2) the CH_4 content of the fines and the fraction of mineral grains in the 400 mesh residue which show an amorphous coating. Unfortunately, this comparison is restricted to two samples. However, the fines with the high CH_4 content have the higher fraction of these grains. Conversely, the fines with a low CH_4 content have a much lower fraction of the grains. The amorphous coating of a grain is highly disordered in comparison with the crystalline interior; it extends to a depth of 200–800 Å, and has been tentatively attributed to solar wind implantation (BORG *et al.*, 1971; DRAN *et al.*, 1970). A high latent track density ($> 10^{10}$ tracks/cm^2), attributable to solar suprathermal ions, is also found in these grains. It has also been suggested that the grains with an amorphous coating represent cosmic dust irradiated in space and accreted by the

Table 3. Correlation of the quantities of CH_4 and CD_4 released by DF etch from lunar fines with various measurements relating to surface exposure.

Sample No.	CH_4(μg/g)	CD_4(μg/g)	^{36}Ar (10^{-8} cc/g) (PEPIN, 1971)	Fraction of grains with amorphous coating in 400 mesh residue (BORG *et al.*, 1971)	Fraction of particles with track density $\geqslant 10^8$/cm^2 (BHANDARI *et al.*, 1971)	Average track density (10^6) of particles with $< 10^8$/cm^2 (BHANDARI *et al.*, 1971)	Cosmic ray exposure age (10^6 yr.) (BHANDARI *et al.*, 1971)
Apollo 11 bulk fines	*ca.* 5‡	*ca.* 22‡	38,000†	67%‖	> 0.9§	*	> 100§
12042	2.1	15	28,000	n.d.	0.92	54.9	160
12001	2.2	12	n.d.	n.d.	0.68	48.2	80
12037	1.2	3	n.d.	n.d.	0.52	16	20
12032	0.28	1.7	4,200	15%	0.17	13	10
12033	0.11	0.8	2,800	n.d.	0.03	5.6	100**

* lowest track density observed (10084) = 1×10^7 (LAL *et al.*, 1970).
† 10084, HINTENBERGER *et al.* (1970); PEPIN *et al.* (1970).
n.d. not determined.
‡ 10,086 D.
§ 10,084: 10,085 LAL *et al.* (1970).
‖ 10,084.
** total in situ exposure age.

moon (BARBER *et al.*, 1970); (3) The CH_4 content and the cosmic ray track density (and, hence, possibly "cosmic ray exposure age"). BHANDARI *et al.* (1971) have examined a large number of individual grains of pyroxene and feldspar selected from several samples of Apollo 12 fines. The fossil cosmic ray tracks occur within 10 μ of the surface of a grain. Table 3 lists the fraction of particles having a track density $\geq 10^8$ tracks/cm^2, the average track density for the remaining particles, and the corresponding exposure age. A high CH_4 abundance is observed in samples of fines (e.g., 12042) having a high frequency of particles with track density $\geq 10^8$ tracks/cm^2, a high average track density for the remaining particles and a correspondingly long exposure age. A low CH_4 abundance is observed (e.g., 12032) when the other parameters are also low. Similar figures are not available for Apollo 11 fines. However, LAL *et al.* (1970) have reported the following observations for 10084 and 10085 fines: (a) 90% of the particles have track densities $> 10^8$ tracks/cm^2; (b) the minimum track density is 1×10^7 tracks/cm^2; and (c) the cosmic ray exposure age is several hundred million years. The CH_4 abundance (5 μg/g) for 10,086 bulk fines is the highest so far observed for any sample of fines.

Cosmic ray exposure ages and solar wind exposure ages need not be directly related. BHANDARI *et al.* (1971) have suggested that grains having a track density of at least 5×10^8 tracks/cm^2 have been exposed "unshielded" on the lunar surface. Grains having such track densities could also represent particles abraded from rock surfaces which had an "unshielded" exposure. In contrast, it is thought that grains with track densities in the range 10^6–10^7 have been located within the regolith at depths between 0.05 and 10 cm. Unshielded exposure to cosmic rays indicates a greater likelihood of exposure to the solar wind, i.e., grains with track density $\geq 10^8$ are more likely to have been exposed to solar wind. The correlations outlined under (1), (2), and (3) are explicable in terms of an extralunar origin for the CH_4, possibly as a result of solar wind activity. If indeed solar wind implantation is the major source of the CH_4 observed in the Apollo 11 and 12 fines, it should be located well within the outer 1000 Å of an individual grain.

A method of shallow etching of lunar fines should provide information about the location of the hydrocarbon gases in relation to the surfaces of the grains. NaOH has been used to reveal the tracks left by cosmogenic heavy nuclei which impinged on silicate grains in meteorites (LAL *et al.*, 1968). The rate of etch of silicate minerals will depend both on the mineralogy and on the track densities present. Etching of these minerals is slow at elevated temperature (*ca.* 160°C); the rate of etch of lunar fines at ambient must be even slower. In comparison with acid treatment, where gas evolution is vigorous and the residue is an off-white flocculent material, alkali treatment affords no visible gas evolution and the samples appear unaltered. The use of NaOD as etching reagent allows indigenous hydrocarbons and hydrolysis products to be distinguished without appreciably altering the slow rate of etch. NaOD treatment of both Apollo 11 fines and a size-differentiated sample of these fines (200–240 mesh) showed that $>60\%$ of the CH_4 released by acid etch can be released by this reagent (Fig. 5). The quantity of CD_4 concurrently released is small. Although we are unable at present to measure accurately the depth of etch, these preliminary data indicate that a significant proportion of the CH_4 may be located at the surface of the fines. An NaOD etch of mineral concentrates having a narrow size-distribution and a known rate of etch should confirm the surface location of the CH_4.

We have previously outlined a number of synthetic processes which could lead to the formation of hydrocarbons from lunar, meteoritic, or solar carbon as a result of solar wind bombardment of the lunar surface (ABELL *et al.*, 1970a, 1970b). There are several distant laboratory analogies for such reactions. Irradiation of diamond crystals with protons or deuterons (0.7–1.5 MeV) affords products with carbon-hydrogen or carbon-deuterium bonds, respectively (ZELLER and DRESCHOFF, 1968). Protons produced by microwave discharge react with soot (HARRIS and TICKENER, 1947) or graphite (SHAHIN, 1962) to produce C_1 and C_2 hydrocarbons. Similarly protons or tritium atoms produced from a hot tungsten filament afforded C_1–C_4 hydrocarbons after reaction with lampblack, graphite or diamond dust (GILL *et al.*, 1967). The conditions prevailing in these reactions are somewhat different from those which might be expected during a solar wind synthesis; however, it is noteworthy that methane and other low molecular weight saturated and unsaturated hydrocarbons were always observed among the products obtained from the interaction of hydrogen atoms and various forms of carbon.

Indigenous gases in lunar basalt 12022

Gas chromatographic analysis was carried out on the gases released by in vacuo crushing of two interior chips of the crystalline rock 12022; CH_4 was the only gas observed and was present in concentrations just in excess of the detection limit of the analytical method. Acid etch (DF) of the crushed sample released (Table 2) methane as CH_4 and ethane as C_2H_6. No CD_4or C_2D_6 was observed. It is unlikely that the CH_4 and C_2H_6 could be contaminants arising from adhering fines; fully deuterated species would also be released from this source of contamination, if present. No data are at present available for the isotopic composition of the acetylene and ethylene released; however, it appears that the CH_4 and C_2H_6 are of primordial origin. If this is the case, there may be a small primordial component in the fines, although the extent of losses of this component during formation of the fines is unknown. BLOCH *et al.* (1971) have examined the gases released from a number of crystalline rocks (12021, 12063, 12075, 10057) using an inefficient in vacuo crushing procedure. The only carbon-containing species observed was CH_4 at concentrations ranging from 0.01 to 1.0 ng/g. BLOCH *et al.* suggest that the CH_4 is representative of the magmatic gas phase (primordial). Application of the same crushing procedure to a number of breccias released 10^2–10^3 times the concentration of trapped gas as did the crystalline rocks. CH_4, C_2H_4, C_2H_6, C_3H_6, and C_3H_8 were among the species observed. Although the crushing procedure used is mild, it is not impossible that hydrolysis reactions take place. Synthetic products would be recognised by D_2O exchange prior to crushing. Glow discharges in gas emissions have been proposed as explanations for transient lunar phenomena. If magmatic hydrocarbon gases are indeed observed in a range of igenous rocks, their presence may explain the carbon-containing species in the spectrum observed (KOZYREV, 1959) during such an event.

Carbides in lunar fines

Mineralogical observations have indicated the presence of cohenite in the Apollo 11 fines (FRONDEL *et al.*, 1970; ANDERSON *et al.*, 1970). Acid hydrolysis (DF) of cohenite liberates the same C_1 and C_2 deuterocarbons (with the exception of acetylene) and in similar relative proportions, as do the Apollo 11 fines. Cohenite must be partly responsible for the hydrocarbon species liberated from the fines; the same conclusion was reached by CHANG *et al.* (1970), using undeuterated acid. However, hydrolysis of cohenite yields only 1–2% of hydrocarbons. Assuming similar yields from lunar carbides, the quantities of deuterocarbons released from the lunar fines by deuterated acids would require carbon contents greatly in excess of those observed. A significant proportion of the carbide in the fines is thus likely to be extremely finely divided. At present there is no mineralogical evidence for the presence of carbide in Apollo 12 fines. DF etch (Table 2) shows that carbides are present in all samples of Apollo 12 fines studied. Their presence should be sought by mineralogical examination. No CD_4 or C_2D_6 was observed among the products of DF etch of the olivine basaltic rock 12022. The isotopic composition of the acetylene and ethylene released is unknown and their source (indigenous gases or carbide hydrolysis products) remains uncertain. More data from a variety of crystalline rocks are clearly necessary; however, the absence of CD_4 and C_2D_6 from the gases released from rock 12022 is in agreement with an extralunar origin for the carbide in the fines giving rise to these species. An extralunar origin could indicate contributions from meteoritic impact, solar wind implantation, or both.

A number of pieces of evidence must be considered in relation to these origins. (1) Meteoritic cohenite is found almost exclusively in coarse octahedrites containing 6–8% by wt. nickel (BRETT, 1967). All lunar petrographers have observed the presence of metallic iron without nickel in the Apollo 11 crystalline rocks, but in the fines and breccias some of the iron is associated with nickel (JEDWAB and HERBOSCH, 1970). FRONDEL *et al.* (1970) have reported that the nickel content of the iron–nickel grains and inclusions ranges from 3 to 14.6%, most particles being in the range 6–7%. GOLDSTEIN and YAKOWITZ (1971) have measured the nickel and cobalt contents of a number of iron particles and inclusions in Apollo 12 fines. Approximately 15% of the metal inclusions had nickel and cobalt contents in the known range for meteoritic composition; four of the six individual metal grains studied were of meteoritic origin. We infer that there are in the lunar fines, iron–nickel grains and inclusions of meteoritic origin, with iron–nickel ratios in the range necessary for cohenite to be present. Thus, both Apollo 11 and 12 fines contain meteoritic iron grains and inclusions with nickel contents compatible with those observed in meteorities containing cohenite. (2) The two values for the $\delta^{13}C$ value (+5, +14 per mil) for the total hydrocarbons released by HCl etch of 12023 fines (CHANG *et al.*, 1971) are outside the range of the $\delta^{13}C$ values (−4 to −8 per mil) reported for meteoritic carbide. The hydrocarbons released from this sample also contain an indigenous component (*ca.* 25%, Table 2) and the conversion of carbide to hydrocarbons may not be quantitative. Despite these considerations, the disparity in $\delta^{13}C$ values indicates that the carbide in lunar samples is different from carbide of known meteoritic origin. (3) The carbon accounted for as CD_4 and C_2D_6 from acid etch of Apollo 11 and Apollo 12 fines (Tables 1 and 2) approximately parallels the total carbon abundances of the samples (MOORE *et al.*, 1970; MOORE *et al.*, 1971). Similar observation can be made for the indigenous hydrocarbons (see above). The parallelism observed in each case can only be approximate because of the variations in the measured CD_4/CH_4 ratios. In addition, this relationship between CD_4 and total carbon content appears to be valid for the size-differentiated fines: a surface area-correlated phenomenon could be responsible for a fraction of the carbide in the fines. (4) The quantities of CD_4, like those of CH_4, released by DF etch of Apollo 11 and Apollo 12 fines, appear to correlate with the ^{36}Ar abundances, cosmic ray exposure ages and extent of amorphous surface coating, measured for the same samples (Table 3). The trend appears to be valid although as already indicated, the CD_4/CH_4 ratios vary between samples. These observations suggest that the carbide, like the CH_4, may have a component resulting from exposure on the lunar surface.

In summary, (1) suggests that the presence in lunar fines of iron/nickel particles of meteoritic origin is compatible with a contribution of meteoritic carbide. The microscopic carbide grains observed in Apollo 11 samples (FRONDEL *et al.*, 1970); ANDERSON *et al.*, 1970; ADLER *et al.*, 1970) probably arise from this source. Isotope ratio measurements, (2) above, suggest that overall, lunar carbide in Apollo 12 fines is different from meteoritic carbide. However, the "hydrogen stripping" process proposed by BERGER (1970) and KAPLAN and SMITH (1970) can be invoked to explain the ^{13}C enrichment (CHANG *et al.*, 1971). Considerations (3) and (4) suggest that the lunar carbide may have a contribution arising from solar wind implantation. Thus, the data at present available point towards a solar wind contribution and a meteoritic contribution to lunar carbide. Carbon atoms implanted by the solar wind into metal grains and inclusions could generate sub-microscopic moieties having the stoichiometry of carbides (cf. CHANG *et al.*, 1971). The necessary metal particles are available from both meteorites and lunar igneous rocks (GOLDSTEIN and YAKOWITZ, 1971; REID *et al.*, 1970). The major portion (80%) of the metal inclusions in the Apollo 12 fines originates in the igneous rocks (GOLDSTEIN and YAKOWITZ, 1971). Rock 12022 does not contain hydrolysable carbide. If this result proves to be typical of other rocks, then definitive evidence for carbide or carbidelike materials arising from solar wind implantation could be provided by DF or DCl etch of abraded material containing metal inclusions, selected from the fines.

Other gases

The quantities of CO observed among the gases released by DF etch of both Apollo 11 and Apollo 12 fines are of the order of 1–3 μg/g. However, the reaction of one sample of Apollo 11 fines with DCl at elevated temperature (120°C for 90 hr) afforded CO at a concentration of 65 μg/g. Similar experiments on Apollo 12 fines (12023 and 12037) released CO in quantities <1 μg/g. We are unable at present to explain this one anomalously high result. At present insufficient data are available for comparisons to be drawn between CO content and other parameters such as total carbon content, particle size, exposure age, etc. ORÓ *et al.* (1971) have suggested that the CO_2 released from Apollo 11 fines by heating at temperatures up to 700°C may be ascribed to terrestrial contamination. The CO_2 observed in our mass spectrometric analyses could not be completely accounted for by residual air in the analytical system. However, it could have arisen through adsorption on to the surfaces of the fines during the handling procedures at Houston and Bristol. We hope to resolve soon the ambiguities in the CO_2 measurements made from gas chromatographic analysis of the products of acid etch. At present we are unable to decide whether the CO_2 is present in the lunar fines as such, is released from carbonates by reaction with acid or is adsorbed contamination.

CONCLUSIONS

All of the Apollo 11 and Apollo 12 samples of fines examined contain indigenous methane and carbide; indigenous ethane has also been detected in several samples. Comparison of the methane content with a number of other parameters indicates that solar wind activity may play a major role in the synthesis of the methane. Synthesis from the solar wind necessarily requires that the methane be located at the surface of the fines and shallow etching at ambient temperature with alkali appears to confirm this surface location. Analysis of interior chips of an olivine basaltic rock (12022) suggests that small quantities of primordial methane are present, which could also contribute to some extent to the methane in the fines. No acid-hydrolysable carbide was detected in the interior of the rock. This observation requires extension to other lunar rocks, but this preliminary result indicates an extralunar origin for the carbide in the fines. The data available at present are in agreement with a dual extralunar origin for the carbide, that is, an origin from both meteoritic impact and solar wind implantation of carbon.

It appears that the chemistry of the carbon on the Moon may be very dependent on the processes operating on the lunar surface (ABELL *et al.*, 1970a, 1970b). More detailed studies will allow an understanding of the nature of these processes. As we have indicated, it is necessary to integrate the findings obtained from a study of the carbon compounds with the analyses of other elements and with physical measurements such as cosmic ray track density, rate of erosion, etc. The importance of such a study is that the processes involved should be applicable to appropriate planets and other solid bodies in this and other solar systems. Information about these processes cannot be gained from analysis of terrestrial rocks, although laboratory experiments designed to simulate their operation will complement the results from detailed analysis of lunar samples.

Acknowledgments—We thank the Science Research Council and the Natural Environment Research Council for financial assistance, and the National Aeronautics and Space Administration for a sub-contract covering organic geochemical studies (NGL–05–003–003), made through the University of California, Berkeley. The award of Fellowships from the following bodies is gratefully acknowledged: The Science Research Council (C.T.P.), and the Petroleum Research Fund of the American Chemical Society (P.I.A., P.H.C.). We also thank Dr. M. Maurette of the Centre de Spectrométrie de Mass du CNRS, Professor P. H. Fowler of the University of Bristol, Dr. R. O. Pepin of the University of Minnesota, Dr. R. Walker of Washington University, St. Louis, for helpful discussions, Mr. B. Mays of the University of Bristol for assistance, and Dr. W. R. Smith of the Cabot Corporation, Billerica, Mass. for a generous gift of Graphon.

References

Abell P. I., Draffan G. H., Eglinton G., Hayes J. M., Maxwell J. R., and Pillinger C. T. (1970a) Organic analysis of the returned Apollo 11 lunar sample. *Proc. Apollo 11 Lunar Sci. Conf., Geochim. Cosmochim. Acta* Suppl. 1, Vol. 2, pp. 1757–1773. Pergamon.

Abell P. I., Eglinton G., Maxwell J. R., Pillinger C. T., and Hayes J. M. (1970b) Indigenous lunar methane and ethane. *Nature* **226**, 251–252.

Adler I., Walter L. S., Lowman P. D., Glass B. P., French B. M., Philpotts J. A., Heinrich K. J. F., and Goldstein J. I. (1970) Electron microprobe analysis of Apollo 11 lunar samples. *Proc. Apollo 11 Lunar Sci. Conf., Geochim. Cosmochim. Acta* Suppl. 1, Vol. 1, pp. 87–92. Pergamon.

Anderson A. T., Jr., Crew A. V., Goldsmith J. R., Moore P. B., Newton J. C., Olsen E. J., Smith J. V., and Wyllie P. J. (1970) Petrologic history of moon suggested by petrography, mineralogy, and crystallography. *Science* **167**, 587–589.

Barber D. J., Hutcheon I., and Price P. B. (1971) Extralunar Dust in Apollo Cores? *Science* **171**, 372–374.

Berger R. (1970) Reaction of carbon and sulphur isotopes in Apollo 11 samples with solar hydrogen atoms. *Nature* **226**, 738–739.

Bhandari N., Bhat S., Lal D., Rajagopalan G., Tamhane A. S., and Venkatavaradan V. S. (1971) Fossil track studies in lunar materials, II. The near surface and postdepositional exposure history of regolith components at the Apollo 12 site including the double core 25,28. Second Lunar Science Conference (unpublished proceedings).

Bloch M., Fechtig H., Funkhouser J., Gentner W., Jessberger E., Kirsten T., Müller O., Newkun G., Schneider E., Steinbrunn F., and Zähringer J. (1971) Meteorite impact craters, crater simulations, and the meteorite flux in the early solar system. Second Lunar Science Conference (unpublished proceedings).

Borg J., Dran J. C., Durrieu L., Jouret C., and Maurette M. (1970) High voltage electron microscope studies of fossil nuclear particle tracks in extra-terrestrial matter. *Earth Planet. Sci. Lett.* **8**, 379–386.

Borg J., Durrieu L., Jouret C., and Maurette M. (1970) The ultramicroscopic irradiation record of micron-sized lunar dust grains. Second Lunar Science Conference (unpublished proceedings).

Brett R. (1967) Cohenite: Its occurrence and proposed origin. *Geochim. Cosmochim. Acta* **31**, 143–159.

Burlingame A. L., Calvin M., Han J., Henderson W., Reed W., and Simoneit B. R. (1970) Study of carbon compounds in Apollo 11 lunar samples. *Proc. Apollo 11 Lunar Sci. Conf., Geochim. Cosmochim. Acta* Suppl. 1, Vol. 2, pp. 1779–1791. Pergamon.

Chang S., Smith J. W., Kaplan I., Lawless J., Kvenvolden K. A., and Ponnamperuma C. (1970) Carbon compounds in lunar fines from Mare Tranquillitatis, IV. Evidence for oxides and carbides. *Proc. Apollo 11 Lunar Sci. Conf., Geochim. Cosmochim. Acta* Suppl. 1, Vol. 2, pp. 1857–1869. Pergamon.

Chang S., Kvenvolden K. A., Lawless J., Ponnamperuma C., and Kaplan I. R. (1971) Carbon in an Apollo 12 sample: Concentration, isotopic composition, pyrolysis products and evidence for indigenous carbides and methane. Second Lunar Science Conference (unpublished proceedings).

Di Corcia A. and Bruner F. (1970) The use of high efficiency packed columns for gas-solid chromatography, II. The semi-preparative separation of isotopic mixtures. *J. Chromatog.* **49**, 139–145.

Dran J. C., Durrieu L., Jouret C., and Maurette M. (1970) Habit and texture studies of lunar and meteoritic materials with a 1 MeV electron microscope. *Earth Planet. Sci. Lett.* **9**, 391–400.

Eberhardt P., Geiss J., Graf H., Grögler N., Krähenbühl, U., Schwaller H., Schwarzmüller J., and Stettler A. (1970) Trapped solar wind noble gases, exposure age and K/Ar-age in Apollo 11 lunar fine material. *Proc. Apollo 11 Lunar Sci. Conf., Geochim. Cosmochim. Acta* Suppl. 1, Vol. 2, pp. 1037–1070. Pergamon.

Frondel C., Klein C., Jr., Ito J., and Drake J. E. (1970) Mineralogical and chemical studies of Apollo 11 lunar fines and selected rocks. *Proc. Apollo 11 Lunar Sci. Conf., Geochim. Cosmochim. Acta* Suppl. 1, Vol. 1, pp. 445–474. Pergamon.

Gill P. S., Toomey R. E., and Moser H. C. (1967) Reactions of hydrogen and tritium atoms with carbon at 77°K. *Carbon* **5**, 43–66.

Goldstein J. I. and Yakowitz H. (1971) Metallic inclusions and metal particles in the Apollo 12 lunar samples. Second Lunar Science Conference (unpublished proceedings).

Hare P. E., Harada K., and Fox S. W. (1970) Analyses for amino acids in lunar fines. *Proc. Apollo 11 Lunar Sci. Conf., Geochim. Cosmochim. Acta* Suppl. 1, Vol. 2, pp. 1799–1803. Pergamon.

Harris G. M. and Tickner A. W. (1947) Reaction of hydrogen atoms with solid carbon. *Nature* **160**, 871.

Henderson W., Kray W. C., Newman W. A., Reed W. E., Burlingame A. L., Simoneit B. R., and Calvin M. (1971) Study of carbon compounds in Apollo 11 and Apollo 12 returned lunar samples. Second Lunar Science Conference (unpublished proceedings).

Heymann D. and Yaniv A. (1970) Inert gases in the fines from the Sea of Tranquillity. *Proc. Apollo 11 Lunar Sci. Conf., Geochim. Cosmochim. Acta* Suppl. 1, Vol. 2, pp. 1247–1259. Pergamon.

Hintenberger H., Weber H. W., Voshage H., Wänke H., Begemann F., and Wlotzka F. (1970) Concentrations and isotopic abundances of the rare gases, hydrogen and nitrogen in Apollo 11 lunar matter. *Proc. Apollo 11 Lunar Sci. Conf., Geochim. Cosmochim. Acta* Suppl. 1, Vol. 2, pp. 1269–1282. Pergamon.

Hodgson G. W., Bunnenberg E., Halpern B., Peterson E., Kvenvolden K. A., and Ponnamperuma C. (1970) Carbon compounds in lunar fines from Mare Tranquillitatis, II. Search for porphyrins. *Proc. Apollo 11 Lunar Sci. Conf., Geochim. Cosmochim. Acta* Suppl. 1, Vol. 2, pp. 1829–1844. Pergamon.

Jedwab J. and Herbosch A. (1970) Tentative estimation of the contribution of Type 1 carbonaceous meteorites to the lunar soil. *Proc. Apollo 11 Lunar Sci. Conf., Geochim. Cosmochim. Acta* Suppl. 1, Vol. 2, pp. 551–559. Pergamon.

Johnson J. H., Knipe R. H., and Gordan A. S. (1970) Chemical Reactions in a Toepler Pump. *Can. J. Chem.* **48**, 3604–3605.

Kaplan I. R. and Smith J. W. (1970) Concentration and isotopic composition of carbon and sulphur in Apollo 11 lunar samples. *Science* **167**, 541–543.

KOZYREV N. A. (1959) Volcanic activity on the Moon. *Pokrohy Mat. Fys. Astron.* **4,** 704–708 (*Chem. Abstr.* **57**, 6932)

KVENVOLDEN K. A., CHANG S., SMITH J. W., FLORES J., PERING K., SAXINGER C., WOELLER F., KEIL K., BREGER I. A., and PONNAMPERUMA C. (1970) Carbon compounds in lunar fines from Mare Tranquillitatis, I. Search for molecules of biological significance. *Proc. Apollo 11 Lunar Sci. Conf., Geochim. Cosmochim. Acta* Suppl. 1, Vol. 2, pp. 1813–1828. Pergamon.

LAL D., MACDOUGALL D., WILKENING L., and ARRHENIUS G. (1970) Mixing of lunar regolith and cosmic ray spectra: Evidence from particle track studies. *Proc. Apollo 11 Lunar Sci. Conf., Geochim. Cosmochim. Acta* Suppl. 1, Vol. 3, pp. 2295–2303. Pergamon.

LAL D., MURALLI A. V., RAJAN R. S., TAMHANE A. S., LORIN J. C., and PELLAS P. (1968) Techniques for proper revelation and viewing of etch tracks in meteoritic and terrestial minerals. *Earth Planet. Sci. Lett.* **5**, 111–119.

LSPET (Lunar Sample Preliminary Examination Team) (1970) Preliminary examination of lunar samples from Apollo 12. *Science* **167**, 1325–1339.

MEINSCHEIN W. G., JACKSON T. J., MITCHELL J. M., CORDES E., and SHINER V. J. JR. (1970) Search for alkanes of 15–30 carbon atom length in lunar fines. *Proc. Apollo 11 Lunar Sci. Conf., Geochim. Cosmochim. Acta* Suppl. 1, Vol. 2, pp. 1875–1877. Pergamon.

MOORE C. B., GIBSON E. K., LARIMER J. W., LEWIS C. F., and NICHIPORUK W. (1970) Total carbon and nitrogen abundances in Apollo 11 lunar samples and selected achondrites and basalts. *Proc. Apollo 11 Lunar Sci. Conf., Geochim. Cosmochim. Acta* Suppl. 1, Vol. 2, pp. 1375–1382. Pergamon.

MOORE C. B., LEWIS C. F., DELLES F. M., GOOLEY R. C., and GIBSON E. K., JR. (1971) Total carbon and nitrogen abundances in Apollo 12 lunar samples. Second Lunar Science Conference (unpublished proceedings).

MURPHY Sister M. E., MODZELESKI V. E., NAGY B., SCOTT W. M., YOUNG M., DREW C. M., HAMILTON P. B., and UREY H. C. (1970) Analysis of Apollo 11 lunar samples by chromatography and mass spectrometry: Pyrolysis products, hydrocarbons, sulfur, amino acids. *Proc. Apollo 11 Lunar Sci. Conf., Geochim. Cosmochim. Acta* Suppl. 1, Vol. 2, pp. 1879–1890. Pergamon.

MURPHY R. C., PRETI G., NAFISSI V M. M., and BIEMANN K. (1970) Search for organic material in lunar fines by mass spectrometry. *Proc. Apollo 11 Lunar Sci. Conf., Geochim. Cosmochim. Acta* Suppl. 1, Vol. 2, pp. 1891–1900. Pergamon.

ORÓ J., UPDEGROVE W. S., GIBERT J., MCREYNOLDS J., GIL-AV E., IBANEZ J., ZLATKIS A., FLORY D. A., LEVY R. L., and WOLF C. J. (1970) Organogenic elements and compounds in types C & D lunar samples from Apollo 11. *Proc. Apollo 11 Lunar Sci. Conf., Geochim. Cosmochim. Acta* Suppl. 1, Vol. 2, pp. 1901–1920. Pergamon.

ORÓ J., FLORY D. A., GIBERT J. W., MCREYNOLDS J., LICHTENSTEIN H. A., and WIKSTROM S. (1971) Abundances and distribution of organogenic elements and compounds in Apollo 12 lunar samples. Second Lunar Science Conference (unpublished proceedings).

PEPIN R. O., NYQUIST L. E., PHINNEY D., and BLACK D. C. (1970) Rare gases in Apollo 11 lunar material *Proc. Apollo 11 Lunar Sci. Conf., Geochim. Cosmochim. Acta* Suppl. 1, Vol. 2, pp. 1435–1454. Pergamon.

PEPIN R. O. (1971) Personal communication.

REID A. M., MEYER C., JR., HARMON R. S., and BRETT R. (1970) Metal grains in Apollo 12 igneous rocks. *Earth Planet. Sci. Lett.* **9**, 1–5.

RHO J. H., BAUMAN A. J., YEN T. F., and BONNER J. (1970) Fluorometric examination of the returned lunar fines from Apollo 11. *Proc. Apollo 11 Lunar Sci. Conf., Geochim. Cosmochim. Acta* Suppl. 1, Vol. 2, pp. 1929–1932. Pergamon.

SHAHIN M. M. (1962) Reaction of elementary carbon and hydrogen in high frequency discharge. *Nature* **195**, 992.

ZELLER E. J. and DRESCHOFF G. (1968) Chemical reactions resulting from heavy particle bombardment of diamond. *Z. Naturforsch.* **23a**, 953–954.

Proceedings of the Second Lunar Science Conference, Volume 2, pp. 1865–1874.
The M.I.T. Press, 1971.

Lunar pigments: Porphyrin-like compounds from an Apollo 12 sample

GORDON W. HODGSON
University of Calgary, Calgary 44, Canada

EDWARD BUNNENBERG and B. HALPERN
Stanford University, Stanford, California 94305

and

ETTA PETERSON, KEITH A. KVENVOLDEN, and CYRIL PONNAMPERUMA
NASA Ames Research Center, Moffett Field, California 94035

(*Received 19 February 1971; accepted in revised form 18 March 1971*).

Abstract—Porphyrin-like pigments were found in the analysis of a sample of lunar soil (12023) collected on the Apollo 12 mission at a sampling point well removed from the lunar landing vehicle. Fluorescence detection was based on an excitation band at 390 nm which had maximum intensity at an emission wavelength of 625 nm. Abundance of the pigments was estimated to be about 5×10^{-5} μg/g. This sample appeared to be free of rocket exhaust products, whereas a sample of soil (12001) taken from near the lunar module showed considerable amounts of fluorescent pigments attributed to the exhaust of the rocket engine.

INTRODUCTION

THE PRESENCE OF carbonaceous matter and specific organic compounds in meteorites (HAYES, 1967), in comets (WURM, 1963; SWINGS and HASER, 1956), and interstellar space (SNYDER *et al.*, 1969; CHEUNG *et al.*, 1968; DONN, 1971; JOHNSON, 1967) indicates that organic matter is generated extraterrestrially (ANDERS, 1968). Lunar fines were examined for organic compounds (KVENVOLDEN *et al.*, 1970b; ABELL *et al.*, 1970; ORÓ *et al.*, 1970) and evidence was obtained for the presence of porphyrins in the soils returned by Apollo 11 from the Sea of Tranquillity (HODGSON *et al.*, 1970). These pigments were thought, however, to be due in part to a novel synthesis of porphyrin-like compounds from unsymmetrical dimethylhydrazine taking place in the firing of the retrorocket of the lunar module (LM). In that investigation the question as to whether porphyrins were indigenous to the lunar surface was not answered. The present study afforded a further opportunity to attempt to resolve the question through the availability of two samples of lunar fines from widely spaced locations visited during the extravehicular activity of the Apollo 12 mission. One sample (12001,27) was collected near the LM, and was designated in the present study as the *near* sample; the other was taken at a location most remote from the landing point, 440 m from the LM and 500 m from the site of Surveyor 3. It was a portion of sample 12023 which had been packaged in the lunar environment sample container (LESC). This sample was taken from a trench 20 cm below the lunar surface. It was designated in the present study as the *remote* sample. Thus the likelihood of contamination from the retrorocket was high in the case of the *near* sample and low for the *remote* sample. The *near* sample exhibited fluorescent pigments characteristic of rocket exhaust similar to the findings in Apollo 11, but was apparently free of the porphyrin-like pigments present in the Apollo 11 sample; the *remote* sample was free of rocket contaminants and showed the presence of porphyrin-like compounds.

ANALYTICAL METHODS

The methods used in the present study were substantially the same as those used in the Apollo 11 search for porphyrins (HODGSON *et al.*, 1970). The overall organization of the analysis of the lunar samples was the same as before. The description and weights of samples used are listed in Table 1 along with the blank samples included in the study: Pueblito de Allende meteorite, basaltic lava from Hawaii, and a tektite from Thailand. The lunar samples were extracted directly in benzene–methanol (9:1), and the organic extracts were subjected to spectral analysis involving fluorescence, absorption, and magnetic circular dichroism (MCD) spectrometry as previously described by HODGSON *et al.* (1970). This was followed with analytical demetallation for porphyrins (HODGSON *et al.*, 1969), after which free-base recovery and remetallation procedures were undertaken. Improvements in the methods centered on a sharp increase in sensitivity developed in the present study for the MCD examination and on more extensive remetallation of the free-base pigments. Attempts were also made to determine if any fluorescent organic pigments were embedded within the mineral grains of the lunar fines.

Blank analyses involving solvents and reagents were consistently negative. Sample sizes for the blank determinations ranged from 5 to 10 g, and results for all of these were below detection limits. Detection limits for porphyrins were redetermined from those reported earlier (HODGSON *et al.*, 1970) by using deuteroporphyrin IX dimethyl ester in free-base form in 6 N HCl in such a manner that the results would apply directly to the lunar analysis samples. Excitation spectral responses were clearly discernible above background noise for 0.05 ng of free-base porphyrins, and considering all the analytical factors involved in analyses of lunar samples, the detection limits for such samples were taken to be 0.1 ± 0.05 ng. Work with the authentic porphyrin was done after all of the lunar and blank analyses had been completed to avoid possible contamination of the analytical system. Detection limits were determined for the improved MCD method using the magnesium complex of deuteroporphyrin under the same conditions as were used for the analysis of the lunar extracts. The MCD detection limit for metalloporphyrins in general was determined to be 3 ng.

RESULTS AND DISCUSSION

Near lunar sample

Direct spectrofluorometry of the extract from the sample of lunar fines taken from near the LM showed nearly the same type of response as that for the Apollo 11 samples, namely a strong broad excitation band at 450 nm for emission at 660 nm. Partitioning between ether and hydrochloric acid resulted in all of the fluorescing pigments remaining in the organic phase and, accordingly, there was no indication of free-base porphyrins at this stage. A weak excitation band observed at about 390 nm

Table 1. Samples analyzed

No.	Description	Identification	Wt. Analyzed, g.
1.	Lunar, Apollo 12—close-in to lunar module	12001,27	5.0147
2.	Lunar, Apollo 12—remote from lunar module	ARC 12023.03	10.111
3.	Pueblito de Allende—interior of large fragment	NASA 31-4A	10.00
4.	Basalt—Kilauea eruption, Hawaii, Aloi-Alae, February, 1970	NASA 040270	10.40
5.	Tektite—N.E. Thailand	NASA 040770	5.05

in the Apollo 11 analyses was not clearly detected in the present determination. Neither absorption nor MCD spectrometry showed the presence of porphyrins. The 450/660 nm fluorescence (excitation at 450 nm for emission at 660 nm) was probably related to rocket exhaust as suggested in the Apollo 11 analyses (HODGSON *et al.*, 1970).

Methanesulfonic acid (MSA) was used to demetallate any porphyrin complexes present in the extract of the *near* sample. Porphyrins did not appear above the background fluorescence, in contrast to the results for Apollo 11 in which porphyrin-like pigments were observed to demetallate at this stage with a threshold of about 25% MSA in ether. Regular recovery procedures were used to search for free-base pigments in the products of the MSA reaction, but little if any indication of such pigments was obtained.

Remote lunar sample

Direct spectrofluorometry of the organic matter extracted from the *remote* sample showed no discernible excitation in the region from 350 to 500 nm for emission in the red end of the visible spectrum. There was no indication of free-base porphyrins, nor of fluorescing pigments associated with rocket exhaust. Figure 1 shows the marked

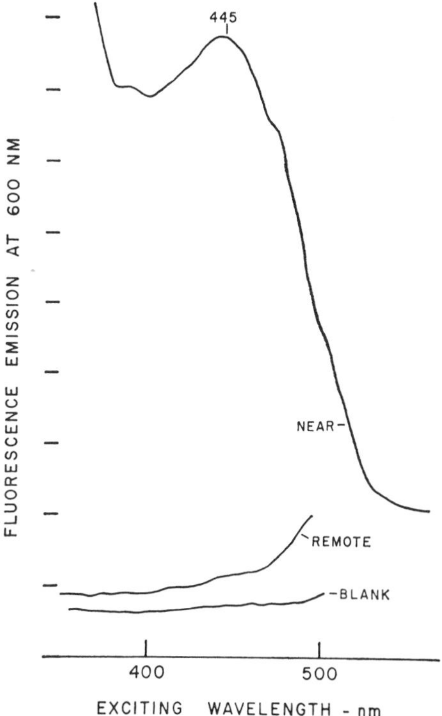

Fig. 1. Direct fluorescence examination of lunar extracts from Apollo 12: *near* refers to sample of lunar soil collected from near the landing module, with 445-nm fluorescence due to exhaust products of descent engine; *remote*, sample from trench dug about 500 meters away.

difference between the extracts of this sample and those of the sample taken from near the lunar module. Absorption spectrophotometry showed nothing. MCD spectrometry, after data aquisition and reduction by computer, did not show bands that could be attributed to metal complexes of porphyrins.

The very low background of fluorescence in the extract of the *remote* sample made it possible to detect the onset of a weak excitation band in the MSA demetallation analysis. This band emerged at about 420–425 nm for emission in the 600–700 nm range. The threshold for the reaction was about 50% MSA, and the intensity of the band increased moderately during the latter part of the analysis. Recovery of presumed free-base pigments was accomplished in the usual manner after the concentration of MSA reached 70%. Free-base recovery showed no specific fluorescence in ether solution, probably because of extensive quenching commonly observed during this stage of the analysis. Extraction of the ether solution with 6 N HCl revealed a fluorescent pigment exciting at 390 nm for emission in the usual porphyrin range.

Analyses based on fluorescence are exceedingly sensitive, and extreme care must be exercised to avoid (a) contaminating the sample, and (b) misinterpreting the data that are obtained. The contamination aspect in the present study was controlled by elaborate precautions taken during the conduct of the analyses which were carried out in a "clean room" using solvents and glassware that had been rigorously cleaned and protected from subsequent contamination. Numerous blank analyses were carried out to test various steps in the process. No indications of any contamination were observed.

In addition, to check the possibility that the samples had picked up contaminants in the lunar receiving laboratory at Houston, a supply of ignited sand from the same supply as used for analytical blanks in the Apollo 11 determinations for porphyrins was exposed in the Houston transfer cabinets in such a manner as to reflect any relevant foreign matter encountered by the Apollo 12 samples. As shown in Fig. 2, there was no indication of porphyrin-like material in the Houston blank. Other contaminants were clearly present, as noted by other workers also, but there was no evidence whatever for the presence of contaminating porphyrin-like pigments.

The other factor, that of misinterpreting spectral data obtained at very low levels, was resolved in a similar manner. In fact, the several laboratory blanks as well as the Houston blank gave direct empirical answers to the question as to whether the signals observed were real or instrumental artifacts as suggested by RHO *et al.* (1971). The simple fact that the lunar sample gave a positive signal while all the blanks gave negative signals augurs strongly for confidence in the data. To strengthen the interpretation of the data, however, a novel cross-plotting approach was developed to demonstrate the direct relationship between fluorescence emission in the red and specific fluorescence excitation in the blue end of the spectrum for the pigments recovered in the lunar analysis. Such a relationship is demanded by porphyrins, and Fig. 3 illustrates the relationship as commonly observed for excitation-emission curves when the concentrations are high enough to observe both maxima directly. In real situations, such as in the case of geochemical samples, background fluorescence is commonly troublesome. Under extreme conditions the emission band may be completely obscured by the background while the excitation bands remain discernible. The specificity of the

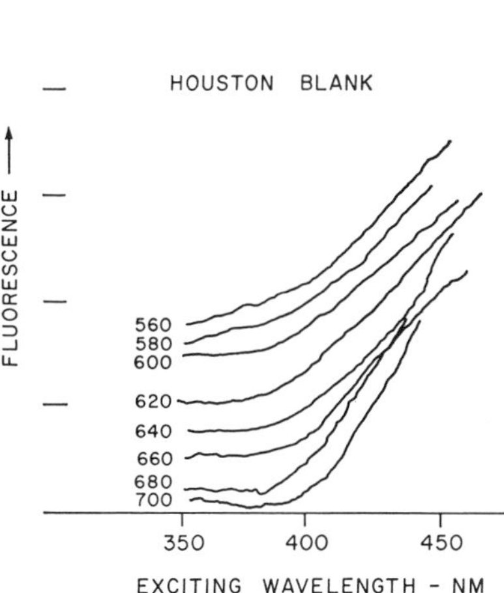

Fig. 2. Excitation scans of fluorescence emission for sand exposed in transfer cabinets in lunar receiving laboratory in Houston. Sand was extracted and demetallated in the same manner as lunar samples; instrumental parameters were same as for lunar samples. There was no indication of porphyrin excitation.

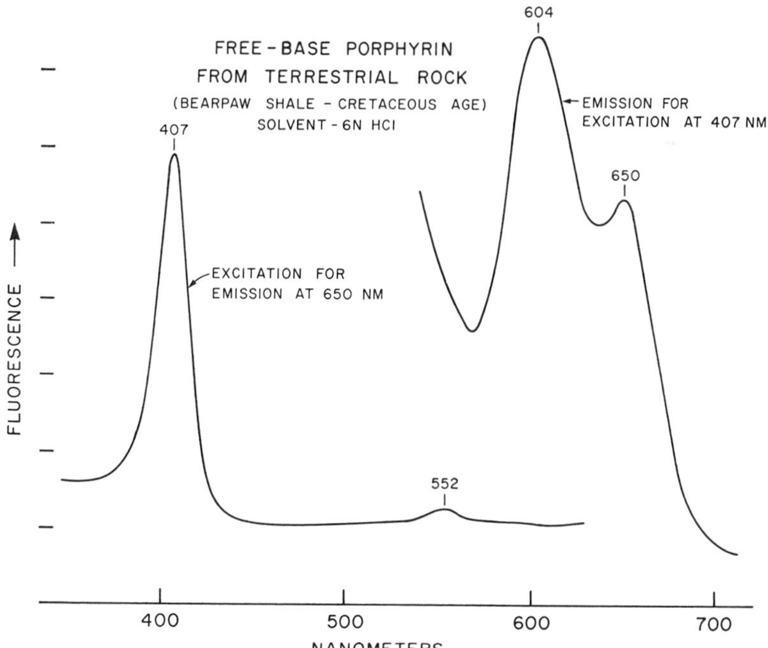

Fig. 3. Typical fluorescence of free-base porphyrin from terrestrial shale. Sample size was 1200 ng; instrument parameters set at very insensitive level.

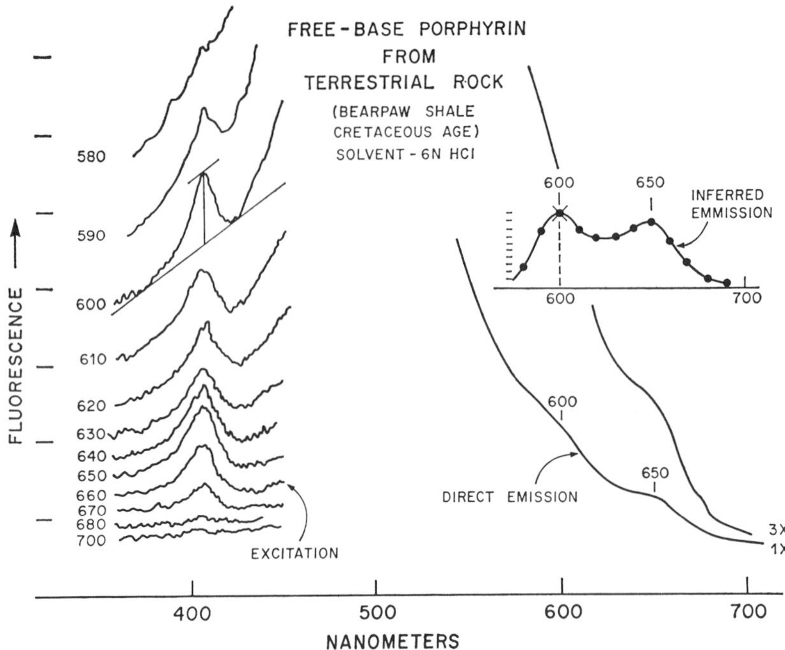

Fig. 4. Fluorescence of 1 ng free-base porphyrin from terrestrial shale (Bearpaw shale, as in Fig. 3) with cross-plotting to illustrate capability for suppressing fluorescent background.

excitation bands under these circumstances is, however, open to question, and they can become confused with instrumental artifacts if care is not exercised. In the present study a novel procedure was adopted to re-establish the specificity. The principle was to note that a porphyrin under these conditions exhibits excitation bands only for a limited region of the emission spectrum. It is possible to generate a family of excitation curves with emission wavelength as the parameter, with the result that an authentic porphyrin shows positive excitation features only for those curves whose emission wavelength is in the region commonly observed for porphyrins, e.g., from about 580 to 700 nm. The final step in the evaluation is to plot the intensity of excitation as a function of emission wavelength and in this manner construct an emission spectrum corrected for all background emission. Figure 4 illustrates this approach for porphyrins recovered in nanogram amounts from a terrestrial shale (sample BH-293 described by HODGSON *et al.*, 1968), in good agreement with the spectra obtained for the same sample in microgram amounts previously shown in Fig. 3. The agreement between the cross-plotted data and the direct-scan data shows clearly how this method can be applied to porphyrins at very low levels in samples with appreciable background fluorescence.

The cross-plotting approach was used for evaluating the excitation signal exhibited by the pigments resulting from the demetallation analysis of the *remote* Apollo 12 sample. The family of excitation curves and the constructed emission curve for these

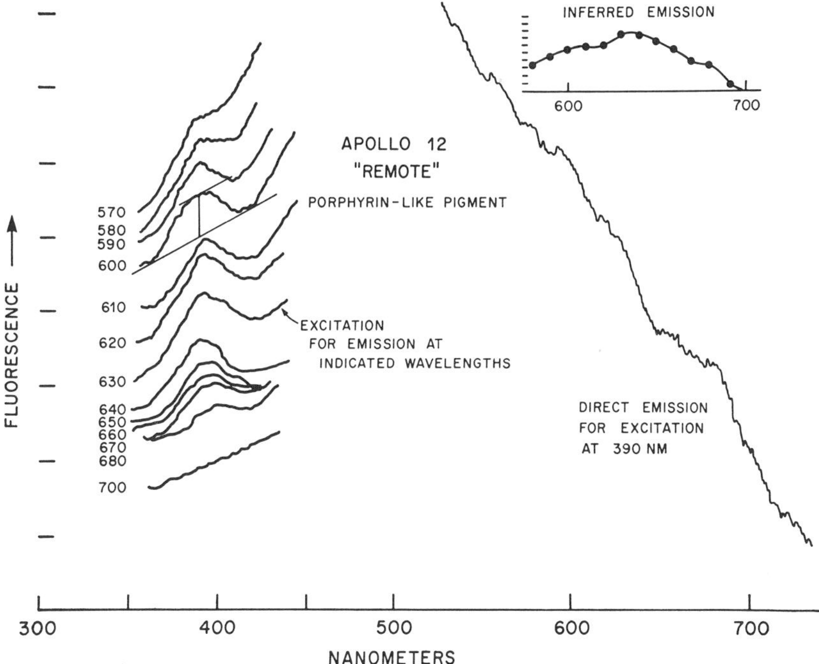

Fig. 5. Fluorescence of porphyrin-like pigments from analysis of Apollo 12 *remote* sample. Note the similarity with naturally occurring porphyrins of terrestrial shale (Fig. 4).

pigments are shown in Fig. 5, and the resemblance to those of the authentic terrestrial porphyrin in Fig. 4 is very close. It seems difficult to attribute such data to instrumental artifacts; on the contrary, it is reasonable to designate the lunar pigments as porphyrins. Until other analyses are available, however, the lunar pigments will be referred to only as porphyrin-like.

Figure 6 shows that the Apollo 12 porphyrin-like compounds are distinct from the rocket-generated pigments of Apollo 11 and have a family resemblance with authentic deuteroporphyrins.

The abundance of the porphyrin-like pigments of the *remote* Apollo 12 sample was estimated on an assumption that the signal was due to a compound similar to deuteroporphyrin IX dimentyl ester, a readily available authentic porphyrin. On this basis, the porphyrin-like material recovered from the *remote* sample was estimated to be about 0.5 ng, corresponding to 5×10^{-5} μg/g of lunar soil. The amount detected was about 5 times greater than the detection limit of the method.

The pigments recovered from the *remote* sample were subjected to remetallation procedures (contact with metal acetates in glacial acetic acid at 100°C for about 10 min) involving copper and zinc to give nonfluorescing and fluorescing complexes, respectively, after the manner of authentic porphyrins. Although the amounts of

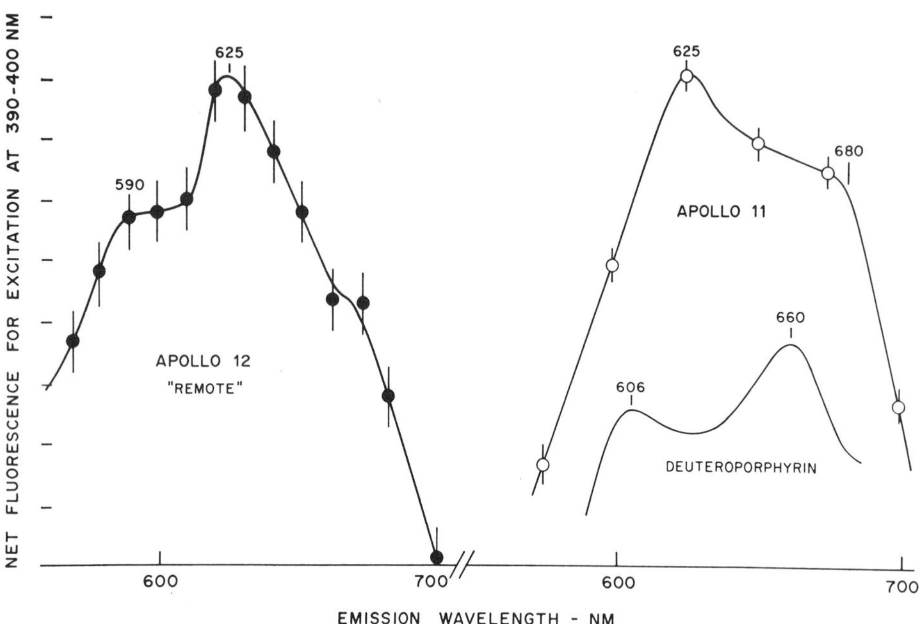

Fig. 6. Comparison of porphyrin-like pigments of Apollo 12 and Apollo 11 with authentic deuteroporphyrin showing common fluorescence emission in the range 580 to 680 nm but with significant differences between individual curves.

presumed free-base porphyrins available for the test were very small, complete quenching was observed for the copper complex, and only minor quenching for the zinc complex.

Table 2 summarizes the several fluorescent pigments encountered in the search for porphyrins in lunar samples and in related samples from the Apollo 11 and Apollo 12 missions. Four classes of pigments were detected: A, B, C, and D. Pigment A comprised organic substances fluorescing at 450/600 nm. They were present in *near* Apollo 11 soil and rocket exhaust, and also in the *near* sample of Apollo 12. These compounds were not porphyrins. The other three groups of pigments resembled porphyrins. B behaved like a free-base porphyrin, and C was probably the same pigment complexed with metals. Both were present in the several samples analyzed from the Apollo 11 mission. Pigment D (detected in Apollo 12 *remote*) appeared also to be a metal complex of a porphyrin, but it was distinguishable from C since (1) the MSA demetallation threshold was 50% contrasted with 25% for C, and (2) that the pairs of emission bands were not only different in position but different in relative intensity.

In order to determine whether the porphyrin-like pigments are surficial or embedded, the *near* sample was exhaustively extracted in the as-received state and then re-extracted after crushing the mineral grains with a mortar and pestle. In addition, the fines were etched with hydrofluoric acid. Neither approach appeared to release additional pigments. Crushing of the *remote* sample also failed to release more fluorescent pigments. It is concluded that the fluorescing organic pigments for the

Table 2. Summary of presence of fluorescent pigments in lunar samples from Apollo 11* and Apollo 12 and related blanks

	Fluorescent Pigments			
	A	B	C	D
Excitation-emission	450/600 nm	390/600 nm	390/625 nm	390/625 nm
Demetallation threshold	0% MSA	0% MSA	25% MSA	50% MSA
Pigment characteristics	not porphyrin	porphyrin-like	porphyrin-like	porphyrin-like
Solvent partition test	organic	free-base	complexed	complexed
Apollo 11 10,086 (i) fines	PRESENT†	PRESENT	PRESENT	P/M
(ii) fines	PRESENT	PRESENT	PRESENT	P/M
Apollo 11 Trap A, exhaust	PRESENT	PRESENT	PRESENT	P/M
Trap E, exhaust	PRESENT	PRESENT	PRESENT	P/M
Apollo 12 Basalt blank	BDL	BDL	BDL	BDL
Apollo 12 Lunar dust, close-in	PRESENT	DL	DL	P/M
Apollo 12 Tektite blank	BDL	BDL	BDL	BDL
Apollo 12 Lunar dust, remote	BDL	BDL	BDL	PRESENT
Apollo 12 Allende blank	BDL	BDL	BDL	BDL

* Apollo 11 data summarized from Hodgson *et al.* (1970).
† Notes: PRESENT—pigment is present at levels significantly above detection limit.
P/M —pigment may be present but masked by other pigments.
BDL —below detection limits.
DL —pigment possibly present, but only near detection levels—detection very uncertain.

most part are not contained within impermeable mineral grains of the lunar surface. Thus, they are present either (1) on the surface of the mineral grains, or (2) in porous particles readily penetrated by organic solvents. In either event, the organic matter appears to have been emplaced in a manner not directly related to the formation of the grains themselves. Condensation of organic matter of extraterrestrial origin, lunar or extralunar, has possibly taken place; perhaps the pigments are present on foreign particles incorporated into the lunar surface as micrometeorites (Keays *et al.*, 1970) or as interplanetary dust. While the organic content of extraterrestrial dust grains is not known, porphyrin-like pigments have been observed in carbonaceous chondrites (Hodgson and Baker, 1969) and the generation of organic matter in interstellar space in general is now clearly indicated (Snyder *et al.*, 1969; Kvenvolden *et al.*, 1970a).

References

Abell P. I., Draffan C. H., Eglinton G., Hayes J. M., Maxwell J. R., and Pillinger C. T. (1970) Organic analysis of the returned Apollo 11 lunar sample. *Proc. Apollo 11 Lunar Sci. Conf., Geochim. Cosmochim. Acta* Suppl. 1, Vol. 2, pp. 1757–1773. Pergamon.

Anders E. (1968) Chemical processes in the early solar system, as inferred from meteorites. *Acct. Chem. Research* **1**, 289–298.

Cheung A. C., Rank D. M., Townes C. H., Thomson D. D., and Welch W. J. (1968) Detection of NH_3 molecules in the interstellar medium by their microwave emission. *Phys. Rev. Lett.* **21**, 1701–1705.

Donn B. (1971) Organic molecules in space. In *Exobiology* (editor C. Ponnamperuma) (in press).

Hayes J. M. (1967) Organic constituents of meteorites—a review. *Geochim. Cosmochim. Acta* **31**, 1395–1440.

Hodgson G. W., and Baker B. L. (1969) Porphyrins in meteorites: Metal complexes in Orgueil, Murray, Cold Bokkeveld and Mokoia carbonaceous chondrites. *Geochim. Cosmochim. Acta* **33**, 943–958.

HODGSON G. W., BUNNENBERG E., HALPERN B., PETERSON E., KVENVOLDEN K. A., and PONNAMPERUMA C. (1970) Carbon compounds in lunar fines from Mare Tranquillitatis—II. Search for porphyrins. *Proc. Apollo 11 Lunar Conf., Geochim. Cosmochim. Acta.* Suppl. 1, Vol. 2, pp. 1829–1844. Pergamon.

HODGSON G. W., PETERSON E., BAKER B. L. (1969) Trace porphyrin complexes: Fluorescence detection by demetallation with methanesulfonic acid. *Mikrochim. Acta*, 805–814.

HODGSON G. W., HITCHON B., TAGUCHI K., BAKER B. L., and PEAKE E. (1968) Geochemistry of porphyrins, chlorins, and polycyclic aromatics in soils, sediments and sedimentary rocks. *Geochim. Cosmochim. Acta* **32**, 737–772.

JOHNSON F. M. (1967) Diffuse interstellar lines and chemical characterization of interstellar dust. In *Interstellar Grains* (editors J. M. Greenberg and T. P. Roark), NASA SP–140, 269 pp., pp. 229–240.

KEAYS R. R., GANAPATHY R., LAUL J. C., ANDERS E., HERZOG G. F., and JEFFRY P. M. (1970) Trace elements and radioactivity in lunar rocks: Implications for meteorite infall, solar-wind flux, and formation conditions of Moon. *Science* **167**, 490–493.

KVENVOLDEN K., LAWLESS J., PERING K., PETTERSON E., FLORES J., KAPLAN I. R., MOORE C., and PONNAMPERUMA C. (1970a) Evidence for extraterrestrial amino acids and hydrocarbons in the Murchison meteorite. *Nature* **228**, 923–926.

KVENVOLDEN K. A., CHANG S., SMITH J. W., FLORES J., PERING K., SAXINGER C., WOEHLER F., KEIL K., BREGER I. A., and PONNAMPERUMA C. (1970b) Carbon compounds from Mare Tranquillitatis—I. Search for molecules of biological significance. *Proc. Apollo 11 Lunar Sci. Conf., Geochim. Cosmochim. Acta* Suppl. 1, Vol. 2, pp. 1813–1828. Pergamon.

ORÓ J., UPDEGROVE W. S., GILBERT J., MCREYNOLDS J., GIL-AV E., IBANEZ J., ZLATKIS A., FLORY D. A., LEVY R. L., and WOLF C. J. (1970) Organogenic elements and compounds in type C and D lunar samples from Apollo 11. *Proc. Apollo 11 Lunar Sci. Conf., Geochim. Cosmochim. Acta* Suppl. 1, Vol. 2, pp. 1901–1920. Pergamon.

RHO J. H., BAUMAN A. J., BONNER J. F., and YEN T. F. (1971) Absence of porphyrins in Apollo 12 lunar samples. Second Lunar Science Conference (unpublished proceedings).

SNYDER L. E., BUHL D., ZUCKERMAN B., PALMER P. (1969) Microwave detection of interstellar formaldehyde. *Phys. Rev. Lett.* **22**, 679–681.

SWINGS P., and HASER L. (1956) Atlas of representative cometary spectra, University of Liège Astrophysical Institute, Louvain.

WURM K. (1963) The physics of comets. In *The Moon, Meteorites and Comets* (editors B. M. Middlehurst and G. P. Kuiper), Vol. IV, The Solar System, pp. 573–617, Chicago University Press.

Proceedings of the Second Lunar Science Conference, Volume 2, pp. 1875–1877.
The M.I.T. Press, 1971.

Absence of porphyrins in an Apollo 12 lunar surface sample

JOON H. RHO and A. J. BAUMAN
Jet Propulsion Laboratory, California Institute of Technology,
Pasadena, California 91103
TEH FU YEN
University of Southern California, Los Angeles, California 90033
and
JAMES BONNER
California Institute of Technology, Pasadena, California 91109

(*Received 24 February 1971; accepted in revised form 5 April 1971*)

Abstract—No porphyrins were found in 15 g of the Apollo 12 lunar fines from the Ocean of Storms under the conditions in which porphyrins would have been detected had they been present in amounts as small as 10^{-14} mole. An instrumental artifact at 600, 630, 680 nm which resembled porphyrin peaks was observed in the control sample in which no porphyrins were present. This was produced by the interaction of grating anomalies with light scattering materials, and was associated with a definite plane of polarization. When the data from the grating monochromator were corrected, or when prism monochromator data were used, we found no fluorescence attributable to the presence of porphyrins.

We have found, however, in the organic phase of the lunar sample extract, the presence of species which fluoresced at 365 to 380 nm when activated at 300 nm. The corresponding aqueous phase of the sample extract also contained a material which exhibits a fluorescence maximum at 415 nm. All fluorescence attributable to organic materials in the lunar sample was also found in the LRL sand blank in equivalent amounts.

INTRODUCTION

THE PRESENCE of porphyrins in lunar samples would suggest that life might once have existed on the Moon. We have, therefore, carefully examined Apollo 12 lunar fines for the presence of porphyrins by the most sensitive fluorometric technique available. The lunar returned sample (LRS) that we studied was collected 440 m from the lunar module in a Lunar Environment Sample Container (LESC) at the near rim of Sharp Crater (MSC–01512, Lunar Sample Information Catalog for Apollo 12, p. 33; January 13, 1970).

ANALYTICAL METHODS

A 15 g sample (LRS No. 12023) was Soxhlet-extracted for 24 h sequentially with 200 ml of benzene-methanol (9:1::v:v) and with methanol, then each extract was examined separately for fluorescent residues as with the Apollo 11 sample (RHO *et al.*, 1970).

The Soxhlet extract of a sample of 200-mesh optical quartz identical in weight to that of the LRS was used as a solvent blank. A 10-g sand blank which had been exposed to the environment of Vacuum Chamber F-201 at the MSC was similarly extracted and both blanks extracted were examined fluorometrically.

The solvents used were of fluorometric grade (Harleco) and all glassware was cleaned with hot aqua regia then rinsed with triple distilled water and fluorometric grade methanol just before use.

All sample operations were carried out in a dry box under a positive pressure of high purity argon.

In each case, the extracts were taken to dryness and the residues were redissolved in a benzene-water two-phase partition system in order to remove fluorescence-quenching impurities such as ferric ion from the benzene phase. As metallated porphyrins are frequently nonfluorescent, the benzene portion was first treated with methanesulfonic acid (MSA) to demetallate porphyrins present and was then neutralized with saturated sodium acetate and repartitioned with aqueous phase. This aqueous phase was discarded and any demetallated free base porphyrins in the benzene fraction were extracted by means of a 6 *N* HCl wash. In this system one would find free base porphyrins in the first benzene phase and demetallated porphyrins resulting from MSA treatment in the HCl wash.

FLUOROMETRIC EXAMINATION OF SAMPLE EXTRACTS

Fluorometric examination of the LRS benzene phase showed the presence of species which fluoresced at 365 to 380 mm when activated at 300 nm, as shown in Fig. 1. This peak was found in equivalent intensity in the sand blank benzene phase. The corresponding aqueous phases for both the LRS and the sand blank showed

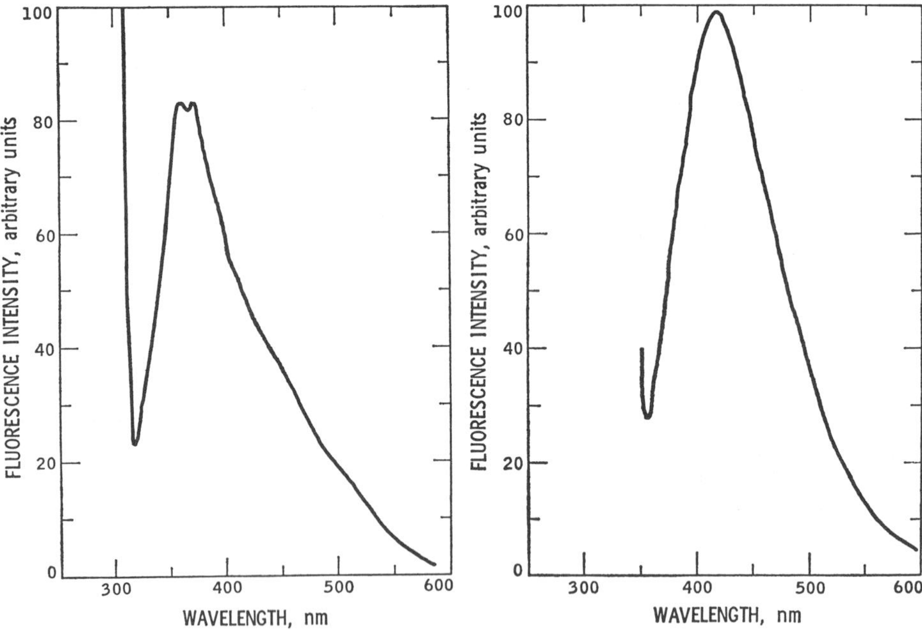

Fig. 1. Fluorescence spectrum of benzene-methanol extract of Apollo 12 lunar sample. The extract in benzene was treated with methane-sulfonic acid in order to demetallate porphyrins and then partitioned with water. The spectrum of the benzene phase of the extract was obtained at an excitation of 300 nm.

Fig. 2. Fluorescence spectrum of methanol extract of Apollo 12 lunar sample. The sample was excited at 330 nm.

identical fluorescence behavior at 415 nm when activated at 330 to 340 nm, as shown in Fig. 2.

We also checked for the presence of anomalous artifact peaks which may often be found in the grating fluorescence spectra of samples which contain finely particulate solids (KING and HERCULES, 1963). We compared the spectral transmission properties of monochromators with 600 grooves/mm and 1200 grooves/mm with those of a prism instrument. An Aminco spectrophotofluorometer with a 600 grooves/mm grating, a Turner Spectro 210 fluorometer with a 1200 grooves/mm grating, and a Perkin–Elmer prism fluorometer were used. In each case we were chiefly interested in the porphyrin fluorescence region.

As we reported elsewhere (RHO, 1971) the anomalies are associated with polarization effects and they appear in a definite wavelength region which depends on the optical properties of individual grating monochromators. The Turner Spectro 210 fluorometer, for example, showed several anomalous spectral regions in which the diffracted energy varied sharply relative to that at other wavelengths.

Grating anomalies of this kind have been described by LOEWEN (1970) and by PALMER (1952). This type of anomaly distorts the spectrum in general and also causes the appearance of artifact peaks.

Artifacts which resemble porphyrin peaks can be produced by the interaction of grating anomalies with almost any fluorescing system containing light scattering material. They are associated with a definite plane of polarization and although this phenomenon is not well known, it appears to be quite general with grating instruments. An understanding of the grating anomaly phenomenon is important because the artifact peaks produced can be mistakenly assumed as evidence of trace organic compounds in any samples containing light-scattering materials.

CONCLUSIONS

When the data obtained with the grating monochromator for the Apollo 12 LRS extract were corrected, or when proper prism monochromator data were used, we found no fluorescence attributable to the presence of porphyrins. Porphyrins, whether indigenous or caused by contamination, would have been detected had they been present in the 15-g LRS sample to the extent of 10^{-14} mole. Indeed, all fluorescence attributable to organic materials in the LRS was also found in the LRL sand blank in equivalent amounts.

Acknowledgments—We thank John R. Thompson for technical assistance. This investigation was carried out at the Jet Propulsion Laboratory, California Institute of Technology, under NASA Contract NAS 7–100.

REFERENCES

KING R. M. and HERCULES D. M. (1963) Correction for anomalous fluorescence peaks caused by grating transmission characteristics. *Anal. Chem.* **35**, 1099–1100.

LOEWEN E. G. (1970) Anomalous energy distribution. In *Diffraction Grating Handbook* (editor E. G. Loewen), pp. 34–36. Bausch & Lomb, Inc.

PALMER C. H., JR. (1952) Parallel diffraction grating anomalies. *J. Opt. Soc. Amer.* **42**, 269–276.

RHO J. H. (1971) Grating anomalies on porphyrin spectra. *Geochim. Cosmochim. Acta* **35**, 743–748.

RHO J. H., BAUMAN A. J., YEN T. F., and BONNER, J. (1970) Fluorometric examination of a lunar sample. *Science* **167**, 754–755.

Proceedings of the Second Lunar Science Conference, Volume 2, pp. 1879–1889.
The M.I.T. Press, 1971.

The search for organic compounds in various Apollo 12 samples by mass spectrometry

G. PRETI, R. C. MURPHY, and K. BIEMANN
Department of Chemistry, Massachusetts Institute of Technology,
Cambridge, Massachusetts 02139

(Received 22 February 1971; accepted in revised form 30 March 1971)

Abstract—Five different Apollo 12 samples were analyzed for organic compounds. Two types of experiments were performed: direct heating in the ion source of a high resolution mass spectrometer up to temperatures of 900°C; and pyrolysis-gas chromatography-low resolution mass spectrometry. The results of the experiments using high resolution mass spectrometry indicated lunar organic material at the ppb level of concentration in all samples. The pyrolysis studies indicated a much higher level of concentration but this result is presently thought to be due to thermal synthesis in the course of these experiments.

INTRODUCTION

THE VARIOUS ORGANIC analyses of the Apollo 11 lunar surface samples presented a fairly consistent picture. Even though a complete carbon balance could not be achieved for the 100–200 ppm of carbon in the samples, most of the carbon could be accounted for as inorganic carbon or as oxides of carbon released by heat treatment of the sample (BURLINGAME *et al.*, 1970; CHANG *et al.*, 1970).

The most predominant indigenous organic compounds were the simple hydrocarbon gases such as methane and ethane; however, at most these hydrocarbons represented only a few ppm of carbon (ABELL *et al.*, 1970). With general survey techniques, more complex compounds such as benzene and other aromatic hydrocarbons (at concentrations up to 1 ppm) evolved during heating the lunar material to 400° or 700°C (MURPHY, SR. M. E. *et al.*, 1970; MURPHY, R. C. *et al.*, 1970; ORÓ *et al.*, 1970). With very sensitive, specific techniques, amino acids were reported after acid hydrolysis of the aqueous extract of lunar fines at 50 ppb (HARE *et al.*, 1970; MURPHY, SR. M. E. *et al.*, 1970). The unambiguous identification of these as indigenous organic compounds was complicated by terrestrial contamination as well as the possibility that they were artifacts produced by the reactions of simpler molecules under the seemingly innocuous treatments.

Five different specimens from Apollo 12 were obtained for analysis in this laboratory. The samples provided an opportunity to investigate any differences in the organic compounds, albeit very low in concentration, in relation to their original location at the Apollo 12 site. The following were analyzed: a sample from the surface near the lunar module and supposedly exposed to the LEM rocket (12001,22); a sample sealed in the Lunar Environment Sample Container (12023,11); three samples of the long core tube (10, 21, and 30 cm depths—12028, 164, 171, and 178, respectively); and a light gray material (suspected to be of volcanic origin) from the Bench crater (12032,6).

Rather than concentrating on a quantitative analysis of the simpler forms of carbon—CO, CO_2, CH_4 (an area which several others are exploring)—our investigation of the lunar samples was aimed at the detection of more complex organic species. Our particular interest lay in the detection of compounds containing heteroatoms such as nitrogen, oxygen, sulfur, and phosphorous, because such molecules are important in living systems on earth. The use of high resolution mass spectrometry in such an investigation allowed not only the detection of organic molecules, but also the assignment of elemental compositions to the various ions. In addition to the high-resolution mass spectrometry experiment, a pyrolysis-gas chromatography-mass spectrometry experiment was devised. This latter experiment used a very sensitive low resolution mass spectrometer to record the mass spectra of pure compounds produced upon rapid heating of the lunar sample and separated by the gas chromatograph. In such an experiment, mass spectra are obtained that are more easily interpretable in terms of molecular structure; however, no information concerning the elemental composition is obtained. Use of both methods of analysis on the same lunar sample can provide a means to correlate the elemental composition data from the high resolution experiment with the mass spectra of the individual compounds obtained from the pyrolysis-gas chromatograph-mass spectrometer experiment.

PROCEDURE

(1) *Direct heating under high vacuum*

As described previously (MURPHY, R. C. *et al.*, 1970), the samples were placed in a small bulb mounted on the ion source of the high resolution mass spectrometer (CEC 21–110B). The sample heater was, however, modified by replacing the tungsten heating wire inside the ceramic oven with an Aerorod heater (American Standard) 76 × 0.173 cm, which was coiled to fit inside the oven and to make physical contact with a quartz sample bulb when it is in position. This new heater was capable of raising the sample to 900°C in less than one minute. The same procedures (e.g., venting the ion source housing with dry nitrogen; taking consecutive 3 min exposures; and using perfluoroalkane as a mass standard) were used as reported previously, except the upper temperature limit for sample heating was 900°C. Unless stated otherwise, the samples were heated to 600°C.

(2) *Pyrolysis in a helium atmosphere*

A commercial Hamilton Multi-Purpose Sampling System was used for the pyrolysis. After a procedural blank run, the lunar sample (100–500 mg) was positioned in the unit and purged with a helium flow of 30 ml/min. The sample was dropped into the pyrolyzer oven, which was maintained at 700°C. The heated transfer line diverted the helium stream that swept over the sample into the injection port of a gas chromatograph, which was part of the previously described gas chromatograph-mass spectrometer-computer system (HITES and BIEMANN, 1968). The sample was pyrolyzed for 8 min while the gas chromatographic column (183 × 0.32 cm, 3% OV-17 on Gas Chrom Q) was heated from 0° to 30°C. After the pyrolysis, the transfer line was removed and the helium flow rerouted as in a normal gas chromatographic run. The column temperature was programmed from 30° to 250°C, at 12°/min. The mass spectrometer scanned continuously (every 4 sec from mass 10 to 600) and the computer recorded, digitized, and stored the mass spectra which were then converted to mass and intensity data.

RESULTS

(1) *Direct heating under high vacuum*

The analyses of the core tube samples were particularly interesting. The organic ions (above mass 28.0000) produced by heating the uppermost core sample (12028,164) to 600°C are listed in Fig. 1. This format was designed to facilitate the comparison of the mass spectra obtained from one sample at different, gradually increasing temperature ranges. In Fig. 1, the numbers directly to the left of the letter X represent the nominal mass (e.g., m/e 28), while the entries under the elemental headings (CH, CHN, CHS, CHO, and CHNO) are the intensities and elemental composition of ions at that nominal mass in each of the exposures (X, A, B, C, D, and E). At m/e 28, upon heating the 10-cm deep core sample from 50° to 275°C (exposure A), two ions were produced—one ion C_2H_8 (intensity of 534) and the other CH_2N (intensity of 368). The intensities listed in this figure are unnormalized and are more representative of the optical density of the ions on the photographic plate, rather than their relative abundance. This tabular arrangement of the data of successive spectra allows a more meaningful comparison of the abundances of the same ions which had been recorded on different exposures but on the same photographic plate. Direct comparison of the three core-tube samples can be made because all three samples were analyzed using the same photographic plate. Test exposures of known amounts of nonane permitted estimation of the total amount of material represented in these spectra. For the data in Fig. 1, the concentration of material is less than 10 ppb.

Inspection of Fig. 1 indicates that the ions which are produced during heating of the sample are not present in the background, spectrum X (for a procedural blank see bottom of Fig. 4). The ions of highest mass, C_7H_7 and C_6H_6, are due to the tropylium ion and benzene, respectively, with the remaining hydrocarbon ions being smaller than C_5H_7. The heteroatoms containing ions have no more than four carbon atoms attached and are therefore from very low molecular weight species; for example, pyrrole (C_4H_5N), dimethylamine (C_2H_7N), and acetaldehyde (C_2H_4O). HCNO was also produced. There were no other organic ions in this sample; however, there were other ions due to SO_2, SO, H_2S, CO_2, CO, CH_4, NH_3, HCN, and the rare gases; but inorganic ions and ions below m/e 28.0000 are not listed in this representation.

Each temperature range corresponds to a 3 min exposure during the heating period, and the ion abundances in Fig. 1 thus indicate that the evolution of organics reaches a maximum between the temperatures of 275° to 525°C (spectrum B). This temperature range for the release of the largest fraction of the material is typical for all the lunar samples investigated with the high resolution mass spectrometer. This is shown in Fig. 2, which is a graphical summary of the combined elemental composition data from the deepest core tube sample (12028,178). The bar graph shows the summed intensities of various component ions in relation to the temperature of the sample. The ions produced by heating this deep core sample suggested the presence of benzene (C_6H_6), toluene (C_7H_8), pyridine (C_5H_5N), methylpyridine (C_6H_7N), benzonitrile (C_7H_5N), phenol (C_6H_6O), as well as lower molecular weight species. The C_3H_7S ion is reminiscent of the contaminant diisopropyldisulfide found in the Apollo 11 samples (MURPHY *et al.*, 1970; GIBERT *et al.*, 1971) and the C_3H_7NO species may represent traces of *N*,*N*-dimethylformamide, a common solvent.

A comparison of the total amounts of organic material released in the three different core tube samples (Fig. 3) shows only minor variations. However, most of the ion abundance making up this type of bar graph is from low mass ions. The ions of higher mass, which often are more informative, are listed in Table 1. These data

		CH	CHN	CHS	CHO	CHNO
28	X					
	A	534 2/ 4	368 1/ 2			
	B	344 2/ 4	322 1/ 2			
	C	1289 2/ 4	1124 1/ 2			
	D	121 2/ 4				
	E					
29	X					
	A	348 2/ 5				
	B	283 2/ 5	53 1/ 3			
	C	484 2/ 5				
	D					
	E					
30	X					
	A				416 1/ 2	
	B		198 1/ 4		268 1/ 2	
	C				591 1/ 2	
	D					
	E					
31	X					
	A				365 1/ 3	
	B		78 1/ 5		190 1/ 3	
	C					
	D					
	E					
32	X					
	A					
	B				68 1/ 4	
	C					
	D					
	E					
37	X					
	A					
	B	302 3/ 1				
	C	649 3/ 1				
	D					
	E					
38	X					
	A	231 3/ 2				
	B	336 3/ 2	224 2/ 0			
	C	1063 3/ 2	354 2/ 0			
	D					
	E					
39	X					
	A	1012 3/ 3				
	B	635 3/ 3	371 2/ 1			
	C	2213 3/ 3	900 2/ 1			
	D	204 3/ 3+				
	E	101 3/ 3+				

		CH	CHN	CHN2	CHO
51	X				
	A	519 4/ 3			
	B	393 4/ 3	111 3/ 1		
	C	859 4/ 3			
	D				
	E				
52	X				
	A	198 4/ 4			
	B	287 4/ 4	202 3/ 2		
	C	425 4/ 4			
	D				
	E				
53	X				
	A				
	B	150 4/ 5	180 3/ 3		
	C				
	D				
	E				
54	X				
	A				
	B	20 4/ 6	62 3/ 4		
	C				
	D				
	E				
55	X				
	A				
	B	195 4/ 7	50 3/ 5		139 3/ 3
	C	86 4/ 7			
	D				
	E				
56	X				
	A				
	B	134 4/ 8			
	C				
	D				
	E				
57	X				
	A				
	B	50 4/ 9			
	C				
	D				
	E				
58	X				
	A				
	B				
	C				158 3/ 6
	D				
	E				

m/e	Exposure	CH	CHN	CHS	CHO	CHNO
40	X					
	A		254 2/ 2			
	B	370 3/ 4	486 2/ 2		201 2/ 0	
	C	1068 3/ 4	1368 2/ 2		882 2/ 0	
	D					
	E					
41	X					
	A	935 3/ 5	546 2/ 3		166 2/ 1	
	B	530 3/ 5	549 2/ 3		443 2/ 1	
	C	1502 3/ 5	1741 2/ 3		1845 2/ 1	
	D	47 3/ 5	48 2/ 3			
	E					
42	X					
	A				1003 2/ 2	
	B	335 3/ 6	461 2/ 4		613 2/ 2	108 1/ 0
	C	502 3/ 6	732 2/ 4		2192 2/ 2	
	D				244 2/ 2	
	E					
43	X					
	A	223 3/ 7			1495 2/ 3	382 1/ 1
	B	340 3/ 7	232 2/ 5		599 2/ 3	444 1/ 1
	C	407 3/ 7	113 2/ 5		1601 2/ 3	833 1/ 1
	D					
	E					
44	X					
	A		158 2/ 6		521 2/ 4	
	B		343 2/ 6	160 1/ 0	424 2/ 4	
	C		552 2/ 6		878 2/ 4	
	D					
	E					
45	X					
	A					
	B		101 2/ 7			
	C					
	D					
	E					
49	X					
	A					
	B	48 4/ 1				
	C					
	D					
	E					
50	X					
	A	424 4/ 2				
	B	367 4/ 2				
	C	889 4/ 2				
	D					
	E					

m/e	Exposure	CH	CHN	CHN2	CHO
63	X				
	A				
	B	27 5/ 3			
	C				
	D				
	E				
65	X				
	A				
	B	28 5/ 5			
	C				
	D				
	E				
66	X				
	A				
	B	50 5/ 6			
	C				
	D				
	E				
67	X				
	A				
	B		75 4/ 5		
	C				
	D				
	E				
69	X				
	A				
	B				
	C				
	D				
	E				
77	X				
	A				
	B	38 6/ 5-			
	C				
	D				
	E				
78	X				
	A	1130 6/ 6			
	B	376 6/ 6-			
	C	846 6/ 6			
	D				
	E				
91	X				
	A				
	B	161 7/ 7			
	C	185 7/ 7			
	D				
	E				

Fig. 1. Summary of all carbon-containing ions above m/e 27 from core tube sample 12028,164 (104 mg). Exposure X, 50°C; A, 50–275° C; B, 275–525°C; C, 525–575°C; D, 575–600°C; and E, 600°C.

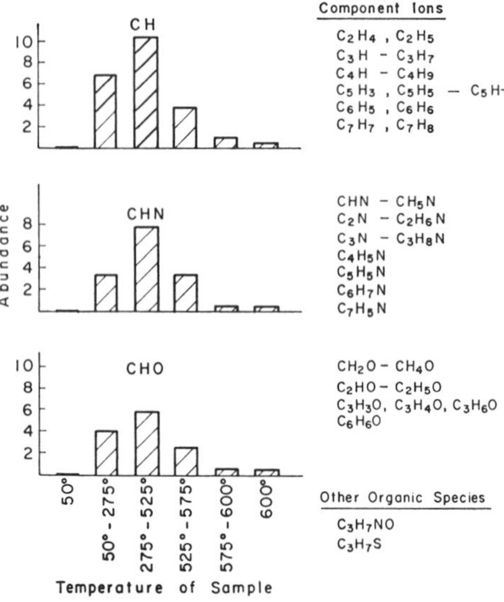

Fig. 2. Summary of the production of hydrocarbon ions (CH), oxygen-containing ions (CHO), and nitrogen-containing ions (CHN) during heating core tube sample 12028,178 (120 mg) from 50 to 600°C in the ion source of a high-resolution mass spectrometer. (See text for discussion of relative abundance and total amount.)

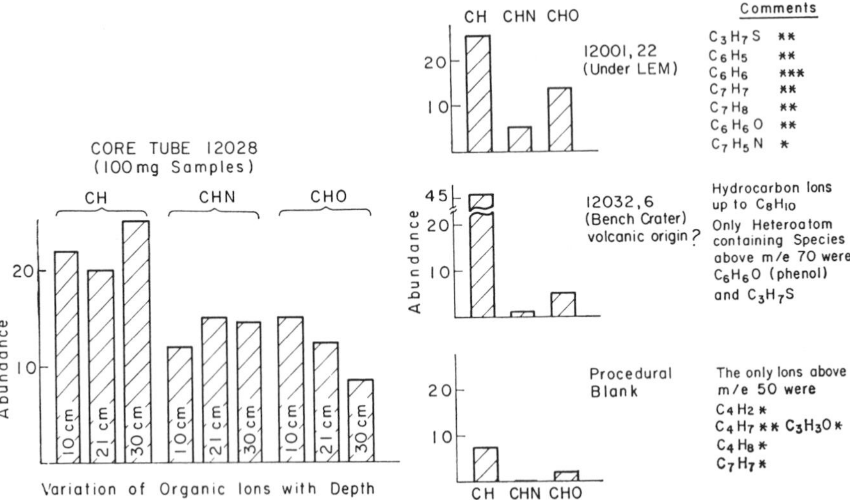

Fig. 3. Summary of the total amount of ion types (CH, CHO, or CHN) produced during the heating of core tube samples 12028,164, 171, and 178 (104, 114, and 120 mg, respectively) from 50 to 600°C in the ion source of a high-resolution mass spectrometer.

Fig. 4. Summary of ions produced from heating 145 mg of sample 12001,22 (LEM) and 115 mg of sample 12032,6 (Bench Crater) from 50 to 600°C in the ion source of a high-resolution mass spectrometer.

Table 1. Core tube samples (approx. 100 mg sample).
Ions above m/e 70

m/e	El. comp.	Abundance*	Comments
12028,164 (10 cm)			
77	C_6H_5	*	phenyl ion
78	C_6H_6	***	benzene
91	C_7H_7	**	tropylium ion
12028,171 (21 cm)			
70	C_4H_6O	*	
72	C_4H_8O	*	
73	C_3H_7NO	***	
75	C_3H_7S	*	
77	C_6H_5	**	phenyl
78	C_6H_6	***	benzene
79	C_5H_5N	*	pyridine
91	C_7H_7	**	tropylium
92	C_7H_8	*	toluene
12028,178 (30 cm)			
73	C_3H_7NO	**	
75	C_3H_7S	*	
77	C_6H_5	***	phenyl
79	C_5H_5N	**	pyridine
91	C_7H_7	**	tropylium
92	C_7H_8	**	toluene
93	C_6H_7N	**	methyl pyridine
94	C_6H_6O	*	phenol
103	C_7H_5N	**	benzonitrile

* Abundances estimated to be in the ppb range.

indicate an increase in different types of larger organic molecules with depth. This evidence is certainly not conclusive because it is based on only one single analysis of three samples. The Preliminary Examination Team (PET) examination of the core tube samples, in fact, did not reveal such a variation (BURLINGAME *et al.*, 1971); however, the different handling of the samples, particularly the heat sterilization of the PET samples, could account for these differences if the results reported here are in fact more representative of the actual conditions on the lunar surface. The extremely small amounts of organic material discussed here would preclude any such trend being observed by total carbon analysis. The material giving rise to these spectra is less than 0.01% of the total carbon content of these samples.

The heat evolution of organic material in samples 12001,22 (fines near the LEM) and 12032,6 (Bench crater) are shown in Fig. 4. A comparison of the LEM sample with the sample from Apollo 11 indicated a similar distribution of ions below m/e 78. However, there are no hydrocarbon ions larger than C_7H_7; no oxygen-containing ions larger than C_3H_6O except for phenol (C_6H_6O); and a complete absence of ions containing more than one oxygen or nitrogen atom in sample 12001,22. These results most certainly reflects the lower level of contamination in Apollo 12 relative to Apollo 11. The results from the PET also indicated that the level of organic contamination was reduced compared to the Apollo 11 samples (LSPET, 1970).

Perhaps the most striking feature of the data from the LEM sample (12001,22) is that there does not appear to be any significant increase in those ions attributed to the rocket-exhaust products (Simoneit *et al.*, 1969) over those ions in the samples which definitely were not exposed to the rocket plume. This can be seen by comparing the relative CH, CHN, and CHO abundances of the core tubes (Fig. 3) and the LEM sample (Fig. 4). One would expect a larger relative ratio of CHN to CH or CHO in the LEM sample, since the rocket burned nitrogenous fuel. However, just the opposite is true; that is, the core tube samples appear to be enriched in nitrogen-containing molecules. These facts would suggest one of two possibilities: either the sample 12001,22 is not representative of a sample contaminated by rocket exhaust, or that contamination due to the LEM rocket is minimal (possibly because the directly exposed surface layer was blown away). The samples from Bench Crater (12032,6) shown in Fig. 4, and Head Crater (12033,11) produced a relatively large amount of hydrocarbon ions (especially between m/e 28–55) as compared with other samples. There were no nitrogen-containing ions above m/e 43.

The organic material present in the returned lunar material showed some variation with respect to sample location. As mentioned above, this variation is evident from the larger abundance of nitrogen-containing ions in the core samples rather than the LEM sample, as well as the large hydrocarbon abundances in the Bench and Head Crater samples. Unfortunately, the levels of organic material detected in these experiments were in the low part per billion range and at such low concentrations the problems of contamination, as evidenced by C_3H_7NO and C_3H_7S, become severe. The interpretation of the results, therefore, must be made with reservations.

By the nature of the experiment, the ions observed in the high resolution mass spectra are mainly fragments of the components of a mixture. The identity of these components can only be inferred from the type of ions observed, particularly from those which correspond to a molecular ion. In an effort to generate, separate, and characterize these components by their individual mass spectra, an attempt was made to vaporize the organic material out of the inorganic matrix into the helium stream used as the carrier gas of a gas chromatograph. The gas chromatograph is directly coupled to a low resolution mass spectrometer and a computer for recording the mass spectra (Hites and Biemann, 1968). It was expected that these spectra could then be correlated with the high resolution spectra obtained in the experiments described above, enabling the elemental composition data to be used as an aid in the interpretation of the conventional spectra where necessary. Such an experiment, commonly called pyrolysis-GC-MS, had indeed been carried out in some Apollo 11 samples (Murphy, Sr. M. E. *et al.*, 1970). The investigators found a series of aromatic hydrocarbons and thiophenes in appreciable quantities but could not exclude the possibility that they are produced by thermal synthesis from smaller molecules. If this were the case, the correlation of the data obtained in such an experiment with the high resolution data would not be possible. Nevertheless, such "pyrolysis" experiments were performed on some of our Apollo 12 samples. The results would either permit such a correlation or provide further support of thermal synthesis of larger molecules from aggregations of indigenous carbon sources. Indeed, the experiments described below seem to indicate the latter.

(2) *Pyrolysis in a helium atmosphere*

The results of the most productive "pyrolysis" experiment, involving a portion of the Bench Crater sample (120 mg of 12032,6) is shown in Fig. 5. The total amount of material represented in the gas chromatogram is estimated to be as much as 25 ppm. These results, except for the absence of thiophene and the presence of compounds emerging after biphenyl, indicated the production of organic compounds as previously published (MURPHY, SR. M. E. *et al.*, 1970). However, the reproducibility of these results was quite poor and other identical experiments with the same sample (12032,6) and other Apollo 12 samples produced much lower quantities of material. In fact, only three compounds could be consistently found in duplicate runs—benzene, toluene, and napthalene.

These results are in direct contradiction with the related experiments described at the beginning of this paper involving the stepwise heating of the sample into the ion source of the high resolution mass spectrometer. An additional experiment with the high resolution instrument indicated that heating the lunar sample from 600 to 900°C did not produce aromatic molecules, but rather produced large quantities of carbon monoxide. The only difference between these two experiments which could explain this dichotomy (discounting contamination) would be the pressure in the system where the samples were heated. In the mass spectrometer the pressure was in the 10^{-6} torr range (the mean free path of molecules is several tens of meters); in the Hamilton pyrolysis oven the pressure is 760 torr or greater when the sample is heated (the mean free path of molecules is approximately 10^{-5} cm). Since large quantities of carbon monoxide and hydrogen evolve at the pyrolysis temperatures, it seems likely that the aromatic compounds observed during the atmospheric pyrolysis are synthesized by the reaction of CO and H_2 with the lunar sample as a catalyst.

With a simple, empirical approach, this hypothesis was tested by slowly passing

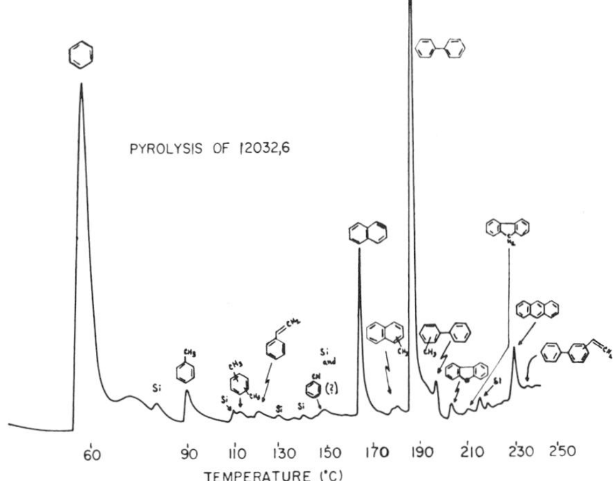

Fig. 5. Gas chromatogram of the products formed from 120 mg of sample 12032,6 (Bench Crater) upon heating in a pyrolysis oven.

mixtures of carbon monoxide and hydrogen over a previously pyrolyzed lunar sample which was at 700°C. In two experiments using a carbon monoxide to hydrogen ratio of 1:1 and 1:9 (w/w), traces of benzene were indeed observed. We believe that thermal synthesis of larger molecules from indigenous carbon sources occurs during the abrupt heating of the lunar material at atmospheric pressure. This conclusion is based on the following data: the results of the above model experiments, the irreproducibility of the yields in the pyrolyses of lunar material, and the failure to detect any of these larger molecules (listed in Fig. 5) upon the more gentle vaporization to the same temperature into the high vacuum of the high resolution mass spectrometer. Further experiments are in progress to substantiate these conclusions, but they are somewhat hampered by the lack of necessary amounts of fresh lunar material of various mineralogical and physical characteristics, which may well have an effect on the catalytic activity leading to difficulties in reproducing the results.

Conclusions

Even if one could exclude the possibility of contamination, the level of organic material in the Apollo 12 lunar samples is quite low—a total of 10 ppb or below (excluding simple hydrocarbon gases). Since the probable structures of the compounds present (e.g., benzene, toluene, pyridine, benzonitrile [or an isomer], etc.) are rather simple, one might speculate that they have been synthesized in a rather direct manner from their component atoms introduced by the solar wind. Another pathway might involve the utilization of the thermal energy produced by meteor impact causing reactions of indigenous carbon and hydrogen. The pyrolysis experiment points out that great care must be exercised in interpreting the results of seemingly mild treatment (e.g., mere heating) of the lunar material.

Acknowledgments—We are indebted to Dr. Norman Mancuso and Mr. Edward Ruiz for assistance with the computer evaluation of the data. This work was supported in part by NASA Research Contract NAS9–8099.

References

Abell P. I., Eglinton G., Maxwell J. R., Pillinger C. T., and Hayes J. M. (1970) Indigenous lunar methane and ethane. *Nature* **226**, 251–252.

Burlingame A. L., Calvin M., Han J., Henderson W., Reed W., and Simoneit B. R. (1970) Study of carbon compounds in Apollo 11 lunar samples. *Proc. Apollo 11 Lunar Sci. Conf., Geochim. Cosmochim. Acta* Suppl. 1, Vol. 2, pp. 1779–1791. Pergamon.

Burlingame A. L., Hauser J. S., Simoneit B. R., Smith D. H., Biemann K., Mancuso N., Murphy R. C., Flory D. A., and Reynolds M. A. (1971) Preliminary organic analysis of the Apollo 12 cores. Submitted for publication.

Chang S., Smith J. W., Kaplan I., Lawless J., Kvenvolden K. A., and Ponnamperuma C. (1970) Carbon compounds in lunar fines from Mare Tranquillitatis. IV. Evidence for oxides and carbides. *Proc. Apollo 11 Lunar Sci. Conf., Geochim. Cosmochim. Acta* Suppl. 1, Vol. 2, pp. 1857–1869. Pergamon.

Gibert J., Flory D., and Oró J. (1971) Identity of a common contaminant of Apollo 11 lunar fines and Apollo 12 York meshes. *Nature* **229**, 33–34.

Hare P. E., Harada K., and Fox S. W. (1970) Analyses for amino acids in lunar fines. *Proc. Apollo 11 Lunar Sci. Conf., Geochim. Cosmochim. Acta* Suppl. 1, Vol. 2, pp. 1799–1803. Pergamon.

HITES R. A. and BIEMANN K. (1968) A mass spectrometer-computer system particularly suited for gas chromatography of complex mixtures. *Anal. Chem.* **40**, 1217–1221.

LSPET (Lunar Sample Preliminary Examination Team) (1970) Apollo 12 preliminary science report (NASA SP–235), National Aeronautics and Space Administration, Washington, D.C.

MURPHY R. C., PRETI G., NAFISSI V. M. M., and BIEMANN K. (1970) Search for organic material in lunar fines by mass spectrometry. *Proc. Apollo 11 Lunar Sci. Conf., Geochim. Cosmochim. Acta* Suppl. 1, Vol. 2, pp. 1891–1900. Pergamon.

MURPHY SR. M. E., MODZELESKI V. E., NAGY B., SCOTT W. M., YOUNG M., DREW C., HAMILTON P., and UREY H. C. (1970) Analysis of Apollo 11 lunar samples by chromatography and mass spectrometry: Pyrolysis products, hydrocarbons, sulfur, amino acids. *Proc. Apollo 11 Lunar Sci. Conf., Geochim. Cosmochim. Acta* Suppl. 1, Vol. 2, pp. 1879–1890. Pergamon.

ORÓ J., UPDEGROVE W. S., GIBERT J., MCREYNOLDS J., GIV-AV E., IBANEZ J., ZLATKIS A., FLORY D. A., LEVY R. L., and WOLF C. J. (1970) Organogenic elements and compounds in types C and D lunar samples from Apollo 11. *Proc. Apollo 11 Lunar Sci. Conf., Geochim. Cosmochim. Acta* Suppl. 1, Vol. 2, pp. 1901–1920. Pergamon.

Proceedings of the Second Lunar Science Conference, Volume 2, pp. 1891–1899.
The M.I.T. Press, 1971.

Preliminary organic analysis of the Apollo 12 cores

A. L. BURLINGAME,* J. S. HAUSER, B. R. SIMONEIT, and D. H. SMITH†
Space Sciences Laboratory, University of California, Berkeley, California 94720

K. BIEMANN, N. MANCUSO, and R. MURPHY
Department of Chemistry, Massachusetts Institute of Technology, Cambridge, Massachusetts 02139

and

D. A. FLORY‡ and M. A. REYNOLDS
NASA-Manned Spacecraft Center, Houston, Texas 77058

(*Received 22 February 1971; accepted in revised form 30 March 1971*)

Abstract—A computer-coupled, high-sensitivity mass spectrometer has been used to estimate the amount and kind of organic matter in the double core (12025 and 12028) from Apollo 12. The concentrations of total organic matter in the core samples ranged from 0.2 to 2.2 ppm. The bulk of this organic matter is derived from contamination. The major volatile compounds present were hydrocarbons of varying degrees of unsaturation displaying ions to the m/e 250 range. Several organic polymers were found in various samples (e.g., polystyrene in 12028,4 and 12025,1 and Teflon in 12025,2). Most of the samples evolve CO_2 and lesser amounts of SO_2 at higher temperatures (500°C) during the analyses. In two samples (12028,2 and 12028,1) where the hydrocarbon background was low, there was an indication (based on the observations of m/e 78 and 91 late in the heating cycle) that pyrolysis of indigenous organic matter may take place. The level of such indigenous organic matter was estimated to be approximately 100 ppb (excluding methane, CO, and CO_2). Cadmium was found present in samples 12028,5 and 12028,15.

INTRODUCTION

THE PRELIMINARY ORGANIC analysis experiment was designed to yield information on the amount and kind of volatilizable organic material in returned lunar samples. This information was used to assist the Lunar Sample Analysis Planning Team (LSAPT) in the distribution of samples to various principal investigators and in compilation of the Lunar Sample Catalog. The data presented in this report are those obtained from the preliminary analysis of samples from the double core tube. These data, which were not included in the Apollo 12 catalog prepared by the Lunar Sample Preliminary Examination Team (LSPET, 1970b), are reported to complete the Apollo 12 preliminary organic analysis results.

EXPERIMENTAL TECHNIQUES

The equipment employed for these studies includes a combined low resolution mass spectrometer-computer system equipped with a special inlet system for direct volatilization-pyrolysis of

* John Simon Guggenheim Memorial Fellow, 1970–1971.

† Present address: School of Chemistry, Geochemistry Unit, University of Bristol, Bristol BS8 ITS, England.

‡ Present address: Department of Biophysical Sciences, University of Houston, Houston, Texas 77004.

the lunar samples into the mass spectrometer. This combined system has been previously described (LSPET, 1969a, b, 1970a, b). A more detailed description for the operation of the computer system is available (SMITH *et al.*, 1971).

The analytical procedure involves direct insertion of the lunar sample into an oven maintained at 500°C and simultaneous recording of mass spectra with 10 sec scans (16.4 sec per total cycle) for approximately 30 min. It is estimated that the sample reaches the oven temperature in one to two minutes. The standard operating conditions of the mass spectrometer and the method of calibration of the system is the same as described earlier (LSPET, 1969a, b, 1970a, b), with the exception that only one hundred repetitive scans were recorded in the Apollo 12 analyses. The sample handling and transfer procedures have been described briefly (LSPET, 1970a, b). The samples were not heat sterilized and were transferred for analysis after quarantine. For these samples a dry nitrogen glove box was installed in the gas analysis laboratory (GAL) to prevent laboratory contamination of the samples. All samples were transferred in the glove box from the LRL stainless steel and aluminum sample containers to the nickel capsule. The same nickel capsule was used for all mass spectral analyses.

The samples analyzed were taken from the double core tube (sample numbers 12025 and 12028). Stratification and morphological change vs. depth of the double core tube has been discussed (LSPET, 1970a). The core tubes were opened in the Bioprep area and 100–200 mg samples were taken at the depths indicated in Table 1. The surface material was removed prior to sampling. The samples were transferred to the GAL in stainless steel and aluminum sample containers. The Bioprep cabinetry was monitored using sintered Ottawa sand supplied by NASA-Ames. Analytical results related to the monitoring have been discussed in more detail by SIMONEIT and FLORY (1971). The control sand had an organic contamination level of 0.4 ppm (12 ng/cm^2), and the sand which was exposed for 48 days in the Bioprep cabinet where the cores were opened had an organic contamination level of 1.1 ppm (40 ng/cm^2). Major contamination resulted from hydrocarbons and Teflon.

RESULTS

The results of the analyses are summarized in Table 1, with an indication of the major types of organic material present. It should stressed that the organic contents noted in Table 1 have not been corrected for contributions from carbon-containing molecules such as CO and CO_2. The m/e values corresponding to such species have been deleted from the summarized data to permit easier interpretation of the higher molecular weight organic ions (LSPET, 1969b). (The standard deleted m/e values are 14, 16 to 20, 23, 26–28, 32, 35 to 37, 40, and 44 (LSPET, 1970b).) The strong evolution of CO_2 from Apollo 12 samples analyzed previously, which represents a significant

Table 1. Apollo 12 double core preliminary organic analysis

Sample number	Depth (cm)	Organic content (ppm)	Comments (major constituents)
12025,2	1.9	0.4	hydrocarbons and Teflon
12025,1	8.0	0.9	hydrocarbons, Teflon, and polystyrene
12028,2	11.0	0.2	low hydrocarbons and possibly benzene
12028,15	13.3	2.2	hydrocarbons, SO_2, m/e 73, Cd, and possibly chlorobenzene
12028,3	16.8	0.6	hydrocarbons, SO_2, and Teflon
12028,4	23.0	1.5	hydrocarbons and polystyrene
12028,5	33.6	~1.3	hydrocarbons, SO_2, Cd, and m/e 73
12028,1*	bottom (~41)	0.8	hydrocarbons, low SO_2, and m/e 58

* LSPET, 1970b.

amount of the total carbon in the samples, has been reported (LSPET, 1970a). ORÓ *et al.* (1971) have presented evidence that carbon dioxide evolution in this temperature range (ambient to 500°C) is probably due to adsorption from the terrestrial atmosphere.

Generally, the results indicate that most of the volatile material is terrestrial contamination and appears to be hydrocarbons and polymers such as Teflon and polystyrene that was depolymerized during the sample heating. All samples released varying amounts of sulfur dioxide (SO_2, m/e 64). The m/e 64 ion was attributed to SO_2, rather than to S_2, since plots of ion current vs. scan numbers (HITES and BIEMANN, 1970) for m/e 64 and m/e 48 (i.e., SO^+ fragment from SO_2) were identical. There appears to be no obvious correlation of amount or type of organic matter evolved with regards to depth or particle size and color.

Sample 12025,2. This sample is characterized by a relatively low organic content consisting primarily of hydrocarbons and Teflon depolymerization products. The data of Fig. 1(a) show, from the sum plot (SMITH *et al.*, 1971), the rapid evolution of material on insertion of the sample into the oven (at scan 4). The mass spectrum of scan 4, Fig. 1(c), illustrates that the material evolved consists of saturated and unsaturated alkanes with m/e values up to at least 230. This spectrum is characteristic of the early volatilization of surface-adsorbed material in all samples. Decomposition products of Teflon later in the analysis are illustrated by the plot, Fig. 1(b), of m/e 100 ($C_2F_4^+$) vs. scan number (HITES and BIEMANN, 1970) and spectrum number 100, Fig. 1(d). The typical fragments observed are m/e 31 (CF^+), m/e 50 (CF_2^+), m/e 69 (CF_3^+) m/e 81($C_2F_3^+$), m/e 100 ($C_2F_4^+$), and m/e 131 ($C_3F_5^+$).

Sample 12025,1. This sample displayed a moderate amount of contamination, consisting of hydrocarbons and Teflon decomposition products as above. For all samples, the hydrocarbon contamination (SIMONEIT and FLORY, 1971) consisted mainly of the more saturated alkane series: C_nH_{2n+2}, C_nH_{2n}, C_nH_{2n+2}, for $n = 2$ to 10 (some samples to 17), with minor amounts of the series C_nH_{2n-4} to C_nH_{2n-8} for various values of n ranging from as low as 4 to a maximum of 18. In addition, a strong evolution of what are most probably decomposition products of polystyrene was observed. These products are discussed in connection with sample 12028,4.

Sample 12028,2. This sample appeared to be the least contaminated of this series. Some selected mass spectral data are shown in Fig. 2. As indicated by the plot of m/e 57 ($C_4H_9^+$) vs. scan number, Fig. 2(b), the hydrocarbon contaminants are quickly volatilized. This initial evolution of hydrocarbon contaminants yields peaks to only m/e 150. The second maximum in the sum plot (SMITH *et al.*, 1971) of Fig. 2(a) is due mainly to the ion currents from m/e 66, m/e 78 [cf. plot of m/e 78 and scan number in Fig. 2(c)], m/e 81, and m/e 93, which increase simultaneously. This could indicate a possible common source. The peak at m/e 78 is tentatively interpreted as being benzene, since the group of peaks around m/e 52 also increase in intensity (see scan 71 in Fig. 2). The peaks at m/e 66, 81, and 93 (see scan 71 in Fig. 2) could indicate aromatic nitrogen heterocycles. However, without accurate mass measurement and subsequent assignment of elemental compositions, it is not possible to make further suggestions.

Sample 12028,15. This sample exhibited the highest level of volatilizable material in

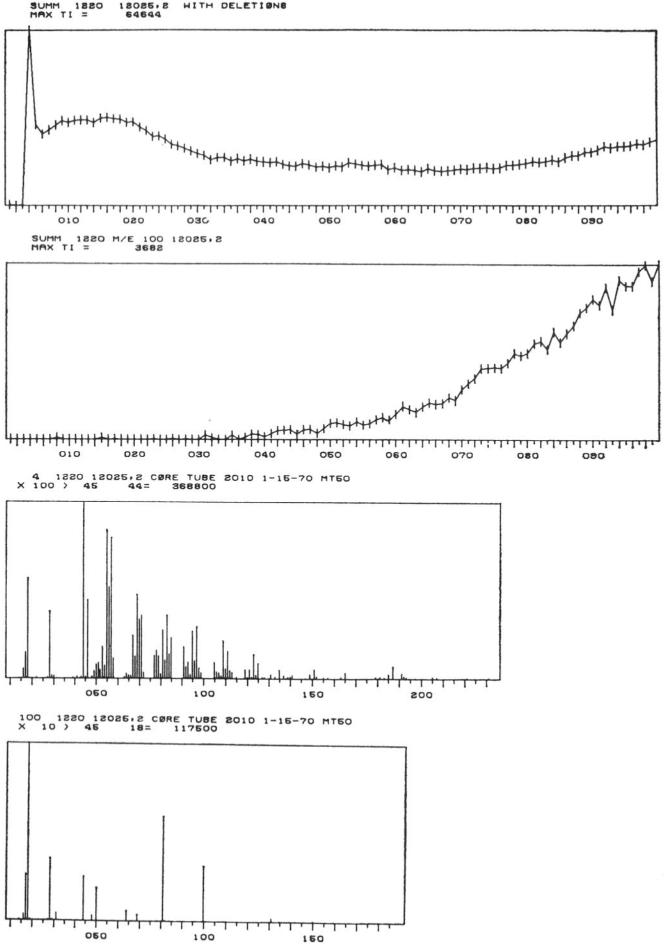

Fig. 1. Mass spectral data for core sample 12025,2. *Top*: Sum plot of total ion current vs. scan number (with deletions). *Second:* Plot of m/e 100 intensity vs. scan number. *Third:* Mass spectrum scan 4 (normalized to m/e 44, intensity scale expanded by 100 above m/e 45). *Fourth:* Mass spectrum scan 100 (normalized to m/e 18, intensity scale expanded by 10 above m/e 45).

this series. The usual hydrocarbon distribution extends to m/e 250, and these data also show a strong release of SO_2. Furthermore, the peak at m/e 73 is intense from scans 10 to 40. The identity of the component responsible for this peak has not been established, but a species such as the trimethylsilyl ion $[(CH_3)_3Si^+]$ is a possibility (Silicone rubber is a probable source—SIMONEIT and FLORY, 1971). A minor constituent of interest is indicated by a group of peaks at m/e 110 to 114 appearing in later scans (cf. scans 50 and 75 in Fig. 3). This is most probably cadmium metal. (The terrestrial isotope distribution of cadmium is m/e 106 (1.22%); 108 (0.88%); 110 (12.39%); 111 (12.75%); 112 (24.07%); 113 (12.26%); 114 (28.86%); and 116

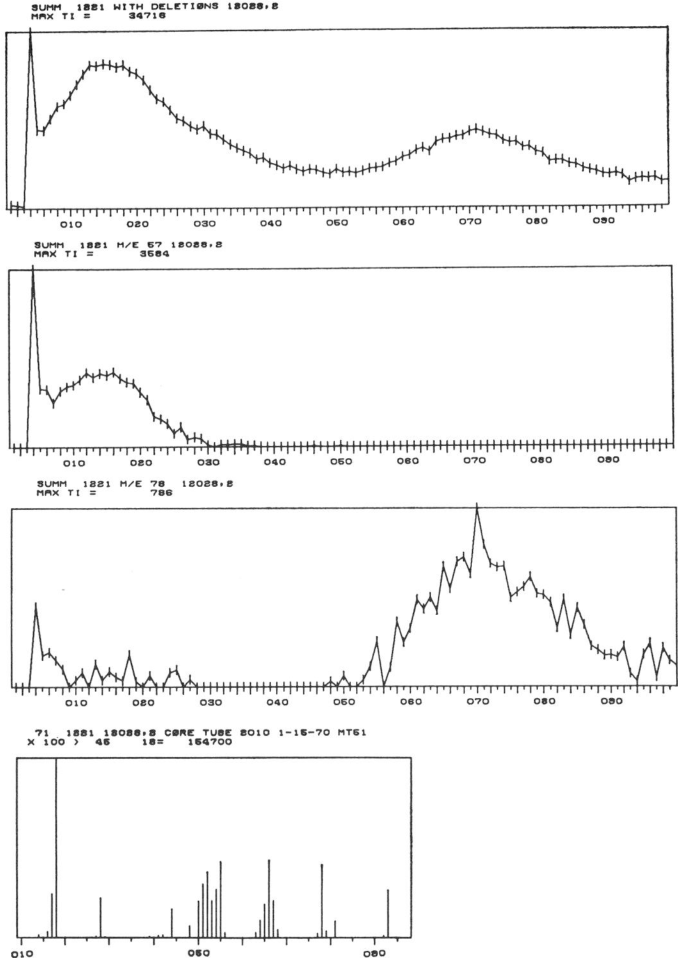

Fig. 2. Mass spectral data for core sample 12028,2. *Top:* Sum plot of total ion current vs. scan number (with deletions). *Second:* Plot of m/e 57 intensity vs. scan number. *Third:* Plot of m/e 78 intensity vs. scan number. *Fourth:* Mass spectrum scan 71 (normalized to m/e 18, intensity scale expanded by 100 above m/e 45).

(7.58%).) The observed total ion current of 1.7×10^5 from 45 mg of sample represents an approximate estimate of 240 ppb Cd (based on *n*-tetracosane calibration—LSPET, 1970b). GANAPATHY *et al.* (1970) observed 22 ppm Cd in sample 12028,66.

Sample 12028,3. This is a relatively uncontaminated sample. The data indicate mainly the usual hydrocarbons, ranging to approximately m/e 150, SO_2, and then Teflon decomposition products later in the analysis.

Sample 12028,4. This sample showed relatively high contamination with hydrocarbons and polystyrene, and it liberated a significant amount of SO_2. The summary plot presented in Fig. 4(a) shows, apart from the initial maximum at scan 4, two substantial

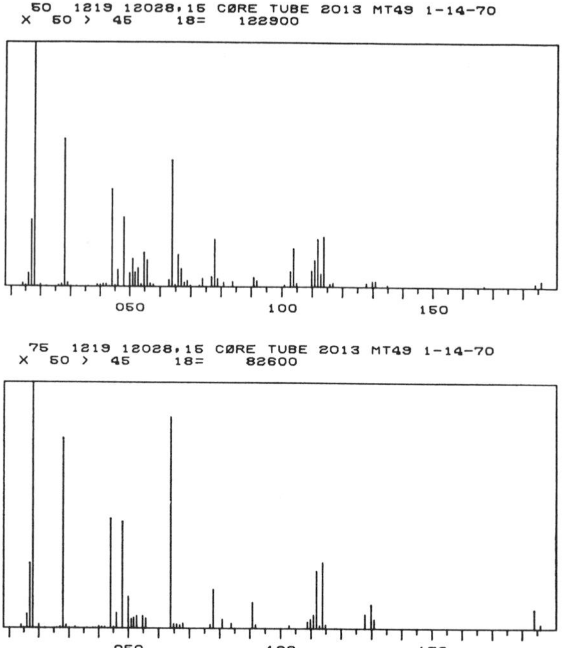

Fig. 3. Mass spectral data for core sample 12028,15. *Top:* Mass spectrum scan 50 (normalized to m/e 18, intensity scale expanded by 50 above m/e 45). *Bottom:* Mass spectrum scan 75 (normalized to m/e 18, intensity scale expanded by 50 above m/e 45).

maxima peaking at scans 15 and 41. A hydrocarbon envelope extending to approximately m/e 310 and SO_2 are responsible for the maximum at scan 15, as shown in Fig. 4(c). The other maximum is due to ions including m/e 104 [see sum m/e 104 plot of Fig. 4(b)], m/e 78, m/e 91, and m/e 207. A representative spectrum, scan 41, [shown in Fig. 4(d)], illustrates what are most probably the decomposition products of polystyrene. The peak at m/e 104 is possibly styrene ($C_8H_8^+$), and m/e 207 can be thought of as a 1,3-diphenylbutene fragment (Structure I).

$\cdot +$

$(C_{16}H_{15})$

I

Sample 12028,5. This is a relatively heavily contaminated sample. The data indicate mainly the usual hydrocarbons to about m/e 260, m/e 73, and the isotopes of cadmium metal [see, for example, scan 50 in Fig. 5(d)]. The total ion current for the cadmium

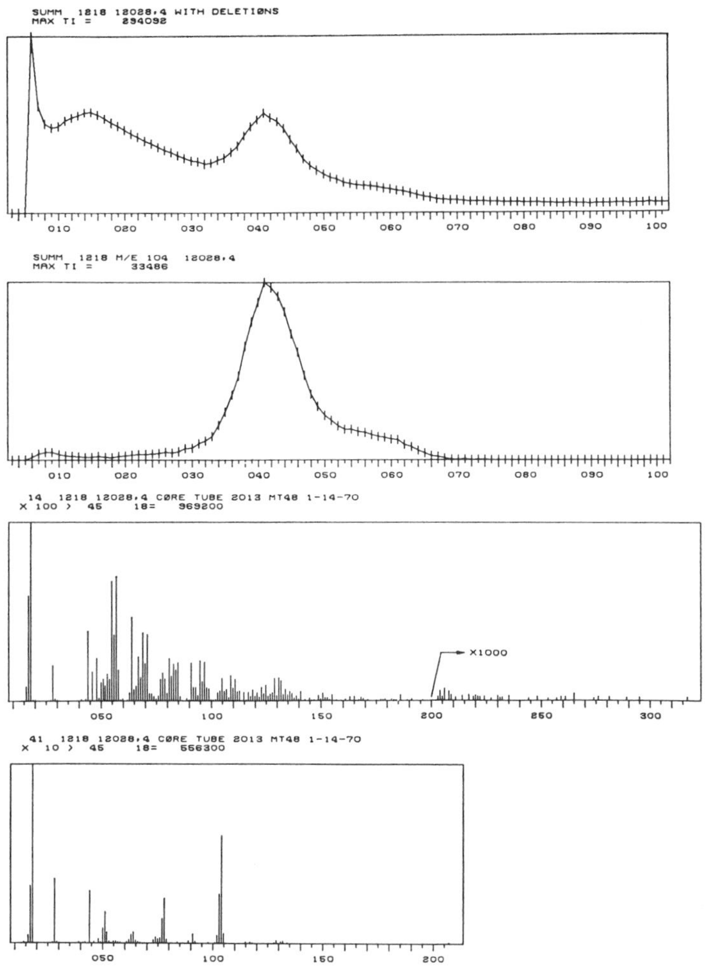

Fig. 4. Mass spectral data for core sample 12028,4. *Top:* Sum plot of total ion current vs. scan number (with deletions). *Second:* Plot of m/e 104 intensity vs. scan number. *Third:* Mass spectrum scan 14 (normalized to m/e 18, intensity scale expanded by 100 above m/e 45 and by 1000 above m/e 200). *Fourth:* Mass spectrum scan 41 (normalized to m/e 18, intensity scale expanded by 10 above m/e 45).

was 2.5×10^5 from a 150 mg sample representing an approximate estimate of 105 ppb Cd (based on n-tetracosane calibration—LSPET, 1970b). The sum plot of m/e 113 [cf. Fig. 5(c)] indicates the volatilization of the cadmium later during the analysis. Large amounts of SO_2 are evolved from the sample [cf. Fig. 5(a) and (b)].

Sample 12028,1. The data for this sample have been presented earlier (LSPET, 1970b) and were briefly discussed. The peak at m/e 58 which was not discussed previously is now thought to be a contaminant such as trimethylene oxide (Structure II) resulting from surface polymerization of ethylene oxide (a sterilizing agent used at the LRL).

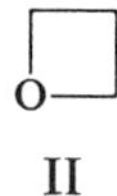

II

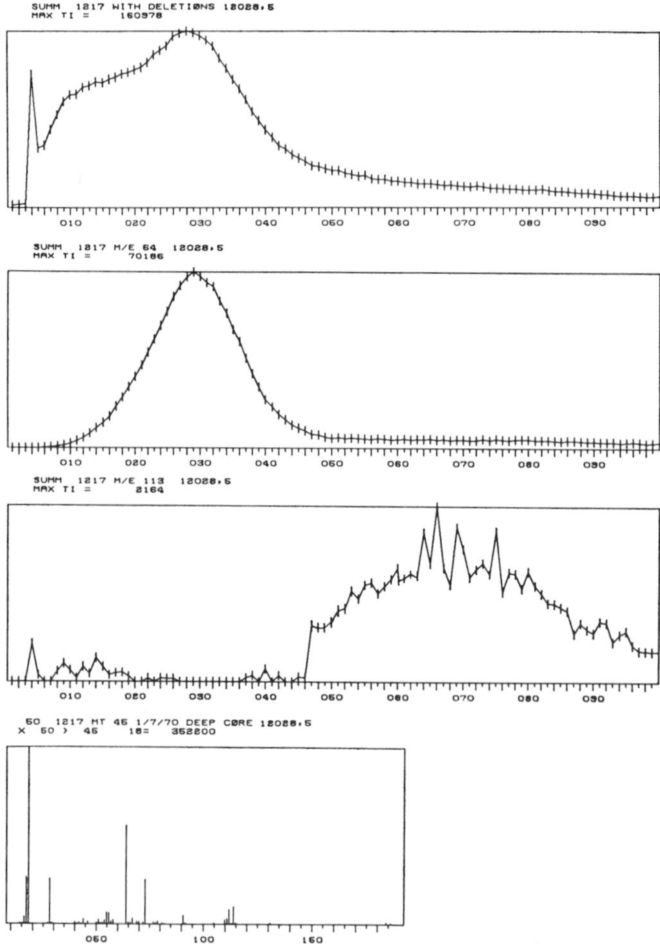

Fig. 5. Mass spectral data for core sample 12028,5. *Top*: Sum plot of total ion current vs. scan number (with deletions). *Second:* Plot of m/e 64 intensity vs. scan number. *Third:* Plot of m/e 113 intensity vs. scan number. *Fourth:* Mass spectrum scan 50 (normalized to m/e 18, intensity scale expanded by 50 above m/e 45).

CONCLUSIONS

The majority of the volatilizable material in the double core tube samples represents contamination. In the case of samples 12028,2 and 12028,1, where the contamination level is relatively low *and* there is no polymeric contaminant (e.g., Teflon) present, there is an indication of the release of organic matter late in the analyses. These data could be interpreted either as pyrolysis products of endogenous organic matter or synthetic products from simpler carbon containing species (PRETI *et al.*, 1971; NAGY *et al.*, 1971). Assuming these data are indicative of the former, the level is estimated to be no greater than 100 ppb, not including the m/e values which were deleted, indicating compounds such as methane, CO, CO_2, etc.

Cadmium was found only in samples 12028,5 and 12028,15 and in none of the others analyzed by LSPET (1970b). This corroborates the results reported by GANAPATHY *et al.* (1970) on sample 12028,66 and may indicate two ejecta levels of cadmium enriched material.

It is evident from the lack of correlation of amount of organic material with depth or sample morphology, that much of the contamination observed is a result of sample handling procedures. It should be emphasized that these samples received minimum exposure and handling between the moon and the mass spectrometric system. Even with this relatively short exposure, some contamination of the samples has occurred. Caution must be exercised in interpretation of subsequent data recorded on these samples because of the many sample handling procedures exercised between the above experiments and the eventual distribution to the principal investigators.

Acknowledgment—We thank the National Aeronautics and Space Administration for financial support (UCB Contract NAS 9–9593 and MIT Contract NAS 9–7381).

REFERENCES

GANAPATHY R., KEAYS R. R., and ANDERS E. (1970) Apollo 12 lunar samples: Trace element analysis of a core and the uniformity of the regolith *Science* **170**, 533–535.

HITES R. A. and BIEMANN K. (1970) Computer evaluation of continuously scanned mass spectra of gas chromatographic effluents *Anal. Chem.* **42**, 855–860.

LSPET (Lunar Sample Preliminary Examination Team) (1969a) Preliminary examination of lunar samples from Apollo 11 *Science*, **165**, 1211–1227.

LSPET (Lunar Sample Preliminary Examination Team) (1969b) *Lunar Sample Information Catalog—Apollo 11*. Lunar Receiving Laboratory, NASA–MSC, August 31.

LSPET (Lunar Sample Preliminary Examination Team) (1970a) Preliminary examination of lunar samples from Apollo 12 *Science* **167**, 1325–1339.

LSPET (Lunar Sample Preliminary Examination Team) (1970b) *Lunar Sample Information Catalog—Apollo 12*. NASA Technical Report R–353.

NAGY B., DREW C. M., HAMILTON P. B., MODZELESKI J. E., MODZELESKI V. E., NAGY L. A., SCOTT W. M., THOMAS J. E., UREY H. C., and WARD R. (1971) Carbon compounds in Apollo 12 lunar samples. Second Lunar Science Conference (unpublished proceedings).

ORÓ J., FLORY D. A., GIBERT J. M., MCREYNOLDS J., LICHTENSTEIN H. A., WIKSTROM S., and GIBSON E. K. JR. (1971) Abundances and distribution of organogenic elements and compounds in Apollo 12 lunar samples. Second Lunar Science Conference (unpublished proceedings).

PRETI G., MURPHY R. C., and BIEMANN K. (1971) The search for organic compounds in various Apollo 12 samples by mass spectrometry. Second Lunar Science Conference (unpublished proceedings).

SIMONEIT B. R. and FLORY D. A. (1971) Apollo 11, 12, and 13 Organic Contamination Monitoring History. NASA Special Publication (in press).

SMITH D. H., OLSEN R. W., WALLS F. C., and BURLINGAME A. L. (1971) Realtime organic mass spectrometry: LOGOS—A generalized laboratory system for high and low resolution, GC–MS and closed-loop applications *Anal. Chem.* (in press).

Proceedings of the Second Lunar Science Conference, Volume 2, pp. 1901–1912.
The M.I.T. Press, 1971.

Study of carbon compounds in Apollo 11 and Apollo 12 returned lunar samples

W. HENDERSON, W. C. KRAY, W. A. NEWMAN, W. E. REED,
B. R. SIMONEIT, and M. CALVIN*
Space Sciences Laboratory, University of California, Berkeley, California 94720

(*Received 22 February 1971; accepted in revised form 31 March 1971*)

Abstract—In order to determine the nature and chemical characteristics of the carbon present in the returned lunar materials, several different experimental approaches have been used. Temperature-dependent release of carbonaceous molecules has been examined from ambient to 1000°C. The results indicate that there are at least two different forms of carbon present in the surface fines in addition to terrestrial organic contaminants.

The release of volatile carbon-containing molecules such as CO, CH_4, and CO_2 during acid dissolution has been investigated further, using mass spectrometry for detection and quantitation. A gas chromatographic method of analysis has also been developed to quantitate the release of CO and C_1–C_4 hydrocarbons. The concentrations of these components has been quantitatively determined as CO, less than 1 ppm; C_1–C_4 total hydrocarbons, up to 38 ppm for Apollo 11 and Apollo 12 samples; and CO_2, up to 40 ppm.

Soluble organic compounds obtained upon solvent extraction of the surface fines have been examined; it appears that the possible indigenous organic compounds amount to less than 1 ppm C (wt/wt fines).

INTRODUCTION

WE HAVE REPORTED previously that, using our techniques, the indigenous organic content of the Apollo 11 fines (10086,A) was < 1.0 ppm and that demineralization of the fines with 20% HF yielded 66 ppm of carbon as carbon monoxide (CO) (BURLINGAME *et al.*, 1970). Using the same detection and quantitation procedure we confirmed the total carbon content found by other groups (MOORE *et al.*, 1970; KAPLAN *et al.*, 1970; ABELL *et al.*, 1970; ORÓ *et al.*, 1970). No complete carbon mass balance for the Apollo 11 fines has as yet been reported.

Since then our efforts have been directed towards confirming the concentration of CO obtained, and to obtaining a mass balance for carbon utilizing three different experimental approaches: (1) organic solvent extraction of lunar fines; (2) Carbon analysis at elevated temperatures; (3) HF demineralization of lunar surface and double core materials, and trace gas analysis in closed systems.

EXPERIMENTAL TECHNIQUES

Organic solvent extraction

All extractions were carried out in a class 100 organic clean-room facility. [Class 100 clean room as specified by Federal Standard 209a (1966) contains less than 1000 particles of 0.3–0.5 μ size and 100 or less particles of 0.5 μ or greater size per cubic foot.] The fines samples were

* Also, Laboratory of Chemical Biodynamics, University of California, Berkeley, California.

extracted in a mixture (3:1) of redistilled benzene and methanol (Mallinckrodt–nanograde purity) following the procedure of HAN *et al.* (1969) and BURLINGAME *et al.* (1970). For samples 12001 and 12023 (8.5 g of each) 50 ml of solvent mixture was used, and for samples 12032 and 12042 (3.5 g of each) only 30 ml was used. After filtration through a fine fritted filter, the extract was concentrated on a rotary evaporator, followed by further concentration under flowing helium. The extracts were introduced into the MS–902 high-resolution mass spectrometry–computer system via a direct introduction probe. The same on-line operating conditions as reported by BURLINGAME *et al.* (1970) were used.

Carbon analysis at elevated temperatures

All samples were analyzed using a modified Model 185 C, H, and N analyzer equipped with an HP–18 helium purifier (Hewlett-Packard, Palo Alto, California). All modifications to enhance stability and sensitivity have been reported elsewhere (JENSEN *et al.*, 1967). Daily calibration was provided with methyl stearate or $CaCO_3$. At levels of 15–18 μg carbon (as carbonate), measured fluctuations in absolute carbon mass calibration were consistently less than $\pm 2\%$ over the reported experimental period, whether measured before or after sample analyses.

Other standards used included argillaceous limestone (Standard Reference Material 1b, National Bureau of Standards), carbon black, calcined (i.e., heated at 550°C for 12 hours in air) graphite, calcined chromium carbide (ROC/RIC Corporation, Sun Valley, California), and calcined iron carbide (Fe–72, Wilshire Chemical Co., Gardena, California). A low level standard and matrix, calcined at 550°C, was prepared from a sample of crushed Table Mountain latite (from near Sonora, California) to which TiO_2 (7% wt/wt) had been added.

All samples were pyrolyzed in a static helium atmosphere. The pyrolysis cavity was maintained at a constant temperature of 1065°C, the present limit of the instrument. The sample was loaded onto a platinum boat and placed into a holding position at the cold ($\sim$60°C) end of the pyrolysis tube under helium. Pyrolysis was accomplished by sliding the boat into the hot cavity. The sample was allowed to remain in the hot cavity for 20 or 60 seconds (cf. Table 2) followed by injection of the products through a combustion tube to convert all released carbon compounds to CO_2. The CO_2 was then measured by a sensitive on-line gas chromatograph using a thermal conductivity detector. Differing pyrolysis times resulted in different terminal temperatures (925°–1000°C., cf. see Table 2) with concomitant release of different amounts of carbon.

Pyrolysis was performed on samples with three different thermal pretreatments: (1) untreated samples were pyrolyzed directly; (2) samples were maintained at 150°C for 12 hours in air to remove most surface adsorbed contaminants and then pyrolyzed; (3) samples were maintained at 550°C for 12 hours in air to remove remaining surface adsorbed materials and all amorphous carbon and then pyrolyzed. When standards were pyrolyzed under these conditions, quantitative recovery was achieved for all organic, amorphous, and carbonate carbon, with nearly quantitative (82%) recovery for graphitic carbon (mixed with an oxidizing catalyst consisting of manganese dioxide and tungsten oxide), but low ($\sim$10%) recovery for carbide carbon (as iron carbide and chromium carbide). The comparative recoveries for these carbides appears anomalous; however, the purity of iron carbide as purchased is questionable in that the composition cannot be balanced stoichiometrically. Therefore, the recovery of carbon from iron carbide is likely to be higher than the 10% shown for the Fe-72 standard.

HF demineralization and trace gas analysis

The previously reported experiment (BURLINGAME *et al.*, 1970) on the release and quantitative recovery of CO from lunar fines was repeated with basically similar apparatus and instrumentation. The modifications introduced were a glass reaction and trap system which was greatly reduced in volume (from that previously used), and the G.E.C.–A.E.I. MS–902 instrument was replaced by a C.E.C. 21–110B high-resolution mass spectrometer (BURLINGAME *et al.*, 1968) which was equipped with a micromanometer and an improved gas inlet system. The following spectrometer conditions were used for calibration and sample: resolution 5000 (static), ionizing voltage 70 eV, and ion source temperature 215°C. The system was calibrated for CO as reported by BURLINGAME *et al.*

(1970), except this time the electron multiplier output voltage was measured for the flat topped CO peak and correlated vs. the CO partial pressure. Lunar fines (10086,A, 5 g) were dissolved as before (BURLINGAME *et al.*, 1970) and the contents of the liquid helium trap were analyzed.

A gas chromatographic system was developed for the quantitative determination of the hydrocarbons and other gases released from lunar materials on solution with acid. The glass reaction system consisted of a dropping funnel connected to a removable reaction vessel via a glass Cajon fitting with a Teflon seal. Gas-tight Teflon stopcocks were used for introduction of HF into the dropping funnel and subsequently to the reaction vessel. The reactor was connected to a welded stainless steel manifold with Nupro valves (all stainless steel) for evacuation prior to a run, for blanketing the reaction with helium and for subsequent transfer of gaseous products for analysis. This transfer was made by flushing the reactor with helium and passing the effluent through a Dry Ice/isopropyl alcohol trap to remove HF and H_2O and then on to a 5 Å molecular sieve trap in liquid nitrogen to concentrate the product gases from the stream of helium. This trap was connected to the gas chromatographic system such that the helium carrier gas could be instantly transferred from a by-pass line to the heated trap via two 3-way valves. This system provided quantitative and reproducible injections of the product gases and the standard gas samples used for calibration.

Two gas chromatographic detection systems were used. Carbon monoxide, methane, and hydrogen were analyzed on a 40′ × $\frac{1}{8}$″ stainless steel column packed with Poropak Q in a constant temperature bath at -22 ± 0.1°C using a micro-thermal conductivity detector (Carle Instruments, Los Angeles). Methane and other hydrocarbons up to C_4 in carbon number were analyzed on a combination stainless steel column (10′ × $\frac{1}{8}$″ Poropak Q and 3′ × $\frac{1}{8}$″ Poropak N), temperature programmed from 50°C to 100°C at 10°C per minute using an Aerograph 665 gas chromatograph with a flame ionization detector. The system was calibrated by injections of known amounts (1–15 μl) of pure gases into the reactor system via a septum followed by a transfer and injection identical to actual run conditions. The minimum detection limits were 0.1 ppm with 100 mg samples using the thermal conductivity detector and 0.1 ppm using 20 mg samples with the flame ionization detector.

All runs with lunar samples, including HF alone and HF plus glass blanks, were performed as follows: Purified 20% HF (HAN *et al.*, 1969) was degassed under helium with ultrasonication and transferred, still under helium, to the dropping funnel. The sample was added to the reaction flask and then heated to 150°C for 5 minutes under vacuum to remove adsorbed gas from the sample. HF was then added to the sample (5 ml of HF solution per 100 mg of sample) under helium and allowed to react for 30 minutes at 50°C. Visual observation indicated that the reaction between the sample and HF had ceased after 20 minutes, often with complete dissolution. All evolved gases were transferred to the molecular sieve/liquid nitrogen trap by flushing with 250 ml of helium. The trap was isolated from the reactor by closing off the inlet valve, followed by heating to 350°C for 30 seconds and then injecting the trap contents into the gas chromatograph by rerouting the helium carrier gas from a bypass line through the trap.

RESULTS AND DISCUSSION

Organic solvent extraction of lunar fines

The concentrations of extractable organic compounds were low as determined by high resolution mass spectrometry only. The main results are shown in Table 1. The levels of contamination from sample handling procedures (SIMONEIT and FLORY, 1971) and from the LM retro-rocket exhaust (SIMONEIT *et al.*, 1969) are low, especially in the Special Environment Sample Container (SESC) sample. Sample 12001 contains the largest amount of extractable material, consisting mainly of the more common saturated hydrocarbons and significant amounts of LM engine exhaust products. Samples 12023, the SESC sample, and 12032 show the lowest amounts of extractable

Table 1. Composition of the benzene-methanol extracts of lunar fines from Apollo 12, as determined by high-resolution mass spectrometry.

Nominal Mass	Species	Blank	Samples			
			12011	12023(SESC)	12032	12042
27	HCN^c	30[a] (4)[b]	3200[a] (7)[b]	20[a] (1)[b]	20[a] (1)[b]	200[a] (8)[b]
30	NO^c	—	3150 (8)	25 (1)	—	3200 (7)
31	HNO^c	—	720 (1)	—	—	1500 (1)
41	CH_3CN^c	—	—	—	—	100 (4)
43	C_2H_3O	1200 (10)	7200 (10)	350 (10)	400 (9)	1500 (10)
	$HCNO^c$	—	700 (1)	—	—	150 (4)
44	$C_2H_6N^c$	—	1380 (4)	—	60 (1)	200 (5)
45	CHO_2	—	5700 (9)	—	—	2500 (9)
	CH_3NO^c	—	100 (1)	20 (1)	—	200 (2)
46	CH_2O_2	—	4500 (9)	—	—	1900 (9)
	CH_4NO^c	—	650 (1)	—	—	300 (2)
57	C_4H^{9d}	5500 (10)	9800 (10)	2000 (10)	2500 (10)	2000 (10)
104	C_8H_8	50 (4)	550 (3)	—	—	100 (6)
149	$C_8H_5O_3$	4000 (10)	5500 (9)	550 (10)	320 (10)	500 (8)

[a] Average ion current (relative).
[b] Occurrence number per 10 scans.
[c] Composition attributable to LM retro-rocket exhaust (SIMONEIT *et al.*, 1969).
[d] Hydrocarbons of the series C_nH_{2n+2} to C_nH_{2n-6}, but mainly the more saturated series, were found in the mass spectral data for $n = 1$ to 14 in the extracts of the samples and the blank.

material observed to date, with essentially no compounds attributable to the LM engine exhaust.

Carbon analysis at elevated temperatures

Results obtained for the temperature/time dependent release of carbon from Apollo 11 samples are summarized in Table 2. Surface adsorbed contaminants released by treatment in air at 150°C for 12 hours amount to 1–2 ppm C. Additional carbon amounting to 16 ppm and 20 ppm for the fines and the rock respectively is released upon treatment at 550 C for 12 hours. Under these conditions, recovery of amorphous carbon and carbonate carbon was 100% and 0%, respectively. The recovery (5 ppm) of extractable organic contaminants (BURLINGAME *et al.*, 1970) accounts for only 25–33% of this material. Therefore, the remainder of this carbon could be surface adsorbed CO_2 and possibly amorphous carbon. Between the 550°C/12 hour treatment and 925°C/20 second pyrolysis, the fines and the rock release additional carbon amounting to 36 ppm and 55 ppm, respectively. Under these conditions carbonate recovery is 100%, graphite recovery is 65%, and carbide recovery, as iron carbide or chromium carbide is less than 10%. Therefore, should carbonate or graphite carbon be present, all of the carbonate, most of the graphite, and only a little of the carbide would be recovered. An additional 24 ppm carbon from the fines results from increasing the terminal pyrolysis temperature and time from 925°C/20 seconds to 1000°C/60 seconds. The latter conditions providing 82% recovery of graphite but only approximately 10% for carbide. Therefore, this 24 ppm could be predominantly graphitic carbon.

These data suggest the existence of several forms of carbon in the lunar fines—adsorbed terrestrial contamination (5 ppm), endogenous organic compounds, adsorbed CO_2 or amorphous carbon (11–15 ppm), graphite ($\geq$24 ppm), and possible

Table 2. Carbon content of lunar samples by total organic carbon determination technique.

Sample	Initial Treatment Temp. (°C)/Time(hr)	Pyrolysis Terminal Temp. (°C)/Time(sec)	Carbon Content (ppm)
Surface fines 10086A	Ambient	925/20[c]	53.9
	150/12[a]	925/20[c]	52.0
	550/12[b]	925/20[c]	36.4
	Ambient	1000/60[d]	78.3
Interior chip Rock 10059	Ambient	925/20[c]	74.0
	150/12[a]	925/20[c]	73.1
	550/12[b]	925/20[c]	53.3

[a] Probably removes most surface adsorbed contaminants in 12 hrs.
[b] Removes amorphous carbon.
[c] 100% recovery of organic compounds, amorphous carbon, and carbonates; 65% recovery of graphite, < 10% recovery of iron and chromium carbide.
[d] 82% recovery of graphite, ~10% recovery of iron and chromium carbide.

carbonate (≤ 36 ppm). In addition, carbide carbon, at least part of which is hydrolyzable (see below), would account for the difference between the total recovered in these experiments (76–80 ppm) and the much higher values (140–200 ppm) previously reported for total carbon content (for example, Moore *et al.*, 1970).

HF demineralization and trace gas analysis

A new experimental system was devised to investigate and quantify the hydrocarbons released from lunar fines, probably by HF acid reaction with carbides (Chang *et al.*, 1971). This system was also designed to investigate CO evolution under the same reaction conditions. A liquid nitrogen trap packed with 5 Å molecular sieve was used to collect effluent components. Quantitative results were obtained using a dual gas chromatographic detection system (microthermal conductivity and flame ionization detectors). The system sensitivity allowed the use of 100 mg aliquots of surface materials for dual detection and 20 mg aliquots using a flame ionization detector alone. The results from these analyses are shown in Table 3, and representative chromatograms are shown in Figs. 1, 2, and 3. The main findings are: (1) CO was not detected by the micro-thermal conductivity system, which has a minimum detection limit of 0.1 ppm carbon as CO per 100 mg of sample; (2) the Apollo 11 fines exhibit a significantly higher production of CH_4 than Apollo 12 fines; (3) the breccia (10059) yields approximately twice as much CH_4 as the fines; (4) there is no correlation of CH_4 production with depth in the core (12028) [the noncorrelation of organic material obtained by pyrolysis-mass spectrometry with depth in the double core was noted (Burlingame *et al.*, 1971)]; (5) the core sample at 11–12 cm depth exhibits significantly more CH_4 than the other core aliquots (cf. Fig. 3) [the sample from the 11 cm depth exhibited an indication of possible indigenous organic matter (Burlingame *et al.*, 1971)]; and (6) large amounts of hydrogen are evolved during hydrolysis.

The total gas mixture from the molecular sieve trap was also analyzed by high

Table 3. Gaseous hydrocarbons released on demineralization of lunar surface materials.

	Carbon Content (ppm C)				
	Total	CH_4	Total C_2	Total C_3	Total C_4
Apollo 11 Fines (10086,A)	21.7	13.2	2.9	4.5	1.1
Apollo 11 Rock (10059) Interior	32.1	23.3	5.3	2.5	0.9
Apollo 11 Rock (10059) Exterior	38.2	24.2	5.7	5.9	2.4
Apollo 12 Fines					
(12001,23)	14.0	10.1	2.1	1.2	0.6
(12023,6)	17.6	11.4	3.0	2.1	1.2
(12032,7)	3.7	1.9	0.7	0.8	0.3
(12033,14)	9.3	3.5	2.5	2.5	0.8
(12042,6)	32.2	13.3	3.6	9.9	5.3
Apollo 12 Core					
(12028,12) 166 (11–12 cm)	19.2	9.3	4.5	3.5	1.9
(12028,24) 172 (20.8–21.8 cm)	7.0	3.7	1.7	1.1	0.5
(12028,29) 175 (25.4–26.1 cm)	8.5	6.4	1.3	0.5	0.3
(12028,33) 179 (28.8–30.0 cm)	9.0	5.6	1.8	1.2	0.4
(12028,35) 184 (30.6–31.2 cm)	7.2	4.2	1.1	1.0	0.9

resolution mass spectrometry (G.E.C.–A.E.I. MS–902). The results shown in Table 4 indicate that the most abundant species are H_2S, CO_2, CO, CS_2, and CH_4. Hydrocarbons in the C_1–C_6 carbon number range and traces of benzene and tropylium ion are present.

There are conflicting results regarding CO liberated on HF treatment from the mass spectrometric experiments on Apollo 11 fines (BURLINGAME *et al.*, 1970), and the mass spectrometric and gas chromatographic experiments reported here. The previous result of 66 ppm carbon as CO from fines sample 10086,A seemed to be corroborated within experimental error by utilizing the CEC 21–110B mass spectrometer and inlet system as described in the experimental section. Hence two possibilities should be considered to explain the carbon monoxide anomalies: (a) instrumental artifacts, and (b) sorption-desorption phenomena on the lunar fines as suggested by ORÓ *et al.* (1971).

There are two possible instrumental sources which could introduce the observed CO signal. First, oxygen in the evolved gases could react with carbon previously deposited on various ion source surfaces. Second, amplifier "ringing" after intense peaks was noted in the high resolution mass spectrometric data, particularly after the large nitrogen peak. The synthesis of CO in the ion source of the MS–902 on hot surfaces such as the filament was investigated by introducing oxygen-18 gas (10 cm^3 at 80 mm) to the source via a doser and a gold leak (Fig. 4). There is indeed a generation of $C^{18}O$ as well as $C^{18}O_2$ and $CO^{18}O$. The $C^{18}O$ intensity ranges from 1 to 13% on the increasing $^{18}O_2$ intensity. The $CO^{18}O$ and $C^{18}O_2$ generated amounts to less than 0.1%

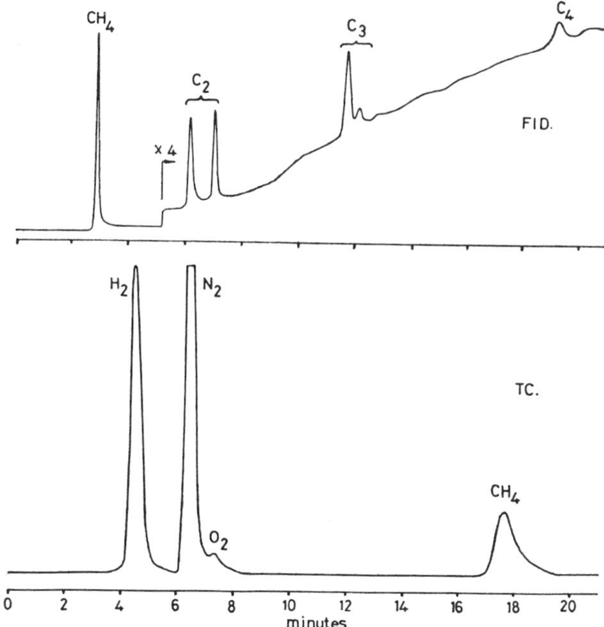

Fig. 1. Gas chromatograms (F.I.D. and T.C.) for Apollo 12 bulk fines 12023 after HF demineralization. (Conditions: (a) F.I.D. combination stainless steel column, 10′ × ⅛″ packed with Poropak Q and 3′ × ⅛″ packed with Poropak N, temperature programmed from 50°–100°C at 10°C per minute, using helium at a flow rate of 40 ml/min. (b) T.C. 40′ × ⅛″ stainless steel column packed with Poropak Q in a constant temperature bath at −22° ± 0.1°C, using helium at a flow rate of 40 ml/min.)

of the $^{18}O_2$ intensity. The "ringing" of the electron-multiplier amplifier amounts to 0.7–1.0% of the nitrogen peak intensity and tails into the CO peak. The net effect is a summed peak area in the high resolution mass spectral calculations.

Correcting for the generation of CO by oxygen, as well as the minor amplifier effect, the level of CO released by HF treatment of fines appears to be less than 1 ppm. This value is corroborated by the gas chromatographic results reported above.

The presence of CO as an artifact, as discussed above, in no way affects the determinations of total carbon as CO by vacuum pyrolysis at 1150°C in this laboratory (BURLINGAME *et al.*, 1970). In contrast to the statements of GIBSON and JOHNSON (1971), the generation of CO by the reaction FeO + C (any form of carbon) → CO + Fe is well known and used in commercial blast furnace processes (other minerals such as SiO_2 analogously form CO). During the vacuum pyrolysis experiments, the oxygen and nitrogen concentrations in the mass spectrometer source were at instrument background levels, thus eliminating the possible synthesis of CO and electron multiplier amplifier "ringing."

The possibility of adsorption of gases on the lunar fines should be considered, since the major LM exhaust products are NH_3, H_2O, HCN, CO, NO, O_2, CO_2, and

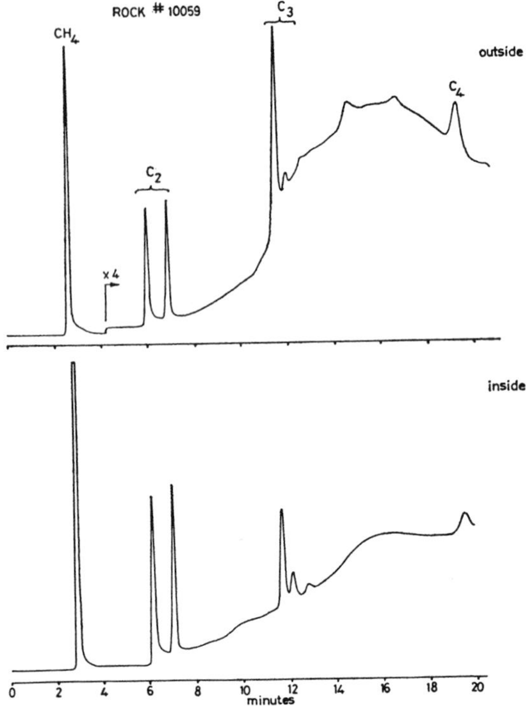

Fig. 2. Gas chromatograms (F.I.D.) after HF demineralization of interior and exterior chips (Breccia 10059). (Conditions as cited in Fig. 1).

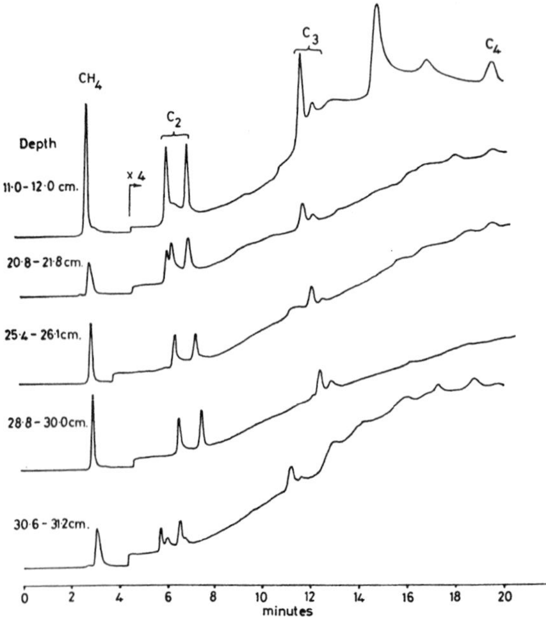

Fig. 3. Gas chromatograms (F.I.D.) of the hydrocarbons from HF demineralization of the 12028 lunar core tube aliquots. (Conditions as cited in Fig. 1).

Table 4. High resolution mass spectral data for the Apollo 11 lunar fines (10086,A) HF demineralization total gases.

Observed Mass	Mass error (mmu*)	Relative Average Intensity†	Elemental Composition
16.03092	1.14	1482	CH_4
27.99498	−0.02	1200	CO
28.00615	0.00	42300	N_2
28.03124	0.06	400	C_2H_4
29.00315	−0.41	354	CHO
29.03896	0.04	158	C_2H_5
30.04695	0.00	85	C_2H_6
31.01834	−0.37	115	CH_3O
31.97227	−0.20	7206	S
31.99001	−0.18	2576	O_2
32.98012	−0.22	6850	HS
33.98800	−0.23	18130	H_2S
40.03164	−0.34	105	C_3H_4
41.03927	−0.15	245	C_3H_5
42.04708	−0.14	131	C_3H_6
43.01684	0.07	6	C_2H_3O
43.05463	−0.07	101	C_3H_7
43.97208	0.02	108	CS^+
43.98988	−0.15	14865	CO_2
44.02638	−0.17	65	C_2H_4O
45.03435	−0.35	88	C_2H_5O
45.99348	−1.35	90	NO_2
46.99627	0.40/−1.83	72	CH_3S/COF
47.96672	0.27	84	SO
52.03084	0.46	65	C_4H_4
53.03920	−0.40	67	C_4H_5
54.04664	0.30	54	C_4H_6
55.05589	−0.21	158	C_4H_7
56.06329	0.13	87	C_4H_8
57.07061	−0.19	69	C_4H_9
59.05062	0.20	112	C_3H_7O
59.96874	−0.05	137	COS
62.97040	0.07	146	CFS
63.94387	0.27	65	S_2
63.96374	0.14	148	SO_2
65.99134	0.37	50	COF_2
67.05514	−0.37	62	C_5H_7
69.06938	−0.08	81	C_5H_9
70.07703	0.98	76	C_5H_{10}
75.94413	0.09	833	CS_2
77.04013	−1.01	73	C_6H_5
78.04474	0.78	98	C_6H_6
91.05414	0.63	40	C_7H_7

* mmu = millimass unit.
† Ten scans averaged and corrected for instrument background.

NO_2 (Simoneit *et al.*, 1969). Some of these gases could have been adsorbed during the lunar landing, then desorbed during long storage of the samples.

It should also be noted that the CO_2 abundance (cf. Table 4) is significantly higher than its atmospheric abundance and therefore cannot be attributed to occluded air. Furthermore, the concentration of CO is six times lower than that of the CO_2. This completely contradicts the previous results obtained using the liquid helium trap and high resolution mass spectrometry (Burlingame *et al.*, 1970). Methane and CO_2 have essentially equal ionization potentials. By taking their respective fragmentation

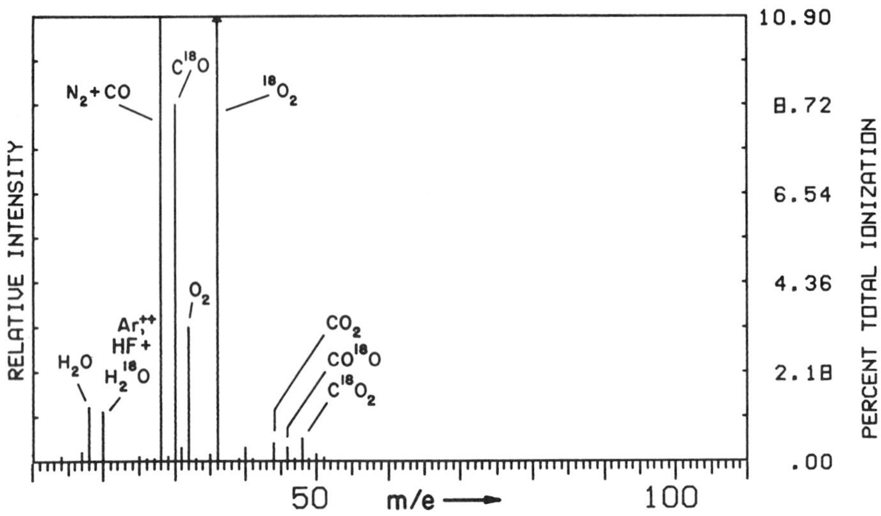

Fig. 4. Low resolution mass spectrum of the products from oxygen-18 gas in the ion source of an MS–902 (the $^{18}O_2$ peak at m/e 36 has a relative intensity of 650, assuming 100 as full scale).

patterns into account and then relating the molecular ion intensities (corrected for both instrument and blank run backgrounds) to the observed CH_4 concentration (GC method), we calculate 62 ppm of CO_2 liberated on HF treatment. (In 70 eV mass spectra of the individual gases, the CH_4 peak is 50% of the total methane ionization and the CO_2 peak is 80% of the total carbon dioxide ionization.) Since we have previously set an upper limit of 36 ppm for CO_2 derived from carbonate in the lunar fines, this suggests that, during the handling procedures at MSC and during the 18 months storage and handling in our laboratories, approximately 15 to 26 ppm of atmospheric CO_2 could have been adsorbed on the surface of the fines. This is supported by the fact that when we first analyzed the evolved gases upon acid dissolution of the fines, the CO_2 concentration found was of the order of 10 ppm (BURLINGAME *et al.*, 1970).

All solution experiments with lunar samples, but none of the standardization or calibration experiments, exhibit a peak in the thermal conductivity chromatographic record which elutes 2 minutes before nitrogen. This peak has been identified as hydrogen on the basis of its retention characteristics and low resolution mass spectrometry. Preliminary quantitation for H_2 indicates a concentration of approximately 4 cc STP/g of lunar fines. Just what proportion of the evolved hydrogen is endogenous to the lunar materials or results from reduction of the acid on contact with those materials is not known at the present time. The use of deuterated acids in dissolution experiments, currently underway in this laboratory, should resolve this question.

FRIEDMAN *et al.* (1970) reported 150 to 450 ppm hydrogen for three breccias (10046,21; 10046,22; and 10060,11), while HINTENBERGER *et al.* (1970) estimate 92 ppm for breccia 10021,20. Therefore, our value of 350 ppm H_2 obtained in HF dissolution experiments, although from lunar fines, corroborates the data obtained in vacuum pyrolysis experiments.

Conclusions

(1) Trace gas analysis by high resolution mass spectrometry appeared to confirm the release of CO from lunar fines by HF dissolution while gas chromatography followed by mass spectrometry did not detect CO above background levels. It was then found that the carbon on the ion source filament reacts with oxygen (approximately a minimum of 1% conversion) yielding CO. Furthermore, the amplifier bounce from the nitrogen peak overshoot makes a minor contribution to the CO peak. Thus, the net amount of CO detected by mass spectrometry of HF dissolution of the fines could be less than 1 ppm. (2) The determinations of total carbon as CO by vacuum pyrolysis at 1150°C in this laboratory are not affected by instrumental artifacts. (3) A range of hydrocarbons, including CH_4, C_2H_4, C_2H_6, C_3H_8, up to C_6 in carbon number are evolved during HF demineralization. These probably are derived from different forms of carbides. (4) Apollo 11 lunar materials yield significantly higher concentrations of hydrocarbons than Apollo 12 materials on HF treatment. (5) There appears to be no correlation of hydrocarbon release upon HF treatment with depth in the Apollo 12 double core (12028), except that sample 12028,12 (closest to the surface) released predominantly more hydrocarbons. (6) Preliminary results from several analytical techniques allow us to tentatively set maximum limits for the categories of carbon present in the lunar fines as follows: endogenous organic carbon ($\ll$1 ppm), surface contamination ($\leq$5 ppm), adsorbed CO_2 (15–26 ppm), carbonate ($\leq$40 ppm) (cf. Gay *et al.*, 1970), graphite (25–50 ppm) (cf. Arrhenius *et al.*, 1970), hydrolyzable carbides ($\leq$25 ppm) nonhydrolyzable carbides and other forms of carbon not accounted for above (10–50 ppm).

Acknowledgments—We acknowledge the contributions of Dr. A. L. Burlingame, Miss M. Petrie, Dr. P. Holland, Miss P. Wszolek, and Mr. F. C. Walls (mass spectrometry); Dr. F. Lindgren (C analyses); Dr. G. Steel and Mr. J. Moss (discussions and technical assistance); and Mr. H. Melling and Mr. D. Aufdenkamp (glass blowing). This work was supported by NASA MSC Contract No. NAS 9–7889.

References

Abell P. I., Draffan C. H., Eglinton G., Hayes J. M., Maxwell J. R., and Pillinger C. T. (1970) Organic analysis of the returned Apollo 11 lunar sample. *Proc. Apollo 11 Lunar Sci. Conf., Geochim. Cosmochim. Acta* Suppl. 1, Vol. 2, pp. 1757–1773. Pergamon.

Arrhenius G., Asunmaa S., Drever J. I., Everson J., Fitzgerald R. W., Frazer J. Z., Fujita H., Hanor J. S., Lal D., Liang S. S., MacDougall D., Reid A. M., Sinkankas J., and Wilkening L. (1970) Phase chemistry, structure and radiation effects in lunar samples. *Science* **167**, 659–661.

Burlingame A. L., Smith D. H., and Olsen R. W. (1968) High resolution mass spectrometry in molecular structure studies, XIV. Real-time data acquisition, processing and display of high resolution mass spectral data. *Anal. Chem.* **40**, 13–19.

Burlingame A. L., Calvin M., Han J., Henderson W., Reed W., and Simoneit B. R. (1970) Study of carbon compounds in Apollo 11 lunar samples. *Proc. Apollo 11 Lunar Sci. Conf., Geochim. Cosmochim. Acta* Suppl. 1, Vol. 2, pp. 1779–1791. Pergamon. Also the references therein.

Burlingame A. L., Hauser J. S., Simoneit B. R., Smith D. H., Biemann K., Mancuso N., Murphy R., Flory D. A., and Reynolds M. A. (1971) Preliminary organic analysis of the Apollo 12 cores. Second Lunar Science Conference (unpublished proceedings).

Chang S., Kvenvolden K., Lawless J., Ponnamperuma C., and Kaplan I. R. (1971) Carbon, carbides, and methane in an Apollo 12 sample. *Science* **171**, 474–477.

Friedman I., O'Neil J. R., Adami L. H., Gleason J. D., and Hardcastle K. (1970) Water hydrogen, deuterium, carbon, carbon-13, and oxygen-18 content of selected lunar material. *Science* **167**, 538–540.

Gay P., Bancroft G. M., and Bown M. G. (1970) Diffraction and Mössbauer studies of minerals from lunar soils and rocks. *Science* **167**, 626–628.

Gibson E. K. and Johnson S. M. (1971) Thermal analysis-inorganic gas release studies of lunar samples. Second Lunar Science Conference (unpublished proceedings).

Han J., Simoneit B. R., Burlingame A. L., and Calvin M. (1969) Organic analysis on the Pueblito de Allende meteorite. *Nature* **222**, 364–365.

Hintenberger H., Weber H. W., Voshage H., Wänke H., Begemann F., Vilscek E., and Wlotzka F. (1970) Rare gases, hydrogen, and nitrogen: Concentrations and isotopic composition in lunar material. *Science* **167**, 543–545.

Jensen L. C., Ewing A. M., Wills R. D., and Lindgren F. T. (1967) The use of the computer in serum lipid and lipoprotein analysis. *J. Amer. Oil Chem. Soc.* **44**, 5–10.

Kaplan I. R., Smith J. W., and Ruth E. (1970) Carbon and sulfur concentration and isotopic composition in Apollo lunar samples. *Proc. Apollo 11 Lunar Sci. Conf., Geochim. Cosmochim. Acta* Suppl. 1, Vol. 2, pp. 1317–1329. Pergamon.

Moore C. B., Gibson E. K., Larimer J. W., Lewis C. F., and Nichiporuk W. (1970) Total carbon and nitrogen abundances in Apollo 11 lunar samples and selected achondrites and basalts. *Proc. Apollo 11 Lunar Science Conf., Geochim. Cosmochim. Acta* Suppl. 1, Vol. 2, pp. 1375–1382. Pergamon.

Oró J., Updegrove W. S., Gibert J., McReynolds J., Gil-Av E., Ibanez J., Zlatkis A., Flory D. A., Levy R. L., and Wolf C. J. (1970) Organogenic elements and compounds in type C and D lunar samples from Apollo 11. *Proc. Apollo 11 Lunar Sci. Conf., Geochim. Cosmochim. Acta* Suppl. 1, Vol. 2, pp. 1901–1920. Pergamon.

Oró J., Flory D. A., Gibert J. M., McReynolds J., Lichtenstein H. A., Wikstrom S., and Gibson E. K., Jr. (1971) Abundances and distribution of organogenic elements and compounds in Apollo 12 lunar samples. Second Lunar Science Conference (unpublished proceedings).

Simoneit B. R., Burlingame A. L., Flory D. A., and Smith I. D. (1969) Apollo lunar module engine exhaust products. *Science* **166**, 733–738.

Simoneit B. R. and Flory D. A. (1971) Apollo 11, Apollo 12, and Apollo 13 organic contamination monitoring history. National Aeronautics and Space Administration, Special Publication (in press).

Proceedings of the Second Lunar Science Conference, Volume 2, pp. 1913–1925.
The M.I.T. Press, 1971.

Abundances and distribution of organogenic elements and compounds in Apollo 12 lunar samples

J. ORÓ, D. A. FLORY, J. M. GIBERT, J. MCREYNOLDS,
H. A. LICHTENSTEIN, and S. WIKSTROM
Departments of Biophysical Sciences and Chemistry, University of Houston,
Houston, Texas 77004

(*Received 22 February 1971; accepted in revised form 25 March 1971*)

Abstract—The Apollo 12 fines have yielded results comparable to our earlier analyses of Apollo 11 samples. The presence of organogenic elements, S (av. 1800 ppm), P (av. 570 ppm), C (230 ppm, one sample, core tube), and of compounds of organogenic elements H_2, H_2S, CO, CO_2, CH_4, C_2H_4, C_2H_6 has been corroborated. Some of these organic compounds were evolved by heating and others were generated by acid treatment. Although traces of benzene, toluene, and other organic substances were detected, no measurable quantities of solvent extractable organic compounds were found. Amino acid derivatives were not detected at a sensitivity of 20 nanograms per component. Most of the CO_2 evolved by heating to 750°C is of terrestrial origin. CO is the major carbon compound liberated at high temperatures. It appears to be endogenous, and its energy of release is approximately 30 kilocalories per mole. In general, the evolved gases result from chemical reactions, chemidesorption processes, or are released from solar wind implantation sites or microvesicles. The ultimate source of these compounds may be primary (lunar), secondary (solar, meteoritic, and cometary), or a combination of both.

INTRODUCTION

OUR PREVIOUS EXAMINATION of Apollo 11 fines and breccia has shown the presence in these samples of organogenic elements (H, C, N, O, S, P) and compounds of these elements (ORÓ *et al.*, 1970a, b, c). This includes appreciable quantities of the oxides of carbon (CO and CO_2), H_2, N_2, H_2S, smaller amounts of light hydrocarbons (CH_4, C_2H_6, C_2H_4, and C_2H_2), and other compounds of carbon (WACHI *et al.*, 1970). No measurable quantities of solvent extractable organic compounds (less than 0.001 ppm) and amino acid derivatives (less than 0.1 ppm) were detected in the bulk fines of the Apollo 11 samples. In some sample aliquots minor traces were observed of compounds such as benzene, toluene, and diisopropyl disulfide. The experimental data indicate these compounds were probably contaminants (GIBERT and ORÓ, 1970; GIBERT *et al.*, 1971). The total carbon content was of the order of 200 ppm and the $^{13}C/^{12}C$ ratio was positive ($\delta^{13}C = +2.6$ to $+18.5$ PDB), atypical of most terrestrial values.

We have continued these studies and generally applied the same analytical techniques to the Apollo 12 and related samples as were used for the Apollo 11 studies. Some modifications in equipment and procedure were adopted and are described here. In this work we have placed particular emphasis on low temperature (up to 770°C) effluent gas analysis and acidolysis in an attempt to learn more about the nature of the volatile, pyrolyzable, or decomposable carbon containing material in the lunar samples.

Organogenic Elements: Spark Source Mass Spectrometric Analysis

By means of spark source mass spectrometry (Oró *et al.*, 1970b; Socha *et al.*, 1970) the most abundant organogenic elements in the Apollo 12 samples (Table 1) were found to be sulfur (av. 1,800 ppm) and phosphorus (av. 570 ppm), in general agreement with the results from Apollo 11 (Oró *et al.*, 1970b) showing also a similar variation range. Carbon could be measured reliably in only one case (core tube, sample 12028,189) because of background interference. However, the value obtained for this sample (230 ppm) is within the range of those obtained for total carbon (70 to 250 ppm) by Moore *et al.* (1971) from double core tube sample 12028 (LSEPT, 1970).

Low Temperature Effluent Gas Analysis

Samples of fines from Apollo 11 (10086,16) and Apollo 12 (12023,9) were studied by an effluent gas analysis technique which was modified to allow determinations of the amounts, temperature release profiles, and binding energies of the volatile components released by heating under vacuum (Oró *et al.*, 1970b). The samples were placed in a stainless steel sample tube connected to the ion source of a quadrupole mass spectrometer (EAI Quad 250) by means of a stainless steel and glass inlet line, evacuated to a pressure of 1 to 3×10^{-8} torr, and pumped overnight prior to heating. An external oven was placed around the sample tube and heated from ambient to 750°C at a linear rate of 3.4°C/min. The maximum temperature was maintained for 2 to 4 hours, or until the evolution of gases subsided. The same sample tube was used for all analyses and was cleaned in acid (6% HF, 40% HNO_3, H_2O) and baked out under vacuum at 750°C prior to loading each sample. The mass spectrometer was used to monitor the cleanliness of the tube at 750°C before cooling and loading the sample. Reproducible background spectra were obtained from a programmed heating of the empty sample tube which approached the level of the vacuum system for all species detected in the samples. The samples were analyzed as received, unsieved, and with the minimum handling necessary.

For the effluent gas analysis, preweighed samples were transferred to the cleaned, baked-out sample tube and attached to the vacuum system immediately so that the sample tube was exposed to the atmosphere for the least amount of time possible. After evacuating the inlet system, as described above, the programmed heating was begun. The pressure of the vacuum system was monitored with a nude Bayard-Alpert ionization gage, and the temperature of the sample tube was measured with a

Table 1. Organogenic element concentrations (in ppm)*

Element	Apollo 12 sample			
	12028,189	12023,9	12032,14	12042,14
S	1600	3500	690	1400
P	190	680	820	590
C	230	—	—	—

* Hydrogen, nitrogen, and oxygen were not measured.

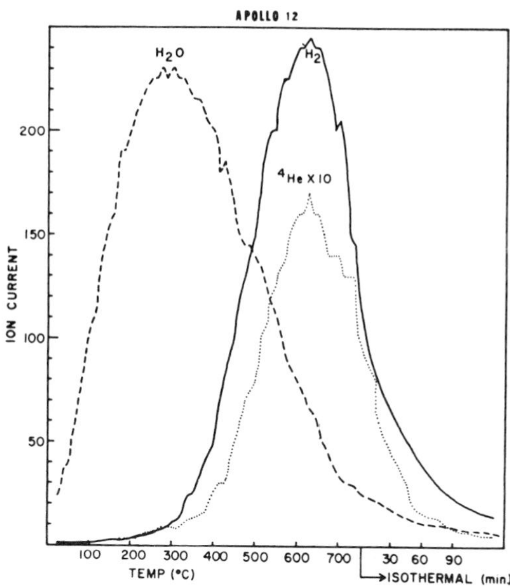

Fig. 1. Effluent gas analysis of water, hydrogen, and helium from room temperature to 750°C at a linear rate of 3.4°C/min using a quadrupole-mass spectrometer. First heating after evacuating the system overnight at 10^{-8} torr.

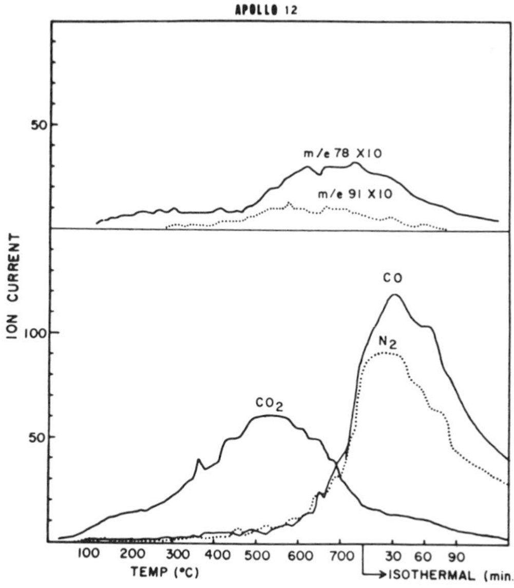

Fig. 2. Same as Fig. 1 for carbon dioxide, carbon monoxide, and nitrogen, as well as benzene (m/e 78) and the tropylium ion (m/e 91) representative of alkyl benzenes.

chromelalumel thermocouple. Mass spectra of the effluent gases were recorded at intervals of 12.5°C throughout the experiment and at 10 minute intervals during the isothermal period. After subtracting the background spectra, the ion intensities of the various volatile species were plotted as a function of temperature and time to obtain a gas evolution profile for each sample. The data are shown in Figs. 1 and 2.

Although the instrument was not capable of resolving N_2 from CO in the m/e 28 peak, it was possible to estimate the relative amounts of these two species using simultaneous equations involving the intensities of the peaks at m/e 44, 28, 16, 14, and 12 and the measured fragmentation patterns of standard gases. Both samples showed the presence of H_2, He, H_2O, N_2, CO, and CO_2, as well as traces of benzene and toluene. The mass spectra of the Apollo 11 fines showed the presence of a contaminant identified as *n*-butane. This particular sample of fines was the remainder of an aliquot taken to another laboratory for analysis, and the history of its handling and exposure to contamination is therefore unknown.

After obtaining a gas evolution profile as described above, the Apollo 12 fines were subjected to a procedure designed to determine the nature and extent of potential contamination resulting from atmospheric exposure. The once-heated sample was allowed to cool under vacuum. A second temperature programming was begun, and the spectra during this second cycle indicated essentially complete removal of all volatile species from the sample. The gas evolution release pattern for this sample is shown in Fig. 3 (note that the H_2O and CO_2 intensity plots are multiplied by 10). After again cooling under vacuum the sample was exposed to the atmosphere for two hours in the sample tube without handling or mixing. At the end of this exposure the tube was again installed on the inlet line and evacuated as before. The gas evolution

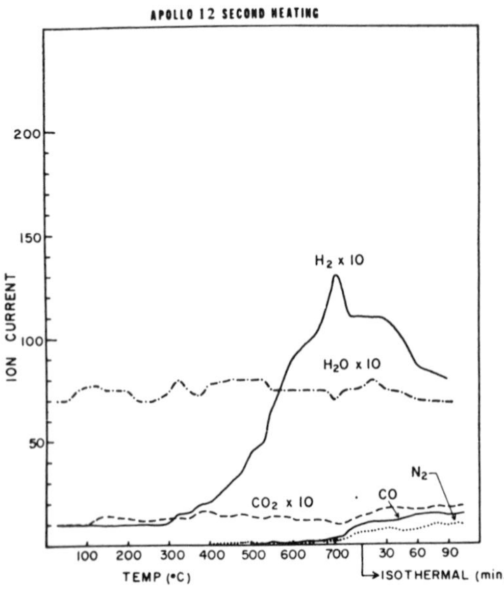

Fig. 3. Same system as for Fig. 1. Second temperature programming after cooling the sample under vacuum.

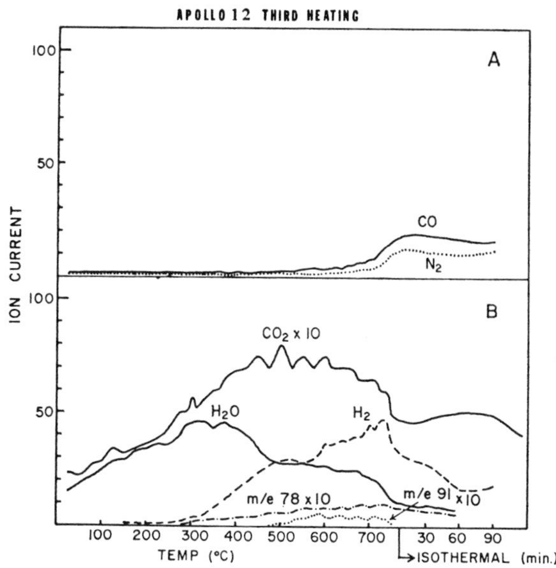

Fig. 4. Same system as for Fig. 1. Third temperature programming after cooling and exposing the sample to the laboratory atmosphere for two hours.

release pattern for the third and final programmed heating is given in Fig. 4 and showed the presence of H_2, H_2O, and CO_2. The evolution profiles of the H_2O and CO_2 were similar to the pattern of these species obtained on the first heating indicating that these species are adsorbed from the atmosphere. The H_2 is emerging at a higher temperature than that evolved initially indicating hydrogen is generated by reaction of atmospheric water with carbides or other solid compounds. MEINSCHEIN (1971) has offered a similar explanation for the generation of H_2 involving the reduction of metallic compounds. The metals thus formed could react with H_2O to yield H_2 and the metal hydroxides or oxides. These hydroxides and oxides could react with carbonic acid and release CO_2 when subsequently heated. A contribution from residual H_2 not completely degassed on the first two heatings also cannot be ruled out.

The ion intensity curves for each species were integrated numerically, and the areas converted into approximate concentration values using experimentally determined instrumental sensitivity factors. Table 2 shows the concentrations of the various volatile species computed by this method. In order to estimate the energy of the transition of a species from the bound state to the vapor state, a modified Clausius-Clapeyron equation was used as described in our earlier papers. These values are shown in Table 3.

It is interesting to note from these data that those components shown to be terrestrially adsorbed have the lowest binding energies. These binding energies give further evidence that the hydrogen (6 ppm) obtained on the third heating is due to the reduction of some of the adsorbed water at temperatures above 400°C. The gases assumed to be of solar origin, H_2 and He, have higher release energies than water consistent with their high energy of implantation. The temperature release profiles for

Table 2. Volatile concentrations (in ppm) Apollo 12 fines (12023,9) heated to 750°C

Component	1st run	2nd run	3rd run
H_2	44.3	1.0	6.33
He	4.14	—	—
H_2O	26.8	1.33	5.68
N_2	16.2	0.16	0.52
CO	137	0.19	0.83
CO_2	17.2	0.53	2.49
Carbon*	65.1	—	1.9
O_2	0	—	0

* Sum of carbon in CO and CO_2.

Table 3. Binding energies of volatile components (kcal/mole).

Component	Apollo 11	Apollo 12 1st run	Apollo 12 3rd run
H_2	10	7.5	n.d.
He	14.5	14	n.d.
H_2O	2.5	3.6	2.0
N_2	n.d.	n.d.	n.d.
CO	37	29	n.d.
CO_2	8.7	5.3	2.7

n.d. = not determined.

these gases match closely those determined by Lord (1958) and Gibson and Johnson (1971). Our values for the concentrations of H_2 and He are in close agreement with Hintenberger *et al.* (1971). Hintenberger *et al.* (1971) also noted that atmospheric contamination by hydrogen (water) and undetermined oxides of carbon occurred upon sieving of partially or totally degassed samples. The results of our experiments, together with the measurements of Epstein and Taylor (1971) on the deuterium content of water in lunar samples conclusively establish terrestrial contamination as the source of the water released upon heating. This is in agreement with the calculations of Anders (1970) and the experiments of Bloch *et al.* (1971).

Our similar observation of terrestrially adsorbed CO_2 amounting to a few parts per million is of particular interest for future organic analyses and carbon isotope ratio measurements. It suggests the need for separating the various carbon-containing compounds before making carbon isotope measurements and shows that there is at least one source of volatile carbon contamination which may influence the measured ratio of indigenous lunar carbon. This could partially explain the negative $\delta^{13}C$ values obtained by Chang *et al.* (1971) for the 500°C volatile fraction of the Apollo 12 fines. This effect will be especially significant for samples with a low concentration of indigenous carbon. The temperature release profiles of the gases found to be adsorbed further show that the methods commonly employed to remove atmospheric contaminants may not be completely effective.

The binding energy for CO is the largest, indicating that this component is strongly bound, chemisorbed, or is released by a chemical reaction involving carbides or other forms of carbon. Other investigators (Abell *et al.*, 1970b; Chang *et al.*, 1971) have made similar suggestions. We did not detect any of saturated or aromatic components reported by PRETI *et al.* (1971) and NAGY *et al.* (1971), with the exception of benzene and toluene. Our experimental conditions should minimize synthesis of higher molecular weight compounds from the primary gases released and, therefore, support PRETI's *et al.* (1971) interpretation that the higher molecular weight compounds are synthesis products. Complex hydrocarbon mixtures (RIESENFELD, 1942; HOERING, 1966; MEINSCHEIN, 1971) could also be synthesized from carbides in the presence of terrestrially adsorbed water upon heating the lunar samples.

The CO/CO_2 ratio for this Apollo 12 fines sample (Table 2) is in line with WELLMAN's (1970) calculations showing CO to be the dominant carbon species under the conditions assumed to have prevailed during the crystallization of the lunar material. This is in agreement with the results obtained for the Apollo 11 fines. The binding energies and temperature release profiles for identical compounds are similar for both the Apollo 11 and Apollo 12 samples (Table 3). For the temperature range covered in these experiments, our results are comparable to those obtained by GIBSON and JOHNSON (1971).

STEPPED VOLATILIZATION GAS CHROMATOGRAPHY–MASS SPECTROMETRY

The study of the volatiles released by stepwise heating from room temperature to 770°C was also carried out by following a method previously described (ORÓ *et al.*, 1970b, c). A 536 mg sample of Apollo 12 fines (sample 12023,9) was evacuated to 10^{-2} torr at 150°C for two hours in a quartz tube on line to an LKB-9000 gas chromatograph-mass spectrometer. The sample was then heated under vacuum at 200°, 400°, 600°, and 775°C. The gases evolved at each step were swept into and separated in a 3m × 2mm 200 mesh Porapak Q gas chromatographic column. Mass spectrometric evidence was obtained for the evolution of carbon monoxide, carbon dioxide, and methane. No gases were detected at 200°C. At 400°C carbon dioxide appeared in about 20 μg/g quantities and CO in comparable amounts. Methane appeared in amounts of the order of 0.8 μg/g. At 600°C the predominant gas was CO, being about 8 times more abundant than CO_2 and methane appeared in very small amounts. At 775°C, CO_2 was found to be on the order of 5.6 μg/g and CO about 7 times more abundant. Methane appeared in only trace amounts. The total amounts of gases released from 400° to 770°C were approximately 35 μg/g of CO_2, 85 μg/g of CO, and 1 μg/g of CH_4. Overall, the release of CO, CO_2, and CH_4 follows very closely earlier data (ORÓ *et al.*, 1970a, b, c).

The predominance of CO at higher temperatures is consistent with the reaction of elemental carbon and carbides with oxygen containing compounds, such as metal oxides and silicates. Comparative studies between Apollo lunar fines and synthetic analogs (composed of troilite, cohenite in a silicate matrix) indicate the general pattern of gas evolution can be duplicated at temperatures above 500°C (GIBSON and JOHNSON, 1971). The evolution of the bulk of methane at lower temperatures ($<$400°C) points towards low binding energy and a surface dependent origin. It appears that this compound could be a reaction product generated by partial reaction of carbides and/or elemental carbon with adsorbed water either before or during heating. This would be consistent with the strong adsorptive properties exhibited by the lunar fines for H_2O, N_2, and CO_2. (See low temperature effluent gas analysis and the second experiment in this section.) Based on the above, the reported indigenousness of a fraction of methane as reported by Abell *et al.* (1970b) and Chang *et al.* (1971) should be questioned.

In a second experiment 359 mg of sample 12023,9 was evacuated to 10^{-1} torr at room temperature for 15 minutes, repressurized with 40 psi of He, and the evolution of gases studied from 25° to 770°C in 200°C steps. N_2 and CO_2 were copiously released at the lower temperatures (25°–500°C). In addition CO, CH_4, C_2H_4, and C_2H_6 were also detected, being more abundant at the higher temperatures. The release of N_2, CO_2, and CO at low temperatures indicates they are adsorbed by the sample during their short terrestrial history (see above), and the additional hydrocarbons found at the higher temperatures indicate that they could be synthesized as suggested by other pyrolysis experiments (e.g., Nagy *et al.*, 1970, 1971; Murphy *et al.*, 1970; Preti *et al.*, 1971).

Gas Evolution by Acid Treatment

A specially designed microreactor was connected to the LKB-9000 gas chromatograph-mass spectrometer as described elsewhere (Oró *et al.*, 1970b, c). The gases generated upon treatment of the sample with 12 *N* HCl were swept into and separated in the same column described above. Semicontinuous mass scans were taken at 15 second intervals to obtain mass spectral data of the eluted components. Figure 5 is a multiple ion plot (mass chromatogram or ion fragmentogram) obtained for a 953 mg aliquot of lunar sample 12023,9, displaying m/e as a function of the time elapsed since injection. The data show acid treatment evolved CH_4, C_2H_4, C_2H_6, and H_2S as well as N_2, CO_2, and CO.

The multiple ion plot provides a more specific gas chromatographic detector, thereby improving the gas chromatographic resolution, which is incomplete due to the relatively long time (30 seconds) required for the carrier gas to sweep out the reaction gases. The availability of multiple ion plots also enables more reliable identification of different components. An evaluation of the relative ratios of m/e 29 (^{15}N, ^{14}N, and ^{13}CO), 16 (O) and 12 (C) shown on the plot suggest the presence of some N_2 and traces of CO (<0.01 ppm). The simultaneous increase of m/e 12 (C) with m/e 16, as well as ions at m/e 14 and 15 not shown on the plot, is also evidence for methane. Similar interpretations were applied for the CO_2, $H_2C{=}CH_2$, and $H_3C{-}CH_3$ assignments. Although no attempt was made to obtain exact quantitative measurements, the approximate amounts of these gases were of the order of a few ppm or less.

The light hydrocarbons appear to be generated by the action of the acid on hydrolyzable carbides and possibly native hydrogen (reaction of acid with free metal) on elemental carbon. The evolution of CO_2 and CO generated by the action of the acid in this manner cannot be distinguished from adsorbed CO_2 and CO released by contact of the acid solution with the grain surfaces. Therefore, evolution of CO_2 and CO by acid treatment cannot be considered conclusive evidence for the presence of carbonates and carbonyls. Other observations on the low temperature release of CO

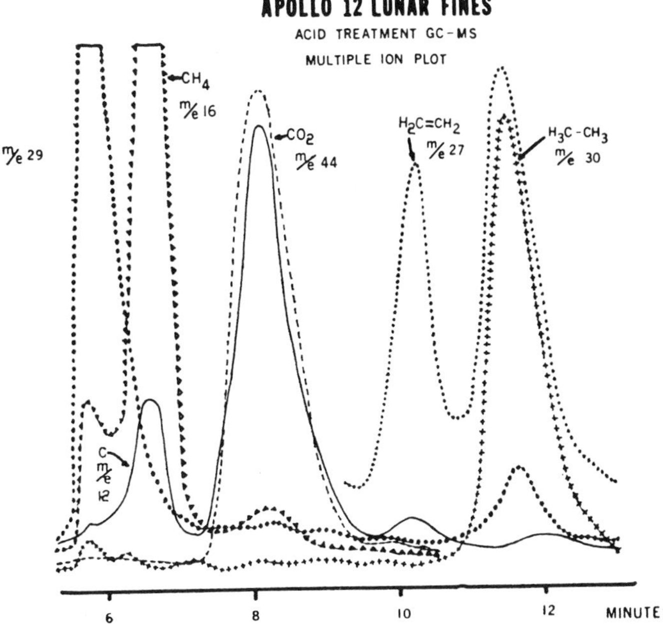

Fig. 5. Multiple ion plot of the gases evolved upon 12N HCl treatment of 0.95 g of sample 12023,9. The intensities corresponding to the masses 12(C^+), 16 (O^+, CH_4^+), 27 ($C_2H_3^+$), 29 ($^{15}N^{14}N$, $^{13}CO^+$), and 30 ($C_3H_6^+$) were obtained at 15 second intervals on an LKB-9000 gas chromatograph–mass spectrometer.

(e.g., Burlingame *et al.*, 1970; Henderson *et al.*, 1971), CH_4, and low molecular weight hydrocarbons (e.g., Abell *et al.*, 1970a, b; Bloch *et al.*, 1971; Chang *et al.*, 1971; Eglinton *et al.*, 1971) by either acid treatment or crushing should be taken into consideration for a more complete discussion of the contamination problem of the gases released at low temperatures.

The indigenousness of methane liberated in ppm amounts by crushing the sample in vacuum or by DCl treatment as reported by Abell *et al.* (1970b) and Chang *et al.* (1971), may be questioned. It appears that the long-term hydrolytic effects of terrestrially adsorbed water on the active sites of hydrolyzable carbon containing species has not been thoroughly considered. The fact that D_2O treatment did not release deuterated hydrocarbons over a 24 hour period (Abell *et al.*, 1970) is not sufficient evidence to disregard the hydrolytic actions of previously adsorbed water. Only 200–500 nanomoles of H_2O are required to produce 100 nanomoles of CH_4 by hydrolysis of common carbides. These quantities of water are present in only 0.25 cc of air at 25°C and 75% relative humidity. The presence of such quantities of water would not, however, insure complete reaction of any carbide present. Factors such as the weakly acidic nature of water, surface polarization phenomena, and the morphological arrangement of the carbide grains within the silicate matrix determine a very slow reaction rate so that most of the carbide remains unreacted even after prolonged contact (days) of the fines with adsorbed water. The addition of much larger quantities of water in the absence of grinding, heating, and other activation phenomena, would not be expected to alter this rate. Consequently the nonrelease of CD_4 or CDH_3 by one day treatment with D_2O is not sufficient to argue against an *in situ* hydrolysis of carbide by terrestrial water. Until additional experimental evidence becomes available, the contribution of methane by the hydrolysis of carbides due to terrestrial exposure (water, etc.) cannot be excluded.

In line with the above interpretation Bloch *et al.* (1971) analyzed at room temperature lunar samples which did not contain any measurable water, CO_2, and CO and found that the amounts of methane produced by crushing were extremely small, i.e., about two orders of magnitude lower than those found by Abell *et al.* (1970b) and Chang *et al.* (1971).

In general, the present work confirms and extends the results obtained with Apollo 11 fines regarding the presence of carbides and sulfides in the lunar fines (Abell *et al.*, 1970a; Oró *et al.*, 1970b, c; Ponnamperuma *et al.*, 1970).

Extractable Organic Compounds

Two aliquots of sample 12023,9, weighing 0.126 g and 1.158 g respectively, were Soxhlet extracted for four hours with 16 cc of a mixture of benzene–methanol (3:1) following the procedure established during our Apollo 11 analyses (Oró *et al.*, 1970b). The extracts were concentrated to a few microliters and aliquots of 1 μl injected into a 120 m × 0.5 mm capillary column coated with SF–96 connected to an F&M 800 gas chromatograph. The extractions did not yield any measurable compounds at the sensitivity limit of 10^{-9} g/g. The sensitivity limit was determined by injection of a n-C_{15} hydrocarbon standard immediately prior to the actual injection of the lunar extracts. The lack of any extractable organic compounds down to this limit correlates with earlier observations on the bulk of Apollo 11 samples (Oró *et al.*, 1970b, c). This is in agreement with the work reported by other investigators (e.g., Lipsky *et al.*, 1971).

Amino Acid Derivatives

A combined 9.2 g portion of Apollo 12 lunar fines (3.0 g 12023,9; 3.1 g 12042,14; 3.1 g 12032,14) was refluxed with 50 ml of triply distilled water for 15 hours at 100°C. The sample was filtered through a sintered glass filter. The filtrate and an equal volume of concentrated HCl were added to a hydrolysis tube, sealed under vacuum, and hydrolyzed for 20 hours at 100°C. The hydrolyzate was evaporated to dryness and the residue washed twice under reduced pressure, brought to a volume of 2 ml with triply distilled water, and passed through 30 ml of ion exchange resin [Dowex 50 × 8(H^+)]. The resin was eluted with 300 ml of distilled water, followed by 300 ml of 2 N NH_4OH.

The latter eluate was evaporated to dryness under reduced pressure, esterified with 3 N HCl-isopropanol for 35 minutes, and eventually reacted with trifluoroacetic anhydride according to established procedures (Roach and Gehrke, 1969; Nakaparksin *et al.*, 1970). The N–TFA-amino acid derivatives were evaporated to dryness, dissolved in chloroform, and injected into a Varian Aerograph 1200 equipped with a flame ionization detector and a 150 m × 0.5 mm stainless steel capillary column coated with N–TFA–L–valyl–L–valine cyclohexyl ester. Chromatography was carried out isothermally at 110°C (GIL-AV *et al.*, 1966). Detection limits were on the order of 20 nanograms for a single peak. Very small peaks with retention times corresponding to the amino acid derivatives were observed, but not significantly higher than the blank. The amounts of the individual components were not sufficient to obtain mass spectrometric evidence. Because of detection limits and the amount of sample analyzed, it is not possible to conclude that amino acids have been detected. We feel that if any meaningful information is to be obtained on amino acids in the lunar fines, sufficient sample must be provided for future analyses to obtain good mass spectra of the amino acid enantiomers.

Acknowledgments—We wish to thank E. K. Gibson, Jr. for helpful discussions. This work was supported by NASA Contract NAS–9–8012.

REFERENCES

ABELL P. I., DRAFFAN C. H., EGLINTON G., HAYES J. M., MAXWELL J. R., and PILLINGER C. T. (1970a) Organic analysis of the returned Apollo 11 lunar sample. *Proc. Apollo 11 Lunar Sci. Conf., Geochim. Cosmochim. Acta* Suppl. 1, Vol. 2, pp. 1757–1773. Pergamon.

ABELL P. I., EGLINTON G., MAXWELL J. R., and PILLINGER C. T. (1970b) Indigenous lunar methane and ethane. *Nature* **226**, 251–252.

ANDERS E. (1970) Water on the moon. *Science* **169**, 1309–1310.

BLOCH M., FECHTIG H., FUNKHOUSER J., GENTNER W., JESSBERGER E., KIRSTEN I., MULLER O., NEUKUM G., SCHNEIDER E., STEINBRUNN F., and ZAHRINGER J. (1971) Active and inert gases in lunar material released by crushing at room temperature and by heating at low temperatures. Second Lunar Science Conference (unpublished proceedings).

BURLINGAME A. L., CALVIN M., HAN J., HENDERSON W., REED W., and SIMONEIT B. R. (1970) Study of carbon compounds in Apollo 11 samples. *Proc. Apollo 11 Lunar Sci. Conf., Geochim. Cosmochim. Acta* Suppl. 1, Vol. 2, pp. 1779–1791. Pergamon.

CHANG S., SMITH J. W., KAPLAN I., LAWLESS J., KVENVOLDEN K. A., and PONNAMPERUMA C. (1970) Carbon compounds in lunar fines from Mare Tranquillitatis, IV. Evidence for oxides and carbides. *Proc. Apollo 11 Lunar Sci. Conf., Geochim. Cosmochim. Acta* Suppl. 1, Vol. 2, pp. 1857–1869. Pergamon.

CHANG S., KVENVOLDEN K., LAWLESS J., PONNAMPERUMA C., and KAPLAN I. R. (1971) Carbon, carbides, and methane in an Apollo 12 sample. *Science* **171**, 474–477.

CHUPKA W. A. and INGHRAM M. G. (1953) Investigation of the heat of vaporization of carbon. *J. Phys. Chem.* **21**, 371–372.

EGLINTON G., ABELL P. I., CADOGAN P. H., MAXWELL J. R., and PILLINGER C. T. (1971) Survey of lunar carbon compounds. Second Lunar Science Conference (unpublished proceedings).

EPSTEIN S. and TAYLOR H. P., JR. (1970) The concentration and isotopic composition of hydrogen, carbon, and silicon in Apollo 11 lunar rocks and minerals. *Proc. Apollo 11 Lunar Sci. Conf., Geochim. Cosmochim. Acta* Suppl. 1, Vol. 2, pp. 1085–1096. Pergamon.

GIBERT J. and ORÓ J. (1970) Gas chromatographic–mass spectrometric determination of potential contaminant hydrocarbons of moon samples. *J. Chromatogr. Sci.* **8**, 295–296.

GIBERT J., FLORY D., and ORÓ J. (1971) Identity of a common contaminant of Apollo 11 lunar fines and Apollo 12 York meshes. *Nature* **229**, 33–34.

GIBSON E. K. and JOHNSON S. M. (1971) Thermal analysis–inorganic gas release studies of lunar samples. Second Lunar Science Conference (unpublished proceedings).

GIL-AV E., FEIBUSH B., and CHARLES-SIGLER R. (1966) Separation of enantiomers by gas-liquid chromatography with an optically active stationary phase. In *Gas Chromatography* (editor A. B. Littlewood) pp. 227–239. Institute of Petroleum.

HENDERSON W., KRAY W. C., NEWMAN W. A., REED W. E., BURLINGAME A. L., SIMONEIT B. R., and CALVIN M. (1971) Study of carbon compounds in Apollo 11 and Apollo 12 returned lunar samples. Second Lunar Science Conference (unpublished proceedings).

HEYMANN D., YANIV A., ADAMS J. A. S., and FRYER G. E. (1970) Inert gases in lunar samples. *Science* **167**, 555–558.

HINTENBERGER H., VOSHAGE H., and SPECHT S. (1971) Heat extraction and mass spectrometric analysis of gases from lunar samples. Second Lunar Science Conference (unpublished proceedings).

HOERING T. C. (1966) Criteria for suitable rocks in precambrian organic geochemistry. *Carnegie Institution Yearbook* **65**, 365.

LORD H. C. (1968) Hydrogen and helium ion implantation into olivine and enstatite: Retention coefficients, saturation concentrations and temperature-release profiles. *J. Geophys. Res.* **73**, 5271–5280.

LSPET (Lunar Sample Preliminary Examination Team) (1970) *Preliminary Science Report* (NASA SP–235). National Aeronautics and Space Administration, Washington, D.C.

MEINSCHEIN W. G. (1971) Personal communication.

MOORE C. B., GIBSON E. K., LARIMER J. W., LEWIS C. F., and NICHIPORUK W. (1970) Total carbon and nitrogen abundances in Apollo 11 Lunar samples and selected achondrites and basalts. *Proc. Apollo 11 Lunar Sci. Conf., Geochim. Cosmochim. Acta* Suppl. 1, Vol. 2, pp. 1375–1382. Pergamon.

MOORE C., LEWIS C. F., DELLES F. M., and GOOLEY R. C. (1971) Total carbon and nitrogen abundance in Apollo 12 samples. Second Lunar Science Conference (unpublished proceedings).

MURPHY R. C., PRETI G., NAFISSI-V M. M., and BIEMANN K. (1970) Search for organic material in lunar fines by mass spectrometry. *Proc. Apollo 11 Lunar Sci. Conf. Geochim. Cosmochim. Acta* Suppl. 1, Vol. 2, pp. 1891–1900. Pergamon.

NAGY B., SCOTT W. M., MODSELESKI V., NAGY L. A., DREW C. M., MCEWAN W. S., THOMAS J. E., HAMILTON P. B., and UREY H. C. (1970) Carbon compounds in Apollo 11 Lunar Samples. *Nature* **225**, 1028–1032.

NAGY B., DREW C. M., HAMILTON P. B., MODZELESKI V. E., NAGY L. A., and UREY H. C. (1971) The organic compounds in the Apollo 12 lunar samples. Second Lunar Science Conference (unpublished proceedings).

NAKAPARKSIN S., GIL-AV E., and ORÓ J. (1970) Study of the racemization of some neutral α-amino acids in acid solution using gas chromatographic techniques. *Anal. Biochem.* **33**, 374–382.

ORÓ J., UPDEGROVE W. S., GIBERT J., MCREYNOLDS J., GIL-AV E., IBANEZ J., ZLATKIS A., FLORY D. A., LEVY R. L., and WOLF C. (1970a) Organogenic elements and compounds in surface samples from the Sea of Tranquillity. *Science* **167**, 765–767.

ORÓ J., UPDEGROVE W. S., GIBERT J., MCREYNOLDS J., GIL-AV E., IBANEZ J., ZLATKIS A., FLORY D. A., LEVY R. L., and WOLF C. J. (1970b) Organogenic elements and compounds in type C and D lunar samples from Apollo 11. *Proc. Apollo 11 Lunar Sci. Conf., Geochim. Cosmochim. Acta* Suppl. 1, Vol. 2, pp. 1901–1920. Pergamon.

ORÓ J., GIBERT J., UPDEGROVE W. S., MCREYNOLDS J., IBANEZ J., GIL-AV E., FLORY D., and ZLATKIS A. (1970c) Gas chromatographic and mass spectrometric methods applied to the analysis of lunar samples from the Sea of Tranquillity. *J. Chromatogr. Sci.* **8**, 297–308.

PRETI G., MURPHY R. C., and BIEMANN K. (1971) The search for organic compounds in various Apollo 12 samples by mass spectrometry. Second Lunar Science Conference (unpublished proceedings).

RIESENFELD E. H. (1942) *Tratado de quimica inorganica* (editor M. Marin). Barcelona, Spain.

ROACH D. and GEHRKE C. W. (1969) Direct esterification of the protein amino acids: Gas-liquid chromatography of N–TFA *n*-butyl esters. *J. Chromatogr. Sci.* **44**, 269–278.

SOCHA A. J., UPDEGROVE W. S., and ORÓ J. (1970) Mass spectrometric analysis of lunar materials by probe technique. Presented at the Eighteenth Annual Conference on Mass Spectrometry and Allied Topics, San Francisco, California, June.

UPDEGROVE W. S. and ORÓ J. (1969) Analysis of the organic matter on the moon by gas chromatography-mass spectrometry: A feasibility study. In *Research in Physics and Chemistry* (editor F. J. Malina), pp. 53–74. Pergamon.

WACHI F. M., GILMARTIN D. E., ORÓ J., and UPDEGROVE W. S. (1970) Differential thermal analysis and gas release studies of Apollo 11 samples. *Icarus*, submitted for publication.

WELLMAN T. R. (1970) Gaseous species in equilibrium in the Apollo 11 holocrystalline rocks during their crystallization. *Nature* **225**, 716–717.

Proceedings of the Second Lunar Science Conference, Volume 2, pp. 1927–1928.
The M.I.T. Press, 1971.

Search for alkanes containing 15 to 30 carbon atoms per molecule in Apollo 12 lunar fines

J. M. MITCHELL, T. J. JACKSON, R. P. NEWLIN, and W. G. MEINSCHEIN
Department of Geology

and

EUGENE CORDES and V. J. SHINER, JR.
Department of Chemistry, Indiana University, Bloomington, Indiana 47401

(*Received 8 February 1971; accepted 1 March 1971*)

Abstract—Benzene-methanol extracts of three samples of lunar fines were obtained on the intact, pulverized, and hydrofluoric acid digested samples. These extracts did not contain detectable quantities of alkanes containing 15 to 30 carbon atoms per molecule. No C_{15} to C_{30} alkane was present in the Apollo 12 samples at concentrations exceeding 1 part per billion by weight.

ALKANES ARE UBIQUITOUS but minor constituents of waxes, fats, and oils of plants and animals (GERARDE and GERARDE, 1962). They are widely used as molecular fossils because they are preferentially preserved relative to other organic materials in sedimentary environments (MEINSCHEIN *et al.*, 1970). The study of alkanes in exobiological research has been reported for a sample from Apollo 11 (MEINSCHEIN *et al.*, 1970). This report deals with the extension of the search for C_{15} to C_{30} alkanes in the Apollo 12 lunar samples.

The samples analyzed were lunar fines #12001,31 (11 g), #12001,32 (9 g), and #12033,6 (5 g) (SHOEMAKER *et al.*, 1970). Except where noted, the procedures followed in the analysis of these samples were identical with those reported in detail for the Apollo 11 sample (MEINSCHEIN *et al.*, 1970). The sensitivity of the gas-liquid chromatograph when using an Apiezon L coated column ($\frac{1}{8}'' \times 10'$) was increased such that 1×10^{-8} g. pristane gave a 10% of full scale deflection. An additional modification of the procedure was the isolation of the mercury head of the extraction bubbler by use of medium sintered glass filters.

Sensitivity checks were run on standard solutions of pristane immediately before and after each gas-liquid chromatographic analysis of a lunar extract. Analysis of each sample consisted of the following three steps: extraction of the intact sample, extraction of the crushed sample after ball milling, and extraction of the sample after its solution in HF (MEINSCHEIN *et al.*, 1970).

No detectable quantities of C_{15} to C_{30} alkanes were found in any of the extracts of the three samples. The lunar fines contained no C_{15} to C_{30} alkanes at concentrations exceeding 1 ppb by weight. In contrast, concentrations of C_{15} to C_{30} alkanes in rocks from the surface of the Earth commonly exceed 100 ppm by weight.

Acknowledgments—The authors express their sincere thanks to the National Aeronautics and Space Administration, who sponsored this work, under contract NAS 9–9974, and supplied the lunar sample. We also deeply appreciate the funds provided by the National Science Foundation, under grant GB–6583, which made possible the establishing of our laboratory for biogeochemical research.

References

Gerarde H. W. and Gerarde D. F. (1962) The ubiquitous hydrocarbons. *Quart. Bull. Assoc. Food Drug Officials U.S.* **26**, 65–89.

Meinschein W. G., Jackson T. J., Mitchell J. M., Cordes E., and Shiner V. J., Jr. (1970) *Proc. Apollo 11 Lunar Sci. Conf., Geochim. Cosmochim. Acta* Suppl. 1, Vol. 2, pp. 1875–1877. Pergamon.

Shoemaker E. M., Batson R. M., Bean A. L., Conrad C., Jr., Dahlen E. H., Goddard E. N., Hart M. H., Larson K. B., Schaber G. G., Schleicher D. L., Sutton R. L., Swann G. A., and Waters A. C., (1970) *Apollo 12 Preliminary Science Report.* NASA Manned Spacecraft Center, Houston, Texas, 139–140.

Proceedings of the Second Lunar Science Conference, Volume 2, pp. 1929–1930.
The M.I.T. Press, 1971.

A search for biogenic structures in the Apollo 12 lunar samples

J. WILLIAM SCHOPF
Department of Geology, University of California, Los Angeles, Los Angeles, California 90024

(*Received 9 February 1971; accepted in revised form 29 March 1971*)

Abstract—Optical and electron microscopic studies of rock chips and dust returned by Apollo 12 and examination of a petrographic thin section of microbreccia have yielded no evidence of lunar organisms. Terrestrial contamination of these samples by particulate organic matter is less than that typical of Apollo 11 samples.

THE GOAL of this study has been to examine samples of lunar rocks and dust returned by Apollo 12 in search of morphologic evidence of present or former life on the moon. The rationale underlying these investigations has been discussed elsewhere (SCHOPF, 1969; 1970).

The Apollo 12 samples investigated in this study consisted of the following: (1) lunar fines (sample 12001,36) from the selected sample container; (2) fines (12032,15; 12033,10; 12037,11; 12042,15) from the documented sample container; (3) fines (12023,16) from the lunar environment sample container; (4) rock chips from the exterior (12034,13) and interior (12034,22) of a microbreccia; and (5) a polished petrographic thin section (12034,33) of a portion of the rock from which chips had been obtained. As a member of the Lunar Sample Preliminary Examination Team for the Apollo 12 mission, I also examined rocks, chips, fines, and the two bioquarantine samples (one of which included portions of drive-tube core sample 12026 and the lowest portion of sample 12028, the deepest core recovered).

Portions of the fines (approximately 0.1 gm per sample) placed on glass microscope slides either as free powder or dispersed in glycerine jelly, were studied with a light microscope at magnifications ranging from 4 to 1500 using normal and polarized transmitted light, phase-contrast optics, and reflected white and ultraviolet light. Similar optical microscopic studies were made of the chips and thin section of the microbreccia. Other samples of the fines and selected glass particles and rocks fragments were coated with a thin gold-palladium film and studied with a scanning electron microscope at magnifications ranging from 30 to 30,000. Previous studies have demonstrated that acid maceration of lunar material does not provide additional significant information (SCHOPF, 1970; CLOUD *et al.*, 1970); this destructive technique was therefore here omitted.

These investigations have yielded no evidence of living, recently dead, or fossil microorganisms. In contrast with material returned by Apollo 11, which contained numerous cellulose fibers and other organic materials (e.g., teflon fragments) of terrestrial origin (SCHOPF, 1970), the samples here studied appear to be generally devoid of particulate organic contaminants. The few contaminants identified included shreds of teflon, presumably derived from sample bags, and minute fragments of aluminum foil and gold-coated Mylar fabric. If organic contamination of other types has been equally minimal, it seems reasonable to suppose that much of the minute amount of total carbon (undifferentiated as to chemical species) detected in the Apollo 12 samples (LSPET, 1970) may be of lunar origin. If this possibility can be further substantiated, perhaps based on analyses of carbon isotopic composition (e.g., KAPLAN and SMITH, 1970; SMITH and KAPLAN, 1970; KVENVOLDEN *et al.*, 1970), uncertainties as to the probable origin(s) and significance of organic substances found to be associated with Apollo 11 samples should be alleviated for Apollo 12.

Acknowledgments—I thank Mrs. Carol Lewis and Mr. Bruce N. Haugh for assistance. Supported by NASA Contract NAS 9-9941.

REFERENCES

CLOUD P., MARGOLIS S. V., MOORMAN M., BARKER J. M., LICARI G. R., KRINSLEY D., and BARNES V. E. (1970) Micromorphology and surface characteristics of lunar dust and breccia. *Science* **167**, 776–777.

KAPLAN I. R., and SMITH J. W. (1970) Concentration and isotopic composition of carbon and sulfur in Apollo 11 lunar samples. *Science* **167**, 541–543.

KVENVOLDEN K., LAWLESS J., PERING K., PETERSON E., FLORES J., PONNAMPERUMA C., KAPLAN I. R., and MOORE C. (1970) Evidence for extraterrestrial amino-acids and hydrocarbons in the Murchison Meteorite. *Nature* **228**, 923–926.

LSPET (Lunar Sample Preliminary Examination Team) (1970) Preliminary examination of lunar samples from Apollo 12. *Science* **167**, 1325–1339.

SCHOPF J. W. (1969) Micropaleontology and extraterrestrial sample studies. *Program, Annual Meeting, Geol. Soc. Amer.*, Atlantic City, p. 200 (abstr.).

SCHOPF J. W. (1970) Micropaleontological studies of lunar samples. *Science* **167**, 779–780.

SMITH J. W., and KAPLAN I. R. (1970) Endogenous carbon in carbonaceous meteorites. *Science* **167**, 1367–1370.

Proceedings of the Second Lunar Science Conference, Volume 2, pp. 1931–1937.
The M.I.T. Press, 1971.

Search for viable organisms in lunar samples: Further biological studies on Apollo 11 core, Apollo 12 bulk, and Apollo 12 core samples

V. I. OYAMA, E. L. MEREK, M. P. SILVERMAN, and C. W. BOYLEN
Exobiology Division, National Aeronautics and Space Administration,
Ames Research Center, Moffett Field, California 94035

(*Received 22 February 1971; accepted in revised form 29 March 1971*)

Abstract—No discernible growth of organisms was obtained from lunar samples either sprinkled directly upon petri dishes containing media or from membrane filters containing fine samples obtained from a gas flotation technique. Colored artifacts similar in appearance to those from earlier Apollo 11 bulk fines appeared in agar culture media around some of the particles. Results of tests of ^{14}C-compound evolution from labeled organics and $^{14}CO_2$ fixation did not indicate biological activity. Although there was production of H_2 when medium was added to lunar samples, the rates and amounts of gas produced contraindicated biological processes.

INTRODUCTION

NO VIABLE ORGANISMS were found in the Apollo 11 lunar material by the biological quarantine team at the Manned Spacecraft Center, Houston (LSPET, 1969). The same results were found in our independent search for viable organisms in a 50 g sample of Apollo 11 bulk fines although iron-containing artifacts superficially resembling microbial clones were seen (OYAMA *et al.*, 1970). We are now reporting the results of similar studies with an Apollo 11 core sample and Apollo 12 bulk and core samples as well as the results from metabolic studies to detect microorganisms. Studies on the effect of lunar samples on terrestrial microorganisms are reported elsewhere (SILVERMAN *et al.*, 1971).

TECHNIQUES AND EXPERIMENTAL RESULTS

Growth studies

All growth and metabolic studies were performed in the Ames Research Center Lunar Biological Laboratory described earlier (OYAMA *et al.*, 1970). Nine different media (Table 1) were used. Five of these media differed from those used previously; the four complex organic media now included carbohydrates and yeast extract (instead of vitamins and nucleic acid bases), and the formose sugars were processed to contain less formaldehyde than those used in the previous tests. Gas mixtures were identical to those used previously.

Two procedures were used for dispensing the samples. A flotation procedure was used first in an attempt to separate organisms from any mineral particles which may have been toxic. The samples were placed on a fine mesh screen which was enclosed in an apparatus designed to blow dry sterile nitrogen vertically through the sample.

Table 1. Media for the attempted isolation of microorganisms from the lunar samples.

Designation	General composition*
1. Complex dilute pH 4	100 mg/l ea. $(NH_4)_2SO_4$, KNO_3, K_2HPO_4, D-ribose, D-fructose, lactate, glycerol, pyruvate, acetate, yeast extract. 50 mg/l ea. D-glucose, L-glucose, L-arabinose, D-arabinose, 20 mg/l ea. 18 DL-amino acids, glycine, 4-OH-L-proline, β-alanine. Final adjustment to pH 4
2. Complex dilute pH 7	Same as 1 but adjusted to pH 7
3. Complex dilute pH 10	Same as 1 but adjusted to pH 10
4. Complex concentrated	Same as 2 with concentrations increased ten times
5. Water agar	
6. Salts	1 g/l ea. $(NH_4)_2SO_4$, KNO_3, K_2HPO_4
7. Sulfur oxidation	Same as 6 adjusted to pH 3.5 and with 0.5 ml of $Na_2S \cdot 9H_2O$ solution saturated with sulfur to precipitate elemental sulfur on surface
8. Formose sugars	Same as 6 with 100 mg/l of formose sugars (SHAPIRA, 1967) passed through an IR 400 ($^-HSO_3$) ion exchange resin to remove formaldehyde
9. Spark discharge	Same as 6 with 50% solution of organics generated by spark discharge (RABINOWITZ *et al.*, 1969).

* Media formulated in synthetic sea salts at 0.1%, 3.0%, or 20% macroelements with ocean concentration of microelements.

Table 2. Media used for the lunar samples for each of the three atmospheres.

	Apollo 11 Core 10005,33	Apollo 12 Bulk 12003,3-7	Apollo 12 Core 12028,205
Sample per plate (mg)	21	123	20
Flotation	1–3L*	1–3L	1–3L
	1–3H	1–3H	1–3H
Direct inoculation	1–9L	1–9L	1–9L
	1–9H	1–9M	1–9H
		1–9H	

* Arabic numbers refer to media (see Table 1), the capital letters refer to sea salt concentration L, 0.1%, M, 3%, H, 20%.

A 1-inch diameter, 0.2 μm membrane filter was positioned in the exhaust end of the tubing to receive particles of low density which were blown free of the sample by the nitrogen. This filter was then placed on the surface of agar media. The duration and flow rate of the nitrogen were those selected previously to yield the maximum number of colonies from terrestrial soils. A settling tower (OYAMA *et al.*, 1970) was then used to dispense the residual sample particles onto the agar surfaces.The sample and media combinations used for the growth studies are listed in Table 2.

All inoculated plates were incubated at 10°C for three weeks. The plates were then incubated at 20°C for three weeks, then at 35°C, and finally at 55°C for the same length of time. No organisms or terrestrial contaminants appeared on any of the plates. A total of 43 nonbiologic artifacts were recorded from 25 of the 189 petri plates inoculated directly with particles. These were of two types: either flat brown or raised structures of various colors similar to those observed previously (OYAMA *et al.*, 1970).

Metabolic experiments

Evolution of ^{14}C-labeled compounds. Lunar samples were distributed to specially designed incubation cells (Fig. 1) that permit removal and renewal of the medium without disturbing the sample. The ^{14}C-labeled components listed in Table 3 were added to media 1, 2, and 3. The initial radioactivity totaled 1.3×10^6 disintegrations per min (DPM)/ml. The quantities of lunar sample and media used are listed in Table 4. At incubation times of 2, 3, 5, and 8 weeks, the sidearms containing 0.5 ml of 40% KOH were removed from the incubation cells and replaced with sidearms containing fresh KOH. The removed sidearms were sterilized in the pass-through autoclave, and 100 μl aliquots were removed for counting radioactivity in a scintillation counter. Control incubation cells contained labeled medium 2 without lunar samples.

We found no evidence that the lunar samples reacted with organic compounds to evolve ^{14}C-compounds (Table 5). Neither the sample size, which ranged from 21 mg to 3680 mg nor the initial pH of the medium influenced the amount of ^{14}C-compounds evolved. More ^{14}C-compounds were evolved from the 6 ml of medium than from 1.5 ml. For each test set, the data in the table show no consistent differences greater than 10% over that of control incubation cells for any period; we ascribe no significance to these differences because of errors introduced by pipetting and counting.

Fixation of CO_2. A set of experiments similar to the ^{14}C-compound evolution experiments was performed using media 1, 2, and 3 with $Na_2^{14}CO_3$ added instead of labeled organic compounds. The specific activity of the $Na_2^{14}CO_3$ was 52 $\mu c/\mu M$, the

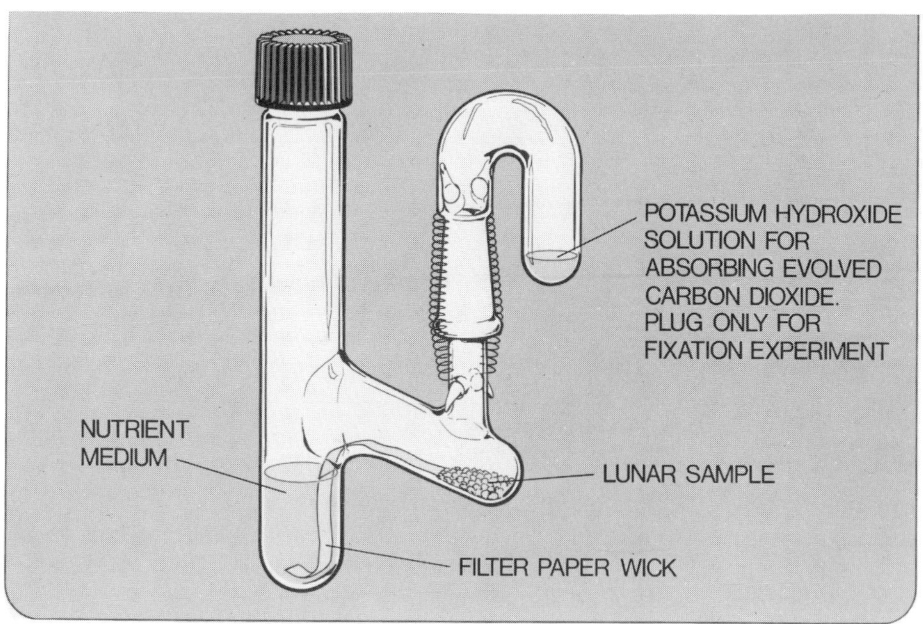

Fig. 1. Special cell for radioactive carbon dioxide evolution and fixation experiments.

Table 3. Radioactive components added to media 1, 2, and 3 for ^{14}C-compound evolution experiments

Labeled Medium Components	Specific Activity $\mu c/\mu M$	μM
D-glucose-UL-^{14}C	7	7
D-arabinose-1-^{14}C	50	0.8
D-ribose-1-^{14}C	9	3.5
D-fructose-UL-^{14}C	210	0.2
DL-glutamic acid-UL-^{14}C	4	10
Glycine-UL-^{14}C	10	4
DL-lysine-1-^{14}C	10	3.8
DL-methionine (Methyl ^{14}C)	31	1.3
DL-alanine-1-^{14}C	52	0.8
Sodium lactate-1-^{14}C	28	1.4
Glycerol-UL-^{14}C	18	2.2
Sodium pyruvate-UL-^{14}C	105	0.4

Table 4. Lunar samples and media added to incubation cells for ^{14}C-compound evolution and $^{14}CO_2$ fixation studies

Lunar Sample Number	mg	Medium ml
12003,3-7 (12 bulk)	3680	6
10005,33 (11 core)	272	1.5
12028,205 (12 core)	21	1.5

Table 5. Evolution of ^{14}C-compounds from lunar samples and labeled organic compounds.

Sample	Medium	DPM/100 μl KOH solution Incubation time, weeks 0–2	2–3	3–4	5–8
Control*	2	520	480	700	865
10005,33	1	550	425	740	840
	2	470	550	620	800
	3	480	530	680	860
12028,205	1	470	550	685	860
	2	490	525	715	860
	3	510	475	705	825
Control†	2	1500	1000	1670	3195
12003,3-7	1	1340	960	1980	3140
	2	1310	1070	1620	3170
	3	1220	1160	1730	3195

* Control with 1.5 ml medium.
† Control with 6 ml medium.

concentration was 0.02 μM, and the initial radioactivity was 5.7×10^6 DPM/ml. The cell (Fig. 1) was used with a cap replacing the KOH sidearm. At intervals the medium was removed from the lower compartment and replaced by fresh medium. The medium removed was acidified with 85 mg of KH_2PO_4/ml of medium, purged for 15 minutes with N_2, and 100 μl aliquots were removed and counted by scintillation.

We found no evidence that the lunar samples incorporated CO_2 into organic

Table 6. Fixation of $^{14}CO_2$ by the lunar samples.

Sample	Medium	DPM/100 μl of medium incubation time, weeks 0–2	2–3	3–5	5–8
Control*	2	22	20	48	40
10005,33	1	20	24	52	45
	2	23	20	49	45
	3	23	25	51	38
12028,205	1	26	21	57	41
	2	29	26	43	38
	3	24	20	51	40
Control†	2	17	19	47	40
12003,3-7	1	13	16	43	33
	2	20	23	46	36
	3	20	23	43	36

* Control with 1.5 ml of medium.
† Control with 6 ml of medium.

compounds (Table 6). The data suggest that neither the size of sample nor the pH of the medium influenced the amount of radioactivity remaining in the medium after acidification and purging.

Gas exchange. Lunar samples 12003,3-7 (0.465 g) and 10005,33 (0.27 g) were incubated with 1 ml of medium 2 in 9.9 ml and 11.4 ml incubation cells respectively. The incubation cells inside the barrier system were individually connected to outside valves. These valves were connected through a manifold to a gas sampling valve of a gas chromatograph (Carle, 1970). At the start of the experiment, the cells were evacuated and filled repeatedly with a mixture of 93 vol % He–7 vol % Kr until the N_2 concentration was reduced to a few parts per million. The gas in the head space was then sampled periodically and analyzed for hydrogen, methane, carbon dioxide, oxygen, nitrogen, nitrous oxide, and an internal standard (krypton).

No gas compositional changes indicating biological activity were detected. No methane, carbon dioxide, or nitrous oxide were detected and the nitrogen concentration remained constant. Hydrogen was produced in both cells, but the rate and amounts suggested nonbiological production. With sample 12003,3-7, the hydrogen appeared early in the incubation period and then the concentration remained constant. The total hydrogen accumulated was 1.02 $\pm$ .32 μ moles over 26 days. Approximately one-third as much hydrogen appeared in sample 10005, 33 over 26 days of incubation. One micromole of hydrogen could have been produced by ten parts per million of elemental Fe in the sample 12003,3-7 if the reaction were $Fe + 2H_2O \rightarrow H_2 + Fe(OH)_2$. Although much more elemental iron, 0.5%, is found in lunar fines, it may be alloyed with nickel (Runcorn *et al.*, 1970), and consequently only a small amount of elemental iron may be capable of this reaction. We are, however, not certain as to the nature of the reaction evolving hydrogen.

Discussion

Thus far, we have tested approximately fifty grams of lunar samples from the sites visited by Apollo 11 and Apollo 12 for the presence of viable microorganisms. Most of these samples consisted of bulk fines; also included, however, was one core sample from Mare Tranquillitatis, recovered from below the surface at a depth no greater than 13.5 cm, and one core sample from Oceanus Procellarum, obtained at a depth of 34.2–35.2 cm. Samples of lunar material have now been examined for the presence of viable microorganisms by us (OYAMA *et al.*, 1970) and by TAYLOR *et al.*, 1970a. Neither group detected viable microorganisms or terrestrial contaminants in any of the samples tested using a wide variety of nutrient media, environments, and techniques, nor did we find evidence of metabolic activity.

Failure to detect microorganisms by culturing raised the question of possible toxicity of the lunar material to microbial life forms. An aqueous extract of one Apollo 11 core sample (10005) was reported to be toxic to selected terrestrial microorganisms (TAYLOR *et al.*, 1970b). However, our independent study using a different portion of the same core failed to detect any inhibitory effects on microbial growth (SILVERMAN *et al.*, 1971). Thus toxicity of the lunar material in this sample cannot be the primary reason for the failure to detect viable microorganisms in our experiments.

In addition to negative results from our attempts to culture viable microorganisms, we also detected no evidence of metabolic activity in the lunar samples. Positive control tests of the ^{14}C-evolution experiment were performed prior to the actual test of lunar material. Under conditions identical to those used in the subsequent lunar tests, terrestrial soils from Death Valley (U.S.A.) and the Atacama Desert (Chile) evolved 300,000 and 105,000 DPM respectively within three days. Background counts were 50 or less for the same soils when sterilized and treated similarly, some 3 orders of magnitude less than the elicited responses. The fact that these soils contain relatively low populations (less than 10^4/gram viable organisms capable of growing either anaerobically or aerobically) coupled with the high levels of net radioactivity evolved demonstrates the high sensitivity of this technique.

Although the medium used in the ^{14}C-evolution experiments was compounded to contain materials having little or no vapor pressure, thereby obviating the capture of volatile components in the KOH trap, radioautodegradation did occur, resulting in radioactive volatile substances being formed. This degradation occurred in the medium with or without the addition of lunar soil, was linear over the eight week incubation period, and was not influenced by the initial pH of the medium.

Similar positive control studies for $^{14}CO_2$ fixation experiments showed that in two days the Death Valley soil fixed 3000 DPM whereas Atacama soil fixed 90,000 DPM. The sterilized soil controls for the terrestrial soil experiment and the medium-only control for the lunar experiment showed that the presumed "pure" ^{14}C-carbonate could not be purged completely of $^{14}CO_2$ after acidification. When pure $^{14}CO_2$ gas was used in previous studies, all of the unreacted $^{14}CO_2$ was removed by purging with N_2 at a pH buffered between 3 and 5 by KH_2PO_4. Although we did not anticipate the carbonate being radioactively impure, the controls without lunar soil sufficed for this experiment. Future experiments on lunar matter will use pure $^{14}CO_2$.

The gas exchange experiment when applied to a variety of terrestrial soils generally results in both qualitative and quantitative changes in the composition of the enclosed atmosphere. All terrestrial soils examined to date that contain anaerobic heterotrophs produce CO_2 and either H_2, N_2, CH_4, and/or N_2O. The production of

H_2 is usually accompanied by or followed by its utilization. Thus for the lunar experiment, the H_2 production by itself cannot be interpreted as a biological phenomenon. In fact, elemental Ca, Cd, Fe, Li, Mg, Zn, and Sb have been demonstrated in our laboratory to react with aqueous media to produce H_2.

The failure to evoke biologic responses has substantiated earlier results. There are several reasons for supporting the hypothesis that the moon is a sterile body (Oyama *et al.*, 1970). All biologic tests to date support this hypothesis, but more extensive testing with additional samples and a wider variety of tests is required.

Acknowledgments—We thank B. J. Berdahl, G. C. Carle, C. C. Johnson, P. J. Kirk, M. E. Lehwalt, A. K. Miyamoto, E. F. Munoz, G. E. Pollock, B. J. Tyson, and O. Whitfield for assistance, and J. W. Schopf for valuable criticisms of the manuscript.

References

Carle G. C. (1970) Gas chromatographic determination of hydrogen, nitrogen, oxygen, methane, krypton, and carbon dioxide at room temperature. *J. Chromatogr. Sci.* **8**, 550–551.

LSPET (Lunar Sample Preliminary Examination Team) (1969). Preliminary examination of lunar samples from Apollo 11. *Science* **165**, 1211–1227.

Oyama V. I., Merek E. L., and Silverman M. P. (1970) A search for viable organisms in a lunar sample. *Proc. Apollo 11 Lunar Sci. Conf., Geochim. Cosmochim. Acta* Suppl. 1, Vol. 2, pp. 1921–1927. Pergamon.

Rabinowitz J., Woeller F., Flores J., and Krebsbach R. (1969) Electric discharge reactions in mixtures of phosphine, methane, ammonia, and water. *Nature* **224**, 796–798.

Runcorn S. K., Collinson D. W., O'Reilly W. O., Battey M. H., Stephenson A., Jones J. M., Manson A. J., and Readman P. W. (1970) Magnetic properties of Apollo 11 lunar samples. *Proc. Apollo 11 Lunar Sci. Conf., Geochim. Cosmochim. Acta* Suppl. 1, Vol. 3, pp. 2369–2387. Pergamon.

Shapira J. (1967) *Closed Life Support System.* NASA SP–134.

Silverman M. P., Munoz E. F., and Oyama V. I. (1971) Effect of Apollo 11 lunar samples on terrestrial microorganisms. *Nature* **230**, 169–170.

Taylor G. R., Ferguson J. K., and Truby C. P. (1970a). Methods used to monitor the microbial load of returned lunar material. *Appl. Microbiol.* **20**, 271–272.

Taylor G., Ellis W. L., Arrendondo M., and Mayhew B. (1970b) Growth response of *Pseudomonas aeruginosa* to the presence of lunar material. *Bacteriol. Proc.*, 42 (in press).

Proceedings of the Second Lunar Science Conference, Volume 2, pp. 1939–1948.
The M.I.T. Press, 1971.

Microbial assay of lunar samples

GERALD R. TAYLOR
Preventive Medicine Division, NASA Manned Spacecraft Center,
Houston, Texas 77058

and

WALTER ELLIS, PRATT H. JOHNSON, KATHRYN KROPP, and THERON GROVES
Brown and Root-Northrop, Houston, Texas 77058

(*Received 22 February 1971; accepted in revised form 25 March 1971*

Abstract—Two drive tube core samples, and two surface samples were assayed for the presence of indigenous microorganisms. Neither terrestrial contaminants nor indigenous biota were demonstrated under a variety of growth conditions. A microbial toxicity was demonstrated by extracts of Apollo 11 core material which had been in contact with complex nutrient media for four months. This toxicity was not demonstrated in Apollo 11 surface material or either of the Apollo 12 samples. The nature of the lethal factor has not yet been determined although the possible role of higher concentrations of ilmenite, nickel, and scandium in the Apollo 11 core material is being investigated.

INTRODUCTION

SERIOUS CONCERN for the possible contamination of extraterrestrial bodies with living microorganisms originated with the concept of Panspermia, proposed by ARRHENIUS in 1908. Since that time the probability of extraterrestrial contamination has increased significantly with the increased use of craft capable of approaching celestial bodies. This problem was recently reviewed by HOTCHIN (1968) who was able to demonstrate the distinct probability of microbial contamination of the moon, especially from fecal discards and other debris. More specifically, TIERNEY (1968) presented conservative calculations which indicated that microorganisms remaining on fragments of a typical U.S. Lunar Probe which has made a hard impact on the moon could be dispersed over a radius of 50 to 60 kilometers.

Several investigators (HAWRYLEWICZ *et al.*, 1968; HAGEN and HAWRYLEWICZ, 1968; CAMERON *et al.*, 1970) have isolated viable microorganisms from unusually harsh environments. Others (VALLENTYNE, 1963; BROCK, 1969) have managed to maintain microorganisms in a viable state after having been subjected to hard vacuum (CAMERON *et al.*, 1970; SILVERMAN *et al.*, 1967), elevated temperatures (UCHINO and DOI, 1967; BROCK, 1967), severe doses of ultraviolet irradiation (SILVERMAN *et al.*, 1967), as well as severe extremes of salinity, pH, hydrostatic pressure, and nutrient concentration (HAWRYLEWICZ *et al.*, 1968, HAGEN and HAWRYLEWICZ, 1968; UCHINO and DOI, 1967; LARSEN, 1967). Most recently, viable microorganisms have been recovered from deep within the Surveyor III TV camera which had resided on the lunar surface for two and one-half years (MITCHELL and ELLIS, 1971). All of these studies support the possibility that having once found their way to the moon, some microorganisms might stand a reasonably good chance of surviving for indefinite periods of time.

METHODS AND RESULTS

As a result of this possibility, a major concern of the first two lunar landing missions was to prevent the introduction, through returned lunar material, of alien microorganisms into the Earth's biosphere; a process termed "back contamination" (MCLANE *et al.*, 1967). To help prevent back contamination, lunar materials returned to Earth after the first and second manned landings were placed under strict quarantine (KEMMERER *et al.*, 1969) and subjected to a comprehensive microbiological examination (TAYLOR *et al.*, 1970b) before sample distribution. Four different samples were tested as part of the Apollo 11 and Apollo 12 protocol. These four samples are identified in Table 1 for reference.

Drive-tube core samples 10004, 10005, and 12026 (ANDERSON, 1970) were obtained by splitting the contents of the respective core tubes along their longitudinal axis and recovering one-half of the contents for study (FRYXELL *et al.*, 1970). The fines represented loose surface material less than one centimeter in diameter, whereas the rock chips were obtained from all three of the main rock types returned (HESS and CALIO, 1970).

Following reduction of the coarse material to a mean particle diameter of 2 μ, culture media were inoculated with an average of 220 mg of lunar material per test condition. Working under strict quarantine, aliquots of each of the four lunar samples were inoculated into three different media. These included: (1) sheep blood agar (BA); (2) Trypticase glucose yeast extract agar (TGY); and, as a control (3) purified agar without nutrients. Other test systems were also employed (TAYLOR, *et al.*, 1970b) but are not germane to the present discussion.

The inoculated test systems were incubated at one of four temperatures (4°C, 24°C, 35°C, and 55°C) while subjected to one of three different gaseous environments (air, 10% CO_2 in air, and N_2). After 21 days of incubation no growth was observed in any of the test systems. The same results were reported by OYAMA *et al.*, (1970) working with different, but analogous, samples.

In order to properly evaluate these results, it was necessary to investigate the possibility of microbial toxicity mediated by the presence of lunar material. For this study, the solids from the lunar sample/agar complexes were separated from the fluids

Table 1. Sources of analyzed lunar material.

Mission	Sample	Description	Percent of total sample	Parent sample number
Apollo 11	Subsurface material	Half of core sample no. 1	54.5	10005
		Half of core sample no. 2	45.5	10004
	Surface material	Chips and fines	12.0	10050
		Chips and fines	61.0	10051
		Chips and fines	26.0	10052
Apollo 12	Subsurface material	Half of core no. 1	80.0	12026
		Bottom portion of core no. 2 (deep core)	20.0	12028
	Surface material	Chips and fines	9.8	12057
		Fines	11.8	12052
		Fines	3.8	12032
		Chips and fines	69.0	12001 and 12003
		Fines	5.7	12033

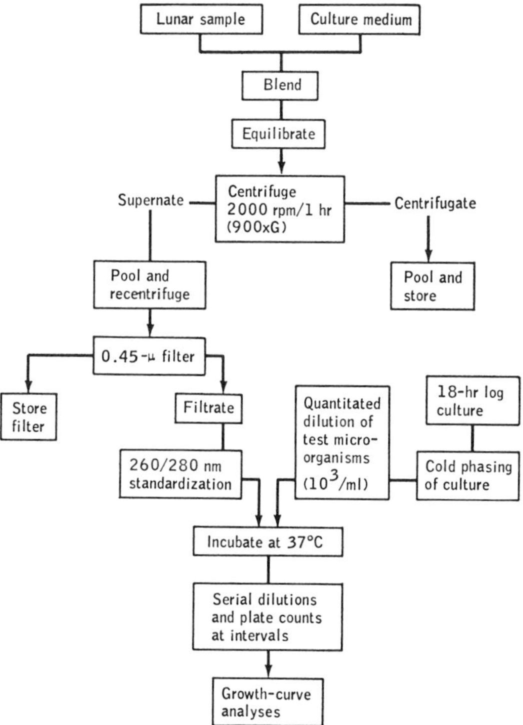

Fig. 1. Lunar sample extraction and test protocol

as shown in Fig. 1. Following quarantine incubation the lunar sample-growth medium complex was blended with the addition of an equal volume of 0.007 *M* sodium phosphate buffer (pH 7.4). Repeated centrifugations, followed by a terminal filtration through a 0.45 μ microbiological filter produced the working fluid (extract) which was free of granular materials.

The resulting extracts were then analyzed for their effects on microbial growth. This was accomplished by adding quantitated cultures of known microorganisms to aliquots of these extracts and analyzing the resulting growth curves as demonstrated in Fig. 1.

Attempts to grow *Pseudomonas aeruginosa* (ATCC 15442) in extracts of Apollo 11 core material that had been in contact with TGY nutrient for four months resulted in the inhibition reported earlier (Taylor *et al.*, 1970a) and shown in Fig. 2. The presence of this extract resulted in the loss of all colony forming units (cfu) within ten hours. Normal growth occurred in extracts of the four-month-old nutrient (TGY) without lunar material, indicating that the toxicity was not due to degradation of the nutrients.

In order to determine if the observed effect was bacteriostatic or bactericidal, the apparently sterile culture was filtered through a microbiological filter with an average pore size of 0.45 μ and washed three times with a solution of phosphate buffered saline. The filter membrane was subsequently placed directly upon a fresh TGY agar

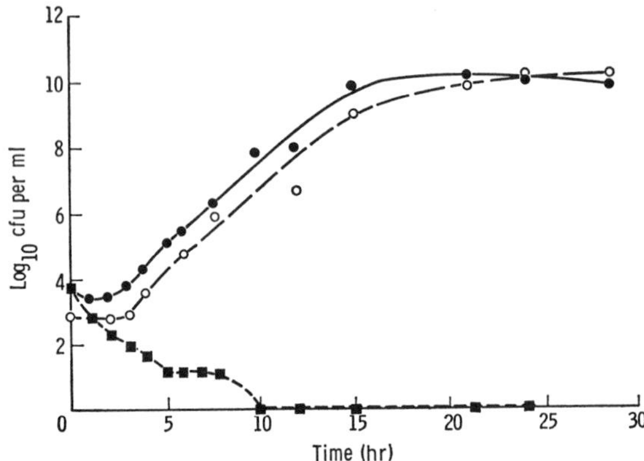

Fig. 2. Growth response of *Pseudomonas aeruginosa* (ATCC 15442) in extract of Apollo 11 core-material/TGY complex: Closed circles, control—Extract of old TGY without lunar material; Open circles, control—Extract of new TGY without lunar material; Squares, experimental—Extract of old TGY with Apollo 11 core material added.

surface and incubated for forty-seven hours at 37°C. The total absence of microbial growth from this membrane indicates that the observed effect is bactericidal. An evaluation of the heat stability of the bactericidal factor indicates that heating an aliquot of the extract for one hour at 80°C produced no discernible change in inhibition. Likewise heating under the same conditions produced no significant toxicity in the control sample.

The toxic effect appears to be limited to the Apollo 11 core sample as extracts of the other three lunar materials, which had been in contact with TGY for approximately the same length of time, did not alter the growth pattern of *P. aeruginosa*. The toxicity of the Apollo 11 core extract was exhibited, however, against all other microbial species tested (*P. aeruginosa*, *Azotobacter vinelandii* (ATCC 12837), and a clinical isolate of *Staphylococcus aureus*).

During quarantine testing of the lunar material, the Apollo 11 core material was added to two other culture media in addition to TGY. One quarantine test system contained Apollo 11 core material which had been in contact with a mixture of 5% sheep blood in blood agar base for four months. When quantitated cultures of *P. aeruginosa* were added to extracts of this complex, a toxic effect indistinguishable from that previously described, was observed. Thus the presence of an inhibitory effect was demonstrated in two completely separate nutrient systems.

The other system consisted of a non-enriched medium in which a suspension of phosphate buffered lunar material had been in contact only with refined agar for four months. TGY and quantitated *P. aeruginosa* cultures were added to extracts of this complex to test for possible toxic effects. The results indicate that there is no demonstrable inhibition elicited by this combination of materials. Similarly, washing the native Apollo 11 core material with phosphate buffer did not result in a bactericidal extract. The data indicate, then, that the toxicity was produced only after prolonged contact of Apollo 11 core material with a complex nutrient solution.

Studies were conducted to determine the presence of certain elements in the lunar material which might contribute to the microbial toxicity noted. The abundance of eighteen elements in dry heat sterilized (160°C for 16 hours) lunar material was measured by a modification of the optical emission spectroscopic method of AHRENS and TAYLOR (1961). Through this modification, the burning of the arc was stabilized permitting triplicate analyses to be obtained from as little as three milligrams of sample. Determinations, conducted on a 3.4 meter Ebert mount spectrograph, were obtained from all four lunar samples. The results of these spectrochemical analyses are summarized in Table 2. The major elements, listed as percent oxides, total nearly one hundred percent. This indicates that the valence state, and therefore the element-to-oxygen ratios, are those shown. Analyses of the major elements were confirmed by atomic absorption spectrophotometry. These determinations, conducted on a dual beam instrument, were performed according to previously reported methods (ANON. 1968).

As the microbial toxicity effect was demonstrated only in extracts of the Apollo 11 core material, the chemical analysis of this sample before extraction was compared with the other three. Compared to the other three, this sample was found to contain slightly higher amounts of calcium, aluminum, and nickel, and a considerably higher amount of scandium. Tests are currently being conducted in this laboratory to determine if the excesses of nickel and scandium could be contributory to the observed toxic effects.

The four heat-sterilized lunar samples were evaluated for their gross relative mineralogical content in an effort to elucidate the toxic agent. Analyses of X-ray diffraction patterns of all four samples indicate (Table 3) that the Apollo 11 core contained more ilmenite and subcalcic augite than any of the other three samples

Table 2. Elemental analysis of heat-sterilized lunar samples. The major elements are reported as oxides and given in weight percent. The minor elements are given in parts per million.

Name	Form	Apollo 11 samples		Apollo 12 samples	
		Core	Fines	Core	Fines
Major elements					
Silicon	SiO_2	43.00	43.00	44.00	49.00
Iron	FeO	14.50	15.00	18.00	15.50
Aluminum	Al_2O_3	14.00	13.00	11.50	11.80
Calcium	CaO	12.00	11.00	11.50	9.80
Magnesium	MgO	7.60	8.60	9.80	10.00
Titanium	TiO_2	7.20	7.50	3.20	2.80
Sodium	Na_2O	0.50	0.50	0.60	0.58
Chromium	Cr_2O_3	0.35	0.34	0.48	0.39
Manganese	MnO	0.19	0.19	0.27	0.23
Potassium	K_2O	0.18	0.12	0.22	0.21
Minor elements					
Zirconium	Zr	380	600	580	550
Nickel	Ni	240	170	210	140
Barium	Ba	160	160	430	300
Yttrium	Y	130	120	170	150
Scandium	Sc	92	73	48	50
Vanadium	V	68	78	160	120
Cobalt	Co	25	22	48	37
Copper	Cu	27	18	34	15

Table 3. Mineral content of lunar material.

Mineral	Percent composition			
	Apollo 11 samples		Apollo 12 samples	
	Core material	Surface material	Core material	Surface material
Clinopyroxene				
Subcalcic augite	35 to 45	15 to 25	0	0
Pigeonite	0	0	30 to 40	35 to 45
Plagioclase	20 to 25	10 to 15	20 to 30	15 to 25
Olivine (Fosterite 65)	1 to 4	3 to 5	5 to 10	3 to 6
Ilmenite	15 to 20	7 to 12	3 to 8	5 to 10
Glass	10 to 15	40 to 50	5 to 15	10 to 20
Tridymite	<1	0	<1	0
Olivine (Fosterite 15)	0	0	<2	0
Unidentified	0	0	0	2 to 3

tested. Analysis of the Apollo 11 core X-ray diffraction pattern indicated that clinopyroxene, plagioclase, ilmenite, olivine, and low tridymite were the major minerals detected. The 221 line of the primary clinopyroxene was 2.998dA ($\pm$0.005dA) indicating that it is a subcalcic augite. The olivine composition, as determined from the 130 line by the method of YODER and SAHAMA (1957), was found to be 65% Fosterite and 35% fayalite. The ilmenite in this sample is stoichiometrically similar to $FeTiO_3$, although, the observed d-space shift to a slightly higher value indicates some ionic substitution has occurred. The plagioclase pattern matches those of calcic bytownite for other lunar samples (LSPET, 1970; STEWART *et al.*, 1970) and is probably due to a similar plagioclase composition. A very weak line at 4.10dA indicated a minor concentration of low tridymite.

In addition to the above studies, in which complete nutrient media were used, portions of the lunar material returned during the Apollo 12 mission (Table 1) were used to inoculate culture media devoid either of organic carbon compounds or of complex nitrogen compounds. Portions of each of the two types of lunar material were inoculated into three different culture media: (1) Burk's nitrogen-free medium (BURK and LINEWEAVER, 1930); (2) 9K basal salts medium with ferrous sulfate as the energy source (SILVERMAN and LUNDGREN, 1959); and (3) 9K basal salts medium with elemental sulfur as the energy source (UNZ and LUNDGREN, 1961). Lunar sample Aliquots of approximately 220 mg of lunar sample each were inoculated into 100 ml of liquid and 10 ml of agar-based culture media. The agar-based media were prepared by adding 1.5% purified agar (BioQuest) to each of the liquid media. To avoid acid hydrolysis of the agar during sterilization, the agar and the acidic (pH 2) 9K basal salts medium were autoclaved separately and combined only after each had cooled to 60°C. Liquid cultures were incubated on a rotary shaker under ambient atmospheric conditions at 25°C and 37°C. The agar-based cultures were incubated at each of twelve different temperature/atmosphere combinations (air, 10% CO_2 in air, or N_2; at 4°C, 25°C, 35°C or 50°C).

After 21 days of incubation, microscopic examination and pH determinations of the fluid media, and stereoscopic examination of the agar plates. gave no indication of microbial growth. Clumps of clear objects with dense centers (Fig. 3) had, however,

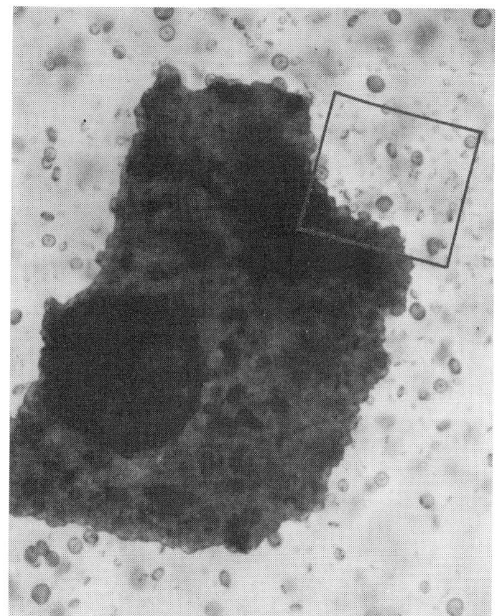

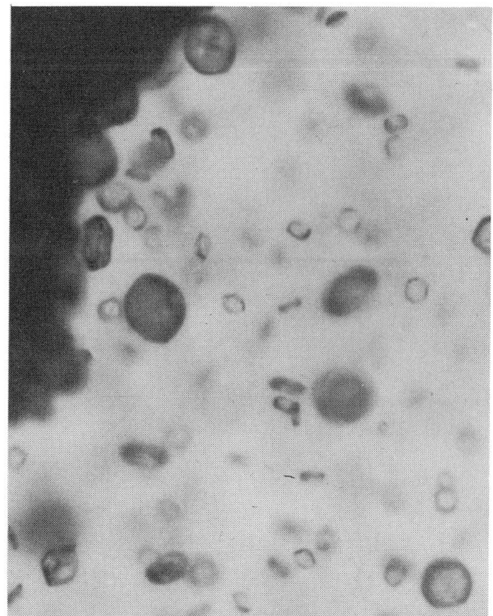

Fig. 3 (left). Object observed on 9K + $FeSO_4$ medium inoculated with core material from Apollo 12. Magnification = 5.3 × (insert shows position of Fig. 4).

Fig. 4 (right). Same as Fig. 3. Magnification = 21 ×.

developed on plates of 9K + $FeSO_4$ which had been inoculated with core material. Microscopic examination of the objects revealed small ovoid objects clustered around particles of lunar material (Fig. 4). Upon prolonged incubation similar objects were observed on many other 9K + $FeSO_4$ plates, including the controls. These objects are considered relevant only to the medium employed, and are presented here to allow comparison with the "artifacts" previously described by Oyama *et al.*, (1970).

Studies were conducted to determine if the lack of microbial growth on these deficient media could be due to some microbial toxicity as reported above for extracts of certain nutrient cultures. For these tests, separate aliquots of the various test systems and appropriate controls were inoculated with well characterized micro-organisms. Test systems containing Burk's medium were inoculated with *Azotobacter vinelandii* (ATCC 12837). Those systems containing 9K medium plus sulfur were inoculated individually with *Thiobacillus thiooxidans* (ATCC 8085) and *Ferrobacillus ferrooxidans* (ATCC 13661). Test systems containing 9K medium plus ferrous sulfate were inoculated with *F. ferrooxidans* only.

The inocula consisted of seven day old cultures which had been washed three times with the specific basal medium minus the energy source. Aliquots of 25 ml were inoculated with 1 ml of the appropriate washed cell suspension. Immediately upon inoculation, and again following incubation on a rotary shaker at 25°C for 14 days, total cell counts were determined using a Petroff-Hauser bacterial counting chamber. The resulting data are shown in Table 4. Comparisons between control and experimental systems indicate that the presence of lunar material did not adversely affect

Table 4.

Microorganisms Tested	Medium Employed	Initial cell Count	Total Cell Count After Incubation for 14 Days with Lunar Material Added as Shown Below		
			None	Surface	Subsurface
Azotobacter vinelandii	Burk's	1.6×10^7	1.5×10^8	1.1×10^8	3.4×10^8
Ferrobacillus ferrooxidans	$9K + FeSO_4$	$< 10^4$	1.6×10^8	1.4×10^8	1.6×10^8
Ferrobacillus ferrooxidans	9K + S	$< 10^4$	$< 10^4$	4.5×10^7	2.7×10^7
Thiobacillus thiooxidans	9K + S	4.8×10^6	4.1×10^8	2.3×10^8	4.0×10^8

the growth rate of any of the species tested. In the case of *F. ferrooxidans*, a three log increase in cell number was recorded in 9K plus sulfur medium to which lunar material had been added. No increase in cell number could be demonstrated in the absence of lunar material.

DISCUSSION

The studies described demonstrate that the fluids extracted from complexes of Apollo 11 lunar sub-surface material, which have incubated in nutrient media for four months, are toxic to all of the bacterial species tested. Fluids extracted from this same Apollo 11 core sample, however, are not toxic if the menstruum is refined agar in phosphate buffer with no nutrient material added. None of the other returned lunar materials from either Apollo 11 or Apollo 12 have exhibited extract toxicity in the systems tested. Chemical analyses of heat-sterilized lunar samples suggest that nickel and scandium are possible contributory factors. Further studies with these two elements are presently underway in this laboratory.

This study was initiated in an effort to better analyze the results of the quarantine microbiological investigation, designed to monitor back contamination of the Earth with microorganisms from the lunar material. The data presented herein demonstrate that extracts from all but one of the lunar sample types were able to support normal microbial growth, indicating that most viable terrestrial microorganisms could have been detected had they occurred in these samples in suitable numbers. The observed toxicity associated with extracts of the Apollo 11 sub-surface material indicates that this sample may, in fact, have assisted the quarantine effort by preventing microbial growth.

SUMMARY

We have reported the total lack of demonstrable, viable microorganisms (either contaminant or indigenous to the lunar material) in any of the four samples tested. These results do not appear to be due to innate bacteriostatic inhibition as three samples supported normal growth of test microorganisms. The fourth, however, elicited a bactericidal response. The nature of the lethal factor has not yet been determined, although the possible role of higher concentrations of ilmenite, nickel, and scandium in the Apollo 11 core material is being investigated.

Acknowledgments—Special acknowledgment is made to: Mary I. Arredondo, for microbiological assistance in this investigation: J. Roger Martin, Weldon B. Nance, James D. Dorsey, A. Dean Bennett, and John H. Allen, for chemical and mineralogical analyses: and Richard Long for production of photomicrographs, all with Brown and Root-Northrop.

References

AHRENS L. H. and TAYLOR S. R. (1961) *Spectrochemical Analysis* (2nd ed.). p. 288. Addison-Wesley.

ANDERSON D. H. (1970) Numbering System for Moon Samples. *Science* **167**, 781.

ANON. (1968) Analytical Methods for Atomic Absorption Spectrophotometry. Perkin-Elmer Corp., Norwalk, Conn.

ARRHENIUS S. A. (1908) *Worlds In The Making*. Harpers Brothers.

BROCK T. D. (1967) Microorganisms adapted to high temperatures. *Nature (Lond.)* **214**, 882–885.

BROCK T. D. (1969) Microbial growth under extreme conditions. *Symp. Soc. for Gen. Microbiol.* **19**, 15–41.

BURK D. and LINEWEAVER H. (1930) The influence of fixed nitrogen on Azotobacter. *J. Bacteriol.* **19**, 389–414.

CAMERON R. E., MORELLI F. A., and CONROW H. P. (1970) Survival of microorganisms in desert soil exposed to five years of continuous very high vacuum. JPL Tech. Rep. 32–1454.

FRYXELL R., ANDERSON D., CARRIER D., GREENWOOD W., and HEIDEN G. (1970) Apollo 11 drive-tube core sample: An initial physical analysis of lunar surface sediment. *Science* **167**, 734–737.

HAGEN C. A. and HAWRYLEWICZ E. J. (1968) Life in extraterrestrial environments. IITRI Rep. No. L6023–15.

HAWRYLEWICZ E. J., HAGEN C. A., TOLKACZ T., ANDERSON B. T., and EWIG M. (1968) Probability of growth (P_G) of viable microorganisms in martian environments. *Life Sciences and Space Research* **7**, 146–156.

HESS W. N. and CALIO A. J. (1970) Summary of scientific results. Apollo 11 Preliminary Science Report. NASA SP No. SP–214.

HOTCHIN J. (1968) The microbiology of space. *J. British Interplanetary Society* **21**, 122–130.

KEMMERER W. W. JR., MASON J. A., and WOOLEY B. C. (1969) Physical, chemical and biological activities at the lunar receiving laboratory. *Bioscience* **19**, 712–715.

LARSEN H. (1967) Biochemical aspects of extreme halophilism. *Adv. Microbial Physiol.* **1**, 97–132.

LSPET (Lunar Sample Preliminary Examination Team) (1970) Preliminary examination of the lunar samples from Apollo 12. *Science* 167, 1325–1339.

MCLANE J. C., KING E. A., FLORY D. A., RICHARDSON K. A., DAWSON J. P., KEMMERER W. W., and WOOLEY B. C. (1967), Lunar receiving laboratory. *Science* **155**, 525–529.

MITCHELL F. J. and ELLIS W. L. (1971) Microbial analysis of the Surveyor III TV camera. Second Lunar Science Conference (unpublished proceedings).

OYAMA V. I., MEREK E. L., and SILVERMAN M. P. (1970) A search for viable organisms in a lunar sample. *Science* 167, 773–775.

SILVERMAN G. J., DAVIS N. S., and BEECHER N. (1967) Resistivity of spores to ultraviolet and radiation while exposed to ultrahigh vacuum or at atmospheric pressure. *Appl. Microbiol.* **15**, 510–515.

SILVERMAN M. P. and LUNDGREN D. G. (1959) Studies on the chemoautotrophic iron bacterium *Ferrobacillus ferrooxidans*, I. An improved medium and a harvesting procedure for securing high cell yields. *J. Bacteriol.* **77**, 642–647.

STEWART D. B., APPLEMAN D. E., HUEBNER J. S., and CLARK J. R. (1970) Crystallography of some lunar plagioclases. *Science* **167**, 634–635.

TAYLOR G. R., ELLIS W. L., ARREDONDO M., and MAYHEW B. (1970a), Growth response of *Pseudomonas aeruginosa* (ATCC 15442) to the presence of lunar material. *Bacterial Proc.* 42 (in press).

TAYLOR G. R., FERGUSON J. K. and TRUBY C. P. (1970b), Methods used to monitor the microbial load of returned lunar material. *Appl. Microbiol.* **20**, 271–272.

TIERNEY M. S. (1968) The chances of retrieval of viable microorganisms deposited on the moon by unmanned lunar probes. Sandia Labs Rep. No. SC–M–68–539.

UCHINO R. and DOI S. (1967) Acido-thermophilic bacteria from thermal waters. *Afric. Biol. Chem.* **31**, 817–822.

UNZ R. F. and LUNDGREN D. G. (1961) A comparative nutritional study of three chemoautotrophic bacteria: *Ferrobacillus ferrooxidans*, *Thiobacillus ferrooxidans* and *Thiobacillus thiooxidans*. *Soil Sci.* **92**, 302–313.

VALLENTYNE J. R. (1963) Environmental biophysics and microbial ubiquity. *Ann. N. Y. Acad. Sci.* **108**, 342–352.

YODER H. S. and SAHAMA T. G. (1957) Olivene X-ray determination curve. *Amer. Mineral.* **42**, 475–491.

Apollo 12 Lunar Sample Inventory

Sample No.	Mass in grams	Description
12001	2216.0	Fines
12002	1529.5	Porphyritic basalt
12003	300.0	Fines and chips
12004	585.0	Porphyritic basalt
12005	482.0	Basalt
12006	206.4	Ophitic basalt
12007	65.2	Basalt
12008	58.4	Porphyritic vitrophyre
12009	468.2	Porphyritic vitrophyre
12010	360.0	Breccia
12011	193.0	Porphyritic basalt
12012	176.2	Porphyritic basalt
12013	82.3	Breccia
12014	159.4	Porphyritic basalt
12015	191.2	Porphyritic vitrophyre
12016	2028.3	Basalt
12017	53.0	Porphyritic basalt
12018	787.0	Porphyritic basalt
12019	462.4	Porphyritic basalt
12020	312.0	Porphyritic basalt
12021	1876.6	Porphyritic basalt
12022	1864.3	Porphyritic basalt
12023	269.3	Fines—LESC
12024	56.5	Fines—GASC
12025	56.1	Fines—double core tube, top
12026	101.4	Fines—first single core tube
12027	80.0	Fines—second single core tube
12028	189.6	Fines—double core tube, bottom
12029	6.5	Fines—in Surveyor scoop
12030	75.0	Fines and chips
12031	185.0	Ophitic basalt
12032	310.5	Fines
12033	450.0	Fines
12034	155.0	Breccia
12035	71.0	Ophitic basalt
12036	75.0	Ophitic basalt
12037	145.0	Fines
12038	746.0	Ophitic basalt
12039	255.0	Ophitic basalt
12040	319.0	Ophitic basalt
12041	24.8	Fines
12042	255.0	Fines
12043	60.0	Ophitic basalt
12044	92.0	Fines
12045	63.0	Porphyritic basalt
12046	166.0	Ophitic basalt
12047	193.0	Ophitic basalt
12048	2.0	Fines
12050	1.0	Fines
12051	1660.0	Ophitic basalt
12052	1866.0	Porphyritic basalt
12053	879.0	Porphyritic basalt
12054	678.0	Basalt
12055	912.0	Porphyritic basalt
12056	121.0	Ophitic basalt
12057	650.0	Fines and chips
12060	20.7	Fines
12061	9.5	Chips
12062	738.7	Ophitic basalt
12063	2426.0	Porphyritic basalt
12064	1214.3	Ophitic basalt
12065	2109.0	Porphyritic basalt
12070	1102.0	Fines
12071	9.16	Chips
12072	103.6	Basalt
12073	407.65	Breccia
12075	232.5	Porphyritic basalt
12076	54.55	Porphyritic basalt
12077	22.63	Basalt

Author Index